COMPUTATIONAL MOLECULAR SPECTROSCOPY

COMPUTATIONAL MOLECULAR SPECTROSCOPY

Edited by

Per Jensen
Bergische Universität—Gesamthochschule Wuppertal, Germany

and

Philip R. Bunker
Steacie Institute for Molecular Sciences,
National Research Council of Canada

JOHN WILEY & SONS, LTD
Chichester • New York • Weinheim • Brisbane • Toronto • Singapore

Copyright © 2000 John Wiley & Sons Ltd,
The Atrium, Southern Gate, Chichester,
West Sussex PO19 8SQ, England

Telephone (+44) 1243 779777

Email (for orders and customer service enquiries): cs-books@wiley.co.uk
Visit our Home Page on www.wileyeurope.com or www.wiley.co.uk

Reprinted May 2003

Other Wiley Editorial Offices

John Wiley & Sons Inc., 111 River Street, Hoboken, NJ 07030, USA

Jossey-Bass, 989 Market Street, San Francisco, CA 94103-1741, USA

Wiley-VCH Verlag GmbH, Boschstr. 12, D-69469 Weinheim, Germany

John Wiley & Sons Australia Ltd, 33 Park Road, Milton, Queensland 4064, Australia

John Wiley & Sons (Asia) Pte Ltd, 2 Clementi Loop #02-01, Jin Xing Distripark, Singapore 129809

John Wiley & Sons Canada Ltd, 22 Worcester Road, Etobicoke, Ontario, Canada M9W 1L1

British Library Cataloguing in Publication Data

A catalogue record for this book is available from the British Library

ISBN 0 471 48998 0

Typeset in 9.5/11.5pt Times by Laser Words, (India) Ltd.
Printed and bound by CPI Antony Rowe, Eastbourne

CONTENTS

LIST OF CONTRIBUTORS

Aleksey B. Alekseyev	Fachbereich 9—Theoretische Chemie Bergische Universität—Gesamthochschule Wuppertal Gaußstr. 20 D-42097 Wuppertal Germany
Wesley D. Allen	Center for Computational Quantum Chemistry Department of Chemistry University of Georgia Athens, Georgia 30602 USA
Ad van der Avoird	Institute of Theoretical Chemistry University of Nijmegen Toernooiveld NL-6525 ED Nijmegen The Netherlands
Timothy A. Barckholtz	Laser Spectroscopy Facility Department of Chemistry The Ohio State University 120 West 18th Avenue Columbus, Ohio 43210 USA
Attila Bérces	Steacie Institute for Molecular Sciences National Research Council of Canada Ottawa, Ontario Canada K1A OR6
John Brown	Physical and Theoretical Chemistry Laboratory South Parks Road Oxford OX1 3QZ UK
Robert J. Buenker	Fachbereich 9—Theoretische Chemie Bergische Universität—Gesamthochschule Wuppertal Gaußstr. 20 D-42097 Wuppertal Germany

Philip R. Bunker — Steacie Institute for Molecular Sciences
National Research Council of Canada
Ottawa, Ontario
Canada K1A OR6

Mark S. Child — Physical and Theoretical Chemistry Laboratory
South Parks Road
Oxford OX1 3QZ
UK

Attila G. Császár — Department of Theoretical Chemistry
Eötvös University
P.O. Box 32
H-1518 Budapest 112
Hungary

Jean Demaison — Laboratoire de Physique de Lasers, Atomes et Molecules
UMR CNRS 249
Bâtiment P5
Université de Lille I
F-59655 Villeneuve d'Ascq
France

R. Benny Gerber — Department of Physical Chemistry and
The Fritz Haber Research Center
The Hebrew University
Jerusalem 91904
Israel
and
Department of Chemistry
University of California
Irvine
California 92697
USA

Jian-ping Gu — Fachbereich 9 — Theoretische Chemie
Bergische Universität — Gesamthochschule Wuppertal
Gaußstr. 20
D-42097 Wuppertal
Germany

Lauri Halonen — Laboratory of Physical Chemistry
P.O. Box 55 (A. I. Virtasen aukio 1)
University of Helsinki
FIN-00014 Helsinki
Finland

Bernd Artur Hess — Lehrstuhl für Theoretische Chemie
Friedrich-Alexander-Universität Erlangen-Nürnberg
Egerlandstr. 3
D-91058 Erlangen
Germany

Gerhard Hirsch — Fachbereich 9 — Theoretische Chemie
Bergische Universität — Gesamthochschule Wuppertal
Gaußstr. 20
D-42097 Wuppertal
Germany

Per Jensen — Fachbereich 9 — Theoretische Chemie
Bergische Universität — Gesamthochschule Wuppertal
D-42097 Wuppertal
Germany

Joon O. Jung — Department of Chemistry
University of California
Irvine
California 92697
USA

Mineo Kimura — School of Allied Health Sciences
Yamaguchi University
Ube, Yamaguchi 755
Japan

Yan Li — Fachbereich 9 — Theoretische Chemie
Bergische Universität — Gesamthochschule Wuppertal
Gaußstr. 20
D-42097 Wuppertal
Germany

Heinz-Peter Liebermann — Fachbereich 9 — Theoretische Chemie
Bergische Universität — Gesamthochschule Wuppertal
Gaußstr. 20
D-42097 Wuppertal
Germany

Jan Makarewicz — A. Mickiewicz University
Faculty of Chemistry
PL-60-780 Poznań
Poland

Christel Maria Marian — GMD Center for Information Technology
Scientific Computing and Algorithms Institute (SCAI)
Schloss Birlinghoven
D-53754 St. Augustin
Germany
and
Institut für Physikalische und Theoretische Chemie
Universität Bonn
Wegelerstrasse 12
D-53115 Bonn
Germany

Terry A. Miller	Laser Spectroscopy Facility Department of Chemistry The Ohio State University 120 West 18th Avenue Columbus, Ohio 43210 USA
Robert Moszynski	Department of Chemistry University of Warsaw Pasteura 1 PL-02-093 Warsaw Poland
Gerald Osmann	Steacie Institute for Molecular Sciences National Research Council of Canada Ottawa, Ontario Canada K1A OR6
Martin J. Packer	Department of Chemistry University of Sheffield Sheffield S3 7HF UK
Roger Rousseau	Steacie Institute for Molecular Sciences National Research Council of Canada Ottawa, Ontario Canada K1A OR6
Kamil Sarka	Department of Physical Chemistry Faculty of Pharmacy Comenius University SK-83232 Bratislava Slovakia
Stephan P. A. Sauer	Chemistry Laboratory IV Department of Chemistry University of Copenhagen Universitetsparken 5 DK-2100 Copenhagen Ø Denmark
Henry F. Schaefer III	Center for Computational Quantum Chemistry Department of Chemistry University of Georgia Athens, Georgia 30602 USA
Tamar Seideman	Steacie Institute for Molecular Sciences National Research Council of Canada Ottawa, Ontario Canada K1A OR6

Jonathan Tennyson	Department of Physics and Astronomy University College London Gower Street London WC1E 6BT UK
John S. Tse	Steacie Institute for Molecular Sciences National Research Council of Canada Ottawa, Ontario Canada K1A OR6
David J. Wales	University Chemical Laboratories Lensfield Road Cambridge CB2 1EW UK
Paul E. S. Wormer	Institute of Theoretical Chemistry University of Nijmegen Toernooiveld NL-6525 ED Nijmegen The Netherlands
Yukio Yamaguchi	Center for Computational Quantum Chemistry Department of Chemistry University of Georgia Athens, Georgia 30602 USA
Dong-Sheng Yang	Department of Chemistry University of Kentucky Lexington Kentucky 40506-0055 USA
David R. Yarkony	Department of Chemistry Johns Hopkins University Baltimore Maryland 21218 USA
Marek Z. Zgierski	Steacie Institute for Molecular Sciences National Research Council of Canada Ottawa, Ontario Canada K1A OR6

FOREWORD

Dennis Salahub

I had been in my new job as Director General of the Steacie Institute for only a few weeks when Phil Bunker asked me if I would like to write a preface for the new book that Per Jensen and he were editing. I immediately said yes, not only because I guessed this was the sort of thing that DGs did for their team, nor because I was flattered at the invitation (I was), but primarily because I knew, even before seeing the Table of Contents, that Phil and Per would have chosen some of the most difficult, and potentially rewarding, problems in theoretical chemistry and would have found the best authors to portray the state of the art. They have certainly done this.

Their book focuses on three areas: quantum chemistry, bound state calculations and dynamics and is written to foster understanding and interaction amongst these fields. Theoretical chemistry has advanced rapidly in recent years so that now the state of the art draws heavily from all three. True progress requires a mixture of theoretical concepts, computational techniques and, often, raw computing power. Jensen and Bunker's book captures a cross section of these activities. It is aimed at both beginning and seasoned researchers. Newcomers will be able to whet their appetites with a sampling of computational molecular spectroscopy as it now stands; veterans will find enough details, and an extensive bibliography, to satisfy their specialist needs.

Following an introductory chapter by the Editors on the Born–Oppenheimer approximation, the opening chapters focus on the quantum chemical calculation of electronic states and potential energy surfaces. Small molecules are treated by Császár, Allen, Yamaguchi and Schaefer, intermolecular interactions by Moszynski, Wormer and van der Avoird. Transition metals are not forgotten (Bérces, Zgierski, and Yang) nor are electronically excited states (Buenker, Hirsch, Li, Gu, Alekseyev and Liebermann). Hess and Marian cover relativistic effects and, lest the fixation on the energy become too strong, Sauer gives an overview of properties calculations. Such a varied canvas of applications requires an equally broad palette of computational techniques: CI, CC, SAPT, DFT...

Once one has a sufficiently accurate surface, a very tough problem, one meets a second challenge of equal magnitude: calculating the rotation–vibration states. In this section, Sarka and Demaison guide the reader through perturbation theory, effective hamiltonians and force constants. Tennyson covers variational calculations and Halonen highly excited states and local modes. In attempting to extend rovibrational calculations to larger molecules one meets a very steep wall (the $3N-6$ catastrophe...) so that more approximate methods are needed. Gerber and Jung show how to do this with a vibrational self-consistent field approach. Even for molecules of relatively modest size, large amplitude, floppy motion requires special techniques (Makarewicz). The final chapter of this section by Wales treats the complex low-energy motion and tunneling splittings in small water clusters.

The Born–Oppenheimer approximation is the mainstay of quantum chemistry but it sometimes breaks down. The penultimate section of the book treats rovibronic states, rotations, vibrations and electronic motion are coupled and understanding the nature of the various couplings and incorporating them into accurate computational approaches remains a great challenge. Yarkony's chapter deals with adiabatic and nonadiabatic representations for diatomic molecules. Jensen,

Osmann and Bunker write on the Renner effect (breaking of symmetry for linear molecules that arises from degeneracy) as does Brown, using an effective Hamiltonian approach. Non-linear molecules in degenerate states also distort and this Jahn–Teller effect is treated, in the presence of spin–orbit coupling, by Barckholtz and Miller. That such intricate and difficult problems as these can now be treated with good accuracy is a testament to the efforts of theoretical chemists, including the authors just mentioned, over the past decades.

One often needs to go beyond eigenstates and worry about the time dependence of molecular (and condensed phase) properties. Indeed, predicting chemical rates on the basis of accurate potential surface calculations (with appropriate non-Born–Oppenheimer corrections when needed) remains the holy grail of theoretical chemistry. The last chapters of the book turn to dynamics. Child writes on semiclassical resonance and wave-packet techniques, Seideman on the superposition of states and Tse and Rousseau provide an entry into very complex systems with a chapter on *ab initio* (DFT) molecular dynamics. These 'on-the-fly' dynamics simulations, in which the potential surface is generated from quantum chemistry as needed, have revolutionized simulations in materials and biological modeling.

Jensen and Bunker have, indeed, convinced leading experts over a broad spectrum of theoretical molecular spectroscopy to contribute to this book. The subjects range from fundamental theory to detailed applications, stressing the applications.

The motto of the Steacie Institute is 'The fundamental things apply.'. Per Jensen and Phil Bunker have edited a volume that, while fundamentally fundamental, provides examples of applications and points out avenues to follow for further, more complex, calculations that will ultimately impinge on real practical issues. If there is a current trend towards simulations of staggering complexity, it should not be forgotten that the ultimate value of the computations depends crucially on the approximations made for the energy surfaces, for eigenstates, and for the dynamics *per se*. I am sure the reader will find such cautionary notes in Jensen and Bunker's book, both explicitly and between the lines, along with a myriad of potential solutions and exciting results.

Bonne lecture!

Dennis Salahub
Director General, Steacie Institute for Molecular Sciences
National Research Council of Canada
Ottawa

PREFACE

The principal aim of this book is to 'bridge the gaps' between traditional quantum chemistry, which is mainly concerned with the calculation of the electronic properties of molecules, theoretical high resolution spectroscopy, traditionally concerned with the calculation of bound rovibronic states, and molecular dynamics which looks at the time dependence of molecular processes. The book is aimed at research workers in all of these fields, and in the border areas between them. The foreword by Dennis Salahub introduces the reader to the 20 chapters that make up this volume.

Editing a book is like a research project. One is eager and enthusiastic at the beginning after one has first developed the idea, one is then often miserable and on the verge of giving up in the middle of the work as unforseen difficulties present themselves, and finally one is happy and relieved when it is completed. During the middle phase one needs constant advice and encouragement. To produce this book we sometimes needed it very much, and thus it is a real pleasure to acknowledge the help given us by Katya Vines of Wiley in this regard. We thank Marion Litz of the Bergische Universität in Wuppertal, Germany, for help with practical matters. It is also necessary to acknowledge the forebearance of the many authors who produced their chapters clearly written in a timely fashion, and then did not complain because of the delays caused by the few who did not.

During the time we worked to produce this book PRB was the holder of an Alexander von Humboldt Senior Award at the Bergische Universität, and he is very grateful for that. Also PJ appreciates the hospitality he received from the Steacie Institute for Molecular Sciences at the National Research Council of Canada in Ottawa. The new Director General of the Steacie Institute for Molecular Sciences, Dennis Salahub, agreed to write a foreword that introduces the chapters, and we thank him for doing such a good job.

Per Jensen
Wuppertal, Germany

Philip R. Bunker
Ottawa, Canada

PART 1

INTRODUCTION

1 THE BORN–OPPENHEIMER APPROXIMATION

Philip R. Bunker
National Research Council of Canada, Ottawa, Canada

and

Per Jensen
Bergische Universität, Wuppertal, Germany

The Born–Oppenheimer (BO) approximation is used in the solution of the rovibronic wave equation, and every chapter of this book implicitly or explicitly involves it; either the BO approximation is assumed or else its breakdown is specifically considered. Thus it is appropriate that the opening chapter explain briefly what this approximation is. Although not always appreciated there are two ways of making the BO approximation: the perturbation theory approach [Born and Oppenheimer (1927); see also Sections 14 and 15, and Appendix VII, of Born and Huang (1954), and Mead (1988)], and the variation theory approach [Born (1951); see also Appendix VII of Born and Huang (1954)]. Before outlining these two approaches we first write out the nonrelativistic rovibronic Hamiltonian in order to introduce the notation used [see, for example, Bunker and Jensen (1998) for the details of the derivation].

We consider a molecule as consisting of l particles, N of which are nuclei and $l - N$ of which are electrons. The derivation of the rovibronic Hamiltonian involves first separating off translation, and then (in order to facilitate the separation of nuclear and electronic motion) referring the coordinates of all particles in the molecule to a (ξ, η, ζ) axis system that has origin at the nuclear center of mass, but with arbitrary space-fixed orientation. Doing this the exact nonrelativistic

Computational Molecular Spectroscopy. Edited by Per Jensen and Philip R. Bunker

(spin-free) rovibronic Hamiltonian is obtained as [see, for example, equations (9–48) of Bunker and Jensen (1998)]:

$$\hat{H}_{\rm rve} = \hat{T}^0_{\rm e} + \hat{T}'_{\rm e} + \hat{T}_{\rm N} + V(\boldsymbol{R}_{\rm N}, \boldsymbol{r}_{\rm elec}), \tag{1}$$

where

$$\hat{T}^0_{\rm e} = -\frac{\hbar^2}{2m_{\rm e}} \sum_{i=N+1}^{l} \nabla_i^2, \tag{2}$$

$$\hat{T}'_{\rm e} = -\frac{\hbar^2}{2M_{\rm N}} \sum_{i,j=N+1}^{l} \nabla_i \cdot \nabla_j, \tag{3}$$

$$\hat{T}_{\rm N} = -\frac{\hbar^2}{2} \sum_{i=2}^{N} \frac{\nabla_i^2}{m_i} + \frac{\hbar^2}{2M_{\rm N}} \sum_{i,j=2}^{N} \nabla_i \cdot \nabla_j, \tag{4}$$

$$\nabla_i^2 = \frac{\partial^2}{\partial \xi_i^2} + \frac{\partial^2}{\partial \eta_i^2} + \frac{\partial^2}{\partial \zeta_i^2}, \tag{5}$$

$$\nabla_i \cdot \nabla_j = \frac{\partial^2}{\partial \xi_i \partial \xi_j} + \frac{\partial^2}{\partial \eta_i \partial \eta_j} + \frac{\partial^2}{\partial \zeta_i \zeta_j}, \tag{6}$$

and

$$V(\boldsymbol{R}_{\rm N}, \boldsymbol{r}_{\rm elec}) = \sum_{r<s=1}^{l} \frac{C_r C_s e^2}{4\pi\varepsilon_0 R_{rs}}. \tag{7}$$

In these equations the mass of each nucleus is m_i, the total mass of all the nuclei is $M_{\rm N}$, the mass of the electron is $m_{\rm e}$, the charge of each particle is $C_r e$ ($C_r = -1$ for the electron), and R_{rs} is the separation of particles r and s. We explicitly show that the potential energy function V depends on the nuclear coordinates $\boldsymbol{R}_{\rm N} = (\xi_2, \eta_2, \zeta_2, \xi_3, \eta_3, \zeta_3, \ldots, \xi_N, \eta_N, \zeta_N)$ and on the electronic coordinates $\boldsymbol{r}_{\rm elec} = (\xi_{N+1}, \eta_{N+1}, \zeta_{N+1}, \xi_{N+2}, \eta_{N+2}, \zeta_{N+2}, \ldots, \xi_l, \eta_l, \zeta_l)$. In these (ξ, η, ζ) coordinates (with the origin at the nuclear center of mass) the kinetic energy is completely separable into an electronic part $\hat{T}_{\rm e} = \hat{T}^0_{\rm e} + \hat{T}'_{\rm e}$, and a nuclear part $\hat{T}_{\rm N}$. We now consider the two ways of making the BO approximation in order to solve the rovibronic wave equation

$$\left[\hat{H}_{\rm rve} - E_{{\rm rve},m}\right] \Psi_{{\rm rve},m}(\boldsymbol{r}_{\rm elec}, \boldsymbol{R}_{\rm N}) = 0. \tag{8}$$

1.1 The perturbation theory approach

Born and Oppenheimer (1927) (called BandO from here on) begin in their introduction with the fundamental idea that the rovibronic Hamiltonian should be expanded in powers of the parameter κ, where κ is given by the fourth root of the ratio of the mass of the electron to the mean nuclear mass M_0, i.e.,

$$\kappa = \left(\frac{m_{\rm e}}{M_{\rm N}/N}\right)^{1/4} = \left(\frac{m_{\rm e}}{M_0}\right)^{1/4}. \tag{9}$$

Born and Heisenberg (1924) had earlier suggested an expansion in $\sqrt{m_{\rm e}/M_0}$. Using their expansion parameter BandO show (as we see below) that, with the electronic energy appearing in zeroth order, the vibrational energy appears in second order, and the rotational energy in fourth order, while the first- and third-order terms disappear.

The second idea in BandO is that the vibrational displacement coordinates are κ times the size of the bond lengths or bond angles. This is expressed by writing

$$\boldsymbol{R}_{\mathrm{N}} = \boldsymbol{R}_{\mathrm{N}}^{0} + \kappa \boldsymbol{U}_{\mathrm{N}}, \tag{10}$$

from which it follows that

$$\frac{\partial}{\partial \boldsymbol{R}_{\mathrm{N}}} = \frac{1}{\kappa}\frac{\partial}{\partial \boldsymbol{U}_{\mathrm{N}}}. \tag{11}$$

In equation (10) $\boldsymbol{R}_{\mathrm{N}}^{0}$ is an as yet unspecified molecular reference configuration, and the rotational coordinates are contained in the $\boldsymbol{R}_{\mathrm{N}}^{0}$ coordinates. BandO show that this reference configuration has to be the equilibrium molecular configuration, and we will see why this is when we repeat their analysis of the first order result below.

The third important part of the development made by BandO, which follows after introducing the vibrational displacement coordinates $\boldsymbol{U}_{\mathrm{N}}$ in equation (10), is that the following order of magnitude relation holds:

$$\frac{\partial \Psi_{\mathrm{rve}}}{\partial \boldsymbol{r}_{\mathrm{elec}}} \sim \frac{\partial \Psi_{\mathrm{rve}}}{\partial \boldsymbol{U}_{\mathrm{N}}}. \tag{12}$$

This equation is necessary in order to relate the orders of magnitude of $\hat{T}_{\mathrm{N}}$ and $\hat{T}_{\mathrm{e}}^{0}$.

The treatment in BandO depends on the following two conditions:

- That the vibrational energy separations be two orders of magnitude smaller than the electronic energy separations, i.e.,

$$\Delta E_{\mathrm{vib}} \sim \kappa^{2}\, \Delta E_{\mathrm{elec}}, \tag{13}$$

and

- That the rotational energy separations be two orders of magnitude smaller than the vibrational energy separations, i.e.,

$$\Delta E_{\mathrm{rot}} \sim \kappa^{2}\, \Delta E_{\mathrm{vib}}. \tag{14}$$

When these conditions are in accord with the experimental observations the BO approximation as treated in BandO will be valid.

To begin the perturbation theory treatment the rovibronic Hamiltonian is written as

$$\hat{H}_{\mathrm{rve}} = \hat{H}_{0} + \hat{T}_{\mathrm{e}}' + \hat{T}_{\mathrm{N}}, \tag{15}$$

where the electronic Hamiltonian is defined as

$$\hat{H}_{0} = \hat{T}_{\mathrm{e}}^{0} + V\left(\boldsymbol{R}_{\mathrm{N}}, \boldsymbol{r}_{\mathrm{elec}}\right). \tag{16}$$

The electronic wave equation

$$\left[\hat{H}_{0} - E_{\mathrm{elec},n}\right]\Phi_{\mathrm{elec},n}(\boldsymbol{r}_{\mathrm{elec}}; \boldsymbol{R}_{\mathrm{N}}) = 0, \tag{17}$$

is supposed solved, with the $\boldsymbol{r}_{\mathrm{elec}}$ as dynamical variables and n as the electronic state quantum number, for $\boldsymbol{R}_{\mathrm{N}} = \boldsymbol{R}_{\mathrm{N}}^{0}$ and for neighbouring molecular configurations around $\boldsymbol{R}_{\mathrm{N}}^{0}$. In the approach of BandO the nth solution of the electronic wave equation is obtained by a perturbation theory expansion about the solution at $\boldsymbol{R}_{\mathrm{N}} = \boldsymbol{R}_{\mathrm{N}}^{0}$ using equation (10) to give:

$$\hat{H}_{0} = \hat{H}_{0}^{(0)} + \kappa \hat{H}_{0}^{(1)} + \kappa^{2}\hat{H}_{0}^{(2)} + \ldots, \tag{18}$$

$$\Phi_{\mathrm{elec},n} = \Phi_{\mathrm{elec},n}^{(0)} + \kappa \Phi_{\mathrm{elec},n}^{(1)} + \kappa^{2}\Phi_{\mathrm{elec},n}^{(2)} + \ldots, \tag{19}$$

and

$$E_{\mathrm{elec},n} = E^{(0)}_{\mathrm{elec},n} + \kappa E^{(1)}_{\mathrm{elec},n} + \kappa^2 E^{(2)}_{\mathrm{elec},n} + \dots, \tag{20}$$

where $\hat{H}_0^{(0)}$, $\Phi^{(0)}_{\mathrm{elec},n}$ and $E^{(0)}_{\mathrm{elec},n}$ are independent of $\boldsymbol{U}_{\mathrm{N}}$ (being appropriate for $\boldsymbol{R}_{\mathrm{N}} = \boldsymbol{R}^0_{\mathrm{N}}$), $\hat{H}_0^{(r)}$ is an operator with respect to the coordinates $\boldsymbol{r}_{\mathrm{elec}}$ and, along with $\Phi^{(r)}_{\mathrm{elec},n}$ and $E^{(r)}_{\mathrm{elec},n}$, it is a homogeneous function of degree r in the $\boldsymbol{U}_{\mathrm{N}}$. In the notation of BandO equations (18)–(20) above are given in their equations (16)–(18), where they use the notation ξ, ζ and V_n for what we call $\boldsymbol{R}_{\mathrm{N}}$, $\boldsymbol{U}_{\mathrm{N}}$ and $E_{\mathrm{elec},n}$. Substituting equations (18)–(20) into equation (17) and equating like powers of κ we obtain:

$$\left[\hat{H}_0^{(0)} - E^{(0)}_{\mathrm{elec},n}\right] \Phi^{(0)}_{\mathrm{elec},n} = 0, \tag{21}$$

$$\left[\hat{H}_0^{(0)} - E^{(0)}_{\mathrm{elec},n}\right] \Phi^{(1)}_{\mathrm{elec},n} = -\left[\hat{H}_0^{(1)} - E^{(1)}_{\mathrm{elec},n}\right] \Phi^{(0)}_{\mathrm{elec},n}, \tag{22}$$

$$\left[\hat{H}_0^{(0)} - E^{(0)}_{\mathrm{elec},n}\right] \Phi^{(2)}_{\mathrm{elec},n} = -\left[\hat{H}_0^{(1)} - E^{(1)}_{\mathrm{elec},n}\right] \Phi^{(1)}_{\mathrm{elec},n} - \left[\hat{H}_0^{(2)} - E^{(2)}_{\mathrm{elec},n}\right] \Phi^{(0)}_{\mathrm{elec},n}, \tag{23}$$

etc., where equation (21) is obtained from the κ^0 coefficients, equation (22) is obtained from the κ^1 coefficients, and equation (23) is obtained from the κ^2 coefficients.

To make an order of magnitude expansion of the complete Hamiltonian $\hat{H}_{\mathrm{rve}}$ we have to determine the orders of magnitude of $\hat{T}'_{\mathrm{e}}$ and $\hat{T}_{\mathrm{N}}$ given in equations (3) and (4) respectively. Comparing the mass coefficient of $\hat{T}'_{\mathrm{e}}$ with that of $\hat{T}^0_{\mathrm{e}}$ it is obvious that $\hat{T}'_{\mathrm{e}}$ is of order of magnitude $\kappa^4 \hat{H}_0$, and so we can express its order of magnitude by writing

$$\hat{T}'_{\mathrm{e}} = \kappa^4 \hat{H}_1^{(0)}. \tag{24}$$

To determine the order of magnitude of $\hat{T}_{\mathrm{N}}$ we have to consider both the size of its mass coefficient and the size of the derivative $\partial^2 \Psi_{\mathrm{rve}}/\partial \boldsymbol{R}_{\mathrm{N}}{}^2$. Using equation (11) for the vibrational part of $\hat{T}_{\mathrm{N}}$ [called $\hat{T}_{\mathrm{N}}(\mathrm{vib})$], which is expressed solely in terms of derivatives of degree two with respect to the displacement coordinates, it follows that

$$\hat{T}_{\mathrm{N}}(\mathrm{vib}) = \hat{T}_{\mathrm{N}}\left(\frac{\partial^2}{\partial \boldsymbol{R}^2_{\mathrm{N}}}\right) \sim \frac{1}{\kappa^2} \hat{T}_{\mathrm{N}}\left(\frac{\partial^2}{\partial \boldsymbol{U}^2_{\mathrm{N}}}\right). \tag{25}$$

Using equation (12), together with a consideration of the sizes of the mass coefficients, there is the order of magnitude relation:

$$\hat{T}_{\mathrm{N}}\left(\frac{\partial^2}{\partial \boldsymbol{U}^2_{\mathrm{N}}}\right) \sim \kappa^4 \hat{H}_0. \tag{26}$$

Substituting equation (26) into equation (25) we obtain the required order of magnitude estimate.

$$\hat{T}_{\mathrm{N}}(\mathrm{vib}) \sim \kappa^2 \hat{H}_0. \tag{27}$$

The order of magnitude of the rotational part of $\hat{T}_{\mathrm{N}}$ (which contains no derivatives with respect to the displacement coordinates) is $\kappa^4 \hat{H}_0$, and the order of magnitude of the rotational–vibration coupling part of $\hat{T}_{\mathrm{N}}$ (which contains derivatives of degree one with respect to the displacement coordinates) is $\kappa^3 \hat{H}_0$. To express these orders of magnitude we write

$$\begin{aligned}\hat{T}_{\mathrm{N}} &= \kappa^2 \hat{H}_{2\mathrm{v}} + \kappa^3 \hat{H}_{2\mathrm{rv}} + \kappa^4 \hat{H}_{2\mathrm{r}} \\ &= \kappa^2 \hat{H}^{(0)}_{2\mathrm{v}} + \kappa^3\left(\hat{H}^{(0)}_{2\mathrm{rv}} + \hat{H}^{(1)}_{2\mathrm{v}}\right) + \kappa^4\left(\hat{H}^{(0)}_{2\mathrm{r}} + \hat{H}^{(1)}_{2\mathrm{rv}} + \hat{H}^{(2)}_{2\mathrm{v}}\right).\end{aligned} \tag{28}$$

This is given in equation (29) of BandO, and the above order of magnitude breakdown of the factors $\hat{H}_{2v}$, $\hat{H}_{2rv}$, and $\hat{H}_{2r}$ is copied from them. Note that $\hat{H}_{2v}^{(0)}$ involves quadratic derivatives with respect to the vibrational displacement coordinates. Thus to fourth order the rovibronic Hamiltonian written as an order of magnitude expansion is

$$\hat{H}_{\text{rve}} = \hat{H}_0^{(0)} + \kappa\hat{H}_0^{(1)} + \kappa^2\left(\hat{H}_0^{(2)} + \hat{H}_{2v}^{(0)}\right) + \kappa^3\left(\hat{H}_0^{(3)} + \hat{H}_{2rv}^{(0)} + \hat{H}_{2v}^{(1)}\right) + \kappa^4\left(\hat{H}_0^{(4)} + \hat{H}_1^{(0)} + \hat{H}_{2r}^{(0)} + \hat{H}_{2rv}^{(1)} + \hat{H}_{2v}^{(2)}\right). \tag{29}$$

To implement perturbation theory for solving the rovibronic wave equation [equation (8)] we expand the eigenfunctions and eigenvalues of $\hat{H}_{\text{rve}}$ as

$$\Psi_{\text{rve},m} = \Psi_{\text{rve},m}^{(0)} + \kappa\Psi_{\text{rve},m}^{(1)} + \kappa^2\Psi_{\text{rve},m}^{(2)} + \ldots, \tag{30}$$

and

$$E_{\text{rve},m} = E_{\text{rve},m}^{(0)} + \kappa E_{\text{rve},m}^{(1)} + \kappa^2 E_{\text{rve},m}^{(2)} + \ldots. \tag{31}$$

As in the perturbation theory treatment of the electronic wave equation in equations (21)–(23) above, we now substitute equations (29)–(31) into equation (8), and equate like powers of κ. If we stop at second order we obtain:

$$\left[\hat{H}_0^{(0)} - E_{\text{rve},m}^{(0)}\right]\Psi_{\text{rve},m}^{(0)} = 0, \tag{32}$$

$$\left[\hat{H}_0^{(0)} - E_{\text{rve},m}^{(0)}\right]\Psi_{\text{rve},m}^{(1)} = -\left[\hat{H}_0^{(1)} - E_{\text{rve},m}^{(1)}\right]\Psi_{\text{rve},m}^{(0)}, \tag{33}$$

and

$$\left[\hat{H}_0^{(0)} - E_{\text{rve},m}^{(0)}\right]\Psi_{\text{rve},m}^{(2)} = -\left[\hat{H}_0^{(1)} - E_{\text{rve},m}^{(1)}\right]\Psi_{\text{rve},m}^{(1)} - \left[\hat{H}_0^{(2)} + \hat{H}_{2v}^{(0)} - E_{\text{rve},m}^{(2)}\right]\Psi_{\text{rve},m}^{(0)}. \tag{34}$$

In zero order, comparing equations (21) and (32), we deduce that

$$E_{\text{rve},m}^{(0)} = E_{\text{elec},n}^{(0)}, \tag{35}$$

and that $\Phi_{\text{elec},n}^{(0)} = \Phi_{\text{elec},n}(\boldsymbol{r}_{\text{elec}}; \boldsymbol{R}_{\text{N}}^0)$ is a solution of the zero-order rovibronic equation (32). The product of this electronic function with an arbitrary function of $\boldsymbol{U}_{\text{N}}$, $\Phi_{\text{nuc}}^{(0)}(\boldsymbol{U}_{\text{N}})$ say, will also be a solution, and thus, in general:

$$\Psi_{\text{rve},m}^{(0)}(\boldsymbol{r}_{\text{elec}}, \boldsymbol{U}_{\text{N}}) = \Phi_{\text{nuc}}^{(0)}(\boldsymbol{U}_{\text{N}})\Phi_{\text{elec},n}^{(0)}(\boldsymbol{r}_{\text{elec}}) = \Phi_{\text{nuc}}^{(0)}(\boldsymbol{U}_{\text{N}})\Phi_{\text{elec},n}\left(\boldsymbol{r}_{\text{elec}}; \boldsymbol{R}_{\text{N}}^0\right). \tag{36}$$

The arbitrariness in the function $\Phi_{\text{nuc}}^{(0)}(\boldsymbol{U}_{\text{N}})$ disappears when we take the treatment to second order.

Multiplying the left hand side of equation (33) on the left by $\Phi_{\text{elec},n}^{(0)*}$ and integrating over $\boldsymbol{r}_{\text{elec}}$ gives zero from equations (21) and (35). Thus multiplying the right hand side on the left by $\Phi_{\text{elec},n}^{(0)*}$ and integrating over $\boldsymbol{r}_{\text{elec}}$ gives zero, so we have

$$\left\langle \Phi_{\text{elec},n}^{(0)} \right| \left(\hat{H}_0^{(1)} - E_{\text{rve},m}^{(1)}\right) \left|\Psi_{\text{rve},m}^{(0)}\right\rangle = \Phi_{\text{nuc}}^{(0)}(\boldsymbol{U}_{\text{N}}) \left\langle \Phi_{\text{elec},n}^{(0)} \right| \left(\hat{H}_0^{(1)} - E_{\text{rve},m}^{(1)}\right) \left| \Phi_{\text{elec},n}^{(0)} \right\rangle = 0, \tag{37}$$

where we integrate only over $\boldsymbol{r}_{\text{elec}}$. However, if we multiply equation (22) by $\Phi_{\text{elec},n}^{(0)}$ and integrate over $\boldsymbol{r}_{\text{elec}}$ we obtain

$$\left\langle \Phi_{\text{elec},n}^{(0)} \right| \left(\hat{H}_0^{(0)} - E_{\text{elec},n}^{(0)}\right) \left| \Phi_{\text{elec},n}^{(1)} \right\rangle = -\left\langle \Phi_{\text{elec},n}^{(0)} \right| \left(\hat{H}_0^{(1)} - E_{\text{elec},n}^{(1)}\right) \left| \Phi_{\text{elec},n}^{(0)} \right\rangle = 0, \tag{38}$$

from which we deduce, by comparing with equation (37), that we must have

$$E^{(1)}_{\text{rve},m} = E^{(1)}_{\text{elec},n}. \tag{39}$$

However $E^{(1)}_{\text{rve},m}$, like $E_{\text{rve},m}$, is a constant independent of $\boldsymbol{U}_{\text{N}}$, whereas $E^{(1)}_{\text{elec},n}$ is a linear function of $\boldsymbol{U}_{\text{N}}$; this implies that we must have

$$E^{(1)}_{\text{elec},n} = 0, \tag{40}$$

and the linear term in equation (20) vanishes. This means that $\partial E_{\text{elec},n}/\partial \boldsymbol{U}_{\text{N}} = 0$, and thus the reference molecular configuration $\boldsymbol{R}^0_{\text{N}}$ has to be the equilibrium molecular configuration for each electronic state n. Setting $E^{(1)}_{\text{elec},n}$ and $E^{(1)}_{\text{rve},m}$ equal to zero in equations (22) and (33), and comparing the two equations, we see that $\Phi^{(0)}_{\text{nuc}}(\boldsymbol{U}_{\text{N}})\Phi^{(1)}_{\text{elec},n}(\boldsymbol{r}_{\text{elec}};\boldsymbol{R}_{\text{N}})$ is a solution of the inhomogeneous equation (33). It is still a solution if we add any solution of the corresponding homogeneous equation so that, in general, the first-order rovibronic wavefunction term is

$$\Psi^{(1)}_{\text{rve},m}(\boldsymbol{r}_{\text{elec}};\boldsymbol{R}_{\text{N}}) = \Phi^{(0)}_{\text{nuc}}(\boldsymbol{U}_{\text{N}})\Phi^{(1)}_{\text{elec},n}(\boldsymbol{r}_{\text{elec}};\boldsymbol{R}_{\text{N}}) + \Phi^{(1)}_{\text{nuc}}(\boldsymbol{U}_{\text{N}})\Phi^{(0)}_{\text{elec},n}(\boldsymbol{r}_{\text{elec}}), \tag{41}$$

where $\Phi^{(1)}_{\text{nuc}}(\boldsymbol{U}_{\text{N}})$ is another, as yet, arbitrary function of $\boldsymbol{U}_{\text{N}}$.

Using equations (36), (40) and (41), the second-order rovibronic equation (34) can be rewritten

$$\begin{aligned}\left[\hat{H}^{(0)}_0 - E^{(0)}_{\text{rve},m}\right]\Psi^{(2)}_{\text{rve},m} &= -\hat{H}^{(1)}_0\left[\Phi^{(0)}_{\text{nuc}}\Phi^{(1)}_{\text{elec},n} + \Phi^{(1)}_{\text{nuc}}\Phi^{(0)}_{\text{elec},n}\right] \\ &\quad - \left[\hat{H}^{(2)}_0 + \hat{H}^{(0)}_{2\text{v}} - E^{(2)}_{\text{rve},m}\right]\Phi^{(0)}_{\text{nuc}}\Phi^{(0)}_{\text{elec},n}.\end{aligned} \tag{42}$$

If we subtract $\Phi^{(1)}_{\text{nuc}}$ times equation (22) and $\Phi^{(0)}_{\text{nuc}}$ times equation (23) from equation (42) we obtain

$$\begin{aligned}&\left[\hat{H}^{(0)}_0 - E^{(0)}_{\text{rve},m}\right]\left[\Psi^{(2)}_{\text{rve},m} - \Phi^{(1)}_{\text{nuc}}\Phi^{(1)}_{\text{elec},n} - \Phi^{(0)}_{\text{nuc}}\Phi^{(2)}_{\text{elec},n}\right] \\ &= -\left[\hat{H}^{(0)}_{2\text{v}} + E^{(2)}_{\text{elec},n} - E^{(2)}_{\text{rve},m}\right]\Phi^{(0)}_{\text{nuc}}\Phi^{(0)}_{\text{elec},n}.\end{aligned} \tag{43}$$

The condition for the solution of this equation is that

$$\left\langle \Phi^{(0)}_{\text{elec},n}\right|\left[\hat{H}^{(0)}_{2\text{v}} + E^{(2)}_{\text{elec},n} - E^{(2)}_{\text{rve},m}\right]\left|\Phi^{(0)}_{\text{nuc}}\Phi^{(0)}_{\text{elec},n}\right\rangle = 0. \tag{44}$$

The integration in equation (44) is over $\boldsymbol{r}_{\text{elec}}$ and $[\hat{H}^{(0)}_{2\text{v}} + E^{(2)}_{\text{elec},n} - E^{(2)}_{\text{rve},m}]\Phi^{(0)}_{\text{nuc}}$ is independent of $\boldsymbol{r}_{\text{elec}}$. Therefore, the condition reduces to

$$\left[\hat{H}^{(0)}_{2\text{v}} + E^{(2)}_{\text{elec},n} - E^{(2)}_{\text{rve},m}\right]\Phi^{(0)}_{\text{nuc}} = 0. \tag{45}$$

Equation (45) determines the vibrational wavefunction $\Phi^{(0)}_{\text{vib}} = \Phi^{(0)}_{\text{nuc}}$ and the second-order correction to the vibronic energy.

It is instructive to stop here in the development and to ponder the result. If we multiply equation (45) by κ^2 then $\kappa^2\hat{H}^{(0)}_{2\text{v}}$ is the vibrational kinetic energy of the nuclei, $\kappa^2 E^{(2)}_{\text{elec},n}$ acts as a potential function for the nuclear vibrational motion (note that it is a quadratic function of $\boldsymbol{U}_{\text{N}}$), and $\kappa^2 E^{(2)}_{\text{rve},m}$ is the second-order correction to the vibronic energy that is derived. At this level of approximation the potential function for the nuclear motion is a quadratic function of the nuclear displacements $\boldsymbol{U}_{\text{N}}$; as a result we will call this the *harmonic Born-Oppenheimer approximation.* In the harmonic BO approximation the vibronic wavefunction is determined only to zero order [from equation (36)] as the product of the vibrational wavefunction $\Phi^{(0)}_{\text{nuc}}(\boldsymbol{U}_{\text{N}})$ [from equation (45)] and the electronic wavefunction $\Phi_{\text{elec},n}(\boldsymbol{r}_{\text{elec}};\boldsymbol{R}^0_{\text{N}})$. From equations (35), (31) and (45) we see that the vibronic eigenvalue is the sum of $E^{(0)}_{\text{elec},n}$ (the electronic energy when the nuclei are at the equilibrium configuration) and $\kappa^2 E^{(2)}_{\text{rve},m}$ where this second-order energy correction is obtained from the harmonic vibration wave equation equation (45). Anharmonicity, and the rotational energies, are of too small an order of magnitude to be considered in the harmonic BO approximation.

The perturbation treatment can be carried to higher order and the fourth-order result is important; this leads to the *quartic BO approximation.* To derive this one uses the expression for $\hat{H}_{\text{rve}}$ given in equation (29). The algebra from BandO involved in the fourth-order development is also given in Appendix VII of Born and Huang (1954), and we summarize it. The rovibronic wavefunction obtained can be written as

$$\begin{aligned}\Psi_{\text{rve},m}(\boldsymbol{r}_{\text{elec}},\boldsymbol{U}_{\text{N}}) &= \left[\Phi^{(0)}_{\text{nuc}}(\boldsymbol{U}_{\text{N}}) + \kappa\Phi^{(1)}_{\text{nuc}}(\boldsymbol{U}_{\text{N}}) + \kappa^2\Phi^{(2)}_{\text{nuc}}(\boldsymbol{U}_{\text{N}})\right]\Phi_{\text{elec},n}(\boldsymbol{r}_{\text{elec}};\boldsymbol{R}_{\text{N}}) \\ &= \widetilde{\Phi}_{\text{rv}}(\boldsymbol{U}_{\text{N}})\Phi_{\text{elec},n}(\boldsymbol{r}_{\text{elec}};\boldsymbol{R}_{\text{N}}),\end{aligned} \tag{46}$$

and the rovibronic energy is obtained by solving an equation like equation (45), but in which the potential function is developed to quartic powers of the nuclear displacement coordinates $\boldsymbol{U}_{\text{N}}$ and the kinetic energy is the rotation–vibration kinetic energy operator $\kappa^2\hat{H}_{2\text{v}} + \kappa^3\hat{H}_{2\text{rv}} + \kappa^4\hat{H}_{2\text{r}}$. The quartic potential function is termed an *effective* potential function since it involves the effect of the electronic kinetic energy term $\kappa^4\hat{H}^{(0)}_1$ [see equation (24)] as well as the effect of the vibrational kinetic energy term acting on the electronic wavefunction.

The important result is that in the (fourth-order) quartic BO approximation the rovibronic wavefunction is still obtained as the product of a rotation–vibration wavefunction $\widetilde{\Phi}_{\text{rv}}$, that depends only on $\boldsymbol{U}_{\text{N}}$, and a single electronic wavefunction, obtained by solving the electronic wave equation at fixed nuclear coordinates. Since the electronic wavefunction is represented as a single function it does not involve coupling with other electronic states, and the quartic BO approximation maintains the adiabatic separation of the electronic from the rotation–vibration degrees of freedom.

Using perturbation theory for an isolated ground electronic state of a diatomic molecule Watson (1980) shows that the eigenvalues can be written as the standard *Dunham expansion* [Dunham (1932)]

$$E_{vJ} = hc\sum_{k=0}^{\infty}\sum_{l=0}^{\infty} Y_{kl}\left(v+\frac{1}{2}\right)^k \left[J(J+1)-\Lambda^2\right]^l, \tag{47}$$

where v is the vibrational quantum number, J is the rotational quantum number, and Λ is the projection of the electronic angular momentum on the molecular axis in units of $\hbar$. To second order the Dunham coefficients are given by

$$Y_{kl} = \mu_C^{(k+2l)/2} U_{kl}\left[1 + m_e\Delta^a_{kl}/M_a + m_e\Delta^b_{kl}/M_b + \text{O}\left(m_e^2/M_i^2\right)\right], \tag{48}$$

where m_e is the electron mass, M_a and M_b are the two atomic masses, and μ_C is the charge-modified reduced mass

$$\mu_C = M_a M_b/(M_a + M_b - Cm_e), \tag{49}$$

where C is the charge number of the molecule, so that the denominator in μ_C is the total mass. An important result of Watson (1980) is that the U_{kl} and the Δ^i_{kl} are nuclear mass independent constants. The correction represented by $\text{O}(m_e^2/M_i^2)$ in equation (48) is too small to be characterized experimentally for an isolated ground electronic state. However, if there are low lying excited electronic states such higher-order terms would become significant. Substituting equation (48) into equation (47) one obtains an equation that can be used in the simultaneous fitting of rotation-vibration energy levels for all isotopomers of a diatomic molecule. The result of such a fitting will be values for the constants U_{kl} and Δ^i_{kl}. Farrenq *et al.* (1991) make such a fitting for the isotopomers of CO.

Bunker and Moss (1980) show for a triatomic molecule how the perturbation theory contact transformation procedure can be applied to eliminate matrix elements between an isolated ground electronic state and the excited electronic states to give an effective rotation–vibration Hamiltonian. This effective rotation–vibration Hamiltonian contains nuclear mass dependent terms that correct for the adiabatic effects, and for the perturbing effect of all the excited electronic states.

1.2 The variation theory approach

In this approach we write the solution to the rovibronic wave equation as

$$\Psi_{\mathrm{rve},m}(\boldsymbol{r}_{\mathrm{elec}}, \boldsymbol{U}_{\mathrm{N}}) = \sum_{n'} \Phi^m_{\mathrm{rv},n'}(\boldsymbol{U}_{\mathrm{N}})\Phi_{\mathrm{elec},n'}(\boldsymbol{r}_{\mathrm{elec}}; \boldsymbol{R}_{\mathrm{N}}), \tag{50}$$

where $\Phi_{\mathrm{elec},n'}$ are a complete set of (orthonormal) solutions to the electronic wave equation (17), and the coefficient functions $\Phi^m_{\mathrm{rv},n'}(\boldsymbol{U}_{\mathrm{N}})$ are to be determined. Substituting this expression into the rovibronic wave equation (8), multiplying on the left by $\Phi^*_{\mathrm{elec},n}$, and integrating over the electronic coordinates $\boldsymbol{r}_{\mathrm{elec}}$, we obtain the following set of coupled equations for the $\Phi^m_{\mathrm{rv},n'}(\boldsymbol{U}_{\mathrm{N}})$:

$$\left[\hat{T}_{\mathrm{N}} + E_{\mathrm{elec},n}(\boldsymbol{R}_{\mathrm{N}}) - E_{\mathrm{rve},m}\right]\Phi^m_{\mathrm{rv},n}(\boldsymbol{R}_{\mathrm{N}}) + \sum_{n'} C_{nn'}\Phi^m_{\mathrm{rv},n'}(\boldsymbol{R}_{\mathrm{N}}) = 0, \tag{51}$$

where

$$C_{nn'} = \left\langle \Phi_{\mathrm{elec},n}\middle|\hat{T}'_{\mathrm{e}} + \hat{T}_{\mathrm{N}}\middle|\Phi_{\mathrm{elec},n'}\right\rangle. \tag{52}$$

Different levels of approximation are achieved depending on how the $C_{nn'}$ are dealt with.

If the $C_{nn'}$ are ignored entirely we have what is traditionally called 'the' BO approximation, which we can summarize by writing:

$$\Psi_{\mathrm{rve},m}(\boldsymbol{r}_{\mathrm{elec}}, \boldsymbol{U}_{\mathrm{N}}) = \Phi^m_{\mathrm{rv},n}(\boldsymbol{U}_{\mathrm{N}})\Phi_{\mathrm{elec},n}(\boldsymbol{r}_{\mathrm{elec}}; \boldsymbol{R}_{\mathrm{N}}), \tag{53}$$

where $\Phi_{\mathrm{elec},n}(\boldsymbol{r}_{\mathrm{elec}}; \boldsymbol{R}_{\mathrm{N}})$ is the solution of the electronic wave equation:

$$\left[-\frac{\hbar^2}{2m_{\mathrm{e}}}\sum_{i=N+1}^{l} \nabla_i^2 + V(\boldsymbol{R}_{\mathrm{N}}, \boldsymbol{r}_{\mathrm{elec}}) - E_{\mathrm{elec},n}(\boldsymbol{R}_{\mathrm{N}})\right]\Phi_{\mathrm{elec},n}(\boldsymbol{r}_{\mathrm{elec}}; \boldsymbol{R}_{\mathrm{N}}) = 0, \tag{54}$$

and $\Phi^m_{\mathrm{rv},n}(\boldsymbol{U}_{\mathrm{N}})$ is the solution of the rotation–vibration wave equation:

$$\left[-\frac{\hbar^2}{2}\sum_{i=2}^{N}\frac{\nabla_i^2}{m_i} + \frac{\hbar^2}{2M_{\mathrm{N}}}\sum_{i,j=2}^{N}\nabla_i\cdot\nabla_j + V_{\mathrm{BO},n}(\boldsymbol{R}_{\mathrm{N}}) - E_{\mathrm{rve},m}\right]\Phi^m_{\mathrm{rv},n}(\boldsymbol{R}_{\mathrm{N}}) = 0, \tag{55}$$

with the Born–Oppenheimer potential energy function $V_{\mathrm{BO},n}(\boldsymbol{R}_{\mathrm{N}})$ given by

$$V_{\mathrm{BO},n}(\boldsymbol{R}_{\mathrm{N}}) = E_{\mathrm{elec},n}(\boldsymbol{R}_{\mathrm{N}}), \tag{56}$$

which is isotopically independent.

If we only ignore off-diagonal elements of $C_{nn'}$ we have 'the adiabatic approximation'. In this approximation the rotation–vibration function is obtained from

$$\left[-\frac{\hbar^2}{2}\sum_{i=2}^{N}\frac{\nabla_i^2}{m_i}+\frac{\hbar^2}{2M_{\rm N}}\sum_{i,j=2}^{N}\nabla_i\cdot\nabla_j+V_{{\rm ad},n}(\boldsymbol{R}_{\rm N})-E_{{\rm rve},m}\right]\Phi^m_{{\rm rv},n}(\boldsymbol{R}_{\rm N})=0, \tag{57}$$

with the adiabatic potential energy function $V_{\rm ad}(\boldsymbol{R}_{\rm N})$ given by

$$\begin{aligned} V_{{\rm ad},n}(\boldsymbol{R}_{\rm N}) &= E_{{\rm elec},n}(\boldsymbol{R}_{\rm N})+C_{nn} \\ &= E_{{\rm elec},n}(\boldsymbol{R}_{\rm N})+\langle\Phi_{{\rm elec},n}|\hat{T}'_{\rm e}+\hat{T}_{\rm N}|\Phi_{{\rm elec},n}\rangle. \end{aligned} \tag{58}$$

The adiabatic potential energy function depends on isotope because of the presence of the nuclear masses in $\hat{T}'_{\rm e}+\hat{T}_{\rm N}$.

The full nonadiabatic calculation would involve making no approximations and including all $C_{nn'}$. This would give rise to an infinite number of coupled equations and each eigenfunction would be the infinite sum of product functions given in equation (50). However, in practice, this complete nonadiabatic calculation is never attempted. Calculations involve either isolated electronic states or situations in which two or three electronic states are close in energy and for which their interaction has to be treated explicitly.

Several of the chapters in this book are concerned with the study of the rotation–vibration energies in an isolated ground electronic state, and the BO approximation is assumed. This means that the rotation–vibration energies are obtained by solving

$$\left[-\frac{\hbar^2}{2}\sum_{i=2}^{N}\frac{\nabla_i^2}{m_i}+\frac{\hbar^2}{2M_{\rm N}}\sum_{i,j=2}^{N}\nabla_i\cdot\nabla_j+V_{\rm BO}(\boldsymbol{R}_{\rm N})-E_{{\rm rv},m}\right]\Phi^m_{\rm rv}(\boldsymbol{R}_{\rm N})=0, \tag{59}$$

where for the ground ($n=0$) electronic state we write

$$V_{\rm BO}(\boldsymbol{R}_{\rm N})=E_{{\rm elec},0}(\boldsymbol{R}_{\rm N})-E_{{\rm elec},0}(\boldsymbol{R}_{\rm N}{}^0) \tag{60}$$

with $\boldsymbol{R}^0_{\rm N}$ being the ground electronic state equilibrium nuclear configuration. However, the chapter by Yarkony and the chapter by Buenker *et al.* are particularly concerned with the situation when two or more electronic potential energy curves $E_{{\rm elec},n}(\boldsymbol{R}_{\rm N})$ come very close for certain internuclear separations in a diatomic molecule. In this case nonadiabatic effects due to the coupling of such close lying states are explicitly considered. Also the chapter by Jensen, Osmann and Bunker on the Renner effect, and the chapter by Barckholtz and Miller on the Jahn–Teller effect, consider situations in polyatomic molecules when two potential energy surfaces are degenerate at a particularly symmetrical nuclear configuration. Nonadiabatic effects have again to be explicitly introduced.

1.3 References

Born, M., 1951, *Nachr. Akad. Wiss. Goettingen* **no. 6**, 1–3.

Born, M., and Huang, K., 1954, *Dynamical Theory of Crystal Lattices*, Oxford Univ. Press, New York.

Born, M., and Heisenberg, W., 1924, *Ann. Phys. (Leipzig)* **74**, 1–31.

Born, M., and Oppenheimer, J. R., 1927, *Ann. Phys. (Leipzig)* **84**, 457–484.

Bunker, P. R., and Jensen, P., 1998, *Molecular Symmetry and Spectroscopy*, 2nd Edition, NRC Research Press, Ottawa.

Bunker, P. R., and Moss, R. E., 1980, *J. Mol. Spectrosc.* **80**, 217–228.

Dunham, J. L., 1932, *Phys. Rev.* **41**, 721–731.

Farrenq, R., Guelachvili, G., Sauval, A. J., Grevesse, N., and Farmer, C. B., 1991, *J. Mol. Spectrosc.* **149**, 375–390.

Mead, C. A., 1988, in *Mathematical Frontiers of Computational Chemical Physics*, D. G. Truhlar, Ed., Springer-Verlag, New York.

Watson, J. K. G., 1980, *J. Mol. Spectrosc.* **80**, 411–421.

PART 2

ELECTRONIC STATES

2 *AB INITIO* DETERMINATION OF ACCURATE GROUND ELECTRONIC STATE POTENTIAL ENERGY HYPERSURFACES FOR SMALL MOLECULES

Attila G. Császár

Eötvös University, Budapest, Hungary

and

Wesley D. Allen, Yukio Yamaguchi, and Henry F. Schaefer III

University of Georgia, Athens GA, USA

Computational Molecular Spectroscopy. Edited by Per Jensen and Philip R. Bunker
© 2000 John Wiley & Sons Ltd

2.1 Introduction

The concept of potential energy (hyper)surfaces (PES) is fundamental to the understanding of most modern branches of chemistry, including, most importantly, almost the whole of spectroscopy and kinetics [Murrell *et al.* (1984), Hirst (1985), Sathyamurthy (1985), Mezey (1987), Truhlar, Steckler, and Gordon (1987), Schatz (1989), Truhlar (1990), and Western (1995)]. Nevertheless, PESs exist only within the so-called Born–Oppenheimer (BO) separation of electronic and nuclear motion [Born and Oppenheimer (1927), Born (1951), and Born and Huang (1954); see also the chapter by Bunker and Jensen in this volume], although adiabatic corrections (see Section 2.3) to the BO–PES relax this strict separation, defining what one might call an adiabatic PES (APES). The BO separation is based on the recognition that in molecular systems motion of the nuclei is much slower than that of the electrons; in most cases, adiabatic separation of these two types of motion is well justified. Under the BO approximation, potential energy hypersurfaces for motion of the relatively massive nuclei arise, describing the variation of the total electronic energy of a chemical system as a function of the position coordinates of the constituent nuclei. Usually, attention is focused on cases where a single BO–PES is sufficiently uncoupled from other surfaces (electronic states) that their interaction may be ignored. Most fundamental among these surfaces are ground electronic state potential energy hypersurfaces, the topic of this review. Although we are not concerned here with nuclear motion, it has to be recognized that in some cases, such as the Jahn–Teller and the Renner effects, motion of the nuclei can affect properties considered to be strictly electronic. In this review we are concerned neither with the important case where several electronic states need to be described equally well (e.g. with processes driven by curve crossings) nor with the evaluation of coupling matrix elements, such as those arising from nonadiabatic BO interactions and spin–orbit couplings.

The spectra of (isolated) molecules are a predominant source of chemical knowledge. Study of non-reactive and reactive processes on a molecular level requires modeling or knowledge of the relevant potential energy hypersurfaces for the system considered. Therefore, much of contemporary experimental physical chemistry, through spectroscopic, scattering, and kinetic studies, is directed toward the elucidation of salient features of potential energy surfaces. There is, for example, clearly an intimate relationship, given by solutions of the Schrödinger equation for nuclear motion, between the PES of a molecule and its ro-vibrational spectra [Wilson, Decius, and Cross (1955), Searles and Nagy-Felsobuki (1993), and Western (1995)]. In the inverse problem [Rydberg (1931), Klein (1932), Rees (1947), Kosman and Hinze (1975), Zhang and Light (1995), Wu and Zhang (1996), and Ho and Rabitz (1996)], one can obtain details of the PES from an analysis of well-resolved rovibrational spectra or perhaps from scattering experiments. While both approaches have been employed extensively to probe intra- and intermolecular potential energy hypersurfaces, spectroscopic techniques possess several advantages over scattering measurements: (a) they can provide results of much higher accuracy; (b) there is no need to average over experimental conditions; and (c) bound-state properties can be calculated much more accurately and easily than scattering cross sections. Generally, experiments, through well-defined modeling approaches, yield parameters in more or less local representations of potential surfaces. The parts of the PES sampled in the energy regime of interest determine the type of parametrization of the PES that is most appropriate for the problem at hand.

Much of modern quantum chemistry is also aimed at understanding given portions or the whole of potential energy hypersurfaces of molecular species or reactive (scattering) systems. The availability of analytic gradients and higher derivative methods [Garrett and Mills (1968), Pulay (1969), Jørgensen and Simons (1986), and Yamaguchi *et al.* (1994)] in standard quantum chemical programs has substantially changed the utilization of quantum chemistry for exploration of PESs: determination of molecular geometries [Helgaker *et al.* (1997a)] and (anharmonic) force fields [Császár (1998)] at various stationary points on surfaces is now commonplace. As more and more complicated systems are being investigated, determination of critical configurations (e.g. reactants, products, and transition structures for chemical reactions) remains a key priority. However, the energy and energy derivative values at a small number of stationary points may not constitute a sufficient description of the PES.

Similarly to experiments, theoretical methods yield the potential values only for some discrete set of configurations. *Ab initio* construction of a hypersurface involves (a) choice of a physically correct and robust electron correlation methodology; (b) application of a highly flexible and still compact basis set; and (c) design of a suitable geometrical grid for the calculations. Independent of the choices made for the basis set and the electron correlation technique, the accuracy of the computational results can vary significantly with varying internuclear separation. It is noted, in this respect, that two kinds of general error criteria are commonplace in describing the accuracy of theoretical calculations of energies and energy differences: chemical and spectroscopic accuracy. These terms are not defined uniquely and are obviously not synonymous with experimental spectroscopic accuracies. There is, however, general agreement that chemical and spectroscopic accuracies mean $\approx 1\,\text{kcal mol}^{-1}$ ($\approx 1\,\text{m}E_\text{h}$) and $\approx 1\,\text{cm}^{-1}$ ($\approx 1\,\mu E_\text{h}$), respectively. The term hyperfine accuracy could perhaps be coined for the next benchmark of $\approx 1\,\text{n}E_\text{h}$ accuracy. Obviously, it is much easier to achieve chemical (spectroscopic) accuracy for relative energies than for absolute energies. Accordingly, in theoretical vibrational spectroscopy, spectroscopic accuracy means prediction of a transition energy with an error less than $1\,\text{cm}^{-1}$, usually by means of significant error cancellations in absolute energies. Spectroscopic accuracy for absolute energies has been achieved only for a few prototypic systems. For example, for He accurate wave function and energy calculations have a long history [Hylleraas (1929), and Hylleraas (1964)]. The best variational result is $E = -2.903\,724\,377\,034\,119\,597\,E_\text{h}$ [Drake and Yan (1994)]. H_2^+ and its isotopomers have been a testing bed for extremely accurate quantum chemical computations [Bukowski *et al.* (1992), and Leach and Moss (1995)]. Chemical accuracy was realized some time ago for H_2 [James and Coolidge (1933), Kołos and Wolniewicz (1964), Wolniewicz (1993), and Kołos and Rychlewski (1993)]. The best variational energy result for $X\,^1\Sigma_g^+\,H_2$ at 0.74143 Å is $E = -1.174\,475\,931\,E_\text{h}$ [Cencek and Kutzelnigg (1996)], of better than spectroscopic accuracy. The best variational results available today for the rovibrational energy levels of H_2 are of hyperfine accuracy [Wolniewicz (1993), and Cencek and Kutzelnigg (1996)], and similar accuracy has recently been achieved for H_3^+. Excellent reviews of laboratory observations and molecular structure computations for H_3^+ have been provided by Oka (1992) and Dalgarno (1994). Although H_3^+, one of the simplest polyatomic molecules, was observed in the laboratory quite a long time ago [Thompson (1912)], its rotation–vibration spectrum has attracted considerable interest only in the last two decades [Oka (1980), Anderson (1992), Röhse *et al.* (1994), Tennyson and Polyansky (1994), Dinelli *et al.* (1995), and Prosmiti, Polyansky, and Tennyson (1997)]. The simplicity of the electronic structure of H_3^+ has made it an almost ideal target for *ab initio* calculations [Anderson (1992)]. A very accurate PES available today is due to Röhse *et al.* (1994), who have published a surface which was constructed using explicitly correlated CISD-R12 wave functions with a nearly saturated basis set. The estimated accuracy of this surface is about $1\,\text{cm}^{-1}$ (only a few μE_h) over its entire range. Even this accuracy has recently been surpassed by Cencek *et al.* (1998), who employed a Gaussian-type geminal (GTG) approach with a large expansion length, and obtained a PES with an absolute error of about $0.02\,\text{cm}^{-1}$ ($0.1\,\mu E_\text{h}$). This accuracy had never before been achieved in quantum chemical calculations for the entire PES of a triatomic molecular system. These *ab initio* computations gave total energies of the equilibrium structure which are lower than the quantum Monte Carlo results of Anderson (1987). Comparisons made with vibrational band origins inferred from experiment suggested that in nuclear motion calculations [Röhse *et al.* (1994), and Jaquet *et al.* (1998)] the BO approximation itself was the major source of the remaining error. Similar conclusions concerning other high accuracy PESs have been made previously [Lie and Frye (1992), and Tennyson and Polyansky (1994)]. The best results available for two-electron systems are nicely reviewed by Klopper (1998) and Sanders (1998). In summary, to achieve spectroscopic accuracy, the basis set limit is to be approached at a very high level of electron correlation treatment and the relativistic, adiabatic and non-adiabatic corrections have to be taken care of. Recently, impressive levels of precision have also been achieved for the three-electron lithium [Yan and Drake (1998) reported $-7.478\,060\,324\,E_h$] and the four-electron beryllium [Komasa, Cencek, and Rychlewski (1995) reported $-14.667\,355\,022\,E_h$] atoms. For molecules comprising elements heavier than H and He, reaching even chemical accuracy for relative energies is still a considerable challenge [see, e.g., studies on the dissociation energy of N_2 by Bauschlicher and Partridge (1994), Császár and Allen (1996), Peterson *et al.* (1997), and Klopper and Helgaker (1998)]. Nevertheless, CCSD(T)-R12 and similar calculations (see Section 2.2.2.5) with an extended and well-designed basis can yield absolute energies with almost chemical accuracy around equilibrium structures [see, e.g., Müller Kutzelnigg, and Noga (1997), and Helgaker *et al.* (1997b)]. *Ab initio* determination of barrier heights [Lee (1997), Császár, Allen, and Schaefer (1998), and Tarczay *et al.* (1999)] of prototypical molecules, closer in spirit to the relative accuracy of calculations covering large portions of PESs, has a current accuracy of about 0.1–0.3 $\mathrm{kcal\,mol^{-1}}$, a considerable achievement but still far from true spectroscopic accuracy.

There are basically three distinct families of PESs most researchers have been studying: hypersurfaces for (a) (ro)vibrational spectra [Searles and Nagy-Felsobuki (1993), Western (1995), and Bürger and Thiel (1997)]; (b) chemical reactions (of usually unsolvated species) [Truhlar, Steckler, and Gordon (1987)]; and (c) intermolecular interactions representing systems of non-covalent chemistry [Le Roy and van Kranendonk (1974), and Truhlar (1990)]. Although these surfaces, obviously, have a lot in common, there are many important and noteworthy differences in their construction due mostly to the different types of experimental information these surfaces are aimed to interpret. For chemical systems, there are four distinct types of (global) PESs available in the literature: empirical, semiempirical, semitheoretical, and theoretical. Empirical hypersurfaces are constructed by fitting a given functional form to available experimental data without the guidance of explicit theoretical data [e.g. Schnupf, Bowman, and Heaven (1992)]. Because of the usually limited range of available experimental data, the accuracy of empirical surfaces may often be questionable. Notable exceptions, however, do exist, including the highly accurate empirical potential, usually denoted as the PJT2, of water [Polyansky, Jensen, and Tennyson (1996)], and a potential for CH_2 [Jensen and Bunker (1988)]. Theoretical PESs are constructed using quantum chemical (preferably *ab initio*) data alone. Because of well recognized inadequacies of most theoretical models (see later) it is still not possible to achieve spectroscopic accuracy with these surfaces except for the smallest systems [Röhse *et al.* (1994)]. The most effective way to obtain high-quality PESs is a combination of theoretical and empirical approaches. The distinction between the resulting types of PESs, which we might call semiempirical and semitheoretical, is not strict. If the final PES is based principally on high-quality *ab initio* data of at least chemical accuracy, adjusted only slightly to reproduce relevant experimental findings, the PES may be called semitheoretical. This is perhaps the best way to obtain a quantitatively accurate PES. This strategy of building a PES has been employed for rovibrational studies [e.g. the water potential of Partridge and Schwenke (1997), which has been further refined by Polyansky *et al.* (1997a) and by Császár *et al.* (1998)], as well as for intermolecular interactions [$(HF)_2$, Klopper, Quack, and Suhm (1998)]. In contrast, a semiempirical PES uses limited results from theoretical studies of average accuracy to guide the empirical construction of the PES.

Since vibrational spectroscopy, including vibrational overtone spectroscopy [Kauppi and Halonen (1995), Lukka, Kauppi, and Halonen (1995), and Halonen (1997)], usually samples a rather limited range of the PES, for the calculation of rovibrational spectra and the related dynamics the PES is needed mostly in the vicinity of a minimum. Therefore, techniques based on power series expansions around a single stationary point can be highly useful. Furthermore, during construction of PESs designed to model rovibrational spectra and dynamics, it is (a) usually not necessary to consider dissociation products or other molecular fragments/isomers; and (b) often an excellent approximation to use a single PES. For example, the PES of the ground electronic state of water has been the testing ground for the many different approaches that can be employed for the calculation of (ro)vibrational states. Consequently, a large number of studies on the local and global PES of water have been published [Hoy, Mills, and Strey (1972), Carter and Handy (1987), Halonen and Carrington (1988), Jensen (1989), Császár and Mills (1997), Polyansky, Jensen, and Tennyson (1996), Partridge and Schwenke (1997), and Császár *et al.* (1998)]. Recently, as a result of drastically increased spectroscopic capabilities for detecting higher-lying bending states [Polyansky *et al.* (1997a)], to the challenges provided by the extremely high density of rovibrational states observed in the recently recorded sunspot spectrum of water [Wallace *et al.* (1995), Oka (1997), and Polyansky *et al.* (1997b)], and to the special role the water molecule plays in the physics and chemistry of the atmospheres of planets and stars, of interstellar medium, and of combustion systems, accurate determination of the ground electronic state PES of water has received renewed interest [Zobov *et al.* (1996), Polyansky, Jensen, and Tennyson (1996), and Császár *et al.* (1998)]. These studies have demonstrated [Zobov *et al.* (1996), and Császár *et al.* (1998)] that even small correction terms accounting for physical effects usually neglected during construction of global PESs, such as the diagonal Born–Oppenheimer correction (DBOC) and relativistic effects, may produce changes on the order of a few cm^{-1} for the rovibrational eigenstates.

In the course of a chemical reaction bonds are broken and formed. The global requirements for these potentials are considerably more stringent, use of a single PES is sometimes less satisfactory [e.g. the $O(^1D) + H_2 \rightarrow OH + H$ reaction involves five surfaces, see Walch and Harding (1988)], and regions have to be sampled where theoretical methods perform quite differently. Nevertheless, for some simplified techniques used to study chemical reactions, it is enough to have a proper representation of the surface close to an intrinsic reaction path [IRP, Fukui (1981)], a minimum-energy reaction path [MEP, Fukui (1970), and Basilevsky (1977)], or some other path [Müller (1980)]. Characterization of an IRP requires both the determination of a saddle point and the calculation of the steepest-descent paths ensuing from it. The construction of an MEP may be hampered by difficulties in the selection of a suitable reaction coordinate. Several other difficulties in computing reaction paths on multidimensional potential energy hypersurfaces have been reviewed [Müller (1980)]. The first quantitative success for a reaction surface occurred for the simplest neutral system, the $H + H_2$ reaction. Reliable energies for 156 nuclear arrangements of the three hydrogen atoms were reported by Siegbahn and Liu (1978). With the energy of the transition state taken to be zero, the *relative* energies were thought to be within 0.1 $kcal\,mol^{-1}$ of the exact BO (clamped nucleus) results. In 1978, this high accuracy was only possible because of special techniques expressly formulated for three-electron systems. Thus, Siegbahn and Liu were able to determine full configuration interaction (FCI) wave functions for H_3 with a relatively large basis set, namely four s functions, three sets of p functions, and one set of d functions on each hydrogen atom. From the quantum mechanical results the authors were able to estimate that the exact position of the saddle point corresponds to $r(\text{H–H}) = 0.9298 \pm 0.0005$ Å and the exact classical barrier height is $9.68 \pm 0.12\,kcal\,mol^{-1}$. Since the pioneering 1978 study, the H_3 system has continued to be of intense theoretical and experimental interest. The most reliable results to date for the H_3 potential surface appear to come from the quantum Monte Carlo (QMC) method. The QMC results of Diedrich and Anderson (1994) give the same H–H saddle point distance as the 1978 study and a classical barrier height of $9.61 \pm 0.01\,kcal\,mol^{-1}$. The absolute energy error in these most reliable H_3 results is said to be only $3\,cm^{-1}$.

Intermolecular surfaces, a prime target of noncovalent chemistry [Finney (1996)], are always multidimensional. In these cases the dimensionality of the problem is one of the greatest obstacles in obtaining quantitative accuracy for PES and consequent property predictions. Therefore, for all but the smallest molecular complexes, reduced dimensionality models are constructed keeping the monomers more or less rigid [e.g. van der Avoird and Wormer (1984), and Cohen and Saykally (1993)]. Prior to about 1985 there existed no quantitatively correct intermolecular PESs. Recently, model potentials of ever-increasing accuracy have been constructed for several weakly bound systems, most notably $(HF)_2$ [Kofranek, Lischka, and Karpfen (1988), and Klopper, Quack, and Suhm (1998)]. The theoretical investigations on $(HF)_2$ have aimed at a quantitative characterization of this prototype of simple hydrogen bonding, resulting in theoretical studies on its potential energy and property hypersurfaces [Kofranek, Lischka, and Karpfen (1988), Truhlar (1990), and Klopper, Quack, and Suhm (1998)] and the related dynamics [Lester (1996)]. These theoretical studies are complemented by very careful and detailed spectroscopic studies. Repeated analytical modeling, quantum treatments of the nuclear dynamics, and empirical refinement of the *ab initio* PES of Kofranek, Lischka, and Karpfen (1988) resulted in the widely employed SQSBDE potential of $(HF)_2$ [Quack and Suhm (1991)]. This semiempirical potential has been further improved by inclusion of information from CCSD(T)-R12 and MP2-R12 calculations employing near-complete basis sets. The resulting semitheoretical PES, labeled as SC-2.9 [Klopper, Quack, and Suhm (1996)] and SO-3 [Klopper, Quack, and Suhm (1998)], reproduces most available experimental results with unprecedented accuracy.

In summary, the ever-increasing accuracy of experimental spectroscopic and scattering investigations, the power of semiclassical [Child (1996)] and quantum dynamical methods (see the chapters by Child and Yarkony in this volume), and of electronic structure techniques [Yarkony (1995), and Schleyer *et al.* (1998)] have all been combined in recent years to focus attention on more and more details of potential energy and property hypersurfaces and on related dynamical questions of larger and larger systems [see, e.g., Bowman, Bittman, and Harding (1986), Bačić and Light (1987), Klepeis *et al.* (1993), Carter, Pinnavaia, and Handy (1995), Atchity and Ruedenberg (1997), Leforestier *et al.* (1997), Qiu and Bačić (1997), and King *et al.* (1998)]. Given all the obstacles for *ab initio* determinations of PESs, described in considerable detail below, the development of methods that use the output of quantum chemical techniques to construct 'global' potential energy hypersurfaces provides still one of the most important contemporary challenges for chemical theory. Nevertheless, the present review focuses on issues related to the pointwise *ab initio* determination of PESs, while techniques employed for the utilization of such information in global parametrizations will be reviewed by other contributors to this book.

2.2 Theoretical armamentarium–A hierarchy of methods

The last three decades have witnessed remarkable advances in the theory and application of computational quantum chemistry. Several excellent introductory and advanced textbooks have been written on the subject [Schaefer (1977), Hehre *et al.* (1986), Náray-Szabó, Surján, and Ángyán (1987), Szabo and Ostlund (1989), Levine (1991), Roos (1992), McWeeny (1992), Yarkony (1995), and Schleyer *et al.* (1998)]. For this reason and because of space limitations, a detailed treatment of modern methods of molecular electronic structure theory is not pursued here; the interested reader is referred to the above and similar volumes. In the following, a necessarily rudimentary and non-technical description of the theoretical methods applied for studies on potential energy hypersurfaces is given to highlight the most important issues which face readers interested in *ab initio* determination of potential energy and property hypersurfaces.

For nearly all systems of chemical interest the exact solution to the (non-relativistic, time-independent) electronic Schrödinger equation cannot be obtained, and thus one must introduce

Series I	Series II	Series III	STO–3G 4–31G 6–31 G* 6–311 + G(2d1f, 2p) … Complete basis DZ TZP QZ2P … Complete basis cc-pVDZ cc-pVTZ cc-pVQZ aug-cc-pCVQZ … Complete basis
HF	HF	HF	
MP2	CISD	CCSD	↓
MP3	CISDT	CCSDT	↓
MP4	CISDTQ	CCSDTQ	↓
MP5	CISDTQP	…	↓
MP6	…		↓
.	.	.	↓
….	.	.	↓
FCI	FCI	FCI	⟶ Exact solution

Figure 2.1. Computational matrix of *ab initio* electronic structure theory indicating quality of one-particle space (basis set) versus quality of the n-particle space (computational method) (for the abbreviations employed see the discussions of this chapter).

approximations. Naturally, it would be highly advantageous to have both upper and lower bounds on calculated properties, such as energies. However, as Bunge (1980) observes for atoms, 'today's rigorous error bounds are too large to be of much use'. Therefore, one is left with a pragmatic approach, where the accuracy of a given approach is judged by comparing the calculated result with experiment.

Basic to the understanding of the hierarchy of methods of electronic structure theory is the computational matrix depicted in Figure 2.1. It clearly shows that there are two fundamental approximations: truncation of the one- and n-particle spaces. Extension of both the one-particle space (atomic basis sets) and n-particle space (many-electron wave function) is needed to achieve results close to the nonrelativistic limit ('exact solution'). The usefulness, quality, and reliability of any given approximation must be assessed from a large number of careful studies. It is highly advantageous if the error introduced by the different approximations (a) is controllable; (b) can be cancelled in some systematic way; (c) is comparable for similar systems and physical situations; and (d) is balanced over a large region of geometrical space.

The electron correlation energy ($\varepsilon_{\text{corr}}$) is defined [Löwdin (1959)] as the energy difference

$$\varepsilon_{\text{corr}} = E_{\text{exact}} - E_{\text{HF}}, \tag{1}$$

where E_{HF} is the electronic energy obtained from a Hartree–Fock (HF) calculation. Obviously, one needs to make a distinction between full correlation energy, defined with respect to the HF and FCI one-particle limits, and the actual correlation energy, defined in a given finite basis set as the FCI − HF energy difference. At minima, the HF method typically recovers some 99 % of the electronic energy, but its performance can deteriorate rapidly when one moves away from equilibrium. Therefore, $\varepsilon_{\text{corr}}$ may vary greatly over a PES. As guaranteed by the variational theorem, the correlation energy is always negative. Fermi correlation arises from the Pauli antisymmetry principle; it is not part of electron correlation as defined above; and it is taken into account already at the HF level. Dynamic correlation [DC; Mok, Neumann, and Handy (1996)] serves to keep electrons apart instantaneously. It is a cumulative effect built up from a myriad of small contributions and usually forms the largest part of $\varepsilon_{\text{corr}}$. It originates primarily from the failure of most reference wave functions to describe the electron–electron cusp, and therefore it is a short-range effect. Non-dynamic correlation (NDC) arises when an electronic state is not adequately described by a single Slater determinant (or configuration state function) of molecular orbitals, often because of near-degeneracy effects. NDC is a long-range effect. Several other conceptual divisions of electron correlation, of less relevance for our treatment, have also been introduced: in–out, left–right, angular, intraatomic and interatomic [Wahl and

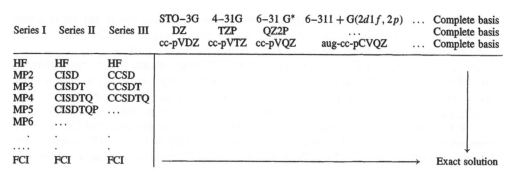

Das (1977), and Mok, Neumann, and Handy (1996)]. The correlation energy per electron pair goes up approximately linearly with the atomic number [Linderberg and Shull (1960)]; for the ground electronic state of He-like ions and H_2, it is approximately $-0.04\,E_h$.

There are two computational strategies possible if NDC is significant. In the single-reference route, NDC is accounted for along with DC merely by sufficiently increasing the highest order of electronic excitations included in coupled-cluster (CC) approaches, perturbation theories (PT), or configuration interaction (CI) techniques. In the multiconfiguration/multireference (MC/MR) route, NDC is accounted for at the start in the zeroth-order wave function, and DC is added subsequently via MR configuration interaction (or less commonly MR CC or PT) schemes. For calculation of truly global PESs, MC/MR approaches are usually more appropriate, provided they are computationally feasible.

Many (global) PESs cover vastly different atomic (bonding) arrangements. Approximate theoretical techniques used to study such PESs should provide comparable accuracy for all subsystems (fragmentation products) investigated. Such considerations lead to the concepts of size consistency [Pople, Binkley, and Seeger (1976), and Crawford and Schaefer (1999)] and size extensivity [Bartlett and Purvis (1978)]. In accord with traditional thermodynamic concepts, a size-extensive method scales correctly with the number of particles in the system, as in the pedagogical example of N non-interacting H_2 molecules [see, for example, McWeeny (1992)], or for an electron gas [Parr and Yang (1989)]. A size-consistent method is one which more specifically leads during molecular fragmentation to a wave function which is multiplicatively separable and an energy which is additively separable. There is some disagreement over the precise definition of size-consistency, and many use the term 'size-extensive' and 'size-consistent' synonymously, despite the fact that in some cases the former property is exhibited but not the latter. Neither property alone ensures that a molecule and its dissociation products are described with precisely the same accuracy. While variational calculations (such as CI) are size-consistent only if an exponentially growing direct-product space of the fragments is employed for their construction, exponential wave functions are natural *ansätze* for multiplicatively separable approximate wave functions. Use of size-extensive techniques is mandatory during determination of PESs involving a change in the number of interacting electrons, and on systems containing different numbers of electrons.

The exact basis-set correlation energy is obtained by full configuration interaction (FCI), by CC calculations with the full cluster operator $\hat{T}$, or by PT carried out to infinite order. Almost all methods of computational quantum chemistry may be viewed as certain approximations to these techniques.

2.2.1 PHYSICALLY-CORRECT REFERENCE WAVE FUNCTIONS

In computations employing methods of molecular electronic structure theory for determination of PESs, it is of considerable importance to determine whether single-configuration and single-reference-based methods, which are applicable with relative ease even for large molecular systems and have a 'black-box' nature, are of practical use, or whether conceptually more involved MC and MR methods need to be applied. Different tests have been developed which allow estimation of non-dynamical electron correlation, in other words the multireference character of a given electronic state at a given geometry. One test involves C_0, the reference CI coefficient in a CI wave function. This diagnostic is, however, of rather limited utility, since C_0 in a limited CI wave function is strongly biased toward the reference function when the molecular orbitals are optimized for this configuration. Some correlation treatments, including methods based on CC theory, are less dependent upon the quality of the reference molecular orbitals. Therefore, one of the simplest and most dependable diagnostics in testing multireference character is the T_1 diagnostic of CC theory [Lee *et al.* (1989), and Lee and Taylor (1989)], based on the norm of the T_1 amplitudes in the CC wave function (see Section 2.2.2.2), as built from the HF reference function. Calculation of T_1 values for a large number of molecules has suggested that closed-shell electronic states with $T_1 < 0.02$ can be adequately described by single-reference electron correlation methods. Analogous diagnostics have also been proposed for open-shell systems [Jayatilaka and Lee (1993)]. Some objections to reliance on such diagnostics have been raised [Watts, Urban, and Bartlett (1995)].

2.2.1.1 Hartree–Fock methods

The simplest standard model that *ab initio* electronic structure theory offers is the Hartree–Fock (HF), mean-field theory [Hartree (1928), Slater (1929, 1930), and Fock (1930)], often called, somewhat misleadingly, self-consistent-field [SCF, Hartree (1928)] theory. Most HF methods are adequately treated in standard textbooks [e.g. Hehre *et al.* (1986), and Szabo and Ostlund (1989)]. An appealing feature of HF theory is that this simple model retains the notion of molecular orbitals (MO) as delocalized one-electron functions describing movement of an electron in an average (effective) field of all the other electrons. In the HF methods the MOs are variationally optimized in order to obtain an energetically 'best' many-electron function of a single-configuration form. Of course, energy optimization does not imply a similar favorableness for properties. The HF model is size-extensive, but the HF wave function is not an eigenfunction of the exact Hamiltonian.

There are several different HF techniques. For closed-shell species the restricted HF (RHF) level is usually a surprisingly adequate approximation. It accounts for a large percentage of the electronic energy of the molecule, and it usually gives adequate reference functions for correlated treatments. In RHF theory it is assumed that all spin-orbitals χ_K are 'pure' space–spin products of the form $\phi_K\alpha$ or $\phi_K\beta$, where ϕ_K denotes the Kth spatial orbital, and α and β are the usual spin functions; thus, pairs of electrons occupy the same spatial orbital. In other words, the RHF energy is optimized with the restriction that the corresponding wave function is an eigenfunction of the total spin (S^2) and one of its components (S_z). At the spin-unrestricted HF (UHF) level spin-orbitals are still restricted to be of product form, but α and β electrons are allowed to occupy different, non-orthogonal spatial functions (different orbitals for different spins, DODS). A considerable weakness of UHF theory is that the UHF wave function is contaminated by eigenfunctions of other spin multiplicity, and thus it is not a pure spin eigenfunction. Similarly to the UHF method, the restricted open-shell HF (ROHF) method is applicable for open-shell molecules; here, however, electrons that are paired with each other are restricted, as in the RHF method, to occupy the same spatial orbital. This restriction leads to an artificial increase in the energy. It may also become necessary to allow for more than one Slater determinant in the HF wave function in order to get the correct spin state. An important variant of the HF method is the generalized HF (GHF) method [Seeger and Pople (1977)]. The result of an HF calculation is the total energy, the wave function consisting of MOs (canonical, localized, or other), and the electron density, from which various properties can be calculated.

As mentioned above, the HF wave function may or may not be a proper zeroth-order representation of the exact wave function. Many transition metal systems with open shells have multireference character even close to equilibrium. Diradical singlets are often at least two-configuration SCF (TCSCF) problems, as are systems with nearly degenerate orbitals in general. A representative list of cases investigated, where one runs into difficulties when using HF theory, includes the lowest or one of the lowest electronic states of Be, Mg, and Ca clusters [Rendell, Lee, and Taylor (1990), and Lee, Rendell, and Taylor (1990)], CuH [Chong *et al.* (1986), and Pou-Amérigo *et al.* (1994)], Cr_2 [Walch *et al.* (1983)], BeO [Scuseria, Hamilton, and Schaefer (1990)], ozone [Yamaguchi *et al.* (1986), Lee, Allen, and Schaefer (1987), Lee and Scuseria (1990); Watts, Stanton, and Bartlett (1991), and Watts and Bartlett (1998)], FOOF [Lee *et al.* (1989)], C_2 [Scuseria, Hamilton, and Schaefer (1990), and Leininger *et al.* (1998)], CN^+ [Scuseria, Hamilton, and Schaefer (1990)], Cl_2O_2 [Rendell and Lee (1991)], and singlet CH_2 [Meadows and Schaefer

(1976)]. Farther away from equilibrium, the HF approximation usually starts to break down, disallowing its use in PES calculations. A notable exception is a Woodward–Hoffmann allowed fragmentation or rearrangement, where the RHF wave function is a proper reference all the way from reactants to products.

Where the exact wave function is dominated by a single configuration, e.g. close to equilibrium structures for most molecular species, determination of the ground-state HF wave function causes no problems. However, if several configurations have comparable importance, e.g. far from equilibrium, several local minima in orbital space can be reached during an SCF optimization. The result obtained depends on several factors, including the initial guess for the form of the MOs and the algorithm used during the energy minimization. In such cases extreme care must be exercised to select the desired solution.

When trying to describe degenerate electronic states, or portions of the potential energy hypersurface where the wave function becomes degenerate, it is mandatory to ensure that the calculations reflect the symmetry properties of the system. HF calculations will result, in many cases, in broken-symmetry solutions (see Table 2.1 and the relevant discussion in Section 2.2.1.3). This disadvantage of the HF model can, of course, be corrected by more elaborate techniques of electronic structure theory (e.g. complete-active-space SCF (CASSCF), Section 2.2.1.2).

Table 2.1. A collection of known examples of spatial symmetry-breaking phenomena.

Species	Reference
Core-hole states of cations	
O_2^+	Bagus and Schaefer (1972), and Ågren, Bagus, and Roos (1981)
N_2^+	Lozes, Goscinski, and Wahlgren (1979)
Li_2^+	Bacskay, Bryant, and Hush (1987)
$C_6H_6^+$	Bigelow and Freund (1982)
Valence-hole states of cations	
Cu clusters	Cox, Benard, and Veillard (1982)
Cu_2^+	Benard (1982)
Ni_2^+ and Ni_4^+	Newton (1982)
Ag_2^+	Benard (1983)
Excited states with equivalent moieties	
CO_2	Engelbrecht and Liu (1983)
CHO · CHO	Gaw and Schaefer (1985)
Cations and ionic species with equivalent centers	
LiO_2	Allen *et al.* (1989)
NaO_2	Horner *et al.* (1991)
C_3^+	Grev, Alberts, and Schaefer (1990)
O_4^+	Lindh and Barnes (1994)
$HOOH^+$	Xie *et al.* (1996)
Doublets with bond fluctuations	
CH_2CCH_2 radical	Paldus and Veillard (1978), Cook (1986), and Szalay *et al.* (1990)
NO_2	Jackels and Davidson (1976), and Crawford *et al.* (1997a)
formyloxyl radical	McLean *et al.* (1985), and Burton *et al.* (1991)
NO_3	Stanton, Gauss, and Bartlett (1992)
Cyclic networks	
N_2S_2	Laidlaw and Benard (1987)
$N_3S_3^-$	Benard, Laidlaw, and Paldus (1985)

2.2.1.2 *Multiconfiguration SCF methods*

When chemical intuition or simple tests indicate a serious breakdown of the HF (single configuration) approximation, it is necessary to turn to multiconfiguration (MCSCF) methods [e.g. CAS, or restricted-active-space (RAS) SCF]. Multiconfiguration SCF methods are to be used when an HF description of the electronic state under consideration is qualitatively incorrect, as when energy hypersurfaces of certain types of chemical reactions are investigated (e.g. a dissociation path for homolytic bond cleavage), when proper space and spin eigenfunctions must be constructed for linear/atomic systems/fragments, or when problematic radicals or transition metals are studied, amongst other things. For a representative list of cases where the HF method provides an incorrect description even in the vicinity of equilibrium structures, see Section 2.2.1.1. In all these instances inclusion of the most important configuration functions in the wave function becomes necessary at the earliest stage, when the molecular orbitals are determined in an SCF procedure. A basic premise of many MCSCF approaches [Frenkel (1934), and Wahl and Das (1977)] is that the important chemical aspects of most molecular systems can be represented by just a limited number of carefully chosen configurations.

While the HF wave function can be written [Jørgensen and Simons (1981)] as

$$|\psi_{\mathrm{HF}}\rangle = |\kappa\rangle = \exp(-\hat{\kappa})|0\rangle, \tag{2}$$

where $|0\rangle$ is a reference configuration of spin orbitals, $\hat{\kappa}$ is an anti-hermitian operator and therefore $\exp(-\hat{\kappa})$ carries out unitary transformations among spin orbitals, an arbitrary MCSCF wave function may be written in the form

$$|\kappa, \boldsymbol{C}\rangle = \exp(-\hat{\kappa})\sum_i C_i|i\rangle = \exp(-\hat{\kappa})\exp\left(-\hat{P}\right)|0\rangle, \tag{3}$$

where the C_i are coefficients of the electronic configurations $|i\rangle$ included in the MCSCF space. The C_i may be viewed as being determined by elements of the anti-Hermitian operator $\hat{P}$, which effects rotations of $|0\rangle$ into its orthogonal complement space of many-electron states. Refinement of the orbital rotation ($\boldsymbol{\kappa}$) and configuration interaction ($\boldsymbol{C}$ or $\boldsymbol{P}$) parameters is performed simultaneously during optimization of the MCSCF wave function by sundry algorithms [Yarkony (1981), Knowles and Werner (1985), and Olsen *et al.* (1988)].

A significant problem with the MCSCF technique, aimed to recover important non-dynamic (static) correlation, is that the selection of the configurations to be included in the wave function is not always straightforward. Selection of individual configurations for an MCSCF treatment has usually been based on physical intuition and/or perturbative estimates. However, the relative importance of configurations can change rapidly over a PES, making their guessing inefficient and problematic and requiring incorporation of the entire union of locally important configurations over all relevant regions of the PES to avert discontinuities in the theoretical treatment. The most rigorous way out is to employ the CASSCF technique [Ruedenberg and Sundberg (1976), Roos, Taylor, and Siegbahn (1980), and Roos (1987)], where the orbitals are divided into three subspaces: inactive, active, and secondary. Inactive and secondary orbitals have fixed occupation numbers of 2 and 0, respectively. Active orbitals have varying occupation: all configurations that are obtained by distribution of the active electrons among the active orbitals in all possible ways, satisfying the relevant symmetry and spin requirements of the total wave function, are included in the wave function. Therefore, within the active space a CASSCF wave function is an FCI wave function. The framework of the CASSCF method is general, its application is straightforward, and it can be employed for all difficult problems of electronic structure theory within certain size limitations. One possible modification of the CASSCF technique is the RASSCF method [Olsen *et al.* (1988), Sanchez de Meras *et al.* (1991), Jensen *et al.* (1989), Sherrill (1996), and Sherrill and Schaefer (1999)], in which the active space is divided into three subspaces (RAS1, RAS2, and RAS3). For the RAS1/RAS3 spaces the number of allowed holes/electrons is selected by the user, whereas the RAS2 space is the same as the active space in CASSCF. A simplified procedure for constructing the CAS space and the CASSCF wave function itself is offered by the so-called UNO-CAS method, developed by Pulay and co-workers [Bofill and Pulay (1989), and Pulay and Hamilton (1988)]. The UNO-CAS method is based on the observation that fractionally occupied UHF natural orbitals are usually good approximations to MCSCF active orbitals. In effect, the UNO-CAS method builds on an inspection of the UHF natural orbital occupation numbers: if an occupation number differs from 2 or 0 by a fixed amount, usually taken to be 0.02, inclusion of the corresponding orbital in the active space is recommended. These orbitals, without further optimization, are employed for the construction of a CAS wave function. Generalized valence bond theory [GVB; Bobrowicz and Goddard (1977)] is a powerful but somewhat specialized technique, which may be viewed as a type of constrained MCSCF method, in which the CI coefficients are fixed by the natural orbital representations of non-orthogonal, variationally optimized orbitals. In essence, these GVB solutions correspond to the best set of orbitals for the superposition of the classic electronic structures of valence bond theory under various types of constraints.

In the (R/C)ASSCF theory the active space has to be kept rather small, comprising no more than about 10 electrons and 15 orbitals. The (R/C)ASSCF electronic energies are normally not accurate enough to be used for the prediction of potential energy or property hypersurfaces. However, a proper choice of active orbitals provides an excellent starting point for a proper multireference dynamical correlation treatment [such as MR-CI, see below, or a low-order perturbational treatment, see Andersson and Roos (1995)].

2.2.1.3 Spin and spatial symmetry breaking

Approximate electronic wave functions are frequently constructed such that they maintain selected spin and spatial symmetry characteristics of the exact wave function. However, these wave functions are not always energetically optimal, and relaxation of enforced symmetry constraints sometimes leads to lower-energy solutions. In such cases, the symmetry-adapted wave function is said to exhibit a symmetry-breaking instability.

In electronic structure theory, symmetry-broken wave functions do not transform as irreducible representations of the full point group of the nuclear framework, but they usually are symmetry species of some nontrivial subgroup thereof. HF wave functions provide the classic example of this behavior in the prediction of the potential energy curve for molecular hydrogen. At long bond distances, spin-restricted (RHF) and spin-unrestricted (UHF) determinants give qualitatively different results, with the latter providing an energetically correct dissociation asymptote at the expense of significant spin impurity and the loss of inversion symmetry. Symmetry-broken wave functions may not be beneficial or even acceptable [Goscinski (1986)]. The question of whether to relax constraints in the presence of an instability has been described as the 'symmetry dilemma' [Löwdin (1963)].

Electronic wave function instabilities were first analyzed in detail by Thouless (1961), Čížek and Paldus (1967), and Paldus and Čížek (1969; 1970), who characterized multiple solutions of the HF equations in terms of the eigenvalues of a Hessian ($\boldsymbol{H}^0$) comprising the second derivatives of the energy with respect to molecular orbital rotations. If all eigenvalues of $\boldsymbol{H}^0$ are positive, the given HF wave function corresponds to a local (perhaps global) minimum on the orbital rotation surface, while one or more negative eigenvalues (λ_-) corresponds to a saddle point or higher-order, unstable stationary point. If the rotations defined by the eigenvector of λ_- involve pairs of orbitals belonging to different irreducible representations of the point group of the nuclear framework, a lower-energy, spatial-symmetry-broken HF wave function exists. In an influential paper, Seeger and Pople (1977) discussed the structure of generalized HF (GHF) theory with emphasis on the stability conditions for release of orbital constraints in the interconnected hierarchy (complex GHF) → (real GHF, complex UHF) → (real UHF, complex RHF) → real RHF. Symmetry breaking in an electronic wave function is also a driving force for symmetry distortions of the nuclear framework. Under these circumstances, unconstrained computations will usually converge to the symmetry-broken solution because of variational collapse. For the determination of potential energy hypersurfaces and many molecular properties obtained by finite-difference procedures (e.g., force constants for non-symmetric vibrations), the existence of these symmetry-breaking instabilities presents a serious obstacle not only for HF theory, but also for correlated methods which use the HF determinant as a reference [Crawford *et al.* (1997b)].

Examples of the difficulties caused by spatial symmetry breaking in HF wave functions are abundant in the literature. A representative list of known spatial symmetry breaking phenomena is given in Table 2.1.

The chemical origins of electronic symmetry breaking can often be explained in valence-bond terms as a competition between orbital size effects and resonance interactions [McLean *et al.* (1985), Allen *et al.* (1989), and Xie *et al.* (1996)]. In doublet instability problems, if an HF determinant is allowed to break symmetry, one of two valence-bond-like solutions will be obtained in which the singly occupied orbital is localized on one of two equivalent centers. Such a wave function may variationally incorporate energy lowering due to orbital size effects by allowing the doubly occupied orbital to be more (or less) diffuse than its singly occupied counterpart. However, in localizing the orbitals, the energy lowering due to the resonance interaction between the valence-bond structures is compromised. On the other hand, the symmetry-restricted determinant best recovers the stabilizing resonance interaction, but its inclusion of the orbital size effect is incomplete. One solution to this problem is to combine the symmetry-broken wave functions; i.e., work with a 2×2 nonorthogonal CI in the space of symmetry-broken solutions. This concept may be used in a more general sense to construct small, conventional MCSCF wave functions which incorporate the physical (resonance) interactions among such nonorthogonal symmetry-broken solutions by doubling the space of active molecular orbitals [Engelbrecht and Liu (1983), McLean *et al.* (1985), Allen *et al.* (1989), and Horner *et al.* (1991)]. In arriving at proper reference wave functioins, traditional MCSCF approaches to symmetry-breaking problems may not be significantly more expensive than single-reference HF methods; however, the addition of dynamic correlation to such treatments by means of multireference CI computations may be hindered by size limitations, and alternative schemes such as multireference perturbation or coupled-cluster theory are often poorly developed for general application. A second option lies in Brueckner-orbital methods [Stanton, Gauss, and Bartlett (1992), and Lindh and Barnes (1994)], such as those based on the coupled-cluster ansatz. Although Brueckner determinants are not *a priori* impervious to symmetry breaking, they appear to have a propensity for preserving symmetry. Some effort has been devoted to this area in recent years, and it is hoped that routine application of such methods to open-shell systems will eventually become more affordable.

It must be made clear that in most cases even high levels of correlation may be unable to overcome inadequacies in the single-determinant reference wave function. The ensuing important questions, namely what general behavior can be expected of force constants computed using correlated wave functions based on unstable reference determinants, and over what range of geometries will correlated wave functions be spuriously affected by reference instabilities, have recently been addressed [Crawford *et al.* (1997b)] in a combined analytic and computational work. The details are, however, outside the scope of the present review.

2.2.2 ELECTRON CORRELATION METHODS

Although the HF description of the electronic structure of most molecular species is surprisingly accurate overall, the HF model does have a weakness of extreme importance for computational quantum chemistry: the electrons are allowed to approach each other more closely than what is allowed by the exact wave function. In other words, compared with the HF model, the exact wave function shows a substantially reduced electron density (a Coulomb hole) around the instantaneous position of each electron. In the wave function models that go beyond the HF (single-configuration) description, techniques of various sophistication are employed to represent the long- and short-range electron–electron interactions, especially the Coulomb hole. While long-range dynamical correlation effects are described adequately by most techniques, it has become clear over the years that it is exceedingly difficult to arrive at an accurate representation of the short-range electron–electron interactions. This difficulty has led to the development of myriad approaches for the treatment of electron correlation.

2.2.2.1 Many-body perturbation theory (MBPT)

As in many branches of physics and chemistry, in computational quantum chemistry it is often expedient to use some form of perturbation theory [PT; Schrödinger (1926), Rayleigh (1927), Kato (1966), Hameka (1981), and Pilar (1990)]. Møller and Plesset (1934) proposed a perturbation treatment of atoms and molecules, equivalent to Rayleigh–Schrödinger perturbation theory, in which the unperturbed Hamiltonian is taken as a sum of one-particle Fock operators; this form of MBPT is called Møller–Plesset (MP) perturbation theory [see, e.g., Szabo and Ostlund (1989), and Hehre *et al.* (1986)]. When MP perturbation theory is carried out to second order (MP2), it defines the simplest method besides density functional theory (DFT) which incorporates electron correlation, and it provides size-extensive energy corrections at low cost.

In the simplest (closed-shell) MP2 case, the HF determinant is coupled only to double substitutions Φ_{ij}^{ab}, as a consequence of Brillouin's theorem, and the first-order correction to the HF wave function is

$$T_2^{(1)}|\Phi_0\rangle = \tfrac{1}{4}\sum_{ijab}\left(t_{ij}^{ab}\right)^{(1)}\left|\Phi_{ij}^{ab}\right\rangle = -\tfrac{1}{4}\sum_{ijab}\frac{(ij\|ab)}{\varepsilon_a+\varepsilon_b-\varepsilon_i-\varepsilon_j}\left|\,\Phi_{ij}^{ab}\right\rangle, \tag{4}$$

where the $(t_{ij}^{ab})^{(1)}$ are amplitudes in the first-order wave function, and the indices ij and ab run over the occupied and virtual subspaces, respectively. This correction results in the following well-known MP2 energy expression:

$$E^{(2)} = \langle\Phi_0|V|T_2^{(1)}\,\Phi_0\rangle_C = \tfrac{1}{4}\sum_{ijab}\left(t_{ij}^{ab}\right)^{(1)}(ij\|ab), \tag{5}$$

where subscript C means that only the connected terms are considered [Čížek (1966), and Čížek (1969)].

The next level of sophistication, MP3, includes second-order doubles (D) corrections, while MP4 is the first method which includes singles, triples, and quadruples (STQ), as well as third-order doubles terms [Pople, Binkley, and Seeger (1976)]. MP4 has found widespread application in theoretical schemes designed to yield bond energies approaching chemical accuracy, such as bond-additivity-corrected fourth-order MP theory, BAC-MP4 [Binkley and Pople (1975), Krishnan, Frisch, and Pople (1980), Ho and Melius (1990), and Melius and Ho (1991)], which uses experimental information to make molecule-dependent empirical corrections. MP4 calculations also form part of the Gaussian-*n* model chemistries [Pople *et al.* (1989)], which use several additivity approximations and small higher-level correction (HLC) terms and usually yield bond energies of chemical accuracy, and of the complete basis set (CBS) model chemistries [see, e.g., Ochterski, Petersson, and Montgomery (1996)]. Complete fifth-order [Laidig, Fitzgerald, and Bartlett (1985), Kucharski, Noga, and Bartlett (1989), and Raghavachari *et al.* (1990)] and sixth-order [Kucharski and Bartlett (1992), and He and Cremer (1996)] theories have also been implemented, although application of these techniques has been restricted to small, benchmark systems. Systematic studies on *intrinsic* errors in MP*n* computations are abundant [e.g., Peterson and Dunning (1995), Peterson and Dunning (1997), Feller and Peterson (1998), and Dunning and Peterson (1998)]. When the underlying HF wave function provides a proper zeroth-order representation, MP2 calculations yield about 90 % of the electron correlation energy. Nevertheless, the MP perturbation treatment may converge poorly, or not converge at all [Handy, Knowles, and Somasundram (1985), Olsen *et al.* (1996), Christiansen *et al.* (1996), and Leininger *et al.* (2000)]. Divergence is associated with the so-called 'intruder state problem' [Schucan and Weidenmüller (1972)]. Several approaches have been advanced to remedy this situation, including construction of an effective Hamiltonian that has certain eigenvalues different from the original Hamiltonian [Malrieu, Durand, and Dausey (1985), and Zaitsevskii and Heully (1992)], resummation of series [e.g., Young, Biedenharn, and Feenberg (1957), and Čížek and Vrscay (1982)], order-dependent mappings [Seznec and Zinn-Justin (1979)], and Λ-transformation [Dietz *et al.* (1993)]. To indicate the current understanding about the quality of low-order MP*n* energy and property results about stationary points on a PES, we quote from Dunning and Peterson (1998), who observed the following for first-row diatomics: 'The perturbation expansions are, in general, only slowly converging and, for HF, N_2, CO, and F_2, appear to be far from convergence at MP4. In fact, for HF, N_2, and CO, the errors in the calculated spectroscopic constants for the MP4 method are larger than those for the MP2 method (the only exception is D_e). The current study, combined with other recent studies, raises serious doubts about the use of Møller–Plesset perturbation theory to describe electron correlation effects in atomic and molecular calculations.' It must also be emphasized that perturbation methods are usually not adequate for calculating global reaction surfaces.

Several schemes for performing extrapolations to the FCI limit based on low-order MP perturbation series have been proposed [Wilson (1980), Pople *et al.* (1983), Laidig, Fitzgerald, and Bartlett (1985), Schmidt, Warken, and Handy (1993), and Cremer and He (1996)]. Representative are the following two formulas for the total basis-set correlation energy ($\varepsilon_{\text{corr}}$) based on nth-order MP terms (ε_n):

$$\varepsilon_{\text{corr}} = \frac{\varepsilon_2 + \varepsilon_3}{1 - \varepsilon_4/\varepsilon_2} \tag{6}$$

and the shifted [2/1] Padé approximant

$$\varepsilon_{\text{corr}} = \frac{\varepsilon_2^2(\varepsilon_4 - \varepsilon_5) + 2\varepsilon_2\varepsilon_3(\varepsilon_4 - \varepsilon_3) + \varepsilon_3^2(\varepsilon_2 - \varepsilon_3)}{(\varepsilon_2 - \varepsilon_3)(\varepsilon_4 - \varepsilon_5) - (\varepsilon_4 - \varepsilon_3)^2}, \tag{7}$$

which is based on the rational extrapolation

$$\varepsilon_2 + \lambda\varepsilon_3 + \lambda^2\varepsilon_4 + \lambda^3\varepsilon_5 + \cdots \approx \frac{a_0 + a_1\lambda}{1 + b_1\lambda + b_2\lambda^2}. \tag{8}$$

Both formulas have been subjected to limited testing [Pople *et al.* (1983), and Laidig, Fitzgerald, and Bartlett (1985)].

Despite being size-extensive, perturbation methods with an RHF reference will suffer from problems due to near-degeneracies [see, e.g., Hirschfelder, Byers Brown, and Epstein (1964), Matcha and King (1967), and Wilson, Jankowski, and Paldus (1985)]. It is therefore obligatory to extend the perturbation approach to handle also multiconfigurational reference functions. Such MC-PT approaches could not be implemented until quite recently [see, e.g., Brandow (1967), Kaldor (1982), Wolinski, Sellers, and Pulay (1987), Wolinski and Pulay (1989), Zarrabian and Paldus (1990), Andersson *et al.* (1990), Hirao (1991), and Andersson, Malmqvist, and Roos (1992)]. Experience shows that large reference spaces are usually necessary to get reasonable results, whereas chemical intuition and simple models suggest that most molecules and complexes can be described by only a few configurations. Moreover, as Zarrabian and Paldus (1990) note, 'MR-MBPT cannot be viewed as a universal approach, which will yield desirable energies for a wide range of molecular geometries when a continuous transition between nondegenerate and highly degenerate situations may occur, as required when determining potential energy surfaces'.

2.2.2.2 Coupled-cluster (CC) methods

The coupled-cluster (CC) method was introduced into molecular electronic structure theory in the 1960s [Čížek (1966); Čížek (1969)]. Efficient single-reference coupled-cluster (SRCC) procedures have been developed which are based on several types of reference wave functions, including closed-shell RHF, and open-shell UHF, ROHF, and quasirestricted HF [QRHF; Rittby and Bartlett (1988)] wave functions [Gauss *et al.* (1991), Scuseria (1991), Watts, Gauss, and Bartlett (1993), Jayatilaka and Lee (1993), and Crawford and Schaefer (1996)], or Brueckner determinants [Handy *et al.* (1989)]. In the vicinity of equilibrium structures these methods have now reached a high degree of sophistication [Paldus (1994), Bartlett (1995, 1997), and Crawford and Schaefer (1999)].

The fundamental equation of CC theory is

$$|\psi\rangle_{\mathrm{CC}} = \mathrm{e}^{\hat{T}}|\Phi_0\rangle, \tag{9}$$

where $|\psi\rangle_{\mathrm{CC}}$ is the correlated molecular electronic wave function based on the exponential CC ansatz and in most applications $|\Phi_0\rangle$ is a normalized HF wave function for the system. Nothing in CC theory is fundamentally limited, however, to the HF choice for the reference function; for example, the QRHF reference is often a preferred choice for open-shell systems. The $\exp(\hat{T})$ operator is defined by its usual Taylor-series expansion

$$\mathrm{e}^{\hat{T}} = 1 + \hat{T} + \frac{\hat{T}^2}{2!} + \ldots = \sum_{k=0}^{\infty} \frac{\hat{T}^k}{k!}, \tag{10}$$

and the global cluster excitation operator $\hat{T}$ is defined as the sum of n-tuple excitation operators $\hat{T}_n$, i.e.

$$\hat{T} = \hat{T}_1 + \hat{T}_2 + \ldots + \hat{T}_n, \tag{11}$$

where the maximum value of n equals the number of electrons (in this case CC becomes equivalent to FCI) but practical restrictions almost always dictate $n \leq 3$. For example, $\hat{T}_1 = \sum_{i,a} t_i^a \{a^+ i\}$ converts the reference Slater determinant into a linear combination of all singly excited Slater determinants according to the variable coefficients t_i^a, called amplitudes. The effect of the $\exp(\hat{T})$ operator is then to express $|\psi\rangle_{\mathrm{CC}}$ as a configuration interaction expansion, including Φ_0 and all possible excitations of electrons from occupied to virtual spin-orbitals, albeit with restrictions on the CI coefficients if $\hat{T}$ is truncated. The aim of a CC calculation is to solve nonlinear equations for CC amplitudes in each $\hat{T}_n$ (as governed by the Schrödinger equation). Once these are determined, the multideterminantal, many-electron wave function is known.

Some highlights of CC theory are as follows: (a) at any given level of excitation/truncation, CC methods are size-extensive; (b) the exponential form of $|\psi\rangle_{\mathrm{CC}}$ ensures that all higher excitations are included in the wave function, although they may be restricted to be simple products of lower excitations; (c) CC *ansätze* are almost always based on determinants rather than configuration state functions (CSFs), because spin adaptation is not straightforward [Li and Paldus (1994, 1995)]; and (d) CC wave functions are not variational. An important practical consequence of (d) is that calculation of energy derivatives of CC wave functions is somewhat costly and requires the setup and solution of additional linear response equations, known as the lambda equations

[Fitzgerald *et al.* (1985), Bartlett (1986), Scheiner *et al.* (1987), Salter, Trucks, and Bartlett (1989), Gauss *et al.* (1991), and Gauss and Stanton (1997)]. The lambda equations utilize a simple and fundamental discovery [Handy and Schaefer (1984)], the so-called Z-vector method, which paved the way to efficient determination of derivatives of almost all non-variational wave functions. As a practical application of the Handy–Schaefer device, the calculation of derivatives of non-variational wave functions has been formulated in a uniform way through effective (also called relaxed) density matrices [Rice and Amos (1985)].

In the usual applications of CC theory, only certain excitation operators are included in the cluster operator. Restricting $\hat{T}$ just to $\hat{T}_2$ results in the CCD method [Čížek (1966)], originally termed coupled-pair many-electron theory (CPMET). Inclusion of $\hat{T}_1$ and $\hat{T}_2$ gives the widely employed CC singles and doubles (CCSD) method [Purvis and Bartlett (1982)]. It is capable of recovering typically 95 % or more of the correlation energy for molecules in the vicinity of their equilibrium structures. Inclusion of $\hat{T}_3$ defines the CCSDT method [Noga and Bartlett (1987), and Scuseria and Schaefer (1988)], which, under favorable circumstances, recovers more than 99 % of the dynamical electron correlation energy. CCSDT is perhaps the most accurate single-reference technique offered by computational quantum chemistry for the treatment of light tri-and tetra-atomic systems in the relative proximity of equilibrium structures. CCSDT applications are becoming increasingly common [e.g., Noga and Bartlett (1987), Scuseria and Schaefer (1988), Scuseria and Lee (1990), Halkier *et al.* (1997), Watts and Bartlett (1998), and Császár, Allen, and Schaefer (1998)]. In the earliest studies, Noga and Bartlett (1987) found that for singly bonded molecules (BH, HF, and H_2O) and for atoms (Ne and F^-) the CCSDT energies agree with FCI energies to chemical accuracy even when bonds are stretched to twice their equilibrium value, while Scuseria and Schaefer (1988) showed that the equilibrium bond lengths and harmonic frequencies of the diatomics HF, OH^-, N_2, and CO agree very well with results obtained from CISDTQ at substantially less cost. Benchmark studies also exist for CCSDTQ [Kucharski and Bartlett (1992)] but not for higher excitation schemes of CC theory.

Simplification in CC theory can be accomplished by restricting the evaluation of (connected) contributions corresponding to higher excitations to certain lead terms. For example, before efficient computational procedures were developed for the calculation of the CC amplitudes [Purvis and Bartlett (1982), and Cullen and Zerner (1982)], it was practical to linearize the exponential CC equations. The linearized coupled-cluster doubles [LCCD; Bartlett and Shavitt (1977), and Purvis and Bartlett (1982)] or the linearized coupled-pair many-electron theory [L-CPMET; Čížek (1969)] is defined as

$$\left\langle \Phi_{ij}^{ab} \right| \hat{H}_N(1 + \hat{T}_2)|\Phi_0\rangle = 0. \qquad (12)$$

In the face of rapidly expanding computational resources, such techniques have lost most of their appeal. In the case of CCSD, linearization can be understood as changing CC coefficients to CI ones and neglecting higher excitations.

Perturbation analysis of the CC hierarchy offers important insights. Although CC methods provide a systematic way of summing up selected terms in the correlation perturbation series to infinite order, a pure correlation order n is not necessarily well defined. The CCSD method is correct through third order in the MP series but is augmented by many higher-order terms. The fourth-order approximation to CCSD is MP4(SDQ). CCSDT is fully correct to fourth order, while CCSDTQ is correct through MP6. The close relationship between MP theory and iterating the CC equations has been discussed widely [Pople, Binkley, and Seeger (1976), and Raghavachari *et al.* (1990)] and exploited frequently [e.g., Kucharski and Bartlett (1992)]. A useful approach is to evaluate computationally intensive, higher-order CC terms by means of lowest-order PT. The resulting terms can be added either iteratively or noniteratively. Such strategies have resulted in

Table 2.2. Force constants for the electronic ground state of water obtained with the aug-cc-pVQZ basis set[a].

Term	RHF	MP2	CCSD	CCSD(T)	Expt.[b]
f^{r}	0.1622	0.0085	0.0307	0.0075	—
f^{α}	−0.0162	0.0011	−0.0017	0.0000	—
f^{rr}	8.606	8.506	8.455	8.437	8.447
$f^{rr'}$	−0.039	−0.135	−0.087	−0.098	−0.102
$f^{r\alpha}$	0.0236	0.260	0.253	0.256	0.263
$f^{\alpha\alpha}$	0.765	0.690	0.713	0.706	0.704
f^{rrr}	−59.11	−58.27	−58.51	−58.51	−56.37
$f^{rrr'}$	−0.0095	−0.074	−0.046	−0.055	−0.276
$f^{rr\alpha}$	−0.084	−0.140	−0.100	−0.107	−0.059
$f^{rr'\alpha}$	−0.533	−0.458	−0.519	−0.516	−0.189
$f^{r\alpha\alpha}$	−0.316	−0.342	−0.314	−0.321	−0.309
$f^{\alpha\alpha\alpha}$	−0.694	−0.706	−0.718	−0.715	−0.751
f^{rrrr}	369.7	362.7	362.9	362.6	320.3
$f^{rrrr'}$	−0.75	−0.44	−0.65	−0.61	1.62
$f^{rrr'r'}$	0.44	0.19	0.36	0.31	1.73
$f^{rrr\alpha}$	−1.23	−1.52	−1.40	−1.44	−2.37
$f^{rrr'\alpha}$	0.81	0.97	0.81	0.80	−6.16
$f^{rr\alpha\alpha}$	−0.26	−0.25	−0.25	−0.25	−0.70
$f^{rr'\alpha\alpha}$	0.51	0.61	0.63	0.63	0.15
$f^{r\alpha\alpha\alpha}$	0.72	0.78	0.77	0.78	1.30
$f^{\alpha\alpha\alpha\alpha}$	−1.01	−0.57	−0.65	−0.62	−0.64

[a] The reference geometry of these calculations corresponds to an empirical estimate of the equilibrium geometry of water, $r(\text{OH}) = 0.95843\,\text{Å}$ and $\angle\text{HOH} = 104.44°$, based on results reported by Carter and Handy (1987), Partridge and Schwenke (1997), and Polyansky, Jensen, and Tennyson (1996). The units of force constants are consistent with energy measured in aJ, distances in Å, and angles in rad.
[b] A set of empirical force constants derived from Polyansky, Jensen, and Tennyson (1996), as reported by Tarczay *et al.* (1999).

a large number of approximate CC techniques with a corresponding large number of acronyms [Bartlett (1995)]. For the case of triple excitations, an important iterative technique is CCSDT-1 [Lee, Kucharski, and Bartlett (1984)], while the most important noniterative technique is CCSD(T) [Raghavachari *et al.* (1990)]. The well-balanced CCSD(T) approach–which includes a noniterative, perturbative correction for the effect of connected triple excitations–is able to provide total and relative electronic energies of chemical accuracy [e.g., Scuseria and Lee (1990), Yarkony (1995), Lee and Scuseria (1995), and Császár, Allen and Schaefer (1998)], as well as excellent results for a wide range of molecular properties [Yarkony (1995), and Bartlett 1997)]. It remains the most popular method for high-accuracy calculations. In single-reference cases and for most properties of general interest, the HF, MP2, CCSD and CCSD(T) methods provide the most useful hierarchy of approximations of increasing accuracy. The force field data presented in Table 2.2 for water show that coupled-cluster methods, especially CCSD(T), result in high-quality force fields, whose accuracy approaches, and in higher orders even surpasses, the precision obtainable in empirical studies.

2.2.2.3 Higher-order and multireference configuration interaction (CI) methods

The technique of configuration interaction [CI; Hylleraas(1928), Weinbaum (1933), Shavitt (1977), and Sherrill and Schaefer (1999)] expresses the approximate wave function either as a linear combination of Slater determinants, which are eigenfunctions of $\hat{S}_z$, the z component of the total spin operator, or as a linear combination of configuration state functions [CSFs, for more details on the form and utilization of CSFs, see Pauncz (1979)], which are eigenfunctions of both $\hat{S}^2$ and $\hat{S}_z$. In conventional CI-type expansions, the determinants/CSFs are antisymmetrized products of one-electron functions. CI theory is covered in practically all the standard textbooks [e.g., Szabo and Ostlund (1989)]. The following equation defines the single-reference CI wave function in general:

$$\psi_{\mathrm{CI}} = \Phi_0 + \sum_{i,a} C_i^a \Phi_i^a + \sum_{\substack{i<j \\ a<b}} C_{ij}^{ab} \Phi_{ij}^{ab} + \cdots, \tag{13}$$

that is in addition to the reference (HF) wave function, Φ_0, the total wave function includes singly-, doubly-, and perhaps more highly-excited determinants/CSFs, Φ_i^a, Φ_{ij}^{ab}, Φ_{ijk}^{abc} ..., wherein the occupied $(ijk\ldots)$ spin orbitals of the reference are replaced by virtual orbitals $(abc\ldots)$. The CI coefficients C_i^a, C_{ij}^{ab}, etc. are then optimized variationally to give the lowest electronic energy, which amounts to solving an eigenvalue problem for the Hamiltonian matrix in the determinant/CSF many-electron basis.

For an n-electron system the full CI (FCI) method is defined as the wave function that includes all possible excitations through order n. With a complete basis set the FCI method would become the exact solution to the non-relativistic electronic Schrödinger equation. It is the FCI method to which all approximate methods of electronic structure theory are to be compared.

The conceptually simple single-reference CI methods, which are variational in nature in contrast to nonvariational CC methods, have been in use from the early days of computational electronic structure theory. If we take the HF wave function as the zeroth-order wave function in PT, then all triple and higher excitations make no contribution to the exact wave function in first order. Accordingly, CI including all single and double excitations, i.e., CISD, became the first standard method for the treatment of electron correlation [see, e.g., Saxe, Schaefer, and Handy (1981)]. Nevertheless, in the last few years low-order CI techniques find less favor among users. It has been shown [Helgaker *et al.* (1997a)], for example, that for prediction of equilibrium geometries, complete basis set (CBS) CISD performs less satisfactorily than any other correlated method, with the possible exception of MP3. CISD is seldomly used today for ground electronic state calculations.

The lack of size-extensivity of truncated CI treatments is another weakness of this technique. It makes studies of many-electron systems problematic, and it is a hindrance for the calculation of potential energy and property hypersurfaces. More or less elaborate ways of trying to correct this disadvantage led to a number of modified CI techniques. The simplest and thus least accurate scheme is the so-called Davidson correction [Langhoff and Davidson (1974)]. Useful estimates of the size-extensivity errors can be obtained from perturbation calculations [see, e.g., Wilson and Silver (1978)]. A more advanced way of handling the lack of size extensivity is the *a priori* correction of the working equations. Chief among the methods resulting from *a priori* correction techniques are the coupled-pair functional (CPF) method [Ahlrichs, Scharf, and Ehrhardt (1985), and Gdanitz and Ahlrichs (1988)], the modified CPF (MCPF) method [Chong and Langhoff (1986)], and the averaged CPF (ACPF) method [Gdanitz and Ahlrichs (1988)]. Several studies have indicated [e.g., Chong *et al.* (1986), and Bauschlicher *et al.* (1989)] that these techniques perform very well, and they can successfully identify configurations to be included in the reference space of multireference electron correlation procedures. Nevertheless, the ACPF method proved to be rather sensitive to the choice of the reference CSFs, which calls for caution in the selection procedure. The QCISD model [Pople, Head-Gordon and Raghavachari (1987)] constitutes a size-extensive version of the CISD model closely related to CCSD. It was first obtained as a modified CISD method in which additional operator terms are added that exactly cancel the terms in CISD that break size extensivity. It is also a truncated version of CCSD in which a large number of

Table 2.3. Number of CSFs and percentage of correlation energy (ε_{corr}) recovered by various CI excitation levels for small chemical systems in a cc-pVDZ basis[a].

	CISD		CISDT		CISDTQ		FCI
Species	No. CSFs	% ε_{corr}	No. CSFs	% ε_{corr}	No. CSFs	% ε_{corr}	No. CSFs
Ne	253	97.16	2289	97.67	11307	99.94	64331
HF	1175	95.77	17199	96.64	133161	99.91	2342800
C_2	1427	80.84	33439	84.11	411225	97.94	27944940
CN^+	2827	80.77	66891	86.63	822041	97.91	55883796
CN	9072	87.78	192697	94.20	2207722	99.35	245880640
N_2	2090	88.93	60842	92.12	969718	99.30	540924024

[a] See Leininger *et al.* (2000).

relatively small terms are omitted. The QCI energies do not correspond to an identifiable wave function. The equations used to determine the QCI amplitudes are linear, as in traditional CI, rather than nonlinear, as in CC. In terms of a perturbation expansion, QCISD and CCSD behave identically through fourth order, and differ only slightly in fifth order.

Highly correlated CI wave functions, such as CISDTQ, going beyond the simple CISD model, can provide very reliable potential energy hypersurfaces. Unfortunately, such wave functions are notoriously difficult to evaluate because of their considerable computational demands. It should also be kept in mind that achieving spectroscopic accuracy is next to impossible in conventional CI treatments. The vast improvement of the CISDTQ wave function over CISDT is a testament to the importance of quadruple excitations in the correct treatment of molecular systems. Through the 'disconnected' terms $\hat{T}_2^n$, CC schemes, even CCSD, allow for quadruple, hextuple, etc. excitations, resulting in a much more robust method than low-order CI, while the cost is comparative since the number of parameters to be optimized is the same for CIn and CCn. The number of configuration state functions and the percentage of correlation energy recovered by various CI excitation levels for selected small molecules is given in Table 2.3. The results presented confirm that while CISD performs relatively poorly in approaching the FCI limit, CISDTQ does a very good job in recovering dynamical electron correlation energy at a substantially reduced cost.

The difference in the convergence behavior among the doubles, triples, and quadruples coefficients/amplitudes with respect to orbital space expansion [Adamowicz and Bartlett (1987), Sherrill and Schaefer (1996), and Klopper *et al.* (1997)] allows for the use of substantially different orbital spaces for their determination. For example, Sherrill and Schaefer (1996) have investigated variational wave functions that incorporate limited triple and quadruple substitutions. The CISD[TQ] method [Grev and Schaefer (1992), Fermann *et al.* (1994), and Sherrill and Schaefer (1996)], which is an *a priori* selection scheme for multireference CISD (MR-CISD) based on natural orbitals, selects as references all single and double substitutions in its active space. It is equivalent to a limited CISDTQ in which no more than two electrons are allowed into external orbitals. If a single reference function dominates, CISD[TQ] provides results almost identical to second-order configuration interaction (SOCI), which distributes electrons in all possible ways as long as no more than two electrons are allowed in external orbitals at once [Blomberg, Siegbahn, and Roos (1980), Saxe, Schaefer, and Handy (1981), Bauschlicher and Partridge (1993), and Sherrill and Schaefer (1999)]. The accuracy of the CISD[TQ] and SOCI approaches can be judged from the data presented in Table 2.4 for a closed- and an open-shell molecule at three distances. The consistency and modest size of the error of SOCI over the three distances is rather remarkable. It is also noted here that Klopper *et al.* (1997) proved that perturbational corrections

Table 2.4. Errors in total energies (in mE_h) relative to FCI for several molecules using a DZP basis set[a].

Molecule	Method	No. CSFs	$E(r_e)$	$E(1.5r_e)$	$E(2r_e)$
$^2B_1$ NH_2	CISD[TQ]	18396	2.897	2.630	4.957
	SOCI	21687	2.853	2.107	1.703
	FCI	2435160	0.000	0.000	0.000
$^1A_1$ H_2O	CISD[TQ]	32361	1.630	2.537	6.867
	SOCI	76660	1.276	1.058	1.020
	CISDTQ	151248	0.397	1.547	6.280
	FCI	6740280	0.000	0.000	0.000

[a] Results are taken from Grev and Schaefer (1992) for $^2B_1$ NH_2, and from Fermann *et al.* (1994) for $^1A_1$ H_2O. The CISD[TQ] and SOCI methods employed CISD natural orbitals. r_e refers to the X–H (X = N/O) bond length; for actual geometries see the original publications.

for triples replacements, (T), in the framework of single-reference coupled cluster theory can be calculated accurately from a reduced (perhaps by as much as 50 %) space of virtual orbitals. A proper choice of the reduced virtual orbital space is critical in the success of this approach [Klopper *et al.* (1997), and Adamowicz and Bartlett (1987)].

Although multireference (MR) CC theories are difficult to formulate, MR-CI methods have been employed for over a quarter of a century [Bagus *et al.* (1973), Buenker and Peyerimhoff (1974), and Buenker and Peyerimhoff (1975)]. The MR-CI technique is treated in detail in the chapter by Buenker *et al.* in this volume. Even today, the MR-CI technique is considered to be the most accurate *ab initio* electron correlation procedure which can be employed for reasonably large molecular systems over an extended range of nuclear configurations. Perhaps the most successful variational method for treating triple and quadruple excitations includes single and double excitations relative to several reference configurations [Bagus *et al.* (1973)], a method called MR-CISD. The motivation of this very reliable approach is the observation that an important quadruple excitation invariably encompasses at least one of the dominant double excitations. Nevertheless, probably most higher excitations are needed only to correct for the lack of size extensivity of the single-reference CISD wave function. In order to avoid introduction of nonphysical kinks in the potential energy and property hypersurfaces determined from MR-CISD computations, use of different reference configurations for different regions of the configuration space covered should be avoided. FCI benchmark studies, although within a rather limited one-particle space, proved that the MR-CISD method based on a CASSCF reference space provides potential energy surfaces which closely parallel the FCI surfaces [Brown, Shavitt, and Shepard (1984), Bauschlicher *et al.* (1986), Bauschlicher and Taylor (1987), and Bauschlicher, Langhoff, and Taylor (1990)].

Like all other truncated CI methods, MR-CISD is not size extensive. The simplest *a posteriori* correction is offered by the MCSCF analog of the Davidson correction [MR-CISD + Q, Davidson and Silver (1977)]:

$$E_{+Q} = (E_{CI} - E_{ref})\left(1 - \sum C_{ref}^2\right), \tag{14}$$

where the sum indicated is over all the configurations included in the reference set, E_{ref} is the corresponding expectation energy value, and E_{CI} is the MR-CISD energy. Several *a priori* corrected MR-CI procedures have been proposed to overcome the problem of size nonextensivity [for a recent review, see Szalay (1997)]. In all these methods the error due to the lack of size extensivity is reduced by an effective inclusion of higher excitation effects. These methods treat the exclusion principle violation (EPV) [see, e.g., McWeeny (1992), and Szalay (1997)] and the redundancy terms [Malrieu, Daudey, and Caballol (1994)] using different approximations. One popular *a posteriori* correction scheme is the multireference averaged coupled pair functional (MR-ACPF) method [Gdanitz and Ahlrichs (1988)], which accounts for the EPV terms in an averaged way. For the generation of a PES the MR-ACPF procedure requires a large amount of fine-tuning so that the reference space is both valid and optimal over the whole configuration range of interest. Another recent proposal for size-extensivity corrections is the multireference averaged quadratic coupled-cluster (MR-AQCC) method [Szalay and Bartlett (1995)].

2.2.2.4 Density functional methods

Density functional theory [DFT; for recent reviews see, for example, Parr and Yang (1989), Labanowski and Andzelm (1991), and Kohn, Becke, and Parr (1996)] was originally developed for investigations in the solid state. However, after introduction of the Kohn–Sham method [Kohn and Sham (1965)], and especially after the development of new, accurate exchange-correlation functionals [e.g., Becke (1988)], DFT techniques have become popular for treating individual molecules. DFT provides a conceptually very different approach from MO theory toward the exact solution of the molecular Schrödinger equation. While in conventional *ab initio* theory the exact Hamiltonian is used and increasingly better approximations for the wave function are sought, in DFT approximate forms for energy terms as a function of the electron density are adopted and parametrized, with no explicit reference to the wave function. Subsequently, the approximate density functionals for the energy are subjected to a variational procedure whereby a precise numerical solution for the electron density profile is found. The theoretical basis and applications of DFT methods employed in molecular quantum chemistry are treated in detail in the chapter by Bérczes, Zgierski, and Yang in this volume. Therefore, no further details are given here. It is only noted that Kohn–Sham DFT offers an affordable scheme for computations on large-scale molecular systems. Strictly speaking, however, most DFT methods employed today, especially the more successful (accurate) ones, are not really *ab initio* techniques, as they contain empirical parameters. The uncertain reliability of present-day DFT schemes is a substantial drawback for their use in determining accurate, global PESs.

2.2.2.5 Explicitly correlated methods

The Coulomb potential in the usual electronic Hamiltonian becomes singular when any two particles i and j coalesce ($r_{ij} = 0$). In order to fulfill the requirements that the wave functions Ψ satisfy the Schrödinger equation and remain bounded at these poles, certain coalescence/cusp conditions need to be satisfied. One must consider both nuclear cusp (when nuclei and electrons coincide) and electronic (Coulomb) cusp (when electrons coincide) conditions [Kato (1957), Roothaan and Weiss (1960), Pack and Byers Brown (1966), and Bingel (1967)].

For an eigenfunction $\Phi(\boldsymbol{r}_1, \boldsymbol{r}_2, \ldots, \boldsymbol{r}_n)$ of the spinless Hamiltonian of an n-electron atom in the clamped-nucleus approximation, Kato (1957) proved rigorously that

$$\left(\frac{\partial\overline{\Phi}}{\partial r_{12}}\right)_{r_{12}=0} = \gamma\Phi(r_{12} = 0), \tag{15}$$

where $\overline{\Phi}$ is Φ averaged spherically about the singularity. The constant γ is equal to $-Z$ (the atomic number) for electron–nucleus coalescence, and to a small numerical value (e.g., 1/2 for singlet coupled electron pairs and 1/4 for triplets) for electron–electron coalescence [Kutzelnigg and Morgan (1992)]. Pack and Byers Brown (1966) extended Kato's treatment, so that the fixed-nucleus approximation of the original treatment could be lifted and the wave function can have a node at the singularity (e.g., in the case of a triplet wave function), without change in the form of the cusp conditions of equation (15).

The nuclear cusp condition is satisfied if the wave function exhibits, close to the nucleus, a simple exponential dependence for the electronic coordinate considered. Therefore, Slater-type functions (STFs) are compatible with the nuclear cusp condition but Gaussian-type functions (GTFs) are not (for more details about STFs and GTFs see Section 2.2.3).

The Coulomb cusp condition leads to a wave function that is continuous but not smooth at $r_{ij} = 0$: the wave function has discontinuous first derivatives for coalescing electrons. It is also well known that the deviation between the exact and an approximate wave function is most significant at short electron–electron distances (in other words, it is exceedingly difficult to arrive at an accurate description of the Coulomb hole around each electron). Straightforward generalization of the electron–electron cusp condition of Kato [Kato (1957)] implies that for an n-electron system and for small r_{12} the (unnormalized) n-electron wave function is linear in r_{12} [Pack and Byers Brown (1966)]. It must be stressed that the electron–electron poles greatly complicate the determination of approximate wave functions and that the Coulomb cusp condition can never be satisfied by traditional CI-type wave functions.

We may impose the correct Coulomb-cusp condition on any determinant-based wave function by multiplying the orbital product expansion by some so-called correlation factor Γ [Löwdin (1959), pp. 303–306 and pp. 316–317]. Methods that employ correlation factors or otherwise make explicit use of the interelectronic coordinates in the wave function are known as explicitly correlated methods. Several choices for the correlation factor have been probed [see, e.g., Hylleraas (1929), Bohm and Pines (1953), Hirschfelder (1963), Hirschfelder, Byers Brown, and Epstein (1964), and Kutzelnigg (1985)]. If a single term Φ_0 dominates in the conventional expansion Φ of the wave function, one can write [Kutzelnigg (1985)]

$$\Psi = \Gamma\Phi = \Gamma(\Phi_0 + \omega) = \Gamma\Phi_0 + \chi, \tag{2.16}$$

and expand χ in the usual manner in a Hilbert space basis. Such methods, which thus employ a simple reference configuration, multiplied linearly by the interelectronic distances within Γ, are referred to as R12 methods [Kutzelnigg (1985), and Klopper (1998)].

Although the presence of the correlation factor Γ may ensure the correct Coulomb cusp behavior, it does not follow *a priori* that the associated improvements in the energy are significant. Numerical experiments showed, however, that addition of even a single linear R12 term to the wave function has a remarkable effect on the energy, reducing the error in the electronic energy substantially at all levels of electron correlation. Still, it appears difficult to converge the FCI-R12 energy of the He atom to within 1 μE_h. Clearly, for errors of the order 1 μE_h and smaller (spectroscopic accuracy), even more flexible wave functions are needed, such as the Hylleraas function for the He atom [Hylleraas (1929)].

Introduction of correlating functions leads to a rather complex representation of the energy. Whereas the energy expression of a determinantal wave function, written in terms of orbital products, contains at most two-electron integrals, the inclusion of two-electron functions in the wave function gives rise to three- and higher-electron integrals in the energy expression. One way forward is the method of Gaussian geminals [Boys (1960), and Singer (1960)], where the factor r_{12} is replaced by $\exp(-\alpha_i r_{12}^2)$ and the exponents α_i are obtained through difficult nonlinear optimizations [see, e.g., Cencek, Komasa, and Rychlewski (1995)]. A successful application of the Gaussian geminal method is a study of water at the level of second-order perturbation theory [Bukowski *et al.* (1995)]. In analogy with fitting STFs with GTFs, Persson and Taylor (1996) studied the approximate replacement

$$r_{12} \approx \sum_{i=1}^{N} b_i \left[1 - \exp\left(-\alpha_i r_{12}^2\right)\right] \tag{17}$$

i.e., use of a basis of Gaussian geminals for the representation of r_{12}. Although Gaussian geminals cannot satisfy the Coulomb cusp condition, experience with them suggests that electron correlation effects are nonetheless described very efficiently. Another approach to the problem of integral evaluation is offered by the linear R12 theory of Kutzelnigg [Kutzelnigg (1985), and Klopper (1998)]. The approximations introduced in linear R12 theory are aimed at eliminating all many-electron integrals by inserting into them the approximate resolution of identity (RI) in terms of the orbital basis. Linear R12 techniques have been implemented within the framework of CI [Kutzelnigg (1985), and Gdanitz (1993)], CC [Noga and Kutzelnigg (1994), and Noga, Klopper, and Kutzelnigg (1997)], and PT theory [Klopper and Kutzelnigg (1987), Klopper and Kutzelnigg (1991), Bearpark and Handy (1992), and Noga, Kutzelnigg, and Klopper (1992)]. The final result of an R12 calculation is the conventional energy, to which an R12 contribution, which may be considered as a correction term for the basis set incompleteness error (BSIE, see Section 2.2.3.6), is added. The CISD-R12 method [Röhse, Klopper, and Kutzelnigg (1993)], when used in conjunction with a contracted $10s8p6d4f$ basis set, provided, for H_3^+, an *ab initio* potential energy hypersurface accurate to 1 cm^{-1}, while a molecule as large as ferrocene has been treated at the MP2–R12 level [Klopper and Lüthi (1996)]. Nevertheless, because of the use of RI, it may be necessary [Tarczay *et al.* (1999)] to use a technique such as intramolecular counterpoise (ICP) correction (see Section 2.2.3.7) to achieve error balance during calculation of relative energies in substantially different regions of the same PES.

Ultimately, the use of correlating functions may provide a very efficient means of generating highly accurate wave functions, as they lead to much more compact and robust expansions than conventional methods of electronic structure theory. Once the remaining technical difficulties are remedied, explicitly correlated methods may well become the preferred methods for highly accurate PES determinations.

2.2.2.6 Full configuration interaction (FCI) extrapolations and benchmarks

Achieving the full configuration interaction (FCI) limit has been a persistent goal of molecular quantum mechanics [for a recent review, see, e.g., Sherrill and Schaefer (1999)]. However, for most chemical systems explicit FCI computations are intractable owing to their factorial growth with respect to the one-particle basis and the number of electrons. Nonetheless, numerous schemes have been developed to estimate the FCI limit from series of explicit, truncated CI computations conjoined with in many cases perturbation theory, and many applications have demonstrated the viability of generating potential energy surface information by such methods. Moreover, explicit FCI computations for small, model systems have proved essential in calibrating diverse electronic structure methods, understanding their physical merits and limitations, and developing improvements [e.g., for prototypical molecules of non-covalent chemistry see Woon (1994), van Mourik and van Lenthe (1995), and Mayer, Vibók, and Valiron (1994)]. The quality of different approximations, both CI and CC at various levels of truncation, vis-à-vis the FCI limit in predicting spectroscopic constants can be judged from data such as those in Table 2.5 for the demanding, multireference diatomic system C_2.

FCI calculations can be performed either with Slater determinants or with configuration state functions [CSFs, Pauncz (1979)]. Determinants offer certain advantages in the context of the highly efficient 'direct CI' methods, which, in contrast to direct SCF, hold integrals on disk, but never explicitly construct or store the Hamiltonian matrix [Roos (1972), Siegbahn (1983), and

Table 2.5. DZP CI and CC spectroscopic constants of $X\,^1\Sigma_g^+\,C_2$[a].

	CISD	CISDTQ	CISDTQPH	CCSD	CCSDT	FCI
r_e	1.2572	1.2650	1.2694	1.2632	1.2675	1.2695
ω_e	1901.6	1842.5	1814.6	1862.3	1829.1	1813.4
$\omega_e x_e$	12.27	13.00	13.28	12.24	12.57	13.19
α_e	0.01600	0.01668	0.01698	0.01620	0.01652	0.01699
f^{rr}	12.7837	12.0014	11.6398	12.2597	11.8265	11.6251
f^{rrr}	−79.45	−75.77	−73.97	−76.34	−74.08	−73.89
f^{rrrr}	376	353	343	365	350	346

[a] All data are taken from Leininger *et al.* (1998). Bond distances (r_e) in Å; harmonic frequencies (ω_e), vibration–rotation interaction (α_e), and anharmonicity ($\omega_e x_e$) constants in cm^{-1}; quadratic (f^{rr}), cubic (f^{rrr}), and quartic (f^{rrrr}) force constants in aJ $Å^{-2}$, aJ $Å^{-3}$, and aJ $Å^{-4}$, respectively.

Knowles and Werner (1988)]. Methods which explore the possibility of elimination of the CI vector itself are called 'superdirect' [Duch (1989), and Duch and Meller (1994)].

Despite the fact that the number of determinants is typically 2–4 times larger than the number of CSFs, all modern, efficient FCI programs, following pioneering work by Handy (1980), are based on Slater determinants. For example, the vectorized FCI algorithm of Knowles and Handy (1984) allowed the evaluation of several important FCI benchmarks [Bauschlicher, Langhoff, and Taylor (1990)]. The unnecessarily large operation count of the Knowles–Handy full CI algorithm has been reduced substantially [Olsen *et al.* (1988), Zarrabian, Sarma, and Paldus (1988), and Harrison and Zarrabian (1989)]. As a result of continuing efficiency improvements, the first converged CI calculation requiring more than one billion Slater determinants was reported in 1996 [Evangelisti *et al.* (1996)], using a $9s2p1d$ CGF basis for Be_2, although the feasibility of iterations toward this milestone had been demonstrated six years earlier [Olsen, Jørgensen, and Simons (1990)]. It is expected that FCI calculations of this size will become common in the next few years on high-end workstations.

Recently there have been reports of passing the one quadrillion mark in FCI extrapolations on a personal computer [Wulfov (1996a,b)]. This work is representative of the immense literature advocating configuration selection schemes for systematic approaches to the FCI limit; see, e.g., the CIPSI approach [Huron, Malrieu, and Rancurel (1973)], the 'selection-plus-perturbation-correction' approach [Harrison (1991)], as well as Feller and Davidson (1989) and Novoa, Mota, and Arnau (1990). For an excellent early review, see Shavitt (1977). It has long been recognized that the FCI Hamiltonian matrix is very sparse and that perturbation theory (PT) can be used to evaluate/estimate contributions for myriad, unimportant configurations. One of the first methods taking advantage of this observation, CIPSI [Huron, Malrieu, and Rancurel (1973)], diagonalizes the Hamiltonian in a subspace of selected determinants, and uses the resulting eigenvector as the zeroth-order wave function in a subsequent PT treatment. The first-order wave function space is increased by the inclusion of determinants based on some energy or wave function threshold η, while the effect of unselected determinants is evaluated by second-order PT. The final energy (E_η) therefore comprises an exact solution within an internal space augmented by perturbational corrections for the residual external space. Most contemporary CI + PT schemes are akin to the original CIPSI approach. In both CIPSI and CI + PT schemes, the FCI limit is estimated by extrapolating E_η sequences to zero threshold. In recent studies [Povill *et al.* (1994)] a self-consistent size-extensive dressing was added to ensure size extensivity of the Hamiltonian matrix. The 'selection-plus-perturbation-correction' approach of Harrison (1991) lacks any *ad hoc* or extrapolation procedures; its accuracy awaits further evaluation.

Another established configuration selection approach is the MRD-CI method [Buenker and Peyerimhoff (1974), and Buenker and Peyerimhoff (1975); see also the chapter by Buenker *et al.* in this volume]. In this treatment the reference space is chosen *a priori* rather than iteratively, based on the observation that the most compact wave functions are obtained by truncating the singles and doubles excitation space and not the reference space. The single and double substitutions are added to the reference space and the computed CI energies are compared with an energy-lowering threshold until convergence. In effect, the multireference excitation space is partitioned into an internal space, subjected to a large CI treatment, and an external space, whose effects are quantified by Brillouin–Wigner PT and extrapolation procedures. These CI + PT schemes, as well as their variants, showed that it is not easy to establish the convergence of the energy. It is expected that convergence of properties can be even more difficult, although there are some indications that this might not always be true [Cave, Xantheas, and Feller (1992), and Wulfov (1996b)].

2.2.3 THE ONE-PARTICLE BASIS

Molecular orbitals can be constructed either numerically or by expansion techniques. While the numerical approach offers remarkable flexibility and accuracy, it is computationally intractable, because of the large number of grid points needed, for all chemical systems but atoms and small linear molecules [see, e.g., Fischer (1977), and Christiansen and McCullough (1977)]. Therefore, for polyatomic systems the only choice is to expand the molecular orbitals in a set of simple analytical one-electron functions. Consequently, all traditional quantum chemical calculations, whether HF, CI, MP, or CC, start with the selection of a one-particle basis set. The accuracy and dependability of any quantum chemical computation depends supremely on this basis set.

A one-center expansion [Coolidge (1932), and Bishop (1967)], despite its many desirable mathematical characteristics (e.g., orthogonality and simple recursive relations), lacks the physical characteristics of electronic structures of chemical systems. Consequently, many-center, atoms-in-molecules (AIM) expansions are much more compact and economical than one-center expansions, in particular for larger molecules. The basic design principles of AIM basis sets for electronic structure calculations are as follows: (a) easy manipulation and evaluation of Hamiltonian matrix elements (e.g., numerical stability, easy differentiation, and fast integration); (b) rapid convergence, requiring only a few terms for a reasonably accurate description of molecular electronic distributions; and (c) an orderly and systematic extension toward completeness.

While exploratory calculations have been performed employing basis functions of different mathematical nature (e.g., ellipsoidal functions [Ebbing (1963)], hydrogenic functions [Hall (1959)], and mixed functions [Silver (1971)] have been tested), there is general agreement that Gaussian-type functions (GTF) and Slater-type (exponential) functions (STFs) are best suited for electronic structure computations. In fact, a large number of hierarchical Gaussian basis sets have been developed for the efficient calculation of energies and molecular properties [Huzinaga (1985), Poirier, Kari, and Csizmadia (1985), Wilson (1987), Almlöf and Taylor (1987), Dunning (1989), Widmark, Malmqvist, and Roos (1990), Almlöf and Taylor (1991), and Almlöf (1995)]. We use the term hierarchical in the sense that these basis sets allow approach to the complete basis set limit (CBS) in a systematic fashion. Perhaps the most successful one in this regard is the family of correlation-consistent basis sets [Dunning (1989)].

The usual choice [Boys (1950)] for the form of basis functions is the Cartesian Gaussian-type function (GTF), whose definition is [McWeeny (1992), p. 526]

$$g^{a,b,c}(x, y, z; \zeta, \boldsymbol{r}_{\mathrm{A}}) = N_a N_b N_c (x - x_{\mathrm{A}})^a (y - y_{\mathrm{A}})^b (z - z_{\mathrm{A}})^c \exp\left(-\zeta |\boldsymbol{r} - \boldsymbol{r}_{\mathrm{A}}|^2\right),$$

with

$$N_\alpha = \left[\frac{(2\alpha-1)!!}{\zeta^a}\left(\frac{\zeta}{\pi}\right)^{1/2}\right]^{1/2} \tag{18}$$

where the N_α ($\alpha = a$, b, c) are normalization constants, a, b, and c are nonnegative integers, the orbital exponents ζ are taken to be positive, and the basis function is centered on atom A at $\boldsymbol{r}_A = (x_A, y_A, z_A)$. The use of the spherical Gaussian-type function, whose definition is

$$S^{n,l,m}(r, \theta, \phi; \varepsilon, \boldsymbol{r}_A) = Nr^{n-1} \exp\left(-\varepsilon r^2\right) Y_{l,m}(\theta, \phi), \tag{19}$$

is also possible [McMurchie and Davidson (1976)], where the spherical polar coordinates $\{r, \theta, \phi\}$ are relative to center $\boldsymbol{r}_A$. Frequently, linear transformations of Cartesian Gaussians are invoked to yield manifolds of pure spherical harmonics ($l = a + b + c$ plus lower contaminants) or real combinations thereof [Schlegel and Frisch (1995)]. Contrary to Slater-type functions (STFs), GTFs neither satisfy the nuclear cusp condition, nor do they decay properly as $\exp(-\zeta r)$ at long range. Nevertheless, GTFs give integrals that can be computed very efficiently, making them superior to STFs and allowing the actual number of primitive functions in the basis to be increased to mitigate any short- and long-range deficiencies. The reason for the ease with which integrals over GTFs may be obtained is related to two important analytic properties of Gaussian distributions: their separability in the Cartesian directions and the Gaussian product rule [Saunders (1975)].

It is computationally more efficient to take each basis function $\chi_r^{a,b,c}$ as a contraction of primitive Cartesian Gaussians centered on the same atom with the same a, b, and c values but different exponents ζ_j:

$$\chi_r^{a,b,c}(x, y, z; \boldsymbol{r}_A) = \sum_j c_{jr} g^{a,b,c}(x, y, z; \zeta_j, \boldsymbol{r}_A). \tag{20}$$

The contraction coefficients c_{jr} and the orbital exponents ζ_j are preoptimized for a given basis set and are held fixed during the actual quantum chemical calculation. The contraction coefficients are usually obtained either from atomic HF calculations or by fitting Gaussians to Slater functions. It is important to distinguish between segmented and general contractions. In the general contraction scheme [Raffenetti (1972)] the contracted basis is a non-sparse transformation of the primitive set, i.e., no restriction is placed on the contractions and all primitive functions of a given angular momentum may contribute to each of several contracted basis functions. In the segmented contraction approach [Taketa, Huzinaga, and O-ohata (1966)], designed to simplify the evaluation of molecular integrals over contracted functions, each primitive function is allowed to contribute to only one contracted orbital (or perhaps two). Although the integrals are evaluated most efficiently over Cartesian Gaussians, to avoid problems with linear dependence, the use of pure spherical harmonics, with supernumerary contaminants removed, should be preferred over the use of Cartesian functions.

At last we note that what is needed in correlation calculations, as opposed to HF calculations, is not only a set of atomic orbitals, that resemble as closely as possible the occupied orbitals of the atomic systems, but also a set of spatially compact, orthogonal virtual orbitals, into which electrons can be excited and hence correlated. Therefore, it is generally expected that optimal basis sets for uncorrelated and correlated calculations will be quite different. Correlated-level calculations demand the use of much better one-electron functions; for example, basis sets of double-zeta quality (i.e., two sets of basis functions for each shell) in the valence region augmented with polarization functions are considered to be of the lowest acceptable quality.

2.2.3.1 Basis set expansions

Indispensable for high-accuracy *ab initio* studies is an understanding of the behavior of finite-order correlation energies with respect to basis set augmentation [see, e.g., Klopper *et al.* (1999)]. The asymptotic behavior of basis set expansions was first examined in the He atom archetype by Schwartz (1962) and by Carroll, Silverstone, and Metzger (1979). The analysis was extended by Kutzelnigg and Morgan (1992) to second- and third-order energies of all possible two-electron atoms and also to MP2 energies of arbitrary n-electron atoms; in particular, coefficients were derived for the leading terms in the $(l+\frac{1}{2})^{-n}$ partial-wave expansion of the atomic correlation energy, where l is a given angular-momentum quantum number. Assuming saturation of the radial basis, the contribution to $E^{(2)}$, the second-order energy, from angular momentum l decays asymptotically as $(l+\frac{1}{2})^{-4}$ for natural parity singlet states, as $(l+\frac{1}{2})^{-6}$ for triplet states, and as $(l+\frac{1}{2})^{-8}$ for unnatural parity singlet states. Clearly, the observed inverse-power decay of correlation energies falls short of the preferable exponential decay of the basis set incompleteness error (BSIE). It must also be noted that (a) l is not a good quantum number for molecules; (b) the partial-wave expansion, i.e., the successive addition of saturated l-shells to the basis, is not a useful practical approach for the design of molecular basis sets; and (c) results obtained from partial-wave expansions form the basis of most energy extrapolation schemes used for molecules (see Section 2.2.3.6). An alternative to the partial-wave expansion is provided by the so-called principal expansion [see, e.g., Klopper *et al.* (1999)], which makes use of the principal quantum number n. In the principal expansion it is assumed that the $\sum_{k=1}^{n} k^2 = \frac{1}{3}n(n+\frac{1}{2})(n+1)$ AO basis functions (1s for $n=1$, 1s2s2p for $n=2$, etc.) are fully optimized, and that systematic sequences of basis sets are defined by increasing values of n. Characteristics of the principal expansion are related to certain hierarchical basis sets of computational quantum chemistry (see Section 2.2.3.5).

In accurate CI expansions of the electronic wave function, by far the most determinants are introduced to describe the short-range correlation of the electrons [see, e.g., Kutzelnigg (1985)]. If one employs instead a wave function that contains r_{ij} terms linearly, where r_{ij} is the interelectronic distance of electrons i and j, much shorter expansions are sufficient. In some sense, we may consider conventional CI, CC, and PT approaches as an attempt to expand the inverse interelectronic distances r_{ij} appearing in the Coulomb potential in products of one-electron functions centered on the nuclei, an expansion which becomes ill behaved near electron coalescence points. Such attempts naturally will lead to very protracted asymptotic decay of the energy and wave function with respect to successive expansion of the one-particle space to higher angular momentum components. Accordingly, in conventional quantum chemical computations, achieving chemical accuracy is predicated on some error cancellation among absolute energies, and spectroscopic accuracy is virtually unattainable.

2.2.3.2 Even-Tempered Basis Sets

For basis sets larger than the minimal STO-kG basis sets [Hehre, Stewart, and Pople (1986), and Stewart (1969)], it is popular to determine orbital exponents by minimization of the total HF energy for the corresponding atom. Such determinations of exponents are tedious, since (a) the nonlinear optimization problem may exhibit multiple solutions; and (b) it must be repeated for each atomic system of interest. One way to circumvent pitfalls and streamline the process is to use orbital exponents which form a geometric progression. This approach [McWeeny (1948)] was explored by Reeves (1963), and was later investigated by several workers, including Raffenetti (1975), Feller and Ruedenberg (1979), Schmidt and Ruedenberg (1979), and Klahn (1985). In these so-called even-tempered (ET) basis sets, the exponents of an m-zeta set for each angular momentum l are constrained to be of the form $\ln \zeta_{klm} = \ln \alpha_{lm}, +k \ln \beta_{lm}$, $k = 1, 2, \ldots, m$.

Advantages of ET basis sets are as follows: (a) for a set value of l and m, they depend only on two parameters, α and β; (b) they allow straightforward generation of hierarchies of basis sets; (c) they may be employed as starting values for an unconstrained optimization of the exponents; (*d*) they allow approach of the CBS limit in a simple and controlled fashion; and (*e*) if β remains greater than unity, no linear dependence in the basis can arise. Related to ET basis sets are the so-called universal basis sets [Silver and Wilson (1978), and Mezey (1979)], in which the exponents are fixed independently of the nuclear charge.

2.2.3.3 Split-valence basis sets

Certain basis sets [such as the Huzinaga–Dunning sets, Huzinaga (1965), Dunning (1970), and Dunning (1971), popular in the 1970s and 1980s] were constructed in a two-step procedure: first a set of primitive functions was deduced from atomic calculations, then a smaller set of contracted functions was generated by taking linear combinations of the primitive atomic functions without exponent reoptimization. A more flexible approach is to optimize the orbital exponents and the contraction coefficients simultaneously, obtaining, in some sense, the best segmented basis set for a fixed number of primitive and contracted functions. Particularly popular among the basis sets designed in this way are the split-valence sets of Pople and co-workers, e.g., the 3-21G, 6-31G, and 6-311G basis sets [Hehre *et al.* (1986)]. Split-valence basis sets are of multiple-ζ character in the valence region [e.g., double-ζ for 6-31G and triple-ζ for 6-311G; see, however, Grev and Schaefer (1989)] but provide only a single-ζ description of the core. The sequences of digits in their designations signify the number of Gaussian primitives in each sp contraction. The split-valence basis sets are often augmented with polarization (*) and diffuse (+) functions on one or more atoms. Polarization functions have angular momentum higher than those of the occupied orbitals in an atomic HF calculation. Therefore, (a) they are not needed for the conceptual bonding description of the atom; and (b) they give extra angular flexibility to the wave function, since polarization functions with exponents similar to those of valence orbitals can be considered as members of a Taylor series expansion helping to describe the effect of displacing the basis functions from the nuclei. An alternative to polarization functions is the use of bond-centered orbitals or floating orbitals [Hurley (1954), Reeves (1963), Frost (1967), and Huber (1980)], the latter being allowed to find optimal positions. Despite their conceptual appeal, floating orbitals find less favor in modern quantum chemistry because of major problems with superposition error [e.g., Bauschlicher (1980), and Martin, François, and Gijbels (1989); see also Section 2.2.3.7]. Diffuse functions are needed for anions as well as to describe longer-range interactions, e.g., hydrogen bonds. While enormous quantities of chemical information have been uncovered via *ab initio* studies based on common split-valence basis sets, their use is not recommended for constructing PESs, as better choices are now available.

2.2.3.4 Atomic natural orbital (ANO) basis sets

Natural (spin)orbitals [Löwdin (1955), and McWeeny and Kutzelnigg (1968)] are eigenfunctions of the one-electron density operator. The associated eigenvalues, known as occupation numbers, are a measure of the contribution of each orbital to the electron density. Atomic natural orbitals (ANOs) are obtained by diagonalizing the one-electron density matrix from a correlated atomic calculation.

The construction of ANO basis sets [Almlöf and Taylor (1987), and Almlöf, Helgaker, and Taylor (1988)], built on the observation that natural orbital occupation numbers provide an excellent criterion for contracting basis functions, proceeds as follows. First, a conventional HF calculation is carried out on the ground electronic state of the atom employing a chosen set of primitive Gaussians. Next, a correlated-level atomic calculation in the uncontracted basis, in reality a CISD computation, is executed. The one-particle density matrix is constructed, and perhaps an average is taken of density matrices for various states of the neutral atom and its ions. Diagonalization of this matrix provides both contraction coefficients for the strongly-occupied orbitals and a truncation of the basis in the weakly-occupied space. Important observations [Almlöf and Taylor (1987), and Almlöf, Helgaker, and Taylor (1988)] on ANO basis sets are:

(1) ANOs selected on the basis of occupation number thresholds and grouped together according to their principal quantum number exhibit similar contraction loss in the energy.
(2) Occupation number criteria [Almlöf and Taylor (1987)] lead to the same contraction of polarization functions as criteria for correlation energy improvement [Jankowski *et al.* (1985)].
(3) Basis functions which are optimal in describing correlation effects in atoms, work similarly well in describing molecular correlation effects.

All ANO sets employ the same underlying primitive set. An important advantage of the procedure is that it allows for hierarchical basis set truncation within the general contraction scheme. The ANO basis sets provide excellent convergence characteristics with respect to the total electronic energy. Although ANOs have been constructed with correlated calculations in mind, their design (they have the correct nodal structure) makes their use appropriate for HF calculations.

Some deficiences and limitations of ANOs are:

(1) They are constructed from a large number of primitive functions, so even in cases of strong contractions, their use is rather expensive. Nevertheless, in modern computational quantum chemistry, calculation of integrals and integral derivatives is not the major bottleneck.
(2) They do not provide enough flexibility in the valence region to allow for highly accurate calculation of some molecular properties. However, the ANO basis sets can either be augmented with diffuse functions, or the most diffuse Gaussians can be left uncontracted.

In addition to the original ANO constructions [Almlöf and Taylor (1987)], ANO basis sets built from average density matrices have been published for first-row [Widmark, Malmqvist, and Roos (1990)] and second-row [Widmark *et al.* (1991)] atoms. In this work CISD calculations were performed for each atom in its ground state, for the corresponding positive and negative ions, for the ground-state atom placed in a small homogeneous electric field, and in some cases for excited atomic states. The ANOs resulting from the average density matrices simultaneously give accurate values for the ionization energy, the electron affinity, and the polarizability of the atoms.

It has been shown [Langhoff, Bauschlicher, and Taylor (1988), and Widmark *et al.* (1991)] that ANO basis sets give only small errors due to basis set superposition error (BSSE) for first- and second-row atoms (for a brief discussion of BSSE, see Section 2.2.3.7). Most of the BSSE obtained is due to missing basis functions with higher angular momentum.

2.2.3.5 Correlation-consistent basis sets

Families of correlation-consistent (cc) basis sets, developed by Dunning and co-workers [Dunning (1989), Kendall, Dunning, and Harrison (1992), and Woon and Dunning (1993)], result from convergence studies of the correlation energy with respect to saturation of both the radial and angular spaces. These basis sets are correlation consistent in the sense that each basis set contains all functions that lower the energy by at least a set amount. It is comforting to know that the use of such energy criteria [Jankowski *et al.* (1985)] during the design of basis sets gives the same pattern of angular momentum functions as found in the case of the ANOs. The resulting cc basis sets are denoted as (aug)-cc-p(C)VXZ (X = D, T, Q, 5, 6, etc.), where (aug) means the addition of diffuse functions, (C) the addition of functions to describe core correlation, and X

is the cardinal number of the basis. The cardinal number also represents the highest spherical harmonic contained in the basis set. The representative equation for the number of contracted functions in cc-pVXZ basis sets, as a function of the cardinal number X, is

$$\tfrac{1}{3}(X+a)\left(X+a+\tfrac{1}{2}\right)(X+a+1)+b, \tag{21}$$

where $a=b=0$ for first-row, $a=1$ and $b=0$ for second-row, and $a=1$ and $b=4$ for third-row atoms. Similar equations hold for the (C) and (aug) correlation-consistent basis set series [Helgaker *et al.* (1997b)].

Unlike the ANO sets, which all employ the same underlying primitive set, each correlation consistent basis set is built around an accurate representation of the atomic ground-state HF orbitals, obtained by carrying out a general contraction of a primitive Gaussian basis, the size of which increases with the cardinal number. For example, the cc-pVDZ basis for B–Ne contains a [2s1p] contraction of a (9s4p) set for the occupied orbitals, as well as a [1s1p1d] set of primitive correlating functions. The occupied [2s1p] contraction is determined variationally in an atomic HF calculation and the correlating [1s1p1d] set is then added so as to maximize its contribution to the correlation energy in an atomic valence CISD calculation. It turns out that the optimized s and p exponents are similar to the most diffuse exponents in the primitive set employed for the occupied HF orbitals. To reduce the computational cost, the exponents for the correlating s and p functions are therefore chosen to be equal to the most diffuse exponents in the primitive set. For cc-pVDZ the underlying primitive set is therefore (9s4p1d). The cc-pVDZ basis set contains too few functions to be useful for accurate molecular calculations. For all atoms, there is a single-zeta representation of the core orbitals and a multiple-zeta representation of the valence orbitals.

Performance of cc basis sets, employing standard electronic wave functions, have been tested in detail for geometric structures of molecules comprising first-row elements [see, e.g., Helgaker *et al.* (1997a)], conformational barriers of prototypical molecules [see, e.g., Császár, Allen, and Schaefer (1998)], as well as dissociation energies [Martin (1996)]. For second-row atoms cc sets lack core-polarization functions [Martin (1999), and Tarczay *et al.* (2000)], thus slowing down their convergence at the HF level. Performance of the different cc-pVXZ basis sets (X = D, T, Q, 5) for the prediction of force constants (potential energy surface derivatives at equilibrium) can be judged from the data presented in Table 2.6. One must note that adequate description

Table 2.6. Basis set effects on the force constants of diatomic molecules at the CCSD(T) level[a].

Molecule	Force constant	cc-pVDZ	cc-pVTZ	cc-pVQZ	cc-pV5Z	expt.[b]
N_2	$f^{(1)}$	−0.5068	−0.0679	−0.0093	0.0209	—
	$f^{(2)}$	26.33	23.59	23.20	23.11	22.94
	$f^{(3)}$	−189.7	−172.7	−169.7	−169.9	−169.6
	$f^{(4)}$	1108.5	1018.9	998.4	991.6	997.6
	$f^{(5)}$	−6114	−6054	−5949	−5606	−5861
	$f^{(6)}$	31017	39481	39117	37074	37975
F_2	$f^{(1)}$	−0.1866	−0.0081	0.0037	0.0147	—
	$f^{(2)}$	4.796	4.816	4.763	4.718	4.703
	$f^{(3)}$	−33.4	−35.8	−34.9	−34.75	−36.39
	$f^{(4)}$	177.6	180.0	177.4	174.5	211.3
	$f^{(5)}$	−952	−843	−903	−902	
	$f^{(6)}$	5714	4642	5272	6192	

[a]Units for the nth-order force constants $f^{(n)}$ are aJ $Å^{-n}$. All calculations have been performed at the experimental equilibrium geometries of $r(N_2) = 1.097685$ Å and $r(F_2) = 1.4119$ Å.
[b] RKR results taken from Allen and Császár (1993).

of multiple bonds, especially triple bonds, requires considerably more extended basis sets than typical single bonds. Another difficulty is the description of atoms with high valence electron density (such as O and F), in which case the use of diffuse functions should be considered.

2.2.3.6 Basis set incompleteness error (BSIE)

For conventional methods of electronic structure theory, convergence of electron correlation energies with the expansion of the one-particle basis set is intrinsically and painfully slow. Fortunately, for energy values the convergence also appears to be rather systematic, making it possible to extrapolate towards the complete basis set (CBS) limit. For properties, the convergence behavior is much less systematic, mostly because of competing terms and error cancellation. For example, in the prediction of geometric parameters of stationary points, corresponding to zero first derivatives with respect to geometric coordinates, the competing effects of nuclear–nuclear, as well as HF and correlation electronic energy first derivatives have to be considered. The convergence properties of these derivatives are quite different, making the apparent extrapolation of geometric parameters rather problematic [Helgaker *et al.* (1997a)].

Based on the use of cc basis sets (aug)-cc-p(C)VXZ, several energy extrapolation formulas have been proposed involving the cardinal number (X) in these series. The assumption that incremental ($X \rightarrow X+1$) lowerings of the total energy lie in a geometric progression is equivalent to extrapolation via the exponential form of Feller (1992), viz.,

$$E_X = E_{\text{CBS}} + a\,\exp(-bX). \tag{22}$$

This form appears operative for HF energies (see Table 2.7) and perhaps properties. As can be seen from Table 2.7, the convergence is in all cases uniform and systematic. Clearly, the cc basis sets provide a convenient framework for the quantitative study of molecular systems containing first-row atoms at the HF level. Apparently, the molecular core orbitals retain atomic forms and are not polarized significantly by chemical bonding. In HF calculations on molecules containing

Table 2.7. RHF total energies (in E_h) obtained with cc basis sets and the corresponding complete basis set (CBS) limits[a].

Basis set	H_2O	NH_3	HNCO	C_2H_6	HCOOH
cc-pVDZ	−76.026719	−56.195630	−167.778298	−79.229596	−188.781864
cc-pVTZ	−76.057074	−56.217862	−167.828765	−79.255132	−188.841301
cc-pVQZ	−76.064703	−56.223091	−167.841576	−79.260385	−188.856117
cc-pV5Z	−76.066957	−56.224713	−167.844527	−79.261721	−188.859785
cc-pV6Z	−76.067267	−56.224921	−167.844906	−79.261883	−188.860214
CBS	−76.067316	−56.224952	−167.844962	−79.261905	−188.860271
aug-cc-pVTZ	−76.060490	−56.220325	−167.831564	−79.260412	−188.844528
aug-cc-pVQZ	−76.065869	−56.223972	−167.842306	−79.265343	−188.857006
aug-cc-pV5Z	−76.067187	−56.224854	−167.844636	−79.266585	−188.859915
aug-cc-pV6Z	−76.067329	−56.224958	—	—	−188.860235
CBS	−76.067346	−56.224972	−167.845281	−79.267002	−188.860275
cc-pCVDZ	−76.027123		−167.779607	−79.230583	−188.782875
cc-pCVTZ	−76.057234	−56.218091	−167.829896	−79.255515	−188.842120
cc-pCVQZ	−76.064815	−56.223188	−167.841979	−79.260585	−188.856522
CBS	−76.067366		−167.845800	−79.261879	−188.861147

[a] Reference geometries have been optimized at the unfrozen-core aug-cc-pVTZ CCSD(T) level for NH_3 and H_2O, and at the unfrozen-core cc-pVTZ CCSD(T) level for HNCO, C_2H_6, and HCOOH [Császár, Allen, and Schaefer (1998)]. Extrapolation to the CBS limit involved the last three total energy values obtained within a given basis set family and the use of equation (22).

first-row atoms, therefore, the use of the smaller valence cc-pVXZ sets is recommended over the cc-pCVXZ sets.

Computational evidence indicates, however, that exponential extrapolations underestimate correlation energy limits [Martin (1996), Helgaker *et al.* (1997a), and Császár, Allen, and Schaefer (1998)]. The partial-wave analyses (see Section 2.2.3.1) suggest alternative extrapolations, but they rigorously apply only to expansions in which the radial space is first saturated for each l, whereas the cc basis sets are built up with radial and angular functions which achieve an energy lowering simultaneously and in a balanced fashion. The well-known slow convergence of electron-correlation effects with basis-set improvement is better accounted for by inverse-power fits. Possible forms [Feller (1992), Martin (1996), Martin and Taylor (1997), Helgaker *et al.* (1997a), Császár, Allen, and Schaefer (1998), Klopper and Helgaker (1998), and Truhlar (1998)] of inverse-power fits tested are

$$E_X = E_{\text{CBS}} + \alpha(X+\beta)^{-\gamma}, \tag{23}$$

where γ is not necessarily an integer,

$$E_X = E_{\text{CBS}} + \sum_{k=3}^{k_{\max}} \alpha_k (X+\beta_k)^{-k}, \tag{24}$$

where $k_{\max}$ is small, or alternatively a polynomial in inverse powers of the number of basis functions. In particular, Helgaker *et al.* (1997b) have proposed equation (23) with $\gamma = 3$ and $\beta = 0$, a simple form which allows E_{CBS} to be estimated from correlation energies for only two basis sets in the correlation-consistent series. In contrast, Martin (1996) has chosen a phenomenological inverse-power form for first-row species,

$$E_X = E_{\text{CBS}} + \frac{a}{(X+1/2)^4} + \frac{b}{(X+1/2)^6}, \tag{25}$$

termed Schwartz4 ($b = 0$) or Schwartz6 ($b \neq 0$). It is yet unclear which form will prove to be of highest accuracy for a wide class of molecular systems. Therefore, further elaborate studies are needed. One hindrance is the substantial cost of these computations. For example, if the goal is to achieve $1\,\text{m}E_\text{h}$ accuracy in the calculation of the CCSD total energy, a cc-pCVXZ basis set with $X = 10$ would be needed for water [Klopper *et al.* (1997)].

Alternative schemes have also been advanced to achieve the CBS limit. Siegbahn, Svensson and Boussard (1995) have proposed a simple parametrized configuration interaction (PCI-X) method which uses a single scale factor to estimate *ab initio* limits from various correlation procedures employing modest basis sets. If it is assumed that a certain one-electron basis recovers X % of the correlation energy, one can simply scale the computed correlation energy with a factor of 100/X. This multiplicative scheme can, in principle, be employed for the calculation of PESs. Another scheme for empirically based extrapolations toward this limit is the complete-basis-set (CBS) model of Petersson and co-workers [Petersson and Braunstein (1985), Petersson *et al.* (1988), Ochterski, Petersson, and Montgomery (1995), and Petersson *et al.* (1998)]. The CBS model chemistries feature basis-set extrapolations based on formulas for the asymptotic convergence of pair natural orbital expansions, progressively higher order correlation treatments with basis sets of decreasing size, and empirical corrections for zero-point energy, spin contamination (for open-shell systems), and remaining correlation effects. Most accurate in the CBS family is the complete basis set-quadratic configuration interaction/atomic pair natural orbital (CBS-QCI/APNO) model. In the Gaussian-2 approach and in its modification [Curtiss *et al.* (1991), Raghavachari and Curtiss (1995), and Curtiss *et al.* (1998)], MP2, MP4, and QCISD(T) energies from basis sets up to spdf in quality are employed with certain additivity approximations, and an empirical additive correction, depending linearly on the number of paired/unpaired electrons, is utilized to account for deviations from the *ab initio* limit. Martin (1992, 1994) has developed another empirical (three-parameter) additive energy correction scheme, which works for molecules having well-defined σ and π bonds and lone electron pairs. In mapping entire PESs, various energy correction schemes may break down if proper account is not taken of the geometric dependence of the correction or if the correction is fundamentally discontinuous in nature.

2.2.3.7 Basis set superposition error (BSSE)

The straighforward way to determine interaction energies is provided by the supermolecule (SM) approach. In the simplest case of two interacting subsystems, A and B, the uncorrected interaction energy E^{AB} of the complex AB is the energy difference

$$\Delta E(\text{R}) = E^{\text{AB}}(\text{AB}; R) - E^{\text{AB}}(\text{AB}; \infty) = E^{\text{AB}}(\text{AB}; R) - E^{\text{A}}(A) - E^{\text{B}}(B), \tag{26}$$

the latter equality holding if the quantum chemical method is size-consistent. In this equation R is characteristic of the AB separation, the letters in parentheses refer to the finite basis set employed in the calculation (N_{A} and N_{B} basis functions on systems A and B, respectively), and the system for which the quantum chemical energy is computed is given in superscript. One considerable advantage of the SM approach is that use can be made of the highly efficient algorithms of widely distributed standard quantum chemical softwares. However, the interaction energy obtained this way is prone to be too negative. This artificial energy lowering is an intrinsic feature of finite-basis calculations on molecular complexes, and it has been termed the basis set superposition error [BSSE; Liu and McLean (1973)]. The size of the BSSE may not be small compared with $\Delta E(R)$. Therefore, accurate and reliable prediction of potential energy hypersurfaces of molecular complexes and chemical reactions is only possible if BSSE is removed *a posteriori* or avoided *a priori*.

There are several strategies for calculation of (almost) BSSE-free molecular properties, including interaction energies of non-covalent molecular complexes. One is the symmetry-adapted perturbation theory [SAPT; Jeziorski, Moszynski, and Szalewicz (1994), and Szalewicz and Jeziorski (1997)]. In SAPT the interaction energies are evaluated directly, and this approach is free of BSSE by construction. The latest developments in SAPT are discussed at length in the chapter by van der Avoird of this book. Another possibility to avoid BSSE is the chemical Hamiltonian approach [CHA, Mayer (1987)]. Localized many-body perturbation theories [LMBPT, Kapuy *et al.* (1988), and Kozmutza and Tfirst (1998)] and local electron-correlation methods [e.g., Saebo, Tong, and Pulay (1993)] open ways to minimize the BSSE content of interaction energies.

Historically, the most common *a posteriori* BSSE correction has been the counterpoise (CP) or Boys–Bernardi (BB) scheme [Jansen and Ross (1969), Boys and Bernardi (1970), Feller (1992), van Duijneveldt, van Duijneveldt-van de Rijdt, and van Lenthe (1994), and van Mourik *et al.* (1998)]. During the calculation on species A resulting in an energy $E^{\text{A}}(\text{AB}; R)$, B is a so-called 'ghost species' with zero nuclear charge, no electrons, and a set of N_{B} basis functions centered at positions where the nuclei of B are in the complex. Vice versa, during the calculation on B resulting in an energy $E^{\text{B}}(\text{AB}; R)$, A is a ghost species. At any given point on the PES of the dimer, the overlap-dependent BSSE, $\delta E^{\text{BSSE}}(R)$, is defined as

$$\delta E^{\text{BSSE}}(R) = E^{\text{A}}(\text{AB}; R) - E^{\text{A}}(\text{A}) + E^{\text{B}}(\text{AB}; R) - E^{\text{B}}(\text{B}), \tag{27}$$

which upon subtraction results in the following CP-corrected PES for the total energy:

$$E^{\text{CP}}(R) = E^{\text{AB}}(\text{AB}; R) + [E^{\text{A}}(\text{A}) - E^{\text{A}}(\text{AB}; R)] + [E^{\text{B}}(\text{B}) - E^{\text{B}}(\text{AB}; R)]. \tag{28}$$

This equation is the starting point of most treatments related to the CP scheme:

(1) Equation (28) can be generalized to the case of an arbitrary number of subsystems [see, e.g., White and Davidson (1990), Turi and Dannenberg (1993), and Valiron and Mayer (1997)]. Calculation of E^{CP} for more interacting subsystems is, however, rather expensive as a large and rapidly growing number of calculations, involving various subsets of the overall union of the basis sets, needs to be performed.
(2) Various derivatives of $E^{CP}(R)$ can easily be calculated [see, e.g., Simon, Duran, and Dannenberg (1996)] from derivatives of the five terms present in equation (28).
(3) When geometries of the subsystems change, as a result of their interaction, the problem of 'monomer geometry relaxation' occurs [Mayer and Surján (1992), and Simon, Duran, and Dannenberg (1996)]. Treatment of this relaxation needs only a slight modification of equation (28).

In research on the prototypical case of $(H_2O)_2$ with cc series of basis sets, Feller (1992) demonstrated that CP-corrected dimerization energies converge monotonically from below the limiting D_e value, whereas their uncorrected counterparts converge monotonically from above. The CP binding energies exhibited smoother convergence characteristics; however, for the aug-cc-pVXZ sets BSSE and BSIE closely cancelled, resulting in an apparently more rapid convergence of the uncorrected D_e values. Although in many complexes studied by high-level *ab initio* calculations (e.g., cc-pVTZ CCSD and above) BSSE may affect already minuscule binding energies by a nonnegligible percentage and may change interfragment separations somewhat, predictions of the basic structure and fragment orientations are not likely to be compromised.

The fully consistent approach for an *a posteriori* BSSE correction would be a decomposition of the interacting system into the smallest meaningful units, which, for chemists, are the atoms. Such a treatment is, however, prohibitively expensive for systems containing at least four atoms. It is noted, also in this respect, that BSSE within a covalently-bound molecule has been investigated. This is of special importance if relative energies are to be determined at a few distinct configurations on the same PES. Let us assume that we are interested in the energy difference at two configurations (e.g., in studies on barrier heights [Jensen (1996), and Tarczay *et al.* (1999)] or on dissociation energies). Traditionally, basis functions are assigned to each nucleus and the difference of the two energies provides the required result. However, in the spirit of the CP method, it must be noted that the two calculations employ different basis sets, since now the positions of the functions change. An intramolecular CP (ICP) basis can be designed for these calculations, which has the same number of functions at the same positions for both calculations. The energy difference obtained with the regular and the ICP basis is denoted the intramolecular BSSE. Jensen (1996) has shown, for example, that even at the cc-pV5Z MP2 level there is a $0.13\,\text{kcal mol}^{-1}$ difference due to ICP for the barrier to linearity of water.

There has been certain suspicion [e.g., Saebo, Tong, and Pulay (1993), Woon (1994), and Davidson and Chakravorty (1994)] that the CP correction, sometimes also termed full counterpoise (FCP), overestimates the BSSE content of $\Delta E(R)$, especially at correlated levels. One popular suggestion [Daudey, Claverie, and Malrieu (1974)] was that a more appropriate correction results if the monomer energies are calculated in a basis that contains the virtual orbital space but not the occupied orbitals of the ghost atoms. However, numerical experiments [van Duijneveldt, van Duijneveldt-van de Rijdt, and van Lenthe (1994)] proved that with reasonable basis sets the CP-corrected energies provide the best PESs. The alternative brute force approach of increasing the size of the basis set until the interaction energy becomes stable to the desired accuracy is slowly convergent and possible only for small systems.

The CHA [Mayer (1987)] eliminates the non-physical terms of the Hamiltonian that are due to BSSE *a priori*; therefore, the CHA can be applied straightforwardly to predict BSSE-free quantities (e.g., potential energy and property hypersurfaces). Furthermore, the CHA allows one to determine a PES by single energy calculations for supermolecules with an arbitrary number of subsystems, a considerable advantage over the CP scheme. The CHA has not yet been developed to treat chemical reactions.

Note finally that (a) as one approaches a complete basis, the difference between the different *a posteriori* BSSE corrections is expected to disappear before the BSSE would actually become zero; and (b) on the whole it appears that for intermolecular interaction energies the differences between the HF-level CHA and CP results are small compared with the errors associated with the use of incomplete basis sets and the neglect of correlation effects.

2.3 *Ab initio* evaluation of small correction terms for potential energy hypersurfaces

Advances in the continuing development and widespread application of quantitatively accurate *ab initio* methods have revealed the necessity of a full understanding of the consequences of small but physically relevant correction terms tacitly neglected in most quantum chemical calculations of potential energy (and property) hypersurfaces. These correction terms include: core–core (CC) and core–valence (CV) electron correlation, adiabatic corrections [the diagonal Born–Oppenheimer correction (DBOC)], non-adiabatic corrections (i.e., interactions among surfaces of multiple electronic states), effects due to special relativity, and electronic radiative corrections. Non-adiabatic corrections are treated in detail in the chapter by Buenker *et al.* in this volume. Therefore, they are not discussed further here. The electronic radiative correction (or Lamb shift) [Pasternack (1938), and Bethe (1947)], arising from the interaction of the electron with the fluctuation of the electromagnetic field in vacuum, predicted by quantum electrodynamics, has been considered for the H atom [Pasternack (1938), Lamb and Rutherford (1947), and Drake and Swainson (1990)], for H_2^+ [Bukowski *et al.* (1992)], and for H_2 [Garcia (1996)]. Since Lamb shifts are expected to be much smaller than relativistic (mass-velocity, Darwin, spin-orbit, and Breit) corrections, they are not discussed further here.

2.3.1 CORE CORRELATION

One of the most common presumptions of computational quantum chemistry concerns the core (or inner shell) orbitals, located closest to the atomic nuclei (1s for first-row atoms, 1s2s2p for second-row atoms, etc.). It is assumed that while core orbitals provide important screening both in atoms and molecules, their change is diminutive upon molecule formation. Therefore, the energetics and structure of common molecular species are determined by interactions of valence (outermost) electrons, spatially much more extended than the tight core orbitals. However, if the goal is to achieve quantitative accuracy in the prediction of a property at any point on a surface, the effects of correlating the core electrons must be taken into account.

The core correlation energy is defined as the difference between the all-electron correlation energy and the valence–valence (VV) correlation energy (usually just called the valence correlation energy). The core correlation energy is often partitioned into two contributions: CC and CV correlation. The CC correlation energy arises from double (or higher) excitations out of the core orbitals. The CV correlation energy results from single excitations out of the core orbitals together with excitations from the valence orbitals. The CV correlation energy represents that part of the core correlation energy which is most sensitive to differential effects (e.g., changes in the geometry of the molecule); therefore, it affects substantially the calculation of spectroscopic constants.

Table 2.8. Core–core, core–valence, and valence–valence correlation energies (in mE_h) in the ground electronic states of the nitrogen and neon atoms[a].

Basis set	Nitrogen			Neon		
	Core–core	Core–valence	Valence–valence	Core–core	Core–valence	Valence–valence
cc-pVDZ	−0.287	−1.259	−90.000	−0.215	−1.626	−187.877
cc-pVTZ	−4.760	−5.223	−114.769	−4.475	−8.031	−265.283
cc-pVQZ	−17.387	−9.983	−121.661	−16.722	−15.872	−326.318
cc-pCVDZ	−29.906	−6.101	−90.600			
cc-pCVTZ	−36.089	−13.089	−116.642			
cc-pCVQZ	−38.854	−15.145	−122.249			

[a]All calculations have been performed at the CISD level employing the MR-AQCC *a posteriori* energy correction scheme of Szalay and Bartlett (1995).

Size-extensive methods should be employed to estimate core correlation effects, because a consistent accuracy needs to be maintained when the number of active electrons is changed. In calculating the core correlation energy, special attention must also be paid to the one-particle basis set [Bauschlicher, Langhoff, and Taylor (1988)], since the usual basis sets are developed to describe only the VV interaction of electrons. The compactness and the large dynamical correlation energy associated with core molecular orbitals requires inclusion in the basis set of higher angular momentum functions, with additional radial and angular nodes and very high exponents. In this way one can obtain a large fraction of the correlation energy and can avoid large basis set superposition errors. Just as the standard correlation-consistent polarized valence basis sets (cc-pV*X*Z) give excellent results for VV correlation, the correlation-consistent polarized CV sets (cc-pCV*X*Z) appear to be an excellent choice for core correlation, giving results very close to traditionally designed CV basis sets [Allen, East, and Császár (1993), Császár and Allen (1996), Martin and Taylor (1997), and Peterson *et al.* (1997)] at a reduced cost. Results obtained with both the cc-pV*X*Z and cc-pCV*X*Z basis sets for the ground electronic states of the nitrogen and neon atoms are collected in Table 2.8. As seen from Table 2.8, in any application where CC correlation effects are important, the cc-pV*X*Z sets are inadequate. For the CV correlation energy, the valence basis sets are able to recover an increasing fraction of the energy. In some applications this fraction may be sufficiently large to make the use of the CV basis sets unnecessary. In one of the earliest studies on this subject, Bauschlicher, Langhoff, and Taylor (1988) studied the Be ^1S–^1P, the C ^3P–^5S, and CH^+ $^1\Sigma^+$–$^1\Pi$ electronic separations, and CH^+ spectroscopic constants, dipole moments and $^1\Sigma^+$–$^1\Pi$ transition dipole moments. The effect of core correlation on atomic excitation energies can be substantial, but excited electronic states are not the focus of our review.

Core correlation in simple one- or two-valence-electron atoms was first discussed in detail by Müller, Flesch, and Meyer (1984). Core correlation results in reduced Coulomb repulsion between core and valence electrons. Accordingly, the valence electrons come closer to the nucleus and become more strongly bound, while the core electrons become more screened by the contracted valence charge. These changes have a substantial effect on properties of the atoms.

For few-electron systems Müller, Flesch, and Meyer (1984) have advocated the use of a model operator to describe the effects of CV correlation. Their core-polarization potential (CPP) is based on a description of the correlation as the effects of the fields generated by the valence electrons (and surrounding cores in a molecule) on a central, polarizable core. The CV correlation is approximated by a model operator, $\hat{V}_{CPP}$, included in the Hamiltonian and projected onto the valence space. This potential is generated by the polarizable core c and is proportional to the resulting field (f_c) on the core from all the valence electrons and other cores and to the induced dipole moment ($\alpha_c f_c$, where α_c is the polarizability of core c) associated with this core. The operator becomes

$$\hat{V}_{\mathrm{CPP}} = -\frac{1}{2}\sum_c \left(\alpha_c f_c \cdot f_c - 2f_c \cdot \mu_c^0 + f_c^0 \cdot \mu_c^0\right) \tag{29}$$

where μ_c^0 is the induced dipole moment of core c and f_c^0 is the static field acting on core c. The last two terms correct for the static core polarization obtained with the basis set used (due to the non-spherical static field in a molecular environment). The instantaneous field on core c, f_c, is obtained as (in atomic units)

$$f_c = \sum_i \frac{\boldsymbol{r}_{ci}}{r_{ci}^3} C(r_{ci}, \rho_c) - \sum_{c' \neq c} \frac{Z_{c'} \boldsymbol{R}_{cc'}}{R_{cc'}^3}, \tag{30}$$

where $C(\boldsymbol{r}_{ci}, \rho_c)$ is a cut-off function introduced to remove contributions from electronic charge penetrating the core. It can be taken as

$$C(r_{ci}, \rho_c) = \left(1 - \exp\left[-(r_{ci}/\rho_c)^2\right]\right)^2, \tag{31}$$

where the cut-off radius, ρ_c, is a parameter. The CPP operator thus contains two parameters, α_e and ρ_c. Note that no extension of the basis set into the core region is needed with the CPP approach.

Many successful applications of the CPP technique and similar approaches based on perturbation theory have been made on alkali metals [Müller and Meyer (1984), Jeung, Malrieu, and Daudey (1982), and Fuentealba *et al.* (1982)], alkaline earths [Partridge *et al.* (1990), and Fuentealba *et al.* (1982)], P_2 [Pettersson and Persson (1993)], and on systems containing copper and silver [Stoll *et al.* (1983), and Pettersson *et al.* (1990)]. A few selected results: (a) for covalently bound molecules core correlation tends to decrease the bond distance as a result of the contraction of the valence charge; (b) for weakly bound alkali dimers the geometric effect is substantial, 0.02–0.2 Å, while the effects on the dissociation energy and vibrational frequency are less dramatic [Müller and Meyer (1984)]; and (c) for the van der Waals-bound Mg dimer, the atomic polarizability dominates, and core correlation results in a reduced binding energy and a small increase in the bond distance [Partridge *et al.* (1990)].

In diatomic model calculations on molecules composed of first-row atoms [Martin (1995), Császár and Allen (1996), Halkier *et al.* (1997), and Peterson *et al.* (1997)] the effect of core correlation has been found to be: (a) equilibrium bond distances of first-row diatomic molecules experience a considerable contraction, 0.002 Å or more for multiple bonds and 0.001 Å for single bonds, reducing the errors in r_e predictions at the CCSD(T) level to less than 0.001 Å; (b) the *direct* effect of CC and CV correlation is a correction function to the valence diatomic potential energy curve which has negative curvature at all bond lengths in the vicinity of the equilibrium position, in other words under the constraint of fixed internuclear distance CC and CV correlation decreases the quadratic force constant such that the frequency is lowered; and (c) CV correlation decreases all higher-order force constants as well. Nevertheless, if the total core correlation effect is considered in a conventional, phenomenological sense as the difference between the $\omega(r)$ values for all-electron and partial-electron treatments, the core correlation can be said to increase the harmonic frequency, in accord with the direction of the shift usually anticipated from the associated changes in r_e. Relevant data for the series B_2–F_2 are given in Table 2.9.

Probably the most detailed treatment of the effect of core correlation on a PES was given by Partridge and Schwenke (1997) for the ground electronic state potential energy hypersurface of water, where core correlation effects were evaluated through a set of averaged coupled pair functional (ACPF) calculations. Partridge and Schwenke (1997) observed that the sign and magnitude of the core correlation effect is strongly dependent on the geometry. For example, the core correlation

Table 2.9. Core correlation effects on the dissociation energy (D_e/kcal mol^{-1}), equilibrium bond length (r_e/Å), and harmonic vibrational frequency (ω_e/cm^{-1}) for the B_2–F_2 series[a].

Molecule	D_e	r_e	ω_e	
			indirect	direct
B_2	1.1	−0.007	+10	
C_2	1.5	−0.004	+13	
N_2	0.9	−0.002	+10	−8
O_2	0.3	−0.002	+6	−6
F_2	−0.1	−0.0015	+2	−3

[a] D_e and r_e data are taken from Peterson *et al.* (1997), while the ω_e data are taken primarily from Császár and Allen (1996) and also from Peterson *et al.* (1997).

shift for the barrier to linearity of water is close to $-100\,\text{cm}^{-1}$. Similarly large core correlation shifts for the appropriate barriers have been observed [Császár, Allen, and Schaefer (1998)] for ammonia ($-60\,\text{cm}^{-1}$) and HNCO ($-80\,\text{cm}^{-1}$). For second-row atoms the effect becomes smaller; e.g., for the barrier to linearity of H_2S it is less than 10 cm^{-1} [Tarczay *et al.* (2000)].

2.3.2 THE DIAGONAL BORN–OPPENHEIMER CORRECTION (DBOC)

In nearly all of conventional quantum chemistry, one works within the adiabatic [Born–Oppenheimer (BO)] *separation* of electronic and nuclear motion [Born and Oppenheimer (1927), Born (1951), Born and Huang (1954), Longuet-Higgins (1961), Pack and Hirschfelder (1970), Kołos (1970), Bardo and Wolfsberg (1978), Combes, Dudos, and Seiler (1981), Sutcliffe (1993), Kutzelnigg (1997), and Bunker and Jensen (1998); see also the chapter by Bunker and Jensen in this volume]. As Kutzelnigg (1997) emphasizes, 'the BO separation is the most important approximation in the quantum theory of molecules and solids'.

After early attempts to treat nuclear and electronic motions separately, Born and Oppenheimer (1927) tried to formulate a hierarchy of approximations in terms of an expansion in powers of $\kappa = (m/M)^{1/4}$, where m is the electronic and M a typical nuclear mass. They observed that through κ^4 in the Hamiltonian and κ^2 in the wave function, 'during nuclear motion the electrons move as though the nuclei were fixed in their instantaneous positions' [Born and Huang (1954)]. The leading term through κ^4 is then what is usually associated with the BO approximation. For example, while the exact electronic ground-state energy of one-electron atoms with nuclear charge Z is (in E_h) $-\mu Z^2/2$ with $\mu = (1 + m/M)^{-1}$, a Taylor expansion in κ yields a BO energy of $-Z^2/2$ and a first-order adiabatic correction of $\kappa^4 Z^2/2$ [Kutzelnigg (1997)]. Therefore, corrections to the BO energy seem to scale as Z^2/M.

The original perturbation treatment [Born and Oppenheimer (1927)] is not only rather complex but it is based on a divergent series. Therefore, modern treatments of the BO separation, following Born (1951), start with the expansion of the total wave function for the simultaneous motion of the electrons and nuclei as

$$\Psi_k(\boldsymbol{r}, \boldsymbol{R}) = \sum_{jvr} c^k_{jvr} \psi^{\mathrm{e}}_j(\boldsymbol{r}; \boldsymbol{R}) \psi^{\mathrm{n}}_{vr}(\boldsymbol{R}), \tag{32}$$

where k is a general index of exact rovibronic wave functions, j enumerates BO electronic states ψ^{e}, which depend parametrically on $\boldsymbol{R}$, v and r index the rovibrational nuclear wave functions ψ^{n}, and the c^k_{jvr} are expansion coefficients. In the usual BO *approximation* (which holds excellently for well-separated potential energy surfaces), only a single term is retained in equation (32), say the ψ_0^e term, while treatments which variationally or perturbationally release the mixing coefficients are called nonadiabatic. The general result for the first-order BO correction (DBOC), diagonal in the electronic state, is [Davydov (1965), Sellers and Pulay (1984), Handy, Yamaguchi, and Schaefer (1986), and Kutzelnigg (1997)]

$$\Delta E^{\mathrm{DBOC}} = \langle \Psi_0^e | \hat{T}_n | \Psi_0^e \rangle = -\frac{1}{2} \langle \Psi_0^e | \sum_A \frac{1}{M_A} \nabla_A^2 | \Psi_0^e \rangle = \frac{1}{2} \sum_A \frac{1}{M_A} \langle \nabla_A \Psi_0^e \mid \nabla_A \Psi_0^e \rangle, \tag{33}$$

where the index A runs over the nuclei. The adiabatic correction thus results from the response of the nuclei of finite mass to the instantaneous positions of the electrons, governed by the motion of the molecular center of mass. The energy $E^{\mathrm{BO}} + \Delta E^{\mathrm{DBOC}}$ provides the best energy possible using a single potential energy surface; the addition of ΔE^{DBOC} to the BO PES constitutes the adiabatic approximation [see, e.g., Pack and Hirschfelder (1970), Pilar (1990), pp. 157–160 and pp. 309–314, and Bunker and Jensen (1998), p. 370]. It appears that Sellers and Pulay (1984) were the first to apply the important simplification, proved rigorously by Kutzelnigg (1997), that it is possible to compute the DBOC correction independently of the coordinate system in which the nuclear kinetic energy operator and the electronic wave function [cf. equation (33)] are expressed. This observation opened up the way for the calculation of DBOC energy corrections for arbitrary molecular systems.

The HF approximation seems to be sufficient to establish the correct order of magnitude of the DBOC in closed-shell molecules [e.g., Cencek *et al.* (1998)]. Sellers and Pulay (1984) showed how to evaluate the DBOC for HF wave functions using numerical procedures. Handy, Yamaguchi, and Schaefer (1986) proposed a simple recipe for the evaluation of the DBOC term for closed-shell HF wave functions via analytic second derivative calculations. Extension of the algorithm to UHF wave functions followed later [Ioannou, Amos, and Handy (1996)]. Yarkony and co-workers [Saxe, Lengsfield, and Yarkony (1985), and Lengsfield and Yarkony (1986)] have given the general theory for the evaluation of the adiabatic correction terms for general configuration interaction (CI) wave functions which are based on multiconfiguration self-consistent-field (MCSCF) reference wave functions.

Even though the theory is well established, there is still only a limited number of studies reporting on the magnitude of the DBOC term for usual chemical problems in the literature. The earliest and perhaps best known example is for the H_2 molecule. In their seminal paper Kołos and Wolniewicz (1964) report that at R_e and for separated atoms the DBOC is 114.588 and 119.532 cm^{-1}, respectively, thus contributing 5 cm^{-1} to the value of D_e (38 297 cm^{-1}). This was an important investigation which revealed an experimental discrepancy in the assignment of spectral lines. Garrett and Truhlar (1985) examined the DBOC for barrier heights in chemical reactions. Using the diatomics-in-molecules (DIM) method, they found that the DBOC correction may have a value as large as 0.2 kcal mol^{-1} on some parts of the $F + H_2$ surface. The DBOC contribution, determined at the HF level, to the barrier of the reaction $F + H_2 \rightarrow HF + H$ is 0.05 kcal mol^{-1} [Ioannou, Amos, and Handy (1996)]. Sellers (1986) found a practically negligible DBOC for the ring puckering potential of oxetane, and somewhat more substantial corrections to the potential functions of H_2, CO, and CO_2. At the DZP RHF level Császár, Allen, and Schaefer (1998) determined DBOC contributions (in cm^{-1}) of −17, −11, and −1 for the conformational barrier heights of H_2O, NH_3, and C_2H_6, respectively. In probably the most serious attempts to investigate the effect of the DBOC term on semi-global PESs of polyatomics, Dinelli *et al.* (1995), Cencek *et al.* (1998), and Zobov *et al.* (1996) determined DBOC correction surfaces for H_3^+ and H_2O, respectively. Some important results of Dinelli *et al.* (1995) for H_3^+ are as follows: (a) the DBOC obtained at the 6s3p RHF level for equilibrium H_2 is 101.2 cm^{-1} as compared with the accurate value of 114.6 cm^{-1} of Kołos and Wolniewicz (1964), thus, RHF theory works reasonably well; (b) the adiabatic correction to the dissociation energy appears to be between 2 and 16 cm^{-1} [Dinelli *et al.* (1995), Lie and Frye (1992), and Röhse *et al.* (1994)]; (c) inclusion of the adiabatic correction greatly improves the calculation of rovibrational energy levels, at least at the level of precision obtainable for this simple and light triatomic system, although the calculations were still far short of the precision of high-resolution experiments. One must note here that for H_2 purely *ab initio* calculations beyond the BO approximation [Schwartz and Le Roy (1987), and Kołos and Rychlewski (1993)] have produced a potential curve more accurate than the effective functions derived from spectroscopic data.

In summary, the DBOC is mass dependent [Bunker (1970), Handy and Lee (1996), and Kutzelnigg (1997)], it is usually small, and its calculation is straightforward if analytic derivative methods are available. For highly accurate constructions of potential energy hypersurfaces, it cannot be neglected, especially if hydrogen or other light atoms are present in the molecular system.

2.3.3 RELATIVISTIC EFFECTS

A relativistic effect is defined as the difference in an observable property that arises from the true velocity of light as opposed to the assumed infinite velocity in traditional treatments of quantum chemistry. The importance of accounting for relativistic effects in quantum chemical calculations of heavy atoms or molecules having such atoms is well recognized. Relativistic phenomena indeed have striking chemical and physical consequences in the lower part of the periodic table (an entire book, Balasubramanian (1997), Part B, has been devoted to chemical applications of relativistic electronic structure theory and to the elucidation of relativistic effects). For general molecular systems, the evaluation of relativistic corrections by numerical methods (finite differences or finite elements), applied successfully for atoms and diatoms [Desclaux (1973), and Sundholm, Pyykkö, and Laaksonen (1987)], is not computationally viable, necessitating the usual expansion in a one-electron basis. The resulting techniques can be classified either as 'fully relativistic' (within the framework of four-component spinors), or 'quasirelativistic' (within the structure of traditional nonrelativistic electron correlation theory) [see, e.g., Pyykkö (1978, 1986, 1993), Pitzer (1979), Pyykkö and Desclaux (1979), Balasubramanian (1997), and Quiney, Skaane, and Grant (1998)]. DBOC corrections are proportional to Z^2, while relativistic corrections may scale up to Z^4. Therefore, for all but the lightest systems relativistic corrections are likely to be more important for accurate PES computations than DBOC corrections, as has been demonstrated for water [Császár *et al.* (1998)].

The initial step in fully relativistic calculations is usually the solution of the four-component Dirac–Hartree–Fock (DHF) equation for the many-electron atom or molecule [Mayers (1957), Grant (1961, 1970), and Desclaux (1975)]. Nevertheless, in many-electron systems there are difficulties related to the proper definition of a relativistic many-electron Hamiltonian. Efficient codes have been written for all-electron DHF calculations [e.g., Dyall *et al.* (1991)]. As in nonrelativistic theory, methods to remedy the mean-field approximation via dynamic electron correlation include the CI, MBPT, and CC approaches. Fully relativistic four-component calculations at the CI level [Visscher *et al.* (1993)] as well as at the MBPT level [Collins, Dyall, and Schaefer (1995)] have been developed for molecules. A fully relativistic Fock-space CCSD method for atoms has been implemented and applied to closed- and open-shell atoms [Eliav, Kaldor, and Ishikawa (1994), and Kaldor (1997)]. The major difference vis-à-vis a nonrelativistic treatment from a computational point of view is that the relativistic functions are complex four-component spinors. The four-component form of relativistic functions substantially multiplies the computational effort compared with nonrelativistic treatments. Moreover, the completeness requirements for the one-particle basis are far more stringent, in part because much more flexibility is required in regions near the nuclear centers. Fully relativistic calculations must also deal with problems of variational collapse to physically undesirable solutions [Schwarz and Wallmeier (1982), and Kutzelnigg

(1984)]. Since we are not aware of the use of fully relativistic methods in the generation of PESs, their properties are not discussed further here.

To handle molecular systems with heavy atoms and a large number of electrons, which is already a serious task in nonrelativistic calculations, much effort has been spent on developing methods which can be considered as approximations to the complete relativistic treatment. The most widely used approach is certainly based on relativistic effective core potentials (RECP) [see, e.g., Gropen (1988) or Schwerdtfeger *et al.* (1989); the ECP method itself was first suggested by Phillips and Kleinman (1959), and was later developed by Bonifacic and Huzinaga (1974) and Melius and Goddard (1974)]. To include the effect of relativity in the effective potentials, use is made of relativistic numerical atomic structure calculations. In order to allow their use in traditional non-relativistic methods, the RECPs are commonly presented in a spin-free or spin-averaged form [see, e.g., Hay and Wadt (1985)].

The Dirac equation [Dirac (1928), Darwin (1928), Gordon (1928), and Swirles (1935)], which is hard to handle computationally in its four-component spinor form, can be brought into a two-component form by appropriate transformations [e.g., FW transformations, Foldy and Wouthuysen (1950)]. The resulting equations usually suffer from approximations and divergent expansions. Modified convergent FW transformations have also been proposed. These include methods based on the Douglas–Kroll approach developed further by Hess [DK, Douglas and Kroll (1974), and Hess (1986)] and the ZORA/CPD scheme [Sadlej and Snijders (1994)]. The resulting equations are usually used in the first order of perturbation theory [the first-order treatment was introduced to an *ab initio* treatment of H_2 by Ladik (1959, 1961)]. At the nonrelativistic limit, the most important terms in the first-order relativistic expression [Itoh (1965), Moss (1973), and Wilson (1984)] include the one-electron mass–velocity (MV) term, the one- and two-electron Darwin (D) terms, and the spin–spin and spin–orbit coupling terms. The effect of spin–orbit coupling on PESs is discussed in the chapter by Buenker *et al.* in this volume. For light atoms, the spin–orbit interaction is not too important for the form of the molecular orbitals; therefore, it is usually neglected. The one-electron mass–velocity–Darwin (MVD) terms have been used to calculate a variety of molecular properties, including relativistic correlation contributions up to CCSD(T), within the framework of first-order perturbation theory [Cowan and Griffin (1976)]. Despite the fact that the Cowan–Griffin (CG) operator neglects the two-electron Darwin interaction and the spin–orbit interaction, it is a remarkably good approximation to the more sophisticated Dirac–Fock Hamiltonian, as is evident from a large number of atomic and molecular calculations [see, e.g., Martin (1983)]. In the limit of a full one-electron basis set, the relativistic direct perturbation theory [DPT; Sewell (1949), Rutkowski (1986), Kutzelnigg (1989), and Klopper (1997)], possibly the most useful scalar (quasirelativistic) approach, results in the same answer as first-order perturbation theory including the one-electron mass–velocity and the one- and two-electron Darwin terms. Recently Klopper (1997) gave a recipe for the calculation of the scalar first-order relativistic corrections in the framework of DPT. If PT in powers of $Z^2\alpha^2$ is used to estimate relativistic corrections, the convergence of the expansion in powers of $Z^2\alpha^2$ may not be fast enough for large Z. The so-called scalar relativistic methods (e.g., DK) usually assume that corrections of higher order than c^{-2} are not important for chemical accuracy, and that the effect of spin–orbit coupling on the shape of the orbitals can be neglected. A few aspects of the terms in first-order PT warrant comment: (a) the MV term corrects the kinetic energy of the system, and it is always negative; (b) the one-electron Darwin (D1) term reduces the Coulomb attraction, and thus it always increases the total energy of the system; (c) the two-electron Darwin (D2) correction term serves to reduce the electronic Coulomb repulsion, it is negative, and it is expected to be diminutive, since it depends on the minuscule probability of two electrons being at the same point in space; and (d) convergence of these quantities with respect to basis set expansion appears to be extremely slow (e.g., convergence of the D2 term was shown [Halkier *et al.* (2000)] to have an X^{-1} dependence,

Table 2.10. Effects of Dirac–Hartree–Fock (DHF), relativistic effective core potential (RECP) and first-order perturbation theory (MVD) calculations on bond length and symmetric stretching frequencies of AH_4 (A = C, Si, Ge, Sn, Pb) molecules[a].

	CH_4	SiH_4	GeH_4	SnH_4	PbH_4
$\Delta^{rel}r_e$					
MVD	−0.00009	−0.00084	−0.0081	−0.0202	−0.0748
RECP			−0.016	−0.038	−0.076
DHF	−0.00013	−0.00066	−0.0070	−0.0206	−0.0733
$\Delta^{rel}\omega_e$					
MVD	−0.05	+0.51	+4.6	+14	+49
RECP			−4	−9	+11
DHF	+0.08	−0.04	+3.0	+10	+27

[a]All data are taken from Dyall *et al.* (1991).

where X refers to the cardinal number of numerically optimized or correlation-consistent basis sets).

Selected results on bond length contractions and harmonic vibrational frequencies due to relativity are collected in Table 2.10. They show the effectiveness of quasirelativistic approaches for the prediction of spectroscopic constants. It is also clear that first-order PT (involving one-electron MVD terms) works remarkably well. The RECP results show poorer agreement with the DHF results for the heavier members of the series [Dyall *et al.* (1991)]. Relativistic corrections to inversion barriers have also been studied extensively [see, e.g., Schwerdtfeger, Laakkonen, and Pyykkö (1992), Császár, Allen, and Schaefer (1998), Tarczay *et al.* (1999), and Aarset *et al.* (2000)]. The size of the relativistic effect for conformational changes in light molecules appears to be integrally related to the extent of attendant sp rehybridization for stereochemically active lone electron pairs. For example, the torsional motions for internal rotation in ethane and formic acid entail no rehybridization, and thus the relativistic corrections are less than 5 cm^{-1}. On the other hand, the inversion barrier of ammonia and the barrier to linearity of water involve substantial changes in the sp character of one and two lone electron pairs, engendering $\approx$23 and $\approx$56 cm^{-1} stabilizations of the s-rich pyramidal and bent forms, respectively. Nevertheless, the relativistic increase of the inversion barrier is much more dramatic in heavier congeners such as BiH_3.

Probably the most serious attempts to investigate the effect of relativistic corrections for the prediction of polyatomic PESs have been made for the ground electronic states of H_3^+ [Cencek *et al.* (1998)] and water [Császár *et al.* (1998)]. There is a slight but significant modulation of the relativistic correction over each PES.

The additivity of relativistic and electron correlation effects has long been questioned. Until recently, few methods were available to assess quantitatively the interplay of these two deviations from non-relativistic mean-field theory in molecular calculations. It is expected that such calculations will have a prominent role in the near future.

Acknowledgements

Much of the original research performed by A.G.Cs. and reported in this review was partially supported by the Hungarian Ministry of Culture and Education (FKFP 0117/1997) and by the Scientific Research Foundation of Hungary (OTKA T024044 and T033074). The research at the Center for Computational Quantum Chemistry at the University of Georgia was supported by the US Department of Energy, Office of Basic Energy Sciences, Fundamental Interactions Branch, Grant. No. DE-FG02-97-00ER14748. A.G.Cs., W.D.A., and H.F.S. acknowledge partial support by a NATO Linkage Grant (CRG.LG 973 892). We would like to thank Professors Matt

Leininger for the provision of the results presented in Table 2.3. The authors wish to thank Professors T. Daniel Crawford, W. Klopper and L. von Szentpály for useful suggestions and corrections regarding earlier versions of the manuscript.

2.4 References

Aarset, K., Császár, A. G., Klopper, W., Allen, W. D., Schaefer III, H. F., and Noga, J., 2000, *J. Chem. Phys.* **112**, 4053–4063.

Adamowicz, L., and Bartlett, R. J., 1987, *J. Chem. Phys.* **86**, 6314–6324.

Ågren, H., Bagus, P. S., and Roos, B. O., 1981, *Chem. Phys. Lett.* **82**, 505–510.

Ahlrichs, R., Scharf, P., and Ehrhardt, C., 1985, *J. Chem. Phys.* **82**, 890–898.

Allen, W. D., Horner, D. A., Dekock, R. L., Remington, R. B., and Schaefer III, H. F., 1989, *Chem. Phys.* **133**, 11–45.

Allen, W. D., Császár, A. G., and Horner, D. A., 1992, *J. Am. Chem. Soc.* **114**, 6834–6849.

Allen, W. D., and Császár, A. G., 1993, *J. Chem. Phys.* **98**, 2983–3015.

Allen, W. D., East, A. L. L., and Császár, A. G., 1993, in *Structures and Conformations of Non-Rigid Molecules*, Laane, J., Dakkouri, M., van der Veken, B., and Oberhammer, H., Eds; Kluwer: Dordrecht, pp. 343–373.

Almlöf, J., and Taylor, P. R., 1987, *J. Chem. Phys.* **86**, 4070–4077.

Almlöf, J. E., Helgaker, T., and Taylor, P. R., 1988, *J. Phys. Chem.* **92**, 3029–3033.

Almlöf, J. E., and Taylor, P. R., 1991, *Adv. Quantum Chem.* **22**, 301–373.

Almlöf, J. E., 1995, in *Modern Electronic Structure Theory, Part I*, Yarkony, D. R., Ed.; World Scientific: Singapore, pp. 110–151.

Amat, G., Nielsen, H. H., and Tarrago, G. *Rotational–Vibration of Polyatomic Molecules*; M. Dekker: New York, 1971.

Anderson, J. B., 1987, *J. Chem. Phys.* **86**, 2839–2843.

Anderson, J. B., 1992, *J. Chem. Phys.* **96**, 3702–3706.

Andersson, K., Malmqvist, P.-Å., Roos, B. O., Sadlej, A. J., and Wolinski, K., 1990, *J. Phys. Chem.* **94**, 5483–5488.

Andersson, K., Malmqvist, P.-Å., and Roos, B. O., 1992, *J. Phys. Chem.* **96**, 1218–1226.

Andersson, K., and Roos, B. O., 1995, in *Modern Electronic Structure Theory, Part I*, Yarkony, D. R., Ed.; World Scientific: Singapore, pp. 55–109.

Atchity, G. J., and Ruedenberg, K., 1997, *Theor. Chem. Acc.* **96**, 176–194.

van der Avoird, A., 2000, in *Computational Molecular Spectroscopy*, Bunker, P. R., and Jensen, P., Eds; Wiley: New York.

van der Avoird, A., and Wormer, P. E. S., 1984, *J. Chem. Phys.* **81**, 1929–1939.

Bačić, Z., and Light, J. C., 1987, *J. Chem. Phys.* **86**, 3065–3077.

Bacskay, G. B., Bryant, G., and Hush, N. S., 1987, *Int. J. Quantum Chem.* **31**, 471–487.

Bagus, P. S., and Schaefer III, H. F., 1972, *J. Chem. Phys.* **56**, 224–226.

Bagus, P. S., Liu, B., McLean, A. D., and Yoshimine, M., 1973, in *Wave Mechanics: The First Fifty Years*, Price, W. C., Chissick, S. S., and Ravensdale, T., Eds; Butterworths: London, pp. 99–118.

Balasubramanian, K. 1997, *Relativistic Effects in Chemistry, Parts A and B*; Wiley: New York.

Bardo, R. D., and Wolfsberg, M., 1978, *J. Chem. Phys.* **68**, 2686–2695.

Bartlett, R. J., and Shavitt, I., 1977, *Chem. Phys. Lett.* **50**, 190–198.

Bartlett, R. J., and Purvis, G. D., 1978, *Int. J. Quantum Chem.* **14**, 561–581.

Bartlett, R. J., 1986, in *Geometrical Derivatives of Energy Surfaces and Molecular Properties*, Jørgensen, P., and Simons, J., Eds; Reidel: Dordrecht, pp. 35–61.

Bartlett, R. J., Cole, S. J., Purvis, G. D., Ermler, W. C., Hsieh, H. C., and Shavitt, I., 1987, *J. Chem. Phys.*, **87**, 6579–6591.

Bartlett, R. J., 1995, in *Modern Electronic Structure Theory*, Part II, Yarkony, D. R., Ed.; World Scientific: Singapore, pp. 1047–1131.

Bartlett, R. J., Ed., 1997, *Recent Advances in Coupled-Cluster Methods*; World Scientific: Singapore.

Basilevsky, M. V., 1977, *Chem. Phys.* **24**, 81–89.

Bauschlicher, C. W., Jr., 1980, *Chem. Phys. Lett.* **74**, 277–279.

Bauschlicher, C. W., Jr., Langhoff, S. R., Taylor, P. R., Handy, N. C., and Knowles, P. J., 1986, *J. Chem. Phys.* **85**, 1469–1474.

Bauschlicher, C. W., Jr., and Taylor, P. R., 1987, *J. Chem. Phys.* **86**, 1420–1424.

Bauschlicher, C. W., Jr., Langhoff, S.R., and Taylor, P.R., 1988, *J. Chem. Phys.* **88**, 2540–2546.

Bauschlicher, C. W., Jr., Langhoff, S. R., Lee, T. J., and Taylor, P. R., 1989, *J. Chem. Phys.* **90**, 4296–4300.

Bauschlicher, C. W., Jr., Langhoff, S. R., and Taylor, P. R., 1990, *Adv. Chem. Phys.* **77**, 103–161.

Bauschlicher, C. W., Jr., and Partridge, H., 1993, *Theor. Chim. Acta* **85**, 255–259.

Bauschlicher, C. W., Jr., and Partridge, H., 1994, *J. Chem. Phys.* **100**, 4329–4335.

Bearpark, M. J., and Handy, N. C., 1992, *Theor. Chim. Acta* **84**, 115–124.

Becke, A. D., 1988, *Phys. Rev. A* **38**, 3098–3100.
Becke, A. D., and Roussel, M. R., 1989, *Phys. Rev. A* **39**, 3761–3767.
Becke, A. D., 1992, *J. Chem. Phys.* **97**, 9173–9177.
Becke, A. D., 1993, *J. Chem. Phys.* **98**, 5648–5652.
Beckel, C. L., and Engelke, R., 1968, *J. Chem. Phys.* **49**, 5199–5200.
Benard, M., 1982, *Theor. Chim. Acta* **61**, 379–385.
Benard, M., 1983, *Chem. Phys. Lett.* **96**, 183–191.
Benard, M., Laidlaw, W. G., and Paldus, J., 1985, *Can. J. Chem.* **63**, 1797–1802.
Bethe, H. A., 1947, *Phys. Rev.* **72**, 339–341.
Bigelow, R. W., and Freund, H.-J., 1982, *J. Chem. Phys.* **77**, 5552–5561.
Bingel, W. A., 1967, *Theor. Chim. Acta* **8**, 54–61.
Binkley, J. S., and Pople, J. A., 1975, *Int. J. Quantum Chem.* **9**, 229–236.
Bishop, D. M., 1967, *Adv. Quantum Chem.* **3**, 25–59.
Blomberg, M. R. A., Siegbahn, P. E. M., and Roos, B. O., 1980, *Int. J. Quantum Chem. Symp.* **14**, 229–247.
Bobrowicz, F. W., and Goddard, W. A. III, 1977, in *Methods of Electronic Structure Theory*, Schaefer, H.F. III, Ed.; Plenum: New York, pp. 79–127.
Bofill, J. M., and Pulay, P., 1989, *J. Chem. Phys.* **90**, 3637–3646.
Bohm, D., and Pines, D., 1953, *Phys. Rev.* **92**, 609–625.
Bonifacic, V., and Huzinaga, S., 1974, *J. Chem. Phys.* **60**, 2779–2786.
Born, M., and Oppenheimer, J. R., 1927, *Ann. Phys.* **84**, 457–484.
Born, M., 1951, *Nachr. Acad. Wiss. Göttingen* **6**, 1.
Born, M., and Huang, K., 1954, *Dynamical Theory of Crystal Lattices*; Oxford University Press: New York.
Bowman, J. M., Bittman, J. S., and Harding, L. B., 1986, *J. Chem. Phys.* **85**, 911–921.
Boys, S. F., 1950, *Proc. R. Soc. London Ser. A* **200**, 542–554.
Boys, S. F., 1960, *Proc. R. Soc. London Ser. A* **258**, 402–411.
Boys, S. B., and Bernardi, F. 1970, *Mol. Phys.* **19**, 553–566.
Brandow, B. H., 1967, *Rev. Mod. Phys.* **39**, 771–828.
Brown, F. B., Shavitt, I., and Shepard, R., 1984, *Chem. Phys. Lett.* **105**, 363–369.
Brown, P. J., and Hayes, E. F., 1971, *J. Chem. Phys.* **55**, 922–926.
Buenker, R. J., and Peyerimhoff, S. D., 1974, *Theor. Chim. Acta* **35**, 33–58.
Buenker, R. J., and Peyerimhoff, S. D., 1975, *Theor. Chim. Acta* **39**, 217–228.
Bukowski, R., Jeziorski, B., Moszynski, R., and Kołos, W., 1992, *Int. J. Quantum Chem.* **42**, 287–319.
Bukowski, R., Jeziorski, B., Rybak, S., and Szalewicz, K., 1995, *J. Chem. Phys.* **102**, 888–897.
Bunge, C. F., 1980, *Phys. Scr.* **21**, 328–334.
Bunker, P. R., 1970, *J. Mol. Spectrosc.* **35**, 306–313.
Bunker, P. R., and Jensen, P., 1998, *Molecular Symmetry and Spectroscopy*, 2nd edition; NRC Research Press: Ottawa.
Bürger, H., and Thiel, W. 1997, Vibration–rotation spectra of reactive molecules: interplay of *ab initio* calculations and high resolution experimental studies, in *Vibration–Rotational Spectroscopy and Molecular Dynamics*, Papousek, D., Ed.; World Scientific: Singapore.
Burton, N. A., Yamaguchi, Y., Alberts, I. L., and Schaefer III, H. F. 1991, *J. Chem. Phys.* **95**, 7466–7478.
Carroll, D. P., Silverstone, H. J., and Metzger, R. M., 1979, *J. Chem. Phys.* **71**, 4142–4163.
Carter, S., and Handy, N. C., 1987, *J. Chem. Phys.* **87**, 4294–4301.
Carter, S., Pinnavaia, N., and Handy, N. C., 1995, *Chem. Phys. Lett.* **240**, 400–408.
Cave, R. J., Xantheas, S. S., and Feller, D., 1992, *Theor. Chim. Acta* **83**, 31–55.
Cencek, W., Komasa, J., and Rychlewski, J., 1995, *Chem. Phys. Lett.* **246**, 417–420.
Cencek, W., and Kutzelnigg, W., 1996, *J. Chem. Phys.* **105**, 5878–5885.
Cencek, W., Rychlewski, J., Jaquet, R., and Kutzelnigg, W., 1998, *J. Chem. Phys.* **108**, 2831–2836.
Child, M. S., 1996, *Molecular Collision Theory;* Dover: Mineola, NY.
Chong, D. P., and Langhoff, S. R., 1986, *J. Chem. Phys.* **84**, 5606–5610.
Chong, D. P., Langhoff, S. R., Bauschlicher, C. W., Jr., Walch, S. P., and Partridge, H., 1986, *J. Chem. Phys.* **85**, 2850–2860.
Christiansen, O., Olsen, J., Jørgensen, P., Koch, H., and Malmqvist, P.-Å., 1996, *Chem. Phys. Lett.* **261**, 369–378.
Christiansen, P. A., and McCullough, E. A., 1977, *J. Chem. Phys.* **67**, 1877–1882.
Čížek, J., 1966, *J. Chem. Phys.* **45**, 4256–4266.
Čížek, J., and Paldus, J., 1967, *J. Chem. Phys.* **47**, 3976–3985.
Čížek, J., 1969, *Adv. Chem. Phys.* **14**, 35–89.
Čížek, J., and Vrscay, E. R., 1982, *Int. J. Quantum Chem.* **31**, 27–68.
Clabo, D. A., Jr., Allen, W. D., Remington, R. B., Yamaguchi, Y., and Schaefer III, H. F., 1988, *Chem. Phys.* **123**, 187–239.
Cohen, R. C., and Saykally, R. J., 1993, *J. Chem. Phys.* **98**, 6007–6030.
Collins, C. L., Dyall, K. G., and Schaefer III, H. F., 1995, *J. Chem. Phys.* **102**, 2024–2031.

Combes, J. M., Duclos, P., and Seiler, R., 1981, *The Born–Oppenheimer Approximation in Rigorous Atomic and Molecular Physics*, NATO Advanced Study Institutes, Vol. 74, Velo, G., and Wightman, A. S., Eds; Plenum: New York, p. 185.
Cook, D. B., 1986, *J. Chem. Soc., Faraday Trans. 2* **82**, 187–199.
Coolidge, A. S., 1932, *Phys. Rev.* **42**, 189–209.
Cowan, R. D., and Griffin, D. C., 1976, *J. Opt. Soc. Am.* **66**, 1010–1014.
Cox, P. A., Benard, M., and Veillard, A., 1982, *Chem. Phys. Lett.* **87**, 159–161.
Crawford, T. D., and Schaefer III, H. F., 1996, *J. Chem. Phys.* **104**, 6259–6264.
Crawford, T. D., Stanton, J. F., Szalay, P. G., and Schaefer III, H. F., 1997a, *J. Chem. Phys.* **107**, 2525–2528.
Crawford, T. D., Stanton, J. F., Allen, W. D., and Schaefer III, H. F., 1997b, *J. Chem. Phys.* **107**, 10626–10632.
Crawford, T. D., and Schaefer III, H. F., 1999, *Rev. Comput. Chem.* **14**, 33–136.
Cremer D., and He, Z., 1996, *J. Phys. Chem.* **100**, 6173–6188.
Császár, A. G., and Allen, W. D., 1996, *J. Chem. Phys.*, **104**, 2746–2748.
Császár, A. G., and Mills, I. M., 1997, *Spectrochim Acta Part A* **53**, 1101–1122.
Császár, A. G., Allen, W. D., and Schaefer III, H. F., 1998, *J. Chem. Phys.* **108**, 9751–9764.
Császár, A. G., Kain, J. S., Polyansky, O. L., and Tennyson, J., 1998, *Chem. Phys. Lett.* **293**, 317–323; **312**, 613–616(E).
Császár, A. G., 1998, Anharmonic molecular force fields, in *The Encyclopedia of Computational Chemistry*, Schleyer, P. v. R., Allinger, N. L., Clark, T., Gasteiger, J., Kollmann, P. A., Schaefer III, H. F., and Schreiner, P. R., Eds, Wiley: Chichester, pp. 13–30.
Cullen, J. M., and Zerner, M. C., 1982, *J. Chem. Phys.* **77**, 4088–4109.
Curtiss, L. A., Raghavachari, K., Trucks, G. W., and Pople, J. A., 1991, *J. Chem. Phys.* **94**, 7221–7230.
Curtiss, L. A., Raghavachari, K., Redfern, P. C., Rassolov, V., and Pople, J. A., 1998, *J. Chem. Phys.* **109**, 7764–7776.
Dalgarno, A., 1994, *Adv. At. Mol. Opt. Phys.* **32**, 57–68.
Darwin, C. G., 1928, *Proc. R. Soc. London, Ser. A* **118**, 654–680.
Daudey, J. P., Claverie, P., and Malrieu, J. P., 1974, *Int. J. Quantum Chem.* **8**, 1–15.
Davydov, A. S., 1965, *Quantum Mechanics*; Addison-Wesley: New York.
Davidson, E. R., and Silver, D. W., 1977, *Chem. Phys. Lett.* **52**, 403–406.
Davidson, E. R., and Chakravorty, S. J., 1994, *Chem. Phys. Lett.* **217**, 48–54.
Desclaux, J. P., 1973, *At. Data Nucl. Data Tables* **12**, 311.
Desclaux, J. P., 1975, *Comp. Phys. Commun.* **9**, 31; (E) 1977, **13**, 71.
Diedrich, D. L., and Anderson, J. B., 1994, *J. Chem. Phys.* **100**, 8089–8095.
Dietz, K., Schmidt, C., Warken, M., and Hess, B. A., 1993, *J. Phys. B* **26**, 1885–1896.
Dinelli, B. M., Le Sueur, C. R., Tennyson, J., and Amos, R. D., 1995, *Chem. Phys. Lett.* **232**, 295–300.
Dirac, P. A. M., 1928, *Proc. R. Soc. London, Ser. A* **117**, 610–624.
Douglas, M., and Kroll, N. M., 1974, *Ann. Phys.* **82**, 89–155.
Drake, G. W. F., and Swainson, R. A., 1990, *Phys. Rev. A* **41**, 1243–1246.
Drake, G. W. F., and Yan, Z.-C., 1994, *Chem. Phys. Lett.* **229**, 486–490.
Duch, W., 1989, *Chem. Phys. Lett.* **162**, 56–60.
Duch, W., and Meller, J., 1994, *Int. J. Quantum Chem.* **50**, 243–271.
Duch, W., and Diercksen, G. H. F., 1994, *J. Chem. Phys.* **101**, 3018–3030.
Van Duijneveldt, F. B., van Duijneveldt-van de Rijdt, J. G. C. M., van Lenthe, J. H., 1994, *Chem. Rev.* **94**, 1873–1885.
Dunning, T. H., Jr., 1970, *J. Chem. Phys.* **53**, 2823–2833.
Dunning, T. H., Jr., 1971, *J. Chem. Phys.* **55**, 716–723.
Dunning, T. H., Jr., 1989, *J. Chem. Phys.* **90**, 1007–1023.
Dunning, T. H., Jr., and Peterson, K. A., 1998, *J. Chem. Phys.* **108**, 4761–4771.
Dyall, K. G., Taylor, P. R., Faegri, K., Jr., and Partridge, H., 1991, *J. Chem. Phys.* **95**, 2583–2594.
Ebbing, D. D., 1963, *J. Chem. Phys.* **36**, 1361–1370.
Eliav, E., Kaldor, U., and Ishikawa, Y., 1994, *Chem. Phys. Lett.* **222**, 82–87.
Ellison, F. O., 1963, *J. Am. Chem. Soc.* **85**, 3540–3544.
Engelbrecht, L., and Liu, B., 1983, *J. Chem. Phys.* **78**, 3097–3106.
Evangelisti, S., Bendazzoli, G. L., Ansaloni, R., Duri, F., and Rossi, E., 1996, *Chem. Phys. Lett.* **252**, 437–446.
Feller, D. F., and Ruedenberg, K., 1979, *Theor. Chim. Acta* **52**, 231–251.
Feller, D., and Davidson, E. R., 1989, *J. Chem. Phys.* **90**, 1024–1030.
Feller, D., 1992, *J. Chem. Phys.* **96**, 6104–6114.
Feller, D., and Peterson, K. A., 1998, *J. Chem. Phys.* **108**, 154–176.
Fermann, J. T., Sherrill, C. D., Crawford, T. D., and Schaefer III, H. F., 1994, *J. Chem. Phys.* **100**, 8132–8139.
Fernley, J. A., Miller, S., and Tennyson, J., 1991, *J. Mol. Spectrosc.* **150**, 597–609.
Finney, J. L., 1996, *Faraday Discuss.* **103**, 1–18.
Fischer, C. F., 1977, *The Hartree–Fock Methods for Atoms*; Wiley: New York.
Fitzgerald, G., Harrison, R., Laidig, W. D., and Bartlett, R. J., 1985, *Chem. Phys. Lett.* **117**, 433–436.

Fock, V., 1930, *Z. Phys.* **61**, 126–148.
Foldy, L. L., and Wouthuysen, S. A., 1950, *Phys. Rev.* **78**, 29–36.
Franke, R., and Kutzelnigg, W., 1992, *Chem. Phys. Lett.* **199**, 561–566.
Frenkel, J., 1934, *Wave Mechanics, Advanced General Theory*; Clarendon Press: Oxford.
Frost, A. A., 1967, *J. Chem. Phys.* **47**, 3707–3713.
Fuentealba, P., Preuss, H., Stoll, H., and von Szentpaly, L., 1982, *Chem. Phys. Lett.* **89**, 418–422.
Fukui, K., 1970, *J. Phys. Chem.* **74**, 4161–4163.
Fukui, K., 1981, *Acc. Chem. Res.* **14**, 363–368.
Garcia, J. D., 1966, *Phys. Rev.* **147**, 66–68.
Garrett, J., and Mills, I. M., 1968, *J. Chem. Phys.* **49**, 1719–1729.
Garrett, B. C., and Truhlar, D. G., 1985, *J. Chem. Phys.* **82**, 4543–4547.
Gauss, J., Lauderdale, W. J., Stanton, J. F., Watts, J. D., and Bartlett, R. J., 1991, *Chem. Phys. Lett.* **182**, 207–215.
Gauss, J., and Stanton, J. F., 1997, *Chem. Phys. Lett.* **276**, 70–77.
Gaw, J. F., and Schaefer III, H. F., 1985, *J. Chem. Phys.* **83**, 1741–1745.
Gdanitz, R. J., and Ahlrichs, R., 1988, *Chem. Phys. Lett.* **143**, 413–420.
Gdanitz, R. J., 1993, *Chem. Phys. Lett.* **210**, 253–260.
Gordon, W., 1928, *Z. Phys.* **48**, 11–14.
Goscinski, O., 1986, *Int. J. Quantum Chem.* **19**, 51–59.
Grant, I. P., 1961, *Proc. R. Soc. London, Ser. A* **262**, 555–576.
Grant, I. P., 1970, *Adv. Phys.* **19**, 747–811.
Grev, R. S., and Schaefer III, H. F., 1989, *J. Chem. Phys.* **91**, 7305–7306.
Grev, R. S., Alberts, I. L., and Schaefer III, H. F., 1990, *J. Phys. Chem.* **94**, 3379–3381.
Grev, R. S., and Schaefer III, H. F., 1992, *J. Chem. Phys.* **96**, 6850–6856.
Gropen, O., 1988, in *Methods of Computational Chemistry*, Vol. 2, Wilson, S. Ed.; Plenum: New York, pp. 109–135.
Halkier, A., Jørgensen, P., Gauss, J., and Helgaker, T., 1997, *Chem. Phys. Lett.* **274**, 235–241.
Halkier, A., Helgaker, T., Klopper, W., and Olsen, J., 2000, *Chem. Phys. Lett.* **319**, 287–295.
Hall, G. G., 1959, *Rep. Prog. Phys.* **22**, 1–32.
Halonen, L., and Carrington, T., Jr., 1988, *J. Chem. Phys.* **88**, 4171–4185.
Halonen, L., 1997, *J. Chem. Phys.* **106**, 7931–7945.
Hameka, H. F., 1981, *Quantum Mechanics*; Wiley: New York.
Handy, N. C., 1980, *Chem. Phys. Lett.* **74**, 280–283.
Handy, N. C., and Schaefer III, H. F., 1984, *J. Chem. Phys.* **81**, 5031–5033.
Handy, N. C., Knowles, P. J., and Somasundram, K., 1985, *Theor. Chim. Acta* **68**, 87–100.
Handy, N. C., Yamaguchi, Y., and Schaefer III, H. F., 1986, *J. Chem. Phys.* **84**, 4481–4484.
Handy, N. C., Pople, J. A., Head-Gordon, M., Raghavachari, K., and Trucks, G. W., 1989, *Chem. Phys. Lett.* **164**, 185–192.
Handy, N. C., and Lee, A. M., 1996, *Chem. Phys. Lett.* **252**, 425–430.
Harrison, R. J., and Handy, N. C., 1983, *Chem. Phys. Lett.* **95**, 386–391.
Harrison, R. J., and Zarrabian, S., 1989, *Chem. Phys. Lett.* **158**, 393–398.
Harrison, R. J., 1991, *J. Chem. Phys.* **94**, 5021–5031.
Hartree, D. R., 1928, *Proc. Cambridge Philos. Soc.* **24**, 89–132.
Hay, P. J., and Wadt, W. R., 1985, *J. Chem. Phys.* **82**, 270–283, 299–310.
He, Z., and Cremer, D., 1996, *Int. J. Quantum Chem.* **59**, 15–29, 31–55, 57–69.
Hehre, W. J., Stewart, R. F., and Pople, J. A., 1969, *J. Chem. Phys.* **51**, 2657–2664.
Hehre, W. J., Radom, L., Schleyer, P. v. R., and Pople, J. A., 1986, *Ab Initio Molecular Orbital Theory*; Wiley: New York.
Helgaker, T., Gauss, J., Jørgensen, P., and Olsen, J., 1997a, *J. Chem. Phys.* **106**, 6430–6440.
Helgaker, T., Klopper, W., Koch, H., and Noga, J., 1997b, *J. Chem. Phys.* **106**, 9639–9646.
Herzberg, G., 1950, *Molecular Spectra and Molecular Structure I–II*; Van Nostrand Reinhold: New York.
Hess, B. A., 1986, *Phys. Rev. A* **33**, 3742–3748.
Hill, R. N., 1985, *J. Chem. Phys.* **83**, 1173–1196.
Hirao, K., 1991, *Chem. Phys. Lett.* **190**, 374–380.
Hirschfelder, J. O., 1963, *J. Chem. Phys.* **39**, 3145–3146.
Hirschfelder, J. O., Byers Brown, W., and Epstein, S. T., 1964, *Adv. Quantum Chem.* **1**, 255–374.
Hirst, D. M., 1985, *Potential Energy Surfaces*; Taylor&Francis: London.
Ho, P., and Melius, C. F., 1990, *J. Phys. Chem.* **94**, 5120–5127.
Ho, T.-S., and Rabitz, H., 1996, *J. Chem. Phys.* **104**, 2584–2597.
Hohenberg, P., and Kohn, W., 1964, *Phys. Rev. B* **136**, 864–871.
Horner, D. A., Allen, W. D., Császár, A. G., and Schaefer III, H. F., 1991, *Chem. Phys. Lett.* **186**, 346–355.
Hoy, A. R., Mills, I. M., and Strey, G., 1972, *Mol. Phys.* **24**, 1265–1290.
Huber, H., 1980, *Theor. Chim. Acta* **55**, 117–126.
Hurley, A. C., 1954, *Proc. R. Soc. London, Ser. A* **226**, 170–178.
Huron, B., Malrieu, J. P., and Rancurel, P., 1973, *J. Chem. Phys.* **58**, 5745–5759.

Huzinaga, S., 1965, *J. Chem. Phys.* **42**, 1293–1302.
Huzinaga, S., 1985, *Comput. Phys. Rep.* **2**, 281–339.
Hylleraas, E. A., 1928, *Z. Phys.* **48**, 469–494.
Hylleraas, E. A., 1929, *Z. Phys.* **54**, 347–366.
Hylleraas, E. A., 1964, *Adv. Quantum Chem.* **1**, 1–33.
Ioannou, A. G., Amos, R. D., and Handy, N. C., 1996, *Chem. Phys. Lett.* **251**, 52–58.
Itoh, T., 1965, *Rev. Mod. Phys.* **37**, 159–165.
Jackels, C. F., and Davidson, E. R., 1976, *J. Chem. Phys.* **64**, 2908–2917.
James, H. M., and Coolidge, A. S., 1933, *J. Chem. Phys.* **1**, 825–835.
Jankowski, K., Becherer, R., Scharf, P., Schiffer, H., and Ahlrichs, R., 1985, *J. Chem. Phys.* **82**, 1413–1419.
Jansen, H. B., and Ross, P., 1969, *Chem. Phys. Lett.* **3**, 140–143.
Jaquet, R., Cencek, W., Kutzelnigg, W., and Rychlewski, J., 1998, *J. Chem. Phys.* **108**, 2837–2846.
Jayatilaka, D., and Lee, T. J., 1993. *J. Chem. Phys.* **98**, 9734–9747.
Jensen, F., 1996, *Chem. Phys. Lett.* **261**, 633–636.
Jensen, J. Aa., Jørgensen, P., Helgaker, T., and Olsen, J., 1989, *Chem. Phys. Lett.* **162**, 355–360.
Jensen, P., and Bunker, P. R., 1988, *J. Chem. Phys.* **89**, 1327–1332.
Jensen, P., 1989, *J. Mol. Spectrosc.* **133**, 438–460.
Jeung, G. H., Malrieu, J. P., and Daudey, J. P., 1982, *J. Chem. Phys.* **77**, 3571–3577.
Jeziorski, B., Moszynski, R., and Szalewicz, K., 1994, *Chem. Rev.* **94**, 1887–1930.
Johnson, B. G., Gill, P. M. W., and Pople, J. A., 1993, *J. Chem. Phys.* **98**, 5612–5626.
Jordan, K. D., Kinsey, J. L., and Silbey, R., 1974, *J. Chem. Phys.* **61**, 911–917.
Jorish, V. S., and Scherbak, N. B., 1979, *Chem. Phys. Lett.* **67**, 160–164.
Jørgensen, P., and Simons, J., 1981, *Second Quantization-Based Methods in Quantum Chemistry*; Academic Press: New York.
Jørgensen, P., and Simons, J., Eds; 1986, *Geometrical Derivatives of Energy Surfaces and Molecular Properties*; D. Reidel: Dordrecht.
Jurgens-Lutovsky, R., and Almlöf, J. E., 1991, *Chem. Phys. Lett.* **178**, 451–454.
Kaldor, U., 1984, *J. Chem. Phys.* **81**, 2406–2410.
Kaldor, U., 1997, in *Recent Advances in Coupled-Cluster Methods*, Bartlett, R. J., Ed.; World Scientific: Singapore, pp. 125–153.
Kapuy, E., Bartha, F., Kozmutza, C., and Bogár, F., 1988, *J. Mol. Struct. (THEOCHEM)* **170**, 59–67.
Kato, T., 1957, *Commun. Pure Appl. Math.* **10**, 151–177.
Kato, T., 1966, *Perturbation Theory for Linear Operators*; Springer: Berlin.
Kauppi, E., and Halonen, L., 1995, *J. Chem. Phys.* **103**, 6861–6872.
Kendall, R. A., Dunning, T. H. Jr., and Harrison, R. J., 1992, *J. Chem. Phys.* **96**, 6796–6806.
King, R. A., Allen, W. D., Ma, B. Y., and Schaefer III, H. F. 1998, *Faraday Discuss.* **110**, 23–50.
Klahn, B., 1985, *J. Chem. Phys.* **83**, 5749–5753.
Klein, O., 1932, *Z. Phys.* **76**, 226–235.
Klepeis, N. E., East, A. L. L., Császár, A. G., Allen, W. D., Lee, T. J., Schwenke, D. W., 1993, *J. Chem. Phys.* **99**, 3865–3897.
Klopper, W., and Kutzelnigg, W., 1987, *Chem. Phys. Lett.* **134**, 17–22.
Klopper, W., and Kutzelnigg, W., 1991, *J. Chem. Phys.* **94**, 2020–2030.
Klopper, W., and Lüthi, H. P., 1996, *Chem. Phys. Lett.* **262**, 546–552.
Klopper, W., Quack, M., and Suhm, M. A., 1996, *Chem. Phys. Lett.* **261**, 35–44.
Klopper, W., 1997, *J. Comput. Chem.* **18**, 20–27.
Klopper, W., Noga, J., Koch, H., and Helgaker, T., 1997, *Theor. Chem. Acc.* **97**, 164–176.
Klopper, W., 1998, *r_{12}-dependent wave functions*, in *The Encyclopedia of Computational Chemistry*, Schleyer, P. v. R., Allinger, N. L., Clark, T., Gasteiger, J., Kollmann, P. A., Schaefer III, H. F., Schreiner, P. R., Eds; Wiley: Chichester, pp. 2351–2375.
Klopper, W., and Helgaker, T., 1998, *Theor. Chem. Acc.* **99**, 265–271.
Klopper, W., Quack, M., and Suhm, M. A., 1998, *J. Chem. Phys.* **108**, 10096–10115.
Klopper, W., Bak, K. L., Jorgensen, P., Olsen, J., and Helgaker, T., 1999, *J. Phys. B* **32**, 103–130.
Knowles, P. J., and Handy, N. C., 1984, *Chem. Phys. Lett.* **111**, 315–321.
Knowles, P. J., and Werner, H.-J., 1985, *Chem. Phys. Lett.* **115**, 259–267.
Knowles, P. J., and Werner, H.-J., 1988, *Chem. Phys. Lett.* **145**, 514–522.
Kofranek, M., Lischka, H., and Karpfen, A., 1988, *Chem. Phys.* **121**, 137–153.
Kohn, W., and Sham, L. J., 1965, *Phys. Rev. A* **140**, 1133–1138.
Kohn, W., Becke, A. D., and Parr, R. G., 1996, *J. Phys. Chem.* **100**, 12974–12980.
Kołos, W., and Wolniewicz, L., 1964, *J. Chem. Phys.* **41**, 3663–3673, 3674–3678.
Kołos, W., 1970, *Adv. Quantum Chem.* **5**, 99–133.
Kołos, W., and Rychlewski, J., 1993, *J. Chem. Phys.* **98**, 3960–3967.
Kołos, W., 1994, *J. Chem. Phys.* **101**, 1330–1332.
Komasa, J., Cencek, W., and Rychlewski, J., 1995, *Phys. Rev. A* **52**, 4500–4507.

Kosman, W. M., and Hinze, J., 1975, *J. Mol. Spectrosc.* **56**, 93–103.
Kozmutza, C., and Tfirst, E., 1998, *Adv. Quantum Chem.* **31**, 231–250.
Krishnan, R., Frisch, M. J., and Pople, J. A., 1980, *J. Chem. Phys.* **72**, 4244–4245.
Kucharski, S. A., Noga, J., and Bartlett, R. J., 1989, *J. Chem. Phys.* **90**, 7282–7290.
Kucharski, S. A., and Bartlett, R. J., 1992, *J. Chem. Phys.* **97**, 4282–4288.
Kutzelnigg, W., 1984, *Int. J. Quantum Chem.* **25**, 107–129.
Kutzelnigg, W., 1985, *Theor. Chim. Acta* **68**, 445–469.
Kutzelnigg, W., 1989, *Z. Phys. D* **11**, 15–28.
Kutzelnigg, W., and Morgan, J. D., 1992, *J. Chem. Phys.* **96**, 4484–4508; **97**, 8821 (E).
Kutzelnigg, W., 1996, *Phys. Rev. A* **54**, 1183–1198.
Kutzelnigg, W., 1997, *Mol. Phys.* **90**, 909–916.
Labanowski, J. K., and Andzelm, J. W., Eds., 1991, *Density-Functional Methods in Chemistry*; Springer: New York.
Ladik, J., 1959, *Acta Phys. Acad. Sci. Hung.* **10**, 271–279.
Ladik, J., 1961, *Acta Phys. Acad. Sci. Hung.* **13**, 123–137.
Laidig, W. D., Fitzgerald, G., and Bartlett, R. J., 1985, *Chem. Phys. Lett.* **113**, 151–158.
Laidlaw, W. G., and Benard, M., 1987, *J. Comput. Chem.* **8**, 727–735.
Lamb, W. E., and Retherford, R. C., 1947, *Phys. Rev.* **72**, 241–243.
Lánczos, C., 1956, *Applied Analysis*; Prentice Hall: Englewood Cliffs, NJ.
Langhoff, S. R., and Davidson, E. R., 1974, *Int. J. Quantum Chem.* **8**, 61–72.
Langhoff, S. R., Bauschlicher, C. W., Jr., and Taylor, P. R., 1988, *J. Chem. Phys.* **88**, 5715–5725.
Leach, C. A., and Moss, R. E., 1995, *Annu. Rev. Phys. Chem.* **46**, 55–82.
Lee, C., Yang, W., and Parr, R. G., 1988, *Phys. Rev. B* **37**, 785–789.
Lee, J. S., 1997, *J. Phys. Chem. A* **101**, 8762–8767.
Lee, T. J., Allen, W. D., and Schaefer III, H. F., 1987, *J. Chem. Phys.* **87**, 7062–7075.
Lee, T. J., Rice, J. E., Scuseria, G. E., and Schaefer III, H. F., 1989, *Theor. Chim. Acta* **75**, 81–98.
Lee, T. J., and Taylor, P. R., 1989, *Int. J. Quantum Chem. Symp.* **23**, 199–207.
Lee, T. J., and Scuseria, G. E., 1990, *J. Chem. Phys.* **93**, 489–494.
Lee, T. J., Rendell, A. P., and Taylor, P. R., 1990, *J. Chem. Phys.* **93**, 6636–6641.
Lee, T. J., and Scuseria, G. E., 1995, in *Quantum Mechanical Electronic Structure Calculations with Chemical Accuracy*, Langhoff, S. R., Ed.; Kluwer: Dordrecht, pp. 47–108.
Lee, Y. S., Ermler, W. C., and Pitzer, K. S., 1977, *J. Chem. Phys.* **67**, 5861–5876.
Lee, Y. S., Kucharski, S. A., and Bartlett, R. J., 1984, *J. Chem. Phys.* **81**, 5906–5912.
Leforestier, C., Braly, L. B., Liu, K., Elrod, M. J., and Saykally, R. J., 1997, *J. Chem. Phys.* **106**, 8527–8544.
Leininger, M. L., Sherrill, C. D., Allen, W. D., and Schaefer III, H. F., 1998, *J. Chem. Phys.* **108**, 6717–6721.
Leininger, M. L., Allen, W. D., Schaefer III, H. F., and Sherrill, C. D., 2000, *J. Chem. Phys.* **112**, 9213–9222.
Lengsfield, B. H., and Yarkony, D. R., 1986, *J. Chem. Phys.* **84**, 348–353.
Le Roy, R. J., and van Kranendonk, J., 1974, *J. Chem. Phys.* **61**, 4750–4769.
Lester, M. I., 1996, *Adv. Chem. Phys.* **96**, 51–102.
Levine, I. N., 1991, *Quantum Chemistry*, 4th edition; Prentice-Hall: London.
Levy, M., 1979, *Proc. Natl. Acad. Sci. USA* **76**, 6062–6065.
Li, X., and Paldus, J., 1994, *J. Chem. Phys.* **101**, 8812–8826.
Li, X., and Paldus, J., 1995, *J. Chem. Phys.* **102**, 2013–2023.
Lie, G. C., and Frye, D., 1992, *J. Chem. Phys.* **96**, 6784–6790.
Linderberg, J., and Shull, H., 1960, *J. Mol. Spectrosc.* **5**, 1–16.
Lindh, R., and Barnes, L. A., 1994, *J. Chem. Phys.* **100**, 224–237.
Liu, B., and McLean, A. D., 1973, *J. Chem. Phys.* **59**, 4557–4558.
Longuet-Higgins, H. C., 1961, *Adv. Spectrosc.* **2**, 429–472.
Lozes, R. L., Goscinski, O., and Wahlgren, U. I., 1979, *Chem. Phys. Lett.* **63**, 77–81.
Löwdin, P.-O., 1955, *Phys. Rev.* **97**, 1474–1489.
Löwdin, P.-O., 1959, *Adv. Chem. Phys.* **2**, 207–322.
Löwdin, P.-O., 1963, *Rev. Mod. Phys.* **35**, 496–501.
Lukka, T., Kauppi, E., and Halonen, L., 1995, *J. Chem. Phys.* **102**, 5200–5206.
Malrieu, J. P., Durand, P., and Daudey, J. P., 1985, *J. Phys. A* **18**, 809–826.
Malrieu, J. P., Daudey, J. P., and Caballol, R., 1994, *J. Chem. Phys.* **101**, 8908–8921.
Martin, J. M. L., François, J. P., and Gijbels, R., 1989, *J. Comput. Chem.* **10**, 152–162, 875–886.
Martin, J. M. L., 1992, *J. Chem. Phys.* **97**, 5012–5018.
Martin, J. M. L., 1994, *J. Chem. Phys.* **100**, 8186–8193.
Martin, J. M. L., 1995, *Chem. Phys. Lett.* **242**, 343–350.
Martin, J. M. L., 1996, *Chem. Phys. Lett.* **259**, 669–678.
Martin, J. M. L., and Taylor, P. R., 1997, *J. Chem. Phys.* **106**, 8620–8623.
Martin, J. M. L., 1999, *J. Chem. Phys.* **108**, 2791–2800.
Martin, R. L., 1983, *J. Phys. Chem.* **87**, 750–754.

Matcha, R. L., and King, S. C., Jr., 1967, *J. Chem. Phys.* **65**, 3355–3356.
Mayer, I., 1987, *Theor. Chim. Acta* **72**, 207–210.
Mayer, I., and Surján, P. R., 1992, *Chem. Phys. Lett.* **191**, 497–499.
Mayer, I., Vibók, A., and Valiron, P., 1994, *Chem. Phys. Lett.* **224**, 166–174.
Mayers, D. F., 1957, *Proc. R. Soc. London, Ser. A* **241**, 93–109.
McLean, A. D., Lengsfield III, B. H., Pacansky, J., and Ellinger, Y., 1985, *J. Chem. Phys.* **83**, 3567–3576.
McMurchie, L. E., and Davidson, E. R., 1976, *J. Comput. Phys.* **26**, 218–231.
McWeeny, R., 1948, Dissertation, University of Oxford.
McWeeny, R., and Kutzelnigg, W., 1968, *Int. J. Quantum Chem.* **2**, 187–203.
McWeeny, R., 1992, *Methods of Molecular Quantum Mechanics*, 2nd edition; Academic Press: London.
Meadows, J. H., and Schaefer III, H. F., 1976, *J. Am. Chem. Soc.* **98**, 4383–4386.
Melius, C. F., and Goddard, W. A., 1974, *Phys. Rev. A* **10**, 1528–1540.
Melius, C. F., and Ho, P., 1991, *J. Phys. Chem.* **95**, 1410–1419.
Mezey, P. G., 1979, *Theor. Chim. Acta* **53**, 183–192.
Mezey, P. G., 1987, *Potential Energy Hypersurfaces*; Elsevier: New York.
Mills, I. M., 1972, in *Molecular Spectroscopy: Modern Research*, Vol. I, Rao, K.N., and Mathews, C. W., Eds; Academic Press: New York, pp. 115–140.
Mok, D. K. W., Neumann, R., and Handy, N. C., 1996, *J. Phys. Chem.* **100**, 6225–6230.
Møller, C., and Plesset, M. S., 1934, *Phys. Rev.* **46**, 618–622.
Moss, R. E., 1973, in *Advanced Molecular Quantum Mechanics*, Chapman and Hall: London.
van Mourik, T., and van Lenthe, J. H., 1995, *J. Chem. Phys.* **102**, 7479–7483.
van Mourik, T., Wilson, A. K., Peterson, K. A., Woon, D. E., and Dunning, T. H., Jr., 1998, *Adv. Quantum Chem.* **31**, 105–135.
Murray, C. W., and Handy, N. C., 1992, *J. Chem. Phys.* **97**, 6509–6516.
Murray, C. W., Laming, G. L., Handy, N. C., and Amos, R. D., 1992, *Chem. Phys. Lett.* **199**, 551–556.
Murrell, J. N., Carter, S., Farantos, S. C., Huxley, P., and Varandas, A. J. C., 1984, *Molecular Potential Energy Surfaces*; Wiley: New York.
Murrell, J. N., Varandas, A. J. C., and Brandao, J., 1987, *Theor. Chim. Acta* **71**, 459–465.
Müller, K., 1980, *Angew. Chem., Int. Ed. Engl.* **19**, 1–13.
Müller, H., Kutzelnigg, W., and Noga, J., 1997, *Mol. Phys.* **92**, 535–546.
Müller, W., Flesh, J., and Meyer, W., 1984, *J. Chem. Phys.* **80**, 3297–3310.
Müller, W., and Meyer, W., 1984, *J. Chem. Phys.* **80**, 3311–3320.
Náray-Szabó, G., Surján, P. R., and Ángyán, J., 1987, *Applied Quantum Chemistry*; Akadémiai Kiadó: Budapest.
Newton, M. D., 1982, *Chem. Phys. Lett.* **90**, 291–295.
Nielsen, H. H., 1951, *Rev. Mod. Phys.* **23**, 90–136.
Noga, J., and Bartlett, R. J., 1987, *J. Chem. Phys.* **86**, 7041–7050; **89**, 3401 (*E*).
Noga, J., Kutzelnigg, W., and Klopper, W., 1992, *Chem. Phys. Lett.* **199**, 497–504.
Noga, J., and Kutzelnigg, W., 1994, *J. Chem. Phys.* **101**, 7738–7762.
Noga, J., Tunega, D., Klopper, W., and Kutzelnigg, W., 1995, *J. Chem. Phys.* **103**, 309–320.
Noga, J., Klopper, W., and Kutzelnigg, W., 1997, CC-R12: An explicitly correlated coupled-cluster theory, in *Recent Advances in Coupled-Cluster Methods*, Bartlett, R. J., Ed.; World Scientific: Singapore, pp. 1–48.
Novoa, J. J., Mota, F., and Arnau, F., 1990, *Chem. Phys. Lett.* **165**, 503–512.
Ochterski, J. W., Petersson, G. A., Montgomery, J. A., Jr., 1995, *J. Am. Chem. Soc.* **117**, 11 299–11 308.
Ochterski, J. W., Petersson, G. A., Montgomery, J. A., Jr, 1996, *J. Chem. Phys.* **104**, 2598–2619.
Oka, T., 1980, *Phys. Rev. Lett.* **45**, 531–534.
Oka, T., 1992, *Rev. Mod. Phys.* **64**, 1141–1149.
Oka, T., 1997, *Science* **277**, 328–329.
Olsen, J., Roos, B. O., Jørgensen, P., and Jensen, H. J. Aa., 1988, *J. Chem. Phys.* **89**, 2185–2192.
Olsen, J., Jørgensen, P., and Simons, J., 1990, *Chem. Phys. Lett.* **169**, 463–472.
Olsen, J., Christiansen, O., Koch, H., and Jørgensen, P., 1996, *J. Chem. Phys.* **105**, 5082–5090.
Pack, R. T., and Byers Brown, W., 1996, *J. Chem. Phys.* **45**, 556–559.
Pack, R. T., and Hirschfelder, J. O., 1970, *J. Chem. Phys.* **52**, 521–534.
Paldus, J., and Čížek, J., 1969, *Chem. Phys. Lett.* **3**, 1–3.
Paldus, J., and Čížek, J., 1970, *J. Chem. Phys.* **52**, 2919–2936.
Paldus, J., and Veillard, A., 1978, *Mol. Phys.* **35**, 445–459.
Paldus, J., 1994, Algebraic approach to coupled cluster theory, in *Relativistic and Electron Correlation Effects in Molecules and Solids*, Malli, G. L., Ed.; Plenum: New York, pp. 207–282.
Parr, R. G., and Yang, W., 1989, *Density-Functional Theory of Atoms and Molecules*; Oxford University Press: New York.
Partridge, H., Bauschlicher, C. W., Jr., Pettersson, L. G. M., McLean, A. D., Liu, B., Yoshimine, M., and Komornicki, A., 1990, *J. Chem. Phys.* **92**, 5377–5383.
Partridge, H., and Schwenke, D. W., 1997, *J. Chem. Phys.* **106**, 4618–4639.
Pasternack, S., 1938, *Phys. Rev.* **54**, 1113.

Pauncz, R., 1979, *Spin Eigenfunctions: Construction and Use*; Plenum: New York.
Perdew, J. P., 1985, *Phys. Rev. Lett.* **55**, 1665–1668.
Perdew, J. P., and Wang, Y., 1986, *Phys. Rev. B* **33**, 8800–8802.
Perdew, J. P., Chevary, J. A., Vosko, S. H., Jackson, K. A., Pederson, M. R., Singh, D. J., and Fiolhais, C., 1992, *Phys. Rev. B* **46**, 6671–6687.
Perdew, J. P., and Wang, Y., 1992, *Phys. Rev. B* **45**, 13244–13249.
Persson, B. J., and Taylor, P. R., 1996, *J. Chem. Phys.* **105**, 5915–5926.
Peterson, K. A., and Dunning, T. H., Jr., 1995, *J. Phys. Chem.* **99**, 3898–3901.
Peterson, K. A., and Dunning, T. H., Jr., 1997, *J. Mol. Struct. (THEOCHEM)* **400**, 93–117.
Peterson, K. A., Wilson, A. K., Woon, D. E., and Dunning, T. H., Jr., 1997, *Theor. Chem. Acc.* **97**, 251–259.
Petersson, G. A., and Braunstein, M., 1985, *J. Chem. Phys.* **83**, 5129–51334.
Petersson, G. A., Bennett, A., Tensfeldt, T. G., Al-Laham, M. A., Shirley, W. A., and Mantzaris, J., 1988, *J. Chem. Phys.* **89**, 2193–2218.
Petersson, G. A., Malick, D. K., Wilson, W. G., Ochterski, J. W., Montgomery J. A., Jr., and Frisch, M. J., 1998, *J. Chem. Phys.* **109**, 10570–10579.
Pettersson, L. G. M., Åkeby, H., Siegbahn, P., and Wahlgren, U., 1990, *J. Chem. Phys.* **93**, 4954–4957.
Pettersson, L. G. M., and Persson, B. J., 1993, *Chem. Phys.* **170**, 149–159.
Phillips, J. C., and Kleinman, L., 1959, *Phys. Rev.* **116**, 287–294.
Pilar, F. L., 1990, *Elementary Quantum Chemistry*; McGraw-Hill: New York.
Pitzer, K. S., 1979, *Acc. Chem. Res.* **121**, 271–275.
Poirier, R., Kari, R., and Csizmadia, I. G., 1985, *Handbook of Gaussian Basis Sets: A Compendium for Ab Initio Molecular Orbital Calculations*; Elsevier: Amsterdam.
Polyansky, O. L., Jensen, P., and Tennyson, J., 1996, *J. Chem. Phys.* **105**, 6490–6497.
Polyansky, O. L., Zobov, N. F., Tennyson, J., Lotoski, J. A., and Bernath, P. F., 1997a, *J. Mol. Spectrosc.* **184**, 35–50.
Polyansky, O. L., Zobov, N. F., Viti, S., Tennyson, J., Bernath, P. F., and Wallace, L. 1997b, *Science* **277**, 346–348.
Pople, J. A., Binkley, J., and Seeger, R., 1976, *Int. J. Quantum Chem. Symp.* **10**, 1–19.
Pople, J. A., Frisch, M. J., Luke, B. T., and Binkley, J. S., 1983, *Int. J. Quantum Chem. Symp.* **17**, 307–320.
Pople, J. A., Head-Gordon, M., and Raghavachari, K., 1987, *J. Chem. Phys.* **87**, 5968–5975.
Pople, J. A., Head-Gordon, M., Fox, D. J., Raghavachari, K., and Curtiss, L. A., 1989, *J. Chem. Phys.* **90**, 5622–5629.
Pople, J. A., Gill, P. M. W., and Johnson, B. G., 1992, *Chem. Phys. Lett.* **199**, 557–560.
Pou-Amérigo, R., Merchan, M., Nebot-Gil, I., Malmqvist, P.-Å. and Roos, B. O., 1994, *J. Chem. Phys.* **101**, 4893–4902.
Povill, A., Rubio, J., Caballol, R., and Malrieu, J. P., 1994, *Chem. Phys. Lett.* **218**, 283–286.
Prosmiti, R., Polyansky, O. L., and Tennyson, J., 1997, *Chem. Phys. Lett.* **273**, 107–114.
Pulay, P. 1969, *Mol. Phys.*, **17**, 197–204.
Pulay, P., Fogarasi, G., Pang, F., and Boggs, J. E., 1979, *J. Am. Chem. Soc.* **101**, 2550–2560.
Pulay, P., and Hamilton, T. P., 1988, *J. Chem. Phys.* **88**, 4926–49333.
Purvis, G. D., and Bartlett, R. J., 1982, *J. Chem. Phys.* **76**, 1910–1918.
Pyykkö, P., 1978, *Adv. Quantum Chem.* **11**, 353–409.
Pyykkö, P., and Desclaux, J. P., 1979, *Acc. Chem. Res.* **12**, 276–281.
Pyykkö, P., 1986, *Relativistic Theory of Atoms and Molecules*, Part I; Springer: Berlin.
Pyykkö, P., 1993, *Relativistic Theory of Atoms and Molecules*, Part II; Springer: Berlin.
Quiney, H. M., Skaane, H., and Grant, I. P., 1998, *Chem. Phys. Lett.* 473–480.
Qiu, Y., and Bačić, Z., 1997, *J. Chem. Phys.* **106**, 2158–2170.
Quack, M., and Suhm, M. A., 1991, *J. Chem. Phys.* **95**, 28–59.
Raffenetti, R. C., 1972, *J. Chem. Phys.* **58**, 4452–4458.
Raffenetti, R. C., 1975, *Int. J. Quantum. Chem. Symp.* **9**, 289–295.
Raghavachari, K., Pople, J. A., Replogle, E. S., and Head-Gordon, M., 1990, *J. Phys. Chem.* **94**, 5579–5586.
Raghavachari, K., and Curtiss, L. A., 1995, in *Modern Electronic Structure Theory, Part. II*, Yarkony, D. R., Ed.; World Scientific: Singapore, pp. 991–1021.
Rayleigh, Lord, 1927, *The Theory of Sound*, Vol. I, London.
Rees, A. L. G., 1947, *Proc. Phys. Soc.* **59**, 998–1008.
Reeves, C. M., 1963, *J. Chem. Phys.* **39**, 1–10.
Rendell, A. P., Lee, T. J., and Taylor, P. R., 1990, *J. Chem. Phys.* **92**, 7050–7056.
Rendell, A. P., and Lee, T. J., 1991, *J. Chem. Phys.* **94**, 6219–6228.
Rice, J. E., and Amos, R. D., 1985, *Chem. Phys. Lett.* **122**, 585–590.
Rittby, M., and Bartlett, R. J., 1988, *J. Phys. Chem.* **92**, 3033–3036.
Roos, B. O., 1972, *Chem. Phys. Lett.* **15**, 153–159.
Roos, B. O., Taylor, P. R., and Siegbahn, P. E. M., 1980, *Chem. Phys.* **48**, 157–173.
Roos, B. O., 1987, in *Ab Initio Methods in Quantum Chemistry*, Lawley, K. P., Ed.; Wiley: Chichester.

Roos, B. O., Ed., 1992, *Lecture Notes in Quantum Chemistry, European Summer School in Quantum Chemistry*, Vol. 58; Springer-Verlag: Berlin.
Roothaan, C. C. J., and Weiss, A. W., 1960, *Rev. Mod. Phys.* **32**, 194–205.
Röhse, R., Klopper, W., and Kutzelnigg, W., 1993, *J. Chem. Phys.* **99**, 8830–8839.
Röhse, R., Kutzelnigg, W., Jaquet, R., and Klopper, W., 1994, *J. Chem. Phys.* **101**, 2231–2243.
Ruedenberg, K., and Sundbarg, K. R., 1976, in *Quantum Science*, Calais, J. L., Goscinski, O., Linderberg, J., and Ohrn, Y., Eds; Plenum: New York, pp. 505–515.
Rutkowski, A., 1986, *J. Phys. B* **19**, 149–158, 3431–3441, 3443–3455.
Rydberg, J. R., 1931, *Z. Phys.* **73**, 376.
Sadlej, A. J., and Snijders, J. G., 1994, *Chem. Phys. Lett.* **229**, 435–438.
Saebo, S., Tong, W., and Pulay, P., 1993, *J. Chem. Phys.* **98**, 2170–2175.
Salter, E. A., Trucks, G. W., and Bartlett, R. J., 1989, *J. Chem. Phys.* **90**, 1752–1766.
Sanchez de Meras, A. M., Jensen, H. J. Aa., Jørgensen, P., and Olsen, J., 1991, *Chem. Phys. Lett.* **186**, 379–385.
Sanders, F. C., 1998, *Adv. Quantum Chem.* **33**, 369–387.
Sathyamurthy, N., 1985, *Comput. Phys. Rep.* **3**, 1–70.
Sato, S., 1955, *J. Chem. Phys.* **23**, 2465–2466.
Saunders, V. R., 1975, in *Computational Techniques in Quantum Chemistry and Molecular Physics*, Diercksen, G. H. F., Sutcliffe, B. T., and Veillard, A., Eds; Reidel: Dordrecht, pp. 347–424.
Saxe, P., Lengsfield, B. H., and Yarkony, D. R., 1985, *Chem. Phys. Lett.* **113**, 159–164.
Saxe, P., Schaefer III, H. F., and Handy, N. C., 1981, *J. Phys. Chem.* **85**, 745–747.
Schaefer III, H. F. Ed., 1977, *Modern Theoretical Chemistry*, Vols. 3 and 4; Plenum: New York.
Schatz, G. C., 1989, *Rev. Mod. Phys.* **61**, 669–688.
Scheiner, A. C., Scuseria, G. E., Rice, J. E., Lee, T. J., and Schaefer III, H. F., 1987, *J. Chem. Phys.* **87**, 5361–5373.
Schlegel, H. B., and Frisch, M. J., 1995, *Int. J. Quantum Chem.* **54**, 83–87.
Schleyer, P. v. R., Allinger, N. L., Clark, T., Gasteiger, J., Kollmann, P. A., Schaefer III, H. F., Schreiner, P. R., Eds, 1998, *The Encyclopedia of Computational Chemistry*; Wiley: Chichester.
Schmidt, C., Warken, M., and Handy, N. C., 1993, *Chem. Phys. Lett.* **211**, 272–281.
Schmidt, M. W., and Ruedenberg, K., 1979, *J. Chem. Phys.* **71**, 3951–3962.
Schnupf, U., Bowman, I. M., and Heaven, M. C., 1992, *Chem. Phys. Lett.* **189**, 487–494.
Schrödinger, E., 1926, *Ann. Phys.* **80**, 437–490.
Schucan, T. H., and Weidenmüller, H. A., 1972, *Ann. Phys.* **73**, 108–135.
Schwartz, C., 1962, *Phys. Rev.* **126**, 1015–1019.
Schwartz, C., and Le Roy, R. J., 1987, *J. Mol. Spectrosc.* **121**, 420–439.
Schwarz, W. H. E., and Wallmeier, H., 1982, *Mol. Phys.* **46**, 1045–1061.
Schwerdtfeger, P., Dolg, M., Schwarz, W. H. E., Bowmaker, G. A., and Boyd, P. D. W., 1989, *J. Chem. Phys.* **91**, 1762–1774.
Schwerdtfeger, P., Laakkonen, L. J., and Pyykkö, P. 1992, *J. Chem. Phys.* **96**, 6807–6819.
Scuseria, G. E., and Schaefer III, H. F., 1988, *Chem. Phys. Lett.* **152**, 382–386.
Scuseria, G. E., Hamilton, T. P., and Schaefer III, H. F., 1990, *J. Chem. Phys.* **92**, 568–573.
Scuseria, G. E., and Lee, T. J., 1990, *J. Chem. Phys.* **93**, 5851–5855.
Scuseria, G. E., 1991, *Chem. Phys. Lett.* **176**, 27–35.
Searles, D. J., and Nagy-Felsobuki, E. I., 1991, in *Vibrational Spectra and Structure*, Vol. 19, Durig, J. R., Ed.; Elsevier: Amsterdam.
Searles, D. J., and Nagy-Felsobuki, E. I., 1991, *Phys. Rev. A* **43**, 3365–3372.
Searles, D. J., and Nagy-Felsobuki, E. I., 1993, *Ab Initio Variational Calculations of Molecular Vibrational–Rotational Spectra*; Springer: Berlin.
Seeger, R., and Pople, J. A., 1977, *J. Chem. Phys.* **66**, 3045–3050.
Sellers, H., and Pulay, P., 1984, *Chem. Phys. Lett.* **103**, 463–465.
Sellers, H., 1986, *Chem. Phys. Lett.* **108**, 339–341.
Sewell, G. L., 1949, *Proc. Cambridge Philos. Soc.* **45**, 631–637.
Seznec, R., and Zinn-Justin, J., 1979, *J. Math. Phys.* **20**, 1398–1408.
Shavitt, I., 1977, in *Methods of Electronic Structure Theory*, Schaefer III, H. F., Ed.; Plenum: New York, pp. 189–275.
Sherrill, C. D., 1996, Ph.D. Thesis, University of Georgia.
Sherrill, C. D., and Schaefer III, H. F., 1996, *J. Phys. Chem.* **100**, 6069–6075.
Sherrill, C. D., and Schaefer III, H. F., 1999, *Adv. Quantum Chem.* **34**, 143–269.
Siegbahn, P. E. M., 1983, in *Methods in Computational Molecular Physics*, Diercksen, G. H. F., and Wilson, S., Eds; D. Reidel: Dordrecht, pp. 189–207.
Siegbahn, P. E. M., and Liu, B., 1978, *J. Chem. Phys.* **68**, 2457–2465.
Siegbahn, P. E. M., Svensson, M., and Boussard, P. J. E., 1995, *J. Chem. Phys.* **102**, 5377–5386.
Silver, D. M., 1971, *J. Chem. Phys.* **55**, 1461–1467.
Silver, D. M., and Wilson, S., 1978, *J. Chem. Phys.* **69**, 3787–3789.

Simon, S., Duran, M., and Dannenberg, J. J., 1996, *J. Chem. Phys.* **105**, 11024–11031.
Singer, K., 1960, *Proc. R. Soc. London, Ser. A* **258**, 412–420.
Slater, J. C., 1929, *Phys. Rev.* **34**, 1293–1322.
Slater, J. C., 1930, *Phys. Rev.* **35**, 210–211.
Stanton, J. F., Gauss, J., and Bartlett, R. J., 1992, *J. Chem. Phys.* **97**, 5554–5559.
Stewart, R. F., 1969, *J. Chem. Phys.* **52**, 431–438.
Stoll, H., Fuentealba, P., Schwerdtfeger, P., Flad, J., von Szentpály, L., and Preuss, H., 1984, *J. Chem. Phys.* **81**, 2732–2736.
Suhm, M. A., 1994, *Chem. Phys. Lett.* **223**, 474–480.
Sundholm, D., Pyykkö, P., and Laaksonen, L., 1987, *Phys. Scr.* **36**, 400–402.
Sutcliffe, B. T., 1993, *J. Chem. Soc. Faraday Trans.* **89**, 2321–2335.
Swirles, B., 1935, *Proc. R. Soc. London, Ser. A* **152**, 625–649.
Szabo, A., and Ostlund, N. S., 1989, *Modern Quantum Chemistry: Introduction to Advanced Electronic Structure Theory*, 1st edition revised; McMillan: New York.
Szalay, P. G., Császár, A. G., Fogarasi, G., and Lischka, H., 1990, *J. Chem. Phys.* **93**, 1246–1256.
Szalay, P. G., and Bartlett, R. J., 1995, *J. Chem. Phys.* **103**, 3600–3612.
Szalay, P. G., 1997, in *Recent Advances in Coupled-Cluster Methods*, Bartlett, R. J., Ed.; World Scientific: Singapore, pp. 81–123.
Szalewicz, K., and Jeziorski, B., 1997, in *Molecular Interactions–From van der Waals to Strongly Bound Complexes*, Scheiner, S., Ed.; Wiley: New York, pp. 3–43.
Taketa, H., Huzinaga, S., and O-ohata, K., 1966, *J. Phys. Soc. Jpn.* **21**, 2313–2324.
Tarczay, G., Császár, A. G., Klopper, W., Szalay, V., Allen, W. D., and Schaefer III, H. F., 1999, *J. Chem. Phys.* **110**, 11971–11981.
Tarczay, G., Császár, A. G., Leininger, M. L., and Klopper, W., 2000, Chem. Phys. Lett. **322**, 119–128.
Tennyson, J., and Polyansky, O. L., 1994, *Phys. Rev. A* **50**, 314–316.
Thompson, J. J., 1912, *Philos. Mag.* **24**, 209–253.
Thouless, D. J., 1961, *The Quantum Mechanics of Many-Body Systems*; Academic Press: New York.
Truhlar, D. G., Steckler, R., and Gordon, M. S., 1987, *Chem. Rev.* **87**, 217–236.
Truhlar, D. G., 1990, in *Proceedings of the NATO Workshop on the Dynamics of Polyatomic van der Waals Complexes*, N. Halberstadt and K. Janda, Eds, NATO Ser. B 227; Plenum: New York, pp. 159–185.
Truhlar, D. G., 1998, *Chem. Phys. Lett.* **294**, 45–48.
Turi, L., and Dannenberg, J. J., 1993, *J. Chem. Phys.* **97**, 2488–2490.
Valiron, P., and Mayer, I., 1997, *Chem. Phys. Lett.* **275**, 46–55.
Varandas, A. J. C., 1990, in *Trends in Atomic and Molecular Physics*, Yanez, M., Ed., Autonomade Madrid: Madrid.
Visscher, L., Saue, T., Nieuwpoort, W. C., Faegri, K., and Gropen, O., 1993, *J. Chem. Phys.* **99**, 6704–6715.
Visscher, L., and Dyall, K. G., 1996, *J. Chem. Phys.* **104**, 9040–9046.
Wadt, W. R., and Hay, P. J., 1985, *J. Chem. Phys.* **82**, 284–298.
Wahl, A. C., and Das, G., 1977, in *Modern Electronic Structure Theory*, Schaefer III, H. F., Ed.; Plenum: New York, pp. 51–78.
Walch, S. P., Bauschlicher, C.W., Jr., Roos, B. O., and Nelin, C. J., 1983, *Chem. Phys. Lett.* **103**, 175–179.
Walch, S. P., and Harding, L. B., 1988, *J. Chem. Phys.* **88**, 7653–7661.
Wallace, L., Bernath, P., Livingstone, W., Hinkle, K., Busler, J., Gou, B. J., and Zhang, K. Q. 1995, *Science* **268**, 1155–1158.
Watts, J. D., Stanton, J. F., and Bartlett, R. J., 1991, *Chem. Phys. Lett.* **178**, 471–474.
Watts, J. D., Gauss, J., and Bartlett, R. J., 1993, *J. Chem. Phys.* **98**, 8718–8733.
Watts, J. D., Urban, M., and Bartlett, R. J., 1995, *Theor. Chim. Acta* **90**, 341–355.
Watts, J. D., and Bartlett, R. J., 1998, *J. Chem. Phys.* **108**, 2511–2514.
Weinbaum, S., 1933, *J. Chem. Phys.* **1**, 593–596.
Western, C. M., 1995, *Chem. Soc. Rev.* **24**, 299–307.
White, J. C., and Davidson, E. R., 1990, *J. Chem. Phys.* **93**, 8029–8035.
Widmark, P. O., Malmqvist, P. Å., and Roos, B. O., 1990, *Theor. Chim. Acta* **77**, 291–306.
Widmark, P. O., Joakim, B., Persson, B. J., and Roos, B. O., 1991, *Theor. Chim. Acta* **79**, 419–432.
Wilson, E. B., Jr., Decius, J. C., and Cross, P. C., 1955, *Molecular Vibrations*; McGraw-Hill: New York.
Wilson, S., and Silver, D. M., 1978, *Mol. Phys.* **36**, 1539–1548.
Wilson, S., 1980, *Int. J. Quantum Chem.* **18**, 905–906.
Wilson, S., 1984, *Electron Correlation in Molecules*; Clarendon Press: Oxford.
Wilson, S., Jankowski, K., and Paldus, J., 1985, *Int. J. Quantum Chem.* **28**, 525–534.
Wilson, S., 1987, *Adv. Chem. Phys.* **67**, 439–500.
Wolinski, K., Sellers, H. L., and Pulay, P., 1987, *Chem. Phys. Lett.* **140**, 225–231.
Wolinski, K., and Pulay, P., 1989, *J. Chem. Phys.* **90**, 3647–3659.
Wolniewicz, L., 1993, *J. Chem. Phys.* **99**, 1851–1868.
Woon, D. E., and Dunning, T. H., Jr., 1993, *J. Chem. Phys.* **98**, 1358–1371.
Woon, D. E., 1994, *J. Chem. Phys.* **100**, 2838–2850.

Wu, Q., and Zhang, J. Z. H., 1996, *Chem. Phys. Lett.* **252**, 195–200.
Wulfov, A. L., 1996a, *Chem. Phys. Lett.* **255**, 300–308.
Wulfov, A. L., 1996b, *Chem. Phys. Lett.* **263**, 79–83.
Xie, Y., Allen, W. D., Yamaguchi, Y., and Schaefer III, H. F., 1996, *J. Chem. Phys.* **104**, 7615–7623.
Yamaguchi, Y., Frisch, M. J., Lee, T. J., Schaefer III, H. F., and Binkley, J. S., 1986, *Theor. Chim. Acta* **69**, 337–352.
Yamaguchi, Y., Osamura, Y., Goddard, J. D., and Schaefer III, H. F., 1994, *A New Dimension to Quantum Chemistry: Analytic Derivative Methods in Ab Initio Molecular Electronic Structure Theory*; Oxford University Press: New York.
Yan, Z.-C., and Drake, G. W. F., 1998, *Phys. Rev. Lett.* **81**, 774–777.
Yarkony, D. R., 1981, *Chem. Phys. Lett.* **77**, 634–635.
Yarkony, D. R., Ed., 1995, *Modern Electronic Structure Theory*, Vols. I–II; World Scientific: Singapore.
Young, R. C., Biedenharn, L. C., and Feenberg, E., 1957, *Phys. Rev.* **106**, 1151–1155.
Zaitsevskii, A. V., and Heully, J. L., 1992, *J. Phys. B* **25**, 603–612.
Zarrabian, S., Sarma, C. R., and Paldus, J., 1988, *Chem. Phys. Lett.* **155**, 183–188.
Zarrabian, S., and Paldus, J., 1990, *Int. J. Quantum Chem.* **38**, 761–778.
Zhang, D. H., and Light, J. C., 1995, *J. Chem. Phys.* **103**, 9713–9720.
Ziegler, T., 1991, *Chem. Rev.* **91**, 651–667.
Zobov, N. F., Polyansky, O. L., Le Sueur, C. R., and Tennyson, J., 1996, *Chem. Phys. Lett.* **260**, 381–387.

3 SYMMETRY ADAPTED PERTURBATION THEORY APPLIED TO THE COMPUTATION OF INTERMOLECULAR FORCES

Robert Moszynski

University of Warsaw, Poland

and

Paul E. S. Wormer and Ad van der Avoird

University of Nijmegen, The Netherlands

Computational Molecular Spectroscopy. Edited by Per Jensen and Philip R. Bunker

3.1 Introduction

The importance of increasing our knowledge of intermolecular potentials hardly needs be emphasized. These potentials determine the properties of non-ideal gases, (pure) liquids, solutions, molecular solids, and the behavior of complex molecular ensembles encountered in biological systems. They describe the so-called non-bonded contributions, as well as the special hydrogen bonding terms, that are part of the force fields used in simulations of processes such as enzyme–substrate binding, drug–receptor interactions, etc.

Let us not start our paper without briefly repeating the well-known fact that the concept of an intermolecular potential is based on the Born–Oppenheimer–or adiabatic–separation of the Schrödinger equation for the electronic and nuclear motions. The solution of the first step (the electronic structure problem) for a number of clamped nuclear coordinates yields the potential surface for the nuclear motions. For an individual molecule the latter are the vibrations, rotations, and translations of this molecule. In an ensemble of interacting molecules there is a hierarchy: the intramolecular forces, which determine the internal vibrations of the molecules, are strong and the intermolecular forces, which determine their relative translational and rotational motions, are much weaker. This chapter is concerned with the relatively weak intermolecular interactions, van der Waals forces and hydrogen bonding, in particular. They play a role in molecular complexes which may be collisional, as in gas phase scattering or in crossed molecular beam experiments, or truly bound, as the van der Waals complexes occurring in high concentrations in cold supersonic nozzle beams, but also–in lower concentrations–in bulk gases.

Intermolecular potentials depend on the intermolecular degrees of freedom, i.e., on the coordinates describing the relative translations and rotations of the molecules in a complex, but also–at least in principle–on the intramolecular coordinates describing the molecular geometries. Because of the hierarchy mentioned above, it is often allowed to make another adiabatic separation, namely between the intramolecular vibrations–with high frequencies–and the intermolecular modes–with much lower frequencies. The latter can then be described with an intermolecular potential averaged over the fast intramolecular vibrations. Or, one may even use the rigid molecule approximation. The intermolecular motions have mostly large amplitudes, because the potential surface corresponding to the weak intermolecular forces is rather flat. Often there are multiple (equivalent or non-equivalent) minima in the potential surface which are accessible through thermal motions of the complex or quantum mechanical tunneling. An appropriate theoretical treatment of the dynamics of such weakly bound complexes requires the knowledge of the full potential surface, not just of the second derivatives–the force constants–at the minimum of the potential, as in the harmonic oscillator model.

Present day methods and computers have evolved to a stage where it is possible to obtain fairly accurate intermolecular potentials from *ab initio* electronic structure calculations. Basically there are two methods: the 'supermolecule' method and symmetry adapted perturbation theory. Supermolecule calculations can be made by standard quantum chemistry program packages. In the following we outline the symmetry adapted perturbation theory (SAPT) for pair and three-body interactions. Although some recent SAPT applications [Korona *et al.* (1999)] are concerned with the interactions between open-shell monomers, we restrict ourselves here to closed-shell systems.

3.2 Symmetry adapted perturbation theories

3.2.1 THE POLARIZATION APPROXIMATION

Before discussing the problem of symmetry adaptation we start with the standard Rayleigh–Schrödinger (RS) perturbation theory and consider the interaction of two closed-shell monomers A and B in their ground states, described by the functions Φ_0^A and Φ_0^B, respectively, which are eigenstates of the respective Hamiltonians H_A and H_B. Introducing $H_0 \equiv H_A + H_B$, we can write the Schrödinger equation for the noninteracting system A–B as

$$H_0\Phi_0 = E_0\Phi_0, \tag{1}$$

where $\Phi_0 = \Phi_0^A\Phi_0^B$ and $E_0 = E_0^A + E_0^B$. The Schrödinger equation for the interacting system A–B takes the form

$$(H_0 - E_0)\Psi = (E_{\text{int}} - V)\Psi, \tag{2}$$

where V is the intermolecular interaction operator that contains all Coulombic interactions between the nuclei and electrons of A and the nuclei and electrons of B. To derive the RS perturbation equations for the dimer wave function Ψ and the interaction energy E_{int} we parametrize the Hamiltonian H with a complex parameter ζ, $H = H_0 + \zeta V$, with the physical value of ζ obviously equal to one. The energy and the wave function become functions of ζ, and can be expanded as power series in ζ

$$\Psi(\zeta) = \Phi_0 + \sum_{n=1}^{\infty} \zeta^n \Phi_{\text{pol}}^{(n)} \qquad \text{and} \qquad E_{\text{int}}(\zeta) = \sum_{n=1}^{\infty} \zeta^n E_{\text{pol}}^{(n)}. \tag{3}$$

The individual corrections $E_{\text{pol}}^{(n)}$ and $\Phi_{\text{pol}}^{(n)}$ are referred to, after Hirschfelder (1967), as the nth-order *polarization energy* and *polarization wave function*, respectively. The nth-order energy is given by

$$E_{\text{pol}}^{(n)} = \langle \Phi_0 | V | \Phi_{\text{pol}}^{(n-1)} \rangle, \tag{4}$$

while the polarization wave functions can be obtained from the following recursion relation

$$\Phi_{\text{pol}}^{(n)} = -\hat{R}_0 V \Phi_{\text{pol}}^{(n-1)} + \sum_{k=1}^{n-1} E_{\text{pol}}^{(k)} \hat{R}_0 \Phi_{\text{pol}}^{(n-k)}, \qquad \text{with } \Phi_{\text{pol}}^{(0)} \equiv \Phi_0. \tag{5}$$

Here we introduced the reduced resolvent

$$\hat{R}_0 = \sum_{k \neq 0} \frac{|\Phi_k\rangle\langle\Phi_k|}{E_k - E_0}, \tag{6}$$

with E_k and Φ_k denoting the excited state eigenvalues and eigen functions of H_0.

Although the equations defining the RS or polarization wave functions and energies are simple, they are not suitable to describe weak intermolecular interactions because the expansions in equation (3) are either divergent or converge much too slowly for $\zeta = 1$. This was shown by extensive theoretical [Claverie (1971), Jeziorski (1978), and Kutzelnigg (1980)] and numerical [Chałasiński, Jeziorski, and Szalewicz (1977), Ćwiok *et al.* (1992a), and Korona, Moszynski, and Jeziorski (1997)] studies. Moreover, the series, if it converges, does not converge to the physical ground state of the dimer (except for one- and two-electron dimers such as H_2^+ and H_2). For three- and four-electron dimers the energy first reaches very quickly the average of the energies of all states including the mathematical, Pauli-forbidden, solutions of the Schrödinger equation corresponding to the same dissociation limit. Then the series converges very slowly to the energy of the mathematical ground state of the Hamiltonian H which corresponds to the fully symmetric, Pauli-forbidden, solution of the Schrödinger equation. Numerical calculations for He_2 [Korona, Moszynski, and Jeziorski (1997)] illustrate these effects. The situation is even more complex when one of the monomers has more than two electrons. It can be shown [Jeziorski (1978), and Kutzelnigg (1980)] that the polarization expansion is divergent in this case, at least for sufficiently large R. These theoretical findings were recently supported by numerical calculations of Adams (1990, 1991, 1992) for the ground state of LiH.

The origins of this pathological convergence pattern of the polarization expansion can be easily understood by considering a hydrogen atom A interacting with a proton B at a large distance R [Jeziorski and Kołos (1982)]. In this case Φ_0 is the $1s_A$ hydrogenic orbital located at proton A. Since the exact ground state wave function Ψ must be symmetric with respect to the reflection in the plane perpendicular to the internuclear axis and passing through its midpoint, the correct form of Ψ at large R is $\Psi \approx 1s_A + 1s_B$. Obviously, $\Phi_0 = 1s_A$ is not a good approximation to Ψ. The component $1s_B$ due to the perturbation V is as large as the unperturbed function itself. Hence, the operator V cannot be considered as a small perturbation, and the polarization theory can only recover the $1s_B$ component of Ψ in very high order, because all polarization wave functions for this system are localized at nucleus A (i.e., they decay exponentially with the distance from nucleus A).

So, although the function Φ_0–and any finite sum of the polarization wave functions–is a poor approximation to the exact wave function Ψ, and the polarization expansion cannot be used to compute the interaction energies of weakly bound complexes, it does provide the correct asymptotic expansion of the interaction energy in the following sense [Jeziorski and Kołos (1982)]

$$E_{\text{int}} = \sum_{n=1}^{N} E_{\text{pol}}^{(n)} + O\left(R^{-\kappa(N+1)}\right), \tag{7}$$

where $\kappa = 2$ if at least one of the interacting molecules has a net charge and $\kappa = 3$ if both molecules are neutral. In addition, the polarization expansion for the wave function, after projection with the operator $\mathcal{A}$ which imposes the permutational symmetry of the exact wave function Ψ–or in the example of H_2^+ the spatial symmetry–gives the correct asymptotic expansion

$$\Psi = \mathcal{A}\Phi_0 + \sum_{n=1}^{N} \mathcal{A}\Phi_{\text{pol}}^{(n)} + O\left(R^{-\kappa(N+1)}\right). \tag{8}$$

Hence, the function $\mathcal{A}\Phi_0$ seems a natural zeroth-order approximation for a perturbation expansion of the interaction energies, but unfortunately it is not an eigenfunction of H_0. One could try to introduce a new partitioning of the Hamiltonian, $H = \widetilde{H}_0 + \widetilde{V}$, such that $\mathcal{A}\Phi_0$ is an eigenfunction of $\widetilde{H}_0$, but no really successful construction of $\widetilde{H}_0$ has been reported to date. On the other hand, one could keep the natural partitioning of the Hamiltonian, $H = H_0 + V$, and modify the perturbation equations in such a way that the function $\mathcal{A}\Phi_0$ can be used as the zeroth-order approximation. Such a modification leads to symmetry adapted perturbation theories (SAPTs).

3.2.2 SYMMETRY ADAPTATION

Since the field of SAPT has been recently reviewed in a number of papers [Jeziorski, Moszynski, and Szalewicz (1994), Szalewicz and Jeziorski (1997), and Jeziorski and Szalewicz (1998)], we will restrict the discussion to two perturbation theories only. That is, we will consider the symmetrized Rayleigh–Schrödinger (SRS) perturbation theory [Jeziorski, Chałasiński, and Szalewicz (1978), and Ćwiok *et al.*, (1992b)] and the Hirschfelder–Silbey (1966) (HS) perturbation theory. The first has been applied with success to two- and three-body (non-additive) interactions of closed shell systems, and the second also to pair interactions involving open-shell monomers.

SAPTs can be divided into two categories. In the first category, corresponding to the so-called *weak symmetry forcing* [Jeziorski, Chałasiński, and Szalewicz (1978), and Jeziorski and Kołos (1977)], the symmetry forcing operator appears only in the energy expressions. The perturbation equations for the wave functions are the polarization equations (4) and (5). Thus far, only these type of theories have been applied to interactions of many-electron systems. In the SRS perturbation theory [Jeziorski, Chałasiński, and Szalewicz (1978), and Ćwiok *et al.*, (1992b)] the usual antisymmetrizer $\mathcal{A}$ appears in the expressions for the energy corrections to the interaction energy in each order (n)

$$E_{\mathrm{SRS}}^{(n)} = N_0 \left[\langle \Phi_0 | V | \mathcal{A}\Phi_{\mathrm{pol}}^{(n-1)} \rangle - \sum_{k=1}^{n-1} E_{\mathrm{SRS}}^{(k)} \langle \Phi_0 | \mathcal{A}\Phi_{\mathrm{pol}}^{(n-k)} \rangle \right], \tag{9}$$

where $N_0 = \langle \Phi_0 | \mathcal{A}\Phi_0 \rangle^{-1}$.

In the second category, referred to as *strong symmetry forcing* [Jeziorski, Chałasiński, and Szalewicz (1978), and Jeziorski and Kołos (1977)], the symmetry operators enter the perturbation equations for the wave functions and complicate significantly their solution. In order to discuss the strong symmetry forcing method of Hirschfelder and Silbey (1966), we must first recall a few group theoretical concepts. The permutation (symmetric) group consisting of $N!$ permutations is denoted by S_N. Its irreducible representations (irreps) are labeled by Young diagrams consisting of N nodes. A Young diagram consisting of n_i rows of length l_i, with $i = 1, \ldots, k$ and $l_1 > l_2 > \ldots > l_k$, is written as $[l_1^{n_1}, l_2^{n_2}, \ldots, l_k^{n_k}]$. The number of nodes satisfies $N = \sum_{i=1}^{k} n_i l_i$. The antisymmetric irrep is labeled by the one-column Young diagram $[1^N]$.

Let the number of electrons on monomer X be N_X, X = A,B. Thus, the group $S_{N_A+N_B}$ is the total permutation group of the dimer; it has the outer product $S_{N_A} \otimes S_{N_B}$ of the monomer groups as a subgroup. The induced product $[1^{N_A}] \otimes [1^{N_B}] \uparrow S_{N_A+N_B}$ is carried by the $\binom{N_A+N_B}{N_A}$ functions obtained by operating with all $(N_A + N_B)!$ permutations in the group $S_{N_A+N_B}$ on the product $\Phi_0^A \Phi_0^B$, where both unperturbed monomer functions are totally antisymmetric. Decomposing the induced product, we find by application of Littlewood's rule of regular application [Hamermesh (1962)]

$$[1^{N_A}] \otimes [1^{N_B}] \uparrow S_{N_A+N_B} = \sum_{\nu=0}^{f} [2^{\nu}, 1^{N_A+N_B-2\nu}], \qquad \text{with } f = \min(N_A, N_B). \tag{10}$$

Note that every irrep in this decomposition has unit multiplicity. The actual decomposition of the induced product may in general be accomplished by diagonal matrix element projection operators (elements of the group algebra of $S_{N_A+N_B}$). If we choose irreps ν adapted to the group chain $S_{N_A} \otimes S_{N_B} \subset S_{N_A+N_B}$, we find by Frobenius' reciprocity theorem [Jansen and Boon (1967)] that the irrep $[1^{N_A}] \otimes [1^{N_B}]$ appears exactly once in each ν. That is, only one of the diagonal matrix element projection operators gives a non-vanishing result. Hence we may as well use character projectors

$$^{\nu}\mathcal{A} = \frac{1}{(N_A + N_B)!} \sum_{P \in S_{N_A+N_B}} \chi^{\nu}\left(P^{-1}\right) P,$$

where $\chi^{\nu}(P)$ is the character of the irrep ν. Each $^{\nu}\mathcal{A}$, where ν appears in the decomposition of equation (10), projects exactly one non-vanishing function out of a function that transforms as $[1^{N_A}] \otimes [1^{N_B}]$.

Note that out of the $f + 1$ functions corresponding to the decomposition of the induced product, cf. equation (10), only the first one, the fully antisymmetric function corresponding to $[1^{N_A+N_B}]$, is physical, the others are unphysical. These $f + 1$ functions are asymptotically degenerate with the physical ground state solution of the Schrödinger equation for the dimer. In the HS theory all these asymptotically degenerate functions are coupled. Jeziorski and Kołos (1977) showed that the HS perturbation equations for the wave functions and energies can be written as

$$^{\nu}E_{\rm HS}^{(n)} = {}^{\nu}N_0 \left[\langle \Phi_0 | V | {}^{\nu}\mathcal{A}F^{(n-1)} \rangle - \sum_{k=1}^{n-1} {}^{\nu}E_{\rm HS}^{(k)} \langle \Phi_0 | {}^{\nu}\mathcal{A}F^{(n-k)} \rangle \right], \tag{11}$$

$$F^{(n)} = -\hat{R}_0 V F^{(n-1)} + \sum_{k=1}^{n} \sum_{\nu=0}^{f} {}^{\nu}E_{\rm HS}^{(k)} \hat{R}_0 {}^{\nu}\mathcal{A}F^{(n-k)}, \qquad F^{(0)} \equiv \Phi_0. \tag{12}$$

It is worth noting that in each order of the SRS and HS theory the energy correction can be separated into polarization and exchange parts

$$E_{\rm SRS/HS}^{(n)} = E_{\rm pol}^{(n)} + E_{\rm exch}^{(n)}. \tag{13}$$

The exchange contributions $E_{\rm exch}^{(n)}$, although different for the SRS and HS theories (except for the first order), vanish exponentially as a function of intermonomer distance in each order, so at large intermonomer separations the SRS and HS results coincide with the RS results.

Extensive numerical calculations for small systems such as H_2^+ [Chałasiński, Jeziorski, and Szalewicz (1977), and Jeziorski, Chałasiński, and Szalewicz (1978)], H_2 [Ćwiok *et al.* (1992b, 1994)], and He_2 [Korona, Moszynski, and Jeziorski (1997)] show that the convergence properties of both the SRS and the HS theories are excellent. Considerably faster convergence of the HS expansion is observed only in very high orders, at the cost of a dramatic increase in the complexity of the theory when compared with the SRS approach. This fact is illustrated in Table 1 of Jeziorski and Szalewicz (1998), where results are reported of numerical calculations [Korona, Moszynski, and Jeziorski (1997)] for two ground state helium atoms at the van der Waals minimum. An inspection of this table shows that the low-order convergence of both expansions is very fast. Indeed, already the second-order calculation reproduces the full configuration interaction (CI) results within 0.1 %. One may note that the differences between the SRS and HS results become important only in high orders. Thus, the simplest SRS theory provides in low orders excellent approximations to the exact interaction energies.

Since in most cases the polarization expansion is divergent, one can expect that for many-electron monomers the SRS expansion will not be strictly convergent. However, the experience gained thus far for large many-electron systems suggests that a second-order SRS calculation correctly accounts for all major polarization and exchange contributions to the interaction energy. In the region of the van der Waals minimum it should be accurate to within a few percent.

3.3 Physical interpretation of the low-order polarization and exchange energies

The polarization and exchange energies through the second order have an appealing physical interpretation. Except for the second-order exchange terms they can also be rigorously related to monomer properties which considerably facilitates their practical evaluation.

3.3.1 ELECTROSTATIC ENERGY

The first-order polarization energy is given by

$$E_{\text{pol}}^{(1)} = \left\langle \Phi_0^{\text{A}} \Phi_0^{\text{B}} \middle| V \middle| \Phi_0^{\text{A}} \Phi_0^{\text{B}} \right\rangle. \tag{14}$$

As shown by Claverie (1978) and by Magnasco and McWeeny (1991) the expression for $E_{\text{pol}}^{(1)}$ can be rewritten in terms of the total charge distributions $\rho_{\text{A}}^{\text{tot}}(\boldsymbol{r})$ and $\rho_{\text{B}}^{\text{tot}}(\boldsymbol{r})$ of the unperturbed monomers,

$$E_{\text{pol}}^{(1)} = \int\int \rho_{\text{A}}^{\text{tot}}(\boldsymbol{r}_1) \frac{1}{|\boldsymbol{r}_1 - \boldsymbol{r}_2|} \rho_{\text{B}}^{\text{tot}}(\boldsymbol{r}_2)\, \text{d}^3\boldsymbol{r}_1\, \text{d}^3\boldsymbol{r}_2, \tag{15}$$

where the total charge distribution for monomer A is given by

$$\rho_{\text{A}}^{\text{tot}}(\boldsymbol{r}) = \sum_{\alpha \in \text{A}} Z_\alpha \delta(\boldsymbol{r} - \boldsymbol{R}_\alpha) - \rho_{\text{A}}(\boldsymbol{r}). \tag{16}$$

Here, and elsewhere in this chapter, atomic units are used. The term containing Dirac's delta $\delta(\boldsymbol{r} - \boldsymbol{R}_\alpha)$ represents the contribution from the positive point charge Z_α at the position $\boldsymbol{R}_\alpha$ of nucleus α, and $-\rho_{\text{A}}(\boldsymbol{r})$ is the electronic charge distribution, given by the diagonal element of the first-order density matrix (normalized to the number of electrons in monomer A).

Equations (15) and (16) show that the first-order polarization energy $E_{\text{pol}}^{(1)}$ represents the energy of the electrostatic interaction of the unperturbed monomers' charge distributions, and hence is referred to as the *electrostatic energy*. At large intermonomer distances R the electrostatic energy can be represented as a sum of classical electrical interactions between the permanent multipole moments of the unperturbed monomers. One should note, however, that the electrostatic energy also contains important short-range components due to the mutual penetration (charge overlap) of the monomers' electron clouds. This short-range part of the electrostatic energy makes a significant contribution to the stabilization energy of van der Waals complexes and cannot be neglected in any accurate calculation of the potential energy surfaces for such systems.

3.3.2 FIRST-ORDER EXCHANGE (HEITLER–LONDON) ENERGY

The first-order energy in the SRS and HS perturbation theories is given by

$$E_{\text{SRS}}^{(1)} = E_{\text{HS}}^{(1)} = \frac{\langle \Phi_0 | V | \mathcal{A} \Phi_0 \rangle}{\langle \Phi_0 | \mathcal{A} \Phi_0 \rangle}, \tag{17}$$

where $\mathcal{A}$ is the antisymmetrizer. In the case that Φ_0 is an exact eigenfunction of H_0 this energy is identical with the so-called Heitler–London energy defined as

$$E_{\text{HL}}^{(1)} = \frac{\langle \mathcal{A}\Phi_0 | H - E_0 | \mathcal{A}\Phi_0 \rangle}{\langle \mathcal{A}\Phi_0 | \mathcal{A}\Phi_0 \rangle}. \tag{18}$$

To separate the exchange and polarization parts of $E^{(1)}_{\rm SRS}$ one has to use the following decomposition of the total antisymmetrizer (van Duijneveldt-van de Rijdt and van Duijneveldt, 1972)

$$\mathcal{A} = \frac{N_{\rm A}!N_{\rm B}!}{(N_{\rm A}+N_{\rm B})!}(1+\mathcal{P})\mathcal{A}_{\rm A}\mathcal{A}_{\rm B}, \tag{19}$$

where $\mathcal{A}_{\rm A}$ and $\mathcal{A}_{\rm B}$ are the antisymmetrizers for the monomers A and B, respectively and $\mathcal{P}$ collects all permutations (with appropriate sign factors) interchanging at least one pair of electrons between the interacting monomers. By inserting equation (19) into equation (17) one finds that

$$E^{(1)}_{\rm SRS} = E^{(1)}_{\rm pol} + E^{(1)}_{\rm exch}, \tag{20}$$

where

$$E^{(1)}_{\rm exch} = \frac{\langle \Phi_0 | V - E^{(1)}_{\rm pol} | \mathcal{P}\Phi_0\rangle}{1+\langle \Phi_0|\mathcal{P}\Phi_0\rangle}. \tag{21}$$

This expression vanishes exponentially at large R since the functions $\Phi_0^{\rm A}$ and $\Phi_0^{\rm B}$ decay exponentially with the distance from the centers of the respective monomers [Ahlrichs (1973)]. The $E^{(1)}_{\rm exch}$ component represents the main exchange contribution to the interaction energy. At the van der Waals minimum it usually accounts for over 90 % of the total exchange effect. The interpretation of $E^{(1)}_{\rm exch}$ is very simple: it is the effect of taking the expectation value of the full Hamiltonian with the simplest possible function ($\mathcal{A}\Phi_0$) representing in zeroth order the resonance tunneling of electrons between all available equivalent minima.

A direct evaluation of $E^{(1)}_{\rm exch}$ is difficult because the multiple electron exchange operators included in $\mathcal{P}$ prevent us from expressing this quantity in terms of monomer properties. For the intermonomer distances corresponding to typical van der Waals minima equation (21) can be greatly simplified by considering only the single electron exchanges [Williams, Schaad, and Murrell (1967), and Jeziorski, Bulski, and Piela (1976)]. Since the resulting approximate value of $E^{(1)}_{\rm exch}$ is quadratic in the intermolecular overlap densities $\rho_{lm}(\boldsymbol{r}) = \psi_l(\boldsymbol{r})\psi_m(\boldsymbol{r})$ (orbital ψ_l on A and ψ_m on B), it is denoted by $E^{(1)}_{\rm exch}(S^2)$,

$$E^{(1)}_{\rm exch}(S^2) = -\langle \Phi_0 | V - E^{(1)}_{\rm pol} | \mathcal{P}_1\Phi_0\rangle, \tag{22}$$

where $\mathcal{P}_1$ denotes the sum of all $N_{\rm A}N_{\rm B}$ transpositions of electrons between the monomers. Equation (22) represents a very good approximation since its error is of the fourth order in the intermonomer overlap densities. It can be shown that $E^{(1)}_{\rm exch}(S^2)$ can be expressed through the one- and two-particle density matrices of the unperturbed monomers [Moszynski *et al.*, (1994a)]

$$E^{(1)}_{\rm exch}(S^2) = \iint \left(\tilde{v}(\boldsymbol{r}_1,\boldsymbol{r}_2) - \frac{1}{N_{\rm A}N_{\rm B}}E^{(1)}_{\rm pol}\right)\rho_{\rm int}(\boldsymbol{q}_1,\boldsymbol{q}_2)\,{\rm d}\boldsymbol{q}_1\,{\rm d}\boldsymbol{q}_2, \tag{23}$$

where

$$\begin{aligned}\rho_{\rm int}(\boldsymbol{q}_1,\boldsymbol{q}_2) = &-\rho_{\rm A}(\boldsymbol{q}_1|\boldsymbol{q}_2)\rho_{\rm B}(\boldsymbol{q}_2|\boldsymbol{q}_1) - \int \Gamma_{\rm A}(\boldsymbol{q}_1\boldsymbol{q}_3|\boldsymbol{q}_1\boldsymbol{q}_2)\rho_{\rm B}(\boldsymbol{q}_2|\boldsymbol{q}_3)\,{\rm d}\boldsymbol{q}_3\\ &- \int \rho_{\rm A}(\boldsymbol{q}_1|\boldsymbol{q}_3)\Gamma_{\rm B}(\boldsymbol{q}_2\boldsymbol{q}_3|\boldsymbol{q}_2\boldsymbol{q}_1)\,{\rm d}\boldsymbol{q}_3\\ &- \iint \Gamma_{\rm A}(\boldsymbol{q}_1\boldsymbol{q}_3|\boldsymbol{q}_1\boldsymbol{q}_4)\Gamma_{\rm B}(\boldsymbol{q}_2\boldsymbol{q}_4|\boldsymbol{q}_2\boldsymbol{q}_3)\,{\rm d}\boldsymbol{q}_3\,dd\boldsymbol{q}_4,\end{aligned} \tag{24}$$

and $\tilde{v}(\boldsymbol{r}_i, \boldsymbol{r}_j)$ is a modified interelectronic interaction potential

$$\tilde{v}(\boldsymbol{r}_i, \boldsymbol{r}_j) = r_{ij}^{-1} - N_{\mathrm{B}}^{-1} \sum_{\beta \in \mathrm{B}} Z_\beta r_{\beta i}^{-1} - N_{\mathrm{A}}^{-1} \sum_{\alpha \in \mathrm{A}} Z_\alpha r_{\alpha j}^{-1} + N_{\mathrm{A}}^{-1} N_{\mathrm{B}}^{-1} \sum_{\alpha \in \mathrm{A}} \sum_{\beta \in \mathrm{B}} Z_\alpha Z_\beta R_{\alpha\beta}^{-1}, \tag{25}$$

defined such that

$$\sum_{i \in \mathrm{A}} \sum_{j \in \mathrm{B}} \tilde{v}(\boldsymbol{r}_i, \boldsymbol{r}_j) = V. \tag{26}$$

Finally, ρ_X and Γ_X, with $X = \mathrm{A}$ or B, are the one- and two-particle density matrices for monomer X defined in the conventional way

$$\rho_X(\boldsymbol{q}_1|\boldsymbol{q}_1') = N_X \int \Phi_0^X(\boldsymbol{q}_1, \boldsymbol{q}_2, \ldots, \boldsymbol{q}_{N_X})^* \Phi_0^X(\boldsymbol{q}_1', \boldsymbol{q}_2, \ldots, \boldsymbol{q}_{N_X})\, \mathrm{d}\boldsymbol{q}_2 \ldots \mathrm{d}\boldsymbol{q}_{N_X} \tag{27}$$

$$\Gamma_X(\boldsymbol{q}_1, \boldsymbol{q}_2|\boldsymbol{q}_1', \boldsymbol{q}_2') = N_X(N_X - 1) \int \Phi_0^X(\boldsymbol{q}_1, \boldsymbol{q}_2, \boldsymbol{q}_3, \ldots, \boldsymbol{q}_{N_X})^* \times \Phi_0^X(\boldsymbol{q}_1', \boldsymbol{q}_2', \boldsymbol{q}_3, \ldots, \boldsymbol{q}_{N_X})\, \mathrm{d}\boldsymbol{q}_3 \ldots \mathrm{d}\boldsymbol{q}_{N_X}, \tag{28}$$

where $\boldsymbol{q}_i = (\boldsymbol{r}_i, s_i)$ denotes the space and spin coordinates of the ith electron. Since theoretical methods for the evaluation of the density matrices ρ_X and Γ_X for many-electron molecules are well developed, equations (23) and (24) enable practical calculations of the first-order exchange energy from accurate electronic wave functions [Moszynski *et al.*, (1994a)].

3.3.3 INDUCTION ENERGY

The second-order polarization energy $E_{\mathrm{pol}}^{(2)}$ is given by

$$E_{\mathrm{pol}}^{(2)} = -\langle \Phi_0 | V \hat{R}_0 V | \Phi_0 \rangle. \tag{29}$$

The induction energy, $E_{\mathrm{ind}}^{(2)}$, is obtained when the reduced resolvent $\hat{R}_0$ [equation (6)] is restricted to terms where one of the monomers is in the ground state and the other in the excited state. The corresponding expression is given by

$$E_{\mathrm{ind}}^{(2)} = E_{\mathrm{ind}}^{(2)}(\mathrm{A} \leftarrow \mathrm{B}) + E_{\mathrm{ind}}^{(2)}(\mathrm{B} \leftarrow \mathrm{A}), \tag{30}$$

where

$$E_{\mathrm{ind}}^{(2)}(\mathrm{A} \leftarrow \mathrm{B}) = -\langle \Phi_0^{\mathrm{A}} | \Omega_{\mathrm{B}} \hat{R}_{0\mathrm{A}} \Omega_{\mathrm{B}} | \Phi_0^{\mathrm{A}} \rangle, \tag{31}$$

and a similar definition holds for $E_{\mathrm{ind}}^{(2)}(\mathrm{B} \leftarrow \mathrm{A})$. Here, Ω_{B} denotes the operator of the electrostatic potential generated by the unperturbed monomer B

$$\Omega_{\mathrm{B}} = \sum_{i \in \mathrm{A}} \omega_{\mathrm{B}}(\boldsymbol{r}_i), \qquad \omega_{\mathrm{B}}(\boldsymbol{r}_i) = \int \frac{1}{r_{ij}} \rho_{\mathrm{B}}^{\mathrm{tot}}(\boldsymbol{r}_j)\, \mathrm{d}^3\boldsymbol{r}_j. \tag{32}$$

The operator $\hat{R}_{0\mathrm{A}}$ is the part of the resolvent [equation (6)] in which B is in its ground state and the sum is over the excited states of A.

Equation (31) has the form of the second-order energy correction for monomer A perturbed by the static electric field generated by the unperturbed monomer B. This field, corresponding to the potential ω_{B}, induces a modification $\Phi_{\mathrm{ind}}^{(1)}(\mathrm{A} \leftarrow \mathrm{B}) = -\hat{R}_{0\mathrm{A}} \Omega_{\mathrm{B}} \Phi_0^{\mathrm{A}}$ in the wave function of monomer A, and the energetic effect of this modification is equal to $E_{\mathrm{ind}}^{(2)}(\mathrm{A} \leftarrow \mathrm{B})$. Thus, the second-order induction energy results from the mutual polarization of the monomers by the static electric fields of unperturbed partners. Asymptotically, at large R, $E^{(2)}_{\rm ind}$ is fully determined by the permanent multipole moments and static multipole polarizabilities of the monomers. At finite R additional information is needed to account for the short-range, penetration part of $E^{(2)}_{\rm ind}$. This information is contained in the short-range part of the electrostatic potentials $\omega_X(\boldsymbol{r})$, $X = \rm A$ or B, and in the polarization propagators of the monomers. The polarization propagator is a molecular property, which fully describes the linear response of a molecule to an arbitrary, spin-independent, external perturbation [Jørgensen and Simons (1981), and Oddershede (1983)]. The use of these propagators for the calculation of molecular properties is treated in the chapter by Sauer. The polarization propagator is defined for an arbitrary frequency ω by

$$\Pi^{ll'}_{kk'}(\omega) = -\langle \Phi^{\rm A}_0 | E^l_k \hat{R}_{\rm A}(-\omega) E^{l'}_{k'} | \Phi^{\rm A}_0 \rangle - \langle \Phi^{\rm A}_0 | E^{l'}_{k'} \hat{R}_{\rm A}(\omega) E^l_k | \Phi^{\rm A}_0 \rangle, \tag{33}$$

where E^l_k is the spin-free unitary group generator (orbital replacement operator), defined by

$$E^l_k = \sum_{\sigma=1}^{2} a^\dagger_{l\sigma} a_{k\sigma} \tag{34}$$

and $a^\dagger_{l\sigma}$ is a creation operator, $l\sigma$ indicates a spin orbital of spin σ, and $a_{l\sigma}$ is the corresponding annihilation operator. Further, $\hat{R}_{\rm A}(\omega)$ is the frequency-dependent resolvent operator defined as $\hat{R}_{\rm A}(\omega) = (H_{\rm A} - E^{\rm A}_0 + \omega)^{-1} Q_{\rm A}$, and $Q_{\rm A} = 1 - |\Phi^{\rm A}_0\rangle\langle\Phi^{\rm A}_0|$. The induction energy $E^{(2)}_{\rm ind}(\rm A \leftarrow B)$ is related to the polarization propagator at $\omega = 0$ by the equation

$$E^{(2)}_{\rm ind}(\rm A \leftarrow B) = \tfrac{1}{2} (\omega_{\rm B})^k_l (\omega_{\rm B})^{k'}_{l'} \Pi^{ll'}_{kk'}(0), \tag{35}$$

where $(\omega_{\rm B})^k_l \equiv \langle \psi_l | \omega_{\rm B} | \psi_k \rangle$ is the matrix element of the electrostatic potential $\omega_{\rm B}(\boldsymbol{r})$ calculated with the one-particle functions (orbitals) ψ_l used to define the operators E^l_k. The Einstein summation convention over repeated lower and upper indices is used in equation (35) and further on in this paper. The spinorbital equivalent of equation (35) has been derived by Magnasco and McWeeny (1991) and by Moszynski, Cybulski, and Chałasiński (1994). Since the electron densities (needed to calculate $\omega_{\rm B}$) and the static propagators can be calculated as the first and second derivatives of the monomer energy with respect to appropriate perturbations, the existing quantum chemical technology for the calculations of analytic first and second derivatives [Salter *et al.* (1987), Trucks *et al.* (1988a, 1988b), Salter, Trucks, and Bartlett (1989), Salter and Bartlett (1989), Koch *et al.* (1990), Handy *et al.* (1985), Harrison *et al.* (1985), Handy *et al.* (1986), and Helgaker, Jørgensen, and Handy (1989)] can directly be employed to study induction interactions in the region where the charge overlap effects play an important role, i.e., in the region of the van der Waals minimum and at shorter distances.

3.3.4 EXCHANGE–INDUCTION ENERGY

The second-order exchange energy in the SRS theory, defined as $E^{(2)}_{\rm exch} = E^{(2)}_{\rm SRS} - E^{(2)}_{\rm pol}$, separates naturally into two contributions: *exchange–induction* and *exchange–dispersion* energies

$$E^{(2)}_{\rm exch} = E^{(2)}_{\rm exch-ind} + E^{(2)}_{\rm exch-disp}. \tag{36}$$

The exchange–induction energy is an energetic effect resulting from the antisymmetrization of the induction wave function, and can be viewed as a coupling between the induction interaction and the electron exchange. At the distances corresponding to the van der Waals wells, it is sufficient to consider only the single-exchange part of the exchange–induction energy. Higher-order terms (in S^2) were computed for the helium dimer by Chałasiński and Jeziorski (1976) and were found to be negligible in the region of the van der Waals minimum. The same two workers [Chałasiński and Jeziorski (1977)] gave the following expression for $E^{(2)}_{\text{exch-ind}}$ in this approximation

$$E^{(2)}_{\text{exch-ind}}\left(S^2\right) = -\langle\Phi_0|\left(V - E^{(1)}_{\text{pol}}\right)\left(\mathcal{P}_1 - \overline{\mathcal{P}}_1\right)|\Phi^{(1)}_{\text{ind}}\rangle, \tag{37}$$

where $\overline{\mathcal{P}}_1 = \langle\Phi_0|\mathcal{P}_1\Phi_0\rangle$, $\Phi^{(1)}_{\text{ind}} = \Phi^{(1)}_{\text{ind}}(\text{A} \leftarrow \text{B})\Phi^B_0 + \Phi^A_0\Phi^{(1)}_{\text{ind}}(\text{B} \leftarrow \text{A})$, and $\Phi^{(1)}_{\text{ind}}(\text{B} \leftarrow \text{A})$ and $\Phi^{(1)}_{\text{ind}}(\text{A} \leftarrow \text{B})$ are the induction wave functions discussed in the preceding section. In the repulsive part of the intermolecular potential the exchange–induction energy quenches a substantial part of the induction contribution and cannot be neglected in any quantitatively accurate calculation.

3.3.5 DISPERSION ENERGY

The second-order dispersion energy $E^{(2)}_{\text{disp}}$ is defined as the difference between the second-order polarization and induction energies, $E^{(2)}_{\text{disp}} = E^{(2)}_{\text{pol}} - E^{(2)}_{\text{ind}}$. One can also use the following direct definition

$$E^{(2)}_{\text{disp}} = -\langle\Phi_0|V\hat{R}_{\text{AB}}V|\Phi_0\rangle, \tag{38}$$

where the operator $\hat{R}_{\text{AB}}$ is that part of $\hat{R}_0$ [equation (6)] which involves only excited states on both A and B. By its very definition the dispersion interaction represents a pure intermolecular correlation effect. It may be viewed as the stabilizing energetic effect of the correlations of instantaneous multipole moments of the monomers. Since the classic work of Casimir and Polder (1948) we know that, asymptotically at large R, the energy of the dispersion interaction can be expressed in terms of the dynamic multipole polarizabilities of the monomers. A powerful generalization of the Casimir and Polder result has been recently reported [Magnasco and McWeeny (1991), Dmitriev and Peinel (1981), McWeeny (1984), Claverie (1986), and Moszynski, Jeziorski, and Szalewicz (1993)]. These authors have shown that the complete dispersion energy, including the charge-overlap effects, can be expressed, via the Casimir–Polder type integral, in terms of the polarization propagators of the isolated monomers

$$E^{(2)}_{\text{disp}} = -\frac{1}{4\pi}v^{k_1m_1}_{l_1n_1}v^{k_2m_2}_{l_2n_2}\int_{-\infty}^{+\infty}\Pi^{l_1l_2}_{k_1k_2}(i\omega)\Pi^{n_1n_2}_{m_1m_2}(-i\omega)\,\text{d}\omega. \tag{39}$$

In the above expression we assumed that k_1, k_2, l_1, l_2 and m_1, m_2, n_1, n_2 label the orbitals of monomers A and B, respectively. We also introduced the following notation for the Coulomb integrals:

$$v^{k_1m_1}_{l_1n_1} \equiv \langle\psi_{l_1}(1)\psi_{n_1}(2)|r^{-1}_{12}|\psi_{k_1}(1)\psi_{m_1}(2)\rangle.$$

Equation (39) is very important since in the region of the van der Waals minimum the charge-overlap contribution to the dispersion energy is always substantial. Moreover, the powerful computational techniques developed in the 1980s to obtain accurate polarization propagators [Oddershede (1983)] can be utilized via equation (39) in the calculations of the dispersion energies at finite distances.

3.3.6 EXCHANGE–DISPERSION ENERGY

The exchange–dispersion energy $E^{(2)}_{\text{exch-disp}}$ is the energetic effect of the antisymmetrization of the dispersion wave function $\Phi^{(1)}_{\text{disp}}(A\cdots B) = \Phi^{(1)}_{\text{pol}} - \Phi^{(1)}_{\text{ind}} = -\hat{R}_{AB}V\Phi_0$, and can be interpreted as a coupling between the dispersion interaction and the electron exchange. In the single-exchange approximation $E^{(2)}_{\text{exch-disp}}$ is given by Chałasiński and Jeziorski (1977)

$$E^{(2)}_{\text{exch-disp}}(S^2) = -\langle\Phi_0|(V - E^{(1)}_{\text{pol}})(\mathcal{P}_1 - \overline{\mathcal{P}}_1)|\Phi^{(1)}_{\text{disp}}(A\cdots B)\rangle. \tag{40}$$

The effect of multiple exchanges has been computed by Chałasiński and Jeziorski (1976) for the He dimer and found to be negligible in the region of the van der Waals minimum. The exchange–dispersion contribution is relatively small, quenching usually only a few percent of the dispersion energy.

3.4 Many-electron formulation of the SRS theory

In principle, the theory reviewed in Sections 3.2 and 3.3 can be applied to interactions of arbitrary systems if the full CI wave functions of the monomers are available, and if the matrix elements of H_0 and V can be constructed in the space spanned by the products of the configuration state functions of the monomers. For the interactions of many-electron monomers the resulting perturbation equations are difficult to solve, however.

A general approach to the *intramonomer* correlation problem is known as the many-electron (or many-body) SAPT method [Jeziorski and Kołos (1977), Szalewicz and Jeziorski (1979), Jeziorski *et al.* (1989), Rybak, Jeziorski, and Szalewicz (1991), and Moszynski, Jeziorski, and Szalewicz (1994)]. In this method the zeroth-order Hamiltonian H_0 is decomposed as $H_0 = F + W$, where $F = F_{\text{A}} + F_{\text{B}}$ is the sum of the Fock operators, F_{A} and F_{B}, of monomers A and B, respectively, and W is the intramonomer correlation operator. The correlation operator can be written as $W = W_{\text{A}} + W_{\text{B}}$, where $W_X = H_X - F_X$, and $X = \text{A}$ or B. The total Hamiltonian can then be represented as $H = F + V + W$. This partitioning of H defines a double perturbation expansion of the wave function and interaction energy. In the SRS theory the wave function is obtained by expanding the parametrized Schrödinger equation as a power series in ζ and λ

$$(F + \zeta V + \lambda W)\Psi(\zeta, \lambda) = E(\zeta, \lambda)\Psi(\zeta, \lambda), \tag{41}$$

where the parameters ζ and λ are introduced to order the double perturbation expansion, and their physical value is equal to unity. Note that $\Psi(0, 0) = \Phi^{\text{HF}}_{\text{A}}\Phi^{\text{HF}}_{\text{B}}$ is the product of the Hartree–Fock determinants of the unperturbed monomers, and $\Psi(0, \lambda) = \Phi^{\text{A}}_0(\lambda)\Phi^{\text{B}}_0(\lambda)$, where $\Phi^{X}_0(\lambda)$ is the eigenfunction of the Hamiltonian $F_X + \lambda W_X$, $X = \text{A}$ or B. The polarization energy corrections are obtained by expanding the function,

$$E(\zeta, \lambda) = \langle\Phi^{\text{A}}_0(\lambda)\Phi^{\text{B}}_0(\lambda)|F + \zeta V + \lambda W|\Psi(\zeta, \lambda)\rangle, \tag{42}$$

while the SRS energy corrections are obtained by expanding

$$\mathcal{E}(\zeta, \lambda) = \frac{\langle\Phi^{\text{A}}_0(\lambda)\Phi^{\text{B}}_0(\lambda)|F + \zeta V + \lambda W|\mathcal{A}\Psi(\zeta, \lambda)\rangle}{\langle\Phi^{\text{A}}_0(\lambda)\Phi^{\text{B}}_0(\lambda)|\mathcal{A}\Psi(\zeta, \lambda)\rangle}. \tag{43}$$

The expansion of equations (42) and (43) leads to the so-called double perturbation expansion of the interaction energy,

$$E_{\text{int}} = \sum_{n=1}^{\infty}\sum_{k=0}^{\infty}\left(E^{(nk)}_{\text{pol}} + E^{(nk)}_{\text{exch}}\right), \tag{44}$$

where the indices n and k denote the orders of $E^{(nk)}_{\text{pol}}$ and $E^{(nk)}_{\text{exch}}$ with respect to the intermolecular interaction and intramonomer correlation, respectively. The nth-order polarization and exchange contributions can be obtained by a direct summation of the expansion (44) over k. Explicit expressions for the individual corrections $E^{(nk)}_{\rm pol}$ and $E^{(nk)}_{\rm exch}$ can be obtained using the techniques of many-body perturbation theory and the coupled-cluster method. See Moszynski *et al.* (1994a), Moszynski, Cybulski, and Chałasiński (1994), Moszynski, Jeziorski, and Szalewicz (1993), Jeziorski, Moszynski, Rybak, and Szalewicz (1989), Rybak, Jeziorski, and Szalewicz (1991), Moszynski, Jeziorski, and Szalewicz (1994), Moszynski *et al.* (1993), and Williams *et al.* (1995) for the details of the derivations of open-ended expressions valid for the interaction of monomers of arbitrary size. The methods to treat the intramonomer correlation effects on the electrostatic, induction, dispersion, and exchange energies are discussed below. These formulas have been implemented in a computer code. See Jeziorski *et al.* (1993) for a description of this program and for the final working equations in the form that they were coded. We refer especially to Table I of this reference for an overview of the different two-body SAPT terms.

We introduce some auxiliary quantities that will arise in the following subsections. The labels a, a' and r, r' indicate occupied and virtual orbitals, respectively, on A. Similarly b, b' and s, s' are occupied and virtual orbitals on B. The labels k, l and m, n designate arbitrary orbitals on A and B, respectively. Two-electron integrals are denoted by $v^{kk'}_{ll'} \equiv \langle \psi_l\psi_{l'}|r_{12}{}^{-1}|\psi_k\psi_{k'}\rangle$ and $\tilde{v}^{km}_{ln} \equiv \langle \psi_l\psi_n|\tilde{v}|\psi_k\psi_m\rangle$, where $\tilde{v}$ contains nuclear attractions, see equation (25). We will also meet $g^{kk'}_{ll'} \equiv \langle \psi_l\psi_{l'}|(2-P_{12})r_{12}{}^{-1}|\psi_k\psi_{k'}\rangle$. The quantities t^a_r and $t^{aa'}_{rr'}$ (one- and two-particle spin-free cluster amplitudes) are in first approximation

$$t^{aa'}_{rr'} = v^{aa'}_{rr'}/\varepsilon^{aa'}_{rr'} \qquad \text{and} \qquad t^a_r = \left(\theta^{a''a}_{r''r'}v^{r''r'}_{a''r} - \theta^{a''a'}_{r''r}v^{r''a}_{a''a'}\right)/\varepsilon^a_r, \tag{45}$$

where $\theta^{aa'}_{rr'} \equiv 2t^{aa'}_{rr'} - t^{a'a}_{rr'}$. The denominators contain orbital energy differences: $\varepsilon^{aa'\cdots}_{rr'\cdots} = \varepsilon_a + \varepsilon_{a'} + \cdots - \varepsilon_r - \varepsilon_{r'} - \cdots$. Often we will decompose terms of order k in W into terms of different order in $W_{\rm A}$ and $W_{\rm B}$, writing

$$E^{(n,k)} = \sum_{l=0}^{k} E^{(n,k-l,l)}. \tag{46}$$

3.4.1 ELECTROSTATIC ENERGY

The intramonomer correlation corrections to the electrostatic energy can be obtained from a simple modification of equation (15) proposed by Moszynski *et al.* (1993)

$$E^{(1)}_{\rm pol} = \iint \rho_{\rm A}(\boldsymbol{r}_1)\tilde{v}(\boldsymbol{r}_1,\boldsymbol{r}_2)\rho_{\rm B}(\boldsymbol{r}_2)\,d\boldsymbol{r}_1\,d\boldsymbol{r}_2. \tag{47}$$

This expression can be easily expanded as a series in the intramonomer correlation. The correction $E^{(1,k)}_{\rm pol}$ is simply given by

$$E^{(1,k)}_{\rm pol} = \sum_{l=0}^{k} \iint \rho^{(l)}_{\rm A}(\boldsymbol{r}_1)\tilde{v}(\boldsymbol{r}_1,\boldsymbol{r}_2)\rho^{(k-l)}_{\rm B}(\boldsymbol{r}_2)\,d\boldsymbol{r}_1\,d\boldsymbol{r}_2, \tag{48}$$

where $\rho^{(l)}_X(\boldsymbol{r})$, $X = $ A or B, is the kth-order term in the Møller–Plesset expansion of the electron density of monomer X. The consecutive correlation corrections to the electron density, $\rho^{(l)}_X(\boldsymbol{r})$, are obtained from the Møller–Plesset expansion of bra and ket in the expectation value $\rho_X(\boldsymbol{r}) = \langle\Phi^X_0|D(\boldsymbol{r})|\Phi^X_0\rangle$ of the electron density operator $D(\boldsymbol{r})$,

$$D(\boldsymbol{r}) \equiv \sum_{i\in X} \delta(\boldsymbol{r}-\boldsymbol{r}_i). \tag{49}$$

Thus, through second-order in the intramonomer correlation,

$$E_{\text{pol}}^{(1)} = \left(\rho_{\text{A}}^{(0)}\right)_k^l \tilde{v}_{ln}^{km} \left(\rho_{\text{B}}^{(0)}\right)_m^n + \left(\rho_{\text{A}}^{(0)}\right)_k^l \tilde{v}_{ln}^{km} \left(\rho_{\text{B}}^{(2)}\right)_m^n + \left(\rho_{\text{A}}^{(2)}\right)_k^l \tilde{v}_{ln}^{km} \left(\rho_{\text{B}}^{(0)}\right)_m^n, \tag{50}$$

where we recall that $\rho_{\text{A}}^{(1)}$ and $\rho_{\text{B}}^{(1)}$ vanish by virtue of Brillouin's theorem. The matrix elements arising here are

$$\left(\rho_{\text{A}}^{(0)}\right)_k^l = 2\delta_k^a \delta_{\text{A}}^l, \qquad \left(\rho_{\text{A}}^{(2)}\right)_{r'}^{r''} = 2\theta_{rr'}^{aa'} t_{aa'}^{rr''}, \qquad \left(\rho_{\text{A}}^{(2)}\right)_{a''}^{a'} = 2\theta_{rr'}^{aa'} t_{aa''}^{rr'}, \qquad \left(\rho_{\text{A}}^{(2)}\right)_r^a = 2t_r^a, \tag{51}$$

with δ_q^p a Kronecker delta. The explicit expressions for $E_{\text{pol}}^{(13)}$ and $E_{\text{pol}}^{(14)}$ are more complicated; they were published a few years ago [Moszynski *et al.* (1993)].

Alternatively one may obtain the perturbation contributions to the density by differentiating with respect to ξ the lth-order Møller–Plesset energy computed with the Hamiltonian $\widetilde{H}_X = H_X + \xi D(\boldsymbol{r})$ at $\xi = 0$. The correlation corrections obtained by the use of the expectation and differentiation method are not equal, since the Møller–Plesset wave function does not fulfill the Hellmann–Feynman theorem. Consequently, the two Møller–Plesset expansions of the electron densities define two intramonomer correlation contributions to the electrostatic energy which will be denoted by $E_{\text{pol}}^{(1,k)}$ and $E_{\text{pol,resp}}^{(1,k)}$, respectively. The corrections $E_{\text{pol,resp}}^{(1,k)}$ computed from the electron densities obtained by differentiation of the perturbed Møller–Plesset energies account for the so-called orbital relaxation or coupled-Hartree–Fock type response [Sadlej (1981)]. Note that $E_{\text{pol,resp}}^{(10)} \equiv E_{\text{pol}}^{(10)}$ since the Hartree–Fock wave functions are fully variational, and fulfill the Hellmann–Feynman theorem. At large intermonomer separations the two corrections behave as

$$E_{\text{pol}}^{(1,k)} \sim \sum_{l=0}^{k} \left(Q^{(l)}\right)_{L_{\text{A}}}^{K_{\text{A}}} \left(Q^{(k-l)}\right)_{L_{\text{B}}}^{K_{\text{B}}} R^{-(L_{\text{A}}+L_{\text{B}}+1)}, \tag{52}$$

where Q_L^K denotes the component K of the first non-vanishing permanent multipole moment of order L. For instance $L = 1$ gives a dipole, $L = 2$ a quadrupole, etc. The nth order correlation contribution to the multipole of monomer X in the expectation value approach [Moszynski *et al.* (1993)] is given by $(Q^{(n)})_{L_X}^{K_X}$, $X = \text{A}$ or B. If we replace the multipole moments by their corresponding response (finite-field) values [Salter *et al.* (1987), Trucks *et al.* (1988a, 1988b), Salter, Trucks, and Bartlett (1989), and Salter and Bartlett (1989)] we obtain the response asymptote.

Moszynski *et al.* (1993) investigated the convergence of the series $\sum_{k=0} E_{\text{pol}}^{(1,k)}$ and $\sum_{k=0} E_{\text{pol,resp}}^{(1,k)}$ for model four-electron dimers. Their results suggest that both expansions truncated at $k = 3$ reproduce the full CI values with a high accuracy. In all cases the response series showed a better convergence rate. Since a single point calculation of $E_{\text{pol,resp}}^{(1,k)}$ is no more computer time demanding than the calculation of $E_{\text{pol}}^{(1,k)}$, it is preferable to use the response approach.

3.4.2 INDUCTION ENERGY

The intramonomer correlation corrections to the induction energy can be obtained from equation (35). This equation can be expanded as a series in the intramonomer correlation, and the correction $E_{\text{ind}}^{(2,k)}$ is given by

$$E_{\text{ind}}^{(2,k)} = \sum_{l=0}^{k} \left[E_{\text{ind}}^{(2,k-l,l)}(\text{A} \leftarrow \text{B}) + E_{\text{ind}}^{(2,k-l,l)}(\text{B} \leftarrow \text{A})\right], \tag{53}$$

with

$$E_{\rm ind}^{(2,k-l,l)}({\rm A}\leftarrow{\rm B})=\frac{1}{2}\sum_{l_1=0}^{l}\left(\Pi^{(k-l)}\right)_{kk'}^{ll'}(0)\left(\omega_{\rm B}^{(l_1)}\right)_{l}^{k}\left(\omega_{\rm B}^{(l-l_1)}\right)_{l'}^{k'},\tag{54}$$

where $(\Pi^{(n)})_{kk'}^{ll'}(0)$, is the nth-order term in the Møller–Plesset expansion of the static polarization propagator of monomer A, and $\omega_{\rm B}^{(n)}$ is defined by the Møller–Plesset expansion of the electron density, cf. equation (32),

$$\omega_{\rm B}^{(n)}(\boldsymbol{r}_1)=u_{\rm B}(\boldsymbol{r}_1)\delta^{(n,0)}+\int\rho_{\rm B}^{(n)}(\boldsymbol{r}_2)r_{12}^{-1}\,{\rm d}\boldsymbol{r}_2,\tag{55}$$

where $u_{\rm B}(\boldsymbol{r}_1)$ is the nuclear potential of B, which contributes only in zeroth ($n=0$) order. By definition $E_{\rm ind}^{(2,k-l,l)}({\rm B}\leftarrow{\rm A})$ contains a propagator of order l on B, while it is of order $k-l$ in the static field due to A.

The uncorrelated induction energy becomes

$$E_{\rm ind}^{(200)}({\rm A}\leftarrow{\rm B})=2\left(\omega_{\rm B}^{(0)}\right)_{a}^{r}\left(\omega_{\rm B}^{(0)}\right)_{r}^{a}/\varepsilon_{r}^{a},\tag{56}$$

and analogously for $E_{\rm ind}^{(200)}({\rm B}\leftarrow{\rm A})$. The first-order correlation contribution is

$$E_{\rm ind}^{(210)}({\rm A}\leftarrow{\rm B})=2\left(\omega_{\rm B}^{(0)}\right)_{a}^{r}g_{ra'}^{ar'}\left(\omega_{\rm B}^{(0)}\right)_{r'}^{a'}/\varepsilon_{r}^{a}\varepsilon_{r'}^{a'}+2\Re\left(\omega_{\rm B}^{(0)}\right)_{a}^{r}g_{rr'}^{aa'}\left(\omega_{\rm B}^{(0)}\right)_{a'}^{r'}/\varepsilon_{r}^{a}\varepsilon_{r'}^{a'}.\tag{57}$$

An analogous equation holds for $E_{\rm ind}^{(201)}({\rm B}\leftarrow{\rm A})$. Note that $E_{\rm ind}^{(211)}$ vanishes. The explicit expression for the second-order intramonomer correlation correction has been published [Szalewicz and Jeziorski (1997), and Moszynski, Cybulski, and Chałasiński (1994)].

Similarly as in the case of the electron density, cf. equations (49)–(51), the consecutive correlation corrections can be obtained from the Møller–Plesset expansion of an appropriate expectation value [Wormer and Hettema (1992)]. Alternatively, one can compute mixed second derivatives by differentiating with respect to ξ and χ the lth-order Møller–Plesset energy computed with the Hamiltonian $\widetilde{H}_X=H_X+\xi E_k^l+\chi E_{k'}^{l'}$ at $\xi=\chi=0$. Again, the correlation corrections obtained using these two methods are not equal, so there are two different intramonomer correlation contributions to the induction energy which will be denoted by $E_{\rm ind}^{(2,k-l,l)}$ and $E_{\rm ind,resp}^{(2,k-l,l)}$, respectively. The corrections $E_{\rm ind,resp}^{(2,k-l,l)}$ account for the so-called orbital relaxation or coupled-Hartree–Fock type response. At large intermonomer separations the two corrections behave as

$$E_{\rm ind}^{(2,k-l,l)}({\rm A}\leftarrow{\rm B})\sim\sum_{l_1=0}^{l}\alpha_{\rm A}^{(k-l)}(0)\left(Q^{(l_1)}\right)_{l_{\rm B}}^{k_{\rm B}}\left(Q^{(l-l_1)}\right)_{l'_{\rm B}}^{k'_{\rm B}}R^{-(l_{\rm B}+l'_{\rm B}+4)}.\tag{58}$$

where $\alpha_X^{(n)}(0)$, $X=$ A or B, denotes the nth-order correlation correction to the static ($\omega=0$) polarizability of monomer X in the standard Møller–Plesset perturbation expansion [Wormer and Hettema (1992)]. The same asymptotic expression is obtained by the finite-field method [Handy *et al.* (1985), Harrison *et al.* (1985), Handy *et al.* (1986)] when we use the corresponding response monomer properties.

No systematic study of the convergence of the expansions $\sum_{k=0}E_{\rm ind}^{(2,k)}$ and $\sum_{k=0}E_{\rm ind,resp}^{(2,k)}$ has been reported in the literature thus far. The results reported by Moszynski, Cybulski, and Chałasiński (1994) show that, except for ionic systems, the leading intramonomer correlation contributions to the induction energy are small, so higher-order terms should be negligible.

3.4.3 DISPERSION ENERGY

The intramonomer correlation corrections to the dispersion energy can be directly obtained from equation (39),

$$E_{\rm disp}^{(2,k)} = -\frac{1}{4\pi}\sum_{l=0}^{k} v_{l_1n_1}^{k_1m_1} v_{l_2n_2}^{k_2m_2} \int_{-\infty}^{+\infty} \left(\Pi^{(l)}\right)_{k_1k_2}^{l_1l_2}(\mathrm{i}\omega)\left(\Pi^{(k-l)}\right)_{m_1m_2}^{n_1n_2}(-\mathrm{i}\omega)\,\mathrm{d}\omega, \tag{59}$$

where $(\Pi^{(n)})_{k_1k_2}^{l_1l_2}(\mathrm{i}\omega)$ is the nth-order correction in the Møller–Plesset perturbation expansion of the frequency-dependent polarization propagator. For two atoms at large distances the correction $E_{\rm disp}^{(2,k)}$ behaves as

$$E_{\rm disp}^{(2,k)} \sim -\frac{C_6^{(k)}}{R^6}, \quad \text{where} \quad C_6^{(k)} = \frac{3}{\pi}\sum_{l=0}^{k}\int_0^{\infty} \alpha_{\rm A}^{(l)}(\mathrm{i}\omega)\alpha_{\rm B}^{(k-l)}(\mathrm{i}\omega)\,\mathrm{d}\omega, \tag{60}$$

and $\alpha_X^{(n)}(i\omega)$ is the nth-order term in the Møller–Plesset perturbation expansion of the frequency-dependent polarizability [Wormer and Hettema (1992)].

The uncorrelated dispersion energy is given by

$$E_{\rm disp}^{(20)} = 4v_{rs}^{ab}v_{ab}^{rs}/\varepsilon_{rs}^{ab}, \tag{61}$$

while the leading intramonomer correlation contribution is

$$E_{\rm disp}^{(210)} = 4t_{rs}^{ab}g_{ar'}^{ra'}t_{a'b}^{r's} + 8\Re\left(t_{rs}^{ab}v_{r'b}^{a's}\theta_{aa'}^{rr'}\right), \tag{62}$$

and analogously for $E_{\rm disp}^{(201)}$. The expressions for the corrections $E_{\rm disp}^{(211)}$ and $E_{\rm disp}^{(220)} + E_{\rm disp}^{(202)}$ are given in the literature [Rybak, Jeziorski, and Szalewicz (1991)]. Note that the first term in equation (94) of this reference should be multiplied by two, and that the definitions of the g and θ quantities slightly differ from our definitions.

The convergence of the intramonomer correlation corrections to the dispersion energy has been studied by Williams *et al.* (1995) for model four-electron and many-electron dimers. For four-electron systems the results of perturbative calculations up to and including $E_{\rm disp}^{(22)}$ were compared with the values from full CI calculations. For many-electron dimers full CI calculations of the dispersion energy are not feasible at present, and the reference results were obtained with the coupled-cluster method restricted to double excitations with an approximate inclusion of the single and triple excitations. In both cases the convergence of the intramonomer correlation series truncated at the second order in W was excellent.

3.4.4 FIRST-ORDER EXCHANGE ENERGY

The intramonomer correlation corrections to the first-order energy can be directly obtained from equation (23) [Moszynski *et al.* (1994a)]

$$E_{\rm exch}^{(1,k)}(S^2) = \iint \tilde{v}(\boldsymbol{r}_1,\boldsymbol{r}_2)\rho_{\rm int}^{(k)}(\boldsymbol{q}_1,\boldsymbol{q}_2)\,\mathrm{d}\boldsymbol{q}_1\,\mathrm{d}\boldsymbol{q}_2$$

$$-\sum_{l=0}^{k}\sum_{l_1=0}^{l} E_{\rm pol}^{(1,k-l)}\iint \rho_{\rm A}^{(l_1)}(\boldsymbol{q}_1|\boldsymbol{q}_2)\rho_{\rm B}^{(l-l_1)}(\boldsymbol{q}_2|\boldsymbol{q}_1)\,\mathrm{d}\boldsymbol{q}_1\,\mathrm{d}\boldsymbol{q}_2, \tag{63}$$

where

$$\rho_{\text{int}}^{(k)}(q_1, q_2) = \sum_{l=0}^{k} \Big[-\rho_{\text{A}}^{(l)}(q_1 \mid q_2)\rho_{\text{B}}^{(k-l)}(q_2 \mid q_1)$$

$$- \int \Gamma_{\text{A}}^{(l)}(q_1q_3 \mid q_1q_2)\rho_{\text{B}}^{(k-l)}(q_2 \mid q_3)\,\text{d}q_3$$

$$- \int \rho_{\text{A}}^{(l)}(q_1 \mid q_3)\Gamma_{\text{B}}^{(k-l)}(q_2q_3 \mid q_2q_1)\text{d}q_3$$

$$- \int\int \Gamma_{\text{A}}^{(l)}(q_1q_3 \mid q_1q_4)\Gamma_{\text{B}}^{(k-l)}(q_2q_4 \mid q_2q_3)\,\text{d}q_3\,\text{d}q_4\Big], \tag{64}$$

further $\rho_X^{(k)}(q_1|q_2)$ and $\Gamma_X^{(k)}(q_1, q_2|q_3, q_4)$, $X = A$ or B, denote the kth-order terms in the Møller–Plesset expansions of the one- and two-particle density matrices [Jeziorski and Moszynski (1993)], cf. equations (27) and (28).

Through first order in the correlation the first-order exchange interaction consists of the following terms:

$$E_{\text{exch}}^{(100)} = -2\left[\tilde{v}_{ab}^{ba} + S_{a'}^{b}\left(2\tilde{v}_{ab}^{aa'} - \tilde{v}_{ab}^{a'a}\right) + S_{b'}^{a}\left(2\tilde{v}_{ab}^{b'b} - \tilde{v}_{ab}^{bb'}\right)\right.$$

$$\left. -2S_{a'}^{b}S_{b'}^{a'}\tilde{v}_{ab}^{ab'} - 2S_{a'}^{b'}S_{b'}^{a}\tilde{v}_{ab}^{a'b} + S_{a'}^{b}S_{b'}^{a}\tilde{v}_{ab}^{a'b'}\right], \tag{65}$$

$$E_{\text{exch}}^{(110)} = -4\Re\left[\theta_{rr'}^{aa'}(\omega_{\text{B}}^{(0)})_{\text{A}}^{r}S_{\text{B}}^{r'}S_{a'}^{b}\right] - 2\theta_{aa'}^{rr'}\tilde{v}_{rb}^{aa'}S_{r'}^{b} - 2\theta_{rr'}^{aa'}\tilde{v}_{ab}^{rr'}S_{a'}^{b}, \tag{66}$$

with a similar equation for $E_{\text{exch}}^{(101)}$. The algebraic expressions for $E_{\text{exch}}^{(111)}$ and $E_{\text{exch}}^{(120)} + E_{\text{exch}}^{(102)}$ are somewhat more complicated, but do exist [Moszynski *et al.* (1994a)].

The convergence of the expansion $\sum_{k=0} E_{\text{exch}}^{(1,k)}$ has been investigated for model four-electron dimers by Moszynski, Jeziorski, and Szalewicz (1994). The results obtained suggested that a truncation of the intramonomer correlation series at $k = 2$ is not a good approximation, and a coupled-cluster method for the calculation of $E_{\text{exch}}^{(1)}$ (including single and double excitations, CCSD) was proposed. The CCSD exchange energy was shown to be an excellent approximation to the full CI results. This is not surprising, as it is well known that CCSD sums all correlation terms with intermediate singly and doubly excited states to infinite order. Also for more-than-four-electron systems the intramonomer correlation expansion truncated at $k = 2$ did not give satisfactory results, so for these larger systems, too, it is preferable to use the coupled-cluster approach. For these systems the CCSD approximations to $E_{\text{exch}}^{(1)}$ were used as reference results [Moszynski *et al.* (1994a)]. Finally we wish to point out that the latest version of the SAPT program [Bukowski *et al.* (1996)] contains a module that computes the first-order exchange for monomers that are correlated at the CCSD level.

3.4.5 SECOND-ORDER EXCHANGE ENERGY

Exchange modifications of induction and dispersion have been considered to date only at the Hartree–Fock level, that is, without intramolecular correlation. See Chałasiński and Jeziorski (1977) for the explicit expressions of $E_{\text{exch-ind}}^{(200)}$ and $E_{\text{exch-disp}}^{(200)}$.

3.5 Comparison with supermolecule Hartree–Fock and Møller–Plesset theories

Apart from the computation of potentials, SAPT can be used as well to interpret the results of supermolecular calculations. Numerous studies [Jeziorska, Jeziorski, and Čížek (1987), Moszynski, Heijmen, and Jeziorski (1996), Chałasiński and Szczęśniak (1988), Cybulski, Chałasiński, and Moszynski (1990), Moszynski *et al.* (1990), Cybulski and Chałasiński (1992), and Moszynski (1992)] have interpreted supermolecule Hartree–Fock and Møller-Plesset perturbation theory interaction energies in terms of physically meaningful contributions. In this section we briefly review the present status of theories connecting the supermolecule Hartree–Fock and Møller–Plesset theories with the SAPT approach.

Although it is not immediately evident whether SAPT should be related to supermolecule results with or without correction for basis set superposition errors (BSSEs), theoretical arguments and numerical results [Korona, Moszynski, and Jeziorski (1997)] show that a perturbation theory expansion in a finite basis can converge only to a supermolecular interaction energy if the latter is corrected by the Boys and Bernardi (1970) counterpoise method. This means that, in agreement with the conclusions reached earlier [Bulski and Chałasiński (1977), Gutowski *et al.* (1986a, 1986b, 1987), Gutowski and Chałasiński (1993), Gutowski *et al.* (1993), and van Duijneveldt, van Duijneveldt-van der Rijdt, and van Lenthe (1994)], the use of the Boys–Bernardi counterpoise correction is fully legitimate. Hence, we will assume in the remainder of this section that the supermolecule interaction energies are BSSE corrected according to the Boys–Bernardi prescription.

3.5.1 HARTREE–FOCK THEORY

An SAPT approach for the calculation of the Hartree–Fock interaction energies has been proposed by Jeziorska, Jeziorski, and Čížek (1987) for the helium dimer, and was generalized to the many-electron case [Moszynski, Heijmen, and Jeziorski (1996)]. The authors of these references developed a basis-set-independent perturbation scheme to solve the Hartree–Fock equations for the dimer, and analyzed the Hartree–Fock interaction energy in terms of contributions related to the many-electron SAPT reviewed in Section 3.4. Specifically, they proposed to replace the Hartree–Fock equations for the canonical orbitals of the dimer by noncanonical equations for orbitals localized on the monomers. Several localizations conditions can be exploited, but Moszynski, Heijmen, and Jeziorski (1996) found it advantageous to employ a generalization of the localization condition used in the (many-electron) Hirschfelder–Silbey (1966) theory. The perturbation expansion of these localized orbitals defines an expansion of the Hartree–Fock interaction energy, $E_{\text{int}}^{\text{HF}}$

$$E_{\text{int}}^{\text{HF}} = E_{\text{pol}}^{(10)} + E_{\text{exch}}^{(10)} + E_{\text{ind,resp}}^{(20)} + E_{\text{exch-ind,resp}}^{(20)} + E_{\text{exch-def,resp}}^{(20)} + \cdots. \tag{67}$$

All terms appearing on the r.h.s. of equation (67) were defined in Sections 3.3 and 3.4, except for the exchange–deformation energy, $E_{\text{exch-def,resp}}^{(20)}$. This contribution, included in Hartree–Fock theory, does not appear in the SRS theory (cf. Section 3); it is a part of the exchange energy that cannot be recovered by perturbation theory with weak symmetry-forcing. For systems with long-range induction interactions this contribution vanishes faster at large intermonomer distances than the exchange–induction energy itself [Certain and Hirschfelder (1970), and Chałasiński and Jeziorski (1973)].

Numerical results for the He–C_2H_2, He–CO, and Ar–HF complexes are summarized in Table 3.1. Consecutive entries in this table represent the contributions appearing on the r.h.s. of equation (67). Also reported are the total nth-order approximations to $E_{\text{int}}^{\text{HF}}$, denoted by $E_{\text{int}}^{\text{HF}}(n)$.

Table 3.1. Comparison of low-order approximations (in μhartree) to the Hartree-Fock interaction energies of the He–C_2H_2, He–CO, and Ar–HF complexes, at the global minima of the potentials.

	He–C_2H_2	He–CO	Ar–HF
$E^{(10)}_{\text{pol}}$	−14.87	−19.99	−192.1
$E^{(10)}_{\text{exch}}$	93.45	99.07	636.2
$E^{\text{HF}}_{\text{int}}(1)$	78.59	79.11	444.7
$E^{(20)}_{\text{ind,resp}}$	−19.15	−5.91	−227.2
$E^{(20)}_{\text{exch-ind,resp}}$	3.51	5.69	157.0
$E^{(20)}_{\text{exch-def,resp}}$	−4.57	−3.89	−31.1
$E^{\text{HF}}_{\text{int}}(2)$	58.37	74.98	342.7
$E^{\text{HF}}_{\text{int}}$	56.52	74.17	332.6

An inspection of Table 3.1 shows that the perturbation expansion (67) converges rapidly. The second-order approximation $E^{\text{HF}}_{\text{int}}(2)$ reproduces the exact Hartree–Fock results to within 3 %. One may note that for some systems the exchange–deformation energy is far from negligible, so the SRS theory does not fully recover the Hartree–Fock interaction energy. In all of the applications, discussed in Section 3.7 below, we used therefore supermolecule Hartree–Fock as a starting point and added only those SAPT terms that are not accounted for by the Hartree–Fock method.

3.5.2 MØLLER–PLESSET THEORY

An analysis similar to that above has not been performed thus far for the Møller–Plesset theories. However, Chalasinski and collaborators [Chałasiński and Szczęśniak (1988), Cybulski, Chałasiński, and Moszynski (1990), Moszynski *et al.* (1990), Cybulski and Chałasiński (1992), and Moszynski (1992)] conjectured that the polarization part of the supermolecule nth-order Møller–Plesset energy, $\Delta E^{\text{MP}n}_{\text{int}}$, $n \leq 4$, is given by

$$\Delta E^{\text{MP}n}_{\text{int}} = E^{(1,n,0)}_{\text{pol,resp}} + E^{(1,0,n)}_{\text{pol,resp}} + E^{(2,n)}_{\text{ind,resp}} + E^{(2,n-2)}_{\text{disp}} + E^{\text{MP}n}_{\text{exch}}$$
$$+ \text{ similar terms higher order in } V. \qquad (68)$$

Here $\Delta E^{\text{MP}n}_{\text{int}}$ denotes the nth-order correlation part of the supermolecule MPn interaction energy. Equation (68) shows that the supermolecule MP2 interaction energy correctly accounts for the leading intramonomer correlation corrections to the electrostatic and induction energies, and for the major part of the dispersion energy. This explains why it could be used with success for several van der Waals and hydrogen-bonded complexes [Chałasiński and Szczęśniak (1994)]. The physical structure of the MPn exchange terms, $E^{\text{MP}n}_{\text{exch}}$, is not well understood. A perturbation theory analysis of the MP2 equations for the pair functions in a localized representation [Moszynski (1992)] suggests that $E^{\text{MP2}}_{\text{exch}}$ accounts for the major part of the uncorrelated exchange–dispersion energy $E^{(20)}_{\text{exch-disp}}$ (the so-called K_1 term [Chałasiński and Jeziorski (1976)]), and for some parts of $E^{(11)}_{\text{exch}}$ and $E^{(12)}_{\text{exch}}$ corrections.

In order to get an idea how well a standard SAPT calculation can reproduce $\Delta E^{\text{MP2}}_{\text{int}}$ Bukowski, Jeziorski, and Szalewicz (1996) defined

$$E^{\text{MP2}}_{\text{SAPT}} = E^{(12)}_{\text{pol,resp}} + E^{(22)}_{\text{ind,resp}} + E^{(20)}_{\text{disp}} + E^{(11)}_{\text{exch}} + E^{(12)}_{\text{exch}} + E^{(20)}_{\text{exch-disp}}, \qquad (69)$$

Table 3.2. SAPT and localized MP2 components of $\Delta E_{\mathrm{int}}^{\mathrm{MP2}}$ for the He dimer. The unit of energy is 1 K.

Component	$R = 4.0$	$R = 5.6$	$R = 7.0$
$\Delta E_{\mathrm{int}}^{\mathrm{MP2}}$	−117.42	−16.003	−3.780
$E_{\mathrm{SAPT}}^{\mathrm{MP2}}$	−128.22	−16.250	−3.792
$\Delta E_{\mathrm{int}}^{\mathrm{MP2}}$(inter)	−132.74	−16.683	−3.812
$E_{\mathrm{disp}}^{(20)} + E_{\mathrm{exch-disp}}^{(20)}$	−140.63	−16.627	−3.848
$\Delta E_{\mathrm{int}}^{\mathrm{MP2}}$(intra)	15.32	0.681	0.032
$E_{\mathrm{pol,resp}}^{(12)} + E_{\mathrm{ind}}^{(22)} + E_{\mathrm{exch}}^{(11)} + E_{\mathrm{exch}}^{(12)}$	12.31	0.378	0.012

and analyzed the performance of this ansatz for the helium dimer at various distances. Note that equation (69) is equivalent to equation (68) except that $E_{\mathrm{exch}}^{\mathrm{MP2}}$ is approximated by the sum $E_{\mathrm{exch}}^{(11)} + E_{\mathrm{exch}}^{(12)} + E_{\mathrm{exch-disp}}^{(20)}$. The results are illustrated in Table 3.2. An inspection of this table shows that the agreement between the supermolecule MP2 interaction energy and the approximate MP2-SAPT results computed from equation (69) is reasonable. Bukowski, Jeziorski, and Szalewicz (1996) also decomposed the MP2 interaction energy $\Delta E_{\mathrm{int}}^{\mathrm{MP2}}$ into an intermonomer part $\Delta E_{\mathrm{int}}^{\mathrm{MP2}}$(inter) and an intramonomer part $\Delta E_{\mathrm{int}}^{\mathrm{MP2}}$(intra). These terms were computed using properly localized MP2 pair functions and Hartree–Fock orbitals. It is interesting to note that the agreement between the sum of the intermonomer correlation terms (dispersion and exchange–dispersion energies) and the localized intermonomer part of $\Delta E_{\mathrm{int}}^{\mathrm{MP2}}$ is very good. The agreement between the localized intramonomer part and the sum of the intramonomer correlation contributions is less satisfactory. Since the electrostatic and induction terms are included in $\Delta E_{\mathrm{int}}^{\mathrm{MP2}}$, the level of disagreement suggests that the first-order exchange terms in MP2 and SAPT are different.

3.5.3 PERFORMANCE OF SAPT IN COMPARISON WITH CCSD(T)

To obtain a further perspective on the accuracy of the SAPT results, we collected in Table 3.3 SAPT and CCSD(T) results for van der Waals and hydrogen-bonded complexes [Milet *et al.*

Table 3.3. Components of the interaction energy (in cm^{-1}) of the $(H_2O)_2$, CO–H_2O, and He–CO_2 complexes and comparison with supermolecule CCSD(T) results computed in the same basis sets [Milet *et al.* (1999)]. Geometries are close to the global minima, but note that the exact values of D_e are somewhat larger.

	$(H_2O)_2$	CO–H_2O	He–CO_2
$E_{\mathrm{pol}}^{(1)}$	−2343	−667.49	−0.85
$E_{\mathrm{exch}}^{(1)}$	2238	713.45	4.81
$E_{\mathrm{ind}}^{(2)}$	−850	−258.57	−0.79
$E_{\mathrm{exch-ind}}^{(2)}$	507	136.37	0.11
$E_{\mathrm{disp}}^{(2)}$	−860	−387.63	−29.72
$E_{\mathrm{exch-disp}}^{(2)}$	119	37.75	0.23
$E_{\mathrm{exch-def}}^{(2)}$	−217	−63.60	−0.18
$E_{\mathrm{int}}^{\mathrm{SAPT}}$	−1476	−489.73	−26.38
$E_{\mathrm{int}}^{\mathrm{CCSD(T)}}$	−1479	−484.87	−25.70

(1999a)]. As representatives we have chosen the water dimer, H_2O-CO, and $He-CO_2$ complexes. The first complex is mainly bound by electrostatic, the second by electrostatic and dispersion, and the third by pure dispersion forces. The binding energy at the minimum varies from $40\,cm^{-1}$ for $He-CO_2$ to $1800\,cm^{-1}$ for the water dimer. An inspection of Table 3.3 shows that the agreement between the SAPT and CCSD(T) results is always excellent. In fact, the discrepancies are smaller than 3 %. It is very gratifying to observe such a good agreement between the two methods, despite substantial cancellations of the attractive and repulsive contributions in the SAPT calculations.

3.6 Non-additive interactions

This section gives an overview of the SAPT of pair-wise nonadditive interactions in trimers. This theory was recently formulated [Moszynski *et al.* (1995b), and Lotrich and Szalewicz (1997a)].

In the earlier sections of this chapter we reviewed the many-electron formulation of the SAPT of two-body interactions. As we saw, all physically important contributions to the potential could be identified and computed separately. We follow the same program for the three-body forces and discuss a triple perturbation theory for interactions in trimers. We show how the pure three-body effects can be separated out and give equations in terms of linear and quadratic response functions. These formulas have a clear, partly classical, partly quantum mechanical interpretation. The exchange terms are also classified; for the explicit orbital formulas we refer to Moszynski *et al.* (1995b).

3.6.1 GENERAL THEORY

We consider the closed-shell systems A, B, and C. The total Hamiltonian of the trimer A–B–C is

$$H = H_0 + \zeta V^{AB} + \eta V^{BC} + \chi V^{CA}, \tag{70}$$

where $H_0 = H_A + H_B + H_C$ is the sum of Hamiltonians of the isolated monomers, the operator V^{XY} collects all Coulombic interactions between electrons and nuclei of monomers X and Y, and the parameters ζ, η, and χ have the physical value unity. The exact interaction energy of the trimer is defined by the asymmetric energy expression

$$E_{int}(\zeta, \eta, \chi) = \langle \Phi_0 |\, \zeta V^{AB} + \eta V^{BC} + \chi V^{CA}\, | \Psi(\zeta, \eta, \chi) \rangle\,, \tag{71}$$

where $\Phi_0 = \Phi_0^A \Phi_0^B \Phi_0^C$ is the ground state eigenfunction of the unperturbed Hamiltonian H_0, and Φ_0^X denotes the ground state wave function of monomer X. The function $\Psi(\zeta, \eta, \chi)$ is the exact solution of the Schrödinger equation of the trimer with Hamiltonian (70).

The interaction energy and wave function are expanded in the usual manner as a power series in the three perturbation parameters ζ, η, and χ,

$$E_{int}(\zeta, \eta, \chi) = \sum_{i,j,k} \zeta^i \eta^j \chi^k E_{pol}^{(ijk)}, \qquad \Psi(\zeta, \eta, \chi) = \sum_{i,j,k} \zeta^i \eta^j \chi^k \Phi_{pol}^{(ijk)}. \tag{72}$$

Hence $E_{pol}^{(ijk)}$ and $\Phi_{pol}^{(ijk)}$ denote the polarization energy and wave function of ith-order in V^{AB}, jth-order in V^{BC}, and kth-order in V^{CA}. The energy and wave function perturbation corrections are solutions of triple RS perturbation equations [Jeziorski and Kołos (1977), and Jeziorski (1974); see Moszynski *et al.* (1995b) for the explicit form of these equations.

As discussed above, the polarization expansion neglects the exchange effects. We saw in the earlier sections that the simple symmetrized SRS perturbation theory shows satisfactory convergence properties (this was also observed recently in a computational study of the convergence properties for the quartet state of H_3 [Korona, Moszynski, and Jeziorski (1996)]) and can be applied in practice to many-electron systems. This weak symmetry forcing may briefly be described by 'perturb first, antisymmetrize later'. That is, we antisymmetrize the solutions $\Phi_{\rm pol}^{(ijk)}$ of the RS equations by the action of the full antisymmetrizer $\mathcal{A}$. Since the antisymmetrized functions no longer satisfy the intermediate normalization conditions, normalization is necessary and the energy $E_{\rm SRS}^{(ijk)}$, $[(i,j,k) \neq (0,0,0)]$ becomes

$$E_{\rm SRS}^{(ijk)} = N_0^{\rm ABC}\left[\langle\Phi_0|V^{AB}|\mathcal{A}\Phi_{\rm pol}^{(i-1,jk)}\rangle + \langle\Phi_0|V^{\rm BC}|\mathcal{A}\Phi_{\rm pol}^{(i,j-1,k)}\rangle + \langle\Phi_0|V^{CA}|\mathcal{A}\Phi_{\rm pol}^{(ij,k-1)}\rangle\right]$$

$$- N_0^{\rm ABC}\sum_{l=0}^{i}\sum_{m=0}^{j}\sum_{n=0}^{k}{}' E_{\rm SRS}^{(lmn)}\langle\Phi_0|\mathcal{A}\Phi_{\rm pol}^{(i-l,j-m,k-n)}\rangle, \tag{73}$$

with the normalization constant $N_0^{\rm ABC} = \langle\Phi_0 \mid \mathcal{A}\Phi_0\rangle^{-1}$. The prime on the summation symbol reminds us that the term with $(lmn) = (ijk)$ is excluded from the summation. The exchange contribution to the interaction energy in each order can now be defined by

$$E_{\rm exch}^{(ijk)} \equiv E_{\rm SRS}^{(ijk)} - E_{\rm pol}^{(ijk)}. \tag{74}$$

3.6.2 PHYSICAL INTERPRETATION OF POLARIZATION EFFECTS

The electrostatic energy $E_{\rm pol}^{(1)}$ is additive, so the first non-additive polarization contribution is given by the second-order term,

$$E_{\rm pol}^{(2)} = E_{\rm pol}^{(110)} + E_{\rm pol}^{(101)} + E_{\rm pol}^{(011)}, \tag{75}$$

where, e.g., $E_{\rm pol}^{(110)}$ is explicitly given by an expression in terms of the linear response function of equation (33) and the electric fields $\omega_{\rm X}$ of equation (32)

$$E_{\rm ind}^{(110)} = (\omega_{\rm A})_m^n(\omega_{\rm C})_{m'}^{n'}\Pi_{nn'}^{mm'}(0). \tag{76}$$

As before, m and n label arbitrary orbitals on B. Similar expressions for $E_{\rm ind}^{(101)}$, etc., can be easily found by a proper permutation of A, B, and C.

Note that equation (75) represents the only second-order non-additive polarization term, there is no non-additive second-order dispersion. The remaining non-additive polarization terms are all of third order at least, i.e., in equation (72): $i + j + k \geq 3$. The three-body induction energy, $E_{\rm ind}^{(3)}$, is defined as that part of $E_{\rm pol}^{(3)}$ that can be obtained by complete neglect of the *intermonomer* correlation effects. The difference $E_{\rm pol}^{(3)} - E_{\rm ind}^{(3)}$ represents all intermonomer correlation effects, and separates into contributions due to pure third-order dispersion interactions and to the coupling of the second-order dispersion interaction with the induction interaction

$$E_{\rm pol}^{(3)} = E_{\rm disp}^{(3)} + E_{\rm ind-disp}^{(3)} + E_{\rm ind}^{(3)}. \tag{77}$$

The dispersion nonadditivity $E_{\rm disp}^{(3)}$ arises from the coupling of intermonomer pair correlations in subsystems $X-Y$ and $Y-Z$ via the intermolecular interaction operator V^{ZX}. This contribution can be expressed as a generalized Casimir–Polder formula

$$E_{\rm disp}^{(3)} = -\frac{1}{2\pi}v_{ln}^{km}v_{n'p}^{m'x}v_{p'l'}^{x'k'}\int_{-\infty}^{+\infty}\Pi_{kk'}^{ll'}(-i\omega)\Pi_{mm'}^{nn'}(i\omega)\Pi_{xx'}^{pp'}(i\omega)\,{\rm d}\omega. \tag{78}$$

The orbitals p and x are on C and, as before, k and l are on A and m, n on B. For the interaction of three spherically symmetric atoms the third-order dispersion nonadditivity contains the famous

Axilrod–Teller–Muto triple-dipole interaction [Axilrod and Teller (1943), and Muto (1943)]. We have implemented the zeroth-order (in the intramolecular correlation) approximation and the random phase approximation (RPA) for the propagators. See Moszynski *et al.* (1995b) for the equation that results when RPA propagators of the monomers are used.

The induction–dispersion contribution, in turn, can be interpreted as the energy of the (second-order) dispersion interaction of monomer X with monomer Y deformed by the electrostatic field of monomer Z. Note that we have six such contributions. In particular, when $X = \mathrm{A}$, $Y = \mathrm{B}$, and $Z = \mathrm{C}$ the corresponding induction–dispersion contribution in terms of response functions is given by

$$E^{(210)}_{\text{ind-disp}} = -\frac{1}{4\pi}(\omega_{\mathrm{C}})^m_n v^{k_1 m_1}_{l_1 n_1} v^{k_2 m_2}_{l_2 n_2} \int_{-\infty}^{+\infty} \Pi^{l l_1 l_2}_{k k_1 k_2}(i\omega)\Pi^{n_1 n_2 n}_{m_1 m_2 m}(-i\omega, 0)\,\mathrm{d}\omega, \tag{79}$$

where the exact quadratic polarization propagator $\Pi^{l l_1 l_2}_{k k_1 k_2}(\omega_1, \omega_2)$ of A is given by [Olsen and Jørgensen (1985)]

$$\begin{aligned}
\Pi^{l l_1 l_2}_{k k_1 k_2}(\omega_1, \omega_2) &= \langle \Phi^{\mathrm{A}}_0 | E^l_k \hat{R}_{\mathrm{A}}(-\omega_1 - \omega_2) \left(E^{l_1}_{k_1} - \rho^{l_1}_{k_1}\right) \hat{R}_{\mathrm{A}}(-\omega_2) E^{l_2}_{k_2} |\Phi^{\mathrm{A}}_0\rangle \\
&+ \langle \Phi^{\mathrm{A}}_0 | E^{l_2}_{k_2} \hat{R}_{\mathrm{A}}(\omega_2) \left(E^{l_1}_{k_1} - \rho^{l_1}_{k_1}\right) \hat{R}_{\mathrm{A}}(\omega_1 + \omega_2) E^l_k |\Phi^{\mathrm{A}}_0\rangle \\
&+ \langle \Phi^{\mathrm{A}}_0 | E^{l_1}_{k_1} \hat{R}_{\mathrm{A}}(\omega_1) \left(E^l_k - \rho^l_k\right) \hat{R}_{\mathrm{A}}(-\omega_2) E^{l_2}_{k_2} |\Phi^{\mathrm{A}}_0\rangle + (1 \leftrightarrow 2),
\end{aligned} \tag{80}$$

and $\rho^l_k = \langle \Phi^{\mathrm{A}}_0 \mid E^l_k \mid \Phi^{\mathrm{A}}_0 \rangle$, and $(1 \leftrightarrow 2)$ denotes three additional terms with all symbols with indices 1 and 2 interchanged (including those with k_1, l_1, and k_2, l_2). The unitary group generator E^l_k is given by equation (34). The corresponding equation can be found for the induction–dispersion energy obtained from the propagator in the random phase approximation (RPA) [Moszynski *et al.* (1997c)]. Similar expressions for $E^{(120)}_{\text{ind-disp}}$, etc., can be easily found by permutation of symbols pertaining to monomers A, B, and C.

The mechanism of the third-order three-body induction interactions is somewhat more complicated. It can be shown that one can distinguish three principal categories. The first mechanism is simply the interaction of permanent moments on monomer C with the moments induced on B by the nonlinear (second-order) effect of the electrostatic potential of monomer A plus contributions obtained by interchanging the roles of monomers A and C. In terms of the quadratic response function the contribution of this mechanism takes the form

$$E^{(210)}_{\text{ind}}(\mathrm{B} \leftarrow \mathrm{A,C}) = \tfrac{1}{2}(\omega_{\mathrm{A}})^m_n (\omega_{\mathrm{A}})^{m'}_{n'} (\omega_{\mathrm{C}})^{m''}_{n''} \Pi^{n n' n''}_{m m' m''}(0, 0). \tag{81}$$

Note that we have six contributions of this type corresponding to six possible permutations of the indices A, B, and C. Again, we can use either the zeroth-order or the RPA approximation for the propagator in this expression.

The second mechanism is the interaction between the multipole moments induced on A and C by the electrostatic potential of the monomer B. The induction energy component corresponding to this particular interaction will be denoted by $E^{(111)}_{\text{ind}}(\mathrm{A} \leftarrow \mathrm{B}; \mathrm{C} \leftarrow \mathrm{B})$, and can be written as

$$E^{(111)}_{\text{ind}}(\mathrm{A} \leftarrow \mathrm{B}; C \leftarrow B) = (\omega_{\mathrm{B}})^k_l (\omega_{\mathrm{B}})^x_p v^{x' k'}_{p' l'} \Pi^{l l'}_{k k'}(0) \Pi^{p p'}_{x x'}(0). \tag{82}$$

Since by definition $E^{(111)}_{\text{ind}}(\mathrm{A} \leftarrow \mathrm{B}; C \leftarrow B) = E^{(111)}_{\text{ind}}(C \leftarrow B; \mathrm{A} \leftarrow \mathrm{B})$, we have three contributions of this kind.

The third mechanism corresponds to the interaction of multipole moments induced in monomers A and B by the electrostatic potentials of monomers B and C, respectively:

$$E^{(210)}_{\text{ind}}(\mathrm{A} \leftarrow \mathrm{B}; \mathrm{B} \leftarrow \mathrm{C}) = (\omega_{\mathrm{B}})^k_l (\omega_{\mathrm{C}})^m_n v^{k' m'}_{l' n'} \Pi^{l l'}_{k k'}(0) \Pi^{n n'}_{m m'}(0). \tag{83}$$

Again we have six contributions of this type corresponding to six possible permutations of the indices A, B, and C.

3.6.3 EXCHANGE EFFECTS

In order to arrive at a closed expression for $E^{(ijk)}_{\text{exch}}$ [cf. equation (74)] the total antisymmetrizer of the trimer is approximated by

$$\mathcal{A} \approx \frac{(N_{\text{A}} + N_{\text{B}} + N_{\text{C}})!}{N_{\text{A}}!N_{\text{B}}!N_{\text{C}}!}(1 + \mathcal{P})\mathcal{A}_{\text{A}}\mathcal{A}_{\text{B}}\mathcal{A}_{\text{C}}, \tag{84}$$

where N_X is the number of electrons of monomer X, $\mathcal{A}_X$ is the antisymmetrizer for monomer X, and the operator $\mathcal{P}$ collects all intermolecular two and three cycle permutations,

$$\mathcal{P} = P^{\text{AB}} + P^{\text{BC}} + P^{\text{CA}} + P^{\text{ABC}}, \tag{85}$$

with

$$P^{XY} = -\sum_{i\in X}\sum_{j\in Y} P_{ij} \qquad \text{and} \qquad P^{XYZ} = \sum_{i\in X}\sum_{j\in Y}\sum_{k\in Z}(P_{ijk} + P_{jik}). \tag{86}$$

This truncation of the antisymmetrizer leads to a more approximate energy expression than equation (73). This more approximate expression is used to derive working equations for the nonadditive exchange terms.

The leading first-order exchange nonadditivity is given by the sum

$$E^{(1)}_{\text{exch}} = E^{(100)}_{\text{exch}} + E^{(010)}_{\text{exch}} + E^{(001)}_{\text{exch}}, \tag{87}$$

where the first term is

$$E^{(100)}_{\text{exch}} = \langle\Phi_0|\, V^{\text{AB}} \left(\mathcal{Q}^{\text{AB}} - \langle\mathcal{Q}^{\text{AB}}\rangle\right) |\Phi_0\rangle \tag{88}$$

and due to the truncation of the antisymmetrizer

$$\mathcal{Q}^{\text{AB}} = \mathcal{P} - P^{\text{AB}} \qquad \text{and} \qquad \langle\mathcal{Q}^{\text{AB}}\rangle = \langle\Phi_0|\, \mathcal{Q}^{\text{AB}}\, |\Phi_0\rangle\,. \tag{89}$$

The other terms of equation (87) follow by permutation of A, B and C. Similarly to the two-body case, see equations (23) and (24), the first-order exchange nonadditivity can be expressed through one- and two-particle density matrices ρ_X and Γ_X of the isolated monomers $X = \text{A, B}$ [Moszynski, Wormer, and van der Avoird (1999)].

In order to consider higher than first-order exchange effects, we write the first-order wave function $\Phi^{(100)}_{\text{pol}}$ appearing in equation (73) as

$$\Phi^{(100)}_{\text{pol}} = \Phi^{(1)}_{\text{ind}}(\text{A} \leftarrow \text{B})\Phi^{\text{B}}_0\Phi^{\text{C}}_0 + \Phi^{\text{A}}_0\Phi^{(1)}_{\text{ind}}(\text{B} \leftarrow \text{A})\Phi^{\text{C}}_0 + \Phi^{(1)}_{\text{disp}}(\text{A}\cdots\text{B})\Phi^{\text{C}}_0, \tag{90}$$

where $\Phi^{(1)}_{\text{ind}}(\text{X} \leftarrow \text{Y})$ is the standard induction wave function corresponding to the polarization of monomer X by monomer Y, and $\Phi^{(1)}_{\text{disp}}(X\cdots Y)$ is the dispersion wave function for the pair X–Y [Jeziorski, Moszynski, and Szalewicz (1994), and Jeziorski and van Hemert (1976)]. Insertion of equation (90) into equation (73) shows that the second-order exchange non-additivity splits into exchange–induction $E^{(2)}_{\text{exch-ind}}$ and exchange–dispersion $E^{(2)}_{\text{exch-disp}}$

$$\begin{aligned} E^{(200)}_{\text{exch-ind}} = {} & \langle\Phi_0|\left(V^{\text{AB}} - \langle V^{\text{AB}}\rangle\right)\left(\mathcal{Q}^{\text{AB}} - \langle\mathcal{Q}^{\text{AB}}\rangle\right)|\Phi^{(1)}_{\text{ind}}(\text{A} \leftarrow \text{B})\Phi^{\text{B}}_0\Phi^{\text{C}}_0\rangle \\ & + \langle\Phi_0|\left(V^{\text{AB}} - \langle V^{\text{AB}}\rangle\right)\left(\mathcal{Q}^{\text{AB}} - \langle\mathcal{Q}^{\text{AB}}\rangle\right)|\Phi^{\text{A}}_0\Phi^{(1)}_{\text{ind}}(\text{B} \leftarrow \text{A})\Phi^{\text{C}}_0\rangle, \end{aligned} \tag{91}$$

$$
\begin{aligned}
E^{(110)}_{\text{exch-ind}} &= \langle\Phi_0|\left(V^{\text{AB}} - \langle V^{\text{AB}}\rangle\right)(\mathcal{P} - \langle\mathcal{P}\rangle) \\
&\quad \times |\Phi_0^{\text{A}}\Phi^{(1)}_{\text{ind}}(\text{B} \leftarrow \text{C})\Phi_0^{\text{C}} + \Phi_0^{\text{A}}\Phi_0^{\text{B}}\Phi^{(1)}_{\text{ind}}(\text{C} \leftarrow \text{B})\rangle \\
&\quad + \langle\Phi_0|\left(V^{\text{BC}} - \langle V^{\text{BC}}\rangle\right)(\mathcal{P} - \langle\mathcal{P}\rangle) \\
&\quad \times |\Phi^{(1)}_{\text{ind}}(\text{A} \leftarrow \text{B})\Phi_0^{\text{B}}\Phi_0^{\text{C}} + \Phi_0^{\text{A}}\Phi^{(1)}_{\text{ind}}(\text{B} \leftarrow \text{A})\Phi_0^{\text{C}}\rangle, \qquad (92)
\end{aligned}
$$

$$
E^{(200)}_{\text{exch-disp}} = \langle\Phi_0|\left(V^{\text{AB}} - \langle V^{\text{AB}}\rangle\right)\left(\mathcal{Q}^{\text{AB}} - \langle\mathcal{Q}^{\text{AB}}\rangle\right)\Phi^{(1)}_{\text{disp}}(\text{A}\cdots\text{B})\Phi_0^{\text{C}}\rangle, \qquad (93)
$$

$$
\begin{aligned}
E^{(110)}_{\text{exch-disp}} &= \langle\Phi_0|\left(V^{\text{AB}} - \langle V^{\text{AB}}\rangle\right)(\mathcal{P} - \langle\mathcal{P}\rangle)\Phi_0^{\text{A}}\Phi^{(1)}_{\text{disp}}(\text{B}\cdots\text{C})\rangle \\
&\quad + \langle\Phi_0|\left(V^{\text{BC}} - \langle V^{\text{BC}}\rangle\right)(\mathcal{P} - \langle\mathcal{P}\rangle)\Phi^{(1)}_{\text{disp}}(\text{A}\cdots\text{B})\Phi_0^{\text{C}}\rangle. \qquad (94)
\end{aligned}
$$

Moszynski *et al.* (1995b) have shown how these expressions can be cast into an explicitly connected commutator form. They also showed that these commutators can be worked out in terms of one- and two-electron integrals.

3.7 Illustrative applications

3.7.1 PAIR POTENTIALS

All the formulas described in Section 3.4 have been implemented in a computer program package [Bukowski *et al.* (1996)] which can routinely be used to compute intermolecular pair potentials. Some of the earliest results, for Ar–H_2 [Williams *et al.* (1993)] and for He–HF [Moszynski (1994b, 1994c)], have already been summarized in previous review articles (Jeziorski, Moszynski, and Szalewicz (1994), Szalewicz and Jeziorski (1997), van der Avoird, Wormer, and Moszynski (1994, 1997), and Moszynski *et al.* 1997a)]. A more recent application of the He–HF potential in a calculation of differential scattering cross sections [Moszynski *et al.* (1996a)] and comparison with experiment shows that this potential is very accurate indeed, also in the repulsive region. Some other results, for Ar–HF [Lotrich *et al.* (1995)] and for the water dimer [Mas *et al.* (1997)], were summarized in the recent review by Jeziorski and Szalewicz (1998). The accuracy of the water pair potential is presently being tested [Groenenboom, van der Avoird, and Wormer (1999)] by a calculation of the various tunneling splittings caused by hydrogen bond rearrangement processes in the water dimer and comparison with high resolution spectroscopic data. Results which were recently obtained by the Nijmegen/Warsaw collaboration are the pair potentials of He–CO [Moszynski *et al.* (1995a), Heijmen *et al.* (1997b)], Ne–CO [Moszynski (1997b)], He–acetylene [Moszynski, Wormer, and van der Avoird (1995)], Ne–acetylene [Bemish *et al.* (1998)], and Ar–CH_4 [Heijmen *et al.* (1999b), and Miller *et al.* (1999)].

A typical feature of all these potentials for weakly interacting systems is that their shape is determined by a subtle balance between the geometry dependence of the repulsive short range interactions and that of the long range forces, which mostly are attractive. As might be expected, the potential surfaces of He–CO and Ne–CO are qualitatively similar, with a nearly T-shaped equilibrium structure for Ne–CO, but a more skew minimum for He–CO, see Figure 3.1. The barrier for moving away from this minimum to the linear CO–Rg geometries ($\theta = 0°$) is very low in both cases, whereas the barriers at the linear Rg–CO geometries ($\theta = 180°$) are much higher. Also the He–acetylene and Ne–acetylene potentials are qualitatively similar, with two minima related by symmetry for a skew structure and two symmetry related minima for a nearly–in the Ne case exactly–linear geometry, see Figure 3.2. The depths of these minima differ only marginally. In the He case the nearly linear structure is the global minimum, whereas in the Ne

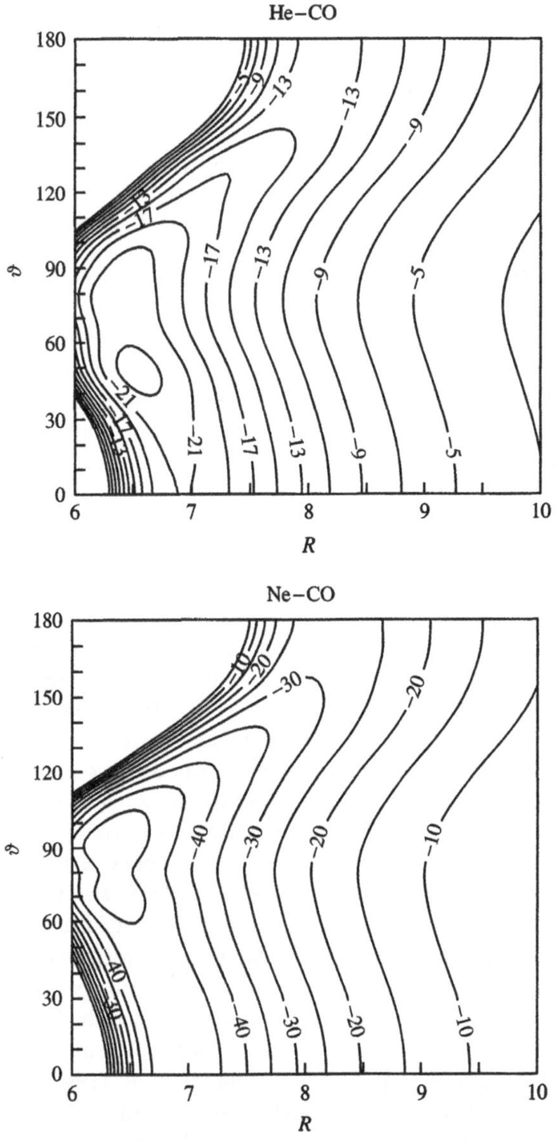

Figure 3.1. Contour plots (in cm^{-1}) of the He–CO and Ne–CO potential surfaces at the CO equilibrium bond length $r_e = 2.132$ bohr. R is the length of the vector $\boldsymbol{R}$ pointing from the CO center of mass to the rare gas atom (Rg) nucleus and θ is the angle between the CO axis and this vector $\boldsymbol{R}$; $\theta = 0°$ at the linear CO–Rg structure.

complex the skew minimum is slightly deeper than the linear one. In accordance with the cigar-like shape of the acetylene molecule, the equilibrium distances R of the (nearly) linear minima are considerably larger than those for the skew minima. Another indication of the subtle balance between the short range repulsion and the long range attraction is that the equilibrium distances in the He complexes are about equal to those of the Ne complexes (sometimes even slightly larger, but this depends on the equilibrium angle θ), in spite of the fact that the He atom is more

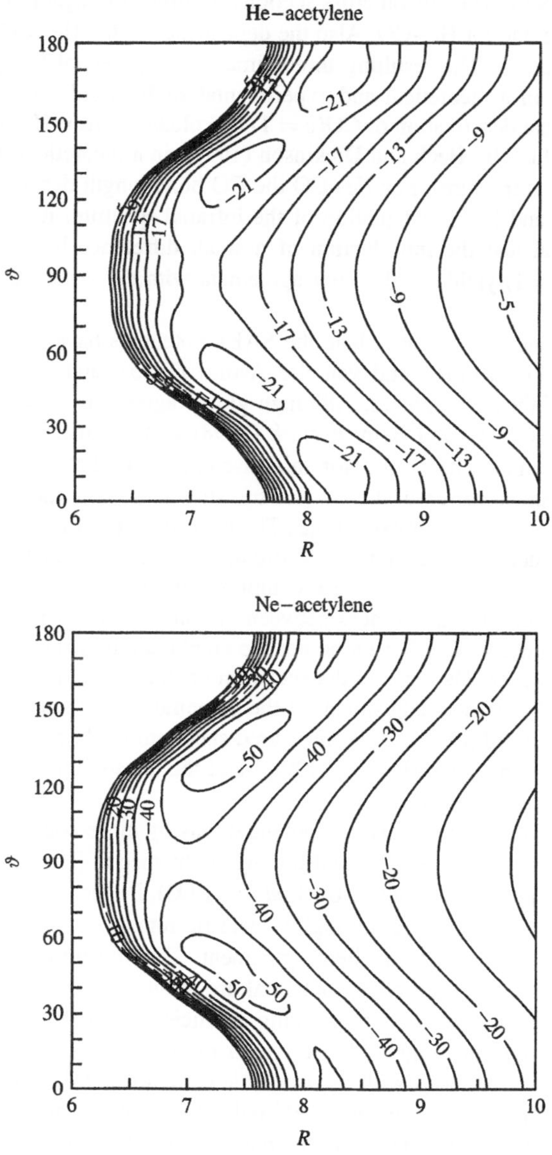

Figure 3.2. Contour plots (in cm^{-1}) of the He–acetylene and Ne–acetylene potential surfaces. R is the length of the vector $\boldsymbol{R}$ pointing from the acetylene center of mass to the rare gas atom (Rg) nucleus and θ is the angle between the molecular axis and this vector $\boldsymbol{R}$.

compact (much less polarizable and with fewer electrons) than the Ne atom. Of course, the wells are much shallower in the case of He.

The pair potentials of He–CO and Ne–CO were applied [Heijmen *et al.* (1997b), and Moszynski *et al.*, (1997b)] in a calculation of the rotationally resolved infrared spectra of these complexes [Chuaqui, Roy, and McKellar (1994), Chan and McKeller (1996)], and by Antonova *et al.* (1999a, 1999b) in a theoretical and experimental study of the state-to-state rotationally inelastic He–CO and Ne–CO collision cross sections. It turned out again that both potentials are accurate, especially the one for He–CO. Also the dependence of the He–CO potential on the CO bond length was computed. The resulting three-dimensional potential [Heijmen *et al.* (1997b)] was successfully used in a theoretical and experimental study [Reid, Simpson, and M. Quiney (1997)] of the vibrational deactivation of CO($v = 1$) by collisions with ^{3}He and ^{4}He. The He–CO potential was also applied, by Heck and Dickinson (1997), in a theoretical study of the transport properties of He/CO gas mixtures. For Ne–CO the CO bond length dependence of the potential was not yet included, and in the calculation of the infrared spectrum it was found that a slight scaling of the potential and the introduction of a small difference between the potentials for CO($v = 0$) and CO($v = 1$) yielded still better agreement with the observed spectrum [Chan and McKeller (1996)].

Also for He–acetylene and Ne–acetylene the SAPT potentials have been applied in *ab initio* calculations [Moszynski, Wormer, and van der Avoird (1995), and Bemish *et al.* (1998)] of the infrared spectra of these complexes. Although global agreement with the spectra measured by Bemish and Miller (1995), and Bemish *et al.* (1998) was obtained, these spectra could not be reproduced line by line. The reason for this discrepancy is that the experimental infrared spectra correspond to an excitation of the asymmetric stretch mode in the acetylene monomer, in combination with the intermolecular excitations. The measured spectra could not yet be assigned, but it could be concluded that the unusual distribution of the lines in these spectra is caused by the fact that the internal rotation in these complexes is just slightly hindered, which makes the spectra sensitive to the tiny differences between the intermolecular potentials for the ground and excited state. For the Ne–C_2HD isotopomer the combination of theory and experiment was entirely successful, however. The lines in the spectrum of Bemish *et al.* (1998) are much sharper in this case. By a small empirical adjustment of the potential for the stretch excited state (which was not explicitly calculated *ab initio*) the measured spectrum could be completely assigned on the basis of the calculations [Bemish *et al.* (1998)] and the unusual distribution of the observed lines was quantitatively reproduced.

An independent test of the He–acetylene potential was by close-coupling calculations of the total differential scattering cross sections and time-of-flight spectra [Heijmen *et al.* (1997c)] and comparison with the experimental data of Buck *et al.* (1993). Perfect agreement was obtained in this comparison. It turned out that rather highly inelastic rotational state-to-state cross sections had to be included to achieve this excellent agreement. This illustrates that both the isotropic and anisotropic parts of the He–acetylene potential are accurate also in the repulsive region. A subsequent calculation of the pressure broadening coefficients and rate constants for rotational (de)excitation of C_2H_2 by collisions with He was a marked success for theory. The pressure broadening coefficients agreed well with the available experimental data, but the calculated rate constants for rotationally inelastic collisions differed substantially from the data extracted from experimentally observed rotational population decays. It was then discovered that these decay rates are not merely determined by single-collision cross sections. After a proper inclusion of multiple collision effects the agreement between theory and experiment becomes very good [Heijmen *et al.* (1999a)].

An example that we describe in somewhat more detail is the pair potential of Ar–CH_4, recently calculated by SAPT [Heijmen *et al.* (1997a)]. A two-dimensional cut that displays most of the interesting features of this three-dimensional intermolecular potential is shown in Figure 3.3. For large R the preferred direction of approach of the Ar atom is along one of the C–H bonds ($\Theta = 55°$), as if the C–H$\cdots$Ar bond were a linear hydrogen bond. At shorter distance, however, also the steric repulsion is the largest for this orientation. The deepest attractive well occurs where the short range repulsion is the weakest, i.e., for the Ar atom in between three C–H bonds ($\Theta = 125°$). In Figure 3.4 one can observe the origin of this behavior. The long range attraction

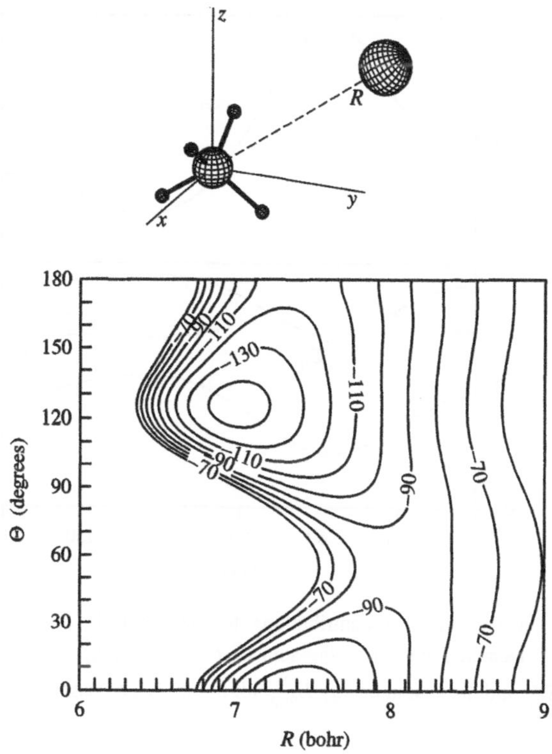

Figure 3.3. Contour plot (in cm^{-1}) of the Ar–CH_4 potential for $\Phi = 0°$, i.e., the Ar atom in the *xz*-plane. The potential depends on R, Θ, and Φ, which are the polar coordinates of the Ar atom in the frame defined at the top of this figure.

is mostly caused by dispersion forces. The attraction caused by induction is much smaller at large R, but increases steeply with decreasing R, when the charge clouds of Ar and CH_4 start to overlap. This latter effect is solely due to penetration, i.e., incomplete screening of the nuclear charges by the electron clouds. The small R behavior is dominated by the short range repulsion, with the contributions of first-order exchange, exchange–induction, and exchange–dispersion in decreasing order of importance. Both the long range attraction and the short range repulsion are largest for Ar along one of the C–H bonds ($\Theta = 55°$), and smallest when the Ar atom approaches one of the faces of the CH_4 tetrahedron ($\Theta = 125°$). Since the long range R^{-n} contributions decrease less rapidly with increasing R than the exponential short range terms this explains the observed behavior, which is typical for a van der Waals complex. Even for hydrogen-bonded complexes one finds such behavior, but in that case the (first-order) electrostatic and (second-order) induction forces are more dominant, and the equilibrium geometry of the complex is often determined by these long range forces.

This Ar–CH_4 potential was used in extensive close-coupling calculations for elastic and rotationally inelastic scattering at various energies. Both the total differential Ar–CH_4 scattering cross sections [Heijmen *et al.* (1998)] and the integral state-to-state cross sections for the rotational transitions of CH_4 induced by collisions with Ar [Heijmen *et al.* (1997a)] agree well with the experimental data of Buck and coworkers [Heijmen *et al.* (1998)] and with the data of Chapman *et al.* (1996).

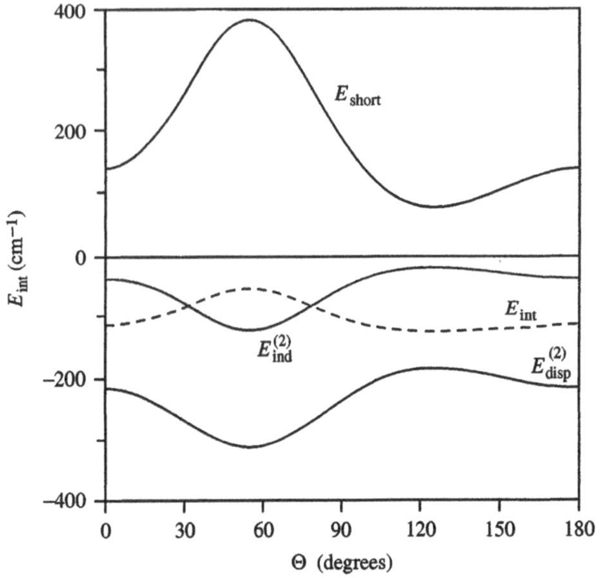

Figure 3.4. Dependence of the different contributions to the Ar–CH_4 potential on the angle Θ at $R = 7.5$ bohr and $\Phi = 0°$. Depicted are E_{short}, the sum of $E^{(1)}_{elec}$, $E^{(1)}_{exch}$, $E^{(2)}_{exch-ind}$, $E^{(2)}_{exch-disp}$, and the long range induction energy $E^{(2)}_{ind}$ and dispersion energy $E^{(2)}_{disp}$, all in cm^{-1}. The dashed line is the total interaction energy.

Another test to which the Ar–CH_4 potential was subjected is the calculation of the infrared spectrum of the Ar–CH_4 complex. This spectrum, in the region of the ν_3 mode of CH_4, was measured by Miller (1994) and presented at the 1994 Faraday Discussion on van der Waals molecules. It shows a lot of detailed structure, with many lines more or less grouped in seven bands, over a range of 40 cm^{-1} around the band origin of the ν_3 mode (3020 cm^{-1}) which could not yet be assigned or understood, however. Recently, the bound levels of Ar–CH_4 for total angular momentum $J = 0, \ldots, 7$ were calculated from the SAPT potential. Also the quasi-bound levels of the complex with the ν_3 mode excited were calculated with the use of the same ground state potential surface, but it was explicitly taken into account that the ν_3 mode is threefold degenerate and the Coriolis coupling between the angular momentum of the ν_3 vibration and the (hindered) internal rotation of the CH_4 subunit inside the complex was included. With the use of the ν_3 transition dipole moment function–it was assumed that the weak interaction with Ar does not affect this transition dipole moment–the infrared spectrum in the region of the ν_3 mode could then be generated completely *ab initio*. This spectrum is shown in Figure 3.5, next to Miller's experimental high resolution spectrum. It is obvious that the agreement is very good and that the structure of the measured spectrum can be fully understood from the *ab initio* calculations. More details are given in two papers [Heijmen *et al.* (1999b), and Miller *et al.* (1999)], where it is shown that even the individual lines in each of the seven bands agree rather well in most cases, and thus could be assigned.

This example, and the previous summaries of the results for other dimers, demonstrate that the pair potentials from *ab initio* SAPT calculations are indeed accurate. Another, more global, comparison with experiment which confirms this finding was made by computations of the (pressure) second virial coefficients of all of these dimers over a wide range of temperatures [Moszynski *et al.* (1998a)].

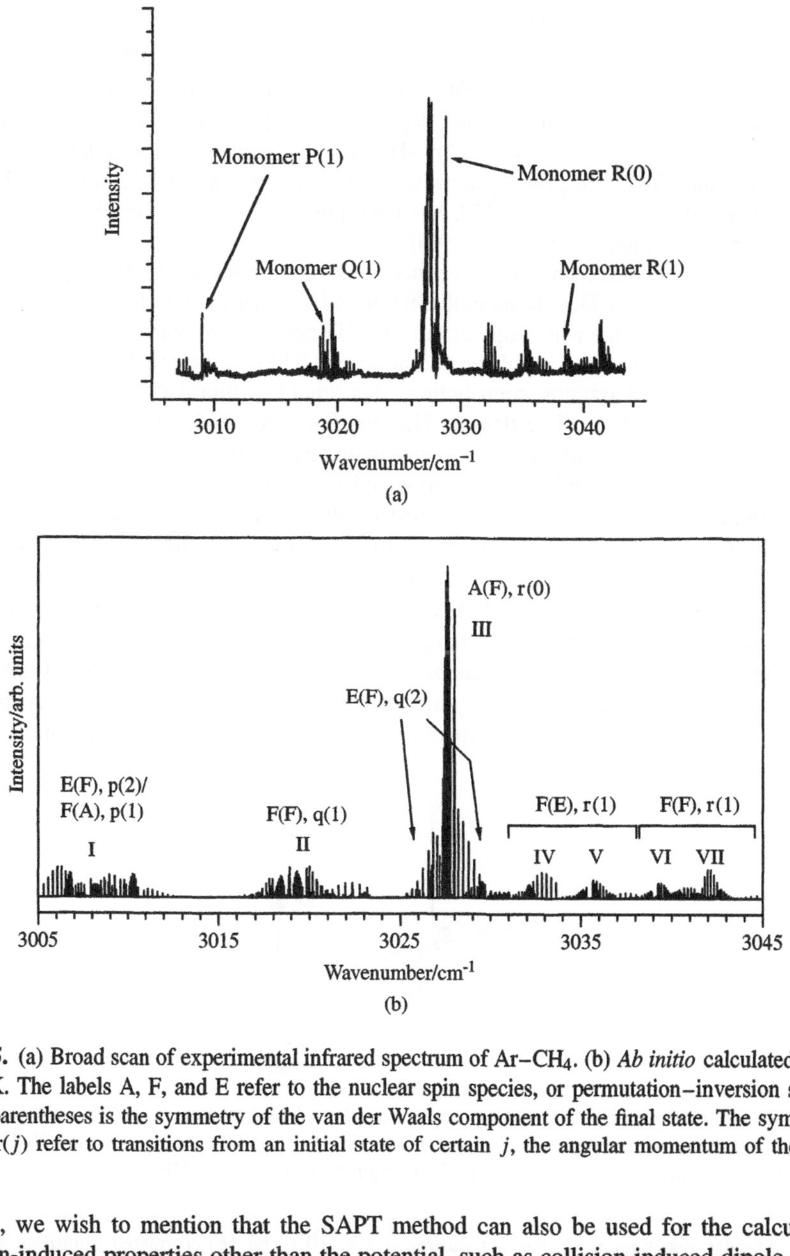

Figure 3.5. (a) Broad scan of experimental infrared spectrum of Ar–CH_4. (b) *Ab initio* calculated spectrum at $T = 1$ K. The labels A, F, and E refer to the nuclear spin species, or permutation–inversion symmetry. Between parentheses is the symmetry of the van der Waals component of the final state. The symbols p(j), q(j), and r(j) refer to transitions from an initial state of certain j, the angular momentum of the methane monomer.

Finally, we wish to mention that the SAPT method can also be used for the calculation of interaction-induced properties other than the potential, such as collision-induced dipole moments and polarizabilities. The way to obtain these properties is by computation of the SAPT interaction energy of a molecular complex A–B in the presence of an external static electric field. (Numerical) differentiation with respect to the components of this field yields the collision-induced dipole moment function–the first derivative–and polarizability–the second derivative. This approach was applied recently to He–He and He–H_2 (Heijmen *et al.*, 1996). The property functions were tested in a full quantum-statistical calculation of the dielectric second virial coefficient and in calculations of the polarized and depolarized Raman spectrum of He gas [Moszynski *et al.* (1996b), and Moszynski, Heijmen, and van der Avoird (1995)].

3.7.2 THREE-BODY INTERACTIONS

Here we present some results obtained with the recently developed SAPT code for three-body interactions [Wormer and Moszynski (1996)]. Since the relative importance of the different pair-wise non-additive interaction contributions depends strongly on the degree of polarity of the constituent molecules we show results for Ar_2–HF, as an example of a non-polar system, and for $(H_2O)_3$, $(H_2O)_4$, and $(H_2O)_5$, as polar systems with strong hydrogen bonds. Another, independent, application of SAPT, to Ar_3 and Ar_2–HF, has been published [Lotrich and Szalewicz (1997b), and Lotrich, Jankowski, and Szalewicz (1998)].

Figure 3.7 shows the various non-additive interaction contributions for Ar_2–HF in the geometries displayed in Figure 3.6. Details are in the literature [Moszynski *et al.* (1998b)]. An inspection of these curves shows that the anisotropy of the Ar_2–HF nonadditive potential results from a delicate balance of the attractive and repulsive components. Unlike in the two-body case there are no leading exchange and polarization contributions that would qualitatively determine the angular shape of the potential. At small angles the Hartree–Fock non-additivity dominates. However, its decomposition shows that all terms up to and including the third order are important. For all geometries considered the second-order induction and exchange–induction non-additivities cancel to a large extent, and $E_{\rm int}^{\rm HF}$ is mostly determined by the Heitler–London energy and the third-order contributions. At larger angles also the third-order induction energy is strongly quenched by its exchange counterpart, and $E_{\rm HL}^{(1)}$ is the dominant contribution. At the correlated level of theory, the Ar_2–HF non-additive potential is dominated by three large contributions: the induction–dispersion, dispersion, and MP2 exchange terms. The first-order exchange–correlation and the third-order exchange–induction–dispersion non-additivities are not yet included in the three-body SAPT code. For the time being they are approximately obtained from the supermolecule MP2 exchange terms. The exchange–dispersion non-additivity is relatively less important, but

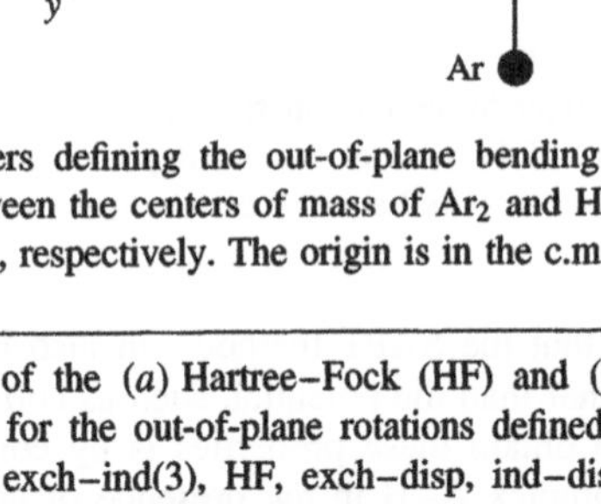

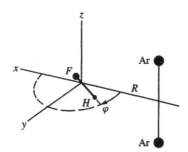

Figure 3.6. Geometrical parameters defining the out-of-plane bending motions of HF within Ar_2–HF. $R = 2.9798$ Å is the distance between the centers of mass of Ar_2 and HF. The Ar–Ar and H–F distances were fixed at 3.826 Å and 0.917 Å, respectively. The origin is in the c.m. of HF.

Figure 3.7. Angular dependence of the (*a*) Hartree–Fock (HF) and (*b*) correlated contributions to the nonadditive potential of Ar_2–HF for the out-of-plane rotations defined in Figure 3.6. The abbreviations HL, ind(2), exch–ind(2), ind(3), exch–ind(3), HF, exch–disp, ind–disp, disp, and MP2–exch refer to the Heitler–London energy $E_{\rm HL}^{(1)}$, which is approximately equal to the first-order exchange interaction, to the second-order induction and exchange–induction energies, $E_{\rm ind,resp}^{(2)}$ and $E_{\rm exch-ind,resp}^{(2)}$, the corresponding third-order contributions, $E_{\rm ind,resp}^{(3)}$ and $E_{\rm exch-ind}^{(3)}$, the Hartree–Fock three-body interaction energy $E_{\rm int}^{\rm HF}$, the second-order exchange-dispersion energy, $E_{\rm exch-disp}^{(2)}$, the third-order induction-dispersion and dispersion energies, $E_{\rm ind-disp,RPA}^{(3)}$, $E_{\rm disp,RPA}^{(3)}$, and the Møller-Plesset exchange energy, $E_{\rm exch}^{\rm MP2}$, respectively. Details are explained in Moszynski *et al.* (1998b).

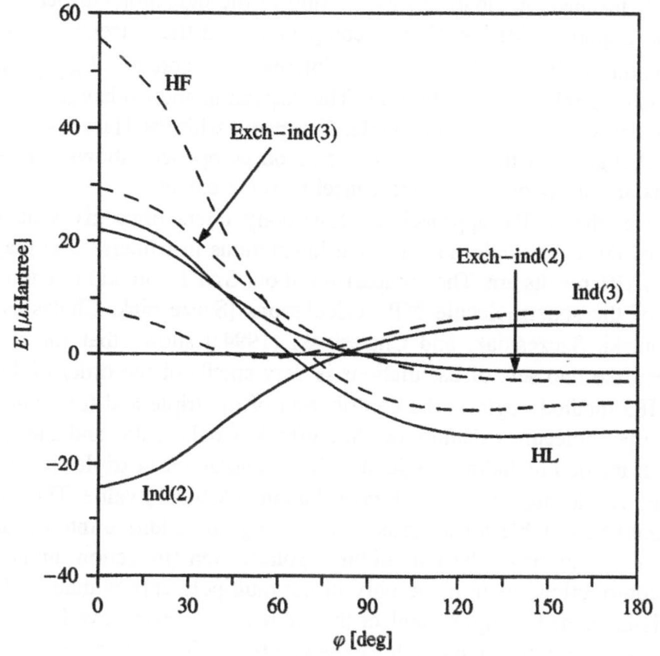
HF
Exch−ind(3)
Exch−ind(2)
Ind(3)
HL
Ind(2)
E [μHartree]
φ [deg]

(a)

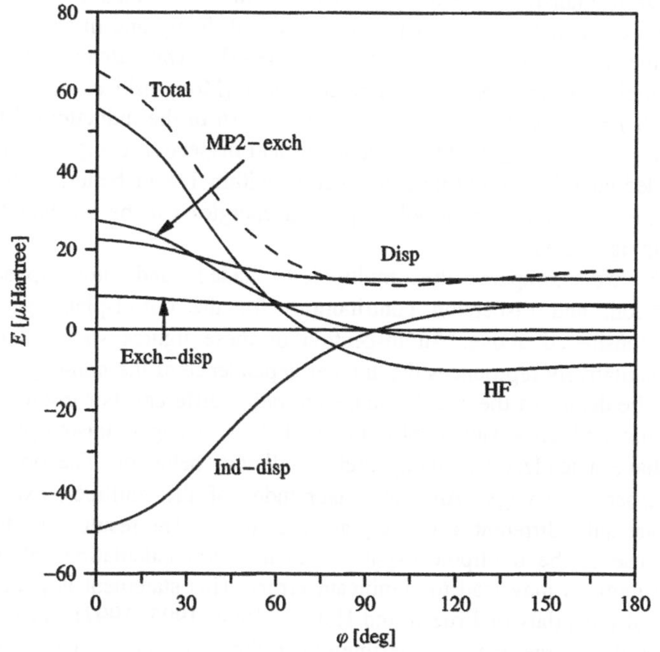
Total
MP2−exch
Disp
Exch−disp
HF
Ind−disp
E [μHartree]
φ [deg]

(b)

not negligible. For instance, at small angles the sum of the induction–dispersion and dispersion terms cancels the important MP2 exchange component, and the correlation contribution to the non-additive potential is given by $E^{(2)}_{\text{exch-disp}}$. It is interesting to note that $E^{(3)}_{\text{ind-disp,RPA}}$ and $E^{\text{MP2}}_{\text{exch}}$ are the most anisotropic correlation contributions. The dispersion and exchange–dispersion nonadditivities are rather flat as functions of the angle. Compared with the Hartree–Fock non-additivity and the MP2 exchange term the induction–dispersion component shows a reverse anisotropy. These three terms are of opposite sign and cancel to some extent.

As shown above, the SAPT approach to three-body interactions gives an insight into the physical origins of the anisotropic non-additive interactions in trimers. One may ask, however, how reliable the SAPT results are. The comparison of our SAPT non-additive interaction energies with the results of the supermolecule MP3 calculations [Szczęśniak, Chałasiński, and Piecuch (1993), and Cybulski, Szczęśniak, and Chałasiński (1994)] shows that for all geometries the difference between the two sets of calculations is very small–of the order of 1 μhartree or less. Of course, the MP3 method neglects the contribution of the triple and (disconnected) quadruple excitations, but supermolecule calculations [Szczęśniak, Chałasiński, and Piecuch (1993)] with the coupled-cluster method including single, double, and noniterative triple excitations [CCSD(T)] show that the MP3 results are only ≈5 % higher than the CCSD(T) value. Thus, the present level of the theory should be suitable for accurately describing non-additive interactions in trimers.

Until recently, calculations of the non-additive polarization (induction, induction–dispersion, and dispersion) contributions were made only in the multipole approximation. It is well known that the multipole-expanded energies neglect the short range charge-overlap (damping) effects. Very little is known about the magnitude of these effects. Theoretical studies of the charge-overlap effects have thus far been restricted to the model H_3 system [O'Shea and Meath (1974, 1976)]. Since our formalism in Section 3.6 enables numerical calculations of the multipole-expanded and non-expanded energies at the same level of theory and in the same basis sets we can investigate the importance of these terms for Ar_2–HF. The calculations in the multipole approximation employed the expressions derived earlier [Moszynski *et al.* (1995b)]. All terms with $\max(l_A, l_B, l_C) = 4$ and $2l_A + l_B + l'_B + l''_B + l_C \leq 16$ in the induction–dispersion energy, and with $2l_A + 2l_B + l_C + l'_C \leq 15$ in the induction and dispersion energies were included. The multipole-expanded energies should be converged to within 1 % or better. Thus, any difference between the non-expanded and multipole-expanded energies can be attributed to the charge-overlap and damping effects.

In Figure 3.8 we compare the multipole-expanded and non-expanded induction, induction–dispersion, and dispersion contributions for the out-of-plane geometries of the Ar_2–HF trimer considered above. An inspection of these figures shows that the multipole approximation qualitatively reproduces the angular dependence of the non-expanded polarization terms. However, the details of the angular shape are rather different. For instance, the multipole-expanded induction and dispersion contributions show a stronger anisotropy, so the charge-overlap effects have a tendency to damp their oscillatory behavior. The opposite is true for the induction–dispersion energy. Also the magnitudes of the multipole-expanded and non-expanded terms are quite different, especially at small angles. The results reported in Figure 3.8 suggest that the use of the multipole expansion in *ab initio* calculations of the non-additive polarization contributions may lead to significant errors. This statement is perhaps not true for the (semi)empirical potentials of Ernesti and Hutson (1994, 1995, 1997), since these potentials effectively include the charge-overlap contributions in the short range exponential terms.

All water clusters $(H_2O)_n$ with $n = 3$, 4, and 5 are cyclic hydrogen-bonded complexes with each monomer acting simultaneously as proton donor and proton acceptor. The trimer has a triangular equilibrium structure with–because of geometry constraints–rather strongly non-linear hydrogen bonds, the tetramer has a square planar system of hydrogen bonds–with much less

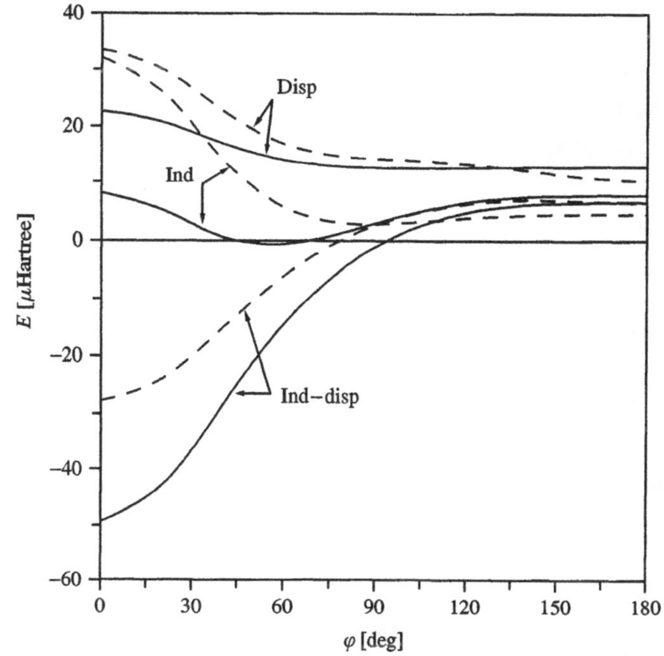

Figure 3.8. Comparison of the multipole-expanded (dashed lines) and nonexpanded (full lines) induction, induction–dispersion, and dispersion energies for the out-of-plane rotations of Ar_2–HF defined in Figure 3.6. The abbreviations ind, ind–disp, and disp refer to $E^{(3)}_{\text{ind,resp}}$, $E^{(3)}_{\text{ind-disp,RPA}}$, and $E^{(3)}_{\text{disp,RPA}}$, respectively.

strain–and the pentamer has a strain free, slightly puckered, pentagonal hydrogen-bonded framework, see Figure 3.9. In all cases the external, non-hydrogen-bonded, protons lie above and below the planes of the hydrogen-bonded 'skeletons' (denoted 'up' and 'down', or u and d). The up–up–down and up–up–down–up–down equilibrium structures of $(H_2O)_3$ and $(H_2O)_5$ have no spatial symmetry, but there are six equivalent structures in the case of the trimer and ten for the pentamer, which are interconnected by up–down flipping motions–quantum mechanical tunneling–so that the effective permutation–inversion symmetry groups are $G_6 \simeq C_{3h}(M)$ and $G_{10} \simeq C_{5h}(M)$, respectively. In addition, there are much slower tunneling processes which break the hydrogen bonds [Liu *et al.* (1994, 1997), and Brown, Keutsch, and Saykally (1998)]. For $(H_2O)_4$ the symmetry of the up–down–up–down equilibrium structure is the point group S_4, and if the slow tunneling to the equivalent down–up–down–up structure is taken into account the effective symmetry group is $C_{4h}(M)$ [Cruzan *et al.* (1997)].

In Table 3.4 we list the various pairwise additive and non-additive interaction energy contributions for the uud, udud, and uudud equilibrium geometries of the water trimer, tetramer, and pentamer [Milet *et al.* (1999b)]. We note that a substantial part of the binding energy originates from the three-body contributions: 17 % for the trimer, 26 % for the tetramer, and 29 % for the pentamer. The dominant three-body term is the second-order induction energy, mainly due to the dipole–induced-dipole interactions, but also the third-order induction energy is important. Hence, if one wishes to include induction effects by iteration [Millot and Stone (1992), and Gregory and Clary (1995)] of the induced dipole moments and the corresponding electric fields, one should proceed with this iteration beyond the first step. The contribution of the third-order induction–dispersion energy is small and, even though the dispersion energy is an important

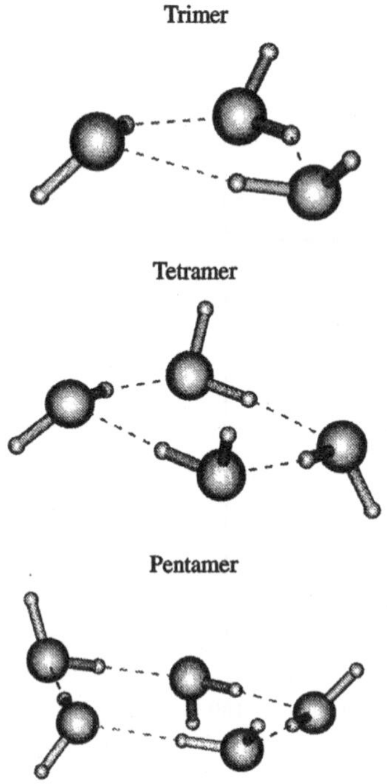

Figure 3.9. Equilibrium structures of the cyclic water clusters $(H_2O)_n$ with $n = 3$, 4, and 5.

Table 3.4. Decomposition of the interaction energy in water clusters.

		Trimer (uud)	Tetramer (udud)	Pentamer (uudud)
2-body	$E^{(1)}_{\text{pol}}$	−26.645	−48.963	−64.090
	$E^{(2)}_{\text{ind}}$	−12.252	−25.364	−34.146
	$E^{(2)}_{\text{disp}}$	−9.121	−16.088	−20.932
	E_{exch}	36.583	72.445	96.634
	$E^{\text{SAPT}}_{\text{int}}$	−11.435	−17.970	−22.534
	$E^{\text{CCSD(T)}}_{\text{int}}$	−11.624	−18.111	−22.201
3-body	$E^{(2)}_{\text{ind}}$	−1.351	−3.169	−4.551
	$E^{(3)}_{\text{ind}}$	−0.688	−1.165	−1.251
	$E^{(3)}_{\text{ind-disp}}$	−0.090	0.026	0.195
	$E^{(3)}_{\text{disp}}$	0.060	0.077	0.045
	E_{exch}	−0.345	−1.958	3.558
	$E^{\text{SAPT}}_{\text{int}}$	−2.414	−6.189	−9.120
	$E^{\text{CCSD(T)}}_{\text{int}}$	−2.371	−6.081	−8.978
4-body	$E^{\text{CCSD(T)}}_{\text{int}}$		−0.562	−1.220
5-body	$E^{\text{CCSD(T)}}_{\text{int}}$			−0.009
Total	$E^{\text{SAPT}}_{\text{int}}$	−13.849	−24.159	−31.654
	$E^{\text{CCSD(T)}}_{\text{int}}$	−13.995	−24.754	−32.481

component of the pair hydrogen bonding energy, the Axilrod–Teller three-body dispersion energy is even smaller. The three-body exchange effects are substantial, however, so one cannot restrict the treatment of non-additive effects in water to the classical induction terms only. Another interesting observation that follows from Table 3.4 is that the pair, three-body, and total SAPT interaction energies agree well with the results of supermolecule CCSD(T) calculations, if these are appropriately corrected for the BSSE. The additional information on the different long and short range components provided by SAPT will be useful if one tries to improve the modeling of the interactions in water and aqueous solutions.

Acknowledgement

We would like to thank Prof. B. Jeziorski for reading the manuscript and useful comments. R. M. thanks the Polish Scientific Research Council (KBN) for support through the University of Warsaw (grant BW-1418/10/98).

3.8 References

Adams, W. H., 1990, *Int. J. Quant. Chem. S* **24**, 531–547.
Adams, W. H., 1991, *Int. J. Quant. Chem. S* **25**, 165–181.
Adams, W. H., 1992, *J. Math. Chem.* **10**, 1–24.
Ahlrichs, R., 1973, *Chem. Phys. Lett.* **18**, 67–68.
Antonova, S., Lin, A., Tsakotellis, A. P., and McBane, G. C., 1999a, *J. Chem. Phys.* **110**, 2384–2390.
Antonova, S., Tsakotellis, A. P., Lin, A., and McBane, G. C., 1999b, *J. Chem. Phys.* **110**, 11742.
Axilrod, B. M., and Teller E., 1943, *J. Chem. Phys.* **11**, 299–300.
Bemish, R. J., and Miller, R. E., 1995, unpublished results.
Bemish, R. J., Oudejans, L., Miller, R. E., Moszynski, R., Heijmen, T. G. A., Korona, T., Wormer, P. E. S., and van der Avoird A., 1998, *J. Chem. Phys.* **109**, 8968–8979.
Boys, S. F., and Bernardi F., 1970, *Mol. Phys.* **19**, 553–566.
Brown, M. G., Keutsch, F. N., and Saykally, R. J., 1998, *J. Chem. Phys.* **109**, 9645–9647.
Buck, U., Ettischer, I., Schlemmer, S., Yang, M., Vohralik, P., and Watts R. O., 1993, *J. Chem. Phys.* **99**, 3494–3502.
Bukowski, R., Jankowski, P., Jeziorski, B., Jeziorska, M., Kucharski, S. A., Moszynski, R., Rybak, S., Szalewicz, K., Williams, H. L., and Wormer, P. E. S., 1996, *SAPT96: An Ab Initio Program for Many-Body Symmetry-Adapted Perturbation Theory Calculations of Intermolecular Interaction Energies*; University of Delaware and University of Warsaw.
Bukowski, R., Jeziorski, B., and Szalewicz K., 1996, *J. Chem. Phys.* **104**, 3306–3317.
Bulski, M., and Chałasiński, G., 1977, *Theor. Chim. Acta* **44**, 399–404.
Casimir, H. B. G., and Polder, D., 1948, *Phys. Rev.* **73**, 360–372.
Certain, P. N., and Hirschfelder, J. O., 1970, *J. Chem. Phys.* **52**, 5992–5999.
Chałasiński, G., and Jeziorski, B., 1973, *Int. J. Quant. Chem.* **7**, 63–73.
Chałasiński, G., and Jeziorski, B., 1976, *Mol. Phys.* **32**, 81–91.
Chałasiński, G., and Jeziorski, B., 1977, *Theor. Chim. Acta* **46**, 277–290.
Chałasiński, G., Jeziorski, B., and Szalewicz K., 1977, *Int. J. Quant. Chem.* **11**, 247–257.
Chałasiński, G., and Szczęśniak, M. M., 1988, *Mol. Phys.* **63**, 205–224.
Chałasiński, G., and Szczęśniak, M. M., 1994, *Chem. Rev.* **94**, 1723–1765.
Chan, M. C., and McKellar, A. R. W., 1996, *J. Chem. Phys.* **105**, 7910–7914.
Chapman, W. B., Schiffman, A., Hutson, J. M., and Nesbitt, D. J., 1996, *J. Chem. Phys.* **105**, 3497–3516.
Chuaqui, C. E., Roy, R. J. L., and McKellar, A. R. W., 1994, **101**, 39–61.
Claverie, P., 1971, *Int. J. Quant. Chem.* **5**, 273–296.
Claverie, P., 1978, in *Intermolecular Interactions: From Diatomics to Biopolymers*, Pullman, B., Ed.; Wiley: New York, pp. 69–305.
Claverie, P., 1986, in *Structure and Dynamics of Molecular Systems*, Daudel, R., Ed.; Reidel: Dordrecht.
Cruzan, J. D., Viant, M. R., Brown, M. G., and Saykally, R. J., 1997, *J. Phys. Chem. A* **101**, 9022–9031.
Ćwiok, T., Jeziorski, B., Kołos, W., Moszynski, R., Rychlewski, J., and Szalewicz, K., 1992a, *Chem. Phys. Lett.* **195**, 67–76.
Ćwiok, T., Jeziorski, B., Kołos, W., Moszynski, R., and Szalewicz, K., 1992b, *J. Chem. Phys.* **97**, 7555–7559.
Ćwiok, T., Jeziorski, B., Kołos, W., Moszynski, R., and Szalewicz, K., 1994, *J. Mol. Struct. (Theochem)* **307**, 135–151.

Cybulski, S. M., and Chałasiński, G., 1992, *Chem. Phys. Lett.* **197**, 591–598.
Cybulski, S. M., Chałasiński, G., and Moszynski, R., 1990, *J. Chem. Phys.* **92**, 4357–4363.
Cybulski, S. M., Szczęśniak, M. M., and Chałasiński, G., 1994, *J. Chem. Phys.* **101**, 10708–10716.
Dmitriev, Y., and Peinel, G., 1981, *Int. J. Quant. Chem.* **19**, 763–769.
Ernesti, A., and Hutson, J. M., 1994, *Faraday Discuss. Chem. Soc.* **97**, 119–129.
Ernesti, A., and Hutson, J. M., 1995, *Phys. Rev. A* **51**, 239–250.
Ernesti, A., and Hutson, J. M., 1997, *J. Chem. Phys.* **106**, 6288–6301.
Gregory, J. K., and Clary, D. C., 1995, *J. Chem. Phys.* **103**, 8924–8930.
Groenenboom, G. C., Mas, E. M., Bukowski, R., Szalewicz, K., Wormer, P. E. S., and van der Avoird, A., 2000, *Phys. Rev. Lett.*, **84**, 4072–4075.
Gutowski, M., and Chałasiński, G., 1993, *J. Chem. Phys.* **98**, 5540–5554.
Gutowski, M., van Duijneveldt, F. B., Chałasiński, G., and Piela, L., 1986a, *Chem. Phys. Lett.* **129**, 325–328.
Gutowski, M., van Duijneveldt, F. B., Chałasiński, G., and Piela, L., 1987, *Mol. Phys.* **61**, 233–247.
Gutowski, M., van Duijneveldt-van der Rijdt, J. G. C. M., van Lenthe, J. H., and van Duijneveldt, F. B., 1993, *J. Chem. Phys.* **98**, 4728–4737.
Gutowski, M., van Lenthe, J. H., Verbeek, J., van Duijneveldt, F. B., and Chałasiński, G., 1986b, *Chem. Phys. Lett.* **124**, 370–375.
Hamermesh, M., 1962, *Group Theory and Its Applications to Physical Problems*; Addison-Wesley: Reading, MA, Section 7–12.
Handy, N. C., Amos, R. D., Gaw, J. F., Rice, J. E., and Simandiras, E. S., 1985, *Chem. Phys. Lett.* **120**, 151–158.
Handy, N. C., Amos, R. D., Gaw, J. F., Rice, J. E., Simandiras, E. S., Lee, T. J., Harrison, R. J., Fitzgerald, G., Laidig, W. D., and Bartlett, R. J., 1986, in *Geometrical Derivatives of Energy Surfaces and Molecular Properties*, Jørgensen, P., and Simons, J., Eds; Reidel: Dordrecht, pp. 179–191.
Harrison, R. J., Fitzgerald, G., Laidig, W. D., and Bartlett, R. J., 1985, *Chem. Phys. Lett.* **124**, 291–294.
Heck, L., and Dickinson, A. S., 1997, *Mol. Phys.* **91**, 31–45.
Heijmen, T. G. A., Korona, T., Moszynski, R., Wormer, P. E. S., and van der Avoird, A., 1997a, *J. Chem. Phys.* **107**, 902–913.
Heijmen, T. G. A., Moszynski, R., Wormer, P. E. S., and van der Avoird, A., 1996, *Mol. Phys.* **89**, 81–110.
Heijmen, T. G. A., Moszynski, R., Wormer, P. E. S., and van der Avoird, A., 1997b, *J. Chem. Phys.* **107**, 9921–9928.
Heijmen, T. G. A., Moszynski, R., Wormer, P. E. S., van der Avoird, A., Buck, U., Ettischer, I., and Krohne, R., 1997c, *J. Chem. Phys.* **107**, 7260–7265.
Heijmen, T. G. A., Moszynski, R., Wormer, P. E. S., van der Avoird, A., Buck, U., Steinbach, C., and Hutson, J. M., 1998, *J. Chem. Phys.* **108**, 4849–4853.
Heijmen, T. G. A., Moszynski, R., Wormer, P. E. S., van der Avoird, A., Rudert, A. D., Halpern, J. B., Martin, J., Gao, W. B., and Zacharias, H., 1999a, *J. Chem. Phys.* **111**, 2519.
Heijmen, T. G. A., Wormer, P. E. S., van der Avoird, A., Miller, R. E., and Moszynski, R., 1999b, *J. Chem. Phys.* **110**, 5639.
Helgaker, T., Jørgensen, P., and Handy, N. C., 1989, *Theor. Chim. Acta* **76**, 227–245.
Hirschfelder, J. O., 1967, *Chem. Phys. Lett.* **1**, 325–329.
Hirschfelder, J. O., and Silbey, R., 1966, *J. Chem. Phys.* **45**, 2188–2192.
Jansen, L., and Boon, M., 1967, *Theory of Finite Groups. Applications to Physics*, North-Holland: Amsterdam, Section 6.3.
Jeziorska, M., Jeziorski, B., and Čížek, J., 1987, *Int. J. Quant. Chem.* **32**, 149–164.
Jeziorski, B., 1974, Ph.D. Thesis, University of Warsaw.
Jeziorski, B. 1978, unpublished result quoted in Kutzelnigg (1980).
Jeziorski, B., Bulski, M., and Piela, L., 1976, *Int. J. Quant. Chem.* **10**, 281–297.
Jeziorski, B., Chałasiński, G., and Szalewicz, K., 1978, *Int. J. Quant. Chem.* **14**, 271–287.
Jeziorski, B., and Kołos, W., 1977, *Int. J. Quant. Chem. S* **12**, 91–117.
Jeziorski, B., and Kołos, W., 1982, in *Molecular Interactions*, Ratajczak, H., and Orville-Thomas, W. J., Eds; Vol. 3; Wiley: New York, pp. 1–46.
Jeziorski, B., and Moszynski, R., 1993, *Int. J. Quant. Chem.* **48**, 161–183.
Jeziorski, B., Moszynski, R., Ratkiewicz, A., Rybak, S., Szalewicz, K., and Williams, H. L., 1993, in *SAPT: A Program for Many-Body Symmetry-Adapted Perturbation Theory Calculations of Intermolecular Interaction Energies*, Vol. B of *Methods and Techniques in Computational Chemistry: METECC-94*, Clementi, E., Ed.; STEF: Cagliari, pp. 79–129.
Jeziorski, B., Moszynski, R., Rybak, S., and Szalewicz, K., 1989, in *Many-Body Methods in Quantum Chemistry*, Vol. 52 of *Lecture Notes in Chemistry*, Kaldor, U., Ed.; Springer: New York, pp. 65–94.
Jeziorski, B., Moszynski, R., and Szalewicz, K., 1994, *Chem. Rev.* **94**, 1887–1930.
Jeziorski, B., and Szalewicz, K., 1998, in *Encyclopedia of Computational Chemistry*, Vol. 2, von Ragué Schleyer, P., Allinger, N. L., Clark, T., Gasteiger, J., Kollman, P. A., III, H. F. S., and Schreiner, P. R., Eds; Wiley: New York, pp. 1376–1398.
Jeziorski, B., and van Hemert, M., 1976, *Mol. Phys.* **31**, 713–729.

Jørgensen, P., and Simons, J., 1981, *Second Quantization-Based Methods in Quantum Chemistry*, Academic Press: New York.
Koch, H., Jørgen, H., Jensen, A., Jørgensen, P., Helgaker, T., Scuseria, G. E., and Schaefer III, H. F., 1990, *J. Chem. Phys.* **92**, 4924–4940.
Korona, T., Jeziorski, B., Moszynski, R., and Diercksen, G. H. F., 1999, *Theor. Chem. Acc.* **101**, 282.
Korona, T., Moszynski, R., and Jeziorski, B., 1996, *J. Chem. Phys.* **105**, 8178–8186.
Korona, T., Moszynski, R., and Jeziorski, B., 1997, *Adv. Quant. Chem.* **28**, 171–188.
Kutzelnigg, W., 1980, *J. Chem. Phys.* **73**, 343–359.
Liu, K., Brown, M. G., Cruzan, J. D., and Saykally, R. J., 1997, *J. Phys. Chem. A* **101**, 9011–9021.
Liu, K., Loeser, J. G., Elrod, M. J., Host, B. C., Rzepiela, J. A., and Saykally, R. J., 1994, *J. Am. Chem. Soc.* **116**, 3507–3512.
Lotrich, V. F., Jankowski, P., and Szalewicz, K., 1998, *J. Chem. Phys.* **108**, 4725–4738.
Lotrich, V. F., and Szalewicz, K., 1997a, *J. Chem. Phys.* **106**, 9668–9687.
Lotrich, V. F., and Szalewicz, K., 1997b, *J. Chem. Phys.* **106**, 9688–9702.
Lotrich, V. F., Williams, H. L., Szalewicz, K., Jeziorski, B., Moszynski, R., Wormer, P. E. S., and van der Avoird, A., 1995, *J. Chem. Phys.* **103**, 6076–6085.
Magnasco, V., and McWeeny, R., 1991, in *Theoretical Models of Chemical Bonding*, Vol. 4, Maksic, Z. B., Ed.; Springer: New York, pp. 133–169.
Mas, E. M., Szalewicz, K., Bukowski, R., and Jeziorski, B., 1997, *J. Chem. Phys.* **107**, 4207–4217.
McWeeny, R., 1984, *Croat. Chem. Acta* **57**, 865–878.
Milet, A., Korona, T., Moszynski, R., and Kochanski, E., 1999a, *J. Chem. Phys.* **111**, 7727.
Milet, A., Moszynski, R., Wormer, P. E. S., and van der Avoird, A., 1999b, *J. Phys. Chem. A* **103**, 6811–6819.
Miller, R. E., 1994, *Faraday Discuss. Chem. Soc.* **97**, 177–178.
Miller, R. E., Heijmen, T. G. A., Wormer, P. E. S., van der Avoird, A., and Moszynski, R., 1999, *J. Chem. Phys.* **110**, 5651–5657.
Millot, C., and Stone, A. J., 1992, *Mol. Phys.* **77**, 439–462.
Moszynski, R., 1992, unpublished results.
Moszynski, R., Cybulski, S. M., and Chałasiński, G., 1994, *J. Chem. Phys.* **100**, 4998–5010.
Moszynski, R., de Weerd, F., Groenenboom, G. C., and van der Avoird, A., 1996a, *Chem. Phys. Lett.* **263**, 107–112.
Moszynski, R., Heijmen, T. G. A., and Jeziorski, B., 1996, *Mol. Phys.* **88**, 741–758.
Moszynski, R., Heijmen, T. G. A., and van der Avoird, A., 1995, *Chem. Phys. Lett.* **247**, 440–446.
Moszynski, R., Heijmen, T. G. A., Wormer, P. E. S., and van der Avoird, A., 1996b, *J. Chem. Phys.* **104**, 6997–7007.
Moszynski, R., Heijmen, T. G. A., Wormer, P. E. S., and van der Avoird, A., 1997a, *Adv. Quant. Chem.* **28**, 119–140.
Moszynski, R., Jeziorski, B., Ratkiewicz, A., and Rybak, S., 1993, *J. Chem. Phys.* **99**, 8856–8869.
Moszynski, R., Jeziorski, B., Rybak, S., Szalewicz, K., and Williams, H. L., 1994a, *J. Chem. Phys.* **100**, 5080–5092.
Moszynski, R., Jeziorski, B., and Szalewicz, K., 1993, *Int. J. Quant. Chem.* **45**, 409–431.
Moszynski, R., Jeziorski, B., and Szalewicz, K., 1994, *J. Chem. Phys.* **100**, 1312–1325.
Moszynski, R., Jeziorski, B., van der Avoird, A., and Wormer, P. E. S., 1994b, *J. Chem. Phys.* **101**, 2825–2835.
Moszynski, R., Korona, T., Heijmen, T. G. A., Wormer, P. E. S., van der Avoird, A., and Schramm, B., 1998a, *Pol. J. Chem.* **72**, 1479–1496.
Moszynski, R., Korona, T., Wormer, P. E. S., and van der Avoird, A., 1995a, *J. Chem. Phys.* **103**, 321–332.
Moszynski, R., Korona, T., Wormer, P. E. S., and van der Avoird, A., 1997b, *J. Phys. Chem. A* **101**, 4690–4698.
Moszynski, R., Rybak, S., Cybulski, S. M., and Chałasiński, G., 1990, *Chem. Phys. Lett.* **166**, 609–614.
Moszynski, R., Wormer, P. E. S., Heijmen, T. G. A., and van der Avoird, A., 1998b, *J. Chem. Phys.* **108**, 579–589.
Moszynski, R., Wormer, P. E. S., Jeziorski, B., and van der Avoird, A., 1994c, *J. Chem. Phys.* **101**, 2811–2824.
Moszynski, R., Wormer, P. E. S., Jeziorski, B., and van der Avoird, A., 1995b, *J. Chem. Phys.* **103**, 8058–8074.
Moszynski, R., Wormer, P. E. S., Jeziorski, B., and van der Avoird, A., 1997c, *J. Chem. Phys.* **107**, 672–673.
Moszynski, R., Wormer, P. E. S., and van der Avoird, A., 1995, *J. Chem. Phys.* **102**, 8385–8397.
Moszynski, R., Wormer, P. E. S., and van der Avoird, A., 1999, to be published.
Muto, Y., 1943, *Proc. Phys. Soc. Jpn.* **17**, 629.
Oddershede, J., 1983, in *Methods in Computational Molecular Physics*, Diercksen, G. H. F., and Wilson, S., Eds; Reidel: Dordrecht, pp. 249–271.
Olsen, J., and Jørgensen, P., 1985, *J. Chem. Phys.* **82**, 3235–3264.
O'Shea, S. F., and Meath, W. J., 1974, *Mol. Phys.* **28**, 1431–1439.
O'Shea, S. F., and Meath, W. J., 1976, *Mol. Phys.* **31**, 515–528.
Reid, J. P., Simpson, C. J. S. M., and Quiney, H. M., 1997, *J. Chem. Phys.* **107**, 9929–9934.
Rybak, S., Jeziorski, B., and Szalewicz, K., 1991, *J. Chem. Phys.* **95**, 6576–6601.
Sadlej, A. J., 1981, *J. Chem. Phys.* **75**, 320–331.
Salter, E. A., and Bartlett, R. J., 1989, *J. Chem. Phys.* **90**, 1767–1773.
Salter, E. A., Trucks, G. W., and Bartlett, R. J., 1989, *J. Chem. Phys.* **90**, 1752–1766.

Salter, E. A., Trucks, G. W., Fitzgerald, G., and Bartlett, R. J., 1987, *Chem. Phys. Lett.* **141**, 61–70.
Szalewicz, K., and Jeziorski, B., 1979, *Mol. Phys.* **38**, 191–208.
Szalewicz, K., and Jeziorski, B., 1997, in *Molecular Interactions: From van der Waals to Strongly Bound Complexes*, Scheiner, S., Ed.; Wiley: New York, pp. 3–43.
Szczęśniak, M. M., Chałasiński, G., and Piecuch, P., 1993, *J. Chem. Phys.* **99**, 6732–6741.
Trucks, G. W., Salter, E. A., Noga, J., and Bartlett, R. J., 1988a, *Chem. Phys. Lett.* **150**, 37–44.
Trucks, G. W., Salter, E. A., Sosa, C., and Bartlett, R. J., 1988b, *Chem. Phys. Lett.* **147**, 359–366.
van der Avoird, A., Wormer, P. E. S., and Moszynski, R., 1994, *Chem. Rev.* **94**, 1931–1974.
van der Avoird, A., Wormer, P. E. S., and Moszynski, R., 1997, in *Molecular Interactions: From van der Waals to Strongly Bound Complexes*, Scheiner, S., Ed.; Wiley: New York, pp. 105–153.
van Duijneveldt, F. B., van Duijneveldt-van der Rijdt, J. G. C. M., and van Lenthe, J. H., 1994, *Chem. Rev.* **94**, 1873–1885.
van Duijneveldt-van de Rijdt, J. G. C. M., and van Duijneveldt, F. B., 1972, *Chem. Phys. Lett.* **17**, 425–427.
Williams, D. R., Schaad, L. J., and Murrell, J. N., 1967, *J. Chem. Phys.* **47**, 4916–4922.
Williams, H. L., Szalewicz, K., Jeziorski, B., Moszynski, R., and Rybak, S., 1993, *J. Chem. Phys.* **98**, 1279–1291.
Williams, H. L., Szalewicz, K., Moszynski, R., and Jeziorski, B., 1995, *J. Chem. Phys.* **103**, 4586–4591.
Wormer, P. E. S., and Hettema, H., 1992, *J. Chem. Phys.* **97**, 5592–5606.
Wormer, P. E. S., and Moszynski, R., 1996, *SAPT3 Package*, Nijmegen.

4 THE STRUCTURE AND ELECTRONIC STATES OF TRANSITION METAL MOLECULES AND CLUSTER COMPOUNDS BY MEANS OF DFT CALCULATIONS AND ZEKE SPECTROSCOPY

Attila Bérces and Marek Z. Zgierski

National Research Council of Canada, Ottawa, Canada

and

Dong-Sheng Yang

University of Kentucky, Lexington KY, USA

Computational Molecular Spectroscopy. Edited by Per Jensen and Philip R. Bunker

4.1 Introduction

Spectroscopy studies the interaction of electromagnetic radiation with atoms, molecules and solids. The data derived from spectroscopy reflect the change in the electronic and molecular structure induced by such interactions. The quantum nature of these interactions requires that the corresponding theoretical models are based on the principles of quantum mechanics. Numerical calculations based on electronic structure theory enable the interpretation of complex spectroscopic data and give significant insight into the fundamental principles of spectroscopy. At the same time, theoretical models need to be validated by comparing the theoretically predicted and experimentally measured observables. In all areas of modern spectroscopy there is an emergence of close collaboration between experiment and theory. This chapter describes such a collaboration to interpret observed zero electron kinetic energy (ZEKE) photoelectron spectra with the help of density functional theory (DFT) calculations. In the next two sections the theoretical background of DFT and that of ZEKE spectroscopy is developed. The theoretical background is followed by the introduction of the experimental technique of ZEKE spectroscopy, examples of its application and interpretation of the observed spectra on a theoretical basis.

4.2 Density functional theory

In this section we develop the mathematical and physical background of DFT. We do not aim to give a complete and mathematically coherent derivation or development of the theory. We leave the proofs of the statements made here to the cited references and give an interpretation of the theory that provides a basis for the critical evaluation of the results.

Some principles of quantum mechanics [Levine (1983); Szabo and Ostlund (1982)] need to be addressed before we discuss DFT. The average energy of an electronic system is described by the time independent Schrödinger equation (in atomic units):

$$\hat{H}\Psi = E\Psi. \tag{1}$$

In equation (1), E is the electronic energy, Ψ is the wave function, and $\hat{H}$ is the Hamiltonian operator for the system of N electrons:

$$\hat{H} = \sum_{i=1}^{N}\left(-\tfrac{1}{2}\nabla_i^2\right) + \sum_{i=1}^{N} v(r_i) + \sum_{i<j}^{N} \frac{1}{r_{ij}}, \tag{2}$$

where the first term is the kinetic energy operator, the last term is the electron–electron electrostatic interaction operator, and $v(r_i)$ is the external potential which in atoms, molecules and solids is the potential due to the nuclei of charges Z_α:

$$v(r_i) = -\sum_{\alpha} \frac{Z_\alpha}{r_{i\alpha}}. \tag{3}$$

The wave function depends on all $3N$ spatial (r) and N spin (s) coordinates of N electrons. The square of the wave function is a probability distribution function but the wave function itself does not have a clear physical meaning. The expectation value of any observable can be calculated from the wave function:

$$\langle \hat{A} \rangle = \frac{\int \Psi^* \hat{A} \Psi \, dx}{\int \Psi^* \Psi \, dx} = \frac{\langle \Psi^* | \hat{A} | \Psi \rangle}{\langle \Psi^* | \Psi \rangle}, \tag{4}$$

where $\hat{A}$ is the operator that corresponds to an observable. The integration $\int dx$ runs over all $3N$ spatial coordinates and involves summation over N spin coordinates.

The exact wave function of multielectron systems cannot be expressed in a simple form and it may contain an infinite number of terms. The Schrödinger equation of many-electron systems can only be solved with some approximations, usually assuming a particular form of the wave function. The simplest wave function that satisfies the Pauli principle, that is antisymmetric with respect to exchange of two electrons, is the Slater determinant composed of orthonormal spin orbitals ψ_i:

$$\Psi = \frac{1}{\sqrt{N!}} \begin{vmatrix} \psi_1(x_1)\psi_2(x_1)\ldots\psi_N(x_1) \\ \psi_1(x_2)\psi_2(x_2)\ldots\psi_N(x_2) \\ \\ \psi_1(x_N)\psi_2(x_N)\ldots\psi_N(x_N) \end{vmatrix} = \frac{1}{\sqrt{N!}} \det|\psi_1\psi_2\ldots\psi_N|. \tag{5}$$

The simplest first principle or *ab initio* method, the Hartree–Fock (HF) method, [Roothaan (1951)] can be derived by substituting the single determinant wave function equation (5) into the Schrödinger equation and subsequently applying the variational principle for the ground state. This procedure yields a set of coupled one-electron equations that can be expressed in a canonical form:

$$\hat{F}\psi_k(r) = \varepsilon_k\psi_k(r), \tag{6}$$

where the operator consists of three terms:

$$\hat{F} = -\tfrac{1}{2}\nabla^2 + v + \hat{g}. \tag{7}$$

The first term is the kinetic energy operator, the second term is the external potential defined in equation (3) and the last term is the electron–electron interaction term which consists of a Coulomb operator and an exchange operator:

$$\hat{g} = \hat{j} - \hat{k}. \tag{8}$$

Here the Coulomb operator is defined as

$$\hat{j}(x_1)f(x_1) = \sum_{k=1}^{N} \int \psi_k^*(x_2)\psi_k(x_2)\frac{1}{r_{12}}f(x_1)\,dx_2, \tag{9}$$

and the exchange operator is

$$\hat{k}(x_1)f(x_1) = \sum_{k=1}^{N} \int \psi_k^*(x_2)f(x_2)\frac{1}{r_{12}}\psi_k(x_1)\,dx_2. \tag{10}$$

Wave function based quantum mechanical methods differ in the form of the wave function. The HF method is one of the so-called independent electron models in which the wave function is composed of one-electron orbitals obtained from a set of coupled one-electron equations (6). Unfortunately, the independent electron model cannot account for the electron correlation energy. Correlation can be incorporated only in methods with wave functions more complicated than the single determinant. Systematic improvements of the HF theory involves the application of increasingly more complicated wave functions but the cost of these calculations scales so rapidly with the size of the system that very accurate methods are restricted to systems of modest size.

An alternative approach to solving the electronic structure problem uses the electron density as the basic variable. The electron density depends only on three spatial coordinates as opposed to the wave function which depends on the spatial and spin coordinates of N electrons. Thus, density-based methods provide the promise of simpler equations. The electron density is determined uniquely by the wave function:

$$\rho(r_1) = N \int \cdots \int |\Psi(x_1, x_2 \ldots, x_N)|^2 \, \mathrm{d}s_1 \, \mathrm{d}x_2 \ldots \mathrm{d}x_N, \tag{11}$$

where s_1 is the spin coordinate of electron 1 and x_2 is the spatial coordinates electron 2. However, there can be different wave functions that yield the same electron density. The consequent uncertainty is eliminated by the first Hohenberg and Kohn theorem that provides the theoretical justification for taking the electron density as a basic variable. This theorem states that the external potential $v(r)$ of an electronic system is determined, within a trivial additive constant, by the system's electron density [Hohenberg and Kohn (1964)]. Since ρ determines both $v(r)$ and the number of electrons, it thus determines also all properties of the ground state including its energy. Consequently, the energy can be expressed as a functional of the electron density:

$$E[\rho] = T[\rho] + V_{\mathrm{ne}}[\rho] + V_{\mathrm{ee}}[\rho], \tag{12}$$

where the first term is the kinetic energy, the second term is the nuclear–electron interaction and the third is the electron–electron interaction. Finding the explicit form of such functional is a long-standing challenge of DFT and has not been solved exactly.

The second Hohenberg–Kohn theorem affords the variational principle for the energy. For any nonnegative trial density $\tilde{\rho}(r)$ that satisfies $\int \tilde{\rho}(r) \, \mathrm{d}r = N$:

$$E_0 \leq E[\tilde{\rho}], \tag{13}$$

where E_0 is the ground state energy. Assuming that $E[\tilde{\rho}]$ is differentiable, the variational principle requires that:

$$\delta \left\{ E[\rho] - \mu \left[\int \rho(r) \, \mathrm{d}r - N \right] \right\} = 0. \tag{14}$$

Equation (14) yields the Euler–Lagrange equation:

$$\mu = \frac{\delta E[\rho]}{\delta \rho(r)}. \tag{15}$$

One of the major stumbling blocks of the direct application of equation (15) to calculate molecular energies is the severe approximations that are necessary for the calculation of the kinetic energy functional. The idea of Kohn and Sham (KS) was to introduce the concept of orbitals to calculate the kinetic energy more accurately; trading simplicity for accuracy. The orbitals of the KS method are required to reproduce the exact molecular density of the system of interest:

$$\rho(r) = \sum_{i=1}^{N} \sum_{s} |\psi_i(r, s)|^2. \tag{16}$$

To introduce this idea, Kohn and Sham referred to a non-interacting reference system for which the exact wave function can be expressed in the form of a single determinant. A system is called non-interacting if its Hamiltonian can be written in a from of the sum of one-electron operators without any explicit electron–electron interaction terms:

$$\hat{H} = \hat{T} + \sum_{i=1}^{N} v(i). \tag{17}$$

The absence of explicit electron–electron interaction terms, however, does not mean that the Hamiltonian does not contain implicit electron–electron interactions. This point is best illustrated using the example of the HF method in which the Hamiltonian can be brought to the form like equation (17), [see Szabo and Ostlund (1982), p. 130] but which clearly contains electron–electron interactions in a form of an averaged potential.

The kinetic energy of a single determinant wave function constructed from the KS orbitals contains the dominant portion of the kinetic energy:

$$T_s[\rho] = \sum_{i=1}^{N} \langle \psi_i | -\tfrac{1}{2}\nabla^2 | \psi_i \rangle. \tag{18}$$

A residual term that arises from the difference between the exact multielectron wave function and the single determinant wave function is treated separately. With this separation the energy functional can be written in the following form:

$$E[\rho] = T_s[\rho] + E_{en}[\rho] + J[\rho] + E_{xc}[\rho], \tag{19}$$

where T_s is given in equation (18). The second term, the electron–nuclear interaction energy is:

$$E_{en}[\rho] = \int v(r)\rho(r)\,dr. \tag{20}$$

The term $J[\rho]$ represents the classical part of the electron–electron interaction term, the Coulomb interaction:

$$J[\rho] = \int \frac{\rho(r_1)\rho(r_2)}{|r_1 - r_2|}\,dr_1\,dr_2. \tag{21}$$

The last term in equation (19), the exchange-correlation energy, contains the difference between T and T_s and the non-classical part of the electron–electron interactions:

$$E_{xc}[\rho] = T[\rho] - T_s[\rho] + E_{ee}[\rho] - J[\rho]. \tag{22}$$

Applying the variational principle for the ground state energy under the constraint that the KS orbitals are orthonormal one obtains the KS equations which in their canonical form read:

$$\hat{h}_{eff}\psi_i = \left[-\tfrac{1}{2}\nabla^2 + v_{eff}\right]\psi_i = \varepsilon_i\psi_i, \tag{23}$$

where the effective potential v_{eff} is defined as:

$$v_{eff}(r) = v(r) + \int \frac{\rho(r')}{|r - r'|}\,dr' + v_{xc}(r). \tag{24}$$

The first term of equation (24) is given in equation (3), the second term is the Coulomb energy functional and the last term is the exchange-correlation potential that can be derived as the functional derivative of equation (22) with respect to the density:

$$v_{xc}(r) = \frac{\delta E_{xc}[\rho]}{\delta\rho(r)}. \tag{25}$$

The exchange-correlation potential is a central term in the KS theory. If one compares the HF equations (6) and the KS equations (23) these two are almost identical except for the treatment of the non-classical (exchange and correlation) part of the electron–electron interactions. While in wave function methods, electron correlation can be included only by going beyond the single determinant wave function, the KS equations include electron correlation within such a formalism. A major difference between HF and KS theories is that the HF determinant is the approximate wave function of the system of interest while the KS determinant is not. To illustrate this point, in principle, the exact electron density and energy can be obtained from the KS equations but since the multielectron wave function cannot be written in a single determinant form, the corresponding KS determinant is not the exact wave function of the system of interest. The KS determinant is only the wave function of an independent electron (or non-interacting) model system that yields the exact density of the real system. Thus the KS determinant and the corresponding orbitals and orbital energies do not have any clear physical significance.

The absence of a wave function in KS theory becomes important when electronic properties are calculated. In wave function-based methods, such as the HF method, one can use equation (4) to calculate properties from the approximate wave function while in the case of the KS method, substituting the KS determinant into equation (4) would give the properties of the non-interacting model system instead of the property of the system of study. In the case of one-electron properties, this difference is insignificant since the density functional of a property and the application of equation (4) yield the same result. However, for any two-electron properties the explicit form of the property functional has to be derived in a formalism of density matrices [Szabo and Ostlund (1982) pp. 9–12]. The most obvious example is the energy which cannot be calculated based on equation (4) using the total energy operator of equation (2) and the KS determinant, but has to be obtained from equation (19).

The exchange-correlation potential is the only term in equation (19) that is not known exactly. While this is the term that provides all the opportunities of KS theory to include correlation in a relatively simple mathematical and computational model, this is also the term that represents the greatest challenge. The exchange-correlation potential is known only in approximate forms and the simplest of these approximations is the local density approximation (LDA). Within LDA, the v_{xc} is calculated as the v_{xc} of the homogenous electron gas. One of the most significant of LDA approaches is the Vosko, Wilk, and Nussair parametrization of v_{xc} [Vosko, Wilk and Nussair (1980)]. One feature of the Hohenberg–Kohn and KS theories is that the spin does not appear in these equations. Explicit spin-dependent terms are included in the exchange correlation functionals within the local spin density (LSD) approximation and its extensions. The spin-unrestricted KS equations are analogs of the unrestricted HF equations. However, the single determinant of unrestricted spin orbitals may not be the wave function of a pure spin state. In wave function-based theories, such as the HF method, the expectation value of the total spin operator can be calculated and the unwanted spin components can be projected out. On the other hand, such a procedure would be meaningless in KS theory since the KS determinant is not the wave function of the system of interest. The significance of spin contamination is still controversial in DFT.

Beyond the LDA, one incorporates terms that arise from the inhomogenous nature of the electron density and the corresponding non-local potential in addition to LDA which is referred to as non-local (NL) DFT, or LDA + NL. The simplest non-local method is the generalized gradient approximation (GGA) which incorporates exchange-correlation potential terms that arise because of the gradient of the electron density [Perdew and Yue (1986)]. Other popular gradient-corrected correlation functionals are those derived by Lee, Yang and Parr (1988) (LYP) and those obtained by Perdew (P86) [Perdew (1986a, 1986b)]. Becke has provided successful gradient-corrected exchange functionals (B88) [Becke (1988a, 1988b, 1988c)].

Since the exchange energy can be calculated exactly from HF theory, Kohn and Sham (1965) suggested that this term is incorporated in the effective potential and only the correlation energy and potential are treated approximately. However, functionals with exact exchange have recently become popular and in practice they contain only a fraction of the exact HF exchange potential.

One of the most popular form of the GGA which contains contribution from exact exchange is Becke's three parameter (B3) exchange potential [Becke (1993)]. The B3 exchange potential became one of the most popular functional combined with the correlation functional of Lee, Yang and Parr (LYP) (1988). Further Perdew and Wong derived both exchange and correlation functionals that are popular; for a recent review of this potential refer to Perdew, Burke and Wang (1996).

The literature of different functionals is extensive and it would be impossible to discuss all functionals here. The quest for the systematic improvement of exchange-correlation functionals brings about many new forms of functionals. One direction is to include higher order terms, such as the Laplacian of the electron density, in the XC functional [Proynov, Vela, and Salahub (1995)]. These methods, however, have not yet gained general recognition. Another extension of the KS method is to calculate excited states [Nagy (1997)]. The variational principle, the basis for the derivation of the KS equations, is valid only for the ground state and the lowest energy states of any given symmetry. Further improvements are necessary for a generally applicable DFT method for excited states.

We use three implementations of the KS theory: Gaussian 94 [Frisch *et al.* (1995)], Amsterdam Density Functional (ADF) [teVelde (1997)] and DeMon [St-Amant and Salahub (1992); Salahub *et al.* (1991)] program systems. All reported ADF calculations used B88 exchange and P86 correlation functionals while all Gaussian calculations used the B3 and P86 combinations. For the complete computational details, the reader should refer to the original references. We tested not only different levels of theories but also different implementations of the same theory combined with different types and sizes of basis sets, different treatment of core electrons and core potential, and different ways of calculating the Coulomb and exchange-correlation potentials. Different implementations of the same level of theory give similar results for the relative energies, geometries and vibrational frequencies. However, the calculated ZEKE spectrum is very sensitive to small differences in the calculated vibrational eigenvectors and geometry shifts upon ionization.

4.3 Introduction to ZEKE photoelectron spectroscopy

4.3.1 BASIC CONCEPT

The fundamental process of photoelectron spectroscopy involves the ionization of molecules (or electron detachment from anions) after irradiating with monochromatic light of sufficiently high photon energy and the measurements of electrons ejected from molecules. By applying the well-known energy conservation relationship, the energy of an ionic state (E_{I}) is the energy difference between the incident light ($h\nu$) and the kinetic energy (E_{kin}) of the emitted electrons, that is $E_{\mathrm{I}} = h\nu - E_{\mathrm{kin}}$. There are two general approaches to probe the ionic states of molecules: the first approach is to measure the photoelectron signal as a function of electron kinetic energy at a fixed photon energy. This approach is sometimes referred to as conventional photoelectron spectroscopy. It is relatively simple to implement experimentally and does not require a tunable light source. Thus, the method has been widely used since the inception of photoelectron spectroscopy in the 1960s, and most of the photoelectron work was performed with this method from the 1960s to the 1980s. However, the method suffers the disadvantage of low spectral resolution (ca $100\,\mathrm{cm}^{-1}$) because of technical difficulties in separating electrons that have small energy differences [Müller-Dethlefs and Schlag (1991)]. The second approach is to measure the photoelectrons of a given kinetic energy as a function of the incident photon energy. The technique that detects electrons with zero (or near zero) kinetic energy is called ZEKE spectroscopy.

ZEKE spectroscopy is reviewed in a recent book by Schlag (1998) and in many review articles, for example Müller-Dethlefs *et al.* (1995); Schlag and Levine (1997); and Müller-Dethlefs and Schlag (1998). Briefly, ZEKE is a threshold photoelectron technique in which electrons are produced when the photon energy of the incident light is scanned across the region of the ionization threshold. There are two ways to generate ZEKE electron signals. First, the ZEKE electrons are produced directly by threshold photoionization of the molecule (or photodetachment of the anion) and then pulled to an electron detector by a small electric pulse. In the meantime, kinetic electrons are also emitted from the states that have lower energy than that of the incident photons. To separate the ZEKE from the kinetic electrons, a time delay is applied between photoionization and electron extraction. During the time delay, the kinetic electrons drift away from the ionization volume or the detection window while the ZEKE electrons remain for extraction and detection. Secondly, ZEKE electrons that are from the ionic core are ejected by a small electric pulsed field, and these are from Rydberg states with high quantum number $n(n > 100)$, $\sim 5\,\mathrm{cm}^{-1}$ below the ionization continuum. High-lying Rydberg states are formed by photoexcitation and stabilized by Stark mixing due to the stray field present in the excitation volume and the inhomogenous electric field formed by nearby prompt ions. Because of the long lifetime (up to tens of microseconds) of Rydberg states, a time delay can be placed between the photoexcitation and field-ionization to discriminate against prompt electrons. Thus, as the photon energy is scanned through the manifold of the Rydberg states, ZEKE electrons are produced a few cm^{-1} below each ionic state. A simple correction for the energy shift induced by the small electric ionization field produces a ZEKE spectrum that is similar to that from direct threshold photoionization. In fact, the delayed field-ionization method is currently the most commonly used ZEKE technique, and it is known as the PFI-ZEKE technique.

The major advantage of the ZEKE technique is its high spectral resolution. Figure 4.1 compares spectra obtained from (a) photoionization efficiency (b) conventional photoelectron, and (c) ZEKE techniques. A single-photon ionization or excitation process was employed in collecting these spectra. The photoionization efficiency spectrum was obtained by collecting the ion signals as the laser wavelength was scanned. The spectrum displays an onset of ionization that can be used to locate the approximate ionization energy of the cluster. The conventional photoelectron spectrum was obtained by collecting the kinetic electron signals with a fixed photon energy. The

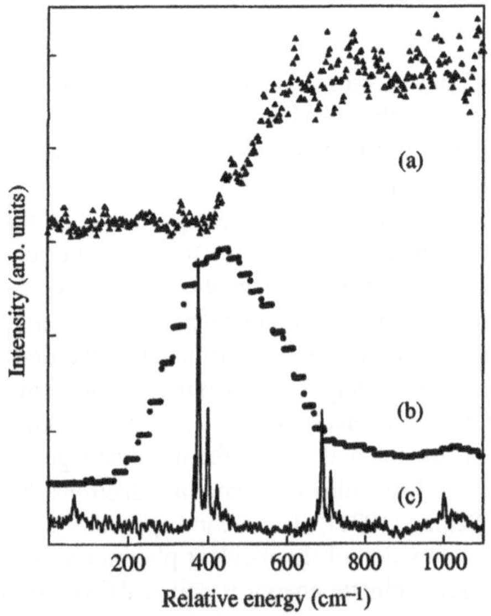

Figure 4.1. Spectra of Nb_3O obtained by (a) threshold photoionization spectroscopy, (b) anion photoelectron spectroscopy, and (c) PFI-ZEKE spectroscopy.

spectrum shows a broad band with a full width at the half maximum (FWHM) of $\sim 40\,\text{meV}$, or $\sim 300\,\text{cm}^{-1}$. In contrast, fully resolved vibrational structure with a FWHM of about $5\,\text{cm}^{-1}$ is revealed in the ZEKE spectrum. This spectrum was obtained by detecting ZEKE electrons from pulsed field ionization of high-lying Rydberg states that were formed while laser wavelengths were scanned across each ionization continuum. Because there is a severe lack of the knowledge about the intermediate states of transition metal-containing molecules, the high ZEKE spectral resolution achieved without resonant state selection is an important advantage over other high resolution techniques, such as resonant two-photon ionization or laser-induced fluorescence each of which require a long-lived excited state. Also, the ionization energy of many transition metal clusters are in a spectral region easily accessible to frequency-doubled tunable dye lasers, making the application of the ZEKE technique fairly routine. Further, the vibrational normal modes of transition metal clusters and complexes are small enough to be populated at room temperature, making it possible to measure vibrational frequencies of the neutral species, in addition to that of the cations. Finally, ZEKE signals are recorded against a null background, and thus the technique is extremely sensitive.

4.3.2 EXPERIMENTAL SETUP AND GENERAL PROCEDURES

Figure 4.2 shows the molecular beam PFI-ZEKE photoelectron spectrometer system schematically. The system consists of two vacuum chambers. The first houses a Smalley-type cluster source [Dietz *et al.* (1981)] and is pumped by a $2200\,\text{l}\,\text{s}^{-1}$ diffusion pump. The second houses the ZEKE spectrometer and is pumped by two $400\,\text{l}/\,\text{s}^{-1}$ turbomolecular pumps. The PFI-ZEKE spectrometer consists of a two-stage extraction assembly, a 34 cm long tube, and a dual microchannel plate detector. The ionization region is well shielded from the high voltages applied in the acceleration region and to the microchannel plate detector. It is also magnetically shielded by a cylindrical, double layer of μ-metal. The spectrometer can also be operated as a two-field, space-focused, Wiley–McLaren time-of-flight mass spectrometer by supplying the appropriate voltages.

Bare metal clusters were produced by laser vaporization of a metal rod (> 99.5 %) in the presence of a pulse of helium gas from a home-built pulsed valve [Proch and Trickl (1989)]. A trace amount of an appropriate reactant (e.g., $\sim 10^{-5}$ of ethylene, nitrogen, oxygen, or dimethyl

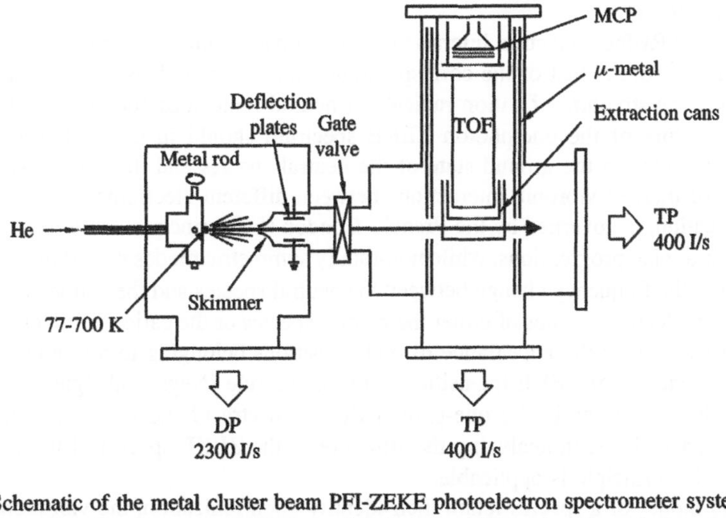

Figure 4.2. Schematic of the metal cluster beam PFI-ZEKE photoelectron spectrometer system. DP, diffusion pump; TP, turbomolecular pump; TOF, time-of-flight tube; MCP, microchannel plate detector.

ether) was doped in helium gas to produce ligated metal clusters and organometallic complexes. The resulting species passed down a clustering tube and were supersonically expanded into the vacuum chamber. The supersonic jet was skimmed a few cm downstream from the exit end of the clustering tube. The clustering tube was maintained at room temperature or cooled by liquid nitrogen. A pair of deflection plates located after the skimmer removed residual charged species from the molecular beam before it entered the second chamber.

Prior to single-photon PFI-ZEKE experiments, photoionization efficiency spectra were recorded to locate the approximate ionization energy of the cluster or complex. Then, with laser energy set above the ionization energy, the experimental conditions, such as the timing and the fluence of both the vaporization and ionization lasers, the backing pressure of the helium gas, and the reactant concentration, were carefully optimized in order to maximize the ratio of the mass peak of the cluster of interest in the time-of-flight mass spectrum to that of all other peaks. This step was crucial for identifying the carrier of the PFI-ZEKE electrons. With the optimized experimental conditions, the cluster or complex was excited to high-lying Rydberg states by a single-photon excitation. After a suitable delay, the high-lying Rydberg states were field ionized by a voltage pulse from a digital delay generator applied to the repeller plate. Typically, a field of $1\ \mathrm{V\,cm^{-1}}$ was applied for 100 ns after a delay of about 3 μs. A DC field, less than $0.1\ \mathrm{V\,cm^{-1}}$, was used to reject electrons from prompt photoionization. The PFI-ZEKE electron signals were capacitively coupled from the microchannel plate detector anode and were amplified by a preamplifier, averaged by a gated integrator, and collected in a laboratory computer.

4.4 Theoretical foundation of ZEKE spectral simulation

The long lifetime of highly excited molecular Rydberg states that lie just below the photoionization threshold plays a crucial role in ZEKE photoelectron spectroscopy [Müller-Dethlefs, Sander, and Schlag (1984), Müller-Dethlefs, and Schlag (1991), and Müller-Dethlefs *et al.* (1995)]. This long lifetime is a result of the heavy mixing of Rydberg states within their quasi-continuum. Therefore, the initially optically populated states with low l and m quantum numbers become quickly 'diluted' in a 'sea' of states with high l and m quantum numbers, which, in turn, are nonradiative and have a long lifetime after initial excitation. It is now clear that the main cause of the mixing of highly excited Rydberg states is the external electric fields produced mainly by the surrounding ions [Vrakking *et al.* (1995)] in agreement with a mechanism suggested first by Chupka (1993a, 1993b).

A highly excited Rydberg electron interacts only marginally with the core of the molecule, thus the structure of the core is that of the corresponding radical cation. This makes ZEKE a perfect tool for studying the structure of cation radicals of neutral molecules and clusters. In particular, the vibronic structure of the one-photon ZEKE spectrum should in the first place reflect the geometry change between the ground state of the neutral species and the cation. This structure, in the absence of marked vibronic interactions between different electronic states in the neutral species and/or cation, is governed by the Franck–Condon (FC) principle. Only totally symmetric modes show vibrational progressions, while non-totally symmetric modes may show $m \rightarrow m \pm 2n$ transitions due to the frequency change between the neutral species and the cation. When vibronic coupling between electronic states of either the neutral species or the cation is important, then one observes in ZEKE spectra the appearance of odd transitions belonging to non-totally symmetric modes and the deviation of ZEKE intensities from the FC rule [Negri and Zgierski (1997)]. We expect that such interactions in the one-photon ZEKE spectra of metal–ligand complexes are of minor importance. Thus, in analyzing the structure of the ZEKE spectra of these species, we assume that the FC principle is applicable.

If the normal modes of a species retain their identity upon ionization, the vibrational structure of the ZEKE spectrum can be described in terms of the displacements $\boldsymbol{Q}^{0+}$ of the normal modes.

These displacements are the projections of the geometry change between the neutral and cation states onto the normal modes of the cation, and they are given by:

$$Q^{0+} = \left[x_0^e - x_+^e\right] M^{1/2} L^+, \tag{26}$$

where x_0^e is the 3N-dimensional vector of the equilibrium Cartesian coordinates of the neutral species, and x_+^e is that of the cation. M is the $3N \times 3N$ diagonal matrix of the atomic masses and L^+ is the matrix relating the normal coordinate q^{0+} to the mass-weighted Cartesian coordinates of the cation.

The vibrational overlap integrals, or FC factors, for a ZEKE spectrum are in this case products of the FC factors of the individual normal modes, and they can be expressed in terms of two dimensionless parameters: a displacement and a distortion parameter. The displacement parameter B_i^{0+} reflects the shift in the equilibrium configuration and it is given by:

$$B_i^{0+} = \left[2\nu_i^0 \nu_i^+ / \hbar \left(\nu_i^0 + \nu_i^+\right)\right]^{1/2} Q_i^{+0}. \tag{27}$$

The distortion parameter, ζ_i^{+0}, relates to the change in force constants:

$$\zeta_i^{0+} = \left(\nu_i^+ - \nu_i^0\right) / \left(\nu_i^+ + \nu_i^0\right). \tag{28}$$

In equation (28), ν_i^0, ν_i^+ are the frequencies of the ith normal mode in the neutral and cationic species, respectively. The vibrational overlap integrals $\langle u_i^0 | v_i^+ \rangle$ can be expressed as follows [Manneback (1951)]:

$$\begin{aligned} \langle 0|0\rangle &= \left[1 - \left(\zeta_i^{0+}\right)^2\right]^{1/4} \exp\left[-\left(B_i^{0+}\right)^2/4\right], \\ \langle 0|1\rangle &= -2^{-1/2} B_i^{0+} \left(1 + \zeta_i^{0+}\right)^{1/2} \langle 0|0\rangle, \\ \langle 1|0\rangle &= 2^{-1/2} B_i^{0+} \left(1 - \zeta_i^{0+}\right)^{1/2} \langle 0|0\rangle, \end{aligned} \tag{29}$$

where $|v_i^0\rangle$ and $|v_i^+\rangle$ are vibrational wave functions of the ith mode in its vth vibrational state of the neutral, and cation species, respectively. Integrals with higher vibrational quantum numbers are then expressed in terms of these basic integrals through the recurrence formulas [Manneback (1951)]:

$$\begin{aligned} (u|v+1) &= (v+1)^{-1/2} \Big[u^{1/2} \left(1 - \left(\zeta_i^{0+}\right)^2\right)^{1/2} (u-1|v) \\ &\quad + \zeta_i^{0+} v^{1/2} (u|v-1) + 2^{-1/2} B_i^{0+} \left(1 - \zeta_i^{0+}\right)^{1/2} (u|v)\Big], \\ (u+1|v) &= (u+1)^{-1/2} \Big[v^{1/2} \left(1 - \left(\zeta_i^{0+}\right)^2\right)^{1/2} (u|v-1) \\ &\quad - \zeta_i^{0+} u^{1/2} (u-1|v) - 2^{-1/2} B_i^{0+} \left(1 + \zeta_i^{0+}\right)^{1/2} (u|v)\Big]. \end{aligned} \tag{30}$$

If any cation normal mode projects onto more than one neutral species normal mode, the above procedure can be only used only for cold spectra [Zgierski (1986)]. Otherwise the mode mixing, called the Dushinsky effect, must be considered explicitly to account for hot bands. The Dushinsky effect is especially important in metal–ligand clusters where low frequency vibrations and hot bands are common. In such a situation, the normal coordinates of the ion, q^+ are related to the normal coordinates of the neutral species, q^0, by:

$$q^+ = Q^{0+} + S q^0, \tag{31}$$

where S is the so-called Dushinsky matrix [Dushinsky (1937)] defined as:

$$S = (L^+)^T L^0. \tag{32}$$

For cases with substantial mode mixing S deviates markedly from a unit matrix. Within the harmonic approximation, it is possible to calculate the FC overlaps in a closed form (without setting $S = 1$) using recurrence relations given by Doktorov, Malkin, and Man'ko (1977).

To simulate a ZEKE spectrum, we calculate the equilibrium geometry and force-field of the lowest electronic state of a given symmetry and spin-multiplicity for the neutral and ionized species, $|0\rangle$ and $|+\rangle$, respectively. The FC overlap integrals, $\langle v^+|v^0\rangle$, for a given $\langle +v^+| \leftarrow \langle 0v^0|$ vibronic transition are then calculated from the equilibrium geometry difference and the Dushinsky matrix S. Here $|v^0\rangle$ is the vibrational wave function in the state $|0\rangle$ with quantum numbers v^0 and $|v^+\rangle$ is a similar function for the v^+ state. The vibrational structure of the spectrum of the $\langle +| \leftarrow \langle 0|$ photoionization process can be represented as:

$$I^{0+}(\Omega) = \Gamma\hbar\Omega Z^0(kT) \sum_{v^0} \exp(-E_{v^0}/kT) \sum_{v^+} |\langle v^+|v^0\rangle|^2 \left[\left(E^0 - E_{v^0} + E_{v^+} - \hbar\Omega\right)^2 + \Gamma^2\right]^{-1}, \tag{33}$$

where $Z^0(kT)$ is the vibrational partition function of the $\langle 0|$ state, and Γ is the (Lorentzian) linewidth assumed the same for all vibronic transitions, E^0 is the ionization threshold energy for the process under investigation, Ω is the frequency of the ionizing laser and E_{v^0} and E_{v^+} are the vibrational energies in the initial and the final state, respectively.

In the case when ionization occurs as a result of a resonant two-photon process, the interpretation is more involved, since it requires *ab initio* calculations of the excited resonant state of the neutral species. An example of such a process is two-photon ionization in the naphthalene molecule [Cockett *et al.* (1993)] in which the first photon prepares an individual vibronic state of the S_1 manifold from which the ZEKE spectrum is measured. One obtains in this case a set of ZEKE spectra from different vibronic levels of the resonant manifold. These contain a wealth of information not only on geometry changes between electronic states of the neutral species and the corresponding cation, but also on the vibronic coupling between the electronic states of the two species. For details we refer to [Negri and Zgierski (1996, 1997)].

4.5 Application of DFT to ZEKE simulation

4.5.1 DIATOMICS: Y_2

There was a long standing controversy as to the nature of the ground electronic state of diyttrium. On the one hand, attempts to obtain the ESR spectrum of Y_2 ended in failure [Knight *et al.* (1983)] indicating an apparent singlet state. On the other hand, elaborate quantum-chemical calculations [Walch and Bauschlicher (1984), and Dai and Balasubramanian (1993, 1995)] predicted the ground state to be $^5\Sigma_u^-$, just as is the case for the isovalent Sc_2 dimer. A similar conclusion was reached by recent density functional calculations with the Gaussian and the ADF programs [Yang *et al.* (1996)] in which also FC factors for various ionization processes, and vibrational frequencies for the cation were obtained. The comparison of various theoretical results is given in Table 4.1. All theoretical methods predict the $^5\Sigma_u^-$ state to be well below the $^1\Sigma_g^+$ state.

Both the Gaussian and the ADF calculation predict nearly equal intensities for the $0 \leftarrow 0$ and $1 \leftarrow 0$ transitions in the $^4\Sigma_g^- \leftarrow {}^5\Sigma_u^-$ photoionization process. The calculated dimensionless displacement parameter B^{0+} is 1.42 and 1.49, from Gaussian and ADF calculations, respectively. On the other hand, the calculations predict a rather large geometry change for the $^2\Pi_u \leftarrow {}^1\Sigma_g^+$

Table 4.1. Summary of experimental and theoretical results on Y_2 and Y_2^+.

Method	Ref.	Parameter	Y_2		Y_2^+	
			$^5\Sigma_u^-$	$^1\Sigma_g^+$	$^4\Sigma_g^-$	$^2\Pi_u$
CASSCF-CL	[a]	T_0 (eV)	0.0	0.87		
		r_e (Å)	3.03	2.74		
		ω_e (cm^{-1})	171	206		
CASSCF-CL	[b]	T_0 (eV)	0.0	0.87	4.92	5.32
		r_e (Å)	3.03	2.76		
		ω_e (cm^{-1})	172	180		
DFT-G92	This work	T_0 (eV)	0.0	0.96	5.14	5.22
		r_e (Å)	2.73	2.76	2.82	2.94
		ω_e (cm^{-1})	214	225	216	204
ADF	This work	T_0 (eV)	0.0	0.29	5.14	5.37
		r_e (Å)	2.94	2.59	2.85	2.84
		ω_e (cm^{-1})	173	207	188	201
Experiment	This work	T_0 (eV)	0.0		4.9756(2)	
		ω_e (cm^{-1})	182(2)		197(2)	

[a]Walch and Bauschlicher (1984).
[b]Dai and Balasubramanian (1983, 1995).

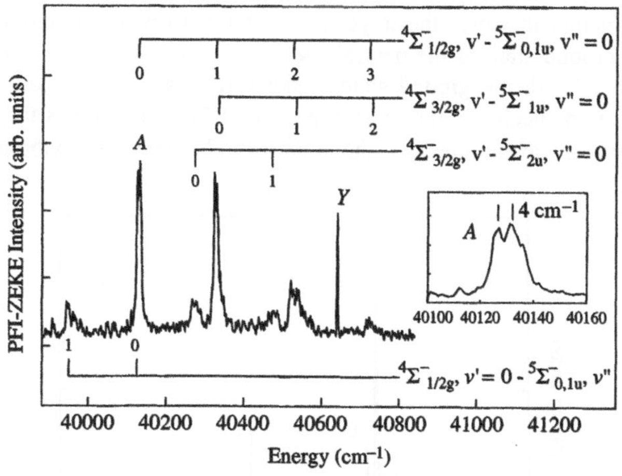

Figure 4.3. PFI-ZEKE spectrum of Y_2.

transition resulting in displacement parameter values of 2.6 (Gaussian) and 3.6 (ADF), and a correspondingly large Stokes' shift for this transition.

A recent PFI-ZEKE experimental study of the ionization process in Y_2 [Yang *et al.* (1996)] resolved the detailed fine structure in the ZEKE spectrum. This spectrum is shown in Figure 4.3. It is evident that the FC progression in the experimental spectrum is short, which lends support to the quintet being the ground electronic state. The main progression with $v^+ = 0$ to 3 is identified as belonging to the $^4\Sigma^-_{1/2,g} \leftarrow {}^5\Sigma^-_{0,1,u}$ overlapping transitions with a small separation between the two spin–orbit split components of the quintet ground state. The separation between the $^5\Sigma^-_{0,u}$ and $^5\Sigma^-_{1,u}$ components is 4 cm^{-1}. The second progression $v^+ = 0$ to 2 with the same relative intensity distribution as the main progression is identified as the $^4\Sigma^-_{3/2,g} \leftarrow {}^5\Sigma^-_{0,1,u}$ transition to the second spin–orbit component of the cation quartet state. The location of the origin of this progression gives the value of 210 cm^{-1} for the spin–orbit splitting in the quartet state of the cation.

The large second-order spin–orbit splitting in the quintet state (the $^5\Sigma^-_{2,u}$ level is 68 cm^{-1} above the $^5\Sigma^-_{0,u}$ level) explains the failure to observe an EPR spectrum. The excellent agreement of the FC structure [Yang *et al.* (1993)] with the observed one identifies beyond any doubt the quintet nature of the ground electronic state of diyttrium.

4.5.2 TRIMER OXIDES: Nb_3O, Zr_3O

The Nb_3O molecule is the first four-atom metal–ligand cluster with an experimentally determined structure [Yang *et al.* (1995)]. The structure was determined by combining DFT, spectral simulation with the analysis of the experimental PFI-ZEKE spectrum. This spectrum is shown in Figure 4.4. at two different temperatures. It shows a short progression with a spacing of 312 cm^{-1} labeled as 0, 1 and 2. The width of each band (about 5 cm^{-1}) is attributed to the unresolved rotational structure. This progression is assigned to the Nb–Nb–Nb bending a_1 mode. It results from a slight change of the Nb–Nb–Nb angle upon ionization (1.2°). The less intense bands marked by ′, ″ and a, b and c belong to transitions from vibrationally excited levels of the neutral species as is evident from the drop of their intensity at lower temperatures.

DFT calculations using Gaussian (at the B3P86/LANL2DZ level) and the deMon-KS program [Salahub *et al.* (1991)] (with Perdew–Wang and Perdew non-local functionals) show that Nb_3O has a planar C_{2v} structure in which the oxygen atom binds to two niobium atoms. These calculations identify the ground state of the neutral species as a doublet of B_1 symmetry [Yang *et al.* (1995)]. On the other hand, the ground state of the cation is a totally-symmetric singlet state. The structure of the two species is shown in Figure 4.5. Table 4.2 shows the comparison of the calculated and observed wavenumbers of the complex. The calculated vibronic structure of the

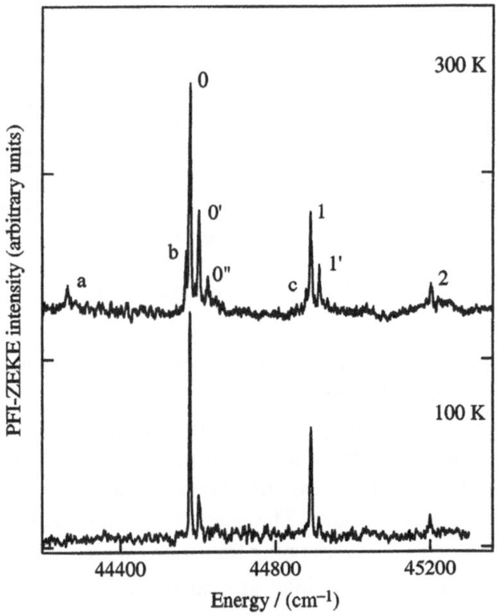

Figure 4.4. PFI-ZEKE spectrum of Nb_3O at 300 and 100 K.

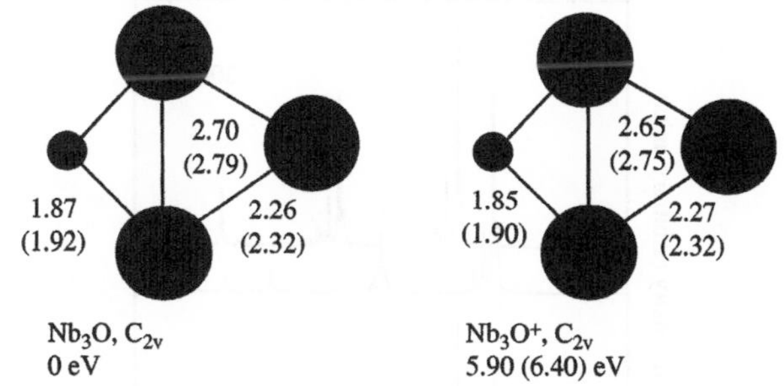

Figure 4.5. Calculated geometries and relative energies for Nb_3O and Nb_3O^+. The deMon-KS DFT code was used, see text. Bond lengths, in Å, are shown. For the planar geometries, Gaussian 92/DFT was also used. Results from these calculations are shown in brackets.

Table 4.2. Calculated[a] vibrational wavenumbers for planar C_{2v} Nb_3O and Nb_3O^+.

Mode	Symmetry	Type[b]	Neutral[a,c]	Ion[a,d]	Δ[c,d]
1	a_1	ν_s Nb–O			
			753	790	−37
			(733)	(764)	(−31)
			710 ± 20[e]		
2	a_1	ν_s Nb–Nb			
			382	392	−10
			(350)	(379)	(−29)
3	a_1	δ_s Nb–Nb–Nb			
			334	337	−3
			(310)	315	(−5)
			320 ± 1[f]	**312 ± 1**[f]	**8 ± 1**[f]
4	b_1	ν_a Nb–O			
			579	625	−46
			(572)	(608)	(−36)
5	b_1	ν_a Nb–Nb			
			238	269	−31
			(233)	(260)	(−27)
					−23 ± 1[f]
6	b_2	π Nb_3O			
			300	289	11
			(286)	(274)	(12)
					11 ± 1[f]

[a]Wavenumbers calculated using the deMon-KS and the Gaussian 92/DFT (shown in brackets) codes. All values in cm^{-1}.
[b]ν_s symmetric stretch; ν_a, asymmetric stretch; δ_s, symmetric bend; π, out of plane deformation.
[c]Experimental values shown as boldface.
[d]Wavenumber difference (neutral–ion) in cm^{-1}.
[e]Alex, S., Green, S. M. E., and Leopold, D. G. (1998).
[f]This work.

ZEKE spectra obtained by the two DFT methods is shown in Figure 4.6. Excellent agreement with the experimentally observed spectrum as to intensity distribution indicates that the geometry change between the neutral and ionized species is very well described by DFT methods. The agreement between the observed and calculated wavenumbers supports the result that the Nb_3O complex is planar with C_{2v} symmetry.

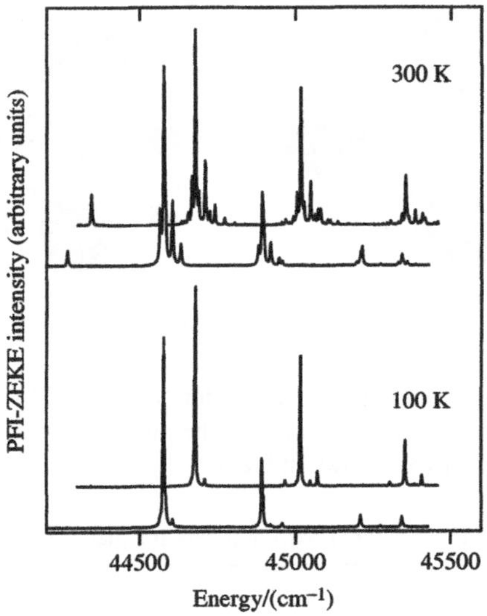

Figure 4.6. Simulations of the PFI-ZEKE spectrum of Nb_3O at 100 and 300 K. The simulations were calculated using the geometry from Gaussian92 (lower curve in each panel) and deMon-KS DFT calculations. The deMon-KS spectra have been offset by $+100\,cm^{-1}$ for clarity. Both spectra were convoluted with a $5\,cm^{-1}$ FWHM Lorentzian line shape to simulate the (rotational) width of the experimental bands.

DFT calculations find another isomer with a trigonal non-planar geometry about 1 eV above the planar conformer [Yang *et al.* (1995)]. For this isomer, however, the calculated wavenumbers do not match the observed wavenumbers, and the calculated change in geometry upon ionization is far too large to reflect the observed short FC progression in the ZEKE spectrum.

The ZEKE spectrum of the Zr_3O cluster looks very similar to that of Nb_3O (see Figure 4.7.) [Yang *et al.* (unpublished)]. It shows a short progression with a wavenumber of $271\,cm^{-1}$ superimposed on hot band sequences and progressions. Two DFT methods (B3P86/LANL2DZ and ADF) predict similar configuration of this complex to that of Nb_3O, i.e., a planar rhomboid with

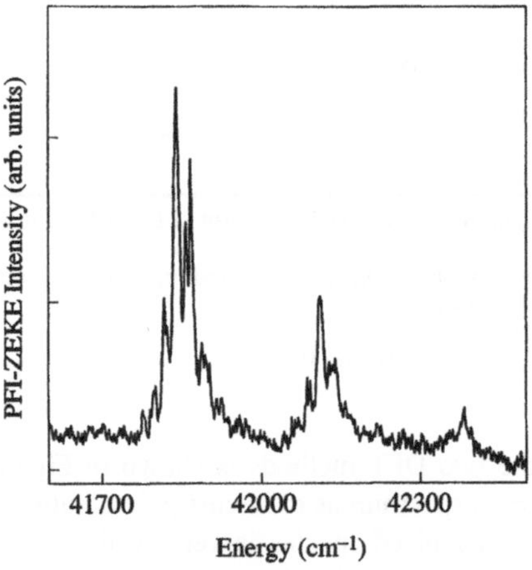

Figure 4.7. PFI-ZEKE spectrum of Zr_3O.

C_{2v} symmetry. The Zr–Zr and Zr–O bond lengths are about 0.1 Å longer than those in the Nb_3O complex. Both DFT methods identify the lowest state of the cation as a quartet of B_1 symmetry with the A_2 doublet state about 0.2–0.3 eV higher. For the neutral species, the quintet state of A_2 symmetry and the triplet state of the same symmetry lie very close. B3P86/LANL2DZ predicts the quintet state to be 0.04 eV below the triplet, while ADF predicts the triplet to lie 0.09 eV below the quintet. Thus the observed ZEKE spectra should belong either to the $^4B_1 \leftarrow {}^5A_2$ transition or to the $^4B_1 \leftarrow {}^3A_2$ transition.

The calculated displacement parameters for the two possible transitions are given in Table 4.3. We see that the calculations properly predict a short progression of the mode with wavenumber of 239 and 227 cm^{-1} by Gaussian and ADF programs, respectively. This mode is identified as the Zr–Zr sym. stretch which leads to the deformation of the Zr–Zr–Zr angle. The calculated wavenumber of the Zr–O–Zr deformation is 40–50 cm^{-1} above that of the Zr–Zr–Zr bending mode. Both DFT methods predict it to have smaller FC activity for the two possible transitions. The Zr–O symmetric stretch with a calculated wavenumber of 650 cm^{-1} in the neutral species and 675 cm^{-1} in the cation, shows weak FC activity only in the $^4B_1 \leftarrow {}^5A_2$ transition. The calculated spectra for this transition at the temperature of 150 K are shown in Figure 4.8. The general structure of these spectra is similar to the observed spectrum, in particular, the Gaussian calculations give a nice reproduction of the sequence band structure of the origin and of the $1 \leftarrow 0$ transitions. What they fail at, is the reproduction of the Zr–Zr symmetric stretch and the Zr–O–Zr deformation wavenumbers which they underestimate by 30–40 cm^{-1}. The experimental spectrum 600–700 cm^{-1} above the origin is too noisy to discern any weak band belonging to the symmetric Zr–O stretch. The calculated ZEKE spectra for the $^4B_1 \leftarrow {}^3A_2$ transition differ substantially for the observed spectrum suggesting the quintet to be the ground state of Zr_3O.

Electronic states of the non-planar, pyramidal configuration of Zr_3O are calculated to lie 0.8 eV above the ground state of the planar C_{2v} structure, similarly to the situation encountered in the Nb_3O cluster.

4.5.3 TRIMER DINITRIDES AND DICARBIDES: Nb_3C_2, Nb_3N_2, Y_3C_2

Following the success of the structure determination of the Nb_3O complex by a combination of density functional calculations and ZEKE spectroscopy, the studies were extended to the following five-atom complexes: Nb_3C_2 [Yang *et al.* (1996b)], Nb_3N_2 [Yang *et al.* (1997)] and Y_3C_2 [Yang, Zgierski, and Hackett (1998)].

Table 4.3. Calculated displacement parameters in totally-symmetric modes, b (Å amu$^{1/2}$ and B (dimensionless), for the two ionization processes in Zr_3O.

$\omega(^4B_1)$	$b[B]$ $^4B_1 \rightarrow {}^5A_2$	$b[B]$ $^4B_1 \rightarrow {}^3A_2$
Gaussian 94		
239	0.263 [0.70]	0.210 [0.56]
288	0.045 [0.13]	0.131 [0.38]
678	0.089 [0.40]	0.021 [0.09]
ADF		
227	0.312 [0.81]	0.354 [0.92]
268	0.188 [0.53]	0.050 [0.14]
676	0.080 [0.36]	0.031 [0.14]

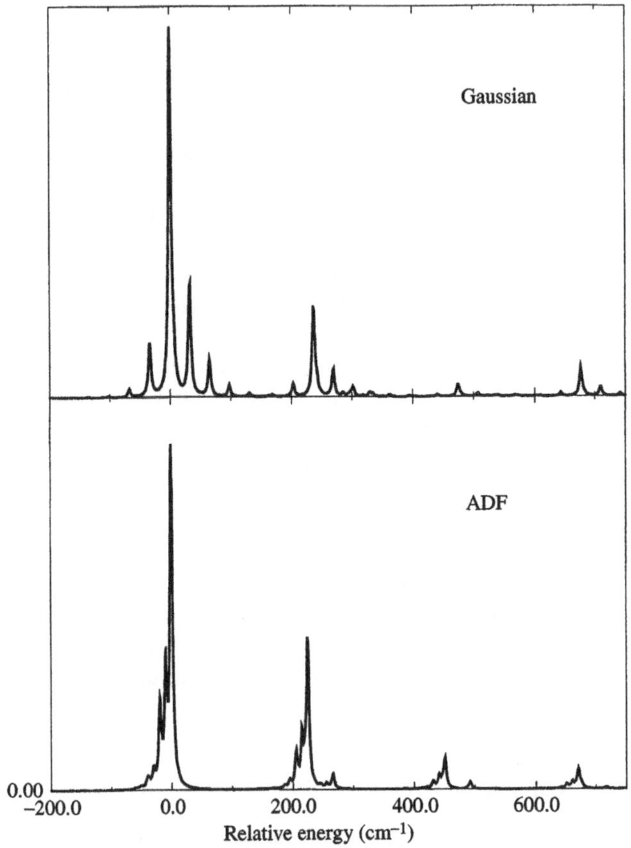

Figure 4.8. Simulated spectrum of Zr_3O.

The experimental ZEKE spectrum of the niobium trimer dicarbide (see Figure 4.9) shows a long FC progression with a 259 cm^{-1} separation which is insensitive to ^{13}C substitution and can be assigned to the symmetric Nb–Nb–Nb deformation mode [Yang *et al.* (1996b)]. Apart from the main progression six secondary progressions with the same spacing were identified as shown in Figure 4.9.

The calculations with three different DFT codes (Gaussian, ADF and deMon-KS-LSDA) indicate that the stable structure of lowest energy in the complex is a trigonal bipyramid of C_{2v} symmetry for the neutral species and of D_{3h} symmetry in the cation. The observed ZEKE transition was assigned to the $^1A_1' \leftarrow {^2A_1}$ transition. Another alternative to the bipyramidal structure (structure **a** in Figure 4.10) is a doubly bridged structure in which two niobium and two carbon atoms lie in one plane, with the third niobium atom forming the apex of a pyramid (structure **b** in Figure 4.10). Both Gaussian and ADF calculations predict the two structures to be almost of equal energy, the bipyramidal structure being more stable by only 0.02 and 0.01 eV, respectively. The two structures differ markedly in their calculated ionization potentials, and that for the doubly bridged structure is predicted to be 0.86 eV higher than that for the double pyramid. Thus, our

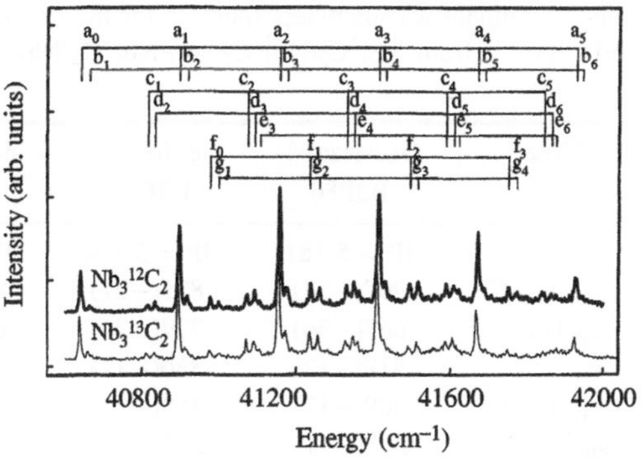

Figure 4.9. PFI-ZEKE spectra of $Nb_3^{12}C_2$ (upper trace) and $Nb_3^{13}C_2$ (lower trace).

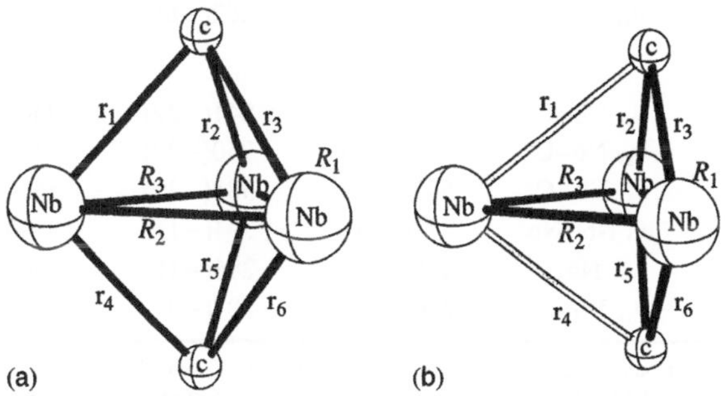

Figure 4.10. Structural alternatives of Nb_3C_2: (a) trigonal bipyramidal, (b) double bridged.

spectrum corresponds to one of the two ionic states which can be observed with different ionizing laser wavelength.

The calculated wavenumbers in the neutral and cation species for the trigonal bipyramid are reproduced in Table 4.4. The comparison between the experimental and simulated ZEKE spectra for this structure is presented in Figure 4.11. The totally symmetric component (under C_{2v}) of the δ Nb_3 e' vibration is assigned to the long progression in the observed ZEKE spectrum. Although the agreement is not perfect and spectra obtained from different methods differ in details, the main structure of the ZEKE spectrum is reproduced well by all calculations confirming the bipyramidal structure. However, this does not preclude the existence of a doubly bridged conformer which could be probed by a higher energy UV laser corresponding to its higher ionization potential. The calculated ZEKE spectrum of the doubly bridged structure differs markedly from that of the bipyramid: the wavenumbers of the active modes are higher and two or three different modes contribute to short FC progressions [Yang *et al.* (1996b)].

The experimental ZEKE spectrum of Nb_3N_2 given in the top panel of Figure 4.12 shows a short progression with a spacing of $257\,cm^{-1}$. This progression is assigned to the symmetric Nb–Nb–Nb deformation similarly to Nb_3C_2. Superimposed on this progression are hot bands belonging to the asymmetric Nb–Nb–Nb deformation. Three different DFT calculations identify the lowest energy structure of the Nb_3N_2 cluster as a doubly bridged structure, similar to the structure **b** of the Nb_3C_2 cluster. Replacement of carbon atoms with nitrogen stabilizes this structure with respect to the bipyramid (**a**) by 1.11 eV, while this stabilization in the cation is 0.69 eV [Zgierski (1998)].

Table 4.4. Calculated harmonic vibrational wavenumbers (cm^{-1})[a] for the trigonal bipyramid structures of $Nb_3^{12}C_2/Nb_3^{12}C_2^+$, isotopic shifts ($cm^{-1}$) from $Nb_3^{12}C_2/Nb_3^{12}C_2^+$ to $Nb_3^{13}C_2/Nb_3^{13}C_2^+$, and ionization potentials (IP, eV) of $Nb_3^{12}C_2$.

Mode	Symmetry[b]	Type[c]	Gaussian94-B3P86	deMon-KS-LSDA[d]	deMon-KS-LSDA[e]	ADF-BP
$Nb_3^{12}C_2$	C_s/C_{2v}		IP = 5.781	IP = 5.164	IP = 5.162	IP = 5.121
ν_1''	a′/a_1	ν_s, Nb–C	807(−37)	835(−31)	826(−31)	798(−30)
ν_2''	a′/b_1	ν_a, Nb–C	667(−34)	714(−26)	693(−25)	659(−24)
ν_3''	a′/a_1	ν_s, Nb–C	519(−17)	530(−17)	539(−18)	529(−18)
ν_4''	a″/b_2	ν_s, Nb–C	509(−17)	469(−15)	518(−17)	525(−18)
ν_5''	a′/a_1	ν_s, Nb–Nb	368(−1)	387(−2)	387(−1)	368(−1)
ν_6''	a′/a_1	δ, Nb_3	241(−1)	238(−2)	256(−1)	266(0)
ν_7''	a″/b_2	δ, Nb_3	86(−1)	94(−1)	65(0)	42(0)
ν_8''	a′/b_1	ν_a, Nb–C	244(−8)	304(−9)	289(−9)	72(−3)
ν_9''	a″/a_2	ν_a, Nb–C	192(−7)	282(−8)	214(−7)	246(−8)
$Nb_3^{12}C_2^+$	D_{3h}					
ν_1^+	a_1'	ν_s, Nb–C	817(−31)	841(−32)	840(−31)	812(−30)
ν_2^+	a_2''	ν_a, Nb–C	709(−26)	750(−27)	749(−27)	727(−26)
$\nu_{3,4}^+$	e′	ν_s, Nb–C	543(−18)	581(−20)	586(−20)	553(−19)
ν_5^+	a_1'	ν_s, Nb–Nb	371(−0)	394(−1)	398(0)	375(−1)
$\nu_{6,7}^+$	e′	δ, Nb_3	270(−1)	286(−1)	290(0)	267(0)
$\nu_{8,9}^+$	e″	ν_a, Nb–C	226(−7)	287(−9)	282(−10)	244(−8)

[a]The numbers in parentheses are isotopic shifts from $Nb_3^{12}C_2/Nb_3^{12}C_2^+$ to $Nb_3^{13}C_2/Nb_3^{13}C_2^+$.
[b]A C_s symmetry was predicted by deMon-KS-LSDA and a C_{2v} by Gaussian94-B3P86 and ADF-BP.
[c]ν_s symmetric stretch; ν_a, asymmetric stretch; δ, deformation.
[d]Using model core potentials for the niobium atoms but all electrons for the carbon atoms. [e]Using model core potentials for the niobium and carbon atoms.

The calculated wavenumbers and ionization potentials of Nb_3N_2 are given in Table 4.5 while the calculated ZEKE spectra are shown in panels B–D of Figure 4.12. Bands denoted by an asterisk (*) and a number sign (#) are associated with the ν_8 and ν_5 modes, respectively. The ADF calculations produce an almost perfect reproduction of the observed spectrum, while the two other methods are not as successful. In particular, they predict rather strong mixing of ν_7 and ν_8 modes upon ionization resulting in shifting part of the displacement along the ν_7 coordinate into the ν_8 coordinate contrary to the experiment. On the other hand, the ADF method retains the identity of the ν_7 upon ionization leading to only one FC progression as observed.

The ZEKE spectrum of the Y_3C_2 cluster [Yang, Zgierski and Hackett (1998)] shows a long main FC progression with a spacing of 86 cm^{-1} belonging to the symmetric Y–Y–Y deformation mode (see Figure 4.13, panel A). This progression is repeated three times starting from different vibronic origins. Two of these are hot band progressions in which the low wavenumber asymmetric Y–Y–Y deformation is excited in the neutral species. The third one starts from one quantum of the Y–Y symmetric stretch. The $^2B_1$ state of the C_{2v} symmetry isomer of Y_3C_2 similar to that of Nb_3C_2 is identified as the ground state by B3P86/LANL2DZ calculations. The $^4E'$ state under D_{3h} symmetry lies 0.295 eV above the neutral ground state. A^1A_1' state of the D_{3h} symmetry isomer is the predicted cation ground state and it is 4.77 eV above the ground state of the neutral species. The lowest triplet, $^3B_1$, and quintet, $^5A_2$, states of the cation have C_{2v} symmetry and are 0.32 and 1.56 eV above the ground state of the cation.

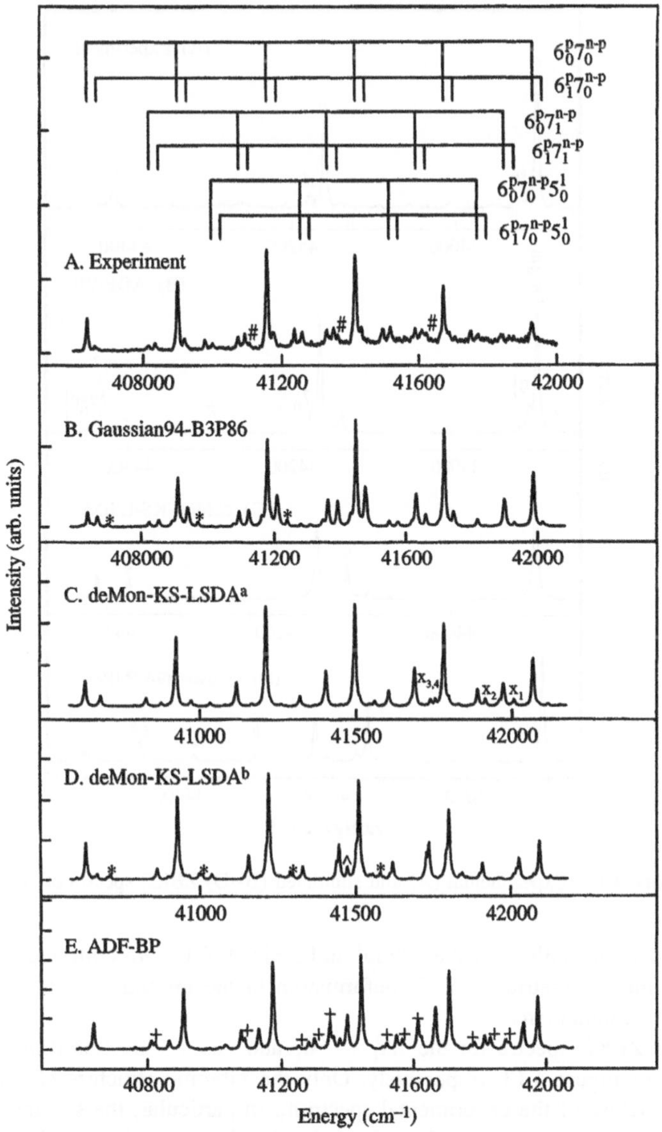

Figure 4.11. Experimental (A) and simulated (B–E) PFI-ZEKE spectra of $Nb_3^{12}C_2$ at 300 K. The spectra were calculated using the trigonal bipyramid geometry from Gaussian94-B3P86 (B), deMon-KS-LSDA (C,D), and ADF-BP (E). In the deMon-KS-LSDA[a] [see Yang *et al.* (1996)] calculations, model core potentials were used for the niobium atoms and all electrons for the carbon. In the deMon-KS-LSDA calculations, model core potentials were used for the niobium and carbon atoms. The spectra were convoluted with a 7 cm^{-1} FWHM Lorentzian line shape to simulate the line width of the experimental bands. Transition energies are reported relative to the energy of the experimental 0–0 transition. The symbol, #, in the experimental spectrum, indicates the features not predicted by theory, while the symbols, *, , +, and $x_{1,2,3,4}$, indicate features not observed by the experiments.

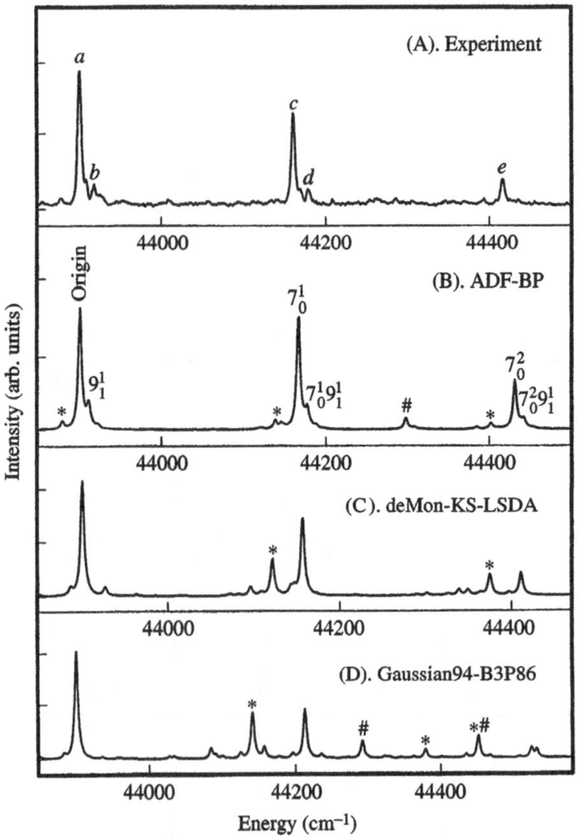

Figure 4.12. Experimental (A) and simulated (B–D) ZEKE spectra of Nb_3N_2.

The calculated wavenumbers of the neutral and ionized Y_3C_2 are collected in Table 4.6. The wavenumber of the symmetric Y–Y–Y deformation in the ground state of the cation is well reproduced by the calculations.

The simulated ZEKE spectra for the $^1A_1' \leftarrow {}^2B_1$ and $^3B_1 \leftarrow {}^2B_1$ transitions are depicted in panels B and C of Figure 4.13, respectively. Only the vibronic structure of the first transition resembles the structure of the experimental spectrum. In particular, the spacing is right and the progressions based on the ν_7 mode and on the ν_9 hot levels are nicely reproduced. However, the calculated progression in the ν_8 mode is too long, indicating an overestimation of the Y–Y–Y angle change upon ionization. The calculated structure of the second transition clearly shows the effect of breaking of the D_{3h} symmetry in the triplet state since two split components of the original e′ mode ν_8 and ν_9 are clearly visible forming rather short FC progressions.

4.5.4 ORGANOMETALLICS: $ZrO(CH_3)_2$

The first organometallic compound studied by a combination of ZEKE spectroscopy and DFT methods is zirconium dimethyl ether, $ZrO(CH_3)_2$ [Yang, Zgierski, and Hackett (unpublished)]. The experimental ZEKE spectrum is shown in the lowest panel of Figure 4.14. The observed ionization potential is 5.28 eV. Three modes can be identified clearly in this spectrum: 248, 450

Table 4.5. Calculated harmonic vibrational wavenumbers (cm^{-1}) and ionization potentials (IP, eV) of Nb_3N_2.

Mode	Symmetry	Type	$Nb_3N_2^+$ ADF-BP	$Nb_3N_2^+$ deMon-KS-LSDA	$Nb_3N_2^+$ Gaussian94-B3P86	Nb_3N_2 ADF-BP	Nb_3N_2 deMon-KS-LSDA	Nb_3N_2 Gaussian94-B3P86
ν_1	a_1	ν_s, Nb–N	773	799	788	743	771	756
ν_2	b_1	ν_a, Nb–N	687	672	694	667	677	666
ν_3	b_2	ν_a, Nb–N	611	648	628	609	640	612
ν_4	a_2	ν_a, Nb–N	466	482	468	449	474	434
ν_5	a_1	ν_s, Nb–Nb	396	400	390	390	401	383
ν_6	b_1	δ_a, Nb_2N_2	318	305	274	310	318	277
ν_7	a_1	δ_s, Nb_3	265	255	311	257	269	255
ν_8	a_1	δ_s, Nb_2N	234	219	239	309	303	328
ν_9	b_2	δ_a, Nb_3	199	193	194	188	193	151
IP						5.60	5.60	6.37

[a] ν_s, symmetric stretch; ν_a, asymmetric stretch; δ_s, symmetric deformation; δ_s, asymmetric deformation.

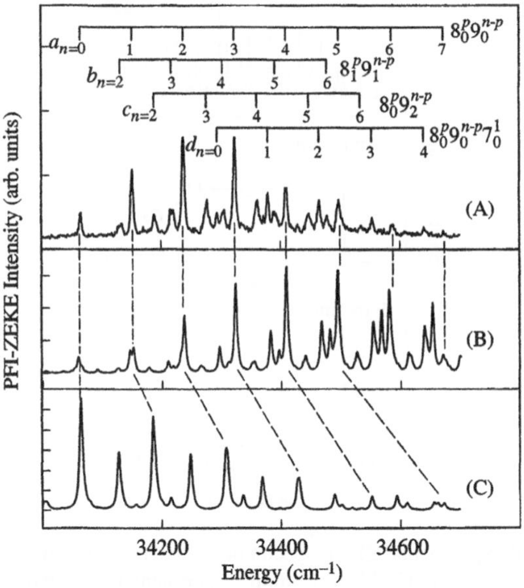

Figure 4.13. PFI-ZEKE spectra of $Y_3^{12}C_2$ (A) and simulations of $^1A_1' \leftarrow {}^2B_1$ (B) and $^3B_1 \leftarrow {}^2B_1$ (C) at 100 K. The simulations were convoluted with a 6 cm^{-1} FWHM Lorentzian line shape to simulate the line width of the experimental bands. Theoretical transition energies are reported relative to the energy of the experimental a_0 band.

and 856 cm^{-1}. They form a very short FC progression which indicates only a small geometry change upon ionization.

The two possible structures of the cluster are an adduct complex and an insertion complex (see Figure 4.14). In the adduct, the Zr atom binds to the oxygen without disrupting any of the CO bonds. In the insertion complex, the Zr atom is inserted between oxygen and carbon. B3P86/LANL2DZ calculations predict [Yang, Zgierski, and Hackett (unpublished)] that the

Table 4.6. B3P86 calculated vibrational wavenumbers[a] (cm^{-1}) for different electronic states of $Y_3^{12}C_2$ ($^2B_1$, $^4E'$) and $Y_3^{12}C_2^+$ ($^1A_1'$, $^3B_1$, $^5A_2$) and isotopic shift (cm^{-1}, in parentheses) from $Y_3^{12}C_2/Y_3^{12}C_2^+$ to $Y_3^{13}C_2/Y_3^{13}C_2^+$.

Mode	Symmetry D_{3h}/C_{2v}	Type	$^2B_1(C_{2v})$	$^4E'(D_{3h})$	$^1A_1'(D_{3h})$	$^3B_1(C_{2v})$	$^5A_2(C_{2v})$
ν_1	e'/b_1	Y–C stre.	594(−20)	632(−21)	604(−20)	676(−24)	836(−33)
ν_2	e'/a_1	Y–C stre.	587(−20)	632(−21)	604(−20)	598(−22)	580(−19)
ν_3	a_1'/a_1	Y–C stre.	522(−18)	575(−21)	557(−20)	529(−18)	519(−21)
ν_4	e''/a_2	Y–C stre.	500(−20)	474(−18)	527(−19)	525(−20)	476(−15)
ν_5	e''/b_2	Y–C stre.	490(−19)	474(−18)	527(−19)	483(−19)	281(−10)
ν_6	a_2''/b_2	YCY def.	379(−14)	380(−14)	418(−15)	378(−14)	248(−10)
ν_7	a_1'/a_1	Y–Y stre.	232(−1)	236(−1)	245(−1)	246(0)	238(0)
ν_8	e'/a_1	Y_3 def.	92(0)	104(0)	87(0)	121(0)	92(0)
ν_9	e'/b_1	Y_3 def.	28(0)	104(0)	87(0)	91(0)	57(0)

Calculated with the Gaussian 94 program.

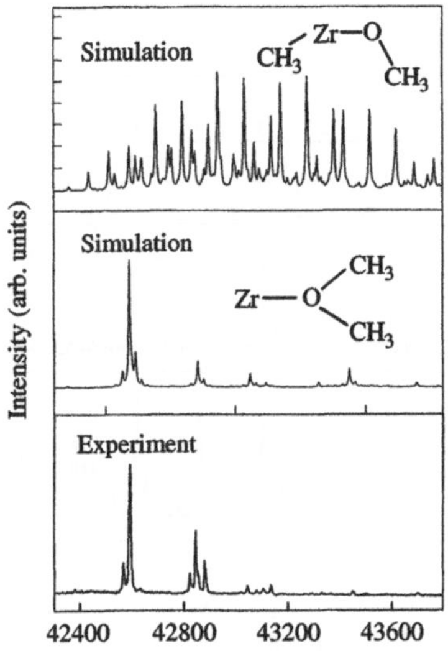

Figure 4.14. Simulated and experimental ZEKE spectra of $ZrO(CH_3)_2$. Top: simulated spectrum of insertion product. Middle: simulated spectrum of adduct. Bottom: experimental PFI-ZEKE spectrum.

insertion complex is 3.44 eV more stable than the adduct. However, the formation of an adduct is a barrierless process, the binding energy of the Zr atom to ether is 1.15 eV on the triplet state potential surface. Calculations show that there is a substantial barrier to formation of the insertion complex from the adduct; it is estimated at 0.58 eV and occurs at the crossing of two potential surfaces correlating with the ground electronic states of the two complexes.

The ground state of the insertion complex, under C_s geometry is the $^3A''$ state, while that of the cation is $^2A'$ with a calculated IP of 6.39 eV. The cation possesses only one totally-symmetric mode near the observed wavenumbers, namely the COZr bend at 239 cm^{-1}. The simulated $^3A'' \rightarrow {}^2A'$ spectrum is characterized by a long progression due to the large geometry change upon ionization. This prediction is inconsistent with the observed spectrum.

The ground state of the adduct (C_{2v} symmetry) is $^3B_1$, while the ground state of the corresponding cation is $^4B_1$ with the calculated IP of 5.54 eV. Three wavenumbers of totally-symmetric modes in the $^4B_1$ cation state match closely the observed ones: 263 cm^{-1}, ZrO stretch, 464 cm^{-1}, COC bend, and 844 cm^{-1}, CO stretch. The simulated vibronic structure of the $^4B_1 \leftarrow {}^3B_1$ transition (middle panel of Figure 4.14) is in an excellent agreement with the experimental spectrum confirming that the adduct is observed in the ZEKE experiment.

This seems to be the first direct observation of an adduct between a transition metal and an organic molecule. Apparently the formation of the complex at low temperature in a jet favors the adduct since no energy barrier has to be overcome.

4.6 Conclusions

The combined DFT and ZEKE spectroscopic studies enables one to determine the structure of transition metal cluster compounds and of a metal–ligand adduct in the gas phase. The theoretical simulated spectrum is very sensitive to the quality of the calculations. We compared several different DFT calculations which all gave similar qualitative structures, vibrational wavenumbers and relative energies. The small quantitative differences in these parameters often translated into more remarkable qualitative differences in the appearance of the simulated ZEKE spectrum. Thus ZEKE simulations provide a very sensitive test to the theoretical results. The positive conclusion is that DFT always predicted the correct qualitative structure; however, the methods were not consistent in the performance and the quantitative differences are sometimes significant. It has varied which method provided the best simulated ZEKE spectrum.

Experimentally the high resolution of the observed ZEKE spectrum, that is, the fully resolved vibrational structure, is the key to the success of this approach. However, without accurate calculations based on first principles it would be almost hopeless to assign the spectrum and to determine structural information from it. Thus this project is a fruitful collaboration between two highly complementary approaches: theoretical simulation and the experimental measurement of ZEKE spectra.

4.7 References

Alex, S., Green, S. M. E., and Leopold, D. G., 1998, unpublished.
Becke, A. D., 1988a, *J. Chem. Phys.* **88**, 1053–1062.
Becke, A. D., 1988b, *J. Chem. Phys.* **88**, 2547–2553.
Becke, A. D., 1988c, *Phys. Rev. A* **38**, 3098–3100.
Becke, A. D., 1993, *J. Chem. Phys.* **98**, 5648–5656.
Chupka, W. A., 1993, *J. Chem. Phys.* **98**, 4520–4530.
Chupka, W. A., 1993, *J. Chem. Phys.* **99**, 5800–5806.
Cockett, M. C. R., Ozeki, H., Okuyama, K., and Kimura, K., 1993, *J. Chem. Phys.* **98**, 7763–7772.
Dai, D. and Balasubramanian, K., 1993, *J. Chem. Phys.* **98**, 7098–7106.
Dai, D., and Balasubramanian, K., 1995, *Chem. Phys. Lett.* **238**, 203–207.
Dietz, T. G., Powers, D. E., Duncan, M. A., and Smalley, R. E., 1981, *J. Chem. Phys.* **74**, 6511–6512.
Doktorov, E. V., Malkin, I. A., and Man'ko, V. I., 1977, *J. Mol. Spectrosc.* **4**, 302–307.
Dushinsky, F., 1937, *Acta Physicochim URSS* **7**, 551–554.
Frisch, M. J., Trucks, G. W., Schlegel, H. B., Gill, P. M. W., Johnson, B. G., Robb, M. A., Cheeseman, J. R., Keith, T., Petersson, G. A., Montgomery, J. A., Raghavachari, K., Al-Laham, M. A., Zakrzewski, V. G., Ortiz, J. V., Foresman, J. B., Peng, C. Y., Ayala, P. Y., Chen, W., Wong, M. W., Andres, J. L., Replogle, E. S., Gomperts, R., Martin, R. L., Fox, D. J., Binkley, J. S., Defrees, D. J., Baker, J., Stewart, J. P., Head-Gordon, M., Gonzalez, C., and Pople, J. A., 1995, *Gaussian 94*, Revision B.3, Gaussian: Pittsburgh, PA.
Hohenberg, P., and Kohn, W., 1964, *Phys. Rev. B* **136**, 864–871.
Knight, L. B., Woodward, R. W., Van Zee, R. J., and Weltner, W., 1983, *J. Chem. Phys.* **79**, 5820–5827.
Kohn, W., and Sham, L. J., 1965, *Phys. Rev. A* **140** 1133–11143.
Lee, C., Yang, W., and Parr, R. G., 1988, *Phys. Rev. B* **37**, 785–791.
Levine, I. N., 1983, *Quantum Chemistry*, 3rd edition; Boston: Allyn and Bacon.

Manneback, C., 1951, Physica Utrecht, **17**, 1001–1010.
Müller-Dethlefs, K., and Schlag, E. W., 1991, *Annu. Rev. Phys. Chem.* **42**, 109–136.
Müller-Dethlefs, K., and Schlag, E. W., 1998, *Angew. Chem. Int. Ed.* **37**, 1346–1374.
Müller-Dethlefs, K., Sander, M., and Schlag, E. W., 1984, *Z. Naturforsch. Teil A* **39**, 1089–1091.
Müller-Dethlefs, K., Schlag, E. W., Grant, E. R., Wang, K, and Mckoy, B. V., 1995, *Adv. Chem. Phys.*, Vol. XC, 1–104.
Nagy, A., 1997, *Adv. in Quantum Chem.*, **29**, 159–178.
Negri, F., and Zgierski, M. Z., 1996, *J. Chem. Phys.* **104**, 3486–3500.
Negri, F., and Zgierski, M. Z., 1997, *J. Chem. Phys.* **107**, 4827–4843.
Perdew, J. P., 1986a, *Phys. Rev. B* **33**, 8822–8824.
Perdew, J. P., 1986b, *Phys. Rev. B* **34**, 7046–7046.
Perdew, J. P., and Yue, W., 1986, *Phys. Rev. B* **33** 8800–8802.
Perdew, J. P., Burke, K., and Wang, Y., 1996, *Phys. Rev. B* **54**, 16533–16543.
Proch, D., and Trickl, T., 1989, *Rev. Sci. Instrum.* **60**, 713–716.
Proynov, E. I., Vela, A., Salahub, D. R., 1995, *Chem. Phys. Lett.* **230**, 419–428.
Roothaan, C. C. J., 1951, *Rev. Mod. Phys.* **23**, 69–89.
Salahub, D. R., Fournier, R., Mlynarski, P., Papai, I., St-Amant, A., and Ushio, J., 1991, in *Density Functional Methods in Chemistry*, Labanowski, J., and Andzelm, J. Eds; Springer: Berlin.
Schlag, E. W., 1998, *ZEKE Spectroscopy*; Cambridge University Press: Cambridge.
Schlag, E. W., and Levine, R. D., 1997, *At. Mol. Phys.* **33**, 159–180.
St-Amant, A., and Salahub, D. R., 1990, *Chem. Phys. Lett.* **169**, 387–397.
Szabo, A., and Ostlund, N. S., 1982, *Modern Quantum Chemistry*; Macmillan: New York.
teVelde, G., 1997, *Amsterdam Density Functional Program Users' Guide*, version 2.3.
Vosko, S. J., Wilk, L., and Nussair, M., 1980, *Can. J. Phys.* **58**, 1200–1211.
Vrakking, M. J. J., Fischer, I., Villeneuve, D. M., and Stolow, A., 1995, *J. Chem. Phys.* **103**, 4538–4550.
Walch, S. P., and Bauschlicher, C. W., 1984, in *Comparison of Ab Initio Quantum Chemistry with Experiment for Small Molecules*, Bartlett, R. J., Ed.; Reidel: Dordrecht, pp. 17–51.
Yang, D-S., Simard, B., Hackett, P. A., Bérces, A., and Zgierski, M. Z., 1996a, *Int. J. Mass Spectrom. Ion Process.*, **159**, 65–74.
Yang, D-S., Zgierski, M. Z., and Hackett, P. A., 1998, *J. Chem. Phys.*, **108**, 3591–3597.
Yang, D-S., Zgierski, M. Z., and Hackett, P. A., unpublished.
Yang, D-S., Zgierski, M. Z., Bérces, A., and Hackett, P. A., unpublished.
Yang, D-S., Zgierski, M. Z., Bérces, A., Hackett, P. A., Martinez, A., and Salahub, D. R., 1997, *Chem. Phys. Lett.* **227**, 71–78.
Yang, D-S., Zgierski, M. Z., Bérces, A., Hackett, P. A., Roy, P-N., Martinez, A. , Carrington Jr., T., Salahub, D. R., Fournier, R., Pang, T., and Chen, C., 1996b, *J. Chem. Phys.* **105**, 10663–10671.
Yang, D-S., Zgierski, M. Z., Rayner, D. M., Hackett, P. A., Martinez, A., Salahub, D. R., Roy, P. N., and Carrington Jr., T., 1995, *J. Chem. Phys.*, **103**, 5335–5342.
Zgierski, M. Z., 1986, *Chem. Phys.* **108**, 61–68.
Zgierski, M. Z., 1998, unpublished results.

5 *AB INITIO* CALCULATIONS OF EXCITED STATE POTENTIAL FUNCTIONS

Robert J. Buenker, Gerhard Hirsch, Yan Li, Jian-ping Gu, Aleksey B. Alekseyev, Heinz-Peter Liebermann
Bergische Universität, Wuppertal, Germany

and

Mineo Kimura
Yamaguchi University, Japan

Computational Molecular Spectroscopy. Edited by Per Jensen and Philip R. Bunker
© 2000 John Wiley & Sons Ltd

5.1 Introduction

When atoms interact with one another their motion is conveniently described in terms of a potential energy function. The clamped-nuclei approximation introduced by Born and Oppenheimer (1927) provides a blueprint of how such entities can be computed in a quantum mechanical formulation [see Slater (1963)]. In essence one defines a Schrödinger equation for electrons moving in the field of nuclei located at fixed positions, and then goes about the task of obtaining solutions, both eigenvalues and eigenvectors, to a suitably high degree of accuracy. This process is then repeated for a series of nuclear conformations covering the coordinate space of interest.

There are many solutions to each electronic Schrödinger equation, however, for ground and various excited states, and one would like to develop computational methods which can obtain a fairly large number of these at a similarly high level of accuracy in each case. The present chapter presents a means of accomplishing this goal with a quite general range of applicability, restricted only by the numbers of electrons and nuclei that need to be considered in a given case. Because of the variation principle there is a tendency to believe that the ground or lowest-energy state of a particular symmetry is fundamentally easier to describe with high accuracy than are its more highly excited states, and for many computational methods this is indeed the case. One technique which has historically enjoyed great success in overcoming the difficulties with describing excited states is configuration interaction (CI), as introduced by Hylleraas (1928, 1930). Accordingly, a many-electron basis of antisymmetrized products of one-electron functions, usually Slater determinants [Slater (1929)] of orthogonal spin orbitals, is employed to construct a matrix representation of the Hamiltonian for a given fixed-nuclei system. A secular equation is then solved to obtain the eigenvalues and eigenvectors of this matrix.

The nature of the electronic Hamiltonian needs to be considered, however, especially when relatively heavy atoms are present, such as antimony, bismuth, lead and various transition metals, lanthanides and actinides, in which case a purely electrostatic nonrelativistic description of the interactions is not sufficiently accurate. Moreover, one should recall that the clamped-nuclei procedure described above is only an approximation and even the concept of potential energy surfaces (PESs) in quantum mechanics represents an oversimplification. Terms are omitted from the global Hamiltonian, in which the coordinates of both the electrons and nuclei are variables and the nuclear kinetic energy operators are included explicitly, which couple the electronic states and lead beyond the simple adiabatic approximation in which nuclei are assumed to move with infinitesimally small velocities along the various potential energy surfaces. Such terms are indispensible in describing the dynamics of atomic and molecular collisions, for example. Particularly instructive in this regard are charge-transfer processes in which the initial channel has an unbalanced electric charge localized on one of the atoms, whereas in the outgoing channel it is removed to a different atom or molecular fragment. The magnitudes of the nonadiabatic couplings between different electronic states determine to a large extent what the reaction probabilities (cross sections) are as well as the temperature dependence of the associated rate constants. When the PESs have barriers or other features of various kinds there is also the possibility of spontaneous nonradiative decay, such as occurs in predissociation phenomena, for example, and also in such cases accurate nonadiabatic couplings between different electronic states are needed in order to obtain a suitably quantitative description of these processes.

In the following it will be shown that it is relatively easy in a CI framework not only to obtain the desired PES of ground and excited electronic states but also to compute the various couplings needed to describe associated dynamical processes. The theoretical aspects described above will be illustrated with many examples, both for diatomic and polyatomic systems, and their relevance to available experimental data will be demonstrated in numerous instances as the true test of the methodology employed.

5.2 Theoretical methods

Within the Born–Oppenheimer method the electronic Hamiltonian $H_{\rm el}$ is defined in atomic units for a set of fixed nuclear coordinates as

$$H_{\rm el} = -\frac{1}{2}\sum_i \nabla_i^2 + \sum_{i,a} \frac{-Z_a}{r_{ia}} + \sum_{i<j} \frac{1}{r_{ij}} + \sum_{a<b} \frac{Z_a Z_b}{R_{ab}}, \tag{1}$$

where the summation indices i and a denote electrons and nuclei, respectively, r_i stands for electronic coordinates and R_a, for nuclear, and Z_a are the nuclear charges. The solutions of the corresponding Schrödinger equation,

$$H_{\rm el}\Psi_i(r, R) = E_i(R)\Psi_i(r, R), \tag{2}$$

are conveniently expanded in terms of Slater determinants D_i, which are antisymmetric products of one-electron functions $u_i(r)$, so that the Pauli principle is automatically satisfied in each case. If the $\{u_i\}$ basis is chosen to be orthonormal (including spin functions), the corresponding Hamiltonian matrix elements can be formed with the help of the Slater–Condon rules [Tinkham (1964)].

The choice of the one-electron basis is critical because only a limited number of such functions can be used in a given application. For this purpose it is convenient to make use of the LCAO–MO method of Hund (1927a, 1927b) and Mulliken (1928a, 1928b, 1929). Accordingly, an atomic orbital (AO) basis is defined for each nuclear center and sometimes also for positions between nuclei (bond functions). In what follows we will simply mention specific examples of such (gaussian) basis sets that have been employed in actual calculations. It is important to have a good representation of the valence electrons (typically an s–p basis) but various high angular momentum functions are also needed to describe polarization effects. One also has to be aware of the existence of special types of states, particularly Rydberg (large-orbit) species, for which relatively diffuse functions (with small exponents) are needed [Buenker, Hirsch and Li (1999)]. Negative-ion states are also known to require small-exponent (semi-diffuse) functions for an adequate description of their electronic structure.

In order to form the $\{u_i\}$ orthonormal basis itself, there are several possibilities. Perhaps the most widely used approach for this purpose is the SCF or Hartree–Fock method [Fock (1930), and Slater (1930)]. The LCAO–MO expansion coefficients are optimized so as to give a minimum in the total energy of a given single determinant, closed- or open-shell, employing methods introduced by Roothaan (1951; 1960) and Hall (1951). A more accurate starting point can be achieved by using various multiconfiguration (MC-SCF) methods, of which one of the most popular is the state-averaged complete active space (CAS-SCF) method [see Shepard (1987) and references cited therein]. In each case the idea is the same, namely to optimize the AO coefficients in accordance with the variation principle until an energy minimum is reached for a given state or mixture of states.

A key fact which needs to be considered in this connection is that there is a direct relationship between the quality of the CI treatment and its sensitivity to the choice of the one-electron functions $\{u_i\}$ which can be formed from a given AO basis. Ideally, if a full CI treatment is carried out, in which all combinations of the $\{u_i\}$ spanning a given AO basis are employed to construct Slater determinants, the actual choice of the $\{u_i\}$ functions themselves is totally immaterial in arriving at the final results of the calculations since they only differ from one another by a unitary transformation. Since such a treatment is almost always impractical, however, the choice of one-electron basis is important, but since the goal must be to come as close as possible to the results which would be obtained at the full CI limit, truncating the CI to the extent that it becomes quite sensitive to the actual choice of $\{u_i\}$ may well mean that the treatment as a whole is inadequate for most purposes. A useful internal check of such calculations involves carrying out the CI for different choices of $\{u_i\}$ for the same AO basis. The results should be nearly independent of the one-electron basis when a high-quality CI has been carried out. The above discussion places the emphasis on a given method's capacity to effect a CI treatment which closely approaches the results at the full CI limit for all electronic states of interest in a given application.

A multireference (MR) single- and double-excitation CI goes a long way toward satisfying the above requirement. This topic has recently been reviewed [Buenker and Krebs (1999)] and so only a brief summary will be given in this work. The MR-CI method affords a great deal of flexibility in the types of electronic states that can be treated simultaneously in the same secular equation. One simply has to choose as reference species all configurations which make major contributions to any one of the electronic states of interest. For continuity of the calculated PES it is highly desirable that the same set of reference configurations be employed for all nuclear conformations. It is generally necessary to run a small set of preliminary calculations in order to arrive at a consistent choice. Wherever possible symmetry should be used to simplify the calculations, because states transforming according to different irreducible representations do not interact and thus can be treated in separate secular equations. For various technical reasons molecular CI calculations are almost always carried out in abelian subgroups of the molecular point group [Buenker and Peyerimhoff (1968)], however.

One of the most basic requirements in calculations of electronically excited states is the ability to accurately describe avoided crossings of neighboring PESs. More than anywhere else, a balanced treatment of different kinds of electronic states is desired to achieve this goal. For many-electron systems it is unfortunately not possible to obtain better than 0.1 eV accuracy for the relative spacings of such states on a general basis. One can certainly do better when the states are quite similar in character, such as for a series of Rydberg states with the same core [Buenker, Hirsch and Li (1999)], but one can improve the theoretical treatment only so far for other less optimum cases, and it is important to be aware of such restrictions when interpreting the results of such calculations.

Avoided crossings can have important effects which cannot be derived from the PES alone. They are often an indication of a breakdown in the Born–Oppenheimer or adiabatic approximation itself. Solution of the electronic Schrödinger equation omits coupling interactions between the various states and it is therefore wise to obtain at least a rough idea of how significant such terms are in a given case. There are three types of nonadiabatic couplings which commonly play a role in spectroscopy or dynamics: radial, rotational and spin–orbit. In the first case the electronic wave functions need to be employed to obtain average values of first and second derivatives with respect to vibrational coordinates such as for the bond stretching motion in diatomic molecules. In the present work a numerical differentiation method is employed which was introduced by Galloy and Lorquet (1977) to compute such integrals [Hirsch *et al.* (1980), and Buenker *et al.* (1982)]. Analytical methods are also available for this purpose [Yarkony (1995); see also Yarkony's contribution to this volume] and explicit comparisons between the two types of results have indicated a high level of agreement between them [Neuheuser, Sukumar and Peyerimhoff (1995)]. Such matrix elements tend to be relatively large in the neighborhood of avoided crossings but they can also be quite important even when the PESs of neighboring states appear to be repelled by one another. All that is required is that the respective electronic wave functions change character as a function of the vibrational coordinate in order for the corresponding derivative matrix elements to be large enough to cause significant nonadiabatic coupling.

In spectroscopic studies it is possible to work in the adiabatic representation by including radial-type couplings explicitly in order to account for vibronic mixing. In simple terms this means that a given rovibrational level cannot be assigned unambiguously to a definite electronic state, but rather must be described as a linear combination involving two or more such states. Radial coupling can also be the cause of nonradiative transitions such as occur in predissociation. They also play an important role in atomic and molecular collisions, in which the radial couplings are often responsible for the system jumping from one channel to another, depending on the relative velocity of the collision partners. Rotational coupling can also produce significant effects in each of the above applications, particularly when symmetry eliminates direct competition with a radial coupling mechanism, for example, in promoting transitions between Σ and Π states in diatomics. These matrix elements are invariably calculated analytically and are also needed for studying magnetic effects or certain dipole-forbidden transitions.

Spin–orbit coupling is important for transitions or reactions involving states of different and in some cases also equal multiplicity. Strictly speaking it is not a nonadiabatic effect but simply a relativistic adjunct to the electronic Hamiltonian. When such terms are included in the treatment adiabatic PESs are obtained in the usual way, but the electronic states to which they correspond are not eigenfunctions of S^2. For light elements an effective means of proceeding is to ignore the spin–orbit effects at an early stage of the calculations and thus to obtain intermediate wave functions for which spin is still a good quantum number. The resulting states can then be coupled via the spin–orbit interaction, and so only in this computational sense are the mixing effects which ensue nonadiabatic. In the present work relativistic effects of various kinds are described in a relatively simple manner through the use of relativistic effective core potentials (RECPs), and such techniques have been shown to be reliable down to the sixth row of the Periodic Table (e.g. for molecules containing atoms such as Pt, Au or Bi) and even beyond.

Finally, it is also important to consider radiative transitions and for this purpose it is necessary to calculate matrix elements for the electric-dipole operator between different electronic states. Clearly all of the quantities mentioned above, not only the potential energy itself, are described by many-dimensional hypersurfaces, each of which is to be calculated point by point according to the Born–Oppenheimer computational scheme. There are thus many uses for the CI wave functions other than to produce the PESs themselves, although in this chapter we will be concentrating mostly on the latter. We turn now to specific examples in which the theoretical methods outlined above have been applied to systems of direct experimental interest.

5.3 Diatomic molecular potentials

The first examples of computed potential curves to be considered are for various diatomic systems. We will begin with applications dealing with collisions of either hydrogen or helium atoms or their ions with heavier atoms or ions. Then a series of calculations for the O_2 and CO molecules will be considered for which the main interest is their spontaneous decay via predissociation. Finally, attention will be directed to diatomics which contain at least one heavy atom such as antimony or bismuth, in which case relativistic interactions have an important effect on the shapes of the potential energy curves.

5.3.1 ATOMIC COLLISIONS AND CHARGE-TRANSFER PROCESSES

The first step in predicting scattering cross sections for atomic collisions is to produce a potential energy diagram for a number of electronic states for the corresponding diatomic molecule. The HeC^+ system is a case in point. Since He^+ ions are produced in large numbers by various radiative processes in the interstellar region, there is the possibility that collisions between them and other atoms such as carbon, oxygen and nitrogen result in the transfer of an electron because of a change in electronic state as we will see below. The adiabatic potential energy curves shown in Figure 5.1 have been calculated with a carbon atom (9s5p1d)/[9s2p1d] contracted basis supplemented with various Rydberg functions [Kimura *et al.* (1994a)]. The corresponding He atom basis is (9s3p1d) contracted to [7s3p1d]. The labelling of the states (B, D and E) is based on their respective dissociation limit. The MR-CI treatment is carried out with configuration selection [Buenker and Peyerimhoff (1974, 1975), and Buenker (1986)] with a threshold of $T = 5.0\,\mu E_h$ and employs the Table CI method [Buenker (1982), and Buenker and Phillips (1985)] for the calculation of Hamiltonian matrix elements.

The initial channel is $He^+(^2S) + C(^3P)$, which produces the $D\,^2\Pi$ and $D\,^2\Sigma^-$ states. Below this lies the charge-transfer asymptote, $He(^1S) + C^+(^2D)$, which corresponds to three molecular states, only the $B\,^2\Delta$ and $B\,^2\Pi$ of which are shown in Figure 5.1. The lowest-energy combination for this system (limit A) is $He(^1S) + C^+(^2P)$, but it is of minor interest in the present application. Instead the metastable channel with an excited $C^+(^2P)$ state is important and its molecular potential curves ($E\,^2\Sigma^-$ and $E\,^2\Pi$) are also shown in Figure 5.1. It can be noted that the B, D and $E\,^2\Pi$ states undergo avoided crossings with one another. As a result there are strong radial couplings between the D and E states and between the D and B states (at slightly smaller R values). Plots of these matrix elements can be found in Figure 5.2 of Kimura *et al.* (1994a). These couplings control the nonradiative electron transfer at relatively high collision energies. The initial $D\,^2\Sigma^-$ crosses the $B\,^2\Pi$ at $R = 2.6\,a_0$ and these states are connected by rotational coupling, which dominates at lower energies. Above 5 eV, the radial coupling of the $E\,^2\Pi$ to $D\,^2\Pi$ takes over. The computed

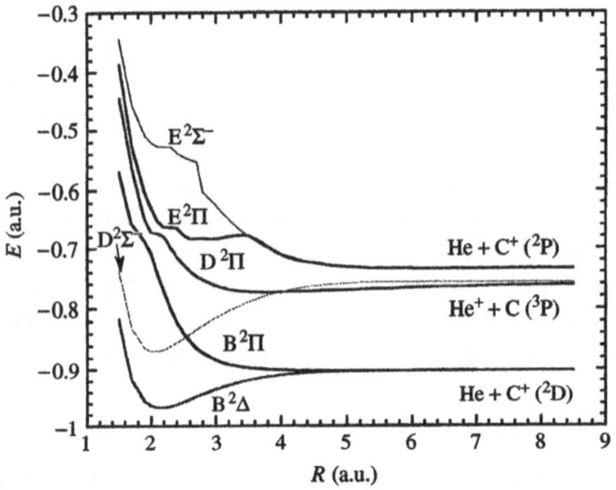

Figure 5.1. Computed adiabatic potential energy curves of HeC^+.

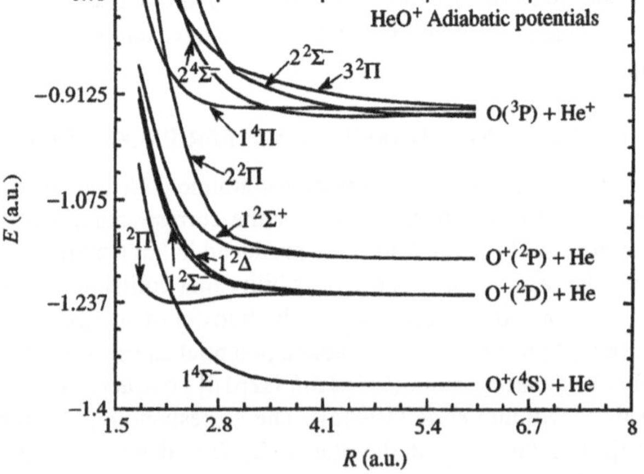

Figure 5.2. Computed adiabatic potential energy curves of HeO^+.

cross sections [Kimura *et al.* (1994a)] for nonradiative charge transfer can therefore be understood to a large extent just by analyzing the potential energy diagram in Figure 5.1, although one must also keep in mind the nature of the nonadiabatic coupling elements between neighboring states in order to have a reasonably complete picture.

The corresponding PES diagram for the HeO^+ system is shown in Figure 5.2. The most interesting process in this case is electron capture from one of the initial He/O^+ states to form $O(^3P) + He^+$. The ground state of O^+ is $^4S^o$ but there are also two low-lying metastable states, $^2D^o$ and $^2P^o$, similarly as for the nitrogen atom [Kimura *et al.* (1994b, 1994c)]. It can be seen from Figure 5.2 that the molecular potential curve of the $^4\Sigma^-$ ground state only rises to the height of the $2\,^4\Sigma^-$ charge-transfer state at very short R values, so the corresponding reaction cross section is expected to be relatively weak, especially at low collision energy. By contrast the $O^+(^2P^o)$ state correlates with a $2\,^2\Pi$ molecular potential curve which undergoes an avoided crossing with the $3\,^2\Pi$ coming from the electron-capture channel near $2.0\,a_0$. Computed radial couplings show that this is a strong interaction [see Figure 5.2a of Kimura *et al.* (1994c)].

On the basis of these data scattering cross sections have been computed at various collision energies [Kimura *et al.* (1994c)] and the results for all three O^+ initial channels are given in Table 5.1. As expected, the $^2P^o$ process dominates at low energy, with $^2D^o$ being 100 times weaker and $^4S^o$ much weaker still. At higher collision energies near 10 keV, however, these distinctions in reaction cross sections become notably smaller.

These results helped to solve an interesting question which arose when two different groups measured these quantities experimentally. In both cases O^+ ions were generated via electron impact ionization. Kusakabe *et al.* (1990) used CO_2 for this purpose and they claimed to have generated $^4S^o$ ground state ions with 21.0 and 24.5 eV electron beams, but a mixture of metastable states with 150 eV electrons. Wolfrum, Schweinzer and Winter (1992) used a similar technique on water with 30, 40 and 130 eV electrons. The measured electron capture cross sections are compared in Figure 5.3 with the theoretical predictions. The lowest-energy results of Kusakabe *et al.* (1990) agree quite well with the calculated cross sections for the $^4S^o$ process, which decrease rather strongly toward lower collision energies. The results obtained by Wolfrum, Schweinzer and Winter (1992) are larger and much less dependent on collision energy and in fact agree quite well with the results of Kusakabe *et al.* (1990) for O^+ ions generated with 150 eV electrons. On this basis Wolfrum, Schweinzer and Winter (1992) concluded that only $^4S^o$ ground state HeO^+ ions are generated at all electron impact energies studied, whereas Kusakabe *et al.* (1990) argued that the measured cross sections corresponding to low-energy electron impact production of O^+ are due to the presence of only $^4S^o$, but that the larger cross sections obtained by both groups were evidence of a mixture of both metastable and ground state species.

Table 5.1. Electron capture cross sections for three initial $He + O^+$ channels. The numbers in brackets denote multiplicative powers of ten.

E(keV)	Cross section (cm^2)		
	$O^+(^4S^o) + He$	$O^+(^2D^o) + He$	$O^+(^2P^o) + He$
0.16	<10 [−20]	2.61 [−18]	2.58 [−16]
0.36	5.62 [−20]	6.62 [−18]	1.71 [−16]
0.64	1.44 [−19]	7.41 [−18]	1.55 [−16]
1.00	5.32 [−19]	1.19 [−17]	1.21 [−16]
1.44	2.88 [−18]	8.30 [−17]	1.15 [−16]
2.56	2.76 [−18]	1.07 [−16]	1.87 [−16]
4.00	1.27 [−17]	2.08 [−16]	3.95 [−16]
9.00	8.92 [−17]	2.68 [−16]	2.17 [−15]

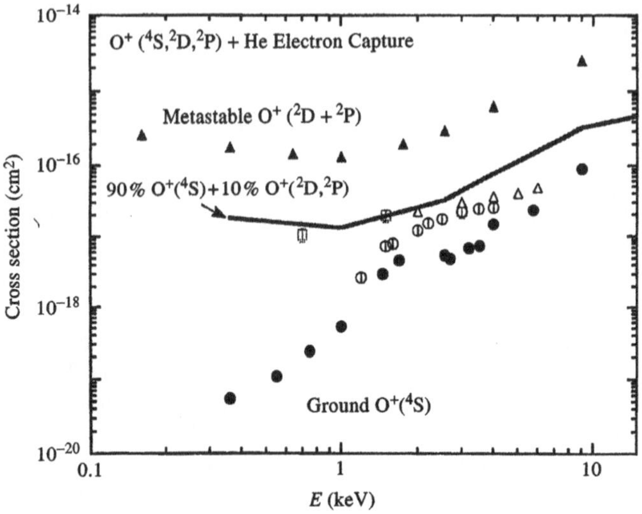

Figure 5.3. Electron capture cross sections for He + O^+ above 100 eV. Calculated: ●, $O^+(^4S)$; ▲, $O^+(^2D, {}^2P)$. Experiment: ○, $O^+(^4S)$; □, capture by mixed ground and metastable $O^+(^2D, {}^2P)$ [Kusakabe *et al.* (1990), and Wolfrum, Schweinzer and Winter (1992)].

The calculations fit in much better with the interpretation of Kusakabe *et al.* (1990). It is possible to obtain a good fit of all the measured cross sections except those obtained using O^+ ions generated with electron impact energies less than 25 eV by taking a mixture of 90 % $^4S^o$ and 10 % ($^2D^o$, $^2P^o$) computed cross sections, for example. Removing the metastable ions from the calculations produces results which are considerably lower, especially below 2 keV, thus bringing the theoretical data into satisfactory agreement with those measured by Kusakabe *et al.* (1990) when they used 21.0 and 24.5 eV electrons to generate the O^+ ions. Furthermore, it should be noted that Hughes and Tiernan (1971) have reported appearance potentials for the $^2D^o$ and $^2P^o$ O^+ ions of 26.5 and 28.3 eV, respectively. On this basis one can conclude that metastable ions were present in all the experiments carried out by Wolfrum, Schweinzer and Winter (1992). As long as attention is restricted to the results of Kusakabe *et al.* (1990) for 150 eV electron impact ionization, there is acceptable agreement between the two sets of experimental results, but discrepancies are noted when electron impact energies are used which are below the above appearance potentials, at which point presumably only $^4S^o$ O^+ ions are present.

Calculations were also carried out for lower collision energies in the meV to eV range for which the radial coupling mechanism is ineffective. Instead charge transfer occurs primarily via the $O^+(^2P^o)$ state by virtue of the spin–orbit interaction between the $1\,^4\Pi$ molecular state emanating from $O(^3P)$ and the $2\,^2\Pi$ of $O^+(^2P^o)$ which undergo a curve crossing near $R = 2.1\,a_0$ (see Figure 5.2). Spin–orbit effects also drive excitation and deexcitation processes between the $^4S^o$ and $^2D^o$ states of the O^+ ion. In this case the pertinent molecular states are $1\,^4\Sigma^-$ and $1\,^2\Pi$, whose potential curves are seen to cross near $R = 2.2\,a_0$. Both of these spin–orbit matrix elements are relatively large (60–70 cm^{-1}) between 3 and 2 a_0 [see Figure 5.2b of Kimura *et al.* (1994c)], but they fall rapidly to zero at larger bond distances. This is because the excitation–deexcitation processes correspond to a dipole-forbidden atomic transition on the one hand, and charge transfer by definition involves states in which the charge imbalance is localized on different atoms on the other (causing the effectively one-electron spin–orbit operator to have vanishing matrix elements at infinite nuclear separation). Earlier calculations of these quantities had assumed constant values

[Augustin *et al.* (1973)] and thus found a rather different energy dependence for the corresponding cross sections than in the present study. Finally, there is a threshold of 1 eV for the electron capture process [Kimura *et al.* (1994c)]. This can be understood from the potential curves in Figure 5.2, which show that the $^4\Pi$ energy must rise by this amount before intersecting the $2\,^2\Pi$ curve which leads down to the O^+ ($^2P^o$) product.

Proton collisions with a variety of atoms have also been studied with the above methods. A potential energy diagram for the HSi^+ system is shown in Figure 5.4. The lowest H^+/Si channel lies fairly high and as a result the ground $H(^2S) + Si^+(^2P^o)$ products cannot be readily accessed because they are considerably more stable. Below 1 eV the calculations [Kimura *et al.* (1997)] indicate that $Si^+(^2D^o)$ is the most probable electron-capture channel from the initial $H^+/Si(^3P)$ state. At 10 eV and beyond there is a clear preference for $Si^+(^4P^o)$ formation, with cross sections as high as 10^{-14} cm^2 having been calculated at 100 eV collision energy. Above 300 eV the most favored products are $H + Si^+(^2P^o)$, indicating that the large associated energy defect is no longer such a critical factor at high energy. Electron-capture cross sections from the $Si(^1D)$ atom are larger than for $Si(^3P)$ at all collision energies studied, but such processes dominate by over an order of magnitude at 1 keV. There has also been a study of SiH^+ photodissociation [Stancil *et al.* (1997)] based on the potentials of Figure 5.4. Potential curves and radial couplings for the CH^+, NH^+ and OH^+ molecules may also be found in Kimura *et al.* (1997). A comparison of different methods for computing scattering cross sections has also been made based on the CH^+ potentials and couplings [Stancil *et al.* (1998)].

Finally, potential curves have recently been obtained for a number of doubly charged systems. Of particular interest in the CH^{2+} system [Gu *et al.* (1998)] is a strongly avoided crossing between its lowest two $^2\Sigma^+$ states (see Figure 5.2 of this reference). The ground state of this symmetry dissociates to $H^+ + C^+(^2P^o)$ and thus has a repulsive Coulomb potential at large R. The $2\,^2\Sigma^+$ state correlates with $H + C^{2+}$, however, and there is thus some polarization of the H atom's electronic charge even at large R values. The calculated potentials indicate a minimum near

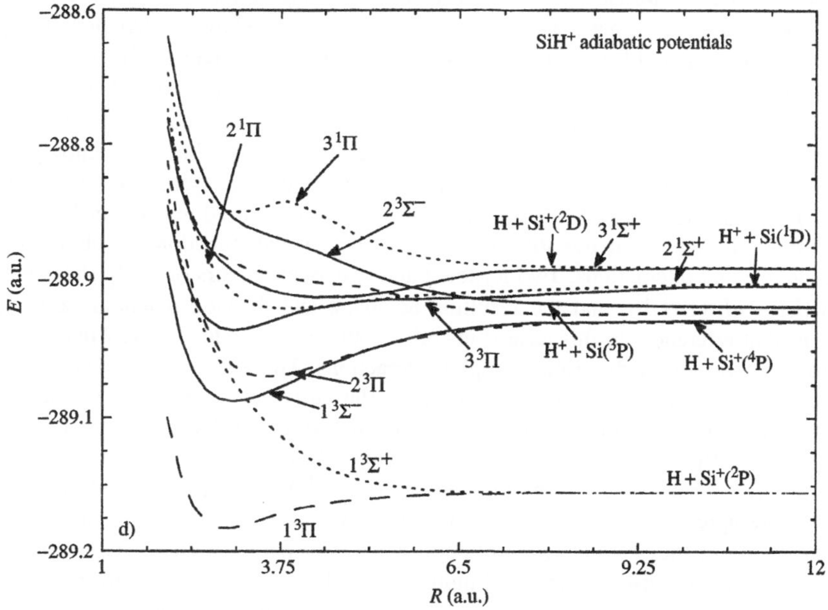

Figure 5.4. Computed adiabatic potential energy curves of SiH^+.

$6.0\,a_0$ for the $2\,^2\Sigma^+$ state, with a well depth of 0.14 eV. Its potential curve is very flat in the neighborhood of the minimum and the corresponding zero-point energy is calculated to be only $180\,\mathrm{cm}^{-1}$. The indication is that several vibrational levels can be held by this well. They are quite stable, however, with computed linewidths of less than $10^{-6}\,\mathrm{cm}^{-1}$. Several AO basis sets have been compared to test these results and it has been found that the well depth only varies by about 0.01 eV from one treatment to the other. There are conflicting experimental data regarding this point [Ast *et al.* (1981); Mathur and Badrinathan (1989); Wetzel *et al.* (1993); Koch *et al.* (1987)], so it is hoped that the present calculations will stimulate further attempts to settle this question.

5.3.2 PREDISSOCIATION IN O_2 AND CO

CI treatments of diatomic molecules with both atoms heavier than He are noticeably more difficult to carry out than those discussed in the last subsection, simply because the number of valence electrons involved is roughly doubled in each case. Nonetheless, systems such as O_2, N_2 and CO can be described with high accuracy with these methods. One of the most sensitive tests for such calculations is predissociation or other nonradiative phenomena. The calculated values for the lifetimes (or linewidths) of resonances depend on a number of key factors such as the height and width of pertinent potential energy barriers and the nature of avoided curve crossings.

There are several interesting examples of such nonradiative transitions in the spectrum of the O_2 molecule. For example, the discrete bands discovered by Tanaka (1952) in the Schumann–Runge continuum are resonances. Lewis *et al.* (1988a, 1988b) have measured photoabsorption cross sections for different O_2 isotopomers and have reported both the energy locations and linewidths of these three levels. To describe these features theoretically potential energy calculations have been carried out at the MR-CI level for three states of $^3\Sigma_u^-$ symmetry, two of Rydberg type and the other, a valence state of π–π^* character which corresponds to the Schumann–Runge upper state. A diabatic transformation was performed on the CI wavefunctions to obtain states of essentially pure Rydberg and valence character and therefore eliminate the avoided crossings in the adiabatic potentials [see Figure 5.1 of Li *et al.* (1992)]. The complex-scaling technique [Balslev and Combes (1971), and Moiseyev (1981)] was used to compute the associated resonances and good agreement with the experimental findings was obtained. Optical oscillator strengths (f values) were also computed for the transitions from the O_2 $X\,^3\Sigma_g^-$ state on this basis.

A somewhat more complicated example of this type is presented by the $^{3,1}\Pi_g$ manifold of O_2 electronic states. Again there is a series of avoided crossings between Rydberg and valence states responsible for the observed phenomena. The REMPI technique has been employed by Sur *et al.* (1986) and Johnson, Long and Hudgens (1987) to study this system and van der Zande *et al.* (1987) and van der Zande, Koot and Los (1988, 1989) have used translational spectroscopy for this purpose. It was again possible to diabatize the CI wavefunctions to obtain a simpler picture of the relevant interactions. Two Rydberg and two valence pairs of $^{3,1}\Pi_g$ states have been calculated [Li *et al.* (1997b)] and the resulting adiabatic and diabatic potentials are shown in Figure 5.1 of that reference. In this case spin–orbit coupling needs to be taken into account for a satisfactory description of the triplet–singlet interactions. There is a bound $^1\Pi_g$ valence state in the problem and the relative position of its potential curve to those of the various Rydberg states must be known accurately in order to obtain satisfactory results for energy locations and linewidths of the perturbed levels. It was necessary to shift this potential upward by $1100\,\mathrm{cm}^{-1}$ so that the difference between its asymptotic energy (^{1}D + ^{1}D) and the computed location of the $C\,^3\Pi_g v = 0$ state agrees with experiment. It was then moved by $0.038\,a_0$ to a smaller R value so that the location of the original crossing point of the two potential curves is unchanged. The theoretical C–X T_0 value itself is obtained quite accurately, only $17\,\mathrm{cm}^{-1}$ in error.

Use of the above potentials and coupling elements in a complex-scaling treatment [Li *et al.* (1997a)] with 2600 vibrational basis functions succeeded in obtaining good quantitative agreement with measured energy level locations and linewidths for the $C\,^3\Pi_g$ and $d\,^1\Pi_g$ states of both $^{16}O_2$ and $^{18}O_2$ isotopomers, as well as explaining trends in these data that had been reported by the experimentalists [Sur *et al.* (1986), Johnson, Long and Hudgens (1987), van der Zande *et al.* (1987), and van der Zande, Koot and Los (1988, 1989)]. For example, it was found that the F_1–F_2 energy separation was consistently smaller than that for F_2–F_3 and this was shown in the calculations to be caused by the fact that the Ω-order of the triplet states is different for the Rydberg and valence states because of the different occupations of the π^* MO in each case. This effect causes the potential curves of the Rydberg $C\,^3\Pi_g$ Ω components to cross the corresponding valence (diabatic) state at notably different R values (longest for F_1, shortest for F_3), hence producing different mixtures of these states in corresponding vibrational levels. Another important effect of this different order of Ω states shows up in the measured linewidths of the corresponding levels (F_1 is typically larger than F_3 by a factor of two). The calculations also demonstrate that there is a double well in the $d\,^1\Pi_g$ potential, again caused by the interaction of a Rydberg and a valence state. This effect shows up in some unusual variations of the rotational constant with v and J in the $d\,^1\Pi_g$ energy levels (Table 5.2).

Another interesting Rydberg–valence interaction occurs for the $^{3,1}\Pi_u$ states of O_2. The calculated CI potential curves for four $^{3,1}\Pi_u$ pairs of states [Li, Hirsch and Buenker (1998)] are shown in Figure 5.5. The spin–orbit interaction has again been included in the calculation of vibrational levels. Recently England *et al.* (1995) and England, Lewis and Ginter (1995) have reported measured values for the level locations and lifetimes of the (4pσ) $^{3,1}\Pi_u$ states, as well as rotational constants and line intensities. The complex-scaling technique was employed in a diabatic representation in order to obtain theoretical predictions of all of these quantities. The T_0 value for the (4pσ) $^3\Pi_u$–X transition is calculated in this treatment to be $800\,cm^{-1}$ in error. The computed energy spacings between corresponding vibrational levels are shown in Table 5.3, along with values of the linewidths. Experimental values are available [England *et al.* (1995), and England, Lewis and Ginter (1995)] for the $\Omega = 0$ and 2 levels of the (4pσ) $^3\Pi_u$ and these are also included in Table 5.3. There is generally good agreement between calculation and experiment in each case. The level spacings are quite irregular, but the observed trends are mirrored in the computed results. The corresponding linewidths are computed to be too large by roughly a factor of two in most cases, but the trends in the measured data are described quite satisfactorily.

The $v = 0$ level has the smallest rotational constant of any of the observed levels for the $^3\Pi_u$ state because it is localized at the outer minimum of this adiabatic potential (Figure 5.5), while $v = 1$ is located in the inner minimum. England *et al.* (1995) and England, Lewis and Ginter (1995) report B_v values of 1.394 ± 0.005 and $1.562 \pm 0.005\,cm^{-1}$, respectively, as compared with calculated values [Li, Hirsch and Buenker (1998)] of 1.397 and $1.542\,cm^{-1}$. Both these vibrational states are strongly influenced by the avoided crossing. There are also measured oscillator strengths [England *et al.* (1995), and England, Lewis and Ginter (1995)] for the (4pσ) $^3\Pi_u$–X vibrational

Table 5.2. Calculated and experimental rotational constants B_{vJ} (in cm^{-1}) of the $v = 0$–$4/J = 1$–15 levels of the $d\,^1\Pi_g(3s\sigma)$ state of $^{16}O_2$.

v	Calculated				Experimental[a]
	$J = 1$	$J = 5$	$J = 10$	$J = 15$	
0	1.66	1.66	1.66	1.66	1.68
1	1.63	1.63	1.63	1.62	1.58
2	1.28	1.22	1.07	0.95	
3	1.00	0.96	0.85	0.76	
4	1.56	1.56	1.56	1.56	1.59

[a] Sur, Friedman and Miller (1991).

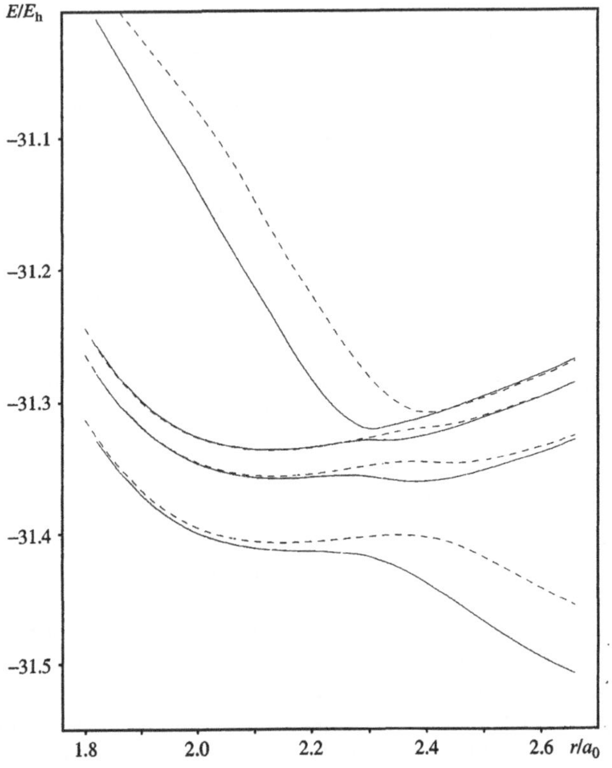

Figure 5.5. Computed adiabatic potential energy curves of four $^3\Pi_u$ (solid lines) and four $^1\Pi_u$ (dashed lines) states of the O_2 molecule.

Table 5.3. Computed energy spacings $\Delta G(v)$ between the vibrational levels $v-1$ and v of the (4pσ) $^{3,1}\Pi_u$ states of $^{16}O_2$ as well as predissociation linewidths $\Gamma(v)$. Experimental results [England *et al.* (1995), and England, Lewis and Ginter (1995)] are given in parentheses; all values are in cm^{-1} [a].

v	$\Delta G(v)$				$\Gamma(v)$		
	$^3\Pi_{0u}$	$^3\Pi_{2u}$	$^3\Pi_{1u}$	$^1\Pi_u$	$^3\Pi_{0,2u}$	$^3\Pi_{1u}$	$^1\Pi_u$
0					3.6 (2.2 ± 0.1)	3.6	0.01
1	656 (750)	661 (756)	647	1496	3.3 (1.6 ± 0.1)	3.4	0.18
2	1246 (1252)	1244 (1248)	1252	870	12.2 (6.4 ± 0.8)	12.0	1.9
3	1166 (1168)	1165 (1168)	1165	1064	2.6 (2.3 ± 0.3)	2.6	1.8
4	1302 (1356)	1302 (1359)	1263		0.003	0.6	
5	1313 (1301)	1310 (1297)			6.0		

[a]For the $v=0$ level the $^3\Pi_{0u}$–$^3\Pi_{2u}$ spin–orbit splitting is calculated to be 175 cm^{-1}, and the energy difference between the $v=0$, $J=1$ rovibrational levels of the $^3\Pi_{1u}$ and $^1\Pi_u$ states is 1020 cm^{-1}.

transitions. The corresponding calculated values are in good agreement with these results, despite the fact that these transitions are quite weak. The $v'=1$ f value is the largest among the (0–3,0) results, both in the calculations and in experiment, with a value of 6.2×10^{-4} calculated and $14.2\pm0.7\times10^{-4}$ measured. For the corresponding $v'=0$ result the agreement is better (4.2×10^{-4} versus $5.8\pm0.3\times10^{-4}$ measured). The corresponding f values for the spin-forbidden

$^1\Pi_u$–X transition fall in the 1.0×10^{-5} to 4.0×10^{-7} range in the calculations but no measured values have been reported.

There have been no measurements of $\Omega = 1$ vibrational levels of the $^3\Pi_u$ state but the $v = 0$, $J = 1$ level of the (4pσ) $^1\Pi_u$ state with which it can interact (Figure 5.5) has been characterized experimentally [England *et al.* (1995), and England, Lewis and Ginter (1995)]. Computed results for the vibrational spacings and linewidths of both electronic states are also contained in Table 5.3. The calculated energy difference between the respective $v = 0$ levels of these two states is 1020 cm^{-1}, which is consistent with the value of ca 1100 cm^{-1} estimated from the observed data. It is hoped that these theoretical findings will aid in future experimental work. The calculated linewidth for the singlet's $v = 0$ level is 0.01 cm^{-1}, which is again consistent with the observed value of <0.1 cm^{-1}. The spin–orbit interaction is expected to have an important influence on the energy gap between respective rovibrational levels of the $^3\Pi_{1u}$ and $^1\Pi_u$ states. The calculations indicate that both the $v = 1$ level of the triplet and the $v = 0$ level of the singlet are located in the inner minimum of their respective PE curves (Figure 5.5) and so there is a large overlap of their vibrational wave functions. Test calculations indicate that the gap between these two levels is lowered from 372 to 342 cm^{-1} when spin–orbit coupling is ignored. This energy separation is about four times that of the spin–orbit matrix element itself. The relatively narrow energy gap and large Franck–Condon factor enhance the spin–orbit effect and cause the $v = 0\,^1\Pi_u$ level to borrow intensity from the $^3\Pi_{1u}$–X$\,^3\Sigma_g^-$ transition. Calculations show that the $v = 0$ singlet level has a transition probability from the O_2 ground state which is 40 times larger than that for $v = 1$ and six times larger than that for $v = 2$. As a result it is quite difficult to detect the latter two levels in absorption studies. There is also a relatively strong interaction between the $v = 4$ triplet level and the $v = 3$ singlet level. The computed energy gap between them is only 122 cm^{-1}, about the same as the spin–orbit matrix element itself. The observed line broadening of the $v = 4$ level of the $^3\Pi_{1u}$ state is thus consistent with its predissociation by the $v = 3$ singlet state as a result of spin–orbit coupling. Finally, it was found that missing lines for the $v = 0$, $J > 1$ levels in the absorption spectrum of the singlet state are probably due to the *L*-uncoupling interaction between this state and the neighboring $^1\Delta_u$ states. Potential curves have also been calculated for the latter states [Li, Hirsch and Buenker (1998)] as well as the corresponding orbital angular momentum matrix elements and these results give strong support to the above conclusion.

The lowest Rydberg states of the CO molecule also provide interesting examples of predissociation effects in diatomic spectra. The $\pi \rightarrow$ 3s and $\pi \rightarrow$ 3pσ B and C states are bound Rydberg states whose potential curves (Figure 5.6) are crossed by various valence states of $\pi \rightarrow \pi^*$ and $\pi^2 \rightarrow \pi^{*2}$ character (D$'$ and C$'$). Cooper and Langhoff (1981) and Cooper and Kirby (1987, 1989) carried out CI calculations for various $^1\Sigma^+$ and $^1\Pi$ states of CO and their computed spectroscopic constants and f values for the corresponding vibrational levels agree well with observed data. Cooper and Kirby (1987, 1989) also found that the $2\,^1\Sigma^+$ potential has a double minimum due to the interaction of the D$'$ valence state with the (3s) B state, thus supporting the experimental interpretation of data reported by Wolk and Rich (1983). The observed predissociation of the B$\,^1\Sigma^+$ state was also considered in the CK calculations, but these results greatly underestimate the observed linewidth [Tchang-Brillet *et al.* (1992)] of the $v = 2$ level (4×10^{-5} versus 0.5 cm^{-1}). Tchang-Brillet *et al.* (1992) were able to deduce empirical potentials which model the B–D$'$ interaction and reproduce the observed vibrational spacings and linewidths of the $v = 0$–2 levels of the B state.

CI calculations have recently been carried out [Li, Buenker and Hirsch (1998)] to study predissociation in both the B and C states of CO. The radial coupling elements including second-derivative results have also been calculated for this purpose. The effective core potentials of Pacios and Christiansen (1985) have been employed in this study along with their (4s4p) uncontracted

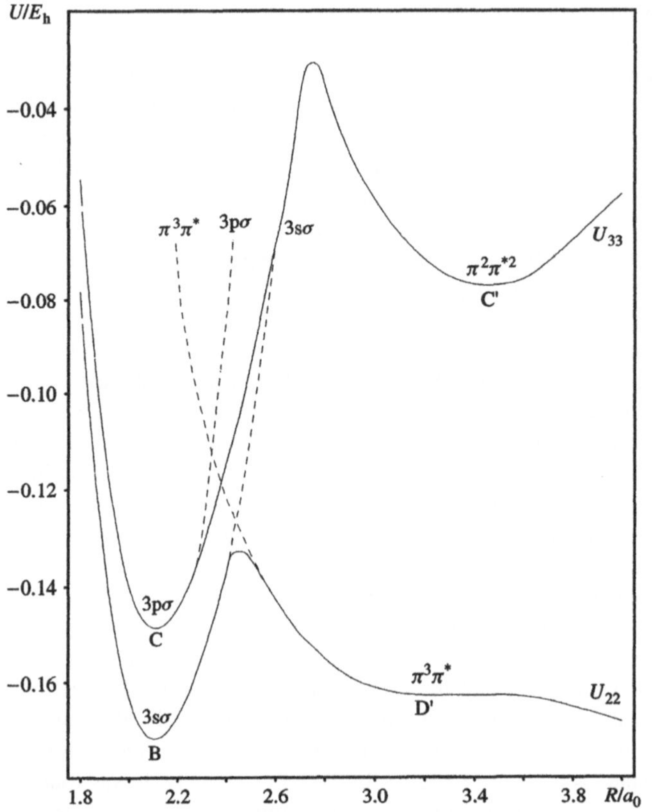

Figure 5.6. Computed adiabatic potential energy curves U_{22} and U_{33} for the second and third $^1\Sigma^+$ states of the CO molecule. The estimated diabatic curves are indicated by dashed lines.

gaussian basis augmented with two sets of d-type polarization functions in addition to the requisite s- and p-type Rydberg orbitals. In contrast to the work described above for the O_2 molecule, it was found in the present case that a diabatic transformation of the original CI results was quite difficult to carry out to a sufficiently high level of accuracy. This is primarily because the valence potential curves cut much more steeply through the Rydberg manifold in CO (Figure 5.6) than for the corresponding O_2 groups of states. It was therefore decided to work directly in the adiabatic representation and to use a numerical procedure to carry out the complex scaling calculations for this system which had already been tested [Li *et al.* (1997a)] on the empirical potentials of Tchang-Brillet *et al.* (1992). This also allows one to use the computed radial coupling elements directly in the calculations, without having to fit these rapidly varying functions in a suitable manner, something which would be quite difficult to do in the present case.

It is interesting to see how the nonadiabatic couplings behave in the critical crossing region (Figure 5.6). Both the diagonal second-derivative corrections to the adiabatic potentials have maxima at the avoided crossing point and become constant within $0.2\,a_0$ on either side of this point. The two coupling terms decrease quickly to zero not far from their peak values near the avoided crossing. The first-derivative coupling has a symmetric appearance and is much smaller than the corresponding second-derivative off-diagonal matrix element, which is also notably less symmetric. Similar behavior is apparent when the diabatic results of Tchang-Brillet *et al.* (1992) are converted to the adiabatic representation, as was done in the tests of the numerical integration procedure [Li *et al.* (1997a)]. As pointed out in the latter study, the second-derivative coupling plays an important role in the predissociation of the B and C states.

The computed vibrational energy spacings and linewidth results are shown in Table 5.4 along with available experimental data. The T_0 value for the B–X transition is calculated to be 87 706 cm^{-1}, which is only 790 cm^{-1} or less than 1 % in error compared with experiment. The gap between the $v = 0$ levels of the B and C states themselves is found to be 5030 cm^{-1}, 27 cm^{-1} higher than that measured. It is predicted that four vibrational levels are possible in the well of the B state, of which $v = 3$ has the broadest width. The computed value for $v = 2$ of 0.1 cm^{-1} is in much better agreement with the observed result of >0.5 cm^{-1} by Tchang-Brillet *et al.* (1992) than in earlier work [Cooper and Kirby (1987, 1989)]. The calculated vibrational spacings are too high by 2–5 %, however, and it is clear that all these discrepancies can be traced to errors in the $2\,^1\Sigma^+$ state's potential, whose energy barrier is both too high and too wide. It is expected that more polarization functions are needed in the AO basis to improve these results significantly, in addition to carrying out the CI at a lower selection threshold.

The description of the C state is relatively good by comparison. The errors in the vibrational spacings are less than 2 % and the observed trend in line broadening of the $v = 0$–4 levels is reproduced as well. Experimental values are available for the $v = 3, 4$ levels, albeit with relatively large error bars. The calculations find Γ values of 4.9 and 8.9 cm^{-1}, each within a factor of two of these estimates (Table 5.4). Lifetime measurements have also been reported for the $v = 0$–2 levels of the C state [Viala *et al.* (1988), Eidelsberg and Rostas (1990), Eidelsberg *et al.* (1990, 1992),

Table 5.4. Comparison between calculated and measured vibrational energy spacings ΔG_v, predissociation linewidths Γ_v and lifetimes τ_v for the B and C $^1\Sigma$ states of the CO molecule.

State	v	ΔG_v (cm^{-1})		Γ_v (cm^{-1})			τ_v(s)	
		this work	exptl.	this work[a]	calc.	exptl.	this work[b]	exptl.
B $^1\Sigma^+$								
	0			0				2×10^{-8} [g]
		2128	2082[c]					
	1			0				3×10^{-8} [g]
		2081	1990[c]					
	2			0.1	4×10^{-5} [e]	>0.5[f]		
		2045	~1900[d]					
	3			23	5×10^{-1} [e]	very diffuse[d]		
C $^1\Sigma^+$								
	0			$<10^{-4}$			5×10^{-8}	10^{-9} [c], 5×10^{-10} [g]
		2100	2146[c]					
	1			0.05			10^{-10}	10^{-9} [c]
		2086	2111[c]					
	2			1.1			5×10^{-12}	10^{-10} [c], 10^{-11} [h]
		2104	2068[c]					
	3			4.9		~2.3[f]		
		2105	2147[c]					
	4			8.9		<5[f]		

[a]For comparison with the experimental data the calculated linewidth values are given for the rotational quantum number $J = 15$.
[b]$\tau(\mathrm{s}) = 2.419 \times 10^{-17} \times [\Gamma(E_h)]^{-1}$.
[c]Eidelsberg and Rostas (1990), and Eidelsberg, *et al.* (1990, 1992).
[d]Baker, Tchang-Brillet and Julilenne (1995).
[e]Cooper and Kirby (1987, 1989).
[f]Tchang-Brillet *et al.* (1992).
[g]Ubachs *et al.* (1994).
[h]Viala *et al.* (1988).

and Ubachs *et al.* (1994)], with roughly order-of-magnitude accuracy. Again the calculations can be said to be in nearly quantitative agreement with these data. The indication is therefore that the calculated nonadiabatic couplings are of high accuracy and that to improve agreement between theory and measurement further it is necessary not only to improve the calculations but also to eliminate errors in the corresponding measured data.

5.3.3 RELATIVISTIC EFFECTS IN MOLECULES WITH HEAVY ATOMS

The role of spin–orbit coupling and other relativistic effects is magnified when the molecular systems contain one or more heavy atoms. There has been greatly increased interest in this field because of the availability of relativistic effective core potentials (RECPs) in the last decade. These quantities allow one to treat only the valence electrons of such heavy atoms as lead and bismuth in CI calculations, greatly reducing the computational expense that would otherwise be required in corresponding all-electron treatments. A key complication is the vector nature of the spin–orbit interaction, which necessitates a reorganization of the overall computational procedure which must be employed relative to that already discussed for lighter systems. The main distinction lies in the fact that spin is no longer a good quantum number for the electronic states of heavy atoms and molecules. To illustrate the effects of adding such relativistic terms to the electronic Hamiltonian, we will now turn our attention to a series of calculations carried out over the past four years for the hydrides and oxides of the Group VA elements, arsenic, antimony and bismuth.

A potential energy diagram for the BiH system including spin–orbit coupling is shown in Figure 5.7. This molecule is isovalent with NH and shares with it a ground π^{*2} electronic configuration. The most stable state is $X\,^3\Sigma^-$, but this splits into a 0^+ and 1 species, which differ in their T_e values by $4917\,cm^{-1}$. The calculated result obtained by using the full-core RECP of Ross *et al.* (1990) in a relativistic version of MRD-CI is $4303\,cm^{-1}$ [Alekseyev *et al.* (1994)]. The corresponding spectroscopic constants for a number of low-lying BiH states are compared with available experimental data in Table 5.5, along with analogous theoretical results obtained by other authors. The vibrational frequency of the $X_2\,^3\Sigma_1^-$ state is observed to be $30\,cm^{-1}$ less than that of the $X_1\,^3\Sigma_{0+}^-$ state. The calculations show the same trend, but only about half as great a decrease. The bond lengths of these states are overestimated by about $0.06\,a_0$ in both cases, but this a clear effect of the RECP employed (with the 5d electrons of Bi included in the core).

The $a2(^1\Delta)$ state is next in energy (Figure 5.7) and it also dissociates to the $Bi\,^4S^o_{3/2}$ ground state. There are also $\Omega = 1$ and 0^- states going to the same asymptote, but their potential curves are found to be repulsive. The location of the a2 state has not yet been determined experimentally, but the calculations place its T_e value at $11\,906\,cm^{-1}$. They also indicate a frequency which is $100\,cm^{-1}$ higher than for X_2 but nearly the same value for its bond length.

The $B0^+$ state is the last of the π^{*2} variety, but it correlates with $^2D^o_{3/2}$ excited bismuth atom state (Figure 5.7). It has a similar bond length as a2, but its frequency is calculated to be $100\,cm^{-1}$ lower. It is interesting to note that without spin–orbit coupling (λ–s representation), the $a\,^1\Delta$ state correlates with $^2D^o$ while the $B\,^1\Sigma^+$ goes to the higher $^2P^o$ asymptote. There is thus a change in the character of the dissociation limit as a result of the addition of relativistic effects. This effect is responsible for a barrier in both the a2 and $B0^+$ potential curves, but that for the former state is more pronounced, which characteristic is consistent with the higher computed a2 frequency compared with those for both $B0^+$ and the X_1 and X_2 states.

The third state of 0^+ symmetry (E) has a much larger bond length and also undergoes an avoided crossing with 0^+(IV). There is therefore a fairly shallow minimum in the $E0^+$ potential (Figure 5.7) which is found to hold two bound levels. Experimentally it is found [Lindgren and

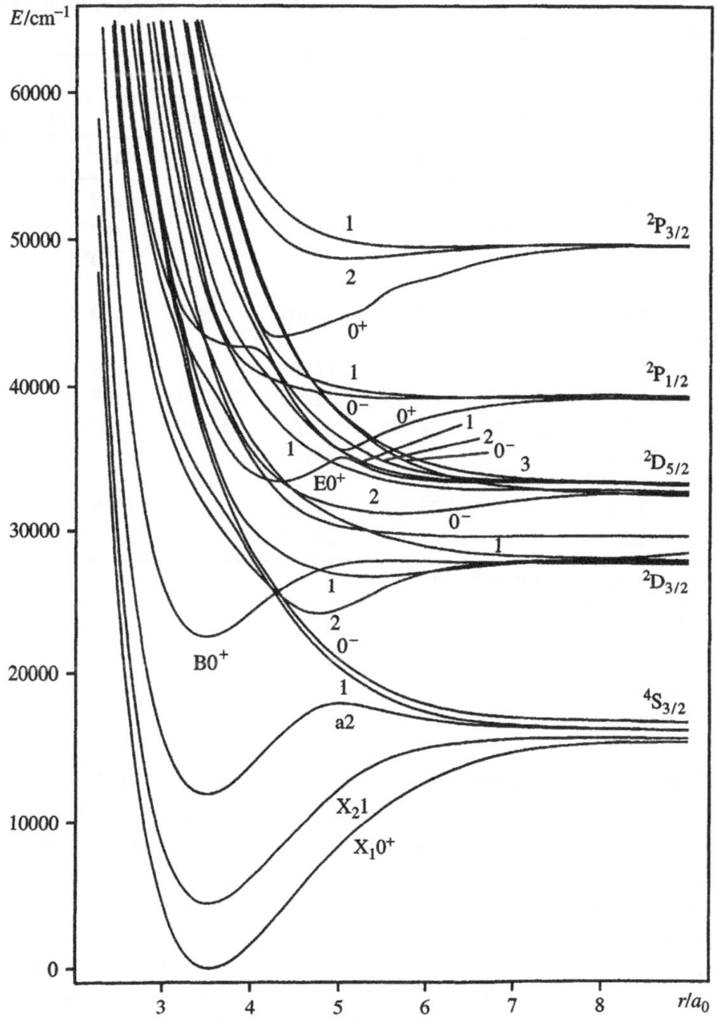

Figure 5.7. Computed potential energy curves for the 23 lowest-lying electronic states of BiH including spin-orbit coupling. The atomic dissociation limits correspond to $H(^2S_{1/2}) + Bi$ (as indicated).

Nilsson (1975)] that the $v = 2$ level is missing rotational lines in its absorption spectrum and this has been interpreted as evidence of a strong predissociation effect caused by the intersection of a repulsive potential curve, all of which fits in quantitatively with the calculations. A barrier height in the order of $2000\,cm^{-1}$ is calculated for the $E0^+$ potential curve, which offers a simple explanation for the fact that no emission bands have been observed for it. Since recombination processes are involved in the experimental studies, it seems clear that such a barrier would prevent formation of the $E0^+$ and hence preclude observation of emission from it. On the basis of this potential curve, it is possible to use experimental data [Lindgren and Nilsson (1975)] to extract the BiH D_e value of 2.28 eV. This result is 0.3 eV greater that that computed directly on the basis of the X_1 potential curve, which is certainly consistent with the level of accuracy expected from the relativistic CI calculations.

Table 5.5. Spectroscopic properties of BiH from various calculations and experiment (bond lengths r_e, transition energies T_e, and vibrational frequencies, ω_e).

State	r_e(Å)			T_e (cm^{-1})			ω_e (cm^{-1})		
	calc.		expt.[a]	calc.		expt.[a]	calc.		expt.[a]
	other[b]	this work		other[b]	this work		other[b]	this work	
$X_1 0^+$	1.90 1.847[e] 1.858[e] 1.869[f]	1.867	1.805 1.80867[d]	0	0	0	1619 1780[e] 1756[e]	1632	1699.5[g]
$X_2 1$	1.89 1.840[e]	1.854	1.791	5737	4303	4917	1630	1618	1669[c]
a2	1.89	1.855		13469	11906		1630	1719	
$B0^+$	1.88	1.851	1.7795	26286	22496	21263	1585	1620	1643[c]
$E0^+$	2.51	2.283	2.1772	38780	33364	32940	1000	1215	1106[c]
0^+(IV)		2.669			35575			2169	
0^+(V)		2.271			43395			1528	
2(II)	2.58	2.521		23450	24100		1529	1013	
1(III)		2.823			26735			559	

[a]Huber and Herzberg (1979) unless otherwise indicated.
[b]Balasubramanian (1986) unless otherwise indicated.
[c]Value quoted is $\Delta G(1/2)$.
[d]Bopegedra, Brazier and Bernath (1989).
[e]Dolg *et al.* (1991).
[f]Dai and Balasubramanian (1990).
[g]Urban, Polomsky and Jones (1989), and Hedderich and Bernath (1993).

Radiative lifetimes have also been calculated on the basis of the relativistic CI treatment. It is found that the X_2 state is metastable, with a lifetime of 16 ms. The a2 state is also relatively stable to radiative decay, but its lifetime is only about 0.5 ms. Because of configuration sharing with the X_1 ground state as a result of the spin–orbit interaction between the $^3\Sigma^-$ and $^1\Sigma^+$ λ–s species, it is found that the $B0^+$ state is much shorter-lived ($\tau = 4.3\,\mu$s). The lifetimes of the next two 0^+ states are 20–30 times shorter still.

Similar calculations have also been carried out for the next heaviest members of this series, SbH [Alekseyev *et al.* (1998b)] and AsH [Alekseyev *et al.* (1998a)]. The available RECPs for arsenic and antimony [Hurley *et al.* (1986); LaJohn *et al.* (1987)] are more accurate than that employed for bismuth hydride, and thus the agreement between theory and experiment is notably better for these lighter systems. Larger basis sets were also employed than in the BiH study. The computed spectroscopic constants for SbH show a slight overestimation of bond lengths by 0.01–0.03 a_0 and a corresponding underestimation of vibrational frequencies by 1–2%. The T_e values of five excited SbH states are known experimentally and the calculated values are found to agree within 0.07 eV in each case. The spin–orbit splittings are much smaller than for BiH, but they are still substantial. The observed X_1–X_2 splitting for SbH is 666.6 cm^{-1}, for example, as compared with 4917 cm^{-1} for BiH.

Two of the four $^3\Pi$ Ω states have been observed [Bollmark and Lindgren (1974)], and the computed T_e values agree with the measured values to within 200 cm^{-1} [Alekseyev *et al.* (1998b)]. The $\Omega = 0^-$, 1 and 2 components of this state are found to be strongly predissociated by the $^5\Sigma^-$ state. The corresponding 0^+ state (A_4) is the most stable of this term and it is found to possess a very unusual potential with a double minimum and a low barrier to dissociation. Vibrational calculations for this state suggest strongly that its $v = 0$ and 2 levels actually correspond to the observed $B0^+$ and $C0^+$ states [Bollmark and Lindgren (1974)]. The very flat and complicated appearance of the A_4 potential produces an irregular vibrational ladder which apparently has led to an incorrect assignment in terms of three separate electronic states. The radiative lifetimes of the $A\,^3\Pi$ states are all calculated in the μs range because of the spin-allowed transitions to the $^3\Sigma^-$ ground state.

The X_1–X_2 splitting for AsH is only $117.6\,cm^{-1}$ [Hensel, Hughes and Brown (1995)], so it is easy to see that the spin–orbit interaction rapidly loses strength as one proceeds to lighter elements. The computed value [Alekseyev *et al.* (1998a)] is $107\,cm^{-1}$. This system has been quite well studied experimentally. The calculations also slightly overestimate the bond lengths and T_e values for this system (by $0.01\,a_0$ and $500\,cm^{-1}$, respectively). Again the $\Omega = 0^-$, 1 and 2 components of the $^3\Pi$ state are found to be strongly predissociated by the $^5\Sigma^-$. The calculated D_e value for the X_1 ground state is 2.75 eV, which is about 0.1 eV lower than inferred from experiment [Dixon and Lamberton (1968), and Berkowitz (1988)]. For this lighter hydride there is also an all-electron treatment in the literature [Matsushita *et al.* (1987)] which takes account of relativistic effects at the CI level. The computed T_e values for the a2 and $B0^+$ states are computed to be 500–$1000\,cm^{-1}$ higher than in the RECP CI study [Alekseyev *et al.* (1998a)] but the agreement for bond lengths of these and the X_1 and X_2 states is quite close.

There have also been relativistic CI calculations of the oxides of arsenic, antimony and bismuth [Alekseyev *et al.* (1995a), and Alekseyev, Liebermann, Buenker *et al.* (1994b, 1995a)]. These molecules are isovalent with NO, which has a $^2\Pi$ ground state followed by a $^4\Pi$ and a series of $^2\Pi$ and $^{4,2}\Sigma^{+/-}$ excited states. Because of spin–orbit coupling the ground state splits into $X_1\,^2\Pi_{1/2}$ and $X_2\,^2\Pi_{3/2}$ states, in that order. This is the regular order of multiplets expected for less than half-filled open shells, consistent with the π^* structure of $X\,^2\Pi$. The $^4\Pi$ state splits into four components, with the computed order for BiO: $3/2 < 1/2 < 5/2 < 1/2$, consistent with its more complicated $\pi^3\pi^{*2}$ configuration. As a result the excited X_2 and a_1 states are adjacent to one another in the electronic spectra of this class of systems, as can be noted from the calculated potential data for SbO. Since antimony is in the fourth row of the Periodic Table, the effect of spin–orbit coupling is only moderately large. For example, the observed X_1–X_2 splitting is $2272\,cm^{-1}$ [Balfour and Ram (1984), and Huber and Herzberg (1979)]. By comparison, the analogous value observed for BiO is $7089\,cm^{-1}$ [Shestakov *et al.* (1998)]. Three other $^2\Pi$ states are derived from the $\pi^3\pi^{*2}$ configuration, as well as a $^2\Phi$. Numerous other low-lying states result from $\sigma \rightarrow \pi^*$ and $\pi \rightarrow \sigma^*$ excitations.

Since the X_2 state has a notably shorter bond length than the $^4\Pi_{3/2}$ state because of the latter's higher occupation of the π^* antibonding MO, the energy difference between these two states decreases quickly towards large R values (Figure 5.8). In BiO this leads to a strongly avoided crossing between the adjacent 3/2 states of $X\,^2\Pi$ and $A\,^4\Pi$ [see Figure 5.2 of Alekseyev *et al.* (1994b)]. Experimentally this is perceived in terms of strong perturbations in the vibrational spectrum of the BiO X_2 state starting at $v = 6$ [Shestakov *et al.* (1998)]. The relativistic CI calculations made it clear that this effect is caused by an interaction of $X\,^2\Pi$ with $A\,^4\Pi$, not with a higher-lying excited $^2\Pi$ state of the same configuration. This was quite unexpected because no such interaction is known for the prototype NO system. Spin–orbit effects mix these two states quite significantly in the much heavier BiO molecule, however, leading to a strongly perturbed $X_2 \rightarrow X_1$ emission spectrum in this case. Similar perturbations have since been observed in the spectra of BiS, BiSe and BiTe, albeit at lower v [Breidohr, Shestakov and Fink (1998)].

Most of the low-lying states of these systems dissociate to their respective ground state atoms. Since the O ground state has ^3P symmetry while that of the nitrogen-like counterparts is $^4S^o$, there are many atomic J limits of nearly equal energy available (Figure 5.8). Since the equilibrium electronic configurations of these states have either one or three open-shell electrons, whereas the corresponding asymptotic wave functions have five open-shell electrons, it is clear that avoided crossings must occur in the associated potential energy curves in passing from their minima to

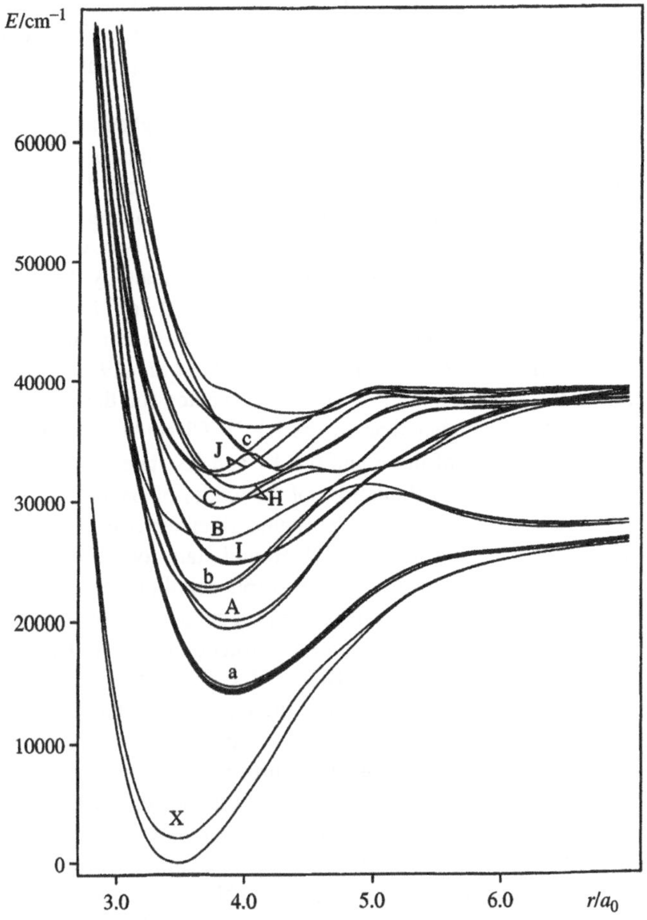

Figure 5.8. Computed potential energy curves for the lowest-lying electronic states of the SbO molecule including spin–orbit coupling.

the separated-atoms limit. These occur at ever shorter bond distances as the degree of excitation increases (Figure 5.8). As a result the lowest $^2\Sigma^+$ state has a fairly large recombination barrier and the lowest $^4\Sigma^+$ state has a broad shoulder in its potential in the 4.5–5.0 a_0 range of R.

A large number of SbO states have been detected experimentally, so it is instructive to compare their measured spectroscopic constants with those calculated in the relativistic CI treatment (Table 5.6). The X_1–X_2 splitting is underestimated by 9 %, whereas the bond length is overestimated by 0.01 a_0 and the frequency is too low by some 60 cm^{-1} (8 %). The errors in the T_e values for the various $^4\Pi$ states are in the 2600–2900 cm^{-1} range. One expects from electron-correlation arguments that a quartet state will be calculated to be too low relative to a doublet with two fewer open shells, but the magnitude of the error is generally in the 0.2 eV range, about 0.15 eV less than is indicated in the SbO measurements [Shestakov and Fink (1996)]. On the other hand, the computed frequencies are only 10 cm^{-1} in error in these cases. The A $^2\Pi$ T_e values are also underestimated in the RCI calculations, but by only about 0.15 eV.

The $^4\Sigma^-$ states arise from a $\sigma \rightarrow \pi^*$ excitation. Since this σ MO is less bonding than the π species, it is to be expected that the bond lengths of these states are notably smaller than for the

Table 5.6. Calculated and experimental spectroscopic properties of SbO (transition energies T_e, bond lengths r_e, and vibrational frequencies ω_e).

State	T_e (cm^{-1})		r_e (Å)		ω_e (cm^{-1})	
	calc.	expt.	calc.	expt.	calc.	expt.
$X_1\,^2\Pi_{1/2}$	0	0	1.8364	1.8258[a]	754	816[a], 819[b]
$X_2\,^2\Pi_{3/2}$	2080	2272[a,b]	1.8358		751	814[b]
$a_1\,^4\Pi_{3/2}$	14207	16862[d]	2.0666		542	558[d]
$a_2\,^4\Pi_{1/2}$	14390	17152[d]	2.0673		541	559[d]
$a_3\,^4\Pi_{5/2}$	14535		2.0678		540	
$a_4\,^4\Pi_{1/2}$	14802	17698[d]	2.0652		545	556[d]
$A_1\,^2\Pi_{3/2}$	19476	20668[a], 20794[b]	2.0627		566	569[a], 570[b]
$A_2\,^2\Pi_{1/2}$	20091	20801[a], 21467[b]	2.0702		546	566[b]
$b_1\,^4\Sigma^-_{1/2}$	22455	24174[d]	1.9581		632	610[d]
$b_2\,^4\Sigma^-_{3/2}$	22836		1.9473		611	
$I_1\,^2\Phi_{7/2}$	24695		2.0750		507	
$I_2\,^2\Phi_{5/2}$	24863		2.0790		500	
$B\,^2\Sigma^+_{1/2}$	26616	26594[a]	1.9418		557	582[a]
$C\,^2\Sigma^-_{1/2}$	29683	29747[a], ~29750[c]	2.0078	1.997[a]	591	571[a], 569[c]
$H_1\,^2\Pi_{3/2}$	30425	30315[a], ~30495[c]	2.0865		495	~546[c]
$H_2\,^2\Pi_{1/2}$	31371		2.0948		501	
$J_1\,^2\Delta_{3/2}$	32348		2.0023		681	
$J_2\,^2\Delta_{5/2}$	32399		1.9944		610	
$c_1\,^4\Sigma^+_{1/2}$	33192		2.2492		533	
$c_2\,^4\Sigma^+_{3/2}$	34041		2.1449		856	
$D\,^2\Pi_{1/2}$		34544[a]		2.073[a]		506[a]

[a]Huber and Herzberg (1979).
[b]Balfour and Ram (1984).
[c]Balfour and Ram (1988).
[d]Shestakov and Fink (1996).

more stable $^2\Pi$ states of $\pi \rightarrow \pi^*$ character, as calculated. One can easily distinguish states of these two types from one another just on the basis of their respective potential curves (Figure 5.8). The $B\,^2\Sigma^+$ arises from a $\pi \rightarrow \sigma^*$ excitation. It undergoes an avoided crossing with b_1, the $^4\Sigma^-$ state of the same Ω. The $C\,^2\Sigma^-$ state also comes from a $\sigma \rightarrow \pi^*$ excitation but it is perturbed by the $H_2\,^2\Pi_{1/2}$ state and thus shows some irregularities in its spectroscopic constants.

Shestakov and Fink (1996) have also measured the radiative lifetimes of a number of the SbO states. The RCI treatment is generally in good agreement with these measurements. The X_2 state's lifetime is apparently too long to be measured in emission experiments, with a value of 0.29 s having been calculated. The observed value for the $a_2\,^4\Pi_{1/2}$ state is about half as large as computed, but the agreement for the shorter-lived $A\,^2\Pi$ states is much closer. Parallel transitions tend to dominate according to the calculations (see Table V of Alekseyev *et al.* (1995a)). Good agreement has also been noted between theory and experiment for the BiO lifetimes [Alekseyev *et al.* (1994b)]. In particular the lifetime of the BiO $^4\Pi_{1/2}$ state is measured to be much shorter (10 μs) than for the corresponding SbO state and this is mirrored in the RCI calculations.

Finally, a parallel treatment of the AsO spectrum [Alekseyev *et al.* (1995b)] has also succeeded in obtaining a high level of agreement with the corresponding measured results (see Table II of that reference), up to about 40 000 cm^{-1} T_e values. One unusual result is the finding that the $B\,^2\Sigma^+$ state has a large amount of Rydberg character. It is the only low-lying AsO state with negative polarity on the arsenic atom. Its potential curve has a minimum which is almost coincident with a maximum in the $A\,^2\Sigma^+$ potential. This causes a break-up in the strong B–X emission intensity pattern at $v' = 0$ and $N' = 21$. On this basis it was possible to obtain an improved estimate of the D_0^0 value for the AsO ground state of 4.22 eV.

5.4 Triatomic molecules and beyond

When there are more than two nuclear centers the problem of calculating potential energy and coupling data is greatly complicated by the increased number of degrees of freedom for internal motion. Most if not all of the computational methods illustrated in the last section for diatomics can be taken over without change for polyatomic systems, however, and so in the present section we will simply discuss the results of such calculations for some problems of current experimental interest.

5.4.1 THE A–X TRANSITION IN HO_2

When a hydrogen atom is bound to molecular oxygen, a bent equilibrium structure results. The electronic configuration of the hydroperoxyl radical can be thought of as having a π^{*3} occupation, but the two components of this MO are no longer degenerate for a bent nuclear conformation. The in-plane 7a′ component becomes more stable than the out-of-plane 2a″ MO, and the energy splitting increases fairly sharply with molecular bending [Buenker and Peyerimhoff (1976)]. As a result, it comes as a surprise to find that the bond angle stays about the same when the first excited state of HO_2 is formed by a 7a′–2a″ excitation. The bending potential curves computed for these two states are shown in Figure 5.9. A large-scale MRD-CI treatment has been employed for this purpose [Gu, Buenker and Hirsch (1998)].

These two electronic states become degenerate in the linear nuclear conformation and thus form a Renner pair [Renner (1934)]. Because both states are bent by roughly 80°, however, this fact has comparatively little consequence on the observed spectrum. The lowering in symmetry caused by bending means that the A–X transition is dipole allowed by the component perpendicular to the plane of the molecule. In the linear geometry this would correspond to a π_x–π_y transition, however, which is forbidden by the dipole selection rules. A calculation of this transition moment at the bent equilibrium geometry of the X state shows that its value is in fact quite small (ca $1.0 \times 10^{-4}\,ea_0$). This result shows clearly that the 7a′ and 2a″ components are still quite similar to their π_x and π_y counterparts to which they correlate in the linear geometry. In other words, the motion of the lone H atom in bending is not sufficient to produce a significant change in this relationship between the two components of the π^* MO.

The above result has some interesting consequences on the appearance of the HO_2 electronic spectrum. First of all, the intensity of this spectrum is quite low, and the upper state is relatively long lived as a result. Because both the A and X states have similar potential curves, the Franck–Condon factors for vibrational transitions are only large for $\Delta v_2 = 0$, and no bending progressions can be observed. Because of the different symmetries of the two states, it is expected from theory that only $\Delta K = 1$ rotational transitions will be allowed [Herzberg (1966)]. It came therefore as a great surprise to find that this was not always the case [Fink and Ramsay (1997)], and that relatively strong $\Delta K = 0$ lines are present in the measured spectrum.

One possible explanation lay in the fact that the two states in question are part of a Renner pair. This means that as a result of a nonadiabatic interaction the rovibrational levels are not actually pure A′ or A″ states, but rather a mixture thereof. Two developments have since all but refuted this interpretation. First, it has been noted experimentally [Fink and Ramsay (1997)] that the $\Delta K = 0$ transitions always occur between rotational states of the same parity. This seemingly rules out dipole-allowed transitions, which require a change in parity in going from initial to final states. Secondly, calculations of the Renner effect [Osmann, Bunker and Jensen (1998)] based on

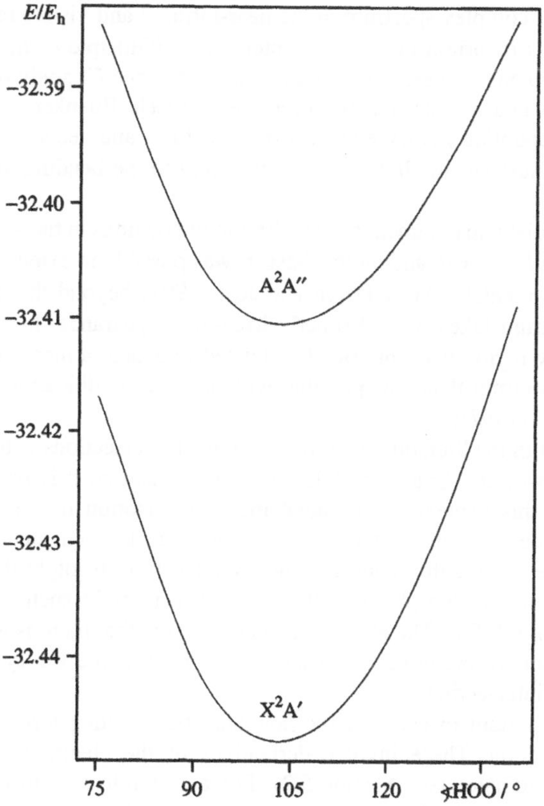

Figure 5.9. Computed bending potential energy curves of the HO_2 molecule at bond distances $r(\text{O–O}) = 2.5133\,a_0$ and $r(\text{O–H}) = 1.833\,a_0$.

the CI potentials and wave functions do not find enough mixing of A and X rotational levels to account for the strength of the $\Delta K = 0$ transitions observed. The correct explanation lies in the fact that the A–X transition is allowed through the parallel component of the magnetic-dipole moment operator (which does have gerade symmetry), even for the linear geometry. Normally, electric-dipole transition moments are several orders of magnitude larger than magnetic-dipole matrix elements, but the weakness of the former quantity in the present case, as discussed above, makes this an exceptional case.

In retrospect, the key to the above solution is the realization that the bent equilibrium geometries of both the A and X states tend to draw attention away from the fact that certain relationships which are true for the linear nuclear conformation maintain their importance even when the atoms do not all lie on the same line. The fact that one can effect the change from linear to bent conformations by simply moving the light H atom while basically leaving the oxygen atoms in their original positions makes this state of affairs more easily understandable.

5.4.2 CONICAL INTERSECTIONS IN NO_2 AND C_2H

Although the Renner effect turned out to be of minor importance in the last example, it emphasizes that polyatomic systems are subject to different types of vibronic interactions than are possible for simple diatomics. Nowhere is this more evident than in the A–X spectrum of the NO_2

molecule. It has a very complex spectrum in the near-infrared and visible range [Hsu, Monts and Zare (1978)] which has its origin in a conical intersection [Gillispie *et al.* (1975)] which occurs between the lowest two NO_2 states. A large-scale multireference CI study of the potential energy surfaces of the NO_2 A and X states has been reported [Hirsch, Buenker and Petrongolo (1991)], in addition to radial coupling matrix elements of both first- and second-derivative type. These are three-dimensional surfaces in all cases, corresponding to the bending and two NO stretching internal coordinates.

Recently a coupled diabatic treatment of the vibronic interactions in the A–X spectrum has been reported [Leonardi *et al.* (1996)], and on this basis it was possible to extend previous assignments of the available experimental data of Delon and Jost (1991) beyond the $10\,000\,\mathrm{cm}^{-1}$ range, at which point the spectrum takes on a distinctly irregular appearance. In addition, the *ab initio* PES in the low-energy region was empirically adjusted to enable a more accurate calculation of the $0{-}10\,000\,\mathrm{cm}^{-1}$ spectrum than was possible without such modification [Tashkun and Jensen (1994); Schryber *et al.* (1997)].

It has been common in the literature to associate conical intersections with the geometric-phase formalism [Berry (1984), and Bunker and Jensen (1998)], and so it is important to discuss the above calculations in this context. In the adiabatic representation it can be shown [Mead and Truhlar (1982)] that the electronic wave functions must change sign upon varying the nuclear geometrical coordinates in a path around a conical intersection. It might therefore be concluded that it is essential to modify the phase of these wavefunctions in such a way as to guarantee this property [Kendrick (1997)]. The above calculations show that there is an alternative method, however, which is actually preferred when attempting to describe the system at or above the energy of the conical intersection.

To see this it is important to consider the nonadiabatic coupling terms in the neighborhood of the conical intersection. These involve derivatives of the electronic wave functions with respect to nuclear coordinates (see Section 5.2). For a diatomic system these are always well defined, but it is in the nature of a conical intersection that these wave functions vary discontinuously as the molecule undergoes a change along an antisymmetric coordinate which destroys the quasi-degeneracy of the interacting states. This means that the radial coupling elements have a singularity along the seams of conical intersections when working in the original adiabatic representation. Adding a geometric phase to the wave functions does not alter this situation, as one continues to employ adiabatic functions which have these undesirable singularities in their associated coupling elements. This procedure is nonetheless quite effective in describing vibronic energies and wave functions near the bottom of the lowest potential well in such a system of states, but this is only because the effects of radial coupling can be safely neglected far away from the conical intersection itself.

The procedure employed to describe the A–X spectrum of NO_2 minimizes the difficulties involved in computing and fitting radial coupling surfaces by performing a diabatic transformation. This has at least two beneficial effects. First, since the diabatic states are by definition slowly varying with nuclear geometry, the radial couplings connecting them are relatively small and easily manageable. The difficulty with this procedure lies mainly in defining a suitable transformation which satisfies this goal, but this has been achieved in the NO_2 work and also in a related study of the C_2H conical intersection [Thümmel *et al.* (1989), and Perić, Buenker and Peyerimhoff (1990a, 1990b)]. Beyond this, however, the diabatization procedure removes the need for adding a geometric phase to the resulting wave functions. They do not change sign as one follows a closed path around the conical intersection because they retain nearly the same composition throughout the entire region. Hence, one can and should employ them without any additional phase other than that which is produced in the diabatization procedure itself.

The best evidence that this analysis does work in practice is in the computed vibronic spectrum obtained for the NO_2 A–X system [Leonardi *et al.* (1996)]. The diabatic approach has allowed for an accurate representation of the NO_2 energy levels throughout the critical 10 000–14 000 cm^{-1} region just above the location of the conical intersection and beyond up to 18 000 cm^{-1}, where the spectrum becomes quite chaotic in appearance [Delon and Jost (1991)].

In NO_2 the conical intersection comes about because in C_{2v} symmetry the $X\,^2B_1$ and $A\,^2A_1$ bending potentials cross one another. When the NO bond lengths are changed slightly in opposite directions, an avoided crossing suddenly occurs between the corresponding adiabatic bending potential curves. The change is discontinuous at the seam of the conical intersection, which is why the nonadiabatic coupling elements have singularities, as discussed above. In C_2H there is an analogous effect for its lowest two electronic states, only in this case it is the bending motion that is the antisymmetric coordinate which is responsible for the sudden change from a crossing to an avoided crossing of the adiabatic potentials. In linear symmetry the two states in question have $^2\Pi$ and $^2\Sigma^+$ symmetry, respectively. As one varies the C–C bond length, the two states change their energetic order, with the $^2\Pi$ state preferring larger values for this quantity. Upon bending out of the linear conformation, one of the $^2\Pi$ components becomes $^2A'$ in the resulting C_s symmetry, just as the $^2\Sigma^+$ state. Thus there is a discontinuous change from Π to Σ^+ character in the crossing region, which again corresponds to a conical intersection in the associated potential surfaces.

In recent work the potential energy and nonadiabatic coupling and electric-dipole moment surfaces have been computed employing a large-scale MRD-CI treatment and a flexible AO basis containing two d polarization functions on each of the carbon atoms [Hirsch, Liebermann and Buenker (1998)]. In order to minimize these coupling terms, especially in the region of the conical intersection itself, a diabatic transformation has been carried out which produces states of nearly pure $^2\Pi$ and $^2\Sigma^+$ character even in highly bent nuclear conformations. This also eliminates the need for adding a geometric phase to the resulting wave functions, as discussed above for the NO_2 molecule. The diabatic potential surfaces are also notably less complicated than the original adiabatic results. A portion of the diabatic PES in which the conical intersection is clearly defined is shown in Figure 5.10 [Hirsch, Liebermann and Buenker (1998)].

In order to check the degree to which the diabatic transformation has been successful in separating out the $^2\Pi$ and $^2\Sigma^+$ character from the adiabatic wave functions, additional calculations have been carried out to evaluate the nuclear derivative matrix elements in the new basis. In some cases it was found that adjustments were necessary to obtain suitably diabatic states and thus to reduce further the magnitudes of the radial couplings between them. Ultimately, the goal is to continue this process until such couplings can be safely neglected in a treatment of the vibronic motion associated with the C_2H conical intersection. It should be emphasized, however, that the diabatic states are still subject to coupling via the electronic Hamiltonian itself as a result of the transformation which has been applied to the original CI wave functions.

An earlier MRD-CI study [Thümmel *et al.* (1989), and Perić, Buenker and Peyerimhoff (1990a, 1990b)] of this type employed a significantly smaller AO basis and did not compute radial couplings for the diabatic states which were generated. Nonetheless it was possible on this basis to demonstrate a high degree of correlation between the computed results and what can be inferred from experimental studies of the C_2H energy levels [Carrick *et al.* (1982), Carrick, Merer and Curl (1983), and Curl, Carrick and Merer (1985)]. The computed T_e value in the latter study is 3300 cm^{-1}, compared with the more recent value obtained above of 3097 cm^{-1} [Hirsch, Liebermann and Buenker (1998)]. It is noteworthy that this energy difference is about one-third of the analogous value for the NO_2 conical intersection. The density of states in the region of the conical section is therefore much lower in C_2H, especially since it possesses a much lighter atom than does NO_2. This fact is quite important in understanding why the perturbations

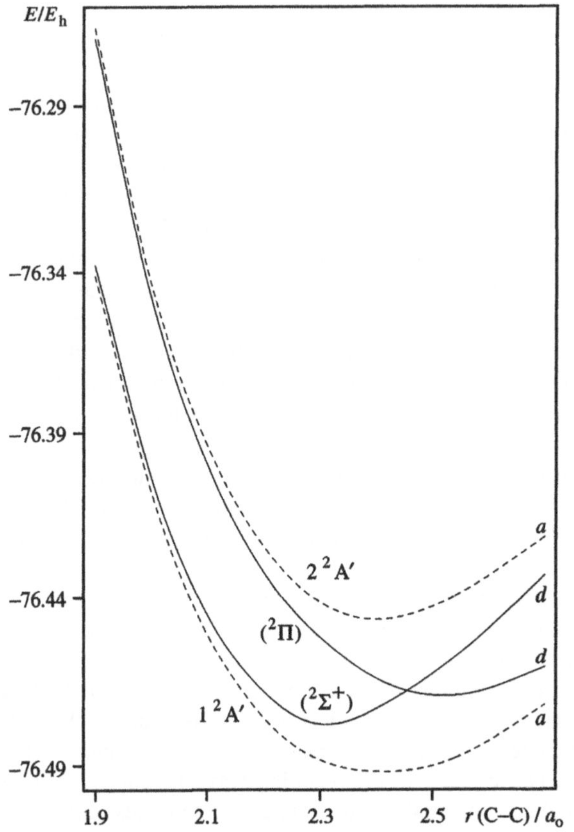

Figure 5.10. Computed stretching potential energy curves of the first two $^2A'$ states of the C_2H molecule at a bond distance r(C–H) $= 2.0\,a_0$ and a bond angle $\theta = 120°$. Dashed curves, adiabatic; solid curves, diabatic with a $(^2\Sigma^+)/(^2\Pi)$ labelling according to their character at linearity.

in the associated electronic spectra are considerably smaller in the C_2H case. As a result it is still possible to assign the C_2H spectrum on the basis of simple mixtures of $^2\Pi$ and $^2\Sigma^+$ levels well beyond the location of the conical intersection [Perić, Buenker and Peyerimhoff (1990b)]. In NO_2 the situation is quite different, with many vibrational functions of both electronic states contributing to the final vibronic wave functions over a wide range of energy. This is an extreme example of the breakdown in the adiabatic approximation upon which much of the theory of molecular spectroscopy is based.

5.4.3 BiOH SPIN–ORBIT EFFECTS

The potential energy surfaces of polyatomic molecules with heavy atoms can also be greatly affected by spin–orbit and other relativistic effects, similarly as for the diatomic systems discussed in Section 5.3.3. A case in point is the BiOH molecule, whose spectrum has recently been detected by Shestakov *et al.* (1998) in the course of their investigation of BiO. Because the triatomic system is isoelectronic with BiF and isovalent with O_2, one expects to find the familiar π^{*2} states lying lowest in its spectrum: a $^3\Sigma^-$ ground state, followed by $^1\Delta$ and $^1\Sigma^+$ excited states. This turns out to be the case but an additional complicating factor is that BiOH has a bent equilibrium conformation and so these states are split into various C_s multiplets.

Multireference CI calculations were first carried out without including the spin–orbit interaction and a three-dimensional potential energy surface was generated [Khandogin *et al.* (1997)]. Diagrams showing various cuts in this PES along all three vibrational coordinates may be found in the original reference. The bending potentials indicate that the $^3A''$ ground state prefers a bond angle of 105°. As expected, the corresponding $^1\Delta$-type states have very similar PESs. The reduced symmetry in the bent conformation causes a splitting in the energy of these two states of about 600 cm^{-1}, with the $^1A'$ component lying lower. This effect can be explained by the latter's interaction with the $2\,^1A'$ state of somewhat greater energy which derives from the $^1\Sigma^+$ state with the same π^{*2} electronic configuration. Beyond this state there is a pair of $^3\Pi$ components, $^3A'$ and $^3A''$. The latter ($2\,^3A''$) is pushed higher in the spectrum because of its interaction with the ground state of the same symmetry. The $^3\Pi$ has a $\pi^*\sigma^*$ configuration and so its PES are notably different than for any of the lower-lying π^{*2} states, particularly along the BiO stretch coordinate. For example, the $^3A'$ state dissociates to the ground state atomic limits, Bi($^4S^o$) + OH($^2\Pi$), and so its potential energy curve along the BiO stretch coordinate crosses those of the $^1\Delta$ and $^1\Sigma^+$ states which correlate with the first excited atomic limit, Bi($^2D^o$) + OH($^2\Pi$).

When spin–orbit is added to the theoretical treatment, a number of interesting effects occur in addition. The $^3A''$ ground state splits apart into three components, not just two as in the case of the analogous diatomic (or linear triatomic) state, $X\,^3\Sigma^-$. The lowest-energy X_1 state has A' symmetry in the C_s double group, which correlates with 0^+ in the linear nuclear arrangement. The corresponding $\Omega = 1$ state is split into two components, $X_2\,A''$ and $X_3\,A'$, albeit by only 29 cm^{-1} in the calculations. The corresponding X_1–X_2 energy splitting is much greater, 5213 cm^{-1} [Khandogin *et al.* (1997)]. This seems to be an underestimate of the observed T_e value, which has been estimated to be about 6200 cm^{-1} [Fink, Shestakov and Setzer (1997)]. The calculations indicate that the analogous splitting in the linear geometry is some 700 cm^{-1} higher. Adding this amount to the above observed estimate gives a value for the X_1–X_2 splitting in linear BiOH of 6900 cm^{-1}, which is in good agreement with the observed BiF splitting of 6768 cm^{-1} [Fink *et al.* (1991)]. Again because of the isoelectronic relationship between BiOH and BiF, one would expect them to exhibit similar properties, especially when the triatomic is in a linear conformation, so this result is strong evidence for the correctness of the experimental assignment of the new spectrum.

The $^1\Delta$ states are not greatly affected by the inclusion of spin–orbit, but a strong interaction is noted between the $^1\Sigma^+$ and $^3\Pi$ A' components. The $3A'$ state at relatively small R values ($3.5\,a_0$) is predominantly $2\,^1A'$, i.e. the $^1\Sigma^+$ state. A contribution of 20 % on a c^2 basis comes from the $^3\Sigma^-$–A' state, but that is quite normal for all π^{*2} systems. As the bond length increases, however, the $^3\Pi$ states become increasingly important, so that by $R = 4.5\,a_0$ they constitute 55 % of the $3A'$, whereas the contribution of the $2\,^1A'$–$^1\Sigma^+$ is now only 20 % ($^3\Sigma^-$ continues to make a similar contribution at this R value). This behavior is very similar to what is found in calculations for isoelectronic BiF [Alekseyev *et al.* (1993)] and it shows up in the corresponding Bi–O stretch potential curve of the $3A'$ state. The latter starts out being parallel to those of the $^3\Sigma^-$ and $^1\Delta$ states, but at large R values it exhibits a shoulder and begins to draw much closer to the $2A'$ curve below it, which is of predominantly $^1\Delta$ character.

Another interesting aspect of the molecular structure of this system is the placement of the H atom relative to the BiO fragment. Additional MRD-CI calculations have also been carried out for the HBiO isomer at the λ–s level of treatment [Khandogin *et al.* (1997)]. It is found that the ground state is no longer $^3A''$, but rather $^1A'$. The latter's T_e value is computed to be 4020 cm^{-1} (with reference to the BiOH $^3A''$ minimal energy). This means it is 2964 cm^{-1} more stable than the analogous state in BiOH. On the other hand, the HBiO $^3A''$ has a T_e value of 11 413 cm^{-1}

according to these calculations. The computed BiO equilibrium bond length in all four states is $2.12\,a_0$, which corresponds to a single bond. This offers a potentially simple way to explain why the BiOH isomer is preferred. Since there is little difference in the BiO bond strengths based on this result, it can be expected that the nature of the X–H bond is decisive in making this determination. That clearly favors BiOH because the O–H bond is definitely more stable than its H–Bi counterpart. The above argument actually works only for the comparison of the two $^3A''$ states, however, since as we have seen, the HBiO $^1A'$ is somewhat more stable than its BiOH counterpart. The $^1A'$ has essentially a closed-shell configuration, and it is much less sensitive to the position of the H atom than the open-shell triplet. This example demonstrates that it is rather difficult to come up with sound qualitative explanations for such energy relationships based on simple MO arguments, a fact which helps to underscore the need for carrying out accurate potential energy calculations to study such questions.

5.4.4 LARGER SYSTEMS

Once one goes beyond triatomics it is extremely difficult to include variations in all internal coordinates in a comprehensive manner in computing potential surfaces. This does not mean that such calculations are of no value for such systems, rather only that one must carefully choose the portions of the PES which have a direct bearing on the structural problem at hand. In the following we will simply mention a few examples of how this can be done in an effective manner.

The calculations discussed in Section 5.3.1 have shown how potential energy curves for a series of electronic states can be combined with coupling terms to predict cross sections and rate constants for atom–atom collisions. For many applications in fields such as astrophysics, radiation therapy and nuclear fusion research, the collision energies of interest are relatively high, often in the keV range. Under these conditions when a proton or helium atom collides with a molecule, it is reasonable to assume that its nuclear conformation does not have sufficient time to change significantly during the time of interaction. One can apply something akin to the Franck–Condon principle to such situations, which means that it is reasonable to freeze the geometry of the molecule in its equilibrium conformation and simply vary the position of the colliding atom relative to it.

Calculations of this type have been carried out for collisions of protons with the small hydrocarbons, methane and ethyne, and the results have recently been reviewed [Buenker *et al.* (1998)]. For CH_4 three different paths of proton approach were explored, along the direction of a CH bond from both sides (C_{3v}) and bisecting a CH_2 angle (C_{2v}). Primary interest lies in charge-transfer reactions but elastic collision cross sections have also been calculated for comparison. Since the ground state of $CH_4{}^+$ is triply degenerate, it is found that its potential curves for proton approach split apart into components, only one of which corresponds to the same irreducible representation in either C_{3v} or C_{2v} as that of the initial state. Because the molecule does not have sufficient time to relax in such high-energy collisions, the methane ion produced is not in its equilibrium conformation. Thus even though its adiabatic IP is nearly the same as that for the H atom, its asymptotic energy is 0.6 eV higher than in the initial channel. The calculated potentials show that the corresponding two potential curves of the same irreducible representation do not really approach each other. It might be anticipated that there is very low probability for charge transfer under the circumstances, but this is not the case. The radial coupling elements connecting these two states are fairly large because, although there is no complete exchange of diabatic character along the reaction path, significant mixing does occur, leading to what is known as Demkov coupling. That such an interaction does occur is already evident from the potential curves, since that of the charge-transfer state of the same symmetry as the initial species is pushed upwards relative to those of the other two components of different symmetry. It is found that the cross section for charge transfer reaches a maximum of $1.0 \times 10^{-15}\,\text{cm}^2$ for the C_{2v} approach, which is more than an order of magnitude greater than for the head-on C_{3v} approach. Qualitatively this result is quite understandable, since the proton has a much freer path to the carbon atom in the former case.

Similar calculations have been carried out for the H^+/C_2H_2 system. In this case the proton was allowed to approach in two directions, along the molecular axis and perpendicular to it along a line to the inversion center. The respective symmetries are $C_{\infty v}$ and C_{2v} and this distinction has a profound effect on the computed cross sections [Buenker *et al.* (1998)]. The initial channel is H^+ plus C_2H_2 in its ground state. The lowest charge-transfer channel lies below this, because the IP of ethyne is so much lower than that of H^+ that the lack of relaxation in the ionic structure still does not alter this relationship. In C_{2v} symmetry these two states are of the same irreducible representation ($^1A_1$), whereas in the linear nuclear arrangement the initial channel is $^1\Sigma^+$ but the charge-transfer state is $^1\Pi$. On the other hand, the first two excited charge-transfer channels both have $^1\Sigma^+$ symmetry for the linear case but only one of them has $^1A_1$ symmetry for the perpendicular approach of the proton. The scattering calculations find that total electron-capture cross section for the C_{2v} approach possesses a minimum near 30 eV and gradually increases with energy to a maximum value of $4.5 \times 10^{-16}\,cm^2$ around 4 keV, while that for the linear approach reaches a maximum of $2.02 \times 10^{-16}\,cm^2$ around 3 keV. The corresponding cross sections for excitation relative to the initial channel are always smaller than for charge transfer, but the difference is notably less in the linear nuclear arrangement. This result is expected based on the symmetry arguments given above.

Another example of a polyatomic molecule for which potential energy calculations have been used successfully to help understand its electronic structure is ethene. The V–N bands in its spectrum have an intensity maximum around 7.66 eV and they are very broad in appearance. Mulliken assigned this as a $\pi \rightarrow \pi^*$ transition in the 1930s [Mulliken (1932, 1933)] and attributed the broadness of the spectrum to progressions in the torsion and C–C stretch vibrations. Petrongolo, Buenker and Peyerimhoff (1982) computed potential energy curves for ethene in the V and N states and concluded on this basis that the C–C stretch was not a primary factor in producing this phenomenon. Instead it was found that an avoided crossing occurs between the $\pi \rightarrow \pi^*$ state and the Rydberg $\pi \rightarrow 3p_y$ species as the molecule undergoes torsion out of the planar conformation. This work has recently been reviewed [Buenker, Hirsch and Li (1999)] and the corresponding potential curves may be found in Figure 5.6 of that reference. This causes a strong vibronic interaction between the torsional levels of these two electronic states.

The calculations also show that the degree of diffuseness in the π^* MO varies rather strongly with torsion. This can be understood in terms of a Rydberg–valence interaction between the diabatic $\pi \rightarrow \pi^*$ and $\pi \rightarrow 3d_\pi$ states of the same symmetry. As torsion occurs out of the planar conformation, the valence state becomes rapidly more stable, reaching a minimum for the perpendicular conformation of the CH_2 groups. The Rydberg state has a potential minimum near 20° torsion but then becomes rapidly less stable. As a result an avoided crossing occurs between these two states and the lower adiabatic state gradually becomes less diffuse as torsion proceeds. The vibronic calculations [Petrongolo, Buenker and Peyerimhoff (1982)] indicate that the transition is distinctly non-vertical as a result of these various effects, especially the conical intersection mentioned first. If the Franck–Condon principle were operative, the calculations indicate that the intensity maximum would occur around 8.0 eV, some 0.4 eV higher than observed, and the broadness of the transition could not be explained. Other symmetric vibrations also help to broaden this spectrum further, but the main effect comes from the avoided crossing in the torsional mode. There is thus a need for isolating the types of vibrational motion which are likely to be responsible for a given experimental effect, so that accurate potential surfaces can be computed for this limited region of coordinate space.

5.5 Conclusions

In order to compute accurate potential energy surfaces for both the ground and excited states of molecules it is necessary to employ a balanced treatment which performs equally well for all types of electronic configurations and nuclear arrangements. It is also advisable to consider various kinds of couplings, radial, rotational and spin–orbit, which drive transitions between the electronic states, both induced and spontaneous. A multireference CI approach has been demonstrated to be quite effective in accomplishing both goals. This allows one to provide a flexible representation of all electronic states of interest, one which can always be upgraded on the basis of the results of a given treatment simply by expanding the list of reference configurations in a succeeding run to include species which have made a significant contribution to one or more of the eigenvectors. Once the final CI wave functions have been generated for a given fixed nuclear conformation, they can be used to calculate the coupling terms mentioned above, as well as quantities such as electric-dipole transition moments which are needed for computing radiative transition probabilities.

The capabilities of the MRD-CI method have been illustrated with the help of results for a large number of diatomic molecules. The first applications have dealt with the scattering of protons and helium atoms off heavier atoms and ions. Subsequent calculations of collision cross sections based on these data have proven quite useful in a variety of fields, including astrophysics, radiation therapy with heavy ions, thin-film etching for semiconductors and nuclear fusion magnetic-confinement experiments. Both differential and total cross sections have been calculated on this basis and the results can be interpreted quite readily on the basis of the associated potential curves and coupling matrix elements. The same information has been employed to study nonradiative decay processes, such as predissociation in the spectra of the O_2 and CO molecules.

For molecules with heavy atoms such as lead and bismuth, it becomes essential to include relativistic effects, both scalar and spin–orbit, in calculating potential curves. The gradual escalation in the importance of these effects as the nuclear charge of the atomic constituents increases has been illustrated with a series of relativistic CI calculations of the hydrides and oxides of the Group VA elements, arsenic through bismuth. The λ–s level of theory becomes gradually less acceptable and it eventually becomes misleading to assign the observed molecular spectra on this basis for systems containing atoms as heavy as bismuth. Spin-forbidden transitions become the rule rather than the exception, and it is necessary to consider spin–orbit effects explicitly in order to have accurate predictions of radiative lifetimes of the excited states of such systems. All the required extensions in the theoretical treatment can be achieved with relative simplicity through the use of effective core potentials (RECPs).

Extending these computational methods to polyatomics is straightforward, but applications are limited by the rapid increase in the number of degrees of freedom with increasing number of the constituent atoms. It is generally feasible to compute full (3D) potential surfaces and couplings for triatomics and a number of examples have been discussed. A key difference between diatomics and polyatomics is the occurrence of conical intersections in the latter case. Adiabatic (CI) wave functions change discontinuously along the seams of such intersections, and as a consequence the radial coupling matrix elements connecting the participating states have singularities in this representation. The basic computational difficulties which thus arise can be minimized by transforming to a diabatic representation and this procedure has been found to be quite effective in analyzing the complex spectra of the A–X transitions of the NO_2 and C_2H molecules. In the neighborhood of the conical intersection vibronic mixing becomes so dominant that it is no longer possible to assign the spectra in terms of integral numbers of vibrational quanta for a particular electronic state. The spectra take on a distinctly irregular appearance and it is not possible to distinguish simple progressions in a particular vibrational mode. It has been pointed out that application of a geometric phase to the CI wavefunctions does not overcome the numerical difficulties associated with the existence of singular radial coupling elements in the adiabatic representation. Rather it has been shown that diabatization eliminates the need for a geometrical phase because the corresponding wave functions are by construction slowly varying with changes in the nuclear conformation, and in particular, they do not change sign upon traversing a closed path around the conical intersection. Finally, calculations of the BiOH isomer have illustrated how relativistic effects can influence the electronic structure of triatomic molecules containing heavy atoms. The lower symmetry which can arise for bent systems creates additional possibilities for the mixing of states of different multiplicity at the $\lambda-s$ level of treatment.

In order to extend these computational methods to the study of general polyatomic molecules it is necessary to restrict the number of vibrational coordinates for which explicit potential energy and coupling results are needed. It becomes more difficult to make meaningful comparisons with experimental data for such large systems as well, and so it becomes essential to reduce the dimensionality of the problem to make substantial progress. For some time to come then, the role of such calculations in the study of large molecules will be to provide quantitative results for relatively small portions of potential surfaces which appear to have a bearing on a given experimental interpretation. That such a restricted role for CI calculations can nonetheless have a positive impact on experimental studies has been illustrated in the work on proton collisions with hydrocarbons and other small molecules, as well as the elucidation of the spectrum of the ethene molecule.

Acknowledgement

This work was supported in part by the Deutsche Forschungsgemeinschaft (grants Bu 450/6, Bu 450/7, Bu 450/9, Bu 450/10, Th 299/4 and Bu 152/12). The financial support of the Fonds der Chemischen Industrie is also gratefully acknowledged.

5.6 References

Alekseyev, A. B., Buenker, R. J., Liebermann, H.-P., and Hirsch, G., 1994a, *J. Chem. Phys.* **100**, 2989–3001.
Alekseyev, A. B., Liebermann, H.-P., Boustani, I., Hirsch, G., and Buenker, R. J., 1993, *Chem. Phys.* **173**, 333–344.
Alekseyev, A. B., Liebermann, H.-P., Buenker, R. J., Hirsch, G., and Li, Y., 1994b, *J. Chem. Phys.* **100**, 8956–8968.
Alekseyev, A. B., Liebermann, H.-P., Buenker, R. J., and Hirsch, G., 1995a, *J. Chem. Phys.* **102**, 2539–2550.
Alekseyev, A. B., Liebermann, H.-P., Hirsch, G., and Buenker, R. J., 1998a, *J. Chem. Phys.* **108**, 2028–2040.
Alekseyev, A. B., Liebermann, H.-P., Lingott, R. M., Bludský, O., and Buenker, R. J., 1998b, *J. Chem. Phys.* **108**, 7695–7706.
Alekseyev, A. B., Sannigrahi, A. B., Liebermann, H.-P., Buenker, R. J., and Hirsch, G., 1995b, *J. Chem. Phys.* **103**, 234–244.
Ast, T., Porter, C. J., Proctor, C. J., and Beynon, J. H., 1981, *Chem. Phys. Lett.* **78**, 439–441.
Augustin, S. D., Miller, W. H., Pearson, P. K., and Schaefer III, H. F., 1973, *J. Chem. Phys.* **58**, 2845–2854.
Baker, J., Tchang-Brillet, W.-Ü. L., and Julienne, P. S., 1995, *J. Chem. Phys.* **102**, 3956–3961.
Balasubramanian, K., 1986, *J. Mol. Spectrosc.* **115**, 258–268.
Balfour, W. J., and Ram, R. S., 1984, *J. Mol. Spectrosc.* **105**, 246–259.
Balfour, W. J., and Ram, R. S., 1988, *J. Mol. Spectrosc.* **130**, 382–388.
Balslev, E., and Combes, J. M., 1971, *Commun. Math. Phys.* **22**, 280–294.
Berkowitz, J., 1988, *J. Chem. Phys.* **89**, 7065–7076.
Berry, M. V., 1984, *Proc. R. Soc. London, Sen., A* **392**, 45–57.
Bollmark, P., and Lindgren, B., 1974, *Phys. Scr.* **10**, 325–330.
Bopegedera, A. M. R. P., Brazier, C. R., and Bernath, P. F., 1989, *Chem. Phys. Lett.* **162**, 301–305.
Born, M., and Oppenheimer, J. R., 1927, *Ann. Phys.* **84**, 457–484.
Breidohr, R., Shestakov, O., and Fink, E. H., 1998, personal communication.
Buenker, R. J., 1982, in *Studies in Physical and Theoretical Chemistry*, Carbó, R., Ed. Vol. 21: *Current Aspects of Quantum Chemistry 1981*; Elsevier: Amsterdam, pp. 17–34.

Buenker, R. J., 1986, *Int. J. Quant. Chem.* **29**, 435–460.
Buenker, R. J., Hirsch, G., and Li, Y., 1999, in *The Role of Rydberg States in Spectroscopy and Photochemistry. Low and High Rydberg States*, Sandorfy, C., Ed.; Kluwer: Dordrecht, pp. 57–91.
Buenker, R. J., Hirsch, G., Peyerimhoff, S. D., Bruna, P. J., Römelt, J., Bettendorff, M., and Petrongolo, C., 1982, in *Studies in Physical and Theoretical Chemistry*, Carbó, R., Ed. Vol. 21: *Current Aspects of Quantum Chemistry 1981*; Elsevier: Amsterdam, pp. 81–97.
Buenker, R. J., and Krebs, S., 1999, in *Recent Advances in Computational Chemistry*, Hirao, K., Ed. Vol. 4: *Recent Advances in Multireference Methods*; World Scientific: Singapore, pp. 1–29.
Buenker, R. J., Li, Y., Hirsch, G., and Kimura, M., 1998, *J. Phys. Chem. A* **102**, 7127–7136.
Buenker, R. J., and Peyerimhoff, S. D., 1968, *Theor. Chim. Acta* **12**, 183–199.
Buenker, R. J., and Peyerimhoff, S. D., 1974, *Theor. Chim. Acta* **35**, 33–58.
Buenker, R. J., and Peyerimhoff, S. D., 1975, *Theor. Chim. Acta* **39**, 217–228.
Buenker, R. J., and Peyerimhoff, S. D., 1976, *Chem. Phys. Lett.* **37**, 208–211.
Buenker, R. J., and Phillips, R. A., 1985, *J. Mol. Struct. Theochem* **123**, 291–300.
Bunker, P. R., and Jensen, P., 1998, *Molecular Symmetry and Spectroscopy*; NRC Canada: Ottawa.
Carrick, P. G., Merer, A. J., and Curl Jr., R. F., 1983, *J. Chem. Phys.* **78**, 3652–3658.
Carrick, P. G., Pfeiffer, J., Curl Jr., R. F., Koestner, E., Tittel, F. K., and Kasper, J. V. V., 1982, *J. Chem. Phys.* **76**, 3336–3337.
Cooper, D. L., and Kirby, K, 1987, *J. Chem. Phys.* **87**, 424–432.
Cooper, D. L., and Kirby, K, 1989, *J. Chem. Phys.* **90**, 4895–4902.
Cooper, D. M., and Langhoff, S. R., 1981, *J. Chem. Phys.* **74**, 1200–1210.
Curl Jr., R. F., Carrick, P. G., and Merer, A. J., 1985, *J. Chem. Phys.* **82**, 3479–3486.
Dai, D., and Balasubramanian, K., 1990, *J. Chem. Phys.* **93**, 1837–1846.
Delon, A., and Jost, R., 1991, *J. Chem. Phys.* **95**, 5686–5718.
Dixon, R. N., and Lamberton, H. M., 1968, *J. Mol. Spectrosc.* **25**, 12–33.
Dolg, M., Kuechle, W., Stoll, H., Preuss, H., and Schwerdtfeger, P., 1991, *Mol. Phys.* **74**, 1265–1285.
Eidelsberg, M., Benayoun, J. J., Viala, Y., and Rostas, F., 1990, *Astron. Astrophys. Suppl.* **90**, 231–282.
Eidelsberg, M., Benayoun, J. J., Viala, Y., Rostas, F., Smith, P. L., Yoshino, K., and Shettel, C. A., 1992, *Astron. Astrophys.* **265**, 839–842.
Eidelsberg, M., and Rostas, F., 1990, *Astron. Astrophys.* **235**, 472–489.
England, J. P., Lewis, B. R., Gibson, S. T., and Ginter, M. L., 1995, *J. Chem. Phys.* **104**, 2765–2771.
England, J. P., Lewis, B. R., and Ginter, M. L., 1995, *J. Chem. Phys.* **103**, 1727–1731.
Fink, E. H., and Ramsay, D. A., 1997, *J. Mol. Spectrosc.* **185**, 304–324.
Fink, E. H., Setzer, K. D., Ramsay, D. A., and Vervloet, M., 1991, *Chem. Phys. Lett.* **179**, 95–102.
Fink, E. H., Shestakov, O., and Setzer, K. D., 1997, *J. Mol. Spectrosc.* **183**, 163–167.
Fock, V., 1930, *Z. Physik* **61**, 126–148.
Galloy, C., and Lorquet, J. C., 1977, *J. Chem. Phys.* **67**, 4672–4680.
Gillispie, G. D., Khan, A. U., Wahl, A. C., Hosteny, R. P., and Krauss, M., 1975, *J. Chem. Phys.* **63**, 3425–3444.
Gu, J.-P., Buenker, R. J., and Hirsch, G., 1998, personal communication.
Gu, J.-P., Hirsch, G., Buenker, R. J., Kimura, M., Dutta, C. M., and Nordlander, P., 1998, *Phys. Rev A.* **57**, 4483–4489.
Hall, G. G., 1951, *Proc. R. Soc., London, Sen. A* **205**, 541–552.
Hedderich, H. G., and Bernath, P. F., 1993, *J. Mol. Spectrosc.* **158**, 170–176.
Hensel, K. D., Hughes, R. A., and Brown, J. M., 1995, *J. Chem. Soc., Faraday Trans.* **91**, 2999–3004.
Herzberg, G., 1966, *Molecular Spectra and Molecular Structure*. Vol. 3: *Electronic Spectra and Electronic Structure of Polyatomic Molecules*; Van Nostrand Reinhold: Princeton, NJ, pp. 197–200.
Hirsch, G., Bruna, P. J., Buenker, R. J., and Peyerimhoff, S. D., 1980, *Chem. Phys.* **45**, 335–347.
Hirsch, G., Buenker, R. J., and Petrongolo, C., 1991, *Mol. Phys.* **73**, 1085–1099.
Hirsch, G., Liebermann, H.-P., and Buenker, R. J., 1998, unpublished results.
Hsu, D. K., Monts, D. L., and Zare, R. N., 1978, *Spectral Atlas of Nitrogen Dioxide from 5530Å to 6480Å*; Academic Press: New York.
Huber, K. P., and Herzberg, G., 1979, *Molecular Spectra and Molecular Structure*. Vol. 4: *Constants of Diatomic Molecules*; Van Nostrand Reinhold: Princeton, NJ.
Hughes, B. M., and Tiernan, T. O., 1971, *J. Chem. Phys.* **55**, 3419–3426.
Hund, F., 1927a, *Z. Physik* **40**, 742–764.
Hund, F., 1927b, *Z. Physik* **42**, 93–120.
Hurley, M. M., Pacios, L. F., Christiansen, P. A., Ross, R. B., and Ermler, W. C., 1986, *J. Chem. Phys.* **84**, 6840–6853.
Hylleraas, E. A., 1928, *Z. Physik* **48**, 469–494.
Hylleraas, E. A., 1930, *Z. Physik* **65**, 209–225.
Johnson, R. D., Long, G. R., and Hudgens, J. W., 1987, *J. Chem. Phys.* **87**, 1977–1981.
Kendrick, B., 1997, *Phys. Rev. Lett.* **79**, 2431–2434.

Khandogin, Y., Alekseyev, A. B., Liebermann, H.-P., Hirsch, G., and Buenker, R. J., 1997, *J. Mol. Spectrosc.* **186**, 22–33.
Kimura, M., Dalgarno, A., Chantranupong, L., Li, Y., Hirsch, G., and Buenker, R. J., 1994a, *Phys. Rev. A* **49**, 2541–2544.
Kimura, M., Gu, J.-P., Hirsch, G., and Buenker, R. J., 1997, *Phys. Rev. A* **55**, 2778–2785.
Kimura, M., Gu, J.-P., Li, Y., Hirsch, G., and Buenker, R. J., 1994b, *Phys. Rev. A* **49**, 3131–3133.
Kimura, M., Gu, J.-P., Liebermann, H.-P., Li, Y., Hirsch, G., Buenker, R. J., and Dalgarno, A., 1994c, *Phys. Rev. A* **50**, 4854–4858.
Koch, W., Liu, B., Weiske, T., Lebrilla, C. B., Drewello, T., and Schwarz, H., 1987, *Chem. Phys. Lett.* **142**, 147–152.
Kusakabe, T., Mizumoto, Y., Katsurayama, K., and Tawara, H., 1990, *J. Phys. Soc. Jpn.* **59**, 1987–1994.
LaJohn, L. A., Christiansen, P. A., Ross, R. B., Atashroo, T., and Ermler, W. C., 1987, *J. Chem. Phys.* **87**, 2812–2824.
Leonardi, E., Petrongolo, C., Hirsch, G., and Buenker, R. J., 1996, *J. Chem. Phys.* **105**, 9051–9067.
Lewis, B. R., Gibson, S. T., Emami, M., and Carver, J. H., 1988a, *J. Quant. Spectrosc. Radiat. Transfer* **40**, 1–13.
Lewis, B. R., Gibson, S. T., Emami, M., and Carver, J. H., 1988b, *J. Quant. Spectrosc. Radiat. Transfer* **40**, 469–477.
Li, Y., Bludský, O., Hirsch, G., and Buenker, R. J., 1997a, *J. Chem. Phys.* **107**, 3014–3020.
Li, Y., Buenker, R. J., and Hirsch, G., 1998, *Theor. Chem. Acc.* **100**, 112–116.
Li, Y., Hirsch, G., and Buenker, R. J., 1998, *J. Chem. Phys.* **108**, 8123–8129.
Li, Y., Honigmann, M., Bhanuprakash, K., Hirsch, G., Buenker, R. J., Dillon, M. A., and Kimura, M., 1992, *J. Chem. Phys.* **96**, 8314–8323.
Li, Y., Petsalakis, I. D., Liebermann, H.-P., Hirsch, G., and Buenker, R. J., 1997b, *J. Chem. Phys.* **106**, 1123–1133.
Lindgren, B., and Nilsson, C., 1975, *J. Mol. Spectrosc.* **55**, 407–419.
Mathur, D., and Badrinathan, C., 1989, *J. Phys. B* **20**, 1517–1525.
Matsushita, T., Marian, C. M., Klotz, R., and Peyerimhoff, S. D., 1987, *Can. J. Phys.* **65**, 155–164.
Mead, C. A., and Truhlar, D. G., 1982, *J. Chem. Phys.* **77**, 6090–6098.
Moiseyev, N., 1981, *Int. J. Quant. Chem.* **20**, 835–842.
Mulliken, R. S., 1928a, *Phys. Rev.* **32**, 186–222.
Mulliken, R. S., 1928b, *Phys. Rev.* **32**, 761–772.
Mulliken, R. S., 1929, *Phys. Rev.* **33**, 730–747.
Mulliken, R. S., 1932, *Phys. Rev.* **41**, 751–758.
Mulliken, R. S., 1933, *Phys. Rev.* **43**, 279–302.
Neuheuser, T., Sukumar, N., and Peyerimhoff, S. D., 1995, *Chem. Phys.* **194**, 45–64.
Osmann, G., Bunker, P. R., and Jensen, P., 1998, personal communication.
Pacios, L. F., and Christiansen, P. A., 1985, *J. Chem. Phys.* **82**, 2664–2671.
Perić, M., Buenker, R. J., and Peyerimhoff, S. D., 1990a, *Mol. Phys.* **71**, 673–691.
Perić, M., Buenker, R. J., and Peyerimhoff, S. D., 1990b, *Mol. Phys.* **71**, 693–719.
Petrongolo, C., Buenker, R. J., and Peyerimhoff, S. D., 1982, *J. Chem. Phys.* **76**, 3655–3667.
Renner, R., 1934, *Z. Phys.* **92**, 172–193.
Roothaan, C. C. J., 1951, *Rev. Mod. Phys.* **23**, 69–89.
Roothaan, C. C. J., 1960, *Rev. Mod. Phys.* **32**, 179–185.
Ross, R. B., Powers, J. M., Atashroo, T., Ermler, W. C., LaJohn, L. A., and Christiansen, P. A., 1990, *J. Chem. Phys.* **93**, 6654–6669.
Schryber, J. H., Polansky, O. L., Jensen, P., and Tennyson, J., 1997, *J. Mol. Spectrosc.* **185**, 234–243.
Shepard, R., 1987, *Adv. Chem. Phys.* **69**, 63–200.
Shestakov, O., Breidohr, R., Demes, H., Setzer, K. D., and Fink, E. H., 1998, *J. Mol. Spectrosc.* **190**, 28–77.
Shestakov, O., and Fink, E. H., 1996, *J. Mol. Spectrosc.* **172**, 215–224.
Slater, J. C., 1929, *Phys. Rev.* **34**, 1293–1322.
Slater, J. C., 1930, *Phys. Rev.* **35**, 210–224.
Slater, J. C., 1963, *Quantum Theory of Molecules and Solids*. Vol. 1: *Electronic Structure of Molecules*; McGraw-Hill: New York.
Stancil, P. C., Havener, C. C., Krstić, P. S., Schultz, D. R., Kimura, M., Gu, J.-P., Hirsch, G., Buenker, R. J., and Zygelman, B., 1998, *Astrophys. J.* **502**, 1006–1009.
Stancil, P. C., Kirby, K., Sannigrahi, A. B., Buenker, R. J., Hirsch, G., and Gu, J.-P., 1997, *Astrophys. J.* **486**, 574–579.
Sur, A., Friedman, R. S., and Miller, P. J., 1991, *J. Chem. Phys.* **94**, 1705–1711.
Sur, A., Ramana, C. V., Chupka, W. A., and Colson, S. D., 1986, *J. Chem. Phys.* **84**, 69–72.
Tanaka, Y., 1952, *J. Chem. Phys.* **20**, 1728–1733.
Tashkun, S. A., and Jensen, P., 1994, *J. Mol. Spectrosc.* **165**, 173–184.
Tchang-Brillet, W.-Ü. L., Julienne, P. S., Robbe, J. M., Letzelter, C., and Rostas, F., 1992, *J. Chem. Phys.* **96**, 6735–6745.
Thümmel, H., Perić, M., Peyerimhoff, S. D., and Buenker, R. J., 1989, *Z. Phys. D.* **13**, 307–316.

Tinkham, M., 1964, *Group Theory and Quantum Mechanics*; McGraw-Hill: New York, pp. 162–167.
Ubachs, W., Eikema, K. S. E., Levelt, P. F., Hogervost, W., Drabbels, M., Meerts, W. L., and ter Meulen, J. J., 1994, *Astrophys. J.* **427**, L55–L58 and references therein.
Urban, R. D., Polomosky, P., and Jones, H., 1989, *Chem. Phys. Lett.* **181**, 485–490.
Viala, Y. P., Letzelter, C., Eidelsberg, M., and Rostas, F., 1988, *Astron. Astrophys.* **193**, 265–272.
Wetzel, T. L., Welton, R. F., Thomas, E. W., Borkman, R. F., and Moran, T. F., 1993, *J. Phys. B* **26**, 49–59.
Wolfrum, E., Schweinzer, J., and Winter, H., 1992, *Phys. Rev. A* **45**, R4218–R4221.
Wolk, G. L., and Rich, J. W., 1983, *J. Chem. Phys.* **79**, 12–18.
Yarkony, D. A., 1995, in *Advanced Series in Physical Chemistry*, Yarkony, D. A., Ed. Vol. 2: *Modern Electronic Structure Theory (Part I)*; World Scientific: Singapore, pp. 642–721.
van der Zande, W. J., Koot, W., Peterson, J. R., and Los, J., 1987, *Chem. Phys. Lett.* **140**, 175–180.
van der Zande, W. J., Koot, W., and Los, J., 1988, *J. Chem. Phys.* **89**, 6758–6770.
van der Zande, W. J., Koot, W., and Los, J., 1989, *J. Chem. Phys.* **91**, 4597–4602.

6 RELATIVISTIC EFFECTS IN THE CALCULATION OF ELECTRONIC ENERGIES

Bernd Artur Hess

Lehrstuhl für Theoretische Chemie, Friedrich–Alexander-Universität Erlangen–Nürnberg, Germany

and

Christel Maria Marian

GMD Center for Information Technology, Scientific Computing and Algorithms Institute, St. Augustin, Germany, and Universität Bonn, Germany

Computational Molecular Spectroscopy. Edited by Per Jensen and Philip R. Bunker

6.1 Introduction

6.1.1 KINEMATIC RELATIVISTIC EFFECTS

Even before quantum theory was given its nowadays generally accepted formulation, Arnold Sommerfeld (1916) in his article 'Zur Quantentheorie der Spektrallinien' discussed the relevance of relativistic effects for the spectroscopy of hydrogen-like atoms. He showed that for a hydrogen-like atom with nuclear charge Z there is a relativistic contribution to the binding energy of the electron proportional to $\alpha^2 Z^2$ in lowest order. This is to a large extent the effect of relativistic kinematics which must be applied to the rapidly moving electrons in the inner shells of heavy atoms.

The effect of relativistic kinematics on the shape of the orbitals can be made plausible by estimating the velocity of the electron in in the 1s orbit of a hydrogen-like atom and using Bohr–Sommerfeld theory of hydogen-like atoms [Pyykkö and Desclaux (1979)]. Adopting the picture that the centrifugal force exerted on the electron should balance the Coulomb attraction,

$$\frac{Ze^2}{4\pi\varepsilon_0 r^2} = \frac{mv^2}{r}, \tag{1}$$

and by using the quantization condition $mvr = n\hbar$, we obtain the familiar expression for the radius of the first Bohr orbit ($n = 1$):

$$r_{\text{Bohr}} = \frac{4\pi\varepsilon_0\hbar^2}{Ze^2 m}. \tag{2}$$

Although *ab initio* calculations are normally carried out in the system of atomic units, in which Planck's constant $\hbar = 1$, the electron mass $m = 1$, the electron charge $e = 1$ as well as the permeability of the vacuum $4\pi\varepsilon_0 = 1$, we include these constants explicitly for convenience.

The transition to relativistic physics is now often performed by noting that the rest mass m of the moving electron measured in the laboratory system (the system of the nucleus which is assumed to be at rest) is increased to the 'relativistic mass'

$$M = m\gamma \equiv \frac{m}{\sqrt{1 - v^2/c^2}}, \tag{3}$$

which enters the expression for the Bohr radius, leading to a value which is smaller than the non-relativistic value by a factor γ. The problem of this argument is an unsatisfactory mixing up of kinematic notions with an invariant property of the particle, namely the mass. The coordinate transformation (i.e., Lorentz transformation) from the system of reference of the moving electron to the laboratory system should only affect the dynamic variables and quantities derived thereof, and not a particle property.

Indeed, modern books [Lindner (1994) and Taylor and Wheeler (1994)] avoid this notion of a 'relativistic mass', in line with a comment of Albert Einstein[1]

> Es ist nicht gut, von der Masse $M = m/\sqrt{1 - v^2/c^2}$ eines bewegten Körpers zu sprechen, da für M keine klare Definition gegeben werden kann. Man beschränkt sich besser auf die 'Ruhe-Masse' m. Daneben kann man ja den Ausdruck für Impuls und Energie geben, wenn man das Trägheitsverhalten rasch bewegter Körper angeben will.[2]

---

[1] Einstein in a letter (19. 6. 1948) to L. Barnett, cited after J. Strnad, *Eur. J. Phys.* **12**, 69–73 (1991); see also Lindner (1994), page 235.

[2] One ought not speak of the mass $M = m/\sqrt{1 - v^2/c^2}$ of a moving body, since no clear definition for M can be given. It is better to confine oneself to the 'rest mass' m. Aside, one might give the expressions for momentum and energy, if one wants to describe the behavior of inertia of rapidly moving bodies.

The argument to obtain the 'relativistic first Bohr radius' should therefore be slightly changed: we make the approximation that at every instance the electron is in an inertial system moving perpendicular to the radius vector which points from the nucleus to the electron and has length r. The relativistic expression for the electric field generated by the electron as measured in the laboratory system (the nucleus) is [Landau and Lifschitz (1981)]

$$E_{\perp} = \frac{e\gamma}{4\pi\varepsilon_0 r^2}. \tag{4}$$

This field multiplied with the charge Ze of the nucleus leads to a concomitant Coulomb force, which should again be balanced by the centrifugal force. The relativistic formulation requires, according to Einstein's remark, the use of the relativistic momentum $p_{\text{rel}} = mv\gamma$. We thus obtain for the force component along the radius vector

$$\frac{Ze^2\gamma}{4\pi\varepsilon_0 r^2} = \frac{L^2}{mr^3} \tag{5}$$

and the quantization condition $|\tilde{L}| = n\hbar$, using the expression for the angular momentum $\tilde{L} = \tilde{r} \times \tilde{p}_{\text{rel}}$. Solving for r, we recover a classical estimate of the 'relativistic' Bohr radius

$$r_{\text{Bohr}}^{\text{rel}} = \frac{1}{\gamma}\frac{4\pi\varepsilon_0\hbar^2}{Ze^2 m}. \tag{6}$$

It can be estimated from the classical expression of the velocity in the first Bohr orbit and also by considering the average radial velocity of an electron in a 1s orbital [Pyykkö and Desclaux (1979)] that the velocity of an inner electron is of the order of Z atomic units of velocity. This can be compared with the velocity of light, which is $1/\alpha \approx 137$ atomic units. For mercury ($Z = 80$) we obtain $v/c \approx 0.58$, and the dilation factor $\gamma \approx 1.23$. Thus we should expect that the 1s shells of 6th row atoms are relativistically contracted by about 20 %, compared with results obtained from non-relativistic theory.

This argument holds in the first place for inner-shell electrons, for which the influence of the relativistic kinematics (the so-called 'direct' relativistic effect) is immediate because of their high velocities. However, the electrons of the valence shell are also affected [Schwarz *et al.* (1989)]: The outer s and to a lesser extent the outer p electrons are close to the nucleus for a certain fraction of their 'revolution time'. In a stationary picture one would say that their orbitals have 'inner tails' or that they are 'core-penetrating orbitals'. For this reason, they also experience the direct relativistic effect. The shells with higher angular momentum, d and f, are not core-penetrating orbitals owing to their large centrifugal barrier. They experience practically exclusively an 'indirect' relativistic effect, which is brought about by the relativistic relaxation of the other shells (in the first place the contraction of s and p orbitals), which will alter the shielding experienced by the d and f electrons. In particular, the contraction of the s and p semi-core (i.e., the s and p shells with the same quantum numbers and about the same spatial extent as the shell in question, albeit with much different energy) will lead to a more effective shielding of the d and f orbitals and thus to an energetic destabilization. Thus, a good rule of thumb is that s and p shells are energetically stabilized (with a concomitant higher ionization potential and electron affinity for ionization or attachment of an electron in these shells) and that d and f shells experience relativistic destabilization.

This destabilization may lead, in turn, to an indirect stabilization of the next higher s and p shells. This situation occurs in the case of the late transition metals and leads to the 'gold maximum' of relativistic effects and the unusually large relativistic effects in the elements of groups 10–12. If the d shell is only weakly occupied, as is the case in the early transition metals, the direct effect on the s and p shells is partly balanced by the indirect effect on those shells, and the relativistic effects are generally much smaller.

6.1.2 FINE STRUCTURE

In the same article, Arnold Sommerfeld (1916) discussed the fine-structure splitting of atomic spectral lines due to spin–orbit coupling, and introduced the fine-structure constant $\alpha \approx 1/137.0359895$ [Cohen and Taylor (1987)], which equals $e^2/\hbar c$ in Gaussian units, $e^2/4\pi\hbar c$ in Heaviside–Lorentz units and $e^2/4\pi\varepsilon_0\hbar c$ in SI units. This quantity can be used as a measure of relativistic effects, since it is essentially the reciprocal of the velocity of light. This can be seen most conveniently by making use of the system of atomic units. Like the kinematic relativistic effects, spin–orbit coupling depends on $Z\alpha$ to the second order, and since the binding energy of a hydrogen-like atom itself grows quadratically with the charge of the nucleus, fine-structure effects grow with the fourth power of the nuclear charge. Therefore, relativistic effects are most important when Z is large, i.e., in heavy atoms and ions and in molecular systems containing such atoms.

It turns out that the spin–orbit coupling, which brings about the fine-structure splittings, is one of the consequences of the entangling of the spin degrees of freedom, which is present in relativistic theories in addition to the effects of spin statistics, namely the Pauli principle. If spin–orbit coupling is large, which is the case in heavy atoms, the distinction between the two effects of the electron spin is blurred to a large extent, and it is mandatory that all multiplets of an atomic configuration are treated on equal footing. This means that the wave function must be described by a superposition (i.e., a configuration interaction wave function) including all spin couplings (Slater determinants) which span the many-particle manifold of interest. This is true in the four-component theories discussed below, as well as in the two-components theories, which are obtained when electronic degrees of freedom and their charge-conjugate ('positron') degrees of freedom are (approximately) decoupled. Indeed, there are problems with the description of multiplets in a restricted multi-configuration Dirac–Fock representation, which have been identified and appreciated only very recently [Kim *et al.* (1998)]. The requirement of accurately describing the coupling of the various fine-structure components is often confused with the question of representation of the basis functions. Thus, it is often claimed that the 'relativistic coupling' is equivalent to jj coupling, while Russell–Saunders-type coupling (*LS* coupling) is taken as synonym for a genuinely non-relativistic approach. While it is indeed true that the jj basis has some technical advantages in the four-component framework, the full manifold of *LS* states is fully equivalent to the jj manifold.

6.2 Methods of relativistic electronic structure theory

6.2.1 FOUR-COMPONENT WAVE FUNCTIONS

The most successful quantum-mechanical equation of motion for a relativistic electron has been derived by Dirac (1928) by means of a linearization of the relativistic expression for the energy of a particle in an external field

$$(\mathcal{E} - V_{\text{ext}})^2 = m^2c^4 + p^2c^2 \equiv p_0^2c^2. \tag{7}$$

A linearization of this equation, i.e., a factorization according to

$$\begin{aligned} c(\gamma_0 p_0 - \gamma_1 p_1 - \gamma_2 p_2 - \gamma_3 p_3)c(\gamma_0 p_0 - \gamma_1 p_1 - \gamma_2 p_2 - \gamma_3 p_3) \\ = c^2(p_0^2 - p_1^2 - p_2^2 - p_3^2) \\ = m^2c^4 \end{aligned} \tag{8}$$

leads to the Dirac equation, which turned out to be a suitable equation for particles with spin 1/2.

Since the factorization is not possible when the γ_μ take values in the field of complex numbers, it is necessary that hypercomplex numbers [Kantor and Solodovnikov (1989)] are considered, and it turns out that an algebra comprising 16 basic entities (formed by the four quantities γ_μ above, six dyadic products $\gamma_\nu\gamma_\mu$, with $\nu < \mu$, four ternary products, one quaternary product of the γ quantities and the real unit) is required to accomplish the factorization. This algebra corresponds to an instance of a four-dimensional Clifford algebra [Porteous (1995)], and the additional degrees of freedom realized by the new algebraic structure may be associated with the spin degrees of freedom for the electron and its charge-conjugated partner.

Using the four Clifford numbers γ_μ, the Dirac equation for a free particle may be immediately read from equation (8):

$$(\gamma_\mu p^\mu - mc)\Psi_{\text{Dirac}} = 0. \tag{9}$$

In this equation, the Einstein summation convention and the Minkowski metric tensor $g_{\mu\nu} = \text{diag}\{1, -1, -1, -1\}$ is employed in forming the scalar product between the four-dimensional quantities γ_μ and p^μ.

The wave function in Dirac's theory of the relativistic electron thus comprises four discrete extra degrees of freedom besides the space–time variable, which correspond to two spin degrees of freedom (spin up/down) both for the electron and its charge-conjugated partner. It is therefore called a four-component spinor. The wave equation is linear in both the space variable, represented by the momentum of the particle and leading to the first derivative with respect to the position upon canonical quantization ($\boldsymbol{p} \to -i\hbar\boldsymbol{\nabla}$), as well as in the time variable represented by the energy E/c. By contrast, the non-relativistic Schrödinger equation is of second order in the momentum and of first order in the time variable.

The coupling between the four components is accomplished by the quantities γ^0 and $\boldsymbol{\gamma} \equiv \gamma^k, k = 1, 2, 3$. They may be represented by 4×4 matrices, which do not depend on the dynamical variables. The requirements of the Clifford algebra does not determine these matrices uniquely. Among several representations which are in practical use, the following 'standard form' is the most important one:

$$\beta \equiv \gamma^0 = \begin{pmatrix} \mathbf{1} & 0 \\ 0 & -\mathbf{1} \end{pmatrix}, \quad \boldsymbol{\alpha} \equiv \gamma^0\boldsymbol{\gamma} = \begin{pmatrix} 0 & \boldsymbol{\sigma} \\ \boldsymbol{\sigma} & 0 \end{pmatrix}. \tag{10}$$

Every entry in the matrices above is to be interpreted as a 2×2 matrix, in particular the $\boldsymbol{\sigma}$ matrices are the familiar Pauli spin matrices

$$\sigma_x = \begin{pmatrix} 0 & 1 \\ 1 & 0 \end{pmatrix}, \quad \sigma_y = \begin{pmatrix} 0 & -\mathrm{i} \\ \mathrm{i} & 0 \end{pmatrix}, \quad \sigma_z = \begin{pmatrix} 1 & 0 \\ 0 & -1 \end{pmatrix}. \tag{11}$$

The 'split notation' employed here is motivated by the presence of two upper and two lower components in the four-component Dirac spinor,

$$\Psi_{\text{Dirac}} = \begin{pmatrix} \psi_L \\ \psi_S \end{pmatrix}. \tag{12}$$

For solutions with positive energy and weak potentials, the lower component is suppressed by a factor $1/c^2$ with respect to the former, and therefore commonly dubbed small component ψ_S, as opposed to the large component ψ_L. It must be stressed that the 'small component' is *not* small in all situations.

A more familiar form of the one-particle Dirac equation for a particle with spin 1/2 in the external potential of the nucleus $V = V_{\text{ext}} = -Z\alpha\hbar c/r$ is obtained in terms of the $\boldsymbol{\alpha}$ matrices defined above. We also subtract the rest mass mc^2 from the energy value, in order to adapt the energy value to the Schrödinger energy scale, an energy $E = 0$ denoting the border between bound electron states and the positive-energy continuum.

$$\begin{aligned} D_{\text{ext}}\Psi_{\text{Dirac}} &= \left(E - mc^2\right)\Psi_{\text{Dirac}}, \\ D_{\text{ext}} &= c\boldsymbol{\alpha}\boldsymbol{p} + (\beta - 1)mc^2 + V. \end{aligned} \tag{13}$$

The spectrum of equation (13) comprises positive as well as negative eigenvalues, and the states with negative energy are stipulated to be occupied. The spectrum of the one-particle equation can thus be partitioned in three regions: Σ_1, the bound states of the electron, which for $Z < 1/\alpha \approx 137$ is confined to the interval $(-2mc^2, 0)$, Σ_2, the positive continuum, describing free electron states in the interval $[0, \infty)$ and the negative continuum Σ_3, $[-\infty, -2mc^2]$, corresponding to the Dirac sea.

For multi-electron systems, an approximate equation is used which is known as the Dirac–Coulomb (DC) equation. The electron–electron interaction is extracted from a perturbation expansion of the quantum eletrodynamics (QED) interaction of the quantized degrees of freedom of the electromagnetic field. In lowest order and in Coulomb gauge, this yields the DC Hamiltonian

$$H_{\text{DC}} = \sum_i D_{\text{ext}}(i) + \alpha\hbar c \sum_{i<j} r_{ij}^{-1}. \tag{14}$$

Here, $D_{\text{ext}}(i)$ denotes the one-particle external-field Dirac operator for the ith electron. This operator does not include the interactions of the transverse degrees of freedom of the electromagnetic field. In Coulomb gauge, they appear in next higher order in α and lead to the so-called Dirac–Coulomb–Breit (DCB) operator. The term which is included now in addition to (14) is called the 'transverse interaction'

$$-\boldsymbol{\alpha}_1\boldsymbol{\alpha}_2 \frac{\exp(\mathrm{i}\omega_{12}r_{12})}{r_{12}} - (\boldsymbol{\alpha}\boldsymbol{\nabla}_1)(\boldsymbol{\alpha}_2\boldsymbol{\nabla}_2)\frac{\exp(\mathrm{i}\omega_{12}r_{12}) - 1}{\omega_{12}^2 r_{12}}, \tag{15}$$

ω_{12} denoting the transition frequency between the orbitals of electrons 1 and 2.

The transverse interaction approximately restores gauge invariance, which is usually lost in approximations unless explicitly taken care of [Grant (1974), Gorceix, Indelicato and Desclaux (1987), Grant (1987), Gorceix and Indelicato (1988), Johnson, Mohr and Sucher (1989), Lindroth and Mårtensson-Pendrill (1989), Lindgren (1990), and Dietz and Hess (1991)]. Moreover, it features important contributions to the spin–orbit interaction [Hess, Marian and Peyerimhoff (1995)], namely the spin–other–orbit part, and is very important for the energy levels of highly ionized systems. For the valence-shell properties of atoms and molecules it can to a good approximation be replaced by the Breit operator, a frequency-independent simplification of the full transverse interaction [Breit (1929), Breit (1930), Breit (1932), and Mann and Johnson (1971)], which can be split into the so-called Gaunt term and a gauge-dependent term, restoring gauge invariance to second order in α

$$H_{\text{Breit}} = H_{\text{Gaunt}} + H_{\text{Gauge}} \tag{16}$$

$$H_{\text{Gaunt}} = -\frac{\alpha\hbar c}{2}\frac{\boldsymbol{\alpha}_1\boldsymbol{\alpha}_2}{r_{12}} \tag{17}$$

$$H_{\text{Gauge}} = -\frac{\alpha\hbar c}{2}(\boldsymbol{\alpha}\boldsymbol{\nabla}_1)(\boldsymbol{\alpha}_2\boldsymbol{\nabla}_2)\frac{1}{r_{12}^3}. \tag{18}$$

The DC Hamiltonian has been used by Swirles [Swirles (1935)] to define the analogue of the Hartree–Fock equation in the framework of Dirac's relativistic one-electron theory. The resulting theory is nowadays called Dirac–Hartree–Fock (DHF) theory and has been implemented for atoms early on [Grant (1961), Grant (1965), Grant (1970), and Desclaux (1975)]. These methods all use some form of the DC or DCB operator. The operators they are based upon are *not* covariant with respect to formulation in a different inertial frame of reference, although the commercial 'fully relativistic' is often used.

Usually, the Dirac Hamiltonian is formulated for a point nucleus. It is, however, advisable to include the structure of the finite nucleus in an approximate way [Visser *et al.* (1987), Dyall and Fægri Jr. (1993), and Chandra and Hess (1994)], and most computer codes make use of some variant of a finite nucleus. By this token, the singularity of the solution of the one-particle point-nucleus Dirac equation can be avoided.

Electron–positron pair creation and other processes described by QED, which has quantized degrees of freedom for *both* the fermions and the electromagnetic field ('radiative corrections'), are usually not included in molecular calculations. Very recently, the size of QED corrections in multi-electron atoms have been calculated for the first time [Pyykkö, Tokman and Labzowsky (1998)]. It was found that their size amounts to roughly 1 % of the relativistic effect in a molecular system, and thus can indeed be neglected unless all other effects (correlation in particular) are calculated to that precision. Earlier results on QED effects were usually based on extrapolations from few-electron systems [Lindgren (1992), Müeller-Nehler and Soff (1994), and Sapirstein (1998)].

Based on the DCB operator, most known methods of quantum-chemical electronic structure determination have been implemented by now also for four-component spinors. This comprises time-honoured poineering work on atoms in the DHF framework, using numerical techniques [Desclaux (1975), and Dyall *et al.* (1989)] and basis set expansion techniques [Kim (1967), Kagawa (1980), and Ishikawa, Baretty and Binning Jr. (1985)], as well as very recent work for molecules in DHF approximation [Dyall, Taylor Jr. and Partridge (1991), Pisani and Clementi (1994), Parpia and Mohanty (1995), and Pisani and Clementi (1995)] and elaborate techniques to treat relativity and correlation on the same footing [Visscher (1995), Lindgren (1996), Dyall (1994), Jensen *et al.* (1996), and Ishikawa and Kaldor (1996)].

Since most of the methods developed for molecules use the linear combination of atomic orbitals (LCAO) expansion technique, special care has to be taken in choosing the appropriate basis set for the small component. From equations (10) and (13) it is obvious that matrix elements of the type $\langle\psi_S|\boldsymbol{\sigma p}|\psi_L\rangle$ have to be evaluated. Since operation of $\boldsymbol{p} = -i\hbar\nabla$ on, e.g., an s function yields a p function, one has to ensure that a p function with appropriate exponent is in the basis set for the small component. The extreme example of using only s functions for the basis set used to expand ψ_S yields a null matrix for the $\boldsymbol{\sigma p}$ operator and illustrates drastically that the matrix representation of $\boldsymbol{\sigma p}$ is grossly wrong unless so-called kinetic balance is maintained. If the matrix representation of $\boldsymbol{\sigma p}$ is too small in norm to balance the attraction by the nuclear potential, 'variational collapse' results. This phenomenon was often confounded with the 'Brown–Ravenhall disease' described below, until the right explanation of the variational collapse problem was given [Lee and McLean (1982), Schwarz and Wechsel-Trakowski (1982), Mark and Schwarz (1982), Schwarz and Wallmeier (1982)] and further analyzed [Stanton and Havriliak (1984), and Dyall, Grant and Wilson (1984)].

The exact relationship between the large and small components can be formalized by writing

$$\phi_S = X\phi_L \tag{19}$$

for any trial function for the small component ϕ_S and for the large component ϕ_L. The operator X is not known in general. If (ϕ_S, ϕ_L) is an exact eigenfunction (ψ_L, ψ_S) of the one-electron Dirac equation (13), it could in principle be determined by expressing the small component in terms of the large component by means of the coupled system of equations resulting from equation (13)

$$\begin{aligned} c\boldsymbol{\sigma p}\,\psi_S + V\psi_L &= E\psi_L, \\ c\boldsymbol{\sigma p}\,\psi_L - 2mc^2\psi_S + V\psi_S &= E\psi_S, \end{aligned} \tag{20}$$

using the expression for ψ_S from the lower equation

$$2mc^2\psi_S = \left(1 + \frac{E-V}{2mc^2}\right)^{-1} c\boldsymbol{\sigma p}\,\psi_L. \tag{21}$$

In the general case, X must fulfill the non-linear equation [Chang, Pélissier and Durand (1986), Heully *et al.* (1986), and Kutzelnigg (1997)]

$$X = \frac{1}{2mc^2}(c\boldsymbol{\sigma p} - [X, V] - X(c\boldsymbol{\sigma p})X). \tag{22}$$

Obviously, the solution of this equation for X is as complex as the solution of the Dirac equation itself, and approximations have to be employed.

For $c \to \infty$, equation (21) can be approximated using

$$\phi_S = \frac{1}{2mc}\boldsymbol{\sigma p}\ \phi_L, \tag{23}$$

which is the non-relativistic limit of an expression valid only for exact eigenfunctions of the Dirac equation. This approximation may be employed to generate the basis set for the small component from the basis set of the large component $\{\chi_L(\alpha, \ell)\} \to \{\chi_S(\alpha, \ell-1), \chi_S(\alpha, \ell+1)\}$ for a basis function with exponent α and angular momentum ℓ. Indeed, this suffices for a valid matrix representation of the kinetic energy, and this relation is therefore dubbed 'kinetic balance' requirement. It was known early on [Stanton and Havriliak (1984), and Kutzelnigg (1997)] that this procedure leads to a lower bound for the variation of the resulting matrix Dirac equation which is *below* the exact value by a term of order $O(c^{-4})$. In the case of a contracted basis set, even more care has to be taken [Ishikawa, Sekino and Binning Jr. (1990), and Matsuoka (1992)]. Satisfactory results are obtained when *atomic balance* is obeyed [Visscher *et al.* (1991)], which amounts to contraction of the spinors with coefficients obtained from atomic calculations with a kinetically balanced uncontracted basis set.

The question of continuum dissolution was first considered in a paper by Brown and Ravenhall (1951). These authors remarked that for a multi-particle state of the DC Hamiltonian (14) there are infinitely many unbound states degenerate with any bound state of this operator. This is easily seen in the case of a two-electron system: Any double excitation of two electrons from an orbital with energy ε resulting in one of the electrons sitting in a continuum orbital with orbital energy $2\varepsilon + mc^2 + \delta$, and the other in a negative-energy continuum orbital with energy $-mc^2 - \delta$ would result in a non-perturbed energy 2ε equal to that of the initial state. Thus, transitions to these doubly excited states and decay of the two-electron bound state into the continuum would result when the Coulomb interaction is switched on. Brown and Ravenhall recommended the use of projectors on positive-energy states bracketing the Coulomb interaction to avoid this decay. The resulting theory is known as no-pair theory, since excitations creating electron–positron pairs are explicitly excluded in the formalism [Mittleman (1971), Mittleman (1972), Buchmüller and Dietz (1980), Sucher (1980), and Mittleman (1981)].

These considerations are not of immediate relevance for DHF calculations, which are not affected by the Brown–Ravenhall disease, since the DHF energy is required to be stationary with respect to single excitations from the DHF Slater determinant only. Double excitations of the Brown–Ravenhall type are not present in the DHF formalism, and thus can do no harm. Correlation is usually implemented by constructing multi-particle states explicitly using positive-energy solutions of the DHF equations, thus trivially implementing the projectors by decreasing the domain of variation. Note that in two-component theories projectors on positive-energy states are trivially implemented as well, since upper and lower components are explicitly decoupled. Thus, one can say that continuum dissolution is not a problem at all, although in the early eighties it was considered a major issue for relativistic electronic structure calculations.

6.2.2 TWO-COMPONENT METHODS

The relativistic formulation requires that the coupling between the charge-conjugated degrees of freedom and the spin degrees of freedom is considered explicitly, which inevitably leads to four-component wave functions for an electron. The four-component 'Dirac-like' theories are unique in the sense that the relation between four components is independent of the kinematics, since α and β are independent of $\boldsymbol{p}, \boldsymbol{r}$. On the other hand, one could ask the question whether there is a unitary transformation that annihilates the coupling between the 'electron-like' and the 'positron-like' degrees of freedom. These wave functions, leading to transformed Hamiltonians, still have formally four components. Since, however, there is no coupling any more between the states of positive energy (the electrons) and the states of negative energy (the positrons), we now have the possibility to focus on the former and work with two-component wave functions only. Needless to say that this is an attractive feature, since the explicit consideration of the small components and the requirement to maintain kinetic balance in the four-component theories requires large computer resources, which could be possibly saved in a two-component theory. There is, however, a price to pay: In the transformed Hamiltonians, the operators will be functions of $\boldsymbol{p}$ (or $\boldsymbol{r}$). There is no way to escape the requirement of implementing the correct relationship between large and small component, equation (19), and in the two-component theories the operators must implement it. While spin–orbit coupling is described in the 'Dirac-like' (four component) representation by a purely algebraic structure (the Clifford algebra of the Dirac matrices), there is a 'space part' of the spin–orbit coupling operator in the decoupled representation.

The decoupled representation is achieved by a unitary transformation

$$H^{\text{decoupled}} = U^{\dagger} D U = \begin{pmatrix} h_{+} & 0 \\ 0 & h_{-} \end{pmatrix} \tag{24}$$

with

$$\begin{aligned} UU^{\dagger} &= 1 \\ U &= \begin{pmatrix} (1+X^{\dagger}X)^{-1/2} & (1+X^{\dagger}X)^{-1/2}X^{\dagger} \\ X(1+XX^{\dagger})^{-1/2} & (1+XX^{\dagger})^{-1/2} \end{pmatrix} \end{aligned} \tag{25}$$

and D denoting a Dirac-type Hamiltonian. The operator X is given by equation (19). Since the transformed large component, now describing electron states only, should be normalized to unity, the equation contains renormalization terms $(1+X^{\dagger}X)^{-1/2}$ to take the change from the Dirac normalization prescription for any four-component wave function Φ

$$\begin{aligned} \langle\Phi|\Phi\rangle &= \langle\phi_{\text{L}}|\phi_{\text{L}}\rangle + \langle\phi_{\text{S}}|\phi_{\text{S}}\rangle \\ &= \langle\phi_{\text{L}}|\phi_{\text{L}}\rangle + \langle X\phi_{\text{L}}|X\phi_{\text{L}}\rangle \end{aligned} \tag{26}$$

into account. Unfortunately, closed-form solutions for equation (19) are known only for a restricted class of potentials [Nikitin (1998)]. A very important special case is, however, the free particle, defined by $V \equiv 0$. In this case, we find a closed-form solution

$$X^{V=0} = \left(mc^2 + \sqrt{m^2c^4 + p^2c^2}\right)^{-1} c\boldsymbol{\sigma p}. \tag{27}$$

This defines the *exact* Foldy–Wouthuysen transformation for the free particle. Note that the square root is not expanded here. Expansion and ensuing truncation of the expansion lead to problems which are discussed below.

An important point is the observation [Kutzelnigg (1997)] that the relationship $\phi_S = X\phi_L$ must be fulfilled for all trial functions in the domain of an approximate $\tilde{H}^{\text{decoupled}}$, which is used to define a variational procedure for approximate solutions of the transformed D. Failure to do so would result in an operator $\tilde{H}^{\text{decoupled}}$ which still may be bounded from below and is thus *variationally stable*, but possibly by a value *below* the exact electronic ground state of D. Thus, the exact solution of D could be attained with an approximate wave function, and go below the exact solution as the wave function approaches the exact one. On the other hand, the four-component Dirac operator is exactly equivalent to the two-component operator h_+, if $\phi_S = X\phi_L$ is ensured for all functions in the domain of h_+. The method is then *variational*, i.e., the lower bound is the exact electronic ground state of D. The same remark on variational stability holds for a matrix representation of the Dirac equation in the four-component formalism, and generally matrix representations obeying approximate kinetic balance prescriptions can be *below* the exact electronic ground state of the Dirac equation, being variationally stable, but not variational.

The various two-component theories known from the literature satisfy $\phi_S = X\phi_L$ to various degrees of accuracy. In the following, we shall discuss methods using an elimination procedure on the one hand, and methods aiming at derivation of transformation operators U on the other hand.

Elimination of the small component

The method of elimination starts from the Dirac equation in the split form, equation (20). The expression for the small component, equation (21), obtained from the second of those equations is inserted into the first one, yielding

$$(V - E)\psi^{L} + \frac{1}{2mc^2}[\boldsymbol{\sigma p}\,\omega(\boldsymbol{r})\boldsymbol{\sigma p}]\psi_{L} = 0, \tag{28}$$

with

$$\omega(\boldsymbol{r}) = \left(1 - \frac{V - E}{2mc^2}\right)^{-1}. \tag{29}$$

This substitution leads to an equation for the large component only, and equation (29) has been used as the basis to formulate energy-dependent, non-hermitean operators [Löwdin (1964), Cowan and Griffin (1976), Wood and Boring (1978), Barthelat, Pélissier and Durand (1980), Karwowski and Kobus (1981), Karwowski and Szulkin (1982), Karwowski and Kobus (1985), and Wood, Grant and Wilson (1985)]. If desired, the spin dependence can be isolated using

$$(\boldsymbol{\sigma u})(\boldsymbol{\sigma v}) = \boldsymbol{uv} + \mathrm{i}\boldsymbol{\sigma}(\boldsymbol{u} \times \boldsymbol{v}). \tag{30}$$

The energy dependence is, however, undesirable, since orbital-dependent Hamiltonians and non-orthogonal orbitals result. The simplest way to arrive at a hermitean, energy-dependent operator is by expanding

$$\left(1 - \frac{V - E}{2mc^2}\right)^{-1} = \sum_{n=0}^{\infty}\left(\frac{V - E}{2mc^2}\right)^{n}. \tag{31}$$

Keeping only the lowest-order term, the non-relativistic Schrödinger equation is recovered. Low-order relativistic corrections can be extracted by keeping the next higher term and eliminating the energy dependence by means of systematic expansion in c^{-2}. This leads to the Pauli Hamiltonian

$$H_{\text{Pauli}} = \frac{p^2}{2m} + V + \frac{1}{4m^2c^2}\left(\frac{-p^4}{2m} + \frac{1}{2}(\Delta V) + \boldsymbol{\sigma}(\nabla V) \times \boldsymbol{p}\right), \tag{32}$$

where the so-called mass–velocity term $-p^4/8m^3c^2$, the Darwin term $\Delta V/4m^2c^2$, and the spin–orbit coupling term $\boldsymbol{\sigma}(\nabla V) \times \boldsymbol{p}/4m^2c^2$ describe relativistic corrections to $O(c^{-2})$. Several problems are connected with this operator: the minus sign of the mass–velocity term yields a strongly attractive term for states with high momentum, and leads to variational collapse in unconstrained variation; the Darwin term degenerates to a highly singular Delta distribution term in the case of the potential of a point-like nucleus; the spin–orbit coupling term leads to variational collapse as well, since it is not bounded below. These problems cannot be remedied by going to higher orders [Morrison and Moss (1980)]. In fact, expansion (31) is invalid for $V - E > 2mc^2$, and this condition occurs certainly in regions close to the nucleus. Operators based on simple expansions of equation (31) in c^{-2} are in general singular and cannot be used for variational calculations. The Pauli operator is therefore defined only for perturbation theory to lowest order. In practical calculations, its expectation values give satisfactory relativistic corrections to the energy up to the first and second transition metal row. Regular expressions for relativistic perturbation theory in higher orders require a special formulation, namely the formalism of direct perturbation theory discussed in the next subsection.

Making use of special features of the matrix representation of the Dirac equation, Dyall has recently worked out a modified elimination of the small component [Dyall (1997, 1998)]. His method takes the proper renormalization mentioned above into account. In particular this normalized variant of the modified elimination of the small component is free from the singularities which plague the classical elimination method.

A very well-studied technique to arrive at regular expansions has been developed in the mid-eighties [Chang, Pélissier and Durand (1986), and Heully *et al.* (1986)]. It is based on rewriting $\omega(\boldsymbol{r})$ in equation (29) and choosing a different expansion parameter. Writing $\omega(\boldsymbol{r})$ as

$$\omega(\boldsymbol{r}) = \frac{2mc^2}{2mc^2 - V}\left(1 + \frac{E}{2mc^2 - V}\right)^{-1} \tag{33}$$

and expanding the term in parentheses is the basis of the so-called *regular approximations*, which were developed by the Amsterdam group [van Lenthe *et al.* (1995), and van Lenthe *et al.* (1996)] to a workable method for electronic-structure calculations.

A truncation of the expansion (33) defines the zero- and first-order regular approximation (ZORA, FORA) [van Lenthe, Baerends and Snijders (1993)]. A particular noteworthy feature of ZORA is that even in the zeroth order there is an efficient relativistic correction for the region close to the nucleus, where the main relativistic effects come from. Excellent agreement of orbital energies and other valence shell properties with the results from the Dirac equation is obtained in this zero-order approximation, in particular in the scaled ZORA variant [van Lenthe, Baerends and Snijders (1994)], which takes the renormalization to the transformed large component approximately into account, using

$$\frac{1}{\sqrt{1 + X^{\dagger}X}} \approx \frac{1}{\sqrt{1 + \langle \phi_L X^{\dagger} | X \phi_L \rangle}} \tag{34}$$

The analysis [van Leeuwen *et al.* (1994)] shows that in regions of high potential the zero-order Hamiltonian reproduces relativistic energies up to an error of order $-E^2/c^2$. On the other hand, in regions where the potential is small, but the kinetic energy of the particle high, the ZORA Hamiltonian does not provide any relativistic correction.

The main disadvantage of the method is its dependence on the zero point of the electrostatic potential, i.e., gauge dependence. This occurs because the potential enters non-linearly (in the denominator of the operator for the energy), so that a constant shift of the potential does not lead to a constant shift in the energy. This deficiency can, however, be approximately remedied by suitable means [van Lenthe, Baerends and Snijders (1994), and van Wüllen (1998)].

Transformation to two components

An alternative to the elimination-type methods is the attempt to achieve the block diagonalization of the Dirac operator according to equation (24) directly. The time-honoured method is the Foldy–Wouthuysen transformation [Foldy and Wouthuysen (1950)]. The idea is to identify 'odd' and 'even' operators in the split form of the Dirac equation, i.e., operators which couple the large and small component, and those which do not. Apart from the even term $(\beta - 1)$, we can identify the even operator $\mathcal{E} = V$ and the odd operator $\mathcal{O} = c\boldsymbol{\alpha}\boldsymbol{p}$, and find

$$[\mathcal{E}, \beta] = 0, \quad \{\mathcal{O}, \beta\} = 0. \tag{35}$$

The braces denote the anticommutator $\{A, B\} = AB + BA$. We now look for a unitary matrix which removes the odd term. The Foldy–Wouthuysen transformation uses the ansatz

$$\begin{aligned} \Phi_1 &= \exp(\mathrm{i}S_1)\Phi, \\ H_1 &= H + \mathrm{i}[S_1, H] + \cdots. \end{aligned} \tag{36}$$

The choice $S_1 = -\mathrm{i}\beta\mathcal{O}/2m$ removes the odd term, but introduces new odd terms of higher order, which are in turn removed by iteration of the transformation: $\Phi_n = \exp(\mathrm{i}S_n)\Phi_{n-1}$. At this point, the resulting operators are again expanded in a power series in c^{-1}. Up to second order, we obtain again the Pauli Hamiltonian, equation (32). While different expressions occur in higher orders, the problems with singular operators are essentially the same as in the case of the elimination of the small component discussed above. Additional problems occur, since the wave functions obtained in the Foldy–Wouthuysen procedure are no longer analytic functions of c^{-1} in the neighborhood of $c^{-1} = 0$ [Kutzelnigg (1989a, 1989b)], as is the case for the Dirac wave function [Titchmarsh (1962)]. This means that the non-relativistic limit is not well defined.

To obtain a valid limiting procedure for $c^{-1} \to 0$, the perturbation theory has to be formulated by considering the non-relativistic limit of the metric (essentially the normalization requirement) and that of the operator itself separately. Both for the metric and for the operator limiting procedures must be defined. This is most conveniently done by formulating the Dirac equation in terms of a scaled small component $c\psi_S$, and a regular perturbation formalism results [Sewell (1949), Titchmarsh (1962), Rutkowski (1986a, 1986b, 1986c), and Kutzelnigg (1989a, 1989b)]. In the more recent literature, this four-component method has been dubbed direct perturbation theory. The second-order results are equivalent to the perturbative results of the Pauli operator in an infinite basis set. In contrast to the singular expansions which are traditionally employed to derive the Pauli operator, direct perturbation theory gives workable and regular results also for higher orders.

Another possibility which has meanwhile proven of considerable practical value is to avoid expansion in reciprocal powers of c throughout, and rather to expand in the coupling strengh $Z\alpha c\hbar$, if closed expressions cannot be obtained [Douglas and Kroll (1974), Hess (1986), and Jansen and Hess (1989)]. The Douglas–Kroll transformation defines a transformation of the external-field Dirac Hamiltonian to two-component form which leads, in contrast to the Foldy–Wouthuysen transformation, to operators which are bounded from below and can be used variationally, similar to the regular approximations discussed above. As in the Foldy–Wouthuysen transformation, it is not possible in the Douglas–Kroll formalism to give the transformation in closed form. It is rather defined by a sequence of unitary transformations $U_0, U_1, \ldots$, the first of which is in fact a free-particle Foldy–Wouthuysen transformation defined by

$$U_0 = A(1+\beta R), \quad U_0^{-1} = (R\beta + 1)A, \tag{37}$$

with

$$A = \sqrt{\frac{E_p + mc^2}{2E_p}} \tag{38}$$

$$R = \frac{c\boldsymbol{\alpha p}}{E_p + mc^2}, \tag{39}$$

$$E_p = c\sqrt{p^2 + m^2c^2}. \tag{40}$$

Applying U_0 to D leads to

$$U_0 D U_0^{-1} = \beta E_p + \mathcal{E}_1 + \mathcal{O}_1 \equiv H_1, \tag{41}$$

with even and odd operators of first order, given by

$$\begin{aligned} \mathcal{E}_1 &= A(V + RVR)A, \\ \mathcal{O}_1 &= \beta A(RV - VR)A. \end{aligned} \tag{42}$$

The following unitary transformation–it turns out that only one more is required to decouple the upper and lower components to sufficient accuracy for chemical applications–is defined by the somewhat unusual parametrization

$$U_1 = \sqrt{1 + W_1^2} + W_1. \tag{43}$$

For any anti-hermitean operator W_1 with $W_1^\dagger = -W_1$, it is easily seen that U_1 is unitary. Performing the transformation through U_1 and expanding the square root in powers of W_1 leads to

$$\begin{aligned} U_1 H_1 U_1^{-1} = \beta E_p - [\beta E_p, W_1] + \mathcal{E}_1 + \mathcal{O}_1 + \tfrac{1}{2}\beta E_p W_1^2 \\ + \tfrac{1}{2} W_1^2 \beta E_p - W_1 \beta E_p W_1 + [W_1, \mathcal{O}_1] + [W_1, \mathcal{E}_1] + \cdots, \end{aligned} \tag{44}$$

where the dots denote terms in higher than second order of W_1. The first-order odd term is now eliminated by equating

$$[\beta E_p, W_1] = \mathcal{O}_1 \tag{45}$$

and solving for W_1. We arrive at a momentum-space integral operator for W_1

$$W_1 \Phi(\boldsymbol{p}) = \int \mathrm{d}^3 p' W_1(\boldsymbol{p}, \boldsymbol{p}')\Phi(\boldsymbol{p}'), \tag{46}$$

with a kernel

$$W_1(\boldsymbol{p}, \boldsymbol{p}') = A(R - R')A' \frac{V(\boldsymbol{p}, \boldsymbol{p}')}{E_{p'} + E_p}, \tag{47}$$

where $V(\boldsymbol{p}, \boldsymbol{p}')$ denotes the Fourier transform of the external potential, and the primed quantities are to be expressed in terms of the variable $\boldsymbol{p}'$.

The final result is

$$H^{\text{decoupled}} \approx \beta E_p + \mathcal{E}_1 - \beta\big(W_1 E_p W_1 + \tfrac{1}{2}[W_1^2, E_p]\big), \tag{48}$$

where the approximation sign denotes equivalence up to second order in the external potential. Higher-order transformations may be devised by definitions similar to equation (43) in order to remove odd terms of higher order in a way similar to the method described above. The performance of the second-order operator was found satisfactory for chemical applications. At this point, a projection to the upper components may be made, with the result that the β matrix becomes the unit matrix, and the $\boldsymbol{\alpha}$ matrices are to be replaced by $\boldsymbol{\sigma}$ matrices.

It has been found convenient to use the Dirac relation, equation (30), to separate the spin–orbit coupling terms. The treatment of the latter is discussed in detail in [Hess, Marian and Peyerimhoff (1995)] and will not be discussed here. For singlet states, it is often convenient to use the spin–averaged approximation and treat the spin–orbit coupling operator in a second step, be it perturbatively or variationally in a spin–orbit configuration interaction procedure with two-component spinors. In this case the R operators are vector quantities $\boldsymbol{P}$, given by

$$\boldsymbol{P} = \frac{c\boldsymbol{p}}{E_p + mc^2}. \tag{49}$$

In most applications (see, however, [Samzow, Hess and Jansen (1992), and Park and Almlöf (1994)]) the Douglas–Kroll transformation of the external potential V is limited to its one-electron part while the two-electron terms are left in their Coulomb form. This leads to the most frequently used spin-averaged 1-component many-electron no-pair Hamiltonian:

$$H_+ = \sum_i E_p(i) + \sum_i V_{\text{eff}}(i) + \sum_{i<j} \frac{1}{r_{ij}}, \tag{50}$$

where

$$\begin{aligned} V_{\text{eff}}(i) = & -A(i)[V(i) + \boldsymbol{P}(i)V(i)\boldsymbol{P}(i)]A(i) \\ & - W_1(i)E_p(i)W_1(i) - \tfrac{1}{2}\left[(W_1(i))^2, E_p(i)\right], \end{aligned} \tag{51}$$

and $V(i)$ is the usual one–electron nuclear attraction operator in the momentum representation.

It is interesting to note that the Hamiltonian of equation (50) is invariant with respect to a change of the electrostatic gauge, i.e., a constant shift in the external potential. It is fairly easy to see that the operators derived from equation (48) satisfy $H_+ \mapsto H_+ + \Delta$ for constant Δ, if

$$V \longmapsto V + \Delta. \tag{52}$$

We can write the effective potential in equation (48) as

$$V^{\text{eff}} \equiv f[V] + g[V] + \cdots, \tag{53}$$

where

$$\begin{aligned} f[V] &= A(V + \boldsymbol{P}V\boldsymbol{P})A, \\ g[V] &= -W_1 E W_1 - \tfrac{1}{2}\big(W_1^2 E - E W_1^2\big). \end{aligned} \tag{54}$$

If the potential is changed according to equation (52),

$$f[V] \longmapsto f[V+\Delta] = f[V] + \Delta. \tag{55}$$

As concerns $g[V]$, we have to consider the kernel $W_1(\boldsymbol{p},\boldsymbol{p}')$:

$$W_1(\boldsymbol{p},\boldsymbol{p}') = A(\boldsymbol{P}-\boldsymbol{P}')A'\frac{\hat{V}(\boldsymbol{p},\boldsymbol{p}')}{E+E'}. \tag{56}$$

Because of the Fourier transform

$$\hat{V}(\boldsymbol{p},\boldsymbol{p}') \longmapsto \hat{V}(\boldsymbol{p},\boldsymbol{p}') + \delta(\boldsymbol{p}-\boldsymbol{p}'), \tag{57}$$

we have

$$g[\Delta]\psi(\boldsymbol{p}) = \int \mathrm{d}^3\boldsymbol{p}'\, A_i(\boldsymbol{P}-\boldsymbol{P}')A'\frac{\delta(\boldsymbol{p},\boldsymbol{p}')}{E+E'}\psi(\boldsymbol{p}') = 0. \tag{58}$$

Douglas–Kroll-transformed Hamiltonians have been used in many quantum-chemical calculations on molecules, density-functional theory [Häeberlen and Röesch (1992)] including implementation of derivatives [Nasluzov and Röesch (1996)], and recently also for calculations of solids [Geipel and Hess (1997), Boettger (1998), and Fehrenbach and Schmidt (1997)]. A numerical analysis of the energy values [Hess (1986), and Molzberger and Schwarz (1996)] and also perturbation theory [Kutzelnigg (1997)] shows that the eigenvalues of the second-order Douglas–Kroll-transformed Hamiltonian for a single particle agrees with the results of the Dirac equation to order c^{-4}. Note that this is the same order in which deviations in the matrix representation of the Dirac equation itself are expected [Stanton and Havriliak (1984), and Kutzelnigg (1997)].

The Douglas–Kroll transformation can be carried to higher orders, if desired [Barysz, Sadlej and Snijders (1997)]. In this way, arbitrary accuracy with respect to the eigenvalues of D can be achieved.

6.2.3 RELATIVISTIC VALENCE-ONLY HAMILTONIANS

Looking for a way to reduce the computational effort connected with a relativistic molecular all-electron calculation, one is led to the use of effective valence-only Hamiltonians. These can be introduced at different computational stages: Right from the beginning, i.e., already in the orbital/spinor optimization step, or after having optimized the molecular orbitals/spinors, for all kinds of interaction or just for the magnetic ones. In either case the interaction between core and valence electrons is described by some kind of potential.

Reducing the number of explicitly treated electrons in the electron correlation step is a common procedure in non-relativistic calculations. In the so-called frozen-core approximation the one-particle space is separated into a core space comprising the strongly bound orbitals and a valence space including everything else. The frozen-core approximation rests on two assumptions: (1) the interaction between core and valence electrons is similar in all considered electronic states; (2) it is sufficient to describe the interaction between core and valence electrons by some mean-field expressions. Deviations from these conditions can partially be counteracted by introducing a semi-empirical core-polarization potential (CPP) [Migdalek and Baylis (1981), and Müller, Flesch and Meyer (1984)]. The frozen-core approximation is not restricted to non-relativistic schemes. In correlation treatments employing four-component spinors as one-particle basis a DHF or DCB mean field is used to decribe the core-valence interaction [Visscher (1995), Lindgren (1996), Dyall (1994), Jensen *et al.* (1996), and Ishikawa and Kaldor (1996)]. In one-component relativistic calculations this interaction is represented by Hartree–Fock Coulomb and exchange potentials or equivalent relativistically modified terms [Hess, Marian and Peyerimhoff (1995), Heinemann *et al.* (1996), and Sjøvoll *et al.* (1998)]. Also the spin–orbit interaction between core and valence electrons can be expressed by means of Coulomb- and exchange-type integrals of the two-electron spin–orbit Hamiltonian for which the summation over the core indices has already been carried out [Blume and Watson (1962, 1963), Langhoff and Kern (1977), Richards, Trivedi and Cooper (1981), Havriliak and Yarkony (1985), Hess, Marian and Peyerimhoff (1995), and Minaev, Vahtras and Ågren (1996)]. Recently, the latter ansatz was extended to define an effective one-electron spin–orbit mean-field Hamiltonian [Hess *et al.* (1996)]. More details will be given below.

The next step in the hierarchy of approximations is to replace the molecular mean field by a sum of atomic mean fields. This idea is based on the observation that molecular core orbitals are fairly localized and closely resemble their atomic counterparts. The one-center approximation allows for an extemely rapid evaluation of spin–orbit mean-field integrals, if the atomic symmetry is fully exploited [Schimmelpfenning (1996)]. Even more efficiency may be gained if also the spin-independent core-valence interactions are replaced by atom-centered potentials. In this case the inner shells do not even emerge in the orbital optimization step and the size of the atomic orbital basis set can be much reduced. Commonly employed atomic effective core potentials are of semi-local or non-local type. Semi-local potentials consist of an analytical (local) radial part and of (non-local) angular projection operators. Their common ground is the Phillips–Kleinman equation [Phillips and Kleinman (1959)]; the familiar pseudo-potentials and the averaged relativistic effective potential (AREP) belong to this class. In non-local potentials the core–valence interaction is partially taken care of by a matrix representation of an effective operator. The method was originally proposed by Huzinaga and Cantu (1971) who introduced non-local core orbital projectors. Nowadays, the most frequently utilized species in this family of core potentials is the *ab initio* model potential (AIMP); in addition to non-local core orbital projectors the AIMP Hamiltonian contains a non-local representation of the exchange operator [Huzinaga *et al.* (1987)]. In the following, we shall use the term relativistic effective core potential (RECP) to denote the superordinate concept rather than a particular type of potential. If a differentiation should become necessary we shall employ special denominations such as PP, AIMP, AREP, etc.

Apart from a few exceptions [Lee, Ermler and Pitzer (1977), Ishikawa and Malli (1981), Dolg (1996, 1997), and Titov and Mosyagin (1999)], all RECP approaches split the potential into a spin-independent (one-component, quasi-relativistic) part and a spin-dependent (spin–orbit coupling) term. This has the advantage that kinematic relativistic corrections can be implemented in every electronic structure code just by modifying the one-electron integrals; special computer programs are required only if magnetic interactions are taken into account. The use of RECPs has a long tradition. Review articles by Krauss and Stevens (1984), Ermler, Ross, and Christiansen (1988), Teichteil (1994), Dolg and Stoll (1995), and Seijo (1998) provide an excellent overview over the various developments until 1998. Here, we shall give only a brief summary of the early work and describe recent developments in some detail. The latter include also the combination of one-component RECP treatments with an effective one-electron spin–orbit mean-field Hamiltonian [Marian and Wahlgren (1996), and Schimmelpfennig *et al.* (1998a, 1998b)].

One-component relativistic effective core potentials

One-component RECPs are extracted either directly from atomic one-component relativistic calculations or they are based on a weighted average of two-component potentials. Among those derived from atomic (one-component) Cowan–Griffin (CG) calculations are the shape-consistent pseudo-potentials by Hay and Wadt (1985) and by Barthelat and Durand (1978), and the AIMPs by Barandiarán, Seijo and Huzinaga (1990, 1991). Most of the energy-adjusted quasi-relativistic pseudo-potentials developed by the Stuttgart–Dresden group [Dolg, Stoll and Preuss (1989), Andrae *et al.* (1990), and Küchle *et al.* (1991)] are based on solutions of the Wood–Boring equation (WB) which differs only slightly from the Cowan–Griffin expression. For large-core

PPs these authors added two important correction terms: firstly an exponential core–nucleus repulsion correction (CNRC) and secondly an empirically parameterized core-polarization potential. In case of core-penetrations the CRNC corrects for deviations of core–core or core–nucleus interactions from a point-charge model [Stoll *et al.* (1983)]. The CPP is of the type introduced by Müller, Flesch and Meyer (1984) for all-electron calculations and accounts for core–valence correlation effects through a fit to an experimental ionization potential. In connection with ECPs it was first used by Fuentealba *et al.* (1982). Teichteil and coworkers [Teichteil, Pélissier and Spiegelmann (1983), and Teichteil and Spiegelmann (1983)] extracted averaged pseudopotentials according to a procedure proposed by Durand and Barthelat (1975); the corresponding atomic all-electron two-component method traces back to Barthelat and Durand (1978). Recently, Wittborn and Wahlgren (1995) and Rakowitz *et al.* [Rakowitz *et al.* (1999), and Rakowitz, Marian and Seijo (1999)] presented spin-free relativistic AIMPs derived from atomic one-component Douglas–Kroll calculations.

One-component quasi-relativistic potentials have frequently been based on the large component of atomic Dirac–Fock spinors. Spin-averaging is achieved by different procedures: one way is to generate different potentials by inverting pseudo-Fock equations for the $j = \ell - 1/2$ and $j = \ell + 1/2$ pseudo-spinors, respectively, and to average the potentials in a second step. This procedure for generating AREPs, proposed by Lee, Ermler and Pitzer (1977), was persued extensively by Ermler, Ross, Christiansen, and coworkers [Pacios and Christiansen (1985), Hurley *et al.* (1986), LaJohn *et al.* (1988), Ross *et al.* (1990), Ermler, Ross and Christiansen (1994), Ross, Gayen and Ermler (1994), Nash, Bursten and Ermler (1997), and Wildman, DiLabio and Christiansen (1997)] and by Stevens and coworkers [Stevens *et al.* (1992), and Cundari and Stevens (1993)]. AREPs are available for nearly the whole periodic table up to element 118. For Cs a corresponding CPP can be found in the work of Marino and Ermler (1993). Basis sets for use with RECPs were reoptimized for Li–Ar [Wallace, Blaudeau and Pitzer (1991)] and for K, Ca, and Ga–Kr [Blaudeau and Curtiss (1997)]. A generalization of the shape-consistent pseudo-spinor transformation approach which allows one to introduce radial nodes in the pseudo-spinors was recently presented by Titov and Mosyagin (1999).

Lately also energy-adjusted one-component pseudo-potentials were parameterized based on results of four-component atomic calculations [Leininger *et al.* (1997), and Dolg (1997)]. In these atomic all-electron reference calculations not only the usual DF Hamiltonian but also the Breit interaction and the leading term of the QED corrections were included. As Leininger *et al.* (1997) point out, different one-component quasi-relativistic potentials are obtained from a fit to averaged energies and from an average of spin-dependent pseudo-potentials.

Valence-only spin–orbit Hamiltonians

Most of the molecular spin–orbit calculations employ some kind of effective spin–orbit Hamiltonian. In order to understand the hierarchy of approximations, let us regard the matrix elements of a pair of determinants interacting via the full one- and two-electron spin–orbit Hamiltonian. For derivation of these formulae it is not important whether the Breit–Pauli spin–orbit operators or their no-pair analog are employed because they are structurally equivalent. The latter reads [Hess, Marian and Peyerimhoff (1995)]

$$H_+^{\mathrm{SO}} = \alpha\hbar c \sum_i \Bigg\{ \sum_I Z_I \frac{A_i}{E_i + mc^2} \boldsymbol{\sigma}_i \left(\frac{\boldsymbol{r}_{iI}}{r_{iI}^3} \times \boldsymbol{p}_i \right) \frac{A_i}{E_i + mc^2}$$

$$- \sum_{j \neq i} \frac{A_i A_j}{E_i + mc^2} \left[\boldsymbol{\sigma}_i \left(\frac{\boldsymbol{r}_{ij}}{r_{ij}^3} \times \boldsymbol{p}_i \right) \frac{A_i A_j}{E_i + mc^2} - 2\boldsymbol{\sigma}_i \left(\frac{\boldsymbol{r}_{ij}}{r_{ij}^3} \times \boldsymbol{p}_j \right) \frac{A_i A_j}{E_j + mc^2} \right] \Bigg\}, \quad (59)$$

where I labels nuclei and i and j electrons. The factors E_i and A_i or A_j, defined in equations (40) and (38), are the same as in the spin-free no-pair Hamiltonian; for the present discussion they are not relevant, however. The matrix element formulae can be derived in a similar way to the Slater–Condon rules for the evaluation of $1/r_{12}$. Differences occur due to the symmetry properties of the angular momentum operators and because of the presence of the spin–other–orbit interaction. The latter requires a symmetrization in the particle indices, doubling the number of two-electron contributions.

(A) Diagonal case: Φ and Φ' are equal.
Because of the symmetry properties of the spin–orbit Hamiltonian there are no diagonal matrix elements in a basis of real (Cartesian) orbitals:

$$\langle \Phi \mid H^{\mathrm{SO}} \mid \Phi' \rangle = 0. \tag{60}$$

(B) Singly excited Slater determinants: Φ and Φ' differ by one pair of spin orbitals i, j.
Contributions come from the one-electron spin–orbit integral and three-index two-electron integrals

$$\langle \Phi | H^{\mathrm{SO}} | \Phi' \rangle = \langle i | H^{\mathrm{SO}}(1) | j \rangle + \frac{1}{2} \sum_k \{ \langle ik | H^{\mathrm{SO}}(2) | jk \rangle - \langle ki | H^{\mathrm{SO}}(2) | jk \rangle - \langle ik | H^{\mathrm{SO}}(2) | kj \rangle \}, \tag{61}$$

where the summation index k runs over common orbitals.

(C) Doubly excited Slater determinants: Φ and Φ' differ by two pairs of spin orbitals ij, kl.
Only four-index two-electron integrals contribute to the spin–orbit coupling matrix element:

$$\langle \Phi | H^{\mathrm{SO}} | \Phi' \rangle = \tfrac{1}{2} \{ \langle ij | H^{\mathrm{SO}}(2) | kl \rangle + \langle ji | H^{\mathrm{SO}}(2) | lk \rangle - \langle ji | H^{\mathrm{SO}}(2) | kl \rangle - \langle ij | H^{\mathrm{SO}}(2) | lk \rangle \}. \tag{62}$$

Except for core excitation processes which rule out the possibility of applying a frozen-core approximation, the indices i and j represent valence orbitals. In the case of doubly excited determinants also the k and l are then in the valence space. The only core contributions occur for singly excited determinants–and they are by far the largest terms in the sum over k in equation (61). By construction, frozen-core orbitals are common to all determinants of a configuration interaction expansion. The summation over the frozen-core orbitals k in equation (61) and the summation over all possible pairs of determinants can therefore be interchanged. Summing up all core contributions to a given index pair i, j one is left with an effective one-electron integral which may be contracted with the proper one-electron term $\langle i|H^{\mathrm{SO}}(1)|j\rangle$. The index k in the sum over three-index two-electron spin–orbit integrals is then restricted to the valence space. At this stage, case (C) contributions remain untouched.

The computational effort can be substantially reduced further, if one manages to get rid of all explicit two-electron terms. In most cases four-index spin–orbit integrals make only a minor contribution to the total spin–orbit coupling matrix element between two configuration interaction wavefunctions and can be neglected. However, in particular for molecules built from light constituents, unacceptable errors result, if also all three-index two-electron valence integrals are set to zero. A much better effective one-electron spin–orbit Hamiltonian is obtained when an average valence orbital occupation–corresponding, e.g., to the Hartree–Fock configuration–is assumed. Equation (61) forms the basis for the definition of the spin–orbit mean-field Hamiltonian, similar to the Coulomb and exchange operators in Hartree–Fock theory [Hess *et al.* (1996)]. To take account of partially filled orbitals, occupation numbers have been explicitly introduced.

Further, the two-electron spin–orbit interaction is averaged over α and β spin orientations before spin-integration. In a matrix representation of basis functions ϕ_a and ϕ_b this effective one-electron spin–orbit mean-field (SOMF) Hamiltonian reads

$$
\begin{aligned}
H_{\text{mf}}^{\text{SO}} = H^{\text{SO}}(1) + \frac{1}{4}\sum_{\mu}\sum_{\nu}\sum_{\substack{M \\ \text{fixed } n_M}} n_M\{ & |\chi_\mu\rangle\langle\chi_\mu M_\alpha|H^{\text{SO}}(2)|\chi_\nu M_\alpha\rangle\langle\chi_\nu| \\
& + |\chi_\mu\rangle\langle\chi_\mu M_\beta|H^{\text{SO}}(2)|\chi_\nu M_\beta\rangle\langle\chi_\nu| - |\chi_\mu\rangle\langle M_\alpha\chi_\mu|H^{\text{SO}}(2)|\chi_\nu M_\alpha\rangle\langle\chi_\nu| \\
& - |\chi_\mu\rangle\langle M_\beta\chi_\mu|H^{\text{SO}}(2)|\chi_\nu M_\beta\rangle\langle\chi_\nu| - |\chi_\mu\rangle\langle\chi_\mu M_\alpha|H^{\text{SO}}(2)|M_\alpha\chi_\nu\rangle\langle\chi_\nu| \\
& - |\chi_\mu\rangle\langle\chi_\mu M_\beta|H^{\text{SO}}(2)|M_\beta\chi_\nu\rangle\langle\chi_\nu|\}.
\end{aligned} \tag{63}
$$

The summation index M runs in principle over all spatial orbitals; occupation numbers n_M are fixed and range from 0 to 2. The extra factor of 1/2 in equation (63) compared with equation (61) comes from the spin-averaging and prevents double counting. Combined with a prescription to set up a correspondence between valence orbitals of all-electron and ECP calculations, the SOMF Hamiltonian may also be employed in RECP calculations [Marian and Wahlgren (1996), and Schimmelpfennig *et al.* (1998a, 1998b)].

Commonest among the approximate spin–orbit Hamiltonians are those derived from RECPs. Semi-empirical local operators such as

$$
H^{\text{SO}}(r) = \frac{\zeta}{r^3}\boldsymbol{LS} \tag{64}
$$

yield molecular results of very limited reliability [Wadt (1982)] and have fallen into disuse. Spin–orbit coupling operators for pseudo-potentials were developed already in the late seventies [Lee, Ermler and Pitzer (1977), and Hafner and Schwarz (1977)]. Commonly used spin-dependent RECPs are based either on the results of four-component [Pacios and Christiansen (1985), Hurley *et al.* (1986), LaJohn *et al.* (1988), Ross *et al.* (1990), Ermler, Ross and Christiansen (1994), Ross, Gayen and Ermler (1994), Nash, Bursten and Ermler (1997), Wildman, DiLabio and Christiansen (1997), Dolg (1997), and Titov and Mosyagin (1999)] or two-component atomic calculations [Teichteil, Pélisser and Spiegelmann (1983), Dolg, Stoll and Preuss (1989), Andrae *et al.* (1990), and Küchle *et al.* (1991)]. In rare cases spin–orbit operators were also derived directly via the gradient of the one-component potential [Seijo (1995)].

Teichteil, Pélissier and Spiegelmann (1983) fit a spin–orbit pseudo-operator such that its action on a pseudo-orbital optimally reproduces the effect of the true spin–orbit operator on the corresponding all-electron orbital. As a reference they used atomic all-electron two-component results obtained with the method by Barthelat and Durand (1978). Ermler, Ross, Christiansen and coworkers [Pacios and Christiansen (1985), Hurley *et al.* (1986), LaJohn *et al.* (1988), Ross *et al.* (1990), Ermler, Ross and Christiansen (1994), Ross, Gayen and Ermler (1994), Nash, Bursten and Ermler (1997), and Wildman, DiLabio and Christiansen (1997)] and Titov and Mosyagin (1999) define a spin–orbit operator as the difference between the ℓ and j dependent pseudo-potentials (REPs) [Ermler, Ross and Christiansen (1988)]

$$
\begin{aligned}
H^{\text{SO}}(r) = \sum_{\ell=1}^{\ell_{\max}} \Delta V_\ell(r) \Bigg(& \frac{\ell}{2\ell+1} \sum_{m_j=-j}^{j} |\ell, j, m_j\rangle\langle\ell, j, m_j| \\
& - \frac{\ell+1}{2\ell+1} \sum_{m'_j=-j'}^{j'} |\ell, j', m'_j\rangle\langle\ell, j', m_{j'}| \Bigg),
\end{aligned} \tag{65}
$$

with

$$j = \ell + 1/2, \quad j' = \ell - 1/2,$$

and

$$\Delta V_\ell(r) = (V_{\ell,\ell+1/2}(r) - V_{\ell,\ell-1/2}(r)).$$

This difference is generally fitted to a linear combination of a few basis functions. According to Pitzer and Winter (1988) this operator (65) can also be written in the form

$$H^{SO}(r) = \sum_{\ell=1}^{\ell_{\max}} \frac{2\Delta V_\ell(r)}{2\ell+1} \boldsymbol{\ell} \cdot \boldsymbol{s} \sum_{m_\ell=-\ell}^{+\ell} |\ell, m_\ell\rangle\langle\ell, m_\ell| \tag{66}$$

$$= \sum_{\ell=1}^{\ell_{\max}} \xi_\ell(r) \boldsymbol{\ell} \cdot \boldsymbol{s} \sum_{m_\ell=-\ell}^{+\ell} |\ell, m_\ell\rangle\langle\ell, m_\ell|, \tag{67}$$

in which its relation to the semi-empirical operator (64) is particularly apparent. Also the Stuttgart and Dresden groups [Dolg, Stoll and Preuss (1989), Andrae *et al.* (1990), Küchle *et al.* (1991), Dolg and Stoll (1995), Leininger *et al.* (1997), and Dolg (1997)] and Seijo (1995) make use of this form to define an effective spin–orbit Hamiltonian. In the former case, the parameters in the expression (67) are fitted directly to atomic fine-structure splittings. Seijo, on the other hand, determines the parameters through least-squares fitting to the radial parts of the numerical Wood–Boring spin–orbit operators and introduces a semi-empirical correction through a change of AO basis set contraction coefficients.

All the above-mentioned procedures to parametrize the spin–orbit interaction for pseudo-potentials have one thing in common: the predominant action of the spin–orbit operator has to be transferred from the region close to the nucleus to the valence region. The reason for this relocation is simply the fact that amplitudes of pseudo-orbitals are small in the proximity of the nucleus. If the original $1/r^3$ dependence is left untouched as, e.g., in the semi-empirical expression (64) huge effective charges may result.

6.3 Case studies

The accurate quantum chemical prediction of spectroscopic properties of molecular systems containing heavy atoms requires both electron correlation and relativistic effects to be included in a calculation. As described in more detail above, a variety of methods have been devised which take these interactions into account–at different levels of approximation. Not unexpectedly, the most sophisticated approaches are also the most expensive ones in terms of computational resources and thus can be applied to small molecules only. On the other hand, the more elaborate relativistic methods constitute a useful tool to gauge more approximate schemes. In the following case studies we shall focus on atoms and molecules, small enough to be treated by any of these methods.

If, in practice, one wants to perform a calculation of molecular spectroscopic properties including spin–orbit coupling, there are several decisions to be made which will in general depend on the molecular size and the required accuracy of spectroscopic parameters.

- The first question concerns the type of Hamiltonian to be used: is it advisable to work with four-component wavefunctions using a (projected) Dirac–Coulomb (–Breit) Hamiltonian or do we want to confine the relativistic treatment to two components employing a Hamiltonian with separate electrostatic and magnetic interaction terms? In the former case, very large AO basis sets are required in order to fulfill the kinetic balance conditions [equation (23)], whereas the latter treatment is more approximate but generally less demanding.

- Closely related to the first point is the question at which stage spin–orbit interaction is allowed to enter. Procedures which include spin–orbit interaction already in the orbital optimization step shall be denoted jj-coupling procedures. All four-component methods and a few two-compenent methods fall into this category. By the term *intermediate coupling* we shall denote cases in which some kind of spin-averaged orbitals are employed but in which the coefficients of individual determinants or configuration state functions (CSFs) are optimized in the presence of spin–orbit coupling. In *LS*-type procedures, finally, electron correlation and spin–orbit interaction are treated independently. Spin–orbit coupling is introduced *a posteriori* at the state-mixing level, i.e., the CSF coefficients are determined in computations with a spin-free Hamiltonian. In the full spin–orbit configuration interaction limit the results of these three approaches are identical, of course, but in practice they never will be.
- Another point concerns the interaction Hamiltonian. Many four-component calculations employ the DC Hamiltonian [equation (14)], while others also take account of the frequency-independent Breit interaction [equation (16)]. Correspondingly, in two-component procedures a decision has to be made whether the two-electron spin–other–orbit interaction is included or not.
- A further question of relevance to both four- and two-component procedures concerns the number of correlated electrons and the type of electron correlation treatment. Commonest are various kinds of configuration interaction calculations, but there are also coupled-pair functional, coupled-cluster, perturbation theory, and density functional theory treatments.
- One of the decisions to be made is whether all electrons or just the outer shells are to be treated explicitly. In the latter case, the interaction between the valence and the core electrons is represented by some ECP. The use of an ECP is certainly an approximation but may save enormous amounts of computer time in the integral evaluation step.
- The following considerations apply only to two-component procedures which explicitly include spin-dependent terms in the interaction Hamiltonian. The question arises, under which circumstances it is necessary to employ the full multi-center one- and two-electron spin–orbit Hamiltonian or alternatively, when can an effective one-electron Hamiltonian be used, probably even in a one-center approximation?

A rigorous test for a method including electron correlation and relativity is provided by studying the ground and excited electronic states of a molecule containing heavy elements. As far as spin–orbit coupling is concerned, it turned out that transition and main group elements put quite different demands on their theoretical treatment and prototypes of each class will be considered here. Further, we shall briefly discuss molecules with light constituents such as carbohydrides. In these molecules the percentage contribution of two-electron spin–orbit terms to a spin–orbit coupling matrix element is much larger–of the order of 50 %–and it is thus not evident how well effective one-electron spin–orbit Hamiltonians will perform. In the following case studies we shall try to exemplify which methods can be used for a given size of molecule and desired accuracy.

6.3.1 MAIN GROUP ELEMENTS: THALLIUM AND THALLIUM HYDRIDE

The molecular bonding in thallium hydride (TlH) is strongly influenced by relativistic effects, both kinematic and magnetic. As a diatomic molecule it is accessible also to the more advanced, expensive relativistic methods and has ever since been a playground for testing the capability of various computational schemes to include relativistic effects in molecular calculations [Pitzer

(1975), Pyykkö and Desclaux (1976), Snijders and Pyykkö (1980), Lee, Ermler and Pitzer (1980), Pyper (1980), Christiansen and Pitzer (1980), Pitzer and Christiansen (1981), Christiansen, Balasubramanian and Pitzer (1982), Wadt (1984), Ramos, Lee amd Malli (1988), Schwerdtfeger (1987), Balasubramanian and Tao (1991), Treboux and Barthelat (1993), Seijo (1995), Dolg *et al.* (1991), Kim, Lee and Lee (1996), van Lenthe, Snijders and Baerends (1996), Rakowitz and Marian (1997), Wildman, DiLabio and Christiansen (1997), Schimmelpfenning *et al.* (1998a), Lee *et al.* (1998), and Han, Bae and Lee (1999)]. Furthermore, the electronic spectrum of TlH proves to be a great challenge for a theoretician, as the excited electronic states give rise to a most puzzling spectrum [Grundström and Valberg (1937), Grundström (1940), Neuhaus and Muld (1959), Larsson and Neuhaus (1963), Ginter and Battino (1965), and Larsson and Neuhaus (1966)]. Unfortunately, only few quantum chemical investigations have been performed which include excited electronic states of TlH [Christiansen, Balasubramanian and Pitzer (1982), Wadt (1984), Treboux and Barthelat (1993), Rakowitz and Marian (1997), and Schimmelpfennig *et al.* (1998a)].

Experimental knowledge about spectroscopic properties of Tl and TlH

Atomic thallium has a ^{2}P ground state with a closed 6s shell and a single electron in the 6p shell. Because of spin–orbit coupling the ^{2}P state splits into a lower $J = 1/2$ and an upper $J = 3/2$ component with a considerable excitation energy of 7793 cm^{-1}. These two ^{2}P components are well separated from the next higher electronic state at 26 478 cm^{-1}, a ^{2}S state with 7s configuration [Moore (1958)].

The first experiment on the electronic spectrum of TlH was conducted as early as 1937 [Grundström and Valberg (1937)]. From the measured line spectrum Grundström and Valberg were able to determine spectroscopic parameters of the electronic ground state–recently refined by Urban *et al.* (1989). It took more than twenty years, however, before the lines were properly assigned to arise from excitations between the TlH $0^+(I)$ ground state and a single excited state with unusually shaped potential energy curve, the $0^+(II)$ state [Neuhaus and Muld (1959), and Ginter and Battino (1965)]. As we shall see later, this unusual shape is caused by spin–orbit interaction between the two 0^+ states. There is ample evidence of further low-lying electronic states; experimentally derived spectroscopic parameters of TlH are reliable, however, only for the low-lying 0^+ states. We shall therefore restrict the comparison with theoretical results to these two.

Qualitative discussion of relativistic effects on the bonding in TlH

Unlike many other heavy element monohydrides such as AuH [Pizlo *et al.* (1993)], TlH experiences a marked destabilization with respect to its dissociation products when relativistic effects are taken into account [Pyykkö and Desclaux (1976), and Lee, Ermler and Pitzer (1980)]. Nevertheless, in relativistic treatments a shorter equilibrium bond length is found accompanied by an increase of the harmonic vibrational force constant [Lee, Ermler and Pitzer (1980), and Snijders and Pyykkö (1980)]. The apparent contradiction between the relativistic bond weakening and bond contraction can be explained in terms of opposite effects of kinematic and magnetic relativistic corrections.

In the dissociation limit the TlH molecule correlates with hydrogen in its ^{2}S ground state and the $^2P_{1/2}$ state of atomic thallium. Since the spin–orbit splitting between the $^2P_{1/2}$ and $^2P_{3/2}$ components of Tl amounts to nearly 1 eV, the jj-coupling picture represents a fair starting point for a qualitative description of the atom. In terms of functions adapted to the molecular symmetry, the $p_{1/2}$ spinor may be written as

$$p_{1/2,m_j=1/2} : \sqrt{\frac{2}{3}}\,(\pi^+,\beta) - \sqrt{\frac{1}{3}}(\sigma,\alpha),$$
$$p_{1/2,m_j=-1/2} : \sqrt{\frac{1}{3}}\,(\sigma^+,\beta) - \sqrt{\frac{2}{3}}\,(\pi^-\alpha)\,. \qquad (68)$$

A simple argument–often put forward in the early days [Pitzer (1979), and Pyykkö and Desclaux (1979)]–explains the weakness of the thallium hydrogen bond by the small amount of σ bonding character in the $6p_{1/2}$ orbital of Tl. Quantitatively, however, the bond strength does not drop to one-third of the strength of a hypothetical 'pure' σ bond as might be expected from the coefficients in equation (68). Calculations reveal a substantial contribution of the $m_j = \pm\frac{1}{2}$ components of the $6p_{3/2}$ spinor

$$p_{3/2,m_j=1/2} \quad : \quad \sqrt{\frac{1}{3}}\,(\pi^+,\beta) + \sqrt{\frac{2}{3}}(\sigma,\alpha)$$
$$p_{3/2,m_j=-1/2} \quad : \quad \sqrt{\frac{2}{3}}\,(\sigma^+,\beta) + \sqrt{\frac{1}{3}}\,(\pi^-\alpha)\,. \qquad (69)$$

to the bonding molecular spinor. [Pyykkö and Desclaux (1976), Lee, Ermler and Pitzer (1980), Christiansen and Pitzer (1980), and Pitzer and Christiansen (1981)] In terms of an *LS* treatment the same qualitative result is obtained, if one starts out with a $^1\Sigma^+$ ground state and allows for spin–orbit coupling with the first excited $^3\Pi_0^+$ state. Upon bond breaking the two *LS* states get closer and closer in energy with the result that their spin–orbit interaction increases. As a consequence, the electronic ground state energy is lowered preferentially at large internuclear separations and the dissociation energy is decreased [Rakowitz and Marian (1997)].

Kinematic relativistic effects, on the other hand, stabilize the bond and lead to a considerable bond length contraction [Lee, Ermler and Pitzer (1980), and Snijders and Pyykkö (1980)]. It was shown by Snijders and Pyykkö that about 90 % of this bond contraction can be obtained perturbatively, i.e., by applying the relativistic kinetic energy operator to a non-relativistic wavefunction.

$^2P_{1/2}$–$^2P_{3/2}$ *splitting in the ground state of atomic thallium*

Let us now turn towards a more quantitative analysis. The first hurdle to get over is the proper description of the dissociation limit, in particular the fine-structure splitting of the ^2P ground state of atomic thallium. In the following, we should like to discuss the results obtained by various authors in the *jj*, *LS*, or *intermediate* coupling schemes. For convenience, these data are collected in Tables 6.1–6.3. To facilitate identification, calculations have been denoted according to their labels in the original work.

The four-component results are displayed in Table 6.1. DHF calculations reproduce the splitting between the two ground state components astonishingly well [Migdalek and Baylis (1981), Ross, Ermler and Christiansen (1986), Küchle *et al.* (1991), Rakowitz and Marian (1996), Eliav *et al.* (1996), and Landau (1999)]. At the single-configuration level it appears to be of minor importance whether separate calculations are carried out for each term, i.e., at optimized level (OL) where core-relaxation is included, or whether spinors are determined from an averaged density matrix, i.e., at the so-called average level (AL). Rakowitz and Marian tested the influence of different nuclear models on the energy. Although total energies differ considerably, the fine-structure splitting was found to vary by less than 1 cm^{-1} [Rakowitz and Marian (1996)]. The frequency-independent Breit (B) interaction [equation (16)]–to a large extent covered by two-electron spin–other-orbit terms in two-component theories–reduces the splitting by 90–100 cm^{-1}.

Table 6.1. Results of jj coupling calculations on the $^2P_{1/2}$–$^2P_{3/2}$ splitting [cm^{-1}] in the ground state of atomic thallium.

Methods	ECP	Correlation		$E(^2P_{3/2}) - E(^2P_{1/2})$
Four-component methods				
DHF(AL)[a]	—		—	7641
DHF(OL)[b,c]	—		—	7684
DHF(OL) + B(var.)[b]	—		—	7597
DHF(OL) + P(pert.)[c]	—		—	7586
DHF(OL) + B(pert.)[d]	—		—	7627
DHF(OL) + B(pert.)[e]	—		—	7600
MCDF(OL)[c]	—	3e$^-$	MC	7841
MCDF(OL) + B(pert.)[c]	—	3e$^-$	MC	7750
DHF(OL) + B(pert.) + CP[d]	—		CP	7816
DHF/RCC1[e]	—	35e$^-$	RCC	7710
DHF + B/RCC1[f]	—	35e$^-$	RCC	7627
DHF/Hermitian CC[f]	—	35e$^-$	RCC	7832
DHF+B/Hermitian CC[f]	—	35e$^-$	RCC	7746
Two-component methods				
REP[b]	Xe4f^{14}5d^{10}		—	7545
REP[g]	Xe4f^{14}		—	7409
WB[a]	—		—	7473
PP, HF, AL[a]	Xe4f^{14}5d^{10}		—	7383
PP, HF, OL[a]	Xe4f^{14}5d^{10}		—	7397
ZORA[h]	—	LDA-DFT		8469
ZORA[h]	—	GGA-DFT		8227
SC RECP 13VE[i]	Xe4f^{14}		—	7528
Wildman SC RECP 13VE[i]	Xe4f^{14}	13e$^-$	MP2	6729
Wildman SC RECP 13VE[i]	Xe4f^{14}	13e$^-$	CCSD	6884
Wildman SC RECP 13VE[i]	Xe4f^{14}	13e$^-$	CCSD(T)	6854
Wildman SC RECP(p) 13VE[i]	Xe4f^{14}		—	8199
Wildman SC RECP(p) 13VE[i]	Xe4f^{14}	13e$^-$	MP2	7088
Wildman SC RECP(p) 13VE[i]	Xe4f^{14}	13e$^-$	CCSD	7249
Wildman SC RECP(p) 13VE[i]	Xe4f^{14}	13e$^-$	CCSD(T)	7212
Schwerdtfeger EC RECP1 3VE[i]	Xe4f^{14}5d^{10}		—	7097
Schwerdtfeger EC RECP1 3VE[i]	Xe4f^{14}5d^{10}	3e$^-$	MP2	6863
Schwerdtfeger EC RECP1 3VE[i]	Xe4f^{14}5d^{10}	3e$^-$	CCSD	6859
Schwerdtfeger EC RECP1 3VE[i]	Xe4f^{14}5d^{10}	3e$^-$	CCSD(T)	6839
Küchle EC RECP2 3VE[i]	Xe4f^{14}5d^{10}		—	7814
Küchle EC RECP2 3VE[i]	Xe4f^{14}5d^{10}	3e$^-$	MP2	7672
Küchle EC RECP2 3VE[i]	Xe4f^{14}5d^{10}	3e$^-$	CCSD	7678
Küchle EC RECP2 3VE[i]	Xe4f^{14}5d^{10}	3e$^-$	CCSD(T)	7668
Leininger EC RECP3 3VE[i]	Xe4f^{14}5d^{10}		—	7465
Leininger EC RECP3 3VE[i]	Xe4f^{14}5d^{10}	3e$^-$	MP2	7318
Leininger EC RECP3 3VE[i]	Xe4f^{14}5d^{10}	3e$^-$	CCSD	7333
Leininger EC RECP3 3VE[i]	Xe4f^{14}5d^{10}	3e$^-$	CCSD(T)	7323
Leininger EC RECP3 13VE[i]	Xe4f^{14}		—	7616
Leininger EC RECP3 13VE[i]	Xe4f^{14}	13e$^-$	MP2	6567
Leininger EC RECP3 13VE[i]	Xe4f^{14}	13e$^-$	CCSD	6573
Leininger EC RECP3 13VE[i]	Xe4f^{14}	13e$^-$	CCSD(T)	6543

[a] Küchle *et al.* (1991).
[b] Ross, Ermler and Christiansen (1986).
[c] Rakowitz and Marian (1996).
[d] Migdalek and Baylis (1981).
[e] Eliav *et al.* (1996).
[f] Landau (1999).
[g] Ross *et al.* (1990).
[h] van Lenthe, Snijders and Baerends (1996).
[i] Han, Bae and Lee (1999).

Table 6.2. Spin–orbit splitting [cm^{-1}] of the ^2P ground state of atomic thallium obtained from *LS*-coupling calculations. Spin–orbit coupling is treated in first-order perturbation theory (PT1), unless indicated otherwise.

Methods	ECP	Correlation		SOC	$E(^2P_{3/2}) - E(^2P_{1/2})$
WB, PT[a]	—		—		6928
PP(3),SCF[b]	$Xe4f^{14}5d^{10}$		—		7021
PP(13),SCF[b]	$Xe4f^{14}$		—		6417
PP(21),SCF[b]	$Kr4d^{10}4f^{14}$		—		6039
Küchle-PP, SCF[c]	$Xe4f^{14}5d^{10}$		—		7014
Ross-RECP, SCF[c]	$Xe4f^{14}5d^{10}$		—		7134
AREP[d]	$Xe4f^{14}5d^{10}$		—		7324
AREP[c,e]	$Xe4f^{14}$		—		7424
PP(3),SCF[b]	$Xe4f^{14}5d^{10}$		CPP		7335
REFCI[f]	—	$3e^-$	Ref.-CI		6406
PP(3),CISD[b]	$Xe4f^{14}5d^{10}$	$3e^-$	CISD		6837
PP(3),CISD[b]	$Xe4f^{14}5d^{10}$	$3e^-$	CISD+CPP		7247
Küchle-PP, FOPT[c]	$Xe4f^{14}5d^{10}$	$3e^-$	CISD		6888
Ross-RECP, FOPT[c]	$Xe4f^{14}5d^{10}$	$3e^-$	CISD		7042
PP, CIPSO[a]	$Xe4f^{14}5d^{10}$	$3e^-$	MRCISD		6850
Küchle-PP, LSC-SO-CI(3)[c]	$Xe4f^{14}5d^{10}$	$3e^-$	MRCISD		6933
Ross-RECP, LSC-SO-CI(3)[c]	$Xe4f^{14}5d^{10}$	$3e^-$	MRCISD		7042
CI3[f]	—	$3e^-$	MRCISD		6221
H^{SOO}_{NP}, CI3[g]	—	$3e^-$	MRCISD		6305
AE MF[h]	—	$3e^-$	MRCISD		6275
ECP MF[h]	$Xe4f^{14}5d^{10}$	$3e^-$	MRCISD		6231
CI3EX[f]	—	$3e^-$	MRCISDT		6224
PP(13),CISD[b]	$Xe4f^{14}$	$13e^-$	CISD		5755
Ross-RECP, FOPT[c]	$Xe4f^{14}$	$13e^-$	CISD		6423
Ross-RECP, LSC-SO-CI(3)[c]	$Xe4f^{14}$	$13e^-$	MRCISD		6423
H^{SOO}_{mf}, CI13(b)[g]	—	$13e^-$	MRCISD		6654
H^{SOO}_{NP}, CI13(b)[g]	—	$13e^-$	MRCISD		6651
CI13[f]	—	$13e^-$	MRD-CI		6255
CI13EX[f]	—	$13e^-$	MRD-CI(T)		6330
CI19EX[f]	—	$19e^-$	MRD-CI(T)		6397
PP(21),CISD[b]	$Kr4d^{10}4f^{14}$	$21e^-$	CISD		6096
Küchle-PP, SOPT, PT2[c]	$Xe4f^{14}5d^{10}$	$3e^-$	CISD		7539
Ross-RECP, SOPT PT2[c]	$Xe4f^{14}5d^{10}$	$3e^-$	CISD		7500
Ross-RECP, SOPT PT2[c]	$Xe4f^{14}$	$13e^-$	CISD		6837
PP(3),MCSCF[b]	$Xe4f^{14}5d^{10}$	$3e^-$	MCSCF	QDPT	7587
PP(3),MCSCF[b]	$Xe4f^{14}5d^{10}$	$3e^-$	MCSCF+CPP	QDPT	7975
PP(13),MCSCF[b]	$Xe4f^{14}$	$3e^-$	MCSCF	QDPT	7567
PP(13),MCSCF[b]	$Xe4f^{14}$	$3e^-$	MCSCF+CPP	QDPT	7768
PP(21),MCSCF[b]	$Kr4d^{10}4f^{14}$	$3e^-$	MCSCF	QDPT	7546
PP(21),MCSCF[b]	$Kr4d^{10}4f^{14}$	$3e^-$	MCSCF+CPP	QDPT	7869
PP, CIPSO[a]	$Xe4f^{14}5d^{10}$	$3e^-$	CISD	QDPT	7627
PP(3),MRCI[b]	$Xe4f^{14}5d^{10}$	$3e^-$	MRCISD	QDPT	7311
PP(3),MRCI[b]	$Xe4f^{14}5d^{10}$	$3e^-$	MRCISD+CPP	QDPT	7808
PP(13),MRCI[b]	$Xe4f^{14}$	$3e^-$	MRCISD+CPP	QDPT	7578
PP(21),MRCI[b]	$Kr4d^{10}4f^{14}$	$3e^-$	MRCISD	QDPT	7298
PP(21),MRCI[b]	$Kr4d^{10}4f^{14}$	$3e^-$	MRCISD+CPP	QDPT	7570
Effective SOCI[i]	$Xe4f^{14}5d^{10}$	$3e^-$	CISD+CPP	H_{eff}	7858

(continued overleaf)

Table 6.2. (*continued*)

Methods	ECP	Correlation		SOC	$E(^2P_{3/2}) - E(^2P_{1/2})$
Küchle-PP, LSC-SO-CI(9)[c]	$Xe4f^{14}5d^{10}$	$3e^-$	MRCISD	QDPT	7698
Ross-RECP, LSC-SO-CI(9)[c]	$Xe4f^{14}5d^{10}$	$3e^-$	MRCISD	QDPT	7538
Ross-RECP, LSC-SO-CI(9)[c]	$Xe4f^{14}$	$13e^-$	MRCISD	QDPT	6862
PP(13),MRCI[b]	$Xe4f^{14}$	$13e^-$	MRCISD	QDPT	5755
PP(21),MRCI[b]	$Kr4d^{10}4f^{14}$	$13e^-$	MRCISD	QDPT	7548
PP(21),MRCI[b]	$Kr4d^{10}4f^{14}$	$21e^-$	MRCISD	QDPT	7810

[a] Küchle *et al.* (1991).
[b] Leininger, *et al.* (1997).
[c] Buenker *et al.* (1998).
[d] Ross, Ermler and Christiansen (1986).
[e] Ross *et al.* (1996).
[f] Rakowitz and Marian (1996).
[g] Wahlgren *et al.* (1997).
[h] Schimmelpfenning *et al.* (1998b).
[i] Vallet, *et al.* (1999).

Table 6.3. Results of intermediate coupling calculations on the $^2P_{1/2}$–$^2P_{3/2}$ splitting [cm^{-1}] in the ground state of atomic thallium.

Method	ECP	Correlation+SOC		$E(^2P_{3/2}) - E(^2P_{1/2})$
H_{NP}^{SOO}, INT(1p)[a]	—	$3e^-$	CAS(val. + 1p)-CI	6625
H_{NP}^{SOO}, INT(f)[a]	—	$3e^-$	CAS(val. + all p)-CI	7537
REP CI[b]	$Xe4f^{14}$	$3e^-$	MRCISD	7440
CASSCF/SOCI/RCI[c]	$Xe4f^{14}$	$3e^-$	MRCISD	6930
PP, DGCI[d]	$Xe4f^{14}5d^{10}$	$3e^-$	MRCISD	7654
Küchle-PP, MR-SO-CI[e]	$Xe4f^{14}5d^{10}$	$3e^-$	MRCISD	7722
Ross-RECP, MR-SO-CI[e]	$Xe4f^{14}5d^{10}$	$3e^-$	MRCISD	7547
WB-AIMP CIDBG[f]	$Xe4f^{14}$	$3e^-$	MRCISD	8457
same, SO-corr. basis[f]	$Xe4f^{14}$	$3e^-$	MRCISD	7473
CI3[g]	—	$3e^-$	MRCISD	7506
CI3EX[g]	—	$3e^-$	MRCISDT	7519
H_{NP}^{SOO}, CI3CV(r)[a]	—	$3e^-$	MRCISD+CP	7784
CI13[g]	—	$13e^-$	MRD-CI	7227
H_{NP}^{SOO}, CI13(b,r)[a]	—	$13e^-$	MRCISD	7656
CI13EX[g]	—	$13e^-$	MRD-CI(T)	7672
CI19EX[g]	—	$19e^-$	MRD-CI(T)	7796

[a] Wahlgren *et al.* (1997).
[b] Christiansen, Balasubramanian and Pitzer (1982).
[c] Balasubramanian and Tao (1991).
[d] Küchle *et al.* (1991).
[e] Buenker *et al.* (1998).
[f] Seijo (1995).
[g] Rakowitz and Marian (1996).

This has to be kept in mind when a comparison with experiment is made because the Breit term is left out in most investigations on Tl. The size of the calculated Breit interaction appears to be insensitive to the kind of treatment: almost equal results are obtained whether the correction is calculated perturbatively or included in the variational spinor determination. Static correlation, as exerted in a valence ($6s^26p$, $6p^3$) multi-configurational self-consistent field treatment, increases the splitting by 200 cm^{-1}, thus improving the results. The same is true for core-polarization (CP). Migdalek and Baylis (1981) tested various empirical CP potentials; the best result was obtained for a potential tailored in a way such that the removal of the 6p electron exactly reproduces the first experimental ionization potential. Although the DF splitting of the ground state is in good agreement with experiment, this is not generally true for other atomic excitation energies of Tl (not shown in the table) which differ from experiment by 2000–4000 cm^{-1} [Eliav *et al.* (1996)]. In general, dynamic correlation will have to be included, if reliable electronic excitation energies are to be obtained. An extensive calculation on the fine-structure splitting of the thallium atomic ground state was presented by Eliav *et al.* (1996), a four-component coupled-cluster (RCC) calculation using a huge basis set comprising 35s,27p,21d,15f,9g,6h,4i Gaussian functions. Since the cluster expansion has to start from a closed-shell state, the Tl^+ ground state was chosen as reference and 35 electrons were correlated, yielding values of 7710 and 7627 cm^{-1} without and with Breit interaction, respectively. Very recently Landau (1999) was able to improve these results employing a newly developed Hermitian coupled cluster approach. Again, Breit interaction reduces the splitting by about 90 cm^{-1} yielding a value of 7746 cm^{-1} in excellent agreement with experiment.

Most of the two-component calculations on Tl employ some kind of RECP. Among the exceptions are WB calculations which were used to parameterize ECPs [Küchle *et al.* (1991)]; this method is restricted to atomic cases, however. The few all-electron treatments which can be extended to molecules [van Lenthe, Snijders and Baerends (1996), Rakowitz and Marian (1996), Wahlgren *et al.* (1997), and Schimmelpfennig *et al.* (1998b)] make either use of the ZORA approximation [equation (33)] or of a one-component no-pair (NP) Douglas–Kroll Hamiltonian [equation (50)] and the corresponding spin–orbit coupling term. As described in more detail above, the RECPs incorporate relativistic effects by adjusting parameters of an analytical potential such that certain properties of relativistic all-electron atomic calculations, e.g., total energies, orbital energies, or orbital shapes are reproduced. In many cases the atomic all-electron reference is a four-component DFC calculation, in other cases a two-component Wood–Boring treatment. It is only recently that people have started to take the Breit interaction into account when parameterizing RECPs [Leininger *et al.* (1997), and Dolg (1997)].

Among the two-component results until very recently only some WB, RECP HF, and ZORA DFT values had been computed according to a jj coupling scheme. The two-component WB and RECP HF results only slightly underestimate the DF values. Compared with experiment they are too small by 300–400 cm^{-1}. The ZORA DFT values, on the other hand, overestimate the experiment by 400–700 cm^{-1} indicating that the 6p spinor is slightly too compact. From these results no definite conclusion can be drawn as to whether the ZORA approximation or the DFT treatment is the cause of this problem. A related study on the hydrogen-like U^{91+} ion [van Lenthe *et al.* (1995)] helps to clarify the situation. Here it is shown that the zeroth-order CPD Hamiltonian (ZORA approximation) tends to overestimate the orbital binding energies and concomitantly also the spin–orbit splittings. Most of the deviations from corresponding DHF values are remedied, however, if also the first-order CPD Hamiltonian (FORA approximation) is included.

Recently, Han, Bae and Lee (1999) tested the performance of commonly used RECPs in various two-component approaches: HF, second-order Møller–Plesset (MP2), coupled-cluster singles and doubles (CCSD), and CCSD(T) which is a CCSD approach augmented by triples in a perturbative way. The RECPs were taken from the works of Wildman, DiLabio and Christiansen (1997), Schwerdtfeger (1987), Küchle *et al.* (1991), and Leininger *et al.* (1997). Unlike the original investigators, Han, Bae and Lee (1999) employ the Gaussian basis functions uncontractedly. Comparing, e.g., the spin–orbit splittings of Küchle *et al.* (1991) (PP, HF, OL, 7397 cm^{-1}) and of Han, Bae and Lee (1999) (EC RECP2 3VE, 7814 cm^{-1}) at the HF level using exactly the same PP it appears that the basis function contraction has an appreciable influence on the outcome of RECP calculations and should not be changed at will. Further, the expansion length of the model spin–orbit Hamiltonian [equation (67)] seems to play an important role. The calculations dubbed SC RECP 13VE and SC RECP(p) 13VE employ both the shape-consistent SPD RECP by Wildman, DiLabio and Christiansen (1997) and an uncontracted basis set. They differ, though, in the maximum angular momentum number used to represent H^{SO}: in the former case the expansion includes up to f projectors whereas ℓ_{max} is restricted to 1 in the latter, indicated by the label (p). In particular the 5d electrons, explicitly contained in the valence space of the Tl atom, do not experience any spin–orbit interaction in that case. In jj coupling scheme this lack of d shell spin–orbit interaction has an indirect impact on the p shell splitting which is increased by about 600 cm^{-1} despite the fact that the 5d shell remains closed in HF calculations. Similar remarks apply to the spin–orbit potentials of the small-core Leininger PPs; in addition to a constant, ℓ-independent shift they include only a term acting on the p orbital space. In jj coupling approaches they should thus be applied only with great care. Because of the appreciable scatter in the spin–orbit splitting [Han, Bae and Lee (1999)] with the various RECPs at the HF level, it seems not meaningful to judge the outcome of their correlation calculations with respect to experiment. It is interesting, however, to note the trends. A second-order perturbative (MP2) treatment of correlation contributions in the valence shell (3e$^-$) decreases the spin-orbit splitting by roughly 150–200 cm^{-1}. Almost no difference is observed when the electron correlation treatment is improved by employing CCSD or CCSD(T) expansions. Adding the $5d_{3/2}$ and $5d_{5/2}$ shells to the set of correlated spinors leads to a drastic reduction of the Tl $^2P_{1/2}$–$^2P_{3/2}$ splitting. The discussion of these electron correlation effects will be deferred as the trends closely resemble those found in *LS*- and intermediate coupling treatments.

Numerous perturbative investigations of the spin–orbit coupling in Tl have been conducted using *LS*-coupled wavefunctions. [Ross, Ermler and Christiansen (1986); Ross *et al.* (1990); Küchle *et al.* (1991); Rakowitz and Marian (1996); Wahlgren *et al.* (1997); Leininger *et al.* (1997); Buenker *et al.* (1998); Schimmelpfennig *et al.* (1998a)]; the results are collected in excerpts in Table 6.2. Compared with a variational solution of the spin-dependent Wood–Boring equation (compare Table 6.1), a first-order perturbation theory (PT1) treatment using one-component WB orbitals underestimates the fine-structure splitting by more than 500 cm^{-1}. Similar results are obtained for expectation values of quasi-relativistic HF wavefunctions when large-core (three valence electrons) RECPs, e.g., [Ross *et al.* (1990), Küchle *et al.* (1991), Leininger *et al.* (1997)] are employed. Before we enter the discussion of how electron correlation effects influence the size of the computed spin–orbit splitting let us have a short look on the influence of different spin–orbit Hamiltonians. A series of tests was performed by Wahlgren *et al.* (1997). As expected, the dominant contribution originates from the (true) one-electron part. The screening by two-electron spin–same-orbit interactions amounts to approximately 250 cm^{-1} in first-order perturbation theory treatments, and the one-electron matrix element is further reduced by the two-electron spin–other-orbit terms. At the PT1 level, the size of the latter interaction is in the order of -30 to -50 cm^{-1}; the spin–other-orbit contribution increases to -70 to -80 cm^{-1} at the spin–orbit configuration interaction (SOCI) level. The size of the spin–other-orbit interaction thus appears to be sensitive to the computational details, in contrast to the Breit interaction in four-component approaches which was found to be approximately equal in variational and perturbative treatments. Wahlgren *et al.*(1997) also tested the performance of the Breit–Pauli Hamiltonian which is often used for light elements. Unlike the regularized all-electron two-component spin–orbit Hamiltonians, the Breit–Pauli (BP) approximation fails badly for an element as heavy as Tl. The spin–orbit interaction is too attractive: BP spin-orbit matrix elements are about 3500 cm^{-1} larger than corresponding no-pair values.

Let us have a closer look now at correlation contributions to the fine-structure splitting of Tl. In accord with the findings at the DF level, (5d)core–valence inter-shell correlation–introduced through an empirical core-polarization potential–enlarges the $^2P_{1/2}$–$^2P_{3/2}$ splitting. Valence-shell correlation, on the other hand, does not improve spin–orbit splittings unless orbital relaxation is allowed for as in an MCSCF procedure. On the contrary, at the PT1 level a complete reference space CI ($6s^26p$, $6p^3$) leads to a decrease of the spin-orbit matrix element. In particular, all-electron CI results deteriorate markedly with respect to the Hartree–Fock expectation values. At this level of spin–orbit treatment, even dynamic correlation introduced through single- (S), double- (D), and triple-excitations (T) from the $6s^26p$, $6p^3$ reference configurations into the virtual space does not change the picture: first-order expectation values of the all-electron no-pair spin–orbit Hamiltonian for correlated *LS*-coupled wavefunctions are too small by $1100–1500\,\mathrm{cm}^{-1}$. The same trend holds true for the small-core RECP calculations. Similar results have been obtained for the lead atom in a recent RECP study [Dolg (1997)] and in an all-electron NP investigation [Kleinschmidt (1999)]. The observed phenomena are thus by no means specific for group 13 elements, but may be generalized to the whole series of heavy main group elements.

An indication as to what is missing in these treatments can be obtained from four-component calculations: the radial expectation value of the Tl $6p_{1/2}$ DHF orbital amounts to $\approx 3.52\,a_0$ and is thus considerably smaller than that of its $p_{3/2}$ pendant ($\approx 4.01\,a_0$) [Desclaux (1973)]. This means that also their charge density distributions and concomitantly their electron correlation contents differ appreciably. There are several possibilities to take account of the spin–orbital relaxation in approaches using one-component (quasi-relativistic) orbitals as a one-particle basis. Küchle *et al.* (1991) and Buenker *et al.* (1998) included spin–orbit interaction with a few ^2P states by means of perturbation theory. It should be mentioned here that the states which have a significant influence on the ground state spin–orbit splitting are energetically high lying and do not have any spectroscopic relevance; they just serve as a basis for a perturbation expansion. Let us begin with a discussion of RECP results where only the 6s and 6p valence electrons were correlated. Evaluating the contributions of two excited ^2P states by a sum-over-states expression at the second-order Rayleigh–Schrödinger perturbation theory (PT2) level already leads to a considerable improvement of the fine-structure splitting. Quasi-degenerate perturbation theory (QDPT), where a small eigenvalue problem is solved in the basis of *LS*-coupled states, yields results of similiar quality as the application of intermediate coupling discussed below. Leininger *et al.* (1997) point out, however, that these methods to account for spin-orbital relaxation are only viable for small p valence basis sets where it is feasible to exhaust fully the p subspace; convergence of the perturbation energy with respect to the number of unperturbed states appears to be very slow if a larger valence basis is employed. Instead, they use symmetrically orthogonalized $6p_{1/2}$ and $6p_{3/2}$ orbitals from atomic SCF calculations to recover most of the relaxation effects. Also at this level of treatment inclusion of 5d core–valence correlation enlarges the splitting, partially compensated by valence correlation effects. Vallet *et al.* (1999) solve this problem in a different way. Very recently they proposed another method which avoids a summation over (highly excited) *LS*-coupled states. They contracted the spin–orbital relaxation effects into an effective Hamiltonian. Preliminary results employing the large-core PPs including CPP terms [Küchle *et al.* (1991)] look very promising.

Exactly the same trends are observed in intermediate coupling schemes (Table 6.3). In a spin–orbit configuration interaction (SOCI) treatment spin–orbital relaxation occurs automatically, if excitations to all virtual orbitals are allowed. One has to be careful, however, in truncated CI expansions where the selection of configurations is based on one-component energies as, e.g., in the MRD-CI approach. As an example, compare the results of the CI13 calculation by Rakowitz and Marian (1996) and the $H_{\mathrm{NP}}^{\mathrm{SOO}}$, CI13(b,*r*) results by Wahlgren *et al.* (1997). Both treatments use a no-pair Hamiltonian, include spin–other-orbit interactions, employ the same AO basis set and CI reference space, and correlate 13 electrons. Nevertheless, the fine-structure splitting computed by Rakowitz and Marian (1996) is smaller by $\approx 400\,\mathrm{cm}^{-1}$ although all single excitations from the $6s^26p^1$ configuration were kept automatically. We attribute this deviation to CI truncation effects. The difference is recovered in the MRD-CI treatment when all single excitations into virtual p orbitals are added to the reference space, from which single and double excitations are then generated and selected after an energy criterion. Removing the 5p shell from the frozen core (19 electron CI) increases the $^2P_{1/2}$–$^2P_{3/2}$ splitting by another $100\,cm^{-1}$.

Let us have a closer look at RECP calculations again. In molecular systems where the $(n-1)$d shell of Tl has appreciable overlap with the valence orbitals of the partner atom it is recommended to include this shell explicitly in the valence space. Several authors tested small-core RECPs with either 13 (5d6s6p) or 21 (5s5p5d6s6p) valence electrons. A comparison of the fine-structure splittings computed as SCF expectation values (Table 6.2) reveals large differences between the various types of RECPs. At this level of treatment the small-core shape-consistent AREPs by Ross and coworkers [Ross, Ermler and Christiansen (1986), and Ross *et al.* (1990)] perform slightly better than the corresponding large-core RECPs. By contrast, in one-component SCF calculations employing the energy-consistent pseudo-potentials (PPs) [Leininger *et al.* (1997)] spin–orbit expectation values considerably deteriorate, if the 5d and the 5s and 5p shells are included explicitly. This is a somewhat strange result and the authors might want to check their parameterization of the spin–orbit model Hamiltonian. Concomitantly spin polarization increases, however, yielding nearly equal fine-structure splittings for all three PPs when orbital relaxation is taken into account. Let us see then how electron correlation affects the results. Again, no systematic behavior is observed for the various RECPs. Taking the small-core RECP [Ross *et al*(1990)] and freezing the 5d shell in its atomic shape Balasubramanian and Feng (1990) find a value of only $6930\,cm^{-1}$ after including spin polarization and valence electron ($3e^-$) correlation; with respect to the SCF value this is a reduction of nearly $500\,cm^{-1}$. Almost the same value is obtained when the 5d shell is added to the valence space and 13 electrons are correlated [Buenker *et al.* (1998)]. Without spin–orbital relaxation this value even drops to $6423\,cm^{-1}$. At all levels of theory–apart from SCF–spin–orbit splittings for the small-core RECP by Ross *et al.* are smaller than the corresponding large-core values by about $600\,cm^{-1}$. The use of this RECP can thus not be recommended. Disastrous results are also obtained for the 13 valence-electron energy-adjusted PP [Leininger *et al.* (1997)], if one tries to include the correlation contributions of the 5d shell explicitly. On the other hand, resonable values are obtained, if only the valence electrons are correlated and the core polarization is added by means of an empirical potential. According to Leininger *et al.* the reason for this failure is connected to the nodal structure of the 6s and 6p pseudo-orbitals: since these are nodeless orbitals smoothly falling off as the radius approaches zero, their amplitudes are small in a region where the 5d orbital has its maximum. In turn this affects the accuracy of the exchange integrals between these orbitals and thus the accuracy of the correlation energy. The error can be considerably reduced, if at least one radial node is reintroduced and the orthogonality tail with respect to the 5s or the 5p orbital, respectively, is built in explicitly. And indeed, Leininger *et al.* (1997) find good agreement with experiment when they employ a 21 valence-electron PP where, in addition to the ns and np shells, also the $(n-1)$s, p, and d shells are treated explicitly. For the same reason, Huzinaga-type RECPs as, e.g., in the *ab initio* model potential (AIMP) by Seijo (1995) should properly describe the exchange interaction because for this kind of core potential all radial nodes are kept in the construction of the valence orbitals. It is interesting to note that Leininger *et al.* (1997) compute nearly the same spin–orbit splitting for either a three valence-electron PP including a CPP ($7808\,cm^{-1}$) or a 21 valence-electron PP ($7810\,cm^{-1}$), both in excellent agreement with experiment. In the latter case also the 5s and 5p shells must be correlated as their correlation contribution to the fine-structure splitting amounts to as much as $250\,cm^{-1}$. It has to be kept in mind, however, that the spin–orbit model potential of the Leininger PPs does not contain any d projector.

Quantum chemical determination of the spectroscopic properties of TlH

Theoretically and experimentially derived spectroscopic parameters of the 0^+ electronic ground state of TlH are collected in Table 6.4. Among the computed values only those which include magnetic interactions may be directly compared with experiment; results of spin-free calculations are therefore displayed in parentheses.

Several authors investigated the influence of kinematic relativistic effects on the spectroscopic properties of TlH by comparing results of non-relativistic and (spin-averaged) quasi-relatvistic ECP calculations, respectively. Unfortunately, the size of the computed relativistic corrections depends considerably on the specific core potential, and so far no non-relativistic all-electron calculation has been published. When only kinematic relativistic effects are taken into account, an appreciable bond contraction by $\approx 10\,\text{pm}$ is observed for the shape-consistent Lee ECPs [Lee, Ermler and Pitzer (1980)] whereas the dissociation energy and the harmonic vibrational frequency hardly change. Less than half of this relativistic bond shortening is found for the Stuttgart pseudopotentials [Dolg *et al.* (1991), and Schwerdtfeger *et al.* (1992)]. Moreover, even without introducing spin–orbit interactions, Dolg *et al.* (1991) observe a slight bond softening. An explanation of the latter effect is readily at hand: on bond formation charge is transferred from thallium to hydrogen; the relativistic energy stabilization is therefore largest in the dissociation limit where three electrons occupy the valence shell of the Tl atom. Unlike the situation in transition metal compounds such as AuH [Pizlo *et al.* (1993)], relativistic and electron correlation effects on the spectroscopic properties of the electronic ground state appear to be nearly independent, however.

In this paragraph we shall analyze the changes of the spectroscopic parameters due to electron correlation. At this point, only trends can be discussed, however, because a comparison of the computed spectroscopic parameters with experiment can only be made after spin–orbit coupling has been included in the theoretical treatment. The most apparent effect of including valence electron correlation at the spin-free level is a considerable increase of the dissociation energy; this bond stabilization is accompanied by a softening of the potential close to the equilibrium bond distance, i.e., the equilibrium internuclear separation increases slightly and the harmonic vibrational frequency becomes smaller. This behavior is typical for sp correlation in main group compounds. Adding 5d inter-shell correlation, either explicitly or by means of an empirical core-polarization potential, further stabilizes the bond. Contrary to the effects of valence shell correlation it leads to a considerable bond contraction. Note, however, that in the molecular case the same restrictions apply to the RECP as discussed earlier for the Tl atom: if the 5d shell is going to be correlated explicitly, one has to make sure that the 6s and 6p orbitals exhibit the proper density in the region where the 5d orbital has its maximum, i.e., they should contain 5s or 5p orthogonality tails instead of using projection operators. Accordingly, results making use of a 13 valence electron RECP of Tl and explicitly correlating 14 electrons of TlH [Kim, Lee and Lee (1996), and Lee *et al.* (1998)] should be considered with some care.

A particularly marked effect of core-valence correlation (not shown in Table 6.4) is observed for the first excited $^3\Pi$ state. At the spin-free level this state is only weakly bound. Correlating just the four valence electrons Rakowitz (1995) obtains a bond length of 212 pm and a dissociation energy of 0.12 eV. Inclusion of 5d inter-shell correlation considerably shortens the equilibrium bond distance to 183 pm. Its dissociation energy increases by 0.11 eV, i.e., by roughly the same amount as that of the electronic ground state thus leaving the adiabatic excitation energy of the $^3\Pi$ state nearly unchanged. Dolg *et al.* (1991) have drawn attention to another point which applies solely for large-core potentials: at short bond distances a neighboring nucleus may overlap with the (large) core of Tl and then the usual point-charge expression for the core–nucleus interaction is no longer a good approximation. Introduction of a core–nucleus repulsion correction potential takes care of this penetration effect and prevents the atoms from approaching too closely.

Table 6.4. Equilibrium distances, dissociation energies, and harmonic vibrational frequencies of the ground state 0^+ (I) of TlH.

	ECP	Correlation	SOC	R_e (pm)	D_e (eV)	ω_e (cm^{-1})
DHF-OCE[a]	—	—	var.	186.7	2.83	1510
NREP3[b]	$Xe4f^{14}5d^{10}$	—	—	(188)	(1.74)	(1410)
AREP3[b]	$Xe4f^{14}5d^{10}$	—	—	(179)	(1.77)	(1380)
REP3[b]	$Xe4f^{14}5d^{10}$	—	var.	179	1.36	1380
NREP13[b]	$Xe4f^{14}5d^{10}$	—	—	(194)	(1.69)	(1380)
REP13[b]	$Xe4f^{14}$	—	var.	184	1.55	1450
NR SCF[c]	$Xe4f^{14}5d^{10}$	—	—	(193.4)	(1.82)	(1439)
QR SCF[c]	$Xe4f^{14}5d^{10}$	—	—	(189.4)	(1.67)	(1357)
QRPP[d]	$Xe4f^{14}5d^{10}$	—	var.	187.9	1.26	1527
REP-KRHF[e]	$Xe4f^{14}$	—	var.	190.7	—	1438
AREP-HF[f]	$Xe4f^{14}$	—	—	(188.9)	(1.70)	(1425)
REP-KRHF[f]	$Xe4f^{14}$	—	var.	186.2	1.18	1454
REP SCF[g]	$Xe4f^{14}$	—	var.	193	0.93	1450
REP MCSCF[g]	$Xe4f^{14}$	$4e^-$ 5det—MCSCF	var.	196	1.66	1330
SR-ZORA[h]	—	LDA-DFT	—	(186.8)	(2.94)	(1390)
ZORA[h]	—	LDA-DFT	var.	190.1	2.39	1370
SR-ZORA[h]	—	GGA-DFT	—	(193.1)	(2.66)	(1320)
ZORA[h]	—	GGA-DFT	var.	190.0	2.10	1320
REP CI[i]	$Xe4f^{14}$	$2e^-$ MRCISD		199	1.81	1300
WB-AIMP CIDBG A[j]	$Xe4f^{14}$	$4e^-$ CISD	—	195.3	2.41	1310
WB-AIMP CIDBG B, $\lambda = 1$[j]	$Xe4f^{14}$	$4e^-$ CISD		193.1	2.40	1325
SOCI–RCI[k]	$Xe4f^{14}$	$4e^-$ MRCISD		195	2.08	—
POL-CI[l]	$Xe4f^{14}5d^{10}$	$4e^-$ CISD+CP	—	(195)	(2.16)	(1139)
POL-CI[l]	$Xe4f^{14}$	$4e^-$ CISD+CP	—	(201)	(2.15)	(1158)
NR MP2[m]	$Xe4f^{14}5d^{10}$	$4e^-$ MP2+CP	—	(192.4)	(2.42)	—
MP2[m]	$Xe4f^{14}5d^{10}$	$4e^-$ MP2+CP	$\Delta^{atom}_{exp.}$	189.9	1.66	—
QCI[m]	$Xe4f^{14}5d^{10}$	$4e^-$ QCI+CP	—	(193.4)	(2.59)	—
QCI[m]	$Xe4f^{14}5d^{10}$	$4e^-$ QCI+CP	$\Delta^{atom}_{exp.}$	191.2	1.85	—
NR CISD+SCC[cn]	$Xe4f^{14}5d^{10}$	$4e^-$ MRCISD	—	(194.4)	(2.40)	(1378)
QR CISD+SCC[cn]	$Xe4f^{14}5d^{10}$	$4e^-$ MRCISD	—	(190.6)	(2.31)	(1294)
QR CISD+SCC+ CPOL[n]	$Xe4f^{14}5d^{10}$	$4e^-$ MRCISD+CP	—	(188.1)	(2.35)	(1438)
QR CISD+SCC+ CPOL+CNRC[cn]	$Xe4f^{14}5d^{10}$	$4e^-$ MRCISD+CP	—	(192.0)	(2.32)	(1359)
CIPSO+CNRC[cn]	$Xe4f^{14}5d^{10}$	$4e^-$ CIPSI	QDPT	190.9	1.97	1417
DGCI+SCC+ CPOL+CNRC[cn]	$Xe4f^{14}5d^{10}$	$4e^-$ MRCISD+CP		189.1	2.09	1346
λ–s/SD[o]	$Xe4f^{14}$	$4e^-$ t-MRCISD	—	(193.6)	(2.42)	(1327)
SO/SD[o]	$Xe4f^{14}$	$4e^-$ t-MRCISD		191.2	1.908	1341
λ–s/SD+D^o_{SO}	$Xe4f^{14}$	$4e^-$ t-MRCISD/MRCIS		191.5	1.885	1340
REP-KRMP2[f]	$Xe4f^{14}$	$14e^-$ MP2	var.	189.8	1.78	1404
AREP-CI[e]	$Xe4f^{14}$	$14e^-$ CISD	—	(193.6)	—	(1371)
AREP-SOCI[e]	$Xe4f^{14}$	$14e^-$ CISD		192.7	—	1381
REP-KRCI[e]	$Xe4f^{14}$	$14e^-$ CISD	var.	191.7	—	1380
AREP-CCSD[f]	$Xe4f^{14}$	$14e^-$ CCSD	—	(192.8)	(2.49)	(1358)
REP-CCSD[f]	$Xe4f^{14}$	$14e^-$ CCSD	var.	190.7	1.99	1371
AREP-CCSD(T)[f]	$Xe4f^{14}$	$14e^-$ CCSD(T)	—	(193.1)	(2.52)	(1351)

Table 6.4. (*continued*)

	ECP	Correlation		SOC	R_e (pm)	D_e (eV)	ω_e (cm^{-1})
REP-CCSD[f]	Xe4f[14]	$14e^-$	CCSD(T)	var.	191.0	2.02	1360
MRD-CI[p]	—	$14e^-$	MRD-CI	—	(191)	(2.53)	(1193)
SOPT[p]	—	$14e^-$	MRD-CI	QDPT	190	2.08	1309
SOCI[p]	—	$14e^-$	t-MRCISD		187	2.15	1388
SOCIEX[p]	—	$14e^-$	t-MRCISD+Ex		186	2.13	1386
Experiment[q]					186.6	2.06	1390.7
Experiment[r]					186.8	—	1391.3

[a] Pyykkö and Desclaux (1976).
[b] Lee, Ermler and Pitzer (1980).
[c] Dolg, *et al.* (1991).
[d] Schwerdtfeger (1987).
[e] Kim, Lee and Lee (1996).
[f] Lee, *et al.* (1998).
[g] Christiansen and Pitzer (1980).
[h] van Lenthe, Snijders and Baerends (1996).
[i] Christiansen, Balasubramanian and Pitzer (1982).
[j] Seijo (1995).
[k] Balasubramanian and Tao (1991).
[l] Wadt (1984).
[m] Schwerdtfeger, *et al.* (1992).
[n] Dolg, *et al.* (1991).
[o] Dilabio and Christiansen (1998).
[p] Rakowitz and Marian (1997).
[q] Grundström and Valberg (1937).
[r] Urban, Bahnmaier, Magg et al. (1989).

How does spin–orbit interaction influence the spectroscopic parameters of the electronic ground state? The most significant effect is seen on the dissociation energy. As discussed qualitatively already above, the molecular ground state correlates with the $^2P_{1/2}$ component of Tl; in first order the energy of this component is located $(2/3) \times 7793\,cm^{-1}$ ($\approx 0.64\,eV$) below the spin-free dissociation limit. However, this atomic level shift is not completely transferred to the dissociation energy. Also in the region of the molecular equilibrium the energy of the electronic ground state is lowered because of spin–orbit coupling. The effect is smaller ($\approx 0.1\,eV$ [Rakowitz (1995)]) though not negligible. Therefore, estimates of a spin–orbit correction to the dissociation energy, which are solely based on the atomic fine-structure splitting, tend to overshoot. While the destabilization of the TlH bond by spin–orbit coupling has been known for nearly two decades now, the bond contraction due to spin–orbit interaction has been recognized only recently [Dolg *et al.* (1991), van Lenthe, Snijders and Baerends (1996), Kim, Lee and Lee (1996), Rakowitz and Marian (1997), DiLabio and Christiansen (1998), and Lee *et al.* (1998)]. For example, at the HF level Lee *et al.* (1998) find a shrinkage of the equilibrium distance by 2.7 pm with respect to the one-component result, if they employ a two-component relativistic pseudo-potential in the self-consistent field procedure. If spin–orbit coupling is taken into account by means of quasi-degenerate perturbation theory, a small bond contraction of $\approx$ 1 pm results [Dolg *et al.* (1991), and Rakowitz and Marian (1997)]. Comparing their results at the spin-averaged (SR-ZORA) and the unrestricted ZORA levels van Lenthe, Snijders and Baerends (1996) observe a bond contraction by 3 pm. Approximately the same difference is found between the results of one-component and spin–orbit CI treatments [Dolg *et al.* (1991)]. A slightly larger bond contraction is reported by Rakowitz and Marian (1997) whereas somewhat smaller shifts are obtained by other authors [Kim, Lee and Lee (1996), Lee *et al.* (1998), and DiLabio and Christiansen (1998)]. The trend is the same, however, in all calculations. The reinforcement of the spin–orbit effects in all-electron variational calculations (intermediate coupling) as compared with a perturbative treatment in the basis of ΛS wavefunctions may be ascribed to spin polarization.

To our knowledge only two theoretical studies have been concerned with the electronically excited states in some detail [Christiansen, Balsubramanian and Pitzer (1982), and Rakowitz and Marian (1997)]. Both use a spin-independent Hamiltonian to optimize the molecular orbitals and they proceed with an intermediate coupling scheme. The latter authors also performed quasi-degenerate perturbation theory calculations including all electronic states which dissociate to the lowest $^2P_{1/2}$ and $^2P_{3/2}$ channels. In addition to the ground state these are the lowest $^{3,1}\Pi$ and $^3\Sigma^+$ states if denoted according to a ΛS picture. Among the low-lying spin-free states only the first excited $^3\Pi$ state is slightly bound whereas $^1\Pi$ and $^3\Sigma^+$ are repulsive. We have already mentioned the large impact of 5d core–valence correlation on the equilibrium bond distance of the $^3\Pi$ state and on the depth of its potential well. Somewhat astonishingly, this kind of correlation does not affect the excitation energy to any major extent. Dynamical valence shell correlation appears to be more important in that respect: truncation of an MRSDCI expansion (selection threshold $10^{-6}\,E_H$) leads to an increase of the adiabatic excitation energy by about 0.2 eV [Rakowitz and Marian (1997)]. Spin–orbit interaction has two major effects on the $^3\Pi$ state. Firstly, it splits the six-fold degenerate potential curve into four components with Ω quantum numbers 0^+, 0^-, 1, and 2, with the $\Omega = 0$ levels being the lowest in energy. The spin–orbit splitting between the components is best reproduced by a spin–orbit CI treatment, for the same reasons that apply to the atomic Tl case. Secondly, it couples states with equal Ω quantum number. We are already familiar with one result of the interaction between the 0^+ states, i.e., the decrease in the ground state dissociation energy. The reverse effect is observed for the upper 0^+ state. Its binding energy is greatly enhanced by spin–orbit coupling because eventually it has to correlate with the upper $^2P_{3/2}$ dissociation limit which lies ≈ 0.32 eV above the spin-free dissociation channel. The strong interaction between the two $\Omega = 0^+$ states leads to a very strange potential energy curve for the upper one, seen both in experiment [Neuhaus and Muld (1959), and Ginter and Battino (1965)] and theory [Christiansen, Balasubramanian and Pitzer (1982), and Rakowitz and Marian (1997)]. Without going into further details we should like to mention that the $\Omega = 1$ components of $^3\Pi$, $^1\Pi$, and $^3\Sigma^+$ also strongly interact with each other and that their potential energy curves undergo substantial changes because of spin–orbit coupling; this is generally to be expected in cases in which electronic excitation energies are of the same size as spin-orbit matrix elements.

Sensitivity of the correlation energy with respect to a truncation of the configuration space may become a serious problem in spin–orbit CI calculations: owing to the presence of a spin–orbit operator, space and spin symmetries can no longer be separated. Although it is possible in some cases to obtain purely real matrix elements through a paricular choice of phase factors [Pitzer and Winter (1988), and Visscher (1996)], the general case yields a complex Hamiltonian matrix. Moreover, it is no longer feasible to choose just one representative of a spin multiplet (e.g., the $m_S = S$ component); by and large all components will have to be included instead. As a consequence the dimension of the Hamiltonian matrix increases by one or two orders of magnitude compared with an ordinary CI calculation. In the particular case of TlH, e.g., the $^1\Sigma^+$ expansion length enters twice (real and imaginary phase), the $^3\Sigma^+$ state comes in with six times its ordinary size (triplet, each with real and imaginary component), the $^1\Pi$ contributes four times its vector length (x and y, both real and imaginary), and finally the $^3\Pi$ adds 24 times the original number of coefficients to the SOCI vectors. Recently, two direct SOCI programs were developed which are capable of coping with several million determinants [Sjøvoll, Gropen and Olsen (1997), and Fleig, Marian and Olsen (1997)]. Both consume a great deal of resources, however, and there are other ways out. In a spin–orbit CI calculation it appears to be of minor importance whether electron correlation and spin–orbit coupling are treated exactly on the same footing: spin–orbit interactions are dominated by single excitations and generally short CI expansion lengths suffice. Electron correlation, on the other hand, is known to converge very slowly with the size of the configuration space. These facts have been exploited in (quasi-degenerate) perturbation theory approaches for a long time where people have combined energies and spin–orbit matrix elements from different quality calculations (see, e.g., [Marian *et al.* (1982), Teichteil, Pélissier and Spiegelmann (1983), and Marian (1991)]). Recently these ideas were transferred to SOCI treatments. Several methods were proposed to achieve this goal, i.e., to simplify spin–orbit CI calculations by decoupling correlation and spin–orbit effects: the spin-free-state-shifted CI technique [Llusar *et al.* (1996), and Rakowitz *et al.* (1998)] employs level shifters to adjust the matrix elements of the spin-free part of the Hamiltonian in a smaller configuration expansion to more accurate values. The SOCIEX method [Rakowitz and Marian (1997)] makes use of a configuration selection and a perturbation theory estimate of the correlation contribution of discarded configurations–in line with the MRD-CI technique for spin-independent cases [Buenker, Peyerimhoff and Butscher (1978)]. DiLabio and Christiansen (1998) proceed in a different way. They carry out high- and low-level one-component CI calculations and run a low-level intermediate coupling CI. The energy shifts of the intermediate coupling expectation values with respect to the corresponding low-level spin-free values are then used to estimate the effect of spin–orbit coupling on the high-level correlated energies. All of these three approximate schemes show a good performance in test calculations.

6.3.2 TRANSITION ELEMENTS: PLATINUM AND PLATINUM HYDRIDE

Also our next example case focuses on a heavy metal with occupied orbitals of principal quantum number six. Relativistic effects are therefore expected to be of the same order of magnitude as those observed for thallium and lead. In the latter elements spin–orbit splitting of electronic states is mainly caused by unfilled p shells and spin–orbital relaxation turned out to be of major concern. This is not the case in low-lying electronic states of transition metals with unpaired s and d electrons only. The radial maxima of the $5d_{3/2}$ and $5d_{5/2}$ components in Pt differ by only $0.03\ a_0$ according to numerical DHF calculations [Desclaux (1973)] which is less than one tenth of the radial splitting between the 6p shell components of Tl. As a consequence spin–orbit coupling in transition metal compounds can be computed very accurately by means of perturbation theory. Thus spin–orbit interaction is not a major difficulty in theoretical spectroscopy of transition metal compounds: the true challenge one has to face is electron correlation.

The spectrum of atomic platinum

Platinum exhibits a closely spaced spectrum of electronic states with orbital occupations $5d^86s^2$, $5d^96s^1$, $5d^{10}$ [Moore (1958)] or admixtures thereof. In an *LS* picture, the d^8s^2 configuration gives rise to five electronic states with ^3F, ^3P, ^1G, ^1D, and ^1S symmetries, the two open shells of d^9s^1 combine to ^3D and ^1D, and finally a d^{10} occupation yields a totally symmetric ^1S state. Spin–orbit coupling is large: a look at the splitting between the various *J* components reveals that Landé's interval rule–which is based on an *LS* picture–is far from being obeyed. *LS* states may serve as a starting point, however. Results of one-component calculations for the lowest terms with d^8s^2, d^9s^1, and d^{10} occupation are displayed in Table 6.5.

'Non-relativistic' platinum has an electronic spectrum similar to that of palladium, i.e., it exhibits a $^1S(d^{10})$ ground state, the first excited state has $^3D(d^9s^1)$ character, and the first d^8s^2 state (^3F) is highly excited. Comparing the excitation energies obtained from a numerical HF treatment [Andrae *et al.* (1990)] and from a Gaussian basis set SCF calculation [Gropen, Almlöf and Wahlgren (1992)] large differences are found in particular for the ^1S state. SCF calculations [Marian (1999)] using the same primitive set of Gaussian type orbitals as Gropen, Almlöf and

Table 6.5. Excitation energies T_e (eV) of low-lying *LS* states of atomic platinum with respect to the $^3D(d^9s^1)$ state.

	ECP	Correlation		Relativity	$^1S(d^{10})$	$^3F(d^8s^2)$
num. HF[a]	—	—		no	−1.41	3.28
SCF, uncontr. basis[b]	—	—		no	−1.42	3.28
SCF[c]	—	—		no	−0.50[d]	2.88–3.04[d]
SCF PT[c]	—	—		MVD	0.62–1.53[d]	0.12–1.19[d]
CG num. HF[e]	—	—		CG	0.90	0.40
num. WB[a]	—	—		WB	0.82	0.42
WB-MEFIT PP SCF[a]	$Xe4f^{14}$	—		RECP	0.81	0.47
SCF, uncontr. basis[b]	—	—		DK	0.87	0.46
SCF AE[f]	—	—		DK	0.92	0.45
SCF AIMP[f]	$Kr4f^{14}$	—		RECP	0.85	0.56
ECP10 HF[g]	$Xe4f^{14}$	—		RECP	1.03	0.46
ECP10 UHF[h]	$Xe4f^{14}$	—		RECP	1.05	0.61
ECP18 UHF[h]	$Kr4d^{10}4f^{14}$	—		RECP	1.22	0.06
ECP10 MP2[h]	$Xe4f^{14}$	$10e^-$	MP2	RECP	0.03	1.13
ECP10 MP4[h]	$Xe4f^{14}$	$10e^-$	MP4	RECP	0.14	0.90
ECP18 MP2[h]	$Kr4d^{10}4f^{14}$	$10e^-$	MP2	RECP	0.56	0.29
ECP18 MP4[h]	$Kr4d^{10}4f^{14}$	$10e^-$	MP4	RECP	0.66	0.17
CI[c]	—	$10e^-$	CI	no	−0.98 to −1.10[d]	2.98–3.28[d]
CI PT[c]	—	$10e^-$	CI	MVD	0.12–1.19[d]	0.07–2.32[d]
RECP (AIMP) CI[c]	$Kr4f^{14}$	$10e^-$	CI	RECP	0.40	0.64
MCPF AE[f]	—	$10e^-$	MCPF	DK	0.25	0.74
MCPF AIMP[f]	$Kr4f^{14}$	$10e^-$	MCPF	RECP	0.14	0.86
no-pair MRD-CI+Dav.[i]	—	$10e^-$	MRD-CI	DK	—	0.70
ECP MRD-CI+Dav.[i]	$Kr4f^{14}$	$10e^-$	MRD-CI	RECP	—	0.71
AIMP MRD-CI+Dav.[i]	$Kr4f^{14}$	$10e^-$	MRD-CI	RECP	—	0.84
no-pair MRD-CI+Dav.[j]	—	$10e^-$	MRD-CI	DK	0.11	0.59

[a] Andrae *et al.* (1990).
[b] Marian (1999).
[c] Gropen, Almlöf and Wahlgren (1992).
[d] Employing different contractions of the primitive Gaussian functions.
[e] Martin and Hay (1981).
[f] Wittborn and Wahlgren (1995).
[g] Basch, Cohen and Topiol (1980).
[h] McMichael Rohlfing, Hay and Martin (1986).
[i] Marian and Wahlgren (1996).
[j] Sjøvoll and Marian (1996).

Wahlgren (1992), but in uncontracted form, clearly show that the deviations are caused by the (over-) contraction of the basis set rather than a deficiency in the primitive basis.

Relativistic stabilization is expected to scale roughly with the number of s electrons. This is found indeed. All all-electron one-component methods which include kinematic relativistic correction terms variationally, i.e., already in the determination of the orbitals, yield more or less identical results: at the HF level the $^3D(d^9s^1)$ state is stabilized by $\approx$2.3 eV with respect to the $^1S(d^{10})$ state, the $^3F(d^8s^2)$ with doubly occupied 6s shell is lowered by $\approx$ 5.1 eV. The HF shifts cannot be used, however, to estimate the kinematic relativistic effects at the correlated level. It is a phenomenon common to all transition elements that kinematic relativistic and electron correlation effects are not additive (see, e.g., [Marian (1990), and Pizlo *et al.* (1993)]). At the correlated level the stabilization with respect to the ^1S state is reduced to $\approx$ 1.2 and $\approx$ 3.5 eV for ^3D and ^3F, respectively. Also most of the RECPs work well; excitation energies

Table 6.6. Spectrum of the platinum atom below 2 eV. Excitation energies (eV) relative to the $J = 3$ for ground state. The correlation methods treated 10 electrons, unless indicated otherwise.

	Correlation	SOC	$^3D_2$ (d^9s^1)	$^3F_4$ (d^8s^2)	$^1S_0$ (d^{10})	$^3P_2$ (d^8s^2)	$^3F_3$ (d^8s^2)	$^3D_1$ (d^9s^1)	$^1D_2$ (d^9s^1)	$^3F_2$ (d^8s^2)
Num. DHF[a]	—	var.	0.159	0.095	2.173	0.846	1.280	1.146	1.868	2.170
Basis set DHF[a]	—	var.	0.163	0.116	2.186	0.856	1.302	1.146	1.874	2.192
Num. DHF+B[a]	—	var.	0.145	0.039	2.151	0.803	1.190	1.109	1.814	2.082
Num. DHF+B[a]	—	var.	0.146	0.045	2.160	0.805	1.195	1.107	1.812	2.087
Basis set CI[a]	CISD	var.	0.145	0.262	1.822	0.926	1.378	1.206	—	—
MCDF aver. spinors[b]	MCDF	var.	0.047	−0.277	1.996	0.677	0.848	1.112	1.650	1.775
MCDF[b,f]	MCDF	var.	0.383	0.090	1.416	1.069	1.170	1.115	1.671	2.073
WB MEFIT MCDF[b,f,g]	MCDF	var.	0.374	0.0005	1.360	0.991	0.985	1.108	1.600	1.922
WB-AIMP CI ^{3}D orbs.[c,h]	DGCI		0.165	0.384	0.942	1.077	1.360	1.626	1.972	2.430
WB-AIMP CI ^{3}F orbs.[c,h]	DGCI		0.086	−0.006	1.252	0.870	1.357	1.230	1.837	2.038
WB-AIMP CI ^{3}AVE orbs.[c,h]	DGCI		0.129	0.157	1.136	0.959	1.363	1.400	1.892	2.206
QDPT S(no shift)[d]	MRCIS	QDPT	0.065	−0.116	0.948	0.959	1.064	1.286	1.747	1.907
SOCI S(no shift)[d]	SOCIS		0.074	−0.086	0.809	0.834	1.101	1.296	1.765	1.943
QDPT S(shift)[d]	MRCIS+shift	QDPT	0.075	0.123	0.792	0.847	1.312	1.285	1.777	2.036
SOCI S(shift)[d]	SOCIS+shift		0.083	0.133	0.809	0.864	1.328	1.296	1.793	2.062
QDPT SD(no shift)[d]	MRCISD	QDPT	0.098	0.066	0.968	0.879	1.204	1.258	1.752	2.012
SOCI SD(no shift)[d]	SOCISD		0.103	0.087	0.945	0.895	1.233	1.269	1.775	2.041
QDPT SD(shift)[d]	MRCISD+shift	QDPT	0.099	0.102	0.766	0.885	1.242	1.258	1.766	2.029
SOCI SD(shift)[d]	SOCISD+shift		0.104	0.114	0.780	0.899	1.262	1.269	1.780	2.054
AE + AMFI[d]	CIPSI 18e$^-$	QDPT	0.091	0.114	0.775	0.825	1.242	1.352	1.686	1.922
ECP + AMFI[d,g]	CIPSI 18e$^-$	QDPT	0.090	0.116	0.780	0.830	1.233	1.358	1.693	1.928
ECP + pseudo op.[d,g]	CIPSI 18e^-	QDPT	0.090	0.144	0.758	0.855	1.059	1.382	1.693	1.928
Experiment[e]			0.096	0.102	0.761	0.814	1.254	1.256	1.673	1.922

[a] Visscher *et al.* (1993).
[b] Dolg (1994).
[c] Seijo (1995).
[d] Schimmelpfennig *et al.* (1998b).
[e] Moore (1958).
[f] Conf. opt. spinors.
[g] ECP Kr $4d^{10}4f^{14}$.
[h] ECP Cd $4f^{14}$.

are similar to those obtained from one-component all-electron calculations. This does not apply to the 18 valence-electron Hay–Wadt potential [Hay and Wadt (1985)], however. Unrestricted HF (UHF) [McMichael Rohlfing, Hay and Martin (1986)] show that this potential is strongly biased towards the $^3F(5d^86s^2)$ state. As we shall see below this tendency also carries over to the correlated level.

In first-row transition metals kinematic relativistic effects are frequently added perturbatively using the mass–velocity and Darwin (MVD) corrections. Trying to proceed in the same way in the third transition metal row badly fails; Gropen, Almlöf and Wahlgren (1992) observe a large dependence of the MVD correction on the type of basis set contraction in the valence region. As shown in Table 6.5, the MVD correction fluctuates by about 1 eV at the SCF level, exceeding by far the variance of corresponding non-relativistic SCF results. The range of excitation energies becomes even wider at the correlated level; e.g., the CI excitation energy of the $^3F(d^8s^2)$ state varies between 0.07 and 2.32 eV for different contraction schemes. We learn from these results that in third-row transition metal compounds a perturbative MVD correction does not yield a reasonable estimate of the kinematic relativistic effects. This criticism does not apply to the Cowan–Griffin and Wood–Boring approaches where the MVD operators are introduced in the HF equations.

Electron correlation follows the typical pattern for transition metals, i.e., dynamic correlation contributions increase with the number of d electrons. Apart from the 18 valence-electron RECP calculations [McMichael Rohlfing, Hay and Martin (1986)] and the MVD PT approach [Gropen, Almlöf and Wahlgren (1992)] all correlation treatments reverse the order of the $^1S(d^{10})$ and $^3F(d^8s^2)$ *LS*-coupled states: computed electronic excitation energies relative to the $^3D(d^9s^1)$ state range from 0.1 to 0.25 eV for the ^1S state and from 0.6 to 0.9 eV for ^3F. Compared with these energy separations the spin–orbit splittings are large; they amount to approximately 1.3 eV in the $^3D(d^9s^1)$ and 1.8 eV in the $^3F(d^8s^2)$ state. A comparison with experiment can thus be made only after spin–orbit coupling has been included.

Visscher *et al.* (1993) investigated the spectrum of atomic platinum with four-component methods, both at the self-consistent field and at the CI level. In the four-component CI pair creation and annihilation processes were excluded (no-pair approximation), thus keeping the number of electrons constant. The authors make a series of interesting observations: first of all, for Pt the differential Breit (B) correction is considerably larger than that for Tl as the size of the Breit interaction strongly depends on the electronic configuration; within the manifold of $5d^86s^2$, $5d^9s^1$, $5d^{10}$ states Breit shifts amount up to 750 cm^{-1}. Replacing the Breit term by its Gaunt (G) part yields nearly identical results. Further, Visscher *et al.* report on difficulties in finding an appropriate one-particle basis set. It turns out this is not a specific feature of four-component approaches but has also been observed in two-component treatments (see below). These difficulties arise because orbital radii vary appreciably from one valence occupation to another. Depending on the electronic configuration of Pt the radial expectation value of the 6s spinor, e.g., varies by as much as 0.5 bohr. Extracting spinors by minimizing the total energy average of all states arising from $5d^86s^2$, $5d^9s^1$, $5d^{10}$ does not solve the problem. This procedure yields a one-particle basis which is biased towards the $5d^86s^2$ states as these form the majority; concomitantly, the $J = 4$ component of the $^3F(d^8s^2)$ state erroneously becomes the ground state. Visscher *et al.* propose to take the relativistic configurational average of $5d^4_{3/2}\,5d^4_{5/2}\,6s^2_{1/2}$, $5d^4_{3/2}\,5d^5_{5/2}\,6s^1_{1/2}$, and $5d^4_{3/2}\,5d^6_{5/2}\,6s^0_{1/2}$ instead; this choice leads to a balanced description of d^9 and d^8 states. With the exception of the $^1S(d^{10})$ state, the relativistic $10e^-$ CISD treatment gives excitation energies in reasonable agreement with experiment. The failure to describe the $^1S(d^{10})$ appropriately may be attributed to the spinor basis; at the time these results were published many of the highly excited spinors had to be deleted from the virtual space due to program limitations. With present day program versions the results could certainly be improved. The same sensitivity with respect to the one-particle basis was observed [Dolg (1994)] in multi-configurational Dirac–Fock calculations. By separately optimizing spinors within the $5d^86s^2$, $5d^96s^1$, and $5d^{10}$ manifolds, respectively, Dolg manages to obtain the appropriate order of states, again except for the $5d^{10}$ state. With respect to the DHF level this state is stabilized by 0.7 eV in the MCDF treatment; for an agreement with experiment there are another 0.65 eV of differential correlation energy to be gained.

In this respect, *LS* and intermediate coupling methods are better off. They can use a larger partition of the underlying basis set to describe electron correlation. Casarrubios and Seijo (1995) investigated the spectrum of the platinum atom employing a AIMP derived from atomic WB results. They performed DGCI calculations using three different one-particle basis sets, namely orbitals generated in CG-AIMP SCF procedures for (a) the $^3D(d^9s^1)$ state, (b) the $^3F(d^8s^2)$, and (c) a weighted average of the $^3D(d^9s^1)$, $^3F(d^8s^2)$, and $^3P(d^8s^2)$ states. Upon an exchange of orbitals they observe considerable shifts of the excitation energies. Certainly, the agreement with experiment is not perfect, but the ^1S excitation energy has improved a lot. Again, the deviations are mainly due to deficiencies in the electron correlation treatment rather than due to spin–orbit effects; very good agreement is observed when the excitation energy of a state is related to the lowest J component with corresponding d occupation.

It appears that in transition metals spin–orbit coupling is not as strongly connected with orbital relaxation and electron correlation as in late main group elements. One reason was already mentioned above: the radial densities of the jj-coupled $d_{3/2}$ and $d_{5/2}$ valence orbitals are similar. The other reason is that in transition metals spin–orbit interaction is large only within a manifold of states exhibiting a common d occupation. For example, spin–orbit matrix elements between the $^3D(d^9s^1)$ and $^3F(d^8s^2)$ states of Pt are almost zero [Sjøvoll and Marian (1996)]. An explanation may readily be given. At the (dominating) one-electron level the mixing of states with different d occupations via spin–orbit interaction would require a coupling of an $s_{1/2}$ electron with a $d_{3/2}$ or $d_{5/2}$ electron which is strictly zero, of course. The amount to which the states can interact via spin–orbit coupling is determined by configuration mixing due to the Coulomb interaction. Let us choose again an example among the states of platinum. In the *LS* coupling picture both the $5d^86s^2$ and $5d^96s^1$ occupations give rise to a ^1D state. The energetically favorable one is dominated by the $5d^96s^1$ configuration and has a large spin-orbit matrix element with the (pure d^9) ^3D state. However, because of CI it also exhibits a non-negligible spin-orbit matrix element with the (pure d^8) ^3F state. The size of the latter matrix element depends on electron correlation effects as these have an influence on the energetic separation of the $^1D(d^9s^1)$ and $^1D(d^8s^2)$ states and thus on their degree of mixing. However, as the ^1D states are the only links, the d^9 and d^8 manifold of states are not strongly interwoven.

Schimmelpfennig *et al.* (1998a) make use of this fact. They employ the spin-free state-shifted method [Llusar *et al.* (1996)] in a way that makes use of empirical corrections. All $5d^96s^1$ states are shifted such that the excitation energy of the $^3D_1$ state relative to the $^1S(5d^{10})$ state in first-order perturbation theory reproduces experiment; similarly the $^3F_4$ component is used as a gauge for the $5d^86s^2$ states. In this way even Douglas–Kroll single-excitation CI treatments, both at the quasi-degenerate perturbation theory and spin–orbit CI levels, yield satisfactory results. Excellent agreement with experiment is found in the spin-free state-shifted Douglas–Kroll MRCISD calculations.

In another publication the authors tested the performance of the SOMF Hamiltonian and a pseudo spin–orbit operator with RECPs of the pseudopotential type. Overall, very good agreement is observed in either case except for the $^3F_3$ component where the pseudo spin–orbit operator approach for unknown reasons appears to have some difficulties.

PtH

Among the $5d^86s^2$, $5d^96s^1$, and $5d^{10}$ electronic states of the platinum atom those with $5d^96s^1$ occupation are best prepared to bind to a single hydrogen atom. Qualitatively the electronic structure of the low-lying bound molecular states are characterized by a doubly occupied σ bond, formed by the 6s and 1s orbitals of Pt and H, respectively, and a single hole in the 5d shell of Pt. In a jj coupling picture one would expect the $5d_{5/2}$ hole states to be clearly preferred in energy over states with one $5d_{3/2}$ spinor unoccupied. A further splitting should then result from the different possible orientations of the hole angular momentum with respect to the electric field along the internuclear axis. Despite its simplicity, this ^{2}D supermultiplet model [Gray *et al.* (1991)] correctly predicts many features in the electronic spectrum of PtH. PtH exhibits an $\Omega = 5/2$ ground state and very low-lying $\Omega = 1/2$ and $\Omega = 3/2$ states, energetically well separated from the two $J = 3/2$ components with Ω quantum numbers 3/2 and 1/2. Of course, the spectrum of PtH can also be explained–and computed–within the ΛS coupling scheme. In this case, one starts out from three ΛS states with $^2\Delta$, $^2\Pi$, and $^2\Sigma^+$ symmetries which undergo strong spin–orbit coupling.

Although none of the $\Omega = 1/2$ states has been detected experimentally so far, there is ample evidence from theoretical investigations [Wang and Pitzer (1983), McMichael Rohlfing, Hay and Martin (1986), Balasubramanian and Feng (1990), Dyall (1993), Visscher *et al.* (1993), Fleig and Marian (1994), Fleig and Marian (1996), Marian and Wahlgren (1996), and Sjøvoll *et al.* (1998)] that the order of the low-lying $\Omega = 1/2$ and $\Omega = 3/2$ states states is reversed with respect to the predictions of the supermultiplet model. This reversal is caused by CI with another $\Omega = 1/2$ state correlating with the d^{10} atomic channel. It is more easily explained in the ΛS coupling picture. The 5d hole may then be characterized as d_δ, d_π, or d_σ. Upon bond formation, the d_δ and d_π holes remain nearly unchanged with respect to their atomic shapes. The singly occupied d_σ orbital, on the other hand, can participate in the σ bonding with hydrogen and is partially filled. As we shall see below, this behavior is clearly reflected in the spin–orbit coupling matrix element between the $^2\Sigma^+$ and $^2\Pi$ state. The extent to which the d_σ orbital participates in the bonding depends on the internuclear separation. The sd hybridization increases as the bond gets shorter, leading to a Mulliken d population of ≈ 9.2 electrons at the equilibrium disctance compared to ≈ 8.9 in the $^2\Pi$ and $^2\Delta$ states. Further, the degree of sd mixing in the $^2\Sigma^+$ state and the atomic $^3D(d^9s^1)$–$^1S(d^{10})$ excitation energy are clearly correlated. This may be seen when comparing PtH with its lighter homologs PdH and NiH. Palladium has a $^1S(4d^{10})$ ground state [Moore (1958)]. Concomitantly, the $^2\Sigma^+$ state becomes the ground state in PdH [Fleig and Marian (1998)]. In the nickel atom, on the other hand, the $^1S(3d^{10})$ state exhibits an excitation energy of ≈ 1.7 eV [Moore (1952)]. As a result, fewer electrons are transferred to the $3d_\sigma$ orbital yielding a $^2\Delta$ ground state [Marian, Blomberg and Siegbahn (1989)]. In this respect platinum is in between. At the (correlated) one-component level its $^1S(5d^{10})$ state is located only slightly above the $^3D(d^9s^1)$ state. This is sufficient for the $^2\Sigma^+$ state to become the ground state of PtH at the spin-free level. In order to obtain this ordering of states it is necessary, however, to include dynamic electron correlation. As shown in Table 6.7, kinematic relativistic effects lead to a considerable bond contraction in the order of 10 pm. They also stabilize the $^2\Delta$ and $^2\Pi$ states slightly with respect to the $^2\Sigma^+$ state because their 6s occupations are somewhat larger. Perturbatively added MVD corrections go in the right direction but do not give the total effect. Apart from the 18 valence-electron RECP calculations [McMichael Rohlfing, Hay and Martin (1986)], which already yielded strange results for the Pt atom, and a small second-order CI approach [Balasubramanian and Feng (1990)] all one-component relativistic correlation treatments agree on a $^2\Sigma^+$ ground state, separated by 0.1–0.3 eV from the first excited $^2\Delta$ and by about 0.7–0.9 eV from the second excited $^2\Pi$ state.

Spin–orbit effects radically change the excitation spectrum. Unlike the situation in TlH, spin–orbit interaction is large not only in the dissociation limit, but persists also in the equilibrium region as PtH is an open-shell molecule. Diagonal and off-diagonal coupling matrix elements involving only the $^2\Delta$ and $^2\Pi$ states are almost independent of the internuclear separation–a prerequisite for the applicability of the ^{2}D supermultiplet model mentioned above. By contrast, the spin–orbit matrix element between the $^2\Sigma^+$ and $^2\Pi$ decreases with decreasing bond distance because of the configuration interaction with the d^{10} state and the concomitant partial filling of the d_σ vacancy [Fleig and Marian (1996)]. Although the resulting spectrum appears to

Table 6.7. Excitation energies T_e (eV) and bond distances r_e (pm) for low-lying ΛS states of PtH. All correlation methods treated 11 electrons.

	ECP	Correlation	Relativity	$^2\Sigma^+$		$^2\Delta$		$^2\Pi$	
				T_e	r_e	T_e	r_e	T_e	r_e
NR-HF[a]	—	—	no	0.00	165	—	—	—	—
PT-HF[a]	—	—	MVD	0.00	158	0.03	159	0.41	167
DK-HF[a]	—	—	DK	0.05	155	0.00	156	0.54	162
DK-HF[b]	—	—	DK	0.05	155	0.00	155	0.53	161
UHF[c]	$Xe4f^{14}$	—	RECP	0.04	159	0.00	160	0.61	168
UHF[c]	$Kr4d^{10}4f^{14}$	—	RECP	0.17	155	0.00	154	0.57	160
CASSCF[d]	$Kr4f^{14}$	CAS-6	RECP	0.13	158	0.00	156	0.54	162
CAS12[b]	—	CAS-12	DK	0.00	154	0.19	155	0.78	161
FOCI[e]	$Xe4f^{14}$	MRCIS	RECP	—	153	—	155	—	163
SOCI[e]	$Xe4f^{14}$	MRCISD	RECP	0.05	—	0.00	—	0.74	—
UHF-MP2[c]	$Xe4f^{14}$	MP2	RECP	0.00	150	0.24	155	0.93	161
UHF-MP4[c]	$Xe4f^{14}$	MP4	RECP	0.00	153	0.17	156	0.86	163
UHF-MP2[c]	$Kr4d^{10}4f^{14}$	MP2	RECP	0.12	148	0.00	149	0.70	153
UHF-MP4[c]	$Kr4d^{10}4f^{14}$	MP4	RECP	0.16	150	0.00	151	0.68	155
MRD-CI $^2\Sigma$ orb.[b]	—	MRD-CI	DK	0.00	150	0.23	151	0.88	157
ACPF[f]	—	ACPF	DK	0.00	149	0.16	150	0.74	158
non-rel. MRSDCI[g]	—	MRSDCI	no	0.00	164	0.72	168	—	—
AIMP-MRSDCI[g]	$Kr4f^{14}$	MRSDCI	RECP	0.00	154	0.34	156	0.69	—
RASCI, opt.orb.[h]	—	MRCISD	DK	0.00	152	0.09	152	0.72	158
RASCI, $^2\Sigma$ orb.[h]	—	MRCISD	DK	0.00	152	0.17	153	0.77	158

[a] Dyall (1993).
[b] Fleig and Marian (1994).
[c] McMichael Rohlfing, Hay and Martin (1986).
[d] Heinemann, Koch and Schwarz (1995).
[e] Balasubramanian and Feng (1990).
[f] Fleig and Marian (1996).
[g] Gropen, Almlöf and Wahlgren (1992).
[h] Sjøvoll *et al.* (1998).

be closer to the jj coupling limit, no difficulties arise when spin–orbit interaction is determined by means of quasi-degenerate perturbation theory, including just the three ΛS states and the coupling elements among them. The reasons are of course the same as those noted in the case of atomic platinum: firstly, only minor changes occur in the radial density distribution whether spin and angular momentum vectors in the d shell are combined in a parallel or anti-parallel fashion and secondly, almost negligible spin–orbit interaction with the nearby quartet states is observed as these exhibit d^8 occupations.

It is interesting to note the influence of electron correlation on the bond distances in PtH. The trends are similar for ΛS and Ω states and will be discussed at the same time. SCF equilibrium distances are too long in either case. When static correlation is added by means of MCSCF or small CI expansions, bond lengths even increase, and eventually shrink upon inclusion of dynamic d shell correlation. (Compare, e.g., the one-component all-electron Douglas–Kroll results at the various levels of correlation treatment.) As a starting point for subsequent MRD-CI and averaged coupled-pair functional (ACPF) runs, Fleig (1993) performed CAS4 orbital optimizations. Herein, the active space is spanned by four σ orbitals, occupied with a total of three or four electrons for the $^2\Sigma^+$ and $\Lambda > 0$ states, respectively. Relative to the SCF results bonds are elongated by roughly 1 pm. By adding the d_δ and d_π and corresponding correlation orbitals to the active space, part of the dynamic d shell correlation is taken into account bringing the bond lengths

Table 6.8. Excitation energies T_e (eV) for low-lying states of PtH. All correlation methods treated 11 electrons.

	Correlation	SOC	5/2 (1)	1/2 (1)	3/2 (1)	3/2 (2)	1/2 (2)
DHF+B[a]	—	var.	0.00	0.34	0.42	1.43	1.55
DHF[b]	—	var.	0.00	0.32	0.41	—	—
DHF+Rel.CI[b]	MRCISD	var.	0.00	0.24	0.44	1.46	1.61
SCF+CI[c] ECP Xe4f^{14}		MRCISD	0.00	0.12	0.34	—	—
SOCI—RCI[d] ECP Xe4f^{14}		MRCIS	0.00	0.19	0.52	1.35	1.35
UHF-MP4[e] ECP Xe4f^{14}	MP4	semi-emp.	0.00	0.31	0.45	1.28	1.57
DK-MRD-CI[f]	MRD-CI	QDPT	0.00	0.17	0.45	1.46	1.54
DK-MRD-CI+SCC[g]	MRD-CI	QDPT	0.00	0.15	0.45	1.46	1.55
AIMP-MRD-CI+SCC[g]	MRD-CI	QDPT	0.00	0.09	0.45	1.48	1.54
DK-RASCI[h]	RASCISD	QDPT	0.00	0.18	0.42	1.45	1.52
DK-SOCI[h]		RASCISD	0.00	0.17	0.42	1.47	1.51
Experiment[i]			0.00	—	0.40	1.45	—

[a] Dyall (1993).
[b] Visscher *et al.* (1993).
[c] Wang and Pitzer (1983).
[d] Balasubramanian and Feng (1990).
[e] McMichael Rohlfing, Hay and Martin (1986).
[f] Fleig and Marian (1994).
[g] Marian and Wahlgren (1996).
[h] Sjøvoll *et al.* (1998).
[i] McCarthy *et al.* (1993).

Table 6.9. Equilibrium bond distances r_e (pm) for low-lying states of PtH. All correlation methods treated 11 electrons.

	Correlation	SOC	5/2 (1)	1/2 (1)	3/2 (1)	3/2 (2)	1/2 (2)
DHF+B[a]	—	var.	155	157	158	158	159
DHF[b]	—	var.	155	157	158	—	—
DHF+Rel.CI[b]	MRCISD	var.	152	153	154	154	156
SCF+CI[c] –ECP Xe4f^{14}		MRCISD	161	159	169	—	—
FOCI-RCI[d] –ECP Xe4f^{14}		MRCIS	155	154	158	159	158
DK-MRD-CI[e]	MRD-CI	QDPT	150	151	154	152	154
DK-MRD-CI+SCC[f]	MRD-CI	QDPT	150	151	153	153	154
AIMP-MRD-CI+SCC[f]	MRD-CI	QDPT	151	151	154	153	155
DK-RASCI[g]	RASCISD	QDPT	153	153	156	155	157
DK-SOCI[g]		RASCISD	153	153	156	155	157
Experiment PtH[h]			152.8	—	152.0	—	—
Experiment PtD[h]			152.4	—	153.2	—	—

[a] Dyall (1993).
[b] Visscher *et al.* (1993).
[c] Wang and Pitzer (1983).
[d] Balasubramanian and Feng (1990).
[e] Fleig and Marian (1994).
[f] Marian and Wahlgren (1996).
[g] Sjøvoll *et al.* (1998).
[h] Gustafsson and Scullman (1989).

back to their SCF values. Further inclusion of dynamic correlation by means of MRCISD or ACPF methods continues the bond length contraction tendency. This behavior is not specific for PtH but rather characteristic of many transition metal compounds. Also the results of Wang and Pitzer (1983) can be interpreted along these lines. They performed very limited spin–orbit CI calculations employing an AO basis of Slater functions. Although they obtain excitation energies in good agreement with experiment or other theoretical treatments, bond lengths are considerably overestimated as dynamic correlation is not included.

A few comments should be given concerning 'experimental' bond lengths of transition metal hydrides. These are commonly extracted from rotationally resolved spectra through a fit to an effective model Hamiltonian. Potential energy curves of transition metal hydrides such as PtH with very close-lying electronic states are frequently coupled by the interaction of rotational and electronic magnetic moments. This interaction–reflected in considerable splittings of even and odd parity sublevels [Fleig and Marian (1996)]–is a particular form of non-adiabatic coupling. In model Hamiltonians this kind of coupling between potential curves is typically taken into account implicitly through the parameters yielding, e.g., different 'effective' bond distances for the hydride and deuteride. To be strictly comparable with theoretically determined values the 'effective' parameters have to be deperturbed.

6.3.3 ON THE PERFORMANCE OF EFFECTIVE SPIN–ORBIT HAMILTONIANS

One-electron approximation

Although most of the spin–orbit calculations for molecules are performed using some kind of effective one-electron spin–orbit Hamiltonian, it is not generally true that two-electron spin–orbit integrals are small. In very light molecules such as H_2 [Langhoff *et al.* (1982)] they even dominate the spin–orbit interaction. With increasing nuclear charge the two-electron contributions become less and less important but not negligible. Experience shows that they make about 50 % of the fine-structure splitting in the case of 2p elements. Their contribution to valence shell spin–orbit interactions drops with increasing nuclear charge to about 10 % for third-row transition metals and to approximately 5 % for 6p elements.

The applicability of effective one-electron spin–orbit Hamiltonians relies on the possibility of incorporating all major two-electron contributions into an effective one-electron operator. This implies that spin–orbit coupling matrix elements between determinants which are doubly excited with respect to one another–and thus require a true two-electron operator for their interaction–make only a small contribution. As long as the underlying physical process is dominated by one-electron terms this will generally be the case. A different situation arises when the one-electron contributions are small, e.g., in the investigation of Auger spectra which originate from double excitations involving at least one core orbital. The other exception we should like to mention is the evaluation of predissociation probabilities between doubly excited states, e.g., $\pi^2 \rightarrow \pi^{*2}$ excitations.

Experience collected so far supports this analysis. In systems containing heavy metals the error introduced through the mean-field approximation is far below 1 % [Hess *et al.* (1996), Marian and Wahlgren (1996), Wahlgren *et al.* (1997), and Schimmelpfennig *et al.* (1998a, 1998b)]. Moreover, it has been shown that the mean-field results for atomic palladium and platinum do not critically depend on the valence d shell occupation entering in equation (63) [Hess *et al.* (1996), and Marian and Wahlgren (1996)]. This means that this kind of effective operator is very well suited for describing d shell excitation processes such as dissociation and spectral excitation.

Only a few tests have been made for light molecules. Here, spin–orbit coupling is much smaller on an absolute scale and the computational savings at this level of approximation are negligible. Tatchen and Marian (1999) investigated the spin–orbit splitting in the $^2\Pi$ electronic ground state of the isoelectronic series HC_6H^+, NC_5H^+, and NC_4N^+. These are linear ions with one hole in the highest occupied delocalized π orbital. Here, the SOMF approximation introduces an error of 1 %. Results on the off-diagonal coupling matrix element between the singlet ground state and the first excited $^3(\pi \rightarrow \pi^*)$ state of ethene indicate that the mean-field approximation is also applicable in this case. Although their spin–orbit interaction amounts to merely a few cm^{-1} the deviations of the mean-field values from matrix elements of the full operator stay below 3 % [Danovich *et al.* (1998)]. On the other hand, neglecting all two-electron valence integrals yields spin–orbit coupling matrix elements with roughly 50 % error.

The fact that the largest two-electron contributions to valence-shell spin–orbit splittings can be incorporated into an effective one-electron operator is also the basis for the great success of effective spin–orbit potentials used in combination with RECPs. The good performance of the semi-local model Hamiltonian [equation (67)] has been demonstrated in many applications; for examples beyond those given in the preceding sections the reader is referred to the excellent reviews by Ermler, Ross, and Christiansen (1988), Dolg and Stoll (1995) and Teichteil (1994). In connection with pseudo-orbitals one should refrain, however, from using empirical spin–orbit operators with a radial $1/r^3$ dependence as, e.g., equation (64) or the Blume–Watson approach [Blume and Watson (1962, 1963)]. The $1/r^3$ dependence puts a large weight on regions close to the nucleus where pseudo-orbitals have very small amplitudes. In order to reproduce experimental or all-electron *ab initio* spin–orbit splittings effective charges then have to be very large (e.g., values between 850 and 1200 were used for Pt [Heinemann, Koch and Schwarz (1995), and Ribbing (1994)] with the consequence that even small variations of the amplitude lead to drastic changes of the calculated magnetic interaction. As described in more detail above, instead spin–orbit pseudo-operators with a different radial dependence are commonly employed [Teichteil, Pélissier and Spiegelmann (1983), Ermler, Ross and Christiansen (1988), Küchle *et al.* (1991), and Dolg (1997)].

One-center approximations

Effective core-potentials rely on the assumption that molecular inner-shell orbitals are atom-like, i.e., that core–valence interactions can be evaluated using atomic core orbitals. Making use of this approximation can save enormous amounts of computer time also in all-electron spin–orbit calculations, if one further assumes that the largest contributions to the two-electron spin–orbit core–valence matrix elements [equation (61)] stem from valence orbitals i and j located at the same center as the core orbital k. In this case all multi-center two-electron spin–orbit integrals can be neglected. If, moreover, spherical symmetry is exploited in the evaluation of the one-center terms [Schimmelpfennig (1996)], the spin–orbit integral processing time can be reduced to the minute scale.

As far as the yet very limited experience can tell, neglect of the multi-center two-electron spin–orbit integrals leads to small errors in molecules containing heavy transition metal atoms, e.g., $2\,\mathrm{cm}^{-1}$ in a total of about $1300\,\mathrm{cm}^{-1}$ in PdCl and Pd_2^+ [Hess *et al.* (1996)]. The small deviations from the multi-center results of PdCl and Pd_2^+ almost disappear if the multi-center one-electron integrals are also neglected [Rakowitz (1999)]. Two reasons for the excellent performance of the one-center approximation in transtion metal compounds may be put forward. Firstly, the open d shell of the transition element exhibits a lower principal quantum number than the outer valence shell; upon bond formation it mostly remains pretty much localized on the metal center. The same argument should hold even more true for lanthanides and actinides. Secondly, the percentage of all two-electron spin–orbit contributions, one- and multi-center, to the total matrix element declines rapidly with increasing nuclear charge. The latter reason supports the assumption that compounds of heavy main group elements with more polarizable open p shells can also be treated using a spin–orbit one-center approximation. To our knowledge, so far no investigation of multi-center spin–orbit contributions has been performed for these kind of molecules.

The tendency of the one- and two-electron multi-center terms to cancel has been observed previously already for light diatomic molecules [Richards, Trivedi and Cooper (1981)]. Here, the cancellation of errors is not always perfect, however. Richards, Trivedi and Cooper (1981) find considerable multi-center contributions in case of NO whereas the one-center approximation appears to work well for a whole series of other diatomics. There are hints that this is a typical feature of delocalized π bonds. As mentioned before, Tatchen and Marian (1999) investigated the spin–orbit splitting in the $^2\Pi$ electronic ground states of HC_6H^+, NC_5H^+, and NC_4N^+ ions which exhibit one hole in the highest occupied delocalized π orbital. Here, the one-center approximation leads to errors of the order of 5 % which may be acceptable for dissociation energies but which are generally too large for spectroscopic purposes.

A recent study of the singlet–triplet coupling in ethene has explicitly checked the geometry dependence of the multi-center spin–orbit terms [Danovich *et al.* (1998)]. In this molecule, the two-electron terms screen the one-electron spin–orbit interaction by as much as 50 %. Total matrix elements are small, of the order of a few wavenumbers. Two kinds of geometry distortions are investigated, a twist of the C–C bond (H–C–C–H torsion) and a syn-pyramidalization of the CH_2 groups. In all cases a partial cancellation of the multi-center one- and two-electron contribution is observed. For the pyramidalization mode it is almost perfect whereas along the torsional coordinate the one-electron multi-center terms are roughly twice as large as the corresponding two-electron terms at all internuclear separations. Moreover, the size of the multi-center terms for the torsional motion depends strongly on the C–C bond distance. For C–C internuclear distances corresponding to a single bond the contribution of the multi-center terms to the total matrix element is about 2 % increasing to 20 % at a typical C–C double-bond distance. It can thus be concluded that multi-center terms cannot be neglected in cases where the total spin–orbit coupling matrix element amounts to just a few cm^{-1} whereas the one-center approximation appears to work excellently for heavy element containing systems.

6.4 Conclusions

In the last decade, relativistic electronic structure methods have been developed to an astounding level of sophistication. We now have at our disposal various tools for the *ab initio* calculation of chemical systems, which are currently being used for analyzing experimental results and giving new insight in various areas of chemistry, including actinide chemistry, coordination chemistry and catalysis. This was made possible by an effective funding in the framework of a programme of the European Science Foundation (ESF) on 'Relativistic Effects in Heavy-Element Chemistry and Physics (REHE)' in the years 1993–1997, and various national programmes, e.g., a corresponding programme with the same name of the Deutsche Forschungsgemeinschaft (DFG), the German Science Foundation.

It is important to have a large variety of methods, which are adapted to the requirements of the problem that is under investigation. While four-component variants are now developed for most of the methods of electronic structure calculation, they are heavily demanding as far as computer resources are concerned. Hence they are presently applied mostly to small systems, for which benchmark results of high accuracy can be performed. Without substantial loss of accuracy, the methods based on transformed Hamiltonians offer a much less computationally demanding treatment of relativistic effects, which can be employed for larger systems including extended systems (surfaces, adsorption on surfaces including heterogeneous catalysis, solids). The working horses of relativistic quantum chemistry are the various valence-only methods, which early on made their way in standard electronic structure programs and nowadays enable highly reliable calculations with very little cost, rendering a standard treatment of relativistic effects a problem which can be considered well under control.

The influence of relativity on the spectroscopic properties of a molecule depends very much on the case. The energy lowering due to kinematic relativistic effects roughly scales with the number of s electrons on the heavy atom(s). This affects both electronic excitation energies and dissociation energies. Kinematic corrections tend to stabilize covalent bonds and often lead to considerable bond length contractions. In more ionic compounds the s electron population at the heavy atom site generally decreases upon bond formation owing to charge transfer. The relativistic energy stabilization is therefore largest in the dissociation limit which in turn results in a bond weakening.

Kinematic relativistic and electron correlation effects are not additive in general. Current one-component methods which include kinematic relativistic correction terms variationally, i.e., already in the determination of the orbitals, yield reliable estimates of their influence, independent of whether all electrons are treated explicitly or RECPs are involved. Adding the mass–velocity and Darwin correction terms perturbationally may be feasible in first-row transition metal compounds, but this procedure fails badly in the third transition metal row and for the sixth-row main group elements. The same remark applies to the Breit–Pauli form of spin–orbit interaction.

Also magnetic interactions which are dominated by spin–orbit coupling affect the electronic spectrum and dissociation energy of a molecule. Particularly strong effects are expected in compounds of heavy elements with unfilled p, d, or f shells. In a molecule spin–orbit splittings are reduced with respect to the dissociation limit because spatial degeneracies are lifted by the external field of the surrounding atoms. In open-shell ground and excited electronic states large spin–orbit interaction may persist, however, also in the equilibrium region.

As far as spin–orbit coupling in heavy element compounds is concerned, transition and main group elements put quite different demands on their theoretical treatment. It appears that in late main group elements spin–orbit coupling is strongly connected with orbital relaxation and electron correlation. This is a consequence of the appreciable difference between the radial densities of the jj-coupled $p_{1/2}$ and $p_{3/2}$ orbitals. The effect is much less pronounced in transition metals where the major spin–orbit coupling stems from open d shells, the components of which exhibit quite similar radial densities. Quasi-degenerate perturbation theory where a small eigenvalue problem is solved in the basis of *LS*-coupled states yields results of similiar quality to that for the application of intermediate coupling in a spin–orbit CI. In either case, one has to be careful, however, in the selection of contributing states or configurations, respectively. It appears to be of minor importance, whether electron correlation and spin–orbit coupling are treated exactly on the same footing: spin–orbit interactions are dominated by single excitations contributing only little to electron correlation. Electron correlation, on the other hand, is known to converge very slowly with the size of the configuration space. Several approximate schemes have been devised recently to cope with these different demands.

Two-electron contributions to the spin–orbit interaction cannot generally be neglected. It is possible, however, to incorporate all major two-electron contributions to valence-shell spin–orbit splittings into an effective one-electron operator. A further reduction of computational expense can be achieved by a spin–orbit mean-field one-center approximation which appears to work excellently for systems containing heavy elements and gives reasonable results even for light open-shell molecules. This is also the basis for the great success of effective spin–orbit potentials used in combination with RECPs.

6.5 References

Andrae, D., Häussermann, U., Dolg, M., Stoll, H., and Preuss, H., 1990, *Theor. Chim. Acta* **77**, 123–141.
Balasubramanian, K., and Feng, P. Y., 1990, *J. Chem. Phys.* **92**, 541–550.
Balasubramanian, K., and Tao, J. X., 1991, *J. Chem. Phys.* **94**, 3000–3010.
Barandiarán, Z., Huzinaga, S., and Seijo, L., 1991, *J. Chem. Phys.* **94**, 3762–3773.
Barandiarán, Z., Seijo, L., and Huzinaga, S., 1990, *J. Chem. Phys.* **93**, 5843–5850.
Barthelat, J. C., and Durand, P., 1978, *Gazz. Chim. Ital.* **108**, 225–236.
Barthelat, J. C., Pélissier, M., and Durand, P., 1980, *Phys. Rev. A*, **21**, 1773–1785.
Barysz, M., Sadlej, A. J., and Snijders, J. G., 1997, *Int. J. Quan. Chem.* **65**, 225–239.
Basch, H., Cohen, D., and Topiol, S., 1980, *Isrl. J. Chem.* **19**, 233–241.
Blaudeau, J.-P., and Curtiss, L. A., 1997, *Int. J. Quant. Chem.* **61**, 943–952.
Blume, M., and Watson, R. E., 1962, *Proc. R. Soc. London, Ser. A*, **270**, 127–143.
Blume, M., and Watson, R. E., 1963, *Proc. R. Soc. London, Ser. A*, **271**, 565–578.
Boettger, J. C., 1998, *Phys. Rev. B*, **57**, 8743–8746.

Breit, G., 1929, *Phys. Rev.* **34**, 553–573.
Breit, G., 1930, *Phys. Rev.* **36**, 383–397.
Breit, G., 1932, *Phys. Rev.* **39**, 616–624.
Brown, G. E., and Ravenhall, D. G., 1951, *Proc. R. Soc. London, Ser. A*, **208**, 552–559.
Buchmüller, W., and Dietz, K., 1980, *Z. Phys. C* **5**, 45–54.
Buenker, R. J., Alekseyev, A. B., Liebermann, H.-P., Lingott, R., and Hirsch, G., 1998, *J. Chem. Phys.* **108**, 3400–3408.
Buenker, R. J., Peyerimhoff, S. D., and Butscher, W., 1978, *Mol. Phys.* **35**, 771–791.
Casarrubios, M., and Seijo, L., 1995, *Chem. Phys. Lett.* **236**, 510–515.
Chandra, P., and Hess, B. A., 1994, *Theor. Chim. Acta* **88**, 182–199.
Chang, C., Pélissier, M., and Durand, P., 1986, *Phys. Scr.* **34**, 394–404.
Christiansen, P. A., Balasubramanian, K., and Pitzer, K. S., 1982, *J. Chem. Phys.* **76**, 5087–5092.
Christiansen, P. A., and Pitzer, K. S., 1980, *J. Chem. Phys.* **73**, 5160–5163.
Cohen, E. R., and Taylor, B. N., 1987, *Rev. Mod. Phys.* **59**, 1121–1148.
Cowan, R. D., and Griffin, D. C., 1976, *J. Opt. Soc. Am.* **66**, 1010–1014.
Cundari, T. R., and Stevens, W. J., 1993, *J. Chem. Phys.* **98**, 5555–5565.
Danovich, D., Marian, C. M., Neuheuser, T., Peyerimhoff, S. D., and Shaik, S., 1998, *J. Phys. Chem. A*, **102**, 5923–5936.
Desclaux, J. P., 1973, *At. Data Nucl. Data Tables*, **12**, 311–406.
Desclaux, J. P., 1975, *Comput. Phys. Commun.* **9**, 31–45.
Dietz, K., and Hess, B. A., 1991, *J. Phys. B*, **24**, 1129–1142.
DiLabio, G. A., and Christiansen, P. A., 1998, *J. Chem. Phys.* **108**, 7527–7533.
Dirac, P. A. M., 1928, *Proc. R. Soc. London, Ser. A* **117**, 610–624.
Dolg, F. M., 1997, Quasirelativistische und relativistische energiekonsistente Pseudopotentiale für quantentheoretische Untersuchungen der Chemie schwerer Elemente. Habilitation, Universität Stuttgart.
Dolg, M., 1994, University of Stuttgart, private communication.
Dolg, M., 1996, *Theor. Chim. Acta* **93**, 141–156.
Dolg, M., Küchle, W., Stoll, H., and Preuss, H., 1991, *Mol. Phys.* **74**, 1265–1285.
Dolg, M., and Stoll, H., 1995, *Handbook on the Physics and Chemistry of Rare Earths*, Gschneider, Jr., K. A., and Eyring, L., Eds, Vol. 22; Elsevier: Amsterdam, chapter 152.
Dolg, M., Stoll, H., and Preuss, H., 1989, *J. Chem. Phys.* **90**, 1730–1734.
Douglas, M., and Kroll, N. M., 1974, *Ann. Phys. (N.Y.)*, **82**, 89–155.
Durand, P., and Barthelat, J. C., 1975, *Theor. Chim. Acta*, **38**, 283–302.
Dyall, K., Grant, I. P., and Wilson, S., 1984, *J. Phys. B*, **17**, 493–503.
Dyall, K. G., 1993, *J. Phys. Chem.* **98**, 9678–9686.
Dyall, K. G., 1994, *Chem. Phys. Lett.* **224**, 186–194.
Dyall, K. G., 1997, *J. Chem. Phys.* **106**, 9618–9626.
Dyall, K. G., 1998, *J. Chem. Phys.* **109**, 4201–4208.
Dyall, K. G., and Fægri, Jr., K., 1993, *Chem. Phys. Lett*, **201**, 27–32.
Dyall, K. G., Grant, I. P., Johnson, C. T., Parpia, F. A., and Plummer, E. P., 1989, *Comput. Phys. Commun.* **55**, 425–456.
Dyall, K. G., Taylor, R. R., Jr., K. F., and Partridge, H., 1991, *J. Chem. Phys.* **95**, 2583–2594.
Eliav, E., Kaldor, U., Ishikawa, Y., Seth, M., and Pyykkö, P., 1996, *Phys. Rev. A*, **53**, 3926–3933.
Ermler, W. C., Ross, R. B., and Christiansen, P. A., 1988, *Adv. Quant. Chem.* **19**, 139–182.
Ermler, W. C., Ross, R. B., and Christiansen, P. A., 1994, *Int. J. Quant. Chem.* **40**, 829–846.
Fehrenbach, G. M., and Schmidt, G., 1997, *Phys. Rev. B*, **55**, 6666–6669.
Fleig, T., 1993, Relativistische ab initio Rechnungen am Platinhydridmolekül, Diploma thesis, University of Bonn.
Fleig, T., and Marian, C. M., 1994, *Chem. Phys. Lett.* **222**, 267–273.
Fleig, T., and Marian, C. M., 1996, *J. Mol. Spectrosc.* **178**, 1–9.
Fleig, T., and Marian, C. M., 1998, *J. Chem. Phys.* **108**, 3517–3521.
Fleig, T., Marian, C. M., and Olsen, J., 1997, *Theor. Chem. Acc.* **97**, 125–135.
Foldy, L. L., and Wouthuysen, S. A., 1950, *Phys. Rev.* **78**, 29–36.
Fuentealba, P., Preuss, H., Stoll, H., and v. Szentpály, L., 1982, *Chem. Phys. Lett.* **89**, 418–422.
Geipel, N. J. M., and Hess, B. A., 1997, *Chem. Phys. Lett.* **273**, 62–70.
Ginter, M. L., and Battino, R., 1965, *J. Chem. Phys.* **42**, 3222–3229.
Gorceix, O., and Indelicato, P., 1988, *Phys. Rev. A*, **37**, 1087–1094.
Gorceix, O., Indelicato, P., and Desclaux, J. P., 1987, *J. Phys. B*, **20**, 639–649.
Grant, I. P., 1961, *Proc. R. Soc. London, Ser. A*, **262**, 555–576.
Grant, I. P., 1965, *Proc. Phys. Soc.* **86**, 523–527.
Grant, I. P., 1970, *Adv. Phys.* **19**, 747–811.
Grant, I. P., 1974, *J. Phys. B*, **7**, 1458–1475.
Grant, I. P., 1987, *J. Phys. B*, **20**, L735–L740.
Gray, J. A., Li, M., Nelis, T., and Field, R., 1991, *J. Chem. Phys.* **95**, 7164–7178.

Gropen, O., Almlöf, J., and Wahlgren, U., 1992, *J. Chem. Phys.* **96**, 8363–8366.
Grundström, B., 1940, *Z. Phys.* **115**, 120–139.
Grundström, B., and Valberg, P., 1937, *Z. Phys.* **108**, 326–337.
Gustafsson, T., and Scullman, R., 1989, *Mol. Phys.* **67**, 981–988.
Häberlen, O. D., and Rösch, N., 1992, *Chem. Phys. Lett.* **199**, 491–496.
Hafner, P., and Schwarz, W. H. E., 1977, *J. Phys. B*, **11**, 217–233.
Han, Y.-K., Bae, C., and Lee, Y. S., 1999, *J. Chem. Phys.* **110**, 8353–8359, 8969–8975.
Havriliak, S. J., and Yarkony, D. R., 1985, *J. Chem. Phys.* **83**, 1168–1172.
Hay, P. J., and Wadt, W. R., 1985, *J. Chem. Phys.* **82**, 299–310.
Heinemann, C., Koch, W., and Schwarz, H., 1995, *Chem. Phys. Lett.* **245**, 509–518.
Heinemann, C., Schwarz, H., Koch, W., and Dyall, K. G., 1996, *J. Chem. Phys.* **104**, 4642–4651.
Hess, B. A., 1986, *Phys. Rev. A*, **33**, 3742–3748.
Hess, B. A., Marian, C. M., and Peyerimhoff, S. D., 1995, *Modern Electronic Structure Theory*, Yarkony, Ed.; World Scientific: Singapore, pp. 152–278.
Hess, B. A., Marian, C. M., Wahlgren, U., and Gropen, O., 1996, *Chem. Phys. Lett.* **251**, 365–371.
Heully, J. L., Lindgren, I., Lindroth, E., Lundquist, S., and Mårtensson-Pendrill, A. M., 1986, *J. Phys. B*, **19**, 2799–2815.
Hurley, M. M., Pacios, L. F., Christiansen, P. A., Ross, R. B., and Ermler, W. C., 1986, *J. Chem. Phys.* **84**, 6840–6853.
Huzinaga, S., and Cantu, A. A., 1971, *J. Chem. Phys.* **55**, 5543–5549.
Huzinaga, S., Seijo, L., Barandiarán, Z., and Klobukowski, M., 1987, *J. Chem. Phys.* **86**, 2132–2145.
Ishikawa, Y., Baretty, R., and Binning, Jr., R. C., 1985, *Chem. Phys. Lett.* **121**, 130–133.
Ishikawa, Y., and Kaldor, U., 1996, *Computational Chemistry–Reviews of Current Trends*, Leszezynski, J., Ed., World Scientific: Singapore, pp. 1–52.
Ishikawa, Y., and Malli, G., 1981, *J. Chem. Phys.* **75**, 5423–5431.
Ishikawa, Y., Sekino, H., and Binning, Jr., R. C., 1990, *Chem. Phys. Lett.* **165**, 237–242.
Jansen, G., and Hess, B. A., 1989, *Phys. Rev. A*, **39**, 6016–6017.
Jensen, J. J. Å., Dyall, K. G., Saue, T., and Jr, K. F., 1996, *J. Chem. Phys.* **104**, 4083–4097.
Johnson, W., Mohr, P., and Sucher, J., 1989, Relativistic Quantum Electrodynamics and Weak Interaction Effects in Atoms, *AIP Conf. Proc.* **189**.
Kagawa, T., 1980, *Phys. Rev. A*, **22**, 2340–2354.
Kantor, I. L., and Solodovnikov, A. S., 1989, *Hypercomplex Numbers*; Springer: Berlin.
Karwowski, J., and Kobus, J., 1981, *Chem. Phys.* **55**, 361–369.
Karwowski, J., and Kobus, J., 1985, *Int. J. Quant. Chem.* **28**, 741–756.
Karwowski, J., and Szulkin, M., 1982, *J. Phys. B*, **15**, 1915–1925.
Kim, M. C., Lee, S. Y., and Lee, Y. S., 1996, *Chem. Phys. Lett.* **253**, 216–222.
Kim, Y. K., 1967, *Phys. Rev.* **154**, 17–39.
Kim, Y. K., Parente, F., Marques, J. P., Indelicato, P., and Desclaux, J. P., 1998, *Phys. Rev. A*, **58**, 1885–1888.
Kleinschmidt, M., 1999, Drehimpulskopplungen in CH/CD und PbH. Eine Studie mit Allelektronen-ab-initio-Methoden, Diploma thesis, University of Bonn.
Krauss, M., and Stevens, W. J., 1984, *Annu. Rev. Phys. Chem.* **35**, 357–385.
Küchle, W., Dolg, M., Stoll, H., and Preuss, H., 1991, *Mol. Phys.* **74**, 1245–1263.
Kutzelnigg, W., 1989a, *Z. Phys. D*, **11**, 15–28.
Kutzelnigg, W., 1989b, *Z. Phys. D*, **15**, 27–50.
Kutzelnigg, W., 1997, *Chem. Phys.* **224**, 203–222.
LaJohn, L. A., Christiansen, P. A., Ross, R. B., Atashroo, T., and Ermler, W. C., 1988, *J. Chem. Phys.* **87**, 2812–2824.
Landau, A., 1999, Poster presented at the European Research Conference Relativistic Quantum Chemistry–Progress and Prospects, Acquafredda di Maratea, Italy.
Landau, L. D., and Lifschitz, E. M., 1981, *Lehrbuch der Theroretischen Physik, Band II: Klassische Feldtheorie*; Akademie-Verlag: Berlin.
Langhoff, S. R., Huo, W. M., Partridge, H., and Bauschlicher, Jr., C. W., 1982, *J. Chem. Phys.* **77**, 2498–2513.
Langhoff, S. R., and Kern, C. W., 1977, *Molecular Fine Structure*; Plenum: New York, pp. 381–437.
Larsson, T., and Neuhaus, H., 1963, *Arkif för Fysik*, **23**, 461–469.
Larsson, T., and Neuhaus, H., 1966, *Arkif för Fysik*, **31**, 299–305.
Lee, H.-S., Han, Y.-K., Kim, M. C., Bae, C., and Lee, Y. S., 1998, *Chem. Phys. Lett.* **293**, 97–102.
Lee, Y. S., Ermler, W. C., and Pitzer, K. S., 1977, *J. Chem. Phys.* **67**, 5861–5876.
Lee, Y. S., Ermler, W. C., and Pitzer, K. S., 1980, *J. Chem. Phys.* **73**, 360–366.
Lee, Y. S., and McLean, A. D., 1982, *J. Chem. Phys.* **76**, 735–736.
Leininger, T., Berning, A., Nicklass, A., Stoll, H., Werner, H.-J., and Flad, H.-J., 1997, *Chem. Phys.* **217**, 19–27.
Lindgren, I., 1990, *J. Phys. B*, **23**, 1085–1093.
Lindgren, I., 1992, *Recent Progress in Many-Body Theories*, Ainsworth, T. L., Ed.; Plenum: New York, pp. 245–276.

Lindgren, I., 1996, *Int. J. Quant. Chem.* **57**, 683–695.
Lindner, A., 1994, *Grundkurs Theoretische Physik*; Teubner: Stuttgart.
Lindroth, E., and Mårtensson-Pendrill, A. M., 1989, *Phys. Rev. A*, **39**, 3794–3802.
Llusar, R., Casarrubios, M., Barandiarán, Z., and Seijo, L., 1996, *J. Chem. Phys.* **105**, 5321–5330.
Löwdin, P. O., 1964, *J. Molec. Spectrosc.* **14**, 131–144.
Mann, J. B., and Johnson, W. R., 1971, *Phys. Rev. A*, **4**, 41–50.
Marian, C. M., 1990, *Chem. Phys. Lett.* **173**, 175–180.
Marian, C. M., 1991, *J. Chem. Phys.* **94**, 5574–5585.
Marian, C. M., 1999, unpublished results.
Marian, C. M., Blomberg, M. R. A., and Siegbahn, P. E. M., 1989, *J. Chem. Phys.* **91**, 3589–3595.
Marian, C. M., Marian, R., Peyerimhoff, S. D., Hess, B. A., Buenker, R. J., and Seger, G., 1982, *Mol. Phys.* **46**, 779–810.
Marian, C. M., and Wahlgren, U., 1996, *Chem. Phys. Lett.* **251**, 357–364.
Marino, M. M., and Ermler, W. C., 1993, *Chem. Phys. Lett.* **206**, 271–277.
Mark, F., and Schwarz, W. H. E., 1982, *Phys. Rev. Lett.* **48**, 673–676.
Martin, R. J., and Hay, P. J., 1981, *J. Chem. Phys.* **75**, 4539–4545.
Matsuoka, O., 1992, *Chem. Phys. Lett.* **195**, 184–188.
McCarthy, M., Field, R., Engleman, Jr., R., and Bernath, P., 1993, *J. Mol. Spectrosc.* **158**, 208–236.
McMichael Rohlfing, C., Hay, P. J., and Martin, R. J., 1986, *J. Chem. Phys.* **85**, 1447–1455.
Migdalek, J., and Baylis, W. E., 1981, *Can. J. Phys*, **59**, 769–774.
Minaev, B., Vahtras, O., and Ågren, H., 1996, *Chem. Phys.* **208**, 299–311.
Mittleman, M. H., 1971, *Phys. Rev. A*, **4**, 893–900.
Mittleman, M. H., 1972, *Phys. Rev. A*, **5**, 2395–2401.
Mittleman, M. H., 1981, *Phys. Rev. A*, **24**, 1167–1175.
Molzberger, K., and Schwarz, W. H. E., 1996, *Theor. Chim. Acta*, **94**, 213–222.
Moore, C., 1952, *Atomic Energy Levels*, Vol. 2, Circular No. 467; NSRDS-NBS, US GPO: Washington, DC.
Moore, C., 1958, *Atomic Energy Levels*, Vol. 3, Circular No. 467; NSRDS-NBS, US GPO: Washington, DC.
Morrison, J. D., and Moss, R. E., 1980, *Mol. Phys.* **41**, 491–507.
Müller, W., Flesch, J., and Meyer, W., 1984, *J. Chem. Phys.* **80**, 3297–3310.
Müller-Nehler, U., and Soff, G., 1994, *Phys. Rep.* **246**, 101–250.
Nash, C. S., Bursten, B. E., and Ermler, W. C., 1997, *J. Chem. Phys.* **106**, 5133–5142.
Nasluzov, V. A., and Rösch, N., 1996, *Chem. Phys.* **210**, 413–425.
Neuhaus, H., and Muld, V., 1959, *Z. Phys.* **153**, 412–422.
Nikitin, A. G., 1998, *J. Phys. A*, **31**, 3297–3300.
Pacios, L. F., and Christiansen, P. A., 1985, *J. Chem. Phys.* **82**, 2664–2671.
Park, C.-Y., and Almlöf, J. E., 1994, *Chem. Phys. Lett.* **231**, 269–276.
Parpia, F. A., and Mohanty, A. K., 1995, *Phys. Rev. A*, **52**, 962–968.
Phillips, J. C., and Kleinman, 1959, *Phys. Rev.* **116**, 287–294.
Pisani, L., and Clementi, E., 1994, *J. Chem. Phys*, **101**, 3079–3084.
Pisani, L., and Clementi, E., 1995, *METECC-95*, Clementi, E., and Corongiu, G., Eds; STEF: Cagliari, pp. 219–241.
Pitzer, K. S., 1975, *J. Chem. Phys.* **63**, 1032–1033.
Pitzer, K. S., 1979, *Acc. Chem. Res.* **12**, 271–276.
Pitzer, K. S., and Christiansen, P. A., 1981, *Chem. Phys. Lett.* **77**, 589–592.
Pitzer, R. M., and Winter, N. W., 1988, *J. Chem. Phys.* **92**, 3061–3063.
Pizlo, A., Jansen, G., Hess, B. A., and von Niessen, W., 1993, *J. Chem. Phys.* **98**, 3945–3951.
Porteous, I. R., 1995, *Clifford Algebras and the Classical Groups*; Cambridge University Press: Cambridge.
Pyper, N. C., 1980, *Chem. Phys. Lett.* **73**, 385–392.
Pyykkö, P., and Desclaux, J. P., 1976, *Chem. Phys. Lett.* **42**, 545–549.
Pyykkö, P., and Desclaux, J. P., 1979, *Acc. Chem. Res.* **12**, 276–281.
Pyykkö, P., Tokman, M., and Labzowsky, L. N., 1998, *Phys. Rev. A*, **57**, R689–R692.
Rakowitz, F., 1995, Relativistische ab initio Berechnungen am Thalliumatom und am Thalliumhydridmolekül, Diploma thesis, University of Bonn.
Rakowitz, F., 1999, Entwicklung, Implementierung und Anwendung effizienter Methoden in der relativistischen Elektronenstrukturtheorie, Dissertation, Universität Bonn.
Rakowitz, F., Casarrubios, M., Seijo, L., and Marian, C. M., 1998, *J. Chem. Phys.* **108**, 7980–7987.
Rakowitz, F., and Marian, C. M., 1996, *Chem. Phys. Lett.* **257**, 105–110.
Rakowitz, F., and Marian, C. M., 1997, *Chem. Phys.* **225**, 223–238.
Rakowitz, F., Marian, C. M., and Seijo, L., 1999, *J. Chem. Phys.* **111**, 10 436–10 443.
Rakowitz, F., Marian, C. M., Seijo, L., and Wahlgren, U., 1999, *J. Chem. Phys.* **110**, 3678–3686.
Ramos, A. F., Lee, S. Y., and Malli, G. L., 1988, *Phys. Rev. A*, **38**, 2729–2739.
Ribbing, C., 1994, University of Stockholm, personal communication.

Richards, W. G., Trivedi, H. P., and Cooper, D. L., 1981, *Spin–orbit Coupling in Molecules*; Clarendon Press: Oxford.
Ross, R. B., Ermler, W. C., and Christiansen, P. A., 1986, *J. Chem. Phys.* **84**, 3297–3300.
Ross, R. B., Gayen, S., and Ermler, W. C., 1994, *J. Chem. Phys.* **100**, 8145–8155.
Ross, R. B., Powers, J. M., Atashroo, T., Ermler, W. C., LaJohn, L. A., and Christiansen, P. A., 1990, *J. Chem. Phys.* **93**, 6654–6669.
Rutkowski, A., 1986a, *J. Phys. B*, **19**, 149–158.
Rutkowski, A., 1986b, *J. Phys. B*, **19**, 3431–3442.
Rutkowski, A., 1986c, *J. Phys. B*, **19**, 3443–3455.
Samzow, R., Hess, B. A., and Jansen, G., 1992, *J. Chem. Phys.* **96**, 1227–1231.
Sapirstein, J., 1998, *Rev. Mod. Phys.* **70**, 55–76.
Schimmelpfennig, B., 1996, *Atomic Spin–orbit Mean-Field Integral Program*, AMFI; Stockholms Universitet.
Schimmelpfennig, B., Maron, L., Wahlgren, U., Teichteil, C., Fagerli, H., and Gropen, O., 1998a, *Chem. Phys. Lett.* **286**, 261–266.
Schimmelpfennig, B., Maron, L., Wahlgren, U., Teichteil, C., Fagerli, H., and Gropen, O. 1998b, *Chem. Phys. Lett.* **286**, 267–271.
Schwarz, W. H. E., van Wezenbeek, E. M., Baerends, E. J., and Snijders, J. G. 1989, *J. Phys. B*, **22**, 1515–1530.
Schwarz, W. H. E., and Wallmeier, H., 1982, *Mol. Phys.* **46**, 1045–1061.
Schwarz, W. H. E., and Wechsel-Trakowski, E., 1982, *Chem. Phys. Lett.* **85**, 94–97.
Schwerdtfeger, P., 1987, *Phys. Scr.* **36**, 453–459.
Schwerdtfeger, P., Heath, G. A., Dolg, M., and Bennett, M. A., 1992, *J. Am. Chem. Soc.* **114**, 7518–7527.
Seijo, L., 1995, *J. Chem. Phys.* **102**, 8078–8088.
Seijo, L., 1998, *Lecture Notes on Effective Core Potentials*; II Escuela Iberoamericana de Química Computacional y Diseño Moelcular, University of Habana: Cuba.
Sewell, G. L., 1949, *Proc. Cambridge Philos. Soc.* **45**, 631–637.
Sjøvoll, M., Fagerli, H., Gropen, O., Almlöf, J., Olsen, J., and Helgaker, T. U., 1998, *Int. J. Quant. Chem.* **68**, 53–64.
Sjøvoll, M., Gropen, O., and Olsen, J., 1997, *Theor. Chem. Acc.* **97**, 301.
Sjøvoll, M., and Marian, C. M., 1996, unpublished results.
Snijders, J., and Pyykkö, P., 1980, *Chem. Phys. Lett.* **75**, 5–8.
Sommerfeld, A., 1916, *Ann. Phys. (Leipzig)*, **51**, 1–94, 125–167.
Stanton, R. E., and Havriliak, S., 1984, *J. Chem. Phys.* **81**, 1910–1918.
Stevens, W. J., Krauss, M., Basch, H., and Jasien, P. G., 1992, *Can. J. Chem.* **70**, 612–630.
Stoll, H., Fuentealba, P., Dolg, M., Flad, J., von Szentpály, L., and Preuss, H., 1983, *J. Chem. Phys.* **79**, 5532–5542.
Sucher, J., 1980, *Phys. Rev. A*, **22**, 348–362.
Swirles, B., 1935, *Proc. R. Soc. London, Ser. A*, **152**, 625–649.
Tatchen, J., and Marian, C. M., 1999, *Chem. Phys. Lett.* **313**, 351–357.
Taylor, E. F., and Wheeler, J. A., 1994, *Physik der Raumzeit*; Spektrum Akademischer Verlag: Heidelberg.
Teichteil, C. H., 1994, *Lecture Notes on Effective Spin–orbit Operators in Multi-configurational Calculations*; Workshop of the European Science Foundation on Relativistic Pseudo- and Model-Potentials, University of Helsinki.
Teichteil, C. H., Pélissier, M., and Spiegelmann, F., 1983, *Chem. Phys.* **81**, 273–282.
Teichteil, C. H., and Spiegelmann, F., 1983, *Chem. Phys.* **81**, 283–296.
Titchmarsh, E. C., 1962, *Proc. R. Soc. London, Ser. A*, **266**, 33–46.
Titov, A. V., and Mosyagin, N. S., 1999, *Int. J. Quant. Chem.* **71**, 359–401.
Treboux, G., and Barthelat, J.-C., 1993, *J. Am. Chem. Soc.* **115**, 4870–4878.
Urban, R. D., Bahnmaier, A. H., Magg, U., and Jones, H., 1989, *Chem. Phys. Lett.* **158**, 443–446.
Vallet, V., Maron, L., Teichteil, C., and Flament, J.-P., 1999, *J. Chem. Phys.* submitted.
van Leeuwen, R., van Lenthe, E., Baerends, E. J., and Snijders, J. G., 1994, *J. Chem. Phys.* **101**, 1271–1281.
van Lenthe, E., Baerends, E. J., and Snijders, J. G., 1993, *J. Chem. Phys.* **99**, 4597–4610.
van Lenthe, E., Baerends, E. J., and Snijders, J. G., 1994, *J. Chem. Phys.* **101**, 9783–9792.
van Lenthe, E., Snijders, J. G., and Baerends, E. J., 1996, *J. Chem. Phys.* **105**, 6505–6516.
van Lenthe, E., van Leeuwen, R., Baerends, E. J., and Snijders, J. G., 1995, *New Challenges in Computational Quantum Chemistry*, Broer, R., Aerts, P. J. C., and Bagus, P. S., Eds; Groningen University Press: Groningen, pp. 93–111.
van Lenthe, E., van Leeuwenand E. J., Baerends, R., and Snijders, J. G., 1996, *Int. J. Quant. Chem.* **57**, 281–293.
van Wüllen, C., 1998, *J. Chem. Phys*, **109**, 392–400.
Visscher, L., 1995, *METECC-95*, Clementi, E., and Corongiu, G., Eds; STEF: Cagliari, pp. 169–218.
Visscher, L., 1996, *Chem. Phys. Lett.* **253**, 20–26.
Visscher, L., Aerts, P. J. C., Visser, O., and Nieuwpoort, W. C., 1991, *Int. J. Quant. Chem. Symp.* **25**, 131–139.
Visscher, L., Saue, T., Nieuwpoort, W. C., Fægri, K., and Gropen, O. 1993, *J. Chem. Phys.* **99**, 6704–6715.
Visser, O., Aerts, P. J. C., Hegarty, D., and Nieuwpoort, W. C., 1987, *Chem. Phys. Lett*, **134**, 34–38.

Wadt, W. R., 1982, *Chem. Phys. Lett.* **89**, 245–248.
Wadt, W. R., 1984, *J. Chem. Phys.* **82**, 284–298.
Wahlgren, U., Sjøvoll, M., Fagerli, H., Gropen, O., and Schimmelpfennig, B., 1997, *Theor. Chem. Acc.* **97**, 324–330.
Wallace, N. M., Blaudeau, J.-P., and Pitzer, R. M., 1991, *Int. J. Quant. Chem.* **40**, 789–796.
Wang, S. W., and Pitzer, K. S., 1983, *J. Chem. Phys.* **79**, 3851–3858.
Wildman, S. A., DiLabio, G. A., and Christiansen, P. A., 1997, *J. Chem. Phys.* **107**, 9975–9979.
Wittborn, C., and Wahlgren, U., 1995, *Chem. Phys.* **201**, 357–362.
Wood, J., Grant, I. P., and Wilson, S. J., 1985, *J. Phys. B*, **18**, 3027–3041.
Wood, J. H., and Boring, A. M., 1978, *Phys. Rev. B*, **18**, 2701–2711.

7 THE *AB INITIO* CALCULATION OF MOLECULAR PROPERTIES OTHER THAN THE POTENTIAL ENERGY SURFACE

Stephan P. A. Sauer
University of Copenhagen, Denmark

and

Martin J. Packer
University of Sheffield, UK

Computational Molecular Spectroscopy. Edited by Per Jensen and Philip R. Bunker

7.1 Introduction

The interaction of a molecule with a weak external electromagnetic field, or the interactions within a molecule involving internal electromagnetic moments such as a nuclear magnetic dipole moment or a nuclear electric quadrupole moment, are all described in terms of so-called 'molecular properties'. These are intrinsic properties of an electronic state of a molecule and are independent of the strength of the external field or internal moments. Molecular properties include the electric dipole moment, the frequency-dependent polarizability tensor, the nuclear magnetic shielding tensor and the indirect nuclear spin–spin coupling tensor among many others. They play an important role in the interpretation of numerous phenomena including the refractive index, the Stark effect, the Kerr effect, and nuclear magnetic resonance spectra. Long-range interactions between molecules can also be understood in terms of molecular electric moments.

Molecular properties are defined classically in Section 7.2, and these definitions are used in Section 7.3 to derive their quantum mechanical analogues. The approach used is semi-classical with only the electrons treated quantum mechanically (and non-relativistically) whereas fields and nuclei are treated classically. Approximate methods which can be used to calculate molecular electromagnetic properties are described in Section 7.4. This exposition is restricted to a discussion of *ab initio* methods and the Born–Oppenheimer approximation is assumed throughout.

7.2 Defining molecular properties

The classical definition of molecular properties is discussed in this section. For that purpose a continuous static or dynamic distribution of charges, defined by a charge density $\rho(\boldsymbol{r})$ and current density $\boldsymbol{j}(\boldsymbol{r})$, is considered. Although in the classical context it is unnecessary to assume a continuous distribution of charges, this is convenient in the light of the quantum mechanical treatment which follows.

7.2.1 PERMANENT AND INDUCED ELECTRIC MOMENTS

Electric charges give rise to an electric field and an electrostatic potential. For a distribution of charges with charge density $\rho(\boldsymbol{r})$, the electrostatic potential $\phi^{\rho}(\boldsymbol{R})$ is given as a superposition of the potentials due to the individual charges

$$\phi^{\rho}(\boldsymbol{R}) = \frac{1}{4\pi\varepsilon_0}\int_{\tau'} \frac{\rho(\boldsymbol{r}')}{|\boldsymbol{R}-\boldsymbol{r}'|}\mathrm{d}\tau'. \tag{1}$$

It is possible to expand $1/|\boldsymbol{R}-\boldsymbol{r}'|$, and thus the potential, in a Taylor series around an origin $\boldsymbol{r}_\mathrm{o}$ within the charge distribution. By evaluating the derivatives of $1/|\boldsymbol{R}-\boldsymbol{r}'|$ and defining the first two electric moments of the charge distribution as

$$q = \int_{\tau'} \rho(\boldsymbol{r}')\,\mathrm{d}\tau', \tag{2}$$

and

$$\mu_\alpha(\boldsymbol{r}_\mathrm{o}) = \int_{\tau'} (r'_\alpha - r_{\mathrm{o},\alpha})\rho(\boldsymbol{r}')\,\mathrm{d}\tau', \tag{3}$$

a multipole expansion of the electrostatic potential is obtained as

$$\phi^{\rho}(\boldsymbol{R}) = \frac{1}{4\pi\varepsilon_0}\left[q\frac{1}{|\boldsymbol{R}-\boldsymbol{r}_\mathrm{o}|} + \mu_\alpha(\boldsymbol{r}_\mathrm{o})\frac{R_\alpha - r_{\mathrm{o},\alpha}}{|\boldsymbol{R}-\boldsymbol{r}_\mathrm{o}|^3} + \cdots\right]. \tag{4}$$

The Greek subscripts α, β, etc. denote vector or tensor components in the molecule-fixed cartesian co-ordinate system. A repeated Greek subscript denotes summation over all three Cartesian components. The zeroth-order electric moment q is the total charge, and the first-order electric moment μ_α is a component of the electric dipole moment vector. The contribution from the higher moments in the series will become negligible as the distance from the origin increases and the potential will then be accurately described by only the charge and dipole moment terms. Convergence of the multipole series for a particular value of $\boldsymbol{R}$ depends on the precise form of the charge distribution.

An important feature of the electric multipole moments is that the first non-vanishing moment of a charge distribution is independent of the choice of origin $\boldsymbol{r}_\text{o}$. However, all higher moments depend on the position of the origin. Thus the charge of an ion and the dipole moment of a neutral molecule are both independent of the choice of origin, whereas the dipole moment of an ion and the quadrupole moment of a neutral molecule are not.

Having obtained the electric multipole moments of a charge distribution, one can calculate the electrostatic potential at any point $\boldsymbol{R}$ from the simple formula in equation (4) instead of evaluating the more complicated expression in equation (1) for each $\boldsymbol{R}$. This explains the importance of the electric multipole moments for the description of intermolecular forces [Buckingham (1967)].

Electric multipole moments also play an important role in the description of interactions between molecules and external electric fields. The potential energy W of a distribution of charges immersed in an external static electric field $\boldsymbol{E}$ is given as

$$W = \int_{\tau'} \rho(\boldsymbol{r}')\phi^E(\boldsymbol{r}')\,\mathrm{d}\tau', \tag{5}$$

where $\phi^E(\boldsymbol{r}')$ is the scalar potential associated with the electric field. A more useful expression can be obtained if one expands the scalar potential in a Taylor series around $\boldsymbol{r}_\text{o}$ and makes use of equations (2) and (3) to obtain

$$W = q\phi^E(\boldsymbol{r}_\text{o}) - \mu_\alpha(\boldsymbol{r}_\text{o})E_\alpha(\boldsymbol{r}_\text{o}) + \cdots. \tag{6}$$

Here E_α denotes components of the electric field vector, which are defined as derivatives of the scalar potential

$$E_\alpha(\boldsymbol{r}_\text{o}) = -\left.\frac{\partial\phi^E(\boldsymbol{r}')}{\partial r'_\alpha}\right|_{\boldsymbol{r}'=\boldsymbol{r}_\text{o}}. \tag{7}$$

From this equation it can be seen that the dipole moment can alternatively be defined as the derivative of the potential energy with respect to the field strength E_α,

$$\mu_\alpha(\boldsymbol{r}_\text{o}) = -\frac{\partial W}{\partial E_\alpha(\boldsymbol{r}_\text{o})}. \tag{8}$$

So far it has been assumed that the distribution of charges is fixed and is not influenced by the external field. However, if the charge distribution can be polarized in the presence of the electric field it will redistribute itself such that the total energy is minimized. As a result, the moments of the charge distribution will change. The field-dependent dipole moment induced by the external field, $\mu_\alpha^{\text{ind}}(\boldsymbol{E})$, is in addition to the field-independent, or 'permanent', moment μ_α^{per}. Traditionally [Buckingham (1967)] the dipole moment of a charge distribution in the presence of an external field $\boldsymbol{E}$ is expanded in the following way:

$$\mu_\alpha(\boldsymbol{E}) = \mu_\alpha^{\text{per}} + \mu_\alpha^{\text{ind}}(\boldsymbol{E}) = \mu_\alpha^{\text{per}} + \alpha_{\alpha\beta}E_\beta + \tfrac{1}{2}\beta_{\alpha\beta\gamma}E_\beta E_\gamma + \tfrac{1}{6}\gamma_{\alpha\beta\gamma\delta}E_\beta E_\gamma E_\delta + \cdots. \tag{9}$$

This equation defines the dipole polarizability tensor $\alpha_{\alpha\beta}$, and the first ($\beta_{\alpha\beta\gamma}$) and second ($\gamma_{\alpha\beta\gamma\delta}$) hyperpolarizability tensors as derivatives of the permanent moments. The energy of the polarizable

Table 7.1. Definitions of tensor components of the electric polarizabilities $\alpha_{\alpha\beta}$ and hyperpolarizabilities $\beta_{\alpha\beta\gamma}$ and $\gamma_{\alpha\beta\gamma\delta}$ as derivatives[a] of components of the field dependent electric dipole moment $\mu_\alpha(\boldsymbol{E})$ or of the energy $W(\boldsymbol{E})$.

	$\mu_\alpha(\boldsymbol{E})$	$W(\boldsymbol{E})$
$\alpha_{\alpha\beta}$	$\frac{\partial}{\partial E_\beta}$	$-\frac{\partial^2}{\partial E_\beta\,\partial E_\alpha}$
$\beta_{\alpha\beta\gamma}$	$\frac{\partial^2}{\partial E_\gamma\,\partial E_\beta}$	$-\frac{\partial^3}{\partial E_\gamma\,\partial E_\beta\,\partial E_\alpha}$
$\gamma_{\alpha\beta\gamma\delta}$	$\frac{\partial^3}{\partial E_\delta\,\partial E_\gamma\,\partial E_\beta}$	$-\frac{\partial^4}{\partial E_\delta\,\partial E_\gamma\,\partial E_\beta\,\partial E_\alpha}$

[a]All derivatives evaluated at zero field in the molecule fixed axis system.

distribution of charges is obtained by inserting equation (9) in equation (8) and integrating:

$$W(\boldsymbol{E}) = W^{(0)} - \mu_\alpha^{\text{per}} E_\alpha - \frac{1}{2}\alpha_{\alpha\beta}E_\alpha E_\beta - \frac{1}{6}\beta_{\alpha\beta\gamma}E_\alpha E_\beta E_\gamma - \frac{1}{24}\gamma_{\alpha\beta\gamma\delta}E_\alpha E_\beta E_\gamma E_\delta + \cdots. \tag{10}$$

Thus, the polarizability and hyperpolarizability tensors can also be defined as derivatives of the energy. The two ways of defining the polarizability and hyperpolarizability tensors are summarized in Table 7.1. By evaluating the derivative in equation (8) at zero field the permanent moment is obtained.

7.2.2 PERMANENT AND INDUCED MAGNETIC MOMENTS

A dynamic system of charges with charge density $\rho(\boldsymbol{r})$ gives rise to a current density $\boldsymbol{j}(\boldsymbol{r})$ given by

$$\boldsymbol{j}(\boldsymbol{r}) = \rho(\boldsymbol{r})\boldsymbol{v}(\boldsymbol{r}), \tag{11}$$

where $\boldsymbol{v}(\boldsymbol{r})$ is the velocity distribution. The vector potential $\boldsymbol{A}^j(\boldsymbol{R})$ due to this current density is given by

$$\boldsymbol{A}^j(\boldsymbol{R}) = \frac{\mu_0}{4\pi}\int_{\tau'} \frac{\boldsymbol{j}(\boldsymbol{r}')}{|\boldsymbol{R}-\boldsymbol{r}'|}\,\mathrm{d}\tau'. \tag{12}$$

Using a Taylor expansion in analogous fashion to Section 7.2.1 one obtains after some manipulations

$$\boldsymbol{A}^j(\boldsymbol{R}) = \frac{\mu_0}{4\pi}\boldsymbol{m}(\boldsymbol{r}_\text{o}) \times \frac{(\boldsymbol{R}-\boldsymbol{r}_\text{o})}{|\boldsymbol{R}-\boldsymbol{r}_\text{o}|^3} + \cdots, \tag{13}$$

for a component of the vector potential, where the first-order magnetic moment $\boldsymbol{m}(\boldsymbol{r}_\text{o})$, the magnetic dipole moment, is defined as

$$\boldsymbol{m}(\boldsymbol{r}_\text{o}) = \frac{1}{2}\int_{\tau'} \mathrm{d}\tau'\,(\boldsymbol{r}'-\boldsymbol{r}_\text{o}) \times \boldsymbol{j}(\boldsymbol{r}') = \frac{1}{2}\int_{\tau'} \mathrm{d}\tau'\,\rho(\boldsymbol{r}')(\boldsymbol{r}'-\boldsymbol{r}_\text{o}) \times \boldsymbol{v}(\boldsymbol{r}'). \tag{14}$$

The absence of a zeroth-order moment in equation (13) reflects the fact that magnetic monopole moments do not exist. Magnetic dipole moments do not play as important a role as their electric counterparts since most electronic states studied are singlet spin states with zero electronic orbital angular momentum, and such states do not possess a permanent magnetic dipole moment. However, nuclei with non-zero spin have a magnetic moment and equation (13) is used to define their vector potential in the following sections.

The potential energy of a distribution of charges immersed in an external magnetic induction $\boldsymbol{B}$ can be expressed in terms of magnetic moments, analogously to the electric field case. In general, the potential energy W of a current distribution in the presence of an external magnetic induction is given by

$$W = -\int_{\tau'} \boldsymbol{j}(\boldsymbol{r}')\cdot\boldsymbol{A}^B(\boldsymbol{r}')\,\mathrm{d}\tau', \tag{15}$$

where $\boldsymbol{A}^B(\boldsymbol{r}')$ is the vector potential associated with the magnetic induction

$$\boldsymbol{B}(\boldsymbol{r}) = \boldsymbol{\nabla}\times\boldsymbol{A}(\boldsymbol{r}). \tag{16}$$

A simpler expression for the potential energy can be obtained by expanding a component of the vector potential $A_\alpha^B(\boldsymbol{r}')$ in a Taylor series around an origin $\boldsymbol{r}_o$. Using the definition of the magnetic dipole moment given in equation (14) and the definition of the vector potential given in equation (16), the expansion of the energy can be written as

$$W = -\boldsymbol{m}(\boldsymbol{r}_o)\cdot\boldsymbol{B}(\boldsymbol{r}_o)+\cdots. \tag{17}$$

From this equation it can be seen that as an alternative to equation (14), the magnetic dipole moment can be defined as the derivative of the potential energy with respect to the field induction B_α,

$$m_\alpha(\boldsymbol{r}_o) = -\frac{\partial W}{\partial B_\alpha(\boldsymbol{r}_o)}. \tag{18}$$

In the presence of a magnetic induction $\boldsymbol{B}$ the energy of the distribution of moving charges changes according to equation (17). A polarizable distribution of charges will adjust itself in order to minimize the energy. This leads to a change in the current density and in the moments of the current density, such that an additional current density $\boldsymbol{j}^{\mathrm{ind}}(\boldsymbol{r})$ and magnetic moment $\boldsymbol{m}^{\mathrm{ind}}$ are induced. An important source of magnetic induction, apart from an external magnetic field, is a neighbouring magnetic moment: in particular the magnetic dipole moment $\boldsymbol{m}^K$ of some nucleus K. The magnetic dipole moment $\boldsymbol{m}(\boldsymbol{B},\boldsymbol{m}^K)$ in the presence of an external magnetic induction and a nuclear magnetic moment can again be expanded in a Taylor series as

$$m_\alpha\left(\boldsymbol{B},\boldsymbol{m}^K\right) = m_\alpha^{\mathrm{per}} + m_\alpha^{\mathrm{ind}}\left(\boldsymbol{B},\boldsymbol{m}^K\right) = m_\alpha^{\mathrm{per}} + \xi_{\alpha\beta}B_\beta - \sigma_{\beta\alpha}^K m_\beta^K + \cdots, \tag{19}$$

where $\xi_{\alpha\beta}$ and $\sigma_{\beta\alpha}^K$ are components of the dipole magnetizability and nuclear magnetic shielding tensor, respectively. The latter is closely related to the chemical shift measured in nuclear magnetic resonance spectroscopy and will be discussed in more detail in Section 7.2.3.

Inserting equation (19) into equation (18) and integrating, the energy of the polarizable charge and current distribution is obtained as

$$W\left(\boldsymbol{B},\boldsymbol{m}^K\right) = W^{(0)} - m_\alpha^{\mathrm{per}}B_\alpha - \tfrac{1}{2}\xi_{\alpha\beta}B_\alpha B_\beta + \sigma_{\beta\alpha}^K m_\beta^K B_\alpha + \cdots. \tag{20}$$

The magnetizability as well as the nuclear magnetic shielding tensor can thus be defined as derivatives of the energy, as was the case for the polarizability and hyperpolarizability tensors. This is summarized in Table 7.2. In order to obtain the permanent, i.e. field-independent, magnetic moment $\boldsymbol{m}$, the derivative in equation (18) has to be evaluated at zero magnetic field.

7.2.3 MOLECULAR ELECTRIC AND MAGNETIC FIELDS

In addition to the moments and polarizabilities which have been considered up to now, the electric and magnetic fields arising from a distribution of charges are also important for describing various

Table 7.2. Definitions of various magnetic properties as derivatives[a] of the perturbed energy $W(\boldsymbol{B}, \boldsymbol{m}^K, \boldsymbol{m}^L)$ or as derivatives[a] of components of the perturbed magnetic dipole moment $m_\alpha(\boldsymbol{B}, \boldsymbol{m}^K)$ and molecular magnetic induction $B^j_\beta(\boldsymbol{R}; \boldsymbol{B}, \boldsymbol{m}^L)$.

	$m_\alpha(\boldsymbol{B}, \boldsymbol{m}^K)$	$B^j_\beta(\boldsymbol{R}; \boldsymbol{B}, \boldsymbol{m}^L)$	$B^j_\beta(\boldsymbol{R}_K; \boldsymbol{B}, \boldsymbol{m}^L)$	$W(\boldsymbol{B}, \boldsymbol{m}^K, \boldsymbol{m}^L)$
$\xi_{\alpha\beta}$	$\frac{\partial}{\partial B_\beta}$	—	—	$-\frac{\partial^2}{\partial B_\beta \, \partial B_\alpha}$
$\sigma_{\beta\alpha}(\boldsymbol{R})$	—	$-\frac{\partial}{\partial B_\alpha}$	—	—
$\sigma^K_{\beta\alpha}$	$-\frac{\partial}{\partial m^K_\beta}$	—	$-\frac{\partial}{\partial B_\alpha}$	$\frac{\partial^2}{\partial m^K_\beta \, \partial B_\alpha}$
$K^L_{\beta\alpha}(\boldsymbol{R})$	—	$-\frac{\partial}{\partial m^L_\alpha}$	—	—
$K^{KL}_{\beta\alpha}$	—	—	$-\frac{\partial}{\partial m^L_\alpha}$	$\frac{\partial^2}{\partial m^K_\beta \, \partial m^L_\alpha}$

[a] All derivatives are evaluated at zero magnetic field and zero nuclear magnetic moment. $\xi_{\alpha\beta}$, magnetizability; $\sigma_{\beta\alpha}(\boldsymbol{R})$, magnetic shielding field; $\sigma^K_{\beta\alpha}$, magnetic shielding; $K^L_{\beta\alpha}(\boldsymbol{R})$, reduced spin–spin coupling field; $K^{KL}_{\beta\alpha}$, reduced spin–spin coupling.

molecular properties. The molecular magnetic induction $\boldsymbol{B}^j(\boldsymbol{R})$ is obtained by application of equation (16) to equation (12) for the vector potential $\boldsymbol{A}^j(\boldsymbol{R})$ of a current distribution

$$\boldsymbol{B}^j(\boldsymbol{R}) = -\frac{\mu_0}{4\pi}\int_{\tau'} \frac{(\boldsymbol{R}-\boldsymbol{r}')\times \boldsymbol{j}(\boldsymbol{r}')}{|\boldsymbol{R}-\boldsymbol{r}'|^3}\mathrm{d}\tau' = -\frac{\mu_0}{4\pi}\int_{\tau'} \rho(\boldsymbol{r}')\frac{(\boldsymbol{R}-\boldsymbol{r}')\times \boldsymbol{v}(\boldsymbol{r}')}{|\boldsymbol{R}-\boldsymbol{r}'|^3}\mathrm{d}\tau'. \tag{21}$$

Although defined everywhere, only the value at a nucleus can be probed experimentally. The interaction of a nuclear magnetic dipole moment $\boldsymbol{m}^K$ with the molecular magnetic induction gives rise to a change in the energy of the distribution of charges

$$W = -\boldsymbol{m}^K \cdot \boldsymbol{B}^j(\boldsymbol{R}_K). \tag{22}$$

Consequently, the molecular magnetic induction can be defined as a derivative of the energy of the distribution of charges with respect to a nuclear magnetic moment

$$B^j_\alpha(\boldsymbol{R}_K) = -\frac{\partial W}{\partial m^K_\alpha}. \tag{23}$$

In Section 7.2.2 the interaction of a charge distribution with an external magnetic induction, $\boldsymbol{B}$, or with a nuclear magnetic moment, $\boldsymbol{m}^L$, was shown to lead to an induced current density $\boldsymbol{j}^{\mathrm{ind}}(\boldsymbol{r})$. According to equation (21) this also gives rise to an induced molecular magnetic induction $\boldsymbol{B}^{\mathrm{j,ind}}$

$$B^j_\alpha\left(\boldsymbol{R}, \boldsymbol{B}, \boldsymbol{m}^L\right) = B^{j,\mathrm{per}}_\alpha + B^{j,\mathrm{ind}}_\alpha\left(\boldsymbol{R}, \boldsymbol{B}, \boldsymbol{m}^L\right) = B^{j,\mathrm{per}}_\alpha - \sigma_{\alpha\beta}(\boldsymbol{R})B_\beta - K^L_{\alpha\beta}(\boldsymbol{R})m^L_\beta + \cdots, \tag{24}$$

where $\sigma_{\alpha\beta}(\boldsymbol{R})$ is the magnetic shielding tensor field [Jensen and Hansen (1999)] and $K^L_{\alpha\beta}(\boldsymbol{R})$ could be called a reduced spin–spin coupling tensor field. The precise definition of both as derivatives of the molecular magnetic induction is shown in Table 7.2. The value of the magnetic shielding tensor field at the position $\boldsymbol{R}_K$ of the nuclear magnetic dipole moment $\boldsymbol{m}^K$ is the well-known nuclear magnetic shielding tensor $\sigma^K_{\alpha\beta} = \sigma_{\alpha\beta}(\boldsymbol{R}_K)$ of nuclear magnetic resonance (NMR) spectroscopy. $K^{KL}_{\alpha\beta} = K^L_{\alpha\beta}(\boldsymbol{R}^K)$ is the reduced indirect nuclear spin–spin coupling tensor, which is related to the indirect nuclear spin–spin coupling tensor $J^{KL}_{\alpha\beta}$ of NMR spectroscopy by

$$J^{KL}_{\alpha\beta} = \frac{\gamma_K}{2\pi}\frac{\gamma_L}{2\pi}hK^{KL}_{\alpha\beta}, \tag{25}$$

where γ_K and γ_L are the gyromagnetic ratios of the two nuclei. The energy of the distribution of charges is obtained as usual by inserting equation (24) into equation (23) and integrating:

$$W\left(\boldsymbol{B}, \boldsymbol{m}^K, \boldsymbol{m}^L\right) = W^{(0)} - m_\alpha^K B_\alpha^{j,\text{per}} + \sigma_{\alpha\beta}^K m_\alpha^K B_\beta + K_{\alpha\beta}^{KL} m_\alpha^K m_\beta^L + \cdots. \tag{26}$$

The nuclear magnetic shielding $\sigma_{\alpha\beta}^K$ and reduced indirect nuclear spin–spin coupling tensors $K_{\alpha\beta}^{KL}$ can therefore be defined as energy derivatives as shown in Table 7.2.

7.2.4 TIME-DEPENDENT FIELDS

The discussion to this point has been limited to static electric and magnetic fields. However, molecules are often exposed to time-dependent fields, as for example in the interaction with electromagnetic radiation. Such interactions must be described by time-dependent molecular properties. The influence of a time-dependent magnetic induction vector $\boldsymbol{B}(\boldsymbol{r}, t)$ and its associated molecular properties have little physical significance and are therefore usually ignored. The electric field for a general polychromatic electromagnetic wave with wave vector $\boldsymbol{k}$ is given by

$$\boldsymbol{E}(\boldsymbol{r}, t) = \int_{-\infty}^{\infty} d\omega\, \boldsymbol{E}^\omega \cos(\boldsymbol{k}\cdot\boldsymbol{r} - \omega t) = \frac{1}{2}\int_{-\infty}^{\infty} d\omega\, \boldsymbol{E}^\omega \left(e^{i(\boldsymbol{k}\cdot\boldsymbol{r}-\omega t)} + e^{-i(\boldsymbol{k}\cdot\boldsymbol{r}-\omega t)}\right). \tag{27}$$

The expansion of the perturbation-dependent electric dipole moment in equation (9) then generalizes to [Buckingham (1967)]

$$\begin{aligned}\mu_\alpha(\boldsymbol{E}(t)) = \mu_\alpha &+ \int_{-\infty}^{\infty} d\omega_1\, \alpha_{\alpha\beta}(-\omega_1;\omega_1) E_\beta^{\omega_1} \cos(\boldsymbol{k}\cdot\boldsymbol{r}_0 - \omega_1 t) \\ &+ \frac{1}{2}\int_{-\infty}^{\infty} d\omega_1 \int_{-\infty}^{\infty} d\omega_2\, \beta_{\alpha\beta\gamma}(-\omega_1-\omega_2;\omega_1,\omega_2) E_\beta^{\omega_1} \cos(\boldsymbol{k}\cdot\boldsymbol{r}_0 - \omega_1 t) E_\gamma^{\omega_2} \\ &\times \cos(\boldsymbol{k}\cdot\boldsymbol{r}_0 - \omega_2 t) + \cdots.\end{aligned} \tag{28}$$

where $\alpha_{\alpha\beta}(-\omega_1;\omega_1)$ and $\beta_{\alpha\beta\gamma}(-\omega_1-\omega_2;\omega_1,\omega_2)$ are components of the frequency-dependent electric dipole polarizability and first hyperpolarizability tensors, respectively.

7.3 Exact quantum mechanical expressions

7.3.1 THE MOLECULAR ELECTRONIC HAMILTONIAN

Molecular properties were defined in the previous section as derivatives of the classical interaction energy of a system with respect to electric and magnetic fields and nuclear moments. Quantum mechanical expressions for these properties will now be derived. Since the properties are defined as energy derivatives, it is necessary to find a quantum mechanical expression for the energy of the system in the presence of a perturbing field.

A complete quantum mechanical treatment would involve quantum electrodynamics and so a number of initial simplifications are required. First, only the molecular electronic structure is treated quantum mechanically. The perturbing fields and nuclear moments are considered to be unaffected by the molecular environment, the so-called minimal coupling approximation. Secondly, the exposition will be restricted to non-relativistic quantum mechanics, i.e. to the Schrödinger equation. However, electron spin perturbations, which can only be derived from a relativistic standpoint, can also be treated by including *ad hoc* spin operators in the Schrödinger Hamiltonian. Finally, the Born–Oppenheimer approximation is applied, yielding a field-free electronic Schrödinger equation

$$\hat{H}^{(0)}\Psi_n^{(0)} = W_n^{(0)}\Psi_n^{(0)}, \tag{29}$$

where $\Psi_n^{(0)}$ is the many-electron wavefunction for state n with electronic energy $W_n^{(0)}$ and the molecular electronic Hamiltonian is defined as

$$\hat{H}^{(0)} = \frac{1}{2m_e}\sum_i \hat{p}_i^2 - \frac{e^2}{4\pi\varepsilon_0}\sum_{iK}\frac{Z_K}{|r_i - R_K|} + \frac{e^2}{4\pi\varepsilon_0}\sum_{i<j}\frac{1}{|r_i - r_j|}. \tag{30}$$

Only the electronic contributions to molecular properties will be obtained from this treatment. Contributions of the fixed nuclear multipoles have to be added afterwards according to the classical expressions given in Section 7.2.

Using minimal coupling, the vector potential enters the mechanical momentum of electron i,

$$m_e\hat{v}_i = \hat{p}_i + e\hat{A}(r_i), \tag{31}$$

and a scalar potential, $-e\phi(r_i)$, is added to the electronic Hamiltonian, yielding

$$\hat{H} = \frac{1}{2m_e}\sum_i\left(\hat{p}_i + e\hat{A}(r_i)\right)^2 - \frac{e^2}{4\pi\varepsilon_0}\sum_{iK}\frac{Z_K}{|r_i - R_K|} + \frac{e^2}{4\pi\varepsilon_0}\sum_{i<j}\frac{1}{|r_i - r_j|} - \sum_i e\phi(r_i). \tag{32}$$

In addition, electron spin is introduced via the Zeeman term of the Breit–Pauli Hamiltonian

$$\hat{H} = \sum_i \frac{g_e e}{2m_e}\hat{s}_i \cdot (\nabla \times A(r_i)). \tag{33}$$

$\hat{s}_i$ is the spin operator of electron i.

From equation (16) it can be seen that the vector potential can be chosen to be divergence free, i.e. $\nabla \cdot A = 0$, which is called the Coulomb gauge. The Hamiltonian can thus be written as

$$\begin{aligned}\hat{H} &= \hat{H}^{(0)} + \hat{H}^{(1)} + \hat{H}^{(2)} \\ &= \sum_i \hat{h}^{(0)}(i) + \sum_{i<j}\hat{g}(i,j) + \sum_i\left(\hat{h}^{(1)}(i) + \hat{h}^{(2)}(i)\right) \\ &= \hat{H}^{(0)} + \frac{e}{m_e}\sum_i \hat{A}(r_i)\cdot\hat{p}_i + \frac{g_e e}{2m_e}\sum_i \hat{s}_i\cdot(\nabla \times A(r_i)) - e\sum_i \phi(r_i) \\ &\quad + \frac{e^2}{2m_e}\sum_i \hat{A}^2(r_i).\end{aligned} \tag{34}$$

$\hat{H}$ is the full Hamiltonian for a molecule in an electromagnetic field, subject to the approximations outlined earlier. $\hat{H}^{(0)}$ is the unperturbed Hamiltonian, $\hat{H}^{(1)}$ includes all operators linear in the perturbing field (first order) and $\hat{H}^{(2)}$ all quadratic terms (second order). Explicit forms for the perturbing operators can be obtained by expressing the scalar and vector potentials in terms of the electric and magnetic fields. The scalar potential of an electric field is, equation (4),

$$\phi^E(r_i) = -(\hat{r}_i - R_O)\cdot E(r_i), \tag{35}$$

where the scalar potential at the origin of the coordinate system, $\phi^E(R_O)$, is set to zero. For the vector potential,

$$A(r_i) = \tfrac{1}{2}B \times (\hat{r}_i - R_{GO}), \tag{36}$$

where R_{GO} is an arbitrary gauge origin. The vector potential of the nuclear magnetic dipole moment of nucleus K is

$$A^K(r_i) = \frac{\mu_0}{4\pi}m^K \times \frac{\hat{r}_i - R_K}{|\hat{r}_i - R_K|^3} \tag{37}$$

(see equation (13)), so that the nuclear magnetic induction is given by

$$\boldsymbol{B}^K(\boldsymbol{r}_i) = \frac{\mu_0}{4\pi}\left(\frac{\left(\boldsymbol{m}^K\cdot(\hat{\boldsymbol{r}}_i-\boldsymbol{R}_K)\right)(\hat{\boldsymbol{r}}_i-\boldsymbol{R}_K)}{|\hat{\boldsymbol{r}}_i-\boldsymbol{R}_K|^5} - \frac{\boldsymbol{m}^K}{|\hat{\boldsymbol{r}}_i-\boldsymbol{R}_K|^3}\right) + \frac{\mu_0}{4\pi}\frac{8\pi}{3}\delta(\boldsymbol{r}_i-\boldsymbol{R}_K)\boldsymbol{m}^K. \tag{38}$$

After some manipulation, the following expressions for the first- and second-order Hamiltonians are obtained:

$$\hat{H}^{(1)} = -(\hat{O}^{lB}_\alpha + \hat{O}^{sB}_\alpha)B_\alpha - \sum_K(\hat{O}^{lm^K}_\alpha + \hat{O}^{sm^K}_\alpha)m^K_\alpha - \hat{O}^E_\alpha E_\alpha, \tag{39}$$

$$\hat{H}^{(2)} = \hat{O}^{BB}_{\alpha\beta}\,B_\alpha B_\beta + \sum_K \hat{O}^{m^K B}_{\alpha\beta}\,m^K_\alpha B_\beta + \sum_{KL}\hat{O}^{m^K m^L}_{\alpha\beta}\,m^K_\alpha m^L_\beta. \tag{40}$$

The interaction operators $\hat{O}$ are thereby defined as

$$\hat{O}^{lB}_\alpha = \sum_i \hat{o}^{lB}_\alpha(i) = -\frac{e}{2m_e}\sum_i\left((\hat{\boldsymbol{r}}_i-\boldsymbol{R}_{\mathrm{GO}})\times\hat{\boldsymbol{p}}_i\right)_\alpha, \tag{41}$$

$$\hat{O}^{sB}_\alpha = \sum_i \hat{o}^{sB}_\alpha(i) = -\frac{g_e e}{2m_e}\sum_i \hat{s}_{i,\alpha}, \tag{42}$$

$$\hat{O}^{lm^K}_\alpha = \sum_i \hat{o}^{lm^K}_\alpha(i) = -\frac{e}{m_e}\frac{\mu_0}{4\pi}\sum_i\left(\frac{\hat{\boldsymbol{r}}_i-\boldsymbol{R}_K}{|\hat{\boldsymbol{r}}_i-\vec{\boldsymbol{R}}_K|^3}\times\hat{\boldsymbol{p}}_i\right)_\alpha, \tag{43}$$

$$\hat{O}^{sm^K}_\alpha = \sum_i \hat{o}^{sm^K}_\alpha(i) = -\frac{g_e e}{2m_e}\frac{\mu_0}{4\pi}\frac{8\pi}{3}\sum_i\delta(\boldsymbol{r}_i-\boldsymbol{R}_K)\hat{s}_{i,\alpha} - \frac{g_e e}{2m_e}\frac{\mu_0}{4\pi}\sum_i\left(\frac{(\hat{\boldsymbol{s}}_i\cdot(\hat{\boldsymbol{r}}_i-\boldsymbol{R}_K))(\hat{r}_{i,\alpha}-R_{K,\alpha})}{|\hat{\boldsymbol{r}}_i-\boldsymbol{R}_K|^5} - \frac{\hat{s}_{i,\alpha}}{|\hat{\boldsymbol{r}}_i-\boldsymbol{R}_K|^3}\right), \tag{44}$$

$$\hat{O}^E_\alpha = \sum_i \hat{o}^E_\alpha(i) = -e\sum_i(\hat{r}_{i,\alpha}-R_{\mathrm{O},\alpha}), \tag{45}$$

$$\hat{O}^{BB}_{\alpha\beta} = \sum_i \hat{o}^{BB}_{\alpha\beta}(i) = \frac{e^2}{8m_e}\sum_i\left((\hat{\boldsymbol{r}}_i-\boldsymbol{R}_{\mathrm{GO}})^2\delta_{\alpha\beta} - (\hat{r}_{i,\alpha}-R_{\mathrm{GO},\alpha})(\hat{r}_{i,\beta}-R_{\mathrm{GO},\beta})\right), \tag{46}$$

$$\hat{O}^{m^K B}_{\alpha\beta} = \sum_i \hat{o}^{m^K B}_{\alpha\beta}(i) = \frac{e^2}{2m_e}\frac{\mu_0}{4\pi}\sum_i\left((\hat{\boldsymbol{r}}_i-\boldsymbol{R}_{\mathrm{GO}})\cdot\frac{(\hat{\boldsymbol{r}}_i-\boldsymbol{R}_K)}{|\hat{\boldsymbol{r}}_i-\boldsymbol{R}_K|^3}\delta_{\alpha\beta} - (\hat{r}_{i,\alpha}-R_{\mathrm{GO},\alpha})\frac{(\hat{r}_{i,\beta}-R_{K,\beta})}{|\hat{\boldsymbol{r}}_i-\boldsymbol{R}_K|^3}\right), \tag{47}$$

$$\hat{O}^{m^K m^L}_{\alpha\beta} = \sum_i \hat{o}^{m^K m^L}_{\alpha\beta}(i) = \frac{e^2}{2m_e}\left(\frac{\mu_0}{4\pi}\right)^2\sum_i\left(\frac{(\hat{\boldsymbol{r}}_i-\boldsymbol{R}_L)}{|\hat{\boldsymbol{r}}_i-\boldsymbol{R}_L|^3}\cdot\frac{(\hat{\boldsymbol{r}}_i-\boldsymbol{R}_K)}{|\hat{\boldsymbol{r}}_i-\boldsymbol{R}_K|^3}\delta_{\alpha\beta} - \frac{(\hat{r}_{i,\alpha}-R_{L,\alpha})}{|\hat{\boldsymbol{r}}_i-\boldsymbol{R}_L|^3}\frac{(\hat{r}_{i,\beta}-R_{K,\beta})}{|\hat{\boldsymbol{r}}_i-\boldsymbol{R}_K|^3}\right). \tag{48}$$

7.3.2 RAYLEIGH–SCHRÖDINGER PERTURBATION THEORY

In Section 7.2 electronic contributions to molecular properties were defined as derivatives of the electronic energy. In order to take derivatives one needs to determine the energy of a system described by the Hamiltonian given in equation (34). In general, however, it is not possible to find exact solutions for this Hamiltonian, i.e. to solve the Schrödinger equation

$$\hat{H}\Psi_0 = W_0\Psi_0. \tag{49}$$

Instead, perturbation or variational methods must be applied to approximate wavefunctions.

In non-degenerate Rayleigh–Schrödinger perturbation theory one assumes that exact energies and wavefunctions of the unperturbed Hamiltonian $\hat{H}^{(0)}$ are known, i.e. that equation (29) has been solved exactly, and that the energies and wavefunctions of the total Hamiltonian, equation (34), can be expanded in a series using the exact solutions of the unperturbed Hamiltonian, $W_n^{(0)}$ and $\Psi_n^{(0)}$, as zeroth-order energies and wavefunctions

$$W_0 = W_0^{(0)} + W_0^{(1)} + W_0^{(2)} + \cdots, \tag{50}$$

$$\Psi_0 = \Psi_0^{(0)} + \Psi_0^{(1)} + \Psi_0^{(2)} + \cdots. \tag{51}$$

First- and second-order corrections to the energy are then given as

$$W_0^{(1)} = \left\langle \Psi_0^{(0)} \middle| \hat{H}^{(1)} \middle| \Psi_0^{(0)} \right\rangle, \tag{52}$$

$$W_0^{(2)} = \left\langle \Psi_0^{(0)} \middle| \hat{H}^{(2)} \middle| \Psi_0^{(0)} \right\rangle + \left\langle \Psi_0^{(0)} \middle| \hat{H}^{(1)} \middle| \Psi_0^{(1)} \right\rangle. \tag{53}$$

Expanding the first-order wavefunction in the complete set of unperturbed wavefunctions $\{\Psi_n^{(0)}\}$

$$\Psi_0^{(1)} = \sum_{n\neq 0} \left|\Psi_n^{(0)}\right\rangle \frac{\left\langle \Psi_n^{(0)} \middle| \hat{H}^{(1)} \middle| \Psi_0^{(0)} \right\rangle}{W_0^{(0)} - W_n^{(0)}}, \tag{54}$$

the second-order correction to the energy becomes

$$W_0^{(2)} = \left\langle \Psi_0^{(0)} \middle| \hat{H}^{(2)} \middle| \Psi_0^{(0)} \right\rangle + \sum_{n\neq 0} \frac{\left\langle \Psi_0^{(0)} \middle| \hat{H}^{(1)} \middle| \Psi_n^{(0)} \right\rangle \left\langle \Psi_n^{(0)} \middle| \hat{H}^{(1)} \middle| \Psi_0^{(0)} \right\rangle}{W_0^{(0)} - W_n^{(0)}}. \tag{55}$$

After insertion of the first- and second-order Hamiltonians, equations (39) and (40), in equations (52) and (55), expressions for the molecular properties defined in equations (8), (18), and (23) and Tables 7.1 and 7.2 are obtained by taking the appropriate derivatives of W_0 or equivalently by comparison with the classical expressions for the interaction energies (10), (20) and (26). The first-order molecular properties are then given as

$$\mu_\alpha^{\text{el}}(\boldsymbol{R}_{\text{O}}) = \left\langle \Psi_0^{(0)} \middle| \hat{O}_\alpha^{E} \middle| \Psi_0^{(0)} \right\rangle, \tag{56}$$

$$m_\alpha^{\text{el}}(\boldsymbol{R}_{\text{GO}}) = \left\langle \Psi_0^{(0)} \middle| \hat{O}_\alpha^{lB} + \hat{O}_\alpha^{sB} \middle| \Psi_0^{(0)} \right\rangle, \tag{57}$$

$$B_\alpha^{j,\text{el}}(\boldsymbol{R}_K) = \left\langle \Psi_0^{(0)} \middle| \hat{O}_\alpha^{lm^K} + \hat{O}_\alpha^{sm^k} \middle| \Psi_0^{(0)} \right\rangle, \tag{58}$$

and the second-order molecular properties are given as

$$P^{(2)} = f_1 \left\langle \Psi_0^{(0)} \middle| \hat{O}_1 \middle| \Psi_0^{(0)} \right\rangle + f_2 \sum_{n\neq 0} \frac{\left\langle \Psi_0^{(0)} \middle| \hat{O}_2 \middle| \Psi_n^{(0)} \right\rangle \left\langle \Psi_n^{(0)} \middle| \hat{O}_3 \middle| \Psi_0^{(0)} \right\rangle}{W_0^{(0)} - W_n^{(0)}}$$

$$+ f_2 \sum_{n\neq 0} \frac{\left\langle \Psi_0^{(0)} \middle| \hat{O}_3 \middle| \Psi_n^{(0)} \right\rangle \left\langle \Psi_n^{(0)} \middle| \hat{O}_2 \middle| \Psi_0^{(0)} \right\rangle}{W_0^{(0)} - W_n^{(0)}}, \tag{59}$$

Table 7.3. Operators and factors for the exact second-order Rayleigh–Schrödinger perturbation theory expressions for molecular properties. See equation (59).

$P^{(2)a}$	f_1	$\hat{O}_1$	f_2	$\hat{O}_2$	$\hat{O}_3$
$\alpha_{\alpha\beta}$	—	—	-1	$\hat{O}^E_\alpha$	$\hat{O}^E_\beta$
$\xi_{\alpha\beta}(\boldsymbol{R}_{\mathrm{GO}})$	-2	$\hat{O}^{BB}_{\alpha\beta}$	-1	$\hat{O}^{lB}_\alpha$	$\hat{O}^{lB}_\beta$
$\sigma^K_{\alpha\beta}(\boldsymbol{R}_{\mathrm{GO}})$	1	$\hat{O}^{m^K B}_{\alpha\beta}$	1	$\hat{O}^{lm^K}_\alpha$	$\hat{O}^{lB}_\beta$
$K^{KL}_{\alpha\beta}$	2	$\hat{O}^{m^K m^L}_{\alpha\beta}$	1	$\hat{O}^{lm^K}_\alpha + \hat{O}^{sm^K}_\alpha$	$\hat{O}^{lm^L}_\beta + \hat{O}^{sm^L}_\beta$

[a] α, electronic dipole polarizability; $\xi_{\alpha\beta}(\boldsymbol{R}_{\mathrm{GO}})$, electronic dipole magnetizability; $\sigma^K_{\alpha\beta}(\boldsymbol{R}_{\mathrm{GO}})$, nuclear magnetic shielding constant; $K^{KL}_{\alpha\beta}$, spin–spin coupling constant.

where the operators $\hat{O}_1$, $\hat{O}_2$ and $\hat{O}_3$ and the factors f_1 and f_2 are collected in Table 7.3. Since non-degenerate perturbation theory is used, the ground state $|\Psi_0^{(0)}\rangle$ is assumed to be a singlet state and to have no orbital degeneracies. This implies that

$$\left\langle \Psi_0^{(0)} \middle| \hat{O}^{sB}_\alpha \middle| \Psi_0^{(0)} \right\rangle = \left\langle \Psi_0^{(0)} \middle| \hat{O}^{sm^K}_\alpha \middle| \Psi_0^{(0)} \right\rangle = \left\langle \Psi_0^{(0)} \middle| \hat{O}^{sB}_\alpha \middle| \Psi_n^{(0)} \right\rangle \left\langle \Psi_0^{(0)} \middle| \hat{O}^{lB}_\alpha \middle| \Psi_0^{(0)} \right\rangle$$
$$= \left\langle \Psi_0^{(0)} \middle| \hat{O}^{lm^k}_\alpha \middle| \Psi_0^{(0)} \right\rangle = 0, \tag{60}$$

and the corresponding terms are not included in Table 7.3. The expressions for the molecular magnetic properties are thus valid only for the case of closed-shell molecules.

7.3.3 STATIC RESPONSE THEORY

It was shown in Section 7.2 that molecular properties such as the dipole polarizability can also be defined as derivatives of the corresponding perturbation-dependent multipole moments. One therefore needs to find quantum mechanical operators for the perturbation-dependent electric and magnetic moments and the molecular magnetic induction. These can be obtained by application of the Hellmann–Feynman theorem, which states that for a hermitian operator $\hat{H}(\lambda)$ depending on a real parameter λ with normalized eigenfunction $\Psi(\lambda)$ and eigenvalue $W(\lambda)$, the derivative of $W(\lambda)$ with respect to λ is given as

$$\frac{\mathrm{d}}{\mathrm{d}\lambda} W(\lambda) = \langle \Psi(\lambda) | \frac{\partial}{\partial \lambda} \hat{H}(\lambda) | \Psi(\lambda) \rangle . \tag{61}$$

Taking the appropriate derivatives, equations (8), (18) and (23), of the Hamiltonian in equation (34) leads to the following expressions for perturbation-dependent operators of the electronic contribution to the electric dipole moment, the magnetic dipole moment and the molecular magnetic induction

$$\hat{\mu}^{\mathrm{el}}_\alpha = \hat{O}^E_\alpha , \tag{62}$$

$$\hat{m}^{\mathrm{el}}_\alpha = \hat{O}^{lB}_\alpha + \hat{O}^{sB}_\alpha - 2\hat{O}^{BB}_{\alpha\beta} B_\beta - \sum_K \hat{O}^{m^K B}_{\beta\alpha} m^K_\beta , \tag{63}$$

$$\hat{B}^j_\alpha(\boldsymbol{R}_K) = \hat{O}^{lm^K}_\alpha + \hat{O}^{sm^K}_\alpha - \hat{O}^{m^K B}_{\alpha\beta} B_\beta - \sum_L \hat{O}^{m^K m^L}_{\beta\alpha} m^L_\beta . \tag{64}$$

Having thus defined operators which are valid in the presence of external and internal perturbations, one can proceed by evaluating expectation values of these operators with the perturbed wavefunctions in equation (51)

$$\langle \Psi_0 | \hat{O} | \Psi_0 \rangle = \left\langle \Psi_0^{(0)} \middle| \hat{O} \middle| \Psi_0^{(0)} \right\rangle + \left\langle \Psi_0^{(1)} \middle| \hat{O} \middle| \Psi_0^{(0)} \right\rangle + \left\langle \Psi_0^{(0)} \middle| \hat{O} \middle| \Psi_0^{(1)} \right\rangle + \cdots . \tag{65}$$

The molecular properties are obtained by taking derivatives of the expectation value, as given in Tables 7.1 and 7.2, or by comparison with the classical expression for the perturbation-dependent moments and fields in equations (9), (19) and (24). One should note that molecular properties such as the polarizability, which were called second-order properties in the context of Rayleigh–Schrödinger perturbation theory, are named linear response functions here and in the following section. For exact wavefunctions the expressions obtained in the previous section, equation (59) and Table 7.3, can be recovered from equation (65) by insertion of the first-order correction to the wavefunction, equation (54). This so-called response theory approach [Olsen and Jørgensen (1985)] can also be generalized to the case of time-dependent perturbations.

7.3.4 TIME-DEPENDENT RESPONSE THEORY

When dealing with time-dependent fields one has to find solutions to the time-dependent Schrödinger equation

$$i\frac{\partial}{\partial t}\Psi_0(t) = \left(\hat{H}^{(0)} + \hat{H}^{(1)}(t)\right)\Psi_0(t), \tag{66}$$

where the time-dependent Hamiltonian $\hat{H}^{(1)}(t)$ can be written as

$$\hat{H}^{(1)}(t) = \hat{V}_{\alpha\ldots}F(t)_{\alpha\ldots}, \tag{67}$$

or can be expressed in terms of its Fourier components

$$\hat{H}^{(1)}(t) = \int_{-\infty}^{\infty} \mathrm{d}\omega\,\hat{H}^{(1)}(\omega)\,\mathrm{e}^{-\mathrm{i}\omega t} = \int_{-\infty}^{\infty} \mathrm{d}\omega\,\hat{V}^{\omega}_{\alpha\ldots}F(\omega)_{\alpha\ldots}\,\mathrm{e}^{-\mathrm{i}\omega t}. \tag{68}$$

The operator $\hat{V}_{\alpha\ldots}$ may depend on coordinates and momenta of the electrons but is independent of time, whereas the time-dependent field $F(t)_{\alpha\ldots}$ does not depend on the electrons. The subscript $\alpha\ldots$ denotes components of a tensor of appropriate rank.

In time-dependent response theory, as described by Zubarev (1974), properties are evaluated as expectation values. The time-dependent expectation value of an operator $\hat{O}$ is thus formally expanded in a series

$$\begin{aligned}\langle\Psi_0(t)|\hat{O}|\Psi_0(t)\rangle &= \left\langle\Psi_0^{(0)}\middle|\hat{O}\middle|\Psi_0^{(0)}\right\rangle + \int_{-\infty}^{\infty}\mathrm{d}t'\left\langle\left\langle\hat{O}^t;\hat{V}^{t'}_{\alpha\ldots}\right\rangle\right\rangle F(t')_{\alpha\ldots}\\ &+\frac{1}{2}\int_{-\infty}^{\infty}\mathrm{d}t'\int_{-\infty}^{\infty}\mathrm{d}t''\left\langle\left\langle\hat{O}^t;\hat{V}^{t'}_{\alpha\ldots},\hat{V}^{t''}_{\beta\ldots}\right\rangle\right\rangle F(t')_{\alpha\ldots}F(t'')_{\beta\ldots}+\cdots,\end{aligned} \tag{69}$$

where $\hat{O}^t = \mathrm{e}^{\mathrm{i}\hat{H}^{(0)}t/\hbar}\hat{O}\,\mathrm{e}^{-\mathrm{i}\hat{H}^{(0)}t/\hbar}$ denotes the Heisenberg representation of operator $\hat{O}$. This representation unitarily transforms the time dependence into the operators so that the wavefunction is time independent. This enables solutions to the time-independent Schrödinger equation to be utilized. The kernels of the integrals are the linear $\langle\langle\hat{O}^t;\hat{V}^{t'}_{\alpha\ldots}\rangle\rangle$ and quadratic response functions $\langle\langle\hat{O}^t;\hat{V}^{t'}_{\alpha\ldots},\hat{V}^{t''}_{\beta\ldots}\rangle\rangle$ in the time domain, where the double bracket notation, $\langle\langle\ldots;\ldots\rangle\rangle$, is used to denote a general response function. The linear response is also called the polarization propagator (see Section 7.4.1.3). Alternatively, by Fourier transforming the response functions to the frequency or energy domain, the expansion can be written as

$$\begin{aligned}\langle\Psi_0(t)|\hat{O}|\Psi_0(t)\rangle &= \left\langle\Psi_0^{(0)}\middle|\hat{O}\middle|\Psi_0^{(0)}\right\rangle + \int_{-\infty}^{\infty}\mathrm{d}\omega_1\,\mathrm{e}^{-\mathrm{i}\omega_1 t}\left\langle\left\langle\hat{O};\hat{V}^{\omega_1}_{\alpha\ldots}\right\rangle\right\rangle_{\omega_1}F(\omega_1)_{\alpha\ldots}+\frac{1}{2}\int_{-\infty}^{\infty}\mathrm{d}\omega_1\,\mathrm{e}^{-\mathrm{i}\omega_1 t}\\ &\times\int_{-\infty}^{\infty}\mathrm{d}\omega_2\,\mathrm{e}^{-\mathrm{i}\omega_2 t}\left\langle\left\langle\hat{O};\hat{V}^{\omega_1}_{\alpha\ldots},\hat{V}^{\omega_2}_{\beta\ldots}\right\rangle\right\rangle_{\omega_1,\omega_2}F(\omega_1)_{\alpha\ldots}F(\omega_2)_{\beta\ldots}+\cdots.\end{aligned} \tag{70}$$

Expressions for the response functions are obtained by standard time-dependent perturbation theory in which the set of exact eigenfunctions $\{\Psi_n^{(0)}\}$ of the unperturbed Hamiltonian $\hat{H}^{(0)}$ are used as basis

$$|\Psi_0(t)\rangle = \frac{\left|\Psi_0^{(0)}\right\rangle + \sum_{n\neq 0} \left|\Psi_n^{(0)}\right\rangle C_n(t)}{\sqrt{1 + \sum_{n\neq 0} C_n^*(t) C_n(t)}}, \tag{71}$$

and the time-dependent coefficients are expanded in a perturbation series

$$\left\langle \Psi_n^{(0)} \middle| \Psi_0(t) \right\rangle = C_n(t) = C_n^{(1)}(t) + C_n^{(2)}(t) + \cdots. \tag{72}$$

The individual terms in the expansion of the coefficients are then determined from the time-dependent Schrödinger equation or a time-dependent variation principle. Evaluating the time-dependent expectation value $\langle \Psi_0(t)|\hat{O}|\Psi_0(t)\rangle$ of an operator $\hat{O}$ with the wavefunction given in equation (71) and collecting terms of the same order allows one to identify the response functions in equation (69). The Fourier transform of the linear response function to the frequency domain gives

$$\left\langle\left\langle \hat{O}; \hat{V}_{\alpha\ldots}^{\omega_1} \right\rangle\right\rangle_{\omega_1} = \sum_{n\neq 0} \left\{ \frac{\left\langle \Psi_0^{(0)} \middle| \hat{O} \middle| \Psi_n^{(0)} \right\rangle \left\langle \Psi_n^{(0)} \middle| \hat{V}_{\alpha\ldots}^{\omega_1} \middle| \Psi_0^{(0)} \right\rangle}{\hbar\omega_1 + W_0^{(0)} - W_n^{(0)}} + \frac{\left\langle \Psi_0^{(0)} \middle| \hat{V}_{\alpha\ldots}^{\omega_1} \middle| \Psi_n^{(0)} \right\rangle \left\langle \Psi_n^{(0)} \middle| \hat{O} \middle| \Psi_0^{(0)} \right\rangle}{-\hbar\omega_1 + W_0^{(0)} - W_n^{(0)}} \right\}. \tag{73}$$

In the limit $\omega \to 0$ this expression becomes equivalent to the second-order energy from Rayleigh–Schrödinger perturbation theory. In this case the property is termed a linear response since it is linear in the perturbing operator $\hat{V}_{\alpha\ldots}^{\omega_1}$, having expanded about the expectation value of operator $\hat{O}$.

Time-dependent response theory can also be developed using the generalized Ehrenfest theorem [Olsen and Jørgensen (1985)] or using the Frenkel variation principle [Frenkel (1934), and Pickup (1992)]. These various approaches, which ultimately produce equivalent results for a variationally optimized wavefunction, are appropriate to different types of quantum mechanical method, as will be illustrated in Section 7.4.

The most important application of time-dependent perturbation theory in the context of molecular properties is to evaluate the time-dependent expectation value of the electric dipole operator $\langle \Psi_0(t)|\hat{\mu}_\alpha^{\mathrm{el}}|\Psi_0(t)\rangle$ in the presence of a monochromatic electromagnetic wave of frequency ω_0. Employing the dipole approximation, which implies setting $\boldsymbol{k}\cdot\boldsymbol{r} = 0$ in equation (27), the perturbation Hamilton operator for the periodic and spatially uniform electric field of the electromagnetic wave is

$$\hat{H}^{(1)}(t) = -\hat{O}_\alpha^E E_\alpha(t) = -\hat{O}_\alpha^E \frac{E_\alpha^{\omega_0}}{2}\left(\mathrm{e}^{-\mathrm{i}\omega_0 t} + \mathrm{e}^{\mathrm{i}\omega_0 t}\right). \tag{74}$$

Comparison with equation (68) shows that the Fourier components of the operator and the field are

$$\hat{V}_{\alpha\ldots}^{\omega} = -\hat{O}_\alpha^E, \tag{75}$$

$$F(\omega)_{\alpha\ldots} = \frac{E_\alpha^{\omega_0}}{2}[\delta(\omega - \omega_0) + \delta(\omega + \omega_0)]. \tag{76}$$

Insertion of these operators into equation (70) yields

$$\langle \Psi_0(t)|\, \hat{\mu}_\alpha^{\mathrm{el}}\, |\Psi_0(t)\rangle = \left\langle \Psi_0^{(0)} \middle| \hat{O}_\alpha^E \middle| \Psi_0^{(0)} \right\rangle + \left(\left\langle\left\langle \hat{O}_\alpha^E; -\hat{O}_\beta^E \right\rangle\right\rangle_{\omega_0} \mathrm{e}^{-\mathrm{i}\omega_0 t} + \left\langle\left\langle \hat{O}_\alpha^E; -\hat{O}_\beta^E \right\rangle\right\rangle_{-\omega_0} \mathrm{e}^{\mathrm{i}\omega_0 t} \right) \frac{E_\beta^{\omega_0}}{2} + \cdots. \tag{77}$$

From the definition of the linear response function in the frequency domain, equation (73), it can be seen that for the hermitian and real operators $\hat{O}^E_\alpha$ and $\hat{O}^E_\beta$ the two response functions $\langle\langle\hat{O}^E_\alpha; -\hat{O}^E_\beta\rangle\rangle_{\omega_0}$ and $\langle\langle\hat{O}^E_\alpha; -\hat{O}^E_\beta\rangle\rangle_{-\omega_0}$ become equal. Thus

$$\langle\Psi_0(t)|\hat{\mu}^{\mathrm{el}}_\alpha|\Psi_0(t)\rangle = \left\langle \Psi_0^{(0)}\middle| \hat{O}^E_\alpha \middle|\Psi_0^{(0)}\right\rangle + \left\langle\left\langle \hat{O}^E_\alpha; -\hat{O}^E_\beta\right\rangle\right\rangle_{\omega_0} E^{\omega_0}_\alpha \cos(\omega_0 t) + \cdots, \tag{78}$$

which can be compared with the classical expansion for a time-dependent dipole moment in equation (28). The frequency-dependent polarizability tensor is then identified as

$$\alpha_{\alpha\beta}(-\omega_0;\omega_0) = -\left\langle\left\langle \hat{O}^E_\alpha; \hat{O}^E_\beta\right\rangle\right\rangle_{\omega_0}. \tag{79}$$

For $\omega_0 = 0$ this reduces to the expression obtained by static response theory or Rayleigh–Schrödinger perturbation theory.

7.4 *Ab initio* methods for the calculation of molecular properties

Ab initio methods for the calculation of molecular electromagnetic properties can be categorized according to several criteria.

- Expressions for the molecular properties can be derived either as derivatives of the electronic energy or as derivatives of molecular electromagnetic moments and fields. This distinction is important for some approximate wavefunctions which do not obey the Hellmann–Feynman theorem [equation (61)]. In general, for an approximate wavefunction $\Phi(\{C_i(\lambda)\})$, which typically depends on a set of parameters $\{C_i(\lambda)\}$ consisting of molecular orbital and configuration coefficients, the derivative of the electronic energy $W(\{C_i(\lambda)\}, \lambda)$ with respect to a real parameter λ in the Hamiltonian $\hat{H}(\lambda)$ is

$$\frac{\mathrm{d}W(\{C_i(\lambda)\},\lambda)}{\mathrm{d}\lambda} = \frac{\partial W(\{C_i(\lambda)\},\lambda)}{\partial\lambda} + \sum_i \left(\frac{\partial(W(\{C_i(\lambda)\},\lambda)}{\partial C_i(\lambda)}\right)\left(\frac{\partial C_i(\lambda)}{\partial\lambda}\right). \tag{80}$$

  If the wavefunction is variationally optimized with respect to all parameters, i.e. $\partial W(\{C_i(\lambda)\}, \lambda)/\partial C_i(\lambda) = 0$, as is the case for a self-consistent field (SCF) and multiconfigurational self-consistent field (MCSCF) wavefunction, the Hellmann–Feynman theorem is satisfied. Truncated configuration interaction (CI) wavefunctions, by contrast, are not variationally optimized with respect to the molecular orbital coefficients. The Hellmann–Feynman theorem is therefore satisfied only in the limit of a full CI wavefunction, when the molecular orbital coefficients are redundant. Similar problems arise for non-variational approaches such as Møller–Plesset perturbation theory (MP) and coupled-cluster (CC). This means that first-order properties obtained as a derivative of the energy for truncated CI or perturbation theory wavefunctions, do not agree with the expectation value of the corresponding operator. It also follows that second- and higher-order properties will also depend on the use of either energy derivatives or derivatives of the expectation value. It is, however, possible to redefine the expectation value for coupled-cluster wavefunctions to remove this discrepancy. Arponen (1983) defined an energy functional which consists of a transition expectation value between the coupled-cluster state and a dual bra state [see equations (122)–(128)], which is stationary with respect to the wavefunction parameters and which therefore satisfies the Hellman–Feynman theorem.
- A distinction can be made between methods which evaluate properties based on perturbation theory, with a partitioning of the Hamiltonian and a series expansion of the wavefunction (Sections 7.4.1–7.4.2) and those which use derivatives of the energy or first-order expectation

values (see Section 7.4.3). The former methods can be further classified according to whether approximate wavefunctions are substituted for the exact wavefunction in the expressions obtained in Sections 7.3.2–7.3.4 or whether perturbation theory is applied directly to approximate wavefunctions. In the case of a variational wavefunction these two alternatives yield identical results–for example, random phase approximation (RPA) and coupled Hartree–Fock (CHF) are formally equivalent [McWeeny (1992)]). Non-variational wavefunctions, on the other hand, give entirely different formulations and results.

- Finally, not all methods are capable of treating time-dependent perturbations to yield frequency-dependent properties. Even though most *ab initio* approaches have now been generalized to the time-dependent case, the pole structure (i.e. values of the perturbing frequency, ω, for which the denominator in equation (73) goes to zero) of MP-based properties can differ (Section 7.4.3.2), giving very different behaviour in the region of resonances.

The following notational conventions will be adopted. The one-electron spatial functions which are solutions to the closed shell Hartree–Fock (HF) equations

$$f(i)\phi_p(i) = \varepsilon_p \phi_p(i) \tag{81}$$

are denoted $\{\phi_p\}$ with Latin indices and are called molecular orbitals. Occupied and unoccupied spatial molecular orbitals are denoted with the indices $i, j, k, \ldots$ and $a, b, c, \ldots$, respectively, while the indices of general spatial molecular orbitals are denoted by $p, q, r, \ldots$ The Fock operator is defined as

$$\hat{F} = \sum_i f(i) = \sum_i \left(\hat{h}^{(0)}(i) + \hat{v}^{\mathrm{HF}}(i)\right), \tag{82}$$

were $\hat{v}^{\mathrm{HF}}(i)$ is an effective one-electron potential, called the HF potential. ε_p is the molecular orbital energy. In the Roothaan–Hartree–Fock approach [Roothaan (1951)] the molecular orbitals are expanded in a basis of one-electron functions, $\{\chi_\mu\}$, denoted by Greek indices, called atomic orbitals (although there is no restriction on their position within the molecule)

$$\phi_p = \sum_\mu \chi_\mu c_{\mu p}, \tag{83}$$

where $\{c_{\mu p}\}$ are the molecular orbital coefficients. One of the most widely-used methods for treating the electron correlation missing in the HF wavefunction is MP perturbation theory [Møller and Plesset (1934), and Pople, Binkley and Seeger (1976)]. The field-free Hamiltonian $\hat{H}^{(0)}$ is partitioned to give the Fock operator $\hat{F}$ and the so-called fluctuation potential $\hat{V}$

$$\hat{H}^{(0)} = \hat{F} + \hat{V} \tag{84}$$

and the wavefunction is expanded in a perturbation series in $\hat{V}$

$$\left|\Psi_0^{(0)}\right\rangle = N\left(|\Phi_{\mathrm{SCF}}\rangle + \left|\Phi^{(1)}\right\rangle + \left|\Phi^{(2)}\right\rangle \ldots\right), \tag{85}$$

where N is a normalization constant and the zeroth-order wavefunction is the single determinant SCF wavefunction, $|\Phi_{\mathrm{SCF}}\rangle$, i.e. the antisymmetrized product of the occupied molecular orbitals $\{\phi_i\}$. The first-order MP correction to the wavefunction $|\Phi^{(1)}\rangle$ consists of determinants which are doubly excited with respect to $|\Phi_{\mathrm{SCF}}\rangle$

$$\left|\Phi^{(1)}\right\rangle = \left|\mathrm{DE}^{(1)}\right\rangle = \frac{1}{4}\sum_{\substack{ai\\bj}} \kappa_{ij}^{ab}\left|\Phi_{ij}^{ab}\right\rangle, \tag{86}$$

while the second-order MP correction to the wavefunction contains singly, doubly, triply and quadruply excited determinants

$$|\Phi^{(2)}\rangle = |\mathrm{SE}^{(2)}\rangle + |\mathrm{DE}^{(2)}\rangle + |\mathrm{TE}^{(2)}\rangle + |\mathrm{QE}^{(2)}\rangle, \tag{87}$$

with e.g.

$$|\mathrm{SE}^{(2)}\rangle = \sum_{ai} \kappa_i^a |\Phi_i^a\rangle. \tag{88}$$

The absence of single excitations in the first-order wavefunction is a consequence of the Brillouin condition on the SCF wavefunction, which is stationary with respect to any mixing of the occupied and virtual orbitals.

7.4.1 APPROXIMATIONS TO EXACT PERTURBATION THEORY

7.4.1.1 Ground state expectation values

In the current context, exact refers to the ground state wavefunction used in Rayleigh–Schrödinger perturbation theory. Computational development from this theory requires approximate wavefunctions to be substituted for the exact wavefunction. According to equations (52) or (65) first-order properties can be evaluated as ground state expectation values. An alternative expression can be obtained by introducing the one-electron density matrix in the basis of the molecular $\{\phi_p\}$ or atomic orbitals $\{\chi_\mu\}$

$$D_{pq} = \int \mathrm{d}\boldsymbol{r}_1\, \mathrm{d}\boldsymbol{r}_1'\, \mathrm{d}s_1\, \mathrm{d}s_1'\, \phi_p^*(\boldsymbol{r}_1)\gamma(\boldsymbol{x}_1, \boldsymbol{x}_1')\phi_q(\boldsymbol{r}_1'), \tag{89}$$

$$D_{\mu\nu} = \int \mathrm{d}\boldsymbol{r}_1\, \mathrm{d}\boldsymbol{r}_1'\, \mathrm{d}s_1,\, \mathrm{d}s_1'\, \chi_\mu^*(\boldsymbol{r}_1)\gamma(\boldsymbol{x}_1, \boldsymbol{x}_1')\chi_\nu(\boldsymbol{r}_1'), \tag{90}$$

where

$$\gamma(\boldsymbol{x}_1, \boldsymbol{x}_1') = N \int \mathrm{d}\boldsymbol{x}_2 \ldots \mathrm{d}\boldsymbol{x}_N\, \Psi_0^{(0)}(\boldsymbol{x}_1, \boldsymbol{x}_2, \ldots, \boldsymbol{x}_N)\Psi_0^{(0)*}(\boldsymbol{x}_1', \boldsymbol{x}_2, \ldots, \boldsymbol{x}_N), \tag{91}$$

and $\boldsymbol{x}_i = \boldsymbol{r}_i s_i$. The expectation value of an operator $\hat{O} = \sum_i \hat{o}(i)$ can thus be written as

$$\left\langle \Psi_0^{(0)} \middle| \hat{O} \middle| \Psi_0^{(0)} \right\rangle = \sum_{pq} D_{pq}\langle \phi_p|\hat{o}|\phi_q\rangle = \sum_{\mu\nu} D_{\mu\nu}\langle \chi_\mu|\hat{o}|\chi_\nu\rangle. \tag{92}$$

For a closed-shell SCF wavefunction $|\Phi_{\mathrm{SCF}}\rangle$, for example, the density matrix in the atomic orbital basis is given as

$$D_{\mu\nu}^{\mathrm{SCF}} = 2\sum_i c_{\mu i}^* c_{\nu i}. \tag{93}$$

$|\Psi_0^{(0)}\rangle$ is therefore replaced by the SCF ground state wavefunction, although generalized density matrices can also be defined for correlated wavefunctions (Section 7.4.3) to give correlated expectation values.

7.4.1.2 The sum-over-states method

The sum-over-states method for the calculation of second- or higher-order properties is based on equations (59) and (73). The main task is thus to obtain a set of excitation energies $W_n^{(0)} - W_0^{(0)}$ and transition moments $\langle \Psi_0^{(0)}|\hat{O}|\Psi_n^{(0)}\rangle$ with the appropriate operator $\hat{O}$ or alternatively a ground state wavefunction $|\Psi_0^{(0)}\rangle$ and a set of excited state wavefunctions $|\Psi_n^{(0)}\rangle$ from which the excitation energies and transition moments can be calculated.

Approximating the excitation energies $W_n^{(0)} - W_0^{(0)}$ by molecular orbital energy differences $\varepsilon_a - \varepsilon_i$ and the transition moments $\langle\Psi_0^{(0)}|\hat{O}|\Psi_n^{(0)}\rangle$ by matrix elements of $\hat{O}$ in the molecular orbital basis $\langle\phi_i|\hat{o}|\phi_a\rangle$ gives the uncoupled HF approximation of Dalgarno (1959), which played an important role in the early days of calculations of molecular properties.

Nowadays the sum-over-states method is mostly used in three cases. The first is benchmark studies of two-electron systems using explicitly correlated wavefunctions [see e.g. Bishop (1994)]. The second is the study of hyperpolarizabilities of larger systems using semi-empirical methods. Finally, it is used in the analysis of contributions to a molecular property from excitations between individual, typically localized, molecular orbitals [see e.g. Hansen and Bouman (1985), and Packer and Pickup (1995)]. The latter is normally done at the level of the RPA, for which excitation energies and transition moments can easily be obtained.

7.4.1.3 Polarization propagator methods

Propagator methods can be used to study both ionization and number-conserving excitation processes [Linderberg and Öhrn (1973)]. The two-time polarization propagator [Zubarev (1974)] is formally equivalent to the linear response function defined in Section 7.3.4. The aim of using the polarization propagator in quantum chemistry is to introduce approximations to the exact representation of the linear response function given in equation (73). Computationally convenient expressions can be obtained by taking time derivatives of the linear response and then Fourier transforming back to the frequency domain. This equation-of-motion approach gives

$$\hbar\omega\langle\langle\hat{O};\hat{V}^{\omega}_{\alpha\cdots}\rangle\rangle_{\omega} = \left\langle\Psi_0^{(0)}\right|\left[\hat{O},\hat{V}^{\omega}_{\alpha\cdots}\right]\left|\Psi_0^{(0)}\right\rangle + \left\langle\left\langle\hat{O};\left[\hat{H}^{(0)},\hat{V}^{\omega}_{\alpha\cdots}\right]\right\rangle\right\rangle_{\omega}. \tag{94}$$

The equality holds only for exact eigenfunctions of the Hamiltonian and can be confirmed by substituting equation (73) for the linear response functions. Iterating on this equation leads to a so-called moment expansion of the polarization propagator

$$\hbar\omega\langle\langle\hat{O};\hat{V}^{\omega}_{\alpha\cdots}\rangle\rangle_{\omega} = \left\langle\Psi_0^{(0)}\right|\left[\hat{O},\hat{V}^{\omega}_{\alpha\cdots}\right]\left|\Psi_0^{(0)}\right\rangle + \left\langle\Psi_0^{(0)}\right|\left[\hat{O},\sum_{n=1}^{\infty}\left(\frac{1}{\hbar\omega}\right)^n(\hat{H}^{(0)})^n\hat{V}^{\omega}_{\alpha\cdots}\right]\left|\Psi_0^{(0)}\right\rangle, \tag{95}$$

where the superoperator $\hat{H}^{(0)}$ operates on another operator to produce a commutator:

$$\hat{H}^{(0)}\hat{V}^{\omega}_{\alpha\cdots} = \left[\hat{H}^{(0)},\hat{V}^{\omega}_{\alpha\cdots}\right]. \tag{96}$$

Using an inner projection technique with a complete set of excitation and de-excitation operators $\{h_n\}$, arranged as a column vector $\boldsymbol{h}$ or as a row vector $\tilde{\boldsymbol{h}}$, an exact matrix representation of the polarization propagator is finally obtained [Pickup and Goscinski (1973)]

$$\begin{aligned}\langle\langle\hat{O};V^{\omega}_{\alpha\cdots}\rangle\rangle_{\omega} = {} & \langle\Psi_0^{(0)}|\left[\hat{O},\tilde{\boldsymbol{h}}\right]|\Psi_0^{(0)}\rangle\langle\Psi_0^{(0)}|[\boldsymbol{h}^{\dagger},\hbar\omega\tilde{\boldsymbol{h}}] - [\boldsymbol{h}^{\dagger},[\hat{H}^{(0)},\tilde{\boldsymbol{h}}]]|\Psi_0^{(0)}\rangle^{-1} \\ & \times\langle\Psi_0^{(0)}|[\boldsymbol{h}^{\dagger},\hat{V}^{\omega}_{\alpha\ldots}]|\Psi_0^{(0)}\rangle.\end{aligned} \tag{97}$$

Completeness of the set of operators $\{h_n\}$ means that all possible excited states $|\Psi_n^{(0)}\rangle$ of the system are generated by operating on $|\Psi_0^{(0)}\rangle$,

$$h_n\left|\Psi_0^{(0)}\right\rangle = |\Psi_n^{(0)}\rangle. \tag{98}$$

This more general expression for the polarization propagator reduces to equation (73) if one chooses the operators $\{h_n\}$ to be $\{|\Psi_n^{(0)}\rangle\langle\Psi_0^{(0)}|, |\Psi_0^{(0)}\rangle\langle\Psi_n^{(0)}|\}$. This equation is exact as long as a complete set of excitation and de-excitation operators $\{h_n\}$ is used and the reference state $|\Psi_0^{(0)}\rangle$ is an eigenfunction of the unperturbed Hamiltonian. Approximate polarization propagator methods are obtained by truncating the set of operators and by using an approximate reference state $|\Psi_0^{(0)}\rangle$. MCSCF and MP perturbation theory wavefunctions are commonly employed as approximate reference states. The inner projection technique therefore allows the formally exact equation of motion in equation (94) to be approximated by variational or perturbation wavefunctions.

In polarization propagator approximations based on MP perturbation theory the reference state is approximated by the wavefunction in equation (85) and the complete set of operators $\boldsymbol{h}$ for an N-electron SCF ground state consists of all possible single excitation and de-excitation operators, $\boldsymbol{h}_2 = \{\boldsymbol{q}^\dagger, \boldsymbol{q}\}$, all possible double excitation and de-excitation operators $\boldsymbol{h}_4 = \{\boldsymbol{q}^\dagger\boldsymbol{q}^\dagger, \boldsymbol{q}\boldsymbol{q}\}$, up to all possible N-tuple excitation operators, with respect to the ground state $|\Phi_{\mathrm{SCF}}\rangle$ [Dalgaard (1979), and Jørgensen and Simons (1981)]. The matrix form of the polarization propagator, equation (97), can thus be written as

$$\langle\langle\hat{O};\hat{V}^{\omega}_{\alpha\ldots}\rangle\rangle_\omega = \left(\langle\Psi_0^{(0)}|[\hat{O},\tilde{\boldsymbol{h}}_2]|\Psi_0^{(0)}\rangle \quad \langle\Psi_0^{(0)}|[\hat{O},\tilde{\boldsymbol{h}}_4]|\Psi_0^{(0)}\rangle\ldots\right)$$
$$\times\left[\hbar\omega\begin{pmatrix}\boldsymbol{S}^{[2]}_{22} & \boldsymbol{S}^{[2]}_{24} & \cdots\\ \boldsymbol{S}^{[2]}_{42} & \boldsymbol{S}^{[2]}_{44} & \cdots\\ \vdots & \vdots & \ddots\end{pmatrix} - \begin{pmatrix}\boldsymbol{E}^{[2]}_{22} & \boldsymbol{E}^{[2]}_{24} & \cdots\\ \boldsymbol{E}^{[2]}_{42} & \boldsymbol{E}^{[2]}_{44} & \cdots\\ \vdots & \vdots & \ddots\end{pmatrix}\right]^{-1}$$
$$\times\begin{pmatrix}\langle\Psi_0^{(0)}|[\boldsymbol{h}_2^\dagger,\hat{V}^{\omega}_{\alpha\ldots}]|\Psi_0^{(0)}\rangle\\ \langle\Psi_0^{(0)}|[\boldsymbol{h}_4^\dagger,\hat{V}^{\omega}_{\alpha\ldots}]|\Psi_0^{(0)}\rangle\\ \vdots\end{pmatrix}, \tag{99}$$

where

$$\boldsymbol{S}^{[2]}_{ij} = \langle\Psi_0^{(0)}|[\boldsymbol{h}_i^\dagger,\tilde{\boldsymbol{h}}_j]|\Psi_0^{(0)}\rangle, \tag{100}$$
$$\boldsymbol{E}^{[2]}_{ij} = \langle\Psi_0^{(0)}|[\boldsymbol{h}_i^\dagger,[\hat{F}+\hat{V},\tilde{\boldsymbol{h}}_j]]|\Psi_0^{(0)}\rangle. \tag{101}$$

A series of approximations of increasing order n is obtained by requiring that all the matrix elements $\boldsymbol{S}^{[2]}_{ij}$, $\boldsymbol{E}^{[2]}_{ij}$ as well as $\langle\Psi_0^{(0)}|[\hat{O},\tilde{\boldsymbol{h}}_i]|\Psi_0^{(0)}\rangle$ are evaluated through order n.

For a first-order polarization propagator approximation (FOPPA) it is only necessary to keep the $\boldsymbol{h}_2$ operator set for an SCF reference state, $|\Phi_{\mathrm{SCF}}\rangle$. This approximation is better known as time-dependent HF [McLachlan and Ball (1964)] (Section 7.4.2.1) or the RPA [Rowe (1968)] and can also be derived as the linear response of an SCF wavefunction, as described in Section 7.4.2.2. The RPA polarization propagator is then given as

$$\langle\langle\hat{O};\hat{V}^{\omega}_{\alpha\ldots}\rangle\rangle_\omega = \left(\langle\Phi_{\mathrm{SCF}}|\,[\hat{O},\tilde{\boldsymbol{q}}^\dagger]\,|\Phi_{\mathrm{SCF}}\rangle\;\langle\Phi_{\mathrm{SCF}}|\,[\hat{O},\tilde{\boldsymbol{q}}]\,|\Phi_{\mathrm{SCF}}\rangle\right)\begin{pmatrix}\boldsymbol{X}\\ \boldsymbol{Y}\end{pmatrix} \tag{102}$$

$$\begin{pmatrix}\boldsymbol{X}\\ \boldsymbol{Y}\end{pmatrix} = \left[\hbar\omega\begin{pmatrix}\boldsymbol{1} & \boldsymbol{0}\\ \boldsymbol{0} & -\boldsymbol{1}\end{pmatrix} - \begin{pmatrix}\boldsymbol{A}^{(0,1)} & \boldsymbol{B}^{(1)}\\ \boldsymbol{B}^{(1)} & \boldsymbol{A}^{(0,1)}\end{pmatrix}\right]^{-1}\begin{pmatrix}\langle\Phi_{\mathrm{SCF}}|\,[\boldsymbol{q},\hat{V}^{\omega}_{\alpha\ldots}]\,|\Phi_{\mathrm{SCF}}\rangle\\ \langle\Phi_{\mathrm{SCF}}|\,[\boldsymbol{q}^\dagger,\hat{V}^{\omega}_{\alpha\ldots}]\,|\Phi_{\mathrm{SCF}}\rangle\end{pmatrix} \tag{103}$$

where the $\boldsymbol{A}^{(0,1)}$ and $\boldsymbol{B}^{(1)}$ matrices are defined as

$$\boldsymbol{A}^{(0,1)} = \langle\Phi_{\mathrm{SCF}}|\left[\boldsymbol{q},\left[\hat{F}+\hat{V},\tilde{\boldsymbol{q}}^{\dagger}\right]\right]|\Phi_{\mathrm{SCF}}\rangle \tag{104}$$

$$\boldsymbol{B}^{(1)} = \langle\Phi_{\mathrm{SCF}}|\left[\boldsymbol{q},\left[\hat{F}+\hat{V},\tilde{\boldsymbol{q}}\right]\right]|\Phi_{\mathrm{SCF}}\rangle . \tag{105}$$

The second-order polarization propagator approximation (SOPPA) of Nielsen, Jørgensen and Oddershede (1980) was historically defined to be second order in matrix elements which contain only $\boldsymbol{h}_2$ operators. This implies that $\boldsymbol{E}_{22}^{[2]}$, $\boldsymbol{S}_{22}^{[2]}$, $\langle\Psi_0^{(0)}|[\hat{O},\tilde{\boldsymbol{h}}_2]|\Psi_0^{(0)}\rangle$ and $\langle\Psi_0^{(0)}|[\boldsymbol{h}_2^{\dagger},\hat{V}_{\alpha\ldots}^{\omega}]|\Psi_0^{(0)}\rangle$ are evaluated through second-order, $\boldsymbol{E}_{24}^{[2]}$, $\boldsymbol{S}_{24}^{[2]}$, $\boldsymbol{E}_{42}^{[2]}$, $\boldsymbol{S}_{42}^{[2]}$, $\langle\Psi_0^{(0)}|[\hat{O},\tilde{\boldsymbol{h}}_4]|\Psi_0^{(0)}\rangle$ and $\langle\Psi_0^{(0)}|[\boldsymbol{h}_4^{\dagger},\hat{V}_{\alpha\ldots}^{\omega}]|\Psi_0^{(0)}\rangle$ are evaluated through first order and $\boldsymbol{E}_{44}^{[2]}$ and $\boldsymbol{S}_{44}^{[2]}$ only through zeroth order [Oddershede, Jørgensen and Yeager (1984)]. An analysis of the matrix elements shows that besides the first-order MP wavefunction $|\Phi^{(1)}\rangle$ only $|SE^{(2)}\rangle$ is required from the second-order wavefunction [equation (88)].

A complete third-order polarization propagator approximation has also been derived but only parts have been implemented [Geertsen, Eriksen and Oddershede (1991)]. However, two other SOPPA-like methods have been used widely. Both methods are based on the assumption that replacement of the first-order MP doubles correlation coefficients in equation (86) and of the second-order MP singles correlation coefficient in equation (88) by coupled-cluster singles and doubles amplitudes will give improved results. In the second-order polarization propagator with coupled-cluster singles and doubles amplitudes–SOPPA(CCSD)–method of Sauer (1997) this is done in all matrix elements, whereas in its precursor, the coupled-cluster singles and doubles polarization propagator approximation (CCSDPPA) [Geertsen, Eriksen and Oddershede (1991)], this was not the case.

In the multiconfigurational polarization propagator approximation of Yeager and Jørgensen (1979), normally called multiconfigurational random phase approximation (MCRPA), the set of operators contains state transfer operators $\{\boldsymbol{R}^{\dagger},\boldsymbol{R}\}$ in addition to the non-redundant single excitation and de-excitation operators. The state transfer operators are defined as

$$R_n^{\dagger} = |\Psi_n\rangle\langle\Psi_{\mathrm{MCSCF}}|, \tag{106}$$

where $|\Psi_n\rangle = \sum_i |\Phi_i\rangle C_{in}$ are the orthogonal complement states of the MCSCF state $|\Psi_{\mathrm{MCSCF}}\rangle = \sum_i |\Phi_i\rangle C_{i0}$ and $\{|\Phi_i\rangle\}$ is the set of configuration state functions. The expression for the polarization propagator in MCRPA can be obtained from equation (99) if one identifies $\boldsymbol{h}_4$ with $\{\boldsymbol{R}^{\dagger},\boldsymbol{R}\}$ and $|\Psi_0^{(0)}\rangle$ with $|\Psi_{\mathrm{MCSCF}}\rangle$. Since this is a variational wavefunction, MCRPA can also be obtained by application of linear response theory (Section 7.4.2.2) or quasienergy derivatives (Section 7.4.3.3) for an MCSCF state.

A feature common to all propagator methods is that the response is given as the product of a 'property gradient' vector $\tilde{\boldsymbol{T}}(\hat{O})$, with an inverse 'Hessian' or principal propagator matrix $(\hbar\omega\boldsymbol{S}-\boldsymbol{E})$, and another 'property gradient' vector $\boldsymbol{T}(\hat{V}_{\alpha\ldots}^{\omega})$. At zeroth order in the fluctuation potential, the property gradients represent transition moments of equation (73) and the diagonal inverse Hessian gives the excitation energies. In actual calculations, however, the inverse is never evaluated. A set of coupled linear equations for a response vector $\boldsymbol{N}(\hat{V}_{\alpha\ldots}^{\omega})$

$$(\hbar\omega\boldsymbol{S}-\boldsymbol{E})\boldsymbol{N}\left(\hat{V}_{\alpha\ldots}^{\omega}\right) = \boldsymbol{T}\left(\hat{V}_{\alpha\ldots}^{\omega}\right) \tag{107}$$

is instead solved iteratively by expanding the response vector in a basis of trial vectors $\{\boldsymbol{b}_i\}$ [Pople *et al.* (1979)]

$$\boldsymbol{N} = \sum_i \boldsymbol{b}_i c_i . \tag{108}$$

In each iteration one has to calculate the linear transformed trial vector $(\hbar\omega \boldsymbol{S} - \boldsymbol{E})\boldsymbol{b}_i$ which can be done directly without ever calculating the $(\hbar\omega \boldsymbol{S} - \boldsymbol{E})$ matrix explicitly [Olsen and Jørgensen (1985), and Packer *et al.* (1996)].

An advantage of propagator methods is that both the transition moments, $\langle \Psi_0^{(0)} | \hat{O} | \Psi_n^{(0)} \rangle$, and the excitation energies, $W_n^{(0)} - W_0^{(0)}$, can be calculated in addition to response properties. This greatly extends the range of application of these methods relative to derivative approaches, which give only response properties. The SOPPA polarization propagator, for example, yields excitation energies correct to second order in the fluctuation potential. These energies are very accurate for molecules with predominantly single determinant ground states [Packer *et al.* (1996)]; in cases where there is appreciable non-dynamical correlation, the MCRPA excitation energies and transition moments are more appropriate.

7.4.2 PERTURBATION THEORY WITH APPROXIMATE WAVEFUNCTIONS

7.4.2.1 Coupled Hartree–Fock and time-dependent coupled Hartree–Fock theory

In the coupled HF method (CHF), probably derived for the first time by Peng (1941), second- and higher-order static properties are obtained by solving the HF equations self-consistently in the presence of the perturbation. The molecular orbitals $\{\phi_i\}$ in the presence of the perturbation, e.g. an electric field E_β, can be expanded in the set of unperturbed molecular orbitals $\{\phi_p^{(0)}\}$

$$\phi_{\beta,i} = \sum_p \phi_p^{(0)} U_{\beta,pi}. \tag{109}$$

The Fock matrices F_{ij}, orbital energies ε_i and the coefficients U_{pi} are then expanded in orders of the perturbation

$$U_{\beta,pi} = U_{pi}^{(0)} + U_{\beta,pi}^{(1)} E_\beta + \cdots, \tag{110}$$

where $U_{pi}^{(0)} = \delta_{pi}$. The first-order coefficients are obtained as

$$U_{\beta,ai}^{(1)} = -\frac{F_{\beta,ai}^{(1)}}{\varepsilon_a^{(0)} - \varepsilon_i^{(0)}}, \tag{111}$$

an equation which must be solved iteratively since the first-order Fock matrix, $F_{\beta,ai}^{(1)}$, depends on $U_{\beta,ai}^{(1)}$. Evaluation of $F_{\beta,ai}^{(1)}$ leads to

$$\sum_{bj} \left(A_{ai,bj}^{(0,1)} \mp B_{ai,bj}^{(1)} \right) U_{\beta,bj}^{(1)} = -\langle \phi_a | \hat{h}_\beta^{(1)} | \phi_i \rangle \tag{112}$$

for real (−) and imaginary (+) perturbation operators, with matrices $\boldsymbol{A}^{(0,1)}$ and $\boldsymbol{B}^{(1)}$ given in equations (104) and (105). The first-order correction to the density matrix is then

$$D_{\beta,\mu\nu}^{(1),\mathrm{SCF}} = 2\sum_i \left(c_{\beta,\mu i}^{(1)*} c_{\nu i} + c_{\mu i}^{*} c_{\beta,\nu i}^{(1)} \right) = 2\left(U_{\beta,\mu\nu}^{(1)*} + U_{\beta,\nu\mu}^{(1)} \right), \tag{113}$$

and the static polarizability can be calculated as

$$\alpha_{\alpha\beta} = \sum_{pq} \langle \phi_p | \hat{o}_\alpha^E | \phi_q \rangle D_{\beta,pq}^{(1)}. \tag{114}$$

Time-dependent HF theory (TDHF) [Langhoff, Epstein and Karplus (1972)] introduces time-dependent molecular orbitals, which are also expanded in the unperturbed orbitals

$$\phi_{\beta,i} = \phi_i^{(0)} + \frac{1}{2}\sum_p E_\beta^\omega \left(U_{\beta,pi}^{(1)}(\omega)\,\mathrm{e}^{\mathrm{i}\omega t} + U_{\beta,pi}^{(1)}(-\omega)\,\mathrm{e}^{-\mathrm{i}\omega t}\right)\phi_p^{(0)} + \cdots. \tag{115}$$

From the time-dependent version of the HF equation, which can be derived from Frenkel's variational principle [Frenkel (1934)], equations for the coefficients $U_{\beta,pi}^{(1)}(\omega)$ and $U_{\beta,pi}^{(1)}(-\omega)$ can be obtained

$$\begin{pmatrix} \boldsymbol{U}_\beta^{(1)}(-\omega) \\ -\boldsymbol{U}_\beta^{(1)}(\omega) \end{pmatrix} = \begin{pmatrix} \boldsymbol{X} \\ \boldsymbol{Y} \end{pmatrix}, \tag{116}$$

where $\boldsymbol{X}$ and $\boldsymbol{Y}$ are defined in equation (103).

As has already been mentioned, the variational nature of the HF wavefunction means that the CHF/TDHF equations are equivalent to the RPA equations. Unlike RPA and its correlated coevals, however, an atomic orbital based solution of the iterative CHF equations cannot give excitation energies and transition moments. Historically, CHF was favoured over RPA since it could be solved in the atomic orbital basis, rather than requiring a transformation to the molecular orbital basis. The need for an inverse Hessian in RPA/SOPPA also restricted the size of systems which could be studied. However, the use of direct atomic orbital driven methods for RPA response properties up to the cubic hyperpolarizability [Ågren *et al.* (1993)] and for SOPPA [Bak *et al.* (2000)], coupled with iterative methods for solving the inverse Hessian, mean that they can now be applied as widely as CHF/TDHF and provide far more information about excited states and properties.

7.4.2.2 Response theory

In the application of response theory to an SCF wavefunction, $|\Psi\rangle = |\Phi_{\mathrm{SCF}}\rangle$, or to an MCSCF wavefunction, $|\Psi\rangle = |\Psi_{\mathrm{MCSCF}}\rangle$, first described by Olsen and Jørgensen (1985), the time-dependent state $|\Psi_0(t)\rangle$ is usually expressed as

$$|\Psi_0(t)\rangle = \mathrm{e}^{\mathrm{i}\kappa(t)}\,\mathrm{e}^{\mathrm{i}S(t)}|\Psi\rangle, \tag{117}$$

where

$$\kappa(t) = \sum_\mu \left[\kappa_\mu(t)q_\mu^\dagger + \kappa_\mu^*(t)q_\mu\right], \tag{118}$$

$$S(t) = \sum_n \left[S_n(t)R_n^\dagger + S_n^*(t)R_n\right], \tag{119}$$

and $S_n(t) = S_n^*(t) = 0$ for the SCF case. The time-dependent parameters, collected in a vector $\gamma(t) = (\kappa_\mu(t), \kappa_\mu^*(t), S_n(t), S_n^*(t))^{\mathrm{T}}$, are then expanded in orders of the perturbation $\hat{H}^{(1)}(t)$

$$\gamma_\mu(t) = \sum_i \gamma_\mu^{(i)}(t). \tag{120}$$

Equations for the coefficients in each order are obtained from a particular form of the time-dependent Schrödinger equation, called the generalized Ehrenfest theorem,

$$\frac{\mathrm{d}}{\mathrm{d}t}\langle\Psi_0(t)|\hat{O}|\Psi_0(t)\rangle + \mathrm{i}\langle\Psi_0(t)|\left[\hat{O}, \hat{H}^{(0)} + \hat{H}^{(1)}(t)\right]|\Psi_0(t)\rangle = 0. \tag{121}$$

Inserting equation (117) for $|\Psi_0(t)\rangle$ and the set of operators $\{q_\mu^\dagger, q_\mu, R_n^\dagger, R_n\}$ for $\hat{O}$ and Fourier transforming to the frequency domain again yields the set of coupled linear equations given in equation (107) [Fuchs, Bonačić-Koutecký and Koutecký (1993)].

Coupled-cluster response functions were derived by Koch and Jørgensen (1990) starting from the time-dependent transition expectation value of Arponen (1983)

$$\langle \Phi_\Lambda(t)|\hat{O}|\Phi_{CC}(t)\rangle, \tag{122}$$

where the time-dependent coupled-cluster state $|\Phi_{CC}(t)\rangle$ and dual or 'lambda' state $\langle\Phi_\Lambda(t)|$ are defined as

$$|\Phi_{CC}(t)\rangle = e^{T(t)}|\Phi_{SCF}\rangle, \tag{123}$$

$$\langle\Phi_\Lambda(t)| = \langle\Phi_{SCF}|(1+\Lambda(t))\,e^{-T(t)}. \tag{124}$$

The time-dependent cluster operator and Λ operator consist of n-tuple excitation, $\tau_\mu^\dagger$, and de-excitation operators, τ_μ, respectively

$$T(t) = \sum_\mu t_\mu(t)\tau_\mu^\dagger, \tag{125}$$

$$\Lambda(t) = \sum_\mu \lambda(t)\tau_\mu, \tag{126}$$

where $\{\tau_\mu^\dagger, \tau_\mu\}$ is a shorthand notation for single $\{q_\mu^\dagger, q_\mu\}$, double $\{q_\mu^\dagger q_\mu^\dagger, q_\mu q_\mu\}$, etc. excitation and de-excitation operators. $t_\mu(t)$ and $\lambda(t)$ are the corresponding time-dependent amplitudes. The amplitudes are then determined from the coupled-cluster time-dependent Schrödinger equations

$$e^{-T(t)} i\frac{d}{dt}|\Phi_{CC}(t)\rangle = e^{-T(t)}\hat{H}|\Phi_{CC}(t)\rangle, \tag{127}$$

$$\left(\frac{d}{dt}\langle\Phi_\Lambda(t)|\right) e^{T(t)} = i\langle\Phi_\Lambda(t)|\hat{H}\,e^{T(t)}. \tag{128}$$

In the presence of a time-dependent perturbation $\hat{H}^{(1)}(t)$, equation (67), the amplitudes $t_\mu(t)$ and $\lambda(t)$ are expanded in a perturbation series $t_\mu(t) = t_\mu^{(0)}(t) + t_\mu^{(1)}(t) + \cdots$ and $\lambda_\mu(t) = \lambda_\mu^{(0)}(t) + \lambda_\mu^{(1)}(t) + \cdots$, yielding a series of equations, which are solved by Fourier transforming to the frequency domain. After insertion of the results for the amplitudes in a perturbation expansion of the time-dependent transition expectation value, $\langle\Phi_\Lambda(t)|\hat{O}|\Phi_{CC}(t)\rangle$, the response functions can be identified by comparison with equation (70).

7.4.3 DERIVATIVE METHODS

7.4.3.1 The finite field method

The finite field method of Cohen and Roothaan (1965) and Pople, McIver and Ostlund (1968) involves numerical evaluation of derivatives of the electronic energy or first- and higher-order properties, P, in the presence of a perturbation operator, $\hat{O}_\alpha^F F_\alpha$. Calculations are performed for various values of field strength F_α. The desired derivative, at zero field strength, can then be obtained either by finite difference or by fitting the calculated values of the energy or property P to a Taylor expansion in the field strength F_α.

In a finite field calculation of the static dipole polarizability α, for example, the perturbation operator, $-\hat{O}_\alpha^E E_\alpha$, is added to the Hamiltonian, $H^{(0)}$, and the electronic energy or the electronic contribution to the dipole moment is calculated for various finite values of the electric field strength E_α. The dipole polarizability is then obtained as the numerical first derivative of the electric field-dependent dipole moment or as the numerical second derivative of the electric field-dependent electronic energy.

The property P for which derivatives are taken need not to be a static property, but could also be a frequency-dependent polarizability $\alpha(-\omega;\omega)$, as done by Jaszuński (1987). Finite field calculations on $\alpha(-\omega;\omega)$ facilitate calculation of $\beta(-\omega;\omega,0)$, $\gamma(-\omega;\omega,0,0)$. Analytical expressions for these higher polarizabilities, using propagators or derivatives, for example, are computationally intensive relative to a finite field evaluation. Once again, however, the propagator/response function formalism would give access to two- and three-photon transition moments which are not available by finite field.

The finite field method is by far the easiest method to implement as long as the perturbation operators are real. Any program for the calculation of the property P can be used, as long as it allows for the inclusion of additional one-electron operators in the Hamiltonian. The finite field method can thus be applied at any level of approximation or correlation and even to approximations for which a wavefunction or a ground state energy is not defined.

Imaginary perturbation operators, such as $\hat{O}_\alpha^{lB}$, $\hat{O}_\alpha^{sB}$, $\hat{O}_\alpha^{lm^K}$ and $\hat{O}_\alpha^{sm^K}$, require the use of complex arithmetic, which prevented a routine usage of the finite field method for the calculation of magnetic properties. Nevertheless, finite field approaches to the calculation of nuclear magnetic shielding constants and nuclear spin–spin coupling constants have been presented by Fukui, Miura and Matsuda (1992), and Fukui *et al.* (1992). In this method the sum-over-states contribution to $\sigma_{\alpha\beta}^K$, for example, is evaluated as a numerical derivative of the expectation value of $\hat{O}_\alpha^{lm^K}$ with respect to B_β, the expectation value having been calculated to second order in electron correlation. Mixed electric/magnetic properties, on the other hand, such as nuclear magnetic shielding polarizabilities can be evaluated as numerical derivatives of nuclear magnetic shielding tensors without complex arithmetic.

A disadvantage of the finite field method lies in the nature of numerical differentiation. Care is required in setting the field strength, in our example for E_α, which must not be too high and in the number of energy evaluations which are used. For higher-order properties or multiple perturbations the method becomes cumbersome, since the number of calculations to be performed increases rapidly.

A variation of this method is the finite point charge method, used by Maroulis and Thakkar (1988), in which the external electric field or field gradient is simulated by an appropriate arrangement of point charges. This method is even simpler to implement, since it requires only the option to include centres with a charge but no basis functions, rather than a modified one-electron Hamiltonian. The finite field method has been widely used but is becoming increasingly obsolete because of the advances in analytical derivative methods: most properties of interest can now be calculated analytically, obviating the need for a finite field calculation.

7.4.3.2 The analytical derivative method

In the analytic derivative method for the calculation of molecular properties, approximate expressions for P within a given approximation are differentiated analytically with respect to the perturbation. It is equally as general as the finite field method and does not suffer from the numerical problems of the latter. However, it is much more difficult to apply to a new type of wavefunction, since expressions for the analytical derivatives have to be derived and implemented. Nevertheless, expressions for first- and second-order properties have been implemented for most *ab initio* methods following the derivation of analytical derivatives with respect to changes in the nuclear coordinates. Explicit expressions can be found in the reviews by e.g. Helgaker and Jørgensen (1988), Amos and Rice (1989), and Gauss and Cremer (1992).

The first derivative of the energy of a system described by the Hamiltonian $\hat{H}^{(0)}+\hat{H}^{(1)}+\hat{H}^{(2)}$, given in equations (34), (39) and (40) with respect to one of the perturbations E_α, B_α and m_α^K or in general F_α can be written for most methods as

$$\left.\frac{\mathrm{d}W(F_\alpha)}{\mathrm{d}F_\alpha}\right|_{F_\alpha=0} = \sum_{pq} D_{pq}\,\langle\phi_p|\,\frac{\partial\hat{h}^{(1)}(1)}{\partial F_\alpha}\,|\phi_q\rangle = \sum_{\mu\nu} D_{\mu\nu}\,\langle\chi_\mu|\,\frac{\partial\hat{h}^{(1)}(1)}{\partial F_\alpha}\,|\chi_\nu\rangle\,. \tag{129}$$

The atomic orbitals χ_μ are here assumed to be independent of the perturbation. For variational wavefunctions, i.e. methods which fulfill the Hellmann–Feynman theorem, this is equivalent to equation (92). However, for non-variational wavefunctions the density matrix is not consistent with the definition in equation (89) and was therefore called the relaxed or response density matrix by Trucks *et al.* (1988). The Hellmann–Feynman theorem can be fulfilled in coupled-cluster theory, however, when the energy and first-order properties are evaluated as transition expectation values, defined in Section 7.4.2.2 [see e.g. Perera, Nooijen and Bartlett (1996)]:

$$\left.\frac{\mathrm{d}\langle\Phi_\Lambda|\hat{H}^{(0)}+\hat{H}^{(1)}|\Phi_{\mathrm{CC}}\rangle}{\mathrm{d}F_\alpha}\right|_{F_\alpha=0} = \langle\Phi_\Lambda|\,\frac{\partial\hat{H}^{(1)}}{\partial F_\alpha}\,|\Phi_{\mathrm{CC}}\rangle\,. \tag{130}$$

The relaxed density matrix can be decomposed into an SCF and a correlation part

$$\boldsymbol{D} = \boldsymbol{D}^{\mathrm{SCF}} + \boldsymbol{D}^{\mathrm{corr}}. \tag{131}$$

The SCF density is given in equation (93) and the correlation contribution consists of two parts

$$\boldsymbol{D}^{\mathrm{corr}} = \boldsymbol{D}^{\mathrm{amp}} + \boldsymbol{D}^{\mathrm{orb}}, \tag{132}$$

where $\boldsymbol{D}^{\mathrm{amp}}$ contains amplitudes or correlation coefficients and $\boldsymbol{D}^{\mathrm{orb}}$, obtained as a solution of the so-called Z-vector equations of Handy and Schaefer (1984), arises because of the relaxation of the orbitals for non-variational wavefunctions. Only the occupied–virtual and virtual–occupied blocks are non-zero in $\boldsymbol{D}^{\mathrm{orb}}$, which is again a result of the Brillouin condition for the SCF ground state. At the level of second-order MP perturbation theory (MP2) it is also the occupied–virtual and virtual–orbital blocks D_{ia} and D_{ai} which differ between a density matrix consistent through second order, which was derived by Jensen *et al.* (1988) from equation (89), and the MP2 relaxed density matrix as given by Gauss and Cremer (1992) or Cybulski and Bishop (1994).

The second derivative of the energy can be written as

$$\left.\frac{\mathrm{d}^2W(F_\beta,F_\alpha)}{\mathrm{d}F_\alpha\,\mathrm{d}F_\beta}\right|_{F_\alpha=F_\beta=0} = \sum_{\mu\nu} D_{\mu\nu}\frac{\partial^2\langle\chi_\mu|\hat{h}_1(1)|\chi_\nu\rangle}{\partial F_\beta\,\partial F_\alpha} + \sum_{\mu\nu} D_{\beta,\mu\nu}^{(1)}\frac{\partial\langle\chi_\mu|\hat{h}_1(1)|\chi_\nu\rangle}{\partial F_\alpha} \tag{133}$$

where $\hat{h}_1(1)$ stands for $\hat{h}^{(0)}(1)+\hat{h}^{(1)}(1)+\hat{h}^{(2)}(1)$ and the atomic orbitals could depend on the perturbation. The derivative of the relaxed density matrix, the so-called first-order relaxed density matrix, in the atomic orbital basis is

$$D_{\beta,\mu\nu}^{(1)} = \frac{\partial D_{\mu\nu}}{\partial F_\beta} = \sum_{pq} c_{\mu p}^{*}\frac{\partial D_{pq}}{\partial F_\beta}c_{\nu q} + \sum_{pq} D_{pq}\left(\frac{\partial c_{\mu p}^{*}}{\partial F_\beta}c_{\nu q} + c_{\mu p}^{*}\frac{\partial c_{\nu q}}{\partial F_\beta}\right). \tag{134}$$

The derivatives of the molecular orbital coefficients $\{c_{\nu q}\}$ are obtained by solving the coupled–perturbed HF equations, which are described in Section 7.4.2.2. The first-order density matrix at the SCF level was given in equation (113). The occupied–occupied and virtual–virtual blocks of the correlated first-order density matrix contain derivatives of the amplitudes or correlation coefficients, which can be obtained by straightforward differentiation of the equations defining the amplitudes. The occupied–virtual and virtual–occupied part requires the solution of the first-order Z-vector equations, i.e. the derivative of the Z-vector equations. Explicit expressions for the relaxed density matrices and first-order relaxed density matrices for many methods can also be found in Helgaker and Jørgensen (1988), Amos and Rice (1989), Gauss and Cremer (1992), and Gauss and Stanton (1995).

7.4.3.3 Time-dependent analytical derivatives

The analytical energy derivative method has been extended to the case of time-dependent perturbations using the pseudo-energy derivative method of Rice and Handy (1991) and the quasi-energy derivative method of Sasagane, Aiga and Itoh (1993). Both methods define frequency-dependent properties as derivatives of the quasi-energy, as defined by Löwdin and Mukherjee (1972) or Kutzelnigg (1992)

$$\tilde{W}(t) = \langle \Psi(t)|\hat{H}^{(0)} + \hat{H}^{(1)}(t) - \mathrm{i}\frac{\partial}{\partial t}|\Psi(t)\rangle, \tag{135}$$

but in the pseudo-energy derivative method (PED) the frequency-dependent polarizability is defined as

$$\alpha_{\alpha\beta}(-\omega;\omega) = -\left.\frac{\partial^2 \tilde{W}(t)}{\partial E^{\omega}_{\beta}\,\partial E^{0}_{\alpha}}\right|_{E=0} = -\left.\frac{\partial^2 \tilde{W}(t)}{\partial E^{0}_{\alpha}\,\partial E^{\omega}_{\beta}}\right|_{E=0}, \tag{136}$$

whereas in the quasi-energy derivative method (QED) it is defined as

$$\alpha_{\alpha\beta}(-\omega;\omega) = -\left.\frac{\partial^2 \tilde{W}(t)}{\partial E^{-\omega}_{\alpha}\,\partial E^{\omega}_{\beta}}\right|_{E=0}. \tag{137}$$

PED and QED expressions for the frequency-dependent polarizability and first hyperpolarizability at the SCF and MP2 level have been derived by Rice and Handy (1991, 1992) and Aiga and co-workers [Aiga, Sasagane and Itoh (1993), and Aiga and Itoh (1996)], whereas QED expressions have also been presented at the coupled-cluster level by Aiga, Sasagane and Itoh (1994) and for second and third hyperpolarizabilities in SCF, MCSCF, full and truncated CI wavefunctions [Sasagane, Aiga and Itoh (1993)]. At the SCF level both methods lead to the time-dependent HF approximation (TDHF). The quasi-energy derivative method for an MCSCF energy was also shown to yield the same expressions as obtained from response theory. However, at the MP2 level the PED and QED methods differ despite the fact that they give identical results at zero frequency. In both methods the Hamiltonian is partitioned in the following way

$$\hat{H}^{(0)} + \hat{H}^{(1)}(t) - \mathrm{i}\frac{\partial}{\partial t} = \hat{F} - \mathrm{i}\frac{\partial}{\partial t} + \hat{V}. \tag{138}$$

The PED method develops from the usual expression for the MP2 closed-shell energy, into which the expansion of the molecular orbitals in the time-dependent field, equation (115), is inserted. In addition, the condition

$$\left\langle \Psi(t) \middle| \frac{\partial \Psi(t)}{\partial E^{0}_{\alpha}} \right\rangle = 0 \tag{139}$$

has to be fulfilled for the first-order MP wavefunction

$$|\Psi(t)\rangle = |\Phi_{\mathrm{SCF}}(t)\rangle + \left|\Phi^{(1)}(t)\right\rangle. \tag{140}$$

In the QED method, by contrast, the derivatives of an MP2 quasienergy Lagrangian are taken, which is variational in the TDHF coefficients, equation (116), and the first-order MP2 amplitudes, as well as in the Lagrangian multipliers for the TDHF coefficients and first-order MP2 amplitudes. A constraint such as equation (139) is not necessary in the QED method as a result of the fact that the second derivative is taken with respect to $E_\alpha^{-\omega}$. The TDHF coefficients have to be obtained by solving the TDHF equations, as in the PED method; the first-order MP2 amplitudes and the Lagrangian multipliers for the TDHF coefficients and first-order MP2 amplitudes are obtained by solving appropriate response equations. Two further differences between the PED and QED method at the MP2 level are that the PED polarizability tensor is not symmetric and that the PED expression contains HF orbital energy differences as poles (excitation energies) whereas the QED [Aiga, Sasagane and Itoh (1993)] method has the TDHF poles and in the latest version, called QED-MP2 [Aiga and Itoh (1996)] also MP2 poles. This difference is potentially very important for low-lying excitation energies, where the error in the HF values is likely to be appreciable.

7.4.4 ILLUSTRATIVE CALCULATIONS

Many different *ab initio* methods for the calculation of molecular properties have been described in the preceding sections, with the purpose of identifying common features of the various approaches. A small selection of computational results is presented in this section, for three typical molecular properties: the nuclear magnetic shielding constant (Table 7.4); the indirect nuclear spin–spin coupling constant (Table 7.5); and the dipole polarizability (Table 7.6). This limited selection is designed to illustrate the kind of accuracy that can be achieved with state-of-the-art *ab initio* methods for small molecules and to point up the differences between some of the methods, in particular the coupled HF method (Section 7.4.2.1), the analytical derivative method at MP2, MP3, MP4, CCSD and CCSD(T) levels (Section 7.4.3.2), the polarization propagator method at the SOPPA and SOPPA(CCSD) levels (Section 7.4.1.3) and the response theory method using CCSD or MCSCF wavefunctions (Section 7.4.2.2). It is not an intention to give a review of the literature on the calculation of molecular properties. Such reviews have recently been published by Bishop (1994) for hyperpolarizabilities, by Helgaker, Jaszuński and Ruud (1999) for NMR parameters and by Ogilvie, Oddershede and Sauer (1999) for rotational *g*-factors.

A few general conclusions can be drawn from these examples. Spin–spin coupling constants in general exhibit such large correlation effects that results at the CHF/RPA level are often not even qualitatively correct. This is not the case for nuclear magnetic shielding constants,

Table 7.4. Comparison of different *ab initio* methods for the calculation of nuclear magnetic shielding constants σ (in ppm)^a.

		CHF	MCSCF	MP2	MP3	CCSD	CCSD(T)	Exp.^{b}
HF	$\sigma^{19}F$	413.6	419.6	424.2	417.8	418.1	418.6	419.7 ± 6^c
H_2O	$\sigma^{17}O$	328.1	336.0	346.1	336.7	336.9	337.9	337 ± 2^c
NH_3	$\sigma^{15}N$	262.3	—	276.5	270.1	269.7	270.7	273.3 ± 0.1^c
CH_4	$\sigma^{13}C$	194.8	198.2	201.0	198.8	198.7	198.9	198.4 ± 0.9^c
F_2	$\sigma^{19}F$	−167.9	−136.6	−170.0	−176.9	−171.1	−186.5	-192.8^c
N_2	$\sigma^{15}N$	−112.4	−52.2	−41.6	−72.2	−63.9	−58.1	-59.6 ± 1.5^c
CO	$\sigma^{13}C$	−25.5	8.2	20.6	−4.2	0.8	5.6	2.8 ± 0.9^c
	$\sigma^{17}O$	−87.7	−38.9	−46.5	−68.3	−56.0	−52.9	-36.7 ± 17.2^c

^{a}Results taken from Gauss (1995), Gauss and Stanton (1995,1996), Ruud *et al.* (1994), and Vaara *et al.* (1998).
^{b}References for the experimental values can be found in the articles given in footnote *a*.
^{c}Experimental data for the equilibrium geometry, i.e. corrected with calculated rovibrational corrections.

Table 7.5. Comparison of different *ab initio* methods for the calculation of indirect nuclear spin–spin coupling constants J (in Hz)[a].

		CHF	MCSCF	SOPPA	SOPPA (CCSD)	Exp.[b]
HF	$^1J^{1}H-^{19}F$	666.9	526.4	539.5	529.4	540[c]
H_2O	$^1J^{1}H-^{17}O$	−103.4	−83.9	−82.4	−80.6	−83.04 ± 0.02[c]
	$^2J^{1}H-^{1}H$	−22.4	−9.6	−9.1	−8.8	−7.8 ± 0.7[c]
CH_4	$^1J^{1}H-^{13}C$	156.9	135.7	126.9	122.3	120.87 ± 0.05[c]
	$^2J^{1}H-^{1}H$	−27.0	−20.8	−15.3	−14.0	−11.878 ± 0.004[c]
N_2	$^1J^{15}N-^{14}N$	−14.9	0.8	2.7	2.1	1.4 ± 0.6[c]
CO	$^1J^{13}C-^{17}O$	−5.7	16.1	20.4	18.6	15.6 ± 0.1[c]
C_2H_2	$^1J^{13}C-^{13}C$	409.5	181.2	189.3	188.7	171.5
	$^1J^{1}H-^{13}C$	411.1	232.1	262.9	253.6	242.40[c]
	$^2J^{1}H-^{13}C$	−49.9	50.1	52.6	51.7	49.2
	$^3J^{1}H-^{1}H$	84.9	10.8	12.2	11.3	10.11[c]

[a]Results taken from Enevoldsen, Oddershede and Sauer (1998), Wigglesworth *et al.* (1998), Kirpekar *et al.* (1997), Åstrand *et al.* (1999), Kaski *et al.* (1998), and Vahtras *et al.* (1993).
[b–c]See footnotes *b–c* of Table 7.4.

Table 7.6. Comparison of different *ab initio* methods for the calculation of static dipole polarizabilities α (in units of $e^2a_0{}^2E_h{}^{-1}$)[a].

	CHF	MP2	MP4	SOPPA	SOPPA (CCSD)	CCSD	Exp.[b]
HF	4.874	5.674	5.637	6.085	5.818	5.724	5.60
H_2O	8.492	9.792	9.866	10.319	9.939	9.824	9.64
NH_3	12.926	14.432	14.411	14.736	14.366	14.411	14.56
CH_4	16.120	16.754	16.704	16.853	16.520	16.709	17.27
F_2	8.593	8.219	8.622	8.903	8.525	8.550	8.38

[a]Results taken from Dalskov and Sauer (1998), and McDowell, Amos and Handy (1995).
[b]References for the experimental values can be found in the articles given in footnote *a*.

apart from notorious cases such as N_2 and CO, or for polarizabilities. Methods based on the application of Møller–Plesset perturbation theory, such as SOPPA and PED-MP2, give similar polarizabilities (Table 7.6) although the latter are consistently smaller. This can be understood on the basis that SOPPA has second-order poles (excitation energies), while PED-MP2 has RPA poles (see Section 7.4.3.3). Since singlet second-order excitation energies are generally lower than those from RPA [Packer *et al.* (1994)] and the excitation energy appears in the denominator of the linear response function, equation (73), SOPPA polarizabilities should exceed the PED-MP2 values, as is observed. Using coupled-cluster-based methods such as CCSD or CCSD(T) derivatives, CCSD response functions or SOPPA(CCSD) improves agreement with experimental values relative to MP2-based methods, as would be expected.

7.5 Vibrational averaging

The *ab initio* methods discussed in the preceding section are capable of producing molecular properties of 'experimental' accuracy, by applying either derivative or perturbation theory methods. However, absolute agreement with experimental values is constrained by the fact that properties are calculated for fixed nuclei, which ignores vibrational corrections to the property. Including these corrections is relatively straightforward for linear molecules but becomes increasingly involved for larger systems. The contribution that these corrections make is not negligible, especially for higher polarizabilities, to the extent that calculated first and second hyperpolarizabilities which do not include any vibrational correction [Bishop (1990)] are of questionable relevance to experiment, even though they may be of value for benchmarking purposes. To take one example, experimentally observable effects such as temperature dependence and isotope shifts of NMR parameters are solely due to differences in these nuclear motion corrections.

The static polarizability will be used to illustrate how these vibrational corrections can be incorporated [Bishop, Cheung and Buckingham (1980)]. A detailed description of corrections to hyperpolarizabilities can be found in the reviews by Bishop (1990, 1994, 1998). In order to incorporate the effects of nuclear motion one requires a Hamiltonian which includes kinetic energy operators for the nuclei. The corresponding eigenfunctions are the so-called vibronic wavefunctions Φ_{nv} and are characterized by electronic, n, and vibrational, v, quantum numbers. In the so-called *clamped-nucleus* treatment, the effect of the perturbation on the electronic and nuclear motion is treated sequentially. The vibronic wavefunction Φ_{0v} is therefore expressed as a product of an electronic wavefunction Ψ_0 and a vibrational wavefunction Θ_v. In the presence of a perturbation, E_β, the electronic wavefunction for the electronic ground state $n = 0$ and the vibrational wavefunction for an arbitrary vibrational state v are then given to first order as

$$\Psi_0 = \Psi_0^{(0)} + \Psi_0^{(1)} = \Psi_0^{(0)} + \sum_{n\neq 0} |\Psi_n^{(0)}\rangle \frac{\langle \Psi_n^{(0)} | \hat{O}_\beta^E E_\beta | \Psi_0^{(0)} \rangle}{W_0^{(0)} - W_n^{(0)}} \tag{141}$$

and

$$\Theta_v = \Theta_v^{(0)} + \Theta_v^{(1)} = \Theta_v^{(0)} + \sum_{n\neq 0} \left|\Theta_{v'}^{(0)}\right\rangle \frac{\left\langle \Theta_{v'}^{(0)} \right| \hat{O}_\beta^E E_\beta \left|\Theta_v^{(0)}\right\rangle}{W_v^{(0)} - W_{v'}^{(0)}}, \tag{142}$$

respectively. An expression for the polarizability can be derived [Bishop, Cheung and Buckingham (1980)] by the static response theory described in Section 7.3.3,

$$\alpha_{\alpha\beta}^{0v} = \frac{\partial}{\partial E_\beta} \langle \Phi_{0v} | \hat{O}_\alpha^E | \Phi_{0v} \rangle^{(1)}, \tag{143}$$

yielding two contributions: the electronic polarizability

$$\alpha_{\alpha\beta}^{\mathrm{e,v}} = -2\langle \Theta_v^{(0)} | \sum_{n\neq 0} \frac{\langle \Psi_0^{(0)} | \hat{O}_\alpha^E | \Psi_n^{(0)} \rangle \langle \Psi_n^{(0)} | \hat{O}_\beta^E | \Psi_0^{(0)} \rangle}{W_0^{(0)}(\{\boldsymbol{R}\}) - W_n^{(0)}(\{\boldsymbol{R}\})} | \Theta_v^{(0)} \rangle, \tag{144}$$

where $W_0^{(0)}(\{\boldsymbol{R}\})$ indicates that the electronic energies are for a given set of nuclear coordinates $\{\boldsymbol{R}\}$, and the vibrational polarizability, sometimes also called atomic polarizability

$$\alpha_{\alpha\beta}^{\mathrm{v}} = -2 \sum_{v'\neq v} \frac{\langle \Theta_v^{(0)} | \left\langle \Psi_0^{(0)} \right| \hat{O}_\alpha^E \left| \Psi_0^{(0)} \right\rangle | \Theta_{v'}^{(0)} \rangle \langle \Theta_{v'}^{(0)} | \left\langle \Psi_0^{(0)} \right| \hat{O}_\beta^E \left| \Psi_0^{(0)} \right\rangle | \Theta_v^{(0)} \rangle}{W_{0v}^{(0)} - W_{0v'}^{(0)}}. \tag{145}$$

The electronic contribution in equation (144) is simply the polarizability as given in Table 7.3 or equation (59) averaged with the vibrational wavefunction $\Theta_v^{(0)}$.

For linear molecules the vibrational wavefunctions can be obtained numerically as solutions of the one-dimensional Schrödinger equation

$$\left\{-\frac{\hbar^2}{2\mu}\left(\frac{\mathrm{d}^2}{\mathrm{d}R^2}+\frac{J(J+1)}{R^2}\right)+V(R)\right\}|\Theta_{v,J}\rangle = W_{v,J}|\Theta_{v,J}\rangle, \tag{146}$$

where J is the rotational quantum number and $V(R)$ is the nuclear repulsion term. The vibrational averaging in the *clamped-nucleus* treatment in equation (144) can then be carried out numerically, if one calculates the polarizability as a function of the internuclear distance R.

For polyatomic molecules the electronic polarizability in the *clamped-nucleus* treatment is frequently expressed as the polarizability evaluated at an equilibrium geometry $\{\boldsymbol{R}_e\}$ plus a so-called zero-point vibrational correction (ZPVC)

$$\alpha_{\alpha\beta}^{\mathrm{e,v}} = \langle\Theta_v^{(0)}|\,\alpha_{\alpha\beta}(\{\boldsymbol{R}\})\,|\Theta_v^{(0)}\rangle = \alpha_{\alpha\beta}(\{\boldsymbol{R}_e\}) + \Delta\alpha_{\alpha\beta}^{\mathrm{ZPVC}}. \tag{147}$$

The latter is usually obtained by perturbation theory. The polarizability is thereby expanded in a power series in the normal coordinates $\{Q_a\}$

$$\alpha_{\alpha\beta}(\{\boldsymbol{R}\}) = \alpha_{\alpha\beta}(\{\boldsymbol{R}_e\}) + \sum_a\left(\frac{\partial\alpha_{\alpha\beta}}{\partial Q_a}\right)Q_a + \frac{1}{2}\sum_{ab}\left(\frac{\partial^2\alpha_{\alpha\beta}}{\partial Q_a\,\partial Q_b}\right)Q_aQ_b + \cdots, \tag{148}$$

and the vibrational wavefunctions $\Theta_v^{(0)}$ are found by solving the vibrational Schrödinger equation with the anharmonic potential

$$V(\{\boldsymbol{R}\}) = V(\{\boldsymbol{R}_e\}) + \frac{1}{2}\sum_a\omega_a^2Q_a^2 + \frac{1}{6}\sum_{abc}K_{abc}Q_aQ_bQ_c + \cdots, \tag{149}$$

where ω_a and K_{abc} are the harmonic vibrational frequencies and cubic force constants, respectively. The potential is normally terminated after the cubic term and the vibrational wavefunctions $\Theta_v^{(0)}$ are then expressed as perturbed harmonic oscillator functions, i.e. they are expanded in the basis of the harmonic oscillator wavefunctions $\{\theta_v\}$

$$|\Theta_v^{(0)}\rangle \approx |\theta_v\rangle - \sum_{v'\neq v}|\theta_v\rangle\frac{\langle\theta_v|\,\frac{1}{6}\sum_{abc}K_{abc}Q_aQ_bQ_c\,|\theta_{v'}\rangle}{\hbar(\omega_{v'}-\omega_v)}. \tag{150}$$

Inserting equations (150) and (148) into equation (147) and using the properties of the harmonic oscillator functions, one obtains for the ZPVC, to first order,

$$\Delta\alpha_{\alpha\beta}^{\mathrm{ZPVC}} = -\frac{\hbar}{4}\sum_a\frac{1}{\omega_a^2}\left(\frac{\partial\alpha_{\alpha\beta}}{\partial Q_a}\right)\left(\sum_b\frac{K_{abb}}{\omega_b}\right) + \frac{\hbar}{4}\sum_a\frac{1}{\omega_a}\left(\frac{\partial^2\alpha_{\alpha\beta}}{\partial Q_a^2}\right), \tag{151}$$

where the first term arises because of the anharmonic term in the potential, equation (149), and the second comes from the non-linear term in the expansion of the polarizability, equation (148). Equivalent expressions for higher vibrational levels have also been derived [Toyama, Oka and Morino (1964)].

The effect of temperature, T, can then be included by Boltzmann averaging the polarizability over several vibrational states of energy E_v

$$\alpha_{\alpha\beta}(T) = \frac{\sum_v \alpha_{\alpha\beta}^{0v}\exp(-E_v/kT)}{\sum_v \exp(-E_v/kT)}. \tag{152}$$

Table 7.7. Calculated ZPVCs to the nuclear magnetic shielding constant (in ppm)[a].

Molecule	Property	Result at R_e	ZPVC	%
HF	$\sigma^{^{19}F}$	419.68	−10.01	2.4
HF	$\sigma^{^{1}H}$	29.01	−0.32	1.1
H_2O	$\sigma^{^{17}O}$	343.94	−9.86	2.9
H_2O	$\sigma^{^{1}H}$	30.97	−0.48	1.6
F_2	$\sigma^{^{19}F}$	−187.84	30.90	16.5
C_2H_2	$\sigma^{^{13}C}$	128.89	−3.78	2.9
C_2H_2	$\sigma^{^{1}H}$	30.45	−0.80	2.6
CO	$\sigma^{^{13}C}$	5.29	−1.82	34.5
CO	$\sigma^{^{17}O}$	−53.5	−4.8	9.0
N_2	$\sigma^{^{15}N}$	−58.7	−3.5	5.9

[a]Results taken from Sundholm, Gauss and Schäfer (1996), Wigglesworth *et al.* (1999), and Wigglesworth *et al.* (2000).

Table 7.8. Calculated ZPVCs to the indirect nuclear spin–spin coupling constant (in Hz)[a].

Molecule	Property	Result at R_e	ZPVC	%
HF	$J^{^{1}H-^{19}F}$	526.4	−26.9	5.1
H_2O	$J^{^{1}H-^{17}O}$	−81.555	3.963	4.9
H_2O	$J^{^{1}H-^{1}H}$	−8.581	0.653	7.6
CH_4	$J^{^{1}H-^{13}C}$	123.846	5.030	4.1
CH_4	$J^{^{1}H-^{1}H}$	−14.450	−0.686	4.7

[a]Results taken from Åstrand *et al.* (1999), and Wigglesworth *et al.* (1997, 1998).

Table 7.9. Calculated ZPVCs to the static dipole polarizability α (in units of $e^2a_0^2E_h^{-1}$) and static dipole second hyperpolarizability $\overline{\gamma}$ (in units of $e^4a_0^4E_h^{-3}$)[a].

Molecule	Property	Result at R_e	ZPVC	%
HF	α	5.48	0.10	1.8
CO	α	13.86	0.05	0.4
N_2	α	11.73	0.05	0.4
CH_4	$\overline{\gamma}$	2152	285	13.2

[a]Results taken from Christiansen, Hättig and Gauss (1998), and Bishop and Sauer (1997).

In Tables 7.7–7.9 some illustrative examples for the ZPVCs to nuclear magnetic shielding constants, σ, indirect nuclear spin–spin coupling constants, J, dipole polarizabilities, α and dipole second hyperpolarizabilities, $\overline{\gamma}$ are collected. The results are taken from the recent literature and were all obtained using correlated *ab initio* methods as described in Section 7.4.

A few general conclusions can be drawn from these examples. ZPVCs are usually larger for properties which describe an interaction with nuclear magnetic moments, such as the nuclear magnetic shielding constant and the spin–spin coupling constant, than for other properties such as dipole polarizabilities. Higher-order properties, such as hyperpolarizabilities, also have larger

ZPVCs than linear response properties. The large ZPVCs to the nuclear magnetic shielding constants of F_2 and CO are two extreme cases, well known in the literature.

7.6 References

Ågren, H., Vahtras, O., Koch, H., Jørgensen, P., and Helgaker, T., 1993, *J. Chem. Phys.* **98**, 6471–6423.

Åstrand, P.-O., Ruud, K., Mikkelsen, K. V., and Helgaker, T., 1999, *J. Chem. Phys.* **110**, 9463–9468.

Aiga, F., Sasagane, K., and Itoh, R., 1993, *J. Chem. Phys.* **99**, 3779–3789.

Aiga, F., Sasagane, K., and Itoh, R., 1994, *Int. J. Quant. Chem.* **51**, 87–97.

Aiga, F., and Itoh, R., 1996, *Chem. Phys.* **251**, 372–380.

Amos, R. D., and Rice, J. E., 1989, *Comput. Phys. Rep.* **10**, 147–187.

Arponen, J. S., 1983, *Ann. Phys.* (N.Y.) **151**, 311–382.

Bak, K. L., Koch, H., Oddershede, J., Christiansen, O., and Sauer, S. P. A., 2000, *J. Chem. Phys.* **112**, 4173–4185.

Bishop, D. M., 1990, *Rev. Mod. Phys.* **62**, 343–374.

Bishop, D. M., 1994, *Adv. Quant. Chem.* **25**, 1–45.

Bishop, D. M., 1998, *Adv. Chem. Phys.* **104**, 1–40.

Bishop, D. M., Cheung, L. M., and Buckingham, A. D., 1980, *Mol. Phys.* **41**, 1225–1226.

Bishop, D. M., and Sauer, S. P. A., 1997, *J. Chem. Phys.* **107**, 8502–8509.

Buckingham, A. D., 1967, *Adv. Chem. Phys.* **12**, 107–142.

Christiansen, O., Hättig, C., and Gauss, J., 1998, *J. Chem. Phys.* **109**, 4745–4757.

Cohen, H. D., and Roothaan, C. C. J., 1965, *J. Chem. Phys.* **43**, S34–S39.

Cybulski, S. M., and Bishop, D. M., 1994, *Intern. J. Quant. Chem.* **49**, 371–381.

Dalgarno, A., 1959, *Proc. R. Soc. London, Ser.* A **251**, 282–290.

Dalgaard, E., 1979, *Int. J. Quant. Chem.* **15**, 169–180.

Dalskov, E. K., and Sauer, S. P. A., 1998, *J. Phys. Chem.* A **27**, 5269–5274.

Enevoldsen, T., Oddershede, J., and Sauer, S. P. A., 1998, *Theor. Chem. Acc.* **100**, 275–284.

Frenkel, J., 1934, *Wave Mechanics Advanced General Theory*; Clarendon Press: Oxford.

Fuchs, C., Bonačić-Koutecký, V., and Koutecký, J., 1993, *J. Chem. Phys.* **98**, 3121–3140.

Fukui, H., Miura, K., and Matsuda, H., 1992, *J. Chem. Phys.* **96**, 2039–2043.

Fukui, H., Miura, K., Matsuda, H., and Baba, T., 1992, *J. Chem. Phys.* **97**, 2299–2303.

Gauss, J., 1995, *Ber. Bunsenges. Phys. Chem.* **99**, 1001–1008.

Gauss, J., and Cremer, D., 1992, *Adv. Quant. Chem.* **23**, 205–299.

Gauss, J., and Stanton, J. F., 1995, *J. Chem. Phys.* **103**, 3561–3577.

Gauss, J., and Stanton, J. F., 1996, *J. Chem. Phys.* **104**, 2574–2583.

Geertsen, J., Eriksen, S., and Oddershede, J., 1991, *Adv. Quant. Chem.* **22**, 167–209.

Handy, N. C., and Schaefer, H. F., 1984, *J. Chem. Phys.* **81**, 5031–5033.

Hansen, A. E., and Bouman, T. D., 1985, *J. Chem. Phys.* **82**, 5035–5047.

Helgaker, T., Jaszuński, M., and Ruud, K., 1999, *Chem. Rev.* **99**, 293–352.

Helgaker, T., and Jørgensen, P., 1988, *Adv. Quant. Chem.* **19**, 183–245.

Jaszuński, M., 1987, *Chem. Phys. Lett.* **140**, 130–132.

Jensen, H. J. Aa., Jørgensen, P., Ågren, H., and Olsen, J., 1988, *J. Chem. Phys.* **89**, 5354.

Jensen, M. Ø., and Hansen, Aa. E., 1999, *Adv. Quant. Chem.* **35**, 193–216.

Jørgensen, P., and Simons, J., 1981, *Second Quantization-Based Methods in Quantum Chemistry*, 1st edition; Academic Press: New York.

Kaski, J., Lantto, P., Vaara, J., and Jokisaari, J., 1998, *J. Am. Chem. Soc.* **120**, 3993–4005.

Kirpekar, S., Enevoldsen, T., Oddershede, J., and Raynes, W. T., 1997, *Mol. Phys.* **91**, 897–907.

Koch, H., and Jørgensen, P., 1990, *J. Chem. Phys.* **93**, 3333–3344.

Kutzelnigg, W., 1992, *Theor. Chim. Acta.* **83**, 263–312.

Langhoff, P. W., Epstein, S. T., and Karplus, M., 1972, *Rev. Mod. Phys.* **44**, 602–644.

Linderberg, J., and Öhrn, Y., 1973, *Propagators in Quantum Chemistry*, Academic Press: London.

Löwdin, P.-O., and Mukherjee, P. K., 1972, *Chem. Phys. Lett.* **14**, 1–7.

Maroulis, G., and Thakkar, A. J., 1988, *J. Chem. Phys.* **88**, 7623–7632.

McDowell, S. A. C., Amos, R. D., and Handy, N. C., 1995, *Chem. Phys. Lett.* **235**, 1–4.

McLachlan, A. D., and Ball, M. A., 1964, *Rev. Mod. Phys.* **36**, 844–855.

McWeeny, R., 1992, *Methods of Molecular Quantum Mechanics*, 2nd edition, Academic Press: New York.

Møller, C., and Plesset, M. S., 1934, *Phys. Rev.* **46**, 618–622.

Nielsen, E. S., Jørgensen, P., and Oddershede, J., 1980, *J. Chem. Phys.* **73**, 6238–6246.

Oddershede, J., Jørgensen, P., and Yeager, D. L., 1984, *Comput. Phys. Rep.* **2**, 33–92.

Ogilvie, J. F., Oddershede, J., and Sauer, S. P. A., 1999, *Adv. Chem. Phys.* **111**, 475–536.

Olsen, J., and Jørgensen, P., 1985, *J. Chem. Phys.* **82**, 3235–3264.

Packer, M. J., Dalskov, E. K., Enevoldsen, T., Jensen, H. J. Aa., and Oddershede, J., 1996, *J. Chem. Phys.* **105**, 5886–5900.

Packer, M. J., and Pickup, B. T., 1995, *Mol. Phys.* **84**, 1179–1192.
Packer, M. J., Dalskov, E. K., Sauer, S. P. A., and Oddershede, J., 1994, *Theor. Chim. Acta* **89**, 323–333.
Peng, H. W., 1941, *Proc. R. Soc. London, Ser.* A **178**, 499–505.
Perera, S. A., Nooijen, M., and Bartlett, R. J., 1996, *J. Chem. Phys.* **104**, 3290–3305.
Pickup, B. T., 1992, *Methods in Computational Chemistry: Theory and Computation of Molecular Properties*, Vol. 5, Wilson, S., Ed.; Plenum: New York, pp. 107–265.
Pickup, B. T., and Goscinski, O., 1973, *Mol. Phys.* **26**, 1013–1035.
Pople, J. A., Binkley, J. S., and Seeger, R., 1976, *Int. J. Quant. Chem. Symp.* **10**, 1–19.
Pople, J. A., Krishnan, R., Schlegel, H. B., and Binkley, J. S., 1979, *Int. J. Quant. Chem. Symp.* **13**, 225–241.
Pople, J. A., McIver, J. W., and Ostlund, N. S., 1968, *J. Chem. Phys.* **49**, 2960–2964.
Rice, J. E., and Handy, N. C., 1991, *J. Chem. Phys.* **94**, 4959–4971.
Rice, J. E., and Handy, N. C., 1992, *Int. J. Quant. Chem.* **43**, 91–118.
Roothaan, C. C. J., 1951, *Rev. Mod. Phys.* **23**, 69–89.
Rowe, D. J., 1968, *Rev. Mod. Phys.* **40**, 153–166.
Ruud, K., Helgaker, T., Kobayashi, R., Jørgensen, P., Bak, K. L., and Jensen, H. J. Aa., 1994, *J. Chem. Phys.* **100**, 8178–8185.
Sasagane, K., Aiga, F., and Itoh, R., 1993, *J. Chem. Phys.* **99**, 3738–3778.
Sauer, S. P. A., 1997, *J. Phys. B: At. Mol. Opt. Phys.* **30**, 3773–3780.
Sundholm, D., Gauss, J., and Schäfer, A., 1996, *J. Chem. Phys.* **105**, 11051–11059.
Toyama, M., Oka, T., and Morino, Y., 1964, *J. Mol. Spectrosc.* **13**, 193–213.
Trucks, G. W., Salter, E. A., Sosa, C., and Bartlett, R. J., 1988, *Chem. Phys. Lett.* **147**, 359–366.
Vaara, J., Lounila, J., Ruud, K., and Helgaker, T., 1998, *J. Chem. Phys.* **109**, 8388–8397.
Vahtras, O., Ågren, H., Jørgensen, P., Helgaker, T., and Jensen, H. J. Aa., 1993, *Chem. Phys. Lett.* **209**, 201–206.
Wigglesworth, R. D., Raynes, W. T., Kirpekar, S., Oddershede, J., and Sauer, S. P. A., 2000, *J. Chem. Phys.* **112**, 736–746.
Wigglesworth, R. D., Raynes, W. T., Sauer, S. P. A., and Oddershede, J., 1997, *Mol. Phys.* **92**, 77–88.
Wigglesworth, R. D., Raynes, W. T., Sauer, S. P. A., and Oddershede, J., 1998, *Mol. Phys.* **94**, 851–862.
Wigglesworth, R. D., Raynes, W. T., Sauer, S. P. A., and Oddershede, J., 1999, *Mol. Phys.* **96**, 1595–1607.
Yeager, D. L., and Jørgensen, P., 1979, *Chem. Phys. Lett.* **65**, 77–80.
Zubarev, D. N., 1974, *Nonequilibrium Statistical Thermodynamics*; Consultants Bureau: New York.

PART 3

ROTATION–VIBRATION STATES

8 PERTURBATION THEORY, EFFECTIVE HAMILTONIANS AND FORCE CONSTANTS

Kamil Sarka

Comenius University Bratislava, Slovakia

and

Jean Demaison

Université de Lille I, Villeneuve d'Ascq, France

Computational Molecular Spectroscopy. Edited by Per Jensen and Philip R. Bunker

8.1 Introduction

In this chapter we shall briefly describe the traditional or more conventional approach to the analysis of vibration–rotation spectra. It will be assumed throughout this chapter that a molecule is in an isolated, nondegenerate, singlet electronic state and the Born–Oppenheimer approximation is adopted. The conventional methods are based on *effective Hamiltonians* for the vibrational state or polyad of interacting vibrational states and on so called *spectroscopic constants* for the states in question. There are several reasons for including such a chapter in the book on computational molecular spectroscopy, the most important among them perhaps being the fact that in spite of dynamic development of *ab initio* methods more than 90 % of spectra are still analyzed by traditional methods using the effective Hamiltonians and spectroscopic constants. There are in principal two reasons for traditional approach, for heavier molecule the *ab initio* methods do not provide as yet accurate enough results and the effective Hamiltonians are also quite helpful in a process of assigning the more complex spectra with resonances and local perturbations.

In vibration–rotational spectroscopy we generally have to deal with several sets of data. The first set represents the experimental data on frequencies and intensities of transitions. We shall denote this set as $\mathcal{E}_b$ if this set describes the data for one band and $\mathcal{E} = (\mathcal{E}_1, \mathcal{E}_2, \mathcal{E}_3, \mathcal{E}_4, \ldots)$ is the set of the data on all bands in a given electronic state.

The second set $\mathcal{X}$ represents the fundamental parameters of the studied molecule, such as the geometric parameters—bond distances, the interbond angles, etc., the potential constants describing the dependence of the potential on the displacements of the molecule from the reference configuration and the parameters characterizing the charge distribution in the molecule, such as a dipole moment, and its derivatives with respect to the displacements.

Finally, the third set represents so called spectroscopic constants such as the rotational constants A_v, B_v, C_v, the centrifugal distortion constants D_v^J, H_v^J, the harmonic frequencies ω_k, the anharmonic corrections x_{kl}, the zeta constants ζ_{kl}^{α}, the doubling constants q_t, etc. Again, we shall denote such set as $\mathcal{S}_v$ if it contains parameters for one vibrational state or a polyad of interacting states.

The dominant part of this book is concerned with the relation

$$\mathcal{X} \Longrightarrow \mathcal{E}, \tag{1}$$

i.e. with finding directly the set of the fundamental molecular parameters $\mathcal{X}$ that would provide optimal agreement with all experimental data $\mathcal{E}$.

In this chapter we provide first a brief account of the theory leading to the spectroscopic constants $\mathcal{S}$ and the effective Hamiltonians. This part is followed by describing how the set of spectroscopic constants $\mathcal{S}_v$ is determined from the experimental data $\mathcal{E}_i$ for one band

$$\mathcal{E}_i \Longrightarrow \mathcal{S}_v. \tag{2}$$

The symbolic equation (2) shows the strong point of the conventional method: the highly sophisticated effective Hamiltonians for the two vibrational states involved in the transitions of the studied band can be used to interpret very accurately the experimental data just for this band, and the spectroscopic constants for the states involved in transitions are determined. This makes it possible to analyze the bands one by one and the size of the problem is reduced considerably. We are obviously paying a price for this advantage because the spectroscopic constants determined are the parameters of the bands involved in transitions rather than universal constants of the studied molecule. On the other hand the spectroscopic constants of the individual states can be accumulated into a set S and the final part of this chapter describes how such a set of all available spectroscopic constants can be converted into a set of the true molecular constants $\mathcal{X}$

$$\mathcal{S} \Longrightarrow \mathcal{X}. \tag{3}$$

Particular attention will be paid to the determination of the force constants.

8.2 From the raw Hamiltonian in the Cartesian coordinates to the effective Hamiltonians and spectroscopic constants

8.2.1 THE OPTIMAL CHOICE OF VIBRATION–ROTATION COORDINATES

As a result of the Born–Oppenheimer approximation we are left with the following Hamiltonian for the motion of nuclei in the Cartesian coordinates

$$\hat{H} = -\sum_{i}^{N} \frac{\hbar^2}{2m_i} \sum_{\alpha=X,Y,Z} \frac{\partial^2}{\partial R_{i\alpha}^2} + V. \tag{4}$$

The Cartesian coordinates $R_{i\alpha}$ describe the complex motion of an atom i in a molecule; however, the complex motion can be decomposed into a superposition of the simpler motions, translation of the whole molecule, rotation of the molecule, and vibrations of the atoms in the molecule about their equilibrium positions. Therefore, it is important to choose an optimal set of coordinates which would allow to the maximum possible extent the separation of the individual motions in the Hamiltonian. We shall introduce such a set of coordinates in several steps. We begin by transformation relating the laboratory–fixed coordinates of the nuclei $\boldsymbol{R}_i$, describing the complex motion, to the rotational and vibrational coordinates

$$\boldsymbol{R}_i = \boldsymbol{R}_0 + S^{-1}(\theta, \phi, \chi)(\boldsymbol{a}_i + \boldsymbol{d}_i). \tag{5}$$

In this equation $\boldsymbol{R}_0$ is a vector containing the three laboratory-fixed coordinates of the center of mass of the molecule, S is a 3×3 rotation matrix of direction cosines relating the orientation of the molecule-fixed axes x, y, z to the orientation of the corresponding laboratory-fixed axes X, Y, Z. We shall use throughout this chapter the definition of the Euler angles given by Wilson, Decius, and Cross (1955), and Papoušek and Aliev (1982). The quantity $\boldsymbol{a}_i$ is a vector containing the three molecule-fixed coordinates of nucleus i at equilibrium, and $\boldsymbol{d}_i$ is a vector containing the three molecule-fixed components of the instantaneous displacement vector of nucleus i from its equilibrium position.

The reader can appreciate that the transformation introduced in equation (5) achieved at least formally the separation of a complex motion described by the laboratory-fixed coordinate $\boldsymbol{R}_i$ of nucleus i into a translation of the whole molecule described by the vector $\boldsymbol{R}_0$, a rotation of the molecule described by the Euler angles, and the vibrations of the nuclei described by the vectors $\boldsymbol{d}_i$. In this chapter we will be concerned only with *rigid* molecules [p. 32 in Bunker and Jensen (1998)] and therefore it is assumed that the vibrational displacements $\boldsymbol{d}_i$ are much smaller than the equilibrium position vectors $\boldsymbol{a}_i$. For the molecules where this is not true, and the amplitudes of certain internal motions of a molecule such as internal rotation or inversion are large, it is best to treat the position vectors $\boldsymbol{a}_i$ as explicit functions of the variables describing the large amplitude motion, and then the small amplitude vibrations are described by the small displacements $\boldsymbol{d}_i$ from such a reference configuration. This problem is addressed in other chapters of this volume.

The transformation described by equation (5) introduces the molecule-fixed axes formally and in order to make it really operational we must specify more precisely the molecule-fixed system of coordinates and also the equilibrium configuration of the molecule. It should be mentioned too that the transformation in equation (5) results in a set of $3N + 6$ coordinates, three components of the vector $\boldsymbol{R}_0$, three Euler angles plus $3N$ displacement coordinates $\boldsymbol{d}_i$, and because the system of N particles is completely described by $3N$ coordinates we have to make six constraints in order to make our variables independent. The most often used approach starts from the classical expression for the kinetic energy of the nuclei T

$$2T = \sum_i^N m_i(\dot{\boldsymbol{R}}_i \cdot \dot{\boldsymbol{R}}_i), \tag{6}$$

where $\dot{\boldsymbol{R}}_i = \mathrm{d}\boldsymbol{R}_i/\mathrm{d}t$ and m_i is the effective mass of the nucleus i. After substitution from equation (5) to equation (6) and after some algebraic manipulation described in detail elsewhere [Wilson, Decius, and Cross (1955), and Papoušek and Aliev (1982)] it can be shown that the translational motion is separated from the other motions exactly if

$$\sum_i^N m_i \boldsymbol{a}_i = 0, \tag{7}$$

and

$$\sum_i^N m_i \boldsymbol{d}_i = 0. \tag{8}$$

We have already anticipated this result, well known from the classical mechanics, in equation (5) where the vector $\boldsymbol{R}_0$ has been defined as the position vector of the center of mass of the molecule. Equation (7) is just the choice of a suitable origin for the reference configuration, while equation (8), called the *translational Eckart condition*, maintains the origin at the center of mass during the vibrations [Aliev and Watson (1985)].

The rotational and vibrational motions cannot be separated completely. The maximum separation, minimizing the Coriolis term that dominates the interaction, is achieved if the following *rotational Eckart condition* with the vector product is satisfied [Eckart (1935), Papoušek and Aliev (1982), Aliev and Watson (1985), Bunker and Jensen (1998)]

$$\sum_i^N m_i(\boldsymbol{a}_i \times \boldsymbol{d}_i) = 0. \tag{9}$$

Calculation of Euler angles following from rotational Eckart conditions has been discussed by Redding and Hougen (1971), and numerical examples are given in Chapter 10 of Bunker and Jensen (1998).

The next step is the transformation of the $3N$ vibrational variables $\boldsymbol{d}_i$ to the normal coordinates Q_k

$$m_i^{1/2} d_{i\alpha} = \sum_{k=1}^{3N-6} l_{i\alpha,k} Q_k \ (\alpha = x, y, z). \tag{10}$$

The transformation is orthogonal so that

$$\sum_{i,\alpha} l_{i\alpha,k} l_{i\alpha,t} = \delta_{kt}. \tag{11}$$

There are two reasons for this transformation.

- To reduce the number of vibration variables to $3N - 6$ independent variables.
- To remove the cross terms from the quadratic term in the expansion of the potential energy.

In the field-free isotropic space the potential energy of a molecule is invariant to translations and rotations and depends only on the distances between the nuclei. For *semirigid molecules* discussed here the potential energy has a deep sharp minimum at the equilibrium nuclear configuration so that the vibrational displacements are restricted to small amplitudes. Therefore the potential energy can be expanded around the equilibrium configuration e

$$V(Q) = V_e + \frac{1}{2}\sum_k \lambda_k Q_k^2 + \frac{1}{6}\sum_{kst} \Phi_{kst} Q_k Q_s Q_t + \frac{1}{24}\sum_{kstu} \Phi_{kstu} Q_k Q_s Q_t Q_u + \cdots . \tag{12}$$

The minimum V_e can be taken as the energy zero, and λ_k, Φ_{kst}, etc. are the corresponding potential energy derivatives. The linear terms vanish, because the first derivatives are zero at the minimum of energy. The transformation in equation (10) is chosen to reduce the quadratic terms in equation (12) to diagonal form. This makes it possible to separate the $(3N - 6)$-dimensional vibrational problem approximately into $3N - 6$ one-dimensional problems, because if the cubic and higher-order terms in the expansion of the potential energy are neglected then the classical vibrational Hamiltonian becomes the sum of the terms corresponding to the individual normal coordinates Q_i and their conjugated momenta P_i

$$H_{\text{classical-vib}} = \frac{1}{2}\sum_{i=1}^{3N-6} (P_i^2 + \lambda_i Q_i^2) \tag{13}$$

Because equation (10) and a $3N \times (3N - 6)$-dimensional transformation matrix l play an important role in relating the spectroscopic constants to the fundamental parameters of a molecule, we give here a useful equation [see equation (3.3.10) in Papoušek and Aliev (1982), or equation (10-138) in Bunker and Jensen (1998)]:

$$l = M^{-1/2}\tilde{B}G^{-1}L, \tag{14}$$

where the matrix l is expressed through the matrices familiar from the Wilson's GF method [Wilson, Decius, and Cross (1955)] for a harmonic force field. B is the matrix of transformation from the Cartesian coordinates to internal coordinates S, M is a $3N \times 3N$ diagonal matrix of atomic masses, G is the kinetic matrix, and the matrix L of transformation from internal coordinates to the normal coordinates is the matrix of eigenvectors of the GF matrix

$$G = BM^{-1}\tilde{B},\ S = LQ,\ GFL = L\Lambda, \tag{15}$$

where Λ is a $(3N - 6, 3N - 6)$ diagonal matrix of λ_k from equation (12).

8.2.2 THE QUANTUM-MECHANICAL HAMILTONIAN

In the previous subsection we have described how the optimal choice of the variables for the vibration–rotation problem has been achieved. The next step is derivation of a correct quantum-mechanical Hamiltonian for the problem. Generally there are two possible ways to address this issue [Aliev and Watson (1985), and Bunker and Jensen (1998)]. We have already briefly touched on one approach starting from the classical expression for the energy, followed by constructing first the classical Hamiltonian and subsequently transforming it to the quantum-mechanical one using the Podolsky 'trick' [Podolsky (1928), and Wilson, Decius, and Cross (1955)] that ensures that the Hamiltonian is Hermitian. The alternative approach starts from the quantum-mechanical Hamiltonian in the laboratory-fixed cartesian coordinates and uses the direct transformation of the momenta [Louck (1976)]. We shall not give here either of the rather lengthy derivations because several excellent presentations of the procedures are available in the literature [Wilson, Decius, and Cross (1955), Papoušek and Aliev (1982), Aliev and Watson (1985), Watson (1968), and Louck (1976)]. The final result after the Watson's simplification [Watson (1968)] can be written in a rather compact form

$$\hat{H}_{\text{vr}} = \frac{\hbar^2}{2} \sum_{\alpha,\beta=x,y,z} \mu_{\alpha\beta}(\hat{J}_\alpha - \hat{\pi}_\alpha)(\hat{J}_\beta - \hat{\pi}_\beta) + \frac{1}{2}\sum_k \hat{P}_k^2 + V(Q) - \frac{\hbar^2}{8}\sum_{\alpha=x,y,z} \mu_{\alpha\alpha}. \tag{16}$$

The translational kinetic energy has been ignored in equation (16) and will be ignored in the rest of this chapter because it is completely separable and not interesting for the effects considered here. The individual symbols used in equation (16) have the following meanings: $(\hbar\hat{J}_\alpha)$ are the components of the total angular momentum of the molecule in a molecule-fixed system of coordinates, and the dimensionless operators $(\hat{J}_\alpha)$ can be expressed through Euler angles

$$\hat{J}_\pm = \hat{J}_x \pm \mathrm{i}\hat{J}_y = \mathrm{i}e^{\mp \mathrm{i}\chi}\left(\frac{1}{\sin\theta}\frac{\partial}{\partial\phi} - \cot\theta\frac{\partial}{\partial\chi} \mp \mathrm{i}\frac{\partial}{\partial\theta}\right), \tag{17}$$

$$\hat{J}_z = -\mathrm{i}\frac{\partial}{\partial\chi}. \tag{18}$$

In many instances it is more advantageous to use instead of the operators $\hat{J}_x$, $\hat{J}_y$ the ladder operators $\hat{J}_\pm$ defined in equation (17).

The components of the vibrational angular operator $(\hbar\hat{\pi}_\alpha)$ are

$$(\hbar\hat{\pi}_\alpha) = \sum_{k,l} \zeta_{kl}^\alpha Q_k \hat{P}_l, \tag{19}$$

where

$$\zeta_{lk}^\alpha = -\zeta_{kl}^\alpha = \sum_i (l_{i\beta,l}l_{i\gamma,k} - l_{i\gamma,l}l_{i\beta,k}), \tag{20}$$

and α, β, and γ are in cyclical order (x, y, z) etc.

$\boldsymbol{\mu}$ is a 3×3 inverse tensor to inertia tensor $\boldsymbol{I}$ and can be decomposed into a matrix product [Aliev and Watson (1985)]

$$\boldsymbol{\mu} = (\boldsymbol{I}'')^{-1}\boldsymbol{I}^{\mathrm{e}}(\boldsymbol{I}'')^{-1}, \tag{21}$$

where $\boldsymbol{I}^{\mathrm{e}}$ is the inertia tensor in the equilibrium configuration

$$I_{\alpha\beta}^{\mathrm{e}} = \sum_i m_i \left[\delta_{\alpha\beta}\sum_\gamma a_{i\gamma}^2 - a_{i\alpha}a_{i\beta}\right]. \tag{22}$$

Normally the axes of the equilibrium configuration are chosen to make $\boldsymbol{I}^{\mathrm{e}}$ diagonal and we will use $I_\alpha = I_{\alpha\alpha}^{\mathrm{e}}$. The components of the tensor $\boldsymbol{I}''$ are

$$I_{\alpha\beta}'' = I_{\beta\alpha}'' = I_{\alpha\beta}^{\mathrm{e}} + \frac{1}{2}\sum_k a_k^{\alpha\beta} Q_k, \tag{23}$$

where

$$a_k^{\alpha\beta} = a_k^{\beta\alpha} = 2\sum_i m_i^{1/2}\left[\delta_{\alpha\beta}\sum_\gamma a_{i\gamma}l_{i\gamma,k} - a_{i\alpha}l_{i\beta,k}\right]. \tag{24}$$

Equation (16) must be replaced for the linear equilibrium configuration by the equation (z axis is the axis of the equilibrium configuration)

$$\hat{H}_{\mathrm{vr}} = \frac{\hbar^2}{2I'}\{(\hat{J}_x - \hat{\pi}_x)^2 + (\hat{J}_y - \hat{\pi}_y)^2\} + \frac{1}{2}\sum_k \hat{P}_k^2 + V(Q), \tag{25}$$

and

$$I' = (I''_{xx})^2/I^{\mathrm{e}}_{xx}. \tag{26}$$

An important property of this isomorphic Hamiltonian [Hougen (1962a), Watson (1970), and Bunker and Jensen (1998)] is that only those eigenvalues that belong to the zero eigenvalue of the operator $(\hat{J}_z - \hat{\pi}_z)$ are physically significant, and the commutation relations between the rotational components $(\hat{J}_\alpha)$ are the same as for nonlinear molecules.

8.2.3 THE EXPANSION OF THE QUANTUM-MECHANICAL HAMILTONIAN

The vibration–rotation Hamiltonians given in equations (16) and (25) are rather compact but, unless a brute-force direct variational approach is applied, they have to be expanded before the subsequent treatment. This is apparent in particular from the form of the tensor $\boldsymbol{\mu}$. It is convenient before expansion to rewrite the rotation–vibration Hamiltonian in a form [Bunker and Jensen (1998)]

$$\hat{H}_{\mathrm{vr}} = \frac{\hbar^2}{2}\sum_\alpha \mu^{\mathrm{e}}_{\alpha\alpha}\hat{J}_\alpha^2 + \frac{1}{2}\sum_{i=1}^{3N-6}(\hat{P}_i^2 + \lambda_i Q_i^2) \tag{27}$$

$$+\frac{\hbar^2}{2}\sum_{\alpha,\beta=x,y,z}(\mu_{\alpha\beta} - \mu^{\mathrm{e}}_{\alpha\beta})\hat{J}_\alpha\hat{J}_\beta \tag{28}$$

$$-\frac{\hbar^2}{2}\sum_{\alpha,\beta=x,y,z}\mu_{\alpha\beta}(\hat{J}_\alpha\hat{\pi}_\beta + \hat{J}_\beta\hat{\pi}_\alpha) \tag{29}$$

$$+\frac{\hbar^2}{2}\sum_{\alpha,\beta=x,y,z}\mu_{\alpha\beta}\hat{\pi}_\alpha\hat{\pi}_\beta + \frac{1}{6}\sum_{kst}\Phi_{kst}Q_kQ_sQ_t$$

$$+\frac{1}{24}\sum_{kstu}\Phi_{kstu}Q_kQ_sQ_tQ_u + \cdots. \tag{30}$$

Following arguments presented by Aliev and Watson (1985) we have ignored here the last term in equation (16), which is small. In the first line is the zeroth-order Hamiltonian consisting of the rigid rotor part and the harmonic oscillator part. The second line, equation (28), is the part respecting nonrigidity of the rotor, and giving corresponding corrections to the rotational tensor $\mu_{\alpha\beta}$. The operator in the third line, equation (29), is the operator of the Coriolis interaction between vibration and rotation and the final part corresponds to vibrational anharmonicity which itself is already the expansion of the potential energy plus the purely vibrational contribution of the vibrational angular momentum.

The expansion of the rotational tensor $\mu_{\alpha\beta}$ is relatively straightforward thanks to equation (21) [Watson (1968)]

$$\mu_{\alpha\beta} = (I_\alpha)^{-1} \left\{ I_\alpha \delta_{\alpha\beta} - \sum_k a_k^{\alpha\beta} Q_k + \frac{3}{4} \sum_{kl\gamma} a_k^{\alpha\gamma} Q_k I_\gamma^{-1} a_l^{\gamma\beta} Q_l - \cdots \right\} (I_\beta)^{-1}. \tag{31}$$

When this equation is substituted to equations (28)–(30) the expanded Hamiltonian becomes a sum of terms, each of which can be expressed as a product of the powers of vibrational operators Q_l, $\hat{P}_l$ and rotational operators $\hat{J}_\alpha$. If we denote such terms as $\hat{H}_{mn}$ where m is the sum of powers of the vibrational operators (Q_l or $\hat{P}_l$) and n is the sum of powers of the rotational operators ($\hat{J}_\alpha$), then the expanded Hamiltonian becomes [Aliev and Watson (1976)]

$$\begin{aligned} \hat{H}_{\mathrm{vr}} &= \hat{H}_{20} + \hat{H}_{30} + \hat{H}_{40} + \cdots \ \text{(vibrational terms)} \\ &+ \hat{H}_{21} + \hat{H}_{31} + \hat{H}_{41} + \cdots \ \text{(Coriolis terms)} \\ &+ \hat{H}_{02} + \hat{H}_{12} + \hat{H}_{22} + \cdots \ \text{(rotational terms).} \end{aligned} \tag{32}$$

Therefore, if the matrix elements of the simple operators Q_l, $\hat{P}_l$, $\hat{J}_\alpha$ in some basis are known, the matrix elements of the terms $\hat{H}_{mn}$ of the expanded Hamiltonian can be easily expressed. It turns out that the matrix elements of the operators Q_l, $\hat{P}_l$, $\hat{J}_\alpha$ can be expressed analytically in the basis of the eigenfunctions of the zeroth-order Hamiltonian–rigid rotor plus harmonic oscillator. We shall discuss now briefly the eigenvalues and eigenfunctions of these operators.

The harmonic oscillator

The term $\hat{H}_{20}$ is a sum of terms

$$\tfrac{1}{2}(\hat{P}_i^2 + \lambda_i Q_i^2), \tag{33}$$

and the eigenvalues of this operator are known [Wilson, Decius, and Cross (1955)]

$$E_i^0/hc = \omega_i(v_i + d_i/2), \quad \omega_i = \frac{\lambda_i^{1/2}}{2\pi c}, \tag{34}$$

$v_i = 0, 1, 2, \ldots$ is a vibrational quantum number, d_i is a degeneracy of the vibration (1 only for asymmetric top, 1 or 2 for linear and symmetric top and 1, 2 or 3 for spherical top). For the one-dimensional harmonic oscillator ($d_i = 1$) the eigenfunction corresponding to the state $v = v_i$ is

$$|v\rangle = \psi_v(q_i) = N_v \exp(-q_i^2/2) H_v(q_i), \tag{35}$$

H_v is the Hermite polynomial and N_v is normalizing factor [Wilson, Decius, and Cross (1955), and Papoušek and Aliev (1982)]. In equation (35) we have used dimensionless normal coordinate q_i defined together with the dimensionless momentum $\hat{p}_i$ as

$$q_i = \frac{\lambda_i^{1/4}}{\hbar^{1/2}} Q_i, \quad \hat{p}_i = -\mathrm{i}\frac{\partial}{\partial q_i} = \frac{1}{\hbar^{1/2}\lambda_i^{1/4}} \hat{P}_i. \tag{36}$$

The introduction of dimensionless vibrational operators q_i, $\hat{p}_i$ is consistent with the dimensionless rotational operators $\hat{J}_\alpha$ introduced in equations (17) and (18) and corresponds to the most often used form of writing the Hamiltonian in the wavenumber units (cm^{-1})

$$\hat{H} = \cdots + X_{mn} f_m(q_l, \hat{p}_l) g_n(\hat{J}_\alpha) + \cdots. \tag{37}$$

It should be understood that in this symbolic equation $\hat{H}$ is actually $\hat{H}/hc$, f_m and g_n are products of degrees m and n in the vibrational and rotational operators respectively, the matrix elements of which are dimensionless, and X_{mn} are the molecular parameters in reciprocal centimeters. With this convention the cumbersome calculation of complicated dimensions of the higher-order terms is avoided.

We shall see later several reasons why it is even more advantageous to use instead of the vibrational operators q_n, $\hat{p}_n$ the ladder operators $\hat{\mathcal{L}}_n^+$, and $\hat{\mathcal{L}}_n^-$ [Makushkin and Tyuterev (1984), Aliev and Watson (1985), and Sarka *et al.* (1997)], defined for the nondegenerate vibrations n by

$$\hat{\mathcal{L}}_n^{\sigma} = q_n - \mathrm{i}\sigma\hat{p}_n, \quad (\sigma = \pm). \tag{38}$$

For the doubly degenerate vibrations $t(d_t = 2)$ where two normal coordinates q_{ta} and q_{tb} describe the vibrations with the same wavenumber ω_t two vibrational quantum numbers are needed to specify the state completely, principal vibrational quantum number v_t and the vibrational angular momentum quantum number l_t. After transformation to the polar coordinates [Papoušek and Aliev (1982)]

$$q_{ta} = \rho\cos\beta, \tag{39}$$

$$q_{tb} = \rho\sin\beta, \tag{40}$$

the corresponding wavefunction is

$$|v, l\rangle = \Psi_{vl} = N_{vl}\,\mathrm{e}^{-\rho^2/2}\rho^{|l|}L^{|l|}_{(v+|l|)/2}(\rho)\,\mathrm{e}^{il\beta}, \tag{41}$$

where N_{vl} is the normalization factor and following ladder operators are useful [Sarka *et al.* (1997)]

$$\hat{\mathcal{L}}_{t,\varepsilon}^{\sigma} = \tfrac{1}{2}[(q_{ta} + \mathrm{i}\varepsilon q_{tb}) - i\sigma(\hat{p}_{ta} + \mathrm{i}\varepsilon\hat{p}_{tb})], \quad (\sigma, \varepsilon = \pm). \tag{42}$$

The commutation and laddering properties of the operators are

$$\left[\hat{\mathcal{L}}_n^{\sigma}, \hat{\mathcal{L}}_n^{\sigma'}\right] = \sigma' - \sigma, \tag{43}$$

$$\left[\hat{\mathcal{L}}_{t,\varepsilon}^{\sigma}, \hat{\mathcal{L}}_{t,\varepsilon'}^{\sigma'}\right] = \tfrac{1}{4}(\sigma' - \sigma)(1 - \varepsilon\varepsilon'), \tag{44}$$

$$\hat{\mathcal{L}}_n^{\sigma}|v_n\rangle = [2v_n + \sigma + 1]^{1/2}|v_n + \sigma\rangle, \tag{45}$$

$$\hat{\mathcal{L}}_{t,\varepsilon}^{\sigma}|v_t, l_t\rangle = [(v_t + \sigma\varepsilon l_t + \sigma + 1)/2]^{1/2}|v_t + \sigma, l_t + \varepsilon\rangle, \tag{46}$$

where the standard symbol $[\hat{A}, \hat{B}]$ has been used for the commutator of two operators $\hat{A}$, $\hat{B}$

$$[\hat{A}, \hat{B}] = \hat{A}\hat{B} - \hat{B}\hat{A}. \tag{47}$$

Note that plus signs of σ and ε in the operator $\hat{\mathcal{L}}$ mean that the action of this operator on a wavefunction increases the principal vibrational quantum number v and the vibrational angular momentum quantum number l_t by one, whereas minus signs of σ and ε imply a decrease of the quantum numbers by one. Another important feature of the ladder operators, apparent from equations (45) and (46) is that, unlike the operators q and $\hat{p}$, they have only one type of matrix elements.

The rigid rotor

The rigid rotor part $\hat{H}_{02}$ can be written in the new convention form introduced in equation (37) as

$$\hat{H}_{02} = \sum_{\alpha=x,y,z} B_{\alpha}^{\mathrm{e}}\hat{J}_{\alpha}^2, \tag{48}$$

where

$$B_\alpha^{e} = \frac{\hbar^2}{2hcI_\alpha}. \tag{49}$$

It is important that we have been able to transform the original Hamiltonian from Cartesian coordinates to the form where the rotational energy is expressed in the terms of total angular momentum components $\hat{J}_\alpha$ because we can use the powerful apparatus developed for this operator. The key operators are the already defined operators $\hat{J}_\pm$, $\hat{J}_z$, the operator of the Z component of the total angular momentum in the laboratory-fixed system of axes $\hat{J}_Z$, and the operator of the square of the total angular momentum $\hat{\boldsymbol{J}}^2$

$$\hat{\boldsymbol{J}}^2 = \hat{J}_x^2 + \hat{J}_y^2 + \hat{J}_z^2. \tag{50}$$

The last three operators have common eigenfunctions $|Jkm\rangle$ which we shall denote as *symmetric top wavefunctions* because they are eigenfunctions of rigid rotor operator of symmetric top molecules

$$|Jkm\rangle = \psi_{Jkm}(\theta, \phi, \chi) = \mathrm{e}^{ik\chi}\,\mathrm{e}^{im\phi}\Theta(\theta), \tag{51}$$

Θ is an associated Legendre polynomial, and

$$k, m = -J, -J+1, \ldots J-1, J.$$

The eigenvalues of the operators are

$$\hat{\boldsymbol{J}}^2|Jkm\rangle = J(J+1)|Jkm\rangle \tag{52}$$

$$\hat{J}_z|Jkm\rangle = k|Jkm\rangle, \tag{53}$$

$$\hat{J}_Z|Jkm\rangle = m|Jkm\rangle. \tag{54}$$

The nondiagonal operators $\hat{J}_\pm$ have following laddering properties that can be used for expressing their matrix elements

$$\hat{J}_\pm|Jkm\rangle = [J(J+1) - k(k\mp 1)]^{1/2}|J, k\mp 1, m\rangle, \tag{55}$$

and we can see again that they have, unlike the operators $\hat{J}_x$, $\hat{J}_y$, only one type of matrix elements. The commutators are

$$[\hat{J}_\pm, \hat{J}_z] = \pm\hat{J}_\pm, \tag{56}$$

$$[\hat{J}_\pm, \hat{J}_\mp] = \mp 2\hat{J}_z. \tag{57}$$

Because of its frequent occurrence in the nondiagonal matrix elements the symbol $F(J,k)$ is often used

$$F(J,k) = [J(J+1) - k(k+1)]^{1/2}. \tag{58}$$

Another important property of the angular momentum operators, which we have seen above, is that they are all diagonal in quantum number J, so that the Hamiltonian matrix consists of a set of noninteracting matrices for the individual values of J.

For spherical top molecules where $B_x^e = B_y^e = B_z^e = B^e$ and for linear molecules where $B_x^e = B_y^e = B^e$, and $B_z^e = 0$ the rigid rotor operator $\hat{H}_{02}$ and its eigenvalues E_{02} take a particularly simple form

$$\hat{H}_{02} = B^e\hat{\boldsymbol{J}}^2, \tag{59}$$

$$E_{02} = B^e J(J+1). \tag{60}$$

For symmetric tops $B_x^e = B_y^e$, and

$$\hat{H}_{02} = B_x^e(\hat{J}_x^2 + \hat{J}_y^2) + B_z^e\hat{J}_z^2 = B_x^e(\hat{J}^2 - \hat{J}_z^2) + B_z^e\hat{J}_z^2, \tag{61}$$

$$E_{02} = B_x^e\{J(J+1) - k^2\} + B_z^e k^2. \tag{62}$$

For asymmetric tops where all three rotational constants B_x^e, B_y^e, B_z^e are different,

$$\hat{H}_{02} = \tfrac{1}{2}(B_x^e + B_y^e)\hat{J}^2 + \tfrac{1}{2}(2B_z^e - B_x^e - B_y^e)\hat{J}_z^2 + \tfrac{1}{4}(B_x^e - B_y^e)(\hat{J}_+^2 + \hat{J}_-^2), \tag{63}$$

and because the last term is nondiagonal, E_{02} cannot be expressed in a closed form, and they have to be obtained by diagonalization of the matrices for the rotational quantum number J in question. This completes our discussion of the zeroth-order Hamiltonian.

The explicit expressions for the individual terms of the expansion in equation (32) are given in a compact form in Table I of Aliev and Watson (1985), and we reproduce here the lower-order terms together with the expressions in the ladder operators where we have used the inverse of the transformation given in equation (38)

$$q_k = \frac{1}{2}(\hat{\mathcal{L}}_k^+ + \hat{\mathcal{L}}_k^-), \quad \hat{p}_k = \frac{i}{2}(\hat{\mathcal{L}}_k^+ - \hat{\mathcal{L}}_k^-), \tag{64}$$

$$\hat{H}_{20} = \frac{1}{2}\sum_k \omega_k(q_k^2 + \hat{p}_k^2) = \frac{1}{2}\sum_k \omega_k(\hat{\mathcal{L}}_k^+\hat{\mathcal{L}}_k^- + 1), \tag{65}$$

$$\hat{H}_{12} = \sum_k R_k q_k = \frac{1}{2}\sum_k R_k(\hat{\mathcal{L}}_k^+ + \hat{\mathcal{L}}_k^-), \tag{66}$$

$$\hat{H}_{21} = \sum_{kl} R_k^l q_k \hat{p}_l$$

$$= \frac{i}{8}\sum_{kl}\{(R_k^l + R_l^k)(\hat{\mathcal{L}}_k^+\hat{\mathcal{L}}_l^+ - \hat{\mathcal{L}}_k^-\hat{\mathcal{L}}_l^-) + (R_k^l - R_l^k)(\hat{\mathcal{L}}_k^+\hat{\mathcal{L}}_l^- - \hat{\mathcal{L}}_k^-\hat{\mathcal{L}}_l^+)\}, \tag{67}$$

$$\hat{H}_{30} = \frac{1}{6}\sum_{klm} k'_{klm} q_k q_l q_m$$

$$= \frac{1}{48}\sum_{klm} k'_{klm}\{(\hat{\mathcal{L}}_k^+\hat{\mathcal{L}}_l^+\hat{\mathcal{L}}_m^+ + \hat{\mathcal{L}}_k^-\hat{\mathcal{L}}_l^-\hat{\mathcal{L}}_m^-) + 3(\hat{\mathcal{L}}_k^+\hat{\mathcal{L}}_l^-\hat{\mathcal{L}}_m^- + \hat{\mathcal{L}}_k^-\hat{\mathcal{L}}_l^+\hat{\mathcal{L}}_m^+)\}. \tag{68}$$

The symbols R_k^l, R_k, k'_{klm} used in equations above are defined by Aliev and Watson (1985) as

$$R_k^l = -2\left(\frac{\omega_l}{\omega_k}\right)^{1/2}\sum_\alpha B_\alpha^e \zeta_{kl}^\alpha \hat{J}_\alpha = -\left(\frac{\omega_l}{\omega_k}\right)R_l^k, \tag{69}$$

$$R_k = \sum_{\alpha\beta} B_k^{\alpha\beta}\hat{J}_\alpha\hat{J}_\beta, \quad B_k^{\alpha\beta} = \hbar^{3/2}\omega_k a_k^{\alpha\beta}/(2\lambda_k^{3/4}I_\alpha I_\beta). \tag{70}$$

Because we shall subsequently use perturbation theory for the treatment of the expanded Hamiltonian in equation (32), it is rather important to assign correct *orders of magnitude* to the individual terms. This problem has been considered by Amat and Nielsen (1962), and Amat, Nielsen, and Tarrago (1971) and an excellent presentation is given by Aliev and Watson (1985), with the conclusion that the order of magnitude of the term $\hat{H}_{mn}$ is

$$\hat{H}_{mn} \simeq \kappa^{m+2n-2} r^m J^n \omega_{\text{vib}}, \tag{71}$$

where r is an abbreviation for q, p, or $\mathcal{L}$ and for low quantum states in the vibrations with $r \simeq 1$ we can ignore the term r^m. The smallness parameter κ is the Born–Oppenheimer expansion parameter of the order of 1/10, Amat and Nielsen (1962) use $\kappa \simeq 1/30$. The value J in equation (71) is the typical value of the J quantum number for the studied spectral interval. If we put $J^n \simeq 1$, equation (71) can be used for the spectra with low rotational quantum numbers such as microwave spectra, but also for the order of magnitude of the coefficients such as X_{mn} in equation (37) or $B_k^{\alpha\beta}$ in equations (66) and (70). In fact by substituting $m = 0$ and $n = 2$, and $J^n \simeq 1$ in equation (71) we find (see also equation (14) of the chapter by Bunker and Jensen in this volume)

$$B_\alpha^{\mathrm{e}} \simeq \kappa^2 \omega_{\mathrm{vib}}, \tag{72}$$

which can be used as a definition of the smallness parameter κ for the studied molecule, where B_α^{e} is a typical value of the rotational constant and ω_{vib} is a typical wavenumber for a vibration. Classification of the coefficients of the terms is more universal because it is independent of the values J, k.

In the regime $J \simeq K \simeq \kappa^{-1}$ corresponding to the range usually studied by infrared spectroscopy

$$\hat{H}_{mn} \simeq \kappa^{m+n-2} \omega_{\mathrm{vib}}, \tag{73}$$

and often $(m + n - 2)$ is called the order of magnitude of $\hat{H}_{mn}$.

It is rather amazing that order of magnitude estimates, which by their very nature are rather crude, play such an important role in a field where an extremely high precision is practically a norm. But they work well for rigid molecules, and in fact, a rigid molecule might be operationally defined as one for which the scheme defined by equation (71) is valid [Aliev and Watson (1985)].

8.2.4 THE CONTACT TRANSFORMATION TO THE EFFECTIVE HAMILTONIAN

Perturbation methods

In the preceding subsection we have obtained the expanded Hamiltonian and showed that its matrix elements are analytical expressions in the basis of the harmonic oscillator and symmetric top wavefunctions. In principle we could set up matrix representation of $\hat{H}_{\mathrm{vr}}$ in this basis set and obtain the eigenvalues and eigenfunctions by numerical diagonalization. Because the matrix blocks for each value of the quantum number J are infinitely large in vibrational quantum numbers, they would have to be truncated in order that the diagonalization, known as approximate *variational calculation*, can be carried out. Such methods are described in the chapters by Tennyson, and by Jensen, Osmann, and Bunker in this volume.

Instead of a variational approach, other approximate techniques can be used. We shall turn our attention to *perturbation methods*. The standard Rayleigh–Schrödinger method has been used by Oka (1967) and Mills (1972). The advantage of this, the perturbation method most often presented in textbooks, is that it gives better insight into the origin of various contributions to the final expressions, and its familiarity. However, because of the extremely high accuracy of the experimental results, which theory tries to match, high orders of the perturbation techniques must be used. The various methods must be judged then by the extent of the required labor. For this reason the contact transformation method is recommended by Aliev and Watson (1985) because their experience was that 'We have found that the Rayleigh–Schrödinger or Bloch methods usually require tedious summations over a large number of intermediate states and that there is a large degree of cancellation in the final algebraic reductions. The reason for these cancellations is that the perturbation formulas are implicit representations of commutators, and a significant reduction in labor is achieved by using the contact transformation method, in which the perturbation formulas are expressed explicitly as commutators that are evaluated as they arise'.

The contact transformation method was first applied by Van Vleck (1929). Shaffer and Nielsen (1939), and Shaffer, Nielsen, and Thomas (1939) applied it to the vibration–rotation problem. The results of the first contact transformation have been reviewed in a well-known *Red Paper* by Nielsen (1951). The extension of the method to higher orders, required by dynamic development of the experimental methods, is described in detail in the book by Amat, Nielsen, and Tarrago (1971), and the method has been again reviewed in an elegant, compact form by Aliev and Watson (1985).

Basics of the contact transformation method

In general, we obtain the eigenvalues and eigenfunctions of the Hamiltonian $\hat{H}$ by constructing the matrix representation of $\hat{H}_{\mathrm{vr}}$ in a basis of the zero-order wavefunctions ϕ, such as $|v\rangle|Jkm\rangle$. By diagonalization we obtain the eigenfunctions ψ as linear combinations of the original basis functions ϕ. This means that we have transformed our original basis set ϕ into another basis set ψ which is the set of eigenfunctions of $\hat{H}_{\mathrm{vr}}$

$$\hat{H}\psi = E\psi. \tag{74}$$

However, the same result can be achieved by transforming the Hamiltonian by a unitary transformation ($\hat{S}$ is Hermitian)

$$\tilde{H} = \mathrm{e}^{\mathrm{i}\hat{S}}\hat{H}\,\mathrm{e}^{-\mathrm{i}\hat{S}} \tag{75}$$

to the Hamiltonian that is diagonal in the original basis set ϕ.

$$\tilde{H}\phi = E\phi, \tag{76}$$

and the eigenvalues E are the same because unitary transformation does not change the eigenvalues, but changes the eigenfunctions

$$\phi = \mathrm{e}^{\mathrm{i}\hat{S}}\psi. \tag{77}$$

This is a formal description of the contact transformation, which up to this point does not seem to bring any advantages, because it simply represents another point of view to the diagonalization process. It should be stressed here that we do not attempt in the contact transformation to diagonalize the Hamiltonian completely, instead we want to bring it to block diagonal form and uncouple the individual vibrational states. If we are able to find algebraically such operators $\hat{S}$ that remove in the transformation of equation (75) the terms of the expanded Hamiltonian such as those in equations (66)–(68), responsible for the interactions among the different vibrational states, the Hamiltonian matrix for the expanded Hamiltonian $\hat{H}^{\mathrm{ex}}$ is transformed from an infinite one to a set of individual matrices for each vibrational state, which can be symbolically described as follows

$$\mathrm{e}^{\mathrm{i}\hat{S}}\begin{bmatrix} \hat{H}^{\mathrm{ex}}_{vv} & \hat{H}^{\mathrm{ex}}_{vv_1} & \hat{H}^{\mathrm{ex}}_{vv_2} & \cdots \\ \hat{H}^{\mathrm{ex}}_{v_1v} & \hat{H}^{\mathrm{ex}}_{v_1v_1} & \hat{H}^{\mathrm{ex}}_{v_1v_2} & \cdots \\ \hat{H}^{\mathrm{ex}}_{v_2v} & \hat{H}^{\mathrm{ex}}_{v_2 1v_1} & \hat{H}^{\mathrm{ex}}_{v_2v_2} & \cdots \\ \vdots & \vdots & \vdots & \ddots \end{bmatrix}\mathrm{e}^{-\mathrm{i}\hat{S}} = \begin{bmatrix} \hat{H}^{\mathrm{eff}}_{v} & 0 & 0 & \cdots \\ 0 & \hat{H}^{\mathrm{eff}}_{v_1} & 0 & \cdots \\ 0 & 0 & \hat{H}^{\mathrm{eff}}_{v_2} & \cdots \\ \vdots & \vdots & \vdots & \ddots \end{bmatrix}. \tag{78}$$

We can see that the result of the contact transformation is that we have now a separate *effective Hamiltonian* $\hat{H}^{\mathrm{eff}}_{v}$ for each vibrational state, or for a polyad of close vibrational states if they are in resonance and this makes it possible to speak about the *spectroscopic constants* such as B_v, D_v^J for a given vibrational state. In addition, the contact transformation brings to the effective Hamiltonian $\hat{H}_v^{\text{eff}}$ new terms. For example, we shall see that, although the original expanded Hamiltonian $\hat{H}_{\text{vr}}$ in equation (32) contained only the angular momentum operators of degree not greater than two, the Hamiltonian $\hat{H}_v^{\text{eff}}$ contains also the terms with the higher powers of these operators.

In order to determine the operators $\hat{S}$ we shall first group the terms of $\hat{H}_{\text{vr}}$ according to their order of magnitude

$$\hat{H}_{\text{vr}} = \hat{H}_0 + \lambda\hat{H}_1 + \lambda^2\hat{H}_2 + \lambda^3\hat{H}_3 + \cdots, \tag{79}$$

where λ is a book-keeping parameter that may be regarded as equal to unity and its only meaning is to show the order of magnitude of the product of two or more operators having various order of magnitudes and group the products accordingly. We postpone for a while the discussion on which terms should be in $\hat{H}_0$, $\hat{H}_1$, $\hat{H}_2$, etc. We shall consider a contact transformation of this operator by operator $\exp(i\lambda\hat{S}_1)$ which is expressed as its Taylor series expansion and the reasons why λ appears in this expression will become clear later

$$e^{i\lambda\hat{S}_1} = 1 + i\lambda\hat{S}_1 - \tfrac{1}{2}\lambda^2\hat{S}_1^2 + \cdots. \tag{80}$$

The transformed Hamiltonian $\tilde{H}$ becomes

$$\tilde{H} = \tilde{H}_0 + \lambda\tilde{H}_1 + \lambda^2\tilde{H}_2 + \cdots, \tag{81}$$

$$\tilde{H}_0 = \hat{H}_0, \tag{82}$$

$$\tilde{H}_1 = \hat{H}_1 + i[\hat{S}_1, \hat{H}_0], \tag{83}$$

$$\tilde{H}_2 = \hat{H}_2 + i[\hat{S}_1, \hat{H}_1] - \tfrac{1}{2}[\hat{S}_1, [\hat{S}_1, \hat{H}_0]]. \tag{84}$$

We recall now the expression for the energy up to the second order in the Rayleigh–Schrödinger perturbation method

$$E_n = E_n^0 + \langle n|\hat{H}_1 + \hat{H}_2|n\rangle + \sum_{m\neq n} \frac{|\langle n|\hat{H}_1|m\rangle|^2}{E_n^0 - E_m^0}. \tag{85}$$

It is apparent from this expression that if after the contact transformation the terms with nondiagonal elements are absent in $\tilde{H}_1$ then we have achieved effectively up to the second order the decoupling described by equation (78). It follows from equation (83) that it is achieved if the operator $\hat{S}_1$ satisfies for $E_n^0 \neq E_m^0$ following equation

$$\langle n|\tilde{H}_1|m\rangle = \langle n|\hat{H}_1|m\rangle - i(E_n^0 - E_m^0)\langle n|\hat{S}_1|m\rangle = 0. \tag{86}$$

The transformed Hamiltonian $\tilde{H}$ can be subjected to the second contact transformation with the operator $\exp(i\lambda^2\hat{S}_2)$ which leaves $\hat{H}_0$ and $\tilde{H}_1$ unchanged and the second-order twice-transformed Hamiltonian $\tilde{H}_2^{(2)}$ becomes

$$\tilde{H}_2^{(2)} = \tilde{H}_2 + i[\hat{S}_2, \hat{H}_0]. \tag{87}$$

This equation can be again used for finding such an operator $\hat{S}_2$ that non-diagonal terms would be absent in $\tilde{H}_2^{(2)}$. It can be shown that with this choice of $\hat{S}_2$ decoupling up to the fourth order of the perturbation theory is achieved.

8.2.5 DETERMINATION OF THE TRANSFORMATION OPERATORS $\hat{S}_i$

In the past, when the vibrational operators q, $\hat{p}$ have been used instead of the ladder operators $\hat{\mathcal{L}}_n^{\sigma}$, the determination of the transformation operators satisfying equation (86) was a rather complex process which is witnessed by the procedure outlined by Papoušek and Aliev (1982). The procedure is considerably simplified by use of the ladder operators. However, we must discuss first which operators should be included in the zeroth-order Hamiltonian $\hat{H}_0$. We take $\hat{H}_0$ to be the harmonic oscillator Hamiltonian $\hat{H}_{20}$ and the rigid-rotor Hamiltonian $\hat{H}_{02}$ will be taken as a part of $\hat{H}_1$. In this way rotational energy differences in the denominators of the perturbation theory are avoided. The presence of $\hat{H}_{02}$ in $\hat{H}_1$ does not cause any problems because it is diagonal in vibrational quantum numbers and therefore becomes immediately part of all effective Hamiltonians $\hat{H}_v^{\text{eff}}$.

With the other terms we use the standard $(m+n-2)$ order of magnitude classification so that $\hat{H}_{mn}$ becomes part of $\hat{H}_{m+n-2}$ in equation (79) which for example gives

$$\hat{H}_1 = \hat{H}_{30} + \hat{H}_{21} + \hat{H}_{12}. \tag{88}$$

We have seen in equations (83) and (87) that equations to be solved for the transformation operators are always of the form

$$\tilde{H}_{mn} = \hat{H}'_{mn} + \mathrm{i}\left[\hat{S}_{mn}, \hat{H}_{20}\right], \tag{89}$$

where $\hat{H}'_{mn}$ is the result of previous transformation and we search for $\hat{S}_{mn}$ that would bring $\hat{H}'_{mn}$ to block-diagonal form $\tilde{H}_{mn}$. The terms $\hat{H}'_{mn}$ can be always written in a form of equation [76] of Aliev and Watson (1985)

$$\hat{H}'_{mn} = \sum C(k, \sigma_k; k', \sigma_{k'}; k'', \sigma_{k''}; \cdots)\hat{\mathcal{L}}_k^{\sigma_k}\hat{\mathcal{L}}_{k'}^{\sigma_{k'}}\hat{\mathcal{L}}_{k''}^{\sigma_{k''}}\ldots, \tag{90}$$

where the coefficients $C(k, \sigma_k; k', \sigma_{k'}; k'', \sigma_{k''}; \cdots)$ are the rotational operators and we have seen examples in equations (66)–(68). Moreover, the commutation relations

$$\left[\hat{\mathcal{L}}_k^{\sigma_k}, \hat{H}_{20}\right] = -\sigma_k\omega_k\hat{\mathcal{L}}_k^{\sigma_k} \tag{91}$$

can be easily deduced from equations (43) and (65) and because $\hat{H}_{20}$ commutes with all rotational operators it can be shown easily that the desired operators $\hat{S}_{mn}$ satisfying equation (89) are

$$\hat{S}_{mn} = -\mathrm{i}\sum^{*}\frac{C(k, \sigma_k; k', \sigma_{k'}; k'', \sigma_{k''}; \cdots)}{(\sigma_k\omega_k + \sigma_{k'}\omega_{k'} + \sigma_{k''}\omega_{k''} + \cdots)}\hat{\mathcal{L}}_k^{\sigma_k}\hat{\mathcal{L}}_{k'}^{\sigma_{k'}}\hat{\mathcal{L}}_{k''}^{\sigma_{k''}}\ldots. \tag{92}$$

The asterisk above the summation sign means that the terms for which the denominator vanishes or nearly vanishes are omitted. It is quite easy to find by using this equation in combination with equations (66)–(68) that

$$\hat{S}_{12} = -\frac{\mathrm{i}}{2}\sum_k R_k(\hat{\mathcal{L}}_k^+ - \hat{\mathcal{L}}_k^-), \tag{93}$$

$$\hat{S}_{21} = \frac{1}{8}\sum_{kl}^{*}\left\{\frac{(R_k^l + R_l^k)}{(\omega_k + \omega_l)}(\hat{\mathcal{L}}_k^+\hat{\mathcal{L}}_l^+ + \hat{\mathcal{L}}_k^-\hat{\mathcal{L}}_l^-) + \frac{(R_l^k - R_k^l)}{(\omega_k - \omega_l)}(\hat{\mathcal{L}}_k^+\hat{\mathcal{L}}_l^- + \hat{\mathcal{L}}_k^-\hat{\mathcal{L}}_l^+)\right\},$$

$$\hat{S}_{30} = \frac{\mathrm{i}}{48}\sum_{klm}^{*}k'_{klm}\left\{\frac{\hat{\mathcal{L}}_k^-\hat{\mathcal{L}}_l^-\hat{\mathcal{L}}_m^- - \hat{\mathcal{L}}_k^+\hat{\mathcal{L}}_l^+\hat{\mathcal{L}}_m^+}{(\omega_k + \omega_l + \omega_m)} + 3\frac{(\hat{\mathcal{L}}_k^-\hat{\mathcal{L}}_l^+\hat{\mathcal{L}}_m^+ - \hat{\mathcal{L}}_k^+\hat{\mathcal{L}}_l^-\hat{\mathcal{L}}_m^-)}{(\omega_k - \omega_l - \omega_m)}\right\}.$$

In this way the transformation operators $\hat{S}_{mn}$ can be determined systematically. These operators can be found in the book by Amat, Nielsen, and Tarrago (1971), in Table IV of Aliev and Watson (1985), and in the book by Makushkin and Tyuterev (1984).

The terms with so called *resonant denominators* where the denominator $(\sigma_k\omega_k + \sigma_{k'}\omega_{k'} + \sigma_{k''}\omega_{k''} + \cdots)$ becomes very small have been excluded from the expression for the operator $\hat{S}_{mn}$ in equation (92) and the reasons for this exclusion are quite clear. The contact transformation method is just another form of perturbation method and vanishing or near vanishing of the resonant denominator implies that the zeroth-order energies of some states are nearly the same, which in the standard Rayleigh–Schrödinger perturbation theory implies a slow convergence, or divergence of the perturbation formulas, and the perturbation method for degenerate states, i.e. explicit diagonalization must be applied. The same is true also for the contact transformation method. We can see that in equation (86), where $\hat{H}_0 = \hat{H}_{20}$, the left-hand side has the order of magnitude $\kappa\omega_{\text{vib}}$, and because the order of magnitude of $E_n^0 - E_m^0$ is ω_{vib} the order of magnitude of $\hat{S}_1$ must be κ, and this is the reason why we have used λ in the operator $\exp(\mathrm{i}\lambda\hat{S}_1)$. The operator $\hat{S}_1$, besides block-diagonalizing $\tilde{H}_1$, contributes to all higher-order terms of the transformed Hamiltonian according to equation [15.5.12] of Papoušek and Aliev (1982)

$$\tilde{H}_n = \hat{H}_n + \sum_{m=0}^{n-1} \frac{\mathrm{i}^{n-m}}{(n-m)!} \underbrace{[\hat{S}_1, [\hat{S}_1, \ldots, [\hat{S}_1, \hat{H}_m] \ldots]]}_{n-m}, \tag{94}$$

and equation (84) is a specific example of this equation.

Let us consider what would happen if the order of magnitude of the expression $E_n^0 - E_m^0$ in equation (86) were $\kappa\omega_{\text{vib}}$, i.e. an order of magnitude smaller than it should be. For example, in the transformation operators presented in equation (93) this could be a case of the Coriolis resonance if $\omega_k \simeq \omega_l$ in the expression for $\hat{S}_{21}$, or the Fermi resonance if $\omega_k \simeq \omega_l + \omega_m$ in the expression for $\hat{S}_{30}$. If the terms with the resonant denominators were not excluded, the operator $\hat{S}_1$ would be one order of magnitude larger than it should be and it would give through equation (94) contributions up to n orders of magnitude larger than it should. This would considerably slow down fast convergence of the Hamiltonian for the medium values of J, k, which is one of its most precious properties, and the higher-order terms would have to be included. Numerical example can be found in Sarka *et al.* (1997), pp. 133–134. The situation is equivalent to the necessity to go to the very high orders if the standard nondegenerate perturbation theory is applied and zeroth-order energies are nearly degenerate.

If the terms with resonant denominators are omitted in equation (92) it means that the interaction block between the corresponding vibrational states cannot be removed and the effective Hamiltonian for the block in equation (78) is actually the effective Hamiltonian for the polyad of states, because interaction blocks with other nonresonant vibrational states have been removed. The near degeneracy of the vibrational levels can be *principal* or *accidental*. Principal resonances are connected with the degenerate vibrations in linear, symmetric top and spherical top molecules. Accidental resonances can occur in all types of molecules and typical examples are the Coriolis and Fermi resonances already mentioned. A more detailed classification of resonances has been given by Aliev and Watson (1985), and for axially symmetric top molecules by Amat and Nielsen (1967).

We shall now briefly touch on the question of an ordering scheme. We have used in equation (88) the so called $(m+n)$ ordering scheme which is useful for values $J \simeq \kappa^{-1}$. Watson (1983) proposed another ordering scheme which he found useful in particular for calculating higher centrifugal terms. In this ordering scheme the operators $\hat{S}_{mn}$ are arranged in the two-dimensional array

$$\begin{aligned}&\hat{S}_{12}, \hat{S}_{13}, \hat{S}_{14}, \hat{S}_{15}, \ldots,\\&\hat{S}_{21}, \hat{S}_{22}, \hat{S}_{23}, \hat{S}_{24}, \ldots,\\&\hat{S}_{30}, \hat{S}_{31}, \hat{S}_{32}, \hat{S}_{33}, \ldots,\\&\hat{S}_{40}, \hat{S}_{41}, \hat{S}_{42}, \hat{S}_{43}, \ldots,\end{aligned} \tag{95}$$

and it is recommended that we start with the lowest value of $m (m = 1)$ and transform to all orders of n before proceeding to the next higher m value. A simple example, demonstrating advantage of this ordering, is presented by Aliev and Watson (1985).

8.2.6 EFFECTIVE HAMILTONIANS, SYMMETRY CONSIDERATIONS AND PHENOMENOLOGICAL TERMS

After determining the transformation operators and choosing the appropriate ordering scheme the remaining algebraic work consists in evaluating the vibrational and rotational commutators in equations such as equation (94). It follows from the commutation relations (43) and (44) that each vibrational commutator reduces the degree of vibrational operators by 2 and from commutation relations (56) and (57) that each rotational commutator reduces the degree of rotational operators by 1.

The final result of the contact transformation is the general effective vibration–rotation Hamiltonian which is presented in Table V of Aliev and Watson (1985) and contains the terms $\tilde{H}_{0,4-8}$, $\tilde{H}_{2,1-4}$, $\tilde{H}_{3,0-2}$, $\tilde{H}_{4,0}$. It is quite fascinating that all these terms could be fitted into one table, but the compactness of the final result is somewhat misleading. The expressions are described in terms of recursive parameters which call again recursive parameters from preceding tables and the result expressed in basic molecular parameters is quite complex. As an illustration one can compare the expression for $\tilde{H}_{23}$ which is one modest line in Table V of Aliev and Watson (1985) with the expressions for molecular parameters η_{tJ}, η_{tK}, given for C_{3v} molecules in Table II of Aliev and Watson (1979), or for the parameter d_t given by Sarka and Demaison (1997). The same comments are applicable to the book by Amat, Nielsen and Tarrago (1972). However, a high degree of complexity must be anticipated in an attempt to describe by complex theory extremely precise experimental results quantitatively. On a practical level these complex formulas are needed only if the spectroscopic constants are used for force field determination, which will be described in the final part of this chapter. On a more principal level these results are important because they give real physical meaning to the determined spectroscopic constants.

We have not used until now symmetry properties of the molecule which can reduce often considerably, especially for molecules with high symmetry, the number of nonvanishing parameters and/or provide the relations among them. Applications of molecular symmetry in spectroscopy are covered quite completely in a recent book by Bunker and Jensen (1998), so a few brief remarks suffice. For example $a_k^{\alpha\beta}$ (or $B_k^{\alpha\beta}$) $\neq 0$ only if the symmetric product $[\Gamma(\hat{J}_\alpha) \times \Gamma(\hat{J}_\beta)]_{\text{sym}}$ contains representation $\Gamma(Q_k)$ and for ζ_{kl}^{α} to be nonvanishing $\Gamma(\hat{J}_\alpha)$ must belong to antisymmetric product $\{\Gamma(Q_k) \times \Gamma(Q_l)\}_{\text{antisym}}$. For definition of symmetric and antisymmetric products see p. 111 in Bunker and Jensen (1998), or footnote on p. 211 of Papoušek and Aliev (1982). For axially symmetric molecules the results have been summarized by Henry and Amat (1960a), and for spherical molecules by Watson (1971). The nonvanishing anharmonic constants and relations among them have been tabulated for axially symmetric molecules by Henry and Amat (1960b).

Symmetry properties of the molecule together with the fundamental physical properties of the operators can be used also for construction of the *phenomenological Hamiltonians* or *phenomenological higher-order terms* in the Hamiltonian. This approach is based upon the premise verified by practice that *each term* that satisfies the fundamental physical and symmetry requirements of being

- invariant to the operation of Hermitian conjugation †,
- invariant under time-reversal operation $T \ldots T^{-1}$ [Watson (1977)],
- totally symmetric in molecular symmetry group,

will be *present in the Hamiltonian*. We shall now demonstrate this approach on a simple example of an isolated nondegenerate vibrational state v.

8.2.7 THE HAMILTONIAN FOR AN ISOLATED NONDEGENERATE VIBRATIONAL STATE–THE WATSONIAN

It has been proposed [p. 352, Bunker and Jensen (1998)], and the proposal is seconded by the authors of this contribution, that the effective rotational Hamiltonian for any molecule is referred to as *Watsonian* in recognition of the contribution made by Dr. J. K. G. Watson to its development.

Several elegant group-theoretical methods can be applied to determine the structure of the phenomenological Watsonian satisfying the requirements given above. Their application, however, requires some knowledge of advanced group theory and we prefer to give here the simple method based on equation (5) in Hougen (1962b). This equation gives the effect of molecular point group operation on Euler angles [see also equation (9.1.2) in Papoušek and Aliev (1982), and Section 12.1 in Bunker and Jensen (1998)] and in combination with equations (17) and (18) can be used to determine the effect of the standard point group operations on angular momentum operators $\hat{J}_\alpha$ [see also equation (8) in Hougen (1962b), equation (19) in Watson (1977), and equation (9.3.1) in Papoušek and Aliev (1982)].

Although we are talking about point group operations it should be clear that the correct symmetry group of the Hamiltonian is the molecular symmetry group (MS) group, which consists of permutations of identical nuclei and such permutations combined with the inversion of the molecule in its center of mass. For the *rigid* molecules considered here the MS group is isomorphic to the point group defined by the equilibrium structure of the molecule, and in the following we use point group notation since this simplifies the discussion. The reader is referred to Section 4.5 of Bunker and Jensen (1998) for more detailed discussion of relations between the two groups.

The results of symmetry operations, together with the effect of Hermitian conjugation † and time reversal given in equations (13)–(15) of Watson (1977), are summarized in Table 8.1, where X is a number or a molecular parameter such as X_{mn} in equation (37) and X^* is its complex conjugate.

Table 8.1. Symmetry properties of rotation operators.

E	$\hat{J}^2$	$\hat{J}_\pm$	$\hat{J}_z$	X
†	$\hat{J}^2$	$\hat{J}_\mp$	$\hat{J}_z$	X*
$T \ldots T^{-1}$	$\hat{J}^2$	$-\hat{J}_\mp$	$-\hat{J}_z$	X*
$C_n(z)$	$\hat{J}^2$	$e^{\mp i2\pi/n}\hat{J}_\pm$	$\hat{J}_z$	X
$S_n(z)$	$\hat{J}^2$	$-e^{\mp i2\pi/n}\hat{J}_\pm$	$\hat{J}_z$	X
$C_2(x)$	$\hat{J}^2$	$\hat{J}_\mp$	$-\hat{J}_z$	X
$\sigma_v(xz)$	$\hat{J}^2$	$-\hat{J}_\mp$	$-\hat{J}_z$	X
$\sigma_h(xy)$	$\hat{J}^2$	$-\hat{J}_\pm$	$\hat{J}_z$	X
i	$\hat{J}^2$	$\hat{J}_\pm$	$\hat{J}_z$	X

When these results together with the reversal rule for the Hermitian conjugate of a product of noncommuting operators

$$(\hat{A}\hat{B})^\dagger = \hat{B}^\dagger \hat{A}^\dagger$$

are applied, it can be found rather easily that the Hamiltonian has the form

$$\hat{H}_v^{\text{eff}} = G_{\text{VL}} + \hat{H}_{\text{diag}} + \hat{H}_{\text{split}}, \tag{96}$$

where G_{VL} is the purely vibrational energy, and

$$\begin{aligned}\hat{H}_{\text{diag}} = {} & \tfrac{1}{2}(B_x^v + B_y^v)\hat{J}^2 + \tfrac{1}{2}(2B_z^v - B_x^v - B_y^v)\hat{J}_z^2 - D_J^v(\hat{J}^2)^2 - D_{JK}^v\hat{J}^2\hat{J}_z^2 - D_K^v\hat{J}_z^4 \\ & + H_J^v(\hat{J}^2)^3 + H_{JK}^v(\hat{J}^2)^2\hat{J}_z^2 + H_{KJ}^v\hat{J}^2\hat{J}_z^4 + H_K^v\hat{J}_z^6,\end{aligned} \tag{97}$$

and the form of $\hat{H}_{\text{split}}$ depends on symmetry. For asymmetric top molecules belonging to orthorhombic point groups (C_{2v}, D_2, D_{2h}) H_J^v, H_{JK}^v, H_{KJ}^v, and H_K^v are replaced by Φ_{600}, Φ_{420}, Φ_{240}, and Φ_{060}, defined in Table 8.1 in Watson (1977), and

$$\begin{aligned}\hat{H}_{\text{split}} = {} & \tfrac{1}{4}(B_x^v - B_y^v)(\hat{J}_+^2 + \hat{J}_-^2) \\ & + \tfrac{1}{2}[T_{202}^v\hat{J}^2 + T_{022}^v\hat{J}_z^2, \hat{J}_+^2 + \hat{J}_-^2]_+ + T_{004}^v(\hat{J}_+^4 + \hat{J}_-^4) \\ & + \tfrac{1}{2}[\Phi_{402}(\hat{J}^2)^2 + \Phi_{222}^v\hat{J}^2\hat{J}_z^2 + \Phi_{042}^v\hat{J}_z^4, \hat{J}_+^2 + \hat{J}_-^2]_+ \\ & + \tfrac{1}{2}[\Phi_{204}^v\hat{J}^2 + \Phi_{024}^v\hat{J}_z^2, \hat{J}_+^4 + \hat{J}_-^4]_+ + \Phi_{006}^v(\hat{J}_+^6 + \hat{J}_-^6),\end{aligned} \tag{98}$$

where the symbol $[A, B]_+$ denotes an anticommutator of two operators A, B

$$[\hat{A}, \hat{B}]_+ = \hat{A}\hat{B} + \hat{B}\hat{A}. \tag{99}$$

For symmetric top molecules of C_{3v} or D_3 symmetry

$$\begin{aligned}\hat{H}_{\text{split}} = {} & \varepsilon^v[\hat{J}_z, \hat{J}_+^3 + \hat{J}_-^3]_+ + \varepsilon_J^v\hat{J}^2[\hat{J}_z, \hat{J}_+^3 + \hat{J}_-^3]_+ \\ & + \varepsilon_K^v[\hat{J}_z^3, \hat{J}_+^3 + \hat{J}_-^3]_+ + h_3^v(\hat{J}_+^6 + \hat{J}_-^6),\end{aligned} \tag{100}$$

and only the last term $h_3^v(\hat{J}_+^6 + \hat{J}_-^6)$ is nonvanishing for molecules possessing D_{3h}, C_{6v}, D_6, or D_{6h} symmetry. For molecules belonging to D_{2d}, C_{4v}, D_4, or D_{4h} point groups

$$\hat{H}_{\text{split}} = d_2^v(\hat{J}_+^4 + \hat{J}_-^4) + h_2^v\hat{J}^2(\hat{J}_+^4 + \hat{J}_-^4) + \eta^v[\hat{J}_z^2, \hat{J}_+^4 + \hat{J}_-^4]_+. \tag{101}$$

For a rather rare group of molecules with fivefold axis C_{5v}, D_5,

$$\hat{H}_{\text{split}} = \varepsilon_{K5}^v[\hat{J}_z, \hat{J}_+^5 + \hat{J}_-^5]_+. \tag{102}$$

For linear molecules $\hat{H}_{\text{split}} = 0$ and in $\hat{H}_{\text{diag}}$ all the terms where $\hat{J}_z$ appear vanish. For spherical top molecules, where the apparatus of irreducible tensors is usually applied

$$\begin{aligned}\hat{H}_{\text{diag}} + \hat{H}_{\text{split}} = {} & B^v\hat{J}^2 - D^v(\hat{J}^2)^2 + H^v(\hat{J}^2)^3 \\ & + D_{4t}^v\hat{\Omega}_4 + H_{4t}^v\hat{J}^2\hat{\Omega}_4 + H_{6t}^v\hat{\Omega}_6,\end{aligned} \tag{103}$$

where $\hat{\Omega}_4$ and $\hat{\Omega}_6$ are the fourth- and sixth-rank tensor operators, explicit expressions for which are given by Watson (1977).

The standard orientation of the axes has been assumed in equations (100)–(102) with x axis in a σ_v plane for C_{nv} point groups, x axis along a C_2 axis for D_{2d}, D_n, D_{nh} groups except for D_3 with y axis along a C_2 axis.

Looking back at Table 8.1, we can draw the following conclusions that explain the structure of the Watsonian

- The Watsonian can be written without loss of generality as the sum of terms

$$\hat{H}_v^{\text{eff}} = \sum_{p,r,t} \mathcal{R}_{prt} (\hat{J}^2)^p [\hat{J}_z^r, \hat{J}_+^t + \hat{J}_-^t]_+. \tag{104}$$

  The Watsonian is Hermitian, if the coefficients $\mathcal{R}_{prt}$ are real. Whereas the operators $\hat{J}_+^t$, $\hat{J}_-^t$ are not Hermitian, their sum is, and the anti-commutator form $[\hat{J}_z^r, \hat{J}_+^t + \hat{J}_-^t]_+$ ensures that the product of two noncommuting Hermitian operators $\hat{J}_z^r$, and $\hat{J}_+^t + \hat{J}_-^t$ is Hermitian.

- This Watsonian with the real coefficients $\mathcal{R}_{prt}$ is invariant to the operation of time reversal $T \ldots T^{-1}$ only if the sum of powers of rotational operators is an even number, i.e.

$$r + t = 0, 2, 4, \ldots, \tag{105}$$

  because $2p$ is always even. If $t = 0$, r must be even and all products $(\hat{J}^2)^p \hat{J}_z^{2n}$ of the commuting operators $\hat{J}^2$, $\hat{J}_z$ are present in $\hat{H}_{\text{diag}}$.

- If the molecule in question belongs to a point group of symmetry that contains the operation of rotation $C_n(z)$, $(n \geq 2)$ or rotation–reflection $S_n(z)$, (n even), the only allowed combination of the $\hat{J}_\pm$ operators in the Watsonian is

$$\hat{J}_+^{sn} + \hat{J}_-^{sn}, \quad s = 0, 1, 2, \ldots,$$

  because only then

$$C_n(z)\hat{J}_\pm = \left(e^{\mp i2\pi/n}\right)^{sn} \hat{J}_\pm C_n(z) = \hat{J}_\pm C_n(z), \tag{106}$$

$$S_n(z)\hat{J}_\pm = \left(-e^{\mp i2\pi/n}\right)^{sn} \hat{J}_\pm S_n(z) = \hat{J}_\pm S_n(z). \tag{107}$$

  If the point group contains both $C_{n1}(z)$ and $S_{n2}(z)$, we take n to be the larger of two integers $n1, n2$. That explains why only the terms $\hat{J}_+^4 + \hat{J}_-^4$ appear in the $\hat{H}_{\text{split}}$ for the group D_{2d} containing both C_2 and S_4 operations and why only the terms $\hat{J}_+^6 + \hat{J}_-^6$ appear in the $\hat{H}_{\text{split}}$ for the group D_3.

These conclusions that can be drawn from Table 8.1 explain completely the structure of the Watsonian for nondegenerate vibrational states. For the molecules with lower than orthorhombic symmetry, such as C_s, C_i, extra terms allowed by lower symmetry in equation (98) can always be removed by an additional rotational contact transformation [Watson (1977)].

It has been shown here how a phenomenological Watsonian can be constructed from principal symmetry properties only. We have taken a simple example and relatively low order terms, so that explicit expressions for most of the coefficients $\mathcal{R}$ of this Watsonian are known and they do not have to be treated as phenomenological parameters. The transformed Watsonian $\hat{H}_v^{\text{eff}}$ can be written in a form similar to equation (37).

$$\hat{H}_v^{\text{eff}} = \cdots + \tilde{X}_{mn} f_m(\hat{\mathcal{L}}_l^+, \hat{\mathcal{L}}_l^-) g_n(\hat{J}^2, \hat{J}_\pm, \hat{J}_z) \ldots, \tag{108}$$

and because we have considered here one isolated nondegenerate vibrational state only, all spectroscopic constants–coefficients of the terms in $\hat{H}_v^{\text{eff}}$ are actually the matrix elements

$$\tilde{X}_{mn}\langle v|f_m(\hat{\mathcal{L}}_l^+, \hat{\mathcal{L}}_l^-)|v\rangle.$$

These coefficients are usually expressed as power series in $(v + d/2)$ [Papoušek and Aliev (1982), and Bunker and Jensen (1998)]

$$G_{\text{VL}} = \sum_r \omega_r \left(v_r + \frac{d_r}{2}\right) + \sum_{r\le r'} x_{rr'} \left(v_r + \frac{d_r}{2}\right)\left(v_{r'} + \frac{d_{r'}}{2}\right) + \sum_{t\le t'} g_{tt'} l_t l_{t'} + \cdots, \tag{109}$$

$$B_\xi^v = B_\xi^e - \sum_r \alpha_r^\xi \left(v_r + \frac{d_r}{2}\right) + \sum_{r\le r'} \gamma_{rr'}^\xi \left(v_r + \frac{d_r}{2}\right)\left(v_{r'} + \frac{d_{r'}}{2}\right) + \sum_{t\le t'} \gamma_{tt'}^\xi l_t l_{t'} + \cdots, \quad (\xi = x, y, z), \tag{110}$$

$$D_J^v = D_J^e + \sum_r \beta_r^J \left(v_r + \frac{d_r}{2}\right) + \cdots, \tag{111}$$

and the analogous expressions can be found for D_{JK}^v, D_K^v, H_J^v, H_{JK}^v, H_{KJ}^v, H_K^v, and for the coefficients in $\hat{H}_{\text{split}}$. We have included in these expressions also the parts depending on vibrational quantum numbers l_t which are equal to zero for a nondegenerate vibrational state, because these expressions are then valid also for degenerate vibrational states or a polyad of interacting states $v, v', \ldots$, where they appear in the matrix elements diagonal in vibrational quantum numbers $\langle v, l\|v, l\rangle$, $\langle v', l'\|v', l'\rangle, \ldots$. Using the conclusions following from symmetry properties of the Hamiltonian presented in Table 8.1 we can easily construct higher-order phenomenological terms in $\hat{H}_{\text{diag}}$ and $\hat{H}_{\text{split}}$, and for example for C_{3v} molecules it can be verified easily that the octic terms in $\hat{H}_{\text{diag}}$, and $\hat{H}_{\text{nondiag}}$ are [Fusina and Carlotti (1989), and Sarka (1989b)]

$$\hat{H}_{\text{diag}} = L_J^v(\hat{J}^2)^4 + L_{JJK}^v(\hat{J}^2)^3\hat{J}_z^2 + L_{JK}^v(\hat{J}^2)^2\hat{J}_z^4 + L_{JKK}^v(\hat{J}^2)\hat{J}_z^6 + L_K\hat{J}_z^8, \tag{112}$$

$$\hat{H}_{\text{split}} = \varepsilon_{JJ}^v\hat{J}^4[\hat{J}_z, \hat{J}_+^3 + \hat{J}_-^3]_+ + \varepsilon_{JK}^v\hat{J}^2[\hat{J}_z^3, \hat{J}_+^3 + \hat{J}_-^3]_+ + \varepsilon_{KK}^v[\hat{J}_z^5, \hat{J}_+^3 + \hat{J}_-^3]_+ + h_{3J}^v\hat{J}^2(\hat{J}_+^6 + \hat{J}_-^6) + h_{3K}^v[\hat{J}_z^2, \hat{J}_+^6 + \hat{J}_-^6]. \tag{113}$$

In this way the effective Watsonian can be expanded by adding the higher-order terms if it is required by the increased accuracy, or range of experimental data. Care must be always taken that the added terms observe the principal symmetries of the Hamiltonian. However, it is rather doubtful that explicit expressions for the higher-order spectroscopic constants will become available, unless computer-aided algorithms are used [Nikitin, Champion, and Tyuterev (1997)], because the amount of algebraic work required grows very rapidly with the increasing order of the terms. For lower-order terms explicit expressions are available. For example, the quartic distortion constants such as D_J^e, D_{JK}^e, D_K^e, etc. can be all expressed as linear combinations of the Wilson–Howard tensor $\tau_{\alpha\beta\gamma\delta}$ [Watson (1977), and Papoušek and Aliev (1982)]

$$\tau_{\alpha\beta\gamma\delta} = -2\sum_k B_k^{\alpha\beta} B_k^{\gamma\delta} \omega_k^{-1}, \tag{114}$$

where $B_k^{\alpha\beta}$ have been defined in equation (70). This tensor can be evaluated if the matrix F of harmonic potential constants has been determined [see equation [120] in Watson (1977)] or vice versa it can be used as an additional information for determination of harmonic force constants. Calculation of the quartic distortion constants is now part of the standard software for the Wilson GF method [Hedberg and Mills (1993)]. Explicit expressions are also available for α_k^ξ parameters [Watson (1977), and Papoušek and Aliev (1982)]

$$-\alpha_k^\xi = \frac{3}{4}\sum_\gamma \frac{(B_k^{\xi\gamma})^2}{B_\gamma} + \frac{2B_\xi^2}{\omega_k}\sum_l (\zeta_{kl}^\xi)^2 \frac{(3\omega_k^2 + \omega_l^2)}{(\omega_k^2 - \omega_l^2)} + \frac{1}{2}\sum_l \frac{k'_{kkl}B_l^{\xi\xi}}{\omega l}, \tag{115}$$

and for all sextic distortion constants [Aliev and Watson (1976), Aliev and Watson (1985), and Papoušek and Aliev (1982)]. For certain types of molecules these expressions are surprisingly simple [see equation (19.2.2) for the H_J constant of linear molecules or equation (18.2.11) for the h_3 constant of pyramidal XY_3 molecules in Papoušek and Aliev (1982)]. The explicit formulas for β_r^J parameters have been published only for linear molecules by Andrade e Silva and Ramadier (1965) and for equilateral X_3 molecule by Watson (1984), although the general formulas have been derived by Sørensen (unpublished). Table 8.2 summarizes information required for the calculation of various spectroscopic constants, or inversely what information can be at least in principle extracted from them.

8.2.8 THE EFFECTIVE HAMILTONIAN FOR DEGENERATE VIBRATIONAL STATES

We have completed our discussion of the effective Hamiltonian for the isolated nondegenerate vibrational states. For asymmetric top molecules where no principal vibrational degeneracies exist, the only complications may present accidental degeneracies. This problem together with the ambiguities of the determined spectroscopic constants has been considered by Perevalov and Tyuterev (1982). For linear, symmetric and spherical top molecules we also have to consider principal vibrational degeneracies. Probably the most frequently studied class of this type are the doubly degenerate vibrations of the C_{3v} molecules, where the purely vibrational energies of the states $|v=1, l=1\rangle$, and $|v=1, l=-1\rangle$ are exactly degenerate. In this case the effective Hamiltonian cannot be written in the form of equation (108) and the vibrational operators must be included explicitly [Watson *et al.* (1998)]

$$\hat{H}_{02} = B\hat{\boldsymbol{J}}^2 + (A-B)\hat{J}_z^2,$$

$$\hat{H}_{21} = -2A\zeta\hat{J}_z\hat{l}_0,$$

$$\hat{H}_{22} = 2q_{22}(\hat{l}_{+2}\hat{J}_-^2 + \hat{l}_{-2}\hat{J}_+^2) + 2q_{12}(\hat{l}_{+2}[\hat{J}_+, \hat{J}_z]_+ + \hat{l}_{-2}[\hat{J}_-, \hat{J}_z]_+),$$

$$\hat{H}_{04} = -D_J\hat{\boldsymbol{J}}^4 - D_{JK}\hat{\boldsymbol{J}}^2\hat{J}_z^2 - D_K\hat{J}_z^4 + \varepsilon[\hat{J}_+^3 + \hat{J}_-^3, \hat{J}_z]_+,$$

$$\hat{H}_{23} = \eta_J\hat{\boldsymbol{J}}^2\hat{J}_z\hat{l}_0 + \eta_K\hat{J}_z^3\hat{l}_0 + d_t(\hat{J}_+^3 + \hat{J}_-^3)\hat{l}_0,$$

Table 8.2. Information contained in the spectroscopic constants.

Information	$\tilde{H}_{mn}$	Parameters
$a_{i\alpha}$	$\tilde{H}_{02}$	B_ξ^e
$a_{i\alpha}$, $l_{i\alpha,k}$	$\tilde{H}_{20}$, $\tilde{H}_{04}$	ω_r, D_J^e, D_{JK}^e, D_K^e, ε^e, T^e, d_2^e
$a_{i\alpha}$, $l_{i\alpha,k}$, k'_{klm}	$\tilde{H}_{22}$, $\tilde{H}_{06}$	α_r^ξ, $\tilde{H}_J^e$, $\tilde{H}_{JK}^e$, $\tilde{H}_{KJ}^e$, $\tilde{H}_K^e$, Φ^e, ε_J^e, ε_K^e, ε_{K5}^e, h_3^e, h_2^e, η^e
$a_{i\alpha}$, $l_{i\alpha,k}$, k'_{klm}, k'_{klmn}	$\tilde{H}_{40}$, $\tilde{H}_{24}$, $\tilde{H}_{08}$	$x_{rr'}$, $g_{tt'}$, β_r^J, β_r^{JK}, β_r^K, L^e, ε_{JJ}^e, ε_{JK}^e, ε_{KK}^e, h_{3J}^e, h_{3K}^e

$$\begin{aligned}
\hat{H}_{24} &= \left[\left(f_{22}^{J}\hat{\boldsymbol{J}}^2 + f_{22}^{K}\hat{J}_z^2\right), \left(\hat{l}_{+2}\hat{J}_-^2 + \hat{l}_{-2}\hat{J}_+^2\right)\right]_+ + \left[\left(f_{12}^{J}\hat{\boldsymbol{J}}^2 + f_{12}^{K}\hat{J}_z^2\right),\right.\\
&\quad \left.\times \left(\hat{l}_{+2}\left[\hat{J}_+, \hat{J}_z\right]_+ + \hat{l}_{-2}\left[\hat{J}_-, \hat{J}_z\right]_+\right)\right]_+ + 2f_{42}\left(\hat{l}_{+2}\hat{J}_+^4 + \hat{l}_{-2}\hat{J}_-^4\right),\\
\hat{H}_{06} &= H_J\hat{\boldsymbol{J}}^6 + H_{JK}\hat{\boldsymbol{J}}^4\hat{J}_z^2 + H_{KJ}\hat{\boldsymbol{J}}^2\hat{J}_z^4 + H_K\hat{J}_z^6 + h_3\left(\hat{J}_+^6 + \hat{J}_-^6\right)\\
&\quad + \varepsilon_J\hat{\boldsymbol{J}}^2\left[\hat{J}_+^3 + \hat{J}_-^3, \hat{J}_z\right]_+ + \varepsilon_K\left[\hat{J}_+^3 + \hat{J}_-^3, \hat{J}_z^3\right]_+,\\
\hat{H}_{25} &= \tau_J\hat{\boldsymbol{J}}^4\hat{J}_z\hat{l}_0 + \tau_{JK}\hat{\boldsymbol{J}}^2\hat{J}_z^3\hat{l}_0 + \tau_K\hat{J}_z^5\hat{l}_0 + \tfrac{1}{2}\left[\left(d_t^J\hat{\boldsymbol{J}}^2 + 2d_t^K\hat{J}_z^2\right), \left(\hat{J}_+^3 + \hat{J}_-^3\right)\right]_+\hat{l}_0.
\end{aligned}$$

The operators $\hat{l}_0, \hat{l}_{+2}, \hat{l}_{-2}$ are identical with the operators

$$\hat{l}_t = \hat{\mathcal{L}}_+^+\hat{\mathcal{L}}_-^- - \hat{\mathcal{L}}_-^+\hat{\mathcal{L}}_+^-, \hat{\mathcal{L}}_+^+\hat{\mathcal{L}}_+^-, \hat{\mathcal{L}}_-^-\hat{\mathcal{L}}_-^+,$$

used in the review paper by Sarka *et al.* (1997). The Hamiltonian can again be constructed easily from the symmetry properties of rotational operators given in Table 8.1 and from the same properties of the vibrational operators $\hat{l}_0, \hat{l}_{+2}, \hat{l}_{-2}$ presented in Table 2, p. 151 of Sarka *et al.* (1997). This effective Hamiltonian has been widely applied to many E bands of C_{3v} molecules [see review paper Sarka *et al.* (1997)]. The explicit expressions for the parameters η_J, η_K have been given by Aliev and Watson (1979) and together with the expressions for the constants q_{22}, q_{12}, d_t they have been summarized recently by Sarka and Demaison (1997). Extension to the $v_t(\mathrm{E}) = 2$ states has been recently discussed by Sarka and Harder (1999). The effective Hamiltonian for Coriolis interacting $v_n(\mathrm{A}_1)$, $v_t(\mathrm{E})$, states have been discussed by Di Lonardo, Fusina, and Johns (1984), Lobodenko *et al.* (1987), and in the review paper of Sarka *et al.* (1997). Practically all possible interactions among the $v = 1$ states of C_{3v} molecules are discussed in a recent paper by Papoušek *et al.* (1998).

The effective Hamiltonian for excited bending vibrational states of linear molecules has been given by Yamada, Birss, and Aliev (1985). For T_d and O_h spherical top molecules Nikitin, Champion, and Tyuterev (1997) derived algorithms that can be implemented on a computer and can be used to model complex interacting states. Recursive procedures are used for generation of the terms, basis functions, commutators, and matrix elements. These algorithms can be applied also to C_{3v} molecules and the method looks very promising.

One brief remark should be given concerning the application of the ladder operators. We have already seen advantage of using these operators in construction of the transformation operator $\hat{S}$ in the contact transformation. Another advantage is ease of constructing operators responsible for a particular local interaction, which may be of relatively higher order, but are important because for several values of quantum numbers J, K the interacting states in a local resonance may be almost degenerate. An excellent example presenting a full discussion of such application to the excited bending states of linear molecule ($HC^{15}NO$) is given by Wagner *et al.* (1993). Another example of local interaction between the states $|v_5 = 1, J = 27, kl = -10\rangle$ and $|v_3 = 1, J = 27, K = 14\rangle$ of $H_3{}^{13}CF$ is described by Papoušek *et al.* (1998).

8.3 Fitting the spectroscopic constants to experimental data

8.3.1 BASIS FUNCTIONS

The contact transformation described in the preceding section symbolically in equation (78) successfully brought the Hamiltonian to the block-diagonal form, and the Hamiltonian is now represented by finite matrices. An additional advantage of the contact transformation is that the matrix elements of the Hamiltonian can be calculated in the simple basis of the harmonic oscillator and symmetric top wavefunctions. Actually, for the nondegenerate vibrational state instead of symmetric top wavefunctions, their linear combinations, obtained by Wang transformation [Watson (1977), and Papoušek and Aliev (1982)]

$$|J, 0^+\rangle = |J, 0\rangle, \tag{116}$$

$$|J, K^+\rangle = 2^{-1/2}\{|J, K\rangle + |J, -K\rangle\}, (K > 0), \tag{117}$$

$$|J, K^-\rangle = 2^{-1/2}\{|J, K\rangle - |J, -K\rangle\}, (K > 0), \tag{118}$$

are used, where $K = |k|$ and this transformation factorizes further $(2J + 1)$-dimensional matrix for a given J into four submatrices [Watson (1977), and Papoušek and Aliev (1982)]. We have ignored here the quantum number m which is the same on both sides of equations. The choice of the proper combinations of the wavefunctions and their phase factors for degenerate vibrational states has been discussed by Yamada (1983), Watson (1991), Bürger *et al.* (1997), and Di Lauro, Lattanzi, and Graner (1990a,b).

The result is that the matrix elements of the effective Hamiltonian $\hat{H}_v^{\text{eff}}$ calculated by using the matrix elements of the basic operators in equations (45), (46), (52), (53) and (55) can be expressed as

$$\left[\hat{H}_v^{\text{eff}}\right]_{ij} = f(\boldsymbol{x}, J, k, v_n, v_t, l_t, \cdots), \tag{119}$$

and they are functions of the quantum numbers and spectroscopic constants $\boldsymbol{x} = B_v, D_J, H_{JK}, q_{22}, \ldots$. The eigenvalues of these matrices are the rotation–vibration energies of the studied vibrational state (or a polyad of states).

8.3.2 LEAST-SQUARES FITTING

In vibration–rotation spectroscopy the so-called inverse problem is typical. The experimental data are the wavenumbers or frequencies of the transitions between the energy levels of a system studied, and the task of the theory is to describe and explain the observed wavenumbers or frequencies. In the process, the spectroscopic constants that contain the important information on fundamental structure parameters of the system are determined. The least-squares fit is applied almost exclusively and the fitted constants minimize the weighted sum of squares of differences between the experimental and calculated values.

$$\sum_{j=1}^{N} w_j \left(y_j^{\text{exp}} - y_j^{\text{calc}}\right)^2,$$

where w_i are the weights describing the relative importance (or accuracy) of the individual data.

The relationship between the observed experimental data and the parameters x_i, $(i = 1 \ldots m)$ can be either linear or nonlinear. In the latter case an iterative procedure is applied. From the set of the constants $\boldsymbol{x}^{(n)}$ obtained in nth iteration cycle we calculate the values y_j^{calc}, differences $\Delta y_j = y_j^{\text{exp}} - y_j^{\text{calc}}$ and the elements of the Jacobian matrix $\boldsymbol{J}$, defined as the derivatives

$$J_{ji} = \left(\partial y_j^{\text{calc}} / \partial x_i\right)_{\boldsymbol{x}=\boldsymbol{x}^{(n)}}. \tag{120}$$

From these values the set of corrections to the parameters $\boldsymbol{\Delta x}$ [Albritton, Schmeltekopf, and Zare (1976)] is calculated

$$\boldsymbol{\Delta x} = \boldsymbol{B}^{-1}\tilde{\boldsymbol{J}}\boldsymbol{W}\,\boldsymbol{\Delta y}, \tag{121}$$

where $\tilde{\boldsymbol{J}}$ denotes the matrix transposed to $\boldsymbol{J}$, and $\boldsymbol{B}$ is the normal matrix

$$\boldsymbol{B} = \tilde{\boldsymbol{J}}\boldsymbol{W}\boldsymbol{J}, \tag{122}$$

$\boldsymbol{W}$ is the diagonal matrix of the weights w_j and the exponent in $\boldsymbol{B}^{-1}$ implies matrix inversion. The set of corrections $\Delta\boldsymbol{x}$ is added to the previous set of parameters $\boldsymbol{x}^{(n)}$ and the whole process is repeated until the set of converged values x_i is obtained. If the problem is linear in the parameters x_i then

$$\boldsymbol{x} = \boldsymbol{B}^{-1}\tilde{\boldsymbol{J}}\boldsymbol{W}\boldsymbol{y}, \tag{123}$$

where $\boldsymbol{y}$ is the column matrix of the experimental data and the values of x_i are obtained in one step without iteration.

For convenience, we will assume that y_j^{exp} are uncorrelated and of equal weight. If the actual observations, say Z_j, have normal error distributions with variances σ_j^2, we can convert to equally weighted observations having unit variance by setting $y_j = Z_j/\sigma_j$ [Lees (1970)]. $\boldsymbol{J}$ is then given by

$$J_{ji} = \frac{1}{\sigma_j}\frac{\partial y_j}{\partial x_i}, \tag{124}$$

and $\boldsymbol{W}$ disappears from equations (121) and (123).

The process of determining the parameters does not always run smoothly as described above. Quite often correlation problems appear that are, from a mathematical point of view traceable to the $\boldsymbol{B}^{-1}$ part of the matrix equations. Only a nonsingular matrix, i.e. a matrix with a nonzero determinant, can be inverted. If the determinant of the matrix $\boldsymbol{B}$ is very small, the inverted matrix may have very large matrix elements. Consequently, $\Delta\boldsymbol{x}$ is large and the iteration process may diverge rather than converge and/or the parameters would be determined with large uncertainties. This is a mathematical description of the problem. However, it is more interesting to know the causes of the above-mentioned singularity or near-singularity of the matrix $\boldsymbol{B}$. These problems have been recently reviewed by Sarka *et al.* (1997), and we will give here only a brief presentation of the most salient points. Generally, the causes of the singularity of the matrix $\boldsymbol{B}$ can be divided into two groups. In the first group are the problems caused by shortage of data while the second group of the problems represent collinearities caused by indeterminacies among the fitted parameters.

8.3.3 SHORTAGE OF DATA

One cause of singularity can be shortage of data. In simple terms it means that we are trying to determine more parameters than our data would allow and the required information on some, usually higher-order parameters, simply is not there. In mathematical terms this weakness is caused by the small length $\|\boldsymbol{J}_i\|$

$$\|\boldsymbol{J}_i\| = \sqrt{\sum_j J_{ji}^2} \tag{125}$$

of $\boldsymbol{J}_i$, the ith column of the matrix $\boldsymbol{J}$ corresponding to the constant x_i causing the problem [Demaison *et al.* (1994a)]. Inspection of equation (124) suggests that there are in principle two ways of increasing the length of $\|\boldsymbol{J}_i\|$, either by reducing σ_j, i.e. by increasing the precision of the measurements, or by increasing $(\partial y_j/\partial x_i)$, which usually depends strongly on J, K for the constants appearing in the diagonal matrix elements of the Hamiltonian, or on $J - K$ for the nondiagonal constants, implying that in general transitions with higher J are required. Thus if this problem occurs the solution either lies completely in the realm of experiment, or alternately the overambitious set of parameters to be determined must be reduced.

In making decisions which spectroscopic constants, and therefore, which corresponding terms, should be included in the fit, it can be useful to apply order of magnitude analysis. If the term in the Hamiltonian is diagonal then its contribution to the energy can be estimated from equations (71) and (72). In order that this contribution is not lost in experimental noise the following relation should be satisfied:

$$\hat{H}_{mn} \simeq \kappa^{m+2n-4} J^n B > \sigma, \tag{126}$$

where σ is the accuracy of the relevant data, B is a typical value of the rotational constant for the studied molecule. The smallness parameter κ can be estimated from equation (72) or from the values of B and D–typical value of the centrifugal distortion constant

$$\kappa \approx \left(\frac{D}{B}\right)^{1/4}, \tag{127}$$

and J^n represents the corresponding dependence of the matrix element in question on the quantum numbers J, K. The most favorable values of J, K contained in experimental data set are used. If the term is nondiagonal then $\hat{H}_{mn}$ on the left-hand side of relation (126) is replaced by $\hat{H}^2_{mn}/\Delta$, where Δ is the distance between the energy levels interacting via considered term and again the most favorable combination of the factors J^n, Δ, corresponding to the data may be used, such as Δ nearest to local resonance. Simple examples are presented in Sarka *et al.* (1997) and more detailed analysis for a linear molecule can be found in Wagner *et al.* (1993).

8.3.4 COLLINEARITY CAUSED BY INDETERMINACIES. REDUCTIONS

Another cause of the singularity or near singularity of the matrix $\boldsymbol{B}$ may be the collinearity which means that the columns of the matrix $\boldsymbol{J}$ are not linearly independent and the following equation is satisfied, or nearly satisfied, for all values of j:

$$c_1 \frac{\partial y_j}{\partial x_1} + c_2 \frac{\partial y_j}{\partial x_2} + \cdots + c_p \frac{\partial y_j}{\partial x_p} = 0, \tag{128}$$

where c_i are constants. The numbering of the parameters x_i has been reordered so that the p parameters involved in the relation (128) are the first ones. It can be shown easily that if equation (128) is satisfied for all values of j, the matrix $\boldsymbol{B}$ is singular. The problem can be demonstrated on a rather simple example.

Consider the equation

$$y = x_1 z + x_2 z^2 + x_3 z(z+1), \tag{129}$$

which one might wish to fit to a set of data points (y_j, z_j) in order to determine the values of the parameters x_1, x_2, and x_3. If the least-squares normal equations are set up, one finds that the matrix $\boldsymbol{B}$ is singular, and its determinant vanishes. Comparison of equations (128) and (129) gives

$$p = 3, c_1 = 1, c_2 = 1, c_3 = -1.$$

The problem that arises can be visualized in a following way. Let us assume that we have independent information, say from *ab initio* calculations, that the 'true' values of the molecular parameters are

$$x_1 = a, x_2 = b, x_3 = d, \tag{130}$$

and we want to compare these values with the values determined from the experimental data by fitting them to equation (129). However, it is easy to see that there is an infinite number of parameter sets, one for each value Δ

$$x_1 = a + \Delta, x_2 = b + \Delta, x_3 = d - \Delta, \tag{131}$$

which all provide the same fit of data for any value of Δ. One could almost say that the poor computer carrying out the fitting cannot decide which of the infinite number of parameters sets is the correct one and the problem is 'ill-conditioned'. When such a problem happens in fitting it is customarily cured by constraining one parameter, usually to the value zero. The constraint solves the problem indeed, as can be seen from equation (128), where after constraint one variable, and therefore also one term on the left-hand side, vanishes, and the equation is not satisfied. The question is what does it mean for the parameters left in a fit. Consider constraint

$$x_3 = 0. \tag{132}$$

This constraint assigns a specific value $\Delta = d$ to thus far arbitrary parameter Δ and inspection of equation (131) shows that the parameters x_1, x_2 determined from the fit with the constraint (132) are not the 'true' molecular parameters a, and b, but the combinations of the 'true' parameters $a + d$, and $b + d$. This point is of extreme importance when the calculated values of the molecular parameters are compared with the values determined from fitting experimental data.

Introduction of the notion of collinearity seemingly did not help us in our search for the causes of singularity of the matrix $\boldsymbol{B}$, because it only transformed the original question to the question 'what then are the causes of collinearity or indeterminacies'?

In the simple example we have presented, the core of the problem could be apparent even to an untrained eye and it could give the impression that collinearity is a trivial problem. Actually, in rotation–vibration spectroscopy collinearity certainly represents a non-trivial problem and we shall see that the indeterminacy problems are caused by the fact that some of the terms in the Hamiltonian are not independent and may be inherently bound together by some relations independent of the data. The problem of indeterminacies among the spectroscopic constants in fitting the experimental data has been known for a long time. Indeed, more than thirty years have passed since Dreizler and Dendl (1965) reported that they were not able to determine all six quartic centrifugal distortion constants for nonplanar dimethylsulfide molecule from the microwave spectra [Dreizler and Rudolph (1965)]. The problem has been explained by Watson (1966,1977) who has shown that the observed indeterminacies were caused by existence of a block-diagonal unitary transformation

$$\tilde{H} = \mathrm{e}^{\mathrm{i}\hat{S}_{vv}} \hat{H}_v^{\mathrm{eff}} \mathrm{e}^{-\mathrm{i}\hat{S}_{vv}}, \tag{133}$$

which transforms the values of parameters–spectroscopic constants $x_i (i = 1 \ldots m)$–in the effective Hamiltonian $\hat{H}_v^{\mathrm{eff}}$ for the vibrational state v in question without changing its eigenvalues.

The operator $\hat{S}_{vv}$ must be block-diagonal, Hermitian, invariant with respect to the molecular symmetry group operations and change sign under the time reversal operation. These restrictive requirements usually limit the number of terms in the operator $\hat{S}$ to just a few, say p terms

$$\mathrm{i}\hat{S} = \sum_{k=1}^{p} s_k (\hat{P}_k - \hat{P}_k^{\dagger}), \tag{134}$$

where $\hat{P}_k(\hat{\mathcal{L}}^+, \hat{\mathcal{L}}^-, \hat{J}_\alpha)$ are generally the products of the vibrational operators $\hat{\mathcal{L}}^+$, $\hat{\mathcal{L}}^-$, and the rotational operators $\hat{J}_\alpha$, $\hat{P}_k^{\dagger}$ is the Hermitian conjugate of $\hat{P}_k$, and the coefficients of these terms $s_k (k = 1 \ldots p)$ are real and must satisfy order-of-magnitude limitations [Watson (1977), and Sarka *et al.* (1997)]. However, within these limitations the coefficients are free to have any value without changing the eigenvalues.

The identical form of the unitary transformation (133) with a block-diagonal operator $\hat{S}_{vv}$ and the contact transformation (75) and (78) with the operator $\hat{S}$ that is not block-diagonal may cause some confusion. It should be stressed that the contact transformation, whose effect is described by equation (78), has been chosen by us for this very positive effect of transforming the Hamiltonian to a set of effective Hamiltonians, whereas the unitary transformation (133) has been so to speak forced on us unwillingly; its effect is negative, it is the source of indeterminacies, but we must study it in order to understand them.

We have seen in the section on contact transformations that actual evaluation of equation (133) is carried out by evaluating commutators and the transformation results in the transformed values $\tilde{x}_i$ of the parameters

$$\tilde{x}_i = x_i + f_i(x_1, \ldots, x_m; s_1, \ldots, s_p). \tag{135}$$

The indeterminacy problem is caused by free parameters s_k [Watson (1966,1977), and Sarka *et al.* (1997)] and comparing equations (131) and (135) shows that the free (arbitrary) parameters s_k play in this more complex problem the same role as does the arbitrary parameter Δ in the simple example and the problem can be solved in the same way by constraining p molecular parameters $\tilde{x}_i$ to the predetermined values, p being the number of the free parameters s_i.

A direct link can be established between the existence of a unitary transformation (133) which uses the operator $\hat{S}_{vv}$ in equation (134) with the free parameters s_i, and the collinearity equation (128) [Sarka, Watson and Pracna (1999)].

Energies E_i which are eigenvalues of the Hamiltonian $\hat{H}_v^{\text{eff}}$, depend on the transformed parameters $\tilde{x}_k$, but are invariant with respect to the unitary transformation (133)

$$\frac{\partial E_i}{\partial s_k} = \sum_{l=1}^{m} \frac{\partial E_i}{\partial \tilde{x}_l}\frac{\partial \tilde{x}_l}{\partial s_k} = 0 \tag{136}$$

for all i and for all the parameters s_k in equation (134). In the fitting of spectroscopic data, the experimental data can be energies, frequencies or wavenumbers of transitions. Because the last two correspond to differences between energies, equation (136) can be generalized to

$$\sum_{l=1}^{m} \frac{\partial y_j}{\partial \tilde{x}_l}\frac{\partial \tilde{x}_l}{\partial s_k} = 0 \tag{137}$$

for all y_j $(j = 1, \ldots n)$, where y_j is the calculated value of the jth experimental quantity, be it energy, frequency or a wavenumber. When equation (137) is compared with equation (128) it becomes apparent that the derivatives $(\partial \tilde{x}_l / \partial s_k)$ in equation (137) can be identified with the constants c_l in equation (128) and the transformed parameters $\tilde{x}_i$ are the parameters x_i in equation (128). Thus we can formulate the following statement. Every parameter s_i in the operator $\hat{S}_{vv}$ causes one independent collinearity described by equation (137) causing in turn the singularity of the matrix $\boldsymbol{B}$ and indeterminacy among the parameters involved in equation (137).

The unitary transformation described by equations (133)–(135) is usually called a reduction because the operator $\hat{S}_{vv}$ may be chosen to eliminate as many terms as possible from the transformed Hamiltonian, which then becomes a reduced Hamiltonian appropriate to the fitting of the spectra. First, the symmetry analysis for the vibrational state or polyad of states in question provides the number of terms and allowed structure of the transformation operator $\hat{S}_{vv}$ for a required level of approximation. Such an analysis can be accomplished in a fashion similar to that used for construction of the phenomenological Hamiltonians, taking into account that the operator $\hat{S}_{vv}$ changes sign under time-reversal operation, or by computer-assisted algebra [Nikitin, Champion and Tyuterev (1997)]. The number of independent terms in this operator equals the number of free parameters and this in turn equals the number of necessary constraints. The operator $\hat{S}_{vv}$ is applied in equation (133) and the expressions are derived for the transformed parameters (135) in terms of the original ones and free parameters of the operator $\hat{S}_{vv}$. These expressions are used in making decisions on which of the parameters are to be constrained and for describing the effect of the constraints on the physical meaning of the remaining parameters.

Thus the reduction method provides the answers to the very practical questions related to spectroscopic data fitting, namely, how many constraints are to be applied and which parameters should be constrained, if ill-conditioned problems are to be avoided? If such constraints are applied, what is the physical meaning of the remaining parameters? For these reasons the reduction theory has been applied to the effective Hamiltonians for various systems of states. The original papers on the reduction of the rotational Hamiltonian of asymmetric rotors [Watson (1966, 1967, 1968)] were followed by a review paper summarizing the results for rotational Hamiltonians of semirigid molecules of all symmetries [Watson (1977)], and Watson's A-reduced Hamiltonian and Watson's S-reduced Hamiltonian have become part of the textbooks on rotation spectroscopy [Papoušek and Aliev (1982), and Bunker and Jensen (1998)]. The special case of the quasi-spherical symmetric top has been treated by Sarka (1989a,b).

The reduction method has also been extended to the Hamiltonians for the rotation-vibration states of asymmetric top molecules [Perevalov and Tyuterev (1981, 1982)], spherical top molecules [Perevalov and Tyuterev (1984), Perevalov, Tyuterev and Zhilinskii (1984), Tyuterev *et al.* (1984), and Nikitin, Champion and Tyuterev (1997)], symmetric top molecules [Nikitin, Champion and Tyuterev (1997), Lobodenko *et al.* (1987), Bürger *et al.* (1997), and Watson *et al.* (1998)], linear molecules [Teffo, Sulakshina and Perevalov (1992), and Teffo, Perevalov and Lyulin (1994)], and to molecules with large-amplitude motions [Sarka and Schrötter (1996), and Tang and Takagi (1993)].

The numerous applications of the reductions to the various states demonstrated repeatedly that the reductions are an appropriate solution for the indeterminacy problems and therefore they should be applied wherever necessary. However, reductions are 'state specific' in the sense that each type of vibrational state, or polyad of vibrational states requires specific reduction. It can be anticipated that computer-assisted algebraic methods [Nikitin, Champion and Tyuterev (1997)], or numerical reductions [Sarka, Watson and Pracna (1999)] will probably play an important role in reductions of complex polyads.

8.3.5 DISTINCTION BETWEEN COLLINEARITY AND SHORTAGE OF DATA

In the fitting process when problems occur it is not *a priori* clear which of the causes described is responsible.

Lees (1970) has shown that diagonalization of the matrix $\boldsymbol{B}$ provides powerful tools for diagnostics. Actually, for numerical reasons that are described in detail in his paper, it is usually the scaled matrix $\boldsymbol{B}^*$ rather than $\boldsymbol{B}$ which is diagonalized [Lees (1970), and Demaison *et al.* (1994a)].

$$\boldsymbol{B}^* = \boldsymbol{CBC}, \tag{138}$$

where $C_{ij} = \delta_{ij}/B_{ii}^{1/2}$. This scaling is equivalent to scaling the columns of the Jacobian matrix $\boldsymbol{J}$ to unit length. The diagonalization of the symmetric and positive semidefinite matrix $\boldsymbol{B}^*$ provides the diagonal matrix Λ of nonnegative eigenvalues λ_i and the orthonormal matrix of eigenvectors $\boldsymbol{V}$

$$\tilde{\boldsymbol{V}}\boldsymbol{B}^*\boldsymbol{V} = \Lambda. \tag{139}$$

The key conclusions of the Lees' paper can be summarized as follows.

- The singularity problems correspond to a small or zero eigenvalues λ_i.
- It is possible to distinguish between the near zero eigenvalues due to shortage of data and those caused by collinearity.

- The first hint can be obtained from inspection of the eigenvector V_i corresponding to the near zero eigenvalue λ_i. If the eigenvector contains only a few nonzero elements it usually implies that it is collinearity that is the cause of the problems, while if there are many nonzero elements it is likely that the problems are due to shortage of data.
- The more powerful test can be performed by hypothetical extension of the data set. It is important to realise that the calculation of the matrix $\boldsymbol{B}$ in equation (122) does not require the actual values of the experimental data y_i^{exp}, except some data needed in nonlinear problems for the reasonable estimates of the parameters x_i. One can then increase the number of data N to include transitions with very high J or K values or even formally include the transitions which would require another experimental technique and for this hypothetical set of data calculate the matrix $\boldsymbol{B}$ and eventually the eigenvalues of $\boldsymbol{B}^*$. If the $\boldsymbol{B}^*$ matrix for the original real data set provided say l suspiciously small eigenvalues λ_i and the same matrix for the hypothetical extended set provided only $p < l$ such eigenvalues, then these p small eigenvalues are caused by collinearity among the parameters, while the remaining $l - p$ small eigenvalues for the real data set are due to shortage of data. Table I in the Lees' paper [Lees (1970)] gives an impressive example of the power of the test.

The test with hypothetical data can also be used in a trial and error way for pointing out the minimum data set required for determination of the desired constants and also for finding out whether the inclusion of additional lower precision data would provide adequate reduction in rms errors of the determined constants.

The method of Lees (1970), although conceptually simple and appealing, has some shortcomings. The first problem is to define precisely what is a small eigenvalue. The second difficulty is that it is possible for an element of the eigenvector V_i to be arbitrary small though the corresponding column of $\boldsymbol{B}^*$ still belongs to the collinear relation [Belsley (1991)]. For these reasons, a great deal of work has been devoted to the problem of detecting the collinearities and, more generally, to the problem of ill-conditioning, [see for instance Watson *et al.* (1984), Femenias (1990), and Grabow, Heineking and Stahl (1992)]. Belsley (1991) has recently reviewed the existing procedures and has concluded that 'none is fully successful in diagnosing the presence of collinearity and variable involvement or in assessing collinearity's potential harm'. He has subsequently proposed a new method which is very easy to use and which is more powerful than the existing ones. Particularly, it allows one to detect whether a collinearity is harmful or not. First the columns of the Jacobian matrix $\boldsymbol{J}$ are scaled to have unit length (each term of the column vector $\boldsymbol{J}_i$ is divided by the norm $\|\boldsymbol{J}_i\|$, see equation (125)). In the following, only the scaled Jacobian matrix will be used, although it will still be named $\boldsymbol{J}$ (this is equivalent to the use of the new variable $x'_i = \|\boldsymbol{J}_i\| \cdot x_i$). Then the singular values of the $\boldsymbol{J}$ matrix are calculated: $\mu_1, \mu_2, \ldots \mu_m$. The singular-value decomposition is a powerful tool to solve a least-squares problem and, more generally, for analyzing linear systems. The singular-value decomposition of $\boldsymbol{J}$ may be written [Brodersen (1990), and Belsley (1991)],

$$\boldsymbol{J} = \boldsymbol{U}\boldsymbol{\Lambda}^{1/2}\tilde{\boldsymbol{V}}, \tag{140}$$

where $\boldsymbol{U}$ is an $N \times m$ matrix with orthonormal columns which are the eigenvectors of $\boldsymbol{J}\tilde{\boldsymbol{J}}$, $\boldsymbol{\Lambda}^{1/2}$ is an $m \times m$ diagonal matrix whose elements are the singular values μ_i of $\boldsymbol{J}$ (i.e. the square roots of the eigenvalues λ_i of $\tilde{\boldsymbol{J}}\boldsymbol{J}$), and $\boldsymbol{V}$ is an $m \times m$ orthonormal matrix whose columns are eigenvectors of $\tilde{\boldsymbol{J}}\boldsymbol{J}$. The mean advantage of this method is that the algorithms that exist for computing the singular-value decomposition are numerically far more stable than those for computing the eigensystem $\tilde{\boldsymbol{J}}\boldsymbol{J}$ (they are faster too). Following the preceding discussion, for each linear dependence among the columns of $\boldsymbol{J}$, there is one small singular value. To determine what small means, Belsley (1991) shows that the degree of ill-conditioning, i.e. the instability of the solution with respect to small changes in either experimental data or in the matrix J, depends on how small the singular value is relative to the maximal singular value. This leads us to define the scaled condition indexes of the matrix J

$$\eta_k = \frac{\mu_{\max}}{\mu_k}, k = 1, \ldots, m. \tag{141}$$

The highest condition index is the condition number $\kappa(J)$. It provides a measure of the potential sensitivity of the solution to errors in the elements of y and J. It is also useful to check whether the Hamiltonian used is fully reduced; when κ is very large, parameters must be removed from the fit until κ becomes acceptably small. This corresponds to procedures which lead to reduced Hamiltonians. The number of near-dependencies is equal to the number of high-scaled condition indexes. Following Belsley (1991), a scaled condition index is high if it is larger than about 30. To determine which parameters are involved in collinearities, Belsley (1991) defines the variance decomposition proportions. The variance–covariance matrix $\Theta(x)$ of the determined parameters is (σ is the estimated standard deviation of the fit)

$$\Theta(x) = \sigma^2 (\tilde{J} J)^{-1} = \sigma^2 V \Lambda^{-1} \tilde{V}, \tag{142}$$

and for the kth parameter x_k it is

$$\mathrm{var}(x_k) = \sigma^2 \sum_j \left(\frac{V_{kj}}{\mu_j} \right)^2. \tag{143}$$

The variance var(x_k) is decomposed into a sum of terms, each depending on one singular value μ_j which appears in the denominator. Thus, a high proportion of the variance of two or more parameters concentrated in components associated with the same small singular value provides evidence that these parameters are highly collinear. It can be simply pointed out using the variance–decomposition proportions whose definition is

$$\pi_{jk} = \left(\frac{V_{kj}}{\mu_j} \right)^2 \Big/ \sum_{i=1}^{m} \left(\frac{V_{ki}}{\mu_i} \right)^2 k, j = 1 \cdots m. \tag{144}$$

Belsley (1991) proposes the following rule: solutions are degraded when two or more variances have at least half of their magnitude (i.e. > 0.5) associated with a high condition index. The last step is to determine which collinearities are harmful, i.e. which parameters cannot be reliably determined in the least-squares fit. It is indeed worth noting that, in a problem with many parameters (which is typical in molecular spectroscopy), it is extremely difficult to avoid collinearities, and only the harmful ones should be eliminated. With that goal, Belsley (1991) defines the signal-to-noise parameter τ. For one parameter, it is simply

$$\tau^2 = \left(\frac{x_i}{s_{x_i}} \right)^2, \tag{145}$$

which is distributed as a noncentral Fisher F distribution and where s_{x_i} is the standard deviation of x_i. The signal-to-noise ratio is adequate (i.e. the corresponding parameter is well determined) if it is greater than a threshold τ_γ^2 which is tabulated in Belsley (1991) (when the number of data is large, a typical value is about 15 for one parameter). The following figure summarizes the different cases which may be encountered:

		Collinearity present? no	yes
Inadequate signal-to-noise?	no	no problems	nonharmful collinearity
	yes	short data	harmful collinearities

Belsley (1991) discusses many examples of this method in his book. An application to the centrifugal distortion analysis may be found in Demaison *et al.* (1994b). Some applications to the determination of geometrical structures of molecules are also presented in Demaison, Wlodarczak and Rudolph (1997). As an illustration, a very simple example will be discussed: the determination of the sextic centrifugal distortion constant H_J in $SiH_3^{79}Br$ [Ceausu *et al.* (1995)]. Table 8.3 collates the results which were obtained for the ground state.

It is to be noted that the experimental value of H_J derived from the microwave spectrum is quite different from the *ab initio* value. Fixing H_J to the *ab initio* value does not significantly degrade the fit because the standard deviation of the fit is still only $\sigma = 48$ kHz whereas the experimental accuracy is estimated to be about 50 kHz. Finally, a fit combining the microwave data and the ground state combination differences (GSCD from infrared data) gives a value for H_J which is a better agreement with the *ab initio* value, although it seems slightely less precisely determined (the Student test $t = H_J/s(H_J)$ is only 7.5 whereas it was 11 with the microwave data alone). There are two condition indexes which are larger than 30 and the analysis of the variance–decomposition proportions allows us to understand the problem:

η_k	B	D_J	D_{JK}	H_J	H_{JK}	H_{KJ}
50	0.777	0.677	0.167	0.563	0.181	0.075
62	0.130	0.322	0.813	0.417	0.811	0.138
sum	0.907	0.999	0.980	0.980	0.992	0.213

As the two condition indexes are of the same order of magnitude, they cannot be analyzed separately, but what should be taken into account is the sum of the variance–decomposition proportions for each parameter. What appears is that there is a near collinearity between B, D_J, D_{JK}, H_J, and H_{JK}. However, all these parameters, except H_J, are well determined if we consider their signal-to-noise parameter. The conclusion is that this collinearity is more harmful for H_J than for other parameters, and adding the GSCD lowers the collinearity. Although H_J is not determined more precisely, it is determined more accurately.

Table 8.3. Spectroscopic constants of $SiH_3{}^{79}Br^a$.

	Unit	Microwave	t	*Ab initio*	Microwave	Microwave + infrared	t
B	MHz	4321.80070 (25)			4321.80210 (18)	4321.80208 (30)	
D_J	kHz	1.86229 (14)			1.863201 (38)	1.86317 (11)	
D_{JK}	Hz	29.5717 (42)			29.5646 (56)	29.5760 (51)	
H_J	mHz	−0.270(42)	11	−0.108	−0.108 (fixed)	−0.122 (16)	7.5
H_{JK}	mHz	24.79 (82)	30	23.3	23.3 (10)	25.4 (10)	25
H_{KJ}	mHz	602.3 (25)	241	497.7	603.1 (34)	602.3 (45)	134
σ	kHz	25			48		

$^a t$ is the Student test parameter.

8.4 Potential constants

8.4.1 METHODS OF CALCULATION

In the framework of the Born–Oppenheimer approximation the electronic energy at a given set of nuclear positions is the potential energy. The equilibrium structure corresponds to the minimum of this potential energy. The potential energy surface governs the displacements of the atoms (i.e. the vibrations), and its knowledge is also important for the determination of the equilibrium structure, for a better understanding of reaction kinetics, particularly for the intramolecular vibration redistribution and the infrared multiphoton dissociation. It is also a convenient model for the study of the local mode behavior. With the exception of diatomic molecules, the complete determination of a potential energy surface from its minimum (the equilibrium structure) up to the dissociation limit is a formidable task. However, for most spectroscopic applications, it is enough to know the surface near its minimum. In that case, the potential may be developed in Taylor series as a function of the nuclear displacement coordinates around the equilibrium structure. The coefficients of this expansion are called the force field. They are usually divided into two parts: the harmonic (or quadratic force field which is the first and most important term of the expansion, the first-order force constants being zero because the expansion is made around the minimum) and the anharmonic force field (the remaining terms of the expansion whose number is theoretically infinite). In the following, we will limit ourselves to the discussion of the force fields of polyatomic molecules (three atoms or more). The determination of the potential function of a diatomic molecule is a rather different problem which is usually solved using the Rydberg–Klein–Rees (RKR) method [Wright (1988)].

In most studies, only the harmonic force field is determined. Although it is not a trivial problem, it is well understood and existing computer programs allow us to obtain the harmonic force constants provided enough experimental data are available (they may also be accurately calculated *ab initio* [Bartlett and Stanton (1994), Lee and Scuseria (1995), and Bürger and Thiel (1997)]. The reader is referred to the recent paper of Hedberg and Mills (1993) which describes in great detail the refinement of harmonic force constants. Their accurate determination requires the use of experimental data which depend only on the harmonic force field, i.e. it is necessary to correct the vibrational data for anharmonicity and to determine equilibrium quartic centrifugal distortion constants. As it is a time consuming and difficult task, it is rarely done and the anharmonicity is generally neglected. Obviously it induces systematic errors into the harmonic force constants. Ideally, to determine a harmonic force field accurately, it is almost necessary to also estimate the anharmonic force field.

The first attempt to determine the anharmonic force field of a polyatomic molecule was made in 1933 by Adel and Dennison (1933) on CO_2. However, although many efforts have been devoted to this task [see Morino (1969), Plíva (1974), Suzuki (1975), and Kuchitsu (1992) for earlier reviews], the potential has been determined for very few molecules [see for instance appendix of Kuchitsu (1992) and Appendix]. Furthermore, as in most cases many assumptions had to be made, the resulting potential might not be reliable.

The first problem to solve is the choice of the parametrization of the potential, i.e. the choice of the coordinates. Although the use of the normal coordinates Q (or the dimensionless normal coordinates q) is at first sight the most convenient, it is generally better first to set up the potential function as a power series expansion in curvilinear internal valence coordinates R (defined as the displacements in internuclear distances r, bond angles α, etc. from their equilibrium values) and then to carry out a non-linear transformation into normal coordinates in terms of which the vibration–rotation Hamiltonian is formulated [Hoy, Mills and Strey (1972)]. For semi-rigid molecules with small amplitudes of vibration, the curvilinear coordinates give a more accurate representation of the potential because atoms move around arcs of circles in the bending and torsion vibrations (i.e. the off-diagonal terms are smaller). Furthermore the parameters are isotope-independent within the Born–Oppenheimer approximation [Hoy, Mills and Strey (1972)]. Finally, they allow the comparison of force constants between related molecules. At this point it should be noted that the truncation of $V(R)$ leads to an infinite series for the corresponding $V(q)$ because the transformation is non-linear:

$$V = \frac{1}{2}\sum_r \omega_r q_r^2 + \frac{1}{6}\sum k'_{rst} q_r q_s q_t + \frac{1}{24}\sum k'_{rstu} q_r q_s q_t q_u + \cdots, \tag{146}$$

$$V = \frac{1}{2}\sum f_{ij} R_i R_j + \frac{1}{6}\sum f_{ijk} R_i R_j R_k + \frac{1}{24}\sum f_{ijkl} R_i R_j R_k R_l + \cdots. \tag{147}$$

The potential constants in normal coordinates, $k'_{rst\ldots}$ can be expressed as a function of the $f_{ijk\ldots}$ and L coefficients which depend only on the quadratic force constants and the molecular structure, see for instance [Papoušek and Aliev (1982), and Allen *et al.* (1996)]. The defining equation of the L coefficients is:

$$R_k = \sum_r L_k^r Q_r + \frac{1}{2}\sum_{r,s} L_k^{rs} Q_r Q_s + \frac{1}{6}\sum_{r,s,t} L_k^{rst} Q_r Q_s Q_t. \tag{148}$$

The L coefficients may be obtained by differentiating the R_k with respect to the normal coordinates. It is interesting to note that all higher L coefficients ($L_k^{rs\cdots}$) may be expressed in terms of the first derivatives of the L coefficients (L_k^r). The expansion of the potential is often truncated after the quartic terms, although in some cases higher-order terms have been determined. The cubic terms give only a second-order perturbation contribution whereas the quartic terms contribute in first order to the vibrational anharmonicities and the anharmonic constants. Higher-order terms give a higher-order contribution and are therefore negligible for most applications.

A Morse-type potential is often profitably used for the stretching modes.

$$V(\Delta r) = D\left[1 - \mathrm{e}^{-a\Delta r}\right]^2. \tag{149}$$

It leads to a simpler potential because all the higher-order terms (from the quartic one on) may be expressed as a simple function of the quadratic and cubic constants. For instance:

$$f_{rrrr} = \frac{7}{9}\frac{f_{rrr}^2}{f_{rr}}. \tag{150}$$

Simons–Parr–Finlan coordinates [Simons, Parr and Finlan (1973), and Simons (1974)],

$$\rho = \frac{R}{R + r_\mathrm{e}}, \tag{151}$$

are also sometimes used because they lead to a more rapid convergence of the potential expansion.

The second difficulty is the choice of the method of calculation. The second-order perturbation theory is often preferred [Hoy, Mills and Strey (1972), and Gaw *et al.* (1990)]. The calculations are at first sight easy because they are computer efficient and the perturbation theory gives closed algebraic expressions between the molecular parameters and the force constants [Mills (1972), Papoušek and Aliev (1982), Aliev and Watson (1976, 1979), Clabo *et al.* (1980), Allen *et al.* (1990), and Sarka and Demaison (1997)]. However, it is only an approximate method because it neglects all the higher-order terms which is sometimes a poor approximation [McCoy and Siebert (1992a,b), and Dateo, Lee, and Schwenke (1994)]. Furthermore, it fails when there is an anharmonic (Fermi, Darling–Dennison, ...) or a Coriolis resonance. In the latter case, terms involving the resonance denominator should be removed from the perturbation sum, and a matrix should be diagonalized to handle the interaction properly. As these resonances are rather common, it makes the perturbation theory rather cumbersome to use. To circumvent this difficulty, the numerical Van Vleck perturbation theory may be used [Chédin (1979)]. It decouples nondegenerate block of states where each block is taken to contain the degenerate and nearly degenerate vibrational states which are coupled by a resonance. This method is computer intensive and the expressions are no longer analytical. It has been used with success on several linear triatomic molecules (see Appendix). In the simple case of triatomic molecules, the vibration–rotation Hamiltonian may be set up using the harmonic oscillator basis functions and numerically diagonalized, as proposed by Whiffen for OCS [Ford, Smith, and Whiffen (1975), Whiffen (1976)] and a few other linear triatomic molecules; see Whiffen (1978), Lacy (1982), Whiffen (1985) and references cited therein. The obvious advantage of the direct diagonalization is that the results are more accurate but it is limited to triatomic molecules. Alternatively, the Hamiltonian of a triatomic molecule may be expanded as a power series of the Morse coordinates and numerically diagonalized, this is the MORBID (Morse oscillator rigid bender internal dynamics) variational approach of Jensen (1988, 1992) [Jensen, Osmann, and Kozin (1997)] where a Morse–cosine series expansion of the potential is used and where the approximate kinetic energy operator is given as a fourth-order Taylor expansion in the stretching coordinates $y_i = 1 - e^{-a_i \Delta r_i}$. The kinetic energy may also be handled exactly and there are many different ways to solve this problem depending on the choice of the coordinates, the expansion bases ...; see Carter, Pinnavaia, and Handy (1995), Fernley, Miller and Tennyson (1991), Choi and Light (1992), Bramley and Carrington (1993), Polyansky, Jensen, and Tennyson (1994) and references cited therein, and also Appendix. A particularly successful variational approach applicable to molecules up to four atoms is that of Carter and Handy [Bramley *et al.* (1993), Carter, Pinnavaia, and Handy (1995), Carter and Handy (1996), and Carter, Handy, and Demaison (1997)]. These variational methods can also be applied to non-rigid molecules (quasilinear molecules, van der Waals complexes, etc.). Most of the existing programs can be used to refine *ab initio* force fields using experimental data.

Usually, to determine the force constants, a nonlinear weighted least-squares method is used. Either the molecular constants (see Table 8.4) or, directly the vibration–rotation energies are fitted. In both cases, the choice of proper weights is important [Foord, Smith and Whiffen (1975), and Brodersen (1991)]. Another difficulty is that the problem is ill-conditioned and often has several minima [Carter and Handy (1996), McCoy and Siebert (1992a,b), Puzzarini *et al.* (1996a,b), and Palmieri *et al.* (1995)]. Finally, during the anharmonic force field refinement, the structure and the harmonic force field are often kept fixed. Although this procedure simplifies the calculations, it may induce important systematic errors in the force field derived. Actually, the main difficulty comes from the number of parameters to determine: the number of anharmonic constants increases rapidly with the number of atoms and the loss of symmetry of the molecule, see Table 8.5. For instance, the cubic force field can be determined from the vibration–rotation constants of one isotopic species only in the case of bent XY_2 molecules [Morino (1969)]. In all other cases, it is necessary to use information from the isotopic species but, taking into account the great number of parameters to refine, it is rarely sufficient because, even when many isotopomers have been studied, the isotopic dependence of the anharmonic contributions is small. This is the reason why many approximations have been tried [Plíva (1974)], the most common (and most reasonable) ones being to calculate the quartic stretching force constants from the cubic ones with the help of the Morse approximation [see equations (149) and (150)] and to neglect the cross-terms associated with both high frequency stretching and other lower frequency modes. The extension of the Urey–Bradley force field has also been attempted [Plíva (1974), and Suzuki (1975)]. However most of these simplifications have not been very successful.

Table 8.4. Main experimental sources of information on the force constants.

Technique	Parameter	Order of force constant	Reference
IR spectroscopy	ω_i harmonic wavenumber[a]	2	c, d
spectroscopy	D quartic centrifugal distortion constants	2	c, d
spectroscopy	ζ Coriolis coupling constant	2	c, d
spectroscopy	Δ inertial defect for a planar molecule	2	c
electron diffraction	l (or u) mean amplitude of vibration	2	d
electron diffraction	δr shrinkage	2	d
IR spectroscopy	ν_i vibrational frequency or	2, 3, 4	c
IR spectroscopy	$x_{ii'}$ vibrational anharmonicity	2, 3, 4	c, e
spectroscopy	α_r^ξ vibration–rotation constants	2, 3	c, e
spectroscopy	$g_{tt'}$ anharmonic constant	2, 3, 4	c, e
spectroscopy	F second-order Coriolis	2, 3	c
spectroscopy	q_t, r_t l-type doubling constants[b]	2, 3	c, e
spectroscopy	H sextic centrifugal distortion constants	2, 3	c
spectroscopy	η_J, η_K diagonal Coriolis coupling[b]	2, 3	f
spectroscopy	d $\Delta K = \pm 3$ matrix element[b]	2, 3	g
spectroscopy	W Fermi resonance constant	3	c
electron diffraction	δr_z isotope effect	2, 3	h
electron diffraction	κ asymmetry parameter	2, 3	i

[a] $\nu_i = \omega_i + x_{ii}(1 + d_i) + (1/2)\sum_{k \neq i} x_{ik} d_k + x_{l_i l_i}$.
[b] For symmetric or linear tops only.
[c] Papoušek and Aliev (1982).
[d] Hedberg and Mills (1993).
[e] Mills (1972).
[f] Aliev and Watson (1979).
[g] Sarka and Demaison (1997).
[h] Hargittai (1988), and Kuchitsu, Nakata and Yamamoto (1988).
[i] Kuchitsu (1967).

Table 8.5. Number of force constants for some molecules[a].

Molecule	Quadratic	Cubic	Quartic
diatomic	1	1	1
linear XY_2	3	3	6
linear XYZ	4	6	9
bent XY_2	4	6	9
bent XYZ	6	10	15
linear XYYX	6	11	25
planar XY_3	5	9	20
tetrahedral XY_4	5	13	33
planar X_2YZ	10	22	48
pyramidal XY_3Z	12	38	102

[a] Collated from Watson (1972), Zhou and Pulay (1989), and Császár (1992).

Currently, the most accurate force fields are obtained with the help of quantum chemical calculations which are able to furnish with good accuracy the cubic and quartic force constants [Lee and Scuseria (1995), Bürger and Thiel (1997), and Császár (1998)]. Actually, the anharmonic force constants can be calculated to higher precision than geometries or quadratic force constants because, for the higher-order force constants, the contribution of the derivatives of the nuclear–nuclear repulsion terms (which may be accurately calculated once the structure is known) become increasingly dominant [Császár (1998)]. The *ab initio* values can then be used to fix the parameters which cannot be accurately determined from the analysis of the experimental data. However, a better method is to refine the *ab initio* force field using experimental information. Most programs based on the variational approach can do that. Different variants exist but an extension of the method originally proposed by Pulay and coworkers [Pulay *et al.* (1983), and Fogarasi and Pulay (1985)] for harmonic force fields to the cubic and quartic potential is easy to use and gives satisfactory results within the framework of the simple second-order perturbation theory. Scaling factors c_i are introduced:

$$f_{ij} = f_{ij}^{0} c_i c_j, \tag{152}$$

$$f_{ijk} = f_{ijk}^{0} c_i c_j c_k, \tag{153}$$

$$f_{ijkl} = f_{ijkl}^{0} c_i c_j c_k c_l, \tag{154}$$

and are least-squares optimized against the experimental data. To avoid convergence problems, the scaling is sequentially performed on the energy terms of increasing order. The usual method is to start from geometry and lowest-order energy terms and, for each order, to keep fixed to the *ab initio* values the higher order constants [Puzzarini *et al.* (1996), and Palmieri *et al.* (1995)]. See Appendix for references to a few applications of this method.

One rapidly growing field is the study of the dynamics of molecular complexes (see Bauder (1996), Leopold *et al.* (1996), and Legon (1997) for a few recent references). A better understanding of intermolecular forces is required to explain properties of the liquid phase. Furthermore, a complex may be a prereactive intermediate, i.e. the first step in a chemical reaction [Legon (1996)]. Nevertheless, the determination of complete intermolecular potential energy surfaces are still rare (see however Hutson (1992) for the determination of the intermolecular potential of the atom–diatom complex Ar · HCl). Generally, the monomers are assumed rigid and unchanged in geometry on dimer formation. With this approximation, the quartic centrifugal distortion constants of the dimer allow us to estimate the intermolecular force constants easily [Legon and Lister (1993)]. It is also possible to obtain the structure and the potential from a direct inversion of the spectrum [Makarewicz and Bauder (1995)].

To conclude this section, two typical examples will be briefly described. The first deals with the experimental determination of the anharmonic (cubic and quartic) force field of a C_{2v} triatomic molecule: chlorine dioxide, ClO_2 [Müller *et al.* (1997)]. It is a good example of what can be done by combining the results of rotational and rovibrational spectroscopies. However, it must be pointed out that, in some particular cases, a much more complete force field (up to sextic terms) could be determined. However, this is limited to light molecules for which a considerable amount of experimental data could be gathered, for instance CO_2 and N_2O, see Appendix. The force field of ClO_2 is completely defined by three curvilinear internal coordinates (two stretching displacements, Δr and $\Delta r'$, and a bending displacement, $\Delta\alpha$) and four quadratic, six cubic and nine quartic force constants. Thus, there are in all 21 unknown molecular parameters to determine: 19 force constants and two structural parameters, the bond distance, r(Cl–O) and the bond angle α(ClOCl). The input data used in the weighted least-squares fit come from two isotopic species ($^{35}ClO_2$ and $^{37}ClO_2$) were

- six ground state rotational and ten quartic centrifugal distortion constants;
- six changes in rotational and ten changes in quartic centrifugal distortion constants for the first excited vibrational states;
- 14 ground state sextic centrifugal distortion constants;
- six wavenumbers of the fundamental frequencies;
- nine anharmonic constants.

The two equilibrium structural parameters were simultaneously determined with the force constants. It was found necessary to correct the rotational and centrifugal distortion constants for the largest higher-order contribution which is due to the bending coordinate (term γ_{22} in equation (110)). As three small off-diagonal quartic constants were not well determined, they were fixed at the corresponding *ab initio* values. The derived force field is given in Table 8.6. Most of the spectroscopic constants are well reproduced by this force field.

The other example deals with the determination of the force field of chlorodifluromethane, $CHClF_2$ [Palmieri *et al.* (1995)]. Although it is still a small molecule (five atoms), it is a low symmetry molecule and, consequently, the number of anharmonic force constants is so huge (69 cubic constants) that there is no hope of determining them using only experimental information. Instead, the harmonic potential has been evaluated by Møller–Plesset theory to second order (MP2) and the cubic and quartic force field was calculated by Hartree–Fock theory only in order to reduce the computational effort. Finally, the potential was scaled to reproduce to very high accuracy the experimental values of the ground state rotational constants, fundamental, overtones, centrifugal distortion and vibration–rotation constants, see equations (152). Table 8.7 gives the scaling factors and Table 8.8 the rms deviation of a few typical parameters before and after scaling. Inspection of Table 8.7 shows that the scaling factors are rather close to unity, indicating that the theoretical prediction is fairly accurate, but they are far enough from unity to justify the scaling procedure. Table 8.8 confirms this result. The interesting conclusion is that the combination of theoretical and experimental methods allow one to obtain results which would be unattainable by each method separately. It must also be emphasized that the *ab initio* method used for $CHClF_2$, namely HF/TZ2P is a relatively cheap method and that it is now possible to use much more sophisticated methods [such as CCSD(T)] with very large basis sets (of quadruple

Table 8.6. Experimental structural parameters (pm, degrees) and force constants (aJ $Å^{-n}$) of OClO[a].

Parameter	Value	Parameter	Value	Parameter	Value
r	146.98234 (120)	$f_{rrr'}$	0.2245 (250)	$f_{rrr'r'}$	16.1 (54)
α	117.40667 (100)	$f_{rr\alpha}$	0.0887 (360)	$f_{rrr\alpha}$	1.908[b]
f_{rr}	7.0354 (83)	$f_{rr'\alpha}$	−0.2066 (93)	$f_{rrr'\alpha}$	6.9 (20)
$f_{rr'}$	−0.2105(65)	$f_{r\alpha\alpha}$	−1.8844 (110)	$f_{rr\alpha\alpha}$	1.232[b]
$f_{r\alpha}$	−0.02660 (410)	$f_{\alpha\alpha\alpha}$	−1.9120 (100)	$f_{rr'\alpha\alpha}$	2.62 (170)
$f_{\alpha\alpha}$	1.39982 (230)	f_{rrrr}	303.2 (62)	$f_{r\alpha\alpha\alpha}$	3.25 (84)
f_{rrr}	−51.815 (110)	$f_{rrrr'}$	0.228[b]	$f_{\alpha\alpha\alpha\alpha}$	5.41 (100)

[a]See Müller *et al.* (1997).
[b]*Ab initio* value.

Table 8.7. Curvilinear internal coordinates and corresponding scaling factors[a] for force constants of $CHClF_2$.

Internal	Scaling factors		
coordinate	Quadratic	Cubic	Quartic
C–F stretch	0.982266	0.920327	1.146928
C–H stretch	0.981208	1.073903	1.249475
C–Cl stretch	1.014052	0.900712	0.744050
HCCl bend	0.978926	1.085966	1.154498
HCF bend	0.995384	1.067965	1.153209
ClCF bend	0.965031	0.948711	1.171661

[a]After Palmieri *et al.* (1995).

Table 8.8. Rms deviations of a few typical parameters of $CHClF_2$ and its isotopomers before and after scaling[a].

Parameter[b]	Unit	Number	Rms deviations	
			ab initio	scaled
B_0	MHz	12	22.6856	0.0374
ν_i, fundamental	cm^{-1}	31	25.9	5.4
α_r^ξ	MHz	12	7.430	4.707
ν_i, overtones	cm^{-1}	26	28.3	4.1

[a]From Palmieri *et al.* (1995).
[b]See definition in Table 8.4.

zeta quality for instance) which give extremely accurate results, at least for small molecules, see for instance the review of Császár (1998) and Appendix.

8.4.2 PRACTICAL APPLICATIONS

Although the determination of potential constants is an important application of spectroscopy, it is not yet widespread because of its many difficulties. On the other hand, the determination of geometrical structures is a flourishing field because approximate methods are available which allow us easily to obtain near-equilibrium structures. This aspect of the problem will not be discussed in detail because it has been recently reviewed [Graner (1992), Van Eijck (1992), Rudolph (1995), and Demaison, Wlodarczak and Rudolph (1997)]. Structural data of molecules have been collected in a series of volumes of group II of the new series of Landolt–Börnstein. Vol. II/23 contains the most recent structures [Graner *et al.* (1995)]. By carefully combining the moments of inertia of the parent molecule with those of isotopically substituted molecules, it is possible to determine a structure which is usually close to the equilibrium structure. A true experimental equilibrium structure is difficult to obtain for polyatomic molecules with more than three atoms because the rotational constants of the vibrational ground state are different from those of the hypothetical vibrationless equilibrium configuration. In principle, it is possible to obtain the equilibrium constants if the rotational constants for all the vibrationally first excited states can be determined, see equation (110). This has to be done for each isotopic species. This is a tedious task which is furthermore complicated by the fact that interactions (anharmonic resonances, Coriolis interactions, etc.) often occur between these vibrational states. Particularly, the correct use of equation (110) assumes that there is no anharmonic resonance between the vibrationally excited states. Although such resonances can be analyzed and unperturbed rotational constants can be derived, this is often a time consuming task requiring the study of many overtone and combination levels [Graner (1992)]. Actually, to avoid the difficult problem of analyzing the anharmonic resonances, the best method is first to determine the cubic force field (see preceding section) to calculate the vibrational corrections. This also makes unnecessary the analysis of the excited states of rare isotopomers. This method was advocated by Hoy, Mills and Strey (1972) and then extensively used by Botschwina and co-workers who showed that it is possible to obtain reliable equilibrium rotational constants by correcting the experimental ground state rotational constants with *ab initio* vibration–rotation interaction constants (α constants) [see for instance Botschwina and Flügge (1991), McCarthy *et al.* (1995) as well as Table 5 of Demaison, Wlodarczak, and Rudolph (1997) and Table 1 of Császár (1998)].

Another use of the molecular parameters is the calculation of thermodynamic quantities. They are also necessary in the theory of chemical kinetics (where the intramolecular vibrational energy redistribution plays an essential part). However, at the moment, one of the most important applications of spectroscopy is the study of properties (temperature, pressure, abundance of species, etc.) of different media such as the Earth's atmosphere, planetary atmospheres, interstellar media, plasmas, etc. In all cases, an accurate knowledge of the spectroscopic parameters (frequencies, intensities, line profile, absorption cross section) is required in order to invert the spectra. In many cases, it is the lack of accurate laboratory data which limits the precision of the inversion. This is particularly important for astrophysics because interstellar gas is a key constituent of galaxies; particularly, it is at the origin of the formation of stars. Most of our knowledge on interstellar media comes from spectroscopy. It allows us to determine the abundance of the different species (atoms, ions, molecules, dust) as well as the temperature and the density of the gas. All we know on the kinematics of interstellar gas (and their galaxies) comes from the Doppler shift of the spectral lines Up to now, about one hundred interstellar molecules have been detected (mainly by millimeter-wave radioastronomy). Most of them are quite difficult to study in the laboratory because they are instable: ions such as HCO^+, radicals such as C_3H, small rings such as SiC_2, isomers such as HNC, long chains such as C_8H, etc. Actually, some of them were first detected by radioastronomy before being studied in the laboratory. Many reviews are devoted to this topic [see for instances Genzel and Stutzki (1989), Wilson and Rood (1994), Herbst (1995), Van Dishoeck (1997), and references cited therein].

The study of planetary atmospheres provides important clues for understanding the origin and evolution of telluric and giant planets. This can be made by detecting and measuring atmospheric constituents, including their spatial distribution and temporal variation. Spectroscopy (mainly in the infrared) is the basic tool used in planetology. After the early detection of CO on Venus [Kakar, Walters and Wilson (1976)], many species have been found in the atmospheres of planets and satellites. Fairly recent reviews may be found in [Encrenaz, Bibring and Blanc (1995), Proceedings ISO (1997)]. In the case of telluric planets, this analysis opens the way to a comparative study of their climatic evolutions. In the case of giant planets, the measurements of elementary and isotopic abundance ratios gives constraints to the models of formation and evolution of these planets. In the case of comets, spectroscopy of the coma (i.e. the region around the core of the comet) opens the way to its chemical composition, therefore to the nature of ices forming its core. It also provides us with information about the coma's thermodynamics (temperature, velocity field) [Crovisier, Leech and Bockelée-Morvan (1997)].

Finally, the spectroscopy of Earth's atmosphere is a rapidly growing field. It is hoped that it will permit a better understanding of the effect of human activities on the evolution of the atmosphere (greenhouse effect due to carbon dioxide and methane, destruction of stratospheric ozone, urban and industrial pollution, etc.). Many different techniques are used: radiometers, grid spectrometers, Fourier transform spectrometers, microwave spectrometers, diode lasers, etc. Sometimes, they are used from the ground, but, more frequently, planes, balloons, or satellites are employed. Here again, it is often the lack of accurate laboratory data which still limits the accuracy of these methods [Flaud, Encrenaz, and Omont (1998), ICAMDATA (1998)].

8.4.3 AVAILABLE COMPUTER PROGRAMS

Short descriptions are given here of a few computer programs often used in anharmonic force field analyses.

INTDER95

INTDER95 is a general program developed by Wesley D. Allen and coworkers (Center for Computational Quantum Chemistry, University of Georgia, Athens, GA 30602) which performs general curvilinear transformations and associated tasks often encountered in the calculation of anharmonic molecular force fields. The capabilities of INTDER95 include:

- Force field transformations between Cartesian and general internal coordinate spaces up to fourth order, including nonstationary references structures. Both forward and reverse transformations are possible; hence, by using intermediate Cartesian coordinates, force fields can be transferred among alternate representations in the internal space [Allen and Császár (1993), and Allen *et al.* (1996)].
- Harmonic frequency and infrared intensity analyses, both in Cartesian and internal coordinate spaces.
- Scaled quantum mechanical (SQM) force field analyses, including optimization of scale factors [Allen, Császár and Horner (1992)].
- Transformation of dipole moment derivatives between Cartesian and internal coordinates.
- Generation of displaced Cartesian geometries along general internal coordinates for use in finite-difference computations of molecular force fields.
- Direct Cartesian projection of any external (translation/rotation) variable dependence out of Cartesian anharmonic force fields.

It is available to anyone free of charge in the form of a FORTRAN listing, a user's manual; and a battery of test/example input and output files.

ANHARM

ANHARM is a set of programs written in FORTRAN by Yukio Yamaguchi and Henry F. Schaefer III (Center for Computational Quantum Chemistry, University of Georgia, Athens, GA 30602, USA). The rotational–vibrational anharmonic constants are determined using the second-order vibrational perturbation theory. It will determine the following vibrational and rotational constants for asymmetric top [Clabo *et al.* (1988)], linear [Allen *et al.* (1990)] and symmetric top molecules:

- Rotational constants (B_e and B_0)
- Vibration–rotation interaction constants (α constants)
- Quartic and sextic centrifugal distortion constants
- Quadratic, cubic and quartic force constants (in normal coordinate space)
- Coriolis (ζ) coupling constants
- Harmonic vibrational frequencies
- Vibrational anharmonic constants ($x_{i,j}$)
- Fundamental vibrational frequencies
- Rotational l-type doubling constants (for linear molecules)
- Vibrational l-type doubling constants (for linear molecules)

Input requires the molecular geometry, quadratic, cubic, and quartic force constants in Cartesian coordinates. By using INTDER95 (see above), the force constants in internal coordinate system may be transformed into Cartesian coordinate system. Currently, there is no stand-alone version of ANHARM. The development of such a version is planed.

SPECTRO

SPECTRO is a program written by A. Willetts, J. F. Gaw, W. H. Green and N. C. Handy (University Chemical Laboratories, Lensfield Road, Cambridge, CB2 1EW, UK) [Gaw *et al.* (1990)]. It uses the geometry and force field of any molecule. Changing to a normal coordinate system it performs a rotational analysis, calculating rotational constants, quartic and sextic centrifugal distortion constants and rotation–vibration interaction constants. It then performs a vibrational analysis finding anharmonic constants and fundamental frequencies. The effects of Fermi and Coriolis resonance can be accounted for. It can also be used to generate vibrational transition intensities, given the molecule's dipole moment field. It also calculates vibrationally averaged interatomic distances. This program uses the perturbation approach and, as input, either Cartesian or internal coordinates may be used.

People who are interested in obtaining copies of this program should apply to the authors.

VIBROT

VIBROT written by G. O. Sørensen (University of Copenhagen, Department of Chemistry, Lab. 5, The H. C. Ørsted Institut, 5, Universitetsparken, DK 2100 Copenhagen ø, Denmark) calculates vibrational frequencies from structure, force constants and atomic masses. The effect of the cubic and quartic anharmonic force constants can be evaluated by second-order perturbation theory and by *direct diagonalization* of the vibrational matrix. Furthermore rotational constants (including centrifugal distortion) can be calculated. If anharmonicity is included, the vibrational dependence of the rotational *and the quartic centrifugal distortion* constants are evaluated by considering second- and fourth-order perturbations respectively and the sextic centrifugal distortion constants are evaluated as well.

The anharmonic force constants may be given either as valence force constants (possibly symmetrized) or as force constants, k, based on the dimensionless normal coordinates, q.

The program has some restrictions in handling symmetric top molecules when anharmonicity is taken into account:

- Presently the centrifugal distortion constants cannot be calculated with the anharmonic contributions included.
- Symmetry can only be used to the extent where an Abelian subgroup of the full symmetry group is considered. However, the symmetry coordinates may well be defined according to the full symmetry.

The program is available for running under Digital's VMS operating system and for running under Windows on a PC. A version for UNIX is presently being prepared. It may be obtained free by anonymous ftp from KL5AXP.KI.KU.DK. However, the author would appreciate to be notified about the results obtained and in particular as to any problems encountered, so that possible bugs may be corrected and the documentation improved.

8.5 Appendix

List of polyatomic molecules for which anharmonic potential constants have been determined using experimental information. The reference for the *ab initio* calculations is also given when available. Only the most recent reference is given. This is an update of the appendix of Kuchitsu (1992). A list of molecules for which anharmonic potential constants have been calculated *ab initio* may be found in Császár (1998).

A BENT XY_2

CH_2	Jensen and Bunker (1988). MORBID.
ClO_2	Müller *et al.* (1997)
Cl_2O	Sugie *et al.* (1995). Cubic.
F_2O	Morino and Saito (1966). Cubic.
	Thiel *et al.* (1988). *Ab initio* CCSD/TZP.

SiF_2	Shoji, Tanaka and Hirota (1973). Cubic.
NH_2	Gabriel *et al.* (1994). Variational.
H_2O	Partridge and Schwenke (1997). *Ab Initio* CASSCF + MRCI and fit to experimental data.
H_2S	Polyansky, Jensen and Tennyson (1996). MORBID in conjunction with exact-kinetic-energy calculations.
	Martin, François and Gijbels (1995). *Ab initio* CCSD(T)/cc-pVTZ.
H_2Se	Jensen and Kozin (1993). MORBID.
H_2Te	Gómez and Jensen (1997). MORBID.
NO_2	Schryber *et al.* (1997). Variational.
SO_2	Carter *et al.* (1982). Variational.
	Martin (1998). *Ab initio* CCSD(T)/cc-pVQZ+1.
O_3	Tyuterev *et al.* (1999).

B LINEAR XY_2

CO_2	Chédin and Teffo (1984). Up to sextic terms, numerical van Vleck transformation.
CS_2	Lacy and Whiffen (1981). Diagonalization.
	Martin, François and Gijbels (1995). *Ab initio* CCSD(T)/cc-pVTZ.

C BENT XYZ

ArHCl	Hutson (1992).
HOBr	Peterson (1997). *Ab initio* CCSD(T)/aug-cc-pVQZ.
BrNO	Degli Esposti *et al.* (1995).
	Palmieri *et al.* (1997). *Ab initio* MRCI scaled cubic.
HCO	Serrano-Andrés, Forsberg and Malmquist (1998).
HOCl	Halonen and Ha (1988). *Ab initio* CASPT2. Peterson (1997). *Ab initio* CCSD(T)/aug-cc-pVQZ.
ClNO	Cazzoli *et al.* (1983). Cubic, four constants fixed at *ab initio* values.
	Martin, François, and Gijbels (1994). *Ab initio* CCSD(T)/cc-pVTZ.
FNO	Degli Esposti, Cazzoli and Favero (1985). Cubic, three constants fixed at *ab initio* values.
	Martin, François and Gijbels (1994). *Ab initio* CCSD(T)/aug-cc-pVTZ.
FSN	Degli Esposti, Cazzoli and Favero (1988). Cubic, some constants fixed at *ab initio* values.
HNO	Hirota (1986). Cubic, some constants fixed at *ab initio* values.
	Dateo, Lee and Schwenke (1994). *Ab initio* CCSD(T)/cc-pVTZ.

D LINEAR XYZ

HBO	Kawashima, Endo and Hirota (1989). Cubic, one constant fixed at zero.
HBS	Turner and Mills (1982). Cubic, Morse approximation for the quartic terms.
BrCN	Degli Esposti *et al.* (1996). Four cubic, other cubic and quartic constants fixed at *ab initio* CASSCF/cc-pVTZ.
ClCN	Lee *et al.* (1995). *Ab initio* CCSD(T)/cc-pVTZ.
FCN	Degli Esposti *et al.* (1982). Lee *et al.* (1995). *Ab initio* CCSD(T)/cc-pVTZ.
HCN	Carter, Mills and Handy (1993). Variational.
HNC	Creswell and Robiette (1978). Cubic, Morse approximation for stretch quartic.

HCO^+	Puzzarini *et al.* (1996). *Ab initio* scaled CASSCF-MRCI/cc-pVQZ.
HCP	Puzzarini *et al.* (1996). *Ab initio* scaled CASSCF/cc-pVQZ.
	Koput and Carter (1997). *Ab initio* CCSD(T)/cc-pV5Z.
ICN	Degli Esposti *et al.* (1998). Five cubic and three quartic constants, other constants fixed at *ab initio* CASSCF/ccpVDZ.
OCS	Martin, François and Gijbels (1995). *Ab initio* CCSD(T)/cc-pVTZ.
N_2O	Teffo and Chédin (1989). Up to sextic terms, numerical Van Vleck transformation.
	Yan, Xian and Xie (1997). Variational.

E LINEAR XYYZ

HCCH	Bramley *et al.* (1993). Variational. Martin, Lee and Taylor (1998). *Ab initio* CCSD(T)/cc-pVQZ.
HCCF	Persson, Taylor and Martin (1998). *Ab initio* CCSD(T)/ANO5432.

F PYRAMIDAL XY_3

AsF_3	Smith (1978). Cubic.
NF_3	Tarroni *et al.* (1996). *Ab initio* scaled SCF/6-311G**.
PF_3	Small and Smith (1978). Cubic.
NH_3	Špirko and Kraemer (1989). Nonrigid invertor Hamiltonian used.
	Martin, Lee and Taylor (1992). *Ab initio* CCSD(T)/cc-pVQZ.

G PLANAR XY_3

BF_3	Pak and Woods (1997). *Ab initio* CCSD(T)/cc-pVTZ.
$CF_3{}^+$	Pak and Woods (1997). *Ab initio* CCSD(T)/cc-pVTZ.
$CH_3{}^+$	Kraemer and Špirko (1991). *Ab initio* + experimental nonrigid inverter Hamiltonian.
$NH_3{}^+$	Kraemer and Špirko (1992). *Ab initio* + experimental nonrigid inverter Hamiltonian.
SO_3	Dorney, Hoy and Mills (1973). Cubic.
	Martin (1999). *Ab initio* CCSD(T)/cc-pVTZ+1.

H TETRAHEDRAL XY_4

CF_4	Brodersen (1991). Cubic, numerical Van Vleck transformation.
CH_4	Mourbat, Aboumajd and Loëte (1998). Cubic.
	Lee, Martin and Taylor (1995). *Ab initio* CCSD(T)/cc-pVTZ.
	Venuti, Halonen and Della Valle (1999). *Ab initio* fitted.
SiH_4	Rotger, *et al.* (1998). *Ab initio* MP2/TZ2Pf.

I PLANAR X_2YZ

H_2CO	Carter, Handy and Demaison (1997). Variational.
H_2CS	Carter and Handy (1998). Variational.

J C_{3v} XY_3Z

SiH_3Br	Klatt, Willetts and Handy (1996) 272. *Ab initio* MP2/3-21G*
CHF_3	Klatt *et al.* (1996). *Ab initio* SCF/DZP.
	Maynard, Wyatt and Iung (1995). *Ab initio* MP2/6-311G**.
CH_3F	Egawa *et al.* (1987).
	Cubic, most constants fixed at *ab initio* MP2/6-311G**.

K OTHERS

H_3^+	Dinelli, Polyansky and Tennyson (1995). Variational.
	Watson (1995).
$CHClF_2$	Palmieri *et al.* (1995). *Ab initio* scaled SCF/TZ2P.
CH_2ClF	Puzzarini, Tarroni and Palmieri (1997). *Ab initio* scaled MP2.
CH_2F_2	Amos *et al.* (1991). *Ab initio* MP2/6-31G**.

8.6 References

Albritton, D. L., Schmeltekopf, A. L., and Zare, R. N., 1976, in *Molecular Spectroscopy: Modern Research*, Vol. II, Narahari Rao, K., Ed.; Academic Press: New York.

Adel, A., and Dennison, D. M., 1933, *Phys. Rev.*, **43**, 716–723; **44**, 99–105.

Aliev, M. R., and Watson, J. K. G., 1976, *J. Mol. Spectrosc.*, **61**, 29–52.

Aliev, M. R., and Watson, J. K. G., 1979, *J. Mol. Spectrosc.*, **75**, 150–160.

Aliev, M. R., and Watson, J. K. G., 1985, in *Molecular Spectroscopy: Modern Research*, Vol. III, Narahari Rao, K., Ed.; Academic Press: San Diego, CA, pp. 1–67.

Allen, W. D., Yamaguchi, Y., Császár, A. G, Clabo, D. A., Remington, R. B., and Schaefer, H. F., 1990, *Chem. Phys.* **145**, 427–466.

Allen, W. D., Császár, A. G., and Horner, D. A., 1992, *J. Am. Chem. Soc.* **114**, 6834–6849.

Allen, W. D., and Császár, A. G., 1993, *J. Chem. Phys.* **98**, 2983–3015.

Allen, W. D., Császár, A. G., Szalay, V., and Mills, I. M., 1996, *Mol. Phys.* **89**, 1213–1221.

Amat, G., and Nielsen, H. H., 1962, *J. Chem. Phys.*, **36**, 1859–1865.

Amat, G., and Nielsen, H. H., 1967, *J. Mol. Spectrosc.*, **23**, 359–364.

Amat, G., Nielsen, H. H., and Tarrago, G., 1971, *Rotation–Vibration of Polyatomic Molecules*; Dekker: New York.

Amos, R. D., Handy, N. C., Green, W. H., Jayatilaka, D., Willetts, A., and Palmieri, P., 1991, *J. Chem. Phys.* **95**, 8323–8336.

Andrade e Silva, M. H., and Ramadier, J., 1965, *J. Phys. (Orsay, Fr.)*, **26**, 246–248.

Bartlett, R. J., and Stanton, J. F., 1994, in *Rev. Comp. Chem.*, Vol. V, Lipkowitz, K. B., and Boyd, D. B., Eds; VCH: New York, pp. 65–127.

Bauder, A., 1996, in *Low Temperature Molecular Spectroscopy*, Fausto, R., Ed.; Kluwer: Amsterdam, pp. 291–309.

Belsley, D. A., 1991, Conditioning Diagnostics; Wiley: New York.

Botschwina, P., and Flügge, J., 1991, *Chem. Phys. Lett.* **180**, 589–593.

Bramley, M. J., and Carrington, T., 1993, *J. Chem. Phys.* **99**, 8519–8541.

Bramley, M. J., Carter, S., Handy, N. C., and Mills, I. M., 1993, *J. Mol. Spectrosc.* **157**, 301–336.

Brodersen, S., 1990, *J. Mol. Spectrosc.* **142**, 122–128.

Brodersen, S., 1991, *J. Mol. Spectrosc.* **145**, 331–351.

Bunker, P. R., and Jensen, P., 1998, *Molecular Symmetry and Spectroscopy*; NRC Research Press: Ottawa.

Bürger, H., and Thiel, W., 1997, in *Vibration–Rotational Spectroscopy and Molecular Dynamics*, Papoušek, D., Ed.; World Scientific; Singapore, pp. 56–115.

Bürger, H., Cosléou, J., Demaison, J., Gerke, C., Harder, H., Mäder, H., Paplewski, M., Papoušek, D., Sarka, K., and Watson, J. K. G., 1997, *J. Mol. Spectrosc.*, **182**, 34–49.

Carter, S., Mills, I. M., Murrell, J. N., and Varandas, A. J. C., 1982, *Mol. Phys.* **45**, 1053–1066.

Carter, S., Mills, I. M., and Handy, N. C., 1993, *J. Chem. Phys.* **99**, 4379–4390.

Carter, S., Pinnavaia, N., and Handy, N. C., 1995, *Chem. Phys. Lett.* **240**, 400–408.

Carter, S., and Handy, N. C., 1996, *J. Mol. Spectrosc.* **179**, 65–72.

Carter, S., Handy, N. C., and Demaison, J., 1997, *Mol. Phys.* **90**, 729–737.

Carter, S., and Handy, N. C., 1998, *J. Mol. Spectrosc.* **192**, 263–267.

Cazzoli, G., Degli Esposti, C., Palmieri, P., and Simeone, S., 1983, *J. Mol. Spectrosc.* **97**, 165–185.

Ceausu, A., Graner, G., Bürger, H., Mkadmi, E. B., Cosléou, J., and Lessari, A. G., 1995, *J. Mol. Spectrosc.* **172**, 16–33.

Chédin, A., 1979, *J. Mol. Spectrosc.* **76**, 430–491.
Chédin, A., and Teffo, J.-L., 1984, *J. Mol. Spectrosc.* **107**, 333–342.
Choi, S. E., and Light, J. C., 1992, *J. Chem. Phys.* **97**, 7031–7054.
Clabo, D. A., Allen, W. D., Yamaguchi, Y., Remington, R. B., and Schaefer, H. F., 1988, *Chem. Phys.* **123**, 187–239.
Creswell, R. A., and Robiette, A. G., 1978, *Mol. Phys.* **36**, 869–876.
Crovisier, J., Leech, K., and Bockelée-Morvan, D., 1997, *Science*, **275**, 1904–1907.
Császár, A., 1992, *J. Phys. Chem.* **96**, 7898–7904.
Császár, A., 1998, in *Encyclopedia of Computational Chemistry*, Ragué Schleyer, P.v. Ed.; Wiley: New York.
Dateo, C. E., Lee, T. J., and Schwenke, D. W., 1994, *J. Chem. Phys.* **101**, 5853–5859.
Degli Esposti, C., Favero, P. G., Serenellini, S., and Cazzoli, G., 1982, *J. Mol. Struct.* **82**, 221–236.
Degli Esposti, C., Cazzoli, G., and Favero, P. G., 1985, *J. Mol. Spectrosc.* **109**, 229–238.
Degli Esposti, C., Cazzoli, G., and Favero, P. G., 1988, *J. Mol. Struct.* **190**, 327–342.
Degli Esposti, C., Tamassia, F., Cazzoli, G., and Kisiel, Z., 1995, *J. Mol. Spectrosc.* **170**, 582–600.
Degli Esposti, C., Tamassia, F., Puzzarini, C., Tarroni, R., and Zelinger, Z., 1996, *Mol. Phys.* **88**, 1603–1620.
Degli Esposti, C., Bizzocchi, L., Tamassia, F., Puzzarini, C., Tarroni, R., and Zelinger, Z., 1998, *Mol. Phys.* **93**, 95–106.
Demaison, J., Bocquet, R., Chen, W. D., Papoušek, D., Boucher, D., and Bürger, H., 1994a, *J. Mol. Spectrosc.*, **166**, 147–157.
Demaison, J., Cosléou, J., Bocquet, R., and Lesarri, A. G., 1994b, *J. Mol. Spectrosc.* **167**, 400–418.
Demaison, J., Wlodarczak, G., and Rudolph, H.-D., 1997, in *Advances in Molecular Structure Research*, Vol. 3, Hargittai, M., and Hargittai, I., Eds; JAI Press: Greenwich, CT, pp. 1–51.
Di Lauro, C., Lattanzi, F., and Graner, G., 1990a, *J. Mol. Spectrosc.*, **143**, 111–136.
Di Lauro, C., Lattanzi, F., and Graner, G., 1990b, *Mol. Phys.*, **71**, 1285–1302.
Di Lonardo, G., Fusina, L., and Johns, J. W. C., 1984, *J. Mol. Spectrosc.*, **104**, 282–301.
Dinelli, B. M., Polyansky, O. L., and Tennyson, J., 1995, *J. Chem. Phys.* **103**, 10 433–10 438.
Dorney, J., Hoy, A. R., and Mills, I. M., 1973, *J. Mol. Spectrosc.* **45**, 253–260.
Dreizler, H., and Dendl, D., 1965, *Z. Naturforsch.*, **20a**, 30–37.
Dreizler, H., and Rudolph, H. D., 1965, *Z. Naturforsch.*, **20a**, 749–751.
Eckart, C., 1935, *Phys. Rev.* **47**, 552–558.
Egawa, T., Yamamoto, S., Nakata, M., and Kuchitsu, K., 1987, *J. Mol. Struct.* **156**, 213–228.
Encrenaz, Th., Bibring, J.-P., and Blanc, M., 1995, *The Solar System*; Springer: Berlin.
Femenias, J. L., 1990, *J. Mol. Spectrosc.* **144**, 212–223.
Fernley, J. A., Miller, S., and Tennyson, J., 1991, *J. Mol. Spectrosc.* **150**, 597–609.
Flaud, J.-M., Encrenaz, Th., and Omont, A., 1998, *J. Chim. Phys.* **95**, 1771–1803.
Fogarasi, G., and Pulay, P., 1985, in *Vibrational Spectra and Structure*, Durig, J. R, Ed.; Elsevier: Amsterdam, pp. 125–219.
Foord, A., Smith, J. G., and Whiffen, D. H., 1975, *Mol. Phys.* **29**, 1685–1704.
Fusina, L., and Carlotti, M., 1989, *J. Mol. Spectrosc.* **130**, 371–381.
Gabriel, W., Chambaud, G., Rosmus, P., Carter, S., and Handy, N. C., 1994, *Mol. Phys.* **81**, 1445–1461.
Gaw, J. F., Willetts, A., Green, W. H., and Handy, N. C., 1990, in *Advances in Molecular Vibrations and Collision Dynamics*, Bowman, J. M, Ed.; JAI Press: Greenwich, CT.
Genzel, R., and Stutzki, J., 1989, *Annu. Rev. Astron. Astrophys.* **27**, 41–86.
Gómez, P. C., and Jensen, P., 1997, *J. Mol. Spectrosc.* **185**, 282–289.
Grabow, J.-U., Heineking, N., and Stahl, W., 1992, *J. Mol. Spectrosc.* **152**, 168–173.
Graner, G., 1992, in *Accurate Molecular Structures*, Domenicano, A., and Hargittai, I., Eds; Oxford University Press: Oxford, pp. 65–94.
Graner, G., Hirota, E., Iijima, T., Kuchitsu, K., Ramsey, D. A., Vogt, J., and Vogt, N., 1995, *Structure Data of Free Polyatomic Molecules*, Landolt–Börnstein, Vol. II/23; Springer: Berlin.
Halonen, L., and Ha, T.-K., 1988, *J. Chem. Phys.* **88**, 3775–3779.
Hargittai, I., 1988, in *Stereochemical Applications of Gas-Phase Electron Diffraction*, Hargittai, I., and Hargittai, M., Eds; VCH: Weinheim, pp. 1–54.
Hedberg, L., and Mills, I. M., 1993, *J. Mol. Spectrosc.*, **160**, 117–142.
Henry, L., and Amat, G., 1960a, *Cah. Phys.*, **14**, 230–256.
Henry, L., and Amat, G., 1960b, *J. Mol. Spectrosc.*, **5**, 319–325.
Herbst, E., 1995, *Ann. Rev. Phys. Chem.* **46**, 27–53.
Hirota, E., 1986, *Nippon Kagaku Kaishi* 1438–1445.
Hougen, J. T., 1962a, *J. Chem. Phys.*, **36**, 519–534.
Hougen, J. T., 1962b, *J. Chem. Phys.*, **37**, 1433–1441.
Hoy, A. R., Mills, I. M., and Strey, G., 1972, *Mol. Phys.* **24**, 1265–1290.
Hutson, J. M., 1992, *J. Phys. Chem.* **96**, 4237–4247.
ICAMDATA-First International Conference, 1998, *Atomic and Molecular Data and their Applications*, Mohr, P. J., and Wiese, W. L., eds; *AIP Conf. Proc.* **434**.

Jensen, P., 1988, *J. Mol. Spectrosc.* **128**, 478–501.
Jensen, P., 1992, in *Methods in Computational Molecular Physics*, Wilson, S., and Diercksen, G. H. F., Eds; Plenum: New York.
Jensen, P., and Bunker, P. R., 1988, *J. Chem. Phys.* **89**, 1327–1332.
Jensen, P., and Kozin, I. N., 1993, *J. Mol. Spectrosc.* **160**, 39–57.
Jensen, P., Osmann, G., and Kozin, I., 1997, in *Vibration–Rotational Spectroscopy and Molecular Dynamics*, Papoušek, D., Ed.; World Scientific: Singapore, pp. 298–351.
Kakar, R. K., Walters, J. W., and Wilson, W. J., 1976, *Science* **191**, 379–380.
Kawashima, Y., Endo, Y., and Hirota, E., 1989, *J. Mol. Spectrosc.* **133**, 116–127.
Klatt, G., Willetts, A., and Handy, N. C., 1996, *Chem. Phys. Lett.* **249**, 272–278.
Klatt, G., Willetts, A., and Handy, N. C., Tarroni, R., and Palmieri, P., 1996, *J. Mol. Spectrosc.* **176**, 64–74.
Koput, J., and Carter, S., 1997, *Spectrochim. Acta A* **53**, 1091–1100.
Kozin, I. N., and Jensen, P., 1994, *J. Mol. Spectrosc.* **163**, 483–505.
Kraemer, W. P., and Špirko, V., 1991, *J. Mol. Spectrosc.* **149**, 235–241.
Kraemer, W. P., and Špirko, V., 1992, *J. Mol. Spectrosc.* **153**, 276–284.
Kuchitsu, K., 1967, *Bull. Chem. Soc. Jpn.* **40**, 504–510.
Kuchitsu, K., Nakata, M., and Yamamoto, S., 1988, in *Stereochemical Applications of Gas-Phase Electron Diffraction*, Hargittai, I., and Hargittai, M., Eds; VCH: Weinheim, pp. 227–263.
Kuchitsu, K., 1992, in *Accurate Molecular Structures*, Domenicano, A., and Hargittai, I., Eds; Oxford University Press: Oxford, pp. 14–46.
Lacy, M., 1982, *Mol. Phys.* **45**, 253–258.
Lacy, M., and Whiffen, D. H., 1981, *Mol. Phys.* **43**, 1205–1217.
Lee, T. J., Martin, J. M. L., Dateo, C. E., and Taylor, P. R., 1995, *J. Phys. Chem.* **99**, 15 858–15 863.
Lee, T. J., Martin, J. M. L., and Taylor, P. R., 1995, *J. Chem. Phys.* **102**, 254–261.
Lee, T. J., and Scuseria, G. E., 1995, in *Quantum Mechanical Electronic Structure Calculations with Chemical Accuracy*, Langhoff, S. R., Ed.; Vol. 13; Kluwer: Dordrecht, pp. 47–108.
Lees, R. M., 1970, *J. Mol. Spectrosc.*, **33**, 124–136.
Legon, A. C., 1997, in *Optical, Electric and Magnetic Properties of Molecules*, Clary, D. C., and Orr, B. J., Eds; Elsevier: Amsterdam, pp. 191–197.
Legon, A. C., 1996, *J. Chem. Soc. Chem. Commun.* (Review) 109–116.
Legon, A. C., and Lister, D. G., 1993, *Chem. Phys. Lett.* **204**, 139–144.
Leopold, K. R., Fraser, G. T., Novick, S. E., and Klemperer, W., 1994, *Chem. Rev.* **94**, 1807–1827.
Lobodenko, E. I., Sulakshina, O. N., Perevalov, V. I., and Tyuterev, V. G., 1987, *J. Mol. Spectrosc.* **126**, 159–170.
Louck, J. D., 1976, *J. Mol. Spectrosc.* **61**, 107–137.
Makarewicz, J., and Bauder, A., 1995, *Mol. Phys.* **84**, 853–878.
Makushkin, J. S., and Tyuterev, V. G., 1984, *Perturbation Methods and Effective Hamiltonians in Molecular Spectroscopy*; Nauka: Novosibirsk (in Russian).
Martin, J. M L., 1998, *J. Chem. Phys.* **108**, 2791–2800.
Martin, J. M L., 1999, *Spectrochim. Acta A* **55**, 713–722.
Martin, J. M L., Lee, T. J., and Taylor, P. R., 1992, *J. Chem. Phys.* **97**, 8361–8371.
Martin, J. M L., François, J.-P., and Gijbels, R., 1994, *J. Phys. Chem.* **98**, 11394–11400.
Martin, J. M L., François, J.-P., and Gijbels, R., 1995, *J. Mol. Spectrosc.* **169**, 445–457.
Martin, J. M L., Lee, T. J., and Taylor, P. R., 1998, *J. Chem. Phys.* **108**, 676–691.
Maynard, A. T., Wyatt, R. E., and Iung, C., 1995, *J. Chem. Phys.* **103**, 8372–8390.
McCarthy, M. C., Gottlieb, C. A., Thaddeus, P., Horn, M., and Botschwina, P., 1995, *J. Chem. Phys.* **103**, 7820.
McCoy, A. B., and Siebert, E. L., 1992a, *Mol. Phys.* **77**, 697–708.
McCoy, A. B., and Siebert, E. L., 1992b, *J. Chem. Phys.* **97**, 2938–2947.
Mills, I. M., 1972, in *Molecular Spectroscopy: Modern Research*, Narahari Rao K., and Matthews, C. W., Eds, Vol. I; Academic Press: New York, pp. 115–140.
Morino, Y., and Saito, S., 1966, *J. Mol. Spectrosc.* **19**, 435–453.
Morino, Y., 1969, *Pure Appl. Chem.* **18**, 323–338.
Mourbat, A., Aboumajd, A., and Loëte, M., 1998, *J. Mol. Spectrosc.* **190**, 198–212.
Müller, H. S. P., Sørensen, G. O., Birk, M., and Friedl, R. R., 1997, *J. Mol. Spectrosc.* **186**, 177–188.
Nielsen, H. H., 1951, *Rev. Mod. Phys.*, **23**, 90–136.
Nikitin, A., Champion, J. P., and Tyuterev, V. G., 1997, *J. Mol. Spectrosc.* **182**, 72–84.
Oka, T., 1967, *J. Chem. Phys.*, **47**, 5410–5426.
Pak, Y., and Woods, R. C., 1997, *J. Chem. Phys.* **106**, 6424–6429.
Palmieri, P., Tarroni, R., Hühn, M. M., Handy, N. C., and Willetts, A., 1995, *Chem. Phys.* **190**, 327–344.
Palmieri, P., Puzzarini, C., Tarroni, R., and Mitrushenkov, A., 1997, *Spectrochim. Acta A* **53**, 1139–1151.
Papoušek, D., and Aliev, M. R., 1982, *Molecular Vibrational–Rotational Spectra*, Elsevier: Amsterdam.
Papoušek, D., Winnewisser, M., Klee, S., *et al.*, 1998, *J. Mol. Spectrosc.* **192**, 220–227.
Partridge, H., and Schwenke, D. W., 1997, *J. Chem. Phys.* **106**, 4618–4639.
Perevalov, V. I., and Tyuterev, V. G., 1981, *Opt. Spectrosc.* **51**, 354–357.

Perevalov, V. I., and Tyuterev, V. G., 1982, *J. Mol. Spectrosc.* **96**, 56–76.
Perevalov, V. I., and Tyuterev, V. G., and Zhilinskii, B. I., 1984, *J. Mol. Spectrosc.* **103**, 147–159.
Perevalov, V. I., and Tyuterev, V. G., 1984, *Chem. Phys. Lett.* **104**, 455–461.
Persson, B. J., Taylor, P. R., and Martin, J. M. L., 1998, *J. Phys. Chem. A* **102**, 2483–2492.
Peterson, K. A., 1997, *Spectrochim. Acta A* **53**, 1051–1064.
Plíva, J., 1974, in *Critical Evaluation of Chemical and Physical Structural Information*, Lide, R. R., Jr. and Paul, M. A., Eds, National Academy of Sciences: Washington, DC, pp. 289–311.
Podolsky, B., 1928, *Phys. Rev.* **32**, 812.
Polyansky, O. L., Jensen, P., and Tennyson, J., 1994, *J. Chem. Phys.* **101**, 7651–7657.
Polyansky, O. L., Jensen, P., and Tennyson, J., 1996, *J. Mol. Spectrosc.* **178**, 184–188.
Proceedings First ISO Workshop on Analytical Spectroscopy, 1997, ESA SP-419.
Pulay, P., Fogarasi, G., Pongor, G., Boggs, J. E., and Vargha, A., 1983, *J. Am. Chem. Soc.* **105**, 7037–7047.
Puzzarini, C., Tarroni, R., Palmieri, P., Carter, S., and Dore, L., 1996, *Mol. Phys.* **87**, 879–898.
Puzzarini, C., Tarroni, R., Palmieri, P., Demaison, J., and Senent, M. L., 1996, *J. Chem. Phys.* **105**, 3122–3141.
Puzzarini, C., Tarroni, R., and Palmieri, P., 1997, *Spectrochim. Acta A* **53**, 1123–1131.
Redding, R. E., and Hougen, J. T., 1971, *J. Mol. Spectrosc.* **37**, 366–370.
Rotger, M., Boudon, V., Lavorel, B., Sommer, S., Bürger, H., Breidung, J., Thiel, W., Bétrencourt, M., and Deroche, J.-C., 1998, *J. Mol. Spectrosc.* **192**, 294–308.
Rudolph, H.-D., 1995, in *Advances in Molecular Structure Research*, Hargittai M., and Hargittai, I., Eds, Vol. 1; JAI Press: Greenwich, CT, pp. 63–114.
Sarka, K., 1989a, *J. Mol. Spectrosc.* **133**, 461–466.
Sarka, K., 1989b, *J. Mol. Spectrosc.* **134**, 354–361.
Sarka, K., and Demaison, J., 1997, *J. Mol. Spectrosc.* **185**, 194–196.
Sarka, K., and Schrötter, H. W., 1996, *J. Mol. Spectrosc.*, **179**, 195–204.
Sarka, K., Papoušek, D., Demaison, J., Mäder, H., and Harder, H., 1997, in *Advances in Physical Chemistry*, Vol. *Vibration–Rotational Spectroscopy and Molecular Dynamics*, Papoušek, D., Ed.; World Scientific: Singapore.
Sarka, K., and Harder, H., 1999, *J. Mol. Spectrosc.* **197**, 254–261.
Sarka, K., Watson, J. K. G., and Pracna, P., 1999, unpublished results.
Schryber, J. H., Polyansky, O. L., Jensen, P., and Tennyson, J., 1997, *J. Mol. Spectrosc.* **185**, 234–243.
Serrano-Andrés, L., Forsberg, N., and Malmquist, P.-Å., 1998, *J. Chem. Phys.* **108**, 7202–7216.
Shaffer, W. H., and Nielsen, H. H., 1939, *Phys. Rev.* **56**, 188–202.
Shaffer, W. H., Nielsen, H. H., and Thomas, L. H., 1939, *Phys. Rev.* **56**, 895–907, 1051–1059.
Shoji, H., Tanaka, T., and Hirota, E., 1973, *J. Mol. Spectrosc.* **47**, 268–274.
Simons, G. R., Parr, R. G., and Finlan, J. M., 1973, *J. Chem. Phys.* **59**, 3229–3234.
Simons, G., 1974, *J. Chem. Phys.* **61**, 369–374.
Small, C. E., and Smith, J. G., 1978, *J. Mol. Spectrosc.* **73**, 215–233.
Smith, J. G., 1978, *J. Mol. Spectrosc.* **35**, 461–475.
Sørensen, G. O., unpublished results.
Špirko, V., and Kraemer, W. P., 1989, *J. Mol. Spectrosc.* **133**, 331–344.
Sugie, M., Ayabe, M., Takeo, H., and Matsumura, C., 1995, *J. Mol. Struct.* **352/353**, 259–265.
Suzuki, I., 1975, in *Applied Spectroscopy Reviews*, Brame, E. G., Jr., Ed., Vol. 9; Dekker: New York, pp. 249–301.
Tang, J., and Takagi, K., 1993, *J. Mol. Spectrosc.*, **168**, 487–498.
Tarroni, R., Palmieri, P., Senent, M. L., and Willetts, A., 1996, *Chem. Phys. Lett.* **257**, 23–30.
Teffo, J.-L., and Chédin, A., 1989, *J. Mol. Spectrosc.* **135**, 389–409.
Teffo, J.-L., Sulakshina, O. N., and Perevalov, V. I., 1992, *J. Mol. Spectrosc.* **156**, 48–64.
Teffo, J.-L., Perevalov, V. I., and Lyulin, O. M., 1994, *J. Mol. Spectrosc.*, **168**, 390–403.
Thiel, W., Scuseria, G., Schäfer, H. F., and Allen, W. D., 1988, *J. Chem. Phys.* **89**, 4965–4975.
Turner, P., and Mills, I. M., 1982, *Mol. Phys.* **46**, 161–170.
Tyuterev, V. G., Champion, J. P., Pierre, G., and Perevalov, V. I., 1984, *J. Mol. Spectrosc.* **105**, 113–138.
Tyuterev, V. G., Tashkun, S., Jensen, P., Barbe, A., and Cours, T., 1999, *J. Mol. Spectrosc.* **198**, 371–375.
Van Dishoeck, E. F., 1997, *Molecules in Astrophysics: Probes and Processes*, Symp. UAI **178**.
Van Eijck, B. P., 1992, in *Accurate Molecular Structures*, Domenicano, A., and Hargittai, I., Eds; Oxford University Press: Oxford, pp. 47–64.
Van Vleck, J. H., 1929, *Phys. Rev.* **33**, 467–506.
Venuti, E., Halonen, L., and Della Valle, R. G., 1999, *J. Chem. Phys.*, **110**, 7339–7347.
Wagner, G., Winnewisser, B. P., Winnewisser, M., and Sarka K., 1993, *J. Mol. Spectrosc.*, **162**, 82–119.
Watson, J. K. G., 1966, *J. Chem. Phys.* **45**, 1360–1361.
Watson, J. K. G., 1967, *J. Chem. Phys.* **46**, 1935–1949.
Watson, J. K. G., 1968, *J. Chem. Phys.* **48**, 181–185.
Watson, J. K. G., 1970, *Mol. Phys.* **19**, 465–487.
Watson, J. K. G., 1971, *J. Mol. Spectrosc.* **39**, 364–379.
Watson, J. K. G., 1972, *J. Mol. Spectrosc.* **41**, 229–230.
Watson, J. K. G., 1977, in *Vibrational Spectra and Structure*, Durig, J., Ed., Vol. 6; Elsevier; Amsterdam, pp. 1–89.

Watson, J. K. G., 1983, *J. Mol. Spectrosc.* **101**, 83–93.
Watson, J. K. G., 1984, *J. Mol. Spectrosc.* **103**, 350–363.
Watson, J. K. G., 1991, *J. Mol. Spectrosc.* **145**, 130–141.
Watson, J. K. G., 1995, *Chem. Phys.* **190**, 291–300.
Watson, J. K. G., Foster, S. C., McKellar, A. R. W., Bernath, P., Amano, T., Pan, F. S., Crofton, M. W., Altman, R. S., and Oka, T., 1984, *Can. J. Phys.* **62**, 1875–1885.
Watson, J. K. G., Gerke, G., Harder, H., and Sarka, K., 1998, *J. Mol. Spectrosc.* **187**, 131–141.
Whiffen, D. H., 1976, *Mol. Phys.* **31**, 989–1000.
Whiffen, D. H., 1978, *Spectrochim. Acta A* **34**, 1173–1192.
Whiffen, D. H., 1985, *J. Mol. Spectrosc.* **111**, 62–65.
Wilson, E. B., Jr., Decius, J. C., and Cross, P. C., 1955, Molecular Vibrations, McGraw-Hill; New York.
Wilson, T. L., and Rood, T. L., 1994, *Annu. Rev. Astron. Astrophys.* **32**, 191–226.
Wright, J. S., 1988, *J. Chem. Soc. Faraday Trans. II* **84** 219–226.
Yamada, K. M. T., 1983, *Z. Naturforsch. A* **38a**, 821–834.
Yamada, K. M. T., Birss, F. W., and Aliev, M. R., 1985, *J. Mol. Spectrosc.* **112**, 347–356.
Yan, G., Xian, H., and Xie, D., 1997, *Chem. Phys. Lett.* **271**, 157–162.
Zhou, X.-F., and Pulay, P., 1989, *J. Comput. Chem.* **10**, 935–938.

9 VARIATIONAL CALCULATIONS OF ROTATION–VIBRATION SPECTRA

Jonathan Tennyson

University College London, UK

9.1 Introduction

Perturbation theory, as described by Sarka and Demaison in this volume, has for a long time been the theoretical bedrock of high resolution molecular spectroscopy. However, there are many problems for which perturbation theory struggles to find accurate solutions. Most standard perturbative treatments of the rotational and vibrational spectra of chemically bound molecules are based on the assumption of small amplitude vibrational motion. There are many situations where this is a

Computational Molecular Spectroscopy. Edited by Per Jensen and Philip R. Bunker

poor assumption: molecules containing hydrogen or multiple symmetry related minima in their potential energy surface and associated tunnelling splittings are typical examples. Furthermore, as a molecule approaches dissociation it must always undergo large amplitude motion. It follows that it is not a question of if, but when, perturbation theory will break down.

Variational methods have been developed over the last twenty years to address directly cases for which perturbation theory is not reliable. These methods are more complete in the sense that they aim at a direct and full solution of the nuclear motion problem. Variational methods can usually solve both the vibrational and rotational problems within the same framework which adds considerably to their scope and flexibility. However these advantages also carry a price: except

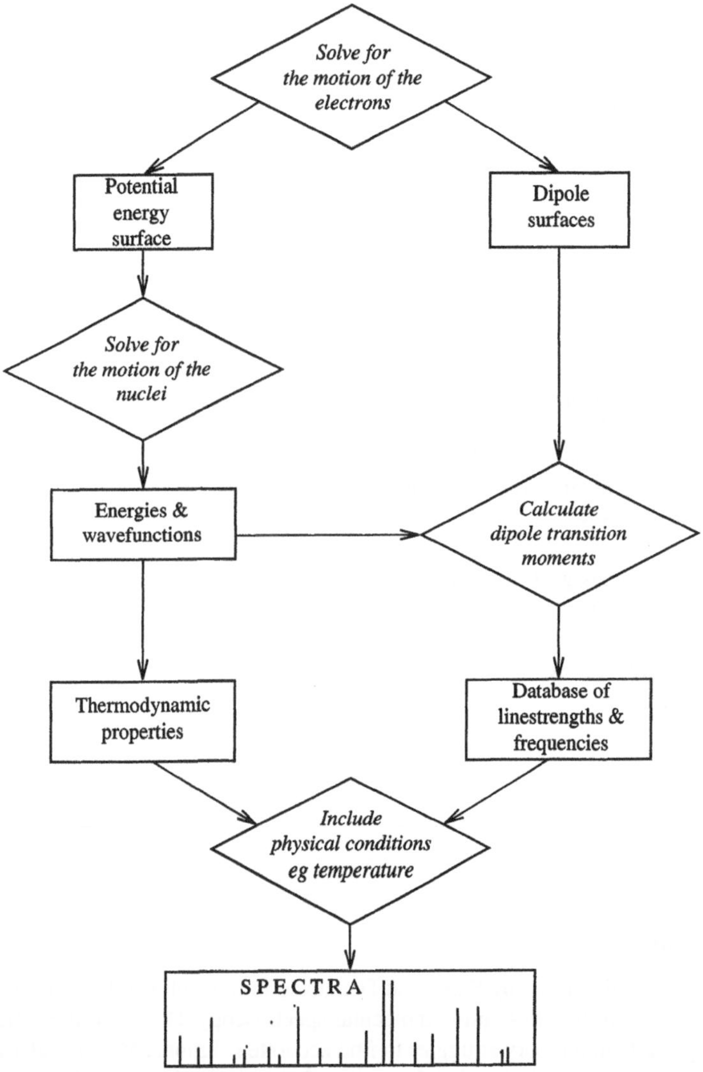

Figure 9.1. Flow diagram depicting the steps involved in generating the rotation–vibration spectrum of a molecule from first principles. Diamonds represent different steps in the calculation, while rectangles represent data transfers.

for diatomic systems for which rotation–vibration energy levels and spectra can be computed essentially exactly for a given potential by direct integration of the Schrödinger equation [Le Roy (1996)], variational calculations are much more computationally intensive than calculations based on perturbation theory. This computational expense also scales rapidly with the number of vibrational modes of the molecule being studied.

For this reason most of the work using variational methods has concentrated on triatomic molecules, although calculations on tetra-atomics are becoming more common and the first attempts at treating penta-atomic systems are beginning to appear [Dunn, Boggs and Pulay (1987), Iung and Leforestier (1992), and Carter, Shnider and Bowman (1999)]. As it is not possible to do justice to all methods here, I will concentrate on methods developed for triatomics and point to possible extensions to larger systems. Similarly there are many methods which have been used only to study the vibrational states of rotationless molecules. These will largely be neglected as not being suitable for generating actual spectra.

There are a number of steps involved in generating the vibration–rotation spectrum of a molecule from first principles. These steps are summarized in Figure 9.1, which assumes a complete *ab initio* procedure. This includes a series of electronic structure calculations to generate the potential energy (and dipole) surface used as the basis for the nuclear motion calculations. In practice, with the sole exception of H_3^+ [Polyansky and Tennyson (1999)], *ab initio* potentials for polyatomic molecules are not accurate enough to give results whose accuracy is even close to that obtained in a typical high resolution spectroscopy experiment. Many calculations therefore replace the *ab initio* stage of the calculation with potentials determined by other means such as by fitting to experimental data.

Given a potential energy surface, the major step in Figure 9.1 is determining the energy levels and associated wavefunctions for that surface. This is the main topic of this chapter, although there will also be some discussion of use of these wavefunctions to give electric dipole transition intensities and hence to generate complete synthetic spectra.

9.2 Hamiltonians

Unlike the electronic structure problems discussed elsewhere in this volume, there is no unique Hamiltonian for studying the nuclear motion problem. This arises from the need to identify and separate out the centre-of-mass motion of the molecule, which gives a continuous spectrum, prior to attempting to determining any rotation–vibration energy levels.

For an N-atomic molecule, there are $3N$ coordinates specifying the position of each atom relative to some arbitrary laboratory-fixed origin. Removal of the three centre-of-mass motion coordinates leaves $3N - 3$ coordinates. These coordinates, which are usually described as space fixed, are widely used for calculations on Van der Waals complexes [Le Roy and Carley (1980), and Hutson (1991)]. Space-fixed coordinates have the disadvantage that they do not distinguish between vibrational and rotational motion of the system. To do this requires a second transformation to body-fixed coordinates. This transformation involves fixing or embedding a rotational axis system within the framework of the molecule. To do this requires defining three rotational angles, two for linear molecules, which link the space-fixed axes to the body-fixed ones. In fact there are situations were even for non-linear molecules it can be advantageous to define less than three rotation angles [see Brocks *et al.* (1983) for example], but these will not be pursued here.

Sutcliffe (1982) gave a general prescription for developing body-fixed Hamiltonians for various choices of internal coordinates based upon application of the chain rule. In principle this prescription is easy to follow but in practice there are two pitfalls. Firstly, the algebra can get very messy. This problem has been alleviated by the use of computer algebra programs [Handy (1987), and Colwell and Handy (1997)]. Secondly, the act of embedding the body-fixed axes *always* leads to regions of the Hamiltonian where it is singular. This is an unavoidable consequence of the mathematical procedure involved. Singularities arise whenever the value of one coordinate causes the Hamiltonian not to depend on some other coordinate. For example, in polar coordinates a vector of length zero is the same for all orientations.

How serious any singularity is depends on the coordinates and embedding chosen, and the nature of the problem to be tackled. For example many of the Hamiltonians discussed below are singular as an atom–atom distance, or something similar to it, approaches zero. Such singularities are usually not important as the potential is very strongly repulsive in this region. Conversely bent molecule Hamiltonians are usually singular for geometries in which the molecule goes linear. As many bent molecules, such as water, sample linear geometries at relatively low energy, such a singularity can be important. In this case strategies have been developed to deal with this problem [Hougen, Bunker and Johns (1970), Tennyson and Sutcliffe (1982), and Jensen 1983].

As the detailed form for the different Hamiltonians used for variational studies of nuclear motion all differ, it is not possible, or desirable, to give them all here. However within the Born–Oppenheimer approximation these Hamiltonians all have the same basic structure:

$$\hat{H} = \hat{K}_V + \hat{K}_{VR} + V, \tag{1}$$

where $\hat{K}_V$ is the vibrational kinetic energy operator, and $\hat{K}_{VR}$ is the vibration–rotation kinetic energy, which is null unless the molecule is rotationally excited. V is the potential energy surface. Both the potential and the vibrational kinetic energy operator are functions of the $3N - 6$ internal coordinates only. This means that the rotationless ($J = 0$) Hamiltonian depends on the internal coordinate system chosen, but is independent of the various ways of defining the body-fixed axis system discussed below. This represents a significant simplification for vibration only calculations.

9.3 Coordinate systems

The computer requirements of a variational calculation are largely determined by the size of the basis set required, which in turn depends on the coordinate system chosen. Although the choice of coordinate is perhaps not quite as crucial as it is for the SCF methods described by Gerber and Jung in this volume, they are still important for the efficiency of a calculation. Even for triatomics it is possible to define unlimited different sets of coordinates [Zúñiga, Bastida and Requena (1997)], so it is helpful to break these into different classes.

9.3.1 NORMAL COORDINATES

Perhaps the most obvious coordinates for treating the vibrational motion, and the ones which were used for the earliest variational calculations [see Carney, Sprandel and Kern (1978)], are normal coordinates. These coordinates are those obtained from diagonalizing the standard harmonic force field problem for the molecule about its equilibrium geometry [Wilson, Decius and Cross (1955)]. The body-fixed Hamiltonian expressed in these coordinates was originally derived, in its simplest form, by Watson (1968), who showed that the Hamiltonian contained terms dependent directly on the inverse of the moment-of-inertia tensor. This term, generally called the Watson term, is singular for linear geometries.

Even more serious than problems with the Watson term was the realization that for large amplitude motion, standard straight line extrapolations of normal coordinates often leave the true domain of the problem. This means that these coordinates are now rarely used for triatomic problems. However, the use of normal coordinates for larger systems is actively being pursued [Carter, Bowman and Handy (1998)]. One advantage in this case is that it is often possible to calculate all the matrix elements for these coordinates analytically [see Dunn, Boggs and Pulay (1987), for example].

9.3.2 INTERNAL COORDINATES: STRETCHES ONLY

One of the conceptually simplest ways of representing the internal structure of a molecule is via atom–atom distances or related stretching coordinates. For three- or four-atom molecules these coordinates might seem natural since the number of interatomic distances coincides with the number of vibrational degrees of freedom. However, there is a severe technical problem with the use of coordinates based directly on atom–atom distances to do with integration ranges. The fact that one atom–atom distance alone cannot go to infinity leads to coupled integration ranges. The coupling of these ranges makes it hard to design efficient algorithms for evaluating matrix elements in these coordinates, although Watson (1995) performed successful calculations on the H_3^+ molecule using these coordinates in a grid representation, which he simply truncated at linear geometries.

One set of stretching only coordinates, used originally by Pekeris (1958), have been proposed which avoid the coupling of integration ranges. These are so-called perimetric coordinates. For a three-atom system perimetric coordinates transform atom–atom distances (r_1, r_2, r_3) into a set defined by $q_1 = -r_1 + r_2 + r_3$, and cyclic permutations thereof. This transformation decouples the integration ranges but thus far use of perimetric coordinates for rotation–vibration calculations has been limited [Sutcliffe (1992)].

9.3.3 INTERNAL COORDINATES: ONE DISTANCE PLUS ANGLES

As it is not possible to define a structure using only angles, any coordinate system must contain at least one radial coordinate. Hyperspherical coordinates [see Manz and Schor (1986), for example] use a single radial coordinate, generally known as the hyperradius, which represents the size of the molecule as a whole. The different structures for a given hyperradius are represented by two angles for a triatomic, five for a tetratomic and so forth.

There has been a considerable amount of theoretical work on hyperspherical coordinates. However these coordinates have often proved difficult to use in actual numerical applications, particularly those with rotational excitation or well-localized motions. This means that even for systems such as H_3^+, whose symmetry is naturally represented in these coordinates [Whitnell and Light (1989), Carter and Meyer (1994), and Wolniewicz and Hinze (1994)], other less symmetric coordinate systems have been widely favoured.

9.3.4 INTERNAL COORDINATES: DISTANCES AND ANGLES

Most successful methods have used a mixture of angular and radial coordinates. For triatomic systems this means two radial coordinates and one angle or (r_1, r_2, θ). A natural form of these coordinates, which have been widely exploited by Carter, Handy and co-workers [Carter and Handy (1986a, 1986b)] are bond length, bond angle coordinates. For a molecule such as water these correspond to normal parameters used to represent the bonding pattern in the molecule.

Bond length, bond angle coordinates suffer from the disadvantage that they are not orthogonal. Orthogonal coordinates are ones for which the kinetic energy operator is diagonal, i.e. contain no terms with mixed differential operators. Hamiltonians expressed in orthogonal coordinates are therefore considerably simpler. This is an important consideration for discrete variable representations (DVRs) where the kinetic energy operator provides the off-diagonal coupling, but is less important in basis set approaches [Carter and Handy (1986a, 1986b)].

The most widely used orthogonal coordinate systems are Jacobi and Radau; these are illustrated in Figure 9.2. Jacobi coordinates are also called atom–diatom scattering coordinates and are appropriate for Van der Waals dimers such as $Ar{-}H_2$. However, they have proved robust and have been widely used for other systems less obvious systems, such as H_3^+.

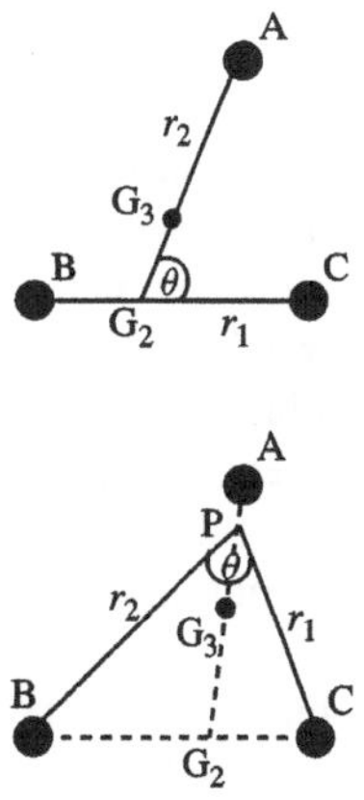

Figure 9.2. Orthogonal coordinates for the ABC triatomic molecule involving two stretches (r_1, r_2) and one angle (θ): (a) Jacobi or atom–diatom scattering coordinates; (b) Radau coordinates. G_2 and G_3 are the diatom (BC) and triatom centres-of-masses respectively. In Radau coordinates the canonical point, P, is defined such that $(PG_2)^2 = AG_3 \times AG_2$ [Hutson (1994)].

Radau coordinates are close to bond length, bond angle coordinates when the central atom is heavy. They were originally proposed for the planetary three-body problem [Radau (1868)] but have been widely used for molecules, such as water, which have a heavy central atom.

A number of workers [Sutcliffe and Tennyson (1991), and Zúñiga, Bastida and Requena (1997)] have developed very general methods of generating (r_1, r_2, θ) coordinate systems. Schwenke (1996) has also shown how orthogonal coordinates and the corresponding Hamiltonians can be derived in a relatively straightforward manner by successive addition of Jacobi or Radau coordinates or even mixtures of the two. Colwell and Handy (1997) extended their bond length, bond angle approach to tetra-atomic systems for which they have derived a number of different Hamiltonians depending on the bonding of the molecule in question: linear, as in acetylene, or branched, as in formaldehyde, for example.

9.3.5 BODY-FIXED AXES

Within the normal coordinate Hamiltonian discussed above it is usual to define the body-fixed axes using the second Eckart (1935) condition. This condition minimizes the interaction between vibrational and rotational motion, the Coriolis coupling, for each geometry. This condition is thus the optimal one for defining rotational motion. Unfortunately, the use of the Eckart embedding in conjunction with the standard normal coordinate Hamiltonian, for which it was originally proposed, suffers from the problems involved with using normal coordinates and discussed above.

While considerable effort has been expended on exploring suitable internal coordinate systems for vibrational calculations, less work has been done exploring possible axis embeddings. Indeed for some time the only general scheme was due to Sutcliffe and Tennyson (1991). Recently, however, Wei and Carrington (1997a, 1997b) have proposed a method of combining the virtues of the Eckart (1935) embedding of the axes, which minimize the Coriolis coupling between vibrational and rotational motion, with the advantages of the orthogonal internal coordinates discussed above. It is too early to say how Wei and Carrington's Hamiltonians perform in practical calculations.

Figure 9.3 illustrates some of the axes embeddings possible with Jacobi and Radau coordinates. Hamiltonians have been derived for other possible orientations of the axes [Sutcliffe and Tennyson (1991)] but most of these suffer from profound problems with singularities which have so far prevented their use. It should be noted that an inappropriate axis embedding will not only lead to greatly increased coupling between the vibrational and rotational motions of the system, but can also lead to loss of symmetry in the system. Furthermore, calculation of vibrational band

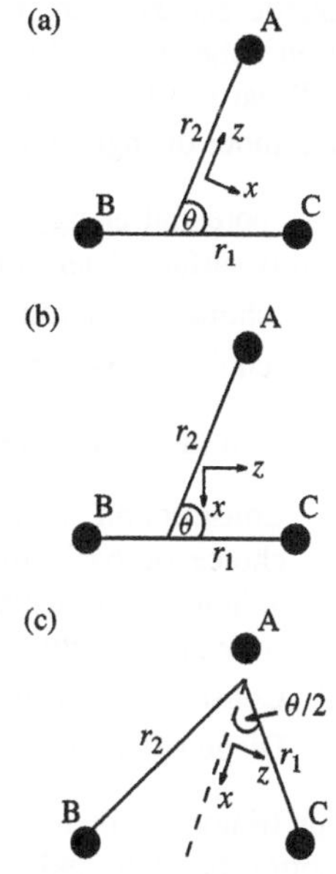

Figure 9.3. Some possible axis embeddings for orthogonal triatomic coordinates. (a) Jacobi coordinates, r_2 embedding; (b) Jacobi coordinates, r_1 embedding; (c) Radau coordinates, bisector embedding. In all cases the y-axis is perpendicular to the plain of the diagram.

intensities using rotationless wavefunctions, a useful method for obtaining intensity estimates cheaply, also depends on the axis system used. Since vibrational band intensities are an approximation which rely on separation of vibrational and rotational motion, they should be calculated using Eckart axes which minimize these interactions [Le Sueur *et al.* (1992)].

9.4 Vibrational motion

For diatomic molecules the vibrational Schrödinger equation can be solved by direct integration [Le Roy (1996)]. For larger molecules, however, there is no single procedure appropriate for all cases. Accurate, non-perturbative methods of solving the vibrational problem for both triatomic and larger molecules have received considerable attention. Not all these methods are based upon the variational principle. One example, direct solution of the problem using quantum Monte Carlo (QMC) procedures, is discussed by Wales elsewhere in this volume.

Another class of methods used for both vibration and rotation–vibration problems are based on generalization of the close coupled equations of Arthurs and Dalgarno (1960). These methods, as encapsulated in the program BOUND [Hutson (1993)], use standard basis set expansions to reduce the problem to solving for one, key interaction coordinate which is attacked using numerical integration procedures adapted from those used for diatomics. These procedures find particular application for Van der Waals dimers [Hutson (1991)] where the interaction coordinate clearly operates on a different energy scale to the internal coordinate(s) of the interacting molecules.

In some ways similar in spirit to the close-coupling methods are the various non-rigid or semi-rigid bender approaches developed by Bunker and co-workers [for a review see Jensen (1983)]. These approaches focus on the coupling between a large amplitude bending motion and rotation, for which very accurate solutions are obtained. In these approaches the other coordinates are treated in a less complete manner. The MORBID (Morse-oscillator rigid bender internal dynamics) method of Jensen (1988), described in more detail in Sections 15.4.7 and 15.4.8 of Bunker and Jensen (1998), can be thought of as the natural successor of the non-rigid bender approach for triatomic systems. The MORBID method has many similarities to the variational approaches discussed below but differs in that there is some approximation in the nuclear motion kinetic energy used. For this reason the more standard variational procedures have sometimes been dubbed exact kinetic energy or EKE methods.

What can be thought of as the standard method for performing a variational calculations has a number of ingredients. In particular it requires:

- A potential energy surface for the system under consideration. In most cases the accuracy of this surface determines the accuracy of the final calculation.
- A choice of coordinates which then defines the Schrödinger equation to be solved.
- A choice of basis functions to be used to represent the vibrational motions in each coordinate.

Given these ingredients, the computational strategy involves the following steps.

(1) Construction of matrix elements for various terms in the Hamiltonian. Depending on the choice of basis functions, some of these matrix elements may be evaluated analytically although it is usually necessary to use numerical procedures to compute matrix elements over the potential.
(2) Construction of the Hamiltonian matrix from the individual matrix elements.
(3) Diagonalization of the Hamiltonian matrix.

For triatomics there are public domain programs which use this strategy to solve the nuclear motion problem and, if required, generate rotation–vibration spectra [Tennyson, Miller and Le Sueur (1993), and Tennyson, Henderson and Fulton (1995)]. For triatomics the computer time required is usually dominated by the final matrix diagonalization. For this reason various strategies have been developed to reduce the size of this matrix. The most common of these involves preconditioning or contracting the final basis by constructing and diagonalizing reduced dimension problems. The idea of such procedures is that fewer basis functions are then required to give converged results.

For molecules with more than three atoms, numerical quadrature can become a computationally very demanding problem. It is in evaluating the potential energy matrix element that this problem is the worst: this involves simultaneous numerical quadrature in all $3N-6$ vibrational coordinates for an N-atomic problem. With M (typically 10–30) quadrature points in each coordinate this leads to M^{3N-6} evaluations of the potential energy surface. One strategy to avoid this problem is to write the potential as a separable or multinomial expansion [Romanowski, Bowman and Harding (1985), Carter and Handy (1996), and Rosenstock *et al.* (1998), for example]. In this form the numerical integrals can all be expressed as products of one-dimensional quadratures. However, this procedure clearly constrains the form of the potential energy surface and the range of any calculation that can be performed using it.

9.4.1 BASIS SET VERSUS GRID METHODS

A number of groups have worked on variational procedures to solve the vibrational problem. Broadly these procedures divide into two camps: those employing (polynomial) basis functions and those employing grid-based procedures. The procedures are closely related but have some subtle differences.

For a triatomic problem in (r_1, r_2, θ) coordinates a basis set representation of the wavefunction of the ith state might be:

$$|i\rangle = \sum_{j,m,n} c^i_{j,m,n} P_j(\theta) Q_m(r_1) R_n(r_2), \tag{2}$$

where, for example, the angular motion might be expanded in terms of Legendre polynomials, and the stretching motion might be represented by Morse or harmonic oscillator functions, which in turn can be expressed in terms of Laguerre and Hermite polynomials, respectively. With preconditioned basis sets, the functions used in equation (2) may be some predetermined linear combination of these polynomials. The calculation involves determining the expansion coefficient matrix, $\boldsymbol{c}$, for each wavefunction. This is done by diagonalizing the real symmetric ('secular') matrix given by:

$$H_{s,s'} = \langle P_j Q_m R_n | \hat{H} | P_{j'} Q_{m'} R_{n'} \rangle, \tag{3}$$

where s is a compound index running over the basis functions used in the expansion; $\hat{H}$ is the Hamiltonian operator whose derivation is discussed above and the integration runs over the vibrational coordinates. These integrals are often referred to as matrix elements.

The variational principle is often stated in a form which deals only with the upper bound to the lowest energy level of a given system. However, it has long been known [MacDonald (1933)] that the matrix formulation of the problem gives an upper bound for excited states of the problem as well. In fact the ith eigenvalue of the problem represents an upper bound to the ith exact solution to the problem. This property is clearly important for rotation–vibration calculations, where one is usually interested in many states of a given system.

Of course knowing something is an upper bound to a desired solution does not mean that it is close to that solution. To be confident that one has obtained a reliable solution to a given problem it is necessary to demonstrate variational convergence. The usual way to do this is to increase the number of terms in the expansion (2) until the results do not change by more than the desired accuracy. It is usually found that lower lying energy levels converge more rapidly than more highly excited ones but one has to be careful because, particularly in multidimensional problems, different levels often converge at markedly different rates. A further complication is that it is sometimes not computationally feasible to perform calculations which conclusively demonstrate convergence.

Procedures which employ basis functions are sometimes labelled FBR for finite basis representation. This name is a tacit acknowledgement that, at least in principle, basis set expansions for vibrational motion involve truncating infinite series of functions. FBR procedures also require a means of evaluating matrix elements numerically. If the basis sets employed are polynomials then it is usual to use Gaussian quadrature schemes based upon these polynomials [Stroud and Secrest (1966)]. Within this method it is possible to perform the numerical integrals to arbitrary accuracy, within the usual limitations of a finite precision computer. For a given potential energy surface and an EKE Hamiltonian, the FBR yields results which are strictly variational upper bounds to the various states of the system. FBR procedures are particularly efficient at obtaining high accuracy results for a few low-lying rotation–vibration states of a molecule.

The most commonly used grid-based method is the so-called discrete variable representation (DVR) [Bačić and Light (1989)]. Formally the DVR is obtained as a transformation from a corresponding FBR,

$$\begin{aligned} H^{\mathrm{DVR}}_{t,t'} &= H_{\alpha,\beta,\gamma,\alpha',\beta',\gamma'} \\ &= \sum_{j,j'} \sum_{m,m'} \sum_{n,n'} T_j^{\alpha} T_{j'}^{\alpha'} T_m^{\beta} T_{m'}^{\beta'} T_n^{\gamma} T_{n'}^{\gamma'} \langle P_j Q_m R_n | \hat{H} | P_{j'} Q_{m'} R_{n'} \rangle, \end{aligned} \tag{4}$$

where the transformation matrices $\boldsymbol{T}$ are defined in terms of the Gaussian quadrature points and weights of the (polynomial) basis functions. For an M-point Gaussian quadrature the quadrature points, $\{\alpha\}$, are defined as the zeroes of Mth order function P_M and the transformation is given by:

$$T_j^{\alpha} = \sqrt{w_\alpha} P_j(\alpha), \tag{5}$$

where w_α is the weight associated with quadrature point α.

Equation (4) demonstrates the close similarity between the FBR and DVR methods. However, the two methods are only identical in the case where the number of basis functions used to represent the vibrational motions in a particular coordinate is equal to the number of Gaussian quadrature points used for evaluating numerical integrals in that coordinate. In practice one would not usually choose to use this few quadrature points in an FBR calculation: it is half the number that can, by the theorems of Gaussian quadrature, be shown to yield exact integrals in most practical applications [Stroud and Secrest (1966)]. This compromise on the accuracy of the integrals used in DVR procedures mean that DVR-based methods, although they often behave variationally, are not actually strictly variational [Wei (1997)].

The loss of strict variational character is a disadvantage, so what advantages does the DVR have? By the so-called quadrature approximation DVR methods are diagonal in the matrix elements of the potential energy operator [Dickinson and Certain (1968)]:

$$\sum_{j,j'}\sum_{m,m'}\sum_{n,n'} T_j^{\alpha} T_{j'}^{\alpha'} T_m^{\beta} T_{m'}^{\beta'} T_n^{\gamma} T_{n'}^{\gamma'} \langle P_j Q_m R_n | V | P_{j'} Q_{m'} R_{n'} \rangle \simeq \delta_{\alpha,\alpha'} \delta_{\beta,\beta'} \delta_{\gamma,\gamma'} V(\alpha, \beta, \gamma). \tag{6}$$

The potential diagonal property makes the Hamiltonian matrix very sparse. It is possible to take advantage of this sparseness directly by using iterative matrix diagonalization procedures to obtained the desired solutions of the Hamiltonian matrix [Bramley and Carrington (1993)]. It is also possible to exploit this property to adapt a DVR grid to a particular potential using a so-called potential-optimized DVR [Echave and Clary (1992)].

Conversely the potential diagonal property makes the DVR particularly good, possibly optimally good, for designing diagonalization and truncation procedures [Bačić and Light (1989)]; these result in a Hamiltonian matrix containing a representation of a much higher fraction of well-converged states than can usually be obtained with an FBR-based procedure. This means that the DVR is particularly effective for problems where a large number of states are of interest. It would appear that which of the two DVR procedures is most effective is dependent on the system under study, and in particular how good the coordinates used are at representing the vibrational motions of this system. With a good match, the diagonalization and truncation procedure performs best, but with a poor match direct diagonalization is computationally more efficient [Bramley and Carrington (1994)].

Finally a grid-based procedure can be used as a means of avoiding singularities in the Hamiltonian by simply dropping the region in the vicinity of the singularity from the calculation [Tennyson and Sutcliffe (1992)]. Of course, this is only useful if the singular regions lie at high enough energy not to effect the energy levels of interest.

9.4.2 DIAGONALIZATION

Diagonalizing the final Hamiltonian (or secular) matrix is a crucial computational step in the above procedures. It is the ability to diagonalize this matrix which usually dictates the level at which it is possible to tackle a problem. Indeed the rapid growth in the application of variational methods at the expense of methods based on perturbation theory is a direct consequence of the increase in computer power which has made such diagonalizations feasible even for fairly challenging systems on standard desktop computers.

In all cases discussed above the matrices that need to be diagonalized are real and symmetric. Usually one is only interested in the k lowest solutions of an N-dimensional matrix. The ratio between k and N is important for determining which diagonalization procedure to use.

Standard real symmetric matrix diagonalization procedures which yield either all eigenvalues or all eigenvalues and eigenvectors have a memory requirement which scales as N^2 and computer time requirement proportional to N^3.

The time requirement of iterative procedures, such as Lanczos diagonalizers, scale as $k \times N^2$ for matrices with few zero elements. However, if the matrix is sparse, it is not necessary either to store or to process the entire matrix. This can lead to considerable saving in computer resources, although, as discussed above, it would appear that this situation is somewhat case dependent. It is even possible to formulate these iterative procedures using a so-called direct approach which means that the Hamiltonian matrix is never actually explicitly constructed [Roy and Carrington (1996)]. Such procedures have been used to diagonalize to obtain the lowest few, $k = 50$, eigenvalues from a very large matrices, $N > 2 \times 10^6$ [Lehoucq *et al.* (1998)].

The diagonalization procedures, discussed above, implicitly assume that one is interested in obtaining k energy levels from the lowest upwards. This assumption is compatible with the behaviour of variational procedures, which generally converge the lowest energy levels first. However, for many applications one is only interested in certain high-lying states of system. A number of methods have been proposed for obtaining only a few results of interest from a large matrix [Roy and Carrington (1995)]. The most successful of these appear to be filter diagonalization which uses semi-classical procedures to project out energy levels in the energy region of interest [Mandelshtam and Taylor (1997)] and the pseudo-spectral method [Antikainen, Friesner and Leforestier (1995)] which replaces the direct diagonalization of the matrix with consideration of a matrix element similar to that used to compute linestrengths, see equation (10) below.

9.5 Rotational motion

There are many variational procedures available for treating the purely vibrational problem. Only a fraction of these procedures have been adapted to treat rotationally excited molecules. This is perhaps somewhat surprising as it is not possible to model nuclear motion spectra correctly without the explicit treatment of rotational excitation.

In some ways the rotational excitation problem is simpler than the vibrational one. In body-fixed coordinates, the rotational motion is a function of the three Euler angles, (α, β, γ), which link the space-fixed axis system to the body-fixed one. Using these coordinates, the rotation matrices, $D^J_{kM}(\alpha, \beta, \gamma)$ [Brink and Satchler (1993), and Zare (1988)] form a complete and finite representation for the rotational motion. In the rotation matrices, k represents the projection of J onto the space-fixed z-axis and M is the projection of J onto the body-fixed z-axis. In the absence of a magnetic field, the dependence on M can be ignored. This means that the rotational motion of a molecule with total angular momentum quantum number J can be entirely represented by $2J + 1$ of these functions with $k = -J, -J + 1, \ldots, 0, \ldots, J - 1, J$. Further simplification can be achieved by use of symmetry:

$$|J, k, p\rangle = \frac{1}{\sqrt{2}}(|J, k, M\rangle + (-1)^p|J, -k, M\rangle), \qquad k > 0, p = 0, 1, \tag{7}$$
$$|J, k, p\rangle = |J, k = 0, M\rangle, \qquad p = 0,$$

where $|J, k, M\rangle$ are the wavefunctions of a rotating rigid symmetric top. These symmetric top functions are closely related to the rotation matrices, [see Zare (1988), equation (3.125)], but have the advantage that they are normalized. In equation (7), p is the Wang symmetry which gives the parity of the rotational functions under spatial inversion symmetry as $(-1)^{J+p}$. Use of this symmetry means that it is possible to separate the rotational basis into two non-interacting portions of dimension J (for $p = 1$) and $J + 1$ (for $p = 0$).

This much is common to all variational treatments of rotational motion. Difficulties arise because it is necessary to treat rotational motion at the same time as vibrational motion, significantly increasing what is often already a large vibrational problem. Superficially it might appear that the addition of rotational motion would result in problems whose usage of computer time would scale as $(J+1)^3 \times N^3$.

The solution to the problem is to solve the vibrational problem first and then use these solutions as the vibrational part of the basis to solve the rotational problem. The simplest way of doing this is to use the solutions of the rotationless ($J = 0$) vibrational problem for this purpose [Chen, Maessen and Wolfsberg (1985)]. However, this method fails for molecules for which a smooth transition from bent to linear geometries is important. There are many examples of these including Van der Waals dimers and floppy but strongly bound systems such as water and H_3^+.

For such molecules it is necessary to solve 'vibrational' problems which include the diagonal part of the rotational kinetic energy operator. In this step of the calculation the projection of J onto the body-fixed z-axis, k, is assumed to be a good quantum number. The second step of the calculation relaxes this assumption but produces a Hamiltonian matrix with a particularly simple block off-diagonal structure [Tennyson and Sutcliffe (1986)]. It should be noted that this two-step procedure places much more importance on a suitable choice of body-fixed axes than direct solution of the full rotation–vibration problem: with a good choice of axes rather few solutions to the 'vibrational' problem are needed to converge the results of the full rotation–vibration problem. The reduction in the number of 'vibrational' states needed is of course equivalent to minimizing the Coriolis interaction between the vibrational states. This is exactly what the Eckart embedding, discussed above, strives to do.

The structure of the Hamiltonian matrix inherent in these two-step procedures can be directly exploited using iterative diagonalization procedures which also lead to a substantial (approximately J-fold) reduction in the memory requirement for the calculation. These two-step procedures have been used to obtain reliable results for highly rotationally excited molecules [Miller and Tennyson (1988)].

Of course the standard spectroscopic method of representing rotational motion is via rotational constants. Although there are well-documented cases where such methods based on perturbation theory do not work [Polyansky (1985)], this representation is compact and reliable for many systems, particularly those undergoing only small amplitude vibrational motion. There are two methods of obtaining rotational constants from a variational calculation. The first is to determine the constants using energy levels obtained from the procedures discussed above. The second is to recognize that, at least in principle, the rotational constants represent the expectation values of various operators. It is straightforward to use vibrational wavefunctions, determined from variational calculations, to evaluate these expectation values as a function of vibrational state.

Table 9.1 compares results obtained with the two methods for the molecule HO_2, which is sufficiently floppy to be regarded as a poor candidate for such a treatment. It is clear that the two

Table 9.1. Comparison of rotational constants, in cm^{-1}, for several vibrational states of HO_2 computed from expectation values of the $J = 0$ vibrational wavefunctions or from the $J = 1$ and $J = 0$ calculated energy levels [Brandão, Rio and Tennyson (1999)].

State	From expectation values			From energy levels		
	A	B	C	A	B	C
(000)	19.75	1.113	1.053	19.74	1.117	1.053
(001)	19.71	1.099	1.039	19.70	1.103	1.039
(010)	20.37	1.107	1.048	20.36	1.117	1.049
(100)	19.02	1.115	1.049	19.02	1.119	1.053
(200)	18.32	1.115	1.045	18.31	1.122	1.052

methods give reasonably similar results. On balance, the expectation value method of evaluating rotational constants is probably the method of choice because (a) it is computationally cheaper as it does not require solving any problems for rotationally excited molecules and (b) the results are more stable, as they are less likely to be distorted by accidental resonances caused by interactions between rotational levels belonging to different vibrational states.

9.6 Symmetry considerations

For a molecule containing no like atoms, the only rigorous symmetry that needs to be considered for the rotation–vibration problem is the total rotational quantum number J and the parity p, defined above. One secular matrix needs be constructed and diagonalized for each (J, p) of interest.

For molecules containing identical atoms it is also necessary to consider the effects of permutation symmetry [Bunker and Jensen (1998)]. It is desirable to include this symmetry explicitly in the nuclear motion calculation for a number of reasons. Inclusion of a symmetry block factorizes the secular matrix into a number of smaller matrices which can be diagonalized separately resulting in a considerable saving of computational resources. Furthermore the results of these separate diagonalizations are automatically symmetry labelled. This helps not only with identifying vibrational states, but is particularly important when generating synthetic spectra, as transitions involving different symmetry species are weighted by different statistical weights, g_{ns}, due to the effects of nuclear spin. In particularly extreme examples, such as H_3^+ or CO_2, statistical weights of zero may occur resulting in the complete absence of certain lines from the spectrum.

Although it is desirable to include the full permutation symmetry of the molecule under consideration in the calculation, this is not always achieved in practice because of other technical and computational considerations. Thus the majority of published variational rotation–vibration calculations on H_3^+, for example, have used Jacobi coordinates despite the fact that it is not possible to represent the full symmetry of H_3^+ using these coordinates.

To exploit the symmetry of a particular molecule it is necessary to use coordinates in which it is possible to represent the effects of the various symmetry operations. Thus for an AB_2 molecule, interchanging the two identical atoms involves changing θ to $-\theta$ in Jacobi coordinates but involves interchanging r_1 and r_2 in Radau coordinates. It is possible to represent this symmetry in both these coordinate systems.

The next step is to define basis functions for the relevant coordinate(s), which can be separated according to symmetry type. Thus for the AB_2 molecule in Jacobi coordinate Legendre polynomials, $P_j(\cos\theta)$ might make an appropriate choice, since functions with even j represent symmetries which are even with respect to interchange of the identical atoms and odd j are odd with respect to this interchange. The secular problem can thus be split according to the parity of j.

When the symmetry operation mixes coordinates, as in the AB_2 system in Radau coordinates example, there are two possible approaches. One option is to redefine the coordinates so that the symmetry behaviour depends only on a single coordinate [Whitnell and Light (1988)]. In the Radau case, or indeed bond length, bond angle coordinates, this transformation is:

$$r = \frac{1}{\sqrt{2}}(r_1 + r_2), \qquad (8)$$

$$q = r_1 - r_2,$$

and the symmetry is now carried by the parity of coordinate q.

An alternative approach is to symmetrize the functions (or grid points) directly:

$$|m,n,s\rangle = \frac{1}{\sqrt{2}}(H_m(r_1)H_n(r_2) + (-1)^s H_n(r_1)H_m(r_2)), \qquad m > n+s, s = 0, 1;$$
$$|m,n,s\rangle = H_m(r_1)H_n(r_2), \qquad m = n, s = 0. \tag{9}$$

In this approach, functions $|m,n,s\rangle$ with $s = 0$ are even and those with $s = 1$ are odd. The two approaches both work although there are some subtle differences over their domains of validity [Polyansky, Tennyson and Zobov (1999)].

9.7 Transition intensities

In nearly all cases of interest, transitions involving changes in rotational and/or vibrational state are driven by electric dipoles. To calculate the probability of such a transition occurring using the wavefunctions derived from variational calculations, it is also necessary to have information on the dipole of the molecule as a function of its internal coordinates. These dipole surfaces can be computed by most *ab initio* electronic structure packages. Usually the dipole is represented by a separate surface for each of its (Cartesian) components. This is necessary as the dipole, unlike the potential, is a vector property, with both a magnitude and a direction.

The computation of rotation–vibration dipole transition intensities is relatively straightforward, if in some cases quite expensive, given a suitable set of rotation–vibration wavefunctions and appropriate dipole surfaces. However the topic has not received a great deal of attention and there are relatively few program suites which include calculation of transition intensities as a routine part of the calculation. This situation perhaps arises because many experiments are not suitable for measuring either absolute or relative intensities and therefore do not yield information for comparison with theory. However, nearly all actual applications of spectroscopy rely on detailed knowledge of transition intensities (see below). Furthermore, as experimental measurements of transition intensities are often only accurate to about 10 %, there are grounds for believing that for simple systems it is often possible to calculate both absolute and relative transition intensities more accurately than they have been measured. Indeed in some cases, such as the spectra of molecular ions, theory remains the only source of transition intensity data.

The linestrength of a dipole transition linking state $|i\rangle$ to state $|f\rangle$ is given by Bunker and Jensen (1998) [equation (14-7)] as:

$$S(f-i) = \sum_{a,b} \sum_{\lambda=-1}^{+1} |\langle f_b|\mu_\lambda|i_a\rangle|^2, \tag{10}$$

where μ_λ is one of the three components of the space-fixed electric dipole expressed at the molecular centre of mass. The first summation in equation (10) runs over not only any degeneracy in the vibrational wavefunctions but also the projection of the rotational functions onto the space-fixed axes, M_i and M_f. Summing over these leads to factors of $(2J_i+1)(2J_f+1)$ in the final expression [Bunker and Jensen (1998)].

The Einstein A_{if} coefficient for spontaneous emission can be obtained from the linestrength:

$$A_{if} = \frac{1}{(2J_f+1)} \frac{64\pi^4\omega_{if}^3}{3h} S(f-i), \tag{11}$$

where is $\omega_{if} = E_f - E_i$ is the transition frequency between the states. A_{if} can be used to generate both emission and adsorption intensities, although certain applications may also require further information to characterize the line profile of a particular transition.

The matrix element in equation (10) involves integration only over the vibrational (or internal) and rotational coordinates [Miller, Tennyson and Sutcliffe (1989), and Schwenke (1996)]. These matrix elements contain all the rigorous dipole selection rules for the system. From analysis of the rotational wavefunctions one finds that electric dipole transitions are strictly forbidden unless:

$$\begin{aligned} \Delta J &= J_f - J_i = \pm 1, \qquad p_i = p_f; \\ \Delta J &= J_f - J_i = 0, \qquad p_i \neq p_f. \end{aligned} \tag{12}$$

Unless the molecule contains identical atoms there are no further selection rules; however, for molecules with permutation symmetry the additional selection rules can be summarized by saying that transitions only occur between states with the same nuclear spin. Thus in the case of water para ($g_{ns} = 1$) states undergo transitions to para states and ortho ($g_{ns} = 3$) to ortho only.

Any other 'rules' governing transitions such as those obtained from the analysis of harmonic oscillators which suggest that overtone bands, combination bands or difference bands should all be forbidden, are not given by analysing equation (10). These rules are therefore approximate and indicate which will be the strong transitions. Indeed the weak forbidden transitions play an important role in spectroscopy.

9.8 Applications

There are a large number of studies which have used variational methods to compute rotation–vibration. Therefore it is only possible to consider here illustrative examples which give the range of problems that variational calculations can be used for. For this purpose only calculations on the single molecule water ($H_2{}^{16}O$) will be considered, and only a very selective subset of these.

One of the original motivations for developing variational methods was to test potential energy surfaces derived from *ab initio* electronic structure calculations [see Whitehead and Handy (1976), for example]. This motivation remains [Partridge and Schwenke (1997), and Kedziora and Shavitt (1997)], although the accuracy of modern electronic structure calculations has led to studies investigating where the 'standard' non-relativistic, Born–Oppenheimer models fails. For example, studies have analysed the failure of the Born–Oppenheimer approximation [Zobov *et al.* (1996)] and the contribution of electronic relativistic effects [Csaszar *et al.* (1998)]. In all these studies, data from high resolution spectroscopy experiments such as the compilation available from the atmospheric database HITRAN [Rothman *et al.* (1998)], are used as the arbiter of accuracy of the underlying (effective) potential energy surface. This match is made via high accuracy nuclear motion calculations.

The realization that it is not possible to calculate *ab initio* potential energy surfaces of many electron molecules to spectroscopic accuracy has led to an alternative strategy. Comparisons between the calculations and experimental data, usually either transition frequencies or energy levels derived from transition frequencies, are used to refine the potential energy surface. This procedure, which results in a spectroscopically determined, effective potential energy surface, is summarized in Figure 9.4. Spectroscopically determined potentials usually rely on high accuracy *ab initio* calculations as a starting point. Of course these calculations only give the potential energy surface at a grid of points, so these are normally interpolated by fitting to some suitable function [see Murrell *et al.* (1984), for example]. This fitted potential is then used as a basis for variational calculations, the results of which can be compared with experiment. The potential is then adjusted and the procedure repeated until convergence is achieved.

The usual technique used to optimize the potential is least-squares fitting for which sophisticated packages are available [Law and Hutson (1997)]. In practice it is desirable for the fitting procedure not to take too many iterations, as each step is fairly computationally expensive. This

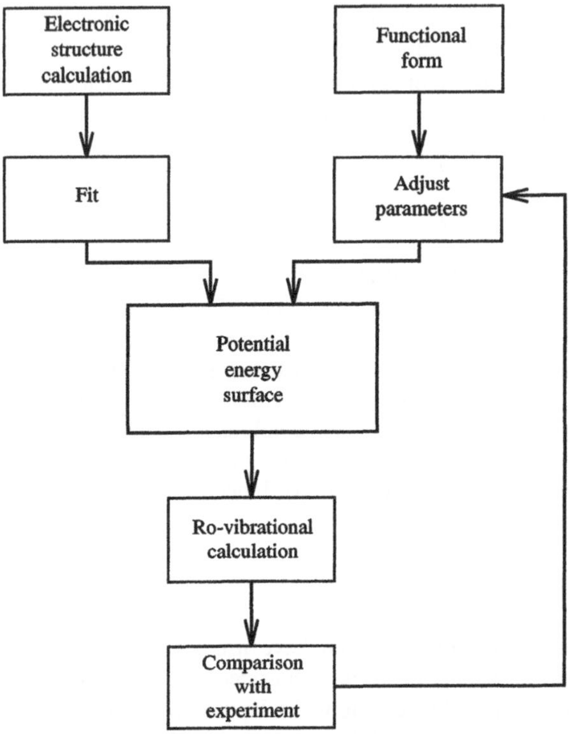

Figure 9.4. Flow diagram depicting the steps involved in fitting an (effective) potential energy surface to experimental data, usually rotation–vibration spectra or energy levels derived from such spectra.

is achieved because the derivatives of the potential, with respect to changing all the various constants that are used to parameterize it, can be obtained at only modest computation cost using the Hellmann–Feynman theorem. This is done by evaluating the expectation value:

$$\frac{\partial}{\partial c}\langle i|\hat{H}|i\rangle = \langle i|\frac{\partial V}{\partial c}|i\rangle \tag{13}$$

for each state of interest, where V is the present approximation to the potential and c is any parameter that one might wish to vary. In equation (13), the derivative of the kinetic energy operator with respect to c is assumed to be zero since, in the EKE approach, the kinetic energy operators are parameter free.

This method of potential optimization has proved highly effective and a number of high accuracy, spectroscopically determined potentials have been derived in this manner for water. The best presently available is due to Partridge and Schwenke (1997) which reproduces a wide range of vibrational and rotational experimental data with a standard deviation of only 0.25 cm^{-1}. It should be noted, however, that a potential energy surface derived in this manner will not correspond to the results of an exact, Born–Oppenheimer electronic structure calculation, as the process of fitting will always attempt to compensate for any effect which influences the spectrum, such as Born–Oppenheimer breakdown, which is not included in the calculation by any other means.

A major reason for performing electronic structure and nuclear motion calculations is to predict molecular spectra, to aid their observation or to help interpret observations of difficult spectra. Traditionally, spectral analysis, the assignment of quantum numbers to observed transitions, has relied heavily on perturbation theory. This situation is changing as variational calculations are able to address problems not amenable to perturbative analysis. A recent example is the assignment of a highly congested spectrum of water recorded in sunspots [Polyansky *et al.* (1997)].

As has already been mentioned, it is sometimes difficult to measure transition intensities even for situations where transition frequencies can be measured to high accuracy. However, applications of high resolution spectroscopy, whether it be the monitoring of a combustion process via infrared emission, modeling the transmission of sunlight through our atmosphere or the study of a species in the interstellar medium, rely heavily on a knowledge of the intensity and temperature dependence of individual transitions. A match of one or more line frequencies may provide proof that a particular molecule is present in the object being studied, but it is only via the intensity that information on the amount present and the physical conditions such the (effective) temperature can be obtained. Furthermore, unlike frequencies which are routinely measured with an accuracy much better than 1 part in a million, even good intensity measurements are often only reliable to around 10 %.

Variational calculations can therefore act as a provider of intensity information even when the frequencies have been measured. One interesting finding is that while spectroscopically determined potential energy surfaces yield much more accurate results than *ab initio* ones for a molecule such as water, the reverse is true for dipole surfaces [Lynas-Gray, Miller and Tennyson (1995)].

Some studies require huge quantities of spectral data. For example, many millions of water transitions are required to model the transport of radiation through super-heated steam. This data is important for modeling situations as diverse as the atmospheres of cool stars and the exhausts of hydrogen burning rockets. In the case of water, many years of laboratory measurements have yielded only a very small fraction of the number of the lines required. For instance the database HITRAN [Rothman *et al.* (1998)] contains just over 30 000 water transitions, sufficient for room temperature models but completely inadequate at high temperature. A number of groups have used variational calculations to compute water linelists appropriate for high temperature studies. The most extensive of these [Partridge and Schwenke (1997)] contains 300 million transitions, the individual measurement of which is clearly beyond any reasonable laboratory campaign.

The radiative transport problems are not the only ones which can benefit from the large quantity of data that can be obtained from variational rotation–vibration calculations. Partition functions, which can be used to determine thermodynamic properties such as specific heats and entropies, can be determined by explicit summation of energy levels. Particularly for hot species, large numbers of energy levels may be required for an accurate computation of the partition function. Variational calculations can be used to compute these properties up to temperatures where other methods fail [Harris *et al.* (1998)].

Calculations, such as those used to compute partition functions at high temperatures, require information on essentially all the bound states of the molecule. Studying bound states up to dissociation naturally links into studies of behaviour just above the dissociation limit. This is, of course, the natural domain of chemical reaction theory, but processes such as photodissociation

$$H_2O + h\nu \rightarrow HO + H \tag{14}$$

or the chemical reaction

$$H_2 + O \rightarrow HO + H \tag{15}$$

can occur on the same (ground state) potential energy surface as that used for studying the spectroscopy of water. Furthermore at low energy and high resolution, these continuum processes often show pronounced structures as a function of energy. These structure are due to quasi-bound states (or resonances) lying in the continuum. It is necessary to adapt the variational methods used here to study these states which do not have an exact energy but are characterized by an energy plus a width or natural lifetime.

A final example of the application of variational calculations has a motivation somewhat different from those discussed above. Chaos theory has seen a major rise in the latter half of the twentieth century. However, this theory only rigorously applies to systems obeying classical mechanics. There has been considerable work in the area of quantum chaology—the study of quantal systems in energy regimes where their classical counterparts are chaotic. Molecular vibrations form a particularly good paradigm for the study of quantum chaology. Anharmonically coupled oscillators are well known to be one of the simplest classically chaotic systems and of course molecular vibrations are generally represented as anharmonically coupled oscillations. So far studies of highly excited states of water have concentrated only on the coupling between the stretching modes [Cho and Child (1994)]. Variational calculations, particularly ones which produce both energy levels and wavefunctions, can provide a rich source of data for testing out ideas on possible manifestations of quantum chaos.

9.9 References

Arthurs, A. M., and Dalgarno, A., 1960, *Proc. R. Soc. London, Ser. A* **256**, 540–551.
Antikainen, J., Friesner, R., and Leforestier, C., 1985, *J. Chem. Phys.*, **102**, 1270–1279.
Bačić, Z., and Light, J. C., 1989, *Annu. Rev. Phys. Chem.*, **40**, 469–498.
Bramley, M. J., and Carrington, T., Jr., 1993 *J. Chem. Phys.*, **99**, 8519–8541.
Bramley, M. J., and Carrington, T., Jr., 1994, *J. Chem. Phys.*, **101**, 8494–8507.
Brandão, J., Rio, C. M. A., and Tennyson, J., 1999, unpublished.
Brink D. M., and Satchler, G. R., 1993, *Angular Momentum*, 3rd edition; Clarendon Press: Oxford.
Brocks, G., van der Avoird, A., Sutcliffe, B. T., and Tennyson, J., 1983, *Mol. Phys.*, **50**, 1025–1043.
Bunker, P. R., and Jensen, P., 1998, *Molecular Symmetry and Spectroscopy*, 2nd edition; NRC Research Press: Ottawa.
Carney, G. D., Sprandel, L. L., and Kern, C. W., 1978, *Adv. Chem. Phys.*, **37**, 305–79.
Carter, S., Bowman, J. M., and Handy, N. C., 1998, *Theor. Chem. Accounts*, **100**, 191.
Carter, S., Shnider, H. M., and Bowman, J. M., 1999, *J. Chem. Phys.*, **110**, 8417–8423.
Carter, S., and Handy, N. C., 1986a, *Mol. Phys.*, **57**, 175–185.
Carter, S., and Handy, N. C., 1986b, *Comput. Phys. Commun.*, **5**, 115–172.
Carter, S., and Handy, N. C., 1996, *J. Mol. Spectrosc.*, **179**, 65–72.
Carter, S., and Meyer, W., 1994, *J. Chem. Phys.*, **100**, 2104–2117.
Chen, C.-L., Maessen, B., and Wolfsberg, M., 1985, *J. Chem. Phys.*, **83**, 1795–1807.
Cho, S. W., and Child, M. S., 1994, *Mol. Phys.*, **81**, 447–465.
Colwell, S. M., and Handy, N. C., 1997, *Mol. Phys.*, **92**, 317–330.
Csaszar, A. G., Kain, J. S., Polyansky, O. L., Zobov, N. F., and Tennyson, J., *Chem. Phys. Lett.*, **293**, 317–323.
Dickinson, A. S., and Certain, P. R., 1968, *J. Chem. Phys.*, **49**, 4209–4211.
Dunn, K. M., Boggs, J. E., and Pulay P., 1987, *J. Chem. Phys.*, **86**, 5088–5093.
Echave, J., and Clary, D. C., 1992, *Chem. Phys. Lett.*, **190**, 339–349.
Eckart, C., 1935, *Phys. Rev*, **47**, 552.
Handy, N. C., 1987, *Mol. Phys.*, **61**, 207–223.
Harris, G. J., Viti, S., Mussa, H. Y., and Tennyson, J., 1998, *J. Chem. Phys.*, **109**, 7197–7204.
Hougen, J. T, Bunker, P. R., and Johns, J. W. C., 1970, *J. Mol. Spectrosc.*, **34**, 136–172.
Hutson, J. M., 1991, *Adv. Molecular Vibrations and Collision Dynamics*, **1A**, 1–45.
Hutson, J. M., 1993, BOUND: A program for calculating bound-state energies for weakly bound molecular complexes, version 5, Distributed via Collaborative Computational Project No. 6 of the Science and Engineering Research Council, on Heavy Particle Dynamics.
Hutson, J. M., 1994, *Comput. Phys. Commun.*, **84**, 1–18.
Iung, C., and Leforestier C., 1992, *J. Chem. Phys.*, **97**, 2481–2489.
Jensen, P., 1983, *Comput. Phys. Commun.*, **1**, 1–55.
Jensen, P., 1988, *J. Mol. Spectrosc.*, **128**, 478–501.
Kedziora, G. S., and Shavitt, I., 1997, *J. Chem. Phys.*, **106**, 8733–8745.
Law, M. M., and Hutson, J. M., 1997, *Comput. Phys. Commun.*, **102**, 252–268.
Lehoucq, R. B., Gray, S. K., Zhang, D.-H., and Light, J. C., 1998, *Comput. Phys. Commun.*, **109**, 15–26.
Le Roy, R. J., 1996, University of Waterloo Chemical Physics Research Report CP-555R, 1.
Le Roy, R. J., and Carley, J. S., 1980, in *Potential Energy Surfaces*, Lawley K. P., Ed.; Wiley: New York, pp. 353–420.

Le Sueur, C. R., Miller, S., Tennyson, J., and Sutcliffe, B. T., 1992, *Mol. Phys.*, **76**, 1147–1156.
Lynas-Gray, A. E., Miller, S., and Tennyson, J., 1995, *J. Mol. Spectrosc.*, **169**, 458–467.
MacDonald, J. K. L, 1933, *Phys. Rev.*, **43**, 830.
Mandelshtam, V. A., and Taylor, H. S., 1997, *J. Chem. Phys*, **106**, 5085–5090.
Manz, J., and Schor, H. H. R, 1986, *J. Phys. Chem.*, **91**, 1813.
Miller, S., and Tennyson, J., 1988, *Chem. Phys. Lett.*, **145**, 117–120.
Miller, S., Tennyson, J., and Sutcliffe, B. T., 1989, *Mol. Phys.*, **66**, 429–456.
Murrell, J. N., Carter, S., Farantos, S. C., Huxley, P., and Varandas, A. J. C., 1984, *Molecular Potential Energy Functions*; Wiley: Chichester.
Partridge, H., and Schwenke, D. W., 1997, *J. Chem. Phys.*, **106**, 4618–4639.
Pekeris, C. L., 1958, *Phys. Rev.*, **112**, 1649.
Polyansky, O. L., 1985, *J. Mol. Spectrosc.*, **112**, 79–87.
Polyansky, O. L., and Tennyson, J., 1999, *J. Chem. Phys.*, **110**, 5056–5064.
Polyansky, O. L., Tennyson, J., and Zobov, N. F., 1999, *Spectrachem. Acta A*, **55**, 659–693.
Polyansky, O. L., Zobov, N. F., Viti, S., Tennyson, J., Bernath, P.F., and Wallace, L., 1997, *Science*, **277**, 346–349.
Radau, R., 1868, *Ann. Sci. Ecole Normale Superior*, **3**, 311.
Rosenstock, M., Rosmus, P., Reinsch, E. A., Treutler, O., Carter, S., and Handy, N. C., 1998, *Mol. Phys.*, **93**, 853–865.
Romanowski, H., Bowman, J. M., and Harding, L., 1985, *J. Chem. Phys.*, **82**, 4155–4165.
Rothman, L. S., Rinsland, C. P., Goldman, A., Massie, S. T., Edwards, D. P., Flaud, J.-M., Perrin, A., Camy-Peyret, C., Dana, V., Mandin, J.-Y., Schroeder, J., McCann, A., Gamache, R. R., Wattson, R. B., Yoshino, K., Chance, K. V., Jucks, K. W., Brown, L. R., Nemtchinov, V., and Varanasi, P., 1998, *J. Quant. Spectrosc. Radiat. Transfer*, **60**, 665–710.
Roy, P. N., and Carrington, T. Jr., 1995, *J. Chem. Phys.*, **103**, 5600–5612.
Roy, P. N., and Carrington, T. Jr., 1996, *Chem. Phys. Lett.*, **257**, 98–104.
Schwenke, D. W., 1996, *J. Phys. Chem*, **100**, 2867–2884.
Stroud, A. H., and Secrest, D., 1966, *Gaussian Quadrature Formulas*; Prentice-Hall: London.
Sutcliffe, B. T., 1982, in *Current Aspects of Quantum Chemistry*, Carbo, R., Ed., Vol. 21; Elsivier: Amsterdam, p. 9.
Sutcliffe, B. T., 1992, *Mol. Phys.*, **75**, 1233–1236.
Sutcliffe, B. T., and Tennyson, J., 1991, *Int. J. Quant. Chem.*, **39**, 183–196.
Tennyson, J., Henderson, J. R., and Fulton, N. G., 1995, *Comput. Phys. Commun.*, **86**, 175–198.
Tennyson, J., Miller, S., and Le Sueur, C. R., 1993, *Comput. Phys. Commun.*, 1993, **75**, 339–364.
Tennyson, J., and Sutcliffe, B. T., 1982, *J. Chem. Phys.*, **77**, 4061–4072.
Tennyson, J., and Sutcliffe, B. T., 1986, *Mol. Phys.*, **58**, 1067–1085.
Tennyson, J., and Sutcliffe, B. T., 1992, *Int. J. Quant. Chem.*, **42**, 941–952.
Watson, J. K. G., 1968, *Mol. Phys.*, **15**, 479–490.
Watson, J. K. G., 1995, *Chem. Phys.*, **190**, 291–300.
Wei, H., 1997, *J. Chem. Phys.*, **106**, 6885–6900.
Wei, H., and Carrington, T. Jr, 1997a, *J. Chem. Phys.*, **107**, 2813–2818.
Wei, H., and Carrington, T. Jr, 1997b, *J. Chem. Phys.*, **107**, 9493–9501.
Whitehead, R. J., and Handy, N. C., 1976, *J. Mol. Spectrosc.*, **59**, 459–469.
Whitnell, R. M., and Light, J. C., 1988, *J. Chem. Phys.*, **89**, 3674–3680.
Whitnell, R. M., and Light, J. C., 1989, *J. Chem. Phys.*, **90**, 1774–1786.
Wilson, E. B. Jr, Decius, J. C., and Cross, P. C., 1955, *Molecular Vibrations: The Theory of Infrared and Raman Vibrational Spectra*, Dover Publications: New York.
Wolniewicz, L., and Hinze, J., 1994, *J. Chem. Phys.*, **101**, 9817–9829.
Zare, R. N., 1988, *Angular Momentum*; Wiley: New York.
Zobov, N. F., Polyansky, O. L., Le Sueur, C. R., and Tennyson, J., 1996, *Chem. Phys. Lett.*, **260**, 381–387.
Zúñiga, J., Bastida, A., and Requena, A., 1997, *J. Chem. Soc., Faraday Trans.*, **93**, 1681–1690.

10 HIGHLY EXCITED STATES AND LOCAL MODES

Lauri Halonen
University of Helsinki, Finland

10.1 Introduction

Traditional vibration–rotation spectroscopy is largely formulated for semirigid molecules in the ground and in low lying vibrational states [Herzberg (1945), Mills (1972), Papousek and Aliev (1982), and Aliev and Watson (1985)]. A cornerstone of this approach is the application of perturbation theory and contact transformation methods to provide physical interpretation of spectroscopic parameters. Fundamental concepts such as molecular structure, potential energy surface, and dipole moment surface can be related to these parameters. Surprisingly, it has been found that even results derived using high order perturbation theory can be physically meaningful.

Computational Molecular Spectroscopy. Edited by Per Jensen and Philip R. Bunker

This was gratifying at the time when the computational methods available were not adequate for more computer time consuming approaches such as large-scale variational calculations. However, the foundation of this theory, the concept of the normal mode coordinate, bears the seeds for the limitation in many practical applications. Normal coordinates do not necessarily provide the best starting point for general vibrational theory because they are rectilinear by definition.

A simple application of standard vibration–rotation theory to high overtone spectroscopy of polyatomic molecules fails to give realistic results owing to the high density of states at high vibrational excitations. Perturbation theory solutions are not a good choice, as the possibility of strong vibration–rotation interactions is high. It was a surprise that at least in some cases the overtone spectra seemed to be more regular than expected. For example, in benzene, it was found that the overtone spectrum consisted of strong absorption features separated by about 3000 cm^{-1} [Martin and Kalantar (1968), and Bray and Berry (1979)]. This could be explained by a model where at high excitations the molecule is assumed to consist of independent CH bond oscillators. This surprising behaviour seemed to contradict what had been expected because of the high density of states in benzene with about 7.3×10^7 vibrational states per 1 cm^{-1} at 15 000 cm^{-1}. Another type of observation is the drastic change of the rotational fine structure of some stretching vibrational bands in GeH_4 [Zhu, Thrush and Robiette (1988)] and in SnH_4 [Zhan *et al.* (1995)] as the vibrational energy increases. The experimental spectra show that the spectral structure of a spherical top becomes that of a symmetric top. This again indicates that bond oscillators become independent at high excitation. In the present application, this would mean that the bond length of one of the independent oscillators would be longer than in the case of the other ones. This results in a dynamical symmetry change at the excited state, i.e. T_d becomes C_{3v}. These examples can be explained with extended models based on standard vibration–rotation theory but another approach, called the local mode model, gives a clearer physical picture. This model is based on the use of internal valence displacement coordinates.

The original local mode model consists of coupled anharmonic bond oscillators where, unlike in normal coordinate based theories, harmonic kinetic and potential energy coupling may exist. The inclusion of bending vibrations such as valence angle bending oscillators may extend this theory. This model can be regarded as a simplification of a general theory which is based on exact vibration–rotation Hamiltonians expressed in terms of curvilinear internal valence coordinates and on full variational methods to obtain eigenvalues. One outcome of this development is the mathematical equivalence of the local and normal mode models once the appropriate cubic and quartic anharmonic coupling terms have been included in the normal mode model. If the stretching vibrational states of many symmetrical hydrogen-containing molecules are treated by the traditional method, it is necessary to include quartic coupling terms in the Hamiltonian. This makes the normal mode model less attractive than the local mode model where the zeroth-order picture is a group of uncoupled anharmonic bond and valence angle bending oscillators. It has been found that the local mode model is particularly useful in describing stretching anharmonic vibrational states involving hydrogen and deuterium stretches but it is less useful for strongly harmonic vibrations. Of course, by extending the model by allowing harmonic and anharmonic kinetic and potential energy couplings between the various oscillators, the model can be made to agree well with experimental data. In summary, the strength of the local mode approach is smaller anharmonic coupling than in the normal mode model.

10.2 Internal coordinate Hamiltonian models for vibrational energies and absorption intensities

10.2.1 GENERAL VIBRATIONAL HAMILTONIAN

Curvilinear internal coordinates, or more precisely curvilinear internal valence displacement coordinates as used in this context, are defined mathematically in Hoy, Mills and Strey (1972). Interest in the present contribution is mainly centered around bond and valence angle displacement coordinates. The exact vibrational Hamiltonian of polyatomic molecules can be written in terms of these coordinates as [Meyer and Günthard (1968), and Pickett (1972)]

$$\hat{H} = \tfrac{1}{2}\hat{\boldsymbol{p}}_r^T \boldsymbol{g}(\boldsymbol{r})\hat{\boldsymbol{p}}_r + V'(\boldsymbol{r}) + V(\boldsymbol{r}), \tag{1}$$

where $\hat{p}_{r_j} = -\mathrm{i}\hbar\partial/\partial r_j$ is the momentum conjugate to the internal valence coordinate r_j (which can be a stretching or bending displacement coordinate) and the eigenfunctions of $\hat{H}$ are normalized so that $\int \psi^*(\boldsymbol{r})\psi(\boldsymbol{r})\,\mathrm{d}r_1\,\mathrm{d}r_2\ldots = 1$. The matrix $\boldsymbol{g}(\boldsymbol{r})$, Wilson's $\boldsymbol{g}$ matrix, is a function of internal coordinates [Wilson, Decius and Cross (1955)]. The quantity $V(\boldsymbol{r})$ is the potential energy function, which is isotope independent because the coordinate system is geometrically defined. The pseudopotential term $V'(\boldsymbol{r})$ is of pure quantum mechanical kinetic energy origin and it does not involve derivative operators. The derivation of the vibrational Hamiltonian is outlined next [Meyer (1985, 1986)]. Readers who wish to proceed directly to the application of this Hamiltonian to local mode problems should continue at the beginning of Section 10.2.2.

The derivation of equation (1) begins with the classical Hamiltonian in the Cartesian representation

$$H = T + V = \sum_{\substack{i=1,\ldots,N \\ \xi=x,y,z}} \frac{1}{2} m_i \dot{x}_{i\xi}^2 + V(\boldsymbol{x}) = \sum_{\substack{i=1,\ldots,N \\ \xi=x,y,z}} \frac{1}{2m_i} p_{x_{i\xi}}^2 + V(\boldsymbol{x}), \tag{2}$$

where $x_{i\xi}$ is the Cartesian coordinate of the ith nucleus with the mass m_i, $\xi(= x, y, \text{or } z)$ is the index for the three Cartesian components, $p_{x_{i\xi}} = \partial L/\partial \dot{x}_{i\xi} = \partial(T - V)/\partial \dot{x}_{i\xi} = \partial T/\partial \dot{x}_{i\xi}$ is the canonical momentum of $x_{i\xi}$, and N is the number of nuclei. The momentum can be expressed as

$$p_{x_{i\xi}} = \frac{\partial L}{\partial \dot{x}_{i\xi}} = \sum_k \frac{\partial L}{\partial \dot{t}_k}\frac{\partial \dot{t}_k}{\partial \dot{x}_{i\xi}} = \sum_k \frac{\partial L}{\partial \dot{t}_k}\frac{\partial t_k}{\partial x_{i\xi}} = \sum_k p_k \frac{\partial t_k}{\partial x_{i\xi}}, \tag{3}$$

where the summation is over three translational, three rotational, and a set of internal coordinates (all of these are denoted by t_k). If the tensor element $g^{(kk')}$ is defined as

$$g^{(kk')} = \sum_{i\xi} \frac{1}{m_i}\frac{\partial t_k}{\partial x_{i\xi}}\frac{\partial t_{k'}}{\partial x_{i\xi}}, \tag{4}$$

then the classical Hamiltonian can be written as

$$H = \frac{1}{2}\sum_{k,k'} p_k g^{(kk')} p_{k'} + V(\boldsymbol{t}). \tag{5}$$

It is not obvious how to find the quantum operator corresponding to the classical momentum p_k. The operator $\hat{p}_k = -\mathrm{i}\hbar\partial/\partial t_k$ is not the correct answer. We start from the Hamiltonian in the Cartesian representation, equation (2), where $\hat{p}_{x_{j\xi}} = -\mathrm{i}\hbar\partial/\partial x_{j\xi}$ in quantum mechanics. A general kinetic energy matrix element can be expressed as

$$\langle a|\hat{T}|b\rangle = \frac{\hbar^2}{2}\int \mathrm{d}\tau \sum_{i\xi} \frac{1}{m_i}\frac{\partial \psi_{xa}}{\partial x_{i\xi}}\frac{\partial \psi_{xb}}{\partial x_{i\xi}}, \tag{6}$$

when the hermitian property of momentum has been taken into account. The volume element $\mathrm{d}\tau$ is given in the Cartesian representation. In terms of the generalized coordinates $\boldsymbol{t}$, it becomes

$$\mathrm{d}\tau = g(\boldsymbol{t})^{1/2}\,\mathrm{d}t_1\,\mathrm{d}t_2\,\mathrm{d}t_3\ldots\mathrm{d}t_n, \tag{7}$$

where

$$g = \begin{vmatrix} g_{11} & g_{12} & \cdots & g_{1n} \\ g_{21} & g_{22} & \cdots & g_{2n} \\ \cdots & \cdots & \cdots & \cdots \\ g_{n1} & g_{n2} & \cdots & g_{nn} \end{vmatrix}, \tag{8}$$

and

$$g_{kk'} = \sum_{i\xi} m_i \frac{\partial x_{i\xi}}{\partial t_k} \frac{\partial x_{i\xi}}{\partial t_{k'}}. \tag{9}$$

Using the chain rule

$$\frac{\partial \psi_{xi}}{\partial x_{j\xi}} = \sum_k \frac{\partial \psi_{xi}}{\partial t_k} \frac{\partial t_k}{\partial x_{j\xi}} (i = a \text{ or } b),$$

and equations (4) and (7), the kinetic energy matrix element becomes

$$\begin{aligned} \langle a|\hat{T}|b\rangle &= \frac{\hbar^2}{2} \int \mathrm{d}t_1\, \mathrm{d}t_2\, \mathrm{d}t_3 \ldots \mathrm{d}t_n \sum_{kk'} \frac{\partial \psi_{xa}}{\partial t_k} g^{1/4} g^{(kk')} g^{1/4} \frac{\partial \psi_{xb}}{\partial t_{k'}} \\ &= -\frac{\hbar^2}{2} \int \mathrm{d}t_1\, \mathrm{d}t_2\, \mathrm{d}t_3 \ldots \mathrm{d}t_n \sum_{kk'} \psi_{xa} \frac{\partial}{\partial t_k} g^{1/4} g^{(kk')} g^{1/4} \frac{\partial}{\partial t_{k'}} \psi_{xb}. \end{aligned} \tag{10}$$

The last equality is obtained by partial integration. The normalization condition is [Podolsky (1928)]

$$\int \mathrm{d}t_1\, \mathrm{d}t_2 \ldots \mathrm{d}t_n g^{1/2} \psi_x^*(t) \psi_x(t) = \int \mathrm{d}t_1\, \mathrm{d}t_2 \ldots \mathrm{d}t_n \psi_t^*(t) \psi_t(t) = 1. \tag{11}$$

Thus,

$$\psi_x(t) = g^{-1/4} \psi_t(t), \tag{12}$$

and the kinetic energy matrix element takes the form

$$\langle a|\hat{T}|b\rangle = -\frac{\hbar^2}{2} \int \mathrm{d}t_1\, \mathrm{d}t_2\, \mathrm{d}t_3 \ldots \mathrm{d}t_n \psi_{ta} \left(\sum_{kk'} g^{-1/4} \frac{\partial}{\partial t_k} g^{1/4} g^{(kk')} g^{1/4} \frac{\partial}{\partial t_{k'}} g^{-1/4} \right) \psi_{tb}. \tag{13}$$

This shows that the quantum mechanical Hamiltonian is

$$\begin{aligned} \hat{H} &= -\frac{\hbar^2}{2} \sum_{kk'} g^{-1/4} \frac{\partial}{\partial t_k} g^{1/4} g^{(kk')}(t) g^{1/4} \frac{\partial}{\partial t_{k'}} g^{-1/4} + V(t) \\ &= -\frac{\hbar^2}{2} \sum_{kk'} \frac{\partial}{\partial t_k} g^{(kk')} \frac{\partial}{\partial t_{k'}} + V'(t) + V(t) = \frac{1}{2} \sum_{kk'} \hat{p}_k g^{(kk')} \hat{p}_{k'} + V'(t) + V(t), \end{aligned} \tag{14}$$

where the pseudopotential term is

$$V'(t) = \frac{\hbar^2}{8} \sum_{kk'} \left(\frac{\partial g^{(kk')}}{\partial t_k} \frac{\partial \ln g}{\partial t_{k'}} + g^{(kk')} \frac{\partial^2 \ln g}{\partial t_k\, \partial t_{k'}} + \frac{1}{4} g^{(kk')} \frac{\partial \ln g}{\partial t_k} \frac{\partial \ln g}{\partial t_{k'}} \right), \tag{15}$$

and the momenta $\hat{p}_k$ are defined as $\hat{p}_k = -\mathrm{i}\hbar\partial/\partial t_k$.

Translational motion can be separated from the other degrees of freedom. The rotational and vibrational motion cannot be separated but in the present application we are only dealing with the pure vibrational problem. For this reason, the momenta conjugate to the rotational degrees of freedom are set equal to zero and equation (1) is obtained.

10.2.2 LOCAL MODE HAMILTONIAN FOR TWO COUPLED IDENTICAL ANHARMONIC OSCILLATORS

The general vibrational Hamiltonian in equation (1) can be easily applied to semirigid molecules. The $\boldsymbol{g}$ matrix elements can be obtained with equation (4). Many of them have been tabulated in Wilson, Decius and Cross (1955), where a useful extension of equation (4) called the s vector method is described for the derivation of these elements. For water-type XY_2 molecules, the Hamiltonian is

$$\hat{H} = (\hat{p}_{r_1} \quad \hat{p}_{r_2} \quad \hat{p}_\theta) \begin{pmatrix} g^{(r_1r_1)} & g^{(r_1r_2)} & g^{(r_1\theta)} \\ g^{(r_1r_2)} & g^{(r_2r_2)} & g^{(r_2\theta)} \\ g^{(r_1\theta)} & g^{(r_2\theta)} & g^{(\theta\theta)} \end{pmatrix} \begin{pmatrix} \hat{p}_{r_1} \\ \hat{p}_{r_2} \\ \hat{p}_\theta \end{pmatrix} + V'(r_1, r_2, \theta) + V(r_1, r_2, \theta), \tag{16}$$

where the $\boldsymbol{g}$ matrix elements are

$$\begin{aligned} g^{(r_1r_1)} &= g^{(r_2r_2)} = \mu_X + \mu_Y = \mu^{-1} = \frac{1}{m_X} + \frac{1}{m_Y}, \\ g^{(\theta\theta)} &= \mu_Y\left(\frac{1}{R_1^2} + \frac{1}{R_2^2}\right) + \mu_X\left(\frac{1}{R_1^2} + \frac{1}{R_2^2} - \frac{2\cos\alpha}{R_1R_2}\right), \\ g^{(r_1r_2)} &= \mu_X \cos\alpha, \\ g^{(r_1\theta)} &= -\frac{\mu_X \sin\alpha}{R_2}, \\ g^{(r_2\theta)} &= -\frac{\mu_X \sin\alpha}{R_1}. \end{aligned} \tag{17}$$

In these equations, r_i (i is 1 or 2) is the bond displacement coordinate of the ith bond, θ is the valence angle displacement coordinate, R_i (i is 1 or 2) is the instantaneous bond length of the ith bond, α is the instantaneous valence angle, and $m_X = 1/\mu_X$ and $m_Y = 1/\mu_Y$ are the masses of the X and Y atoms, respectively. The pseudopotential term V' takes the form [Halonen and Carrington (1988)]

$$V' = \frac{\hbar^2 \cos^3\alpha}{4m_X R_1 R_2 \sin^2\alpha} - \frac{\hbar^2}{8\mu}\left(\frac{1}{R_1^2} + \frac{1}{R_2^2}\right)(1 + \mathrm{cosec}^2\alpha). \tag{18}$$

Note that usually in equation (16), for example, $g^{(r_1r_1)}$ would be written as $G_{r_1r_1}$ [Wilson, Decius and Cross (1955)]. However, in this text, we wish to use the unconventional notation in order to emphasize the origin of the $\boldsymbol{g}$ matrix elements.

A simple stretching vibrational local mode Hamiltonian is obtained from equation (16) by constraining the valence angle displacement coordinate to its equilibrium value, i.e. to zero. The conjugate momentum $\hat{p}_\theta$ disappears in this case. The pseudopotential term is neglected (its effect is often small for well-bent triatomic molecules [Halonen and Carrington (1988)]). The potential energy surface is approximated as a sum of two Morse oscillator functions [Morse (1929)] for the bond stretching oscillators that are coupled by a bilinear harmonic term. Thus, the Hamiltonian is

$$\hat{H} = \left(\tfrac{1}{2}g^{(rr)}\hat{p}_{r_1}^2 + D_e y_1^2\right) + \left(\tfrac{1}{2}g^{(rr)}\hat{p}_{r_2}^2 + D_e y_2^2\right) + g_e^{(rr')}\hat{p}_{r_1}\hat{p}_{r_2} + f_{rr'}r_1r_2, \tag{19}$$

where $g^{(rr)} = g_e^{(rr)} = g^{(r_1r_1)} = g^{(r_2r_2)}$ and $g_e^{(rr')} = g_e^{(r_1r_2)}$. The subscript e indicates that the $\boldsymbol{g}$ matrix element in question is calculated at the equilibrium structure. The Morse variable y_i is defined as

$y_i = 1 - \exp(-a_r r_i)$, where the index $i = 1$ or 2. The parameters D_e and a_r are the Morse dissociation energy and steepness parameter, respectively. The parameter $f_{rr'}$ is a harmonic force constant that describes the strength of the potential energy coupling between the bond oscillators. If a realistic asymptotic limit at large stretching displacements is required, then the potential energy coupling term should be replaced by $f_{rr'}a_r^{-2}y_1 y_2$ [Sage and Williams (1983), and Lehmann, Scherer and Klemperer (1983)]. By expanding both of the y_i variables as Taylor series and retaining the first terms, the coupling term in equation (19) is recovered.

The eigenvalues of the Hamiltonian in equation (19) can be obtained variationally by using a suitable basis set. However, it turns out that not all couplings between states are significant. The close lying states can be grouped together and the couplings between different groups can be neglected in the first approximation or better can be taken into account by using second-order perturbation theory. A simple block diagonal model called a harmonically coupled anharmonic oscillator model (HCAO) is obtained [Child and Lawton (1981), Mortensen, Henry and Mohammadi (1981), and Child and Halonen (1984)].

In the HCAO model, the Morse oscillator energy level expression is used as matrix elements for the two terms in parentheses on the right-hand side of equation (19). In wavenumber units, it is ($i = 1$ or 2) [Morse (1929)]

$$E_i = \omega_r \left(v_{r_i} + \tfrac{1}{2}\right) + x_{rr} \left(v_{r_i} + \tfrac{1}{2}\right)^2, \tag{20}$$

where v_{r_i} is the vibrational quantum number, $\omega_r = \hbar(2a_r^2 D_e g^{(rr)})^{1/2}/hc = \hbar(f_{rr} g^{(rr)})^{1/2}/hc$ is the harmonic wavenumber and $x_{rr} = -a_r^2\hbar^2 g^{(rr)}/2hc$ is the bond anharmonicity parameter. c is the speed of light. The parameter f_{rr} is the harmonic stretching force constant, i.e.

$$f_{rr} = \left(\frac{\partial^2 V}{\partial r_1^2}\right)_e = \left(\frac{\partial^2 V}{\partial r_2^2}\right)_e = 2a_r^2 D_e.$$

In the HCAO model, the matrix elements of the coupling terms in equation (19) are approximated by harmonic oscillator formulas. These are obtained by employing bond product harmonic oscillator basis functions $|v_{r_1} v_{r_2}\rangle = |v_{r_1}\rangle |v_{r_2}\rangle$. Results for the harmonic oscillator are such that there exist both $\Delta v = 0$ and ± 2 matrix elements [Wilson, Decius and Cross (1955)]. The quantity $v = v_{r_1} + v_{r_2}$ is the total stretching quantum number. The $\Delta v = \pm 2$ couplings are taken into account by second order perturbation theory. This is justified because of the large energy separation of the interacting states. The final matrix elements take the forms

$$\begin{aligned} \langle v_{r_1} v_{r_2} | \hat{H}/hc | v_{r_1} v_{r_2}\rangle &= \omega_r'(v_{r_1} + v_{r_2} + 1) + x_{rr}\left[\left(v_{r_1} + \tfrac{1}{2}\right)^2 + \left(v_{r_2} + \tfrac{1}{2}\right)^2\right], \\ \langle (v_{r_1} + 1)(v_{r_2} - 1) | \hat{H}/hc | v_{r_1} v_{r_2}\rangle &= \lambda_r[(v_{r_1} + 1)v_{r_2}]^{1/2}, \\ \langle (v_{r_1} - 1)(v_{r_2} + 1) | \hat{H}/hc | v_{r_1} v_{r_2}\rangle &= \lambda_r[v_{r_1}(v_{r_2} + 1)]^{1/2}, \end{aligned} \tag{21}$$

where

$$\omega_r' = \omega_r - \tfrac{1}{8}\omega_r\left(g_e^{(rr')}/g^{(rr)} - f_{rr'}/f_{rr}\right)^2, \tag{22}$$

and

$$\lambda_r = \tfrac{1}{2}\omega_r\left(g_e^{(rr')}/g^{(rr)} + f_{rr'}/f_{rr}\right). \tag{23}$$

The second term on the right-hand side of equation (22) is the perturbation theory contribution mentioned above. Its effect is small and it can often be neglected. Although the Hamiltonian in equation (19) together with the approximations introduced gives the theoretical foundation, the matrix elements in equations (21) define the HCAO model.

The Hamiltonian matrix is in the block diagonal form because only states with the same total stretching quantum number v are coupled. The different blocks can be factorized further by using symmetrized basis functions

$$\begin{aligned}
|v_{r_1}v_{r_2}\mathrm{A}_1\rangle &= \frac{1}{\sqrt{2}}(|v_{r_1}v_{r_2}\rangle + |v_{r_2}v_{r_1}\rangle), \qquad v_{r_1} > v_{r_2},\\
|v_{r_1}v_{r_2}\mathrm{B}_2\rangle &= \frac{1}{\sqrt{2}}(|v_{r_1}v_{r_2}\rangle - |v_{r_2}v_{r_1}\rangle), \qquad v_{r_1} > v_{r_2},\\
|v_{r_1}v_{r_2}\mathrm{A}_1\rangle &= |v_{r_1}v_{r_1}\rangle, \qquad v_{r_1} = v_{r_2},
\end{aligned} \tag{24}$$

where the $C_{2\mathrm{v}}$ point group symbols are used. On the right-hand side of equations (24), in $|v_{r_i}v_{r_j}\rangle$, v_{r_i} and v_{r_j} refer to the first and to the second bond oscillator, respectively. When zero-point energy is subtracted from the diagonal elements, the matrices up to $v = 5$ are

$$\begin{aligned}
&H(v=1,\mathrm{A}_1) = \omega_r' + 2x_{rr} \pm \lambda_r,\\
&H(v=2,\mathrm{A}_1) = \begin{pmatrix} 2\omega_r' + 6x_{rr} & 2\lambda_r \\ 2\lambda_r & 2\omega_r' + 4x_{rr} \end{pmatrix}, \qquad H(v=2,B_2) = 2\omega_r' + 6x_{rr},\\
&H(v=3,\mathrm{A}_1) = \begin{pmatrix} 3\omega_r' + 12x_{rr} & \sqrt{3}\lambda_r \\ \sqrt{3}\lambda_r & 3\omega_r' + 8x_{rr} + 2\lambda_r \end{pmatrix},\\
&H(v=3,\mathrm{B}_2) = \begin{pmatrix} 3\omega_r' + 12x_{rr} & \sqrt{3}\lambda_r \\ \sqrt{3}\lambda_r & 3\omega_r' + 8x_{rr} - 2\lambda_r \end{pmatrix},\\
&H(v=4,\mathrm{A}_1) = \begin{pmatrix} 4\omega_r' + 20x_{rr} & 2\lambda_r & 0 \\ 2\lambda_r & 4\omega_r' + 14x_{rr} & \sqrt{12}\lambda_r \\ 0 & \sqrt{12}\lambda_r & 4\omega_r' + 12x_{rr} \end{pmatrix},\\
&H(v=4,\mathrm{B}_2) = \begin{pmatrix} 4\omega_r' + 20x_{rr} & 2\lambda_r \\ 2\lambda_r & 4\omega_r' + 14x_{rr} \end{pmatrix},\\
&H(v=5,\mathrm{A}_1) = \begin{pmatrix} 5\omega_r' + 30x_{rr} & \sqrt{5}\lambda_r & 0 \\ \sqrt{5}\lambda_r & 5\omega_r' + 22x_{rr} & \sqrt{8}\lambda_r \\ 0 & \sqrt{8}\lambda_r & 5\omega_r' + 18x_{rr} + 3\lambda_r \end{pmatrix},\\
&H(v=5,\mathrm{B}_2) = \begin{pmatrix} 5\omega_r' + 30x_{rr} & \sqrt{5}\lambda_r & 0 \\ \sqrt{5}\lambda_r & 5\omega_r' + 22x_{rr} & \sqrt{8}\lambda_r \\ 0 & \sqrt{8}\lambda_r & 5\omega_r' + 18x_{rr} - 3\lambda_r \end{pmatrix}.
\end{aligned} \tag{25}$$

The remarkable property of this model is the fact that the eigenvalue structure depends only on three parameters, the harmonic wavenumber ω_r', the anharmonicity parameter x_{rr}, and the bond oscillator coupling parameter λ_r'. The model has been applied to stretching vibrational overtone data of H_2S taken from Vaittinen *et al.* (1997). The observed vibrational term values have been used as data in a calculation where the model parameters have been optimized by the non-linear least-squares method. The fit together with observed and calculated vibrational term values are given in Table 10.1 and the model parameters with standard errors in Table 10.2. The result is pleasing taking into account the simplicity of the model.

If we define the concept of an overtone manifold as a set of energy levels for which the total stretching quantum number v is constant, then within a specified overtone manifold the energy level structure is determined by the ratio $\xi = -x_{rr}/\lambda_r$ [Child and Lawton (1981)]. The large magnitude of the stretching anharmonicity over the magnitude of the coupling strength is called

Table 10.1. $H_2{}^{32}S$ stretching vibrational energy level fit[a].

$v_{r_1} v_{r_2} \Gamma$	$v_1 v_3$	$\nu_{obs}(cm^{-1})$	$(o-c)(cm^1)$
$10A_1$	10	2614.41	0.84
$10B_2$	01	2628.46	1.19
$20A_1$	20	5144.99	1.43
$20B_2$	11	5147.22	1.72
$11A_1$	02	5243.10	0.33
$30A_1$	30	7576.38	1.94
$30B_2$	21	7576.55	2.00
$21A_1$	12	7752.26	−0.74
$21B_2$	03	7779.32	−0.99
$40A_1$	40	9911.02	2.05
$40B_2$	31	9911.02	2.05
$50A_1$	50	12149.46	1.40
$50B_2$	41	12149.46	1.40
$41A_1$	32	12524.63	−3.88
$41B_2$	23	12525.20	−3.69
$60A_1$	60	14291.12	−0.68
$60B_2$	51	14291.12	−0.68

[a] $o-c$ denotes observed minus calculated value. Data are taken from Vaittinen *et al.* (1997) and Vaittinen, Campargue, and Flaud (1998).

Table 10.2. Stretching vibrational parameters (cm^{-1}) of $H_2{}^{32}S$[a].

Parameter	Value
ω_r'	2715.76 (48)
x_{rr}	−47.671 (88)
λ_r	−6.85 (62)

[a] The numbers in parentheses are one-standard errors in the least significant digit. All data in Table 10.1 were given a unit weight in the non-linear least-squares optimization of the model parameters.

the near local mode limit region. On the other hand, the region where the magnitude of the coupling strength is large compared with the magnitude of the bond anharmonicity is called the near normal mode limit case.

10.2.3 LOCAL MODE HAMILTONIAN FOR THREE AND FOUR COUPLED IDENTICAL ANHARMONIC OSCILLATORS

The model described above for two identical anharmonic oscillators can be extended to three and four identical oscillators in XH_3- and XH_4-type C_{3v} and T_d molecules. For example, the Hamiltonian in equation (19) becomes

$$\hat{H} = \sum_{i=1}^{n} \left(\frac{1}{2} g^{(rr)} \hat{p}_{r_i}^2 + D_e y_i^2 \right) + \sum_{i<j} \left(g_e^{(rr')} \hat{p}_{r_i} \hat{p}_{r_j} + f_{rr'} r_i r_j \right), \tag{26}$$

where i and j are indices over the identical bond oscillators. n is the number of the identical bond oscillators. The symmetrization of the bond product basis set can be carried out with projection operators [Cotton (1990), and Bunker and Jensen (1998)] or symmetrized promotion operators with vector coupling coefficients [Halonen and Child (1983)]. For the XH_3 case, the results is [Halonen and Child (1983)]

$$
\begin{aligned}
|v_{r_1}v_{r_1}v_{r_1}\mathrm{A}_1\rangle &= |v_{r_1}v_{r_1}v_{r_1}\rangle,\\
|v_{r_1}v_{r_2}v_{r_2}\mathrm{A}_1\rangle &= \frac{1}{\sqrt{3}}(|v_{r_1}v_{r_2}v_{r_2}\rangle + |v_{r_2}v_{r_1}v_{r_2}\rangle + |v_{r_2}v_{r_2}v_{r_1}\rangle),\\
|v_{r_1}v_{r_2}v_{r_3}\mathrm{A}_1\rangle &= \frac{1}{\sqrt{6}}(|v_{r_1}v_{r_2}v_{r_3}\rangle + |v_{r_1}v_{r_3}v_{r_2}\rangle + |v_{r_2}v_{r_1}v_{r_3}\rangle,\\
&\quad + |v_{r_3}v_{r_1}v_{r_2}\rangle + |v_{r_2}v_{r_3}v_{r_1}\rangle + |v_{r_3}v_{r_2}v_{r_1}\rangle),\\
|v_{r_1}v_{r_2}v_{r_3}\mathrm{A}_2\rangle &= \frac{1}{\sqrt{6}}(|v_{r_1}v_{r_2}v_{r_3}\rangle - |v_{r_1}v_{r_3}v_{r_2}\rangle - |v_{r_2}v_{r_1}v_{r_3}\rangle,\\
&\quad + |v_{r_3}v_{r_1}v_{r_2}\rangle + |v_{r_2}v_{r_3}v_{r_1}\rangle - |v_{r_3}v_{r_2}v_{r_1}\rangle),\\
|v_{r_1}v_{r_2}v_{r_2}\mathrm{E_a}\rangle &= \frac{1}{\sqrt{6}}(2|v_{r_1}v_{r_2}v_{r_2}\rangle - |v_{r_2}v_{r_1}v_{r_2}\rangle - |v_{r_2}v_{r_2}v_{r_1}\rangle),\\
|v_{r_1}v_{r_2}v_{r_2}\mathrm{E_b}\rangle &= \frac{1}{\sqrt{2}}(|v_{r_2}v_{r_1}v_{r_2}\rangle - |v_{r_2}v_{r_2}v_{r_1}\rangle),\\
|v_{r_1}v_{r_2}v_{r_3}\mathrm{E_a}\rangle &= \frac{1}{\sqrt{12}}(2|v_{r_1}v_{r_2}v_{r_3}\rangle + 2|v_{r_1}v_{r_3}v_{r_2}\rangle - |v_{r_2}v_{r_1}v_{r_3}\rangle,\\
&\quad - |v_{r_3}v_{r_1}v_{r_2}\rangle - |v_{r_2}v_{r_3}v_{r_1}\rangle - |v_{r_3}v_{r_2}v_{r_1}\rangle),\\
|v_{r_1}v_{r_2}v_{r_3}\mathrm{E_b}\rangle &= \frac{1}{2}(|v_{r_2}v_{r_1}v_{r_3}\rangle + |v_{r_3}v_{r_1}v_{r_2}\rangle - |v_{r_2}v_{r_3}v_{r_1}\rangle - |v_{r_3}v_{r_2}v_{r_1}\rangle),\\
|v_{r_1}v_{r_2}v_{r_3}\mathrm{E_a}\rangle &= \frac{1}{2}(-|v_{r_2}v_{r_1}v_{r_3}\rangle + |v_{r_3}v_{r_1}v_{r_2}\rangle - |v_{r_2}v_{r_3}v_{r_1}\rangle + |v_{r_3}v_{r_2}v_{r_1}\rangle),\\
|v_{r_1}v_{r_2}v_{r_3}\mathrm{E_b}\rangle &= \frac{1}{\sqrt{12}}(-2|v_{r_1}v_{r_2}v_{r_3}\rangle + 2|v_{r_1}v_{r_3}v_{r_2}\rangle - |v_{r_2}v_{r_1}v_{r_3}\rangle,\\
&\quad + |v_{r_3}v_{r_1}v_{r_2}\rangle + |v_{r_2}v_{r_3}v_{r_1}\rangle - |v_{r_3}v_{r_2}v_{r_1}\rangle),
\end{aligned}
\qquad (27)
$$

where $v_{r_1} \neq v_{r_2} \neq v_{r_3} \neq v_{r_1}$ and for XH_4 in Halonen (1986). On the right-hand side of equations (27) in $|v_{r_i}v_{r_j}v_{r_k}\rangle$, v_{r_i}, v_{r_j}, and v_{r_k} refer to the first, second, and third bond oscillators, respectively. In the HCAO model, symmetrized Hamiltonian matrices are obtained from these basis functions and from the extension of the matrix elements in equations (21). When the zero-point energy has been subtracted from the diagonal elements in the case of $v = 1$, 2 and 3, the result for XH_3 is [Halonen and Child (1983)]

$$
\begin{aligned}
H(v=1,\mathrm{A}_1) &= \omega_r' + 2x_{rr} + 2\lambda_r,\\
H(v=1,\mathrm{E}) &= \omega_r' + 2x_{rr} - \lambda_r,\\
H(v=2,\mathrm{A}_1) &= \begin{pmatrix} 2\omega_r' + 6x_{rr} & 2\sqrt{2}\lambda_r \\ 2\sqrt{2}\lambda_r & 2\omega_r' + 4x_{rr} + 2\lambda_r \end{pmatrix},
\end{aligned}
$$

$$H(v=2,\mathrm{E})=\begin{pmatrix}2\omega_r'+6x_{rr} & -\sqrt{2}\lambda_r\\ -\sqrt{2}\lambda_r & 2\omega_r'+4x_{rr}-\lambda_r\end{pmatrix},\tag{28}$$

$$H(v=3,\mathrm{A}_1)=\begin{pmatrix}3\omega_r'+12x_{rr} & \sqrt{6}\lambda_r & 0\\ \sqrt{6}\lambda_r & 3\omega_r'+8x_{rr}+3\lambda_r & 2\sqrt{3}\lambda_r\\ 0 & 2\sqrt{3}\lambda_r & 3\omega_r'+6x_{rr}\end{pmatrix},$$

$$H(v=3,\mathrm{A}_2)=3\omega_r'+8x_{rr}-3\lambda_r,$$

$$H(v=3,\mathrm{E})=\begin{pmatrix}3\omega_r'+12x_{rr} & \sqrt{6}\lambda_r & 0\\ \sqrt{6}\lambda_r & 3\omega_r'+8x_{rr} & -\sqrt{3}\lambda_r\\ 0 & -\sqrt{3}\lambda_r & 3\omega_r'+8x_{rr}\end{pmatrix},$$

and for XH_4 [Halonen and Child (1982)]

$$H(v=1,\mathrm{A}_1)=\omega_r'+2x_{rr}+3\lambda_r,$$

$$H(v=1,\mathrm{F}_2)=\omega_r'+2x_{rr}-\lambda_r,$$

$$H(v=2,\mathrm{A}_1)=\begin{pmatrix}2\omega_r'+6x_{rr} & 2\sqrt{3}\lambda_r\\ 2\sqrt{3}\lambda_r & 2\omega_r'+4x_{rr}+4\lambda_r\end{pmatrix},$$

$$H(v=2,\mathrm{E})=2\omega_r'+4x_{rr}-2\lambda_r,$$

$$H(v=2,\mathrm{F}_2)=\begin{pmatrix}2\omega_r'+6x_{rr} & 2\lambda_r\\ 2\lambda_r & 2\omega_r'+4x_{rr}\end{pmatrix},$$

$$H(v=3,\mathrm{A}_1)=\begin{pmatrix}3\omega_r'+12x_{rr} & 3\lambda_r & 0\\ 3\lambda_r & 3\omega_r'+8x_{rr}+4\lambda_r & 2\sqrt{6}\lambda_r\\ 0 & 2\sqrt{6}\lambda_r & 3\omega_r'+6x_{rr}+3\lambda_r\end{pmatrix},\tag{29}$$

$$H(v=3,\mathrm{E})=3\omega_r'+8x_{rr}+\lambda_r,$$

$$H(v=3,\mathrm{F}_1)=3\omega_r'+8x_{rr}-3\lambda_r,$$

$$H(v=3,\mathrm{F}_2)=\begin{pmatrix}3\omega_r'+12x_{rr} & \sqrt{6}\lambda_r & \sqrt{3}\lambda_r & 0\\ \sqrt{6}\lambda_r & 3\omega_r'+8x_{rr}-\lambda_r & \sqrt{2}\lambda_r & 0\\ \sqrt{3}\lambda_r & \sqrt{2}\lambda_r & 3\omega_r'+8x_{rr}+2\lambda_r & -2\sqrt{2}\lambda_r\\ 0 & 0 & -2\sqrt{2}\lambda_r & 3\omega_r'+6x_{rr}-\lambda_r\end{pmatrix}.$$

In Figure 10.1(a) and (b) for spherical tops, a correlation diagram is shown which gives reduced energy levels

$$\varepsilon=\frac{E-\overline{E}}{\left(\lambda_r^2+x_{rr}^2\right)^{1/2}}\tag{30}$$

as a function of the local mode parameter $\xi=-x_{rr}/\lambda_r$ for $v=1-5$ [Halonen and Child (1982)]. In equation (30), E is the energy of the vibrational level in question, and $\overline{E}$ is the mean energy of a particular overtone manifold. The striking feature is the large change in the nature of the energy level structure in going from one limit to the other. Another interesting aspect is the effective decoupling of the so-called local mode pair of states $|v000\mathrm{A}_1\rangle$ and $|v000\mathrm{F}_2\rangle$ from the rest of the overtone manifold in question as the stretching vibrational energy increases. This is a result of the fact that the energy level separation of the interacting states $|v000\rangle$ and $|(v-1)100\rangle$ grows linearly in v but the coupling between the states grows as a square root of v as v increases. The same conclusion applies to the states $|(v-1)100\mathrm{A}_1\rangle$ and $|(v-1)100\mathrm{F}_2\rangle$ and so on but the decoupling occurs more slowly, i.e. it appears at higher quantum levels (see Figure 10.1). The

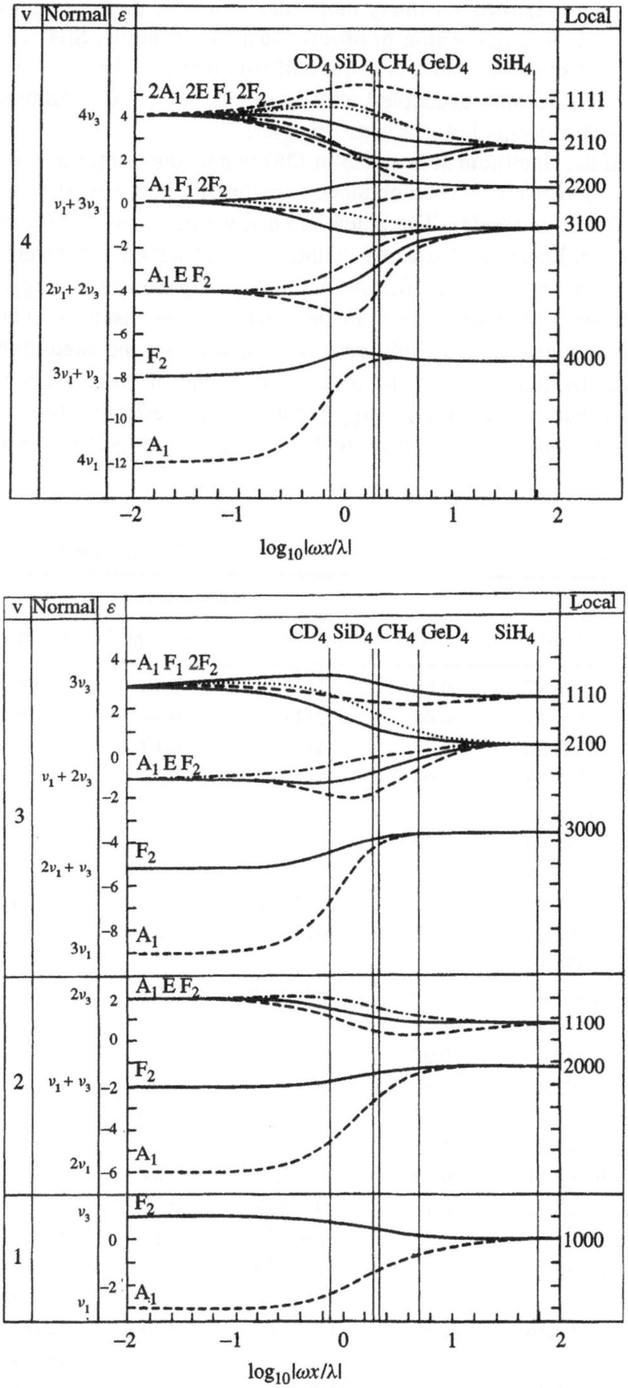

Figure 10.1. Correlation diagram of XH_4-type tetrahedral molecules. $\omega x = -x_{rr}$. (a) $v = 1$–3 and (b) $v = 4$. Reproduced by permission of Taylor & Francis from Halonen and Child (1982).

local mode parameter ξ is known for many molecules. Some of these are placed in appropriate positions in Figure 10.1. It is interesting to observe that, for example, SiH_4 is close to the local mode limit but CH_4, on the other hand, is about half way between the two limits. Therefore, the bond oscillators in SiH_4 become decoupled already at low stretching vibrational excitations. In CH_4, the decoupling also occurs but at higher energies.

The eigenvalues of the Hamiltonian in equation (26) (where the last term inside the last sum has been replaced by $f_{rr'}a_r^{-2}y_iy_j$) have been obtained variationally in this work with the symmetrized basis set for SiH_4, GeH_4, and SnH_4. The potential energy parameters D_e, a_r, and $f_{rr'}$ have been optimized with the non-linear least-squares method. The observed values taken from Halonen (1998) and the fits are given in Table 10.3 and the potential parameters are given in Table 10.4. It is pleasing that with a three-parameter model such a good result is obtained in all three cases. However, if vibrational data of fully deuterated species are included in the fits, the results are less encouraging. Discrepancies of the order of a couple of wavenumbers appear in some cases. This indicates that the potential energy parameters are effective because of interactions between stretches and bends. This aspect of the local mode theory will be discussed later in this article.

Table 10.3. Stretching vibrational term values and fits for $^{28}SiH_4$, $^{74}GeH_4$, and $^{116}SnH_4$[a].

		$^{28}SiH_4$		$^{74}GeH_4$		$^{116}SnH_4$	
$v_{r_1}v_{r_2}v_{r_3}v_{r_4}\Gamma$	v_1v_3	$\nu_{obs}(cm^{-1})$	$(o-c)(cm^{-1})$	$\nu_{obs}(cm^{-1})$	$(o-c)(cm^{-1})$	$\nu_{obs}(cm^{-1})$	$(o-c)(cm^{-1})$
$1000A_1$	10	2186.87	0.88	2110.71	0.18	1908.10	−0.09
$1000F_2$	01	2189.19	−0.05	2111.14	0.54	1905.83	−0.21
$2000A_1$	20	4308.38	−0.87	4153.55	0.10	3752.75	−0.28
$2000F_2$	11	4309.35	−0.07	4153.83	0.38	3752.66	−0.34
$1100A_1$	02	4374.56	0.09				
$1100F_2$	02	4378.40	0.77				
1100E	02	4380.28	1.06				
$3000A_1$	30	6361.98	−0.98	6128.58	−0.03	5539.07	−0.28
$3000F_2$	21	6362.08	−0.89	6128.58	−0.03	5539.04	−0.31
$2100A_1$	12	6496.13	0.21				
$2100F_2$	12	6497.48	0.19	6263.67[b]	−0.92		
2100E	12	6500.30	1.69				
$2100F_2$	03	6500.60	0.07				
$2100F_1$	03	6502.88	1.15				
$4000A_1$	40	8347.87	−1.25				
$4000F_2$	31	8347.87	−1.25				
$5000A_1$	50	10267.11	−0.73	9875.78	−0.04		
$5000F_2$	41	10267.11	−0.73	9875.78	−0.04		
$6000A_1$	60			11647.23	−0.63	10538.18	0.58
$6000F_2$	51	12118.3	−0.84	11647.23	−0.63	10538.16	0.56
$7000A_1$				13352.66	0.46	12083.78	0.34
$7000F_2$	61	13905.18	2.17	13352.66	0.46	12083.77	0.33
$8000A_1$						13568.63	−0.52
$8000F_2$	71	15625.4[b]	5.93	15000[b]	11.17	13568.63	−0.52
$9000F_2$	81	17266.6[b]	−1.98	16574[b]	16.24		

[a]Data taken from Halonen (1998) where references to original references are found.
[b]Not included in the least-squares optimization of the potential energy parameters given in Table 10.4.

Table 10.4. Stretching vibrational potential energy parameters of SiH_4, GeH_4, and SnH_4[a].

Parameter	SiH_4	GeH_4	SnH_4
D_e (aJ)	0.749 22 (63)	0.695 81 (33)	0.638 79 (27)
a_r (Å^{-1})	1.394 75 (69)	1.413 04 (42)	1.334 82 (37)
$f_{rr'}$ (aJ Å^{-2})	0.032 34 (37)	0.012 53 (38)	0.007 89 (35)

[a]The vibrational term value data used in the least squares optimization are given in Table 10.3. All data included were given a unit weight. The numbers in parentheses are one-standard errors in the least significant digit.

10.2.4 INFRARED ABSORPTION INTENSITIES IN THE SIMPLE LOCAL MODE APPROACH

As mentioned in the introduction, the overtone intensity patterns of many molecules such as benzene are unusually simple according to traditional vibration–rotation spectroscopy. How can this be understood in the light of the local mode interpretation of vibrational states? The additional concept needed is the bond dipole model [Mecke (1950), and Schek, Jortner and Sage (1979)] where the molecular dipole function is assumed to be a sum of bond dipole functions, i.e.

$$\mu(\boldsymbol{R}) = \sum_i \mu_i(R_i)\boldsymbol{e}_i, \tag{31}$$

where R_i is the instantaneous bond length of the ith bond and $\boldsymbol{e}_i$ is a unit vector along the ith bond. For simplicity, we use the HCAO model for the vibrational problem and apply it to bent XH_2 systems. In the case of overtone transitions, the lower state is the ground vibrational state which is the product state $|i\rangle = |00\rangle = |0\rangle|0\rangle$. On the other hand, the upper state $|f\rangle$ (the final state) is expressed as a linear combination of local mode product states $|v_{r_1}v_{r_2}\rangle = |v_{r_1}\rangle|v_{r_2}\rangle$ according to the HCAO model. The coefficients of these linear combinations are obtained as eigenvectors of the HCAO model Hamiltonian matrices. The infrared absorption intensities are proportional to the square of the transition moment integral $\mu_{fi} = \langle f|\mu(\boldsymbol{R})|i\rangle$. Because of the orthonomality of the basis functions and the form of dipole moment function, equation (31), the only non-zero matrix elements are of the type $\langle v0|\mu|00\rangle$ or $\langle 0v|\mu|00\rangle$. This means that, in the case of upper states such as $|v_{r_1}v_{r_2}\mathrm{A}_1\rangle$ or $|v_{r_1}v_{r_2}\mathrm{B}_2\rangle$ with $v_{r_1} > 0$ and $v_{r_2} > 0$, the intensity comes from the mixing of these states with $|v0\mathrm{A}_1\rangle$ and $|v0\mathrm{B}_2\rangle$, respectively, where $v = v_{r_1} + v_{r_2}$. Near the local mode limit such as in the case of high hydrogen stretching excitations, this mixing is small. Consequently, transitions from the ground vibrational states to states of the type $|v_{r_1}v_{r_2}\mathrm{A}_1\rangle$ or $|v_{r_1}v_{r_2}\mathrm{B}_2\rangle$ with both $v_{r_1} > 0$ and $v_{r_2} > 0$ are very weak. On the other hand, transitions to the local mode type states $|v0\mathrm{A}_1\rangle$ and $|v0\mathrm{B}_2\rangle$ with $v = v_{r_1} + v_{r_2}$ can be strong when compared with other overtone transitions. This is what is observed, for example, in the benzene overtone spectrum. Transitions from the ground state to the $|v00000\mathrm{E}_{1u}\rangle$ states are the strongest absorption bands observed in the overtone spectrum. Table 10.5 contains intensity calculations for benzene, where an extension of the simple HCAO model has been used [Halonen (1982)].

It is more difficult to give a good theoretical argument for the mathematical form of the bond dipole moment function than for the form of the diatomic potential energy surface. A reasonable starting point would be a function with correct asymptotic limits although this is not the whole answer. One form, which fulfills this requirement, is the often-used Mecke function [Mecke (1950)] for one bond dipole, i.e.

$$\mu(R) = \mu_0 R^m \,\mathrm{e}^{-R/R^*}, \tag{32}$$

where μ_0 and R^* are adjustable parameters, R^* gives the bond length of the maximum dipole, and m is often taken to be an integer ≥ 1. Analytic matrix elements with the Morse basis

Table 10.5. Benzene (C_6H_6) intensity calculation[a].

$v_{r_1}v_{r_2}v_{r_3}v_{r_4}v_{r_5}v_{r_6}\Gamma$	ν_{obs} (cm^{-1})	ν_{calc} (cm^{-1})	I_{obs}	I_{calc}
$100000E_{1u}$	3053	3039.8	[1][b]	[1][b]
$200000E_{1u}$	5957	5955.6	2.1×10^{-2}	3.3×10^{-2}
$110000E_{1u}$		6066.6		1.0×10^{-4}
		6078.9		9.3×10^{-5}
$300000E_{1u}$	8763	8760.1	1.7×10^{-3}	1.7×10^{-3}
$210000E_{1u}$		8983.4		6.1×10^{-7}
		8987.0		1.0×10^{-8}
		8993.1		3.0×10^{-7}
		8996.3		1.7×10^{-7}
		9001.3		1.4×10^{-6}
$111000E_{1u}$		9105.3		1.7×10^{-9}
		9106.0		1.7×10^{-9}
		9116.4		1.6×10^{-9}
$400000E_{1u}$	11442	11448.4	1.3×10^{-4}	1.2×10^{-4}
$500000E_{1u}$	14024	14021.1	1.3×10^{-5}	1.1×10^{-5}

[a]The experimental data are from Halonen (1982) where references to original experimental work can be found. The bond dipole function was approximated by the Mecke function [equation (32)] with $m = 1$ and $R^* = 4$ Å. The vibrational term values and eigenfunctions were obtained with a simple extension of the HCAO model where van Vleck perturbation theory was used to take into account nonresonance kinetic energy interactions between CH and CC stretching vibrations. For more details of the model see Halonen (1982).
[b]Scaled values.

are available for some m values [Schek, Jortner and Sage (1979)]. However, because these are complicated it might be easier to calculate the required transition moment matrix elements with the Gauss–Laguerre numerical integration method. The μ_0 parameter is less interesting because just relative intensities of different overtone bands are often dealt with. The R^* parameter can be obtained from experimental intensity data or from electronic structure calculations. A Taylor series expansion of the dipole moment function is also a possibility. The dipole moment surfaces are improved in both parametrization forms by including cross terms between bond dipole contributions.

In addition to the form of the dipole moment surface, another point of concern is the lower and the upper state eigenfunction. Particularly, if one is interested in the very weak non-local mode bands (for example, in XH_2-type molecules, in bands with upper states such as $|v_{r_1}v_{r_2}\rangle$ with both v_{r_1} and v_{r_2} different from zero being the major contribution in the eigenfunction), it is advisable to go beyond the HCAO vibrational model. It is probably best to treat the vibrational problem variationally (or the appropriate part of the vibrational problem) and then use the lower and upper state eigenfunctions in calculating transition moment integrals.

The choice of molecular axes has also received some attention [Lawton and Child (1980), Le Sueur *et al.* (1992), and Kjaergaard *et al.* (1994)]. It has been shown that particularly in weak overtone transitions it is important to tie the molecular axes to the molecular frame with Eckart conditions [Eckart (1935), and Bunker and Jensen (1998)], which ensures the maximum separation of vibration and rotation. This is important in calculating the vibrational transition moment integrals.

10.2.5 FERMI RESONANCES AND LOCAL MODES IN BENT XH_2 MOLECULES

The inclusion of bending vibrations is an essential step in making the local mode approach useful for many practical applications. A simple mathematical formulation of this in the spirit of the HCAO model is obtained by starting from the exact vibrational Hamiltonian given in equation (1).

The pseudopotential term is neglected and the g matrix elements are expanded as Taylor series around the equilibrium configuration. This produces a Hamiltonian in an infinite series form, which is practical only when it is truncated at a desired point. The simplest Hamiltonian of a polyatomic molecule contains coupled harmonic (or better anharmonic) bending oscillators in a similar fashion as the local mode model contains coupled stretching oscillators. However, this idea, which could be called the zeroth order model in the present context, is often not enough. Interactions between stretches and bends should be considered. In the first approximation, there are two types of couplings, the harmonic couplings and the anharmonic couplings called Fermi resonances.

In XY_2-type bent molecules, the operator parts of harmonic coupling terms between stretches and bend are such as $\hat{p}_{r_i}\hat{p}_\theta$ and $r_i\theta$, where $i = 1$ or 2. The Hamiltonian with the coupling approximations discussed is (see equation (16))

$$\hat{H} = \left(\tfrac{1}{2}g^{(rr)}\hat{p}_{r_1}^2 + D_e y_1^2\right) + \left(\tfrac{1}{2}g^{(rr)}\hat{p}_{r_2}^2 + D_e y_2^2\right) + g_e^{(rr')}\hat{p}_{r_1}\hat{p}_{r_2} + f_{rr'}r_1 r_2 + \tfrac{1}{2}g_e^{(\theta\theta)}\hat{p}_\theta^2 + \tfrac{1}{2}f_{\theta\theta}\theta^2 + g_e^{(r\theta)}(\hat{p}_{r_1} + \hat{p}_{r_2})\hat{p}_\theta + f_{r\theta}(r_1 + r_2)\theta, \quad (33)$$

where $g_e^{(\theta\theta)} = 2[m_X + m_Y(1 - \cos\alpha_e)]/m_X m_Y R_e^2$ and $g_e^{(r\theta)} = -\sin\alpha_e/m_X R_e$ are kinetic energy coefficients evaluated at the equilibrium geometry [Wilson, Decius and Cross (1955)]. The parameter R_e is the equilibrium XY bond length and α_e is the equilibrium valence angle. The harmonic force constants are defined in the usual way, i.e.

$$f_{\theta\theta} = \left(\frac{\partial^2 V}{\partial\theta^2}\right)_e \text{ and } f_{r\theta} = \left(\frac{\partial^2 V}{\partial r_1 \partial\theta}\right)_e = \left(\frac{\partial^2 V}{\partial r_2 \partial\theta}\right)_e.$$

If desired, anharmonic pure bending terms can be added in a straightforward manner.

The eigenvalues of equation (33) can be obtained variationally using, for example, product basis functions with Morse eigenfunctions for stretches and harmonic oscillator eigenfunctions for the bend such as $|v_{r_1}v_{r_2}, v_\theta\rangle = |v_{r_1}\rangle|v_{r_2}\rangle|v_\theta\rangle$. The stretching part can be symmetrized as shown in equation (24). A simple block diagonal model is obtained by employing similar approximations as in the HCAO model. The stretch–bend coupling terms are such that in the harmonic approximation they couple states with $\Delta v = \Delta v_{r_1} + \Delta v_{r_2} = \pm 1$ and $\Delta v_\theta = \pm 1$ or ∓ 1, where v_θ is the bending quantum number. Both v and v_θ remain good quantum numbers. Thus, the coupling term does not couple states with similar energies in many molecules because usually the bending wavenumbers are about half of their stretching counterparts. For example, in water, the symmetric and antisymmetric stretching fundamentals ν_1 and ν_3 are at 3657 and 3756 cm^{-1}, respectively, and the bending fundamental ν_2 is at 1595 cm^{-1}. In the first approximation, these coupling terms do not have any effect on the vibrational energy level structure. However, a better model is obtained by employing second-order perturbation theory. The end result is such that the HCAO model is recovered but the harmonic wavenumber and the coupling parameter are modified. Specifically the model becomes

$$\begin{aligned}
&\langle v_{r_1}v_{r_2}, v_\theta|\hat{H}/hc|v_{r_1}v_{r_2}, v_\theta\rangle \\
&\quad = \omega_r'(v_{r_1} + v_{r_2} + 1) + x_{rr}\left[\left(v_{r_1} + \tfrac{1}{2}\right)^2 + \left(v_{r_2} + \tfrac{1}{2}\right)^2\right] + \omega_\theta'\left(v_\theta + \tfrac{1}{2}\right), \\
&\langle (v_{r_1}+1)(v_{r_2}-1), v_\theta|\hat{H}/hc|v_{r_1}v_{r_2}, v_\theta\rangle = \lambda_r'[(v_{r_1}+1)v_{r_2}]^{1/2}, \\
&\langle (v_{r_1}-1)(v_{r_2}+1), v_\theta|\hat{H}/hc|v_{r_1}v_{r_2}, v_\theta\rangle = \lambda_r'[v_{r_1}(v_{r_2}+1)]^{1/2},
\end{aligned} \quad (34)$$

where

$$\omega_r' = \omega_r - \frac{1}{8}\omega_r\left(g_e^{(rr')}/g^{(rr)} - f_{rr'}/f_{rr}\right)^2 + \frac{1}{4}\left(\frac{d_{r\theta}(1)^2}{\omega_r - \omega_\theta} - \frac{d_{r\theta}(2)^2}{\omega_r + \omega_\theta}\right),$$

$$\omega_\theta' = \omega_\theta - \frac{1}{2}\left(\frac{d_{r\theta}(1)^2}{\omega_r - \omega_\theta} + \frac{d_{r\theta}(2)^2}{\omega_r + \omega_\theta}\right), \tag{35}$$

$$\lambda_r' = \lambda_r + \frac{1}{4}\left(\frac{d_{r\theta}(1)^2}{\omega_r - \omega_\theta} - \frac{d_{r\theta}(2)^2}{\omega_r + \omega_\theta}\right),$$

and

$$\omega_\theta = \frac{1}{2\pi c}\left(f_{\theta\theta}g_e^{(\theta\theta)}\right)^{1/2},$$

$$d_{r\theta}(1) = (\omega_r\omega_\theta)^{1/2}\left[g_e^{(r\theta)}/\left(g^{(rr)}g_e^{(\theta\theta)}\right)^{1/2} + f_{r\theta}/(f_{rr}f_{\theta\theta})^{1/2}\right], \tag{36}$$

$$d_{r\theta}(2) = (\omega_r\omega_\theta)^{1/2}\left[-g_e^{(r\theta)}/\left(g^{(rr)}g_e^{(\theta\theta)}\right)^{1/2} + f_{r\theta}/(f_{rr}f_{\theta\theta})^{1/2}\right].$$

The bilinear harmonic coupling between stretches and bends often gives rise to small effects in XH_2-type hydrides but there is another coupling called Fermi resonance, which is usually more important particularly at higher overtones. Fermi coupling is due to cubic Hamiltonian terms. In XY_2-type molecules, it connects the symmetric stretching vibrational state ν_1 with the bending overtone $2\nu_2$, where normal model labels are used for the interacting states. These states are quite widely separated in H_2X (X = O, S, Se, and Te). For example, in water, ν_1 is at 3657 cm^{-1} and $2\nu_2$ is at 3152 cm^{-1}. Thus, the resonance is weak but the key point is the observation that, because of large stretching anharmonicity, the interacting states at higher stretching energies can get close resulting in strong resonance. It is clear from the above that, if we consider harmonic oscillator matrix elements giving dominant contributions, then the Fermi resonance terms are one-power operators in stretching coordinates and momenta and two-power operators in the bending coordinate and momentum. Using this rule, the Fermi Hamiltonian operator is derived from equation (16) for XY_2-type molecules by expanding the g matrix elements and the potential energy function as a Taylor series. The result is [Halonen and Carrington (1988)]

$$H_{\text{Fermi}} = \frac{1}{2}\left(\frac{\partial g^{(\theta\theta)}}{\partial r}\right)_e (r_1 + r_2)\hat{p}_\theta^2 + \frac{1}{2}\left(\frac{\partial g^{(r\theta)}}{\partial \theta}\right)_e (\hat{p}_{r_1} + \hat{p}_{r_2})(\hat{p}_\theta\theta + \theta\hat{p}_\theta)$$
$$+ \frac{1}{2}f_{r\theta\theta}(r_1 + r_2)\theta^2, \tag{37}$$

where

$$\left(\frac{\partial g^{(\theta\theta)}}{\partial r}\right)_e = \left(\frac{\partial g^{(\theta\theta)}}{\partial r_1}\right)_e = \left(\frac{\partial g^{(\theta\theta)}}{\partial r_2}\right)_e = -\frac{2}{R_e^3}[\mu_Y + \mu_X(1 - \cos\alpha_e)],$$

$$\left(\frac{\partial g^{(r\theta)}}{\partial \theta}\right)_e = \left(\frac{\partial g^{(r_1\theta)}}{\partial \theta}\right)_e = \left(\frac{\partial g^{(r_2\theta)}}{\partial \theta}\right)_e = -\frac{\mu_X\cos\alpha_e}{R_e}, \tag{38}$$

$$f_{r\theta\theta} = \left(\frac{\partial^3 V}{\partial r_1 \partial\theta^2}\right)_e = \left(\frac{\partial^3 V}{\partial r_2 \partial\theta^2}\right)_e.$$

The operator in equation (37) is such that in resonance situations both $v_{r_1} + v_{r_2}$ and v_θ lose their meaning as quantum numbers but $v = 2(v_{r_1} + v_{r_2}) + v_\theta$ still remains a good quantum label. This is seen from the following Fermi resonance matrix elements [Kauppi and Halonen (1992)]

$$\begin{aligned}
\langle (v_{r_1}-1)v_{r_2}, (v_\theta+2)|\hat{H}_{\text{Fermi}}/hc|v_{r_1}v_{r_2}, v_\theta\rangle &= \tfrac{1}{4}\phi_{r\theta\theta}\left[\tfrac{1}{2}v_{r_1}(v_\theta+1)(v_\theta+2)\right]^{1/2},\\
\langle (v_{r_1}+1)v_{r_2}, (v_\theta-2)|\hat{H}_{\text{Fermi}}/hc|v_{r_1}v_{r_2}, v_\theta\rangle &= \tfrac{1}{4}\phi_{r\theta\theta}\left[\tfrac{1}{2}(v_{r_1}+1)v_\theta(v_\theta-1)\right]^{1/2},\\
\langle v_{r_1}(v_{r_2}-1), (v_\theta+2)|\hat{H}_{\text{Fermi}}/hc|v_{r_1}v_{r_2}, v_\theta\rangle &= \tfrac{1}{4}\phi_{r\theta\theta}\left[\tfrac{1}{2}v_{r_2}(v_\theta+1)(v_\theta+2)\right]^{1/2},\\
\langle v_{r_1}(v_{r_2}+1), (v_\theta-2)|\hat{H}_{\text{Fermi}}/hc|v_{r_1}v_{r_2}, v_\theta\rangle &= \tfrac{1}{4}\phi_{r\theta\theta}\left[\tfrac{1}{2}(v_{r_2}+1)v_\theta(v_\theta-1)\right]^{1/2},
\end{aligned} \tag{39}$$

where

$$\phi_{r\theta\theta} = -\frac{\hbar^2}{hc}\left[\left(\frac{\partial g^{(\theta\theta)}}{\partial r}\right)_{\rm e}\alpha_r^{-1/2}\alpha_\theta - 2\left(\frac{\partial g^{(r\theta)}}{\partial \theta}\right)_{\rm e}\alpha_r^{1/2}\right] + (hc)^{-1}f_{r\theta\theta}\alpha_r^{-1/2}\alpha_\theta^{-1}, \tag{40}$$

and

$$\begin{aligned}
\alpha_r &= 4\pi^2 c\omega_r/\left(hg^{(rr)}\right) = \left(f_{rr}/g^{(rr)}\right)^{1/2}/\hbar,\\
\alpha_\theta &= 4\pi^2 c\omega_\theta/\left(hg_{\rm e}^{(\theta\theta)}\right) = \left(f_{\theta\theta}/g_{\rm e}^{(\theta\theta)}\right)^{1/2}/\hbar.
\end{aligned} \tag{41}$$

The vibrational model defined by the matrix elements in equations (34) and (39) is often too simple for practical calculations. More matrix elements of higher-order terms are needed. These arise by applying second-order perturbation theory to cubic operators and first-order perturbation theory to quartic operators. These operators are not given explicitly in this context but for many applications the diagonal matrix element can be improved when compared with equations (34). It takes the form

$$\begin{aligned}
G(v_{r_1}v_{r_2}, v_\theta) = \langle v_{r_1}v_{r_2}, v_\theta|\hat{H}/hc|v_{r_1}v_{r_2}, v_\theta\rangle &= \omega_r'(v_{r_1}+v_{r_2}) + x_{rr}[v_{r_1}(v_{r_1}+1) + v_{r_2}(v_{r_2}+1)]\\
&\quad + \omega_\theta' v_\theta + x_{\theta\theta}v_\theta(v_\theta+1) + x_{r\theta}\left[v_\theta(v_{r_1}+v_{r_2}+1) + \tfrac{1}{2}(v_{r_1}+v_{r_2})\right],
\end{aligned} \tag{42}$$

where the energy has been measured from the ground vibrational state. In the case of $x_{\theta\theta}$ and $x_{r\theta}$, it is possible to derive theoretical expressions similar to equation (40) [Kauppi and Halonen (1992)]. As an example of the Hamiltonian matrices, results are given for both $v = 2$ and 4. When the zero-point energy is subtracted from the diagonal elements, the result for $v = 2$ is

$$\begin{aligned}
H(v=2;\mathrm{A}_1) &= \begin{pmatrix} G(10,0)+\lambda_r' & \frac{\sqrt{2}}{4}\phi_{r\theta\theta} \\ \frac{\sqrt{2}}{4}\phi_{r\theta\theta} & G(00,2) \end{pmatrix}\\
&= \begin{pmatrix} \omega_r' + 2x_{rr} + \frac{1}{2}x_{r\theta} + \lambda_r' & \frac{\sqrt{2}}{4}\phi_{r\theta\theta} \\ \frac{\sqrt{2}}{4}\phi_{r\theta\theta} & 2\omega_\theta + 6x_{\theta\theta} + 2x_{r\theta} \end{pmatrix},\\
H(v=2;\ \mathrm{B}_2) &= (G(10,0)-\lambda_r') = \left(\omega_r' + 2x_{rr} + \frac{1}{2}x_{r\theta} - \lambda_r'\right),
\end{aligned} \tag{43}$$

and for $v = 4$ it is

$$H(v=4;A_1) = \begin{pmatrix} G(20,0) & 2\lambda'_r & \frac{\sqrt{2}}{4}\phi_{r\theta\theta} & 0 \\ 2\lambda'_r & G(11,0) & \frac{\sqrt{2}}{4}\phi_{r\theta\theta} & 0 \\ \frac{\sqrt{2}}{4}\phi_{r\theta\theta} & \frac{\sqrt{2}}{4}\phi_{r\theta\theta} & G(10,2)+\lambda'_r & \frac{\sqrt{3}}{2}\phi_{r\theta\theta} \\ 0 & 0 & \frac{\sqrt{3}}{2}\phi_{r\theta\theta} & G(00,4) \end{pmatrix}$$

$$= \begin{pmatrix} 2\omega'_r + 6x_{rr} + x_{r\theta} & 2\lambda'_r & \frac{\sqrt{2}}{4}\phi_{r\theta\theta} & 0 \\ 2\lambda'_r & 2\omega'_r + 4x_{rr} + x_{r\theta} & \frac{\sqrt{2}}{4}\phi_{r\theta\theta} & 0 \\ \frac{\sqrt{2}}{4}\phi_{r\theta\theta} & \frac{\sqrt{2}}{4}\phi_{r\theta\theta} & \omega'_r + 2x_{rr} + \lambda'_r + 2\omega_\theta + 6x_{\theta\theta} + \frac{9}{2}x_{r\theta} & \frac{\sqrt{3}}{2}\phi_{r\theta\theta} \\ 0 & 0 & \frac{\sqrt{3}}{2}\phi_{r\theta\theta} & 4\omega_\theta + 20x_{\theta\theta} + 4x_{r\theta} \end{pmatrix},$$

$$H(v=4;B_2) = \begin{pmatrix} G(20,0) & \frac{\sqrt{2}}{4}\phi_{r\theta\theta} \\ \frac{\sqrt{2}}{4}\phi_{r\theta\theta} & G(10,2)-\lambda'_r \end{pmatrix}$$

$$= \begin{pmatrix} 2\omega'_r + 6x_{rr} + x_{r\theta} & \frac{\sqrt{2}}{4}\phi_{r\theta\theta} \\ \frac{\sqrt{2}}{4}\phi_{r\theta\theta} & \omega'_r + 2x_{rr} - \lambda'_r + 2\omega_\theta + 6x_{\theta\theta} + \frac{9}{2}x_{r\theta} \end{pmatrix}. \qquad (44)$$

The model is applied to the vibrational data of CH_2Cl_2 taken from Halonen (1988). Experimental CH stretching and CH_2 bending energy levels have been used as data in a least-squares calculation where the model parameters have been optimized. Table 10.6 gives the model parameters with uncertainties and Table 10.7 shows the quality of the fit. The standard deviation of the fit obtained, 1.93 cm^{-1}, is reasonable taking into account the fact that the CH stretching and bending states have been assumed to be decoupled from the other vibrational modes. The model has been applied to H_2S and SO_2 [Kauppi and Halonen (1992), and Halonen (1998)] and to H_2O [Kjaergaard *et al.* (1994)]. Variational calculations with models that are extensions of equations (33) and (37) have been performed for H_2X ($X = O$, S, and Se) including many symmetrical isotopic species [Halonen and Carrington (1988)]. The model has been extended by including rotational motion using perturbation theory [Kauppi and Halonen (1990)]. Potential energy surfaces have been obtained which are in good agreement with *ab initio* calculations and with other more sophisticated models.

Triatomic molecules are such that effectively exact vibrational calculations can be performed for them even for high vibrational excitations. Rotational motion can also be included [Carter and Handy (1987), Tennyson and Sutcliffe (1982), Kauppi (1994), and Wei and Carrington (1992)]. A different type of approach is Jensen's MORBID model where interactions between vibrations and

Table 10.6. CH_2Cl_2 parameters (cm^{-1})[a].

Parameter	Values
ω_r'	3156.91 (160)
x_{rr}	−61.54 (52)
λ_r'	−28.92 (68)
ω_θ'	1465.90 (157)
$x_{\theta\theta}$	−5.54 (63)
$x_{r\theta}$	−20.20 (76)
$\theta_{r\theta\theta}$	70.13 (196)

[a]Numbers in parentheses are one-standard errors in the least significant digit.

Table 10.7. Observed and calculated vibrational term values (cm^{-1}) of CH_2Cl_2[a].

$v_1v_2v_6\Gamma$	Obs	$o-c$
$010A_1$	1435.0	0.38
$020A_1$	2853.6	−0.19
$100A_1$	2997.7	−1.48
$001B_2$	3055.0	2.35
$030A_1$	4260.0	1.61
$011B_2$	4466.9	−0.16
$120A_1$	5828.0	5.00
$021B_2$	5860.4	−0.33
$200A_1$	5912.0	−1.41
$101B_2$	5935.4	1.35
$002A_1$	6072.3	1.13
$031B_2$	7237.0	−2.37
$210A_1$	7322.0	−4.10
$111B_2$	7341.7	−0.17
$012A_1$	7467.0	−0.12
$041B_2$	8603.0	1.16
$300A_1$	8678.0	0.64
$201B_2$	8683.7	−0.37
$220A_1$	8748.0	3.73
$121B_2$	8752.4	−0.65
$022A_1$	8841.0	0.73
$102A_1$	8926.0	8.39
$003B_2$	9008.0	−6.51

[a]The weights used in the least squares optimization of the model parameters are given in Halonen (1988). v_1, v_2, and v_6 denote symmetric CH stretching, CH_2 scissoring, and antisymmetric CH stretching normal mode quantum labels.

rotations are minimized with Eckart conditions [Jensen (1988)]. The bending vibration is treated as a large amplitude motion and the stretching part is expanded as a power series in the Morse variable y_i. The eigenvalues are obtained variationally. Numerically integrated basis functions are used for the bend and Morse oscillator eigenfunctions for the stretches. This model works well even for highly excited angular momentum states.

10.2.6 FERMI RESONANCES AND LOCAL MODES IN PYRAMIDAL XH_3 MOLECULES

The extension of the Fermi resonance–local mode model to pyramidal XY_3 molecules such as NH_3, PH_3, AsH_3, and SbH_3 is possible [Lukka, Kauppi and Halonen (1995) and Lummila *et al.* (1996)]. The Hamiltonian consists of a local mode stretching vibrational part, a bending part which is similar to the stretching part, harmonic bilinear coupling terms between stretches and bends, Fermi resonance terms, and possible non-resonance cubic coupling and higher-order coupling terms, i.e.

$$\begin{aligned}
\hat{H} = {} & \sum_{i=1}^{3}\left(\frac{1}{2}g^{(rr)}\hat{p}_{r_i}^2 + D_e y_i^2\right) + \sum_{i<j}(g_e^{(rr')}\hat{p}_{r_i}\hat{p}_{r_j} + f_{rr'}r_i r_j) + \sum_{i=1}^{3}\left(\frac{1}{2}g_e^{(\theta\theta)}\hat{p}_{\theta_i}^2 + \frac{1}{2}f_{\theta\theta}\theta_i^2\right) \\
& + \sum_{i<j}(g_e^{(\theta\theta')}\hat{p}_{\theta_i}\hat{p}_{\theta_j} + f_{\theta\theta'}\theta_i\theta_j) + \hat{H}(\text{bend})_{\text{anh}} + \sum_{i=1}^{3}(g_e^{(r\theta)}\hat{p}_{r_i}\hat{p}_{\theta_i} + f_{r\theta}r_i\theta_i) \\
& + \sum_{i\neq j}(g_e^{(r\theta')}\hat{p}_{r_i}\hat{p}_{\theta_j} + f_{r\theta'}r_i\theta_j) + \sum_{i\neq j}\left[\left(\frac{\partial g^{(\theta\theta')}}{\partial r}\right)_e r_i\hat{p}_{\theta_i}\hat{p}_{\theta_j} + \left(\frac{\partial g^{(r\theta)}}{\partial \theta'}\right)_e \theta_i\hat{p}_{r_j}\hat{p}_{\theta_j}\right. \\
& \left. + \frac{1}{2}\left(\frac{\partial g^{(r\theta')}}{\partial \theta'}\right)_e \hat{p}_{r_i}(\theta_j\hat{p}_{\theta_j} + \hat{p}_{\theta_j}\theta_j) + f_{r\theta\theta'}r_i\theta_i\theta_j + \frac{1}{2}\left(\frac{\partial g^{(\theta\theta)}}{\partial r'}\right)_e r_i\hat{p}_{\theta_j}^2 + \frac{1}{2}f_{r\theta'\theta'}r_i\theta_j^2\right] \\
& + \sum_{i\neq j\neq k\neq i,\, j<k}\left[\left(\frac{\partial g^{(\theta\theta')}}{\partial r''}\right)_e r_i\hat{p}_{\theta_j}\hat{p}_{\theta_k} + f_{r\theta'\theta''}r_i\theta_j\theta_k\right] \\
& + \sum_{i}\left[\frac{1}{2}\left(\frac{\partial g^{(r\theta)}}{\partial \theta}\right)_e \hat{p}_{r_i}(\theta_i\hat{p}_{\theta_i} + \hat{p}_{\theta_i}\theta_i) + \frac{1}{2}f_{r\theta\theta}r_i\theta_i^2\right] + \hat{H}(\text{stretch–bend})_{\text{anh}},
\end{aligned} \tag{45}$$

where the operator $\hat{H}(\text{bend})_{\text{anh}}$ contains anharmonic bending terms and the operator $\hat{H}(\text{stretch–bend})_{\text{anh}}$ anharmonic stretch–bend coupling terms excluding the Fermi resonance operator. The notation is a simple extension of that employed for bent triatomics. The $\boldsymbol{g}$ matrix elements and their derivatives with respect to internal coordinates are given as [Lukka, Kauppi and Halonen (1995)]

$$g^{(rr)} = g_e^{(rr)} = \mu_X + \mu_Y,$$

$$g_e^{(rr')} = \mu_X \cos\alpha_e,$$

$$g_e^{(\theta\theta)} = 2(\mu_X + \mu_Y - \mu_X\cos\alpha_e)/R_e^2,$$

$$g_{\rm e}^{(\theta\theta')} = [-\mu_{\rm X} + 4\mu_{\rm X}\cos\alpha_{\rm e} + 2\mu_H\cos\alpha_{\rm e} - 3\mu_{\rm X}\cos(2\alpha_{\rm e})]\sec^2\left(\tfrac{1}{2}\alpha_{\rm e}\right)/4R_{\rm e}^2,$$

$$g_{\rm e}^{(r\theta)} = -2\mu_{\rm X}\cos\alpha_{\rm e}\tan\left(\tfrac{1}{2}\alpha_{\rm e}\right)/R_{\rm e},$$

$$g_{\rm e}^{(r\theta')} = -\mu_{\rm X}\sin\alpha_{\rm e}/R_{\rm e},$$

$$\left(\frac{\partial g_{\rm e}^{(\theta\theta)}}{\partial r'}\right)_{\rm e} = 2(-\mu_{\rm X} - \mu_{\rm Y} + \mu_{\rm X}\cos\alpha_{\rm e})/R_{\rm e}^3, \tag{46}$$

$$\left(\frac{\partial g_{\rm e}^{(\theta\theta')}}{\partial r}\right)_{\rm e} = -\mu_{\rm X}(1 + 2\cos\alpha_{\rm e})\tan^2\left(\frac{1}{2}\alpha_{\rm e}\right)/R_{\rm e}^3,$$

$$\left(\frac{\partial g_{\rm e}^{(\theta\theta')}}{\partial r''}\right)_{\rm e} = (-\mu_{\rm X} - \mu_{\rm Y} + \mu_{\rm X}\cos\alpha)\cos\alpha_{\rm e}\sec^2\left(\frac{1}{2}\alpha_{\rm e}\right)/R_{\rm e}^3,$$

$$\left(\frac{\partial g_{\rm e}^{(r\theta)}}{\partial \theta}\right)_{\rm e} = -\mu_{\rm X}\cos\alpha_{\rm e}\sec^2\left(\frac{1}{2}\alpha_{\rm e}\right)/R_{\rm e},$$

$$\left(\frac{\partial g_{\rm e}^{(r\theta)}}{\partial \theta'}\right)_{\rm e} = 2\mu_{\rm X}\sin^2\left(\frac{1}{2}\alpha_{\rm e}\right)/R_{\rm e},$$

$$\left(\frac{\partial g_{\rm e}^{(r\theta')}}{\partial \theta'}\right)_{\rm e} = -\mu_{\rm X}\cos\alpha_{\rm e}/R_{\rm e}.$$

The quantities $\mu_{\rm X} = 1/m_{\rm X}$ and $\mu_{\rm Y} = 1/m_{\rm Y}$, where $m_{\rm X}$ and $m_{\rm Y}$ are the masses of the X and Y nuclei, respectively. The parameters $R_{\rm e}$ and $\alpha_{\rm e}$ are the equilibrium bond length and valence angle, respectively.

Using similar approximations as in the corresponding XH_2 case, a block diagonal model is obtained with the product basis

$$\begin{aligned} |v_{r_1}v_{r_2}v_{r_3}, v_{\theta_1}v_{\theta_2}v_{\theta_3}\rangle &= |v_{r_1}v_{r_2}v_{r_3}\rangle|v_{\theta_1}v_{\theta_2}v_{\theta_3}\rangle = |v_{r_1}v_{r_2}v_{r_3}\rangle_{\rm s}|v_{\theta_1}v_{\theta_2}v_{\theta_3}\rangle_{\rm b} \\ &= |v_{r_1}\rangle|v_{r_2}\rangle|v_{r_3}\rangle|v_{\theta_1}\rangle|v_{\theta_2}\rangle|v_{\theta_3}\rangle, \end{aligned} \tag{47}$$

which is an obvious extension of the notation used earlier. The diagonal matrix elements are

$$\begin{aligned} G(v_{r_1}v_{r_2}v_{r_3}, v_{\theta_1}v_{\theta_2}v_{\theta_3}) &= \langle v_{r_1}v_{r_2}v_{r_3}, v_{\theta_1}v_{\theta_2}v_{\theta_3}|\hat{H}/hc|v_{r_1}v_{r_2}v_{r_3}, v_{\theta_1}v_{\theta_2}v_{\theta_3}\rangle \\ &= \omega_r'(v_{r_1} + v_{r_2} + v_{r_3}) + x_{rr}(v_{r_1}^2 + v_{r_1} + v_{r_2}^2 + v_{r_2} + v_{r_3}^2 + v_{r_3}) \\ &\quad + \omega_\theta'(v_{\theta_1} + v_{\theta_2} + v_{\theta_3}) + x_{\theta\theta}(v_{\theta_1}^2 + v_{\theta_1} + v_{\theta_2}^2 + v_{\theta_2} + v_{\theta_3}^2 + v_{\theta_3}) \\ &\quad + x_{r\theta'}(v_{r_2}v_{\theta_1} + v_{r_3}v_{\theta_1} + v_{r_1}v_{\theta_2} + v_{r_3}v_{\theta_2} + v_{r_1}v_{\theta_3} + v_{r_2}v_{\theta_3} \\ &\quad + v_{r_1} + v_{r_2} + v_{r_3} + v_{\theta_1} + v_{\theta_2} + v_{\theta_3}) \\ &\quad + x_{r\theta}\left[v_{r_1}v_{\theta_1} + v_{r_2}v_{\theta_2} + v_{r_3}v_{\theta_3} + \tfrac{1}{2}(v_{r_1} + v_{r_2} + v_{r_3} + v_{\theta_1} + v_{\theta_2} + v_{\theta_3})\right] \end{aligned} \tag{48}$$

Theoretical expressions for the coefficients are obtained using both first and second order perturbation theory. This leads to the following expressions

$$\begin{aligned}
\omega_r' &= \omega_r - \frac{1}{4}\omega_r\left(g_{\rm e}^{(rr')}/g^{(rr)} - f_{rr'}/f_{rr}\right)^2 \\
&\quad + \frac{1}{4}\left(\frac{d_{r\theta}(1)^2 + 2d_{r\theta'}(1)^2}{\omega_r - \omega_\theta} - \frac{d_{r\theta}(2)^2 + 2d_{r\theta'}(2)^2}{\omega_r + \omega_\theta}\right) \\
\omega_\theta' &= \omega_\theta - \frac{1}{4}\omega_\theta\left(g_{\rm e}^{(\theta\theta')}/g_{\rm e}^{(\theta\theta)} - f_{\theta\theta'}/f_{\theta\theta}\right)^2 \\
&\quad - \frac{1}{4}\left(\frac{d_{r\theta}(1)^2 + 2d_{r\theta'}(1)^2}{\omega_r - \omega_\theta} + \frac{d_{r\theta}(2)^2 + 2d_{r\theta'}(2)^2}{\omega_r + \omega_\theta}\right)
\end{aligned} \tag{49}$$

where

$$\begin{aligned}
d_{r\theta'}(1) &= (\omega_r\omega_\theta)^{1/2}\left[g_{\rm e}^{(r\theta')}/\left(g^{(rr)}g_{\rm e}^{(\theta\theta)}\right)^{1/2} + f_{r\theta'}/(f_{rr}f_{\theta\theta})^{1/2}\right] \\
d_{r\theta'}(2) &= (\omega_r\omega_\theta)^{1/2}\left[-g_{\rm e}^{(r\theta')}/\left(g^{(rr)}g_{\rm e}^{(\theta\theta)}\right)^{1/2} + f_{r\theta'}/(f_{rr}f_{\theta\theta})^{1/2}\right]
\end{aligned} \tag{50}$$

The quantities $d_{r\theta}(1)$ and $d_{r\theta}(2)$ are defined in equation (36). The anharmonicity parameter x_{rr} is defined as before. The other anharmonicity parameters are functions of harmonic, cubic, and quartic force constants.

The local mode off-diagonal matrix elements are of the type

$$\langle (v_{r_1}-1)(v_{r_2}+1)v_{r_3},\, v_{\theta_1}v_{\theta_2}v_{\theta_3}|\hat{H}/hc|v_{r_1}v_{r_2}v_{r_3},\, v_{\theta_1}v_{\theta_2}v_{\theta_3}\rangle = \lambda_r'[v_{r_1}(v_{r_2}+1)]^{1/2} \tag{51}$$

and

$$\langle v_{r_1}v_{r_2}v_{r_3},\, (v_{\theta_1}-1)(v_{\theta_2}+1)v_{\theta_3}|\hat{H}/hc|v_{r_1}v_{r_2}v_{r_3},\, v_{\theta_1}v_{\theta_2}v_{\theta_3}\rangle = \lambda_\theta'[v_{\theta_1}(v_{\theta_2}+1)]^{1/2}, \tag{52}$$

with

$$\begin{aligned}
\lambda_r' &= \frac{1}{2}\omega_r\left(g_{\rm e}^{(rr')}/g^{(rr)} + f_{rr'}/f_{rr}\right) \\
&\quad + \frac{1}{4}\left(\frac{2d_{r\theta}(1)d_{r\theta'}(1) + d_{r\theta'}(1)^2}{\omega_r - \omega_\theta} - \frac{2d_{r\theta}(2)d_{r\theta'}(2) + d_{r\theta'}(2)^2}{\omega_r + \omega_\theta}\right), \\
\lambda_\theta' &= \frac{1}{2}\omega_r\left(g_{\rm e}^{(\theta\theta')}/g_{\rm e}^{(\theta\theta)} + f_{\theta\theta'}/f_{rr}\right) \\
&\quad + \frac{1}{4}\left(\frac{2d_{r\theta}(1)d_{r\theta'}(1) + d_{r\theta'}(1)^2}{\omega_r - \omega_\theta} - \frac{2d_{r\theta}(2)d_{r\theta'}(2) + d_{r\theta'}(2)^2}{\omega_r + \omega_\theta}\right).
\end{aligned} \tag{53}$$

The Fermi resonance operator gives rise to matrix elements that are of the type

$$\begin{aligned}
&\langle (v_{r_1}-1)v_{r_2}v_{r_3},\, (v_{\theta_1}+2)v_{\theta_2}v_{\theta_3}|\hat{H}_{\rm Fermi}/hc|v_{r_1}v_{r_2}v_{r_3},\, v_{\theta_1}v_{\theta_2}v_{\theta_3}\rangle \\
&\quad = d_{r\theta\theta}[v_{r_1}(v_{\theta_1}+1)(v_{\theta_1}+2)]^{1/2}, \\
&\langle (v_{r_1}-1)v_{r_2}v_{r_3},\, v_{\theta_1}(v_{\theta_2}+2)v_{\theta_3}|\hat{H}_{\rm Fermi}/hc|v_{r_1}v_{r_2}v_{r_3},\, v_{\theta_1}v_{\theta_2}v_{\theta_3}\rangle \\
&\quad = d_{r\theta'\theta'}[v_{r_1}(v_{\theta_2}+1)(v_{\theta_2}+2)]^{1/2},
\end{aligned}$$

$$\langle (v_{r_1}-1)v_{r_2}v_{r_3}, (v_{\theta_1}+1)(v_{\theta_2}+1)v_{\theta_3}|\hat{H}_{\text{Fermi}}/hc|v_{r_1}v_{r_2}v_{r_3}, v_{\theta_1}v_{\theta_2}v_{\theta_3}\rangle \tag{54}$$
$$= d_{r\theta\theta'}[v_{r_1}(v_{\theta_1}+1)(v_{\theta_2}+1)]^{1/2},$$
$$\langle (v_{r_1}-1)v_{r_2}v_{r_3}, v_{\theta_1}(v_{\theta_2}+1)(v_{\theta_3}+1)|\hat{H}_{\text{Fermi}}/hc|v_{r_1}v_{r_2}v_{r_3}, v_{\theta_1}v_{\theta_2}v_{\theta_3}\rangle$$
$$= d_{r\theta'\theta''}[v_{r_1}(v_{\theta_2}+1)(v_{\theta_3}+1)]^{1/2},$$

where

$$\begin{aligned}
d_{r\theta\theta} &= \frac{\hbar^2}{2\sqrt{2}hc}\left(\frac{\partial g^{(r\theta)}}{\partial\theta}\right)_{\text{e}}\alpha_r^{1/2} + \frac{1}{4\sqrt{2}hc}f_{r\theta\theta}\alpha_r^{-1/2}\alpha_\theta^{-1},\\
d_{r\theta'\theta'} &= -\frac{\hbar^2}{4\sqrt{2}hc}\left(\frac{\partial g^{(\theta\theta)}}{\partial r'}\right)_{\text{e}}\alpha_r^{-1/2}\alpha_\theta + \frac{\hbar^2}{2\sqrt{2}hc}\left(\frac{\partial g^{(r\theta')}}{\partial\theta'}\right)_{\text{e}}\alpha_r^{1/2} + \frac{1}{4\sqrt{2}hc}f_{r\theta'\theta'}\alpha_r^{-1/2}\alpha_\theta^{-1},\\
d_{r\theta\theta'} &= -\frac{\hbar^2}{2\sqrt{2}hc}\left(\frac{\partial g^{(\theta\theta')}}{\partial r}\right)_{\text{e}}\alpha_r^{-1/2}\alpha_\theta + \frac{\hbar^2}{2\sqrt{2}hc}\left(\frac{\partial g^{(r\theta)}}{\partial\theta'}\right)_{\text{e}}\alpha_r^{1/2} + \frac{1}{2\sqrt{2}hc}f_{r\theta\theta'}\alpha_r^{-1/2}\alpha_\theta^{-1},\\
d_{r\theta'\theta''} &= -\frac{\hbar^2}{2\sqrt{2}hc}\left(\frac{\partial g^{(\theta\theta')}}{\partial r''}\right)_{\text{e}}\alpha_r^{-1/2}\alpha_\theta + \frac{1}{2\sqrt{2}hc}f_{r\theta'\theta''}\alpha_r^{-1/2}\alpha_\theta^{-1},
\end{aligned} \tag{55}$$

where α_r and α_θ are defined in equations (41).

The model is such that the label $v = 2(v_{r_1}+v_{r_2}+v_{r_3}) + v_{\theta_1}+v_{\theta_2}+v_{\theta_3}$ remains a good quantum number. The Hamiltonian matrices are in block diagonal form where only states with the same quantum number v are coupled. In spite of this, the matrix dimensions quickly become large as the energy increases. Employing symmetrized basis functions can further reduce the dimensions. This is achieved by symmetrizing the bending basis functions as in equation (27) for stretches. Final symmetrized basis functions are obtained by standard group theoretical methods from the symmetrized stretching and bending basis functions. A complication occurs when two E functions are combined to produce A_1, A_2, and E functions. Symmetrized product functions are obtained by using projection operators or vector coupling coefficients [Halonen and Child (1983)]. When $v = 2$, the A_1 basis functions are (s and b refer to stretches and bends, respectively)

$$\begin{aligned}
|100,000;\text{A}_1\rangle &= |100\text{A}_1\rangle_\text{s}|000A_1\rangle_\text{b} = \frac{1}{\sqrt{3}}(|100\rangle_\text{s}+|010\rangle_\text{s}+|001\rangle_\text{s})|000\rangle_\text{b},\\
|000,200;\text{A}_1\rangle &= |000\text{A}_1\rangle_\text{s}|200A_1\rangle_\text{b} = \frac{1}{\sqrt{3}}|000\rangle_\text{s}(|200\rangle_\text{b}+|020\rangle_\text{b}+|002\rangle_\text{b}),\\
|000,110;A_1\rangle &= |000A_1\rangle_\text{s}|110\text{A}_1\rangle_\text{b} = \frac{1}{\sqrt{3}}|000\rangle_\text{s}(|110\rangle_\text{b}+|101\rangle_\text{b}+|011\rangle_\text{b}),
\end{aligned} \tag{56}$$

and the Hamiltonian matrix in wavenumber units is given as

$$H(v=1;\text{A}_1)$$
$$= \begin{pmatrix} G(100,000)+2\lambda'_r & \sqrt{2}(d_{r\theta\theta}+2d_{r\theta'\theta'}) & d_{r\theta'\theta''}+2d_{r\theta\theta'}\\ \sqrt{2}(d_{r\theta\theta}+2d_{r\theta'\theta'}) & G(000,200) & 2\sqrt{2}\lambda'_\theta\\ d_{r\theta'\theta''}+2d_{r\theta\theta'} & 2\sqrt{2}\lambda'_\theta & G(000,110)+2\lambda'_\theta \end{pmatrix}$$
$$= \begin{pmatrix} \omega'_r+2x_{rr}+x_{r\theta}+\frac{1}{2}x_{r\theta'}+2\lambda'_r & \sqrt{2}(d_{r\theta\theta}+2d_{r\theta'\theta'}) & d_{r\theta'\theta''}+2d_{r\theta\theta'}\\ \sqrt{2}(d_{r\theta\theta}+2d_{r\theta'\theta'}) & 2\omega'_\theta+6x_{\theta\theta}+2x_{r\theta}+x_{r\theta'} & 2\sqrt{2}\lambda'_\theta\\ d_{r\theta'\theta''}+2d_{r\theta\theta'} & 2\sqrt{2}\lambda'_\theta & 2\omega'_\theta+4x_{\theta\theta}+2x_{r\theta}+x_{r\theta'}+2\lambda'_\theta \end{pmatrix}. \tag{57}$$

The E_a functions are

$$
\begin{aligned}
|100,000;\mathrm{E_a}\rangle &= |100\mathrm{E_a}\rangle_s|000\mathrm{A_1}\rangle_b = \frac{1}{\sqrt{6}}(2|100\rangle_s - |010\rangle_s - |001\rangle_s)|000\rangle_b, \\
|000,200;\mathrm{E_a}\rangle &= |000\mathrm{A_1}\rangle_s|200\mathrm{E_a}\rangle_b = \frac{1}{\sqrt{6}}|000\rangle_s(2|200\rangle_b - |020\rangle_b - |002\rangle_b), \\
|000,110;\mathrm{E_a}\rangle &= |000\mathrm{A_1}\rangle_s|110\mathrm{E_a}\rangle_b = \frac{1}{\sqrt{6}}|000\rangle_s(2|011\rangle_b - |101\rangle_b - |110\rangle_b),
\end{aligned}
\tag{58}
$$

and the corresponding E_b functions take the forms

$$
\begin{aligned}
|100,000;\mathrm{E_b}\rangle &= |100\mathrm{E_b}\rangle_s|000\mathrm{A_1}\rangle_b = \frac{1}{\sqrt{2}}(|010\rangle_s - |001\rangle_s)|000\rangle_b, \\
|000,200;\mathrm{E_b}\rangle &= |000\mathrm{A_1}\rangle_s|200\mathrm{E_b}\rangle_b = \frac{1}{\sqrt{2}}|000\rangle_s(|020\rangle_b - |002\rangle_b), \\
|000,110;\mathrm{E_b}\rangle &= |000\mathrm{A_1}\rangle_s|110\mathrm{E_b}\rangle_b = \frac{1}{\sqrt{2}}|000\rangle_s(|101\rangle_b - |110\rangle_b).
\end{aligned}
\tag{59}
$$

These symmetrized basis functions give rise to two identical E Hamiltonian blocks, which are given as

$$
\begin{aligned}
H(v=1;\mathrm{E}) &= \begin{pmatrix} G(100,000)-\lambda'_r & \sqrt{2}(d_{r\theta\theta}-d_{r\theta'\theta'}) & d_{r\theta'\theta''}-d_{r\theta\theta'} \\ \sqrt{2}(d_{r\theta\theta}-d_{r\theta'\theta'}) & G(000,200) & -\sqrt{2}\lambda'_\theta \\ d_{r\theta'\theta''}-d_{r\theta\theta'} & -\sqrt{2}\lambda'_\theta & G(000,110)-\lambda'_\theta \end{pmatrix} \\
&= \begin{pmatrix} \omega'_r+2x_{rr}+x_{r\theta}+\frac{1}{2}x_{r\theta'}-\lambda'_r & \sqrt{2}(d_{r\theta\theta}-d_{r\theta'\theta'}) & d_{r\theta'\theta''}-d_{r\theta\theta'} \\ \sqrt{2}(d_{r\theta\theta}-d_{r\theta'\theta'}) & 2\omega'_\theta+6x_{\theta\theta}+2x_{r\theta}+x_{r\theta'} & -\sqrt{2}\lambda'_\theta \\ d_{r\theta'\theta''}-d_{r\theta\theta'} & -\sqrt{2}\lambda'_\theta & 2\omega'_\theta+4x_{\theta\theta}+2x_{r\theta}+x_{r\theta'}-\lambda'_\theta \end{pmatrix}.
\end{aligned}
\tag{60}
$$

The Fermi resonance model described has been applied to the vibrational data of arsine (AsH_3) [Lukka, Kauppi and Halonen (1995)] and stibine (SbH_3) [Lummila *et al.* (1996)]. In these papers, the model is parametrized in terms of potential energy parameters (force constants) as shown for example in equations (55) for Fermi resonance coefficients. The initial force constants were taken from an *ab initio* study [Breidung and Thiel (1996)] and some of them were optimized with the non-linear least-squares method using experimental vibrational term values as data. The potential energy parameters of AsH_3 and SbH_3 are given in Table 10.8 and the corresponding fits are in Table 10.9. A five-dimensional model has been published for ammonia [Kauppi and Halonen (1995)]. The inversion motion has been excluded but its effect has been taken into account by second-order perturbation theory.

10.2.7 FERMI RESONANCES AND LOCAL MODES IN TETRAHEDRAL XH_4 MOLECULES

Tetrahedral XH_4-type molecules bring a new aspect to the theory presented. All six valence angle bending coordinates are not independent, that is all of them cannot be increased simultaneously.

Table 10.8. AsH_3 and SbH_3 surface parameters[a].

Parameter	AsH_3		SbH_3	
	Fermi	*Ab initio*	Fermi	*Ab initio*
f_{rr} (aJ Å^{-2})	2.876[b]	2.829	2.293[b]	2.243
$f_{rr'}$ (aJ Å^{-2})	−0.0010 (12)	−0.0097	−0.00419 (85)	−0.0037
f_{rrr} (aJ Å^{-3})	−13.088[b]	−14.192	−9.656[b]	−9.887
f_{rrrr} (aJ Å^{-4})	46.32[b]	54.40	31.62[b]	35.20
$f_{\theta\theta}$ (aJ)	0.6470 (30)	0.642	0.5798 (17)	0.5703
$f_{\theta\theta'}$ (aJ)	−0.03344 (47)	−0.027	−0.01617 (29)	−0.0117
$f_{r\theta}$ (aJ Å^{-1})		0.0147		0.0137
$f_{r\theta'}$ (aJ Å^{-1})		0.0617		0.0407
$f_{r\theta'\theta'}$ (aJ Å^{-1})	−0.2341[c]	−0.2341	−0.2001[c]	−0.2001
$f_{r\theta\theta'}$ (aJ Å^{-1})	0.0662[c]	0.0662	0.0504[c]	0.0504
$f_{r\theta'\theta''}$ (aJ Å^{-1})	0.0686[c]	0.0686	0.0562[c]	0.0562
$f_{r\theta\theta}$ (aJ Å^{-1})	0.0403[c]	0.0403	0.0308[c]	0.0308
$f_{rr\theta'\theta'}$ (aJ Å^{-2})	−3.63 (37)	0.291	−2.94 (18)	0.267

[a]The surfaces have been taken from Lukka, Kauppi, and Halonen (1995) (Fermi, AsH_3), Lummila *et al.* (1996) (Fermi, SbH_3), and Breidung and Thiel (1996) (*ab initio*). The *ab initio* surfaces contain more cubic force constants than given here.
[b]Calculated from ω_r and x_{rr}.
[c]Transferred from the appropriate *ab initio* surface.

This so-called redundancy can be expressed mathematically as [Simanouti (1949)]

$$\begin{vmatrix} 1 & \cos\alpha_{12} & \cos\alpha_{13} & \cos\alpha_{14} \\ \cos\alpha_{12} & 1 & \cos\alpha_{23} & \cos\alpha_{24} \\ \cos\alpha_{13} & \cos\alpha_{23} & 1 & \cos\alpha_{34} \\ \cos\alpha_{14} & \cos\alpha_{24} & \cos\alpha_{34} & 1 \end{vmatrix} = 0, \tag{61}$$

where α_{ij} is the valence angle between the ith and jth bond. It is possible to expand the left-hand side and obtain the redundancy relation up to any order in bending displacement coordinates. Apart from the first-order term, these are somewhat complicated expressions. In fact, in going beyond the harmonic approximation, it becomes impractical to express one of the bending displacement coordinates in terms of the others. Instead, symmetrized curvilinear internal coordinates are commonly used. There are four coordinates for the stretches [Gray and Robiette (1979)]

$$\begin{aligned} S_1 &= \tfrac{1}{2}(r_1 + r_2 + r_3 + r_4), \\ S_{3x} &= \tfrac{1}{2}(r_1 - r_2 + r_3 - r_4), \\ S_{3y} &= \tfrac{1}{2}(r_1 - r_2 - r_3 + r_4), \\ S_{3z} &= \tfrac{1}{2}(r_1 + r_2 - r_3 - r_4), \end{aligned} \tag{62}$$

and five for bends

$$\begin{aligned} S_{2a} &= \frac{1}{\sqrt{12}}(2\theta_{12} - \theta_{13} - \theta_{14} - \theta_{23} - \theta_{24} + 2\theta_{34}), \\ S_{2b} &= \frac{1}{2}(\theta_{13} - \theta_{14} - \theta_{23} + \theta_{24}), \\ S_{4x} &= \frac{1}{\sqrt{2}}(\theta_{24} - \theta_{13}), \end{aligned} \tag{63}$$

Table 10.9. AsH_3 and SbH_3 vibrational term value fits.

		AsH_3		$^{121}SbH_3$	
$v_{r1}v_{r2}v_{r3}, v_{\theta 1}v_{\theta 2}v_{\theta 3}; \Gamma$	$v_2 v_4^{l_4}$	ν_{obs} (cm^{-1})	$(o-c)$(cm^{-1})	ν_{obs} (cm^{-1})	$(o-c)$(cm^{-1})
000,100; A_1	10^0	906.75	1.97	782.24	1.95
000,100; E	$01^{\pm 1}$	999.23	0.31	827.85	1.05
000,110; A_1	20^0	1806.15	−1.59	1559.0	−0.67
000,200; E	$11^{\pm 1}$	1904.12	0.53		
000,200; A_1	02^0	1991.00	−6.25	1652.7	−0.77
000,110; E	$02^{\pm 2}$	2003.48	5.92		
100,000; A_1	00^0	2115.16	0.14	1890.50	0.82
100,000; E	00^0	2126.42	0.34	1894.50	−0.31
100,100; A_1	10^0	3013	0.64	2661	−0.61
100,100; E	$01^{\pm 1}$	3102	−0.16	2705	0.50
200,000; A_1	00^0	4166.77	0.76	3719.93	0.42
200,000; E	00^0	4167.94	0.78	3719.86	0.08
110,000; A_1	00^0	4237.70	−1.06		
110,000; E	00^0	4247.53	−1.14		
200,100; A_1	10^0	5057	5.48	4545	4.87
200,100; E	10^0	5057	5.11	4513	−5.54
110,100; A_1	10^0	5128	2.67		
110,100; E	10^0	5128	−4.42		
200,100; A_1	$01^{\pm 1}$	5158	0.42		
200,100; E	$01^{\pm 1}$	5158	1.16		
300,000; A_1	00^0	6136.32	−0.63	5480.29	0.02
300,000; E	00^0	6136.31	−0.69	5480.24	−0.04
210,000; A_1	00^0	6275.83	−1.54	5607	−0.45
210,000; E	00^0	6282.36	−1.15		
210,000; E	00^0	6294.71	−1.29		
111,000; A_1	00^0	6365.96	−2.62		
400,000; A_1	00^0	8028.97	9.64	7173.80	−0.93
400,000; E	00^0	8028.97	9.67	7173.78	−0.95
310,000; A_1	00^0	8249.52	−3.01		
310,000; E	00^0	8258.38	−2.17		
600,000; A_1	00^0			10358	−10.93
600,000; E	00^0			10358	−10.93
510,000; A_1	00^0			10691.5	−2.80
510,000; E	00^0			10691.5	−2.80
700,000; A_1	00^0			11843.5	−17.14
700,000; E	00^0			11843.5	−17.14

[a] $o-c$ denotes observed minus calculated value. The calculated values are from Lukka, Kauppi, and Halonen (1995) and Lummila *et al.* (1996). The experimental data are from these same sources, where references to the original experimental work are found, and from Ulenikov *et al.* (1996) and from Lin *et al.* (1998).

$$S_{4y} = \frac{1}{\sqrt{2}}(\theta_{23} - \theta_{14}),$$

$$S_{4z} = \frac{1}{\sqrt{2}}(\theta_{34} - \theta_{12}),$$

where θ_{ij} is the valence angle displacement coordinate between the bonds i and j. The coordinate S_1 spans the A_1, (S_{2a}, S_{2b}) spans the E, and (S_{3x}, S_{3y}, S_{3z}) and (S_{4x}, S_{4y}, S_{4z}) span the F_2 irreducible representation in the T_d point group.

The vibrational Hamiltonian in terms of the symmetrized coordinates can be derived starting from the general internal coordinate Hamiltonian given in equation (1). For this purpose, transformations are needed from the unsymmetrized internal coordinates and momenta to the symmetrized ones as explained in Hoy, Mills and Strey (1972). Transformation from the six unsymmetrized valence angle displacement coordinates to the five symmetrized coordinates defined in equation (63) is non-linear. Retaining terms up to the second order, the result for coordinates is [Halonen (1997)]

$$
\begin{aligned}
r_1 &= \frac{1}{2}(S_1 + S_{3x} + S_{3y} + S_{3z}),\\
r_2 &= \frac{1}{2}(S_1 - S_{3x} - S_{3y} + S_{3z}),\\
r_3 &= \frac{1}{2}(S_1 + S_{3x} - S_{3y} - S_{3z}),\\
r_4 &= \frac{1}{2}(S_1 - S_{3x} + S_{3y} - S_{3z}),\\
\theta_{12} &= \frac{2}{\sqrt{12}}S_{2a} - \frac{1}{\sqrt{2}}S_{4z} + \frac{1}{24\sqrt{2}}\left(S_{2a}^2 + S_{2b}^2\right) - \frac{1}{8\sqrt{2}}\left(S_{4x}^2 + S_{4y}^2 + S_{4z}^2\right),\\
\theta_{13} &= -\frac{1}{\sqrt{12}}S_{2a} + \frac{1}{2}S_{2b} - \frac{1}{\sqrt{2}}S_{4x} + \frac{1}{24\sqrt{2}}\left(S_{2a}^2 + S_{2b}^2\right) - \frac{1}{8\sqrt{2}}\left(S_{4x}^2 + S_{4y}^2 + S_{4z}^2\right),\\
\theta_{14} &= -\frac{1}{\sqrt{12}}S_{2a} - \frac{1}{2}S_{2b} - \frac{1}{\sqrt{2}}S_{4y} + \frac{1}{24\sqrt{2}}\left(S_{2a}^2 + S_{2b}^2\right) - \frac{1}{8\sqrt{2}}\left(S_{4x}^2 + S_{4y}^2 + S_{4z}^2\right),\\
\theta_{23} &= -\frac{1}{\sqrt{12}}S_{2a} - \frac{1}{2}S_{2b} + \frac{1}{\sqrt{2}}S_{4y} + \frac{1}{24\sqrt{2}}\left(S_{2a}^2 + S_{2b}^2\right) - \frac{1}{8\sqrt{2}}\left(S_{4x}^2 + S_{4y}^2 + S_{4z}^2\right),\\
\theta_{24} &= -\frac{1}{\sqrt{12}}S_{2a} + \frac{1}{2}S_{2b} + \frac{1}{\sqrt{2}}S_{4x} + \frac{1}{24\sqrt{2}}\left(S_{2a}^2 + S_{2b}^2\right) - \frac{1}{8\sqrt{2}}\left(S_{4x}^2 + S_{4y}^2 + S_{4z}^2\right),\\
\theta_{34} &= \frac{2}{\sqrt{12}}S_{2a} + \frac{1}{\sqrt{2}}S_{4z} + \frac{1}{24\sqrt{2}}\left(S_{2a}^2 + S_{2b}^2\right) - \frac{1}{8\sqrt{2}}\left(S_{4x}^2 + S_{4y}^2 + +S_{4z}^2\right),
\end{aligned}
\tag{64}
$$

and similarly for the conjugate momenta

$$
\begin{aligned}
\hat{p}_{r_1} &= \frac{1}{2}(\hat{P}_1 + \hat{P}_{3x} + \hat{P}_{3y} + \hat{P}_{3z}),\\
\hat{p}_{r_2} &= \frac{1}{2}(\hat{P}_1 - \hat{P}_{3x} - \hat{P}_{3y} + \hat{P}_{3z}),\\
\hat{p}_{r_3} &= \frac{1}{2}(\hat{P}_1 + \hat{P}_{3x} - \hat{P}_{3y} - \hat{P}_{3z}),\\
\hat{p}_{r_4} &= \frac{1}{2}(\hat{P}_1 - \hat{P}_{3x} + \hat{P}_{3y} - \hat{P}_{3z}),\\
\hat{p}_{\theta_{12}} &= \frac{2}{\sqrt{12}}\hat{P}_{2a} - \frac{1}{\sqrt{2}}\hat{P}_{4z},\\
\hat{p}_{\theta_{13}} &= -\frac{1}{\sqrt{12}}\hat{P}_{2a} + \frac{1}{2}\hat{P}_{2b} - \frac{1}{\sqrt{2}}\hat{P}_{4x},
\end{aligned}
\tag{65}
$$

$$\hat{p}_{\theta_{14}} = -\frac{1}{\sqrt{12}}\hat{P}_{2a} - \frac{1}{2}\hat{P}_{2b} - \frac{1}{\sqrt{2}}\hat{P}_{4y},$$

$$\hat{p}_{\theta_{23}} = -\frac{1}{\sqrt{12}}\hat{P}_{2a} - \frac{1}{2}\hat{P}_{2b} + \frac{1}{\sqrt{2}}\hat{P}_{4y},$$

$$\hat{p}_{\theta_{24}} = -\frac{1}{\sqrt{12}}\hat{P}_{2a} + \frac{1}{2}\hat{P}_{2b} + \frac{1}{\sqrt{2}}\hat{P}_{4x},$$

$$\hat{p}_{\theta_{34}} = \frac{2}{\sqrt{12}}\hat{P}_{2a} + \frac{1}{\sqrt{2}}\hat{P}_{4z}.$$

The kinetic energy can be symmetrized up to the fourth order with these equations. This is best done with computer algebra programs.

The model with the block diagonal Hamiltonian matrix structure similar to the XH_2 and XH_3 cases has been applied to methane overtone data [Venuti, Halonen and Della Valle (1999)]. All the kinetic and potential energy terms up to the fourth order have been included. Second order Van Vleck perturbation theory has been employed to remove non-resonance inter-block couplings. An *ab initio* surface [Lee, Martin and Taylor (1995)] has been the starting point of a nonlinear least squares optimization of the force constants. Experimental vibrational term values of the isotopic species $^{12}CH_4$, $^{13}CH_4$, $^{12}CD_4$, and $^{13}CD_4$ up to $10\,000\,\mathrm{cm}^{-1}$ have been used as data. The quality of the fit in $^{12}CH_4$ and $^{12}CD_4$ can be judged from Table 10.10. The potential energy parameters obtained together with *ab initio* values for comparison purposes are shown in Table 10.11. It is obvious that the results are pleasing for this nine-dimensional model. This gives confidence that it may be possible to extend this approach to larger molecular systems. Another interesting outcome concerns the significance of redundancy. On the whole redundancy contributions are small but the effect is not totally negligible in the case of quartic bending force constants [Venuti, Halonen and Della Valle (1999)].

10.3 Local mode and Fermi resonance models in the normal coordinate approach

In the conventional vibration–rotation theory based on the normal mode, the vibrational energy level structure of asymmetric, symmetric top, and linear molecules is modeled by the expression [Herzberg (1945)]

$$G = \sum_s \omega_s \left(v_s + \frac{d_s}{2}\right) + \sum_{s\le s'} x_{ss'} \left(v_s + \frac{d_s}{2}\right)\left(v_{s'} + \frac{d_{s'}}{2}\right) + \sum_{t\le t'} g_{tt'} l_t l_{t'}, \tag{66}$$

where the indices s and s' span all normal modes and the indices t and t' span doubly degenerate modes. Doubly degenerate modes are counted only once in all summations. The harmonic wavenumber ω_s refers to the sth normal mode, $x_{ss'}$ and $g_{tt'}$ are the anharmonicity parameters, d_s is the degeneracy of the sth mode, v_s is the usual vibrational quantum number, and l_t is the vibrational angular momentum quantum number. The harmonic wavenumbers are obtained from the harmonic force field in the standard way [Wilson, Decius and Cross (1955)]. The anharmonicity parameters are functions of the harmonic, cubic, and quartic force constants [Mills (1972, 1974)].

When applied to the stretching vibrations of water, equation (66) becomes

$$G = \omega_1\left(v_1 + \tfrac{1}{2}\right) + \omega_3\left(v_3 + \tfrac{1}{2}\right) + x_{11}\left(v_1 + \tfrac{1}{2}\right)^2 + x_{33}\left(v_3 + \tfrac{1}{2}\right)^2 + x_{13}\left(v_1 + \tfrac{1}{2}\right)\left(v_3 + \tfrac{1}{2}\right), \tag{67}$$

where the indices 1 and 3 refer to the symmetric and antisymmetric stretch, respectively. However, this result does not describe well many of the stretching vibrational overtone and combination

Table 10.10. Methane vibrational term value fit[a].

$v_1v_2v_3v_4\Gamma$	$^{12}CH_4$ ν_{obs} (cm^{-1})	$^{12}CH_4$ $(o-c)$ (cm^{-1})	$^{12}CD_4$ ν_{obs} (cm^{-1})	$^{12}CD_4$ $(o-c)$ (cm^{-1})
$0001F_2$	1310.76	0.26	997.87	0.03
0100E	1533.33	−0.69	1091.65	−0.53
$0002A_1$	2587.04	−0.41	1965.55	0.10
$0002F_2$	2614.26	0.16	1990.47	0.65
0002E	2624.62	0.21	1996.83	0.30
$0101F_2$	2830.32	−0.04	2083.40	−0.09
$0101F_1$	2846.08	−0.25	2090.88	−0.15
$1000A_1$	2916.48	−1.46	2101.37	3.16
$0010F_2$	3019.49	−0.46	2260.08	1.33
$0200A_1$	3063.65	−0.29	2182.23	−0.20
0200E	3065.14	−0.67	2182.59	−0.66
$0003F_2$	3870.49	−1.06		
$0003A_1$	3909.18	−0.85		
$0003F_1$	3920.50	−0.13		
$0003F_2$	3930.92	−0.11		
0102E	4105.15	2.72		
$0102F_1$	4128.57	0.46		
$0102A_1$	4132.99	1.04		
$0102F_2$	4142.86	−0.54		
0102E	4151.22	−0.11		
$0102A_2$	4161.87	−0.13		
$1001F_2$	4223.46	−3.05		
$0011F_2$	4319.21	−0.49		
0011E	4322.15	0.14		
$0011F_1$	4322.58	−0.22		
$0011A_1$	4322.72	−4.47		
$0201F_2$	4348.77	0.32		
$0201F_1$	4363.31	−0.11		
$0201F_2$	4379.10	0.17		
1100E	4446.41	7.87		
$0110F_1$	4537.57	−0.90		
$0110F_2$	4543.76	−0.42		
0300E	4592.03	0.59		
$0300A_2$	4595.32	−0.04		
$0300A_1$	4595.56	0.19		
$0012F_2$	5587.98	−0.66		
$0012F_2$	5623.01	5.50		
$0012F_2$	5628.40	−1.96		
$2000A_1$	5790.25	−3.30		
$0111F_2$	5819.72	−1.77		
$0111F_2$	5826.65	−17.41		
$1200A_1$	5968.09	0.23		
$0020F_2$	6004.69	−0.18		
0020E	6043.87	−0.96		
$0210F_2$	6054.64	−0.37		
$0210F_2$	6065.32	0.36		
$1210F_2$	8906.78	0.32		
$0030F_1$	8947.95	2.94		
$1400A_1$	8975.34	−1.86		
$0030F_2$	9045.82	0.07		

[a]Data and fits are from Venuti, Halonen, and Della Valle (1999), where references to original experimental papers can be found. v_i quantum numbers refer to a Hamiltonian model with basis sets expressed in terms of symmetrized internal coordinates.

Table 10.11. Methane potential energy parameters[a].

Parameter	Experimental fit	*Ab initio*
F_{11} (aJ Å^{-2})	5.4335 (34)	5.46865
F_{3x3x} (aJ Å^{-2})	5.4178 (60)	5.36602
F_{3x4x} (aJ Å^{-1})	−0.2478 (77)	−0.21036
F_{2a2a} (aJ)	0.57924 (47)	0.57919
F_{4x4x} (aJ)	0.53134 (55)	0.53227
F_{111} (aJ Å^{-3})	−14.1127 (53)	−15.17114
F_{13x3x} (aJ Å^{-3})	−15.654 (50)	−15.49772
F_{3x3y3z} (aJ Å^{-3})	−16.338 (55)	−15.57546
F_{13x4x} (aJ Å^{-2})	0.0657 (65)	0.06598
F_{2a3x3x} (aJ Å^{-2})	0.183 (17)	0.17803
F_{3x3y4z} (aJ Å^{-2})	−0.227 (21)	−0.21811
F_{12a2a} (aJ Å^{-1})	−0.2953 (91)	−0.25438
F_{14x4x} (aJ Å^{-1})	−0.2574 (87)	−0.22556
F_{2a3x4x} (aJ Å^{-1})	−0.0943 (36)	−0.09002
F_{3x4y4z} (aJ Å^{-1})	−0.0917 (84)	−0.09616
F_{2a2a2a} (aJ)	0.1016 (78)	0.09116
F_{2a4x4x} (aJ)	0.1724 (62)	0.17165
F_{4x4y4z} (aJ)	0.309 (15)	0.34391
F_{1111} (aJ Å^{-4})	37.4537 (63)	37.41710
F_{113x3x} (aJ Å^{-4})	39.8499 (11)	39.80537
$F_{13x3y3z}$ (aJ Å^{-4})	41.0028 (23)	40.64855
$F_{3x3x3x3x}$ (aJ Å^{-4})	41.077 (18)	41.04703
$F_{3x3x3y3y}$ (aJ Å^{-4})	40.6256 (61)	41.14033
F_{112a2a} (aJ Å^{-2})	−0.0126 (12)	−0.01264
F_{114x4x} (aJ Å^{-2})	0.0580 (59)	0.05929
$F_{12a3x4x}$ (aJ Å^{-2})	0.0213 (21)	0.02136
$F_{13x4y4z}$ (aJ Å^{-2})	−0.0209 (20)	−0.02093
$F_{2a2a3x3x}$ (aJ Å^{-2})	−0.374 (31)	−0.37363
$F_{2a2b3x3x}$ (aJ Å^{-2})	−0.1044 (99)	−0.10194
$F_{3x3x4x4x}$ (aJ Å^{-2})	0.0977 (98)	0.09940
$F_{3x3x4y4y}$ (aJ Å^{-2})	−0.242 (23)	−0.24415
$F_{2a3x3y4z}$ (aJ Å^{-2})	0.269 (27)	0.27554
$F_{3x3y4x4y}$ (aJ Å^{-2})	−0.00524 (52)	−0.00523
$F_{2a2a2a2a}$ (aJ)	0.157 (12)	0.17308
$F_{2a2a4x4x}$ (aJ)	0.2821 (69)	0.28687
$F_{2a2a4z4z}$ (aJ)	0.0180 (17)	0.01744
$F_{4x4x4x4x}$ (aJ)	0.426 (33)	0.49876
$F_{4x4x4y4y}$ (aJ)	0.6871 (95)	0.70977

[a]Experimental surface is from Venuti, Halonen, and Della Valle (1999) and *ab initio* results are from Lee, Martin, and Taylor (1995).

states of water. It is necessary to include Darling–Dennison resonance coupling terms [Darling and Dennison (1940)], which in the present application are defined by the following matrix elements [Mills and Robiette (1985)]

$$\begin{aligned}\langle v_1+2, v_3-2|\hat{H}/hc|v_1, v_3\rangle &= \tfrac{1}{4}K_{11,33}[(v_1+1)(v_1+2)v_3(v_3-1)]^{1/2},\\ \langle v_1-2, v_3+2|\hat{H}/hc|v_1, v_3\rangle &= \tfrac{1}{4}K_{11,33}[v_1(v_1-1)(v_3+1)(v_3+2)]^{1/2},\end{aligned} \tag{68}$$

where $|v_1, v_3\rangle = |v_1\rangle|v_3\rangle$ and $K_{11,33}$ is a parameter that describes the strength of this interaction. The label $v = v_1 + v_3$ is a good quantum number in the model defined by equation (67) with $\langle v_1, v_3|\hat{H}/hc|v_1, v_3\rangle = G$ and by equations (68). Thus, only states with the same v are coupled.

Additional insight into the traditional model including Darling–Dennison resonance coupling is obtained by considering the vibrational Hamiltonian operator

$$\begin{aligned} H/hc_0 &= \tfrac{1}{2}\omega_1\left(\hat{p}_1^2+q_1^2\right)+\tfrac{1}{2}\omega_3\left(\hat{p}_3^2+q_3^2\right)+\tfrac{1}{4}x_{11}\left(\hat{p}_1^2+q_1^2\right)^2+\tfrac{1}{4}x_{33}\left(\hat{p}_3^2+q_3^2\right)^2 \\ &\quad+\tfrac{1}{4}x_{13}\left(\hat{p}_1^2+q_1^2\right)\left(\hat{p}_3^2+q_3^2\right)+K_{11,33}\left[\left(\hat{q}_1^+\right)^2\left(\hat{q}_3^-\right)^2+\left(\hat{q}_1^-\right)^2\left(\hat{q}_3^+\right)^2\right], \end{aligned} \tag{69}$$

where q_1 and q_3 are the dimensionless normal coordinates associated with the symmetric and antisymmetric stretch, respectively. The quantities $\hat{p}_1 = -\mathrm{i}\partial/\partial q_1$ and $\hat{p}_3 = -\mathrm{i}\partial/\partial q_3$ are their conjugate momentum operators. The dimensionless normal coordinates q_s are related to the usual normal coordinates Q_s as $q_s = (2\pi c\omega_s/\hbar)^{1/2}Q_s$ where $s = 1$ or 3 [Wilson, Decius and Cross (1955) and Papousek and Aliev (1982)]. $\hat{q}_s^{\pm} = \frac{1}{2}(q_s \mp \mathrm{i}\hat{p}_s)$. Using product harmonic oscillator basis functions $|v_1, v_3\rangle = |v_1\rangle|v_3\rangle$, which are expressed in terms of q_1 and q_3 and which are consistent with the harmonic wavenumbers ω_1 and ω_3, the matrix elements given in both equations (67) and (68) are obtained. It is obvious that the Darling–Dennison resonance is due to quartic anharmonic resonance terms in the normal mode picture. Note that in the full Hamiltonian model the theoretical expression of the K coefficient is a function of the cubic surface by second-order perturbation theory and a function of the quartic surface by the first-order perturbation theory [Mills and Robiette (1985)]. It is interesting to observe that in the internal coordinate local mode model the most important coupling between the stretching vibrational states is due to bilinear harmonic terms but in the normal mode model it is due to quartic terms.

The block diagonal model described above contains more parameters than the local mode model defined by equations (21). If all these normal mode parameters were independent, then the two models would not give identical eigenvalues. However, if certain special relations called x–K relations between the normal mode parameters are fulfilled, then this will be the case [Lehmann (1983)]. Because of the definition of normal coordinates, the vibrational harmonic Hamiltonian of XH_2-type molecules is separable when expressed in terms of dimensionless normal coordinates q_1 and q_3 and their conjugate momenta. The harmonic bond oscillators are coupled by bilinear harmonic coupling terms in the internal coordinate representation. The parameter λ_r describes the strength of this coupling. In the case of stretching fundamentals, we denote the normal mode basis functions as $|1, 0\rangle$ and $|0, 1\rangle$ (where the first label is v_1 and the second one is v_3) and the local mode basis functions as $|10\rangle$ and $|01\rangle$ (where the first label is v_{r_1} and the second one is v_{r_2}). In these bases, using harmonic oscillator formulas for both diagonal and off-diagonal terms, the normal mode Hamiltonian matrix $\boldsymbol{N}$ and the local mode Hamiltonian matrix $\boldsymbol{L}$ for the stretching fundamentals are related by the unitary transformation

$$\boldsymbol{N} = \begin{pmatrix} \omega_1 & 0 \\ 0 & \omega_3 \end{pmatrix} = \boldsymbol{D}^{\dagger}\boldsymbol{L}\boldsymbol{D} = \boldsymbol{D}^{-1}\boldsymbol{L}\boldsymbol{D} = \begin{pmatrix} \dfrac{1}{\sqrt{2}} & \dfrac{1}{\sqrt{2}} \\ \dfrac{1}{\sqrt{2}} & -\dfrac{1}{\sqrt{2}} \end{pmatrix} \begin{pmatrix} \omega_r' & \lambda_r \\ \lambda_r & \omega_r' \end{pmatrix} \begin{pmatrix} \dfrac{1}{\sqrt{2}} & \dfrac{1}{\sqrt{2}} \\ \dfrac{1}{\sqrt{2}} & -\dfrac{1}{\sqrt{2}} \end{pmatrix}. \tag{70}$$

The $\boldsymbol{D}$ matrix defines a transformation between local mode shift operators $\hat{a}_i$ ($i = 1$ or 2) and normal mode shift operators $\hat{A}_s$ ($s = 1$ or 3) as

$$\begin{aligned} \hat{A}_s^+ &= \sum_i D_{is}\hat{a}_i^+, \\ \hat{A}_s^- &= \sum_i D_{is}^*\hat{a}_i^-, \end{aligned} \tag{71}$$

where the $\hat{a}_i^+$ and $\hat{a}_i^-$ operators are defined as

$$\hat{a}_1^+|v_{r_1}v_{r_2}\rangle = (v_{r_1}+1)^{1/2}|(v_{r_1}+1)v_{r_2}\rangle, \hat{a}_2^+|v_{r_1}v_{r_2}\rangle = (v_{r_2}+1)^{1/2}|v_{r_1}(v_{r_2}+1)\rangle,$$
$$\hat{a}_1^-|v_{r_1}v_{r_2}\rangle = v_{r_1}^{1/2}|(v_{r_1}-1)v_{r_2}\rangle, \hat{a}_2^-|v_{r_1}v_{r_2}\rangle = v_{r_2}^{1/2}|v_{r_1}(v_{r_2}-1)\rangle. \quad (72)$$

It follows from equations (72) that $\hat{a}_i^+$ and $\hat{a}_i^-$ obey the commutation relations $[\hat{a}_i^-, \hat{a}_j^+] = \delta_{ij}$ and $[\hat{a}_i^+, \hat{a}_j^+] = [\hat{a}_i^-, \hat{a}_j^-] = 0$. Because of the unitarity of the $\boldsymbol{D}$ matrix, the $\hat{A}_s^{\pm}$ operators also obey similar commutation relations. Therefore, they can be regarded as shift operators with equations such as (72) being fulfilled, i.e.

$$\hat{A}_1^+|v_1, v_3\rangle = (v_1+1)^{1/2}|v_1+1, v_3\rangle, \hat{A}_1^-|v_1, v_3\rangle = v_1^{1/2}|v_1-1, v_3\rangle,$$
$$\hat{A}_3^+|v_1, v_3\rangle = (v_3+1)^{1/2}|v_1, v_3+1\rangle, \hat{A}_3^-|v_1, v_3\rangle = v_3^{1/2}|v_1, v_3-1\rangle. \quad (73)$$

The local mode stretching vibrational Hamiltonian can be expressed in terms of the local mode shift operators as

$$\hat{H}/hc = \omega'\left(\hat{a}_1^+\hat{a}_1^- + \hat{a}_2^+\hat{a}_2^- + 1\right) + x_{rr}\left[\hat{a}_1^+\hat{a}_1^-\left(\hat{a}_1^+\hat{a}_1^- + 1\right) + \hat{a}_2^+\hat{a}_2^-\left(\hat{a}_2^+\hat{a}_2^- + 1\right) + \tfrac{1}{2}\right]$$
$$+ \lambda_r\left(\hat{a}_1^+\hat{a}_2^- + \hat{a}_2^+\hat{a}_1^-\right), \quad (74)$$

and the normal mode vibrational Hamiltonian can be written in terms of the normal mode shift operators as

$$\hat{H}/hc = \omega_1\left(\hat{A}_1^+\hat{A}_1^- + \tfrac{1}{2}\right) + \omega_3\left(\hat{A}_3^+\hat{A}_3^- + \tfrac{1}{2}\right) + x_{11}\left[\hat{A}_1^+\hat{A}_1^-(\hat{A}_1^+\hat{A}_1^- + 1) + \tfrac{1}{4}\right]$$
$$+ x_{33}\left[\hat{A}_3^+\hat{A}_3^-(\hat{A}_3^+\hat{A}_3^- + 1) + \tfrac{1}{4}\right] + x_{13}\left(\hat{A}_1^+\hat{A}_1^-\hat{A}_3^+\hat{A}_3^- + \tfrac{1}{2}(\hat{A}_1^+\hat{A}_1^- + \hat{A}_3^+\hat{A}_3^-) + \tfrac{1}{2}\right)$$
$$+ \tfrac{1}{4}K_{11,33}\left(\hat{A}_1^+\hat{A}_1^+\hat{A}_3^-\hat{A}_3^- + \hat{A}_3^+\hat{A}_3^+\hat{A}_1^-\hat{A}_1^-\right). \quad (75)$$

Using equations (71), it can be shown that the Hamiltonians in equations (74) and (75) give identical eigenvalues if the so-called x–K relations

$$\omega_1 = \omega_r' + \lambda_r,$$
$$\omega_3 = \omega_r' - \lambda_r, \quad (76)$$
$$x_{11} = x_{33} = x_{13}/4 = K_{11,33}/4 = x_{rr}/2$$

are fulfilled.

The Fermi resonance problem in XH_2-type molecules can be modeled in a similar fashion as the pure stretching Hamiltonians. In the normal mode approach, the Fermi effect is due to cubic potential energy term of the form $\hat{H}_{\text{Fermi}}/hc = \frac{1}{2}\phi_{122}q_1q_2^2$, where

$$\phi_{122} = \frac{1}{hc}\left(\frac{\partial^3 V}{\partial q_1\,\partial q_2^2}\right)_{\text{e}}$$

is the potential energy coefficient (a cubic force constant in the dimensionless normal coordinate representation) which describes the strength of the resonance. When harmonic oscillator formulas are used, the matrix elements of this cubic term are

$$\langle v_1-1, v_2+2, v_3|\hat{H}_{\text{Fermi}}/hc|v_1, v_2, v_3\rangle = \tfrac{1}{4}\phi_{122}\left[\tfrac{1}{2}v_1(v_2+1)(v_2+2)\right]^{1/2},$$
$$\langle v_1+1, v_2-2, v_3|\hat{H}_{\text{Fermi}}/hc|v_1, v_2, v_3\rangle = \tfrac{1}{4}\phi_{122}\left[\tfrac{1}{2}(v_1+1)v_2(v_2-1)\right]^{1/2}, \quad (77)$$

where v_2 is the bending quantum number.

The Fermi resonance coupling is such that $v = 2(v_1 + v_3) + v_2$ remains a good quantum number. Thus, a block diagonal model similar to that described in Section (10.2.5) is obtained. Using the shift operator technique described above, both the normal and the internal coordinate Hamiltonians are shown to give identical eigenvalues provided that, in addition to the x–K relations in equations (76), the equations $\omega_2 = \omega_\theta$, $x_{12}^* = x_{23} = x_{r\theta}$, $x_{22}^* = x_{\theta\theta}$, and $\phi_{122} = \sqrt{2}\phi_{r\theta\theta}$ are obeyed. The stars in some of the anharmonicity parameters indicate that Fermi resonance contributions should be excluded if standard theoretical perturbation theory formulas are used to intepret these parameters.

Normal coordinate-based Fermi resonance–local mode models can be extended to XH_3 and XH_4 (and larger) molecules [Halonen (1998)]. Relations between the parameters of both models can be derived using the technique described above.

10.4 Molecular rotations and local modes

The stretching vibrational energy level structure with near degeneracies in molecules close to the local mode limit is unusual according to the standard vibration–rotation theory based on normal coordinates. It turns out that the rotational energy levels associated with these vibrational states also contain unusual features. As an example, Figure 10.2 shows parts of the stretching vibrational infrared spectra of a tetrahedral molecule, monoisotopic stannane, $^{120}SnD_4$ [Halonen *et al.* (1998)]. A large change in the appearance of the spectral structure in going from the fundamental region ($1000A_1/F_2$) to overtones ($2000A_1/F_2$) and ($3000A_1/F_2$) is evident. The second overtone spectrum looks like part of a parallel band of a symmetric top molecule. Some of the spectral features have been marked with the symmetric top K quantum number. The appearance of these spectra can be qualitatively understood by observing that near the local mode limit vibrational energy has been localized in particular bond oscillators for periods of time longer than typical rotational periods. This has resulted in a dynamical symmetry change from T_d to

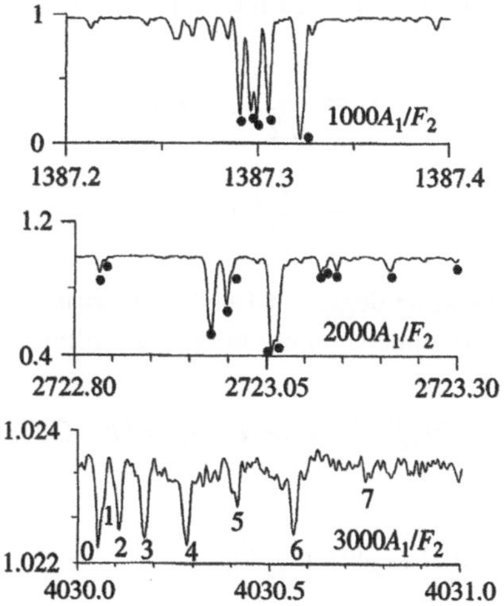

Figure 10.2. $^{120}SnD_4$ infrared spectrum showing rotational local mode effects. Spectra are taken from M. Halonen, L. Halonen, H. Bürger, W. Jerzembeck, 1998, Vibrational energy localization in the stretching vibrational ($1000A_1/F_2$), ($2000A_1/F_2$) and ($3000A_1/F_2$) band systems of $^{120}SnD_4$, *Journal of Chemical Physics* **108** (22) 9285–9290. Copyright 1998 American Institute of Physics.

C_{3v} when the excited bond has become longer than the three unexcited ones. This can be explained quantitatively as described below.

We take an XH_2-type bent molecules as an example and we start by applying the usual vibration–rotation spectroscopic theory to the rotational energy level structure of close lying vibrational states $|v0A_1/B_2\rangle$. The appropriate vibration–rotation Hamiltonian is [Halonen and Robiette (1986), and Halonen (1987)]

$$\begin{aligned}\hat{H}/hc = {} & G_v + B_e^{(x)}\hat{J}_x^2 + B_e^{(y)}\hat{J}_y^2 + B_e^{(z)}\hat{J}_z^2 - \overline{q}_1^2\left(\alpha_1^{(x)}\hat{J}_x^2 + \alpha_1^{(y)}\hat{J}_y^2 + \alpha_1^{(z)}\hat{J}_z^2\right) \\ & - \overline{q}_3^2\left(\alpha_3^{(x)}\hat{J}_x^2 + \alpha_3^{(y)}\hat{J}_y^2 + \alpha_3^{(z)}\hat{J}_z^2\right) + d_{13}\overline{q}_1\overline{q}_3\left[\hat{J}_x, \hat{J}_z\right]_+ \\ & - 2B_e^{(y)}\zeta_{13}^{(y)}\left(\sqrt{\frac{\omega_3}{\omega_1}}q_1\hat{p}_3 - \sqrt{\frac{\omega_1}{\omega_3}}\hat{p}_1 q_3\right)\hat{J}_y. \end{aligned} \tag{78}$$

where $[\ ]_+$ denotes an anticommutator. The vibrational operators ($i = 1$ and 3 for the symmetric and antisymmetric stretch, respectively) are defined as

$$\begin{aligned}\overline{q}_i^2 &= \tfrac{1}{2}\left(q_i^2 + \hat{p}_i^2\right), \\ \overline{q}_1\overline{q}_3 &= \tfrac{1}{2}\left(q_1 q_3 + \hat{p}_1\hat{p}_3\right).\end{aligned} \tag{79}$$

In equation (78), G_v is the vibrational term value, B_e coefficients are equilibrium rotational constants, and the α parameters describe the vibrational dependence of the rotational constants in the usual way, i.e. $\alpha_1^{(\xi)} = B_0 - B_{\nu_1}$ and $\alpha_3^{(\xi)} = B_0 - B_{\nu_3}$, where B_0 is the ground state rotational constant and the subindices ν_1 and ν_3 refer to the stretching fundamentals. The term in d_{13} (H_{22} term, α resonance term) and in $\zeta_{13}^{(y)}$ (H_{21} term, Coriolis term) couple rotational states of different stretching vibrational states [Aliev and Watson (1985)]. The rotational energy level patterns near the local mode limit can be investigated by making a coordinate transformation to the internal coordinate representation. For this symmerized internal coordinates are defined as

$$\begin{aligned}S_1 &= \frac{1}{\sqrt{2}}(r_1 + r_2), \\ S_3 &= \frac{1}{\sqrt{2}}(r_1 - r_2).\end{aligned} \tag{80}$$

In the present application, the bending degree of freedom is assumed to be completely decoupled from the stretches. Therefore, the dimensionless normal coordinates can be expressed as ($i = 1$ or 3)

$$q_i = \left(2\pi c\omega_i/\hbar G_e^{(ii)}\right)^{1/2} S_i = \alpha_i^{1/2} S_i = \left(F_{ii}/G_e^{(ii)}\hbar^2\right)^{1/4} S_i, \tag{81}$$

where

$$\begin{aligned}G_e^{(11)} &= g^{(rr)} + g_e^{(rr')}, \\ G_e^{(33)} &= g^{(rr)} - g_e^{(rr')}, \\ F_{11} &= f_{rr} + f_{rr'} = 2a_r^2 D_e + f_{rr'}, \\ F_{33} &= f_{rr} - f_{rr'} = 2a_r^2 D_e - f_{rr'}.\end{aligned} \tag{82}$$

The transformed Hamiltonian is

$$
\begin{aligned}
\hat{H}/hc = {} & G_v + B_e^{(x)}\hat{J}_x^2 + B_e^{(y)}\hat{J}_y^2 + B_e^{(z)}\hat{J}_z^2 - \frac{1}{2}\left(\overline{r}_1^2 + 2\overline{r}_1\overline{r}_2 + \overline{r}_2^2\right)\left(\alpha_1^{(x)}\hat{J}_x^2 + \alpha_1^{(y)}\hat{J}_y^2 + \alpha_1^{(z)}\hat{J}_z^2\right) \\
& - \frac{1}{2}\left(\overline{r}_1^2 - 2\overline{r}_1\overline{r}_2 + \overline{r}_2^2\right)\left(\alpha_3^{(x)}\hat{J}_x^2 + \alpha_3^{(y)}\hat{J}_y^2 + \alpha_3^{(z)}\hat{J}_z^2\right) + \frac{1}{2}d_{13}\left(\overline{r}_1^2 - \overline{r}_2^2\right)\left[\hat{J}_x, \hat{J}_z\right]_+ \\
& - \hbar^{-1} B_e^{(y)}\zeta_{13}^{(y)}\left[\sqrt{\frac{\omega_3}{\omega_1}}(r_1\hat{p}_1 + r_2\hat{p}_1 - r_1\hat{p}_2 - r_2\hat{p}_2)\right. \\
& \left. - \sqrt{\frac{\omega_1}{\omega_3}}(r_1\hat{p}_1 - r_2\hat{p}_1 + r_1\hat{p}_2 - r_2\hat{p}_2)\right]\hat{J}_y.
\end{aligned} \tag{83}
$$

In the above equation ($i = 1$ or 2 and $i < j = 2$),

$$
\begin{aligned}
\overline{r}_i^2 &= \tfrac{1}{2}\left(\alpha r_i^2 + \hbar^{-2}\alpha^{-1}\hat{p}_{r_i}^2\right), \\
\overline{r}_i\overline{r}_j &= \tfrac{1}{2}\left(\alpha r_i r_j + \hbar^{-2}\alpha^{-1}\hat{p}_{r_i}\hat{p}_{r_j}\right).
\end{aligned} \tag{84}
$$

A single scaling factor $\alpha = \alpha_r = 4\pi^2 c\omega_r/(hg^{(rr)}) = (f_{rr}/g^{(rr)})^{1/2}/\hbar$ is used instead of two different scaling factors α_1 and α_3 [see equation (81)] because in the zeroth-order model $\omega = \omega_1 = \omega_3$ [Mills and Robiette (1985)].

The operators of the type $\overline{r}_1\overline{r}_2$, $r_1\hat{p}_2$ and $r_2\hat{p}_1$ couple rotational states of different local mode states in the Hamiltonian given in equation (83). This coupling disappears in the rovibrational local mode limit [Lukka and Halonen (1994)]. It is possible to achieve this limit if the coefficients fulfill special relations called α relations. These are [Lehmann (1991), and Lukka and Halonen (1994)]

$$
\begin{aligned}
\alpha_1^{(x)} &= \alpha_3^{(x)}, \\
\alpha_1^{(y)} &= \alpha_3^{(y)}, \\
\alpha_1^{(z)} &= \alpha_3^{(z)}, \\
\zeta_{13}^{(y)} &= 0.
\end{aligned} \tag{85}
$$

These results imply that in the rovibrational local mode limit the appropriate rotational constants of the stretching fundamentals are equal and Coriolis coupling between the rotational states of symmetric and antisymmetric stretching states disappears. Note that the rotational resonances due to the term in d_{13} may be significant depending on the magnitude of the resonance coefficient.

By averaging the internal coordinate Hamiltonian in equation (83) with the unsymmetrized local mode wavefunction $|v_r0\rangle$ in the rovibrational local mode limit [Child and Zhu (1991)], an effective rotational Hamiltonian is obtained for this state in the harmonic approximation as

$$
\langle v_r 0|\hat{H}/hc|v_r 0\rangle = G_v + B_v^{(x)}\hat{J}_x^2 + B_v^{(y)}\hat{J}_y^2 + B_v^{(z)}\hat{J}_z^2 + \tfrac{1}{2}v_r d_{13}\left[\hat{J}_x, \hat{J}_z\right]_+, \tag{86}
$$

The rotational constant $B_v^{(\xi)} = B_e^{(\xi)} - (v_r + 1)\alpha^{(\xi)}$, where $\alpha^{(\xi)} = \alpha_1^{(\xi)} = \alpha_3^{(\xi)}$. The rotational part in the matrix representation is

$$
\hat{H}_{\text{rot}}/hc = (\hat{J}_x \quad \hat{J}_y \quad \hat{J}_z)\begin{pmatrix} B_v^{(x)} & 0 & \frac{1}{2}v_r d_{13} \\ 0 & B_v^{(y)} & 0 \\ \frac{1}{2}v_r d_{13} & 0 & B_v^{(z)} \end{pmatrix}\begin{pmatrix} \hat{J}_x \\ \hat{J}_y \\ \hat{J}_z \end{pmatrix}. \tag{87}
$$

Diagonalization of the 3×3 matrix on the right-hand side provides a transformation to a form where the rotational Hamiltonian does not contain rotational cross terms. The transformed Hamiltonian is

$$\hat{H}_{\text{rot}}/hc = B_v^{(x')}\hat{J}_{x'}^2 + B_v^{(y')}\hat{J}_{y'}^2 + B_v^{(z')}\hat{J}_{z'}^2, \tag{88}$$

where

$$\begin{aligned}
B_v^{(x')} &= \tfrac{1}{2}\left(B_v^{(x)} + B_v^{(z)}\right) - \tfrac{1}{2}\left[\left(B_v^{(x)} - B_v^{(z)}\right)^2 + v_r^2 d_{13}^2\right]^{1/2}, \\
B_v^{(y')} &= B_v^{(y)}, \\
B_v^{(z')} &= \tfrac{1}{2}\left(B_v^{(x)} + B_v^{(z)}\right) + \tfrac{1}{2}\left[\left(B_v^{(x)} - B_v^{(z)}\right)^2 + v_r^2 d_{13}^2\right]^{1/2},
\end{aligned} \tag{89}$$

and

$$\begin{aligned}
\hat{J}_{x'} &= \cos\theta \hat{J}_x + \sin\theta \hat{J}_z, \\
\hat{J}_{y'} &= \hat{J}_y, \\
\hat{J}_{z'} &= -\sin\theta \hat{J}_x + \cos\theta \hat{J}_z,
\end{aligned} \tag{90}$$

with

$$\theta = \frac{1}{2}\arctan\frac{v_r d_{13}}{B_v^{(x)} - B_v^{(z)}} \quad \text{and} \quad -\frac{\pi}{4} < \theta < \frac{\pi}{4}.$$

A single asymmetric top rotational Hamiltonian is obtained with the rotational constants related to the parameters of the coupled problem as given above. Thus, in this special case (in the rovibrational local mode limit), the rotational energy level structure of the degenerate local mode states $|v_r 0A_1\rangle$ and $|v_r 0B_2\rangle$ becomes that of an asymmetric rotor. The result is not limited to the use of harmonic oscillator matrix elements but it is necessary to assume that the rovibrational local mode limit relations between parameters are obeyed. Otherwise, even in the case of pure local mode states, there remains rovibrational coupling between different states.

Real molecules do not exactly obey the assumptions of the rovibrational local mode limit. However, it can be shown that effective vibration–rotation parameters of close lying local mode states $|v_r 0\text{A}_1/\text{B}_2\rangle$ called a local mode pair of states obey analogous α relations as the stretching energy increases. For example, the vibrational dependence of the rotational constants of pure stretching states is normally expressed in the conventional vibration–rotation theory as

$$B_i^{(\xi)} - B_0^{(\xi)} = -\alpha_1^{(\xi)} v_1 - \alpha_3^{(\xi)} v_3, \tag{91}$$

where $B_0^{(\xi)}$ is the rotational constant of the ground vibrational state and $\xi = x$, y, or z. However, this is not adequate for a local mode pair of states, for example, because of strong Darling–Dennison resonance effects. Instead equation (91) becomes

$$B_{v_r 0\Gamma}^{(\xi)} - B_0^{(\xi)} = -\alpha_1^{(\xi)}\left(\langle v_r 0\Gamma|\overline{q}_1^2|v_r 0\Gamma\rangle - \tfrac{1}{2}\right) - \alpha_3^{(\xi)}\left(\langle v_r 0\Gamma|\overline{q}_3^2|v_r 0\Gamma\rangle - \tfrac{1}{2}\right), \tag{92}$$

where the symmetry label Γ is A_1 or B_2 corresponding to the two local mode pair of states. The wavefunctions in equation (92) are calculated using the Darling–Dennison resonance model as discussed in Section 10.3. The vibrational dependence of the rotational constants can also be expressed as

$$\begin{aligned}
B_{v_r 0\Gamma}^{(\xi)} - B_0^{(\xi)} = &-\tfrac{1}{2}\alpha_1^{(\xi)}\left(\langle v_r 0\Gamma|\left(\overline{r}_1^2 + 2\overline{r}_1\overline{r}_2 + \overline{r}_2^2\right)|v_r 0\Gamma\rangle - \tfrac{1}{2}\right) \\
&- \tfrac{1}{2}\left(\alpha_3^{(\xi)}\langle v_r 0\Gamma|\left(\overline{r}_1^2 - 2\overline{r}_1\overline{r}_2 + \overline{r}_2^2\right)|v_r 0\Gamma\rangle - \tfrac{1}{2}\right),
\end{aligned} \tag{93}$$

where the wavefunctions are calculated, for example, with the HCAO model. At high vibrational energies, the states $|v_r0\Gamma\rangle$ of near local mode molecules become pure local mode states, i.e. $|v_r0\Gamma\rangle = (1/\sqrt{2})(|v_r0\rangle \pm |0v_r\rangle)$. For these high energy states, the matrix elements of the operators of type $\overline{r}_1\overline{r}_2$ disappear in the harmonic approximation. Thus, the rotational constants of both A_1 and B_2 local mode states are equal. This conclusion is not limited to the harmonic approximation. It holds for anharmonic coupling as well. If the harmonic approximation is adopted, then the effective rotational constants for the local mode pair of states in the local mode limit can be expressed as

$$B^{(\xi)}_{v_r0\Gamma} = B^{(\xi)}_0 - \tfrac{1}{2}\left(\alpha^{(\xi)}_1 + \alpha^{(\xi)}_3\right) v_r. \tag{94}$$

This result should be compared with that in equation (91), which holds for the case of the harmonic normal mode limit.

Similar arguments to those given above can be used to investigate rovibrational coupling between the rotational states of the local mode pair of states $|v_r0A_1/B_2\rangle$. The result is such that the Coriolis coupling disappears but the α resonance remains significant. If the effective d parameter is defined as

$$d_{\text{eff}} = d_{13}\langle v_r0A_1|\overline{q}_1\overline{q}_3|v_r0B_2\rangle = d_{13}\langle v_r0A_1|\left(\overline{r}_1^2 - \overline{r}_2^2\right)|v_r0B_2\rangle, \tag{95}$$

then in the local mode limit for the local mode pair of states

$$d_{\text{eff}} = v_r d_{13}. \tag{96}$$

Note that the discussion has been limited to the $|v_r0A_1/B_2\rangle$ states because transitions from the ground state to these states are often the strongest in intensity (see Section 10.2.4). The treatment can also be extended to other states, i.e. $|(v_r-1)1A_1/B_2\rangle$, $|(v_r-2)2A_1/B_2\rangle$ etc. Tables 10.12 and 10.13 show results when the models have been applied to vibrational dependence of the rotational, d and Coriolis constants of $H_2{}^{32}S$. In the first approach, the HCAO vibrational wavefunctions with equations (92) and (95) have been used with the vibrational parameters as given in Table 10.2 to produce the Model I results. Model II results are obtained with the local mode limit formulas equations (94) and (96).

There are interesting dynamical effects worth mentioning in the context of local mode pair states [Lehmann (1991, 1992)]. There are two time-dependent phenomena to be considered. The speed of the rotational period compared with the speed the stretching vibrational excitation has changed the bond. If the latter lasts much longer, then the dynamical symmetry changes from C_{2v} to C_s. Thus, the rovibrational Hamiltonian is that of an asymmetric rotor with a single vibrational mode. If on the other hand the rotational period is long, then the appropriate rovibrational model consists of two asymmetric top Hamiltonians with rovibrational interaction terms. Another dynamical effect is the observation of the quenching of tunneling splitting (local mode splitting) due to molecular rotation. This can be understood by thinking that in a rotating molecule it is necessary for the angular momentum to reorient when the vibrational excitation swaps the bond.

10.5 Conclusion

It is likely that the use of internal coordinates in overtone spectroscopy will become popular. Many spectroscopists might find it hard to abandon the usage of the more familiar normal coordinates. This is understandable because a rather complete and general picture of molecular vibrations and rotations has emerged from the traditional approach. Thus, one might well ask why should one then make a change. There are several reasons for this. First, the author hopes that this

Table 10.12. Observed and calculated vibrational term values and rotational constants (cm^{-1}) of stretching states in $H_2{}^{32}S^a$.

$v_1 v_3$	$v_{r_1} v_{r_2} \Gamma$		ν	ΔA	ΔB	ΔC
10	10A$_1$	Obs.	2614.41	0.1596	0.1237	0.0698
		Model I	2613.57			
01	10B$_2$	Obs.	2628.46	0.2178	0.0789	0.0544
		Model I	2627.27			
20	20A$_1$	Obs.	5144.99	0.3592	0.2144	0.1289
		Model I	5143.56	0.3613	0.2150	0.1285
		Model II		0.3774	0.2026	0.1242
11	20B$_2$	Obs.	5147.22	0.3765	0.2012	0.1235
		Model I	5145.50	0.3774	0.2026	0.1242
		Model II		0.3774	0.2026	0.1242
02	11A$_1$	Obs.	5243.10	0.3868	0.1956	0.1183
		Model I	5242.77	0.3935	0.1902	0.1199
40	40A$_1$	Obs.	9911.02	0.7453	0.4028	0.2542
		Model I	9908.97	0.7492	0.4095	0.2499
		Model II		0.7548	0.4052	0.2484
31	40B$_2$	Obs.	9911.02	0.7453	0.4028	0.2542
		Model I	9908.97	0.7492	0.4095	0.2499
		Model II		0.7548	0.4052	0.2484
50	50A$_1$	Obs.	12149.46	0.8852	0.5462	0.3196
		Model I	12148.06	0.9383	0.5105	0.3119
		Model II		0.9435	0.5065	0.3105
41	50B$_2$	Obs.	12149.46	0.8852	0.5462	0.3196
		Model I	12148.06	0.9383	0.5105	0.3119
		Model II		0.9435	0.5065	0.3105
32	41A$_1$	Obs.	12524.60	0.8937	0.5386	0.3154
		Model I	12528.48	0.9293	0.5175	0.3143
		Model II		0.9435	0.5065	0.3105
23	41B$_2$	Obs.	12525.20	0.9002	0.5331	0.3131
		Model I	12528.89	0.9345	0.5134	0.3129
		Model II		0.9435	0.5065	0.3105
60	60A$_1$	Obs.	14291.12	0.9228	0.8053	0.3942
		Model I	14291.80	1.1272	0.6117	0.3739
		Model II		1.1323	0.6078	0.3726
51	60B$_2$	Obs.	14291.12	0.9228	0.8053	0.3942
		Model I	14291.80	1.1272	0.6117	0.3739
		Model II		1.1322	0.6078	0.3726

[a]Experimental data are taken from Vaittinen *et al.* (1997) and Vaittinen, Campargue, and Flaud (1998). ΔA etc. denote $\Delta B^{(\xi)}$.

article in part has convinced the readers of the better physical description obtained by using internal coordinates. Second, the curvilinear nature of the internal coordinates should make series expansions (if needed) converge rapidly. Third, many of the molecular parameters expressed in terms of internal coordinate-based Hamiltonians are often directly related to Born–Oppenheimer potential energy surface force constants. The methane example given in this article shows that many isotopic species can be treated simultaneously.

Table 10.13. Observed and calculated d and Coriolis coefficients of stretching states in $H_2{}^{32}S^a$.

$v_{r_1}v_{r_2}$		$\|d\|$	ζ
10	Obs.	0.2961	0.0092
20	Obs.	0.5713	0.0002
	Model I	0.5864	0.0026
	Model II	0.5922	0.0
40	Obs.	1.1393	0.0
	Model I	1.1830	0.0
	Model II	1.1844	0.0
50	Obs.	1.3601	0.0
	Model I	1.4795	0.0
	Model II	1.4805	0.0
41	Obs.	0.8191	0.0
	Model I	0.8830	0.0
	Model II	0.8883	0.0
60	Obs.	1.4636	0.0
	Model I	1.7759	0.0
	Model II	1.7766	0.0

[a]See the footnote in Table 10.1. ζ constants are obtained from the references given in Vaittinen *et al.* (1997).

Acknowledgement

I wish to thank the Rector of the University of Helsinki, Academy of Finland, and the European Commission (TMR-network contract number FMRX-CT96-0088) for financial support.

10.6 References

Aliev, M. R., and Watson, J. K. G., 1985, in *Molecular Spectroscopy: Modern Research*, Vol. III, Rao, K. N., Ed.; Academic Press: New York.
Bray, R. G., and Berry, M. J., 1979, *J. Chem. Phys.* **71**, 4909–4922.
Breidung, J., and Thiel, W., 1996, *J. Mol. Spectrosc.* **169**, 166–180.
Bunker, P. R., and Jensen, P., 1998, *Molecular Symmetry and Spectroscopy*, 2nd edition; NRC Research Press: Ottawa.
Carter, S., and Handy, N. C., 1987, *J. Chem. Phys.* **87**, 4294–4301.
Child, M. S., and Halonen, L., 1984, *Adv. Chem. Phys.* **57**, 1–58.
Child, M. S., and Lawton, R. T., 1981, *Faraday Discuss. Chem. Soc.* **71**, 273–285.
Child, M. S., and Zhu, Q., 1991, *Chem. Phys. Lett.* **184**, 41–44.
Cotton, F. A., 1990, *Chemical Applications of Group Theory*, 3rd edition; Wiley: New York.
Darling, B. T., and Dennison, D. M., 1940, *Phys. Rev.* **57**, 128–139.
Eckart, C., 1935, *Phys. Rev.* **47**, 552–558.
Gray, D. L., and Robiette, A. G., 1979, *Mol. Phys.* **37**, 1901–1920.
Halonen, L., 1982, *Chem. Phys. Lett.* **87**, 221–225.
Halonen, L., 1986, *J. Mol. Spectrosc.* **120**, 175–184.
Halonen, L., 1987, *J. Chem. Phys.* **86**, 588–596.
Halonen, L., 1988, *J. Chem. Phys.* **88**, 7599–7603.
Halonen, L., 1997, *J. Chem. Phys.* **106**, 831–845.
Halonen, L., 1998, *Adv. Chem. Phys.* **104**, 41–179.
Halonen, L., and Carrington, T., Jr., 1988, *J. Chem. Phys.* **88**, 4171–4185.
Halonen, L., and Child, M. S., 1982, *Mol. Phys.* **46**, 239–255.
Halonen, L., and Child, M. S., 1983, *J. Chem. Phys.* **79**, 559–570.
Halonen, L., and Child, M. S., 1983, *J. Chem. Phys.* **79**, 4355–4362.
Halonen, M., Halonen, L., Bürger, H., and Jerzembeck, W., 1998, *J. Chem. Phys.* **108**, 9285–9290.

Halonen, L., and Robiette, A. G., 1986, *J. Chem. Phys.* **84**, 6861–6871.

Herzberg, G., 1945, *Molecular Spectra and Molecular Structure, II Infrared and Raman Spectra of Polyatomic Molecules*; Van Nostrand: New York.

Hoy, A. R., Mills, I. M., and Strey, G., 1972, *Mol. Phys.* **24**, 1265–1290.

Jensen, P., 1988, *J. Mol. Spectrosc*, **128**, 478–501.

Kauppi, E., 1994, *Chem. Phys. Lett.* **229**, 661–666.

Kauppi, E., and Halonen, L., 1990, *J. Phys. Chem.* **94**, 5779–5785.

Kauppi, E., and Halonen, L., 1992, *J. Chem. Phys.* **96**, 2933–2941.

Kauppi, E., and Halonen, L., 1995, *J. Chem. Phys.* **103**, 6861–6872.

Kjaergaard, H. G., Henry, B. R., Wei, H., Lefebvre, S., Carrington, T., Jr., Mortensen, O. S., and Sage, M. L., 1994, *J. Chem. Phys.* **100**, 6228–6239.

Lawton, R. T., and Child, M. S., 1980, *Mol. Phys.* **40**, 773–792.

Lee, T. J., Martin, J. M. L., and Taylor, P. R., 1995, *J. Chem. Phys.* **102**, 254–261.

Lehmann, K. K., 1983, *J. Chem. Phys.* **79**, 1098–1098.

Lehmann, K. K., 1991, *J. Chem. Phys.* **95**, 2361–2370.

Lehmann, K. K., 1992, *J. Chem. Phys.* **96**, 7402–7409.

Lehmann, K. K., Scherer, G. J., and Klemperer, W., 1983, *J. Chem. Phys.* **79**, 1369–1376.

Le Sueur, C. R., Miller, S., Tennyson, J., and Sutcliffe, B. T., 1992, *Mol. Phys.* **76**, 1147–1156.

Lin, H., Ulenikov, O. N., Yurchinko, S., Wang, X., and Zhu, Q., 1998, *J. Mol. Spectrosc.* **187**, 89–96.

Lukka, T., and Halonen, L., 1994, *J. Chem. Phys.* **101**, 8380–8390.

Lukka, T., Kauppi, E., and Halonen, L., 1995, *J. Chem. Phys.* **102**, 5200–5206.

Lummila, J., Lukka, T., Halonen, L., Bürger, H., and Polanz, O., 1996, *J. Chem. Phys.* **104**, 488–498.

Martin, T. E., and Kalantar, A. H., 1968, *J. Chem. Phys.* **49**, 235–243.

Mecke, R., 1950, *Z. Electrochem.* **54**, 38–42.

Meyer, R., 1985/1986, *Intramolekulare Bewegung*, lecture notes; Laboratorium für Physikalische Chemie, ETH: Zürich.

Meyer, R., and Günthard, Hs. H., 1968, *J. Chem. Phys.* **49**, 1510–1520.

Mills, I. M., 1972, in *Molecular Spectroscopy: Modern Research*, Vol. I, Rao, K. N., and Mathews, C. W., Eds; Academic Press: New York.

Mills, I. M., 1974, in *A Specialist Periodical Report, Theoretical Chemistry*, Vol. I, Dixon, R. N., Ed.; Chemical Society: London.

Mills, I. M., and Robiette, A. G., 1985, *Mol. Phys.* **56**, 743–765.

Morse, P. M., 1929, *Phys. Rev.* **34**, 57–64.

Mortensen, O. S., Henry, B. R., and Mohammadi, M. A., 1981, *J. Chem. Phys.* **75**, 4800–4808.

Papousek, D., and Aliev, M. R., 1982, *Molecular Vibrational–Rotational Spectra*; Elsevier: Amsterdam.

Pickett, H. M., 1972, *J. Chem. Phys.* **56**, 1715–1723.

Podolsky, B., 1928, *Phys. Rev.* **32**, 812–816.

Sage, M. L., and Williams, J. A. III, 1983, *J. Chem. Phys.* **78**, 1348–1358.

Schek, I., Jortner, J., and Sage, M. L., 1979, *Chem. Phys. Lett.* **64**, 209–212.

Simanouti, T., 1949, *J. Chem. Phys.* **17**, 245–248.

Tennyson, J., and Sutcliffe, B. T., 1982, *J. Chem. Phys.* **77**, 4061–4072.

Ulenikov, O. N., Sun, F., Wang, X., and Zhu, Q., 1996, *J. Chem. Phys.* **105**, 7310–7315.

Vaittinen, O., Biennier, L., Campargue, A., Flaud, J.-M., and Halonen, L., 1997, *J. Mol. Spectrosc.* **184**, 288–299.

Vaittinen, O., Campargue, A., and Flaud, J.-M., 1998, *J. Mol. Spectrosc.* **190**, 262–268.

Venuti, E., Halonen, L., and Della Valle, R., 1999, *J. Chem. Phys.* **110**, 7339–7347.

Wei, H., and Carrington, T., Jr., 1992, *J. Chem. Phys.* **97**, 3029–3037.

Wilson, E. B., Decius, J. C., and Cross, P. C., 1955, *Molecular Vibrations*; McGraw-Hill: New York.

Zhan, X., Halonen, M., Halonen, L., Bürger, H., and Polanz, O., 1995, *J. Chem. Phys.* **102**, 3911–3918.

Zhu, Q., Thrush, B. A., and Robiette, A. G., 1988, *Chem. Phys. Lett.* **150**, 181–183.

11 THE VIBRATIONAL SELF-CONSISTENT FIELD APPROACH AND EXTENSIONS: METHOD AND APPLICATIONS TO SPECTROSCOPY OF LARGE MOLECULES AND CLUSTERS

R. Benny Gerber[†,‡] **and Joon O. Jung**[‡,1]

† *The Hebrew University, Jerusalem, Israel*
‡ *University of California, Irvine CA, USA*

---

[1] Present address: Chemistry Department, Columbia University, New York, NY 10027, USA.

Computational Molecular Spectroscopy. Edited by Per Jensen and Philip R. Bunker

11.1 Introduction

High-resolution vibrational spectroscopy has long been amongst the most important and useful tools for exploring the dynamics and interaction potentials of small polyatomic systems. The recognized sensitivity of vibrational spectroscopy to the potential energy surfaces has, of course, been central to the success of the field. The power and wide applicability of the approach depend on two preconditions that are satisfied for many small polyatomic systems: detailed, well-resolved experimental data is at hand for the cases of interest, and theoretical tools for calculation and quantitative interpretation are available. For examples of striking progress and success of combined experimental and computational spectroscopy in the area of small van der Waals and other weakly-bound clusters see, e.g., the reviews by Nesbitt (1994), Miller (1990), Saykally (1989), Hutson (1990), and Bačić and Miller (1996). The experimental challenge of measuring high-resolution vibrational spectroscopy for large molecular systems is far more difficult, because of the rotational congestion of the initial states and other problems. Nevertheless, there has been great progress on this issue, and well-resolved data for larger clusters and polyatomic molecules are increasingly becoming available. Much of the progress involves advances in nozzle jet spectroscopy [e.g. Nesbitt and Field (1996), Yang *et al.* (1995), Liu *et al.* (1996), and Liu, Cruzan and Saykally (1996)] and in molecular spectroscopy in superfluid helium droplets [e.g. Hartmann *et al.* (1995, 1996), and Huisken, Kaloudis and Kulicke (1996)]. It seems, therefore, very timely and desirable to pursue theoretical methods for the quantitative interpretation and calculation of vibrational spectroscopy of large anharmonic systems. Most existing methods for accurate calculations of vibrational states of coupled anharmonic degrees of freedom are, however, geared towards and inherently restricted to systems of several modes only. Among the tools that proved successful in rigorous computational studies of vibrational energy levels and wavefunctions in small polyatomic systems and clusters are grid methods, close-coupling algorithms and variational techniques, [Hutson (1990), Bačić and Light (1989), Yang and Peet (1990), Halberstadt *et al.* (1992), Zuniga *et al.* (1992), and Cohen and Saykally (1990)]. Perhaps the most widely used technique at present is the discrete variable representation (DVR) method [Bačić and Light (1986), Light, Hamilton and Liu (1985), Henderson, Tennyson and Sutcliffe (1993), Bromley *et al.* (1994), Mandelshtam and Taylor (1997), and Wright and Hutson (1999)]. State-of-the-art applications to systems such as H_3^+ [Mandelshtam and Taylor (1997)], or Ar_3 [Wright and Hutson (1999)], show the power of the method for vibrational states of small molecules, but also bring out the enormous difficulties that will be encountered, owing to both computer memory and speed limitations, in any effort to extend approaches of this type to substantially larger systems. A method that scales very well with system size as the number of particles increases is the diffusion quantum Monte Carlo (DQMC) algorithm, and related techniques such as the variational quantum Monte Carlo approach [e.g. Anderson (1975), Suhm and Watts (1991), Buch (1992), Barnett and Whaley (1993), and Bačić *et al.* (1992)]. DQMC, introduced by Anderson for electronic structure calculations, is a very effective method for obtaining the vibrational ground state of large anharmonic systems, including the extreme case of quantum clusters [McMahon, Barnett and Whaley

(1993), McMahon, Barnett and Whaley (1996), Cheng and Whaley (1996), Blume *et al.* (1996), Vegiri *et al.* (1994), and Broude and Gerber (1996)]. However, in general, the method is limited to calculations of the ground state. In specific cases it is possible to employ the method also for excited states [Anderson (1976), and Coker and Watts (1986)], and applications to spectroscopy were made on this basis [Sandler, Buch and Sadlej (1996), Gregory and Clary (1995), and Blume *et al.* (1997)]. However, a general extension of DQMC for excited vibrational states is not yet at hand. [We shall discuss in the review application of DQMC to fundamental vibrational excitations, based on combining it with vibrational self-consistent field (VSCF) calculations]. Specific cases where the application to excited states could be made involved mostly, so far, particular types of transitions and small systems.

The approach that seems at present to provide the most effective tool for the spectroscopy of large polyatomic systems is the VSCF approximation and its generalizations [e.g. Bowman (1978, 1986), Gerber and Ratner (1979, 1988)]. The present article reviews the recent methodological developments, with focus on the applications to large, 'floppy' polyatomic systems. In particular, VSCF calculations of the spectroscopy of large hydrogen-bonded clusters, ion–molecule clusters, and biological molecules such as peptides, sugars and proteins are discussed. Recent extensions of VSCF and improved algorithms are outlined, with emphasis on their large-molecule applications. The role of the method is examined in the context of interpretation of spectra and the relation to anharmonic vibrational dynamics; of testing potential functions against experiment, and of developing improved potential functions for large systems by fitting of data. In Section 11.2 we present the methods, their physical basis and the computational aspects involved. Section 11.3 presents applications of the methods. Concluding comments and suggestions for future developments are brought in Section 11.4.

11.2 Methods

VSCF is essentially the application of the Hartree approximation, known for electronic structure theory from the early stages of quantum mechanics, to the problem of coupled anharmonic vibrations. It is thus somewhat surprising that the method was introduced only in the late 1970s [Bowman (1978), Carney, Sprandel and Kern (1978), Cohen, Greita and McEachran (1979), and Gerber and Ratner (1979)]. The early applications of the method were to model problems of two coupled anharmonic modes. The approaches of Bowman (1978) and of Carney, Sprandel and Kern (1978) were fully quantum mechanical, while Gerber and Ratner (1979) employed a semiclassical VSCF theory. As numerical examples showed, the computational time saving gained by the additional semiclassical approximation is insignificant, although at the same time the semiclassical treatment hardly affects the VSCF error. It thus seems that in general the full quantum VSCF is preferable, and only in special applications does the greater mathematical simplicity of semiclassical VSCF offer significant advantages. One such example is the direct inversion of vibrational spectroscopic data to obtain the multidimensional potential energy surface of a polyatomic system. Such an explicit inversion transform of the complete (vib-rotational) spectrum is formally possible in the framework of semiclassical VSCF [Gerber, Roth and Ratner (1981)], but as far as we are aware, not for full quantum VSCF. The earliest extensions of VSCF, to incorporate correlations between the modes, used a configuration interaction (CI) approach, that is a treatment in which the vibrational wavefunction is described by a finite sum of separable SCF terms [Bowman, Christoffel and Tobin (1979), and Ratner, Buch and Gerber (1980)]. The CI-VSCF method was applied extensively and very successfully over the years to accurate spectroscopic calculations for small polyatomic systems. Very efficient codes were written for such applications, especially by Bowman's group, and CI-VSCF is still definitely a state-of-the-art method, especially for highly accurate calculations. Developments of VSCF with a view to large system applications are much more recent. Extensions of VSCF to include correlations between the different modes that scale well with the number of coupled degrees of freedom for large systems have been proposed recently. The perturbation theory correlation corrections to VSCF [Jung and Gerber (1996), and Norris *et al.* (1996)], and DQMC combined with VSCF [Broude, Jung and Gerber (1999)] seem feasible for systems of hundreds or even thousands of modes, and realistic applications for fairly large polyatomic systems were already made. This is the aspect of VSCF-based methods and applications that will be emphasized mostly in the present review.

11.2.1 THE VSCF APPROXIMATION

The brief description given here follows closely that of Jung and Gerber (1996). We consider, for simplicity, a system of total angular momentum $J = 0$, and neglect all rotational coupling effects. Then, we assume the minimum energy configuration of the system is available, and that the normal-mode displacement coordinates from that minimum are computed. The vibrational Schrödinger equation in mass-weighted normal-mode coordinates $Q_1, \ldots, Q_N$ can be written [Wilson, Decius and Cross (1955)]:

$$\left[-\frac{1}{2}\sum_{j=1}^{N}\frac{\partial^2}{\partial Q_j^2} + V(Q_1, \ldots, Q_N)\right]\Psi_n(Q_1, \ldots, Q_N) = E_n\Psi_n(Q_1, \ldots, Q_N), \tag{1}$$

where V is the potential function of the system, and N the total number of vibrational modes. Although normal modes are chosen to describe the vibrations of the system, the harmonic approximation is not made, and the anharmonicity of the potential function $V(Q_1, \ldots, Q_N)$ is fully retained. The choice of coordinates in VSCF is, however, physically very important and will be discussed later. The VSCF approximation is based on the ansatz:

$$\Psi_n(Q_1, \ldots, Q_n) = \prod_{j=1}^{N}\psi_j^{(n)}(Q_j), \tag{2}$$

which, using the variational principle, leads to the single-mode VSCF equations [Gerber and Ratner (1988), and Bowman (1986)]:

$$\left[-\frac{1}{2}\frac{\partial^2}{\partial Q_j^2} + \overline{V}_j^{(n)}(Q_j)\right]\psi_j^{(n)}(Q_j) = \varepsilon_j^{(n)}\psi_j^{(n)}(Q_j), \tag{3}$$

where the effective potential $\overline{V}_j^{(n)}(Q_j)$ for the mode Q_j is given by:

$$\overline{V}_j^{(n)}(Q_j) = \left\langle \prod_{l\neq j}^{N}\psi_l^{(n)}(Q_l)\middle| V(Q_1, \ldots, Q_N)\middle| \prod_{l\neq j}^{N}\psi_l^{(n)}(Q_l)\right\rangle. \tag{4}$$

Equations (3) and (4) for the single-mode wavefunctions, energies and effective potentials must be solved self-consistently. A variety of methods for solving the single-mode Schrödinger equation (3) can be employed. Both ground and excited states of equation (3), and therefore both ground and excited SCF states of the total system can be obtained. The VSCF approximation for the total energy is given by:

$$E_n = \sum_{j=1}^{N}\varepsilon_j^{(n)} - (N-1)\left\langle \prod_{j=1}^{N}\psi_j^{(n)}(Q_j)\middle| V(Q_1, \ldots, Q_N)\middle| \prod_{j=1}^{N}\psi_j^{(n)}(Q_j)\right\rangle. \tag{5}$$

The main computational difficulty in solving equations (3) and (4) for large systems is due to the need to evaluate numerically the multidimensional integrals of equation (4). As we shall see below, at least for low-lying excited states (e.g. fundamentals), the computational effort required can be reduced greatly by simplifying approximations that work very well for many types of realistic systems. This leads to efficient algorithms for solving the VSCF equations, with reasonable numerical effort even for systems of the order of 10^3 coupled modes. Before discussing the computational aspects we consider the conceptually and practically important question of choice of coordinates for VSCF.

11.2.2 CHOICE OF COORDINATES FOR VSCF

The accuracy of VSCF depends on the choice of coordinates. The quality of a separability approximation must depend on the coordinates that are being mutually separated. For the ground state and for low-lying excited states, the choice of normal modes is expected to work very well, and this is indeed supported by many applications. In the harmonic limit, normal modes provide rigorous separability. For energy levels near the bottom of the potential well, it seems reasonable that the same coordinates should remain a good choice for VSCF. For extremely anharmonic cases, VSCF in normal coordinates may be inadequate even for the ground state. The experience of the authors is that the *only* realistic exceptions of this type are 'quantum clusters', i.e. $(He)_n$ or $(H_2)_n$, etc. The anharmonicity is extreme in these cases even at the zero-point energy, to an extent that normal modes become physically irrelevant. Horn, Gerber and Ratner (1989) found that VSCF in normal coordinates fails completely for the quantum clusters $Xe(He)_2$, I_2He. Physical considerations suggest the choices of hyperspherical and ellipsoidal coordinates, respectively in these two cases, and the results of VSCF with these modes are excellent [Horn, Gerber and Ratner (1989)]. In this case, there is a large separation of frequencies between the vibrational modes, and frequency-scale separation between different degrees of freedom is favorable for the accuracy of VSCF. Interestingly, even for Ar_3, the choice of hyperspherical coordinates gives a better VSCF approximation than the normal-mode treatment, but at least for the low-lying states also the normal-mode VSCF results are reasonable [Horn *et al.* (1991)]. For all but the quantum clusters, the issue of choosing good coordinates for VSCF arises only for states of considerable vibrational excitation, as for the lowest-lying states the normal-mode choice works very well. However, for highly excited states, VSCF in normal modes quite typically breaks down. The only general solution for such states is to include correlation effects, i.e. to go beyond the VSCF approximation. For special types of highly excited states, with guidance from physical considerations, it proved possible in some cases to introduce 'good' coordinates, for which VSCF gives reasonable results. Examples are the bending excitations for HCN [Bačić, Gerber and Ratner (1986)] and of $Ar–CO_2$ [Horn, Gerber and Ratner (1993)]. The use of spheroidal (ellipsoidal) coordinates for the bending excitation of HCN, for instance, is motivated by the shape of the equipotential lines of the potential surface. With these modes, VSCF seems to give good results also for highly excited states. While the choice of coordinate systems such as hyperspherical or spheroidals may be physically well motivated in many cases, such coordinates are not computationally convenient because of the structure of the kinetic energy operator. Indeed, in these cases the VSCF averaging for obtaining the effective single-mode Hamiltonians affects also the kinetic energy operator terms. Except for certain few-atoms systems, the use of such 'good modes' does not seem a practical option.

Roth, Gerber and Ratner (1983) compared local (or bond) coordinates with normal modes in VSCF calculations for several molecules. Normal coordinates were found to be better in nearly all cases, even for systems such as D_2O. This was attributed to the fact that collective modes provided better mutual averaging of the potential field, since they are more delocalized [Ratner and Gerber (1986), and Gerber and Ratner (1988)]. A very interesting idea proposed by Thompson and Truhlar (1982), by Lefebvre (1983) and by Moiseyev (1983), is to determine optimal coordinates for VSCF through the use of the variational principle: a family of coordinate systems is introduced by applying a coordinate rotation mapping to given, fixed cartesian modes. The variational principle then yields equations for the optimal coordinate rotation. So far, this approach was used mostly in illustrating calculations, but it may have very useful applications.

In summary, the most useful coordinates for VSCF are obviously those which are both physically motivated, thus yielding good separability, and are computationally convenient. A case of wide applications is obviously normal coordinates when used for low-lying excitations. To a lesser extent, local (bond) coordinates can be used. There are possibilities of very successful choices for certain specific cases, and the possibility of further progress towards variationally determined optimal modes seems open. A new direction which seems very promising is in the context of application of VSCF to tunneling spectroscopy, for multidimensional double-well potentials: Jung *et al.* (unpublished) have proposed to use for such tunneling states normal-mode displacements about the transition-state (barrier top) configuration.

11.2.3 CALCULATION OF THE EFFECTIVE POTENTIALS IN VSCF

The calculation of the single-mode effective potentials, equation (4), which involves in general the evaluation of multidimensional integrals, is the key to the computational effectiveness of the method. We present briefly several approaches to the evaluation of the integrals, following mostly the discussion in Jung and Gerber (1996):

Power series expansion of the potential

A very advantageous approach, in certain cases, is to expand the potential function in powers of the normal modes. The expansion can be written:

$$V(Q_1, \ldots, Q_N) = \sum_{m_1,\ldots,m_N} V_{m_1,\ldots,m_N} (Q_1)^{m_1} \ldots (Q_N)^{m_N}. \tag{6}$$

This obviously leads to a simplification in which only the evaluation of one-dimensional integrals is required in order to obtain the effective potential $\overline{V}_j^{(n)}(Q_j)$. These potentials are then given by

$$\overline{V}_j^{(n)}(Q_j) = \sum_{m_1,\ldots,m_N} V_{m_1,\ldots,m_N} \prod_{l \neq j}^{N} F_{m_l}^{(l),(n)} (Q_j)^{m_j}, \tag{7}$$

where

$$F_{m_l}^{(l),(n)} = \left\langle \psi_l^{(n)}(Q_l) \middle| (Q_l)^{m_l} \middle| \psi_l^{(n)}(Q_l) \right\rangle. \tag{8}$$

Although the number of one-dimensional integrals grows with the size of the system, if the order of the expansion is modest the computational task is feasible even for systems of thousands of modes. In many applications to molecular systems, a fourth-order polynomial approximation for the potential field was used and proved successful [e.g. Romanowski, Bowman and Harding (1985) Norris *et al.* (1996), Roitberg *et al.* (1995), and Jung and Gerber (1996)]. Such an approximation made possible the calculation of the wavefunctions and the IR absorption spectrum for a large system such as the protein BPTI [Roitberg *et al.* (1995)]. In a recent study by the present authors, sixth-order and fourth-order polynomial representations were both used in calculations for the cluster $(Ar)_{13}$ and the close agreement of the two results supports the validity of the approach in this case [Jung and Gerber (1996)]. When expansions to higher orders are required, however, the method loses it usefulness. The coefficients $V_{m_1,\ldots,m_N}$ were obtained in nearly all applications hitherto by differentiation of the potential function at the equilibrium configuration.

When the potential is given in terms of simple analytic functions in the coordinates of the particles involved, the derivatives can be obtained analytically. In general, this is not the case, and numerical differentiation of the potential is required. The accuracy of high-order numerical derivatives is poor, limiting the application of this approach in practice to cases where a fourth-order expansion is sufficient. Determination of a polynomial form of equation (6) by a fitting procedure was recently attempted with some success for $(Ar)_{13}$ [Jung and Gerber (1996)]. For very anharmonic systems, however, such as small water clusters $(H_2O)_n$, the polynomial expansion method breaks down completely, at least from a practical point of view. The power series for the potential either diverges, or requires extremely high powers for acceptable accuracy, to the point of being unfeasible [Jung and Gerber (1996)]. The difficulty is illustrated in Figures 11.1 and 11.2 for the water dimer, $(H_2O)_2$. The results are from Jung and Gerber (1996), computed with the RKW-2 potential of Reimers, Watts and Klein (1982). Figure 11.1 shows the VSCF potential for mode 12, a soft torsional mode of $(H_2O)_2$. The failure of the expansion is obvious from the flat bottom of the potential. As happens for a number of torsional modes in floppy hydrogen-bonded clusters, in the experience of the present authors, the potential is roughly 'square well' like rather than harmonic. Figure 11.2 shows the coupling potential between modes 11 and 12 in $(H_2O)_2$. The quartic polynomial approximation fails completely for this interaction.

Approximation of pairwise interactions between normal modes, and grid integration for the VSCF potentials

Experience of recent years shows that for many realistic empirical *and* computed *ab initio* potential functions, one can represent the potential to excellent accuracy by assuming only pairwise interactions between normal modes [e.g. Jung and Gerber (1996), and Chaban, Jung and Gerber

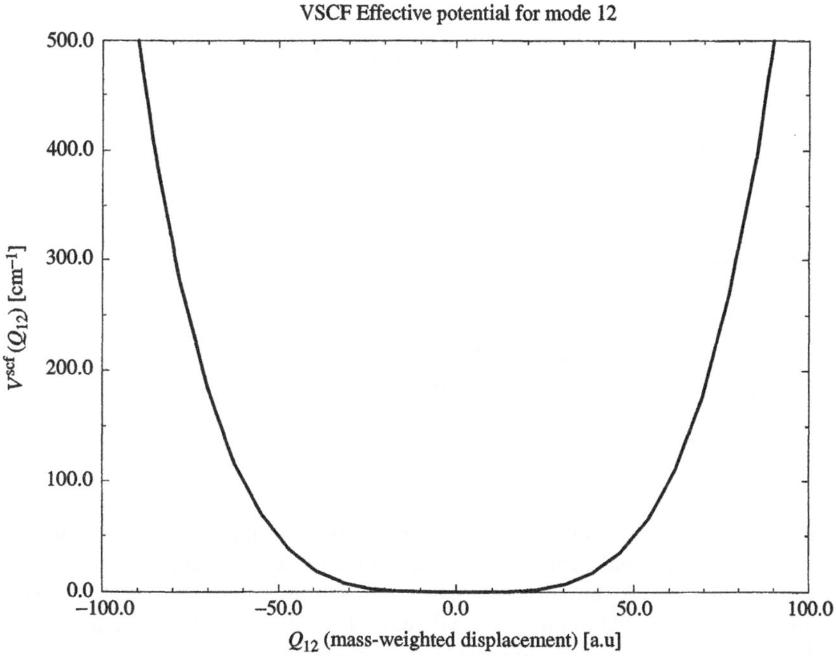

Figure 11.1. The VSCF potential for a torsional mode of $(H_2O)_2$. The normal mode here is Q_{12} [Jung and Gerber (1996)].

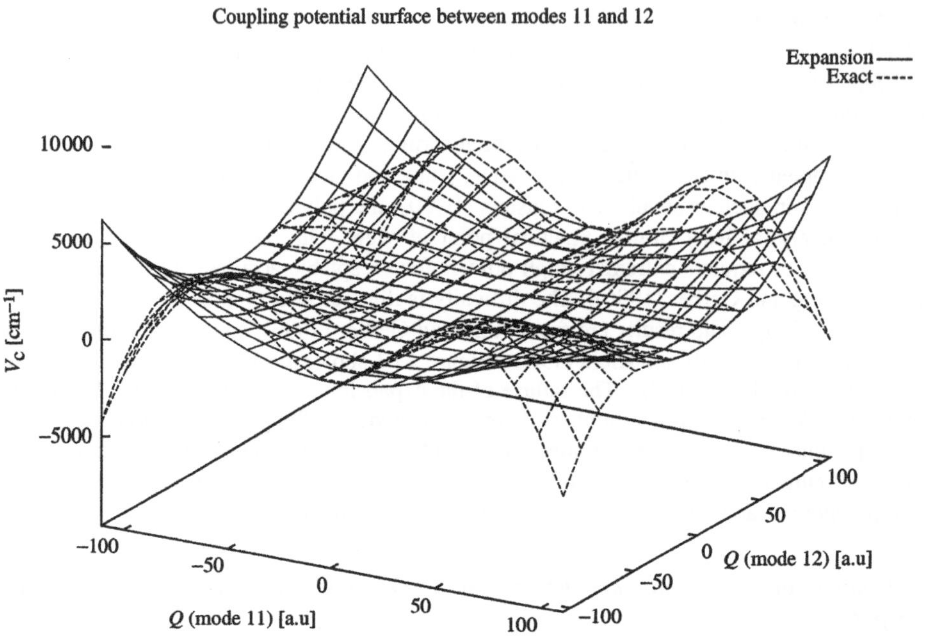

Figure 11.2. The coupling potential between normal modes Q_{11} and Q_{12} of $(H_2O)_2$. The correct coupling is compared with the quartic potential approximation [Jung and Gerber (1996)].

(submitted)]. In this framework, the potential function of the system can be written:

$$V(Q_1, \ldots, Q_N) = \sum_j^N V_j^{\text{diag}}(Q_j) + \sum_j \sum_{i>j} W_{ji}^{\text{coup}}(Q_j, Q_i), \tag{9}$$

where the 'diagonal' potential functions $V_j^{\text{diag}}(Q_j)$ are defined by

$$V_j^{\text{diag}}(Q_j) = V(O, \ldots, Q_j, \ldots, O), \tag{10}$$

and the pairwise interactions are:

$$W_{ji}^{\text{coup}}(Q_j, Q_i) = V(O, \ldots, Q_j, \ldots, Q_i, \ldots, O). \tag{11}$$

With the pairwise coupling approximation (9), the calculation of the effective single-mode VSCF potential $V_j^{(n)}(Q_j)$ requires only one-dimensional integrals of the type:

$$V_j^{\text{coup},(n)}(Q_j) = \sum_{i \neq j} \left\langle \psi_i^{(n)}(Q_i) \middle| W_{ji}^{\text{coup}}(Q_j, Q_i) \middle| \psi_i^{(n)}(Q_i) \right\rangle. \tag{12}$$

These integrals can be readily evaluated using grid methods, combined with a routine technique for solving the one-dimensional Schrödinger equation for the $\psi_i(Q_i)$. Grids up to 32×32 were used in previous calculations, but in most cases 8×8 grids are sufficient. It is possible to extend the method, and to retain computational feasibility, also for cases which include three-mode correlations. Also, it is possible to introduce hybrid approaches, which combine expansion treatments for some of the modes, with grid treatments of the rest. All this provides a sufficient arsenal for treating most realistic cases.

Monte Carlo integration for computing the VSCF potentials

A general numerical method for computing the VSCF potentials is by Monte Carlo integration [Koonin (1986)] of the multidimensional integrals of equation (4). Based on available experience we estimate that potential terms involving many modes are unlikely to be significant. A scheme for the evaluation of integrals of several dimensions is therefore sufficient. The Monte Carlo random integration method has satisfactory scaling with increasing number of dimensions [Koonin (1986)]. Several considerations are favorable for the application of the method in VSCF calculations: (i) the factors $|\psi_l(Q_l)|^2$ that appear in the integrand of equation (4) are all positive. For low-lying vibrational states, which are the ones of interest here, the wavefunctions have few maxima and minima; (ii) also the potential $V(Q_l, \ldots, Q_N)$ is not a very oscillatory function for configurations in the vicinity of the equilibrium point, which are the ones pertinent to the low-lying states. A Monte Carlo integration is effective for smooth integrands having no or only a few oscillations.

A Monte Carlo integration code for computing the VSCF potentials was written in the framework of the present study. However, so far it was used only in test calculations against the other methods. It is clear that for low-dimensional integrals the Monte Carlo method is not competitive. However, it is numerically feasible for calculations of the VSCF in several dimensions, it is quite general, and it is therefore the method of choice when the approaches of the previous subsections are inapplicable.

11.2.4 THE CONFIGURATION INTERACTION (CI) EXTENSION OF VSCF

CI-VSCF was the first extension proposed for VSCF [Bowman, Christoffel and Tobin (1979), Ratner, Buch and Gerber (1980), and Thompson and Truhlar (1980)]. Developed mostly by the group of J. Bowman, it is a very effective and accurate algorithm for the vibrational spectroscopy of polyatomic molecules. In the CI-VSCF method, the eigenfunctions of the vibrational Schrödinger equation (1) are expanded in a basis of VSCF states:

$$\psi_n(Q_1, \ldots, Q_N) = \sum_m C_m^n \prod_{j=1}^{N} \phi_j^{(m)}(Q_j), \tag{13}$$

where the $\phi_j^{(m)}(Q_j)$ are eigenstates of the VSCF Hamiltonian for mode Q_j, and state m as in equation (3). Note that different VSCF states are not orthogonal. The energies E_n are therefore solutions of the secular equation:

$$\det[H - ES] = 0, \tag{14}$$

where H is the full vibrational Hamiltonian and:

$$H_{m,m'} = \left\langle \prod_{j=1}^{N} \phi_j^{(m)}(Q_j) \middle| H \middle| \prod_{j'=1}^{N} \phi_{j'}^{(m')}(Q_{j'}) \right\rangle, \tag{15}$$

$$S_{m,m'} = \left\langle \prod_{j=1}^{N} \phi_j^{(m)}(Q_j) \middle| \prod_{j'=1}^{N} \phi_{j'}^{(m')}(Q_{j'}) \right\rangle. \tag{16}$$

It is possible to use, in the CI expansion (13), not proper VSCF states but *virtual* VSCF states, all obtained as solutions from the VSCF Hamiltonians for a fixed vibrational effective potential. The virtual VSCF states are orthogonal, and S becomes the unit matrix. When the basis used is sufficiently large, CI-VSCF becomes a rigorous algorithm for solving the vibrational eigenvalue problem for coupled-mode systems, with spectroscopic accuracy (assuming an exact potential, of course). In the present state of the art, given the limitations of available potentials, an accuracy of $\sim 1\,\mathrm{cm}^{-1}$ in solving the vibrational eigenvalue for a small polyatomic molecule is, in general, amply sufficient. Such very high accuracies with the CI-VSCF algorithm have been accomplished by Bowman and coworkers in several applications to small polyatomic systems [e.g. Romanowski, Bowman and Harding (1985), Carter, Culik and Bowman (1997), Carter and Bowman (1998), and Carter, Bowman and Harding (1997)]. Moreover, Bowman and coworkers were able to incorporate rotational coupling effects (including Coriolis coupling) in CI-VSCF treatments of the vibrational state calculations. With these generalizations [Carter and Bowman (1998), and Carter, Culik and Bowman (1998)], the method can be applied to the full Watson Hamiltonian [Watson (1968)] of the vibrational problem. This is essential, at least for small polyatomics, for quantitative comparisons with experimental data, but plays a lesser role in larger systems. Applications for systems such as HO_2, H_2O, H_2, CN, HCO and a nine-model of CO absorbed on Cu(100) [Carter and Bowman (1998), and Carter, Culik and Bowman (1997)] demonstrate the accuracy of the method. The disadvantages of CI-VSCF are first that for some systems and states the CI-VSCF expansion may converge slowly, making it computationally costly. Second, and more important, the computational effort in CI-VSCF is expected to scale as N^5 or worse, with the number of modes N (an estimate for the case of pairwise interactions only between the modes). The method is a very powerful tool for small or intermediate size systems, but its application for large systems will be difficult.

11.2.5 CORRELATION CORRECTIONS FOR VSCF BY PERTURBATION THEORY

This approach, developed by Jung and Gerber (1996) and Norris *et al.* (1996) is based on the fact that at least for low-lying vibrational excited states the VSCF itself is already of very good accuracy. This suggests that correlation effects for such states are relatively small, and can be treated by perturbation theory. The approach is analogous to the very familiar Møller–Plesset method for electronic structure calculations, where perturbation theory is used to introduce correlation effects and improve on the Hartree–Fock approximation [McWeeny (1992), Pople, Binkley and Seeger (1976), and Frisch, Head-Gordon and Pople (1990)]. In this approach [Jung and Gerber (1996)], the full Hamiltonian is written as:

$$H = H^{\mathrm{SCF},(n)} + \Delta V(Q_1, \ldots, Q_N), \tag{17}$$

where

$$H^{\mathrm{SCF},(n)} = \sum_j \overline{H}_j^{(n)}(Q_j), \tag{18}$$

and $\overline{H}_j^{(n)}$ is the SCF Hamiltonian for the mode Q_j, in the state (n)

$$\overline{H}_j^{(n)}(Q_j) \equiv -\frac{1}{2}\frac{\partial^2}{\partial Q_j^2} + \overline{V}_j^{(n)}(Q_j). \tag{19}$$

ΔV in equation (17) is

$$\Delta V(Q_1, \ldots, Q_N) = V(Q_1, \ldots, Q_N) - \sum_{j=1}^{N} \overline{V}_j^{(n)}(Q_j). \tag{20}$$

The correlation effects are all included in ΔV. Expecting the SCF part of the Hamiltonian to be dominant, and the correlation effects to be relatively small, the effect of ΔV is treated by perturbation theory. Using standard second-order perturbation theory, the energy expression to second order in ΔV is

$$E_n^{\mathrm{MP2}} = E_n^{\mathrm{VSCF}} + \sum_{m \neq n} \frac{\left| \left\langle \prod_{j=1}^{N} \psi_j^{(n)}(Q_j) \right| \Delta V \left| \prod_{j'=1}^{N} \psi_{j'}^{(m)}(Q_{j'}) \right\rangle \right|^2}{E_n^{(0)} - E_m^{(0)}} \tag{21}$$

We use the label MP2 in equation (21), since it has become common to use the title Møller–Plesset perturbation theory in any perturbation approach where the zeroth-order Hamiltonian employed is the SCF one. All the wave functions $\psi_j^{(m)}$ and the energy $E_m^{(0)}$ in equation (21) are calculated from the VSCF Hamiltonian $H^{\mathrm{SCF},(n)}$, corresponding to the state (n). E_n^{VSCF} is the SCF expectation energy, given by equation (5) and

$$E_m^{(0)} = \sum_{j=1}^{N} \varepsilon_j^{(n),m}. \tag{22}$$

The superscript (n) in equation (22) indicates that the levels are calculated from the Hamiltonian $\overline{H}_j^{(n)}(Q_j)$, while m is the label of the state computed. In equation (22), $\varepsilon_j^{(n),m}$ is thus the mth SCF energy level of jth mode, computed from the Hamiltonian $\overline{H}_j^{(n)}(Q_j)$. Higher-order corrections, to the energy (and the wave functions) can be computed with a reasonable effort, but tests indicate that these corrections are probably very small, and not necessary at the level of accuracy for which we aim here. The integrals in equation (21) for the correlation corrections become simplified when the potential can be treated by the polynomial representation approximation of Section 11.2.3 above (in which case the whole expression can be reduced to one-dimensional integrals). Likewise, equation (21) becomes much simpler when the pairwise mode–mode coupling representation is employed (in this case the energy correction involves at most two-dimensional integrals). The calculations in the case of $(H_2O)_n$, $n = 2$–5 [Jung and Gerber (1996)], included up to four possible excited states for each mode, in each computation for a correlation energy correction. It was verified that this number of states suffices for convergence of the perturbation series, equation (21). Monte Carlo calculations of the integrals were not pursued for $(H_2O)_n$, since for these systems the pairwise mode–mode coupling approximation is quite adequate. We estimate that with suitable adaptations Monte Carlo calculations should be feasible for the correlation corrections in the systems studied, though these would definitely be computationally much more demanding. Finally, we note that difficulties with the correlation corrections for excited states can arise in cases of near degeneracy $E_m^{(0)} \sim E_n^{(0)}$. This can be handled as follows. A small CI calculation is first carried out in the subspace of the nearly degenerate SCF states only [McWeeny (1992)]. This is essentially equivalent to applying degenerate perturbation theory to this subspace. Then, the split levels and the corresponding states could be coupled to other states with which they are not degenerate, by ordinary perturbation theory. Perturbation corrections to fourth order are quite feasible for systems of the order of 100 modes at least, but so far have proved unnecessary in realistic cases. Consider, for instance, water clusters $(H_2O)_n$ for $n = 2, 3$. Correlation corrections to VSCF are on the average, of the order of $25\,\mathrm{cm}^{-1}$, for the stretching modes, and less than $20\,\mathrm{cm}^{-1}$ for the bending modes. Our estimates suggest that the residual errors at the MP2 level are, on the average, less than $\sim 15\,\mathrm{cm}^{-1}$ for these systems and states. This perturbation theoretic approach, to which we refer as correlation-corrected VCSF (CC-VSCF) is computationally very efficient, at least provided that excited state degeneracies do not occur, and that second-order perturbation suffices. Such CC-VSCF calculations should be feasible for systems of many hundreds of coupled modes at least. An interesting and important point is that the correlation effects are generally smaller for larger systems, at least on the average. Consider the quantity:

$$R = \frac{1}{N}\left[\sum_{j=1}^{N} r(j)\right] \times 100, \tag{23}$$

where $r(j)$ is the ratio of the MP2 correction to the jth fundamental excitation. N is the number of modes. Figure 11.3, from Jung and Gerber, shows the dependence of R, in the case of $(H_2O)_n$, on cluster size. Clearly, the correlation errors decrease on the average with cluster size. This suggests that the Møller–Plesset perturbation series should converge better for larger systems. It is reasonable to assume that the MP2 level will suffice for most applications to large systems. Finally, we note that by the experience of the present authors, the vibrational Møller–Plesset series seems to converge better than the electronic one, although this is an impression hard to quantify. We believe that the absence of exchange and Pauli principle issues in the vibrational case are favorable for the convergence and certainly for efficiency of the method. In summary, while CI-VSCF may be the best algorithm available for small molecular systems, since it is essentially numerically exact, the perturbation-corrected VSCF seems to us the obvious method of choice for intermediate and large systems (more than 5–6 atoms). It can be applied for large systems (at least up to hundreds of coupled modes) in terms of computational efficiency, and by all indications, provide an adequate level of accuracy. Extension of the present CC-VSCF algorithms to include rotational coupling effects was not pursued yet, and is a very desirable goal.

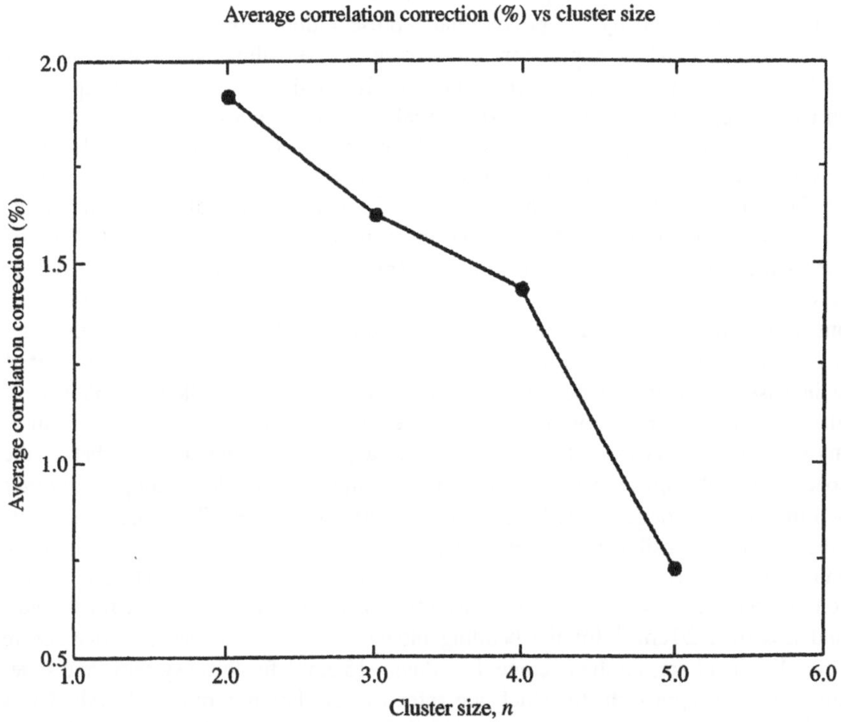

Figure 11.3. Average correlation-correction percentage versus cluster size n for $(H_2O)_n$. The 'correlation-correction percentage' for the vibrational frequencies is defined in the text.

11.2.6 COMBINED DIFFUSION QUANTUM MONTE CARLO–VIBRATIONAL SELF-CONSISTENT FIELD (DQMC–VSCF)

The DQMC method and several methods related to it are a very powerful computational tool for obtaining the vibrational ground state of polyatomic systems [e.g. Suhm and Watts (1991), Buch (1992), Barnett and Whaley (1993), and Bacic *et al.* (1992)]. In general, the method cannot be used for excited states. An important exception is when the node positions of the excited state can be estimated. Several authors have shown how, in specific cases, this can be done by various physical considerations: Coker and Watts (1986), Sandler, Buch and Sadlej (1996), Gregory and Clary (1995), Blume *et al.* (1997), and others, with elegant applications to spectroscopic calculations. However, in general, a practical procedure for computing the nodal positions of excitations in complicated systems is as yet unavailable. Broude, Jung and Gerber (1999) have proposed the use of VSCF to estimate the nodal position for a DQMC calculation. With this, DQMC can be applied, quite generally, for excited state calculations for which VSCF provides a reasonable first approximation. Another advantage of the combined DQMC–VSCF is that the VSCF calculation provides an assignment for the modes and the transitions. This could be a difficulty with DQMC, at least for complicated collective excitations. The method is very recent, and only applied in test calculations for Ar_3 and Ar_{13}, so far [Broude, Jung and Gerber (1999)]. Its expected accuracy is still hard to assess, though from the test case of Ar_3. DQMC–VSCF seems roughly equivalent in accuracy to CC-VSCF with MP2 corrections. However, computational considerations suggest the method may have an important advantage for very large systems. Both DQMC and VSCF are almost ideally computationally parallelizable, and so should the combined method be. That could very greatly enhance the computational performance of the method in large system applications. CC-VSCF (with MP2 corrections) is much faster than DQMC–VSCF for small systems, but is less efficiently parallelizable. We estimate that with high-performance state-of-the-art parallel computers, DQMC-VSCF may become the faster method for systems of hundreds of modes, or larger.

11.2.7 *AB INITIO* VIBRATIONAL SPECTROSCOPY: ELECTRONIC STRUCTURE CALCULATIONS COMBINED WITH VSCF

Hitherto, direct calculations of vibrational spectroscopy from a first-principles point of departure, i.e. starting from electronic structure calculations, were possible only in a few restricted cases. Many available electronic structure algorithms provide second derivatives of the potential energy surface at equilibrium. However, this only corresponds to harmonic frequencies. Efforts to compute anharmonic vibrational states from electronic structure input used fitting procedures to generate analytic potential surfaces from computed points on the potential energy surface [e.g. Skokov, Peterson and Bowman (1998), and Bigwood, Milam and Gruebele (1998)], or pursued calculations of higher-order derivatives of the potential energy surface to compute a polynomial force field [Maslen *et al.* (1992), Bludsky *et al.* (1995), and Persson, Taylor and Martin (1998)]. These approaches are confined to a few degrees of freedom only. Recently, Chaban, Jung and Gerber (submitted) proposed a method based on combining electronic calculations with VSCF, and with CC-VSCF (using perturbation theoretic correlation corrections to second order). The approximation used is the restriction of the potential energy surface to pairwise coupling only between normal modes, equation (9). Assuming that each pair of modes $\{Q_i, Q_j\}$ requires S grid points for the vibrational calculation, the total number of points on the potential surface that need to be computed from the electronic structure code is $SN(N-1)/2$, where N is the number of vibrational modes. The method was successfully applied to H_2O, $Cl^-(H_2O)$, $(H_2O)_2$ [Chaban, Jung and Gerber (submitted)]. It was implemented with Møller–Plesset electronic structure theory (at MP2 level) from the GAMESS electronic structure code [Schmidt *et al.* (1993)]. With a sufficient basis set, the results obtained were very good, but short of spectroscopic accuracy.

Electronic structure calculations cannot, in the present state of the art, give potential surfaces of spectroscopic accuracy for systems such as $(H_2O)_2$ and $Cl^-(H_2O)_2$–the results seem comparable with those of the better empirically fitted potentials for such systems. The results for $Cl^-(H_2O)$, $(H_2O)_2$, ar shown in Tables 11.1 and 11.2 respectively. Estimates suggest that the method should be applicable to systems of 10–15 atoms. The importance of the method lies, in our view, in the fact that it makes it possible to compare electronic structure calculations (both method and basis set) against accurate vibrational spectroscopy. This may become one of the most useful and demanding tests of electronic structure approximations.

11.2.8 VSCF AND THE FITTING OF POTENTIAL ENERGY SURFACES TO SPECTROSCOPIC DATA

Determination of potential energy surfaces from spectroscopic data is one of the main goals of methods for calculation of vibrational states. For diatomic molecules, the celebrated semi-classical RKR method provides an inversion of the rovibrational energy levels to yield the potential energy

Table 11.1. Fundamental excitation frequencies for $Cl^-(H_2O)$: *ab initio* vibrational spectroscopy versus experiment.

Excited mode number	*Ab initio* harmonic frequency (cm^{-1})	*Ab initio* VSCF frequency (cm^{-1})	*Ab initio* corrected VSCF frequency (cm^{-1})	Experimental frequency[a] (cm^{-1})
1	3919	3723	3735	3690
2	3470	3130	3151	3130
3	1693	1662	1633	
		3305	3286	3283 (bend overtone)
4	782	764	694	
5	329	392	325	
6	195	197	194	210

[a] From Chaban, Jung and Gerber (submitted).

Table 11.2. Fundamental excitation frequencies for $(H_2O)_2$: *ab initio* vibrational spectroscopy versus experiment (frequencies are given in cm^{-1}).

Mode	Description	Harmonic	VSCF	Corrected VSCF	Experimental[a]
1	Accept asym	4003	3763	3724	
2	Donor asym (free OH)	3958	3733	3745	3735
3	Accept sym	3869	3690	3647	
4	Donor sym (bonded OH)	3768	3560	3565	3601
5	Donor bend	1646	1612	1605	1619
6	Accept bend	1618	1565	1564	1601
7		676	769	743	
8		360	549	520	
9		212	450	409	
10	O–O stretch	189	260	161	150
11		167	416	570	
12		138	545	412	

[a] From Chaban, Jung and Gerber (submitted).

curve by an explicit transform [Schutte (1976)]. The method has been extensively applied, and is extremely useful. Gerber, Roth and Ratner (1981) showed that within semiclassical VSCF theory, an inversion transform can be given for obtaining polyatomic potential energy surfaces from rovibrational spectra. Mathematically, the result is an extension of the RKR procedure for diatomics. Realistically, the result requires, however, knowledge of the complete vibrational energy spectrum, which is rarely available. Indeed, the larger the molecule, the more limited is, in general, the part of the spectrum available. This severely restricts the applicability of the method. An application was given to the stretching-modes spectroscopy of CO_2, in which case a very accurate two-mode potential surface was obtained [Roth, Ratner and Gerber (1981)]. Later, this work was extended to provide an inversion of the full stretching–bending potential surface of CO_2 from the vibration–rotation spectra [Romanowski, Gerber and Ratner (1988), and Romanowski, Ratner and Gerber (1988)]. In fact, in this inversion, corrections beyond the semiclassical VSCF level were introduced in a second step. There is strong evidence that the CO_2 potential energy surface determined in this effort is highly accurate. We believe there is some future scope for such inversion methods, when a very extensive set of data is at hand. A direction that, in our view, deserves more practical attention is the fitting of potential surfaces to spectroscopic data, based on using parametrized potential functions, even if the data are very incomplete. As we shall see in the next section, there are already numerous applications of VSCF and CC-VSCF along these lines, but it should be desirable to optimize the fitting procedure and integrate it with VSCF.

11.3 Applications to large-molecule spectroscopy

The most interesting applications of VSCF and its extensions are for highly anharmonic systems, involving large amplitude motions and strong coupling between modes. The focus here will be mostly on hydrogen-bonded clusters and on 'floppy' molecules such as proteins that do fall in the above category.

11.3.1 SPECTROSCOPY AND ITS RELATION TO VIBRATIONAL DYNAMICS OF $(H_2O)_n$, $n = 2–8$

VSCF and CC-VSCF (including correlation effects between different modes to second order in perturbation theory) were applied for calculation of the full vibrational fundamental excitation spectra of water clusters $(H_2O)_n$ in the range 2–8 [Jung and Gerber (1996), and Jung and Gerber (to be published)]. The results provide insight into the highly anharmonic vibrational dynamics of water clusters, and on the reflection of the latter in the properties of the vibrational spectra. Most of the conclusions here are drawn from results obtained with the potential functions of Coker and Watts (1982) and of Reimers, Watts and Klein (1982), but qualitatively they seem to be valid also for other state-of-the-art potentials (Jung and Gerber (to be published)). The fundamental frequency spectrum of $(H_2O)_n$ consists of three bands: intramolecular stretching mode transitions, above 3000 cm^{-1}; intramolecular bending modes, around 1600 cm^{-1}, and intermolecular modes, below 1000 cm^{-1}. There are important anharmonic effects for all three groups of transitions, but for some of the intermolecular modes the effects are huge. Consider, for instance, the dimer $(H_2O)_2$. The 12 normal modes for this system are shown in Figure 11.4. An important property is the excitation frequency versus the mode number. This is shown in Figure 11.5, where the modes are enumerated in order of decreasing *harmonic* frequency. As Figure 11.5 shows, the true (VSCF and CC-VSCF) behavior is, for some of the soft modes, qualitatively different from the harmonic one. The anharmonic frequency in VSCF, or in VSCF-MP2, can differ for some of the modes by a factor of 3–4(!) from the corresponding harmonic value. The modes of the most extreme anharmonic behavior are shearing and torsional modes. Typically for these modes the anharmonic frequency is much larger than the harmonic one. The interpretation is obvious from Figure 11.3.

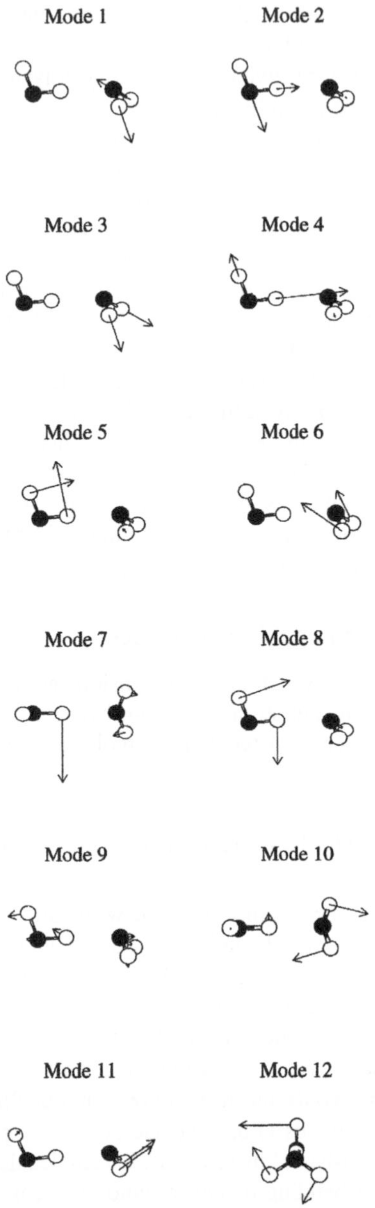

Figure 11.4. The normal mode displacements for $(H_2O)_2$.

The effective potential from these modes deviates far from the harmonic model, and is, for some region of the mode Q_j, roughly similar to a box or *square well* in shape. Very strong anharmonic effects are found also for larger clusters, but they decrease with cluster size. Figure 11.6 shows the excitation energy versus mode number for the fundamental spectrum of $(H_2O)_5$, and clearly the deviations from harmonic behavior are much less extreme. The anharmonic effects for the intramolecular stretching modes are not large as a percentage of the harmonic frequency, but their absolute magnitudes can be of the order of $300\,\mathrm{cm}^{-1}$ (!). Generally for the intramolecular

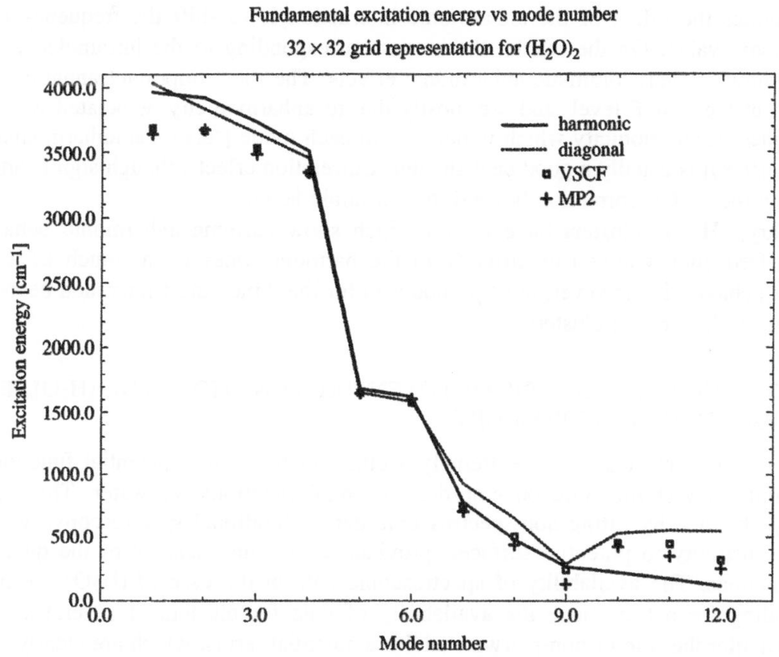

Figure 11.5. The excitation energy versus mode number for the fundamental transitions of $(H_2O)_2$. Results are shown for the harmonic, 'diagonal anharmonic', VSCF and CC-VSCF (MP2) approximations. From Jung and Gerber (1996).

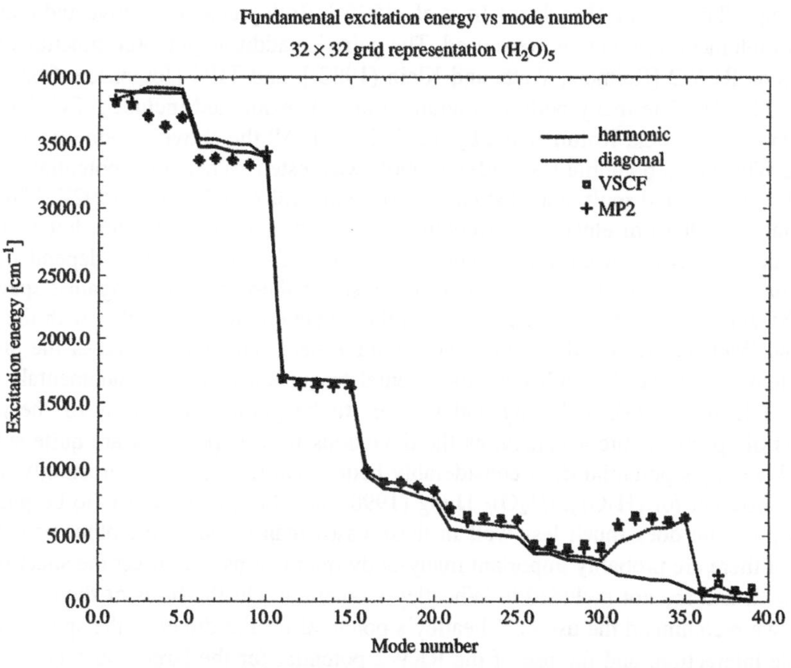

Figure 11.6. The fundamental excitation energies of $(H_2O)_5$ versus mode number. Results are shown for the harmonic, 'diagonal' anharmonic, VSCF and CC-VSCF (MP2) approximations. From Jung and Gerber (1996).

stretching modes the effect of the anharmonic interaction is to shift the frequency to the red of the harmonic value. On the whole, the modes corresponding to the intramolecular bending excitations show the least dramatic anharmonic effects. The most important anharmonic effects are included at the VSCF level, and are mostly due to anharmonicity associated with coupling between modes. Anharmonicity which is intrinsic to each mode ['diagonal anharmonicity', Jung and Gerber (1996)] is usually a good deal smaller. Correlation effects, though significant, are less dramatic than the VSCF correction beyond the harmonic level.

In summary, $(H_2O)_n$ clusters have modes which show extreme anharmonic behavior. The anharmonic frequency values can differ from the harmonic ones by as much as a factor of 3–4(!). This behavior is, however, most pronounced for the dimer and trimer, and becomes more moderate for the larger ring clusters.

11.3.2 TESTS AND COMPARISONS OF POTENTIAL FUNCTIONS FOR $(H_2O)_n$ BY VIBRATIONAL SPECTROSCOPY

Vibrational spectroscopy can be an extremely useful tool for testing potential functions. There is, for example, a vast literature on empirical potential functions for water. These potentials were mostly obtained by fitting non-spectroscopic data. Vibrational spectroscopy, with its well established sensitivity to potential surfaces, provides a rigorous measure of the quality of the potential functions. The availability of spectroscopic data in the case of $(H_2O)_n$ seems particularly appealing for a test, since the availability of data for clusters of several sizes should probe in particular the role of non-pairwise additive potential terms, which are clearly one of the important issues for condensed-phase systems in general, and for water in particular. Jung and Gerber (to be published) and Jung (1998) carried out tests of several state-of-the-art potential functions for $(H_2O)_n$, $n = 2–8$, against experiment. So far, these tests employed the stretching-mode data of Huisken, Kaloudis and Kulicke (1996), obtained in measurements of the cluster spectra in droplets of superfluid helium [Frochtenicht *et al.* (1996)]. Both pairwise additive and non-additive (polarizable) interaction potentials were used. The pairwise additive potential functions that were studied include RKW2 [Reimers, Watts and Klein (1982)]; and TIP4 [Jorgensen, Chandrasekhar and Madura (1983)). The many-body non-additive models examined include NCC [Niesar *et al.* (1990)] and CKL [Cieplek, Kollman and Lybrand (1990)]. All the above are empirical potentials. In addition, Wheatley's potential [Wheatley (1996)] was tested. Wheatley's potential is based on the method of distributed multipoles [Stone (1996), and Millot and Stone (1992)]. This interaction potential is built from electronic structure calculations for the monomer, but is also fitted to experimental properties of the dimer. The above potentials do not carry a dependence on the intramolecular vibrations of the monomer. In the tests of these potentials against spectroscopy [Jung (1998), Jung and Gerber (to be published)] they were employed together with the potential of Coker and Watts (1982) for the intramolecular force field. The comparison of the predictions of these potentials for $(H_2O)_2$ with the experimental frequencies for the fundamentals is shown in Table 11.3 [Jung (1998), and Jung and Gerber (to be published)]. None of the potentials gives results of spectroscopic accuracy, as the deviations from experiment are quite substantial. However, Wheatley's potential does considerably better than the others. Interestingly, the situation is very different for $(H_2O)_3$, $(H_2O)_5$ [Jung (1998), and Jung and Gerber (to be published)]; Wheatley's potential does much less well in these cases than some of the other potentials. We conclude that there are probably important many-body interactions that affect the spectroscopy of $(H_2O)_n$, $n > 2$, and are not included in Wheatley's potential. On the basis of the tests that were carried out, we recommend the use of Wheatley's potential for the dimer as the spectroscopically most reliable interaction, and the use of the RKW2 potential for the larger clusters.

It is of interest to compare the predictions of the different potentials for the spectroscopy with each other. Jung (1998) and Jung and Gerber (to be published) carried out VSCF calculations of

Table 11.3. Fundamental excitation energies of water dimer: comparison of predictions of several empirical potentials with experiment (all in cm^{-1}).

Mode	RWK2	TIP4	Wheatley	NCC	CKL	Experiment[a]
1	3683.72	3716.14	3704.61	3865.63	3994.64	3714
2	3663.55	3710.72	3697.57	3889.42	4002.75	3698
3	3528.81	3546.72	3555.49	3627.49	3582.08	3626
4	3337.89	3350.05	3522.74	3422.80	3365.70	3548
5	1652.98	1680.00	1662.53	1643.72	1645.90	1618
6	1578.47	1593.72	1609.98	1602.99	1548.15	1600
7	750.40	802.68	681.99	802.90	893.17	
8	506.53	589.73	557.86	564.61	693.09	
9	239.61	306.87	276.62	187.01	348.03	150
10	452.08	447.14	456.05	508.16	541.53	
11	445.57	500.57	418.95	526.87	443.85	
12	317.45	463.36	445.46	517.59	613.65	

[a] From Jung (1998). The RWK2, TIP4, Wheatley, NCC and CKL potentials are briefly discussed in the text.

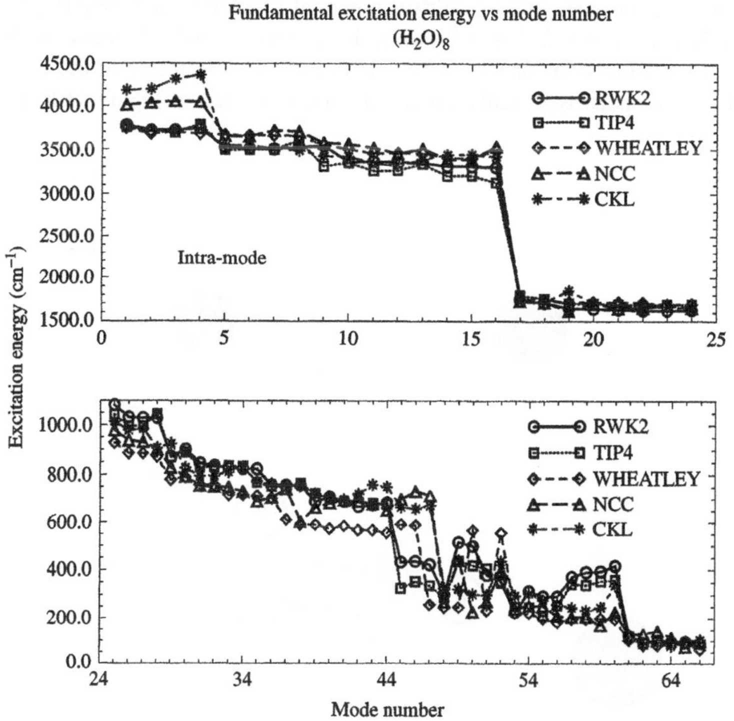

Figure 11.7. The fundamental excitation energies of $(H_2O)_8$ versus mode number. Results for five different empirical potentials are compared.

the complete fundamental spectra of each of the clusters $(H_2O)_n$, $n = 2–8$, for the five potential functions discussed above. It is interesting to note that all these different potentials predict the same equilibrium structures for $(H_2O)_n$, $n = 2–8$, in nearly all cases. This is *not* the case for the vibrational spectra. Figure 11.7 gives the excitation frequencies versus mode number for $(H_2O)_8$, for the RKW2, TIP4, Wheatley, NCC and CKL potentials. Very large differences are seen. The differences are substantial also for the stiff stretching, but are large above all for major part of the soft intermolecular mode spectra. For some of the transitions (especially shearing and torsional modes) corresponding frequencies for different potentials can differ by as much as a factor of 3! Obvious, *at most* one of these potentials can be of acceptable accuracy, since the mutual differences in the predicted spectra are so large. This also shows that vibrational spectra is a far more demanding property to predict than equilibrium structures.

11.3.3 FITTING OF IMPROVED POTENTIALS FOR $I^-(H_2O)_n$, $CL^-(H_2O)_n$, $I^-(H_2O)_nAr_m$

VSCF calculations were used in several cases for fitting potential functions to spectroscopic data. A most recent example is for ion–water complexes. Important experimental progress has led to extensive data on the spectra in the O–H stretching mode frequency range for a number of negative ions in complexes with water, such as $Cl^-(H_2O)_n$ and $I^-(H_2O)_n$ [Choi *et al.* (1998), Ayotte *et al.* (1998), and Ayotte and Johnson (1997)]. A remaining problem for analysis of the spectra is that the ion clusters are formed relatively hot (typically a temperature in the range of 80–150 K is estimated). This can be overcome by formation of mixed ion–water–argon clusters, the temperature of which is low [Ayotte *et al.* (1999)]. Using the spectroscopic data, Jung, Gerber, Ayotte *et al.* (to be published) fitted empirical potential functions for $I^-(H_2O)_n$, $I^-(H_2O)_n\ Ar_m$, with $n = 1, 2, 3$, and $m = 1, 2, 3, 4$. Likewise, Jung, Gerber and Okamura (to be published) fitted a new potential function for $Cl^-(H_2O)_n$, $n = 1$–4. The structure of the cluster $I^-(H_2O)_2\ Ar_3$ predicted by the spectroscopically fitted potential is shown in Figure 11.8. In these cases,

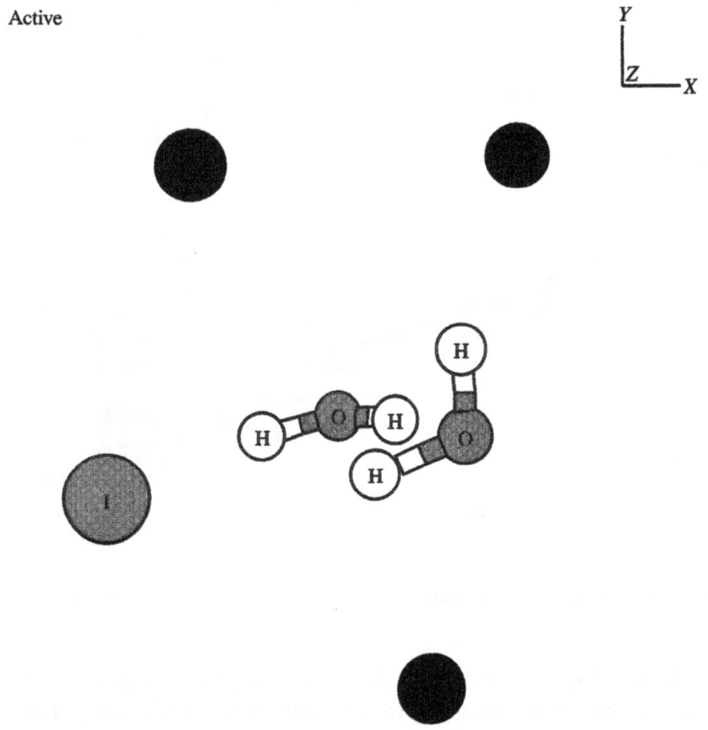

Figure 11.8. Minimum energy structure of $I^-(H_2O)_2\ Ar_3$, predicted from spectroscopically fitted potentials [From Jung (1998)].

the departure point was available empirical potential functions that included also nonadditive polarization effects [Perera and Berkowitz (1993), Dang and Garrett (1993), Stuart and Berne (1996), and Dang *et al.* (1991)]. These potentials, in their exact original form, gave frequencies that did not compare well with experiment. Trial and error fitting by Jung (1998), by Jung, Gerber, Ayotte *et al.* (to be published) and by Jung, Gerber and Okamura (to be published) led to much better agreement. In addition to some changes in the partial charges, the main modification was in using improved additive atom–atom van der Waals (dispersion) potentials, that appear in the empirical force field mentioned. The pairwise atom–atom van der Waals forces in the existing potential functions were modeled by Lennard-Jones functions which appear not be sufficiently realistic. Good *ab initio* potentials for $Cl^-(H_2O)_n$ were computed by Xantheas (1996) and by others. The relative spectroscopic merits (at the rigorous anharmonic level) of the empirical and *ab initio* potential remain to be assessed. A tentative impression is that the state-of-the-art *ab initio* calculations are more accurate than empirical (but *not* spectroscopic) fitted potentials. We stress that fitting efforts so far were limited, and by using much more flexible parameterized potential functions, enormous progress can be made, admittedly at the expense of great effort, now that excellent spectroscopic data is becoming available.

11.3.4 IMPROVED POTENTIAL FOR GLUCOSE BY FITTING OF VIBRATIONAL SPECTRA

Gregurick *et al.* (1999) obtained a new potential field for glucose, by fitting spectroscopic data, using the VSCF method. The approach used an AMBER type force field from the MOIL package of Elber *et al.* (1995). Parameters in this force field, atomic partial charges and several torsional potential coefficients, were varied to fit the experimental vibrational spectra for α-D-glucose. The AMBER force field with the previous parameters is in poor agreement with the measured vibrational frequencies. For the new potential, the disagreement between the calculated and the experimental levels is $3.3\,cm^{-1}$ for α-glucose and $5.1\,cm^{-1}$ for β-glucose. The determined potential function, combined with VSCF calculations supports suggestions that COH bending motion is strongly coupled to methylene and methane modes. The spectroscopically derived force field was tested also for thermodynamic properties, and gave very satisfactory results. In particular, it proved successful in yielding good results for the anomeric effect of glucose in water, a notoriously difficult property to predict. In summary, a spectroscopically derived potential seems to describe very well the spectroscopic, as well as the structural and thermodynamic properties of glucose. The data for the fitting came from spectroscopy of crystalline glucose, but in the VSCF calculations, intermolecular forces were neglected. When high resolution data for gas-phase glucose becomes available, this may be used to fit an even more accurate potential. The absence of a good source for glucose–water interaction at the present stage is a severely limiting factor for predicting the solution behavior of glucose. Such data may become available soon from jet experiments. In any case, this example presents the usefulness of VSCF in fitting large-molecule spectroscopic data, and it shows the merits of spectroscopically derived potentials also for other properties.

11.3.5 PEPTIDES AND PEPTIDE–WATER COMPLEXES: SPECTROSCOPY AND STRUCTURE

Several peptides are known to crystallize and for the crystalline phase well-resolved IR and Raman spectroscopy can often be obtained. For example trialanine (Ala_3) is known to crystallize in both parallel (P) and antiparallel (AP) β sheet configurations. The structure, IR and Raman spectra for Ala_3 for both configurations is given by Qian, Bandekar and Krimm (1991), who also analyzed the frequency spectra at the harmonic approximation level, using isolated-molecule normal modes. Gregurick *et al.* (1997) studied the spectrum of Ala_3 using the VSCF method, and the AMBER-type force field from the MOIL package [Elber *et al.* (1994)]. Crystal packing effects and interactions were neglected. For many of the 'stiff' vibrations good agreement was found between the frequencies computed from the anharmonic force fields and the experimental ones. However, the remaining differences between the calculated and experimental frequencies call for a substantial revision of the force field. Important discrepancies with experiments were noted in the amide III mode region (1356–1200 cm^{-1}), which corresponds to an NH in-plane bend plus a CN stretch; for the skeletal modes (1200–850 cm^{-1}), in which most NC, CC and CH stretch-bond frequencies occur, for the amide V modes (725–675 cm^{-1}) which are characterized by the out-of-plane NH bend plus a CN torsion and, most severely, for the low frequency modes (below 600 cm^{-1}). Frequencies of highly anharmonic modes, such as methyl group torsions, appear in this region. The differences are a sufficient basis for carrying out a fitting of a much improved potential, but this was not yet attempted.

Peptide–water interactions are of great importance, but suitable spectroscopic data for extracting potentials is not yet available. There are good prospects that such data will be obtained soon from jet experiments. In anticipation of such experiments, Gregurick *et al.* (1997) carried out VSCF calculations on the spectra of a cluster of a blocked di-L-serine molecule with H_2O. It was found that the water molecule induces shifts of up to 50 cm^{-1} in the frequencies of the transitions associated with the peptide modes. The presence of water gives rise, of course, also to the additional transitions. The sensitivity of the transitions to the site of the H_2O is such, that an experiment should be able to determine the conformer structure from the measured frequencies.

11.3.6 PREDICTION OF PROTEIN VIBRATIONAL SPECTRA

Calculations of protein vibrational spectra at this stage have a predictive character, and their significance lies in the fact that they may help greatly in guiding future experiments. Preliminary calculations of this type, using the AMBER-type force field from the MOIL code, were carried out for the protein BPTI [Roitberg *et al.* (1995), and Roitberg, Gerber and Ratner (1997)]. In these studies, a classical molecular dynamics simulation of the protein in water was carried out as a first step, to determine how many water molecules interact strongly with the protein. Using an energetic criterion, a hydration layer of 196 molecules that interact strongly with the BPTI was identified. Then, the minimum energy structure of BPTI plus the hydration water was determined, normal modes of the whole system with respect to the minimum structure were computed, and finally a VSCF calculation of the vibrational spectrum was carried out. In the VSCF calculation, the anharmonic potential function was approximated by a quartic force field in the normal modes. The system has over 3500 modes, many mutually coupled quite strongly. Even at the ground state, the protein with its hydration water layer, is a strongly anharmonic system. Figure 11.9 shows the frequencies of the lowest 600 modes [Roitberg, Gerber and Ratner (1997)], calculated at the harmonic level, with 'diagonal anharmonicity' only, and at the VSCF level. The modes were numbered in this case in increasing order with harmonic frequency.Clearly, the anharmonic effects are very large, and the anharmonic coupling between the modes makes an essential contribution, 'diagonal anharmonicity' effects being insufficient. One important conclusion from these studies is that, if proteins can be cooled to their lowest-energy structure, the resulting vibrational spectrum will be well defined and assignable. While the low frequencies give a congested spectrum, at higher frequencies fairly well-resolved transitions are expected [Roitberg *et al.* (1995)]. On the basis of these results, it should be possible to assign and analyze protein vibrational spectra when experimental data become available, and to attempt fitting of improved force fields. The flexibility of protein spectroscopy and force field determination suggested by the calculations is exciting, but there are many assumptions involved. In particular, it is not clear whether the protein can be cooled to a single, global minimum structure. If that is not the case and local minimum structures will appear at the experimental conditions, their effects on the spectrum may be large and remain to be studied.

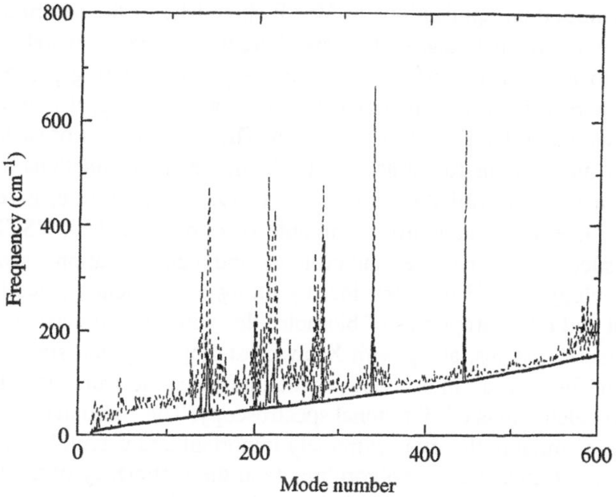

Figure 11.9. The 600 lowest fundamental frequencies of the hydrated BPTI protein, versus mode number. Solid dark line, harmonic approximation; thin line, 'diagonal anharmonicity'; broken line, VSCF.

Despite this and other complications, the promise of protein vibrational spectroscopy is great, as the limitations of the available force fields are quite severe, and they must be held in doubt. Proteins and their force fields are clearly a frontier challenge for large-molecule spectroscopy.

11.4 Conclusions

The VSCF method and its extensions were described here mostly as a tool for interpreting large-molecule spectroscopy. At present, it may well be the only tool of broad applicability for this purpose. The algorithms and version of the method available already have a wide range of applications to realistic systems. At the same time, it is clear that the field for methodological extension and improvements is wide open. There is a strong motivation for pursuing computationally more efficient and simpler methods for introducing correlation effects between modes, beyond the presently available arsenal of CI-VSCF; perturbation-theoretic corrections for VSCF; and the combined DQMC–VSCF approach. VSCF with some measure of correlation effects included, seems to be the level of treatment sufficient for useful analysis of large-molecule spectroscopy in the present state-of-the-art of potential fitting. At such a level of treatment, any (corrected) VSCF method errors are negligible compared with the errors of the best state-of-the-art potentials for such systems. This seems to be the case also for potentials from good *ab initio* calculations. Methods such as CC-VSCF can be used to test the accuracy of *ab initio* potential surfaces against experimental spectroscopy since, based on the evidence at hand, errors of the vibrational energy level are considerably smaller than errors of the electronic structure methodology for the potential surface.

So far, VSCF was employed mostly for fundamental transitions. Applications to more highly excited states are desirable. This was already achieved for small molecules, but is a very realistic and important goal also for large ones. This is important for making contact with overtone and combination-mode spectroscopy, with vibration–rotation tunneling spectroscopy, and also with Raman spectroscopies. The information from such measurements covers a much larger range of configurations than do fundamental transitions. The applications should be straightforward, though for more highly excited states larger correlation corrections to VSCF are expected.

One of the most important applications to VSCF should be to the determination of potential functions by fitting experimental data. As we saw, there were already several useful studies along this line. In order to increase the effectiveness of the potential fitting procedure, to allow for more flexible and more general parametrized functions in the fitting, it is desirable to integrate the VSCF method with an effective fitting algorithm. The development of such an approach, and the pursuit of many applications to extraction of reliable potential functions for large molecules is bound to prove a most fruitful direction. With a view to the future, given these advances of method and the increased availability of suitable spectroscopic data, VSCF methods can be applied to systems such as large peptides and proteins, and make vibrational spectroscopy a major tool of structural biology. VSCF may provide, by fitting experiments, potentials so accurate as to predict structural and other properties of biomolecules with great precision. It may become an important tool for these systems, along with X-ray crystallography and structural NMR.

Another important future goal, in our view, is the integrated electronic structure–VSCF calculations, i.e. *ab initio* calculations of vibrational spectroscopy. The computation of potential energy surfaces by electronic structure theory is extremely important and useful, but it is essential to test the accuracy of the electronic structure algorithms (and the sufficiency of the basis sets used) as directly as possible against demanding experiment. Vibrational spectroscopy is clearly among the most useful tests, and VSCF methods provide the means for comparison. This is another major challenge for the method in the years to come.

Acknowledgements

RBG thanks Professor M. A. Ratner for many helpful discussions, throughout a long-standing cooperation on this topic. We thank Drs. Galina Chaban, Jeremy Harvey, Nicholas Wright, Pavel Jungwirth and Susan Gregurick for very useful comments. We thank Prof. J. Bowman for important remarks on VSCF, and for providing us with his CI-VSCF code. We are grateful to Professor Martin Gruebele and to Professor Nancy Makri for their helpful comments.

The work at UC Irvine was supported by a grant from the Petroleum Research Fund, administered by the American Chemical Society (to RBG). Work at The Hebrew University was supported by the Israel National Science Foundation (to RBG), and by the US–Israel Binational Science Foundation (to RBG).

11.5 References

Anderson, J. B., 1975, *J. Chem. Phys.* **63**, 1499–1503.
Anderson, J. B., 1976, *J. Chem. Phys.* **65**, 4121–4127.
Ayotte, P., Barley, C. G., Weddle, G. H., and Johnson, M. A., 1998, *J. Phys. Chem. A*, **102**, 3067–3071.
Ayotte, P., and Johnson, M. A., 1997, *J. Chem. Phys.* **106**, 811–814.
Ayotte, P., Weddle, G. H., Kim, J., Kelley, J., and Johnson, M. A., 1999, *J. Chem. Phys. A*, **103**, 443–447.
Bačíc, Z., Gerber, R. B., and Ratner, M. A., 1986, *J. Phys. Chem.* **90**, 3606–3612.
Bačić, Z., Kennedy-Mandziuk, M., Moskowitz, J. W., and Schmidt, K. E., 1992, *J. Chem. Phys.* **97**, 6472–6480.
Bačić, Z., and Light, J. C., 1986, *J. Chem. Phys.* **85**, 4594–4604.
Bačić, Z., and Light, J. C., 1989, *Annu. Rev. Phys. Chem.* **40**, 469–498.
Bačić, Z., and Miller, R. E., 1996, *J. Phys. Chem.* **100**, 12 945–12 959.
Barnett, R. N., and Whaley, K. B., 1993, *J. Chem. Phys.* **99**, 9730–9744.
Bigwood, R., Milam, B., and Gruebele, M., 1998, *Chem. Phys. Lett.* **287**, 333–341.
Bludský, O., Bak, K. L., Jørgensen, P., and Špirko, V., 1995, *J. Chem. Phys.* **103**, 10 170–10 115.
Blume, D., Lewerenz, M., Huisken, F., and Kaloudis, M., 1996, *J. Chem. Phys.* **105**, 8666–8683.
Blume, D., Lewerenz, M., Niyaz, P., and Whaley, K. B., 1997, *Phys. Rev. E*, **55**, 3664–3675.
Bowman, J. M., 1978, *J. Chem. Phys.* **68**, 608–610.
Bowman, J. M., Christoffel, K., and Tobin, F., 1979, *J. Phys. Chem.* **83**, 905–912.
Bowman, J. M., 1986, *Acc. Chem. Res.* **19**, 202–208.
Bromley, M. J., Tramp, J. W., Carrington, T., and Corey, G. C., 1994, *J. Chem. Phys.* **100**, 6175–6194.
Broude, S., and Gerber, R. B., 1996, *Chem. Phys. Lett.* **258**, 416–429.

Broude, S., Jung, J. O., and Gerber, R. B., 1999, *Chem. Phys. Lett.* **258**, 437–442.
Buch, V., 1992, *J. Chem. Phys.* **97**, 726–729.
Carney, G. D., Sprandel, L. I., and Kern, C. W., 1978, *Adv. Chem. Phys.* **37**, 305.
Carter, S., Bowman, J. M., and Harding, L. B., 1997, *Spectrochim. Acta A*, **53**, 1179–1188.
Carter, S., and Bowman, J. M., 1998, *J. Chem. Phys.* **108**, 4397–4404.
Carter, S., Culik, S., and Bowman, J. M., 1997, *J. Chem. Phys.* **107**, 10 458–10 469.
Chaban, G. M., Jung, J. O., and Gerber, R. B., *J. Chem. Phys.*, submitted.
Cheng, E., and Whaley, K. B., 1996, *J. Chem. Phys.* **104**, 3155–3175.
Choi, J. H., Kuwata, K. T., Cao, Y. B., and Okumura, M., 1998, *J. Phys. Chem. A*, **102**, 503–507.
Cieplak, P., Kollmann, P., and Lybrand, T., 1990, *J. Chem. Phys.* **92**, 6755–6760.
Cohen, R., and Saykally, R. J., 1991, *Annu. Rev. Phys. Chem.* **42**, 369–392.
Coker, D. F., and Watts, R. O., 1986, *Mol. Phys.* **58**, 1113–1123.
Dang, L. X., and Garrett, B. C., 1993, *J. Chem. Phys.* **99**, 2972–2977.
Dang, L. X., Rice, J. E., Caldwell, J., and Kollman, P. A., 1991, *J. Am. Chem. Soc.* **113**, 2481–2486.
Elber, R., Roitberg, A., Simmerling, C., Goldstein, R., Li, H., Verkhivker, G., Keasar, C., Zhang, J., and Ulitsky, A., 1994, *Comput. Phys. Commun.* **91**, 159–189.
Frisch, M. J., Head-Gordon, M., and Pople, J. A., 1990, *Chem. Phys. Lett.* **166**, 275–280.
Frochtenicht, R., Kaloudis, M., Koch, M., and Huisken, F., 1996, *J. Chem. Phys.* **105**, 6128–6140.
Gerber, R. B., and Ratner, M. A., 1979, *Chem. Phys. Lett.* **68**, 195–198.
Gerber, R. B., and Ratner, M. A., 1988, *Adv. Chem. Phys.* **70**, 97–132.
Gerber, R. B., Roth, R. M., and Ratner, M. A., 1981, *Mol. Phys.* **44**, 1335–1353.
Gregory, J. K., and Clary, D. C., 1995, *J. Chem. Phys.* **103**, 8924–8930.
Gregurick, S. K., Fredj, E., Elber, R., and Gerber, R. B., 1997, *J. Phys. Chem. B*, **101**, 8595–8606.
Gregurick, S. K., Liu, J. H. -Y., Brant, D. A., and Gerber, R. B., 1999, *J. Phys. Chem. B*, **103**, 3476–3488.
Halberstadt, N., Serna, S., Roncero, O., and Janda, K. C., 1992, *J. Chem. Phys.* **97**, 341–354.
Hartmann, M., Miller, R. E., Toennies, J. P., and Vilesov, A., 1995, *Phys. Rev. Lett.* **75**, 1566–1569.
Hartmann, M., Miller, R. E., Toennies, J. P., and Vilesov, A., 1996, *Science* **272**, 1631–1634.
Henderson, J. P., Tennyson, J., and Sutcliffe, B. T., 1993, *J. Chem. Phys.* **98**, 7191–7203.
Horn, T. R., Gerber, R. B., and Ratner, M. A., 1993, *J. Phys. Chem.* **97**, 3151–3158.
Horn, T. R., Gerber, R. B., Valentini, J. J., and Ratner, M. A., 1991, *J. Chem. Phys.* **94**, 6728–6736.
Huisken, F., Kaloudis, M., and Kulicke, A., 1996, *J. Phys. Chem.* **106**, 17–28.
Hutson, J. M., 1990, *Annu. Rev. Phys. Chem.* **41**, 123–154.
Jorgensen, W. L., Chandrasekhar, J., and Madura, J. D., 1983, *J. Chem. Phys.* **79**, 926–935.
Jorgensen, W. L., and Severance, D. L., 1993, *J. Chem. Phys.* **99**, 4223–4235.
Jung, J. O., 1998, Ph.D. Thesis, University of California, Irvine, CA.
Jung, J. O., and Gerber, R. B., 1996, *J. Chem. Phys.* **105**, 10 332–10 347.
Jung, J. O., and Gerber, R. B. (to be published).
Jung, J. O., Gerber, R. B., Ayotte, P., and Johnson, M. A. (to be published).
Jung, J. O., Gerber, R. B., and Okamura, M., (to be published).
Jung, J. O., Harvey, J. N., Chaban, G. M., and Gerber, R. B. (unpublished).
Koonin, S. E., 1986, *Computational Physics*; Addison-Wesley; California, Chapter 8.
Lefebvre, R., 1983, *Int. J. Quant. Chem.* **23**, 543–547.
Light, J. C., Hamilton, I. P., and Lill, J. V., 1985, *J. Chem. Phys.* **82**, 1400–1409.
Liu, K., Brown, M. G., Cruzan, J. D., and Saykally, R. J., 1996, *Science*, **271**, 62–64.
Liu, K., Cruzan, J. D., and Saykally, R. J., 1996, *Science*, **271**, 929–933.
Liu, K., Brown, M. G., and Saykally, R. J., 1997, *J. Phys. Chem. A*, **101**, 8895–9010.
Mandelshtam, V. A., and Taylor, H. S., 1997, *Faraday Trans. Chem. Soc.* **93**, 847–860.
Maslen, P. E., Handy, N. C., Amos, R. D., and Jayatilaka, P., 1992, *J. Chem. Phys.* **97**, 4233–4254.
McMahon, M. A., Barnett, R. N., and Whaley, K. B., 1993, *J. Chem. Phys.* **99**, 8816–8829.
McWeeny, R., 1992, *Methods of Molecular Quantum Mechanics*, 2nd edition; Academic Press: San Diego, CA.
Miller, R. E., 1990, *Acc. Chem. Res.* **23**, 10–26.
Millot, C., and Stone, A. J., 1992, *Mol. Phys.* **77**, 439–462.
Moiseyev, N., 1983, *Chem. Phys. Lett.* **98**, 233–238.
Nesbitt, D. J., 1994, *Annu. Rev. Phys. Chem.* **45**, 367–399.
Nesbitt, D. J., and Field, R. W., 1996, *J. Phys. Chem.* **100**, 12 735–12 756.
Niesar, U., Corongiu, G., Clementi, E., Kneller, G. R., and Bhattacharya, 1990, *J. Phys. Chem.* **94**, 7949–7956.
Norris, L. S., Ratner, M. A., Roitberg, A. E., and Gerber, R. B., 1996, *J. Chem. Phys.* **106**, 11 261–11 267.
Perera, L., and Berkowitz, M. L., 1993, *J. Chem. Phys.* **99**, 4222–4224.
Persson, B. J., Taylor, P. R., and Martin, J. M. L., 1998, *J. Phys. Chem. A*, **102**, 2483–2492.
Pople, J. A., Binkley, J. S., and Seeger, R., 1976, *Int. J. Quant. Chem.* **10**, 1.
Qian, W., Bandekar, J., and Krimm, S., 1991, *Biopolymers*, **31**, 193–210.
Ratner, M. A., Buch, V., and Gerber, R. B., 1980, *Chem. Phys.* **53**, 345–356.
Ratner, M. A., and Gerber, R. B., 1986, *J. Phys. Chem.* **90**, 20–29.

Reimers, J. R., Watts, R. O., and Klein, M. J., 1982, *Chem. Phys.* **64**, 95–114.
Roitberg, A., Gerber, R. B., Elber, R., and Ratner, M. A., 1995, *Science*, **268**, 1319–1322.
Roitberg, A., Gerber, R. B., and Ratner, M. A., 1997, *J. Phys. Chem.* **101**, 1700–1706.
Romanowski, H., Bowman, J. M., and Harding, L. B., 1985, *J. Chem. Phys.* **82**, 4155–4169.
Romanowski, H., Gerber, R. B., and Ratner, M. A., 1988, *J. Chem. Phys.* **88**, 6757–6767.
Romanowski, H., Ratner, M. A., and Gerber, R. B., 1988, *Comput. Phys. Commun.* **51**, 161–171.
Roth, R. M., Gerber, R. B., and Ratner, M. A., 1983, *J. Phys. Chem.* **87**, 2376–2382.
Roth, R. M., Ratner, M. A., and Gerber, R. B., 1984, *Phys. Rev. Lett.* **52**, 1288–1291.
Sandler, P., Buch, V., and Sadlej, J., 1996, *J. Chem. Phys.* **105**, 10 387–10 397.
Saykally, R. J., 1989, *Accts. Chem. Res.* **22**, 295–300.
Schmidt, M. N., Baldridge, K. K., Boatr, J. A., Elbert, S. T., Gordon, M. S., Jensen, J. H., Matsunaga, N., Nguyet, K. A., Sa, S., Windus, T. L., Dupuis, M., and Montgomery, J. A., 1993, *J. Comput. Chem.* **14**, 1347–1363.
Schutte, C. H. J., 1976, *Theory of Molecular Spectroscopy*; North-Holland: Amsterdam.
Skokov, S., Peterson, K. A., and Bowman, J. M., 1998, *J. Chem. Phys.* **109**, 2662–2671.
Stone, A. J., 1996, *The Theory of Intermolecular Forces*; Clarendon Press: Oxford.
Stuart, S. J., and Berne, B. J., 1996, *J. Phys. Chem.* **100**, 11 934–11 943.
Suhm, M. A., and Watts, R. O., 1991, *Phys. Rep.* **204**, 293–329.
Thompson, T. C., and Truhlar, D. G., 1980, *Chem. Phys. Lett.* **75**, 87–91.
Thompson, T. C., and Truhlar, D. G., 1982, *J. Chem. Phys.* **76**, 1790–1798.
Vegiri, A., Alexander, M. H., Gregurick, S., McCoy, A. B., and Gerber, R. B., 1994, *J. Chem. Phys.* **101**, 2577–2591.
Watson, J. K. G., 1968, *Mol. Phys.* **15**, 479–490.
Wheatley, R. J., 1996, *Mol. Phys.* **87**, 1083–1116.
Wilson, E. B. Jr., Decius, J. C., and Cross, P. C., 1955, Molecular Vibrations; McGraw-Hill: New York.
Wright, N. J., and Hutson, J. M., 1999, *J. Chem. Phys.* **110**, 902–911.
Xantheas, S. S., 1996, *J. Phys. Chem.* **100**, 9703–9713.
Yang, W., and Peet, A. C., 1990, *J. Chem. Phys.* **92**, 522–526.
Yang, X., Kerstel, E. R. Th., Scoles, G., Bemish, R. J., and Miller, R. E., 1995, *J. Chem. Phys.* **103**, 8828–8838.
Zuniga, J., Bastida, A., Requena, A., Halberstadt, N., and Beswick, J. A., 1993, *J. Chem. Phys.* **98**, 1007–1017.

12 THE CALCULATION OF ROTATION–VIBRATION ENERGIES FOR MOLECULES WITH LARGE AMPLITUDE VIBRATIONS

Jan Makarewicz

A. Mickiewicz University Poznań, Poland

Computational Molecular Spectroscopy. Edited by Per Jensen and Philip R. Bunker

12.1 Introduction

Spectroscopy provides information on the structure, potential energy surface and internal dynamics of molecules. The most accurate information is extracted mainly from high resolution rovibrational spectra. Such spectra require a detailed assignment and analysis based on theoretical models describing molecular motions from the point of view of quantum mechanics. However, the exact quantum mechanical treatment of molecular motions is practically impossible because of the complexity of the multi-dimensional Schrödinger equation for molecular quantum states. To solve this equation, approximate methods have been developed which simplify the problem and offer approximate models of molecular motions.

Most important for the description of molecular motions is the approximate *separation* method, first proposed by Born and Heisenberg (1924), even before quantum wave mechanics emerged. They proposed a hierarchic multi-step separation of the electronic, vibrational, rotational and translational molecular motions, using perturbation theory. In order to separate the electronic and the nuclear motions, they introduced the concept of the molecular frame, i.e., the molecular axis system attached to the nuclei of the molecule. The molecular frame was necessary to describe the electronic motion relative to the moving nuclei, independently of the orientation of the nuclear frame in the space.

The hierarchic scheme of Born and Heisenberg was adapted for calculation of the quantum molecular states by Born and Oppenheimer (1927). In their method it was possible to define the coupled harmonic oscillator model obtained by assuming that the electronic energy can be represented by a quadratic function of the nuclear displacements. In the Born–Oppenheimer theory, also the rigid rotor model appeared as a consequence of the hierarchic separation scheme. The main idea of this theory was a reduction of the Schrödinger equation depending on the multi-dimensional configuration space of the molecule, to equations of lower dimensions describing dynamical subsystems in the molecule.

Independently, Slater (1927) proposed a similar idea of the hierarchic separation of the electronic motions in the helium atom. His method was not limited by perturbation theory, so it was more general and could potentially be applied to other systems with strong interaction between various types of motion.

A rigorous theory of the adiabatic separation of the electronic and nuclear motions in the diatomic molecule [Born and Flügge (1933)] was later developed and extended to polyatomic molecules by Born and Huang (1954). According to this theory, electrons can be considered as fast particles which adjust immediately to the instantaneous configurations of the slow nuclei. The electrons sense only the positions, but not the momenta of the nuclei, so in the first approximation, the nuclear momenta are neglected. As a result, the electronic motion is described by the reduced electronic Hamiltonian without nuclear momentum operators. The eigenfunctions of the electronic Hamiltonian (electronic wavefunctions), and the eigenvalues (electronic energies) can be calculated for many possible nuclear configurations. In this way, the potential energy surface (PES) can be constructed for a given electronic state and then used to define an effective adiabatic nuclear Hamiltonian which governs the nuclear motions. The adiabatic nuclear Hamiltonian can be improved by taking into account the nonadiabatic corrections.

The dynamical nuclear subsystem can be further partitioned into a set of the internal motions and the rotational motions. In many molecules, inside the subsystem of the internal motions, high frequency small amplitude vibrations (SAVs) and low frequency large amplitude motions (LAMs) can be distinguished. For each of these subsystems, the effective Hamiltonian can be derived, step by step, by applying the adiabatic separation method.

The PES plays a crucial role in the analysis of the molecular structure and its dynamics. The PES parameters appear in the definition of the effective rovibrational Hamiltonian. The global minimum in the PES determines the equilibrium molecular structure, if other local minima are of much higher energy. The curvature of the PES at the global minimum, defined by the force constants, determines the forms and frequencies of the harmonic normal modes. If the rovibrational Hamiltonian is further reduced to the rotational Hamiltonian, its parameters, such as the rotational constants and the centrifugal distortion constants, depend also on the equilibrium structure and the force constants.

The standard theory of the rovibrational states of semirigid molecules [Eckart (1935), Wilson and Howard (1936), Elyashevich (1938), Wilson (1939), Darling and Dennison (1940), Nielsen (1951), Kivelson and Wilson (1952, 1953), Wilson, Decius and Cross (1955), Watson (1968, 1970), Gordy and Cook (1970), Amat, Nielsen and Tarrago (1971), Hoy, Mills and Strey (1972), and Papoušek and Aliev (1982)], in which the nuclei execute SAVs near the equilibrium structure, provides analytic expressions for the effective Hamiltonian parameters, called *spectroscopic constants*, as functions of the equilibrium structure and of the PES characteristics. If the values of these constants are known, the spectrum of a given molecule can be interpreted or predicted in the unexplored range. The spectroscopic constants can be estimated from the PES calculated by *ab initio* methods of quantum chemistry, which helps in the interpretation of spectra. However, such calculations do not reach spectroscopic accuracy, except for some small molecules.

A reverse procedure, namely, determination of the spectroscopic constants from the molecular spectra, is commonly used in spectroscopy to estimate the molecular structure and the shape of the PES. Much information about molecules has been obtained in this way.

However, this approach does not work properly for nonrigid molecules exhibiting LAMs. For example, internal motions in weakly bound molecular complexes, such as van der Waals or hydrogen-bonded complexes, usually cannot be viewed as SAVs because in their PES's several local minima separated by low potential barriers are found. Thus, the LAMs cannot be considered as displacements from a single global minimum. The LAMs are delocalized and the tunneling effect generates a complicated energy level pattern difficult to analyze. The LAMs strongly interact with the other mode motions, especially with the rotation. This interaction cannot be adequately treated using the standard effective Hamiltonians, so rich information about the LAM dynamics cannot be derived from the spectra.

To describe the dynamics of the LAMs and their strong coupling with the rotation, *semirigid models* have been proposed, in which all SAVs are frozen out and one or a few LAMs together with the overall rotation are treated as dynamical degrees of freedom. The first such models were developed for a single internal rotation LAM in molecules in which a rigid part can rotate about a fixed axis relative to other part [Nielsen (1932), Koehler and Dennison (1940), Burkhard and Dennison (1951), Burkhard (1953), Burkhard and Irvin (1955), Hecht and Dennison (1957), Lin and Swalen (1959), Quade and Lin (1963), Quade (1966, 1967), Knopp and Quade (1968), Bauder *et al.* (1968), Hirota (1970), Durig, Craven and Harris (1972), Gut *et al.* (1978), Liu and Quade (1991), Groner (1992)]. The studies of the bending LAM started from a crude model [Thorson and Nakagawa (1960)]. Then, the rigorous theory of Hougen, Bunker and Johns (1970) for the bending motion in a triatomic molecule was soon applied to other molecules, other types of LAMs and extended to two or three LAMs [Bunker and Stone (1972), Papoušek, Stone and Špirko (1973), Špirko, Stone and Papoušek (1973), Moule and Rao (1973), Kręglewski (1978), Bunker (1980, 1983), Wierzbicki, Koput and Kręglewski (1983), Koput and Wierzbicki (1983), Kręglewski and Jensen (1984), Kręglewski (1984, 1993), and Koput (1984, 1986)].

In the semirigid LAM models, the influence of the SAVs on the LAM dynamics is completely neglected. Some interaction between these two different kinds of the motions can be accounted for using more realistic *flexible models*. Such models introduced for the analysis of the the ring-puckering [Blackwell and Lord (1972), Laane (1972), Carreira, Lord and Malloy (1979), and Harthcock and Laane (1985)], the internal rotation [Ribeaud, Bauder and Günthard (1972), Nösberger, Bauder and Günthard (1974)] and the bending LAM [Bunker and Landsberg (1977),

Bunker, Landsberg and Winnewisser (1979), Jensen (1983a), Brown, Godfrey and Kleibömer (1985), Ross (1988, 1993), and Ross, Niedenhoff and Yamada (1994)], assumed that the molecular structure is deformed adiabatically when the LAMs are executed. Such structural deformations can be calculated by *ab initio* methods which allow the molecular geometry optimization when one or more internal LAM coordinates are varied. The *ab initio* calculations were often used to predict the flexible molecular structures needed in the treatment of the flexible models [Brown, Godfrey and Champion (1987), Champion, Godfrey and Bettens (1991,1992), Jensen, Bunker and Karpfen (1991), Jensen *et al.* (1992), Pracna, Winnewisser and Winnewisser (1993), Koput (1995, 1996), and Mekhtiev, Godfrey and Szalay (1996)].

To calculate the energy levels of floppy molecules, Meyer (1979) developed a very general computer program applicable to any type of one or two LAMs in an arbitrary nonlinear molecule. However, only states with the low angular momentum quantum numbers $J = 0$, 1 and 2 can be treated by this program which seriously limits its applicability.

Very recently East and Bunker (1997) developed a computer program for the solution of the flexible models for an arbitrary nonlinear molecule with a single LAM. However, this program cannot be easily extended to two LAMs, because is based on the Numerov–Cooley method [Cooley (1961)] which needs a large number of the integration points to solve the differential eigenvalue problem.

In the flexible models, the SAVs are not treated as dynamical degrees of freedom. They are allowed to adjust adiabatically to varying LAM coordinates, so only the static part of the interaction of the SAVs with the LAMs is accounted for in such models. In more advanced *nonrigid models*, the influence of the SAVs on the LAM dynamics is treated by employing various variants of the perturbation theory. The effective nonrigid Hamiltonian for molecules containing an internal rotor of C_{3v} symmetry was first derived by Kirtman (1962) who treated the interaction of the SAVs with the internal rotation, using Van Vleck's (1929) perturbation theory.

The analysis of the SAV–internal rotation–rotation interaction in molecules with symmetric and asymmetric internal rotors [Bunker (1967), and Quade (1975, 1980, 1998)] was further extended to other types of LAMs [Hoy and Bunker (1974, 1979), Špirko, Stone and Papoušek (1976), Papoušek and Špirko (1976), Sørensen (1978), Bunker, (1983), Jensen (1983a), Špirko (1983), Bunker and Jensen (1986), Guan and Quade (1986), Jensen and Bunker (1988), and Jensen and Kraemer (1988)].

In spectroscopic practice, very often phenomenological effective nonrigid Hamiltonians are postulated because a general form of the interaction terms is typical [see, e.g., Coudert (1992, 1994, 1997)] and some of them can be selected on the basis of symmetry, as was done for the methanol molecule [De Lucia, *et al.* (1989), Tang and Takagi (1993), and Xu and Hougen (1995)].

Recently, a general numerical method for the treatment of the effective phenomenological Hamiltonian for an arbitrary type of a single LAM coupled to the rotation in nonlinear molecules was proposed by the author [Makarewicz (1996)]. In this chapter the separation of various types of the molecular motions is discussed systematically from the point of view of adiabatic theory. In Section 12.2, analytical and numerical methods of the derivation of the molecular rovibrational Hamiltonian are considered. A general form of the nonadiabatic effective Hamiltonians is discussed in Section 12.3; it is obtained with second-order perturbation theory. Section 12.4 presents the effective nonadiabatic nuclear Hamiltonian of a polyatomic molecule in an isolated electronic state. In Section 12.5 the separation of the high frequency SAVs from the LAM modes coupled to the rotation is discussed. A fully numerical approach to the effective nonrigid Hamiltonian applicable to more than one LAM is presented. Limitations of the nonrigid models, due to bifurcations of the minimum energy path on the potential energy surface, are considered. The method of the solution of the multi-dimensional rovibrational eigenvalue problem is presented in

Section 12.6. Applications of the effective LAM–rotation Hamiltonian to molecular complexes are outlined in Section 12.7 and Section 12.8 gives a summary.

12.2 Molecular Hamiltonian

12.2.1 ANALYTICAL DERIVATION OF THE KINETIC ENERGY OPERATOR

The calculation of molecular spectra is based on finding the eigensolutions of a molecule's Hamiltonian operator $\hat{H}$, which is composed of two parts, the kinetic energy operator $\hat{T}$ and the potential energy function V. In a free molecule unperturbed by external fields, V is the Coulomb interaction which keeps together all particles, the electrons and nuclei of a molecule. This interaction depends only on interparticle distances, so it is invariant with respect to the uniform translation of the whole molecule. The operator $\hat{T}$ is also invariant, so the translational motion can be exactly separated from the rest of the molecular motions.

This separation can easily be performed in the Cartesian coordinate system formed by the components of the particle position vectors with the origin in a laboratory fixed frame. For a molecule composed of N_p particles (N nuclei and N_e electrons) $3N_p$ Cartesian coordinates are needed to describe all its possible motions. The translational motion can be described by the molecular center of mass vector with three Cartesian components. The remaining $3N_p - 3$ independent coordinates can be defined as linear combinations of the all $3N_p$ coordinates, constrained in such a way that $3N_p - 3$ new coordinates will be invariant under three independent translations in the x, y and z directions. A further exact separation of the molecular motions is impossible, although an approximate separation may be achieved by choosing properly new $3N_p - 3$ generalized coordinates. This means that in the new coordinates the Hamiltonian may approximately be written as a sum of independent sub-Hamiltonians, each a function of its own subset of the coordinates. The Hamiltonian terms coupling the coordinates of different non-intersecting subsets should be small. In such a case the determination of the eigenfunctions of $\hat{H}$ is greatly simplified, because they can be factorized, i.e., represented as products of the eigenfunctions of the sub-Hamiltonians. In this way, the dimension of the problem is greatly reduced.

To separate approximately the electronic and the nuclear motions, the molecular frame (MF) has to be defined. The MF is specified by the orientation of the Cartesian axis system attached to the moving nuclei. These MF axes should be defined such that the rotation of the whole set of the nuclei, without a variation of the internuclear distances, will be reflected in the rotation of the MF axes. The Euler angles $\Omega = (\theta_E, \varphi_E, \chi_E)$ commonly used to describe this rotation define the instantaneous orientation of the MF axes relative to the laboratory fixed axes. More complicated nuclear motions can be viewed as a combination of the rotation and the internal nuclear motions.

The MF frame allows for a natural and simple definition of the generalized electronic coordinates. They are taken as Cartesian components $r_{i\alpha}(\alpha = x, y, z)$ of the N_e electron position vectors from an origin attached to the nuclei. These components are defined with respect to the MF axis system, so they do not change under the uniform translation and rotation of the electron and nuclear position vectors about the laboratory fixed axes.

The last and most important step in defining the generalized coordinates is the choice of $3N - 6$ internal coordinates $\boldsymbol{q} = (q_1, q_2, \ldots, q_{3N-6})$ for N nuclei in the molecule. For convenience, the word 'internal' will be often replaced by the word 'vibrational', although the internal motion does not always have the character of a vibration. These coordinates have to be invariant under uniform translation and rotation. This requirement guarantees that the potential energy function V, which is also invariant under these operations, can be expressed solely by $\boldsymbol{q}$ and $\boldsymbol{r} = (\boldsymbol{r}_1, \boldsymbol{r}_2, \ldots, \boldsymbol{r}_{N_e})$. In this way, the Euler angles are eliminated from V and this simplifies the solution of the rovibrational problem. The transformation from the initial Cartesian coordinates $\boldsymbol{R}^{\mathrm{L}}$, relative to the laboratory fixed axis system, of all particle position vectors to the generalized coordinates $\boldsymbol{q}^{\mathrm{G}} = (\boldsymbol{R}_{\mathrm{cm}}, \Omega, \boldsymbol{q}, \boldsymbol{r},)$, where $\boldsymbol{R}_{\mathrm{cm}}$ is the molecular center of mass vector, induces the transformation of the kinetic energy operator $\hat{T}(\boldsymbol{R}^{\mathrm{L}}) \rightarrow \hat{T}(\boldsymbol{q}^{\mathrm{G}})$ and the potential energy function $V(\boldsymbol{R}^{\mathrm{L}}) \rightarrow V(\boldsymbol{q}^{\mathrm{G}})$. Naturally, the kinetic energy of the pure translational motion of the molecule is exactly separable, so it will be further neglected. The function $V(\boldsymbol{q}^{\mathrm{G}})$ can be derived much more easily than the operator $\hat{T}(q^{\mathrm{G}})$, since V is independent of the Euler angles Ω. The derivation of $\hat{T}(\boldsymbol{q}^{\mathrm{G}})$ is much more difficult and will be discussed below.

A general form of the kinetic energy operator $\hat{T}(\Omega, \boldsymbol{q}, \boldsymbol{r})$ was derived by Webster, Huang and Wolfsberg (1981) for an arbitrary chosen MF axis system and arbitrary internal coordinates $\boldsymbol{q}$. The form of their operator $\hat{T}(\Omega, \boldsymbol{q}, \boldsymbol{r})$ was complicated because the origin of the electronic vectors $\boldsymbol{r}$ was defined in a general way. However, if the nuclear center of mass is taken as such an origin, this operator can be easily built from the purely nuclear kinetic energy operator $\hat{T}(\Omega, \boldsymbol{q})$ and the recipe of the construction of $\hat{T}(\Omega, \boldsymbol{q}, \boldsymbol{r})$ from $\hat{T}(\Omega, \boldsymbol{q})$ will be presented in Section 12.4. Thus, the effort of the derivation of the quantum mechanical molecular Hamiltonian can be focused on the derivation of the operator $\hat{T}(\Omega, \boldsymbol{q})$. This operator should be expressed in the coordinates $\boldsymbol{q}$ optimally suited for solving the rovibrational eigenvalue problem. In most of the solution methods, the Hamiltonian matrix has to be computed in some functional basis set defined on the space of $\boldsymbol{q}$ variables. Thus the choice of $\boldsymbol{q}$ affects the efficiency of the solution strategy.

The optimal coordinates should satisfy a few conditions. They should span the whole configuration space accessible for the nuclei of the molecule and the boundary conditions defining this space should be simple to permit easy computation of the Hamiltonian matrix. Desirable are coordinates compatible with the symmetry of the molecule. Such coordinates simplify the partitioning of the Hamiltonian matrix into smaller independent sub-matrices. Good coordinates should minimize the kinetic and the potential interaction terms, making the total Hamiltonian nearly separable.

It is difficult to satisfy the above conditions even for small molecules. The coordinates optimal for the kinetic energy operator are often not acceptable for the potential energy function. For example, Cartesian coordinates are ideal for $\hat{T}$ because it separates into one-particle operators. However, they are inappropriate for V. The widely used rectilinear normal coordinates, defined together with the Eckart MF axes [Eckart (1935)] are optimally adapted to SAVs in the molecule. The exact rovibrational operator $\hat{T}(\Omega, \boldsymbol{q})$ expressed in these coordinates was worked out a long time ago by Wilson and Howard (1936), Darling and Dennison (1940) and then simplified by Watson (1968, 1970). The normal coordinates satisfy all the criteria discussed if the internal motions are executed with small amplitudes around the equilibrium nuclear configuration. However, LAMs such as the inversion or internal rotation cannot be conveniently described by normal coordinates because of their curvilinear character. As a consequence of this property, the boundary conditions for a single periodic motion cannot be defined using only one normal coordinate. For SAVs, the representation of the potential energy function V as a quadratic polynomial form of the normal coordinates is adequate. This function can be transformed to a diagonal form with complete elimination of the intermode coupling terms. For LAMs, the function V usually has several local minima, so a polynomial expansion around only one local minimum is practically useless. Curvilinear coordinates have to be employed in a proper description of the LAMs. The familiar valence coordinates, widely used in molecular modeling have many attractive properties, so they are good candidates for the nuclear internal coordinates.

Valence coordinates describe the configuration of N nuclei in the molecule in terms of only three types of geometric parameters:

- $N-1$ distances (bond lengths) r_i, between nuclei p and q,
- $N-2$ interbond angles $\theta_{(ij)}$, between the bonds $i = (pq)$ and $j = (qs)$

- $N - 3$ dihedral (torsional) angles τ_k between two planes spanned by the bonds (pq) and (qs), and the bonds (qs) and (st).

Naturally, these coordinates are rotationally and translationally invariant. The instantaneous nuclear configurations can be conveniently specified by the so called Z-matrix [Frisch, (1990)] which has the advantage that the Cartesian coordinates of the nuclei can be calculated from a fairly simple formula. The valence coordinates specified by the Z-matrix are not necessarily related to chemical bonds in the molecule, because r_i may represent the distances between arbitrary pairs of the nuclei. The Z-matrix specifies which $N - 1$ distances from the all possible $N(N-1)/2$ distances between the nuclei are treated as the internal coordinates and play the role of the 'bonds'.

The potential energy function V with multiple minima can be adequately represented in valence coordinates using the Fourier expansions in the angular coordinates θ and τ, and various type expansion, for example the Morse expansion, in the stretching coordinate r. In many cases V appears to be approximately separable.

The quantum mechanical kinetic energy operator $\hat{T}$ in valence coordinates was derived in first days of quantum mechanics by Hylleraas (1928) for the helium atom with assumed infinite mass of the nucleus. He directly transformed the kinetic energy operator $\hat{T}(\boldsymbol{R}^{\mathrm{L}})$ from the Cartesian coordinates in the laboratory fixed axis system to three Euler angles describing the overall rotation of the atom, and three internal valence coordinates r_1, r_2 and θ. This *direct transformation* method has been used later to derive the rovibrational Hamiltonians of triatomic molecules in internuclear distances [Diehl *et al.* (1961)] in atom–diatom Jacobi coordinates [Ruder (1968), and Tennyson and Sutcliffe (1982)], and in valence coordinates [Bardo and Wolfsberg (1977), and Carter and Handy (1982)]. The direct transformation method has been further developed by Sutcliffe (1982, 1992). In order to make feasible difficult algebraic derivations, Handy employed computer algebra programs and obtained a rovibrational kinetic energy operator in valence coordinates for four-atomic molecules [Handy (1987), and Bramley, Green and Handy (1991)] and the pure vibrational operator $\hat{T}_{\mathrm{V}}(\boldsymbol{q})$ for some penta-atomic molecules [Császár and Handy (1995a)] and sequentially bounded polyatomic molecules [Császár and Handy (1995b)]. Handy's method, which employs computer algebra requires an individual treatment of each molecule and is practically useful only when the generalized coordinates $\boldsymbol{q}^{\mathrm{G}}$ can be expressed as analytic functions of the Cartesian coordinates $\boldsymbol{R}^{\mathrm{L}}$.

An alternative derivation method for the kinetic energy operator $\hat{T}$ was proposed by Podolsky (1928). In his method, the classical Hamiltonian in the Cartesian coordinates $\boldsymbol{R}^{\mathrm{L}}$ is set up and then is transformed to the generalized coordinates $\boldsymbol{q}^{\mathrm{G}}$. Finally the classical Hamiltonian $H(\boldsymbol{q}^{\mathrm{G}})$ is converted to the quantum mechanical operator $\hat{H}(\boldsymbol{q}^{\mathrm{G}})$. This method was used in numerous works on the molecular Hamiltonian. It is very convenient when the approximate LAM semirigid or flexible models of the internal motions have to be considered.

Exact Hamiltonians were also derived in this way, for example the above-mentioned rovibrational Hamiltonian in rectilinear normal coordinates. The exact operator $\hat{T}$ for a four-atomic molecule, expressed in terms of the lengths and angles, defined for various sets of Jacobi interparticle vectors, was obtained by Büttner and Ruder (1969). More general coordinates, depending on two external parameters were formulated for triatomic molecules and a general rovibrational Hamiltonian was derived by Makarewicz (1988). These parameters could be varied to adapt the internal coordinates to a given type of internal motion in a molecule. The generalized coordinates included the Jacobi, Radau and valence coordinates. This rovibrational Hamiltonian was latter used by Sutcliffe and Tennyson (1991) and Schwenke (1992) in variational calculations of the triatomic rovibrational states.

The starting point in Podolsky's method is the classical Lagrangian form of the kinetic energy T, expressed by the Cartesian velocity vector $\dot{\boldsymbol{R}}^{\mathrm{L}}$ components. Eckart (1935) transformed T to the generalized velocities of the internal motions $\dot{\boldsymbol{q}}$ and the angular velocity column vector $\dot{\boldsymbol{\omega}} = (\omega_x, \omega_y, \omega_z)^{\mathrm{T}}$ describing the rotation of the molecular frame. The translational velocity was separated out, so the kinetic energy could be written as a quadratic form

$$T = \frac{1}{2}[\dot{\boldsymbol{\omega}}^{\mathrm{T}}, \dot{\boldsymbol{q}}^{\mathrm{T}}] \begin{bmatrix} \boldsymbol{K}_{\mathrm{R}} & \boldsymbol{K}_{\mathrm{C}} \\ \boldsymbol{K}_{\mathrm{C}}^{\mathrm{T}} & \boldsymbol{K}_{\mathrm{V}} \end{bmatrix} \begin{bmatrix} \dot{\boldsymbol{\omega}} \\ \dot{\boldsymbol{q}} \end{bmatrix}. \tag{1}$$

The kinetic energy matrix $\boldsymbol{K}$ is divided into submatrices describing the contributions of the generalized velocities to the total kinetic energy. The rotational submatrix $\boldsymbol{K}_{\mathrm{R}}$ is defined as the familiar inertia tensor with the components

$$K_{\alpha\beta} = I_{\alpha\beta} \equiv \sum_n m_n(\delta_{\alpha\beta}\boldsymbol{r}'_n \cdot \boldsymbol{r}'_n - \alpha'_n\beta'_n). \tag{2}$$

The vibrational submatrix $\boldsymbol{K}_V$ is defined by its elements

$$K_{kl} = \sum_n m_n(\mathrm{d}\boldsymbol{r}'_n/\mathrm{d}q_k) \cdot (\mathrm{d}\boldsymbol{r}'_n/\mathrm{d}q_l), \tag{3}$$

and $\boldsymbol{K}_{\mathrm{C}}$ corresponding to the Coriolis interaction between the internal velocities and angular velocities is given by

$$K_{\alpha k} = \sum_n m_n(\boldsymbol{r}'_n \times \mathrm{d}\boldsymbol{r}'_n/\mathrm{d}q_k)_\alpha, \tag{4}$$

where α and β label the Cartesian components x, y and z of the nuclear position vectors $\boldsymbol{r}'_k$ measured with respect to the MF axis system placed at the nuclear center of mass.

The generalized momenta $\boldsymbol{p}$ conjugate to $\boldsymbol{q}$ and total angular momentum vector $\boldsymbol{J} = (J_x, J_y, J_z)^{\mathrm{T}}$ can be determined using the known form of T given in equation (1). As a result, T can be expressed by $\boldsymbol{p}$ and $\boldsymbol{J}$ in the Hamiltonian form

$$T = \frac{1}{2}[\boldsymbol{J}^{\mathrm{T}}, \boldsymbol{p}^{\mathrm{T}}] \begin{bmatrix} \boldsymbol{G}_{\mathrm{R}} & \boldsymbol{G}_{\mathrm{C}} \\ \boldsymbol{G}_{\mathrm{C}}^{\mathrm{T}} & \boldsymbol{G}_{\mathrm{V}} \end{bmatrix} \begin{bmatrix} \boldsymbol{J} \\ \boldsymbol{p} \end{bmatrix}, \tag{5}$$

where the matrix $\boldsymbol{G}$, called also the tensor $\boldsymbol{G}$, is the inverse of the matrix $\boldsymbol{K}$ from equation (1).

According to Podolsky's rule, a correct quantum mechanical operator $\hat{T}$ can be obtained by replacing $\boldsymbol{p}$ in equation (5) by $\hat{\boldsymbol{p}} = -i\hbar\partial/\partial\boldsymbol{q}$ and $\boldsymbol{p}^{\mathrm{T}}$ by $\hat{\boldsymbol{p}}_g^{\mathrm{T}} = -i\hbar g^{-1/2}(\partial/\partial\boldsymbol{q})^{\mathrm{T}} g^{1/2}$ where g is the determinant of the $\boldsymbol{K}$ matrix. Because g is a function of $\boldsymbol{q}$, the differential operator $-i\hbar\partial/\partial\boldsymbol{q}$ does not commute with g so $\hat{\boldsymbol{p}}$ and $\hat{\boldsymbol{p}}_g$ are not the same operators. The function $g^{1/2}$ is a part of the Jacobian, $J(\boldsymbol{q}, \boldsymbol{\Omega}) = g^{1/2} \sin\theta_{\mathrm{E}}$, of the transformation from the Cartesian to the coordinates $\boldsymbol{q}$ and Euler angles $\boldsymbol{\Omega} = (\theta_{\mathrm{E}}, \varphi_{\mathrm{E}}, \chi_{\mathrm{E}})$. This Jacobian determines also the volume element $\mathrm{d}V = J(\boldsymbol{q}, \Omega)\,\mathrm{d}\boldsymbol{q}\,\mathrm{d}\boldsymbol{\Omega}$ used in the integration over the space of the variables $\boldsymbol{q}$ and $\boldsymbol{\Omega}$. The *standard normalization* condition for the wavefunctions in this space is

$$\int |\Psi|^2 J(\boldsymbol{q}, \boldsymbol{\Omega})\,\mathrm{d}\boldsymbol{q}\,\mathrm{d}\boldsymbol{\Omega} = 1. \tag{6}$$

The operator $\hat{T}$ is composed of the vibrational ($\hat{T}_{\mathrm{V}}$), Coriolis ($\hat{T}_{\mathrm{C}}$) and the rotational ($\hat{T}_{\mathrm{R}}$) parts, which can easily be obtained from equation (5) and Podolsky's rule. For example

$$\hat{T}_{\mathrm{V}} = \frac{1}{2}\hat{\boldsymbol{p}}_g^{\mathrm{T}} G_{\mathrm{V}} \hat{\boldsymbol{p}} = -\frac{1}{2}\hbar^2 \left(\sum_{k,l} G_{\mathrm{V}}^{kl} \partial^2/\partial q_k\,\partial q_l + \sum_k h^k \partial/\partial q_k \right), \tag{7}$$

where

$$h^k = \sum_l \left(\partial G_V^{kl}/\partial q_l + \frac{1}{2} G_V^{kl} \partial \ln g/\partial q_l \right). \tag{8}$$

The form of the kinetic energy operator $\hat{T}$ depends on the normalization convention used for the wavefunction. Very often, a unit weight function is used, instead of $J(\boldsymbol{q}, \boldsymbol{\Omega})$, to avoid a difficult integration, if $g^{1/2}$ is a complicated not factorizable function of $\boldsymbol{q}$. Then, the Jacobian is absorbed in the modified wavefunction and the kinetic energy operator, which changes its linear terms h^k and takes an additional nondifferential term called a 'pseudopotential'. The weight function can be, in principle, arbitrary. For example, only a part of the Jacobian can be absorbed in the wavefunction and the kinetic energy operator [Handy (1987)]. An appropriate normalization convention should yield a simple and economic form of the operator $\hat{T}$ with minimum number of terms of a simple structure. Such an operator allows us to economize the calculation of the matrix elements in a variational or perturbational solution to the eigenvalue problem.

Recently, Makarewicz and Skalozub (1999) have found that the standard normalization convention provides a clear form of the vibrational operator $\hat{T}_V$ in valence coordinates. They used the facts that general formulae for the elements of the matrix $\boldsymbol{G}_V$ for these coordinates are known [Decius (1948)] and the function $g^{1/2}$ can be determined without the necessity of calculating the determinant $|\boldsymbol{K}|$.

The valence coordinates are the local polar–spherical coordinates for each of N nuclei, except for first three nuclei, which define only two vectors parametrized by two bond lengths r_1, r_2 and one bond angle $\theta_{(12)}$. Hence, the transformation from the Cartesian to the valence coordinates generates the Jacobian

$$g^{1/2} = \mu r_1^2 \prod_{i=2}^{N-1} r_i^2 \prod_{(ij)=1}^{N-2} \sin \theta_{(ij)}, \tag{9}$$

where μ is a constant factor.

The determined linear terms h^k in equation (8) appeared to be independent of a structure of the Z-matrix corresponding to a given molecule. Each bond length r_i connecting the pth and qth nuclei generates the corresponding linear term h^i given by

$$h^i = 2(\mu_i r_i)^{-1}, \tag{10}$$

where $\mu_i^{-1} = m_p^{-1} + m_q^{1}$.

Each angle $\theta_{(ij)}$ between the ith bond pq and jth bond qs generates the corresponding term $h^{(ij)}$ given by

$$h^{(ij)} = G_V^{(ij)(ij)} \cot \theta_{(ij)} - 2 \sin \theta_{(ij)} (m_q r_i r_j)^{-1}, \tag{11}$$

where $G_V^{(ij)(ij)}$ is the diagonal $\boldsymbol{G}_V$ matrix element for the angle $\theta_{(ij)}$

$$G_V^{(ij)(ij)} = \left(\mu_i r_i^2\right)^{-1} + \left(\mu_j r_j^2\right)^{-1} - 2 \cos \theta_{(ij)} (m_q r_i r_j)^{-1}, \tag{12}$$

with $\mu_i^{-1} = m_p^{-1} + m_q^{-1}$ and $\mu_j^{-1} = m_q^{-1} + m_s^{-1}$.

The linear terms h^k for the torsional angles vanish. As a result, $\hat{T}_V$ has a very simple form:

$$\hat{T}_V = -\frac{\hbar^2}{2} (\hat{T}_r + \hat{T}_\theta + \hat{T}_\tau + \hat{T}_{\text{int}}), \tag{13}$$

where

$$\hat{T}_r = \sum_{i=1}^{N-1} \frac{1}{\mu_i}\left(\frac{\partial^2}{\partial r_i^2} + \frac{2}{r_i}\frac{\partial}{\partial r_i}\right), \tag{14}$$

$$\hat{T}_\theta = \sum_{(ij)=1}^{N-2}\left[G_{\rm V}^{(ij)(ij)}\left(\frac{\partial^2}{\partial\theta_{(ij)}^2} + \cot\theta_{(ij)}\frac{\partial}{\partial\theta_{(ij)}}\right) - 2(m_q r_i r_j)^{-1}\sin\theta_{(ij)}\frac{\partial}{\partial\theta_{(ij)}}\right], \tag{15}$$

$$\hat{T}_t = \sum_{k=1}^{N-3} G_{\rm V}^{\tau_k\tau_k}\frac{\partial^2}{\partial\tau_k^2}, \tag{16}$$

and

$$\hat{T}_{\rm int} = \sum_{k\neq l}^{3N-6} G_{\rm V}^{q_k q_l}\frac{\partial^2}{\partial q_k\,\partial q_l}. \tag{17}$$

$G_{\rm V}^{\tau_k\tau_k}$ are the diagonal torsional elements and $G_{\rm V}^{q_kq_l}$ are the off-diagonal elements of the matrix $\boldsymbol{G}_{\rm V}$ tabulated by Decius (1948). The above formulae valid for an arbitrary polyatomic molecule specify the operator $\hat{T}_{\rm V}$ with the simple linear differential terms of a uniform structure independent of a specific structure of the Z-matrix defining the valence coordinates.

Although the valence coordinates have many advantages, they have an essential drawback. They are typical *local* coordinates which are not adapted to the global symmetry of a molecule. Even for simple highly symmetric molecules such as ammonia or methane, the valence coordinates do not reflect their symmetries. Experience has shown that there is no single coordinate set optimal for all molecules. Thus, there is a need for feasible methods of determination of the rovibrational Hamiltonian for internal coordinates adapted to a particular molecule.

12.2.2 NUMERICAL DETERMINATION OF THE KINETIC ENERGY OPERATOR

Podolsky's method is well suited for numerical determination of the $\boldsymbol{G}$ matrix and the determinant g. Since in general g is not factorizable function, it is preferable to use the unit weight function in the volume element for the integration of the matrix elements, which gives the modified form of the operator $\hat{T}$ with the absorbed Jacobian:

$$\hat{T} = \tfrac{1}{2}\left(\hat{\boldsymbol{p}}^{\rm T}\boldsymbol{G}_{\rm V}\hat{\boldsymbol{p}} + \hat{\boldsymbol{p}}^{\rm T}\boldsymbol{G}_{\rm C}\hat{\boldsymbol{J}} + \hat{\boldsymbol{J}}^{\rm T}\boldsymbol{G}_{\rm C}\hat{\boldsymbol{p}} + \hat{\boldsymbol{J}}^{\rm T}\boldsymbol{G}_{\rm R}\hat{\boldsymbol{J}}\right) + U, \tag{18}$$

where the pseudopotential U is defined by the derivatives of $\boldsymbol{G}$ and g [Meyer and Günthard (1968)]

$$U = \tfrac{1}{2}\hbar^2\left\{\boldsymbol{d}^{\rm T}\boldsymbol{G}_{\rm V}\boldsymbol{d} - (\partial/\partial\boldsymbol{q})^{\rm T}\boldsymbol{G}_{\rm V}\boldsymbol{d}\right\}, \tag{19}$$

where

$$\boldsymbol{d} = \tfrac{1}{4}\partial\ln g/\partial\boldsymbol{q}, \tag{20}$$

and the operator $\partial/\partial\boldsymbol{q}$ acts only within the braces $\{\ldots\}$. To determine $\boldsymbol{G} = \boldsymbol{K}^{-1}$ and g from equations (2)–(4), the Cartesian components, with respect to the MF axis system, of the nuclear vectors $\boldsymbol{r}'$ must be known as functions of $\boldsymbol{q}$. In the direct derivation method, the more complicated function $\boldsymbol{R}^{\rm L} \to (\boldsymbol{R}_{\rm cm}, \boldsymbol{\Omega}, \boldsymbol{q})$ must be specified and differentiated.

The set of the nuclear position vectors $\boldsymbol{r}'$ as functions of $\boldsymbol{q}$ defines the dynamical model of the molecule. In the LAM *semirigid models*, only a small number, L, of the LAM coordinates $\tau = (\tau_1, \tau_2, \ldots, \tau_{\rm L})$, selected from the whole set of the internal coordinates $\boldsymbol{q}$, are treated as dynamical variables. The rest of the coordinates $\boldsymbol{s} = (q_{L+1}, q_{L+2}, \ldots, q_{3N-6}) = (s_1, s_2, \ldots, s_M)$, where $L + M = 3N - 6$, are fixed at the value of s_0. This means that when the LAMs are executed, most of the molecular structural parameters described by s_0 remain 'frozen'. As a result, $\boldsymbol{r}'$ is the function of the LAM coordinates only:

$$\boldsymbol{r}' = \boldsymbol{R}(\tau, s_0). \tag{21}$$

In the *flexible models*, the molecular structure represented by s is allowed to follow the variations of the LAMs, so the s coordinates become functions of τ, i.e., $s = S(\tau)$. These functions can be determined from the optimization of the molecular geometry for variable LAMs. The optimization yields the minimum of the potential $V(\tau, s)$ at a given τ. In the following, the pseudopotential will be included in $V(\tau, s)$. From the minimum condition

$$\partial V(\tau, s)/\partial s = 0, \tag{22}$$

the optimum (equilibrium) value $s_e(\tau)$ of s at a given τ can be found. In this way, the function $S(\tau) = s_e(\tau)$ describing the relaxation of the molecular structure, caused by the LAMs, is established. Simultaneously, the *relaxed* PES for the LAMs, $V_e(\tau) = V(\tau, s_e(\tau))$, (the minimum energy path for a single LAM) is obtained.

The *nonrigid models* need more information on the PES. They approximate the internal dynamics as a combination of the LAMs and the accompanying SAVs executed around the flexible molecular structure. The SAVs are modeled by coupled harmonic oscillators, for which the force constants

$$f_{ij}(\tau) = \left[\partial^2 V(\tau, s)/\partial s_i\, \partial s_j\right]_{s=s_e(\tau)} \tag{23}$$

depend of the LAM coordinates. The PES represented as an expansion truncated at quadratic terms,

$$V(\tau, s) \approx V_e(\tau) + \tfrac{1}{2}(s - s_e(\tau))^{\mathrm{T}} f (s - s_e(\tau)), \tag{24}$$

approximates locally the global PES better than the potential $V_e(\tau)$ of the flexible model. Such an expansion of the PES is possible only when equation (22) has a unique solution, corresponding to the flexible structure $s_e(\tau)$. This condition is not always satisfied and this fact will be discussed in Section 12.5.3

In order to set up the rovibrational Hamiltonian, the matrix $\boldsymbol{G}$ and the determinant g must be known as functions of τ (for the LAM rigid and flexible models) or τ and s (for the nonrigid models). They are necessary for the calculation of the Hamiltonian matrix elements. They can be computed efficiently using the numerical integration methods which need the values of the integrated function on a discrete set of points. Thus, $G(\tau, s)$ have to be computed on some grid points. The idea of a total numerical calculation of $G(\tau, s)$ was first proposed by Meyer (1979). A numerical algorithm of this method consist of the steps:

- Define the MF axis system attached to the molecule.
- Choose the origin of the nuclear position vectors $\boldsymbol{r}_n$ and express their MF Cartesian components as functions of the internal coordinates $\boldsymbol{q} = (\tau, s)$.
- Determine the vector of the center of mass of the nuclei $\boldsymbol{r}_{\mathrm{CMN}} = \sum_n m_n \boldsymbol{r}_n / \sum_n m_n$ and then calculate the new vectors $\boldsymbol{r}'_n = \boldsymbol{r}_n - \boldsymbol{r}_{\mathrm{CMN}}$ which are shifted to the center of mass of the nuclei as the origin.
- For a given point $\boldsymbol{q}_0$ calculate numerically all derivatives $\partial r'_{n\alpha}/\partial q_k$.
- Set up the kinetic energy matrix $\boldsymbol{K}$, according to equations (2)–(4), using $r'_{n\alpha}$ and $\partial r'_{n\alpha}/\partial q_k$.
- Calculate the determinant $g = |\boldsymbol{K}|$ and $\boldsymbol{G}$ as the inverse of $\boldsymbol{K}$.

The two first steps depend on the molecule and its dynamical model which can be defined in a separate subroutine. The remaining steps can be carried out automatically. The above calculations have to be repeated for all grid points. Additionally, the first derivatives of g and $\boldsymbol{G}$ can also be computed numerically at the same grid. The accuracy of the calculated derivatives can be controlled using the interpolation formulae of increasing order.

12.3 Adiabatic and nonadiabatic effective Hamiltonians

A molecule can be considered as a system comprising an ordered sequence of dynamical subsystems: electrons and nuclei. The internal nuclear motions are faster than the overall rotation, so they can be treated as a well-defined dynamical subsystems. Within this subsystem we can distinguish the high frequency SAVs, for example the stretching and bending vibrations of light atoms, and low frequency LAMs, for example the internal rotation, inversion, ring puckering, etc. The electrons are the most 'important' because they generate an effective potential for the rest of molecular motions, so they constitute the first subsystem in the hierarchy of the molecular subsystems. The effective Hamiltonians for each of the subsystems can be derived using a multi-step adiabatic procedure [Makarewicz (1985)], which is almost equivalent to the usual adiabatic separation method applied successively step-by-step to dynamical subsystems. Thus, let us consider the electrons described by the set of generalized coordinates $\boldsymbol{q}_1$ and conjugate momenta $\boldsymbol{p}_1$, and the nuclei described by coordinates $\boldsymbol{q}_2$ and conjugate momenta $\boldsymbol{p}_2$. The approximate *adiabatic wavefunction* is represented by the product

$$\Psi^{A}(\boldsymbol{q}_1, \boldsymbol{q}_2) = \Psi_1^{A}(\boldsymbol{q}_1, \boldsymbol{q}_2)\Psi_2^{A}(\boldsymbol{q}_2), \tag{25}$$

where the electronic adiabatic wavefunction $\Psi_1^{A}(\boldsymbol{q}_1, \boldsymbol{q}_2)$ fulfils the normalization condition

$$\langle \Psi_1^{A}(\boldsymbol{q}_1, \boldsymbol{q}_2) | \Psi_1^{A}(\boldsymbol{q}_1, \boldsymbol{q}_2) \rangle_{\boldsymbol{q}_1} = 1 \tag{26}$$

on the whole subspace of the coordinates $\boldsymbol{q}_2$. The integration in equation (26) runs only over the $\boldsymbol{q}_1$ coordinates.

Now, the upper bound of the true energy E of the total system can be calculated from the variational principle

$$\begin{aligned} E \leq E^{A} &\equiv \left\langle \Psi^{A}(\boldsymbol{q}_1, \boldsymbol{q}_2) \left| \hat{H}(\boldsymbol{q}_1, \hat{\boldsymbol{p}}_1, \boldsymbol{q}_2, \hat{\boldsymbol{p}}_2) \right| \Psi^{A}(\boldsymbol{q}_1, \boldsymbol{q}_2) \right\rangle_{\boldsymbol{q}_1\boldsymbol{q}_2} \\ &= \left\langle \Psi_2^{A}(\boldsymbol{q}_2) \left| \hat{H}_2^{A}(\boldsymbol{q}_2, \hat{\boldsymbol{p}}_2) \right| \Psi_2^{A}(\boldsymbol{q}_2) \right\rangle_{\boldsymbol{q}_2}, \end{aligned} \tag{27}$$

where the *effective adiabatic Hamiltonian* for the second subsystem is given by

$$\hat{H}_2^{A}(\boldsymbol{q}_2, \hat{\boldsymbol{p}}_2) \equiv \langle \Psi_1^{A}(\boldsymbol{q}_1, \boldsymbol{q}_2) | \hat{H}(\boldsymbol{q}_1, \hat{\boldsymbol{p}}_1, \boldsymbol{q}_2, \hat{\boldsymbol{p}}_2) | \Psi_1^{A}(\boldsymbol{q}_1, \boldsymbol{q}_2) \rangle_{\boldsymbol{q}_1}, \tag{28}$$

where $\hat{H}$ is the Hamiltonian of the total system. The operator $\hat{H}$ is averaged over the wavefunction $\Psi_1^{A}(\boldsymbol{q}_1, \boldsymbol{q}_2)$ on the space of the $\boldsymbol{q}_1$ coordinates. Hence, $\hat{H}_2^{A}$ is the operator which acts in the space of functions of the $\boldsymbol{q}_2$ coordinates. The eigenfunctions of $\hat{H}_2^{A}$ can be calculated from the equation

$$\left[\hat{H}_2^{A}(\boldsymbol{q}_2, \hat{\boldsymbol{p}}_2) - E^{A}\right] \Psi_2^{A}(\boldsymbol{q}_2) = 0, \tag{29}$$

because the variational principle for equation (29) leads to equation (27). The form of $\hat{H}_2^{A}$, and consequently the accuracy of the adiabatic approximation, depends on the chosen function Ψ_1^{A}. According to the adiabatic principle [Born and Huang (1954)], this function is assumed to be the eigenfunction of the operator $\hat{H}_1^A$ which is obtained by dropping the operators $\hat{\boldsymbol{p}}_2$ from the total Hamiltonian $\hat{H}$,

$$\hat{H}_1^A(\boldsymbol{q}_1, \hat{\boldsymbol{p}}_1, \boldsymbol{q}_2) \equiv \hat{H}(\boldsymbol{q}_1, \hat{\boldsymbol{p}}_1, \boldsymbol{q}_2, \boldsymbol{0}). \tag{30}$$

This assumption reflects the idea that the first subsystem is not sensitive to the momenta of the second one but feels only its instantaneous configuration $\boldsymbol{q}_2$. Thus, the coordinates $\boldsymbol{q}_2$ are treated as parameters, not as dynamical variables, in the equation which defines Ψ_1^A:

$$\left[\hat{H}_1^A(\boldsymbol{q}_1, \hat{\boldsymbol{p}}_1, \boldsymbol{q}_2) - E_1^A(\boldsymbol{q}_2)\right] \Psi_1^A(\boldsymbol{q}_1, \boldsymbol{q}_2) = 0. \tag{31}$$

This equation has a set of solutions which will be labeled by the index n. For each given $E_{1n}^A(\boldsymbol{q}_2)$ and $\Psi_{1n}^A(\boldsymbol{q}_1, \boldsymbol{q}_2)$ there exists a set of partner eigenvalues $E_{2n,k}^A$ and eigenfunctions $\Psi_{2n,k}^A(\boldsymbol{q}_2)$ of the second subsystem, labeled by the additional index k. The set $\{\Psi_{1n}^A \Psi_{2n,k}^A\}_{n,k}$ of the adiabatic wavefunctions can be used to calculate more accurate *nonadiabatic* energy levels and wavefunctions of the total Hamiltonian. When we are interested in a series of the states n, k with a fixed n, which belong to the only one partner sub-state Ψ_{1n}^A, then we can use perturbation theory. The nondiagonal matrix elements of the perturbation operator $\Delta\hat{H} \equiv \hat{H} - \hat{H}_1^A$ in the adiabatic basis set contribute to the second-order nonadiabatic energy correction

$$\Delta E_{n,k}^{NA} = \sum_{m,l \neq n,k} \left(E_{n,k}^A - E_{m,l}^A\right)^{-1} \left\langle \Psi_{2n,k}^A \right| \left\langle \Psi_{1n}^A \right| \Delta\hat{H} \left| \Psi_{1m}^A \right\rangle_{q_1} \left| \Psi_{2m,l}^A \right\rangle_{q_2}$$

$$\left\langle \Psi_{2m,l}^A \right| \left\langle \Psi_{1m}^A \right| \Delta\hat{H} \left| \Psi_{1n}^A \right\rangle_{q_1} \left| \Psi_{2n,k}^A \right\rangle_{q_2}. \tag{32}$$

The labels $\boldsymbol{q}_1$ and $\boldsymbol{q}_2$, indicating the integration variables, will subsequently be omitted.

Since the differences between the excited energy levels of the first subsystems (electrons) are usually much larger than the differences between the excited levels of the second subsystem (the nuclei of the molecule) we can replace $(E_{n,k}^A - E_{m,l}^A)$ by the their simpler approximate values $(E_{n,0}^A - E_{m,0}^A)$ where $k = 0$ refers to the ground state of the second subsystem. This assumption is not valid if the energies $E_{n,0}^A$ and $E_{m,0}^A$ are close to each other, i.e., are generated by close lying energy functions $E_{1n}^A(\boldsymbol{q})$ and $E_{1m}^A(\boldsymbol{q})$. Then the perturbation theory should be extended by constructing the effective nonadiabatic Hamiltonian for the two series of states belonging to Ψ_{1n}^A and Ψ_{1m}^A, respectively.

The above approximation and the closure relation $\Sigma_l |\Psi_{2m,l}^A\rangle\langle\Psi_{2m,l}^A| = \hat{1}$, leads to the simple result

$$\Delta E_{n,k}^{NA} = \left\langle \Psi_{2n,k}^A \left| \Delta\hat{H}_{2n}^{NA} \right| \Psi_{2n,k}^A \right\rangle, \tag{33}$$

where the nonadiabatic operator is given by

$$\Delta\hat{H}_{2n}^{NA} = \sum_{m \neq n} \left(E_{n,0}^A - E_{m,0}^A\right)^{-1} \left\langle \Psi_{1n}^A \left| \Delta\hat{H} \right| \Psi_{1m}^A \right\rangle \left\langle \Psi_{1m}^A \left| \Delta\hat{H} \right| \Psi_{1n}^A \right\rangle. \tag{34}$$

Thus, in order to calculate approximately the nonadiabatic energy levels of the nth well-isolated state of the first subsystem, the effective adiabatic Hamiltonian $\hat{H}_{2n}^A$ for the second subsystem should be replaced in equation (29) by *the effective nonadiabatic* Hamiltonian $\hat{H}_{2n}^A + \Delta\hat{H}_{2n}^{NA}$. The above scheme will be employed for the derivation of the effective Hamiltonians for dynamical subsystems in a molecule, although a more sophisticated formalism of contact transformations can be used to solve this problem [Bunker and Moss (1977, 1980)]. This formalism yields the same Hamiltonian terms with some additional small contributions which are inessential in our discussion.

12.4 Separation of the electronic and nuclear motions and the effective nuclear Hamiltonian

High resolution rovibrational spectra provide much evidence for the breakdown of the Born–Oppenheimer approximation. The effective nonadiabatic rovibrational Hamiltonian for a diatomic molecule in an isolated electronic state was first derived by Bunker and Moss (1977). Recently, new relationships between the mass-dependent adiabatic and nonadiabatic corrections to the Born–Oppenheimer approximation have been obtained for diatomic molecules by Herman and Ogilvie (1998). Theoretical investigations of the nonadiabatic effect in polyatomic molecules have been limited mainly to triatomic molecules. Bunker and Moss (1980) derived the total electron–nuclear Hamiltonian for a symmetric triatomic molecule, in the coordinates used in the model of Hougen, Bunker and Johns (1970) and obtained the effective nuclear Hamiltonian for a closed shell electronic state, including second-order nonadiabatic corrections. Such corrections with additional hyperfine interaction terms have also been derived for a polyatomic semirigid molecule by Michelot (1982). Recently, Teffo (1993) investigated the nonadiabatic effects in linear semirigid molecules.

The accuracy of the empirical PESs of the triatomic molecules H_3^+ [Dinelli, Polyansky and Tennyson (1995)], H_2O [Polyansky, Jensen and Tennyson (1994)] and H_2S [Polyansky, Jensen and Tennyson (1996)] obtained from the fit to high resolution spectra achieved such a high level that the effects of the breakdown of the Born–Oppenheimer approximation needed to be considered explicitly.

In order to derive a general effective nonadiabatic nuclear Hamiltonian describing the overall rotation and internal molecular motions which include the SAV modes and the LAMs in a polyatomic molecule, the total Hamiltonian must be known. Its form depends on the definition of the MF axis system, on chosen coordinates describing the internal nuclear motions and on electronic coordinates measured relative to the MF axis system. The first molecular Hamiltonian, with included relativistic effects, was presented by Howard and Moss (1970). They used rectilinear normal coordinates for the nuclei, the MF axes defined by Eckart conditions and Cartesian electronic position vectors measured from the center of mass of the nuclei. Their Hamiltonian is optimal for semirigid molecules, but not for floppy molecules. A more general electronic–nuclear Hamiltonian for arbitrary MF axes and curvilinear internal coordinates $\boldsymbol{q}$, and the electronic coordinates mentioned above, can be obtained directly using the simple recipe formulated by Makarewicz and Bauder (1995):

- First, for chosen nuclear curvilinear coordinates $\boldsymbol{q}$ and the MF axes, derive the nuclear kinetic energy operator in the form of equation (18), ignoring the electrons.
- Replace in equation (18) the nuclear angular momentum operator $\hat{\boldsymbol{J}}$ by $\hat{\boldsymbol{J}} - \hat{\boldsymbol{j}}$ where now $\hat{\boldsymbol{J}}$ is treated as the total angular momentum operator for the *nuclei and electrons*, and $\hat{\boldsymbol{j}} = (\hat{j}_x, \hat{j}_y, \hat{j}_z)$ is the total angular momentum operator of the electrons only. Both $\hat{\boldsymbol{J}}$ and $\hat{\boldsymbol{j}}$ are measured with respect to the MF axes.
- Add to a modified equation (18) the electronic contribution to the Hamiltonian, defined by the kinetic energy operator and the nuclear–electron potential energy function

$$\hat{H}_e(\boldsymbol{r}, \hat{\boldsymbol{p}}_e, \boldsymbol{q}) = \tfrac{1}{2}\hat{\boldsymbol{p}}_e^{\mathrm{T}} \boldsymbol{G}_e \hat{\boldsymbol{p}}_e + V(\boldsymbol{r}, \boldsymbol{q}), \tag{35}$$

  where the electronic kinetic energy tensor $\boldsymbol{G}_e$ is independent of $\boldsymbol{r}$ and $\boldsymbol{q}$.

The result is:

$$\hat{H} = \frac{1}{2}\left[\hat{p}_e^{\mathrm{T}} G_e \hat{p}_e + \hat{p}^{\mathrm{T}} G_V \hat{p} + \hat{p}^{\mathrm{T}} G_C (\hat{J} - \hat{j}) + (\hat{J} - \hat{j})^{\mathrm{T}} G_C \hat{p} + (\hat{J} - \hat{j})^{\mathrm{T}} G_R (\hat{J} - \hat{j})\right] + U + V. \tag{36}$$

According to the adiabatic principle, the adiabatic Hamiltonian $\hat{H}_e^{\mathrm{A}}$ for the electronic subsystem can be obtained by removing the operators $\hat{p}$ and $\hat{J}$ from the Hamiltonian of equation (36), which gives

$$\hat{H}_e^{\mathrm{A}}(r, \hat{p}_e, q) = \tfrac{1}{2}\left[\hat{p}_e^{\mathrm{T}} G_e \hat{p}_e + \hat{j}^{\mathrm{T}} G_R(q) \hat{j}\right] + V(r, q). \tag{37}$$

Usually, the second term in equation (37) is not included in the electronic Hamiltonian and is shifted, at the second stage of the adiabatic separation, to the effective nuclear Hamiltonian $\hat{H}_n^{\mathrm{A}}$. However, the operator $\hat{H}_n^{\mathrm{A}}$ is simpler for the full form of $\hat{H}_e$ given by (37). To determine $\hat{H}_n^{\mathrm{A}}$ for the nth electronic *nondegenerate* state, the operators of the total Hamiltonian must be averaged with the electronic wavefunction $\Psi_{1n}^{\mathrm{A}} \equiv \Psi_n^{\mathrm{e}}(r, q)$, which is a real eigenfunction of $\hat{H}_e^{\mathrm{A}}$ with the eigenvalue $E_{1n}^{\mathrm{A}} \equiv E_n^{\mathrm{e}}(q)$. For example, the averaged nuclear kinetic energy operator takes the form

$$\langle \Psi_n^{\mathrm{e}} | \hat{p}^{\mathrm{T}} G_V \hat{p} \, | \Psi_n^{\mathrm{e}} \rangle_r = \hat{p}^{\mathrm{T}} G_V \hat{p} + \langle \{ \Psi_n^{\mathrm{e}} | \hat{p}^{\mathrm{T}} G_V \hat{p} \, | \Psi_n^{\mathrm{e}} \} \rangle_r \equiv \hat{p}^{\mathrm{T}} G_V \hat{p} + \{ \hat{p}^{\mathrm{T}} G_V \hat{p} \}_{nn}, \tag{38}$$

where the operators act within the braces $\{\ldots\}$. As a result, $\hat{H}_n^{\mathrm{A}}$ is obtained in a form analogous to that in equation (18):

$$\hat{H}_n^{\mathrm{A}} = \tfrac{1}{2}(\hat{p}^{\mathrm{T}} G_V \hat{p} + \hat{p}^{\mathrm{T}} G_C \hat{J} + \hat{J}^{\mathrm{T}} G_C \hat{p} + \hat{J}^{\mathrm{T}} G_R \hat{J}) + U + V_n^{\mathrm{A}} \tag{39}$$

where the adiabatic potential energy function $V_n^{\mathrm{A}}(q)$ is defined by the electronic energy and the adiabatic corrections:

$$V_n^{\mathrm{A}}(q) = E_n^{\mathrm{e}}(q) + \tfrac{1}{2}\{\hat{p}^{\mathrm{T}} G_V \hat{p}\}_{nn} - \{\hat{j}^{\mathrm{T}} G_C \hat{p}\}_{nn}. \tag{40}$$

The nonadiabatic second-order corrections to the adiabatic effective Hamiltonian $\hat{H}_n^{\mathrm{A}}$ can be calculated from the nondiagonal matrix elements:

$$\Delta \hat{H}_{nm} = \tfrac{1}{2} A_{nm} + B_{nm}^{\mathrm{T}} \hat{p} + C_{nm}^{\mathrm{T}} \hat{J}, \tag{41}$$

where

$$A_{nm} = \{\hat{p}^{\mathrm{T}} G_V \hat{p}\}_{nm} - \left\{\hat{j}^{\mathrm{T}} G_C \hat{p}\right\}_{nm} - \left\{\hat{p}^{\mathrm{T}} G_C \hat{j}\right\}_{nm}, \tag{42}$$

$$B_{nm}^{\mathrm{T}} = p_{nm}^{\mathrm{T}} G_V - j_{nm}^{\mathrm{T}} G_C, \tag{43}$$

$$C_{nm}^{\mathrm{T}} = p_{nm}^{\mathrm{T}} G_C - j_{nm}^{\mathrm{T}} G_R, \tag{44}$$

and $p_{nm} = \{\hat{p}\}_{nm}, j_{nm} = \{\hat{j}\}_{nm}$.

Now, the nonadiabatic operator $\Delta \hat{H}_n^{\mathrm{NA}}$ of equation (34) can be written in a clear form owing to the useful properties of the matrix elements. The elements B_{nm} and C_{nm} are antisymmetric, i.e., $B_{nm} = -B_{mn}$ and $C_{nm} = -C_{mn}$, and the elements A_{mn} fulfil the relation

$$A_{mn} = A_{nm} - 2\{\hat{p}^{\mathrm{T}} B_{nm}\}. \tag{45}$$

This leads to the final form of the nonadiabatic operator

$$\Delta \hat{H}_n^{NA} = \tfrac{1}{2}(\hat{p}^{\mathrm{T}} \Delta G_V \hat{p} + \hat{p}^{\mathrm{T}} \Delta G_C \hat{J} + \hat{J}^{\mathrm{T}} \Delta G_C \hat{p} + \hat{J}^{\mathrm{T}} \Delta G_R \hat{J}) + \Delta V. \tag{46}$$

The nonadiabatic contributions to the internal (V), Coriolis (C) and the rotation (R) kinetic energy in the nth electronic state are defined through the corrections to the $\boldsymbol{G}$ tensor :

$$\Delta G_{\mathrm{V}}^{ij} = -2B^{i} \circ B^{j}, \tag{47}$$

$$\Delta G_{\mathrm{C}}^{i\alpha} = -2B^{i} \circ C^{\alpha}, \tag{48}$$

$$\Delta G_{\mathrm{R}}^{\alpha\beta} = -2C^{\alpha} \circ C^{\beta}, \tag{49}$$

where the symbol $X \circ Y$ stands for

$$X \circ Y = \sum_{m \neq n} \left(E_{n,0}^{\mathrm{A}} - E_{m,0}^{\mathrm{A}}\right)^{-1} X_{nm} Y_{nm}.$$

The additional contribution to the potential energy is given by

$$\Delta V^{\mathrm{NA}} = \frac{1}{4} A \circ A - \frac{1}{2} A \circ \sum_{i} \{\hat{p}_i B^{i}\} + \frac{1}{2} \sum_{i} B^{i} \circ \{\hat{p}_i A\} - \sum_{i,j} B^{i} \circ \{\hat{p}_i \hat{p}_j B^{j}\}. \tag{50}$$

It is clear, that the nuclear nonadiabatic effective Hamiltonian $\hat{H}_n^{\mathrm{A}} + \Delta \hat{H}_n^{\mathrm{NA}}$ preserves the general form of the adiabatic one. The kinetic energy tensor $\boldsymbol{G}$ is modified as a result of the coupling between the total electronic angular momentum $\boldsymbol{j}$ and the nuclear linear momenta $\boldsymbol{p}$. This coupling influences the rotational and vibrational motion, changes the Coriolis interaction, and generates an additional coupling between the internal motions.

For a diatomic molecule the Coriolis term $\boldsymbol{G}_{\mathrm{C}}$ is zero and $\Delta \boldsymbol{G}_{\mathrm{C}}$ vanishes because of the cylindrical symmetry of the electronic states. For this case, the nonadiabatic Hamiltonian takes the form similar to that given by Bunker and Moss (1977). However, in their formulae the variable electronic energies $E_n^{\mathrm{e}}(\boldsymbol{q})$ appear, whereas the constant energies $E_{n,0}^{\mathrm{A}}$ enter in our formulae (47)–(49). Herman and Ogilvie (1998) who analyzed the nonadiabatic effective Hamiltonian for a diatomic molecule argued that using constant energies $E_{n,0}^{\mathrm{A}}$ does not lead to serious errors.

The effective nonadiabatic Hamiltonian of Michelot (1982), derived in the rectilinear normal coordinates and the Eckart MF axes, can be obtained from the above formulae after neglecting the matrix $\boldsymbol{p}_{nm}$. This is justified when the electronic wavefunction does not depend on the nuclear coordinates. Such a simplification is reasonable for semirigid molecules, but not for molecules exhibiting LAM modes.

12.5 The effective LAM–rotation Hamiltonian

12.5.1 THE ADIABATIC LAM–ROTATION HAMILTONIAN

The idea of the adiabatic separation can be extended to the nuclear degrees of freedom. The efficiency of the adiabatic method is based on the possibility of separating the high-frequency fast SAV modes from the other low-frequency motions while retaining the ability to describe properly the main features of the interaction between these two types of motions. This separation allows a reduction of the whole dynamical problem to partial problems of lower dimensionality. As the first step of the separation, the dynamical nuclear coordinates have to be partitioned into the coordinates $\boldsymbol{s}$ of the SAV modes and the coordinates τ of the low-frequency LAMs, and three rotational coordinates (Euler angles). We do not assume that the SAV motions behave as harmonic vibrations. The adiabatic separation is valid also for anharmonic vibrations, for example, for the stretching vibrations of X–H bonds [Carrington (1987)]. Such vibrational modes are well described as weakly coupled Morse oscillators. The valence coordinates $\boldsymbol{s}$ are suitable for anharmonic modes, so the coordinates $\boldsymbol{s}$ will be assumed as curvilinear.

The rovibrational Hamiltonian for the system of the SAVs, LAMs and the overall rotation can be defined in terms of the nonadiabatic kinetic energy tensor $\boldsymbol{G}^{\mathrm{NA}}(\boldsymbol{s},\tau)$

$$\hat{H}_{\mathrm{VR}}(\boldsymbol{s},\hat{\boldsymbol{p}}_s,\tau,\hat{\boldsymbol{p}}_\tau,\hat{\boldsymbol{J}}) = \frac{1}{2}\left[\hat{\boldsymbol{J}}^{\mathrm{T}},\hat{\boldsymbol{p}}_\tau^{\mathrm{T}},\hat{\boldsymbol{p}}_s^{\mathrm{T}}\right]\begin{bmatrix} \boldsymbol{G}^{RR} & \boldsymbol{G}^{R\tau} & \boldsymbol{G}^{Rs} \\ \boldsymbol{G}^{\tau R} & \boldsymbol{G}^{\tau\tau} & \boldsymbol{G}^{\tau s} \\ \boldsymbol{G}^{sR} & \boldsymbol{G}^{s\tau} & \boldsymbol{G}^{ss} \end{bmatrix}\begin{bmatrix} \hat{\boldsymbol{J}} \\ \hat{\boldsymbol{p}}_\tau \\ \hat{\boldsymbol{p}}_s \end{bmatrix} + U + V, \tag{51}$$

which is a function of both types of the coordinates $\boldsymbol{s}$ and τ. All components of this tensor, similarly as the potential energy function, include the nonadiabatic contributions, discussed in the previous section, but the index NA is omitted.

The vibrational adiabatic Hamiltonian for the 'fast' subsystem of the SAVs can be easily obtained removing the momenta $\hat{\boldsymbol{p}}_\tau$ and the angular momentum $\hat{\boldsymbol{J}}$ operators from the total rovibrational Hamiltonian $\hat{H}_{\mathrm{VR}}$. As a result, the eigenvalue equation for the SAV modes is obtained in the form

$$\left[\tfrac{1}{2}\hat{\boldsymbol{p}}_s^{\mathrm{T}}\boldsymbol{G}^{ss}\hat{\boldsymbol{p}}_s + V(\boldsymbol{s},\tau) - E_\nu^{\mathrm{A}}(\tau)\right]\Psi_\nu^{\mathrm{A}}(\boldsymbol{s},\tau) = 0, \tag{52}$$

where the pseudopotential is included in $V(\boldsymbol{s},\tau)$. The adiabatic energy $E_\nu^{\mathrm{A}}(\tau)$ and the eigenfunction $\Psi_\nu^{\mathrm{A}}(\boldsymbol{s},\tau)$ of the SAVs, in the state assigned the index ν, depend on the LAM coordinates.

The rest of the kinetic energy operator is partly recovered at the second stage of the adiabatic method in the averaged form. The effective adiabatic Hamiltonian for the LAM modes and the rotation contains only the terms describing these two types of motions, given by

$$\hat{H}_\nu^{\mathrm{A}} = \tfrac{1}{2}\left(\hat{\boldsymbol{p}}_\tau^{\mathrm{T}}\boldsymbol{G}_\nu^{\tau\tau}\hat{\boldsymbol{p}}_\tau + \hat{\boldsymbol{p}}_\tau^{\mathrm{T}}\boldsymbol{G}_\nu^{\tau R}\hat{\boldsymbol{J}} + \hat{\boldsymbol{J}}^{\mathrm{T}}\boldsymbol{G}_\nu^{R\tau}\hat{\boldsymbol{p}}_\tau + \hat{\boldsymbol{J}}^{\mathrm{T}}\boldsymbol{G}_\nu^{RR}\hat{\boldsymbol{J}}\right) + V_\nu^{\mathrm{A}}(\tau), \tag{53}$$

where

$$\boldsymbol{G}_\nu^{ab}(\tau) \equiv \left\langle \Psi_\nu^{\mathrm{A}}(\boldsymbol{s},\tau)\right| \boldsymbol{G}^{ab}(\boldsymbol{s},\tau)\left|\Psi_\nu^{\mathrm{A}}(\boldsymbol{s},\tau)\right\rangle_{\boldsymbol{s}} \quad \text{for } a,b = \tau, R \tag{54}$$

is the mean value of the subtensor $\boldsymbol{G}^{ab}$, averaged in the state ν of the SAVs.

The effective LAM potential energy function,

$$V_\nu^{\mathrm{A}} = E_\nu^{\mathrm{A}} + \tfrac{1}{2}\left[\{\hat{\boldsymbol{p}}_\tau^{\mathrm{T}}\boldsymbol{G}^{\tau\tau}\hat{\boldsymbol{p}}_\tau\}_{\nu\nu} + \{\hat{\boldsymbol{p}}_\tau^{\mathrm{T}}\boldsymbol{G}^{\tau s}\hat{\boldsymbol{p}}_s\}_{\nu\nu} + \{\hat{\boldsymbol{p}}_s^{\mathrm{T}}\boldsymbol{G}^{\tau s}\hat{\boldsymbol{p}}_\tau\}_{\nu\nu}\right], \tag{55}$$

contains the energy $E_\nu^{\mathrm{A}}(\tau)$ of the SAV modes and the adiabatic corrections which emerge as a result of the dependence of the SAV wavefunction on the LAM coordinates. The symbol $\{\hat{A}\}_{\nu\mu}$ introduced in equation (55) is defined as a matrix element

$$\left\{\hat{A}\right\}_{\nu\mu} \equiv \left\langle\left\{\Psi_\nu^{\mathrm{A}}(\boldsymbol{s},\tau)\left|\hat{A}(\boldsymbol{s},\tau)\right|\Psi_\mu^{\mathrm{A}}(\boldsymbol{s},\tau)\right\}\right\rangle_{\boldsymbol{s}}, \tag{56}$$

where $\hat{A}$ operates only within the braces $\{\ldots\}$.

In the adiabatic Hamiltonian, only the overall rotation energy, the LAM kinetic and the potential energy, and the energy of the Coriolis interaction between the LAM and the rotation motions are taken into account. The Coriolis interaction generated by the SAVs is completely neglected. The kinetic interaction of the SAV modes with the LAM modes is only partly taken into account through a small adiabatic correction in the last two terms of equation (55). The neglected interactions can be recovered in the nonadiabatic Hamiltonian which is derived in the Appendix. However, if these interactions are large, the perturbation expansion of the effective Hamiltonian may be slowly convergent. Thus, the minimization of $\boldsymbol{G}^{Rs}$ and $\boldsymbol{G}^{s\tau}$ is desirable to avoid the convergence problems.

The adiabatic separation of SAVs and LAM modes breaks down in the case of closely lying coupled SAV states. This coupling can be described by an effective Hamiltonian constructed for a given set of the SAV states. The two-state semirigid bender Hamiltonian for a quasilinear molecule has been worked out by Jensen (1983a, 1984) for the special case of large amplitude bending states superimposed on a doubly degenerate small amplitude bending state. The multi-state effective purely rotational Hamiltonians are commonly used in the interpretation of the rovibrational spectra of resonating vibrational states [Papoušek and Aliev (1982), and Makushkin and Tyuterev (1984)]. A phenomenological effective rotational Hamiltonian has also been proposed for molecules with LAM states delocalized over symmetrically equivalent multiple local minima in the PES [Hougen (1981, 1985), Ohashi and Hougen (1987), Oda, Ohashi and Hougen (1990), and Coudert and Hougen (1990)]. An analogous effective LAM–rotation Hamiltonian for the nonadiabatically coupled SAV states can be constructed using the matrix elements derived in the Appendix.

12.5.2 MINIMIZATION OF THE CORIOLIS AND INTERMODE COUPLINGS

The kinematic interaction of the SAV and LAM modes can be reduced by redefining the LAM coordinates. Hougen, Bunker and Johns (1970) used the conditions of Eckart (1935) and Sayvetz (1939) to eliminate G^{Rs} and $G^{\tau s}$ along the single LAM coordinate τ, for the equilibrium values of the *rectilinear* SAV coordinates. For curvilinear s coordinates, Quade (1975) proposed a transformation of a single LAM variable τ, to cancel $G^{s\tau}$ along a new LAM coordinate at fixed values s_0 of the SAV coordinates. The restriction of a constant s_0 is not convenient for the flexible models. Thus, let us consider a more general transformation W from (τ, s) to new variables (τ', s'), defined by

$$\tau' = \tau + W(\tau)(s - s_e(\tau)), \quad (57)$$

$$s' = s. \quad (58)$$

Under this transformation $G^{\tau s}$ changes to

$$G^{\tau' s'} = [1 + (\partial W/\partial\tau)(s - s_e(\tau)) - W(\partial s_e(\tau)/\partial\tau)]G^{\tau s} + WG^{ss}. \quad (59)$$

The new subtensor $G^{\tau' s'}$ vanishes at $s = s_e(\tau)$ if the matrix W fulfils the condition

$$W = -G^{\tau s}[G^{ss} - (\partial s_e(\tau)/\partial\tau)G^{\tau s}]^{-1}. \quad (60)$$

which can easily be solved numerically, if the flexible model is specified by the functional form of $s_e(\tau)$. In this way, the LAM–SAV interaction can be removed on the relaxed PES for few LAMs (on the minimum energy path, for a single LAM).

Additionally, the coupling among the SAV modes can be canceled in the harmonic approximation by introducing the *curvilinear normal coordinates* Q defined by the transformation

$$s - s_e(\tau) = L(\tau)Q. \quad (61)$$

The matrix $L(\tau)$ can be determined in a standard way at each τ such that the subtensor G^{QQ} corresponding to $G^{ss}(\tau, s_e(\tau))$ and the matrix of the force constants f in equation (24) will be diagonal. As a result, the PES describing the SAVs reduces to the sum of one-mode harmonic potentials, parametrically dependent on τ, and the global PES factorizes as

$$V(\tau, s) \cong V_e(\tau) + \frac{1}{2}\sum_i \omega_i^2(\tau)Q_i^2, \quad (62)$$

which yields the analytical formula for the adiabatic energy

$$E_v^A(\tau) \cong V_e(\tau) + \hbar\sum_i \omega(\tau)\left(v_i + \frac{1}{2}\right). \quad (63)$$

This fact simplifies the determination of the eigenfunctions of the LAM modes.

To remove on the relaxed PES the Coriolis coupling generated by the SAVs, the transformation of the MF axis can be performed employing the numerical method of Makarewicz and Bauder (1995). In this method, the initial MF axes are rotated to new MF axes, in order to minimize the selected Coriolis interaction terms $\boldsymbol{G}^{\alpha q}$, using a modified Nielsen transformation [Nielsen (1932), Lin and Swalen (1959), Pickett (1972), Szalay (1984), and Guan and Quade (1986)]. In our case the subtensor $\boldsymbol{G}^{Rs}$ has to vanish at $s = s_e(\tau)$. The cancellation of the three tensor components $\boldsymbol{G}^i = [G^{xi}, G^{yi}, G^{zi}]$ for the ith SAV mode needs three successive rotations about the x, y and z MF axes, represented by the product of the three rotation matrices

$$\boldsymbol{S}(\boldsymbol{q}) = \boldsymbol{S}_z(\varepsilon_z(\boldsymbol{q}))\boldsymbol{S}_y(\varepsilon_y(\boldsymbol{q}))\boldsymbol{S}_x(\varepsilon_x(\boldsymbol{q})). \tag{64}$$

The corresponding rotation angles $\varepsilon_x(\boldsymbol{q})$, $\varepsilon_y(\boldsymbol{q})$ and $\varepsilon_z(\boldsymbol{q})$, depend on the internal coordinates $\boldsymbol{q}$. Since the Coriolis tensor $\boldsymbol{G}^{Rq}$ transforms according to the rule

$$\boldsymbol{G}'^{Rq} = \boldsymbol{S}\left(\boldsymbol{G}^{Rq} + \boldsymbol{\omega}\boldsymbol{G}^{qq}\right) \tag{65}$$

where

$$\boldsymbol{\omega}_i = \begin{bmatrix} \partial\varepsilon_x/\partial q_i \\ \partial\varepsilon_y/\partial q_i \\ \partial\varepsilon_z/\partial q_i \end{bmatrix}, \tag{66}$$

so the Coriolis interaction terms for the ith internal coordinate can be removed if ω_i satisfies the condition

$$\omega_i = \boldsymbol{w}_i(\boldsymbol{q}) \equiv -\left[\boldsymbol{G}^{Rq}(\boldsymbol{G}^{qq})^{-1}\right]_i. \tag{67}$$

The vector $\boldsymbol{w}_i$ represents the internal angular momentum generated by the ith vibrational mode. Not all the internal angular momenta can be cancelled on the whole space of the internal coordinates $\boldsymbol{q}$. However, the SAV–Coriolis interaction terms can be removed for $\boldsymbol{s} = \boldsymbol{s}_e(\tau)$ because the rotation angles

$$\varepsilon_\alpha^i = w_{\alpha i}(\tau)(s_i - s_{e,i}(\tau)) \tag{68}$$

solve equation (65) for the ith selected SAV mode. The Coriolis interaction for all SAV modes can be eliminated on the relaxed PES by a successive sequence of rotations for each of the SAV modes.

The two transformations, $\boldsymbol{L}$ and $\boldsymbol{W}$, allow the elimination of the main part of the intermode SAV and the SAV–LAM couplings. The successive rotations of the MF axes permit the minimization of the SAV–Coriolis interaction. The information necessary for these transformations can be inferred from the molecular geometry, optimized along the LAM coordinates, and the force constant matrix $\boldsymbol{f}(\tau)$.

12.5.3 LIMITATIONS OF THE NONRIGID SAV–LAM MODEL

The minimization of the kinematic couplings can be performed for an arbitrary molecule, if a unique relaxed PES, defined by the function $\boldsymbol{S}(\tau) = s_e(\tau)$, can be determined from its total PES. However, even for simple molecules with only one LAM, multiple minimum energy paths (MEPs) may arise. To illustrate this fact let us consider the bending LAM in the water molecule.

We choose the valence coordinates, two bond lengths r_1 and r_2 treated as the coordinates of the SAV modes, and the bond angle θ as the LAM coordinate. The optimization of r_1 and r_2 for variable bending angle was carried out using the GAUSSIAN 94 program package [Frisch *et al.* (1995)]. The Møller–Plesset perturbation theory to second order and the basis set 6-311++G(3df,3pd) were employed to calculate the electronic energy. Such a method provided an accuracy sufficient for our purpose. The optimized bond lengths $r_1(\theta)$ and $r_2(\theta)$ as functions

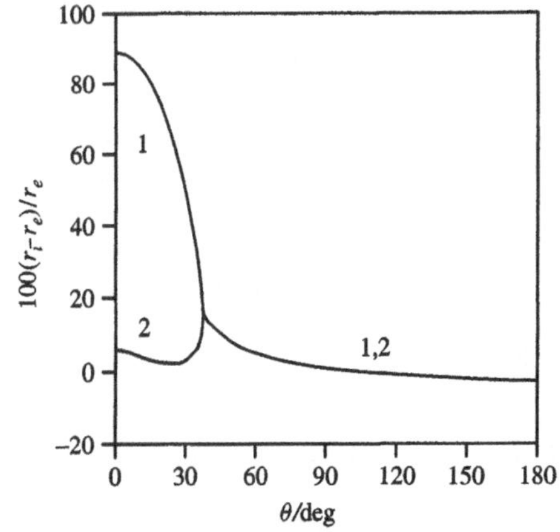

Figure 12.1. The optimized bond lengths $r_i(\theta)$ ($i = 1$ and 2) of the H_2O molecule as functions of the bending angle.

of the bending angle θ specify the flexible structure of the molecule. These functions are plotted in Figure 12.1.

The critical angle θ_c of 37.5° separates two regions in the interval from 0° to 180°. For $\theta \geq \theta_c$, the optimized bond lengths are equal and the point group symmetry of the flexible structure of H_2O is C_{2v}. However, at the critical point θ_c two functions $r_i(\theta)$ split into two different branches. This *bifurcation* results from the fact that asymmetric configuration of the molecule at small bending angle is more favorable than the symmetric one, owing to a strong Coulomb repulsion of two protons. In the extreme case of zero bending angle, the molecule becomes linear HHO. Naturally, the molecule has two equivalent asymmetric configurations, because of permutation symmetry exchanging the two protons. Thus, the equation $\partial V(r_1, r_2, \theta)/\partial\theta = 0$ has two solutions for $\theta < \theta_c$; one with $r_1(\theta) < r_2(\theta)$ and another one with $r_1(\theta) > r_2(\theta)$, so the MEP obtained bifurcates at $\theta = \theta_c$.

It is convenient to describe the asymmetric molecular configurations by the antisymmetric coordinate $r_- = r_1 - r_2$. The PES of H_2O optimized with respect to the symmetric coordinate $r_+ = (r_1 + r_2)/2$, is presented in Figure 12.2 as a function of r_- and θ.

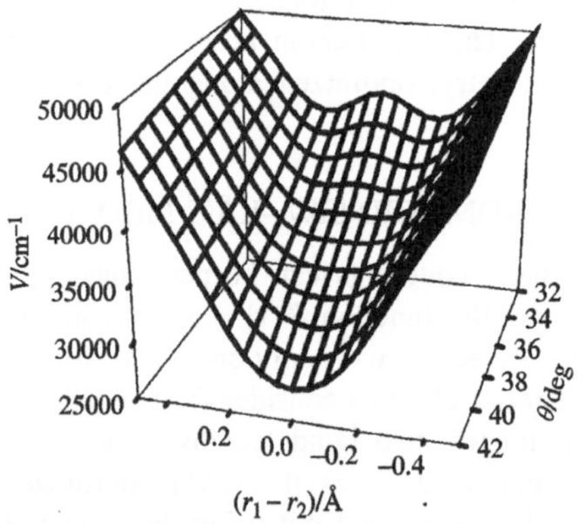

Figure 12.2. The PES of H_2O, as a function of the antisymmetric stretching r_- and the bending θ coordinates, calculated for the optimized symmetric stretching coordinate r_+. The surface exhibits the bifurcation of a single valley into two symmetrically equivalent valleys.

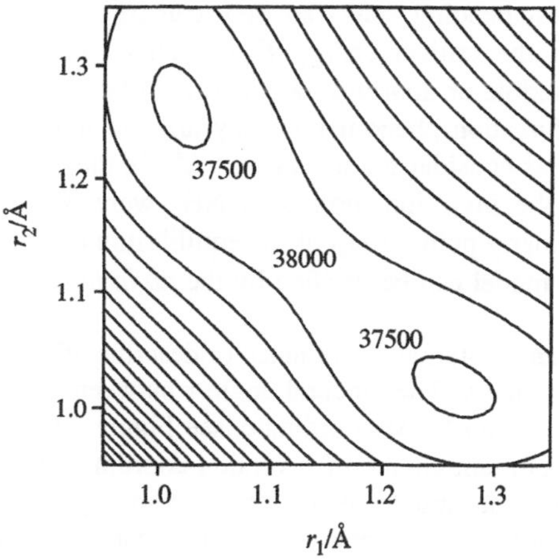

Figure 12.3. The PES of H_2O, as function of the stretching coordinates, calculated for the bending angle $\theta = 35°$. The contours are plotted in step of $500\,cm^{-1}$.

The single valley floor bifurcates at $\theta < \theta_c$ into two equivalent valleys. At the bifurcation point, the curvature of the PES along the antisymmetric coordinate r_- is zero which means that the frequency $\omega(\theta_c)$ of the antisymmetric stretching mode is zero. For $\theta < \theta_c$ the PES along the r_- coordinate is strongly anharmonic and exhibits two local minima. The character of the PES as function of the stretching coordinates for $\theta < \theta_c$ is illustrated in Figure 12.3.

The double well in the PES for $\theta < \theta_c$ changes the character of the antisymmetric stretching mode which becomes a strongly anharmonic LAM; thus it cannot be described by the SAV model and the simple equation (63) for the adiabatic energy is no longer valid.

The one-dimensional nonrigid LAM–SAV model should be substantially modified or extended to the two-dimensional model including explicitly the antisymmetric stretching mode as a LAM. Note that in the general adiabatic theory strongly anharmonic SAV modes are allowed. However, the adiabatic equation (52) cannot be simply solved within the framework of the SAV model and a more general solution method has to be employed. In such a case the flexible model should be abandoned or modified. We cannot use only one branch of the MEP to minimize the Coriolis coupling, because we then obtain the *symmetry-broken* LAM Hamiltonian which is not invariant under the nuclear permutation–inversion symmetry group.

The flexible model can be modified by imposition of an additional optimization constraint to preserve the proper symmetry of the LAM Hamiltonian. The constraint $r_1(\theta) = r_2(\theta)$ prevents symmetry breaking and yields a single *symmetry-constrained* MEP. However, the constrained minimum really corresponds to a maximum of the full PES for $\theta < \theta_c$. For this reason the potential energy function $V_s(\theta)$ calculated on the constrained MEP is higher than the analogous potential $V_e(\theta)$ calculated on the symmetry-broken branch of the unconstrained MEP.

In the nonrigid model based on the constrained MEP, the expansion of equation (62) is inadequate for $\theta < \theta_c$ since then $\omega_-^2(\theta)$ is negative. To describe properly the double well in the PES depending on the antisymmetric stretching coordinate r_-, the quartic terms should be included in the expansion (62). Naturally, the Coriolis coupling of the mode r_- and its coupling with the bending LAM cannot be minimized in the important region of the two valleys which appear in the PES for $\theta < \theta_c$. These terms can be minimized only around the symmetry-constrained MEP.

The bifurcation in H_2O emerges at very high energy region of the PES. The energy $E(\theta_c)$ calculated at the critical bending angle is $33\,900\,cm^{-1}$. This region is not sampled by the bending wavefunctions in the lowest bending states so they can be accurately described by the SAV nonrigid model. However, the critical energy $E(\theta_c)$ can be much lower for other triatomic molecules for which the flexible or nonrigid models should be applied with a caution.

A more complicated bifurcation phenomenon may arise in four-atomic molecules exhibiting LAMs. As an example, let us consider the inversion LAM in the ammonia molecule. All models used hitherto to describe the inversion motion in NH_3 were based on the 'umbrella' model assuming that the C_{3v} symmetry point group of the equilibrium configuration is preserved during the inversion motion. This model can be verified by the *ab initio* calculation of the MEP for the inversion.

We use the standard valence internal coordinates to describe all possible molecular configurations not constrained by symmetry. The dihedral angle τ between the plane H_1NH_2 and the plane H_1NH_3 is chosen as the inversion LAM coordinate, see Figure 12.4.

The angle τ is not the optimum choice of inversion coordinate for the ammonia molecule. It is more convenient for pyramidal molecules of lower symmetry, for example NH_2D. However, the bifurcation occurs also when the internal coordinates adapted to the permutational symmetry of NH_3 are used.

The three bond lengths r_i and two bond angles θ_j are treated as the flexible parameters optimized for each value of the dynamical variable τ. The electronic energy of NH_3 was calculated using the same method as for H_2O. The optimized interbond angles as functions of the inversion coordinate τ are plotted in Figure 12.5. The redundant angle θ_3 between the bonds NH_2 and NH_3 was not optimized but calculated from the remaining coordinates.

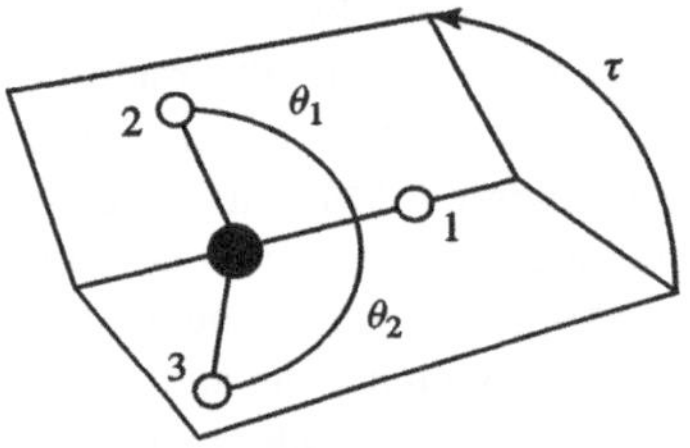

Figure 12.4. The internal valence coordinates for the ammonia molecule. The black (white) circles represent the N (H) nuclei.

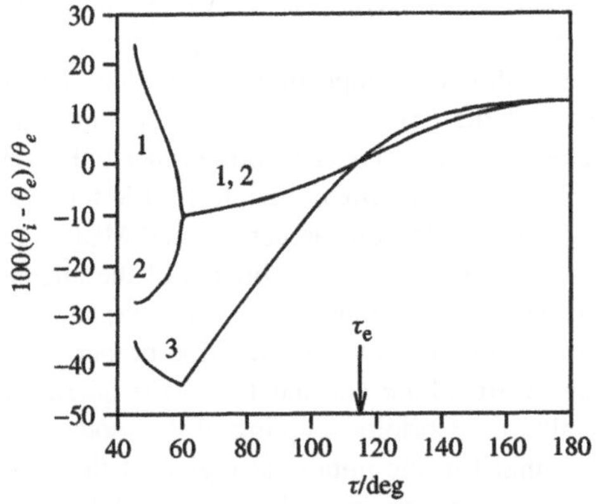

Figure 12.5. The optimized bond angles θ_i ($i = 1$, 2 and 3) of the ammonia molecule as functions of the inversion coordinate τ.

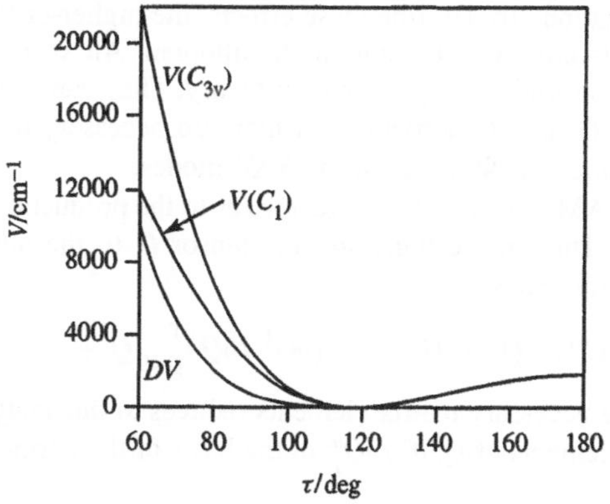

Figure 12.6. The inversion potential energy function V of the ammonia molecule calculated along the symmetry-broken MEP (C_1), and the symmetry-constrained MEP (C_{3v}). DV is the difference $V(C_{3v}) - V(C_1)$.

The planar configuration at $\tau = 180°$ corresponding to the saddle point in the PES and the equilibrium configuration at $\tau = 114.2°$ have the C_{3v} symmetry, see Figure 12.5. The remaining optimized configurations exhibit lower symmetry, C_s or C_1. The MEP starting from the saddle point is split into two symmetry-broken MEPs which meet at the global minimum and then split again. At $\tau = 60°$ one of the branch bifurcates into two new branches. At $\tau = \tau_c = 45.5°$ the MEPs truncate. The optimization process becomes divergent owing to a catastrophe which will be not discussed here. The flexible model cannot be defined for the inversion angle $\tau < \tau_c$. The optimized bond lengths behave in an analogous way. The origin of the bifurcation, a strong Coulomb repulsion of the nuclei at short internuclear distances, is the same as in the triatomic molecules. A similar effect has been recently found in some molecules with internal rotation [Cioslowski, Scott and Radom (1997)].

To avoid the bifurcation problem, a flexible model obtained from the symmetry-constrained geometry optimization can be defined. However, it gives the one-dimensional inversion potential function, shown in Figure 12.6, too high in comparison with the potential obtained from the flexible model based on the symmetry-broken MEP. Thus the expansion of the PES for the symmetry-constrained nonrigid model needs to include large quartic terms in the SAV coordinates.

The examples discussed above show that the SAVs which follow adiabatically the LAM mode may lose the character of harmonic oscillations with small amplitudes around the MEP. At some critical values of the LAM coordinate, the MEP may bifurcate. Then, the correct flexible model cannot be defined because of the symmetry breaking. The optimization of the flexible geometry must be symmetry constrained to obtain an acceptable model. In the bifurcation region the constrained MEP does not follow the true minimum in the PES. Hence, the expansion of the PES must be extended in order to describe properly the multiple valleys which substantially modify the character of the SAVs.

12.5.4 SEMIEMPIRICAL EFFECTIVE LAM–ROTATION HAMILTONIAN

The eigenvalues and eigenfunctions of the adiabatic Hamiltonian $\hat{H}_v^{A}$ are sensitive to the excitations of the SAV modes, since the form of the effective tensor $\boldsymbol{G}_v$ and the potential V_v^{A} depend upon the quantum numbers of the SAV modes. Such an effect is observable in the rovibrational spectra in the region of the LAM transition frequencies. Usually, these frequencies are shifted when the SAV states are excited. However, other important effects, such as the centrifugal distortion of the rotating molecule due to the SAV modes cannot be accounted for in the framework of the adiabatic approximation. To describe these effects, the higher-order interaction terms have to be included into the effective LAM–rotation Hamiltonian. All such terms are derived in the Appendix using first- and second-order perturbation theory. They can be calculated *ab initio* from the kinetic energy tensor $\boldsymbol{G}$ and its derivatives, which are necessary to evaluate various matrix elements with the eigenfunctions $\Psi_\nu^{\mathrm{A}}(\boldsymbol{s},\tau)$ of the SAV modes.

In the nonrigid SAV–LAM model, $\Psi_\nu^{\mathrm{A}}(\boldsymbol{s},\tau)$ factorizes to the product of the one-mode harmonic oscillator wavefunctions. Thus, if we transform the tensor $\boldsymbol{G}$ to the normal coordinates $\boldsymbol{Q}$ and then expand it into a Taylor series

$$G^{kl}(\tau,\boldsymbol{Q}) \approx G_{\mathrm{e}}^{kl}(\tau) + \left(\partial G^{kl}/\partial \boldsymbol{Q}\right)_{Q=0}^{\mathrm{T}} \boldsymbol{Q} + \cdots, \tag{69}$$

then the calculation of the necessary matrix elements reduces to the analytic evaluation of simple one-dimensional matrix elements of $Q_i^n \partial^m/\partial Q_j^m$ in the basis of the harmonic oscillator wavefunctions.

In the case of bifurcation, the function $\Psi_\nu^{\mathrm{A}}(\boldsymbol{s},\tau)$ is given approximately by a product of one-mode basis functions which must be computed numerically for an anharmonic potential. The one-mode approximation reduces the problem to a numerical evaluation of the one-dimensional matrix elements. Thus, a fully numerical construction of the effective nonadiabatic Hamiltonian is feasible, if the Taylor expansions of the PES and the $\boldsymbol{G}$ tensor are known.

The quantum chemistry standard *ab initio* methods provide analytic gradients and second derivatives of the PES. However, they are not available for highly correlated levels of theory. Thus, it is difficult to determine a detailed form of the PES for polyatomic molecules, except for triatomic molecules. Also, the accurate rovibrational tensor $\boldsymbol{G}(\tau,\boldsymbol{s})$ is not known, because the nonadiabatic electronic–nuclear interaction contributing to $\boldsymbol{G}(\tau,\boldsymbol{s})$ (see equations (47)–(49)) cannot be treated theoretically at present. However, in the adiabatic approximation, $\boldsymbol{G}(\tau,\boldsymbol{s})$ can be determined numerically for an arbitrary dynamical LAM model. If the main features of the PES and the tensor $\boldsymbol{G}$ are known or can be predicted from some simple models, then their form can be refined using the spectral data. Such a *semiempirical* methodology can be directly applied to the effective nonadiabatic LAM–rotation Hamiltonian which contains the effective potential energy function $V_\nu(\tau)$ and two types of the kinetic energy terms:

- the quadratic terms such as $\hat{p}_k G^{kl} \hat{p}_l$, with the effective tensor $\boldsymbol{G}_\nu(\tau)$, which includes the adiabatic and small nonadiabatic contributions presented in the Appendix. The adiabatic part of $\boldsymbol{G}_\nu(\tau)$ is the mean value of the total rovibrational tensor $\boldsymbol{G}(\tau,\boldsymbol{s})$ averaged in the state ν of the SAVs; see equations (53)–(54).
- the quartic terms such as $\hat{p}_k^4$, $\hat{p}_k^3\hat{J}_\alpha$, $\hat{p}_k^2\hat{J}_\alpha^2$, $\hat{p}_k\hat{J}_\alpha^3$, $\hat{J}_\alpha^4$. The full set of these operators is given in the Appendix.

Some selected terms of the Hamiltonian can be treated empirically and fitted to the observed spectra. The effective tensor $\boldsymbol{G}_\nu(\tau)$ can also be refined through a modification of the function $s_{\mathrm{e}}(\tau)$ specifying the flexible model. In this way, only the part $\boldsymbol{G}_{\mathrm{e}}(\tau)$ independent of the state ν of the SAV modes can be improved. The v-dependent adiabatic and nonadiabatic contributions to $\boldsymbol{G}_\nu(\tau)$ can be fitted by use of correction functions $\boldsymbol{F}_\nu(\tau)$ which modify the approximate tensor $\boldsymbol{G}_{\mathrm{e}}(\tau)$ obtained from the flexible model:

$$G_\nu^{kl}(\tau) \cong G_{\mathrm{e}}^{kl}(\tau) F_\nu^{kl}(\tau). \tag{70}$$

It is convenient to assume $F_\nu^{kl}(\tau)$ in the form of an expansion into a reasonably chosen set of *expansion basis functions* $f_m^6(\tau)$

$$F_\nu^{kl}(\tau) = \sum_m D_{\nu,m}^{kl} f_m^6(\tau). \tag{71}$$

The advantage of such an expansion is efficiency of the fitting procedure due to the linearity of the $D^{kl}_{v,m}$ parameters. The functions $f_m(\tau)$ should be adapted to a given LAM model. For example, the periodic functions $\sin(m\tau)$ and $\cos(m\tau)$ are appropriate for the treatment of the internal rotation LAM, whereas polynomials are adequate for aperiodic LAMs. The effective potential $V_v(\tau)$ and a dipole moment function, necessary for the calculation of the transition intensities, can also be expanded into analogous series defined by their own expansion basis functions.

When the dynamical LAM model cannot be established owing to a lack of data on the structure and internal dynamics, the tensor $\boldsymbol{G}_v(\tau)$ and the potential $V_v(\tau)$ can be treated as purely empirical functions fully specified by the expansion functions. Then $\boldsymbol{G}_e$ is formally replaced by unity in equation (70).

The quartic Hamiltonian terms have to be selected and treated in a similar fashion. Three types of these terms have been taken into account [Makarewicz (1996)]: the standard centrifugal distortion terms with pure rotational operators such as $\hat{J}^2\hat{J}_z^2$, the Coriolis terms containing operators $\hat{p}$ and $\hat{p}^3$, and the vibration–rotation terms proportional to $\hat{p}^2$ and $\hat{p}^4$. Additional terms have been included in the Hamiltonian to describe the interaction of the nuclear quadrupole with the angular momentum, responsible for the hyperfine quadrupole structure of the rovibrational spectra.

12.6 Rovibration eigenvalue problem

12.6.1 GENERAL SOLUTION STRATEGY

The rovibrational molecular states are most often calculated by diagonalizing the Hamiltonian matrix in a properly chosen functional basis set. The effective LAM–rotation Hamiltonian can be written in a general form

$$\hat{H}(\tau,\hat{p},\hat{\boldsymbol{J}}) = \hat{H}_{\mathrm{V}}(\tau,\hat{p}) + \sum_M \hat{V}_M(\tau,\hat{p})\hat{R}_M(\hat{\boldsymbol{J}}), \tag{72}$$

where $\hat{H}_{\mathrm{V}}(\tau,\hat{p})$ is the pure vibrational Hamiltonian, $\hat{R}_M(\hat{\boldsymbol{J}})$ are rotational operators and $\hat{V}_M(\tau,\hat{p})$ are associated vibrational operators. Some of the vibrational operators may become singular for linear configurations of a molecule, when one of the moments of inertia vanishes. Here, we consider a nonlinear molecule for which the singularities can be ignored and its wavefunctions can be constructed in the form of a direct product of the vibrational eigenfunctions $\Phi_N(\tau)$ and rotational basis functions. The molecular states can approximately be assigned by the set N of the vibrational quantum numbers and the set r of the rotational quantum numbers labeling the rotational eigenfunctions of the effective rotational Hamiltonian

$$\hat{H}^N_{\mathrm{rot}}(\hat{\boldsymbol{J}}) = \sum_M \langle\Phi_N(\tau)|\,\hat{V}_M(\tau,\hat{p})\,|\Phi_N(\tau)\rangle_\tau\,\hat{R}_M(\hat{\boldsymbol{J}}). \tag{73}$$

The mean values $\langle\Phi_N|\hat{V}_M|\Phi_N\rangle$ of the vibrational operators play the role of the rotational constants of the Nth vibrational state. The set of the rotational eigenfunctions $\{S_r\}$ of a given vibrational state N can be combined with the set $\{\Phi_{N'}(\tau)\}$ to span the rovibrational basis set used to calculate accurate wavefunctions of the rotational multiplet in the Nth vibrational state. In this approach, the *local basis set* and the corresponding Hamiltonian matrix is built for each rotational multiplet in a given vibrational state. Such a method is convenient when the number of necessary vibrational states is small. For a large number of the vibrational states to be calculated, only one rotational basis set, determined from the ground vibrational state, may be used to construct only one *global basis set* for all considered states. The global basis set is larger than the local one, but needs to be built only once.

To determine the rotational basis functions from the equation

$$\left[\hat{H}^N_{\text{rot}}(\hat{\boldsymbol{J}}) - E_r\right] \boldsymbol{S}_r = 0, \tag{74}$$

we used the linear combinations of the standard symmetric-top functions $S_{J,K}$,

$$S^{\sigma}_{J,K} = (S_{JK} + \sigma S_{J,-K})\mathrm{e}^{i\pi(2K+1-\sigma)/4}/\sqrt{2}, \tag{75}$$

for $\sigma = \pm 1$ and for $K = 1$ to J, and $S^0_{J,0} = S_{J0}$ for $K = 0$. The phase factors of S_{JK} are chosen such that the matrix elements of the operators $\hat{J}_x$ ($\hat{J}_y$ and $\hat{J}_z$) are imaginary (real). As a consequence, the matrix elements of the all operators $\hat{J}_\alpha$ in the basis set $\{S^{\sigma}_{J,K}\}$ are imaginary. This choice guarantees that all the matrix elements of the operators $\hat{V}_M\hat{R}_M$ are real when the vibrational basis functions are real. After solving equation (74), the matrix elements $\langle S^{\sigma}_{J,K}|\hat{R}_M|S^{\sigma'}_{J,K'}\rangle$ are transformed to $\langle \boldsymbol{S}_r|\hat{R}_M|\boldsymbol{S}_{r'}\rangle$ which are then combined with the vibrational matrix elements, in order to set up the total Hamiltonian matrix $\boldsymbol{H}$. The diagonalization of $\boldsymbol{H}$ yields the rovibrational wavefunctions

$$\Psi^J = \sum_{N,r} C(N,r)\Phi_N \boldsymbol{S}_r \tag{76}$$

and the corresponding energy levels, for each value of the angular momentum quantum number J.

12.6.2 SCF–DVR METHOD

The critical point in the above procedure is the solution of the vibrational equation

$$\left[\hat{H}_{\mathrm{V}}(\tau, \hat{\boldsymbol{p}}) - E_N\right] \Phi_N(\tau) = 0 \tag{77}$$

for coupled LAMs and the calculation of the vibrational matrix elements as multi-dimensional integrals. We used the self-consistent field (SCF) method to generate optimum one-dimensional vibrational basis functions which formed a direct product basis set

$$\Theta_n(\tau) = \varphi_{n(1)}(\tau_1)\varphi_{n(2)}(\tau_2)\ldots\varphi_{n(L)}(\tau_L). \tag{78}$$

The vibrational SCF method has been extensively reviewed in the literature [Bowman (1978), Roth, Gerber and Ratner (1983), Makarewicz, Wierzbicki and Koput (1985), Bowman (1986), Ratner and Gerber (1986), Ratner, Gerber and Buch (1987), Ratner *et al.* (1991), Jelski, Haley and Bowman (1996), Roitberg, Gerber, and Ratner (1997), and Carter, Culik and Bowman (1998)] so we present only the main idea of the method. It is based on the approximate wavefunction of equation (78) which is factorized into single-mode functions determined from coupled Hartree equations. The effective SCF Hamiltonian for the kth mode is obtained by averaging the total L-mode vibrational Hamiltonian over the remaining L-1 modes. Thus, (L-1)-dimensional integrals must be calculated to determine the SCF Hamiltonians and then the kth SCF eigenfunctions must be calculated from the equation of a general form

$$\left[-\frac{\hbar^2}{2}\partial/\partial\tau_k G_k(\tau_k)\partial/\partial\tau_k + V(\tau_k) - e_{k,n(k)}\right] \varphi_{n(k)}(\tau_k) = 0. \tag{79}$$

The functions $\varphi_n(\tau)$, with the label k subsequently omitted, were constructed as linear combinations of primitive basis functions $f_m(\tau)$ satisfying proper boundary conditions. For the periodic problems we employ Meyer's (1979) orthogonal basis functions $f^M_m(\tau)$ localized at equidistant points $\{\tau(m)\}$ from the interval $(-\pi, +\pi)$. The matrix elements of an arbitrary function, for example the potential function $V(\tau)$ are approximated by the simple formula

$$\langle f_m^M(\tau)|\, V(\tau)\, |f_{m'}^M(\tau)\rangle \cong V(\tau(m))\delta_{mm'}, \tag{80}$$

which needs evaluation of $V(\tau(m))$ at only one point. In this basis, the matrix of V is diagonal. This property allows also a simple calculation of the matrix of the kinetic energy operator:

$$\langle f_m^M(\tau)|\, \hat{p}_\tau G(\tau)\hat{p}_\tau\, |f_{m'}^M(\tau)\rangle \cong \sum_i p_{mi} p_{im'} G(\tau(i)), \tag{81}$$

where the matrix elements p_{mi} of the momentum operator are defined by some analytic expression.

For the nonperiodic boundary conditions we propose the primitive functions in the form

$$f_m(\tau) = |\mathrm{d}y/\mathrm{d}\tau|^{1/2}\, w(y)^{1/2} P_m(y), \tag{82}$$

where $P_m(y)$ are the normalized *classical orthogonal polynomials* with the weight function $w(y)$. These polynomials include Hermite, Laguerre, Jacobi, Legendre and other polynomials as special cases [Abramovitz and Stegun (1972)]. We assume, that y is some monotonic function of τ. The function $y(\tau)$ may depend on some nonlinear parameters which make flexible basis functions $f_m(\tau)$. For example, if we choose $P_m(y)$ as Laguerre polynomials and y as the exponential function $y = A\exp(-B(r - r_e))$ then the resulting functions $f_m(r)$ become the same as those proposed by Tennyson and Sutcliffe (1982) to solve the radial problems for the stretching coordinate. For $y = A/r$ the functions $f_m(r)$ of Kauppi (1994) are obtained.

The set of functions $\{f_m(\tau)\}$ is orthonormal, since from equation (82) it follows that

$$\int_a^b f_m(\tau) f_{m'}(\tau)\, \mathrm{d}\tau = \int_{y(a)}^{y(b)} w(y) P_m(y) P_{m'}(y)\, \mathrm{d}y = \delta_{mm'}. \tag{83}$$

Naturally, $y(a)$ and $y(b)$ must correspond to the appropriate boundaries for a given set of polynomials. For example, $y(a) = -1$ and $y(b) = +1$ for Jacobi polynomials, $y(a) = -\infty$ and $y(b) = +\infty$ for Hermite polynomials, etc.

To calculate the kinetic energy matrix elements $\langle m|\hat{T}|m'\rangle$ we need an explicit expression for the derivatives $\mathrm{d}f_m(\tau)/\mathrm{d}\tau$. Using the recurrence relation

$$\mathrm{d}P_m/\mathrm{d}y = a_m(y)P_m(y) + b_m(y)P_{m-1}(y) \tag{84}$$

for the polynomials $P_m(y)$ [Abramovitz and Stegun (1972)] we obtain the simple relation

$$\mathrm{d}f_m(\tau)/\mathrm{d}\tau = w^{1/2}\, |\mathrm{d}y/\mathrm{d}\tau|^{1/2}\, (\mathrm{d}y/\mathrm{d}\tau)[(g(y) + A_m(y))P_m(y) + b_m(y)P_{m-1}(y)], \tag{85}$$

where

$$g(y) = \tfrac{1}{2}(\mathrm{d}^2 y/\mathrm{d}\tau^2)/(\mathrm{d}y/\mathrm{d}\tau)^2 \tag{86}$$

depends only on the assumed function $y(\tau)$, and

$$A_m(y) = \tfrac{1}{2}\mathrm{d}\ln w/\mathrm{d}y + a_m(y) \tag{87}$$

depends only on the type of the polynomial.

We can easily generate a variety of primitive basis functions and calculate the matrix elements $\langle m|\hat{T}|m'\rangle$ and $\langle m|V|m'\rangle$ using Gaussian quadratures adapted to a given type of the polynomials P_m. So, the one-dimensional equations (79) can be efficiently solved.

We solve the SCF equations only for the ground vibrational state. As a result, the set of the functions $G_k(\tau_k)$ and $V_k(\tau_k)$ is determined for L modes τ_k, $k = 1, 2, \ldots, L$. Then, solving the equation (79) for the excited states, with unchanged $G_k(\tau_k)$ and $V_k(\tau_k)$, we generate one-dimensional basis sets. From these sets, the L-dimensional product basis (78) can be formed, but its size is usually too large. To select conveniently the SCF basis functions, we choose the cutoff energy $E_{\max}$ and include the SCF basis function Θ_n in the set, if

$$\sum_{k=1}^{L} W(k) e_{n(k)} < E_{\max}, \tag{88}$$

where the weight factors $W(k)$ are introduced for balancing properly the resulting basis set.

Now the matrix elements $\langle \Theta_n | \hat{H}_{\mathrm{V}} | \Theta_{n'} \rangle$ as L-dimensional integrals must be calculated to solve accurately the vibrational eigenequation (77) beyond the SCF approximation. This is a very time consuming process which can be optimized by reduction of the number of the Gaussian integration points. This reduction was achieved by determination of a new set of the integration points adapted to the SCF solutions. It is well known that the Gausian quadrature points can be obtained from the diagonalization of the coordinate matrix $\langle m | \tau | m' \rangle$ calculated for the classical orthogonal polynomials [Harris, Engerholm and Gwinn (1965), and Dickinson and Certain (1968)]. By analogy, we diagonalize the matrix $\langle n | \tau | n' \rangle$ calculated for the SCF basis functions $\varphi_n(\tau)$ of each kth mode, to obtain the integration points which should be optimally adapted to the SCF basis sets. The diagonalization of the coordinate matrix yields a new basis set with the functions $\chi_i(\tau)$ localized around the obtained integration points $\tau(i)$, so they can be treated as functions of a discrete variable representation (DVR); for reviews see Bačić and Light (1989), and Corey and Tromp (1995). Thus, the one-dimensional integrals involving these functions can be approximated by a formula analogous to equation (80). A similar approach for determination of the potential-optimized DVR basis functions was proposed by Echave and Clary (1992).

The functions $\varphi_n(\tau)$ are related to the SCF–DVR basis functions $\chi_i(\tau)$ through the orthogonal transformation

$$\varphi_n(\tau) = \sum_i c(n, i) \chi_i(\tau), \tag{89}$$

whose matrix $\boldsymbol{c}^{\mathrm{T}}$ is obtained from the diagonalization of the matrix $\langle n | \tau | n' \rangle$. Hence, the integral involving an arbitrary function $F(\tau)$ and the SCF basis functions φ_n can be replaced by the usual sum

$$\begin{aligned} \langle n | F | n' \rangle &= \sum_{i,i'} c(n, i) c(n', i') \langle i | F | i' \rangle \cong \sum_{i,i'} c(n, i) c(n', i') F(\tau(i)) \delta_{ii'} \\ &= \sum_i c(n, i) c(n', i) F(\tau(i)). \end{aligned} \tag{90}$$

In practice, this approximation gives very accurate results. This fact allows the reduction of the computation time needed to evaluate the matrix elements $\langle \Theta_n | \hat{V}_M | \Theta_{n'} \rangle$. After solving the vibrational problem, they are transformed to $\langle \Phi_N | \hat{V}_M | \Phi_{N'} \rangle$ in order to set up the matrix $\boldsymbol{H}$ of the total rovibrational Hamiltonian. The diagonalization of $\boldsymbol{H}$ yields the rovibrational energy levels, and the eigenfunctions which are used to calculate the hyperfine quadrupole structure of the spectrum and the intensities of the rovibrational transitions.

12.7 Applications

The numerical determination of the effective Hamiltonian for the LAMs coupled to the rotation and the solution of the eigenvalue problem are applicable to any nonlinear molecule and any kind of the LAM modes. In the computer program which performs these calculations [Makarewicz (1996)] the separate subroutines specifying the dynamical LAM model of a molecule have to be provided by the user.

To define the LAM model, the Cartesian components of the nuclear position vectors, in the MF axis system, must be defined as functions of the LAM coordinates τ and of the flexible structural parameters $\boldsymbol{S}(\tau, \boldsymbol{P}^S)$, i.e. $\boldsymbol{r} = \boldsymbol{R}(\tau, \boldsymbol{S}(\tau, \boldsymbol{P}^S))$. The functions $\boldsymbol{S}(\tau, \boldsymbol{P}^S)$ may be determined from the *ab initio* optimization of the molecular geometry, or may be assumed as empirical functions with some nonlinear parameters $\boldsymbol{P}^S$ fitted to the spectral data.

The expansion basis functions $f_m^G(\tau, \boldsymbol{P}^G)$ for the correction of the tensor $\boldsymbol{G}$ [see equations (70) and (71)] and analogous functions $f_m^V(\tau, \boldsymbol{P}^V)$ for the expansion of the potential energy function $V(\tau)$ depend on the LAMs considered, so they have to be defined by the user. The functions f_m^G are also used for the expansion of the dipole moment function and may also contain some fitted parameters $\boldsymbol{P}^G$. A general scheme of the input parameters is shown in Figure 12.7. The output is the rovibrational energy levels and wavefunctions, expanded into linear combinations of the vibrational Φ_N and rotational S_r basis functions with the expansion coefficients $C(N, r)$. In order to assign the approximate vibrational and rotational quantum numbers to a given state, the diagonal elements of the vibrational and rotational population matrices

$$P_N = \sum_r |C(N, r)|^2 \text{ and } P_r = \sum_N |C(N, r)|^2$$

are calculated. The $N'(r')$ corresponding to the maximum element of $P_{N'}(P_{r'})$ is treated as an approximate vibrational (rotational) quantum number. The N' can also be further decomposed to obtain information on the contributions of the one-mode SCF functions $\varphi_{n(k)}$ to the N'th vibrational state. Some states may exhibit a mixed character due to a strong rovibrational interaction. Then they cannot be assigned in this way, and only the ordering label of the eigenvectors indicates the state.

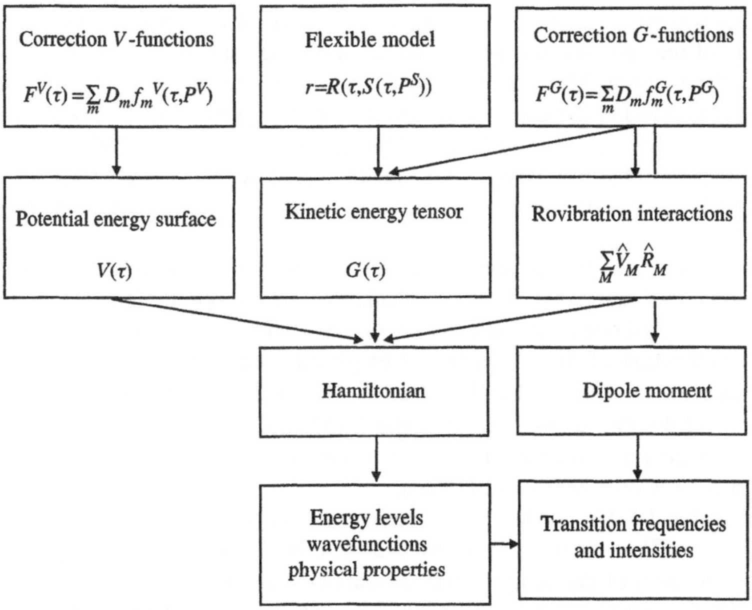

Figure 12.7. The input parameters of the effective rovibrational Hamiltonian and the calculated spectral data.

For a series of rotational states belonging to a given vibrational state N, the effective rotational constants are calculated as average values $\frac{1}{2}\langle\Phi_N|G^{\alpha\beta}|\Phi_N\rangle$ of the rotational components of the tensor $\boldsymbol{G}$. The analogous values $\langle\Phi_N|\hat{V}_M|\Phi_N\rangle$ provide information on the Coriolis and centrifugal interactions produced by the SAV modes.

From the rovibrational energy levels and wavefunctions the transition frequencies and their intensities are calculated. For a molecule with a single quadrupolar nucleus the hyperfine quadrupole structure of the spectrum is calculated by perturbation theory.

The program is useful for the assignment of the rovibrational spectra observed in several excited states of the LAM mode. The correct assignment of such spectra is difficult when the rotational transitions are separately fitted with Watson's rotational Hamiltonian for each exited LAM state. For closely lying LAM states with similar sets of rotational constants, the assignment may be ambiguous. A global fit of the rotational transitions in all observed LAM states with the use of the effective LAM–rotation Hamiltonian is very helpful in resolving the ambiguities.

The LAM excited states in floppy molecules contain the information about a large portion of the PES. If the local minima in the PES are separated by low potential energy barriers, the tunneling motion causes the splitting of the LAM energy levels. The analysis of the splitting provides precious data on the tunneling paths and the potential barriers. The splitting patterns observed in molecules with low symmetry are very complicated owing to the interaction of the LAMs with the rotation. The program presented is very helpful in analysis of the observed splittings in the rovibrational spectra of floppy molecules, especially molecular complexes with a hydrogen bond or van der Waals complexes.

An example of such analysis of the microwave spectra of the van der Waals complex formed by furan and CO is given in a recent work of Brupbacher, Makarewicz and Bauder (1998). The rovibrational transitions in the ground and the first excited state of the internal rotation of CO with respect to furan was first analyzed separately using a rotational Watson Hamiltonian. This standard model was able to predict crudely only the unsplit spectrum, but the fit was very poor. The spectrum was successfully interpreted by employing the one-dimensional effective LAM Hamiltonian for the internal rotation. The influence of the other van der Waals modes on the internal rotation was partly taken into account by an empirical correction of the tensor $\boldsymbol{G}$.

The analysis of the split spectral lines becomes more difficult in the presence of quadrupole coupling with the rotation. The hyperfine quadrupole structure of the rotational spectra in the ground and excited states of the internal rotation in N-methylpyrrole and its argon complexes has been recently measured and analyzed [Huber, Makarewicz and Bauder (1998)] using the flexible model for the internal rotation. The hyperfine quadrupole structure of many rotational transitions could not be predicted with the standard model of a semirigid rotor perturbed by the quadrupole interaction.

In molecular complexes, the geometry relaxation accompanying LAMs is often complicated and cannot be predicted intuitively. *Ab initio* calculations are indispensable in determination of the flexible models. The hydrogen-bonded complex of water with carbon dioxide is a good example of a fruitful combination of the *ab initio* and empirical methodologies in the investigation of the LAMs [Makarewicz, Ha and Bauder (1993)]. The *ab initio* calculation of one- and two-dimensional cuts of the PES along intermolecular LAM coordinates helped to select an adequate flexible model for the internal rotation, which was later refined empirically using the measured rovibrational spectra [Columberg *et al.* (1998)].

A three-dimensional LAM model of a water–formaldehyde hydrogen-bonded complex was determined *ab initio*. The calculated structure and the relaxed PES allowed the prediction of the splitting in the spectrum of the complex [Ha, Makarewicz and Bauder (1993)].

In van der Waals complexes, the interaction of the intramolecular SAV modes with the intermolecular LAM modes is rather weak. For this reason, the structural relaxation of the monomers can be neglected and the simple semirigid LAM models, with empirically corrected tensor $\boldsymbol{G}$, describe properly the dynamics of the van der Waals intermolecular modes. Using such models, three-dimensional PESs were fitted to pure rotational spectra of the complexes formed by rare gas atoms with aromatic ring molecules, such as benzene [Brupbacher, Makarewicz and Bauder (1994)], pyridine [Makarewicz and Bauder (1995)], 2,5-dihydrofuran [Maris *et al.* (1997)], 2,3-dihydrofuran [Caminati *et al.* (1997b)], and pyridazine [Caminati *et al.* (1997a)].

The quality of the PES obtained could be verified only for the Ar–benzene complex for which the frequencies of the van der Waals modes were determined by Kim and Felker (1997) from the Raman spectra. The empirical PES was tested by Riedle and van der Avoird (1996) who established that this PES predicts observable vibrational frequencies with accuracy better than $1\,\mathrm{cm}^{-1}$. The rotational constants calculated for the excited vibrational states were also in very good agreement with the experimental ones. Recently, Sussmann, Neuhauser and Neusser (1995) and Neuhauser *et al.* (1998) measured the vibrational overtones of the van der Waals modes in Ar–benzene, using coherent ion dip spectroscopy, and confirmed the theoretical predictions.

12.8 Conclusions

The adiabatic method has been used in the approximate treatment of the rovibrational dynamics of a polyatomic molecule. This method assumes that the molecular degrees of freedom can be considered as a series of dynamical subsystems ordered according to their contribution to the total energy. The electronic motions, the high-frequency SAVs and the low-frequency LAMs coupled to the overall rotation have been approximately separated. Starting from the total molecular Hamiltonian, the effective adiabatic Hamiltonians for each of these systems have been derived. The nonadiabatic effect was accounted for within the second-order perturbation theory.

The flexible and nonrigid models describing the interaction of the SAVs with the LAMs have been analyzed. To construct these models, the minimum energy path (MEP) for the LAM mode should be determined. We have shown that at some critical value of the LAM coordinate this path may bifurcate into equivalent symmetry-broken branches. In the bifurcation region the SAV modes lose the character of small amplitude vibrations around a single local branch of the MEP and become strongly anharmonic motions. In such a case the nonrigid model must be defined for the symmetry-constrained MEP, and the PES must be expanded up to quartic terms in the coordinates of the SAV modes, to describe properly multiple valleys in the PES.

A convenient numerical method of the minimization of the Coriolis coupling and the interactions between the SAV and LAM modes has been presented. This method allows a construction of the optimal LAM–rotation effective Hamiltonian, but it requires a knowledge of the multidimensional PES of a molecule. A less demanding method is based on a semiempirical effective Hamiltonian with high-order interaction terms fitted to the observed spectra. These terms and additional corrections to the kinetic $\boldsymbol{G}$ tensor take into account the interaction of the SAVs with the LAMs and the rotation, and allow the determination of realistic PES describing the LAM modes.

To solve the eigenvalue problem for the multidimensional effective Hamiltonian, we proposed a computational method using SCF basis functions and the SCF-adapted DVR integration points. This efficient method could be employed in the assignment of the rovibrational spectra of the floppy molecules, and in the fitting of their PESs describing the coupled LAMs.

The presented methodology based on the adiabatic separation of the molecular motions is valid only for an isolated electronic state and for the SAV state well separated from the remaining similar states. In the case of resonating SAV states, the nonadiabatic coupling changes qualitatively the character of the interaction between the SAV and LAM modes. This interaction can be described by an effective Hamiltonian constructed for a set of closely lying SAV states. Work on the extension of the effective LAM–rotation Hamiltonian for a single isolated SAV state to a more general Hamiltonian for a set of interacting SAV states is in progress.

12.9 Appendix

To derive the nonadiabatic LAM–rotation Hamiltonian, the matrix elements of the kinetic energy operators must be calculated for the states ν and μ of the SAV modes. The operators of the vibrational kinetic energy

$$\hat{T}^{\mathrm{V}} = \tfrac{1}{2}\,(\hat{\boldsymbol{p}}_\tau^{\mathrm{T}}\boldsymbol{G}^{\tau\tau}\hat{\boldsymbol{p}}_\tau + \hat{\boldsymbol{p}}_\tau^{\mathrm{T}}\boldsymbol{G}^{\tau s}\hat{\boldsymbol{p}}_s + \hat{\boldsymbol{p}}_s^{\mathrm{T}}\boldsymbol{G}^{s\tau}\hat{\boldsymbol{p}}_\tau)\,, \tag{91}$$

the Coriolis interaction for the LAM modes

$$\hat{T}^{R\tau} = \tfrac{1}{2}\left(\hat{\boldsymbol{p}}_\tau^{\mathrm{T}}\boldsymbol{G}^{\tau R}\hat{\boldsymbol{J}} + \hat{\boldsymbol{J}}^{\mathrm{T}}\boldsymbol{G}^{R\tau}\hat{\boldsymbol{p}}_\tau\right), \tag{92}$$

the Coriolis interaction for the SAV modes

$$\hat{T}^{Rs} = \tfrac{1}{2}\left(\hat{\boldsymbol{p}}_s^{\mathrm{T}}\boldsymbol{G}^{sR}\hat{\boldsymbol{J}} + \hat{\boldsymbol{J}}^{\mathrm{T}}\boldsymbol{G}^{Rs}\hat{\boldsymbol{p}}_s\right) \tag{93}$$

and the rotation

$$\hat{T}^{R} = \tfrac{1}{2}\hat{\boldsymbol{J}}^{\mathrm{T}}\boldsymbol{G}^{RR}\hat{\boldsymbol{J}} \tag{94}$$

give the following matrix elements:

$$\left(\hat{T}^{\mathrm{V}}\right)_{\nu\mu} = \tfrac{1}{2}\left(\hat{\boldsymbol{p}}_\tau^{\mathrm{T}}\boldsymbol{G}_{\nu\mu}^{\tau\tau}\hat{\boldsymbol{p}}_\tau + A_{\nu\mu} + \boldsymbol{B}_{\nu\mu}^{\mathrm{T}}\hat{\boldsymbol{p}}\right), \tag{95}$$

$$\left(\hat{T}^{R\tau}\right)_{\nu\mu} = \tfrac{1}{2}\left(\hat{\boldsymbol{p}}_\tau^{\mathrm{T}}\boldsymbol{G}_{\nu\mu}^{\tau R}\hat{\boldsymbol{J}} + \hat{\boldsymbol{J}}^{\mathrm{T}}\boldsymbol{G}_{\nu\mu}^{R\tau}\hat{\boldsymbol{p}}_\tau + \hat{\boldsymbol{J}}^{\mathrm{T}}H_{\nu\mu}^{R}\right), \tag{96}$$

$$\left(\hat{T}^{Rs}\right)_{\nu\mu} = \tfrac{1}{2}\left(\{\hat{\boldsymbol{p}}_\tau^{\mathrm{T}}\boldsymbol{G}^{sR}\}_{\nu\mu}\hat{\boldsymbol{J}} + \hat{\boldsymbol{J}}^{\mathrm{T}}\{\boldsymbol{G}^{Rs}\hat{\boldsymbol{p}}_s\}_{\nu\mu}\right), \tag{97}$$

and

$$\left(\hat{T}^{R}\right)_{\nu\mu} = \tfrac{1}{2}\hat{\boldsymbol{J}}^{\mathrm{T}}\boldsymbol{G}_{\nu\mu}^{RR}\hat{\boldsymbol{J}}, \tag{98}$$

where

$$A_{\nu\mu} = \{\hat{\boldsymbol{p}}_\tau^{\mathrm{T}}\boldsymbol{G}^{\tau\tau}\hat{\boldsymbol{p}}_\tau\}_{\nu\mu} + \{\hat{\boldsymbol{p}}_\tau^{\mathrm{T}}\boldsymbol{G}^{\tau s}\hat{\boldsymbol{p}}_s\}_{\nu\mu} + \{\hat{\boldsymbol{p}}_s^{\mathrm{T}}\boldsymbol{G}^{s\tau}\hat{\boldsymbol{p}}_\tau\}_{\nu\mu}, \tag{99}$$

$$\boldsymbol{B}_{\nu\mu}^{\mathrm{T}} = \{\boldsymbol{G}^{\tau\tau}\hat{\boldsymbol{p}}_\tau\}_{\nu\mu}^{\mathrm{T}} - \{\boldsymbol{G}^{\tau\tau}\hat{\boldsymbol{p}}_\tau\}_{\mu\nu}^{\mathrm{T}} + \{\hat{\boldsymbol{p}}_s^{\mathrm{T}}\boldsymbol{G}^{s\tau}\}_{\nu\mu} - \{\hat{\boldsymbol{p}}_s^{\mathrm{T}}\boldsymbol{G}^{s\tau}\}_{\mu\nu}, \tag{100}$$

and

$$\boldsymbol{H}_{\nu\mu}^{R} = \{\boldsymbol{G}^{R\tau}\hat{\boldsymbol{p}}_\tau\}_{\nu\mu} - \{\boldsymbol{G}^{R\tau}\hat{\boldsymbol{p}}_\tau\}_{\mu\nu}. \tag{101}$$

The second-order perturbation corrections to the adiabatic Hamiltonian can be calculated from the general formula of equation (39). The antisymmetry of the matrix $\boldsymbol{B}^{\mathrm{T}}$ and the relation

$$A_{\mu\nu} = A_{\nu\mu} - \dot{B}_{\nu\mu}, \tag{102}$$

where

$$\dot{B}_{\nu\mu} = \{\hat{\boldsymbol{p}}^{\mathrm{T}}\boldsymbol{B}\}_{\nu\mu} = \sum_k \{\hat{p}_k B^k\}_{\nu\mu}, \tag{103}$$

simplify the final expressions.

The pure vibrational part of the nonadiabatic Hamiltonian can be obtained for the state ν from

$$\Delta\hat{H}_\nu^{\rm V} = T^{\rm V} \circ T^{\rm V} \equiv \sum_{\mu\neq\nu} \left(E_{\nu,0}^{\rm A} - E_{\mu,0}^{\rm A}\right)^{-1} \left(\hat{T}^{\rm V}\right)_{\nu\mu} \left(\hat{T}^{\rm V}\right)_{\mu\nu}, \qquad (104)$$

which yields

$$\Delta\hat{H}^{\rm V} = \tfrac{1}{4}\left(\hat{P}_2 \circ \hat{P}_2 + \hat{\boldsymbol{p}}_\tau^{\rm T}\, \Delta\boldsymbol{G}_1^{\tau\tau}\hat{\boldsymbol{p}}_\tau + \hat{\boldsymbol{p}}_\tau^{\rm T}\, \Delta\boldsymbol{G}_2^{\tau\tau}\hat{\boldsymbol{p}}_\tau + \Delta V\right), \qquad (105)$$

where the index ν is omitted. The corrections to the LAM component of the adiabatic tensor $\boldsymbol{G}^{\rm A}$ become

$$\Delta G_1^{kl} = 2G^{kl} \circ A - B^k \circ B^l + \left\{\hat{B} \circ G^{kl}\right\}_- \qquad (106)$$

and

$$\Delta G_2^{kl} = \sum_j \left(\{\hat{p}_j B^k\} \circ G^{jl} + \{\hat{p}_j B^l\} \circ G^{jk}\right), \qquad (107)$$

where the indices j, k and l label the LAM modes.

The operator $\hat{B}$ is given by

$$\hat{B}_{\nu\mu} = \boldsymbol{B}_{\nu\mu}^{\rm T}\hat{\boldsymbol{p}} = \sum_k B_{\nu\mu}^k \hat{p}_k, \qquad (108)$$

and the 'commutator' $\{\hat{B} \circ F\}_-$ defines the difference

$$\left\{\hat{B} \circ F\right\}_- \equiv \left\{\hat{B} \circ F - F \circ \dot{B}\right\}. \qquad (109)$$

The nonadiabatic contribution to the potential energy function can be written in the form

$$\Delta V = A \circ A + \left\{\hat{B} \circ A\right\}_- - \left\{\hat{B} \circ \dot{B}\right\} + \left\{\hat{P}_2 \circ (A - \dot{B})\right\}. \qquad (110)$$

The nonadiabatic interaction modifies the LAM kinetic energy tensor ($\boldsymbol{G}^{\rm NA} = \boldsymbol{G}^{\rm A} + \frac{1}{2}\Delta\boldsymbol{G}$) and the potential energy function, an creates an additional Hamiltonian term $\frac{1}{4}\hat{P}_2 \circ \hat{P}_2$, quartic in the LAM momenta, with

$$\hat{P}_2 = \hat{\boldsymbol{p}}_\tau^{\rm T}\boldsymbol{G}^{\tau\tau}\hat{\boldsymbol{p}}_\tau. \qquad (111)$$

The contribution of the vibrational and the Coriolis interactions to the effective Hamiltonian

$$\Delta\hat{H}_\nu^{\rm VC} = \hat{T}^{\rm V} \circ \left(\hat{T}^{R\tau} + \hat{T}^{Rs}\right) + \left(\hat{T}^{R\tau} + \hat{T}^{Rs}\right) \circ \hat{T}^{\rm V} \qquad (112)$$

can be calculated from

$$\Delta\hat{H}^{\rm VC} = \frac{1}{4}\left(\sum_\alpha (\hat{P}_2 \circ \hat{C}^\alpha + \hat{C}^\alpha \circ \hat{P}_2)\hat{J}_\alpha + \sum_{k\alpha} (\hat{p}_k\, \Delta G^{k\alpha} + \Delta G^{k\alpha}\, \hat{p}_k)\hat{J}_\alpha\right), \qquad (113)$$

where

$$\hat{C}_{\nu\mu}^\alpha = \sum_k \left(\hat{p}_k G_{\nu\mu}^{k\alpha} + G_{\nu\mu}^{\alpha k}\hat{p}_k\right), \quad \alpha = x, y, z \qquad (114)$$

and

$$\Delta G^{k\alpha} = B^k \circ \left(H^\alpha - S^\alpha\right) + \hat{E}^k \circ \left(H^\alpha - S^\alpha\right) + G^{k\alpha} \circ \left(A - \dot{B}\right). \qquad (115)$$

The matrix elements of S^α and $\hat{E}^k$ are defined by

$$S^\alpha_{\nu\mu} = \sum_s \{\hat{p}_s G^{s\alpha} + G^{s\alpha}\hat{p}_s\}_{\nu\mu} \tag{116}$$

and

$$\hat{E}^k_{\nu\mu} = \sum_l G^{lk}_{\nu\mu}\hat{p}_k,$$

with the indices k and l running over the LAM modes and the index s running over the SAV modes.

The nonadiabatic operator arising from the vibrational and the rotational interactions

$$\Delta\hat{H}^{\mathrm{VR}} = \hat{T}^{\mathrm{V}} \circ \hat{T}^{\mathrm{R}} + \hat{T}^{\mathrm{R}} \circ \hat{T}^{\mathrm{V}} \tag{117}$$

is reduced to

$$\Delta\hat{H}^{\mathrm{VR}} = \sum_{\alpha,\beta}[\hat{J}_\alpha\, \Delta G_1^{\alpha\beta}\,\hat{J}_\beta + \hat{J}_\alpha(\hat{P}_2 \circ G^{\alpha\beta} + G^{\alpha\beta} \circ \hat{P}_2)\hat{J}_\beta]/2 \tag{118}$$

with the rotational tensor correction

$$\Delta G_1^{\alpha\beta} = A \circ G^{\alpha\beta} + \left\{\hat{B} \circ G^{\alpha\beta}\right\}_- . \tag{119}$$

The Coriolis interaction terms generate three nonadiabatic corrections. The first one is created by the Coriolis interaction of the SAVs:

$$\Delta H^{RsRs} = \hat{T}^{Rs} \circ \hat{T}^{Rs} = \tfrac{1}{4}\sum_{\alpha,\beta}\hat{J}_\alpha\, \Delta G_2^{\alpha\beta}\,\hat{J}_\beta \tag{120}$$

with

$$\Delta G_2^{\alpha\beta} = S^\alpha \circ S^\beta.$$

This nonadiabatic interaction perturbs only the rotational components of $\boldsymbol{G}$ tensor.

The second correction, which emerges from the nonadiabatic Coriolis interaction between the SAV and the LAM modes, contributes to the rotational and the Coriolis kinetic energy:

$$\begin{aligned}\Delta H^{RsR\tau} &= \hat{T}^{Rs} \circ \hat{T}^{R\tau} + \hat{T}^{R\tau} \circ \hat{T}^{Rs} \\ &= \frac{1}{4}\sum_{\alpha,\beta}\left(\Delta G_3^{\alpha\beta}\left[\hat{J}_\alpha, \hat{J}_\beta\right]_+ + \sum_k(\hat{p}_k\,\Delta G_1^{k\alpha\beta} + \Delta G_1^{k\alpha\beta}\hat{p}_k)\left[\hat{J}_\alpha, \hat{J}_\beta\right]_-\right)\end{aligned} \tag{121}$$

because $[\hat{J}_\alpha, \hat{J}_\beta]_+ \equiv \hat{J}_\alpha\hat{J}_\beta + \hat{J}_\beta\hat{J}_\alpha$ is a typical quadratic rotational operator, and $[\hat{J}_\alpha, \hat{J}_\beta]_- \equiv \hat{J}_\alpha\hat{J}_\beta - \hat{J}_\beta\hat{J}_\alpha = -i\hbar e_{\alpha\beta\gamma}\hat{J}_\gamma$, where $e_{\alpha\beta\gamma}$ is the Levi–Civita antisymmetric tensor, is coupled to the linear momenta $\hat{p}_k$. The nonadiabatic corrections to the rotational and Coriolis components of $\boldsymbol{G}$ are

$$\Delta G_3^{\alpha\beta} = H^\alpha \circ S^\beta + \left\{\hat{D}^\alpha \circ S^\beta\right\} \tag{122}$$

and

$$\Delta G_1^{k\alpha\beta} = G^{k\alpha} \circ S^\beta, \tag{123}$$

where

$$\hat{D}^\alpha_{\nu\mu} = \sum_k G^{\alpha k}_{\nu\mu}\hat{p}_k. \tag{124}$$

The third correction, from the LAM–Coriolis interaction, contributes to the rotational and the Coriolis kinetic energy:

$$\Delta H^{R\tau R\tau} = \hat{T}^{R\tau} \circ \hat{T}^{R\tau} = \frac{1}{4}\sum_{\alpha,\beta}(\Delta G_4^{\alpha\beta}[\hat{J}_\alpha, \hat{J}_\beta]_+$$

$$+\sum_k \left(\hat{p}_k \Delta G_2^{k\alpha\beta} + \Delta G_2^{k\alpha\beta}\hat{p}_k\right)[\hat{J}_\alpha, \hat{J}_\beta]_-) + \Delta\hat{T}^R, \tag{125}$$

with the rotational tensor correction

$$\Delta G_4^{\alpha\beta} = \left\{\hat{D}^\alpha \circ D^\beta\right\} + \left\{\hat{D}^\beta \circ D^\alpha\right\} + \left\{\hat{D}^\alpha\right\} \circ D^\beta - \sum_k \left[\{\hat{p}_k H^\alpha\} \circ G^{\beta k} + \{\hat{p}_k H^\beta\} \circ G^{\alpha k}\right], \tag{126}$$

where

$$D^\alpha_{\nu\mu} = \sum_k \{\hat{p}_k G^{\alpha k}\}_{\nu\mu}, \tag{127}$$

and with the Coriolis tensor correction

$$\Delta G_2^{k\alpha\beta} = G^{k\alpha} \circ (D^\beta - H^\beta) - G^{k\beta} \circ (D^\alpha - H^\alpha). \tag{128}$$

The additional perturbation term $\Delta\hat{T}^R$ describes the influence of the LAM momenta on the rotational motion:

$$\Delta\hat{T}^R = \sum_{\alpha,\beta}(D^\alpha + \hat{D}^\alpha) \circ \hat{D}^\beta \left[\hat{J}_\alpha, \hat{J}_\beta\right]_+. \tag{129}$$

The nonadiabatic perturbation generated by the rotation and the Coriolis interactions of the SAV and the LAM modes yields the correction

$$\Delta H^{Rs\tau} = (\hat{T}^{Rs} + \hat{T}^{R\tau}) \circ \hat{T}^R + \hat{T}^R \circ (\hat{T}^{Rs} + \hat{T}^{R\tau})$$

$$= \frac{1}{4}\sum_{\alpha,\beta,\gamma}(\Delta G_1^{\alpha\beta\gamma}\left[\hat{J}_\alpha, \hat{J}_\beta\hat{J}_\gamma\right]_- + \Delta\hat{G}_2^{\alpha\beta\gamma}\left[\hat{J}_\alpha, \hat{J}_\beta\hat{J}_\gamma\right]_+), \tag{130}$$

with

$$\Delta G_1^{\alpha\beta\gamma} = (S^\alpha + H^\alpha + \hat{D}^\alpha) \circ G^{\beta\gamma} \tag{131}$$

and

$$\Delta\hat{G}_2^{\alpha\beta\gamma} = 2\sum_k \hat{p}_k G^{k\alpha} \circ G^{\beta\gamma}\hat{p}_k. \tag{132}$$

The comutator $[\hat{J}_\alpha, \hat{J}_\beta\hat{J}_\gamma]_-$ can be reduced to a simpler form

$$\left[\hat{J}_\alpha, \hat{J}_\beta\hat{J}_\gamma\right]_- = \left[\hat{J}_\alpha, \hat{J}_\beta\right]_- \hat{J}_\gamma + \hat{J}_\beta, \left[\hat{J}_\alpha\hat{J}_\gamma\right]_- = -i\hbar(e_{\alpha\beta\delta}\hat{J}_\delta\hat{J}_\gamma + e_{\alpha\gamma\delta'}\hat{J}_\beta\hat{J}_{\delta'}), \tag{133}$$

which means that $\Delta G_1^{\alpha\beta\gamma}$ contributes to the rotational $\boldsymbol{G}$ tensor. The last term in equation (130) specifies the perturbed Coriolis interaction.

The nonadiabatic influence of the rotational motion on the SAV modes represents the familiar centrifugal distortion effect described by the operator

$$\Delta H^{RR} = \hat{T}^R \circ \hat{T}^R = \frac{1}{4}\sum_{\alpha,\beta,\gamma,\delta} \Delta G^{\alpha\beta\gamma\delta}\hat{J}_\alpha\hat{J}_\beta\hat{J}_\gamma\hat{J}_\delta, \tag{134}$$

with

$$\Delta G^{\alpha\beta\gamma\delta} = G^{\alpha\beta} \circ G^{\gamma\delta}, \tag{135}$$

which, however, is not a constant value, as in the standard theory. All the derived second-order nonadiabatic corrections ΔG are functions of the LAM coordinates.

12.10 References

Abramowitz, M., and Stegun, I. A., 1972, *Handbook of Mathematical Functions*; Dover Publications: New York.

Amat, G., Nielsen, H. H., and Tarrago, G., 1971, *Rotation Vibration of Polyatomic Molecules*; M. Dekker: New York.

Bačić Z., and Light, J. C., 1989, *Annu. Rev. Phys. Chem.* **40**, 469–498.

Bardo, R. D., and Wolfsberg, M., 1977, *J. Chem. Phys.* **67**, 593–603.

Bauder, A., Mathier, E., Meyer, R., Ribeaud M., and Günthard Hs. H., 1968, *Mol. Phys.* **15**, 597–614.

Blackwell C. S., and Lord, R. C., 1972, in *Vibrational Spectra and Structure*, Vol 1, Durig, J. R., Ed.; Dekker: New York, pp. 1–24.

Born, M., and Heisenberg, W., 1924, *Ann. Phys.* (Leipzig), **74**, 1–31.

Born, M., and Flügge, S., 1933, *Ann. Phys.* (Leipzig), **16**, 768–780.

Born, M., and Huang, K., 1954, *Dynamical Theory of Crystal Lattices*; Clarendon Press: Oxford.

Born, M., and Oppenheimer, J. R., 1927, *Ann. Phys.* (Leipzig), **84**, 457–484.

Bowman, J. M., 1978, *J. Chem. Phys.* **68**, 608–610.

Bowman, J. M., 1986, *Acc. Chem. Res.* **19**, 202–231.

Bramley, M. J., Green, W. H., and Handy, N. C., 1991, *Mol. Phys.* **73**, 1183–1209.

Brown, R. D., Godfrey, P. D., and Kleibömer, M., 1985, *J. Mol. Spectrosc.* **114**, 257–273.

Brown, R. D., Godfrey, P. D., and Champion, R., 1987, *J. Mol. Spectrosc.* **123**, 93–125.

Brupbacher, Th., Makarewicz, J., and Bauder, A., 1994, *J. Chem. Phys.* **101**, 9736–9746.

Brupbacher, Th., Makarewicz, J., and Bauder, A., 1998, *J. Chem. Phys.* **108**, 3932–3939.

Bunker, P. R., 1967, *J. Chem. Phys.* **47**, 718–739.

Bunker, P. R., 1980, *J. Mol. Spectrosc.* **80**, 422–437.

Bunker, P. R., 1983, *Annu. Rev. Phys. Chem.* **34**, 59–75.

Bunker, P. R., and Jensen, P., 1986, *J. Mol. Spectrosc.* **118**, 18.

Bunker, P. R., and Jensen P, 1998, *Molecular Symmetry and Spectroscopy*, 2nd edition; NRC Research Press: Ottawa.

Bunker, P. R., and Landsberg, B. M., 1977, *J. Mol. Spectrosc.* **67**, 374–385.

Bunker, P. R., Landsberg, B. M., and Winnewisser, B. P., 1979, *J. Mol. Spectrosc.* **74**, 9–25.

Bunker, P. R., and Moss, R. E., 1977, *Mol. Phys.* **33**, 417–424.

Bunker, P. R., and Moss, R. E., 1980, *J. Mol. Spectrosc.* **80**, 217–230.

Bunker, P. R., and Stone, J. M. R., 1972, *J. Mol. Spectrosc.* **41**, 310–332.

Burkhard, D. G., 1953, *J. Chem. Phys.* **21**, 1541–1549.

Burkhard, D. G., and Dennison, D. M., 1951, *Phys. Rev.* **84**, 408–417.

Burkhard, D. G., and Irvin, J. C., 1955, *J. Chem. Phys.* **23**, 1405–1414.

Büttner, W., and Ruder, H., 1969, *Z. Naturforsch.* **23a**, 1163–1171.

Caminati, W., Millemaggi, A., Favero, P. G., and Makarewicz, J., 1997a, *J. Phys. Chem.* **101**, 9272–9275.

Caminati, W., Favero, P. G., Melandri, S., and Makarewicz, J., 1997b, *Mol. Phys.* **91**, 663–667.

Carreira, L. A., Lord, R. C., and Malloy, T. B., 1979, in *Topics in Current Chemistry*, Vol. 2; Springer: Heidelberg, pp. 1–95.

Carrington, T. Jr, 1987, *J. Chem. Phys.* **86**, 2207–2223.

Carter, S., Culik, S. J., and Bowman, J. M., 1998, *J. Chem. Phys.* **107**, 10458–10469.

Carter, S., and Handy, N. C., 1982, *Mol. Phys.* **47**, 1445–1455.

Champion, R., Godfrey, P. D., and Bettens, F. L., 1991, *J. Mol. Spectrosc.* **147**, 488–495; 1992, *J. Mol. Spectrosc.* **155**, 18–24.

Cioslowski, J., Scott, A. P., and Radom, L., 1997, *Mol. Phys.* **91**, 413–420.

Columberg, G., Bauder, A., Heineking, N., Stahl, W., and Makarewicz, J., 1998, *Mol. Phys.* **93**, 215–228.

Cooley, J. W., 1961, *Math. Comput.* **15**, 363–374.

Corey, G. C., and Tromp, J. W., 1995, *J. Chem. Phys.* **103**, 1812–1820.

Coudert, L. H., 1992, *J. Mol. Spectrosc.* **154**, 427–442; 1994, *J. Mol. Spectrosc.* **165**, 406–425; 1997, *J. Mol. Spectrosc.* **181**, 246–273.

Coudert, L. H., and Hougen, J. T., 1990, *J. Mol. Spectrosc.* **139**, 259–277.

Császár, A. G., and Handy, N. C., 1995a, *Mol. Phys.* **86**, 959–979.

Császár, A. G., and Handy, N. C., 1995b, *J. Chem. Phys.* **102**, 3962–3966.

Darling, B. T., and Dennison, D. M., 1940, *Phys. Rev.* **57**, 128–139.

Decius, J. C., 1948, *J. Chem. Phys.* **16**, 1025–1034.
De Lucia, F. C., Herbst, E., Anderson, T., and Helminger, P., 1989, *J. Mol. Spectrosc.* **134**, 395–411.
Dickinson, A. S., and Certain, P. R., 1968, *J. Chem. Phys.* **49**, 4209–4213.
Diehl, H., Flügge, S., Schröder, U., Völkel, A., and Weiguny, A., 1961, *Z. Phys.* **162**, 1–14.
Dinelli, B. M., Polyansky, O. L., and Tennyson, J., 1995, *J. Chem. Phys.* **103**, 10433–10438.
Durig, J. R., Craven, S. M., and Harris, W. C., 1972, in *Vibrational Spectra and Structure*, Durig, J. R., Ed., Vol 1; Dekker: New York, pp. 73–177.
East, A. L. L., and Bunker, P. R., 1997, *J. Mol. Spectrosc.* **183**, 157–162.
Echave, J., and Clary, D. C., 1992, *Chem. Phys. Lett.* **190**, 225–230.
Eckart, C., 1935, *Phys. Rev.* **47**, 552–558.
Elyashevich, M. A., 1938, *Trudy Gosudarstvenogo Opticheskogo Instituta (SSSR)* **12**, No.106.
Frisch, M. J., 1990, *Gaussian 90 User's Guide and Programmer's Reference*; Gaussian: Pittsburgh, PA, p. 71.
Frisch, M. J., Trucks, G. W., Schlegel, H. B., Gill, P. M. W., Johnson, B. G., Robb, M. A., Cheeseman, J. R., Keith, T., Peterson, G. A., Montgomery, J. A., Raghavachari, K., Al-Laham, M. A., Zakrzewski, V. G., Ortiz, J. V., Foresman, J. B., Challacombe, M., Peng, C. Y., Ayala, P. Y., Chen, W., Wong, M. W., Martin, R. L., Fox, D. J., Binkley, J. S., Defrees, D. J., Baker, J., Stewart, J. P., Head-Gordon, M., Gonzales C., and Pople, J. A., 1995, *GAUSSIAN 94*; Gaussian: Pittsburgh, PA.
Gordy, W., and Cook, R. L., 1970, *Microwave Molecular Spectra*; Interscience: New York.
Groner, P., 1992, *J. Mol. Spectrosc.* **156**, 164–189.
Guan, Y., and Quade, C. R., 1986, *J. Chem. Phys.* **84**, 5624–5638.
Gut, M., Meyer, R., Bauder A., and Günthard, Hs. H., 1978, *Chem. Phys.* **31**, 433–446.
Ha, T.-K., Makarewicz J., and Bauder, A., 1993, *J. Phys. Chem.* **97**, 11415–11419.
Handy, N. C., 1987, *Mol. Phys.* **61**, 207–223.
Harris, D. O., Engerholm, G. G., and Gwinn, W. D., 1965, *J. Chem. Phys.* **43**, 1515–1517.
Harthcock, M. A., and Laane, J., 1985, *J. Phys. Chem.* **89**, 4231–4240.
Hecht, K. T., and Dennison, D. M., 1957, *J. Chem. Phys.* **26**, 48–69.
Herbst, E., Messer, J. K., De Lucia, F. C., and Helminger, P., 1984, *J. Mol. Spectrosc.* **108**, 42–57.
Herman, R. M., and Ogilvie, J. F., 1998, *Adv. Chem. Phys.* **103**, 187–215.
Hirota, E., 1970, *J. Mol. Spectrosc.* **34**, 516–527.
Hougen, J. T., 1981, *J. Mol. Spectrosc.* **89**, 296–327; 1985, *J. Mol. Spectrosc.* **114**, 395–426.
Hougen, J. T., Bunker, P. R., and Johns, J. W. C., 1970, *J. Mol. Spectrosc.* **34**, 136–172.
Howard, B. J., and Moss, R. E., 1970, *Mol. Phys.* **19**, 433–450.
Hoy, A. R., and Bunker, P. R., 1974, *J. Mol. Spectrosc.* **52**, 439–456; 1979, *J. Mol. Spectrosc.* **74**, 1–8.
Hoy, A. R., Mills, I. M., and Strey, G., 1972, *Mol. Phys.* **24**, 1265–1290.
Huber, S., Makarewicz, J., and Bauder, A., 1998, *Mol. Phys.* **95**, 1021–1043.
Hylleraas, E. A., 1928, *Z. Phys.* **48**, 469–494.
Jelski, D., Haley, R. H., and Bowman, J. M., 1996, *J. Comput. Chem.* **17**, 1645–1652.
Jensen, P., 1983a, *Comp. Phys. Rep.* **1**, 1–55.
Jensen, P., 1983b, *J. Mol. Spectrosc.* **101**, 422–439.
Jensen, P., 1984, *J. Mol. Spectrosc.* **104**, 59–71.
Jensen, P., and Bunker, P. R., 1988, *J. Chem. Phys.* **89**, 1327.
Jensen, P., Bunker, P. R., Epa, V. C., and Karpfen, A., 1992, *J. Mol. Spectrosc.* **151**, 384–395.
Jensen, P., Bunker, P. R., and Karpfen, A., 1991, *J. Mol. Spectrosc.* **148**, 385–390.
Jensen, P., and Kraemer, W. P., 1988, *J. Mol. Spectrosc.* **129**, 172–185.
Kauppi, E., 1994, *Chem. Phys. Lett.* **229**, 661–666.
Kim, W., and Felker, P. M., 1998, *J. Chem. Phys.* **107**, 2193–2204.
Kirtman, B., 1962, *J. Chem. Phys.* **11**, 2516–2539.
Kivelson, D., and Wilson, E. B., Jr., 1952, *J. Chem. Phys.* **20**, 1575–1579; 1953, *J. Chem. Phys.* **21**, 1229–1236.
Knopp, J. V., and Quade, C. R., 1968, *J. Chem. Phys.* **48**, 3317–3324.
Koehler, J. S., and Dennison, D. M., 1940, *Phys. Rev.* **57**, 1006–1021.
Koput, J., 1984, *J. Mol. Spectrosc.* **106**, 12–21.
Koput, J., 1986, *J. Mol. Spectrosc.* **118**, 448–458.
Koput, J., 1995, *J. Phys. Chem.* **99**, 15874–15879.
Koput, J., 1996, *Chem. Phys. Lett.* **259**, 661–668.
Koput, J., and Wierzbicki, A., 1983, *J. Mol. Spectrosc.* **99**, 116–132.
Kręglewski, M., 1978, *J. Mol. Spectrosc.* **72**, 1–19; 1984, *J. Mol. Spectrosc.* **105**, 8–23.
Kręglewski, M., 1993, in *Structures and Conformations of Non-Rigid Molecules*, Laane, J., Ed.; Kluwer: Dordrecht.
Kręglewski, M., and Jensen, P., 1984, *J. Mol. Spectrosc.* **103**, 312–320.
Laane, J., 1972, in *Vibrational Spectra and Structure*, Durig, J. R., Ed., Vol. 1; M. Dekker: New York, pp. 25–50.
Lin, C. C., and Swalen, J. D., 1959, *Rev. Mod. Phys.* **31**, 841–890.
Liu, M., and Quade, C. R., 1991, *J. Mol. Spectrosc.* **146**, 238–251.
Makarewicz, J., 1985, *Theor. Chim. Acta* **68**, 321.
Makarewicz, J., 1988, *J. Phys. B* **21**, 1803–1819.

Makarewicz, J., 1996, *J. Mol. Spectrosc.* **176**, 169–179.
Makarewicz, J., and Bauder, A., 1995, *Mol. Phys.* **84**, 853–878.
Makarewicz, J., Ha, T.-K., and Bauder, A., 1993, *J. Chem. Phys.* **99**, 3694–3699.
Makarewicz, J., and Skalozub, A., 1999, *Chem. Phys. Lett.* **306**, 352–356.
Makarewicz, J., Wierzbicki, A., and Koput, J., 1985, *Chem. Phys.* **97**, 311–319.
Makushkin, J. S., and Tyuterev, Vl. G., 1984, *Perturbation Methods and Effective Hamiltonians in Molecular Spectroscopy*; Nauka: Novosibirsk (in Russian).
Maris, A., Melandri, S., Caminati, W., Favero, P. G., and Makarewicz, J., 1997, *J. Chem. Phys.* **107**, 5714–5719.
Mekhtiev, M., Godfrey, P. D., and Szalay, V., 1996, *J. Mol. Spectrosc.* **180**, 42.
Meyer, R., 1979, *J. Mol. Spectrosc.* **76**, 266–300.
Meyer, R., and Günthard, Hs. H, 1968, *J. Chem. Phys.* **49**, 1510–1520.
Michelot, F., 1982, *Mol. Phys.* **45**, 971–1001.
Moule, D. C., and Rao, Ch. V. S., 1973, *J. Mol. Spectrosc.* **45**, 120–141.
Neuhauser, R., Braun, J., Neusser, H. J., and van der Avoird, A., 1998, *J. Chem. Phys.* **108**, 8408.
Nielsen, H. H., 1932, *Phys. Rev.* **10**, 445–456.
Nielsen, H. H., 1951, *Rev. Mod. Phys.* **23**, 90–136.
Nösberger, P., Bauder, A. and Günthard, Hs. H., 1974, *Chem. Phys.* **4**, 196–219.
Ohashi, N., and Hougen, J. T., 1987, *J. Mol. Spectrosc.* **121**, 474–501.
Oda, M., Ohashi, N., and Hougen, J. T., 1990, *J. Mol. Spectrosc.* **142**, 57–84.
Papoušek D., and Aliev, M. R., 1982, *Molecular Vibrational Rotational Spectra*; Academia: Prague.
Papoušek, D., Stone, J. M. R., and Špirko, V., 1973, *J. Mol. Spectrosc.* **48**, 17–37.
Papoušek, D., and Špirko, V., 1976, in *Topics in Current Chemistry*, Vol. 68; Springer: Berlin, pp. 59–102.
Pickett, H. M., 1972, *J. Chem. Phys.* **56**, 1715–1723.
Podolsky, B., 1928, *Phys. Rev.* **32**, 812–816.
Polyansky, O. L., Jensen, P., and Tennyson, J., 1994, *J. Chem. Phys.* **101**, 7651–7657.
Polyansky, O. L., Jensen, P., and Tennyson, J., 1996, *J. Mol. Spectrosc.* **178**, 184–188.
Pracna, P., Winnewisser, M., and Winnewisser, B. P., 1993, *J. Mol. Spectrosc.* **162**, 127–141.
Quade, C. R., 1966, *J. Chem. Phys.* **44**, 2512–2523; 1967, *J. Chem. Phys.* **47**, 1073–1090; 1975, *J. Chem. Phys.* **65**, 700–705, 1980, *J. Chem. Phys.* **73**, 2107–2114.
Quade, C. R., 1998, *J. Mol. Spectrosc.* **188**, 190–199.
Quade, C. R., and Lin, C. C., 1963, *J. Chem. Phys.* **38**, 540–550.
Ratner, M. A., and Gerber, R. B., 1986, *J. Phys. Chem.* **90**, 20–30.
Ratner, M. A., Gerber, R. B., and Buch, V., 1987, in *Stochasticity and Intermolecular Redistribution of Energy*, Mukamel, S., and Lefebvre, R., Eds.; Reidel: Boston, MA, pp. 57–80.
Ratner, M. A., Gerber, R. B., Horn, T. R., and Williams, C. J., 1991, in *Advances in Molecular Vibrations and Collision Dynamics*, Vol. 1A, Bowman, J. M., and Ratner, M. A., Eds.; JAI Press: Creenwich, CT, pp. 215–253.
Ribeaud, M., Bauder, A., and Günthard, Hs. H., 1972, *Mol. Phys.* **23**, 235–248.
Riedle, E., and van der Avoird., 1996, *J. Chem. Phys.* **104**, 882–898.
Roitberg, A. E., Gerber, R. B., and Ratner, M. A., 1997, *J. Phys. Chem.* **101**, 1700–1706.
Ross, S. C., 1988, *J. Mol. Spectrosc.* **132**, 48–79.
Ross, S. C., 1993, *J. Mol. Spectrosc.* **161**, 102–108.
Ross, S. C., Niedenhoff, M., and Yamada, K. M., 1994, *J. Mol. Spectrosc.* **164**, 432–444.
Roth, R. M., Gerber, R. B., and Ratner, M. A., 1983, *J. Phys. Chem.* **87**, 2376–2382.
Ruder, H., 1968, *Z. Naturforsch.* **23a**, 579–596.
Sayvetz, A., 1939, *J. Chem. Phys.* **7**, 383–389.
Schwenke, D. W., 1992, *Comput. Phys. Commun.* **70**, 1–14.
Slater, J. C., 1927, *Proc. Natl Acad. Sci.* **13**, 423–426.
Sørensen, G. O., 1978, in *Topics in Current Chemistry*, Vol. 82; Springer: Berlin, pp. 97–175.
Sussmann, R., Neuhauser, R., and Neusser, H. J., 1995, *J. Chem. Phys.* **103**, 3315–3324.
Sutcliffe, B. T., 1982, in *Current Aspects of Quantum Chemistry*, Carbo, R., Ed.; Elsevier: New York.
Sutcliffe, B. T., 1992, in *Methods of Computation Chemistry*, Wilson, S., Ed.; Plenum: New York.
Sutcliffe, B. T., and Tennyson, J., 1991, *Int. J. Quantum. Chem.* **39**, 183.
Špirko, V., 1983, *J. Mol. Spectrosc.* **101**, 30–47.
Špirko, V., Stone, J. M. R., and Papoušek, D., 1973, *J. Mol. Spectrosc.* **48**, 38–46.
Špirko, V., Stone, J. M. R., and Papousek D., 1976, *J. Mol. Spectrosc.* **60**, 159–178.
Szalay, V., 1984, *J. Mol. Spectrosc.* **102**, 13–32.
Tang, J., and Takagi, K., 1993, *J. Mol. Spectrosc.* **161**, 487–498.
Teffo, J.-L., 1993, *Mol. Phys.* **78**, 1493–1512.
Tennyson, J., and Sutcliffe, B. T., 1982, *J. Chem. Physc.* **77**, 4061–4072.
Thorson, W., and Nakagawa, I., 1960, *J. Chem. Phys.* **33**, 994–1004.
Van Vleck, J. H., 1929, *Phys. Rev.* **33**, 467–506.
Watson, J. K. G., 1968, *Mol. Phys.* **15**, 479–490; 1970, *Mol. Phys.* **19**, 465–487.

Webster, F., Huang M.-J., and Wolfsberg, M., 1981, *J. Chem. Phys.* **75**, 2306–2313.
Wierzbicki, A., Koput, J., and Kręglewski, M., 1983, *J. Mol. Spectrosc.* **99**, 102–115.
Wilson, E. B., Jr., 1939, *J. Chem. Phys.* **7**, 1047–1052.
Wilson, E. B., Jr., Decius, J. C., and Cross, P., 1955, *Molecular Vibrations*; McGraw-Hill: New York.
Wilson, E. B., Jr., and Howard, J. B., 1936, *J. Chem. Phys.* **4**, 260–268.
Xu, L. H., and Hougen, J. T., 1995, *J. Mol. Spectrosc.* **169**, 396–409.

13 REARRANGEMENTS AND TUNNELING SPLITTINGS OF SMALL WATER CLUSTERS

David J. Wales
University of Cambridge, UK

13.1 Introduction

Small water clusters are currently an extremely active research area for both theory and experiment. The ubiquity of water as an almost universal solvent throughout chemistry and biochemistry is often cited as the motivation for such studies. In truth, however, there are many fundamental issues still to resolve in our understanding of even the prototypical water dimer, and the implications of this work for our understanding of intermolecular forces [see e.g. Maitland *et al.* (1981), Nelson, Fraser and Klemperer (1987), Hutson (1990), Cohen and Saykally (1991), and Stone (1996)] and hydrogen bond rearrangement dynamics [Ohmine and Tanaka (1993)] are equally profound.

The trigger for the new wave of research in this field was undoubtedly the advent of far-infrared vibration–rotation tunneling (FIR-VRT) spectroscopy where resolution of order 1 MHz

Computational Molecular Spectroscopy. Edited by Per Jensen and Philip R. Bunker

is now possible [Cohen and Saykally (1990), Liu, Cruzan and Saykally 1996, Paul *et al.* (1997), and Saykally and Blake (1993)]. New data have now been obtained for the water dimer, trimer, tetramer, pentamer, and hexamer. Microwave and submillimeter spectra have also been obtained for water dimer isotopomers by Karyakin *et al.* (1995). The experimental frequency range and the high resolution pose a new challenge to theory because these systems possess large amplitude vibrations which probe regions of the potential energy surface in saddle point regions, away from the local minima. To explain the tunneling splitting patterns requires a detailed understanding of the underlying rearrangement mechanisms and the resulting reaction graph, which summarizes how the various minima are connected. Hence all the transition states which result in observable tunneling splittings should be calculated, along with the corresponding pathways.

Theoretical treatments can be loosely divided into a number of categories: characterization of rearrangement mechanisms and tunneling splitting patterns, high accuracy *ab initio* calculations and decomposition of the many-body contributions to the energy [Pastor and Ortega-Blake (1993), Xantheas (1994), and Hodges, Stone and Xantheas (1997)], development of torsional potentials, quantum nuclear dynamics of the torsional hydrogens using the discrete variable representation (DVR), quantum nuclear dynamics of the intermolecular degrees of freedom using the diffusion Monte Carlo (DMC) approach [Gregory and Clary (1996a), Gregory *et al.* (1997), Gregory and Clary (1998), and Buch, Sandler and Sadlej (1998)] and other methods [Fredj, Gerber and Ratner (1996), and Jung and Gerber (1996)] and investigation of intermolecular potentials.

The present overview also includes some results that have not yet been published. Unconstrained pathways are characterized for water dimer in Section 13.5, showing, for example, that the acceptor tunneling mechanism is not a simple rotation about the local C_2 axis of the acceptor monomer. In Section 13.8 a complete analysis of the possible tunneling splitting patterns in water pentamer is provided for all possible combinations of flip and bifurcation mechanisms. Finally, two-dimensional nuclear dynamics calculations are described for the water hexamer in Section 13.9, which suggest that delocalization over different cage isomers may occur in the ground or low-lying vibrational states.

We consider the molecular symmetry group which must be employed to label the energy levels of these non-rigid molecules in Section 13.2, the classification of different rearrangement mechanisms in Section 13.3 and the characterization of stationary points and rearrangement paths with both *ab initio* and empirical potentials in Section 13.4.

13.2 The molecular symmetry group

The observation of quantum tunneling effects in high resolution spectra provides indirect information about the corresponding rearrangement mechanisms. To make use of this information we must employ group theory to classify the energy levels of such systems. Point groups are useful when the molecule under consideration is rigid on the appropriate experimental timescale. However, molecular potential energy surfaces (PESs) generally contain local minima corresponding to permutational isomers of any given stationary point. We will adopt the nomenclature of Bone *et al.* (1991) where a 'structure' is understood to mean a particular molecular geometry and a 'version' is a particular labeled permutational isomer of a given structure.

Tunneling splittings may be observed when rovibronic wavefunctions localized in the potential wells corresponding to different versions interfere with each other. For example, the ammonia molecule displays doublet splittings because pairs of permutational isomers are interconverted by the inversion mechanism in which the molecule passes through a planar transition state [Dennison and Hardy (1932), and Bell (1980)].

The energy levels of such non-rigid molecules can be classified using the complete nuclear permutation–inversion (CNPI) group which is the direct product of the group containing all possible permutations of identical nuclei and the inversion group [Bunker and Jensen (1998)]. The latter group contains only the identity operation, E, and the operation of inversion of all particle coordinates through the space-fixed origin, E^*, which commutes with the permutations. The CNPI group is a true symmetry group of the full molecular Hamiltonian in the absence of external fields, and its elements are generally referred to as permutation–inversion (PI) operations. The order of this group, h_{CNPI}, increases factorially with the number of atoms of the same element, and rapidly becomes difficult to use. Fortunately, Longuet-Higgins (1963) showed that it not necessary to consider the whole CNPI group, but rather only the PIs which correspond to tunneling splittings that are resolvable for a given experiment. The corresponding rearrangement mechanisms are said to be 'feasible'. The resulting subgroup of the CNPI group is known as the molecular symmetry (MS) group.

For rigid molecules the appropriate MS group, $\mathcal{G}_{\mathrm{RM}}$, of order h_{RM}, contains only PIs which are always feasible because they involve only a barrierless reorientation of the reference version. $\mathcal{G}_{\mathrm{RM}}$ is isomorphic to the usual rigid molecule point group [Hougen (1962, 1963, 1986), and Bunker and Jensen (1998)]. Pure permutation operations in $\mathcal{G}_{\mathrm{RM}}$ must produce a labeled framework which can be rotated back onto the reference version, and can therefore be mapped onto the proper operations of the point group. $\mathcal{G}_{\mathrm{RM}}$ can only contain operations involving E^* if the rigid molecule point group contains improper operations, for otherwise the equilibrium framework is chiral, and no permutation will be able to combine with E^* to give a version that is superimposable on the reference. Furthermore, a PI in $\mathcal{G}_{\mathrm{RM}}$ must combine E^* with a permutation which produces the enantiomer of the reference version. Hence, operations involving E^* in $\mathcal{G}_{\mathrm{RM}}$ can be mapped onto improper operations of the point group. Each such point group operation provides a permutation that can be combined with E^* to give a superimposable version if we imagine the point group operations to act upon the nuclear labels. These permutations each produce the enantiomer of the reference version, and then combination with E^* changes the handedness back again. For planar molecules the operation corresponding to reflection in the molecular plane corresponds to the identity permutation, and so this point group operation maps onto E^*. For linear molecules the correspondence between $\mathcal{G}_{\mathrm{RM}}$ and the point group is a homomorphism because operations in classes with an infinite number of members in the point group all correspond to the same permutation or PI in $\mathcal{G}_{\mathrm{RM}}$.

For a rigid molecule each rovibronic level is $h_{\mathrm{CNPI}}/h_{\mathrm{RM}}$-fold degenerate because there are $h_{\mathrm{CNPI}}/h_{\mathrm{RM}}$ distinct versions on the PES [Bone *et al.* (1991)] and each one supports an identical stack of energy levels.

If there exists a feasible rearrangement with a finite barrier then the MS group is enlarged from $\mathcal{G}_{\mathrm{RM}}$. In general, a given mechanism will not link all the possible versions of a given structure but rather the versions will be partitioned into a number of closed sets with equivalent connectivities within each set. If we consider a representative set of versions all of which can be interconverted by repeated application of the feasible rearrangement, then the corresponding wavefunction must be a linear combination of the localized functions from members of this set. The delocalized wavefunctions can be found by solving a secular problem, just as in the linear combination of molecular orbitals approach to electronic structure.

The enlarged MS group includes the new PI and all its products with the PIs of $\mathcal{G}_{\mathrm{RM}}$. Further feasible mechanisms are included in a similar fashion. The resulting MS group is a subgroup of the CNPI group obtained by removing all the PIs for a given reference version which are not feasible. The order of the MS group, h_{MS}, must satisfy $h_{\mathrm{CNPI}}/h_{\mathrm{MS}} = M$, where M is an integer. The versions are then divided into M disconnected sets. Each set contains $h_{\mathrm{MS}}/h_{\mathrm{RM}}$ distinct versions [Bone *et al.* (1991)] which can be interconverted by one or more feasible operations.

If the wavefunctions decay rapidly in the classically forbidden regions of the PES between minima then it may be sufficient to consider only nearest-neighbor interactions. The resulting splitting pattern is then determined largely by symmetry. This Hückel-type approximation is not always appropriate [Bunker (1996), and Kolbuszewski and Bunker (1996)], but has proved to be useful for water clusters. The connectivity of the reaction graph, the associated MS group and the splitting pattern can then be found automatically by a computer program once a minimal set of generator permutation-inversions is known [Wales (1993a)].

13.3 Classification of rearrangement mechanisms

We define a *degenerate* rearrangement mechanism as one which links permutational isomers of the same structure via a single transition state [Leone and v. R. Schleyer (1970)]. We also follow Murrell and Laidler's (1968) definition of a transition state as a stationary point with a single imaginary frequency. The largest tunneling splittings are expected for degenerate rearrangements (where the localized states are in resonance) with low barriers, short paths and small effective masses. There are two classes of degenerate rearrangement: in a symmetric degenerate rearrangement (SDR) the two sides of the path are related by a symmetry operation, and in an asymmetric degenerate rearrangement (ADR) they are not [Nourse (1980)]. Most of the rearrangements discussed in the following sections are ADRs.

Recent experimental [Liu *et al.* (1996c)] and theoretical [Sabo *et al.* (1996), Sorenson, Gregory and Clary (1996)] studies of water trimer isotopomers indicate that significant tunneling splittings still occur in these species, even though certain rearrangements are no longer degenerate. Within the Born–Oppenheimer approximation the isotopomers have the same PES, but the rovibronic states are not in resonance between different wells. We will also see in the following sections that tunneling splittings are observed for water tetramer and water hexamer where no suitable degenerate rearrangements have been found theoretically. In each case the tunneling may be the result of a series of steps mediated by true transition states, or perhaps even higher index saddles, where the end points are true permutational isomers. We suggest the term 'indirect tunneling' to describe this situation and distinguish it from 'nondegenerate tunneling' where the interaction would be between versions with different structures.

13.4 Geometry optimizations and potentials

Eigenvector-following provides a powerful technique for locating minima and transition states and calculating reaction pathways [Pancíř (1975), Cerjan and Miller (1981), Simons *et al.* (1983), O'Neal, Taylor and Simons (1984), Banerjee *et al.* (1985), and Baker (1986)]. The precise algorithms employed in the present work have been described in detail elsewhere [Wales (1994), and Wales and Walsh (1996)]. The optimization step for eigendirection i is

$$h_i = \frac{\pm 2F_i}{|b_i|(1+\sqrt{1+4F_i^2/b_i^2})}, \qquad \left\{ \begin{array}{l} +\text{for walking uphill} \\ -\text{for walking downhill} \end{array} \right\},$$

where b_i is the corresponding eigenvalue and F_i is the component of the gradient in this direction. Here the eigenvectors and eigenvalues refer to the Hessian matrix of second derivatives of the energy. For each eigendirection the true eigenvalue is compared with an estimate based upon the current and previous gradients and the previous step. From this comparison a 'trust radius' is dynamically adjusted for each direction, and steps outside the trust radius are scaled down. Analytic first and second derivatives of the energy were used at every step.

In the *ab initio* calculations the derivatives were all generated by the CADPAC program [Amos *et al.* (1995)], and Cartesian coordinates were used throughout. Pathways were calculated by taking small displacements of order 0.01 a_0 away from a transition state parallel or antiparallel to the transition vector, and then employing eigenvector-following energy minimization to find the associated minimum. (The atomic unit of distance, a_0, is $5.291\,772 \times 10^{-11}$ m). The pathways obtained by this procedure have been compared with steepest-descent paths and pathways that incorporate a kinetic metric [Banerjee and Adams (1992)] in previous work–the mechanism is usually found to be represented correctly [Walsh and Wales (1997), and Merrill and Gordon (1998)].

Calculations employing rigid body intermolecular potentials were performed using the ORIENT3 program [Popelier, Stone and Wales (1994), Wales, Popelier and Stone (1995), and Wales, Stone and Popelier (1995)], which contains the same optimization package adapted for center-of-mass/orientational coordinates. This program can treat intermolecular potentials including Stone's distributed multipoles [Stone (1981), and Stone and Alderton (1985)] and polarizabilities [Stone (1985), and Le Sueur and Stone (1993)]. Calculations for the water hexamer in Section 13.9 were performed using the latest version of the relatively sophisticated ASP-W2 potential of Millot and Stone [Millot and Stone (1992), Millot *et al.* (1998)] and the much simpler but widely used TIP4P form [Jorgensen (1981), and Jorgensen *et al.* (1983)].

In the *ab initio* Hartree–Fock (HF) calculations for water dimer two basis sets were considered. The smaller double-ζ [Dunning (1970), and Huzinaga (1965)] plus polarization (DZP) basis employed polarization functions consisting of a single set of p functions on each hydrogen atom (exponent 1.0) and a set of six d functions on each oxygen atom (exponent 0.9) to give a total of 26 basis functions per monomer. The larger basis set, denoted DZP + diff, includes the above DZP functions with an additional diffuse s function on each hydrogen atom (exponent 0.0441) and diffuse s and p functions on each oxygen atom (exponents 0.0823 and 0.0651, respectively) [Fowler and Schaefer (1995)], to give 32 basis functions per monomer. Finally, we performed some calculations with the aug-cc-pVDZ basis [Dunning (1989), and Kendall, Dunning and Harrison 1992) with 43 basis functions per monomer. Correlation corrections were obtained through both second-order Møller–Plesset (MP2) theory [Møller and Plesset (1934)] and density functional theory (DFT). In the DFT calculations we employed the Becke (1988) nonlocal exchange functional and the Lee–Yang–Parr correlation functional [Lee, Yang and Parr (1988)], together referred to as BLYP; derivatives of the grid weights were not included and the core electrons were not frozen. Numerical integration of the BLYP functionals was performed using grids between the CADPAC 'MEDIUM' and 'HIGH' options. The 'MEDIUM' grids were not accurate enough to give the right number of negative Hessian eigenvalues, whereas the 'HIGH' grids contained more points than necessary. CADPAC actually uses different sized grids for different parts of the calculation [Amos *et al.* (1995)]; in the present work these grids contained 14 386 and 97 008 points for water dimer after removal of those with densities below the preset tolerances. Calculations were deemed to be converged when the root-mean-square gradient fell below 10^{-6} atomic units. This threshold is sufficient to reduce the six 'zero' normal mode frequencies to less than $1\,\text{cm}^{-1}$ in the HF and MP2 calculations. Since derivatives of the grid weights were not included, the largest of the six 'zeros' can be as large as $20\,\text{cm}^{-1}$ for the DFT stationary points.

Three additional parameters are useful in describing the rearrangement mechanisms. The first is the integrated path length, S, which was calculated as a sum over eigenvector-following steps, m:

$$S \approx \sum_m |\boldsymbol{Q}(m+1) - \boldsymbol{Q}(m)|, \tag{1}$$

where $\boldsymbol{Q}(m)$ is the $3n$-dimensional position vector for n nuclei in Cartesian coordinates at step m. The second is the distance between the two minima in nuclear configuration space, D:

$$D = |\boldsymbol{Q}(s) - \boldsymbol{Q}(f)|, \tag{2}$$

where $Q(s)$ and $Q(f)$ are the $3n$-dimensional position vectors of the minima at the start and finish of the path. The third is the moment ratio of displacement [Stillinger and Weber (1983)], γ, which gives a measure of the cooperativity of the rearrangement:

$$\gamma = \frac{n\sum_i [Q_i(s) - Q_i(f)]^4}{\left(\sum_i [Q_i(s) - Q_i(f)]^2\right)^2}, \tag{3}$$

where $Q_i(s)$ is the position vector of atom i in starting minimum s, etc., and n is the number of atoms. If every atom undergoes the same displacement in one Cartesian component then $\gamma = 1$, while if only one atom has one non-zero component then $\gamma = n$.

13.5 Rearrangements of water dimer

There have been many experimental investigations of the water dimer [Fraser (1991)] following the original work of Dyke, Mack and Muenter (1977). The dimer rovibronic energy levels were first classified in terms of PI group theory by Dyke (1977), and Coudert and Hougen applied their internal axis approach to the intermolecular dynamics using an empirical potential [Hougen (1985), Coudert and Hougen (1988), and Coudert and Hougen (1990)]. Leforestier *et al.* (1997) have recently performed coupled nuclear dynamics calculations for all six intermolecular degrees of freedom. Smith *et al.* (1990) performed *ab initio* calculations and identified three true transition states for $(H_2O)_2$; they also performed constrained calculations of the donor-tunneling pathway. Many *ab initio* calculations have been performed for $(H_2O)_2$ [see e.g. Kim *et al.* (1995), Mas and Szalewicz (1996), and Famulari *et al.* (1998b, 1998a)] and a review is also available [Scheiner (1994)]. DMC techniques have also been applied [Gregory and Clary (1994), Gregory and Clary (1995a), and Gregory (1998)]. In this section we will describe unconstrained pathway calculations for all three rearrangements; the energetics of the various stationary points at different levels of theory are summarized in Table 13.1 and counterpoise-corrected [Boys and Bernardi (1970)] binding energies (including monomer relaxation [(Xantheas (1996)] are given in Table 13.2.

There are a total of $2 \times 2! \times 4!/2 = 48$ distinct versions of the water dimer global minimum on the PES; we divide by two because the rigid molecule point group has order two [Bone *et al.* (1991)]. However, mechanisms which involve the making and breaking of covalent bonds lie too high in energy to give rise to observable tunneling splittings. The largest possible MS group when covalent bond-breaking is not feasible has order $2 \times 2! \times (2!)^2 = 16$, where the first factor accounts for the inversion operation, the second factor accounts for permutation of the two oxygen nuclei, and the last term accounts for the permutation of the two hydrogen (or deuterium) atoms within each monomer. This group is denoted $\boldsymbol{G}_{16}$ and is isomorphic to the point group D_{4h} [Dyke (1977)]. Since the equilibrium geometry has C_s symmetry the maximum number of

Table 13.1. Energies (hartree) and point groups at various levels of theory for the water dimer global minimum and the transition states for acceptor tunneling, donor–acceptor interchange and donor tunneling.

	Min (C_s)	Acceptor	Don–acc (C_i)	Donor (C_{2v})
DZP/HF	−152.102 057	−152.101 274 (C_s)	−152.100 688	−152.099 832
DZP + diff/HF	−152.107 971	−152.107 282 (C_s)	−152.106 287	−152.105 946
DZP + diff/BLYP	−152.87 817	−152.87 701 (C_1)	−152.87 591	−152.87 535
DZP + diff/MP2	−152.540 770	−152.539 673 (C_1)	−152.538 743	−152.538 149
aug-cc-pVDZ/BLYP	−152.86 416	−152.86 322 (C_1)	−152.86 230	−152.86 091

Table 13.2. Counterpoise-corrected binding energies in millihartree at various levels of theory for the water dimer global minimum and the transition states for acceptor tunneling, donor–acceptor interchange and donor tunneling.

	Minimum	Acceptor	Donor–acceptor	Donor
DZP/HF	7.42	6.88	5.73	5.13
DZP + diff/HF	6.92	6.38	5.82	5.07
DZP + diff/BLYP	7.75	6.64	6.03	4.79
DZP + diff/MP2	7.45	6.65	6.46	5.28
aug-cc-pVDZ/BLYP	6.47	5.53	4.71	3.34

distinct versions that can be interconverted without breaking covalent bonds is $16/2 = 8$ (see Section 13.2).

In the equilibrium C_s geometry one 'donor' monomer acts as a single hydrogen-bond donor to the other 'acceptor' molecule [Odutola and Dyke (1980)]. The largest tunneling splitting is due to a mechanism in which the two hydrogen atoms of the acceptor monomer are effectively interchanged, for which Smith *et al.* (1990) reported a barrier of $206\,\text{cm}^{-1}$. In the present work a planar transition state of C_s symmetry was found for both basis sets in the HF calculations, but this changed to an out-of-plane C_1 structure when correlation corrections were applied, in agreement with Smith *et al.* (1990). The pathway is shown in Figure 13.1, and corresponds to a 'methylamine-type' process [Tsuboi *et al.* (1964)] rather than a direct rotation about the local C_2 axis of the acceptor monomer. The path is represented by nine snapshots where the first and last frames are the two minima, the middle frame is the transition state and three additional frames on each side of the path were selected to best illustrate the mechanism. All the pathways in the present work were visualized using Mathematica [Wolfram (1996)].

The above mechanism is in agreement with the analysis of Pugliano *et al.* (1993) for the ground state acceptor tunneling path. The generator PI corresponding to the labeling scheme of

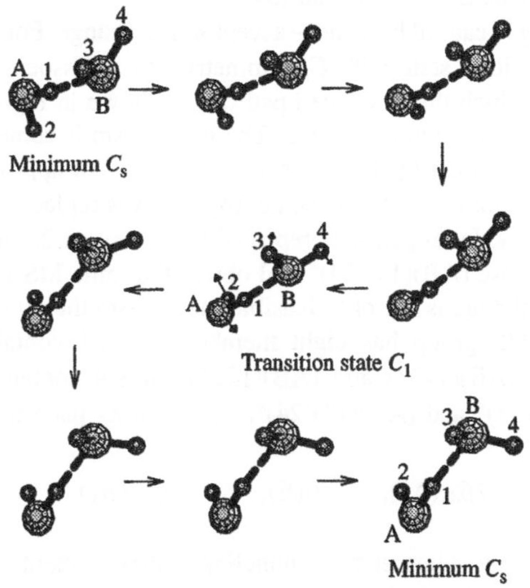

Figure 13.1. Acceptor tunneling path for $(H_2O)_2$ calculated at the DZP + diff/BLYP level. For this path $S = 4.2\,a_0$, $D = 2.8\,a_0$ and $\gamma = 1.3$.

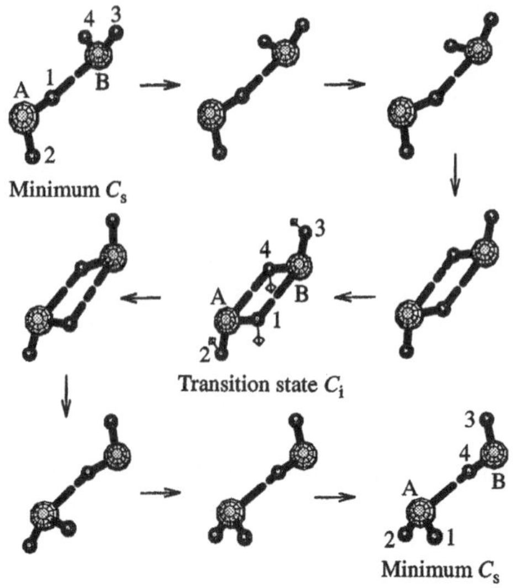

Figure 13.2. Donor–acceptor interchange tunneling mechanism for $(H_2O)_2$ calculated at the DZP + diff/BLYP level. For this path $S = 5.4\,a_0$, $D = 4.4\,a_0$ and $\gamma = 1.4$.

Figure 13.1 is (34), and it connects the versions in pairs. Hence each rigid-dimer rovibrational level is split into two by this process. Experimentally the ground state splitting due to acceptor tunneling is 2.47 cm^{-1} [Zwart *et al.* (1991)], in good agreement with a five-dimensional treatment of the nuclear dynamics by Althorpe and Clary (1995), which yielded a value of 2.34 cm^{-1}. The MS group for the rigid dimer labeled according to Figure 13.1 contains only the identity, E, and the PI (34)*, where hydrogens three and four change places and all coordinates are inverted through the space-fixed origin. The appropriate MS group when the generator (34) operation is feasible contains the elements E, E^*, (34) and (34)*.

The next largest splitting is caused by donor–acceptor interchange. For this process Smith *et al.* (1990) found a cyclic transition state with C_i symmetry and an associated barrier of 304 cm^{-1}. In this rearrangement, for which the calculated pathway is shown in Figure 13.2, the roles of the donor and acceptor monomers are interchanged. The mechanism is somewhat different from the 'geared internal rotation' assumed by Karyakin *et al.* (1995). An appropriate generator for the labeling scheme of Figure 13.2 is (AB) (1423), i.e. oxygen A is replaced by oxygen B, hydrogen 1 is replaced by hydrogen 4, hydrogen 4 is replaced by hydrogen 2, etc. This generator is not unique–one could also choose (AB)(14)(23)* and obtain the same MS group.

If donor–acceptor interchange is the only feasible mechanism then the versions are connected in sets of four and the MS group has eight members: class 1 contains E, class 2 contains (12)(34), class 3 contains (AB)(1423) and (AB)(1324), class 4 contains (34)* and (12)*, and class 5 contains (AB)(14)(23)* and (AB)(13)(24)*. The splitting pattern in the simplest Hückel-type approximation is:

$$2\beta_{da}(A_1), \qquad 0(E), \qquad -2\beta_{da}(B_1), \tag{2}$$

where β_{da} is the donor–acceptor interchange tunneling matrix element and we have labeled the levels according to appropriate irreducible representations of the $\boldsymbol{G}_8$ group which is a subgroup of $\boldsymbol{G}_{16}$ [Dyke (1977)]. When this process and acceptor tunneling are both feasible the MS group has order 16, i.e. the largest MS group possible without breaking covalent bonds, and it is isomorphic to D_{4h} [Dyke (1977)]. The versions are then connected in sets of eight and the splitting pattern is:

$$\beta_a + 2\beta_{da}(A_1^+), \quad \beta_a(E^+), \quad \beta_a - 2\beta_{da}(B_1^+),$$
$$-\beta_a + 2\beta_{da}(A_2^-), \quad -\beta_a(E^-), \quad -\beta_a - 2\beta_{da}(B_2^-),$$

where β_a is the acceptor–tunneling matrix element. Experimentally, the tunneling splittings due to donor–acceptor interchange are about a factor of 5 smaller than those associated with acceptor tunneling.

The third process which is generally presumed to have an observable effect on the energy level diagram is donor tunneling, but this can only lead to energy level shifts rather than further splittings because G_{16} is the largest MS group possible without breaking covalent bonds. Smith *et al.* (1990) calculated a barrier of 658 cm^{-1} for this mechanism, and the corresponding pathway found in the present work is shown in Figure 13.3. An appropriate generator PI for the labeling scheme of Figure 13.3 is (12)(34), and on its own this process would simply lead to doublet splittings with versions linked in pairs and an MS group of order 4. However, when combined with the other two mechanisms the eigenvalues of the corresponding secular problem (assuming a Hückel-type approximation) are:

$$\beta_a + 2\beta_{da} + \beta_d(A_1^+), \quad \beta_a - \beta_d(E^+), \quad \beta_a - 2\beta_{da} + \beta_d(B_1^+),$$
$$-\beta_a + 2\beta_{da} + \beta_d(A_2^-), \quad -\beta_a - \beta_d(E^-), \quad -\beta_a - 2\beta_{da} + \beta_d(B_2^-).$$

This pattern is in complete agreement with those obtained in detailed calculations [Hougen (1985), Coudert and Hougen (1988), Coudert and Hougen (1990), Althorpe and Clary (1995), and Leforestier *et al.* (1997)]. The tunneling levels are no longer in plus–minus pairs because donor tunneling introduces odd-membered rings into the reaction graph.

The DZP + diff/BLYP energy profiles for all three rearrangements are shown in Figure 13.4. A maximum step size of 0.1 a_0 was used for the left-hand side of the acceptor–tunneling path. This value was increased to 0.15 a_0 for all the other paths, resulting in profiles that are not as smooth, but a faster execution time.

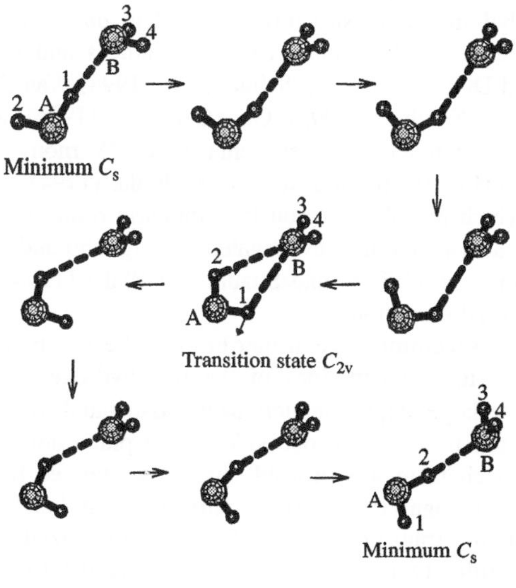

Figure 13.3. Donor tunneling bath for $(H_2O)_2$ calculated at the DZP + diff/BLYP level. For this path $S = 5.8\,a_0$, $D = 4.5\,a_0$ and $\gamma = 2.2$.

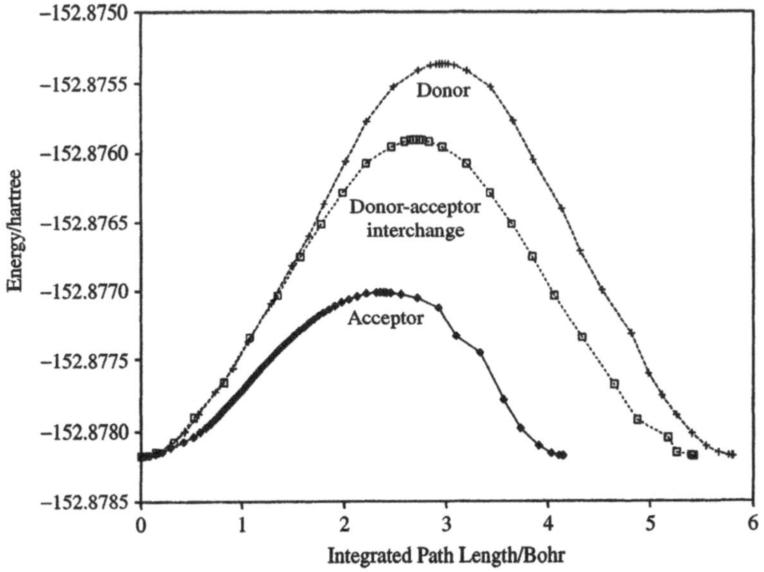

Figure 13.4. Energy versus the integrated path length, S, for the three degenerate rearrangement mechanisms of the water dimer described in Section 13.5.

13.6 Water trimer

Recent interest in the water trimer began with the FIR-VRT experiment of Pugliano and Saykally (1992) who reported spectra for $(D_2O)_3$ characteristic of an oblate symmetric rotor, with each line split into a regularly spaced quartet with a spacing of roughly 6 MHz (2×10^{-4} cm^{-1}). Accurate values for the vibrationally averaged rotational constants revealed a large negative inertial defect, indicative of extensive out-of-plane motion of the non-hydrogen-bonded hydrogens. However, a cyclic, asymmetric global minimum structure is found in *ab initio* calculations [Bene and Pople (1973), Mó and Yáñez (1992), Wales (1993b), Xantheas and Dunning (1993a), Schütz *et al.* (1993), Xantheas and Dunning (1993b), Schütz *et al.* (1994), Van Duijneveldt-Van de Rijdt and Van Duijneveldt (1995), Xantheas (1995), González *et al.* (1996), and Schütz, Rauhut and Werner (1998)], in agreement with earlier experimental results [Vernon *et al.* (1982)]. Subsequent experiments [Liu *et al.* (1994a, 1994b), and Suzuki and Blake (1994)] assigned new transitions for $(H_2O)_3$ and $(D_2O)_3$ which revealed rigorously symmetric rotor structure, in contrast to the strongly perturbed band that was initially investigated by Pugliano and Saykally (1992). All the transitions reported by Liu *et al.* (1994b) and Suzuki and Blake (1994) show a regular quartet splitting of every rovibrational transition.

The oblate symmetric top spectrum of the trimer has now been explained by vibrational averaging over large amplitude torsional motions of the free hydrogens on the timescale of the FIR-VRT experiment. These large amplitude motions are associated with a facile 'flip' rearrangement where a free hydrogen moves from one side of the plane defined by the three oxygen atoms to the other. This mechanism was probably first characterized for an empirical potential by Owicki, Shipman and Scheraga (1975). An *ab initio* pathway has been presented by Wales (1993b) and the corresponding transition state was also characterized by Fowler and Schaefer (1995). Wales (1993b) identified two other degenerate rearrangement mechanisms for the trimer with rather larger barriers than the flip, and suggested that one of them (christened the 'donor' or 'bifurcation' pathway) might be responsible for the quartet splittings observed experimentally.

In the associated transition state one monomer acts as a double donor to a neighbor which acts as a double acceptor in a configuration similar to that of the 'donor tunneling' transition state in the water dimer, discussed in Section 13.5 (Figure 13.3).

The single flip mechanism links each permutational isomer to two others in cyclic sets of six, giving a secular problem analogous to the π system of benzene [Wales (1993b), and Bürgi *et al.* (1995)] with splitting pattern:

$$2\beta_f(A_1), \qquad \beta_f(E_2), \qquad -\beta_f(E_1), \qquad -2\beta_f(A_2),$$

where β_f is the tunneling matrix element for the flip. The MS group has order six and is isomorphic to C_{3h} [Pugliano and Saykally (1992)].

The rearrangement responsible for the regular quartet splittings observed experimentally probably involves the bifurcated transition state [Wales (1993b)]. When both the bifurcation mechanism and the single flip are feasible the MS group has order 48 [Wales (1993b)] and is identical to the one deduced by Bone *et al.* (1991) for acetylene trimer. Liu *et al.* (1994b) reported that their new spectra were consistent with the G_{48} group suggested by the mechanisms characterized by Wales (1993b). However, there are six distinct ways for the bifurcation to occur with accompanying flips of neighboring monomers, and these give two distinct splitting patterns [Walsh and Wales (1996)]. The most recent experiments and theory show that some of the quartets are further split by Coriolis coupling [der Avoird, Olthof and Wormer (1996), and Olthof *et al.* (1996)]. Entirely regular quartets are found to be associated with generators containing E^* [Walsh and Wales (1996), and Olthof *et al.* (1996)].

A more detailed account of the above mechanisms and group theory can be found elsewhere [Walsh and Wales (1996), and Wales (1998)]. There have also been a number of studies of the nuclear dynamics [Sabo *et al.* (1995), Klopper and Schütz (1995), Gregory and Clary (1995a, 1995b), Sabo *et al.* (1996), Wales (1996a), and Guiang and Wyatt (1998)]. The analysis of different bifurcation and flip mechanisms is considered in more detail in Section 13.8 for the water pentamer.

13.7 Water tetramer

A symmetric doublet splitting of 5.6 MHz ($1.9 \times 10^{-4}\,cm^{-1}$) has been reported in two FIR-VRT experiments for $(D_2O)_4$ [Cruzan *et al.* (1996a, 1996b)] and a splitting of 2260 MHz ($0.075\,cm^{-1}$) has been observed in a transition around $67.9\,cm^{-1}$ for $(H_2O)_4$ [Cruzan *et al.* (1997)]. The cyclic global minimum of the tetramer has S_4 symmetry [Bene and Pople (1970), Lentz and Scheraga (1973), Kistenmacher *et al.* (1974), Kim *et al.* (1986), Xantheas and Dunning (1993a), and Xantheas (1995)] with the unbound terminal hydrogens alternating above and below the ring. Hence this system lacks the 'frustration' inherent in the trimer and pentamer, where two of the terminal hydrogens are forced to be adjacent. Schütz and coworkers found no true transition states corresponding to degenerate rearrangements in the torsional space of the tetramer [Schütz *et al.* (1995), and Engkvist *et al.* (1997)]. A more systematic survey revealed many more non-degenerate rearrangement mechanisms and just one true degenerate rearrangement of the global minimum, which disappeared in correlated calculations [Wales and Walsh (1997)]. The predicted splitting pattern for the latter mechanism is much more complex than the experimentally observed symmetric doublets, and so it cannot explain the latter results. Gregory and Clary (1996b) conducted a DMC study of $(H_2O)_4$ and $(D_2O)_4$ which included estimates of tunneling splittings without specifying the mechanism. A four-dimensional torsional potential energy surface has recently been fitted [Graf and Leutwyler (1998)] enabling reduced dimensionality DVR calculations to be carried out [Sabo *et al.* (1998)].

The lack of a degenerate rearrangement mechanism which would produce doublet splittings in the above calculations makes the observed tunneling splitting pattern far harder to explain than for the water dimer and trimer. However, Loerting, Liedl and Rode (1998) have recently found a true transition state for concerted proton transfer in $(H_2O)_4$. This process would produce symmetric doublet splittings, but their magnitude appears to be too small to account for the experimental observations.

For an overall quadruple flip [Cruzan *et al.* (1996a)] the MS group is isomorphic to C_{2v}. The effective generator PI is the same for three possible paths [Dalton and Nicholson (1975)]: (1) a concerted quadruple flip (via an index four saddle), (2) true transition states and local minima, or (3) index two saddles and local minima [Schütz *et al.* (1995), and Wales and Walsh (1997)]. The existence of the 'indirect tunneling' paths (see Section 13.3) with lower barriers than the concerted quadruple flip is a consequence of the Murrell and Laidler (1968) theorem. Since versions of the global minimum are linked in pairs by each of these mechanisms they would all give rise to doublet splittings, and cannot be distinguished experimentally. The experimental assignment of the tunneling mechanism as a 'concerted flipping motion' [Cruzan *et al.* (1997)] should probably be relaxed to an effective quadruple flip. In fact it was also suggested that broadened lines in the $(H_2O)_4$ spectrum may contain 'unresolved bifurcation tunneling splittings' [Cruzan *et al.* (1997)]. However, no degenerate single bifurcation rearrangement has been found in calculations. Instead there appear to exist a wide variety of non-degenerate rearrangements, which could lead to enlargement of the MS group via 'indirect tunneling' in a variety of ways [Wales and Walsh (1997)].

Four-dimensional DVR calculations in the torsional space by Sabo *et al.* (1998) suggest that interconversion occurs by 'a sequence of double O–H flips'. Their results are fairly consistent with analogous calculations in our group using different empirical potentials, although we see a larger weighting for sequences of single flips. Of course, these reduced dimensionality calculations do not admit the concerted proton transfer mechanism [Loerting, Liedl and Rode (1998)] or any of the 'indirect tunneling' pathways involving bifurcation [Wales and Walsh (1997)], and include a number of fairly drastic simplifications.

The present picture for water tetramer is therefore still rather confused. It is probably worth pointing out once again that the experiments measure sums or differences of tunneling splittings and involve vibrational excitations which could have a significant effect upon the nuclear dynamics.

13.8 Water pentamer

The water pentamer exhibits a number of similarities to the trimer due to the frustrated cyclic global minimum [Xantheas and Dunning (1993a), and Xantheas (1995)]. The first FIR-VRT results for $(D_2O)_5$ did not reveal any tunneling splittings [Liu *et al.* (1996b)]. However, analogs of the single flip and bifurcation mechanisms, described above for the trimer, have been found along with a number of pathways connecting higher energy minima [Wales (1996b), and Wales and Walsh (1996)]. Both the flip and the bifurcation mechanism considered in isolation produce an MS group of order 10 isomorphic to C_{5h} with each version connected to two others in a cyclic reaction graph containing 10 versions. The predicted splitting pattern in the simplest Hückel approximation is the same as for the π-system of 10-annulene (cyclodecapentaene):

$$2\beta(\mathrm{A}'), \qquad \phi\beta(\mathrm{E}_2''), \qquad \phi^{-1}\beta(\mathrm{E}_1'), \qquad -\phi^{-1}\beta(\mathrm{E}_1''), \qquad -\phi\beta(\mathrm{E}_2'), \qquad -2\beta(\mathrm{A}''),$$

where β is the appropriate tunneling matrix element, $\phi = (\sqrt{5}+1)/2$ is the golden ratio and $\phi^{-1} = (\sqrt{5}-1)/2 = 1/\phi$. The symmetry species in parentheses are appropriate if the generator corresponds to the operation S_5 of C_{5h}. The nuclear spin weights are almost identical for each irreducible representation in both $(H_2O)_5$ and $(D_2O)_5$ [Wales and Walsh (1996)].

If both the flip and bifurcation mechanisms are feasible then the MS group increases in dimension to order 320 and the splitting pattern becomes rather complicated [Wales and Walsh (1996)]. Qualitative estimates for the magnitude of the splittings suggested that the flip should lead to observable effects in both $(H_2O)_5$ and $(D_2O)_5$, while bifurcation tunneling might produce splittings at the limit of current resolution [Wales and Walsh (1996)]. Subsequent experimental results for $(D_2O)_5$ were interpreted in terms of a pseudorotation scheme for the flip rearrangement, but no bifurcation splittings were resolved [Liu *et al.* (1997), and Cruzan *et al.* (1998)]. However, the spectrum of $(H_2O)_5$ does appear to have resolvable structure due to bifurcation tunneling, and has been interpreted in terms of the nuclear spin weights predicted for the G_{320} group [Brown, Keutsch and Saykally (1998)].

In view of the new experimental results for $(H_2O)_5$ a more detailed analysis of the possible rearrangements and group theory is warranted [Wales (1997)]. The labeling scheme and assignment of generator PIs is illustrated for two possible bifurcation + flip combinations in Figure 13.5. The labeling scheme in the figure enables us to identify all the possible bifurcation/flip processes which correspond to degenerate rearrangements using the letters A–E to label the five monomers. Here we exclude processes where the bifurcating monomer also flips. The bifurcation can occur at any of the five monomers, and in each case the frustrated pair of neighboring free hydrogens can appear on any of the five edges. This gives a total of 25 bifurcation/flip combinations. Five of the resulting generators are self-inverse while the other 20 fall into non-self-inverse pairs to give a total of 15 distinct mechanisms (Table 13.3). For the same labeling scheme the generator for the single flip is (ABCDE) (13 579) (246 810)* with inverse (AEDCB) (19 753) (210 864)*.

Previously we have identified two splitting patterns for different bifurcation/flip combinations in the pentamer, and labeled these A and B. In a systematic analysis a third splitting pattern, denoted type C, emerges, as shown in Table 13.4. All 15 combinations result in the same MS group, G_{320}, and nuclear spin weights described previously [Wales and Walsh (1996)]. The C-type pattern occurs for the five generators of the form $(ij)^*$; for the trimer such generators also

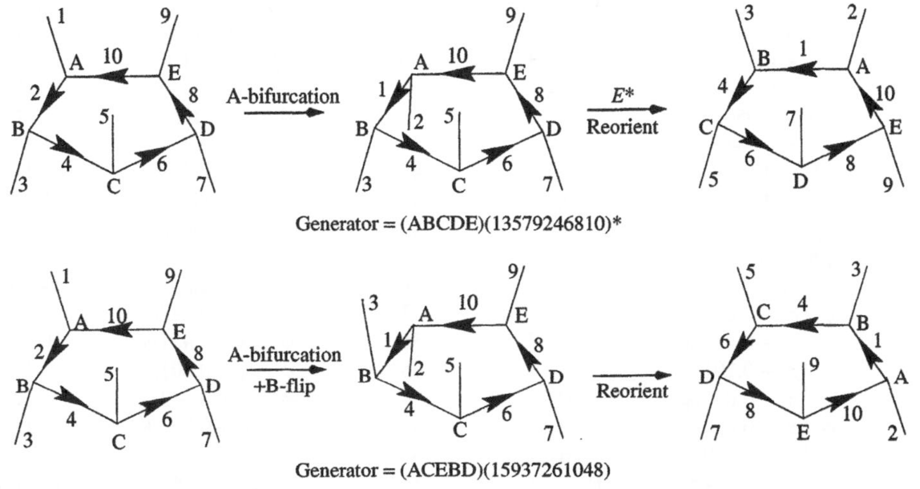

Figure 13.5. The effect of A-bifurcation and A-bifurcation + B-flip on a reference version of the water pentamer. The arrows indicate the direction of hydrogen-bond donation. The generator PI is deduced by putting the structure in coincidence with the reference, with an intervening inversion if necessary.

Table 13.3. Generators and splitting patterns for the 15 distinct bifurcation + flip combinations considered for water pentamer in Section 13.8. See also Figure 13.5.

Description	Generator	Inverse	Pattern
A-bif	(ABCDE)(13579246810)*	E-bif	A
A-bif + BCD-flips	(AEDCB)(19753210864)	B-bif + CDE-flips	A
E-bif + BCD-flips	(ABCDE)(13571024689)	D-bif + ABC-flips	A
D-bif + BCE-flips	(ABCDE)(13581024679)	C-bif + ABD-flips	A
B-bif + ACD-flips	(AEDBC)(19754210863)	C-bif + BCD-flips	A
A-bif + B-flip	(ACEBD)(15937261048)	D-bif + E-flip	B
A-bif + BC-flips	(ADBEC)(17395284106)*	C-bif + DE-flips	B
E-bif + D-flip	(ADBEC)(17310628495)	B-bif + A-flip	B
E-bif + CD-flips	(ACEBD)(17310628495)*	C-bif + AB-flips	B
D-bif + EC-flips	(ACEBD)(15938261047)*	B-bif + AC-flips	B
A-bif + BCDE-flips	(12)*	self-inverse	C
E-bif + ABCD-flips	(910)*	self-inverse	C
D-bif + ABCE-flips	(78)*	self-inverse	C
B-bif + ACDE-flips	(34)*	self-inverse	C
C-bif + ABDE-flips	(56)*	self-inverse	C

Table 13.4. Splitting pattern type C for five of the possible pentamer bifurcation + flip combinations.

Level	Energy	Symmetry	Level	Energy	Symmetry
1	$2\beta_f + \beta_b$	A_1^+	56	$-2\beta_f - \beta_b$	A_1^-
2	$2\beta_f + 0.603\beta_b$	H_6^+	55	$-2\beta_f - 0.603\beta_b$	H_6^-
3	$2\beta_f + 0.206\beta_b$	H_1^+	54	$-2\beta_f - 0.206\beta_b$	H_1^-
4	$2\beta_f + 0.203\beta_b$	H_2^+	53	$-2\beta_f - 0.203\beta_b$	H_2^-
5	$2\beta_f - 0.194\beta_b$	H_3^+	52	$-2\beta_f + 0.194\beta_b$	H_3^-
6	$2\beta_f - 0.197\beta_b$	H_4^+	51	$-2\beta_f + 0.197\beta_b$	H_4^-
7	$2\beta_f - 0.597\beta_b$	H_5^+	50	$-2\beta_f + 0.597\beta_b$	H_5^-
8	$2\beta_f - \beta_b$	A_2^+	49	$-2\beta_f + \beta_b$	A_2^-
9	$\phi\beta_f + \beta_b$	$E_4^- \oplus H_5^-$	48	$-\phi\beta_f - \beta_b$	$E_4^+ \oplus H_5^+$
10	$\phi\beta_f + 0.848\beta_b$	H_3^-	47	$-\phi\beta_f - 0.848\beta_b$	H_3^+
11	$\phi\beta_f + 0.451\beta_b$	$H_1^- \oplus H_4^-$	46	$-\phi\beta_f - 0.451\beta_b$	$H_1^+ \oplus H_4^+$
12	$\phi\beta_f + 0.204\beta_b$	H_5^-	45	$-\phi\beta_f - 0.204\beta_b$	H_5^+
13	$\phi\beta_f + 0.050\beta_b$	H_2^-	44	$-\phi\beta_f - 0.050\beta_b$	H_2^+
14	$\phi\beta_f - 0.044\beta_b$	H_4^-	43	$-\phi\beta_f + 0.044\beta_b$	H_4^+
15	$\phi\beta_f - 0.196\beta_b$	H_6^-	42	$-\phi\beta_f + 0.196\beta_b$	H_6^+
16	$\phi\beta_f - 0.444\beta_b$	$H_2^- \oplus H_3^-$	41	$-\phi\beta_f + 0.444\beta_b$	$H_2^+ \oplus H_3^+$
17	$\phi\beta_f - 0.846\beta_b$	H_1^-	40	$-\phi\beta_f + 0.846\beta_b$	H_1^+
18	$\phi\beta_f - \beta_b$	$E_2^- \oplus H_6^-$	39	$-\phi\beta_f + \beta_b$	$E_2^+ \oplus H_6^+$
19	$\phi^{-1}\beta_f + \beta_b$	$E_1^+ \oplus H_6^+$	38	$-\phi^{-1}\beta_f - \beta_b$	$E_1^- \oplus H_6^-$
20	$\phi^{-1}\beta_f + 0.847\beta_b$	H_2^+	37	$-\phi^{-1}\beta_f - 0.847\beta_b$	H_2^-
21	$\phi^{-1}\beta_f + 0.451\beta_b$	$H_1^+ \oplus H_4^+$	36	$-\phi^{-1}\beta_f - 0.451\beta_b$	$H_1^- \oplus H_4^-$
22	$\phi^{-1}\beta_f + 0.201\beta_b$	H_6^+	35	$-\phi^{-1}\beta_f - 0.201\beta_b$	H_6^-
23	$\phi^{-1}\beta_f - 0.042\beta_b$	H_3^+	34	$-\phi^{-1}\beta_f + 0.042\beta_b$	H_3^-
24	$\phi^{-1}\beta_f - 0.053\beta_b$	H_1^+	33	$-\phi^{-1}\beta_f + 0.053\beta_b$	H_1^-
25	$\phi^{-1}\beta_f - 0.199\beta_b$	H_4^+	32	$-\phi^{-1}\beta_f + 0.199\beta_b$	H_5^-
26	$\phi^{-1}\beta_f - 0.444\beta_b$	$H_2^+ \oplus H_3^+$	31	$-\phi^{-1}\beta_f + 0.444\beta_b$	$H_2^- \oplus H_3^-$
27	$\phi^{-1}\beta_f - 0.847\beta_b$	H_4^+	30	$-\phi^{-1}\beta_f + 0.847\beta_b$	H_4^-
28	$\phi^{-1}\beta_f - \beta_b$	$E_3^+ \oplus H_5^+$	29	$-\phi^{-1}\beta_f + \beta_b$	$E_3^- \oplus H_5^-$

$\phi = (\sqrt{5} + 1)/2$

correspond to one particular splitting pattern [Walsh and Wales (1996)]. The A-type pattern is found for the bifurcation accompanied by zero or three flips and the B-type pattern is found for the bifurcation accompanied by one or two flips (Table 13.3). All three patterns include four sets of ten lines and two sets of eight; they all involve irregular submultiplets. In the C pattern every energy level λ has a partner with energy $-\lambda$ (and opposite parity with respect to E^*), implying that there are no odd-membered rings in the reaction graph. The splittings within each manifold with a constant coefficient of β_f (the tunneling matrix element for the flip) also obey an approximate pairing rule.

13.9 Water hexamer

Liu *et al.* (1996a) identified a VRT band of $(H_2O)_6$ at 83 cm^{-1} on the basis of an isotope mixture test. In contrast to smaller water clusters the lowest energy isomer of the hexamer is probably not cyclic, and the four lowest energy structures found by Tsai and Jordan (1993) lie within an energy range of only 100 cm^{-1}. Kim and Kim (1998) have recently reviewed the evolution of theoretical work on the hexamer [see e.g. Mhin *et al.* (1991, 1994), Lee, Chen and Fitzgerald (1994), Krishnan, Jensen and Burke (1994), Pedulla, Vila and Jordan (1996), and Pedulla, Kim and Jordan (1998)]. The most accurate calculations conducted so far suggest that a 'cage' structure lies lowest, followed closely by 'prism' and 'book' forms (Figure 13.6). DMC calculations were used to find the vibrationally averaged rotational constants for each structure using an empirical potential, and the best match was found for the cage structure. On this basis the spectrum was assigned to the cage [Liu *et al.* (1996a)].

Calculations reveal numerous isomers of all three morphologies illustrated in Figure 13.6 with different arrangements of the hydrogen bonds. The remaining discussion will concentrate on the cage and the results of calculations using the ASP-W2 and TIP4P potentials described in Section 13.4. The ASP-W2 form should be very similar to that employed in the DMC calculations of nuclear dynamics by Liu *et al.* (1996a). However, we note that the cage isomer is not the lowest in energy for either of these potentials, although inclusion of zero-point energy can alter the ordering [Liu *et al.* (1996a)]. In fact, for the ASP-W2 potential the cyclic structure of the water pentamer is also not the global minimum, but nevertheless, the lowest energy rearrangements of this isomer are quite well reproduced [Wales and Walsh (1996)].

For the ASP-W2 potential only the first-order induction energy was considered, since iterating this term to convergence is time consuming and was found to make no qualitative difference to results for the water pentamer in previous work [Wales and Walsh (1996)]. There are then four isomers of the cage structure shown in Figure 13.6 differing only in the position of the free hydrogen atoms of the two terminal, single-donor, single-acceptor monomers (Figure 13.7). No low energy degenerate rearrangements of these isomers were found, but transition states were located for non-degenerate single flip and bifurcation mechanisms [Wales (1996a)]. For every single flip process there is an analogous bifurcation which links different permutational isomers of the same structures. Details of the paths are given in Table 13.5 and the single flip and

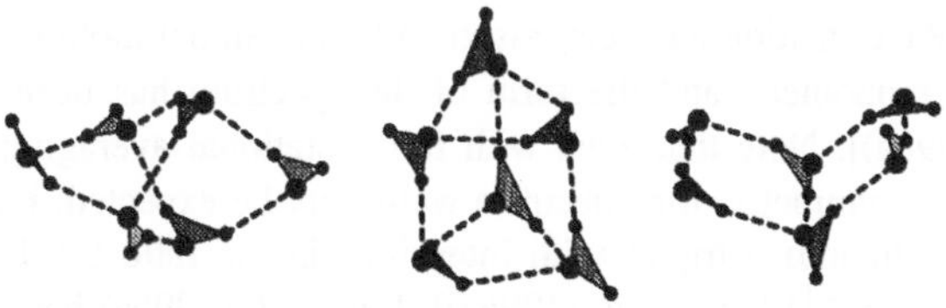

Figure 13.6. Cage, prism and book forms of $(H_2O)_6$.

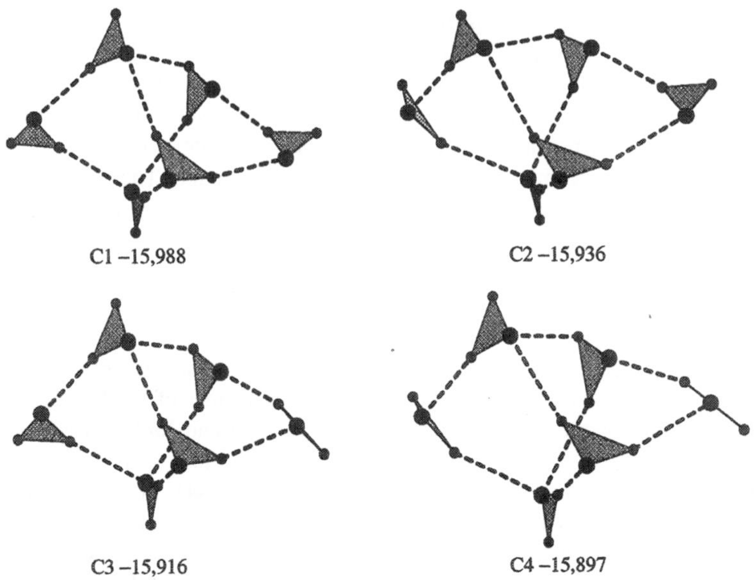

Figure 13.7. Isomers of the cage structure for $(H_2O)_6$ calculated using the ASP-W2 potential with binding energies in cm^{-1}.

Table 13.5. Rearrangement mechanisms which interconvert cage isomers of $(H_2O)_6$ calculated with the ASP-W2 potential. The energies are in cm^{-1}. Min_l is the lower minimum, Δ_l is the higher barrier, TS is the transition state and Δ_2 is the smaller barrier corresponding to the higher minimum Min_2. S is the integrated path length in Å, D is the displacement between minima in Å and γ is the cooperativity index. All these quantities are defined in Section 13.4.

Min_l	Δ_l	TS	Δ_2	Min_2	S	D	γ	Description
C1	560	−15 428	508	C2	2.8	2.1	13.7	single flip
C1	641	−15 347	589	C2	2.9	2.2	7.6	bifurcation
C1	323	−15 664	251	C3	1.7	1.3	15.4	single flip
C1	1135	−14 853	1063	C3	3.2	2.1	8.5	bifurcation
C2	325	−15 612	285	C4	1.7	1.3	16.2	single flip
C2	1058	−14 878	1018	C4	3.4	2.1	8.6	bifurcation
C3	524	−15 391	506	C4	2.7	2.0	14.1	single flip
C3	650	−15 266	631	C4	2.7	2.2	7.9	bifurcation

bifurcation pathways linking cage isomers C1 and C2 are illustrated in Figures 13.8 and 13.9. The other paths are omitted for brevity. Some of the corresponding stationary points have been characterized in *ab initio* calculations by Gregory and Clary (1997).

The experimental VRT transition has been assigned to a torsional motion of one of the single-donor, single-acceptor monomers, and the form of the spectrum has been described as 'near-prolate' [Liu *et al.* (1996a)]. Note that even with full vibrational averaging over the four cage isomers in Figure 13.7 a symmetric top spectrum would not be expected. Experimentally, every line was found to be split into a triplet with intensities in the ratio 9:6:1 and equal spacings of 1.92 MHz ($6.4 \times 10^{-5}\,cm^{-1}$) [Liu *et al.* (1996a)]. Liu *et al.* (1996a) have explained how this pattern might emerge in terms of hypothetical degenerate rearrangements of the cage which they

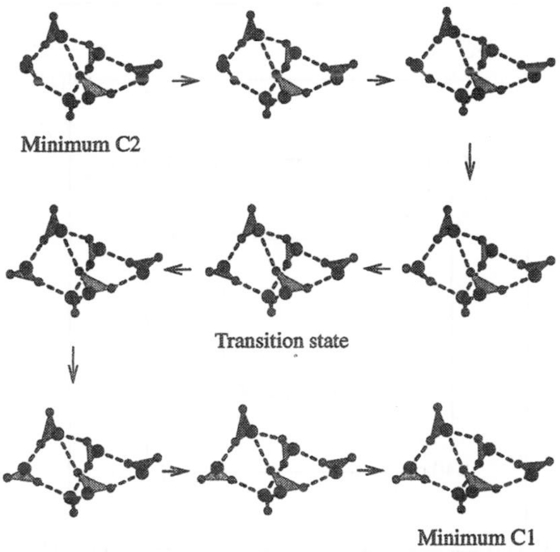

Figure 13.8. Single flip mechanism which interconverts C1 and C2 for the ASP-W2 potential.

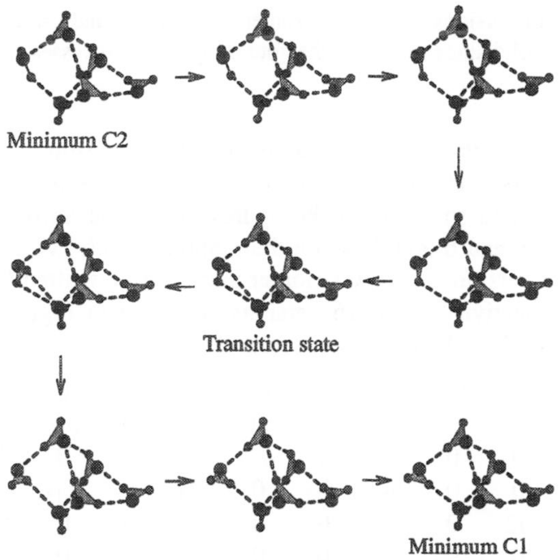

Figure 13.9. Bifurcation mechanism which interconverts C1 and C2 for the ASP-W2 potential.

assumed must interchange the hydrogens of two monomers in almost equivalent environments. This could lead to a doublet of doublets where the middle lines are essentially superimposed. The resulting nuclear spin weights can then reproduce the observed intensity pattern.

Liu *et al.* (1996a) additionally suggested that the two monomers in question might be the two terminal molecules which we have found to undergo flip and bifurcation rearrangements above. However, no direct degenerate rearrangements of the cage isomers have been found to date. Since a combination of sequential bifurcation and flip rearrangements would achieve the desired effect we will now consider the splittings that might result in more detail. The eight

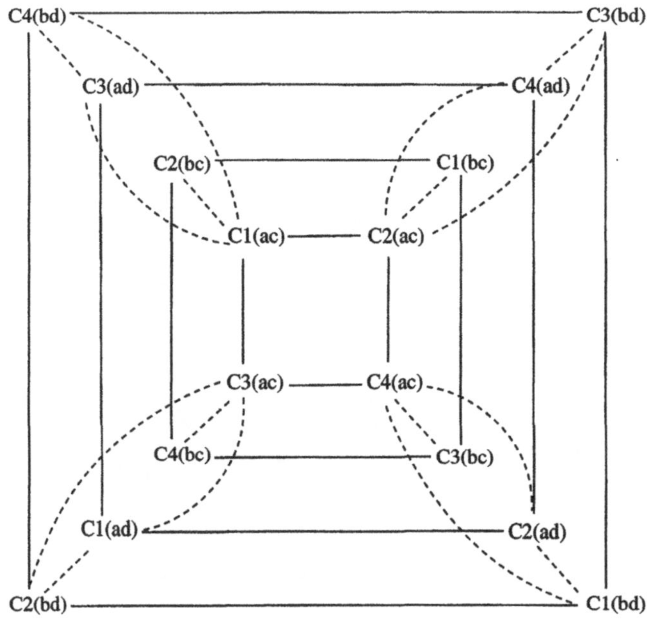

Figure 13.10. Connectivity of versions of the cage isomers C1, C2, C3 and C4 according to the rearrangements listed in Table 13.5. Solid lines represent the four single flip processes and dashed lines the four bifurcations.

single flip and bifurcation processes described above for the cage isomers link 16 versions: four of each isomer. To distinguish between versions of each isomer we need only specify which of the two hydrogens is free, and so we label the hydrogens on the terminal monomer with two double-acceptor neighbors a and b and those on the monomer with two double-donor neighbors c and d. The four relevant versions of the C1 isomer may then be written C1(ac), C1(bc), C1(ad) and C1(bd). The interconnectivity of all 16 versions is shown in Figure 13.10. If we make a Hückel-type approximation then the resulting secular determinant is:

$$\begin{pmatrix}
0 & 0 & 0 & 0 & f_{12} & b_{12} & 0 & 0 & f_{13} & 0 & b_{13} & 0 & 0 & 0 & 0 & 0 \\
0 & 0 & 0 & 0 & b_{12} & f_{12} & 0 & 0 & 0 & f_{13} & 0 & b_{13} & 0 & 0 & 0 & 0 \\
0 & 0 & 0 & 0 & 0 & 0 & f_{12} & b_{12} & b_{13} & 0 & f_{13} & 0 & 0 & 0 & 0 & 0 \\
0 & 0 & 0 & 0 & 0 & 0 & b_{12} & f_{12} & 0 & b_{13} & 0 & f_{13} & 0 & 0 & 0 & 0 \\
f_{12} & b_{12} & 0 & 0 & 52 & 0 & 0 & 0 & 0 & 0 & 0 & 0 & f_{24} & 0 & b_{24} & 0 \\
b_{12} & f_{12} & 0 & 0 & 0 & 52 & 0 & 0 & 0 & 0 & 0 & 0 & 0 & f_{24} & 0 & b_{24} \\
0 & 0 & f_{12} & b_{12} & 0 & 0 & 52 & 0 & 0 & 0 & 0 & 0 & b_{24} & 0 & f_{24} & 0 \\
0 & 0 & b_{12} & f_{12} & 0 & 0 & 0 & 52 & 0 & 0 & 0 & 0 & 0 & b_{24} & 0 & f_{24} \\
f_{13} & 0 & b_{13} & 0 & 0 & 0 & 0 & 0 & 72 & 0 & 0 & 0 & f_{34} & b_{34} & 0 & 0 \\
0 & f_{13} & 0 & b_{13} & 0 & 0 & 0 & 0 & 0 & 72 & 0 & 0 & b_{34} & f_{34} & 0 & 0 \\
b_{13} & 0 & f_{13} & 0 & 0 & 0 & 0 & 0 & 0 & 0 & 72 & 0 & 0 & 0 & f_{34} & b_{34} \\
0 & b_{13} & 0 & f_{13} & 0 & 0 & 0 & 0 & 0 & 0 & 0 & 72 & 0 & 0 & b_{34} & f_{34} \\
f_{24} & 0 & b_{24} & 0 & f_{24} & 0 & b_{24} & 0 & f_{34} & b_{34} & 0 & 0 & 91 & 0 & 0 & 0 \\
0 & f_{24} & 0 & b_{24} & 0 & f_{24} & 0 & b_{24} & b_{34} & f_{34} & 0 & 0 & 0 & 91 & 0 & 0 \\
b_{24} & 0 & f_{24} & 0 & b_{24} & 0 & f_{24} & 0 & 0 & 0 & f_{34} & b_{34} & 0 & 0 & 91 & 0 \\
0 & b_{24} & 0 & f_{24} & 0 & b_{24} & 0 & f_{24} & 0 & 0 & b_{34} & f_{34} & 0 & 0 & 0 & 91
\end{pmatrix}$$

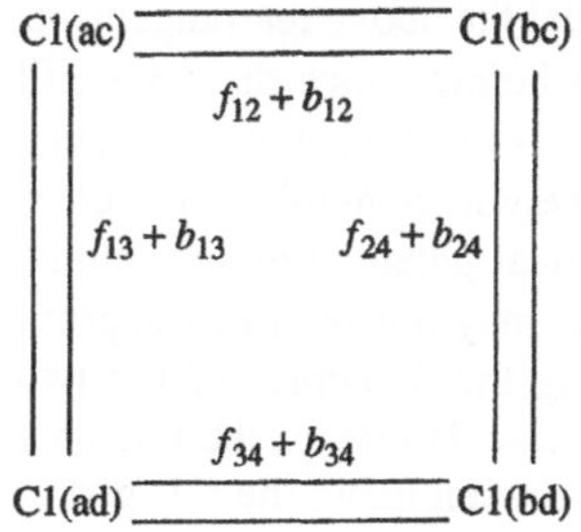

Figure 13.11. Effective reaction graph for the four versions of cage isomer C1 that are interconnected indirectly by flip and bifurcation rearrangements.

where f_{ij} and b_{ij} are the appropriate tunneling matrix elements between C_i and C_j. We can simplify this problem by focusing on one of the isomers, say C1, and considering an effective generator [Dalton and Nicholson (1975)] corresponding to a flip and a bifurcation (in either order). The reaction graph can then be represented as shown in Figure 13.11, where the double lines indicate that there are two routes between each pair of versions depending upon whether the flip or bifurcation comes first. The MS group contains the elements E, (ab), (bc) and (ab)(bc) and is isomorphic to C_{2v}. If the four connections are each represented by the same tunneling matrix element β then the energy level pattern will be identical to that obtained in the Hückel treatment of the π system of butadiene, i.e.

$$2\beta(A_1), \qquad 0(A_1, B_1), \qquad -2\beta(B_1), \tag{5}$$

where the correspondence between the MS group elements and the operations of C_{2v} has been chosen as (cd) $\equiv C_2$, (ab) $\equiv \sigma_v(xz)$ and (ab)(cd) $\equiv \sigma'_v(yz)$. The accidental degeneracy of the A_2 and B_1 states would be broken at higher resolution. The relative nuclear spin weights for rovibronic states are easily shown to be 9:3:3:1 for $(H_2O)_6$ and 4:2:2:1 for $(D_2O)_6$ corresponding to $A_1 : A_2 : B_1 : B_2$. If the accidental degeneracy is unresolved then the relative intensities of the three triplet components would be 9:6:1 for $(H_2O)_6$ and 4:4:1 for $(D_2O)_6$.

For the TIP4P potential the situation is slightly different because there are only two cage isomers rather than four. The single-donor, single-acceptor molecule that has two double-donor neighbors exhibits only one torsional minimum intermediate between the two states found in C1–C4 above. The other single-donor, single-acceptor has the same two torsional states as in C1–C4, and undergoes single flip and bifurcation rearrangements as before. For the single-donor, single-acceptor molecule with two double-donor neighbors a direct degenerate rearrangement corresponding to a bifurcated transition state was found. The pathways are summarized in Table 13.6 where we label the two isomers C1′ and C2′. Illustrations of these rearrangements are omitted for brevity. A degenerate bifurcation rearrangement of the C2′ isomer could not be located, despite starting a number of transition state searches from points around the expected geometry.

Table 13.6. Rearrangement mechanisms which interconvert cage isomers of $(H_2O)_6$ calculated with the TIP4P potential. The energies are in cm^{-1}. Min_1 is the lower minimum, Δ_1 is the higher barrier, TS is the transition state, and Δ_2 is the smaller barrier corresponding to the higher minimum Min_2. S is the integrated path length in Å, D is the displacement between minima in Å and γ is the cooperativity index. All these quantities are defined in Section 13.4.

Min_1	Δ_1	TS	Δ_2	Min_2	S	D	γ	Description
C1′ (−16533)	14	−16519	2	C2′ (−16521)	1.3	1.2	14.1	single flip
C1′ (−16533)	825	−15708	813	C2′ (−16521)	3.6	2.2	8.6	bifurcation
C1′ (−16533)	1361	−15172	1361	C1′ (−16533)	4.5	2.2	8.7	bifurcation

Although the topology of the PES is different for the TIP4P potential the above rearrangements could produce the same tunneling splittings and intensity pattern as before, since there are still pathways linking all four versions of each isomer. Unfortunately there is insufficient experimental information to distinguish the two possibilities. Because of the relatively large number of degrees of freedom and the uncertainties associated with the choice of empirical potential more accurate theoretical treatments of the dynamics will not be easy. However, it may not be a bad approximation to neglect relaxations of the rest of the cage in considering the dynamics of the two single-donor, single-acceptor monomers. Two-dimensional quantum calculations of the torsional dynamics of these two molecules were performed under this approximation using the C1′ geometry and the DVR [Bačić and Light (1989)]. These calculations follow the three-dimensional DVR calculations of Sabo *et al.* (1995) for $(H_2O)_3$ in employing a model Hamiltonian in which only rotation about the hydrogen-bonded O–H bond is permitted:

$$\hat{\mathcal{H}} = -B_{\text{eff}}\left(\frac{\partial^2}{\partial\phi_1^2} + \frac{\partial^2}{\partial\phi_2^2}\right) + V(\phi_1, \phi_2), \tag{6}$$

where ϕ_1 and ϕ_2 are the two torsional angles in question and the value of B_{eff} was taken to be $19.63\,\text{cm}^{-1}$ for H_2O and $9.82\,\text{cm}^{-1}$ for D_2O. For each torsional degree of freedom the basis functions were chosen as:

$$\psi(\phi) = \frac{1}{\sqrt{2\pi}}e^{im\phi}, \qquad m = 0, \pm 1, \ldots, \pm N, \tag{7}$$

giving a total of $(2N+1)^2$ basis functions for a given value of N. The DVR grid points are then uniformly spaced in each coordinate:

$$\phi^j = \frac{2\pi j}{2N+1}, \qquad j = 1, 2, \ldots, 2N+1. \tag{8}$$

Hence we obtain a direct product DVR with grid points (ϕ_1^j, ϕ_2^k), and using the known analytic form for the kinetic energy operator [Colbert and Miller (1992)] we obtain the Hamiltonian matrix elements:

$$H_{cd}^{ab} = T_{ca}\delta_{db} + T_{db}\delta_{ca} + V(\phi_1^c, \phi_2^d)\delta_{ac}\delta_{db}, \tag{9}$$

where

$$T_{ca} = B_{\text{eff}}(-1)^{c-a}\begin{cases} N(N+1)/3, & a = c, \\ \dfrac{\cos[\pi(c-a)/(2N+1)]}{2\sin^2[\pi(c-a)/(2N+1)]}, & a \neq c. \end{cases} \tag{10}$$

The lowest eigenvalues were converged to an accuracy better than $0.1\,\text{cm}^{-1}$ for both $(H_2O)_6$ and $(D_2O)_6$ at a value $N = 14$; an iterative Lanczos matrix diagonalization procedure was employed.

The lowest eight eigenvectors for $(H_2O)_6$ along with assignments are shown in Figure 13.12 for ASP-W2 and Figure 13.13 for TIP4P. In both cases the range of the torsional angles has been restricted to exclude regions where the amplitude is essentially zero. The center of each surface corresponds to the geometry with both free hydrogens in intermediate positions. Not surprisingly, the wavefunctions for the TIP4P potential are all delocalized over torsional space, and can be classified according to the number of nodes in the ϕ_1 and ϕ_2 directions. The three lowest energy wavefunctions for the ASP-W2 potential are localized in the wells corresponding to cage isomers C1, C2 and C3. However, the fourth, seventh and eighth functions are delocalized over two isomers in each case. For this potential the wavefunctions are described in terms of localized functions in the four different wells, e.g. C1(1,0) is the function localized in the C1 well with

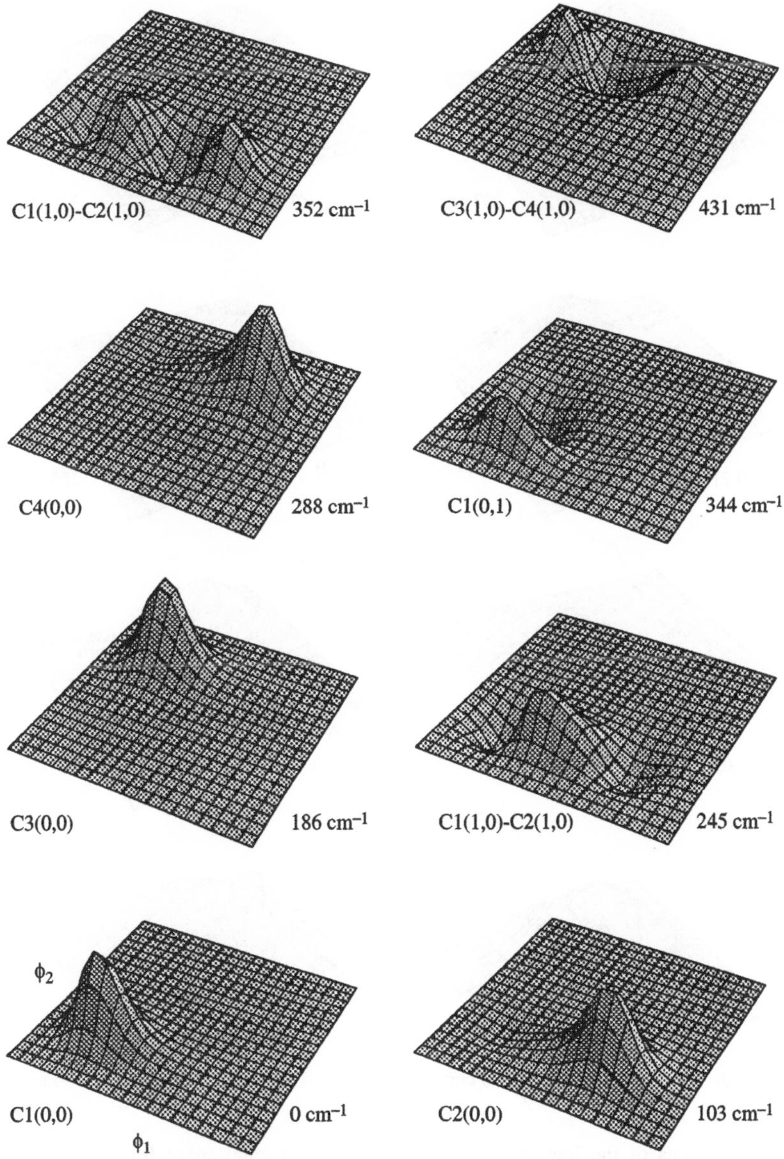

Figure 13.12. Amplitudes of the lowest eight eigenvectors in torsional space found by two-dimensional DVR calculations for $(H_2O)_6$ with the ASP-W2 potential. The interval between grid lines is 12.41° in both directions.

one node in the ϕ_1 direction and none in the ϕ_2 direction. The results for $(D_2O)_2$ are omitted for brevity, and can be obtained from the author on request.

The above MS group analysis leads to the same result as that obtained by Liu *et al.* [Liu *et al.* (1996a), and Liu, Brown and Saykally (1997)], who considered the consequences of hypothetical direct permutations (ab) and (cd). These authors have argued against the involvement of more than one structure in the tunneling splittings observed. Their reasoning provides strong evidence

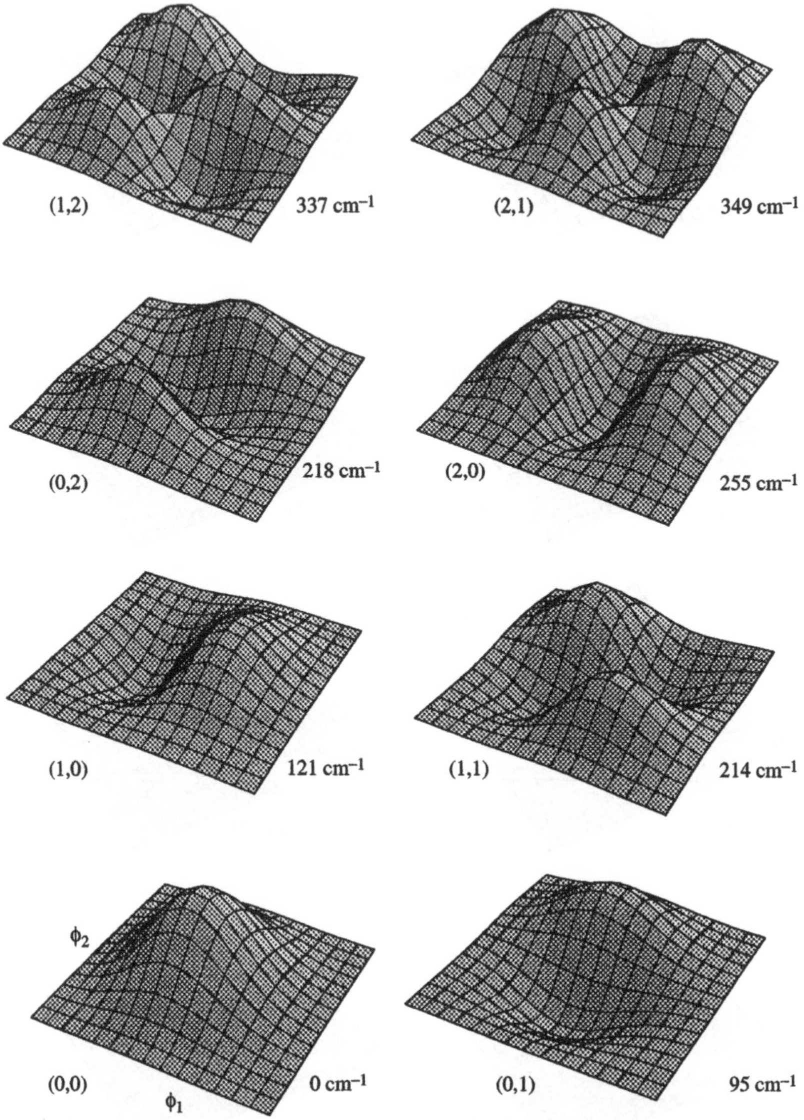

Figure 13.13. Amplitudes of the lowest eight eigenvectors in torsional space found by two-dimensional DVR calculations for $(H_2O)_6$ with the TIP4P potential. The interval between grid lines is 12.41° in both directions.

against any contribution from tunneling between inequivalent structures, but not against the 'indirect tunneling' scenario described above. No single step process corresponding to the 'C_2' mechanism proposed by Liu, Brown and Saykally (1997) has been found, despite extensive searches. Since 'indirect tunneling' has been observed in water trimer isotompomers [Liu *et al.* (1996c), Sabo *et al.* (1996), and Sorenson, Gregory and Clary (1996)], and perhaps in water tetramer, it does not seem unreasonable to invoke it here too. For the latter framework to be consistent we would expect to find similar splitting patterns for all four cage isomers, although the magnitude of the splitting may be different for each one.

The delocalization observed in our two-dimensional DVR calculations provides some evidence that the four low-lying cage isomers are all involved. In DMC calculations Gregory and Clary (1997) found that the ground state wavefunction for the lowest energy cage isomer was essentially localized, in agreement with the present DVR calculations using the ASP potential. These results can only be consistent with the experimentally observed tunneling splittings if delocalization occurs in vibrationally excited states. If the experimentally observed vibrational transition does indeed correspond to the torsional motion of a single-donor, single-acceptor monomer then the splitting in the excited state might be significantly different from the ground state.

13.10 Conclusions

At present, trying to write an overview of the literature concerning small water clusters is an attempt to hit a moving target. The rapid progress which has been made recently is particularly evident in the combined experimental and theoretical interpretation of the spectra of water trimer. A simultaneous fit of 554 rovibrational transitions in five different bands for $(D_2O)_3$ using a model Hamiltonian will soon appear [Viant *et al.* (1999)], and represents the culmination of work which began with the first FIR-VRT results [Pugliano and Saykally (1992)] and mechanistic assignments [Wales (1993b)]. Discovery of the flip and bifurcation rearrangements in water pentamer [Wales (1996b), and Wales and Walsh (1996)] provided theoretical predictions of tunneling splitting patterns which have now been substantially verified (Brown, Keutsch and Saykally (1998)].

Unfortunately, quantitative nuclear dynamics calculations for rearrangements beyond simple torsional processes such as the flip are currently lacking. The number of dimensions is too great for a DVR treatment, and the definition of nodal surfaces creates problems for DMC calculations of excited states. Knowledge of the corresponding transition state or MS group does not define these surfaces [Klein and Pickett (1976)]. which can be rather complicated even in the restricted torsional space [Sabo *et al.* (1998)]. The most serious difficulties in interpretation clearly arise for the tetramer and hexamer, where tunneling splittings are observed but corresponding degenerate rearrangements cannot be found theoretically. 'Indirect tunneling' processes, i.e. degenerate rearrangements mediated by more than one transition state, may be involved. However, novel mechanisms, such as concerted proton transfer [Loerting, Liedl and Rode (1998)] may yet be discovered.

Acknowledgements

The author gratefully acknowledges discussions with Prof. R. J. Saykally and his group, particularly concerning their unpublished results.

13.11 References

Althorpe, S. C. and Clary, D. C., 1995, *J. Chem. Phys.* **102**, 4390–4399.

Amos, R. D., Alberts, I. L., Andrews, J. S., Colwell, S. M., Handy, N. C., Jayatilaka, D., Knowles, P. J., Kobayashi, R., Laidig, K. E., Laming, G., Lee, A. M., Maslen, P. E., Murray, C. W., Rice, J. E. Simandiras, E. D., Stone, A. J., Su, M. D., and Tozer, D. J., 1995, *The Cambridge Analytic Derivatives Package Issue 6*, Cambridge.

Baker, J., 1986, *J. Comput. Chem.* **7**, 385–395.

Banerjee, A., and Adams, N. P., 1992, *Int. J. Quant. Chem.* **43**, 855–871.

Banerjee, A., Adams, N., Simons, J., and Shepard, R., 1985, *J. Phys. Chem.* **89**, 52–57.

Bačić, Z., and Light, J. C., 1989, *Annu. Rev. Phys. Chem.* **40**, 469–498.

Becke, A. D., 1988, *Phys. Rev. A* **38**, 3098–3100.

Bell, R. P., 1980, *The Tunnel Effect in Chemistry*; Chapman and Hall: New York.

Bene, J. D., and Pople, J. A., 1970, *J. Chem. Phys.* **52**, 4858–4866.

Bene, J. D., and Pople, J. A., 1973, *J. Chem. Phys.* **58**, 3605–3608.

Bone, R. G. A., Rowlands, T. W., Handy, N. C., and Stone, A. J., 1991, *Mol. Phys.* **72**, 33–73.

Boys, S. F., and Bernardi, F., 1970, *Mol. Phys.* **19**, 553–566.
Brown, M. G., Keutsch, F. N., and Saykally, R. J., 1998, *J. Chem. Phys.* **109**, 9645–9647.
Buch, V., Sandler, P., and Sadlej, J., 1998, *J. Phys. Chem. B.* **102**, 8641–8653.
Bunker, P. R., 1996, *J. Mol. Spectrosc.* **176**, 297–304.
Bunker, P. R., and Jensen, P., 1998, *Molecular Symmetry and Spectroscopy, 2nd edition*; NRC Research Press: Ottawa.
Bürgi, T., Graf, S., Leutwyler, S., and Kopper, W., 1995, *J. Chem. Phys.* **103**, 1077–1084.
Cerjan, C. J., and Miller, W. H., 1981, *J. Chem. Phys.* **75**, 2800–2806.
Cohen, R. C., and Saykally, R. J., 1990, *J. Phys. Chem.* **94**, 7991–8000.
Cohen, R. C., and Saykally, R. J., 1991, *Annu. Rev. Phys. Chem.* **42**, 369–392.
Colbert, D. T., and Miller, W. H., 1992, *J. Chem. Phys.* **96**, 1982–1991.
Coudert, L. H., and Hougen, J. T., 1988, *J. Mol. Spectrosc.* **130**, 86–119.
Coudert, L. H., and Hougen, J. T., 1990, *J. Mol. Spectrosc.* **139**, 259–277.
Cruzan, J. D., Braly, L. B., Liu, K., Brown, M. G., Loeser, J. G., and Saykally, R. J., 1996a, *Science* **271**, 59–62.
Cruzan, J. D., Brown, M. G., Liu, K., Braly, L. B., and Saykally, R. J., 1996b, *J. Chem. Phys.* **105**, 6634–6644.
Cruzan, J. D., Viant, M. R., Brown, M. G., Lucas, D. D., Liu, K., and Saykally, R. J., 1998, *Chem. Phys. Lett.* **292**, 667–676.
Cruzan, J. D., Viant, M. R., Brown, M. G., and Saykally, R. J., 1997, *J. Phys. Chem. A.* **101**, 9022–9031.
Dalton, B. J., and Nicholson, P. D., 1975, *Int. J. Quant. Chem.* **9**, 325–377.
Dennison, D. M., and Hardy, J. D., 1932, *Phys. Rev.* **39**, 938–947.
Dunning, T. H., 1970, *J. Chem. Phys.* **53**, 2823–2833.
Dunning, T. H., 1989, *J. Chem. Phys.* **90**, 1007–1023.
Dyke, T. R., 1977, *J. Chem. Phys.* **66**, 492–497.
Dyke, T. R., Mack, K. M., and Muenter, J. S., 1977, *J. Chem. Phys.* **66**, 498–510.
Engkvist, O., Forsberg, N., Schütz, M., and Karlstrom, G., 1997, *Mol. Phys.* **90**, 277–287.
Famulari, A., Raimondi, M., Sironi, M., and Gianinetti, E., 1998a, *Chem. Phys.* **232**, 289–298.
Famulari, A., Raimondi, M., Sironi, M., and Gianinetti, E., 1998b, *Chem. Phys.* **232**, 275–287.
Fowler, J. E., and Schaefer, H. F., 1995, *J. Am. Chem. Soc.* **117**, 446–452.
Fraser, G. T., 1991, *Int. Rev. Phys. Chem.* **10**, 189–206.
Fredj, E., Gerber, R. B., and Ratner, M. A., 1996, *J. Chem. Phys.* **105**, 1121–1130.
González, L., Mó, O., Yáñez, M., and Elguero, J., 1996, *J. Mol. Struct. (Theochem)* **371**, 1–10.
Graf, S., and Leutwyler, S., 1998, *J. Chem. Phys.* **109**, 5393–5403.
Gregory, J. K., 1998, *Chem. Phys. Lett.* **282**, 147–151.
Gregory, J. K., and Clary, D. C., 1994, *Chem. Phys. Lett.* **228**, 547–554.
Gregory, J. K., and Clary, D. C., 1995a, *J. Chem. Phys.* **102**, 7817–1829.
Gregory, J. K., and Clary, D. C., 1995b, *J. Chem. Phys.* **103**, 8924–8930.
Gregory, J. K., and Clary, D. C., 1996a, *J. Phys. Chem.* **100**, 18 014–18 022.
Gregory, J. K., and Clary, D. C., 1996b, *J. Chem. Phys.* **105**, 6626–6633.
Gregory, J. K., and Clary, D. C., 1997, *J. Phys. Chem. A.* **101**, 6813–6819.
Gregory, J. K., and Clary, D. C., 1998, *in Advances in Molecular Vibrations and Collision Dynamics*, Bowman, J. M., and Bačić, Z., Eds. Vol. 3; JAI Press: Stamford, pp. 311–359.
Gregory, J. K., Clary, D. C., Liu, K., Brown, M. G., and Saykally, R. J., 1997, *Science* **275**, 814–817.
Guiang, C. S., and Wyatt, R. E., 1998, *Int. J. Quant. Chem.* **68**, 233–252.
Hodges, M. P., Stone, A. J., and Xantheas, S. S., 1997, *J. Phys. Chem. A.* **101**, 9163–9168.
Hougen, J. T., 1962, *J. Chem. Phys.* **37**, 1433–1441.
Hougen, J. T., 1963, *J. Chem. Phys.* **39**, 358–365.
Hougen, J. T., 1985, *J. Mol. Spectr.* **114**, 395–426.
Hougen, J. T., 1986, *J. Phys. Chem.* **90**, 562–568.
Hutson, J. M., 1990, *Annu. Rev. Phys. Chem.* **41**, 123–154.
Huzinaga, S. J., 1965, *J. Chem. Phys.* **42**, 1293–1302.
Jorgensen, W. L., 1981, *J. Am. Chem. Soc.* **103**, 335–340.
Jorgensen, W. L., Chandraesekhar, J., Madura, J. W., Impey, R. W., and Klein, M. L., 1983, *J. Chem. Phys.* **79**, 926–935.
Jung, J. O., and Gerber, R. B., 1996, *J. Chem. Phys.* **105**, 10 332–10 348.
Karyakin, E. N., Fraser, G. T., Lovas, F. J., Suenram, R. D., and Fujitake, M., 1995, *J. Chem. Phys.* **102**, 1114–1121.
Kendall, R. A., Dunning, T. H., and Harrison, R. J., 1992, *J. Chem. Phys.* **96**, 6769–6806.
Kim, J., and Kim, K. S., 1998, *J. Chem. Phys.* **109**, 5886–5895.
Kim, J., Lee, J. Y., Lee, S., Mhin, B. J., and Kim, K. S., 1995, *J. Chem. Phys.* **102**, 310–317.
Kim, K. S., Dupuis, M., Lie, G. C., and Clementi, E., 1986, *Chem. Phys. Lett.* **131**, 451–456.
Kistenmacher, H., Lie, G. C., Popkie, H., and Clementi, E., 1974, *J. Chem. Phys.* **61**, 546–561.
Klein, D. J., and Pickett, H. M., 1976, *J. Chem. Phys.* **64**, 4811–4812.
Klopper, W., and Schütz, M., 1995, *Chem. Phys. Lett.* **237**, 536–544.

Kolbuszewski, M., and Bunker, P. R., 1996, *J. Chem. Phys.* **105**, 3649–3653.
Krishnan, P. N., Jensen, J. O., and Burke, L. A., 1994, *Chem. Phys. Lett.* **217**, 311–318.
Le Sueur, C. R., and Stone, A. J., 1993, *Mol. Phys.* **78**, 1267–1291.
Lee, C., Chen, H., and Fitzgerald, G., 1994, *J. Chem. Phys.* **101**, 4472–4473.
Lee, C., Yang, W., and Parr, R. G., 1988, *Phys. Rev. B* **37**, 785–789.
Leforestier, C., Braly, L. B., Liu, K., Elrod, M. J., and Saykally, R. J., 1997, *J. Chem. Phys.* **106**, 8527–8544.
Lentz, B. R., and Scheraga, H. A., 1973, *J. Chem. Phys.* **58**, 5296–5308.
Leone, R. E., and v. R. Schleyer, P., 1970, *Angew. Chem. Int. Ed. Engl.* **9**, 860–890.
Liu, K., Brown, M. G., Carter, C., Saykally, R. J., Gregory, J. K., and Clary, D. C., 1996a, *Nature* **381**, 501–502.
Liu, K., Brown, M. G., Cruzan, J. D., and Saykally, R. J., 1996b, *Science* **271**, 62–64.
Liu, K., Brown, M. G., Cruzan, J. D., and Saykally, R. J., 1997, *J. Phys. Chem. A.* **101**, 9011–9021.
Liu, K., Brown, M. G., and Saykally, R. J., 1997, *J. Phys. Chem. A.* **101**, 8995–9010.
Liu, K., Brown, M. G., Viant, M. R., Cruzan, J. D., and Saykally, R. J. 1996c, *Mol. Phys.* **89**, 1373–1396.
Liu, K., Cruzan, J. D., and Saykally, R. J., 1996, *Science* **271**, 929–932.
Liu, K., Elrod, M. J., Loeser, J. G., Cruzan, J. D., Pugliano, N., Brown, M. G., Rzepiela, J., and Saykally, R. J., 1994a, *Faraday Discuss.* **97**, 35–41.
Liu, K, Loeser, J. G., Elrod, M. J., Host, B. C., Rzepiela, J. A., Pugliano, N., and Saykally, R. J., 1994b, *J. Am. Chem. Soc.* **116**, 3507–3512.
Loerting, T., Liedl, K. R., and Rode, B. M., 1998, *J. Chem. Phys.* **109**, 2672–2679.
Longuet-Higgins, H. C., 1963, *Mol. Phys.* **6**, 445–460.
Maitland, G. C., Rigby, M., Smith, E. B., and Wakeham, W. A., 1981, *Intermolecular Forces*; Clarendon Press: Oxford.
Mas, E. M., and Szalewicz, K., 1996, *J. Chem. Phys.* **104**, 7606–7614.
Merrill, G. N., and Gordon, M. S., 1998, *J. Phys. Chem. A.* **102**, 2650–2657.
Mhin, B. J., Kim, H. S., Kim, H. S., Yoon, C. W., and Kim, K. S. J. N., 1991, *Chem. Phys. Lett.* **176**, 41–45.
Mhin, B. J., Kim, J., Lee, S., Lee, J. Y., and Kim, K. S., 1994, *J. Chem. Phys.* **100**, 4484–4486.
Millot, C., Soetens, J. C., Costa, M. T. C. M., Hodges, M. P., and Stone, A. J., 1998, *J. Phys. Chem. A.* **102**, 754–770.
Millot, C., and Stone, A. J., 1992, *Mol. Phys.* **77**, 439–462.
Mó, O., and Yáñez, M., 1992, *J. Chem. Phys.* **97**, 6628–6638.
Møller, C., and Plesset, M. S., 1934, *Phys. Rev.* **46**, 618–622.
Murrell, J. N., and Laidler, K. J., 1968, *Trans. Faraday Soc.* **64**, 371–377.
Nelson, D. D., Fraser, G. T., and Klemperer, W., 1987, *Science* **238**, 1670–1674.
Nourse, J. G., 1980, *J. Am. Chem. Soc.* **102**, 4883–4889.
Odutola, J. A., and Dyke, T. R., 1980, *J. Chem. Phys.* **72**, 5062–5070.
Ohmine, I., and Tanaka, H., 1993, *Chem. Rev.* **93**, 2545–2566.
Olthof, E. H. T., der Avoird, A. V., Wormer, P. E. S., Liu, K., and Saykally, R. J., 1996, *J. Chem. Phys.* **105**, 8051–8063.
O'Neal, D., Taylor, H., and Simons, J., 1984, *J. Phys. Chem.* **88**, 1510–1513.
Owicki, J. C., Shipman, L. L., and Scheraga, H. A., 1975, *J. Phys. Chem.* **79**, 1794–1811.
Pancíř, J., 1975, *Collect. Czech. Chem. Commun.* **40**, 1112–1118.
Pastor, N., and Ortega-Blake, I., 1993, *J. Chem. Phys.* **99**, 7899–7906.
Paul, J. B., Collier, C. P., Saykally, R. J., Scherer, J. J., and O'Keefe, A., 1997, *J. Phys. Chem. A.* **101**, 5211–5214.
Pedulla, J. M., Kim, K., and Jordan, K. D., 1998, *Chem. Phys. Lett.* **291**, 78–84.
Pedulla, J. M., Vila, F., and Jordan, K. D., 1996, *J. Chem. Phys.* **105**, 11091–11099.
Popelier, P. L. A., Stone, A. J., and Wales, D. J., 1994, *Faraday Discuss.* **97**, 243–264.
Pugliano, N., Cruzan, J. D., Loeser, J. G., and Saykally, R. J., 1993, *J. Chem. Phys.* **98**, 6600–6617.
Pugliano, N., and Saykally, R. J., 1992, *Science* **257**, 1937–1940.
Sabo, D., Bačić, Z., Bürgi, T., and Leutwyler, S., 1995, *Chem. Phys. Lett.* **244**, 283–294.
Sabo, D., Bačić, Z., Bürgi, T., and Leutwyler, S., 1996, *Chem. Phys. Lett.* **261**, 318–328.
Sabo, D., Bačić, Z., Graf, S., and Leutwyler, S., 1998, *J. Chem. Phys.* **109**, 5404–5419.
Saykally, R. J., and Blake, G. A., 1993, *Science* **259**, 1570–1575.
Scheiner, S., 1994, *Annu. Rev. Phys. Chem.* **45**, 23–56.
Schütz, M., Bürgi, T., Leutwyler, S., and Bürgi, H. B., 1993, *J. Chem. Phys.* **99**, 5228–5238.
Schütz, M., Bürgi, T., Leutwyler, S., and Bürgi, H. B., 1994, *J. Chem. Phys.* **100**, 1780.
Schütz, M., Klopper, W., Lüthi, H. P., and Leutwyler, S., 1995, *J. Chem. Phys.* **103**, 6114–6126.
Schütz, M., Rauhut, G., and Werner, H. J., 1998, *J. Phys. Chem. A.* **102**, 5997–6003.
Simons, J., Jorgensen, P., Taylor, H., and Ozment, J., 1983, *J. Phys. Chem.* **87**, 2745–2753.
Smith, B. J., Swanton, D. J., Pople, J. A., Schaefer, H. F., and Radom, L., 1990, *J. Chem. Phys.* **92**, 1240–1247.
Sorenson, J. M., Gregory, J. K., and Clary, D. C., 1996, *Chem. Phys. Lett.* **263**, 680–686.
Stillinger, F. H., and Weber, T. A. 1983, *Phys. Rev. A* **28**, 2408–2416.
Stone, A. J., 1981, *Chem. Phys. Lett.* **83**, 233–239.
Stone, A. J., 1985, *Mol. Phys.* **56**, 1065–1082.

Stone, A. J., 1996, *The Theory of Intermolecular Forces*; Clarendon Press: Oxford.

Stone, A. J., and Alderton, M., 1985, *Mol. Phys.* **56**, 1047–1064.

Suzuki, S., and Blake, G. A., 1994, *Chem. Phys. Lett.* **229**, 499–505.

Tsai, C. J., and Jordan, K. D., 1993, *Chem. Phys. Lett.* **213**, 181–188.

Tsuboi, M., Hirakawa, A. Y., Ino, T., Sasaki, T., and Tamagake, K., 1964, *J. Chem. Phys.* **41**, 2721–2734.

Van der Avoird, A. V., Olthof, E. H. T., and Wormer, P. E. S., 1996, *J. Chem. Phys.* **105**, 8034–8050.

Van Duijneveldt-Van de Rijdt, J. G. C. M., and Van Duijneveldt, F. B., 1995, *Chem. Phys. Lett.* **237**, 560–567.

Vernon, M. F., Krajnovich, D. J., Kwok, H. S., Lisy, J. M., Shen, Y. R., and Lee, Y. T., 1982, *J. Chem. Phys.* **77**, 47–57.

Viant, M. R., Brown, M. G., Cruzan, J. D., Saykally, R. J., Geleijns, M., and van der Avoird, A. 1999, *J. Chem. Phys.* **110**, 4369–4381.

Wales, D. J., 1993a, *J. Am. Chem. Soc.* **115**, 11 191–11 201.

Wales, D. J., 1993b, *J. Am. Chem. Soc.* **115**, 11 180–11 190.

Wales, D. J., 1994, *J. Chem. Phys.* **101**, 3750–3762.

Wales, D. J., 1996a, in *Theory of Atomic and Molecular Clusters–II*, Jellinek, J., Ed.; Springer: Heidelberg, in press [available from the Los Alamos preprint server at URL http://xxx.lanl.gov/abs/physics/9810031].

Wales, D. J., 1996b, *Science* **271**, 925–929.

Wales, D. J., 1997, in *Recent Theoretical and Experimental Advances in Hydrogen-Bonded Clusters*, Xantheas, S., Ed.; Kluwer: Dordrecht, in press [available from the Los Alamos preprint server at URL http://xxx.lanl.gov/abs/physics/9810032].

Wales, D. J., 1998, in *Advances in Molecular Vibrations and Collision Dynamics*, Bowman, J. M., and Bačić, Z., Eds., Vol. 3; JAI Press: Stamford, pp. 365–396.

Wales, D. J., Popelier, P. L. A., and Stone, A. J., 1995, *J. Chem. Phys.* **102**, 5551–5565.

Wales, D. J., Stone, A. J., and Popelier, P. L. A., 1995, *Chem. Phys. Lett.* **240**, 89–96.

Wales, D. J., and Walsh, T. R., 1996, *J. Chem. Phys.* **105**, 6957–6971.

Wales, D. J., and Walsh, T. R., 1997, *J. Chem. Phys.* **106**, 7193–7207.

Walsh, T. R., and Wales, D. J., 1996, *J. Chem. Soc., Faraday Trans.* **92**, 2505–2517.

Walsh, T. R., and Wales, D. J., 1997, *Z. Phys. D* **40**, 229–235.

Wolfram, S., 1996, *The Mathematica Book*, 3rd edition; Wolfram Media/Cambridge University Press: Cambridge.

Xantheas, S. S., 1994, *J. Chem. Phys.* **100**, 7523–7524.

Xantheas, S. S., 1995, *J. Chem. Phys.* **102**, 4505–4517.

Xantheas, S. S., 1996, *J. Chem. Phys.* **104**, 8821–8824.

Xantheas, S. S., and Dunning, T. H., 1993a, *J. Chem. Phys.* **99**, 8774–8792.

Xantheas, S. S., and Dunning, T. H., 1993b, *J. Chem. Phys.* **98**, 8037–8040.

Zwart, E., ter Muelen, J. J., Meerts, W. L., and Coudert, L. H., 1991, *J. Mol. Spectrosc.* **147**, 27–39.

PART 4

ROVIBRONIC STATES AND THE BREAKDOWN OF THE BORN–OPPENHEIMER APPROXIMATION

14 VIBRONIC ENERGIES AND THE BREAKDOWN OF THE BORN–OPPENHEIMER APPROXIMATION IN DIATOMIC MOLECULES: ADIABATIC AND DIABATIC REPRESENTATIONS

David R. Yarkony

Johns Hopkins University, Baltimore MD, USA

Computational Molecular Spectroscopy. Edited by Per Jensen and Philip R. Bunker

© 2000 John Wiley & Sons Ltd

14.1 INTRODUCTION

This review considers the theoretical treatment of electronically nonadiabatic processes in diatomic molecules with (relatively) low atomic number. In this context electronically nonadiabatic processes can be partitioned into two classes: (i) spin-nonconserving processes induced by the spin–orbit interaction and (ii) spin-conserving processes induced by the kinetic energy operator. The latter class of processes, which subsumes the former when the spin–orbit interaction becomes large and cannot be treated perturbatively, is the focus of this review.

Section 14.2 reviews the Born–Huang approximation [Born and Huang (1954)] introducing the concepts of adiabatic, and both rigorous [Smith (1969)] and approximate diabatic [Baer (1975), Cederbaum, Köppel and Domke (1981), and Mead and Truhlar (1982)] bases and their interconversion. The derivative couplings are defined and their role in nonadiabatic processes described. Nonadiabatic processes can be fundamentally different in polyatomic and diatomic molecules. In polyatomic (diatomic) molecules potential energy surfaces (curves) may (may not) exhibit conical intersections, the corresponding adiabatic state wave functions may (may not) exhibit the geometric phase effect [Longuet-Higgins (1961)] and the wave functions cannot (can) be transformed to rigorous diabatic states. [Mead and Truhlar (1982)]. The origin and consequences of these differences are reviewed.

Nonadiabatic interactions may lead to internal conversion and predissociation. Section 14.3 reviews the solution of the nuclear Schrödinger equation within the Born–Huang approximation for diatomic molecules, with emphasis on these classes of phenomena. Based on the ideas presented in Sections 14.2 and 14.3, Section 14.4 presents practical, computational, descriptions of processes in which nonadiabatic effects are essential: the direct predissociation of a nominally bound state produced by nonadiabatic coupling to dissociative states; and the indirect predissociation of an otherwise largely stable state induced by nonadiabatic interactions with other nominally bound states that are in turn efficiently predissociated. In each case rigorous diabatic states are constructed and used to discuss aspects of approximate diabatic states.

14.2 Born–Huang expansion

Within the Born–Huang approximation the total wave function is expressed as:

$$\Psi_k^{\mathrm{T}}(\mathbf{r},\mathbf{R}) = \sum_{I=1}^{N^{\mathrm{a}}} \chi_I^k(\mathbf{R})\Psi_I^{\mathrm{e}}(\mathbf{r};\mathbf{R}) \equiv \boldsymbol{\chi}^k(\mathbf{R})^{\dagger}\boldsymbol{\Psi}^{\mathrm{e}}(\mathbf{r};\mathbf{R}), \tag{1}$$

where vectors (matrices) are written in bold (bold italic) face, $\mathbf{r} \equiv (\mathbf{r}_1, \ldots, \mathbf{r}_{N^{\mathrm{e}}})$ and $\mathbf{R} \equiv (\mathbf{R}_1, \ldots, \mathbf{R}_{N^{\mathrm{nuc}}})$ denote the coordinates of the N^{e} electrons and N^{nuc} nuclei respectively, in a space fixed frame, $\chi^k(\mathbf{R})$ are the nuclear wave functions and $\Psi^{\mathrm{e}}(\mathbf{r};\mathbf{R})$ are the electronic wave functions, both taken to be single-valued functions of $\mathbf{R}$. The $\Psi_I^{\mathrm{e}}(\mathbf{r};\mathbf{R})$, $I = 1, \ldots, N^{\mathrm{a}}$ form a subspace of the total electronic space of dimension N^{CSF}, with $N^{\mathrm{CSF}} \gg N^{\mathrm{a}}$. When no confusion will result the $\mathbf{r}$ and/or $\mathbf{R}$ dependence of a function may be suppressed for clarity.

Inserting expansion (1) into the Schrödinger equation (in atomic units) [Baylis and Drake (1996)]

$$\left[\sum_{\alpha=1}^{N^{\mathrm{nuc}}} \frac{-1}{2M_\alpha}\nabla_\alpha^2 + H^{\mathrm{e}}(\mathbf{r};\mathbf{R}) - E_k\right]\Psi_k^{\mathrm{T}}(\mathbf{r},\mathbf{R}) = 0, \tag{2a}$$

where

$$H^{\mathrm{e}}(\mathbf{r};\mathbf{R}) = H^0(\mathbf{r};\mathbf{R}) + H^{\mathrm{SO}}(\mathbf{r};\mathbf{R}), \tag{2b}$$

$H^0(\mathbf{r};\mathbf{R})$ is the nonrelativistic, coulombic, electronic Hamiltonian and $H^{SO}(\mathbf{r}:\mathbf{R})$ is the spin–orbit Hamiltonian in the Breit–Pauli approximation [Yarkony (1992)] and projecting onto Ψ_I^e gives

$$\left[\sum_{\alpha=1}^{N^{nuc}} \frac{-1}{2M_\alpha}\nabla_\alpha^2 + \left(\nabla_\alpha \cdot \mathbf{f}_\alpha^{e,II}(\mathbf{R}) + \mathbf{f}_\alpha^{e,II}(\mathbf{R}) \cdot \nabla_\alpha\right) + \overline{E}_I^e(\mathbf{R}) - E_k\right] \chi_I^k(\mathbf{R})$$
$$= \sum_{J \neq I}^{N^a} \left(\overline{H}_{IJ}^e(\mathbf{R}) + \sum_{\alpha=1}^{N^{nuc}} \left\{\frac{1}{2M_\alpha}\left[\nabla_\alpha \cdot \mathbf{f}_\alpha^{e,IJ}(\mathbf{R}) + \mathbf{f}_\alpha^{e,IJ}(\mathbf{R}) \cdot \nabla_\alpha\right]\right\}\right) \chi_J^k(\mathbf{R}). \tag{3}$$

In equation (3) ∇_α acts on all quantities to its right,

$$\mathbf{f}_\alpha^{e,JI}(\mathbf{R}) = \left\langle \Psi_J^e(\mathbf{r};\mathbf{R}) \middle| \nabla_\alpha \Psi_I^e(\mathbf{r};\mathbf{R})\right\rangle_r \equiv \left(f_{\alpha x}^{e,JI}, f_{\alpha y}^{e,JI}, f_{\alpha z}^{e,JI}\right) \tag{4}$$

is the first derivative, or derivative, coupling

$$k_\alpha^{e,JI}(\mathbf{R}) = \left\langle \nabla_\alpha \Psi_J^e(\mathbf{r};\mathbf{R}) \cdot \middle| \nabla_\alpha \Psi_I^e(\mathbf{r};\mathbf{R})\right\rangle_r \tag{5a}$$

$$= \sum_{K=1}^{N^{CSF}} \mathbf{f}_\alpha^{e,KJ}(\mathbf{R})^* \cdot \mathbf{f}_\alpha^{e,KI}(\mathbf{R}) \tag{5b}$$

is a second derivative coupling,

$$\overline{H}_{IJ}^e(\mathbf{R}) = H_{IJ}^e(\mathbf{R}) + \sum_{\alpha=1}^{N^{nuc}} \frac{k_\alpha^{e,IJ}(\mathbf{R})}{2M_\alpha}, \tag{6a}$$

where the potential energy surface $E_I^e(\mathbf{R}) \equiv H_{II}^e(\mathbf{R})$ and the potential coupling, $H_{IJ}^e(\mathbf{R})$, are given by:

$$H_{IJ}^e(\mathbf{R}) = \left\langle \Psi_I^e(\mathbf{r};\mathbf{R}) \middle| H^e(\mathbf{r};\mathbf{R}) \middle| \Psi_J^e(\mathbf{r};\mathbf{R})\right\rangle_r \equiv H_{IJ}^{0,e}(\mathbf{R}) + H_{IJ}^{SO,e}(\mathbf{R}). \tag{6b}$$

The left hand side of equation (3) describes a nuclear wave packet $\chi_I^k(\mathbf{R})$ moving on an effective potential energy surface $\overline{E}_I^e(\mathbf{R}) \equiv \overline{H}_{II}^e(\mathbf{R})$, while the right hand side of that equation describes the nonadiabatic transitions.

At this point the choice of Ψ^e is arbitrary except that they be single-valued as functions of **R**. The Ψ^e must be chosen to limit the size of the expansion (1), that is limit N^a, and should, if possible, facilitate the numerical treatment of equation (3). The choice of Ψ^e is a key issue in this work, and it is to this point that we now turn.

14.2.1 ADIABATIC STATES I

To limit the size of N^a it is, at least initially, advantageous to use as the basis of electronic states the real-valued *adiabatic* electronic wave functions $\Psi_I^a(\mathbf{r};\mathbf{R})$, solutions of the electronic Schrödinger equation

$$\left[H^0(\mathbf{r};\mathbf{R}) - E_I^{0,a}(\mathbf{R})\right]\Psi_I^a(\mathbf{r};\mathbf{R}) = 0. \tag{7a}$$

The $\Psi_I^a(\mathbf{r};\mathbf{R})$ in turn will be expressed in the basis of symmetry-adapted configuration state functions (CSFs) [Yarkony (1996b)] as:

$$\Psi_I^a(\mathbf{r};\mathbf{R}) = \sum_{\alpha=1}^{N^{CSF}} c_\alpha^{a,I}(\mathbf{R})\psi_\alpha(\mathbf{r};\mathbf{R}) \tag{7b}$$

In using equation (7a) we tacitly make the, quite reasonable, assumption that the spin–orbit interaction is sufficiently small that its effects can be treated using quasi-degenerate perturbation theory [Yarkony (1988)].

14.2.2 DIABATIC STATES I

To simplify equation (3) it desirable (see below) to introduce a new electronic basis, the diabatic basis $\Psi_I^{\rm d}(\mathbf{r};\mathbf{R})$, $I = 1, \ldots, N^{\rm a}$, for which, to the extent possible, $f_\tau^{{\rm d},IJ} = 0$, for τ any of the $N^{\rm int}$ internal coordinates. The $\Psi^{\rm d}(\mathbf{r};\mathbf{R})$ are obtained as an orthogonal transformation of the adiabatic states

$$\Psi_I^{\rm d}(\mathbf{r};\mathbf{R}) = \sum_{J=1}^{N^{\rm a}} \Psi_J^{\rm a}(\mathbf{r};\mathbf{R}) U_{JI}(\mathbf{R}), \tag{8a}$$

which can be re-expressed in the CSF basis as

$$\Psi_I^{\rm d}(\mathbf{r};\mathbf{R}) = \sum_{\alpha=1}^{N^{\rm CSF}} c_\alpha^{{\rm d},I}(\mathbf{R}) \psi_\alpha(\mathbf{r};\mathbf{R}). \tag{8b}$$

The requirement [Smith (1969)] $f_\tau^{{\rm d},IJ} = 0$ leads to the following set of coupled linear differential equations for $\boldsymbol{U}$:

$$\frac{\partial}{\partial\tau} \boldsymbol{U}_{IJ}(\mathbf{R}) + \sum_{K=1}^{N^{\rm a}} f_\tau^{{\rm a},IK}(\mathbf{R}) \boldsymbol{U}_{KJ}(\mathbf{R}) = \mathbf{0} \tag{9a}$$

for all internal coordinates τ. For diatoms, for which there is only one internal coordinate, Q, equation (9a) is a total differential equation and can be straightforwardly integrated from $Q = \infty$ where $\boldsymbol{U}(Q \to \infty) = \boldsymbol{I}$, a unit matrix, to finite Q using standard numerical integration techniques. Since $f_\tau^{{\rm a},IK} = -f_\tau^{{\rm a},KI}$, $\boldsymbol{U}$ is orthogonal. For polyatomic molecules, as discussed below, the situation is more complicated, since $N^{\rm int} > 1$.

The case $N^{\rm a} = 2$ will be of particular interest. In this case $\boldsymbol{U} = \boldsymbol{U}(\Theta^{f^{{\rm a},IJ}})$ where

$$\boldsymbol{U}(\Theta) = \begin{pmatrix} \cos\Theta & \sin\Theta \\ -\sin\Theta & \cos\Theta \end{pmatrix} \tag{9b}$$

and equation (9a) reduces to

$$\frac{\partial}{\partial\tau} \Theta^{f^{{\rm a},IJ}}(\mathbf{R}) = -f_\tau^{{\rm a},IJ}(\mathbf{R}) \tag{9c}$$

14.2.3 IDIOSYNCRATIC PROPERTIES OF THE DERIVATIVE COUPLING

The boundary condition at $Q \to \infty$ for equation (9) suggests that $f_\tau^{{\rm a},IK}(Q \to \infty) = 0$. This is only approximately true in general and can be understood in terms of the following analysis which illustrates two, perhaps unexpected, properties of the derivative couplings:

(i) $f_\kappa^{{\rm a},JI}$ does not vanish for κ a translation or rotation of the *nuclei*.

(ii) $f_\tau^{{\rm a},JI}(\mathbf{R})$ is origin dependent, that is, $f_\tau^{{\rm a},JI}(\mathbf{R})$ depends on the choice of the coordinates used to describe the location of the molecule as a whole.

To see property (i) let $O(\mathbf{r},\mathbf{R}) = O^{\rm e}(\mathbf{r};\mathbf{R}) + O^{\rm n}(\mathbf{R})$, where $O^{\rm e}(O^{\rm n})$ generates a translation, $O = {\rm i}P_w$, or rotation, $O = {\rm i}L_w$, $W = x, y, z$, of the electrons (nuclei). Here ${\rm i}\mathbf{P} = \sum_{a=1}^{N} \nabla_a$ and

$\mathbf{iL} = \sum_{a=1}^{N} \rho_a \times \nabla_a$, with $N = N^{\mathrm{e}}$ and $\rho = \mathbf{r}$, ($N = N^{\mathrm{nuc}}$ and $\rho = \mathbf{R}$) and we have observed that translations [rotations] of the electrons (nuclei) are generated by the total linear [angular] momentum operator for the electrons(nuclei) [Chapter 5, Tinkham (1964)]. Then

$$[H^0, O] = 0 = [O^{\mathrm{n}}, E_I^0(\mathbf{R})], \tag{10a}$$

where the second equality depends on the fact that E_I^0 depends only on internal coordinates. Then from equation (10a)

$$\begin{aligned} 0 &= \langle \Psi_J^{\mathrm{a}}(\mathbf{r};\mathbf{R}) | [H^0(\mathbf{r};\mathbf{R}), O(\mathbf{r};\mathbf{R})] | \Psi_I^{\mathrm{a}}(\mathbf{r};\mathbf{R}) \rangle \\ &= E_J^{\mathrm{a},0}(\mathbf{R}) \langle \Psi_J^{\mathrm{a}}(\mathbf{r};\mathbf{R}) | O(\mathbf{r};\mathbf{R}) \Psi_I^{\mathrm{a}}(\mathbf{r};\mathbf{R}) \rangle_{\mathbf{r}} - \langle \Psi_J^{\mathrm{a}}(\mathbf{r};\mathbf{R}) | O(\mathbf{r};\mathbf{R}) E_I^{0,\mathrm{a}}(\mathbf{R}) \Psi_I^{\mathrm{a}}(\mathbf{r};\mathbf{R}) \rangle_{\mathbf{r}} \\ &= \left\{ E_J^{0,\mathrm{a}}(\mathbf{R}) - E_I^{0,\mathrm{a}}(\mathbf{R}) \right\} \langle \Psi_J^{\mathrm{a}}(\mathbf{r};\mathbf{R}) | O(\mathbf{r};\mathbf{R}) \Psi_I^{\mathrm{a}}(\mathbf{r};\mathbf{R}) \rangle_{\mathbf{r}}, \end{aligned} \tag{10b}$$

so that

$$-\langle \Psi_J^{\mathrm{a}}(\mathbf{r};\mathbf{R}) | O^{\mathrm{e}}(\mathbf{r};\mathbf{R}) \Psi_I^{\mathrm{a}}(\mathbf{r};\mathbf{R}) \rangle_{\mathbf{r}} = \langle \Psi_J^{\mathrm{a}}(\mathbf{r};\mathbf{R}) | O^{\mathrm{n}}(\mathbf{R}) \Psi_I^{\mathrm{a}}(\mathbf{r};\mathbf{R}) \rangle_{\mathbf{r}} \equiv f_{O^{\mathrm{n}}}^{\mathrm{a},JI}(\mathbf{R}). \tag{10c}$$

Thus the derivative coupling due to a translation (overall rotation) will be nonvanishing provided the matrix elements of total electronic momentum $\langle \Psi_J^{\mathrm{a}}(\mathbf{r};\mathbf{R}) | iP_w^{\mathrm{e}}(\mathbf{r}) \Psi_I^{\mathrm{a}}(\mathbf{r};\mathbf{R}) \rangle_{\mathbf{r}}$ (total electronic orbital angular momentum, $\langle \Psi_J^{\mathrm{a}}(\mathbf{r};\mathbf{R}) | iL_w^{\mathrm{e}}(\mathbf{r};\mathbf{R}) \Psi_I^{\mathrm{a}}(\mathbf{r};\mathbf{R}) \rangle_{\mathbf{r}}$) is nonvanishing.

Property (ii) follows from the fact that $f_{O^{\mathrm{n}}}^{\mathrm{a},JI}(\mathbf{R}) \neq 0$. To illustrate, consider two nuclei constrained to a line. They can be described by the following sets of coordinates $S_0 = (X_1, X_2)$ the atom-centered, space-fixed frame, coordinates; $S_1 = (x = X_2 - X_1, X = X_1 + X_2)$ the stretch and center of mass coordinates; or $S_2 = (x = X_2 - X_1, X' = X_1)$. By the chain rule for S_1, $(f_x^{\mathrm{e},IJ})_X = \frac{1}{2}[(f_{X_2}^{\mathrm{e},IJ})_{X_1} - (f_{X_1}^{\mathrm{e},IJ})_{X_2}]$, while for S_2, $(f_x^{\mathrm{e},IJ})_{X'} = (f_{X_2}^{\mathrm{e},IJ})_{X_1}$. The difference $(f_x^{\mathrm{e},IJ})_{X'} - (f_x^{\mathrm{e},IJ})_X = \frac{1}{2}[(f_{X_2}^{\mathrm{e},IJ})_{X_1} + (f_{X_1}^{\mathrm{e},IJ})_{X_2}] = (f_X^{\mathrm{e},IJ})_x$, which is attributable to the variable held constant when x is changed, is equal to the derivative coupling for a uniform translation of the molecule, $(f_X^{\mathrm{e},IJ})_x$.

14.2.4 COMPARISON OF DIATOMIC AND POLYATOMIC MOLECULES

The principal difference between a diatomic molecule and all larger molecules is that for a diatom there is only one internal coordinate, Q, whereas for a polyatomic molecule there are three or more such coordinates. This rather trivial statement turns out to have profound consequences for both the potential energy curves/surfaces and derivative couplings.

14.2.4.1 Noncrossing rule, conical intersections and the geometric phase effect

14.2.4.1.1 Symmetry, Dimensionality and the Noncrossing Rule

According to the noncrossing rule [von Neumann and Wigner (1929), and Teller (1937)], states I and J of the same symmetry may intersect (conically) on a seam of dimension $N^{\mathrm{int}} - 2$. If the states have different symmetry the seam has dimension $N^{\mathrm{int}} - K - 1$ where K constraints are required to restrict the molecule to geometries for which the states in question are of different symmetry. Alternatively two conditions must be satisfied to have states of the same symmetry intersect while only one condition is required to cause states of different symmetry to intersect. Thus potential energy curves in diatoms—with only one internal degree of freedom—can only cross if the states are of different symmetry. They can, however, almost intersect, producing avoided intersections with large derivative couplings. In polyatomic molecules, however—with at least three internal degrees of freedom—seams of intersection for two states with the same symmetry are possible and are, in fact, not rare events [Yarkony (1996a)].

14.2.4.1.2 Effects of Conical Intersections

Conical intersections modify/complicate nuclear dynamics so that the situation in diatomic molecules can be profoundly different from that in their polyatomic counterparts. In the excited state the conical topography can funnel the molecule into the region of the conical intersection facilitating radiationless decay while on the ground state potential energy surface it produces the geometric phase (sometimes referred to as the Berry phase [Berry (1984)]) effect [Shapere and Wilczek (1989), Mead (1992), and Yarkony (1996a)], which has implications for the form of the adiabatic wave functions that can be used in equation (1).

Since $H^0(\mathbf{r};\mathbf{R})$ is real valued it would appear that $\Psi_I^{\mathrm{a}}(\mathbf{r};\mathbf{R})$ could be chosen real valued. This is the case for diatomic molecules but for polyatomic molecules complications may arise. The geometric phase effect, the signature property of a conical intersection, requires that the *real-valued* adiabatic electronic wave function change sign when transported along a closed loop—a pseudorotation [Berry (1960)] path—surrounding (only) one point of conical intersection [Longuet-Higgins (1961)]. This makes $\Psi_I^{\mathrm{a}}(\mathbf{r};\mathbf{R})$ double valued, as a function of $\mathbf{R}$, that is at any $\mathbf{R}$ the sign of the wave function is either + or − with the appropriate value being path dependent. Single-valuedness is restored through the use of a complex phase factor $e^{iA(\mathbf{R})}$. The single-valued electronic wave function $\tilde{\Psi}_I^{\mathrm{a}}$ required in equation (1) then has the form $\tilde{\Psi}_I^{\mathrm{a}}(\mathbf{r};\mathbf{R}) = e^{iA_I(\mathbf{R})}\Psi_I^{\mathrm{a}}(\mathbf{r};\mathbf{R})$ [Mead and Truhlar (1979)]. The construction of $A_I(\mathbf{R})$ is discussed below and in Refs [Kendrick and Mead (1995), Kendrick and Pack (1996), and Yarkony (1999)].

14.2.5 DIABATIC STATES II: EXISTENCE AND CONSTRUCTION OF APPROXIMATE DIABATIC BASES

At a conical intersection at least one (depending on the coordinates chosen) $f_\tau^{\mathrm{a},IJ}$ is singular. At a narrowly avoided crossing in a diatom $f_Q^{\mathrm{a},IJ}$ is large and sharply peaked. This type of behavior complicates the solution of equation (3) so that it is desirable to replace Ψ^{a} with Ψ^{d} using equations (8) and (9). Use of equations (8) and (9) is routine for diatoms [Parlant *et al.* (1990), Manaa and Yarkony (1994), and Han, Hettema and Yarkony (1995)] It is therefore natural to try to extend them to polyatomic molecules.

14.2.5.1 A necessary condition for the existence of diabatic states

When $N^{\mathrm{int}} > 1$ for (8) to have a solution, the mixed partial derivatives of $\boldsymbol{U}$ $(\mathbf{R})$ must be equal [Hobey and McLachlan (1960), Baer (1975), and Mead and Truhlar (1982)] that is

$$[(\partial^2/\partial\tau_i\,\partial\tau_j) - (\partial^2/\partial\tau_j\,\partial\tau_i)]\mathbf{U} = \mathbf{0} \tag{11a}$$

which is equivalent to

$$0 = \sum_{K>N^{\mathrm{a}}}^{N^{\mathrm{CSF}}} \left[f_{\tau_i}^{\mathrm{a},KJ}(\mathbf{R})f_{\tau_j}^{\mathrm{a},KI}(\mathbf{R}) - f_{\tau_j}^{\mathrm{a},KJ}(\mathbf{R})f_{\tau_i}^{\mathrm{a},KI}(\mathbf{R})\right] \tag{11b}$$

for all pairs (I, J) with $I, J \le N^{\mathrm{a}}$ and all pairs (i, j) with $i, j \le N^{\mathrm{int}}$. When $N^{\mathrm{CSF}} > N^{\mathrm{a}}$ equation (11b) is in general never satisfied, that is $\mathbf{f}^{\mathrm{d},IJ}$ $(\mathbf{R})$ cannot be made $\mathbf{0}$. In this case one must settle for approximate diabatic bases, for which $\mathbf{f}^{\mathrm{d},IJ} \sim \mathbf{0}$.

Although approximate diabatic bases might seem to be relevant only to polyatomic molecules they are in fact relevant in the present context. They are frequently used to treat nonadiabatic processes in diatomic molecules because of their ease of construction and the historic lack of algorithms for determining the derivative couplings, although the later objection is no longer valid [Lengsfield, Saxe and Yarkony (1984), and Saxe, Lengsfield and Yarkony (1985)].

14.2.5.2 Approximate diabatic bases from molecular properties

A variety of approaches for determining approximate diabatic bases exist [Cederbaum, Köppel and Domke (1981), Sidis (1992), and Domcke and Stock (1997)]. As noted previously, for $N^a = 2$, from equation (9b) only the single, **R**-dependent, parameter, $\Theta(\mathbf{R})$ need be determined. A frequently used approach determines Θ by requiring the smoothness of a molecular property. In this approach $\Theta(\mathbf{R}) = \Theta^B(\mathbf{R})$ is determined from

$$2\Theta^B(\mathbf{R}) = \arctan\left[\frac{B_{IJ}(\mathbf{R})}{\Delta B_{IJ}(\mathbf{R})}\right]. \tag{12}$$

where $B_{IJ}(\mathbf{R}) = \langle\Psi_I(\mathbf{r};\mathbf{R})|B^e(\mathbf{r};\mathbf{R})|\Psi_J(\mathbf{r};\mathbf{R})\rangle_\mathbf{r}$, and $B^e(\mathbf{r};\mathbf{R})$ is any hermitian (molecular property) operator for which $B_{IJ}(\mathbf{R})$ and $\Delta B_{IJ}(\mathbf{R}) \equiv [B_{II}(\mathbf{R}) - B_{JJ}(\mathbf{R})]/2$ do not simultaneously vanish. Equation (12) with the choice $B^e = \mu^{e,w}$, where $\mu^{e,w}$ is the wth component electronic dipole moment operator, was originally suggested [Werner and Meyer (1981)] as a means for obtaining diabatic bases in diatomic molecules. Because of its simplicity it is frequently used to obtain approximate diabatic bases in polyatomic molecules.

Since the choice $B^e = \mu^{e,w}$ is well studied in this work we consider, for illustrative purposes,

$$B^e(\mathbf{r};\mathbf{R}) = \mathrm{t}^\dagger(\mathbf{r};\mathbf{R})\mathrm{t}(\mathbf{r};\mathbf{R}) \tag{13a}$$

where the operator $\mathrm{t}(\mathbf{r};\mathbf{R})$ is based on electronic transition moments to a state Ψ_g^a not included in transformation (9c):

$$\mathrm{t}(\mathbf{r};\mathbf{R}) = \sum_{K=I,J}\langle\Psi_g^a(\mathbf{r};\mathbf{R})|\mu^{e,w}|\Psi_K^a(\mathbf{r};\mathbf{R})\rangle_\mathbf{r}\langle\Psi_K^a(\mathbf{r};\mathbf{R})| \equiv \sum_{K=I,J}\mu_g^{a,k}\langle\Psi_K^a(\mathbf{r};\mathbf{R}). \tag{13b}$$

Recently the theoretical foundation for this approach was strengthened with the demonstration that for $B^e(\mathbf{r};\mathbf{R}) = B^e(\mathbf{r})$, transformation (9b) with $\Theta = \Theta^B$ removes all the singularity in the derivative coupling at a conical intersection [Yarkony (1998)]. A similar result is obtained for $B^e(\mathbf{r};\mathbf{R})$ [Yarkony (1999)] Differentiating equation (12) gives:

$$2\frac{\partial}{\partial\tau}\Theta_I^B(\mathbf{R}) = \frac{\Delta B_{IJ}(\mathbf{R})\dfrac{\partial B_{IJ}(\mathbf{R})}{\partial\tau} - B_{IJ}(\mathbf{R})\dfrac{\partial\Delta B_{IJ}(\mathbf{R})}{\partial\tau}}{\Delta B_{IJ}(\mathbf{R})^2 + B_{IJ}(\mathbf{R})^2}. \tag{14}$$

For $f_\tau^{a,IJ}$ sufficiently large, for example near an avoided crossing of states I and J in a diatomic molecule or near a conical intersection in a polyatomic molecule,

$$\frac{\partial B_{IJ}(\mathbf{R})}{\partial\tau} = 2\Delta B_{IJ}(\mathbf{R})f_\tau^{a,JI}(\mathbf{R}), \tag{15a}$$

and

$$\frac{\partial\Delta B_{IJ}(\mathbf{R})}{\partial\tau} = -2B_{IJ}(\mathbf{R})f_\tau^{a,JI}(\mathbf{R}), \tag{15b}$$

so that

$$\frac{\partial}{\partial\tau}\Theta^B(\mathbf{R}) \equiv f_\tau^{IJ,B}(\mathbf{R}) = f_\tau^{a,IJ}(\mathbf{R}). \tag{16}$$

Thus near an avoided crossing in a diatomic molecule and (certainly) near a conical intersection in a polyatomic molecule, $(\partial/\partial\tau)\Theta^B(\mathbf{R}) \equiv f_\tau^{IJ,B}(\mathbf{R})$ provides a good approximation to the actual derivative coupling. For this reason the angle Θ^B and its gradient have also proved useful in analyzing unusual conical intersection seams in Jahn–Teller molecules. See [Sadygov and Yarkony (1999)] and below.

14.2.6 ADIABATIC STATES II

In Section 14.2.4.1.2 it was observed that for polyatomic molecules a complex-valued function $\exp[A_I(\mathbf{R})]$ may be required to make the adiabatic wave function single-valued. In the past $A_I(\mathbf{R})$ has been determined after the locus of the points of conical intersection was determined [Kendrick and Pack (1996)]. That approach has its limitations. Below we consider an alternative approach that does no require *a priori* knowledge of the locus of points of conical intersection. The discussion, which is only relevant to polyatomic molecules, is included here because of its relation to the transformation to approximate diabatic states.

14.2.6.1 Construction of single-valued adiabatic wave functions

The angle $\Theta^B(\mathbf{R})$ in equation (12) can be used for $A_I(\mathbf{R})$ [Yarkony (1999)]. This choice can be understood as follows.

Given a vector-valued function $\mathbf{g}(\mathbf{R})$ define $A_I^g(\mathbf{R})$ by

$$A_I^g(\mathbf{R}) = \int_{\mathbf{R}_0}^{\mathbf{R}} \mathbf{g}(\mathbf{R}) \cdot d\mathbf{R} + A_I^g(\mathbf{R}_0). \tag{17}$$

where $A_I^g(\mathbf{R}_0)$ is an arbitrary offset. Then $A_I^g(\mathbf{R})$ can be used for $A_I(\mathbf{R})$ provided the criteria

$$\kappa(C_\rho^1, \mathbf{g}) \equiv \int_{C_\rho^1} \mathbf{g}(\mathbf{R}) \cdot d\mathbf{R} = \pi \tag{18a}$$

$$\kappa(C_\rho^0, \mathbf{g}) = 0 \tag{18b}$$

hold for *arbitrary* ρ where $C_\rho^1 \equiv C_\rho(\mathbf{R}_x(I,J))$ a circle of radius ρ centered at $\mathbf{R}_x(I,J)$ a point of conical intersection of states I and $J = I + 1$, and $C_\rho^0 \equiv C_\rho(\mathbf{R})$ which does *not* enclose a point of conical intersection. Below $\mathbf{R}_x(I,J)$ will be denoted $\mathbf{R}_x$ when no confusion will result.

The derivative couplings satisfy criteria (18) for infinitesimal loops [Yarkony (1997)], that is for $\varepsilon \to 0$

$$\kappa(C_\varepsilon^1, \mathbf{f}^{a,IJ}) = \pi \quad \text{and} \quad \kappa(C_\varepsilon^0, \mathbf{f}^{a,IJ}) = 0 \tag{19}$$

Here C_ε^1 must have a nonzero projection on the plane for which the intersecting potential energy surfaces exhibit conical behavior. See below. Equation (19) follows from a rigorous perturbative analysis of the vicinity of a conical intersection [Mead (1983), and Yarkony (1997)]. This analysis expresses the derivative coupling in terms of $\mathbf{g}^{IJ}(\mathbf{R}_x)$ and $\mathbf{h}^{IJ}(\mathbf{R}_x)$, the characteristic parameters of the conical intersection at $\mathbf{R}_x(I,J)$ where

$$2g_\tau^{IJ}(\mathbf{R}) = \left[\mathbf{c}^I(\mathbf{R}_x) + \mathbf{c}^J(\mathbf{R}_x)\right]^\dagger \frac{\partial H(\mathbf{R})}{\partial \tau} \left[\mathbf{c}^I(\mathbf{R}_x) - \mathbf{c}^J(\mathbf{R}_x)\right], \tag{20a}$$

$$h_\tau^{IJ}(\mathbf{R}) = \mathbf{c}^I(\mathbf{R}_x)^\dagger \frac{\partial H(\mathbf{R})}{\partial \tau} \mathbf{c}^J(\mathbf{R}_x). \tag{20b}$$

$\mathbf{g}^{IJ}(\mathbf{R}_x)$ and $\mathbf{h}^{IJ}(\mathbf{R}_x)$ define the $g\text{–}h(\mathbf{R}_x)$ plane which contains all the conical part of the intersection. The circle $C_\rho(\mathbf{R}_x(I,J))$ in the $g\text{–}h(\mathbf{R}_x)$ plane can be expressed in terms of polar coordinates ρ–a size coordinate, and θ–a 'shape' coordinate, defined by $x = \rho\cos\theta$ and $y = \rho\sin\theta$, where $\hat{\mathbf{x}} = \mathbf{h}^{IJ}(\mathbf{R}_x)/\|\mathbf{h}^{IJ}(\mathbf{R}_x)\|$, $\hat{\mathbf{y}} = \mathbf{g}^{IJ}(\mathbf{R}_x)^\perp/\|\mathbf{g}^{IJ}(\mathbf{R}_x)^\perp\|$ and $\mathbf{g}^{IJ}(\mathbf{R}_x)^\perp = \mathbf{g}^{IJ}(\mathbf{R}_x) - (\hat{\mathbf{x}} \cdot \mathbf{g}^{IJ}(\mathbf{R}_x))\hat{\mathbf{x}}$, with the origin at $\mathbf{R}_x$.

The key here is that having identified the $g\text{–}h(\mathbf{R}_x)$ plane, the *only* singular term in the derivative coupling is $(1/\rho)\ f_\theta^{IJ}\ (\mathbf{R})$. The leading term in an expansion of $f_\theta^{IJ}\ (\mathbf{R})$ in powers of ρ is:

$$f_\theta^{IJ}(\theta) = \frac{1}{2}\frac{d\lambda}{d\theta}, \tag{21}$$

where the conical parameters, $q(\theta)$ and $\lambda(\theta)$ are given by:

$$q(\theta)^2 = h_x^2 \cos^2\theta + (g_x \cos\theta + g_y \sin\theta)^2 \equiv h^2 \cos^2\theta + g^2 \sin^2(\theta + \beta), \tag{22a}$$

$$\cos\lambda(\theta) = [h/q(\theta)]\cos\theta, \qquad \sin\lambda(\theta) = [g/q(\theta)]\sin(\theta+\beta), \tag{22b}$$

with $l_w = \mathbf{l}^{IJ}(\mathbf{R}_x)\cdot\hat{\mathbf{w}}$ for $w = x$, y and $\mathbf{l}^{IJ} = \mathbf{g}^{IJ}, \mathbf{h}^{IJ}$.

Thus the first part of equation (19) follows from equation (22) since $\kappa(C_\varepsilon^1, \mathbf{f}^{a,IJ}) = \int_0^{2\pi} d\theta(1/2)(d\lambda/d\theta) = \pi$ while the second part reflects the fact that there is no singularity in the absence of a conical intersection.

Equations (17)–(19) suggest the choice $A_I = A_I^{f^{a,IJ}}$ since $A_I^{f^{a,IJ}}$ satisfies criteria (18) for infinitesimal loops. However the choice $A_I^{f^{a,IJ}}$ is not viable. Since equation (11) does not hold, by Stokes' theorem equations (18a) and (18b) do not apply to closed loops of *arbitrary* ρ. Consequently $A_I^{f^{a,IJ}}$ is not an acceptable A_I. However $A_I^{f^{B,IJ}} = \Theta^B$ where $f_\tau^{B,IJ}(\mathbf{R}) \equiv (\partial/\partial\tau)\Theta^B(\mathbf{R})$, is an acceptable A_I, provided the denominator in equation 14 does not vanish. In this case $f_\tau^{B,IJ}(\mathbf{R})$ is the gradient of a scalar. Therefore by Stokes theorem criterion (18b) holds. Criterion (18a) is satisfied since as shown above $f_\tau^{B,IJ}(\mathbf{R}) \to f_\tau^{a,IJ}(\mathbf{R})$ as $\mathbf{R} \to \mathbf{R}_x(I, J)$. Numerical examples of the use of $A_I^{f^{B,IJ}} = \Theta^B$ can be found in Refs [Sadygov and Yarkony (1999), and Yarkony (1999)]

14.3 Nuclear Schrödinger equation

Two equivalent forms of equation (3) will be considered for diatomic molecules. In the adiabatic representation, which is single-valued since $N^{\text{int}} = 1, H_{IJ}^0(Q)$ is diagonal so that equation (3) becomes:

$$\left[\sum_{\alpha=1}^{2} \frac{-1}{2M_\alpha}\nabla_\alpha^2 + \overline{E}_I^a(\mathbf{R}) - E_k\right]\chi_I^k(\mathbf{R}) = \sum_{J\neq I}^{N^a} O_{IJ}(\mathbf{R})\chi_J^k(\mathbf{R}), \tag{23}$$

where

$$O_{IJ}(\mathbf{R}) = \sum_{\alpha=1}^{2}\left\{\frac{1}{2M_\alpha}\left[k_\alpha^{a,IJ}(\mathbf{R}) + \nabla_\alpha\cdot\mathbf{f}_\alpha^{a,IJ}(\mathbf{R}) + \mathbf{f}_\alpha^{a,IJ}(\mathbf{R})\cdot\nabla_\alpha\right]\right\}. \tag{24a}$$

If the diabatic bases is used equation (3) becomes equation (23) with $\overline{E}_I^a$ replaced with $\overline{H}_{II}^d$ and

$$O_{IJ}(\mathbf{R}) = \overline{H}_{IJ}^d(\mathbf{R}). \tag{24b}$$

Equation (24a) clearly looks more formidable than equation (24b) and for that reason an approach based on a diabatic basis and equation (24b) is usually preferred. However, the adiabatic basis and potential energy surfaces may be better suited for the phenomenological description of molecules moving on potential energy surfaces and in the absence of narrowly avoided intersections the adiabatic description can be quite tractable computationally. See Section 14.4.

The character of the solutions to equation (23) is determined by the relation between the total energy E_k and the asymptotic energies $E_I^a(Q \to \infty) = H_{II}^d(Q \to \infty)$. We will be concerned with solutions to equation (23) where, for the energy range under consideration, the electronic states $\Psi_J^{e,c}(\mathbf{r}; Q), J = 1 - M^c$ are closed channels, their potential energy curves support only bound states, and the electronic states $\Psi_S^{e,o}(\mathbf{r}; Q)$, $S = 1 - M^o$ are open channels, with their potential energy curves supporting only dissociative states. In all cases $E_k > E_S^a(Q \to \infty)$ for some open channel S so that the states in question are resonances. Summarized below is the multichannel resonance method [Han and Yarkony (1995, 1996)] used in Section 14.4. This approach is exact in the isolated resonance limit and is equally effective for both broad and narrow resonances.

14.3.1 RESONANCES

14.3.1.1 Radiationless decay/partial widths

The solution to equation (23) is considered for diatomic molecules in a Hund's case (a) representation. In this case

$$\Psi_I^{\mathbf{e}}(\mathbf{r};\mathbf{R})\chi_I^k(\mathbf{R}) \rightarrow \Psi_{I^{2S+1}\Lambda_\Omega}^{\mathbf{e}}(\mathbf{r};Q)\Psi_{J\Omega M}^{\mathrm{rot}}(\theta,\phi)\chi_{I^{2S+1}\Lambda_\Omega}^{k,x}(Q), \tag{25}$$

where $\mathbf{e} = (e, x)$, $x = \mathrm{o}, \mathrm{c}$ and o, c denote the open and closed channels respectively; (Q, θ, ϕ) are the spherical polar coordinates of the internuclear vector $\mathbf{Q}$; and $\Psi_{J\Omega M}^{\mathrm{rot}}(\theta,\phi) = \sqrt{(2J+1)/8\pi^2}\, D_{M\Omega}^{J*}$, with $D_{M\Omega}^{J*}$ is a D-matrix [Zare (1988)]. For notational convenience below we suppress the contributions due to rotational angular momentum. These contributions can readily be, and in Section 14.4 are, included.

Define, in a diabatic basis:

$$\boldsymbol{H}^{\mathrm{d},x}(Q) = T^{\mathrm{N}}\boldsymbol{I} + \left\langle \Psi^{\mathrm{d},x}(\mathbf{r};Q)\middle|H^{\mathrm{e}}(\mathbf{r};Q)\middle|\Psi^{\mathrm{d},x}(\mathbf{r};Q)^{\dagger}\right\rangle_{\mathbf{r}} \tag{26a}$$

for $x = \mathrm{o}$ and c, and

$$\boldsymbol{H}^{\mathrm{d,co}}(Q) = \left\langle \Psi^{\mathrm{d,c}}(\mathbf{r};Q)\middle|H^{\mathrm{e}}(\mathbf{r};Q)\middle|\Psi^{\mathrm{d,o}}(\mathbf{r};Q)^{\dagger}\right\rangle_{\mathbf{r}}. \tag{26b}$$

It is assumed that the $\Psi^{\mathrm{d,o}}$ have been chosen such that $H_{ps}^{\mathrm{o}} \xrightarrow{Q\to\infty} 0$ for $p \neq s$. Then the M^{o} independent solutions $(\chi_J^{r\mathrm{c}}(Q;E), \chi_S^{r\mathrm{o}}(Q;E))$, $r = 1 - M^{\mathrm{o}}$ satisfy, suppressing the d label,

$$\begin{pmatrix} \boldsymbol{H}^{\mathrm{c}} - E\mathbf{I} & \boldsymbol{H}^{\mathrm{co}} \\ \boldsymbol{H}^{\mathrm{oc}} & \boldsymbol{H}^{\mathrm{o}} - E\mathbf{I} \end{pmatrix} \begin{pmatrix} \chi^{\mathrm{c}} \\ \chi^{\mathrm{o}} \end{pmatrix} = \begin{pmatrix} \mathbf{0} \\ \mathbf{0} \end{pmatrix}. \tag{27}$$

Using a Feshbach type analysis [Feshbach (1958, 1962)] the following results are obtained. The kth resonance energies, E_{res}, is obtained from the iterative solution of the fixed point equation

$$E_\beta(E_{\mathrm{res}}) = E_{\mathrm{res}}, \tag{28}$$

where E_β is the βth root of

$$\left[E_\alpha(E)\boldsymbol{I} - \boldsymbol{H}^{\mathrm{eff}}(E)\right]\xi^\alpha(E) = \mathbf{0}, \tag{29a}$$

the $\boldsymbol{H}^{\mathrm{eff}}$ is given by

$$\boldsymbol{H}^{\mathrm{eff}}(E) = \boldsymbol{H}^{\mathrm{c}} + \boldsymbol{H}^{\mathrm{co}}\,{}^{\mathrm{R}}\boldsymbol{G}(E)\boldsymbol{H}^{\mathrm{oc}}, \tag{29b}$$

and β(E) is the root of $\boldsymbol{H}^{\mathrm{eff}}$(E). In addition to the desired resonance states equation (29a) has spurious and free wave solutions so that $\beta \geq k$ and the solution must be checked to ascertain its resonance character [Han and Yarkony (1995)]. In equation (29b) ${}^{\mathrm{R}}\boldsymbol{G}(E)$, which gives $\boldsymbol{H}^{\mathrm{eff}}$ its energy dependence is the standing wave multichannel Green's function, ${}^{\mathrm{R}}\boldsymbol{G}(Q, Q'; E)$ [Hwang and Rabitz (1979), and Singer *et al.* (1987)]. It is an $M^{\mathrm{o}} \times M^{\mathrm{o}}$ matrix that satisfies

$$(E\boldsymbol{I} - \boldsymbol{H}^{\mathrm{o}})^{\mathrm{R}}\boldsymbol{G}(Q, Q'; E) = \delta(Q - Q')\boldsymbol{I}, \tag{30}$$

and ${}^{\mathrm{R}}\boldsymbol{G}(Q', Q; E) = {}^{\mathrm{R}}\boldsymbol{G}^{\dagger}(Q, Q'; E)$ [Hwang and Rabitz (1979)] and can be expressed in terms of the two linearly independent solutions of:

$$(E\boldsymbol{I} - \boldsymbol{H}^{\mathrm{o}})^{\mathrm{R}}\phi(Q) = \mathbf{0} \tag{31}$$

the regular solution, ${}^{\mathrm{R}}\phi(Q, E)$ and irregular solution, ${}^{\mathrm{I}}\phi(Q, E)$. At the origin ${}^{\mathrm{R}}\phi(Q = 0, E) = 0$ while ${}^{\mathrm{I}}\phi(Q = 0, E)$ is singular and asymptotically

$$ {}^{\mathrm{R}}\phi(Q; E) \xrightarrow{Q \to \infty} \sqrt{\frac{2\mu}{\pi k}} \sin(kQ + \delta^{\mathrm{o}}), \tag{32a} $$

$$ {}^{\mathrm{I}}\phi(Q; E) \xrightarrow{Q \to \infty} \sqrt{\frac{2\mu}{\pi k}} \cos(kQ + \delta^{\mathrm{o}}), \tag{32b} $$

where δ_m^0 is background phase shift on the mth dissociative potential energy curve. ${}^{\mathrm{R}}\boldsymbol{G}(Q', Q; E)$ is given by:

$$ {}^{\mathrm{R}}\boldsymbol{G}(Q, Q'; E) = -\pi\, {}^{\mathrm{I}}\phi(Q; E)^{\mathrm{R}}\phi^{\dagger}(Q'; E) \quad \text{for} \quad Q > Q', \tag{33a} $$

$$ {}^{\mathrm{R}}\boldsymbol{G}(Q, Q'; E) = -\pi\, {}^{\mathrm{R}}\phi(Q; E)^{\mathrm{I}}\phi^{\dagger}(Q'; E) \quad \text{otherwise.} \tag{33b} $$

In practice [Han and Yarkony (1995)] the kth resonance energy, E_k^{res}, denoted E_{res} here, is determined as follows. For an initial guess $E_{\mathrm{res}} = E_{\beta_0}$, $\beta_0 = k$ from $\boldsymbol{H}^{\mathrm{c}}$, ${}^{\mathrm{R}}\phi(E)$ and ${}^{\mathrm{I}}\phi(E)$ are obtained by standard discretization techniques and used to construct ${}^{\mathrm{R}}\boldsymbol{G}(Q', Q; E)$ and hence $\boldsymbol{H}^{\mathrm{eff}}(E)$ which is then diagonalized. If the β^{t} 1th eigenvalue of $\boldsymbol{H}^{\mathrm{eff}}(E_{\beta_0})$ is equal to E_{β_0} a solution to equation (28) has been found. If no such β_1 exists the $E_{\beta_1}(E_{\beta_0})$ for which $|E_{\beta_0} - E_{\beta_1}(E_{\beta_0})|$ is minimized is used as the next guess for E_{res} and the sequence is iterated to convergence. As noted above owing to the $\boldsymbol{H}^{\mathrm{co}\,\mathrm{R}}\boldsymbol{G}(E)\boldsymbol{H}^{\mathrm{oc}}$ term in equation (28) β_{n+1} need not equal β_n. We denote the limit point β_∞ as k.

The partial widths of the kth resonance state are given by

$$ \Gamma_s(k)/2 = \pi\left[\gamma_s^k\right]^2 \Big/ \left[1 - \frac{\mathrm{d}}{\mathrm{d}E} E_{\beta_\infty}(E_{\mathrm{res}})\right], \tag{34a} $$

and are related to the total width Γ by

$$ \Gamma(k) = \sum_{s=1}^{M^{\mathrm{o}}} \Gamma_s(k) \equiv 1/[2\pi\tau_k] = 2\pi\gamma^{k\dagger}\gamma^k \Big/ \left[1 - \frac{\mathrm{d}}{\mathrm{d}E} E_{\beta_\infty}(E_{\mathrm{res}})\right], \tag{34b} $$

where τ_κ is the lifetime for radiationless decay of the kth resonance and the vector $\gamma^{k\dagger}(E)$ has components

$$ \gamma^{k\dagger}(E) \equiv \langle \xi^{\beta_\infty \dagger}(Q, E) | \boldsymbol{H}^{\mathrm{co}}(Q) | {}^{\mathrm{R}}\phi(Q, E) \rangle_Q. \tag{35} $$

In equation (34a) $(\mathrm{d}/\mathrm{d}E)E_\beta(E_{\mathrm{res}})$ is evaluated using a centered divided difference, $[E_\beta(E_{\mathrm{res}} + \varepsilon) - E_\beta(E_{\mathrm{res}} - \varepsilon)]/2\varepsilon$. The partial widths provide the branching ratios for the decay of the resonances into the available channels.

The small $\boldsymbol{H}^{\mathrm{co}}$, perturbative, limit will prove quite useful in Section 14.4. In that case $\boldsymbol{H}^{\mathrm{eff}}(E) \to \boldsymbol{H}^{\mathrm{c}}$, so that $\xi^{\beta_\infty} \to \xi^{\beta_0}$, $(\mathrm{d}/\mathrm{d}E)E_\beta(E_{\mathrm{res}}) \to 0$, and equations (29a) and (34b) become:

$$ \left[E_{\beta_0}\boldsymbol{I} - \boldsymbol{H}^{\mathrm{c}}\right]\xi^{\beta_0} = \boldsymbol{0}, \tag{29a$'$} $$

$$ \Gamma(\beta_0) = \sum_{s=1}^{M^0} \Gamma_s(\beta_0) = 2\pi\gamma^{\beta_0\dagger}\gamma^{\beta_0}. \tag{34b$'$} $$

Since equations (29a$'$) and (34b') are just the Golden rule [Rice (1929b, 1929a)] it is seen that this exact isolated resonance theory bears a formal resemblance to the Golden rule to which it reduces when $\boldsymbol{H}^{\mathrm{co}}$ is small.

14.3.1.2 Radiative decay

The spontaneous radiative decay rate depends on the integral, over the width of the resonance, of the square of the matrix element of the electronic dipole moment operator in the space fixed frame between the *energy-normalized* (*incoming* scattered) wave function for the resonance state and the wave function for a bound state. To correctly describe such a process the full rovibronic wave functions must be employed. Again to simplify the presentation we will continue to suppress the rotational dependence.

The matrix element of the resonance state with the vth vibrational level $[\chi_g^v(Q)]$ of the gth bound electronic state $[\Psi_g(\mathbf{r};Q)]$ is required and found to be:

$$T_k^{(2)}(gv) \approx \frac{|\mu(gv,k)|^2}{\left[1 - \frac{\mathrm{d}}{\mathrm{d}E}E_{\beta_\infty}(E_{\mathrm{res}})\right]}. \tag{36a}$$

$$\mu(gv,k) \equiv \left\langle \chi_g^v \middle| \left(\boldsymbol{\mu}^{\mathrm{c}\dagger} + \boldsymbol{\mu}^{\mathrm{o}\dagger\mathrm{R}}\boldsymbol{G}\boldsymbol{H}^{\mathrm{oc}}\right)\xi^{\beta_\infty}\right\rangle_Q \equiv \mu^{\mathrm{c}}(gv,k) + \mu^{\mathrm{o}}(gv,k), \tag{36b}$$

where $\boldsymbol{\mu}^C \equiv \boldsymbol{\mu}_g^{\mathrm{d,c}}$. Note that when $\boldsymbol{H}^{\mathrm{co}}$ is small equations (36a) and (36b) reduce to a standard bound state form $\mu(gv,\beta) \to \langle \chi_g^v | \mu^{\mathrm{c}\dagger}\xi^{\beta_0}\rangle_Q$ Equations (36a) and (36b) have the obvious interpretation of the bound–bound transition moment $\mu^{\mathrm{c}}(gv,k)$, perturbed by contributions from the continuum.

14.4 Applications

14.4.1 Al_2 $(2,3\,^3\Pi_g, v)$ LEVELS–A THREE ELECTRONIC STATE PROBLEM

14.4.1.1 Statement of the problem

The spectroscopy of the Al_2 dimer has been studied in both noble-gas matrices [Abe and Kolb (1983), and Douglas, Hauge and Margrave (1983)] and the gas phase [Cai, Dzugan and Bondybey (1989), Fu *et al.* (1990), and Cai *et al.* (1991)] and has been the object of several theoretical studies. One of the most interesting spectroscopic problems in Al_2 is the fate of the electronically excited $2\,^3\Pi_g$ and $E'3\,^3\Pi_g$ states (see Figure 14.1). The $E'3\,^3\Pi_g \leftarrow X\,^3\Pi_u$ transition is the most intense band observed in resonant two photon ionization spectroscopy yet the fluorescence quantum yield of the $E'3\,^3\Pi_g$ state is minuscule [Fu *et al.* (1990)]. Experimentally observed [Abe and Kolb (1983), and Douglas, Hauge and Margrave (1983)] absorption spectra isolated in Ar, Kr, or Xe matrices in the range 24 400–25 600 cm^{-1} have been assigned on the basis of theoretical calculations to the $2\,^3\Pi_g \leftarrow X\,^3\Pi_u$ transition [Langhoff and Bauschlicher (1990)]. However, this transition is not observed in either laser-induced fluorescence *or* resonant two-photon ionization spectroscopy [Fu *et al.* (1990)].

The lowest state of $^3\Pi_g$ character, the $1\,^3\Pi_g$ state, is repulsive and its skewed shape reflects strong nonadiabatic interactions with the higher $^3\Pi_g$ states, the $2,3\,^3\Pi_g$ states (see Figure 14.1). Thus the $2,3\,^3\Pi_g$ states are embedded in the $1\,^3\Pi_g$ continuum, and coupled to it, and to each other, through the $f_Q^{\mathrm{a},IJ}(Q)$. The vibrational levels of Al_2 in the $2,3\,^3\Pi_g$ electronic states are therefore resonances predissociated by $1\,^3\Pi_g$ state. Although there had been several theoretical studies of the excited electronic states of Al_2 [Basch, Stevens and Krauss (1984), Upton (1986), and Langhoff and Bauschlicher (1990)] none had addressed the question of nonadiabatic radiationless decay in the $2,3\,^3\Pi_g$ states. The work described below [Han, Hettema and Yarkony (1995), and Han and Yarkony (1995)] addressed this deficiency.

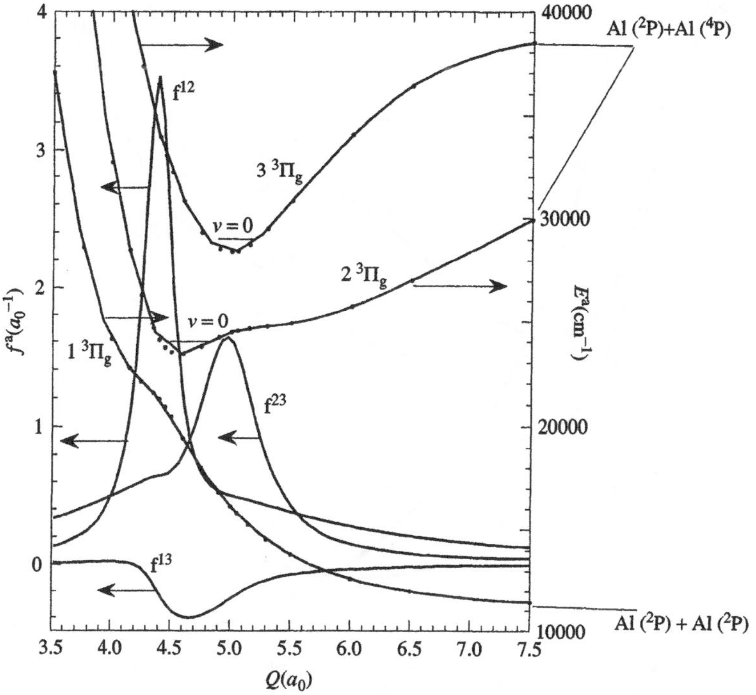

Figure 14.1. Al_2: the potential energy curves for the adiabatic 1,2,3 $^3\Pi_g$ states. The first-derivative nonadiabatic coupling matrix elements $f^{a}_{IJ}(Q)$, $I, J \in 1, 2, 3\,^3\Pi_g$.

14.4.1.2 Electronic structure aspects: diabatic states

Figure 14.1 reports the adiabatic potential energy curves and corresponding $f_Q^{a,IJ}(Q)$ for the $1, 2, 3\,^3\Pi_g$ states determined at the state averaged-multiconfigurational self-consistent field (SA-MCSCF)/second-order configuration interaction (SOCI) level [Yarkony (1996b)]. The peaks in the $f_Q^{a,IJ}$ for $(I, J) = (1\,^3\Pi_g, 2\,^3\Pi_g)$ and $(2\,^3\Pi_g, 3\,^3\Pi_g)$ coincide with the $1\,^3\Pi_g$–$2\,^3\Pi_g$ and $2\,^3\Pi_g$–$3\,^3\Pi_g$ avoided intersections at $Q = 4.4\,a_0$ and $Q = 5.0\,a_0$ respectively. The peak in $f_Q^{a,IJ}(Q)$ for the nonadjacent pair $(I, J) = (1\,^3\Pi_g, 3\,^3\Pi_g)$ is not readily interpreted in terms of avoided intersections. Using equations (8) and (9) the diabatic potential energy curves and potential couplings were computed and are reported in Figure 14.2.

From Figures 14.1 and 14.2 its seen that the diabatic potential energy curves and couplings are, as expected, considerably smoother than their adiabatic counterparts. As noted previously diabatic states are expected to have other properties including configurational or CSF uniformity. See for example [Ruedenberg and Atchity (1993)]. To investigate this question the Q-dependence of the $c_\alpha^{aI}(Q)$ and $c_\alpha^{dI}(Q)$ was determined. The ground state of the Al has the electron configuration $[1s^22s^22p^6]3s^23p$, where [ ] encloses the core electrons. Therefore the lowest three $^3\Pi_g$ states are expected to be linear combinations of CSFs arising from the following electron configurations

$$(1)\ 4\sigma_g^2 4\sigma_u 5\sigma_g^2 2\pi_u, \quad (2)\ 4\sigma_g^2 4\sigma_u 2\pi_u^3, \quad (3)\ 4\sigma_g^2 4\sigma_u^2 5\sigma_g 2\pi_g, \quad (4)\ 4\sigma_g^2 4\sigma_u^2 2\pi_u 5\sigma_u$$

where only the valence molecular orbitals, which are approximated by:

$$4\sigma_{g/u} \sim 3s_1 \pm 3s_2, \quad 5\sigma_{u/g} \sim 3p_{z1} \pm 3p_{z2}, \quad 2\pi_{u/g} \sim 3p_{w1} \pm 3p_{w2}, \quad w = x, y,$$

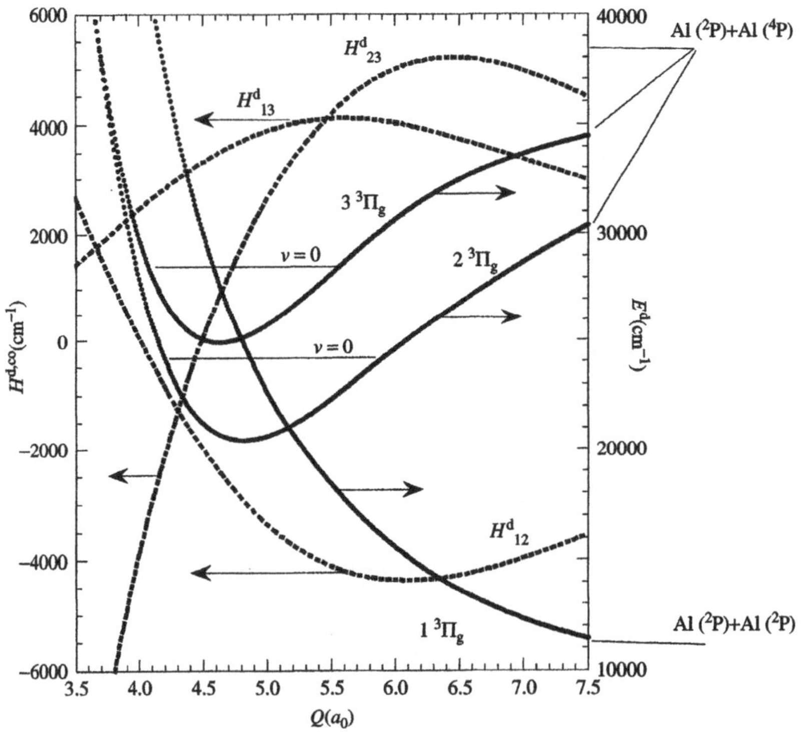

Figure 14.2. Al_2: the diagonal potentials (solid lines) and off-diagonal interaction potentials (dashed lines) for diabatic $1,2,3\,^3\Pi_g$ states. The potentials dashed lines in the diabatic picture of $1,2,3\,^3\Pi_g$ states.

are shown. Figure 14.3 presents the contributions from these CSFs, $c_\alpha^{al}(Q)$ and $c_\alpha^{dl}(Q)$, $\alpha = 1, \ldots, 4$. Note from this figure that while c_α^{al}, $\alpha = 1, \ldots, 4$ change dramatically particularly in the regions of the avoided crossings, the c_α^{dl}, are slowly varying functions of Q. The geometry dependence of the c_α^{dl} quantifies the notion configurational uniformity for the 'exact' diabatic states.

14.4.1.3 Resonance energies and lifetimes

In order to make the connection with the questions posed by the experimental observations, the resonance energies (E_v), radiative decay (τ_v^{rad}) and nonradiative (radiationless or predissociative decay) (τ_v^{nrad}) lifetimes are required. These results are presented in Table 14.1. They not only explain the experimental observations but also provide insights into the mechanism of the predissociation and its theoretical description.

We begin by considering the relative merits of the adiabatic and diabatic state representations. The following observations are relevant. From Figures 14.1 and 14.2 it is seen that min $E^a_{2\,^3\Pi_g}$, the minimum of $E^a_{2\,^3\Pi_g}$ is approximately $3000\,cm^{-1}$ higher than min $E^d_{2\,^3\Pi_g}$, while min $E^a_{3\,^3\Pi_g}$ − min $E^d_{3\,^3\Pi_g} \sim 4000\,cm^{-1}$. From Table 14.1 the energy levels in the adiabatic representation obtained in the golden rule limit, denoted by an additional superscript gr–equation (26a′), and those in the diabatic representation obtained from the exact resonance formulation, denoted by an additional superscript fq, are in excellent accord. In Figures 14.1 and 14.2 the $(2\,^3\Pi_g, v = 0)$ and $(3\,^3\Pi_g, v = 0)$ levels 'fit nicely' in the $2\,^3\Pi_g$ and $3\,^3\Pi_g$ adiabatic potential

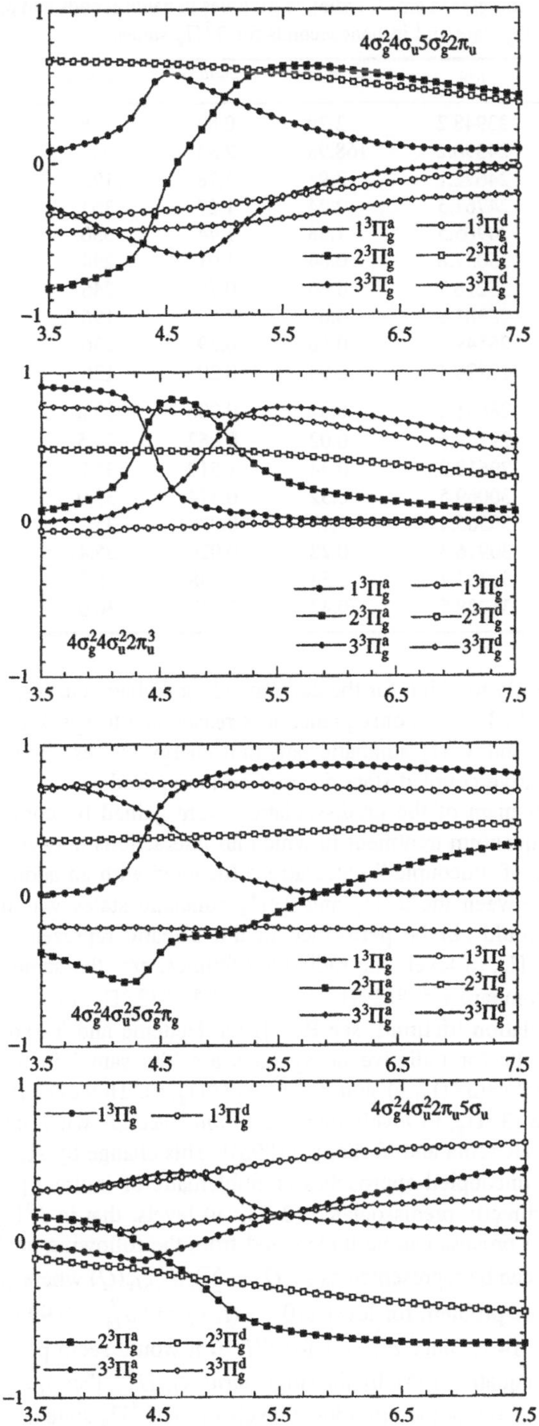

Figure 14.3. Al_2: coefficients, $b_{I\alpha}^{e}$, of four principal CSFs for $e = \text{a, d}$ and $I = 1, 2, 3\,^3\Pi_g$.

Table 14.1. Al_2: resonance energies in cm^{-1}, radiative lifetimes in nanoseconds and predissociative lifetimes (in picoseconds) for the $2\,^3\Pi_g$ state and in nanoseconds for $3\,^3\Pi_g$ states.

$\beta(n, v)$	E^{gr}	E^{fq}	$\tau^{nrad,gr}$	$\tau^{nrad,fq}$	$\tau^{rad,gr}$	$\tau^{rad,fq}$	$\tau^{rad,gr-uc}$
1(2,0)	23949.3	23948.2	2.29	0.83	796	796	793
2(2,1)	24389.6	24390.2	168.98	2.33	337	336	342
3(2,2)	24671.9	24672.1	3.98	2.78	193	191	187
4(2,3)	24761.4	24761.5	2.24	1.67	363	352	372
5(2,4)	24905.4	24905.5	1.38	1.44	330	319	318
6(2,5)	25058.7	25057.8	0.84	1.01	294	279	286
7(2,6)	25217.9	25216.7	0.66	0.75	246	233	238
8(2,7)	25384.1	25381.3	0.61	0.56	196	187	190
9(2,8)	25552.8	25549.7	0.68	0.39	156	151	153
10(2,9)	25724.2	25720.4	0.94	0.20	126	125	125
27(3,0)	28661.6	28661.2	6.37	0.006	35.8	34.8	36.1
31(3,1)	29150.1	29149.6	0.02	0.062	35.5	35.4	35.6
35(3,2)	29618.5	29618.2	0.34	1.512	35.2	35.2	35.2
39(3,3)	30069.7	30069.5	0.32	0.474	35.0	35.1	35.1
43(3,4)	30505.3	30505.1	0.13	0.078	35.1	35.2	35.1
47(3,5)	30926.7	30926.3	0.28	0.024	35.4	35.3	35.2
50(3,6)	31334.1	31333.7	1.54	0.048	35.5	35.6	35.4
54(3,7)	31728.9	31728.5	0.42	0.027	36.0	35.9	35.7

energy curves but are clearly too high for the corresponding diabatic curves, given the vibrational spacing evident in Table 14.1. As a consequence it is reasonable to speak of motion on the individual, see below, $2\,^3\Pi_g$ and $3\,^3\Pi_g$ adiabatic potential energy curves. However, in the diabatic representation a less intuitive coupled state description is required.

Insights into the mechanism of the predissociation were gained by comparing the results of the full coupling or full quantum treatment in which all interactions between the adiabatic states were included with those of 'uncoupled' calculations (denoted with an additional superscript uc) in which the coupling between the $2\,^3\Pi_g$ and $3\,^3\Pi_g$ adiabatic states was neglected. Note that this important analysis could *not* be performed in the diabatic representation for the reasons noted above. For the $(2\,^3\Pi_g, v)$ levels the radiative lifetimes, are 'the same' in the coupled and uncoupled approaches, compare $\tau^{rad,fq}(2\,^3\Pi_g, v)$ and $\tau^{rad,gr-uc}(2\,^3\Pi_g, v)$. A similar conclusion is reached for the predissociation lifetimes, see Ref. [Han, Hettema and Yarkony. (1995)]. For the $(3\,^3\Pi_g, v)$ levels the results for radiative decay rates are 'the same' in the coupled state and uncoupled approaches, $\tau^{rad,fq}(3\,^3\Pi_g, v)$ and $\tau^{rad,gr-uc}(3\,^3\Pi_g, v)$. However the computed predissociation lifetimes of the $(3\,^3\Pi_g, v)$ levels increase to *milliseconds* when coupling to the $2\,^3\Pi_g$ levels is neglected [Han, Hettema and Yarkony. (1995)]. This change by *six* orders of magnitude in the coupled state and uncoupled approaches is emblematic of indirect predissociation of the $(3\,^3\Pi_g, v)$ levels by the directly predissociated $(2\,^3\Pi_g, v)$ levels, that is $3\,^3\Pi_g \sim 2\,^3\Pi_g \sim 1\,^3\Pi_g$. The origin of this indirect process can be understood from the following analysis.

The bound state $\xi^\beta(Q)$ can be represented as $\xi^\beta(Q) = \sum_v d^\beta_{Iv}\zeta_{Iv}(Q)$ where $\zeta_{Iv}(Q)$ is the solution to the single adiabatic state problem for level $\beta(0) = (I, v)$ and $d^\beta_{Iv} \sim 0.99$ enabling the association $\beta \leftrightarrow (I, v)$ see Table 14.1. Since $d^\beta_{Iv} \sim 1$ for $\beta(I, v)$ it would seem possible to approximate $\sum_{Iv} d^\beta_{Iv}\langle\zeta_{Iv}|H^{d,co}_{I,m}|^R\phi_m\rangle_Q$, equation (35), by the single term $\langle\zeta_{Iv}|H^{d,co}_{I,m}|^R\phi_m\rangle_Q$. This is in fact largely appropriate for $\beta(2\,^3\Pi_g, v)$ so that the vibrational levels of the $2\,^3\Pi_g$ state are *directly* predissociated by the $1\,^3\Pi_g$ state, $2\,^3\Pi_g \sim 1\,^3\Pi_g$. Hence the agreement of the coupled and uncoupled state results. However, for $\beta(3\,^3\Pi_g, v)$, $\langle\zeta_{Iv}|H^{d,co}_{I,m}|^R\phi_m\rangle_Q$ is 10^6 larger for $I = 2\,^3\Pi_g$ than for $I = 3\,^3\Pi_g$.

This is a consequence of both the closer proximity and larger derivative coupling of the $1, 2\,^3\Pi_g$ states when compared with the $1, 3\,^3\Pi_g$ states. Thus the above truncation cannot be justified on the basis of the size of d^{β}_{Iv} alone. The vibrational levels of the $3\,^3\Pi_g$ state are predissociated by the small mixing with the (bound but strongly predissociated) levels of the $2\,^3\Pi_g$ adiabatic state, that is, the predissociation is *indirect*, $3\,^3\Pi_g \sim 2\,^3\Pi_g \sim 1\,^3\Pi_g$.

14.4.1.4 Comparison with experiment

The radiative and radiationless decay processes compete to reduce the population of the $\beta(2, 3\,^3\Pi_g, v)$ levels. Based on the lack of observable gas phase, resonant two-photon ionization, and laser-induced fluorescence [Cai *et al.* (1991)] signals for the $2\,^3\Pi_g$ state, Fu *et al.* (1990) concluded that the non-radiative lifetimes of the $\beta(2\,^3\Pi_g, v)$ levels must be shorter than 1 ns. We [Han, Hettema and Yarkony. (1995)] predicted predissociation lifetimes on the order of picoseconds. Fu *et al.* (1990) also concluded that the $\beta(3\,^3\Pi_g, v)$ levels have lifetimes short compared with the time scale of fluorescence but long enough to be observed in the resonant two-photon ionization experiment. We find that the radiative decay rates are generally less than 1–0.1 % of the predissociative rate which are in turn approximately 1000 times *slower* than those of the $\beta(2\,^3\Pi_g, v)$ levels. Thus our predictions are consistent with and provide an explanation for the experimental observations.

14.4.2 MgBr $(1, 2\,^2\Pi, v, J)$ LEVEL–INDIRECT PREDISSOCIATION INDUCED BY MOLECULAR ROTATION

14.4.2.1 Statement of the problem

The $A\,^2\Pi$–$X\,^2\Sigma^+$ system in MgBr had been observed in absorption and laser-induced fluorescence(LIF) [Morgan (1936), and Fethi (1995)]. The $A\,^2\Pi$ manifold is known to have an anomalous v-dependent fine structure splitting constant A^{SO}_v and is strongly predissociated. In analogy with MgCl [Parlant *et al.* (1990), Rostas *et al.* (1990), and Shafizadeh *et al.* (1990)], these observations had been attributed to coupling with the $2\,^2\Pi$ state, a dissociative state having a large fine structure splitting and having a wave function representing qualitatively $Mg(^1S)Br(^2P)$ [Fethi (1995)]. See Figure 14.4. Although the anomalous dependence of the fine structure splitting constant on v is well explained in this model quantitative determination of the predissociation as a function of fine structure level remained elusive. The work described below addressed this issue [Sadygov *et al.* (1997)].

14.4.2.2 Electronic structure aspects

14.4.2.2.1 Nonrelativistic Adiabatic States

The $X\,^2\Sigma^+$ and the two lowest $^2\Pi$ nonrelativistic adiabatic wave functions were obtained at the SA-MCSCF/SOCI level of theory using a (12s9p6d/8s6p2d) basis for magnesium and a (17s14p9d/6s6p5d) basis for bromine. The active space consists of the CSFs arising from the distribution of seven electrons, in the three σ and two π orbitals arising from magnesium $\{3s^2 3p^0\}$ and bromine $\{4p^5\}$ shells. Figure 14.4 reports the adiabatic potential energy curves, ($E^{0,a}_{I^2\Pi}(Q)$, $I = 1, 2$), and the corresponding $f^{a,IJ}_Q(Q)$ for the $1, 2\,^2\Pi$ states. This figure clearly evinces a $1\,^2\Pi$–$2\,^2\Pi$ avoided intersection at $Q_{ax} = 5.03\,a_0$.

The nonrelativistic diabatic potential energy curves ($E^{0,d}_{I^2\Pi}(Q)$, $I = 1, 2$) and interstate couplings [$H^{0,d}_{1^2\Pi,2^2\Pi}(Q)$] were determined using equations (9b) and (9c) and are presented in Figure 14.5.

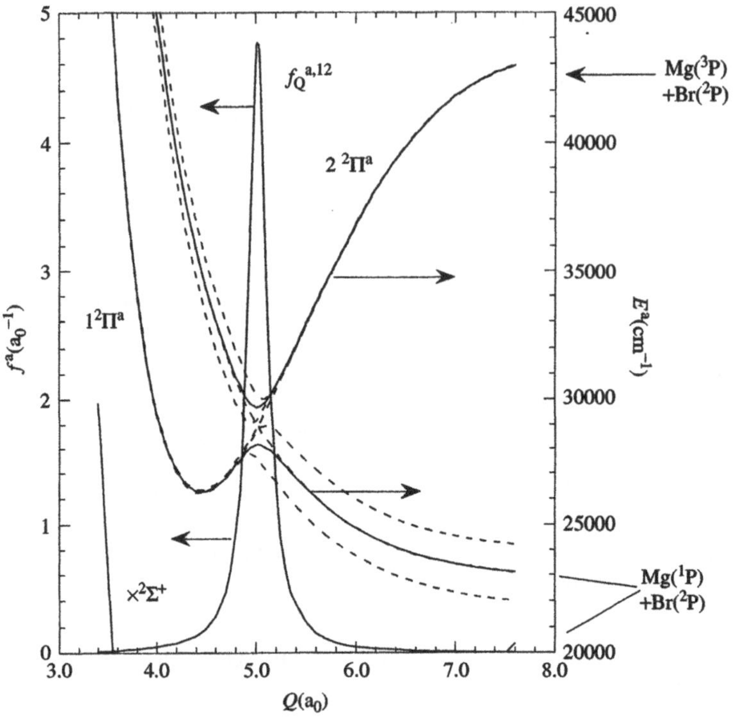

Figure 14.4. MgBr: adiabatic potential energy curves (solid lines) and $f^a_{JI}(R)$ for the nonrelativistic $1, 2\,^2\Pi$ states. Dashed lines are corrections due to the spin–orbit interactions.

From Figures 14.4 and 14.5 $E^{0,a}_{I\,^2\Pi}(Q)$ and $E^{0,d}_{I\,^2\Pi}(Q)$ are seen to be in good accord except in the region of the avoided crossing at Q_{ax}. $E^{0,d}_{1\,^2\Pi}(Q) > E^{0,d}_{2\,^2\Pi}(Q = \infty)$, and as a result all vibrational levels of the bound state are resonances. This conclusion is unaltered by the inclusion of the spin–orbit interaction, see later.

Figure 14.6 reports $f_Q^{B,IJ}(Q)$ and Θ^B equations (12) and (13), with $\Psi_g(\mathbf{r};Q) = \Psi_{X\,^2\Sigma^+}(\mathbf{r};Q)$ and compares them with $f_Q^{a,IJ}(Q)$ and $\Theta^{f^{a,IJ}}$ [equation (9c)]. As anticipated in Section 14.2.4.2.2 agreement between $f_Q^{B,IJ}(Q)$ and $f_Q^{a,IJ}(Q)$ is excellent near the avoided crossing. $\Theta^B(Q = 7)$ was set equal to $\Theta^{f^{a,IJ}}(Q = 7)$. This adjustment, which does not effect $f_Q^{B,IJ}$, reflects the fact that for Q large B_{IJ} and ΔB_{IJ} both approach zero, since $X\,^2\Sigma^+$, $1\,^2\Pi \to$ Mg(^{2}S) + Br(^{2}P) and $2\,^2\Pi \to$ Mg(^{3}P) + Br(^{2}P), so that use of equation (13) becomes problematical.

14.4.2.2.2 Relativistic effects: Spin–orbit interactions for the $^2\Pi$ states

Within the two-state approximation used here, since Ω is a good quantum number, there are six nonvanishing spin–orbit interactions

$$\begin{aligned} H^{so,e}_{I\,^2\Pi_{1\pm1/2},J\,^2\Pi_{1\pm1/2}}(Q) &= \langle \Psi^e_{I\,^2\Pi_{1\pm1/2}}(\mathbf{r};Q) | H^{so}(\mathbf{r};Q) | \Psi^e_{J\,^2\Pi_{1\pm1/2}}(\mathbf{r};Q) \rangle_{\mathbf{r}} \\ &= \mp i \langle \Psi^e_{I\,^2\Pi_y}(\mathbf{r};Q) | H^{so}(\mathbf{r};Q) | \Psi^e_{J\,^2\Pi_x}(\mathbf{r};Q) \rangle_{\mathbf{r}} \\ &\equiv \mp H^{so,e}(Iy, Jx)(Q) \quad \text{for } (I,J) = (1,1), (2,2), (1,2). \end{aligned} \tag{37}$$

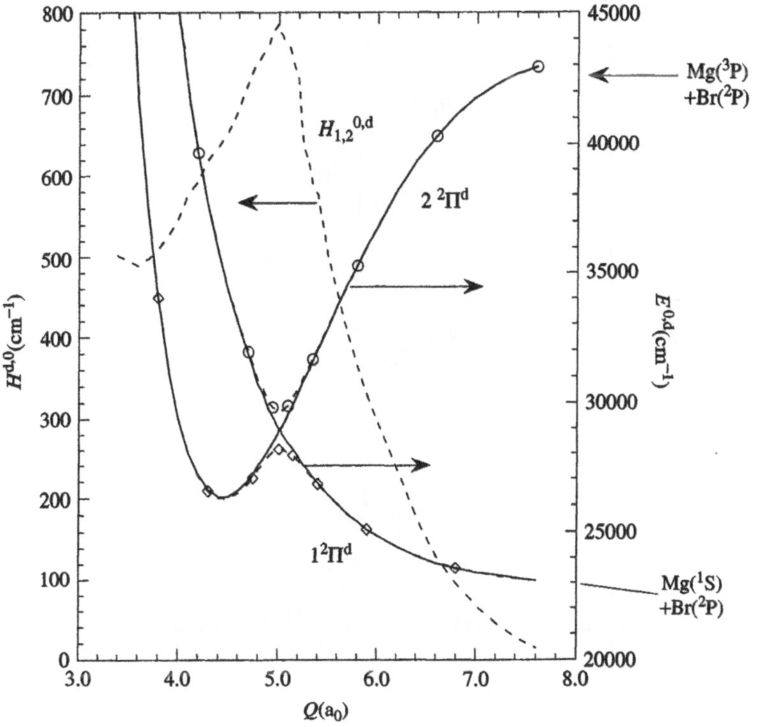

Figure 14.5. MgBr: potential energy curves, $H_{II}^{0,d}$, solid lines, and coupling potential $H_{12}^{0,d}$, dashed line, for diabatic $1, 2\,^2\Pi$ states.

In equation (37) the standard conventions in which a subscript denotes either an Ω value (a number) or the column of an irreducible representation (a letter) have been used.

$H^{so,e}(Ix, Jy)(Q)$, $e = a$ and d, are shown in Figure 14.7. The large changes in the adiabatic matrix elements near Q_{ax} are largely absent in the diabatic basis. However the comparatively small $H^{so,d}(Ix, Jy)$, $I = J = 1$ and $I = 1$, $J = 2$ do exhibit an unexpected Q-dependence This Q-dependence does not reflect a failure of the diabatic transformation. Rather it reflects the heavy atom effect observed in metal–rare gas van der Waals complexes [Sohlberg and Yarkony (1997a, 1997b, 1997c)]. In this situation the $Mg(3p_\pi)$ orbitals of the small spin–orbit splitting state, $Mg^+(3p)Br^-(^1S)$, increase their effective spin–orbit constants through (antibonding) overlap with the bromine. This overlap decreases with increasing Q producing the observed trends, in $H^{so,d}(Ix, Jy)$, $I = J = 1$ and $I = 1$, $J = 2$ near Q_{ax}.

$E^d_{I\,^2\Pi_\omega}(Q)$ are shown in Figure 14.8 together with the vibrational levels of the uncoupled diabatic states $1\,^2\Pi^d_{1/2,3/2}$. For the regular $1\,^2\Pi^d_{1/2}$ manifold $E^d_{1\,^2\Pi_{1/2}}(Q)$ lies only slightly below $E^d_{1\,^2\Pi_{3/2}}(Q)$ while in the inverted $2\,^2\Pi^d_\Omega$ manifold the $E^d_{2\,^2\Pi_{1/2}}(Q)$ is approximately $2200\,cm^{-1}$ above $E^d_{2\,^2\Pi_{3/2}}(Q)$. Consequently $Q_{dx}(^2\Pi_{3/2}) = 4.945\,a_0$, the point where $E^d_{1\,^2\Pi_{3/2}}(Q) = E^d_{2\,^2\Pi_{3/2}}(Q)$, is $1800\,cm^{-1}$ higher than the zero point energy of the $1\,^2\Pi^d_{3/2}$ state, while $Q_{dx}(^2\Pi_{1/2}) = 5.105\,a_0$ and $E^d_{1\,^2\Pi_{1/2}}(Q = 5.105) = 3000\,cm^{-1}$ relative to the zero-point energy of the $1\,^2\Pi^d_{1/2}$ state. From Figure 14.8 it is seen that $E^d_{1\,^2\Pi_{3/2}}(Q_{dx}(^2\Pi_{3/2}))$ [$E^d_{1^2\Pi_{1/2}}(Q_{dx}(^2\Pi_{1/2}))$] is approximately at the energy of $v = 5[8]$ vibrational level. Thus, as a result of the large

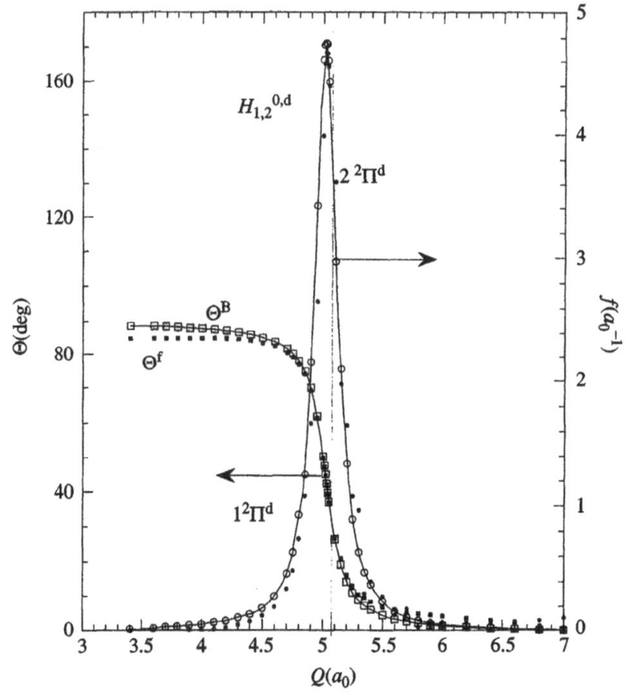

Figure 14.6. MgBr: $f_Q^{a,JI}(Q)$ and $\theta^f(Q)$ compared with $\theta^B(R)$ and $f_Q^{B,JI}(Q)$ for adiabatic $1, 2\,^2\Pi$ states and $B = tt^\dagger$ as described in Section 14.2.5.2.

magnitude of $H^{so,d}(2x, 2y)$ for a given v the $(1\,^2\Pi^d_{3/2}, v)$ level is expected to be much more rapidly predissociated than $(1\,^2\Pi^d_{1/2}, v)$ although as it will be seen below, vibrational levels well below the relevant $E^d_{1\,^2\Pi_\Omega}(Q_{dx}(^2\Pi_\Omega))$ are effectively predissociated.

14.4.2.2.3 Radiative and radiationless decay

Equation (20) was solved using two levels of approximation, the pure Hund's case (a) approximation, and the Hund's (a) approximation plus the S-uncoupling interaction, which is proportional to J and couples the $A\,^2\Pi_{3/2}$ and $A\,^2\Pi_{1/2}$ manifolds [Lefebvre-Brion and Field (1986)]. In the pure Hund's case (a) approximation, the 'two-state case', Ω is a good quantum number so that four electronic manifolds $I\,^2\Pi_\Omega$ $I = 1,2$ $\Omega = 1/2, 3/2$, are block diagonal in Ω. When the S-uncoupling interaction is included, the 'four-state case', all four manifolds are coupled.

Using the above approximations and the resonance theory of Section 14.3 the vibrational energy levels, $E_v(A\,^2\Pi_\Omega)$ and radiative $[\tau^{rad}(A\,^2\Pi_{3/2}, v, J)]$ and radiationless $[\tau^{nrad}(A\,^2\Pi_{3/2}, v, J)]$ lifetimes were obtained and are reported in Table 14.2, for two values of J, 23.5 and 63.5. The $J = 23.5(63.5)$ results are relevant to the P_1 and Q_2 (P_{12} and P_2) band heads. As expected from our previous study of the analogous states in MgCl [Parlant *et al.* (1990)] the $E_v(A\,^2\Pi_\Omega)$ and A_v^{so} are in satisfactory accord with experiment in both the two- and four-state approximations [Sadygov *et al.* (1997)]. The radiative decay rates are also similar the two approaches.

For radiationless decay, however, there are significant differences between the two approaches. These differences, which parallel those observed for Al_2 in the coupled and uncoupled representation, provide valuable insights into the mechanism of the predissociation. In the two state

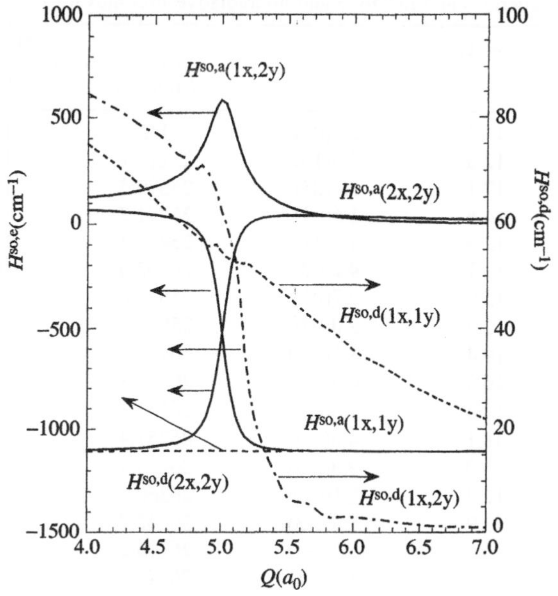

Figure 14.7. MgBr: spin–orbit coupling matrix elements in the adiabatic (solid lines) and diabatic (dotted lines) bases.

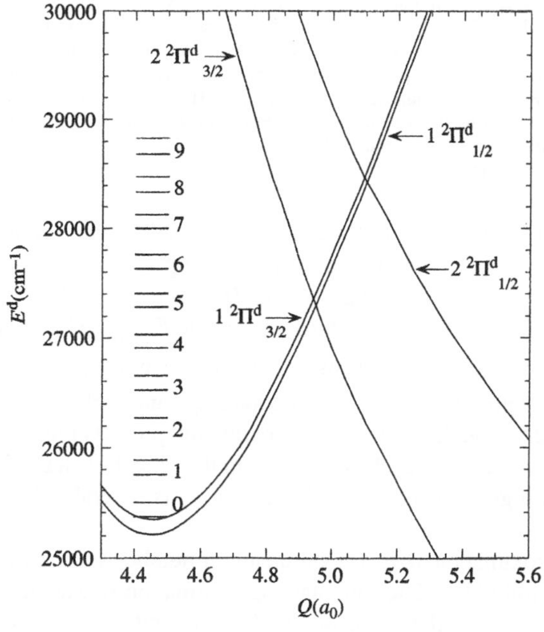

Figure 14.8. MgBr: fine structure splittings of the $1, 2\,^2\Pi_\Omega$ manifolds in the diabatic basis and vibrational levels of the uncoupled $1\,^2\Pi^d_{1/2,3/2}$ states.

Table 14.2. MgBr: energies and radiative and nonradiative lifetimes for $(A\,^2\Pi_\Omega, v, J)$ levels[a].

v	$\Omega = 1/2$			$\Omega = 3/2$		
	$E(\text{cm}^{-1})$	τ^{rad} (ns)	τ^{nrad} (ns)	$E(\text{cm}^{-1})$	τ^{rad} (ns)	τ^{nrad} (ns)
0	25023.4	14.4	2.9(07)[b]	25115.5	11.7	1.7(04)
	25021.6	12.3	3.3(15)	25112.1	12.0	1.2(03)
	25598.3	17.1	5.4(05)	25690.5	10.3	2.4(03)
1	25409.9	14.6	1.6(04)	25491.7	11.5	1.4(01)
	25407.4	12.3	1.1(11)	25486.4	12.0	1.4(00)
	25980.8	17.7	4.2(02)	26062.0	10.1	2.8(00)
2	25790.0	15.0	2.7(01)	25855.8	11.4	4.4(−2)
	25786.8	12.3	2.0(07)	25846.0	12.0	5.4(−3)
	26357.6	19.4	8.3(−1)	26420.8	9.8	1.1(−2)
3	26166.2	16.0	4.5(−2)	26200.8	10.9	3.0(−4)
	26162.2	12.4	1.3(04)	26161.3	11.9	3.6(−5)
	26730.6	18.1	4.6(−3)	26751.4	7.8	9.7(−5)
4	26540.5	13.0	9.9(−3)	26425.9	14.2	1.3(−5)
	26534.3	12.2	2.1(01)	26360.0	12.1	9.0(−6)
	27104.8	13.9	1.9(−3)	26942.7	15.7	1.3(−5)
5	26894.0	13.5	5.9(−4)	26705.2	12.6	1.5(−5)
	26883.6	12.3	1.0(−1)	26740.3	12.1	3.0(−5)
	27457.9	13.1	3.5(−4)	27287.6	11.8	1.5(−5)
6	27379.6	14.8	6.2(−5)	27242.4	12.5	5.3(−4)
	27222.2	12.4	1.4(−3)	27170.5	12.3	3.2(−4)
	27771.6	14.9	4.3(−5)	27604.4	10.1	4.1(−4)

[a] $J = 23.5$ four-state above the $J = 23.5$ two-state results above the $J = 63.5$ four-state results. The $J = 23.5(63.5)$ four-state results are relevant to the P_1 and Q_2 (P_{12} and P_2) band heads. The resonance energies are determined with respect to the zeroth vibrational level of the $X\,^2\Sigma^+$ state which is −2772.213238 au. The minimum of electronic energy is −2772.214049 au. In four-state calculations the $2\,^2\Pi_{1/2}$ and $2\,^2\Pi_{3/2}$ potential energy curves are separated by an additional by 200 cm^{-1} to reproduce the asymptotic value of the Br(^{2}P) fine structure splitting. Also, the $A\,^2\Pi_\Omega$ diabatic curves are shifted down by 400 cm^{-1} to improve the agreement with the experimental data on the vibrational spacings in the $A\,^2\Pi$ manifold. Two-state energies shifted by −400 cm^{-1} to conform better with four-state results below.
[b] Characteristic base 10 is given parenthetically.

approximation all the $(A\,^2\Pi_{3/2},, v)$ levels exhibit observable predissociation in contrast to the $(A\,^2\Pi_{1/2},, v)$ levels for which radiative decay prevails up to $v = 4$. The origin of this difference is found in the spin–orbit couplings. While $H^{\text{so,d}}(1x, 1y)$ and $H^{\text{so,d}}(1x, 2y)$ are modest, <100 cm^{-1}, $H^{\text{so,d}}(2x, 2y) \sim 2200\,\text{cm}^{-1}$ which results in $E^{\text{d}}_{2\,^2\Pi_{3/2}}(Q)$ being shifted down with respect to $E^{\text{d}}_{2\,^2\Pi_{1/2}}(Q)$ by 2200 cm^{-1}. Thus as noted above $E^{\text{d}}_{1\,^2\Pi_{1/2}}(Q_{\text{dx}}(^2\Pi_{1/2}))-E^{\text{d}}_{1\,^2\Pi_{3/2}}(Q_{\text{dx}}(^2\Pi_{3/2})) \sim 1200\text{cm}^{-1}$, approximately three vibrational quanta of the $A\,^2\Pi_{1/2}$ state, Figure 14.8. In accord with this observation the two-state approximation predicts predissociation to start in the $\Omega = 3/2$ channel three vibrational levels lower than in the $\Omega = 1/2$ channel. Equally important is the observation that for a given v and $J\,\tau^{\text{nrad}}(A\,^2\Pi_{3/2}, v, J)$ is several orders of magnitude greater than $\tau^{\text{nrad}}(A\,^2\Pi_{1/2}, v, J)$.

In the two-state approximation Ω is a good quantum number so that mixing of $\Omega = 1/2$ and $\Omega = 3/2$ is rigorously forbidden. The four-state approximation relaxes this segregation. For the $A\,^2\Pi_{1/2}$ resonances even a small mixing of the $A\,^2\Pi_{3/2}$ and $A\,^2\Pi_{1/2}$ states will be important due to the efficient predissociation associated with the $2\,^2\Pi_{3/2}$ continuum. The S-uncoupling interaction does not directly couple the $A\,^2\Pi_{1/2}$ state with the $2\,^2\Pi_{3/2}$ continuum. But there is *indirect* coupling. The $(A\,^2\Pi_{1/2}, v)$ levels are mixed with the $(A\,^2\Pi_{3/2}, v)$ states which are in turn strongly predissociated by the $2\,^2\Pi_{3/2}$ continuum. As a result of this indirect predissociation,

$A\,^2\Pi_{1/2} \sim A\,^2\Pi_{3/2} \sim 2\,^2\Pi_{3/2}$ predissociation starts to compete with the radiative decay for the $(A\,^2\Pi_{1/2},\ v = 2,\ J = 23.5)$ which is more than 2000 cm^{-1} below $E^{d}_{1^2\Pi_{1/2}}(Q_{dr}(^2\Pi_{1/2}))$.

14.4.2.2.4 Comparison with Experiment

The results within the four-state approximation provide a clear picture of the experimental observations. The $J = 23.5$ data are relevant to the low J head (Q_2, P_1) branches while the $J = 63.5$ data are relevant to the high J head (P_2, P_{12}) branches. Note that for the higher J value the mixing of the $A\,^2\Pi_{3/2}$ and $A\,^2\Pi_{1/2}$ states is increased. Experimentally Q_2 and P_2 heads for the $(A\,^2\Pi_{3/2},\ v = 0, 1)$ levels were recorded with the fluorescence from the $(A\,^2\Pi_{3/2},\ v = 1)$ level being weak indicating that this level is partly predissociated. No fluorescence is recorded from the $(A\,^2\Pi_{3/2},\ v > 1)$ levels. This is consistent with the data in Table 14.2 that radiationless decay for the $(A\,^2\Pi_{3/2},\ v,\ J)$ levels becomes preeminent for $v \geq 2$. The P_{12} (high J) and P_1 (low J) heads were observed for the $(A\,^2\Pi_{1/2},\ v = 0, 1)$ levels, while $(A\,^2\Pi_{1/2},\ v = 2)$ level exhibits fluorescence signal only in the P_1 (low J) branch and it is very weak. This is again consistent with the predictions of the four state model. Note that had the two state model been used the experimental findings could not have been understood as solely resulting from the competition of radiationless and radiative decay and would have, *erroneously*, suggested, dynamical bottlenecks in the reaction used to prepare the MgBr.

14.5 Conclusions

The computational tools to describe nonadiabatic processes in diatomic molecules, for which the relativistic effects can be treated perturbatively, are currently available. In addition to the usual techniques for determining potential energy curves, efficient algorithms exist for the determination of derivative couplings, and of spin-orbit interactions within the Breit–Pauli approximation. Diabatic bases are readily determined from the adiabatic states using either a rigorous approach based on the derivative couplings or approximate techniques. The diabatic representation provides computational advantages for the solution of the nuclear Schrödinger which can be achieved to good accuracy. Although not considered explicitly in this work the effects of angular momentum coupling are readily included [Singer and Freed (1985), Singer *et al.* (1987), and Parlant and Yarkony (1999)]. These tools bring considerable predictive power to the field of nonadiabatic processes in diatomic molecules. Equally if not more important are the invaluable mechanistic insights provided only by computational approaches. In this regard we observed that while the diabatic basis is usually superior for numerical studies the adiabatic basis may provide conceptual advantages.

On the other hand nonadiabatic radiationless decay rates frequently turn out to be extremely sensitive to the locations of the curve crossings or avoided crossings [Han *et al.* (1995)]. Here theory and experiment can interact synergistically, with experimental data used to validate a computational treatment which can then be extended beyond the range of the experimental data.

An important area for further research is the extension of current capabilities to the realm of heavy atom diatomics for which a four component description of the electronic wave function is desirable.

Acknowledgments

The preparation of this work and the calculations reported herein were made possible by funds provided by the Air Force Office of Scientific Research, the Department of Energy and the National Science Foundation.

14.6 References

Abe, H., and Kolb, D. M., 1983, *Ber. Bunsenges. Phys. Chem.* **87**, 523–527.

Baer, M., 1975, *Chem. Phys. Lett.* **35**, 112–118.

Basch, H., Stevens, W. J., and Krauss, M., 1984, *Chem. Phys. Lett.* **109**, 212.

Baylis, W. E., and Drake, G. D., 1996, Units and constants in *Atomic, Molecular and Optical Physics Handbook*, Drake, G. L., Ed., AIP: New York, pp. 1–5.

Berry, M. V., 1984, *Proc. R. Soc. London, Ser. A* **392**, 45–57.

Berry, R. S., 1960, *J. Chem. Phys.* **32**, 933–938.

Born, M., and Huang, K., 1954, *Dynamical Theory of Crystal Lattices*; Oxford University Press: Oxford.

Cai, M. F., Carter, C. C., Miller, T. A., and Bondybey, V. E., 1991, *Chem. Phys.* **155**, 233–245.

Cai, M. F., Dzugan, T. P., and Bondybey, V. E., 1989, *Chem. Phys. Lett* **155**, 430–436.

Cederbaum, L. S., Köppel, H., and Domke, W., 1981, *Int. J. Quant. Chem. Symp* **15**, 251–267.

Domcke, W., and Stock, G., 1997, Theory of ultrafast excited state nonadiabatic processes and their spectroscopic detection in real time, *Adv. Chem. Phys.*, Vol. 100, Prigogine, I., and Rice, S. A., Eds; Wiley: New York, pp. 1–168.

Douglas, M. A., Hauge, R. H., and Margrave, J. L., 1983, *J. Phys. Chem.* **87**, 2945–2947.

Feshbach, H., 1958, *Ann. Phys.(NY)* **5**, 357–390.

Feshbach, H., 1962, *Ann. Phys.(NY)* **19**, 287–313.

Fethi, F., 1995, Ph. D. Thesis, Université de Paris-Sud.

Fu, Z., Lemire, G. W., Bishea, G. A., and Morse, M. D., 1990, *J. Chem. Phys.* **93**, 8420–8441.

Han, S., Hettema, H., and Yarkony, D. R., 1995, *J. Chem. Phys.* **102**, 1955–1964.

Han, S., and Yarkony, D., 1995, *J. Chem. Phys.* **103**, 7336–7346.

Han, S., and Yarkony, D., 1996, *Mol. Phys.* **88**, 53–68.

Hobey, W. D., and McLachlan, A. D., 1960, *J. Chem. Phys.* **33**, 1695–1703.

Hwang, J.-T. and Rabitz, H., 1979, *J. Chem. Phys.* **70**, 4609–4621.

Kendrick, B., and Mead, C. A., 1995, *J. Chem. Phys.* **102.**, 4160–4168.

Kendrick, B., and Pack, R. T., 1996, *J. Chem. Phys.* **104**, 7502–7514.

Langhoff, S. R., and Bauschlicher, C. W., 1990, *J. Chem. Phys.* **92**, 1879–1886.

Lefebvre-Brion, H., and Field, R. W., 1986, *Perturbations in the Spectra of Diatomic Molecules*; Academic Press: New York.

Lengsfield, B. H., Saxe, P., and Yarkony, D. R., 1984, *J. Chem. Phys.* **81**, 4549–4553.

Longuet-Higgins, H. C., 1961, *Adv. Spectrosc.* **2**, 429–472.

Manaa, M. R., and Yarkony, D. R., 1994, *J. Chem. Phys.* **100**, 8204–8211.

Mead, C. A., 1983, *J. Chem. Phys.* **78**, 807–814.

Mead, C. A., 1992, *Rev. Mod. Phys.* **64**, 51–85.

Mead, C. A., and Truhlar, D. G., 1979, *J. Chem. Phys.* **70**, 2284–2296.

Mead, C. A., and Truhlar, D. G., 1982, *J. Chem. Phys.* **77**, 6090–6098.

Morgan, F., 1936, *Phys. Rev.* **50**, 503–506.

Parlant, G., Rostas, J., Taieb, G., and Yarkony, D. R., 1990, *J. Chem. Phys.* **93**, 6403–6418.

Parlant, G., and Yarkony, D. R., 1999, *J. Chem. Phys.* **110**, 363–376.

Rice, O. K., 1929a, *Phys. Rev.* **34**, 1451–1462.

Rice, O. K., 1929b, *Phys. Rev.* **33**, 748–759.

Rostas, J., Shafizadeh, N., Taieb, G., Bourguigon, B., and Prisant, M. G., 1990, *Chem. Phys.* **142**, 97–109.

Ruedenberg, K., and Atchity, G. J., 1993, *J. Chem. Phys.* **99**, 3799–3803.

Sadygov, R. G., Rostas, J., Taieb, G., and Yarkony, D. R., 1997, *J. Chem. Phys.* **106**, 4091–4101.

Sadygov, R. G., and Yarkony, D. R., 1999, *J. Chem. Phys.* **110**, 3639–3642.

Saxe, P., Lengsfield, B. H., and Yarkony, D. R., 1985, *Chem. Phys. Lett.* **113**, 159–164.

Shafizadeh, N., Rostas, J., Taieb, G., Bourguigon, B., and Prisant, M. G., 1990, *Chem. Phys.* **142**, 111–122.

Shapere, A., and Wilczek, F., Eds, 1989, *Geometric Phases in Physics*; World Scientific: Singapore.

Sidis, V., 1992, Diabatic potential energy surfaces for charge-transfer processes, in *State-Selected and State-to-State Ion–Molecule Reaction Dynamics Part 2, Theory*, Vol. 82, Baer, M., and Ng, C.-Y., Eds; Wiley: New York, pp. 73–134.

Singer, S. J., and Freed, K. F., 1985, *Adv. Chem. Phys.* **61**, 1–113.

Singer, S. J., Lee, S., Freed, K. F., and Band, Y. B., 1987, *J. Chem. Phys.* **87**, 4762–4778.

Smith, F. T., 1969, *Phys. Rev.* **179**, 111–123.

Sohlberg, K., and Yarkony, D. R., 1997a, *J. Phys. Chem. A* **101**, 3166–3173.

Sohlberg, K., and Yarkony, D. R., 1997b, *J. Chem. Phys.* **107**, 7690–7694.

Sohlberg, K., and Yarkony, D. R., 1997c, *J. Phys. Chem.* **101**, 9520–9524.

Teller, E., 1937, *J. Phys. Chem.* **41**, 109–116.

Tinkham, M., 1964, *Group Theory and Quantum Mechanics*; McGraw-Hill: New York.

Upton, T. H., 1986, *J. Phys. Chem.* **90**, 754–759.

von Neumann, J., and Wigner, E., 1929, *Physik. Z.* **30**, 467–470.

Werner, H. J., and Meyer, W., 1981, *J. Chem. Phys.* **74**, 5802–5807.
Yarkony, D. R., 1988, *J. Chem. Phys.* **89**, 7324–7333.
Yarkony, D. R., 1992, *Int. Rev. Phys. Chem.* **11**, 195–242.
Yarkony, D. R., 1996a, *Rev. Mod. Phys.* **68**, 985–1013.
Yarkony, D. R., 1996b, Molecular structure, in *Atomic Molecular and Optical Physics Handbook*, Drake, G. L., Ed., AIP: New York, pp. 357–377.
Yarkony, D. R., 1997, *J. Phys. Chem. A* **101**, 4263–4270.
Yarkony, D. R., 1998, *J. Phys. Chem. A* **102**, 8073–8077.
Yarkony, D. R., 1999, *J. Chem. Phys.* **110**, 701–705.
Zare, R. N., 1988, *Angular Momentum*; Wiley: New York.

15 THE RENNER EFFECT

Per Jensen[†], Gerald Osmann[‡], and Philip R. Bunker[‡]
[†]*Bergische Universität, Wuppertal, Germany*
[‡]*National Research Council of Canada, Ottawa, Canada*

15.1 Introduction

In the Born–Oppenheimer approximation (Chapter 1) the rovibronic energies and wavefunctions of a molecule in a particular electronic state are calculated in two steps. First the electronic Schrödinger equation is solved *ab initio* for the electronic state under investigation at many molecular geometries, and second the rotation–vibration energies are determined from the rotation–vibration Schrödinger equation. The rotation–vibration Schrödinger equation involves the Born–Oppenheimer potential energy surface determined in the *ab initio* calculation. In principle there are infinitely many Born–Oppenheimer potential energy surfaces, corresponding to the infinitely many electronic states of the molecule, but in the Born–Oppenheimer approximation we consider only one of these surfaces when we solve the rotation–vibration Schrödinger equation. The Born–Oppenheimer approximation will usually fail if the potential energy surface of the electronic state under investigation is not well separated in energy (relative to rotation–vibration energy separations) from all other potential energy surfaces at all accessible molecular geometries.

Computational Molecular Spectroscopy. Edited by Per Jensen and Philip R. Bunker

If two or more potential energy surfaces are close in energy at one or more accessible molecular geometries then the rovibronic energies associated with them usually cannot be accurately obtained by considering only one surface at a time. In this case a more complicated calculation of the rovibronic energies, involving all the near-degenerate potential energy surfaces, is necessary.

The near-degeneracy of electronic potential energy surfaces can be accidental, but at certain (symmetrical) molecular geometries degeneracies are often necessary. In this chapter we consider chain molecules that can vibrate through a linear configuration. For such molecules, it can happen that a pair of Born–Oppenheimer potential energy surfaces have to be degenerate at that linear geometry. In these circumstances the interactions between the rovibronic states associated with the two surfaces are normally significant. Such electronic states are said to exhibit the *Renner effect* since Renner (1934) was the first to consider the effect of this particular type of degeneracy at a linear geometry. Degeneracies forced by symmetry at nonlinear molecular geometries give rise to the Jahn–Teller effect which is discussed elsewhere in this volume.

Figure 15.1 shows bending potentials of the Renner-degenerate $\tilde{X}\,^2A''$ and $\tilde{A}\,^2A'$ electronic states of the hydroperoxyl radical HOO, calculated *ab initio* by Gu, Hirsch, and Buenker (1998). The figure is produced by fixing the bond lengths at their ground ($\tilde{X}$) state equilibrium values and drawing the variation of the Born–Oppenheimer potential energy with $\bar{\rho} = \pi - \angle(\mathrm{HOO})$, the

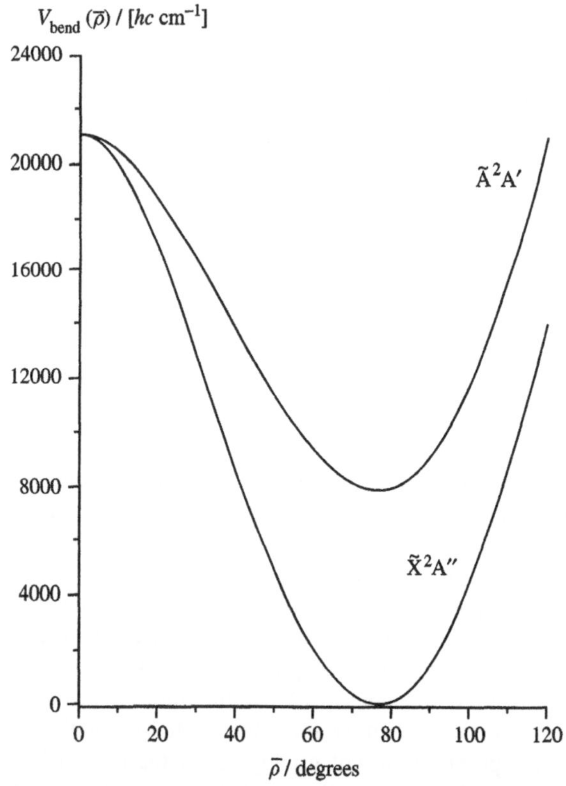

Figure 15.1. Bending potential curves for the lowest electronic states of the hydroperoxyl radical HOO [from Gu, Hirsch and Buenker (1998)]. The abscissa is the bending coordinate $\bar{\rho} = \pi - \angle(\mathrm{HOO})$, the supplement of the bond angle. The bond lengths are held constant.

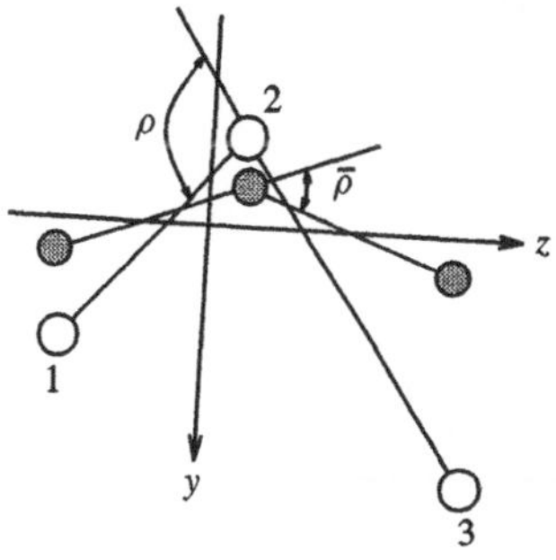

Figure 15.2. The labeling of the nuclei, the definitions of the angles $\overline{\rho}$ and ρ, and the molecule fixed axis system used in the Hougen–Bunker–Johns Hamiltonian for a triatomic molecule. The open circles are the positions of the nuclei in the reference configuration; the shaded circles are their instantaneous positions. The angle $\overline{\rho}$ is the instantaneous value of the supplement of the bond angle and ρ is the corresponding angle in the reference configuration. The (x, y, z) axes form a right-handed axis system where the x axis is perpendicular to the molecular plane.

supplement of the bond angle (see Figure 15.2[1]). At the linear H–O–O configuration, for $\overline{\rho} = 0$, the two potential energy surfaces are exactly degenerate, but at bent geometries, when $\overline{\rho} > 0$, they split into two distinct energies. This qualitative picture is found for any values of the bond lengths; the two surfaces are degenerate at all linear configurations.

Renner degeneracy in a chain molecule is imposed by symmetry, and in order to explain it we must make a brief discourse on molecular symmetry. The vast majority of theoretical work on the Renner effect has been carried out for triatomic molecules, and we limit the discussion of symmetry to this class of molecules. The use of molecular symmetry for simplifying the theoretical description of the electronic structure and the rotation and vibration of molecules is the subject of Bunker and Jensen (1998), to which we refer the reader for an in-depth discussion. The molecular symmetry group (or MS group) is the appropriate symmetry group for labeling the rovibronic (i.e., rotation–vibration–electronic) wavefunctions of a molecule. For a symmetrical ABA triatomic molecule (such as CH_2) the MS group is $\boldsymbol{C}_{2v}$(M) = $\{E, (13), E^*, (13)^*\}$, and for an unsymmetrical triatomic molecule (such as HOO) the MS group is $\boldsymbol{C}_s$(M) = $\{E, E^*\}$. Here, E is the identity operation, (13) is the interchange of the identical nuclei 1 and 3 (see Figure 15.2), and E^* the spatial inversion operation (see Chapter 2 of Bunker and Jensen (1998)). The group $\boldsymbol{C}_{2v}$(M) is isomorphic to the point group $\boldsymbol{C}_{2v} = \{E, C_{2y}, \sigma_{yz}, \sigma_{xy}\}$, and $\boldsymbol{C}_s$(M) is isomorphic to the point group $\boldsymbol{C}_s = \{E, \sigma_{yz}\}$. The point group operation C_{2y} is a rotation of 180° about the molecule fixed y axis (Figure 15.2), whereas σ_{yz} and σ_{xy} are reflections in the yz and xy planes, respectively. In Tables 15.1 and 15.2 we give the character tables of the groups $\boldsymbol{C}_{2v}$(M) and $\boldsymbol{C}_s$(M), respectively. In these tables, we indicate the isomorphisms with the groups $\boldsymbol{C}_{2v}$ and $\boldsymbol{C}_s$, and we give the symmetries of $(\hat{N}_x, \hat{N}_y, \hat{N}_z)$ and $(\hat{L}_x, \hat{L}_y, \hat{L}_z)$. The operator $\hat{N}_\alpha$, $\alpha = x, y, z$, is the component of the rovibronic orbital angular momentum (i.e., the angular momentum resulting from the orbital motion of the nuclei and electrons in the molecule) along the molecule fixed α axis defined in Figure 15.2. For an electronic singlet state $\hat{N}_\alpha = \hat{J}_\alpha$. The operator $\hat{L}_\alpha$ is the analogous component of the electronic orbital angular momentum which results from the orbital motion of the electrons only.

Bunker and Jensen (1998) discuss in their Section 4.5 how the electronic wavefunctions of a molecule can usefully be symmetry labeled in its point group. For bent geometries of HOO the appropriate point group is $\boldsymbol{C}_s$, and for linear geometries it is $\boldsymbol{C}_{\infty v}$, whose character table is given in Table A-19 of Bunker and Jensen (1998). The electronic wavefunctions corresponding to the $\tilde{X}\,^2A''$ and $\tilde{A}\,^2A'$ electronic states of HOO (Figure 15.1) have A″ and A′ symmetry, respectively, in $\boldsymbol{C}_s$ at bent HOO geometries. However, for linear geometries these two wavefunctions span the doubly degenerate irreducible representation Π of $\boldsymbol{C}_{\infty v}$ and, in consequence, the electronic

[1] Together with $\overline{\rho}$, Figure 15.2 also defines the angle ρ and the molecule fixed axis system (x, y, z) used in the Hougen–Bunker–Johns (HBJ) approach [Hougen, Bunker and Johns (1970)] for treating the rovibrational motion of a triatomic molecule. We discuss the HBJ approach below.

Table 15.1. The character table of the group $C_{2v}(M)$.

$C_{2v}(M)$:	E 1	(13) 1	E^* 1	$(13)^*$ 1	
C_{2v}:	E	C_{2y}	σ_{yz}	σ_{xy}	
A_1:	1	1	1	1	:
A_2:	1	1	−1	−1	: $\hat{N}_y, \hat{L}_y$
B_1:	1	−1	−1	1	: $\hat{N}_z, \hat{L}_z$
B_2:	1	−1	1	−1	: $\hat{N}_x, \hat{L}_z$

Table 15.2. The character table of the group $C_s(M)$.

$C_s(M)$:	E 1	E^* 1	
C_s:	E	σ_{yz}	
A′:	1	1	: $\hat{N}_x, \hat{L}_x$
A″:	1	−1	: $\hat{N}_y, \hat{N}_z, \hat{L}_y, \hat{L}_z$

energies of the two states are equal. At linear geometries, HOO (or any other molecule with $C_{\infty v}$ point group symmetry) will have doubly degenerate electronic wavefunctions spanning the irreducible representations Π, Δ, Φ, ... of $C_{\infty v}$. At bent geometries, the appropriate point group is C_s which has the two non-degenerate irreducible representations A′ and A″ given in Table 15.2. Electronic energies that are doubly degenerate at linearity are not required by symmetry to stay doubly degenerate at bent geometries, so they will split into two distinct energies as shown for HOO in Figure 15.1. That is, they will exhibit the Renner effect. From the correlation table of $C_{\infty v} \rightarrow C_s$ [given in Table B-3 of Bunker and Jensen (1998)] we determine that for the symmetries of two electronic states of the bent molecule to correlate with a doubly degenerate symmetry (Π, Δ, Φ, ...) in the linear configuration, these two states must necessarily transform according to the reducible representations $A' \oplus A''$ in C_s. For triatomic molecules with $D_{\infty h}$ symmetry [see Table A-20 of Bunker and Jensen (1998)] at linearity (such as CH_2 with equal bond lengths), obviously electronic states belonging to the doubly degenerate representations Π_g, Π_u, Δ_g, Δ_u, Φ_g, Φ_u, ... of $D_{\infty h}$ exhibit the Renner effect. From the correlation table of $D_{\infty h} \rightarrow C_{2v}$ [Table B-3 of Bunker and Jensen (1998)] we determine that Renner-degenerate electronic states must transform according to $A_1 \oplus B_1$ or $A_2 \oplus B_2$ in C_{2v}. That is, if we denote the point group symmetries of the two electronic states of the bent molecule as Γ_{elec} and Γ'_{elec}, respectively, we have $\Gamma_{elec} \otimes \Gamma'_{elec} = A''$ for C_s molecules and B_1 for C_{2v} molecules. The symmetry of a product

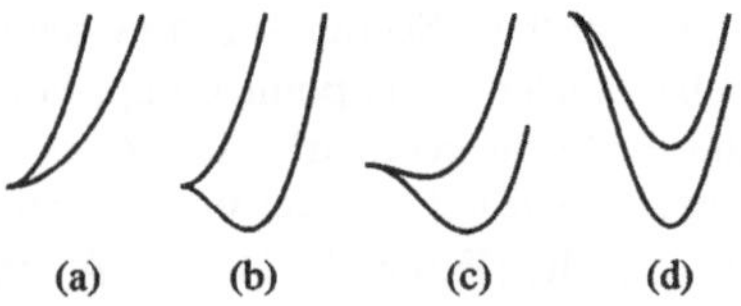

Figure 15.3. Bending potential curves of triatomic molecules. For each of the four diagrams, the abscissa is the bending coordinate $\bar{\rho} = \pi - \angle(ABC)$, the supplement of the bond angle for the triatomic molecule ABC. The bond lengths are held constant. The left hand side of each diagram corresponds to a linear configuration ($\bar{\rho} = 0$). This shows four ways that an electronic degeneracy at the linear nuclear configuration can be resolved when the molecule is bent.

$\Gamma_{\rm elec} \otimes \Gamma'_{\rm elec}$ is defined in Section 6.4 of Bunker and Jensen (1998). Clearly, a double electronic degeneracy at linearity can be resolved in several ways as the molecule bends; this is sketched in Figure 15.3. In the original paper of Renner (1934) the splitting as indicated in Figure 15.3(a) was the only case considered. However, the first observation of the Renner effect [Dressler and Ramsay (1959)] turned out to be the situation depicted in Figure 15.3(c).

15.2 The Hamiltonian

Theoretical work on the Renner effect for triatomic molecules can be broadly divided into two classes: those that use perturbation theory to develop an effective Hamiltonian, and those that use variational theory to diagonalize a rovibronic Hamiltonian that includes the electronic angular momentum. The works of Renner (1934), Pople and Longuet-Higgins [see Pople and Longuet-Higgins (1958), and Pople (1960)], and Brown and co-workers [see Brown (1977), and Brown and Jørgensen (1983)] are examples of the former type, which is treated by Brown in a separate chapter of this book. They constitute what might be considered the classical way of developing a Hamiltonian for the purpose of fitting to observed molecular spectra. The present chapter is concerned with the use of variational calculations involving the diagonalization of the Hamiltonian matrix for the rotation and bending motion of a triatomic molecule in a given bending potential; such work has been undertaken by the schools of Dixon and Duxbury [Barrow, Dixon and Duxbury (1974), Duxbury and Dixon (1981)], Jungen and Merer (1976, 1980)] and Buenker and Peyerimhoff [see Perić, Peyerimhoff and Buenker (1983)]. More recently, variational calculations have been carried out by Handy, Carter and co-workers [Carter and Handy (1984), Carter *et al.* (1990), and Green *et al.* (1991)], by Duxbury, Jungen, and co-workers [Duxbury and Jungen (1988), Duxbury *et al.* (1998a, 1998b)] and by Jensen, Bunker, Kraemer, and co-workers [Jensen *et al.* (1995a); Kolbuszewski *et al.* (1996), and Osmann *et al.* (1997)].

As an example, we outline here the approach by Jensen, Bunker, Kraemer, and co-workers to the theoretical treatment of the Renner interaction between two electronic states of a triatomic molecule. This approach is based on the theoretical model of Hougen, Bunker and Johns (1970), henceforth called HBJ [see also Chapter 15 of Bunker and Jensen (1998)], in the form developed by Jensen (1988a, 1988b, 1988c, 1992, 1994), and on the ideas of Barrow, Dixon and Duxbury (1974). In the HBJ approach, the coordinates for the motion of the nuclei are defined in terms of the molecular reference configuration shown in Figure 15.2. In the reference configuration, the bond angle α varies with time and its supplement, $\rho = \pi - \alpha$, is used as one of the vibrational coordinates, whereas the bond lengths are fixed at the values $r_{12}^{\rm (ref)}$ and $r_{32}^{\rm (ref)}$. Here, $r_{j2}^{\rm (ref)}$ is the reference value of the distance between the 'terminal' nucleus $j = 1$ or 3 and the 'central' nucleus 2; the instantaneous value of this length is called r_{j2}, and the displacement $\Delta r_{j2}^{\rm (ref)} = r_{j2} - r_{j2}^{\rm (ref)}$.

Bunker and Moss (1980) have expressed the translation-free HBJ Hamiltonian $\hat{H}_{\rm rve}$ for nuclear and electronic motion in terms of the following coordinates

- the bending coordinate ρ,
- the stretching coordinates S_1 and S_3 where, for a symmetrical triatomic molecule ABA

$$\begin{aligned} S_1 &= \left(\Delta r_{12}^{\rm (ref)} + \Delta r_{32}^{\rm (ref)}\right)/\sqrt{2}, \\ S_3 &= \left(\Delta r_{12}^{\rm (ref)} - \Delta r_{32}^{\rm (ref)}\right)/\sqrt{2}, \end{aligned} \tag{1}$$

  and for an unsymmetrical triatomic molecule $S_j = \Delta r_{j2}^{\rm (ref)}$, $j = 1$ or 3.
- Euler angles θ, ϕ, and χ which describe the orientation of the molecule fixed (x, y, z) axis system (Figure 15.2) relative to a space fixed axis system [see Chapter 10 of Bunker and Jensen (1998)], and

- the Cartesian electron coordinates (x_k, y_k, z_k), where the index k runs over all electrons in the molecule, in the (x, y, z) axis system.

We have [equations (1)–(4) of Bunker and Moss (1980)]

$$\hat{H}_{\rm rve} = \hat{H}_{\rm e} + \hat{T}'_{\rm e} + \hat{T}_{\rm N}, \tag{2}$$

where

$$\hat{H}_{\rm e} = \frac{1}{2m_{\rm e}} \sum_k \boldsymbol{p}_k^2 + V, \tag{3}$$

$$\hat{T}'_{\rm e} = \frac{1}{2M_{\rm N}} \boldsymbol{P}^2, \tag{4}$$

and

$$\begin{aligned}\hat{T}_{\rm N} = {} & \frac{1}{2}\mu^{1/4} \sum_{j,k=1,3} \hat{P}_j \mu^{-1/2} G_{jk}(S_1, S_3, \rho) \hat{P}_k \mu^{1/4} \\ & + \frac{1}{2}\mu^{1/4} \sum_{\delta,\gamma=x,y,z,\rho} (\hat{N}_\delta - \hat{p}_\delta - \hat{L}_\delta) \mu^{-1/2} \mu_{\delta\gamma}(S_1, S_3, \rho) \\ & \times (\hat{N}_\gamma - \hat{p}_\gamma - \hat{L}_\gamma) \mu^{1/4}. \end{aligned} \tag{5}$$

In $\hat{H}_{\rm rve}$, $m_{\rm e}$ is the electron mass, $\boldsymbol{p}_k$ is an electron momentum, $M_{\rm N}$ is the total nuclear mass, $\boldsymbol{P} = \sum_k \boldsymbol{p}_k$, $\hat{N}_\alpha$ and $\hat{L}_\alpha$ $(\alpha = x, y, z)$ are the α-components of the rovibronic angular momentum and the electronic angular momentum, respectively, as defined above, $\hat{N}_\rho = -i\hbar\partial/\partial\rho$, $\hat{L}_\rho = 0$, $\hat{p}_\alpha$ $(\alpha = x, y, z)$ is a component of the vibrational angular momentum, $\hat{p}_\rho$ can be obtained from the third term of equation (17) in HBJ, V is the electrostatic potential energy function, $\hat{P}_j = -i\hbar\partial/\partial S_j$, and μ is the determinant of the 6×6 matrix having an upper 4×4 block of $\mu_{\delta\gamma}$ (the inverse of $\boldsymbol{I}$ in equation (19) of HBJ) and a lower 2×2 block of G_{jk} (the $\boldsymbol{G}$ matrix).

For isolated electronic states, the $\hat{L}_\alpha$ terms in equation (5) are usually neglected, or their effects are described by perturbation theory as done by Bunker and Moss (1980). This is permissible because, for an electronic state well separated in energy from other electronic states, these terms will contribute little to the rovibronic energies. For Renner-degenerate electronic states, however, some of them are important and must be considered at an early stage of the rovibronic calculation. For simplicity, we rewrite equation (2) as

$$\hat{H}_{\rm rve} = \hat{H}_{\rm e} + \hat{T}'_{\rm e} + \hat{T}_{\rm N}^{\rm (isol)} \tag{6}$$

$$+ \frac{1}{2} \sum_{\delta,\gamma=x,y,z,\rho} \mu_{\delta\gamma} \left(\hat{L}_\delta \hat{L}_\gamma - \hat{N}_\delta \hat{L}_\gamma - \hat{N}_\gamma \hat{L}_\delta\right) \tag{7}$$

$$+ \frac{1}{2} \sum_{\delta,\gamma=x,y,z,\rho} \left(\mu^{1/4} \hat{p}_\delta \mu^{-1/4} \mu_{\delta\gamma} \hat{L}_\gamma + \mu^{-1/4} \mu_{\delta\gamma} \hat{p}_\gamma \mu^{1/4} \hat{L}_\delta\right). \tag{8}$$

In equation (6), $\hat{T}_{\rm N}^{\rm (isol)}$ is the nuclear kinetic energy operator obtained by neglecting all $\hat{L}_\alpha$ terms. In equations (7) and (8) we have used the fact that $\hat{N}_\delta$ depends on the Euler angles θ, ϕ, and χ only, and $\hat{L}_\gamma$ depends on the electronic coordinates (x_k, y_k, z_k) only, so that both of these operators commute with μ and $\mu_{\delta\gamma}$.

We often study Renner-degenerate electronic states with $S > 0$ [i.e., non-singlet states]. For such states, we must add to $\hat{H}_{\rm rve}$ in equation (2) a term $\hat{H}_{\rm SO}$ to account for spin–orbit interaction.

We usually take $\hat{H}_{SO}$ to be given by the 'phenomenological' expression in equation (13-225) of Bunker and Jensen (1998):

$$\hat{H}_{SO} = \frac{hc}{\hbar^2} A_{SO} \hat{\boldsymbol{L}} \cdot \hat{\boldsymbol{S}}, \tag{9}$$

where A_{SO} is a function (given in units of cm^{-1}) depending on the vibrational coordinates of the molecule and $\hat{\boldsymbol{L}}$ and $\hat{\boldsymbol{S}}$ are the total electron angular momentum and total electron spin, respectively. Jensen *et al.* (1995a) consider only the contribution from $\hat{L}_z \hat{S}_z$ to $\hat{\boldsymbol{L}} \cdot \hat{\boldsymbol{S}}$ and parameterize A_{SO} as

$$A_{SO}\left(\Delta r_{12}^{(\mathrm{ref})}, \Delta r_{32}^{(\mathrm{ref})}, \overline{\rho}\right) = \sum_{i=0}^{4} \mathcal{A}_0^{(i)} (1 - \cos \overline{\rho})^i + \sum_j \sum_{i=0}^{3} \mathcal{A}_j^{(i)} (1 - \cos \overline{\rho})^i \Delta r_{j2}^{(\mathrm{ref})}$$
$$+ \sum_{j \le k} \sum_{i=0}^{2} \mathcal{A}_{jk}^{(i)} (1 - \cos \overline{\rho})^i \, \Delta r_{j2}^{(\mathrm{ref})} \, \Delta r_{k2}^{(\mathrm{ref})}, \tag{10}$$

where the indices j and k assume the values 1 or 3, and the $\mathcal{A}_{j\ldots}^{(i)}$ are expansion coefficients.

15.3 Electronic matrix elements

The initial step in obtaining the rovibronic energies for two Renner-degenerate electronic states is normally the *ab initio* solution of the electronic (orbital) Schrödinger equation

$$\hat{H}_e \psi_e(\mathbf{R}_n, \mathbf{r}_e) = V^{(\mathrm{BO})}(\mathbf{R}_n) \psi_e(\mathbf{R}_n, \mathbf{r}_e) \tag{11}$$

at many fixed values of the nuclear coordinates $\mathbf{R}_n$. For triatomic molecules, we normally use the geometrically defined coordinates $\mathbf{R}_n = (r_{12}, r_{32}, \overline{\rho})$ [see Figure 15.2 for the definition of $\overline{\rho}$] at this stage; in the calculation of the rovibronic energies we convert to the HBJ coordinates (S_1, S_3, ρ). In equation (11), $\mathbf{r}_e$ is a shorthand notation for all the electronic coordinates, $\psi_e(\mathbf{R}_n, \mathbf{r}_e)$ is the electronic wavefunction depending parametrically on $\mathbf{R}_n$, and $V^{(\mathrm{BO})}(\mathbf{R}_n)$ is the effective (Born–Oppenheimer) nuclear potential energy function.

The Renner effect in a triatomic molecule occurs when two solutions of the electronic Schrödinger equation (11) are degenerate at linearity. We denote the Born–Oppenheimer potential energy functions for these two solutions as $V_-^{(\mathrm{BO})}(r_{12}, r_{32}, \overline{\rho})$ and $V_+^{(\mathrm{BO})}(r_{12}, r_{32}, \overline{\rho})$, respectively, where $V_-^{(\mathrm{BO})}$ is the lower electronic state of the pair and where the associated electronic wavefunctions are denoted $\psi_e^{(-)}(r_{12}, r_{32}, \overline{\rho}, \mathbf{r}_e)$ and $\psi_e^{(+)}(r_{12}, r_{32}, \overline{\rho}, \mathbf{r}_e)$. At linear geometries

$$V_-^{(\mathrm{BO})}(r_{12}, r_{32}, \overline{\rho} = 0) = V_+^{(\mathrm{BO})}(r_{12}, r_{32}, \overline{\rho} = 0) \tag{12}$$

for all values of r_{12} and r_{32}, whereas at bent geometries ($\overline{\rho} > 0$) the two potential energy surfaces are split apart.

In the calculation of the rovibronic energies for the Renner-degenerate electronic states, we take the rovibronic wavefunctions to have the form

$$\psi_{\mathrm{rve}} = \psi_{\mathrm{vib-rot}}^{(-)}(\theta, \phi, \chi, S_1, S_3, \rho)\, \psi_e^{(-)}(S_1, S_3, \rho, \mathbf{r}_e)$$
$$+ \psi_{\mathrm{vib-rot}}^{(+)}(\theta, \phi, \chi, S_1, S_3, \rho)\, \psi_e^{(+)}(S_1, S_3, \rho, \mathbf{r}_e), \tag{13}$$

where we have transformed the electronic wavefunctions $\psi_e^{(\pm)}$ to depend on the HBJ nuclear coordinates (S_1, S_3, ρ). We calculate the wavefunctions ψ_{evr} and the corresponding eigenvalues E_{rve} by expanding the rotation–vibration factor functions $\psi_{\mathrm{vib-rot}}^{(-)}$ and $\psi_{\mathrm{vib-rot}}^{(+)}$ in equation (13) as linear combinations of suitable rotation–vibration basis functions to be defined below. In this manner, we can obtain the rovibronic energies and corresponding eigenfunctions in a variational calculation, i.e., by diagonalizing a truncated matrix representation of the rovibronic Hamiltonian $\hat{H}_{\text{rve}}$ [equations (6)–(8)] in this basis. Electronic states other than the pair that become degenerate at linearity are neglected.

Clearly, in order to compute the rovibronic matrix elements required for the variational calculation, we need certain integrals over the electronic coordinates $\mathbf{r}_e$. In the rovibronic Hamiltonian given by equations (6)–(8), the terms $\hat{H}_e$ and $\hat{T}'_e$ in equation (6), and the factors of $\hat{L}_\delta$ ($\delta = x, y, z$) in equations (7) and (8) depend on the electronic coordinates. It is simple to obtain the electronic matrix elements of $\hat{H}_e$ since the electronic wavefunctions $\psi^{(-)}_{\text{vib-rot}}$ and $\psi^{(+)}_{\text{vib-rot}}$ are eigenfunctions for it [equation (11)]. That is,

$$\langle \psi_e^{(\sigma)} | \hat{H}_e | \psi_e^{(\sigma)} \rangle_{\text{el}} = V_\sigma^{(\text{BO})}(r_{12}, r_{32}, \overline{\rho}), \tag{14}$$

and

$$\langle \psi_e^{(-)} | \hat{H}_e | \psi_e^{(+)} \rangle_{\text{el}} = 0, \tag{15}$$

where σ = '−' or '+', and the subscript 'el' indicates integration over the electronic coordinates only.

We represent the Born–Oppenheimer potentials $V_\pm^{(\text{BO})}(r_{12}, r_{32}, \overline{\rho})$ as parameterized functions of $(r_{12}, r_{32}, \overline{\rho})$; the values of the parameters are determined by least-squares fitting to the *ab initio* values of $V_\pm^{(\text{BO})}$. Jensen *et al.* (1995a) use the following parameterized functions:

$$\begin{aligned} V_\pm^{(\text{BO})}(r_{12}, r_{32}, \overline{\rho}) &= V_0^{(\pm)}(\overline{\rho}) + \sum_j F_j^{(\pm)}(\overline{\rho}) y_j^{(\text{ref})} + \sum_{j\le k} F_{jk}^{(\pm)}(\overline{\rho}) y_j^{(\text{ref})} y_k^{(\text{ref})} \\ &+ \sum_{j\le k\le m} F_{jkm}^{(\pm)}(\overline{\rho}) y_j^{(\text{ref})} y_k^{(\text{ref})} y_m^{(\text{ref})} \\ &+ \sum_{j\le k\le m\le n} F_{jkmn}^{(\pm)}(\overline{\rho}) y_j^{(\text{ref})} y_k^{(\text{ref})} y_m^{(\text{ref})} y_n^{(\text{ref})}, \end{aligned} \tag{16}$$

where all of the indices j, k, m and n assume the values 1 or 3,

$$y_j^{(\text{ref})} = 1 - \exp\left(-a_j \, \Delta r_{j2}^{(\text{ref})}\right), \tag{17}$$

and the a_j are molecular parameters. The expansion coefficients $F_{jk\ldots}^{(\pm)}(\overline{\rho})$ entering into equation (16) depend on $\overline{\rho}$, and they are chosen to be general cosine expansions:

$$F_{jk\ldots}^{(\pm)}(\overline{\rho}) = f_{jk\ldots}^{(0)} + \sum_{i=1}^{N} f_{jk\ldots}^{(i,\pm)} (1 - \cos\overline{\rho})^i, \tag{18}$$

where the $f_{jk\ldots}^{(0)}$ and $f_{jk\ldots}^{(i,\pm)}$ are expansion coefficients. Note that the parameters $f_{jk\ldots}^{(0)}$ (and the parameters a_j) are common for the two potential energy surfaces; this ensures that the two functions are degenerate at linear configurations where $\overline{\rho} = 0$. The function $F_j^{(\pm)}(\overline{\rho})$ has $N = 4$, $F_{jk}^{(\pm)}(\overline{\rho})$ has $N = 3$, $F_{jkl}^{(\pm)}(\overline{\rho})$ has $N = 2$, and $F_{jklm}^{(\pm)}(\overline{\rho})$ has $N = 1$. Symmetry relations exist between the potential energy parameters for the symmetrical ABA molecule to ensure that the potential energy functions are totally symmetric under the interchange of r_{12} and r_{32}. Finally, $V_0^{(\pm)}(\overline{\rho})$ is parameterized as

$$V_0^{(\pm)}(\overline{\rho}) = \sum_{i=1}^{8} f_0^{(i,\pm)} (1 - \cos\overline{\rho})^i \tag{19}$$

where the $f_0^{(i,\pm)}$ are expansion coefficients.

The term $\hat{T}'_e$ [equation (4)] has a very small contribution to the electronic energy since $M_N \gg m_e$; it is usually neglected and we do not consider its matrix elements here. However, in order to derive the rovibronic matrix elements of the Hamiltonian terms in equations (7) and (8) we require expressions for the electronic matrix elements $\langle\psi_e^{(\sigma)}|\hat{L}_\delta\hat{L}_\gamma|\psi_e^{(\sigma')}\rangle_{el}$ and $\langle\psi_e^{(\sigma)}|\hat{L}_\delta|\psi_e^{(\sigma')}\rangle_{el}$, where σ, σ' = '−' or '+', and $\delta, \gamma = x, y, z$. These electronic integrals are initially expressed as functions of the nuclear coordinates r_{12}, r_{32}, and $\overline{\rho}$; in a later stage of the calculation we convert them to functions of S_1, S_3, and ρ. The coordinates r_{12}, r_{32}, and $\overline{\rho}$ all have positive parity in that they are invariant under the spatial inversion operation E^*. Since any function of these coordinates has positive parity, only those electronic integrals $\langle\psi_e^{(\sigma)}|\hat{L}_\delta\hat{L}_\gamma|\psi_e^{(\sigma')}\rangle_{el}$ and $\langle\psi_e^{(\sigma)}|\hat{L}_\delta|\psi_e^{(\sigma')}\rangle_{el}$ are nonvanishing whose integrand has positive parity. We have discussed above that $\Gamma_{elec} \otimes \Gamma'_{elec} = A''$ for C_s molecules and B_1 for C_{2v} molecules, and it follows that the two electronic wavefunctions $\psi_e^{(-)}$ and $\psi_e^{(+)}$ have opposite parities. From these results, and from the symmetries of the L-components given in Tables 15.1 and 15.2, it would appear that the nonvanishing electronic integrals diagonal in electronic state are $\langle\psi_e^{(\sigma)}|\hat{L}_\delta^2|\psi_e^{(\sigma)}\rangle_{el}$, $\langle\psi_e^{(\sigma)}|\hat{L}_y\hat{L}_z|\psi_e^{(\sigma)}\rangle_{el}$, and $\langle\psi_e^{(\sigma)}|\hat{L}_x|\psi_e^{(\sigma)}\rangle_{el}$, where σ = '−' or '+', and $\delta = x, y, z$. However, we can show that $\langle\psi_e^{(\sigma)}|\hat{L}_x|\psi_e^{(\sigma)}\rangle_{el} = 0$. This result follows from the fact that this integral can be viewed as the expectation value of a Hermitian operator, and as such it must be real. We can choose the electronic wavefunctions $\psi_e^{(\pm)}$ to be real, and in this event it follows from the definition of $\hat{L}_x$ that the integrand of $\langle\psi_e^{(\sigma)}|\hat{L}_x|\psi_e^{(\sigma)}\rangle_{el}$ is purely imaginary. The only resolution of this paradox is for the integral to vanish. The nonvanishing electronic integrals off-diagonal in electronic state are $\langle\psi_e^{(-)}|\hat{L}_x\hat{L}_y|\psi_e^{(+)}\rangle_{el}$, $\langle\psi_e^{(-)}|\hat{L}_x\hat{L}_z|\psi_e^{(+)}\rangle_{el}$, $\langle\psi_e^{(-)}|\hat{L}_y|\psi_e^{(+)}\rangle_{el}$, and $\langle\psi_e^{(-)}|\hat{L}_z|\psi_e^{(+)}\rangle_{el}$.

In principle, we could now express all the nonvanishing electronic integrals as parameterized functions of r_{12}, r_{32}, and $\overline{\rho}$. In practice, however, it is customary to neglect many of these integrals. To appreciate how we select the neglected integrals we consider the factors $\mu_{\delta\gamma}$ entering into equations (7) and (8). These factors (with $\delta, \gamma = x, y, z, \rho$) can be viewed as the elements of a 4×4 matrix $\boldsymbol{\mu}$, and Jensen *et al.* (1995a) expand this matrix as

$$\boldsymbol{\mu} = \boldsymbol{\mu}^{(\mathrm{ref})}(\rho) + \sum_j \boldsymbol{\mu}_j(\rho) y_j^{(\mathrm{ref})} + \sum_{j\le k} \boldsymbol{\mu}_{jk}(\rho) y_j^{(\mathrm{ref})} y_k^{(\mathrm{ref})}$$
$$+ \sum_{j\le k\le m} \boldsymbol{\mu}_{jkm}(\rho) y_j^{(\mathrm{ref})} y_k^{(\mathrm{ref})} y_m^{(\mathrm{ref})} + \sum_{j\le k\le m\le n} \boldsymbol{\mu}_{jkmn}(\rho) y_j^{(\mathrm{ref})} y_k^{(\mathrm{ref})} y_m^{(\mathrm{ref})} y_n^{(\mathrm{ref})}, \quad (20)$$

where $\boldsymbol{\mu}^{(\mathrm{ref})}(\rho)$ is a 4×4 matrix with elements $\mu_{\delta\gamma}^{(\mathrm{ref})}(\rho)$, and $\boldsymbol{\mu}_{jk\ldots}(\rho)$ is a 4×4 matrix with elements $\mu_{\delta\gamma}^{(jk\ldots)}(\rho)$, $\delta, \gamma = x, y, z, \rho$. For a molecule of $\boldsymbol{C}_s$(M) symmetry, the nonvanishing $\boldsymbol{\mu}^{(\mathrm{ref})}(\rho)$ elements are $\mu_{xx}^{(\mathrm{ref})}(\rho)$, $\mu_{yy}^{(\mathrm{ref})}(\rho)$, $\mu_{zz}^{(\mathrm{ref})}(\rho)$, $\mu_{\rho\rho}^{(\mathrm{ref})}(\rho)$, and $\mu_{yz}^{(\mathrm{ref})}(\rho)$, whereas for a $\boldsymbol{C}_{2v}$(M) molecule, only $\mu_{xx}^{(\mathrm{ref})}(\rho)$, $\mu_{yy}^{(\mathrm{ref})}(\rho)$, $\mu_{zz}^{(\mathrm{ref})}(\rho)$, and $\mu_{\rho\rho}^{(\mathrm{ref})}(\rho)$ are nonvanishing. In Figures 15.4 and 15.5 we plot the nonvanishing $\boldsymbol{\mu}^{(\mathrm{ref})}(\rho)$ elements for HOO; the bond lengths are fixed at the values $r_{12}^{(\mathrm{ref})} = r(\mathrm{HO}) = 0.95$ Å and $r_{32}^{(\mathrm{ref})} = r(\mathrm{OO}) = 1.34$ Å. We note that $\mu_{xx}^{(\mathrm{ref})}(\rho)$, $\mu_{yy}^{(\mathrm{ref})}(\rho)$, and $\mu_{yz}^{(\mathrm{ref})}(\rho)$, are almost an order of magnitude smaller than $\mu_{zz}^{(\mathrm{ref})}(\rho)$ and $\mu_{\rho\rho}^{(\mathrm{ref})}(\rho)$. We have $\mu_{\delta\gamma} \approx \mu_{\delta\gamma}^{(\mathrm{ref})}(\rho)$ since the displacements from the reference configuration caused by the stretching vibrations are small. Thus we expect that μ_{xx}, μ_{yy}, and μ_{yz} are also typically an order of magnitude smaller than μ_{zz} and $\mu_{\rho\rho}$. We have considered HOO as an example here but the order-of-magnitude considerations are valid for all triatomic molecules. Also, for all triatomic molecules $\mu_{zz}^{(\mathrm{ref})}(\rho)$ (and hence μ_{zz}) is singular at linear configurations ($\rho = 0$). It can be shown that for ρ near 0 this function is given by

$$\mu_{zz}^{(\mathrm{ref})}(\rho) \approx \frac{\mu_{\rho\rho}^{(\mathrm{ref})}(\rho = 0)}{\rho^2}, \quad (21)$$

where $\mu_{\rho\rho}^{(\mathrm{ref})}(\rho = 0)$ is a positive constant (see Figure 15.4).

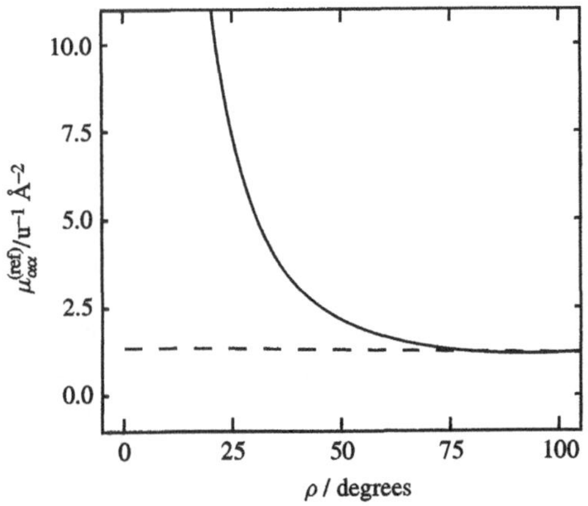

Figure 15.4. The $\boldsymbol{\mu}^{(\mathrm{ref})}(\rho)$-elements $\mu_{zz}^{(\mathrm{ref})}(\rho)$ (solid curve) and $\mu_{\rho\rho}^{(\mathrm{ref})}(\rho)$ (dashed curve) for HOO.

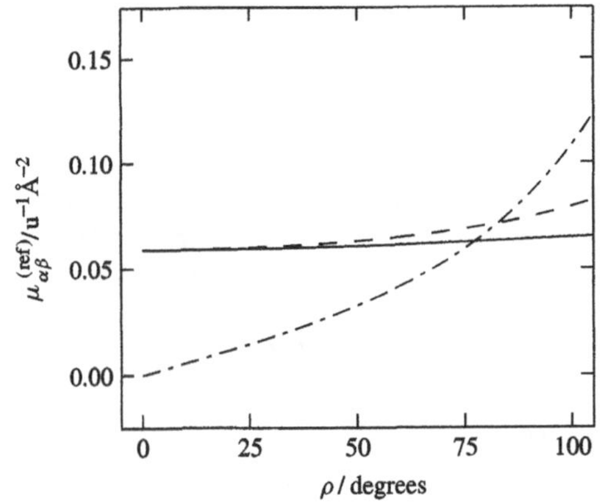

Figure 15.5. The $\boldsymbol{\mu}^{(\mathrm{ref})}(\rho)$-elements $\mu_{xx}^{(\mathrm{ref})}(\rho)$ (solid curve), $\mu_{yy}^{(\mathrm{ref})}(\rho)$ (dashed curve), and $\mu_{yz}^{(\mathrm{ref})}(\rho)$ (dot-dash curve) for HOO.

In theoretical treatments of the Renner effect, it is customary to neglect the terms in equations (7) and (8) associated with the small $\boldsymbol{\mu}^{(\mathrm{ref})}(\rho)$ elements $\mu_{xx}^{(\mathrm{ref})}(\rho)$, $\mu_{yy}^{(\mathrm{ref})}(\rho)$, and $\mu_{yz}^{(\mathrm{ref})}(\rho)$. Since $\hat{L}_\rho = 0$, and $\hat{p}_z$ vanishes by symmetry,[2], only the terms

$$\frac{1}{2}\mu_{zz}(\hat{L}_z^2 - 2\hat{N}_z\hat{L}_z) \tag{22}$$

are considered.[3] The rovibronic matrix elements of these two terms depend on the electronic integrals $\langle\psi_e^{(-)}|\,\hat{L}_z^2|\psi_e^{(-)}\rangle_{\mathrm{el}}$, $\langle\psi_e^{(+)}|\hat{L}_z^2|\psi_e^{(+)}\rangle_{\mathrm{el}}$, and $\langle\psi_e^{(-)}|\hat{L}_z|\psi_e^{(+)}\rangle_{\mathrm{el}}$, and we shall discuss in detail the parameterization of these $(r_{12}, r_{32}, \overline{\rho})$-dependent functions.

---

[2] Formally, $\hat{p}_z$ has the same symmetry as $\hat{N}_z$. This symmetry is B_1 for C_{2v}(M) molecules and A″ for C_s(M) molecules and so $\hat{p}_z$ has negative parity. However, it depends solely on the nuclear coordinates which all have positive parity. This paradox can only be resolved for $\hat{p}_z = 0$.

[3] Note, however, that in the work of Jensen *et al.* (1995a), Kolbuszewski *et al.* (1996), and Osmann *et al.* (1997), the term $-\mu_{yz}\hat{N}_y\hat{L}_z$ is also taken into account.

It is well known that at linear geometries, the electronic Hamiltonian $\hat{H}_e$ commutes with $\hat{L}_z$ so that the eigenfunctions of $\hat{H}_e$ [i.e., the electronic wavefunctions] can be chosen as simultaneous eigenfunctions of $\hat{H}_e$ and $\hat{L}_z$. The eigenvalues of $\hat{L}_z$ are $\pm\hbar\Lambda$, where Λ is a nonnegative integer depending on the symmetry of the electronic wavefunctions in $\boldsymbol{C}_{\infty v}$ or $\boldsymbol{D}_{\infty h}$ as appropriate. We have $\Lambda = 0$ for Σ states [which, of course, do not exhibit the Renner effect], $\Lambda = 1$ for Π states, $\Lambda = 2$ for Δ states, $\Lambda = 3$ for Φ states, and so on. We define an electronic coordinate ζ to be conjugate to $\hat{L}_z$:

$$\hat{L}_z = -i\hbar\frac{\partial}{\partial\zeta}. \tag{23}$$

The eigenfunctions of $\hat{L}_z$ can be written as

$$\psi_e^{(\pm\Lambda)} = \psi_\Lambda \frac{1}{\sqrt{2\pi}} \exp(\pm i\Lambda\zeta), \tag{24}$$

where ψ_Λ depends on the nuclear coordinates and on all electronic coordinates other than ζ. At linearity, these functions are also eigenfunctions of $\hat{H}_e$. However, the particular electronic eigenfunctions $\psi_e^{(-)}$ and $\psi_e^{(+)}$ obtained in an *ab initio* calculation are normally chosen to be real, and therefore they must be identified with the linear combinations

$$\psi_e^{(-)} = \frac{1}{i\sqrt{2}}\left(\psi_e^{(+\Lambda)} - \psi_e^{(-\Lambda)}\right) = \psi_\Lambda \frac{1}{\sqrt{\pi}} \sin(\Lambda\zeta) \tag{25}$$

and

$$\psi_e^{(+)} = \frac{1}{\sqrt{2}}\left(\psi_e^{(+\Lambda)} + \psi_e^{(-\Lambda)}\right) = \psi_\Lambda \frac{1}{\sqrt{\pi}} \cos(\Lambda\zeta). \tag{26}$$

Perić, Peyerimhoff, and Buenker (1983) refer to $\psi_e^{(\pm)}$ as *adiabatic* electronic basis functions, and to $\psi_e^{(\pm\Lambda)}$ as *diabatic* electronic basis functions. In equations (25) and (26), we have arbitrarily assigned $\psi_e^{(-)}$ to the sine function and $\psi_e^{(+)}$ to the cosine function. The assignment could be the other way around. Assuming equations (25) and (26) to be valid we straightforwardly derive that at linear geometries,

$$\langle\psi_e^{(-)}|\hat{L}_z^2|\psi_e^{(-)}\rangle_{el} = \langle\psi_e^{(+)}|\hat{L}_z^2|\psi_e^{(+)}\rangle_{el} = \hbar^2\Lambda^2 \tag{27}$$

and

$$\langle\psi_e^{(-)}|\hat{L}_z|\psi_e^{(+)}\rangle_{el} = i\hbar\Lambda \tag{28}$$

In general, we can always construct electronic wavefunctions $\psi_e^{(-)}$ and $\psi_e^{(+)}$ which satisfy equations (27) and (28). In the event that we must interchange sine and cosine in equations (25) and (26), we can simultenously change the sign [i.e., the phase factor] of $\psi_e^{(-)}$, say, and equations (27) and (28) will still hold.

At nonlinear geometries, the values of $\langle\psi_e^{(-)}|\hat{L}_z^2|\psi_e^{(-)}\rangle_{el}$, $\langle\psi_e^{(+)}|\hat{L}_z^2|\psi_e^{(+)}\rangle_{el}$, and $\langle\psi_e^{(-)}|\hat{L}_z|\psi_e^{(+)}\rangle_{el}$ will differ from those given in equations (27) and (28). Jensen *et al.* (1995a) parameterize these functions as

$$\langle\psi_e^{(\sigma)}|\hat{L}_z^2|\psi_e^{(\sigma)}\rangle_{el} = \hbar^2\left[\Lambda^2 - \sum_{i=1}^{4}\mathcal{L}^{(i;\sigma\sigma)}(1-\cos\bar{\rho})^i\right], \tag{29}$$

$\sigma = -$ or $+$, and

$$\langle\psi_e^{(-)}|\hat{L}_z|\psi_e^{(+)}\rangle_{el} = i\hbar\left[\Lambda - \sum_{i=1}^{4}\mathcal{L}^{(i;-+)}(1-\cos\bar{\rho})^i\right], \tag{30}$$

where the $\mathcal{L}^{(i;\sigma\sigma')}$ are expansion coefficients. These authors neglect the dependence of these matrix elements on the stretching coordinates.

The term $\frac{1}{2}\mu_{zz}\hat{L}_z^2$ in equation (22) gives rise to a diagonal (adiabatic) correction to the Born–Oppenheimer potential energy surfaces. The adiabatic potential energy surfaces are given by

$$V_\sigma^{(\mathrm{ad})} = V_\sigma^{(\mathrm{BO})} + \frac{1}{2}\mu_{zz}\langle\psi_\mathrm{e}^{(\sigma)}|\hat{L}_z^2|\psi_\mathrm{e}^{(\sigma)}\rangle_\mathrm{el}, \tag{31}$$

where $\sigma = +$ or $-$. The adiabatic correction is extremely important since it is singular at linearity (see Figures 1–4 in Kraemer, Jensen and Bunker (1994)).

15.4 The basis functions

We determine the rovibronic energies and wavefunctions in a variational calculation. That is, we diagonalize a matrix representation of the total Hamiltonian

$$\hat{H}_\mathrm{Renner} = \hat{H}_\mathrm{rve} + \hat{H}_\mathrm{SO} \tag{32}$$

in a suitable set of basis functions. There are obviously infinitely many possible choices for these basis functions so, as an example, we outline here the choice made by Jensen *et al.* (1995a). In equation (32), $\hat{H}_\mathrm{rve}$ is given by equations (6)–(8), and $\hat{H}_\mathrm{SO}$ is given by equation (9). The eigenfunctions of $\hat{H}_\mathrm{Renner}$ are the rovibronic wavefunctions given in equation (13), and Jensen *et al.* (1995a) express such a function as

$$|\psi_\mathrm{rve}^{(J,M_J,S,\Gamma_\mathrm{rve})}\rangle = \sum_{\eta=a,b}\sum_{N=|J-S|}^{J+S}\sum_{K=0}^{N}\sum_{v_2^{(\eta)},N_\mathrm{vib}^{(\eta)},\Gamma_\mathrm{vib}^{(\eta)}} c^{(J,M_J,S,\Gamma_\mathrm{rve})}_{\eta,N,K,v_2^{(\eta)},N_\mathrm{vib}^{(\eta)},\Gamma_\mathrm{vib}^{(\eta)}}$$
$$\times\, |v_2^{(\eta)},K\rangle|N_\mathrm{vib}^{(\eta)},\Gamma_\mathrm{vib}^{(\eta)}\rangle|\eta;N,J,S,K,M_J,\tau\rangle, \tag{33}$$

where the $c^{(J,M_J,S,\Gamma_\mathrm{rve})}_{\eta,N,K,v_2^{(\eta)},N_\mathrm{vib}^{(\eta)},\Gamma_\mathrm{vib}^{(\eta)}}$ are expansion coefficients and $|v_2^{(\eta)},K\rangle$, $|N_\mathrm{vib}^{(\eta)},\Gamma_\mathrm{vib}^{(\eta)}\rangle$, and $|\eta;N,J,S,K,M_J,\tau\rangle$ are basis functions. The functions $|v_2^{(\eta)},K\rangle$ describe the bending motion, the functions $|N_\mathrm{vib}^{(\eta)},\Gamma_\mathrm{vib}^{(\eta)}\rangle$ describe the stretching motion, and $|\eta;N,J,S,K,M_J,\tau\rangle$ is an electronic–rotation–spin basis function describing the electronic motion, the rotation of the molecule, and the effects of electron spin. We shall discuss these basis functions in detail below. As indicated in equation (33), J, M_J, S, and Γ_rve are good quantum numbers which we can use to characterize the rovibronic wavefunctions. The quantum number J is associated with the operator $\hat{\boldsymbol{J}}^2$, where the total angular momentum $\hat{\boldsymbol{J}}$ is given by

$$\hat{\boldsymbol{J}} = \hat{\boldsymbol{N}} + \hat{\boldsymbol{S}}, \tag{34}$$

M_J defines the projection of $\hat{\boldsymbol{J}}$ on a space fixed axis, S is associated with the operator $\hat{\boldsymbol{S}}^2$ [see Section 10.3 of Bunker and Jensen (1998)], and Γ_rve is the symmetry of $|\psi_\mathrm{rve}^{(J,M_J,S,\Gamma_\mathrm{rve})}\rangle$ in the MS group. Since all bending basis functions $|v_2^{(\eta)},K\rangle$ are totally symmetric in the MS group (because ρ is totally symmetric), we have for each of the basis functions in equation (33)

$$\Gamma_\mathrm{rve} = \Gamma_\mathrm{vib}^{(\eta)} \otimes \Gamma_\mathrm{ers}, \tag{35}$$

where Γ_ers is the MS group symmetry of the electronic–rotation–spin basis function $|\eta;N,J,S,K,M_J,\tau\rangle$ (Table 15.3). The integer $\tau(=0$ or $1)$ is also a good quantum number. It is defined so that

$$E^*\,|\psi_\mathrm{rve}^{(J,M_J,S,\Gamma_\mathrm{rve})}\rangle = (-1)^\tau\,|\psi_\mathrm{rve}^{(J,M_J,S,\Gamma_\mathrm{rve})}\rangle \tag{36}$$

(i.e., the parity of the rovibronic state is $(-1)^\tau$), so that its value can be derived from Γ_rve.

15.4.1 ROTATION–SPIN BASIS FUNCTIONS

The basis functions $|\eta; N, J, S, K, M_J, \tau\rangle$ in equation (33) are defined in terms of Hund's case (b) basis functions [see, for example, equation (20) of Brown and Howard (1976)] given by

$$|N, J, S, k, M_J\rangle = \sum_{M=-N}^{N} \sum_{M_S=-S}^{S} (-1)^{N-S+M_J} \sqrt{2J+1}$$
$$\times \begin{pmatrix} N & S & J \\ M & M_S & -M_J \end{pmatrix} |S, M_S\rangle |N, k, M\rangle. \quad (37)$$

This function is a simultaneous eigenfunction for $\hat{N}^2$ [with eigenvalue $\hbar^2 N(N+1)$], $\hat{J}^2$ [with eigenvalue $\hbar^2 J(J+1)$], $\hat{S}^2$ [with eigenvalue $\hbar^2 S(S+1)$], $\hat{N}_z$ [with eigenvalue $\hbar k$], and $\hat{J}_Z$, the component of $\hat{J}$ along the space-fixed quantization axis Z [with eigenvalue $\hbar M_J$]. In equation (37), the quantity in parenthesis is a $3j$-symbol [Zare (1988)], $|S, M_S\rangle$ is an electronic spin function quantized along space fixed axes, and $|N, k, M\rangle$ is an eigenfunction for the rigid symmetric rotor [see Chapter 11 of Bunker and Jensen (1998)].

In the work of Jensen *et al.* (1995a), which we outline here, the relative phases of the $|N, k, M\rangle$ functions are chosen so that

$$\hat{N}_{\pm}|N, k, M\rangle = \left[\hat{N}_x \pm i\hat{N}_y\right]|N, k, M\rangle$$
$$= \pm i\hbar\sqrt{N(N+1) - k(k \mp 1)}|N, k \mp 1, M\rangle. \quad (38)$$

This phase choice differs from that of, for example, Bunker and Jensen (1998), who choose the phases to make the matrix elements of $\hat{N}_{\pm}$ real and positive. The $|N, k, M\rangle$ functions used here can be obtained from those used in Bunker and Jensen (1998) by multiplication of these latter functions by a k-dependent phase factor [see equation (50) of Jensen (1988c)]. However, this factor is common to all $|N, k, M\rangle$ functions in equation (37) and so it does not alter the form of this equation.

As discussed by Bunker and Jensen (1998), the electron spin functions $|S, M_S\rangle$ are unaffected by the symmetry operations in the MS group. The transformation properties of the functions $|N, k, M\rangle$ under these operations can be obtained from equations (12-46) and (12-47) of Bunker and Jensen 1998 after allowing for the changes caused by our different phase factors. The elements of the transformation matrices depend on the quantum numbers N [which is called J in Bunker and Jensen (1998)] and k only so that all $|N, k, M\rangle$ functions present in equation (37) have identical transformation properties. The function $|N, J, S, k, M_J\rangle$ has the same transformation properties under a given MS group operation as any one of the functions $|N, k, M\rangle$, $M = -N$, $-N+1, \ldots, N$. Hence, the symmetrized rotation–spin basis functions are given by an expression analogous to equation (7.1) of Jensen (1983):

$$|N, J, S, K, M_J, \tau\rangle = \frac{1}{\sqrt{2}}\left[|N, J, S, K, M_J\rangle + (-1)^{N+K+\tau}|N, J, S, -K, M_J\rangle\right], \quad (39)$$

where $K > 0$ and the integer τ assumes the values 0 or 1. For $K = 0$,

$$|N, J, S, 0, M_J, \tau\rangle = |N, J, S, 0, M_J\rangle \quad (40)$$

and $\tau = N \bmod 2$. The symmetries of the $|N, J, S, K, M_J, \tau\rangle$ functions in the C_{2v}(M) and C_s(M) groups are given in Table 15.3.

There are three equivalent symmetry labeling schemes used for the rovibronic levels of a triatomic molecule exhibiting the Renner effect. One scheme (which we prefer) uses the irreducible representation labels of C_{2v}(M) and C_s(M) as given in Tables 15.1 and 15.2, respectively.

Table 15.3. The symmetry of the rotation–spin basis functions $|N, J, S, K, M_J, \tau\rangle$ in the MS group.

	C_{2v}(M) molecules		
	$K = 0$	K even	K odd
$\tau = 0$	A_1 (N even)	A_1	B_2
$\tau = 1$	B_1 (N odd)	B_1	A_2
	C_s(M) molecules		
	$K = 0$	$K \neq 0$	
$\tau = 0$	A′ (N even)	A′	
$\tau = 1$	A″ (N odd)	A″	

Table 15.4. Correlation between different symmetry labeling schemes.

C_{2v}(M) molecules			C_s(M) molecules		
C_{2v}(M)	($\pm$a/s)	$D_{\infty h}$	C_s(M)	($\pm$)	$C_{\infty v}$
A_1	(+s)	Σ_g^+	A′	(+)	Σ^+
A_2	(−s)	Σ_u^-	A″	(−)	Σ^-
B_1	(−a)	Σ_g^-			
B_2	(+a)	Σ_u^+			

Another scheme uses ($\pm$a/s) labels, and a third scheme uses the non-degenerate irreducible representations of the linear molecule point groups $D_{\infty h}$ and $C_{\infty v}$ (see Chapter 17 of Bunker and Jensen (1998)). Table 15.4 gives the correlation between the three labeling schemes and the character tables for the molecular symmetry groups C_{2v}(M) and C_s(M).

15.4.2 ELECTRONIC–ROTATION–SPIN BASIS FUNCTIONS

In order to obtain rovibronic wavefunctions in the form indicated by equation (13), we use the Born–Oppenheimer electronic (orbital) functions $\psi_e^{(-)}(r_{12}, r_{32}, \overline{\rho}, \mathbf{r}_e)$ and $\psi_e^{(+)}(r_{12}, r_{32}, \overline{\rho}, \mathbf{r}_e)$ as electronic basis functions for constructing the complete rovibronic wavefunction ψ_{rve}. We know that these two wavefunctions form a basis for the representation $A_1 \oplus B_1$ or for $A_2 \oplus B_2$ in C_{2v}, or for $A' \oplus A''$ in C_s. In actual calculations, it is convenient to assume that the electronic wavefunction for the lower component of the Renner pair, $\psi_e^{(-)}$, has A_1 symmetry for C_{2v} molecules and A' symmetry for C_s molecules. Thus, $\psi_e^{(+)}$ has B_1 or A″ symmetry, respectively. With these assumptions, we can determine the MS group symmetry Γ_{rve} of each rovibronic wavefunction ψ_{rve}. In the event that the symmetry of $\psi_e^{(-)}$ is $\Gamma_e^{(-)} \neq A_1$ (or A′), we can straightforwardly determine the actual symmetry of ψ_{rve} by forming the direct product of $\Gamma_e^{(-)}$ and Γ_{rve}.

The 'Renner interaction' terms in equation (22) give rise to large shifts in the unperturbed (Born–Oppenheimer) energies [i.e., the energies resulting from a calculation neglecting the Renner interaction]. It is clearly desirable to find a set of basis functions in which these terms are diagonal. Following the ideas of Barrow, Dixon and Duxbury (1974) [henceforth BDD], we can find such basis functions by obtaining the eigenfunctions for the Hamiltonian

$$\hat{H}_{ez} = \hat{H}_e + \frac{1}{2}\mu_{zz}^{(\mathrm{ref})}(\rho)\left(\hat{N}_z - \hat{L}_z\right)^2 \tag{41}$$

$$= \hat{H}_e + \frac{1}{2}\mu_{zz}^{(\mathrm{ref})}(\rho)\left(\hat{N}_z^2 + \hat{L}_z^2 - 2\hat{N}_z\hat{L}_z\right), \tag{42}$$

where $\mu_{zz}^{(\mathrm{ref})}(\rho)$ is an element of the matrix $\boldsymbol{\mu}^{(\mathrm{ref})}(\rho)$ in equation (20). The Hamiltonian $\hat{H}_{\mathrm{ez}}$ describes the electronic motion and the rotation around the molecule-fixed z axis for a molecule with its bond lengths fixed at the reference values $(r_{12}^{(\mathrm{ref})}, r_{32}^{(\mathrm{ref})})$ and ρ fixed at an arbitrary value. We diagonalize a matrix representation of $\hat{H}_{\mathrm{ez}}$ in the symmetrized electronic–(electron spin)–rotation basis functions $|\psi_{\mathrm{e}}^{(-)}\rangle|N, J, S, K, M_J, \tau\rangle$ and $\mathrm{i}|\psi_{\mathrm{e}}^{(+)}\rangle|N, J, S, K, M_J, \delta_{0\tau}\rangle$. This is done for a grid of ρ values. When $|\psi_{\mathrm{e}}^{(-)}\rangle$ is totally symmetric, these two functions have the same MS group symmetry as $|N, J, S, K, M_J, \tau\rangle$. The Kronecker symbol $\delta_{0\tau}$ is 0 when $\tau = 1$, and 1 when $\tau = 0$. The 2×2 matrix representation of $\hat{H}_{\mathrm{ez}}$ [constructed from equations (14), (15), (29), and (30) where, in equations (29) and (30), we neglect the $\mathcal{L}$-dependent terms at this stage of the calculation] is

$$\begin{Bmatrix} V_0^{(-)}(\rho) + \dfrac{\hbar^2}{2}\mu_{zz}^{(\mathrm{ref})}(\rho)(K^2 + \Lambda^2) & \hbar^2 \mu_{zz}^{(\mathrm{ref})}(\rho) K\Lambda \\ \hbar^2 \mu_{zz}^{(\mathrm{ref})}(\rho) K\Lambda & V_0^{(+)}(\rho) + \dfrac{\hbar^2}{2}\mu_{zz}^{(\mathrm{ref})}(\rho)(K^2 + \Lambda^2) \end{Bmatrix},$$

where the pure bending potential $V_0^{(\sigma)}(\rho)$ is defined in equation (19). This matrix has the eigenfunctions

$$\begin{aligned} |a; N, J, S, K, M_J, \tau\rangle &= \cos[\gamma_K(\rho)]\big|\psi_{\mathrm{e}}^{(-)}\big\rangle|N, J, S, K, M_J, \tau\rangle \\ &\quad + \sin[\gamma_K(\rho)]\mathrm{i}\big|\psi_{\mathrm{e}}^{(+)}\big\rangle|N, J, S, K, M_J, \delta_{0\tau}\rangle \end{aligned} \tag{43}$$

and

$$\begin{aligned} |b; N, J, S, K, M_J, \tau\rangle &= \cos[\gamma_K(\rho)]\mathrm{i}\big|\psi_{\mathrm{e}}^{(+)}\big\rangle|N, J, S, K, M_J, \delta_{0\tau}\rangle \\ &\quad - \sin[\gamma_K(\rho)]\big|\psi_{\mathrm{e}}^{(-)}\big\rangle|N, J, S, K, M_J, \tau\rangle; \end{aligned} \tag{44}$$

these functions enter into equation (33). The corresponding eigenvalues are

$$\begin{aligned} V_\eta^{(K)}(\rho) &= \frac{1}{2}\big(V_0^{(+)}(\rho) + V_0^{(-)}(\rho)\big) + \frac{\hbar^2}{2}\mu_{zz}^{(\mathrm{ref})}(\rho)(K^2 + \Lambda^2) \\ &\quad \mp \frac{1}{2}\sqrt{\big(V_0^{(+)}(\rho) - V_0^{(-)}(\rho)\big)^2 + 4\,\hbar^4 \mu_{zz}^{(\mathrm{ref})}(\rho)^2 K^2 \Lambda^2}, \end{aligned} \tag{45}$$

where the $-$ sign corresponds to $\eta = a$ and the $+$ sign to $\eta = b$. The angle $\gamma_K(\rho)$, which defines the eigenvectors entering into equations (43) and (44), has been discussed already by BDD and by Duxbury and Dixon (1981). The definitions of the functions $|a; N, J, S, K, M_J, \tau\rangle$ and $|b; N, J, S, K, M_J, \tau\rangle$ in equations (43) and (44) require that $-\pi/4 \le \gamma_K(\rho) \le 0$, so that $\cos[\gamma_K(\rho)] > 0$ and $\sin[\gamma_K(\rho)] \le 0$. Obviously for $K = 0$, $\gamma_0(\rho) = 0$ for all ρ. For $K > 0$, $\gamma_K(\rho) \approx -\pi/4$ for ρ near zero, since for such ρ values $V_0^{(-)}(\rho) \approx V_0^{(+)}(\rho)$ and $\mu_{zz}^{(\mathrm{ref})}(\rho)$, being proportional to ρ^{-2}, is large. For ρ near zero we can set $\cos[\gamma_K(\rho)] = -\sin[\gamma_K(\rho)] = 1/\sqrt{2}$, and we can express $|\psi_{\mathrm{e}}^{(-)}\rangle$ and $|\psi_{\mathrm{e}}^{(+)}\rangle$ as given in equations (25) and (26). Some manipulation involving equation (39) shows that in this case

$$\begin{aligned} &|a; N, J, S, K, M_J, \tau\rangle \\ &= -\frac{\mathrm{i}}{\sqrt{2}}\big(\psi_{\mathrm{e}}^{(+\Lambda)}|N, J, S, K, M_J\rangle - (-1)^{N+K+\tau}\psi_{\mathrm{e}}^{(-\Lambda)}|N, J, S, -K, M_J\rangle\big) \end{aligned} \tag{46}$$

and

$$\begin{aligned} &|b; N, J, S, K, M_J, \tau\rangle \\ &= -\frac{\mathrm{i}}{\sqrt{2}}\big(\psi_{\mathrm{e}}^{(-\Lambda)}|N, J, S, K, M_J\rangle - (-1)^{N+K+\tau}\psi_{\mathrm{e}}^{(+\Lambda)}|N, J, S, -K, M_J\rangle\big). \end{aligned} \tag{47}$$

From this result, it is easily shown that in the limit $\rho \to 0$, the functions $|a; N, J, S, K, M_J, \tau\rangle$ and $|b; N, J, S, K, M_J, \tau\rangle$ are simultaneous eigenfunctions for $(\hat{N}_z - \hat{L}_z)^2$ [with eigenvalues $\hbar^2(K - \Lambda)^2$ and $\hbar^2(K + \Lambda)^2$, respectively] and for $\hat{H}_e$ [with the common eigenvalue $V_0^{(-)}(\rho = 0) = V_0^{(+)}(\rho = 0)$]. In this limit, we can say that we describe the Renner interaction in terms of the diabatic electronic basis of Perić, Peyerimhoff and Buenker (1983) [see the discussion after equation (26)].

For large ρ values, where the molecule is well bent, we will normally have $\Delta V_0(\rho) = V_0^{(+)}(\rho) - V_0^{(-)}(\rho) \gg |\hbar^2 \mu_{zz}^{(\mathrm{ref})}(\rho) K \Lambda|$, leading to $\gamma_K(\rho) \approx 0$. In this limit, the functions $|a; N, J, S, K, M_J, \tau\rangle$ and $|b; N, J, S, K, M_J, \tau\rangle$ are approximate eigenfunctions of $\hat{H}_e$ with eigenvalues $V_0^{(-)}(\rho)$ and $V_0^{(+)}(\rho)$, respectively, and we are using the adiabatic electronic basis of Perić, Peyerimhoff and Buenker (1983).

In Figures 15.6–15.9 we show $\gamma_K(\rho)$ functions for the four molecules CuH_2, CH_2^+, CH_2, and HOO. These molecules are examples of the four ways, shown in Figure 15.3, that an electronic degeneracy at the linear nuclear configuration can be resolved when the molecule is bent. The value of $\gamma_K(\rho)$ indicates the extent of the breakdown of the Born–Oppenheimer approximation: It has the value −45° when there is a complete breakdown of the approximation (i.e., when there is a 50–50 mixture of the two Born–Oppenheimer basis states) and the value 0° when there is no mixing of the basis states. The extent of the breakdown of the Born–Oppenheimer

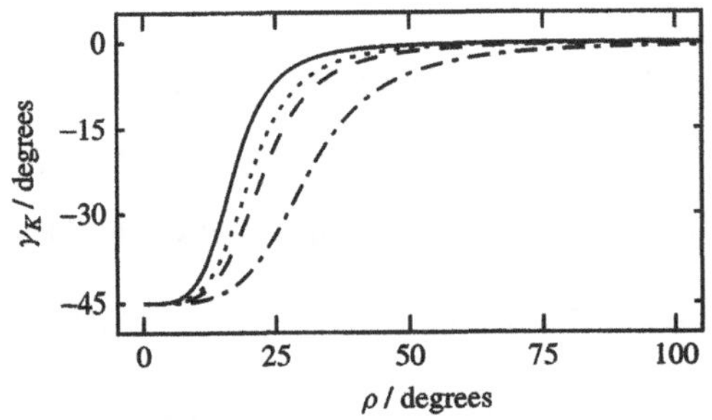

Figure 15.6. The functions $\gamma_K(\rho)$ for the $^2A_2$ and $^2B_2$ electronic states of CuH_2 treated by Lee *et al.* (1984) for $K = 1$ (solid curve), $K = 2$ (dotted curve), $K = 3$ (dashed curve), and $K = 10$ (dot-dash curve). The two electronic states correlate with a $^2\Pi_g$ state at linearity. The bending potential energy curves $V_0^{(\pm)}(\rho)$ of these electronic states are as sketched in Figure 15.3(a).

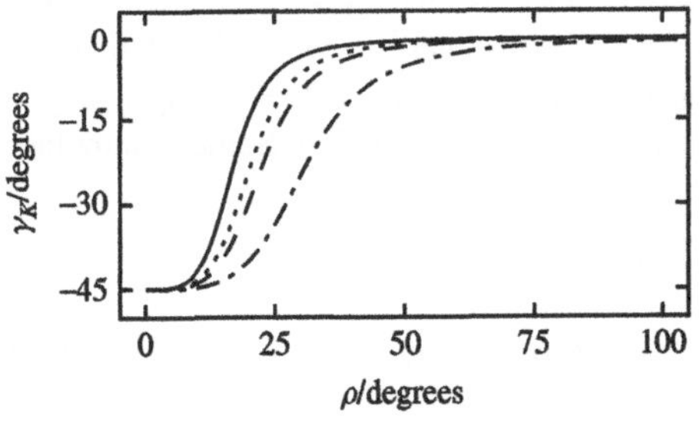

Figure 15.7. The functions $\gamma_K(\rho)$ for the $\tilde{X}\,^2A_1$ and $\tilde{A}\,^2B_1$ electronic states of the molecular ion CH_2^+ for $K = 1$ (solid curve), $K = 2$ (dotted curve), $K = 3$ (dashed curve), and $K = 10$ (dot-dash curve). The two electronic states correlate with a $^2\Pi_u$ state at linearity. The bending potential energy curves $V_0^{(\pm)}(\rho)$ of these electronic states are as sketched in Figure 15.3(b).

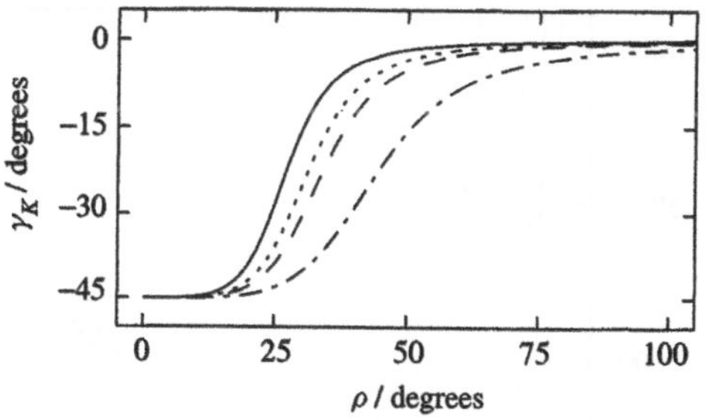

Figure 15.8. The functions $\gamma_K(\rho)$ for the $\tilde{a}\,^1A_1$ and $\tilde{b}\,^1B_1$ electronic states of the methylene radical CH_2 for $K = 1$ (solid curve), $K = 2$ (dotted curve), $K = 3$ (dashed curve), and $K = 10$ (dot-dash curve). The two electronic states correlate with a $^1\Delta_g$ state at linearity. The bending potential energy curves $V_0^{(\pm)}(\rho)$ of these electronic states are as sketched in Figure 15.3(c).

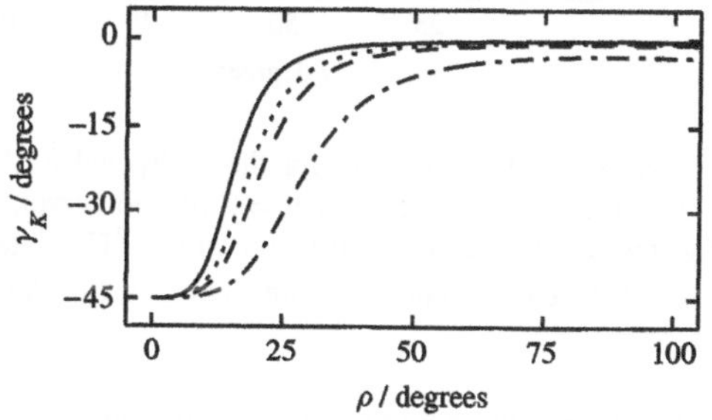

Figure 15.9. The functions $\gamma_K(\rho)$ for the $\tilde{X}\,^2A''$ and $\tilde{A}\,^2A'$ electronic states of HOO for $K = 1$ (solid curve), $K = 2$ (dotted curve), $K = 3$ (dashed curve), and $K = 10$ (dot-dash curve). The two electronic states correlate with a $^2\Pi$ state at linearity. The bending potential energy curves $V_0^{(\pm)}(\rho)$ of these electronic states are as given in Figure 15.1.

approximation depends on the ratio $|\hbar^2\mu_{zz}^{(\mathrm{ref})}(\rho)K\Lambda|/\Delta V_0(\rho)$. This ratio reflects the competition between the Coriolis coupling term $-\mu_{zz}^{(\mathrm{ref})}(\rho)\hat{N}_z\hat{L}_z$, and the Born–Oppenheimer electronic energy separation $\Delta V_0(\rho)$. The former term (i.e., $-\mu_{zz}^{(\mathrm{ref})}(\rho)\hat{N}_z\hat{L}_z$) mixes the Born–Oppenheimer basis states, causes a breakdown of the Born–Oppenheimer approximation, and leads to a $\gamma_K(\rho)$ value significantly less than zero, whereas the latter ($\Delta V_0(\rho)$), when large, prevents significant mixing, upholds the Born–Oppenheimer approximation, and leads to a $\gamma_K(\rho)$ value close to zero. For all triatomic molecules the variation of $\mu_{zz}^{(\mathrm{ref})}(\rho)$ with ρ is qualitatively as given for HOO in Figure 15.4, and it rapidly decreases as ρ increases. Hence, at larger ρ values the Coriolis term, and the breakdown of the Born–Oppenheimer approximation, will become less important. The electronic energy separation $\Delta V_0(\rho)$ in general increases as ρ increases and this will also act in the direction of upholding the Born–Oppenheimer approximation more and more as the molecule is bent. Thus the general trend of $\gamma_K(\rho)$ changing from $-45°$ to $0°$ as ρ increases from zero is to be expected, and it occurs in all the figures. However, it is obvious that as K increases the Coriolis coupling term will become larger; this means that the breakdown of the Born–Oppenheimer approximation will extend to larger ρ values and hence $\gamma_K(\rho)$ will more slowly approach zero as K increases. This also occurs in all the figures. The potential curves of the $\tilde{a}$ and $\tilde{b}$ singlet states of CH_2 are as in Figure 15.3(c) (see Figure 13-7 of Bunker and Jensen (1998)), and for this pair of states the separation $\Delta V_0(\rho)$ remains less than $1000\,\mathrm{cm}^{-1}$ out beyond $\rho = 25°$. Hence

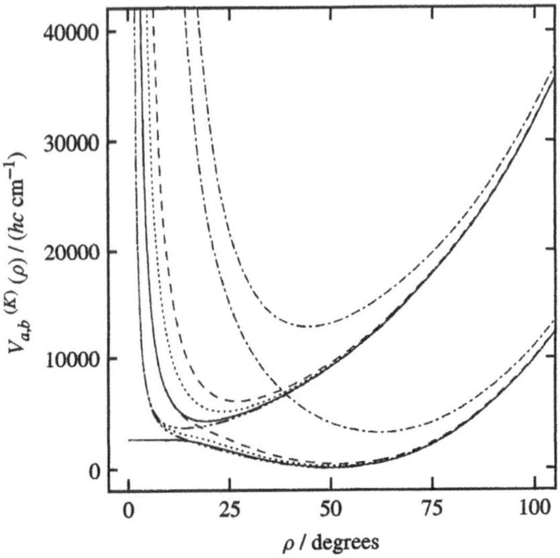

Figure 15.10. The bending potential functions $V_\eta^{(K)}(\rho)$ for the $\tilde{A}\,^2B_1$ and $\tilde{X}\,^2A_1$ electronic states of BH_2 for $K = 0$ (dot-dot-dot-dash curve), $K = 1$ (solid curve), $K = 2$ (dotted curve), $K = 3$ (dashed curve), and $K = 10$ (dot-dash curve). The two electronic states correlate with a $^2\Pi_u$ state at linearity. The bending potential energy curves $V_0^{(\pm)}(\rho)$ of these electronic states are as sketched in Figure 15.3(b).

significant breakdown of the Born–Oppenheimer approximation extends to larger values of ρ than in the other three cases; the $\gamma_K(\rho)$ values for CH_2 reflect this. At very large ρ values for HOO we see from Figure 15.1 that $\Delta V_0(\rho)$ is not diverging so that for $K = 10$ the value of $\gamma_K(\rho)$ is still significantly different from zero even for $\rho = 100°$. However, the value of $\Delta V_0(\rho)$ is obtained by extrapolation of the *ab initio* potential energy curves from their values at smaller bond angles, and this may not be the true situation.

In Figure 15.10 we show the bending potential functions $V_\eta^{(K)}(\rho)$ for BH_2 for several different values of K. For these two states $\Lambda = 1$. We have discussed above that for $\rho \approx 0$, the functions $|a; N, J, S, K, M_J, \tau\rangle$ and $|b; N, J, S, K, M_J, \tau\rangle$ in equations (43) and (44) are eigenfunctions of $(\hat{N}_z - \hat{L}_z)^2$, so in this region, the eigenvalues $V_\eta^{(K)}(\rho)$ of $\hat{H}_{ez}$ [equation (41)] are given by

$$V_\eta^{(K)}(\rho) = \frac{1}{2}\left(V_0^{(-)}(\rho) + V_0^{(+)}(\rho)\right) + \frac{\hbar^2}{2}\mu_{zz}^{(\text{ref})}(\rho)(K \mp \Lambda)^2, \tag{48}$$

where the $-$ (+) sign applies for $\eta = a\,(b)$, and $V_0^{(-)}(\rho) \approx V_0^{(+)}(\rho)$. This result can also be understood by ignoring the first term under the square root in equation (45). For $K = \Lambda = 1$, $V_a^{(1)}(\rho) = (V_0^{(-)}(\rho) + V_0^{(+)}(\rho))/2$ for ρ near zero; this function is nonsingular at $\rho = 0$. For large ρ values, $|a; N, J, S, K, M_J, \tau\rangle$ and $|b; N, J, S, K, M_J, \tau\rangle$ are eigenfunctions of $\hat{H}_e$, and

$$V_\eta^{(K)}(\rho) = V_0^{(\mp)}(\rho) + \frac{\hbar^2}{2}\mu_{zz}^{(\text{ref})}(\rho)\left(K^2 + \Lambda^2\right), \tag{49}$$

where, again, the $-$ (+) sign applies for $\eta = a\,(b)$. This result can be understood by ignoring the second term under the square root in equation (45).

15.4.3 VIBRATIONAL BASIS FUNCTIONS

The bending basis function $|v_2^{(\eta)}, K\rangle$ (which depends on ρ) in equation (33) is an eigenfunction of the bending Hamiltonian

$$\hat{H}_{\text{Bend}}^{(\eta,K)} = \frac{1}{2}\hat{N}_\rho \mu_{\rho\rho}^{(\text{ref})}(\rho)\hat{N}_\rho + \frac{1}{2}\hbar^2 \mu_{\rho\rho}^{(\text{ref})}(\rho)g(\rho)$$
$$+ \frac{1}{2}\left(\mu^{(\text{ref})}\right)^{-1/4}\left[\hat{N}_\rho, \mu_{\rho\rho}^{(\text{ref})}\right]\left[\hat{N}_\rho, \left(\mu^{(\text{ref})}\right)^{1/4}\right] + V_\eta^{(K)}(\rho), \tag{50}$$

i.e., the Hamiltonian for the molecule bending with the bending potential energy $V_\eta^{(K)}(\rho)$ given in equation (45). The bond lengths of the molecule are fixed at the reference values $(r_{12}^{(\text{ref})}, r_{32}^{(\text{ref})})$. The function $g(\rho)$ in equation (50) is given by equation (15-96) of Bunker and Jensen (1998) and $\mu^{(\text{ref})}$ is the determinant of the 4×4 matrix $\boldsymbol{\mu}^{(\text{ref})}(\rho)$. The quantum number $v_2^{(\eta)}$ is the bending quantum number appropriate for a bent molecule, i.e., $v_2^{(\eta)}$ is equal to the number of nodes of the function $|v_2^{(\eta)}, K\rangle$ for ρ between 0 and π. The bending functions are calculated using the Numerov–Cooley numerical integration method as explained in Bunker and Landsberg (1977) and in Section 6 of Jensen (1983). As shown by BDD, we have

$$\left|v_2^{(a)}, K\right\rangle \approx C_a \rho^{|K-\Lambda|+1/2} \text{ and } \left|v_2^{(b)}, K\right\rangle \approx C_b \rho^{|K+\Lambda|+1/2}, \tag{51}$$

where C_a and C_b are constants, for ρ near zero. With this limiting behavior of the bending basis functions, all vibronic integrals required for setting up the matrix representation of the rovibronic Hamiltonian are finite.

We have discussed in connection with equations (46) and (47) how the BDD electronic–rotation–spin basis functions [equations (43) and (44)] effect a smooth change in the electronic basis functions from the diabatic basis [eigenfunctions of $\hat{L}_z$; equation (24)] at $\rho \approx 0$ to the adiabatic basis [eigenfunctions of $\hat{H}_e$; equations (25) and (26)] at large ρ values. The bending basis functions reflect the choice of electronic basis through the term $V_\eta^{(K)}(\rho)$ in equation (50). It is very important for the convergence of the variational calculation producing the rovibronic energies that the bending basis functions are consistent with the diabatic electronic basis at $\rho \approx 0$, i.e., that they vary with ρ as given in equation (51). In the theory developed here, we could introduce bending basis functions associated with the adiabatic electronic basis for all ρ values by setting all $\gamma_K(\rho) = 0$ in the electronic–rotation–spin basis functions of equations (43) and (44). Such bending basis functions are the eigenfunctions of a bending Hamiltonian obtained from that in equation (50) by replacing $V_\eta^{(K)}(\rho)$ by

$$V_\mp^{(K)}(\rho) = V_0^{(\mp)}(\rho) + \frac{\hbar^2}{2}\mu_{zz}^{(\text{ref})}(\rho)\left(K^2 + \Lambda^2\right); \tag{52}$$

they vary as $\rho^{\sqrt{K^2+\Lambda^2}+1/2}$ for $\rho \approx 0$. In an early stage of the work by Jensen *et al.* (1995a), these bending basis functions were used in calculations of the rovibronic energies associated with the $\tilde{a}\,^1A_1$ and $\tilde{b}\,^1B_1$ electronic states of CH_2 from the *ab initio* potential surfaces by Green *et al.* (1991). In these calculations, it proved impossible in practice to obtain converged rovibronic energies with $K > 0$.[4] It was estimated that to obtain an acceptable convergence (on the order of magnitude of 1 cm^{-1}) at least 100 bending basis functions were needed for each K value, and this exceeded the capacity of the computer available.

[4] Since, in the BDD theory, $\gamma_0(\rho) = 0$, there is no difference between the two choices of bending basis set for $K = 0$.

The basis function $|N_{\rm vib}^{(\eta)}, \Gamma_{\rm vib}^{(\eta)}\rangle$ (which depends on $\Delta r_{12}^{(\rm ref)}$ and $\Delta r_{32}^{(\rm ref)}$) of equation (33) is an eigenfunction of the stretching Hamiltonian

$$\hat{H}_{\rm Stretch}^{(\eta)} = \frac{1}{2}\left[\frac{1}{m_1}+\frac{1}{m_2}\right]\hat{\mathcal{P}}_1^2 + \frac{1}{2}\left[\frac{1}{m_3}+\frac{1}{m_2}\right]\hat{\mathcal{P}}_3^2 - \frac{1}{m_2}\cos\rho_e^{(\mp)}\hat{\mathcal{P}}_1\hat{\mathcal{P}}_3$$
$$+\sum_{j\le k} w_{jk}^{(\mp)} y_j^{(\mp)} y_k^{(\mp)} + \sum_{j\le k\le m} w_{jkm}^{(\mp)} y_j^{(\mp)} y_k^{(\mp)} y_m^{(\mp)}$$
$$+\sum_{j\le k\le m\le n} w_{jkmn}^{(\mp)} y_j^{(\mp)} y_k^{(\mp)} y_m^{(\mp)} y_n^{(\mp)}, \qquad (53)$$

where the indices j, k, m, n assume the values 1 and 3, $\hat{\mathcal{P}}_j = -i\hbar\partial/\partial r_{j2}$, m_α is the mass of atom α, and

$$y_j^{(\mp)} = 1 - \exp\left(-a_j\left[r_{j2} - r_{j2}^{(\mp)}\right]\right). \qquad (54)$$

In equations (53) and (54) the quantities are chosen with the (−) superscript to give $\hat{H}_{\rm Stretch}^{(a)}$ and with the (+) superscript to give $\hat{H}_{\rm Stretch}^{(b)}$. The equilibrium bond lengths $r_{12}^{(\sigma)}$ and $r_{32}^{(\sigma)}$, and the equilibrium $\overline{\rho}$ value $\rho_{\rm e}^{(\sigma)}$, are those associated with the Born–Oppenheimer potential energy surface $V_\sigma^{(\rm BO)}$, where σ = '+' or '−'. The parameter a_j is chosen to have the same value as in the $y_j^{(\rm ref)}$ defined in equation (17). The Hamiltonian $\hat{H}_{\rm Stretch}^{(\eta)}$ describes a molecule stretching with its bond angle fixed at the value $\pi - \rho_{\rm e}^{(\mp)}$. The expansion coefficients $w_{jk...}^{(\mp)}$ are determined as described in Section IIa of Jensen *et al.* (1995a). The potential energy functions $V_\mp^{(\rm BO)}(\Delta r_{12}^{(\rm ref)}, \Delta r_{32}^{(\rm ref)}, \overline{\rho})$ can be transformed exactly to expansions in $y_1^{(\mp)}$, $y_3^{(\mp)}$, and $\cos\rho_{\rm e}^{(\mp)} - \cos\overline{\rho}$. In these expansions, we then set $\overline{\rho}$ equal to $\rho_{\rm e}^{(\mp)}$, and the $w_{jk...}^{(\mp)}$ are the coefficients in the resulting expansions in $y_1^{(\mp)}$ and $y_3^{(\mp)}$. The eigenfunctions (and the corresponding eigenvalues, which are our zero-order stretching energies) for $\hat{H}_{\rm Stretch}^{(\eta)}$ are calculated by setting up its matrix representation in a basis of symmetrized Morse oscillator functions as described in Section V of Jensen (1988c). Different Morse oscillator basis functions are used for $\eta = a$ or b, respectively. For $\eta = a$, the Morse oscillator functions are 'centered' at the bond length values $r_{12}^{(-)}$ and $r_{32}^{(-)}$, whereas for $\eta = b$, they are 'centered' at $r_{12}^{(+)}$ and $r_{32}^{(+)}$. The resulting eigenfunctions $|N_{\rm vib}^{(\eta)}, \Gamma_{\rm vib}^{(\eta)}\rangle$ are characterized by $\Gamma_{\rm vib}^{(\eta)}$, their symmetry in the MS group, and by a running index $N_{\rm vib}^{(\eta)}$.

15.5 The calculation of the rovibronic energies

The rovibronic energies can now be calculated as the eigenvalues of the matrix representation of $\hat{H}_{\rm Renner}$ [equation (32)] in the basis of functions $|v_2^{(\eta)}, K\rangle\ |N_{\rm vib}^{(\eta)}, \Gamma_{\rm vib}^{(\eta)}\rangle\ |\eta; N, J, S, K, M_J, \tau\rangle$. The matrix is block diagonal in J and in the MS group symmetry $\Gamma_{\rm rve}$ of the basis functions. The expansion coefficients $c_{\eta,N,K,v_2^{(\eta)},N_{\rm vib}^{(\eta)},\Gamma_{\rm vib}^{(\eta)}}^{(J,M_J,S,\Gamma_{\rm rve})}$ in equation (33) are the eigenvector components of the matrix blocks. We do not discuss here the construction of the matrices; details are given by Jensen *et al.* (1995a).

One might think that with the BDD basis functions of equations (43) and (44), we would obtain matrix representations that are almost diagonal. However, bend–stretch interaction usually produces significant coupling between the basis functions. Also, for the type of bending potential energy functions shown in Figure 15.3(c) [i.e., when both Renner components have relatively small barriers to linearity], certain matrix elements can cause very strong coupling. These matrix elements originate in terms of $\hat{T}_{\rm N}$ [equation (5)] involving the operator $\hat{N}_\rho = -i\hbar\partial/\partial\rho$. In the basis function $|v_2^{(\eta)}, K\rangle\, |N_{\rm vib}^{(\eta)}, \Gamma_{\rm vib}^{(\eta)}\rangle\, |\eta; N, J, S, K, M_J, \tau\rangle$ this operator acts not only on the ρ-dependent factor $|v_2^{(\eta)}, K\rangle$ but also on the coefficients $\cos[\gamma_K(\rho)]$ and $\sin[\gamma_K(\rho)]$ in $|\eta; N, J, S, K, M_J, \tau\rangle$ [see equations (43) and (44)]. As discussed by BDD, by this mechanism the operator $(1/2)\hat{N}_\rho \mu_{\rho\rho} \hat{N}_\rho$ in $\hat{T}_{\rm N}$ gives rise to a pseudopotential term

$$\mathcal{V}_{\eta\eta}^{(K)}\left(\Delta r_{12}^{(\rm ref)}, \Delta r_{32}^{(\rm ref)}, \rho\right) = \frac{\hbar^2}{2}\mu_{\rho\rho}\left(\Delta r_{12}^{(\rm ref)}, \Delta r_{32}^{(\rm ref)}, \rho\right)\left(\frac{\mathrm{d}\gamma_K}{\mathrm{d}\rho}\right)^2, \tag{55}$$

which only contributes to $\Delta K = 0$ matrix elements diagonal in η. Matrix elements off-diagonal in η but diagonal in K have contributions involving integrals over ρ given by

$$\begin{aligned} &\mathcal{V}_{ab}^{(K)}\left(v_2^{(a)}, v_2^{(b)}, \Delta r_{12}^{(\rm ref)}, \Delta r_{32}^{(\rm ref)}\right) \\ &= \frac{\hbar^2}{2}\left\langle v_2^{(a)}, K \left| \frac{\mathrm{d}}{\mathrm{d}\rho}\mu_{\rho\rho}\left(\frac{\mathrm{d}\gamma_K}{\mathrm{d}\rho}\right) + \left(\frac{\mathrm{d}\gamma_K}{\mathrm{d}\rho}\right)\mu_{\rho\rho}\frac{\mathrm{d}}{\mathrm{d}\rho}\right| v_2^{(b)}, K\right\rangle \qquad (56) \\ &= \frac{\hbar^2}{2}\int_0^\pi \mu_{\rho\rho}\left(\rho, \Delta r_{12}^{(\rm ref)}, \Delta r_{32}^{(\rm ref)}\right)\gamma_K'(\rho)\left[\phi_{v_2^{(a)},K}\phi'_{v_2^{(b)},K} - \phi'_{v_2^{(a)},K}\phi_{v_2^{(b)},K}\right]\mathrm{d}\rho. \qquad (57) \end{aligned}$$

Equation (57) is obtained from equation (56) by setting $|v_2^{(\eta)}, K\rangle = \phi_{v_2^{(\eta)},K}(\rho)$, $\mathrm{d}\phi_{v_2^{(\eta)},K}/\mathrm{d}\rho = \phi'_{v_2^{(\eta)},K}(\rho)$, and $\mathrm{d}\gamma_K/\mathrm{d}\rho = \gamma_K'(\rho)$. Contributions to the matrix elements related to those in equations (55) and (56) are obtained from other Hamiltonian terms containing $\hat{N}_\rho$. For example, the term $(1/2)[\hat{N}_\rho\mu_{x\rho} + \mu_{x\rho}\hat{N}_\rho]\hat{N}_x$ gives rise to terms depending on $\mathrm{d}\gamma_K/\mathrm{d}\rho$ in the $\Delta K = 1$ matrix elements.

Barrow, Dixon and Duxbury (1974) were the first to derive the matrix elements in equations (55) and (56) which, in their equation (17), they present in a matrix called $T'_{\rm ev}$. They state that 'since γ_K will be a slowly varying function of ρ, $T'_{\rm ev}$ will be small for all values of ρ'. Jensen *et al.* (1995a) carried out calculations on the $\tilde{\rm a}\,^1{\rm A}_1$ and $\tilde{\rm b}\,^1{\rm B}_1$ electronic states of the CH_2 radical which confirmed this statement. However, in subsequent calculations by the same authors [Jensen *et al.* (1995b)] on the $\tilde{\rm X}\,^2{\rm A}_1$ and $\tilde{\rm A}\,^2{\rm B}_1$ states of $CH_2{}^+$ it was found that some of the matrix elements involving the integral $\mathcal{V}_{ab}^{(K)}(v_2^{(a)}, v_2^{(b)}, \Delta r_{12}^{(\rm ref)}, \Delta r_{32}^{(\rm ref)})$ of equation (56) can reach values of several hundred cm^{-1}. For example, in the $N = 1$ matrices for $CH_2{}^+$, the basis function with $K = 1$ associated with the vibrational ground state in the $\eta = b$ ($\approx \tilde{\rm A}\,^2{\rm B}_1$) electronic state is coupled to the $\eta = a$, $K = 1$ basis functions associated with the bending states $v_2^{(a)} = 0, 1, 2, \ldots, 7$ by matrix elements of −62, 127, 353, 328, 155, −47, −204, and $-285\,\mathrm{cm}^{-1}$, respectively, originating in the operator of equation (56).

Kolbuszewski *et al.* (1996) showed that for the $\tilde{\rm X}\,^2{\rm A}_1$ and $\tilde{\rm A}\,^2{\rm B}_1$ states of $CH_2{}^+$, the large matrix elements were caused by a 'constructive interference' of the functions $\gamma_K'(\rho)$, $\phi_{v_2^{(b)},K}(\rho)$, and $\phi'_{v_2^{(a)},K}(\rho)$ occurring for certain values of K, $v_2^{(a)}$, and $v_2^{(b)}$ [see their Figures 2 and 3]. They concluded that the large integral values are only obtained in the case when the two Renner electronic states both have small (or zero) barriers to linearity. For $CH_2{}^+$, the barrier to linearity in the $\tilde{\rm X}\,^2{\rm A}_1$ state is less than $1000\,\mathrm{cm}^{-1}$, and the $\tilde{\rm A}\,^2{\rm B}_1$ state is linear. When the lower state is strongly bent and the upper state is linear or quasilinear (as in the case of CH_2, where, in the $\tilde{\rm a}\,^1{\rm A}_1$ electronic state, the potential energy barrier to linearity is approximately $8700\,\mathrm{cm}^{-1}$, whereas this barrier is around $700\,\mathrm{cm}^{-1}$ for the $\tilde{\rm b}\,^1{\rm B}_1$ state) constructive interference of the wavefunctions and their derivatives never seems to occur, and the integrals of equation (57) are small as expected by BDD.

Owing to the large matrix elements coupling the $\eta = a$ and b states for $CH_2{}^+$, it was very difficult for Jensen *et al.* (1995a) to obtain converged energies. They had to consider bending basis functions $|v_2^{(\eta),K}\rangle$ with $v_2^{(\eta)} \leq 40$, $\eta = a$ or b, to obtain energies converged to better than $1.0\,\text{cm}^{-1}$ in the final diagonalization. With so many bending basis functions, it becomes unfeasible to calculate high rovibronic energies for higher N since the matrices to be diagonalized become prohibitively large. In calculations on the $\tilde{\text{X}}\,^2\text{A}_1$ and $\tilde{\text{A}}\,^2\text{B}_1$ electronic states of BH_2, Kolbuszewski *et al.* (1996) experienced exactly the same convergence difficulties as described above for CH_2^+, and in order to reduce the size of the matrix blocks to be diagonalized they changed slightly the technique for obtaining the rovibronic energies. Initially, they consider the subsets of basis functions $|v_2^{(\eta)}, K\rangle\ |N_{\text{vib}}^{(\eta)}, \Gamma_{\text{vib}}^{(\eta)}\rangle\ |\eta; K, J, S, K, M_J, \tau\rangle$ obtained for one particular value of $K = 0, 1, 2, \ldots$. The subsets consist exclusively of functions with $N = K$. They then construct the matrix representation of the partial Hamiltonian $\hat{H}_\text{e} + \hat{T}_\text{N}$ [that is, they neglect the spin–orbit coupling in this part of the calculation, see equations (9) and (32)] in the subset of basis functions and diagonalize this matrix block (which they call a K-block), obtaining the eigenfunctions

$$|K, J, S, K, M_J, \tau\rangle_{\text{cont}} = \sum_{\eta=a,b} \sum_{v_2^{(\eta)}} \sum_{N_{\text{vib}}^{(\eta)}, \Gamma_{\text{vib}}^{(\eta)}} c^{(K\tau)}_{\eta, v_2^{(\eta)}, N_{\text{vib}}^{(\eta)}, \Gamma_{\text{vib}}^{(\eta)}} \left|v_2^{(\eta)}, K\right\rangle \left|N_{\text{vib}}^{(\eta)}, \Gamma_{\text{vib}}^{(\eta)}\right\rangle |\eta; K, J, S, K, M_J, \tau\rangle. \tag{58}$$

The elements of the K-block, and hence the eigenvector coefficients $c^{(K\tau)}_{\eta, v_2^{(\eta)}, N_{\text{vib}}^{(\eta)}, \Gamma_{\text{vib}}^{(\eta)}}$ do not depend on the quantum number M_J. Since spin–orbit coupling is neglected, they are also independent of J and S. The eigenvector coefficients are used to form new contracted basis functions

$$|N, J, S, K, M_J, \tau\rangle_{\text{cont}} = \sum_{\eta=a,b} \sum_{v_2^{(\eta)}} \sum_{N_{\text{vib}}^{(\eta)}, \Gamma_{\text{vib}}^{(\eta)}} c^{(K\tau)}_{\eta, v_2^{(\eta)}, N_{\text{vib}}^{(\eta)}, \Gamma_{\text{vib}}^{(\eta)}} \left|v_2^{(\eta)}, K\right\rangle \left|N_{\text{vib}}^{(\eta)}, \Gamma_{\text{vib}}^{(\eta)}\right\rangle |\eta; N, J, S, K, M_J, \tau\rangle \tag{59}$$

for arbitrary values of $N \geq K$. When the K-blocks have been diagonalized, the matrix representation of the complete $\hat{H}_{\text{Renner}}$ (including spin–orbit interaction) is set up in a basis of the contracted functions given by equation (59) and it is diagonalized to yield the final rovibronic energies. However, not all the contracted functions are included in this matrix. It contains only those which, in the diagonalization of the K-block, correspond to eigenvalues lower than an energy threshold E_{limit}.

In the contraction scheme outlined above, the very large $\Delta K = 0$ matrix elements originating in the integral of equation (56) are placed in the K-blocks, whose sizes are independent of K [with the exception that the $K = 0$ blocks are smaller than blocks with $K > 0$ for symmetry reasons; see Jensen *et al.* (1995a)]. In the contracted basis, the off-diagonal matrix elements are not large, and the sizes of the final matrix blocks to be diagonalized (which depend on N and J) can be kept comparatively small.

As an example we consider the $\tilde{\text{X}}\,^2\text{A}_1$ and $\tilde{\text{A}}\,^2\text{B}_1$ electronic states of the BH_2 molecule. These states are degenerate as a $^2\Pi_\text{u}$ state ($\Lambda = 1$) at linearity. Kolbuszewski *et al.* (1996) have calculated *ab initio* the potential energy surfaces for these states. The *ab initio* potential energy surfaces were adjusted slightly to reproduce better the observed rovibronic energies [Herzberg and Johns (1967)]. Figure 15.11 shows the bending potentials obtained from the adjusted potential energy surfaces. The barrier to linearity in the $\tilde{\text{X}}\,^2\text{A}_1$ electronic state is $2666\,\text{cm}^{-1}$. In Figure 15.12, we show the rovibronic energies calculated for the $\tilde{\text{X}}\,^2\text{A}_1$ state by Kolbuszewski *et al.* (1996) using the theoretical model outlined above. The term values in the figure are labeled by the quantum number N, the rovibronic angular momentum quantum number here, and $J = |N - 1/2|$ or $N + 1/2$; we also use the usual quantum numbers K_a, K_c, v_1, $v_2^{(\text{bent})}$ and v_3. The quantum

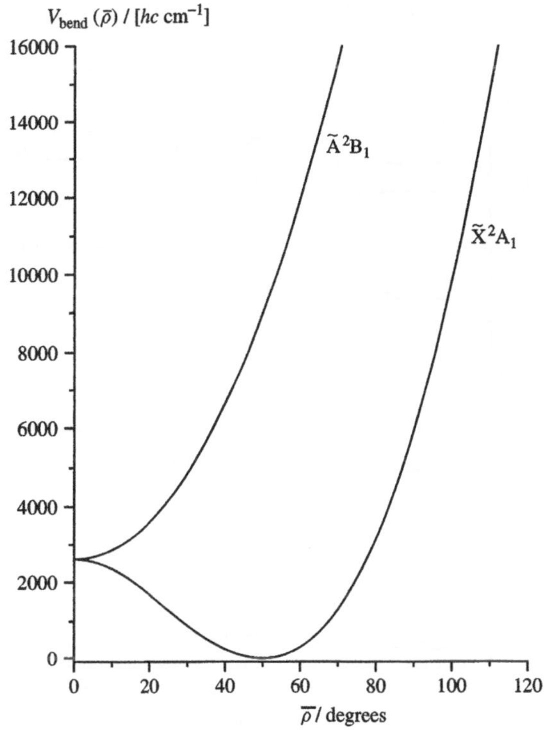

Figure 15.11. Bending potential curves for the lowest electronic states of the BH_2 molecule. The abscissa is the bending coordinate $\bar{\rho} = \pi - \angle(\text{HBH})$, the supplement of the bond angle. The bond lengths are held constant.

number $v_2^{(\text{bent})}$ is that appropriate for a bent triatomic molecule; it is equal to the quantum number $v_2^{(a)}$ defined above. We plot the energies with $N \leq 2$ against their K_a values. On the scale of the figure, the energies with common values of $v_1 v_2^{(\text{bent})} v_3$ [but with different values of $N = K_a$, $K_a + 1, \ldots, 2$, and $J = |N - 1/2|$ or $N + 1/2$] lie close together, and we label them by their value of $v_1 v_2^{(\text{bent})} v_3$. Perhaps the most striking effect of the Renner interaction is the so-called energy 'reordering' of the K_a rotational structure in the lower Renner component close to the barrier to linearity. Figure 15.12 shows that in the $\tilde{X}\,^2A_1$ electronic state of BH_2 energy reordering takes place in the $v_2^{(\text{bent})} = 3$ state, where the $K_a = 1$ energies are below the $K_a = 0$ energies. In their Section 13.4.1, Bunker and Jensen (1998) discuss the reordering phenomenon for the $\tilde{a}\,^1A_1$ and $\tilde{b}\,^1B_1$ electronic states of CH_2 and for the $\tilde{X}\,^2A_1$ and $\tilde{A}\,^2B_1$ states of $CH_2{}^+$.

The energy level structure in the $\tilde{A}\,^2B_1$ electronic state of BH_2 is shown in Figure 15.13; it is that of a linear molecule. The energies increase approximately linearly with K_a, for a given value of (v_1, $v_2^{(\text{bent})}$, v_3), as expected for a linear molecule; there is no energy reordering. The reasons for this are discussed in Section 13.4.1 of Bunker and Jensen (1998) for the analogous case of $\tilde{A}\,^2B_1$ $CH_2{}^+$ (see Figure 13-11 of Bunker and Jensen (1998)). In Figure 15.13, the energies are labeled by $v_1 v_2^{(\text{lin})} v_3$, where $v_2^{(\text{lin})}$ is the bending quantum number appropriate for a linear triatomic molecule. The general expression for this quantum number is

$$v_2^{(\text{lin})} = 2v_2^{(\text{bent})} + |K \mp \Lambda|, \tag{60}$$

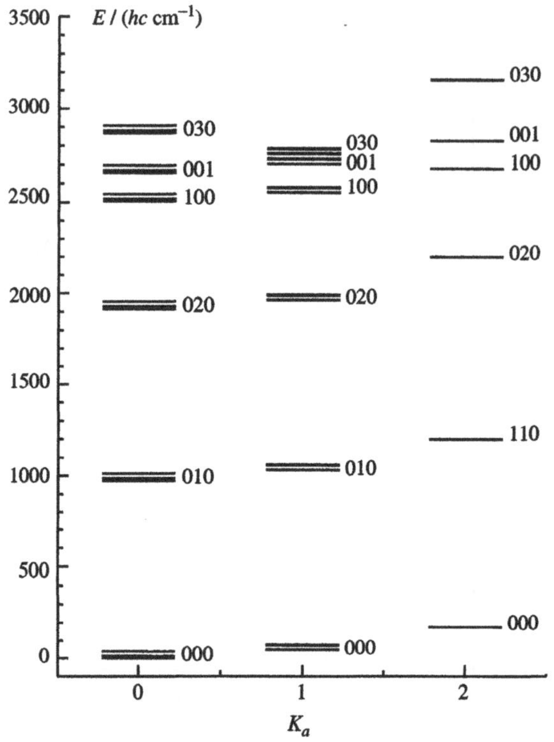

Figure 15.12. Term level diagram for $\tilde{X}\,^2A_1$ BH_2. Energies from Table 4 of Kolbuszewski *et al.* (1996) are labeled by the vibrational quantum numbers $v_1 v_2^{(bent)} v_3$. The energies with common values of $v_1 v_2^{(bent)} v_3$ have different values of $N = K_a, K_a + 1, \ldots, 2$, and $J = |N - 1/2|$ or $N + 1/2$.

where the minus sign applies for the lower Renner component, and the plus sign for the upper component. Thus, for $\tilde{A}\,^2B_1$ BH_2 we have from equation (60)

$$v_2^{(lin)} = 2v_2^{(bent)} + K + 1. \tag{61}$$

Figure 15.13 shows the significance of $v_2^{(lin)}$ in this case: Energies with a common value of $v_2^{(lin)}$, but with different values of $K_a = v_2^{(lin)} - \Lambda$, $v_2^{(lin)} - \Lambda - 2$, $v_2^{(lin)} - \Lambda - 4, \ldots$, 0 or 1, are close in energy.

15.6 Intensity calculations

Osmann *et al.* (1997) have recently extended the theoretical treatment of the Renner effect given above to allow the calculations of the *line strengths* which determine the intensities for rovibronic absorption and emission transitions within and between the two Renner-degenerate electronic states. The line strength of an electric dipole transition in field-free space from an initial state $|\psi_{rve}^{(J'',M_J'',S,\Gamma_{rve}'')}\rangle$ to a final state $|\psi_{rve}^{(J',M_J',S,\Gamma_{rve}')}\rangle$ is

$$S(f \leftarrow i) = \sum_{\varepsilon=-1}^{1} \sum_{M_J'',M_J'} \left| \left\langle \psi_{rve}^{(J',M_J',S,\Gamma_{rve}')} \right| \mu_s^{(1,\varepsilon)} \left| \psi_{rve}^{(J'',M_J'',S,\Gamma_{rve}'')} \right\rangle \right|^2, \tag{62}$$

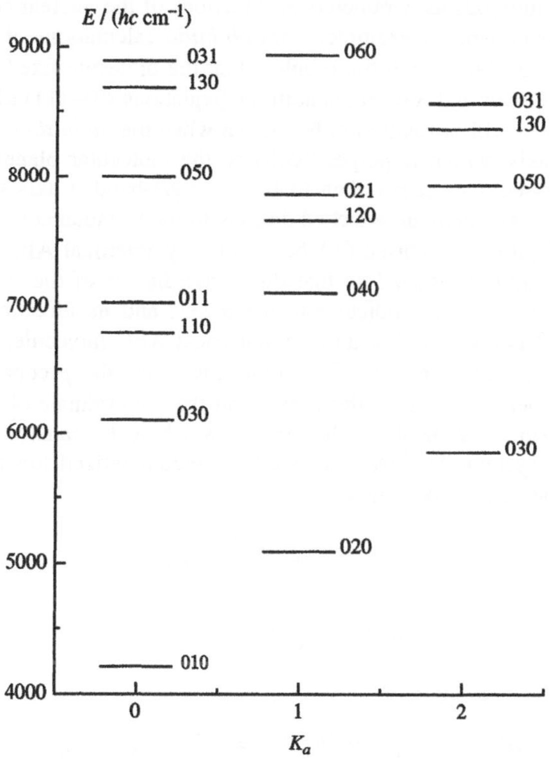

Figure 15.13. Term level diagram for $\tilde{A}\,^2B_1$ BH_2. Energies from Table 5 of Kolbuszewski *et al.* (1996) are labeled by the vibrational quantum numbers $v_1 v_2^{(\mathrm{lin})} v_3$. The energies with common values of $v_1 v_2^{(\mathrm{lin})} v_3$ are effectively degenerate on the scale of the figure; they have $N = K_a$, $J = |N - 1/2|$ or $N + 1/2$, and parities given by the e and f labels [see Section 17.5.1 of Bunker and Jensen (1998)]. The zero of energy is the $J = 1/2$, $N = 0$ level in the vibrational ground state of $\tilde{X}\,^2A_1$ BH_2 (Figure 15.12).

where the irreducible tensor operators $\mu_s^{(1,\varepsilon)}$, $\varepsilon = -1, 0, 1$, are given by

$$\mu_s^{(1,0)} = \mu_Z \tag{63}$$

and

$$\mu_s^{(1,\pm 1)} = \frac{1}{\sqrt{2}}(\mp\mu_X - \mathrm{i}\mu_Y). \tag{64}$$

The (μ_X, μ_Y, μ_Z) are the components of the molecular dipole moment operator in the space-fixed XYZ axis system.

The derivation of the matrix elements entering into equation (62) is discussed by Osmann *et al.* (1997); the general technique for the derivation of line strengths is described in Chapter 14 of Bunker and Jensen (1998). The reader is referred to these publications for details.

From symmetry arguments analogous to those employed for the electronic integrals $\langle\psi_e^{(\sigma)}|\hat{L}_\delta\hat{L}_\gamma|\psi_e^{(\sigma')}\rangle_{\mathrm{el}}$ and $\langle\psi_e^{(\sigma)}|\hat{L}_\delta|\psi_e^{(\sigma')}\rangle_{\mathrm{el}}$ in Section 15.3 above, we can show that there exist a total of five non-vanishing electronic integrals involving the dipole moment components, $\langle\psi_e^{(\sigma)} \mid \mu_y \mid \psi_e^{(\sigma)}\rangle_{\mathrm{el}}$ and $\langle\psi_e^{(\sigma)}|\mu_z|\psi_e^{(\sigma)}\rangle_{\mathrm{el}}$, $\sigma = +$ or $-$, together with $\langle\psi_e^{(-)}|\mu_x|\psi_e^{(+)}\rangle_{\mathrm{el}}$. Here, μ_α ($\alpha = x, y, z$) is the component of the molecular dipole moment along the molecule fixed α axis (Figure 15.2). We aim at obtaining these integrals as parameterized functions of the nuclear coordinates, where the parameter values are obtained, for example, from *ab initio* calculations of the molecular dipole moment. However, we are faced with the problem that the molecule-fixed xyz axis system used here is defined by the Eckart and Sayvetz conditions [equations (9)–(11) of Jensen (1988a)] and the position of these axes will normally not be known when the *ab initio* calculations are carried out (except for the x axis, which is perpendicular to the molecular plane). To circumvent this problem, we employ three axes xqp (taken to form a right-handed axis system with origin in the nuclear center of mass) which are attached directly to the instantaneous nuclear configuration of the molecule [see Figure 1 of Jensen (1988c)]. For a symmetrical ABA molecule, the q axis bisects the bond angle and is directed so that the q coordinates of the 'terminal' nuclei 1 and 3 are positive. The p axis is perpendicular to the q axis and its direction is such that the p coordinate of nucleus 3 is positive. For an unsymmetrical ABC molecule, the p axis is parallel to the line connecting the central nucleus 2 with nucleus 3, and the p coordinate of nucleus 3 is positive. The q axis is perpendicular to the p axis and the q coordinate of nucleus 1 is positive. It is described in considerable detail by Jensen (1988c) how to transform from the xqp axis system to the xyz axis system. That is, when we have parameterized functions representing the nonvanishing dipole moment components

$$\overline{\mu}_q^{(\sigma)} = \langle \psi_e^{(\sigma)} | \mu_q | \psi_e^{(\sigma)} \rangle_{el}, \tag{65}$$

and

$$\overline{\mu}_p^{(\sigma)} = \langle \psi_e^{(\sigma)} | \mu_p | \psi_e^{(\sigma)} \rangle_{el}, \tag{66}$$

with $\sigma = -$ or $+$, and

$$\overline{\mu}_x^{(-+)} = \langle \psi_e^{(-)} | \mu_x | \psi_e^{(+)} \rangle_{el} = \langle \psi_e^{(+)} | \mu_x | \psi_e^{(-)} \rangle_{el}, \tag{67}$$

we can transform to obtain parameterized functions representing the electronic integrals involving the (x, y, z) components of the dipole moment. It can be shown that with the phase choice for the electronic functions $\psi_e^{(\sigma)}$, $\sigma = -$ or $+$, defined by the matrix element of the electronic angular momentum component $\hat{L}_z$ given in equation (28), $\overline{\mu}_x^{(-+)}$ is real as indicated in equation (67).

In the present work, we follow Jensen (1988c) in parameterizing the functions $\overline{\mu}_q^{(\sigma)}$ and $\overline{\mu}_p^{(\sigma)}$ as

$$\begin{aligned}\overline{\mu}_q^{(\sigma)}\left(\Delta r_{12}^{(\text{ref})}, \Delta r_{32}^{(\text{ref})}, \overline{\rho}\right) = \sin\overline{\rho}\,\Big[\mu_0^{(q;\sigma)}(\overline{\rho}) &+ \sum_j \mu_j^{(q;\sigma)}(\overline{\rho})\Delta r_{j2}^{(\text{ref})} + \sum_{j\le k} \mu_{jk}^{(q;\sigma)}(\overline{\rho})\Delta r_{j2}^{(\text{ref})}\Delta r_{k2}^{(\text{ref})} \\ &+ \sum_{j\le k\le m} \mu_{jkm}^{(q;\sigma)}(\overline{\rho})\Delta r_{j2}^{(\text{ref})}\Delta r_{k2}^{(\text{ref})}\Delta r_{m2}^{(\text{ref})} \\ &+ \sum_{j\le k\le m\le n} \mu_{jkmn}^{(q;\sigma)}(\overline{\rho})\Delta r_{j2}^{(\text{ref})}\Delta r_{k2}^{(\text{ref})}\Delta r_{m2}^{(\text{ref})}\Delta r_{n2}^{(\text{ref})}\Big],\end{aligned} \tag{68}$$

and

$$\begin{aligned}\overline{\mu}_p^{(\sigma)}\left(\Delta r_{12}^{(\text{ref})}, \Delta r_{32}^{(\text{ref})}, \overline{\rho}\right) = \mu_0^{(p;\sigma)}(\overline{\rho}) &+ \sum_j \mu_j^{(p;\sigma)}(\overline{\rho})\Delta r_{j2}^{(\text{ref})} + \sum_{j\le k} \mu_{jk}^{(p;\sigma)}(\overline{\rho})\Delta r_{j2}^{(\text{ref})}\Delta r_{k2}^{(\text{ref})}\end{aligned}$$

$$+ \sum_{j\le k\le m} \mu_{jkm}^{(p;\sigma)}(\overline{\rho})\Delta r_{j2}^{(\text{ref})}\Delta r_{k2}^{(\text{ref})}\Delta r_{m2}^{(\text{ref})}$$

$$+ \sum_{j\le k\le m\le n} \mu_{jkmn}^{(p;\sigma)}(\overline{\rho})\Delta r_{j2}^{(\text{ref})}\Delta r_{k2}^{(\text{ref})}\Delta r_{m2}^{(\text{ref})}\Delta r_{n2}^{(\text{ref})}, \tag{69}$$

with $\sigma = +$ or $-$ and $j, k, m, n = 1$ or 3. The angle-dependent coefficients are given by

$$\mu_{jk\ldots}^{(w;\sigma)}(\overline{\rho}) = \sum_{i=0}^{N} w_{jk\ldots}^{(i;\sigma)}(1 - \cos\overline{\rho})^i. \tag{70}$$

with $w = p$ or q. The function $\mu_0^{(w;\sigma)}(\overline{\rho})$ has $N = 8$, $\mu_j^{(w;\sigma)}(\overline{\rho})$ has $N = 4$, $\mu_{jk}^{(w;\sigma)}(\overline{\rho})$ has $N = 3$, $\mu_{jkm}^{(w;\sigma)}(\overline{\rho})$ has $N = 2$ and $\mu_{jkmn}^{(w;\sigma)}(\overline{\rho})$ has $N = 1$.

It can be shown that the function $\overline{\mu}_x^{(-+)}$ [equation (67)] is equal to zero for linear configurations with $\overline{\rho} = 0$. Furthermore, *ab initio* calculations of this quantity for small values of $\overline{\rho}$ indicate that it is an odd function of $\overline{\rho}$, exactly as $\overline{\mu}_q^{(\sigma)}$. Hence we parameterize $\overline{\mu}_x^{(-+)}$ by analogy with equation (68):

$$\overline{\mu}_x^{(-+)}\left(\Delta r_{12}^{(\text{ref})}, \Delta r_{32}^{(\text{ref})}, \overline{\rho}\right) = \sin\overline{\rho}\Bigg[\mu_0^{(x;-+)}(\overline{\rho}) + \sum_j \mu_j^{(x;-+)}(\overline{\rho})\Delta r_{j2}^{(\text{ref})} + \sum_{j\le k} \mu_{jk}^{(x;-+)}(\overline{\rho})\Delta r_{j2}^{(\text{ref})}\Delta r_{k2}^{(\text{ref})}$$
$$+ \sum_{j\le k\le m} \mu_{jkm}^{(x;-+)}(\overline{\rho})\Delta r_{j2}^{(\text{ref})}\Delta r_{k2}^{(\text{ref})}\Delta r_{m2}^{(\text{ref})}$$
$$+ \sum_{j\le k\le m\le n} \mu_{jkmn}^{(x;-+)}(\overline{\rho})\Delta r_{j2}^{(\text{ref})}\Delta r_{k2}^{(\text{ref})}\Delta r_{m2}^{(\text{ref})}\Delta r_{n2}^{(\text{ref})}\Bigg], \tag{71}$$

with

$$\mu_{jk\ldots}^{(x;-+)}(\overline{\rho}) = \sum_{i=0}^{N} x_{jk\ldots}^{(-+)}(1 - \cos\overline{\rho})^i \tag{72}$$

where the number of summation terms N is given exactly as for equation (70).

For symmetrical molecules of type ABA, relations exist between the expansion coefficients in equations (70) and (72), so that the functions $\overline{\mu}_q^{(\sigma)}$, $\sigma = -$ or $+$, and $\overline{\mu}_x^{(-+)}$ are unchanged under the interchange of the identical A nuclei, whereas the functions $\overline{\mu}_p^{(\sigma)}$ are antisymmetric under this operation.

In Figure 15.14, we show as an example a simulation of the $\tilde{\text{A}}(0, 0, 1) \leftarrow \tilde{\text{X}}(0, 0, 0)$ rovibronic band of the HOO molecule. The band was been simulated for a maximum N-value of 10 at the absolute temperature $T = 296$ K. In the calculation we used bending basis functions $|v_2^{(\eta)}, K\rangle$, $\eta = a$ or b (see above), with $v_2^{(a)} \le 30$ and $v_2^{(b)} \le 10$. The stretching Hamiltonians $\hat{H}_{\text{Stretch}}^{(a)}$ and $\hat{H}_{\text{Stretch}}^{(b)}$ were diagonalized in basis sets of Morse oscillator eigenfunctions $|n_1\rangle|n_3\rangle$ [Jensen *et al.* (1995a)] with $n_1 + n_3 \le 13$, and we retained the stretching eigenfunctions $|N_{\text{vib}}^{(\eta)}, \Gamma_{\text{vib}}^{(\eta)}\rangle$ with $N_{\text{vib}}^{(a)} \le 30$ and $N_{\text{vib}}^{(b)} \le 10$ in setting up the Hamiltonian matrices for the contraction step. Finally, we employed $E_{\text{limit}} = 17\,000\ \text{cm}^{-1}$. For the calculation of the intensities only transitions with lower state energy $E'' < 3000\ \text{cm}^{-1}$ were considered in order to reduce the computational expense.

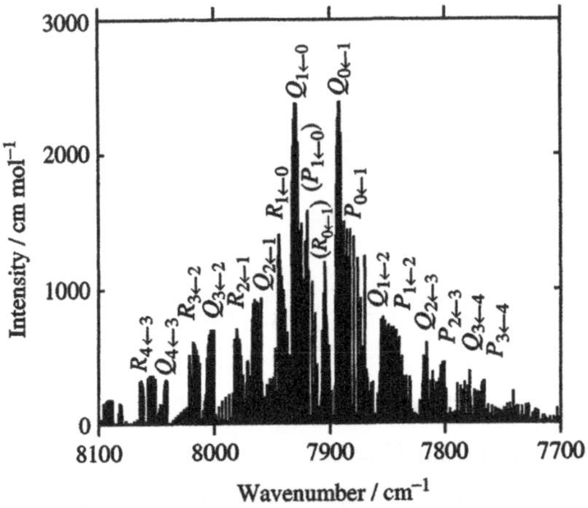

Figure 15.14. Simulation of the $\tilde{A}(0,0,1) \leftarrow \tilde{X}(0,0,0)$ rovibronic band of the HOO molecule at $T = 296$ K. The maximum N value is 10. The branches are labeled by $\Delta N_{K'_a \leftarrow K''_a}$. Note that the spectrum is displayed with the wavenumber increasing towards the left, a format preferred by electronic spectroscopists.

15.7 Other theoretical methods

At present, the theoretical approach by Jensen, Bunker, Kraemer, and co-workers outlined in the preceding sections is the only one that allows the intensities of individual rovibronic transitions to be calculated from wavefunctions describing all rotational and vibrational motions of a triatomic molecule.[5] However, as mentioned in Section 15.2, there exist several alternative variational techniques for calculating the rovibronic *energies* in Renner-degenerate electronic states of triatomic molecules.

The two methods first proposed by Barrow, Dixon and Duxbury (1974) and by Jungen and Merer (1976, 1980), respectively, are closely related. In their original form, both of these methods neglect the stretching vibrations of the triatomic molecule. The Jungen and Merer (JM) approach is based on the Hougen–Bunker–Johns treatment of molecular rotation and vibration, and the BDD treatment is based on a similar treatment of the bending vibration and K-type rotation [i.e., the rotation about the molecule fixed z axis chosen as the axis of least moment of inertia] for a triatomic molecule by Freed and Lombardi (1966). The BDD methods involves a pre-diagonalization procedure leading to the basis functions in the form given in equations (43) and (44). By this procedure, we obtain basis functions in which the most important Renner interaction terms are diagonal. Jungen and Merer (1976, 1980) make an analogous pre-diagonalization, but they start with a different electronic basis set. The matrix representation of the $\hat{L}_z$ operator in the electronic basis set $\psi_e^{(-)}$ and $\psi_e^{(+)}$ is

$$\left\{ \begin{matrix} 0 & \langle \psi_e^{(-)} | \hat{L}_z | \psi_e^{(+)} \rangle_{el} \\ \langle \psi_e^{(-)} | \hat{L}_z | \psi_e^{(+)} \rangle_{el}^* & 0 \end{matrix} \right\};$$

the two diagonal elements vanish by symmetry as discussed in Section 15.3. The off-diagonal elements are purely imaginary by our phase choice in equation (30). As usual, we neglect all electronic states other than $\psi_e^{(-)}$ and $\psi_e^{(+)}$, and in this approximation we have that the eigenfunctions

[5] Jungen and Merer (1980) and Perić, Peyerimhoff and Buenker (1985) have discussed the calculation of intensities by means of models neglecting the stretching vibrations.

for $\hat{L}_z$ are given by

$$\psi_{\rm e}^{(\pm L)} = \frac{1}{\sqrt{2}}\left(\psi_{\rm e}^{(-)} \pm {\rm i}\psi_{\rm e}^{(+)}\right), \tag{73}$$

where

$$\hat{L}_z\psi_{\rm e}^{(\pm L)} = \pm{\rm i}\left\langle\psi_{\rm e}^{(-)}\middle|\hat{L}_z\middle|\psi_{\rm e}^{(+)}\right\rangle_{\rm el}^{*}\psi_{\rm e}^{(\pm L)}. \tag{74}$$

The basis functions $|\psi_{\rm e}^{(+L)}\rangle|N, J, S, K, M_J, \tau\rangle$ and $|\psi_{\rm e}^{(-L)}\rangle|N, J, S, K, M_J, \tau\rangle$ are eigenfunctions for $\hat{N}_z - \hat{L}_z$ with eigenvalues $K\hbar \mp {\rm i}\langle\psi_{\rm e}^{(-)}|\hat{L}_z|\psi_{\rm e}^{(+)}\rangle_{\rm el}^{*}$. Jungen and Merer (1976, 1980) use basis functions of this type, which they refer to as the ℓ basis [since they define $\ell = K \pm \Lambda$]. It is now straightforward to show by means of equations (14) and (15) that the matrix representation of the operator $\hat{H}_{\rm ez}$ [equation (42)] in these basis functions has the diagonal elements H_{--} and H_{++} where

$$H_{\mp\mp} = \frac{1}{2}\left[V_0^{(-)}(\rho) + V_0^{(+)}(\rho)\right] + \frac{\hbar^2}{2}\mu_{zz}^{(\rm ref)}(\rho)(K \mp \Lambda)^2 \tag{75}$$

and the off-diagonal elements

$$H_{+-} = H_{-+} = \frac{1}{2}\left[V_0^{(-)}(\rho) - V_0^{(+)}(\rho)\right]. \tag{76}$$

In equations (75) and (76), a subscript '$-$' refers to $|\psi_{\rm e}^{(-L)}\rangle|N, J, S, K, M_J, \tau\rangle$ and a subscript '$+$' to $|\psi_{\rm e}^{(+L)}\rangle|N, J, S, K, M_J, \tau\rangle$, and we have used equation (28) to approximate $\langle\psi_{\rm e}^{(-)}|\hat{L}_z|\psi_{\rm e}^{(+)}\rangle_{\rm el}^{*}$. Jungen and Merer (1976, 1980) do not express the eigenfunctions of $\hat{H}_{\rm ez}$ in terms of the ℓ basis functions by analogy with equations (43) and (44), and use these as basis functions. Instead, they start by obtaining energies and numerical bending wavefunctions using one of the potential energy curves, usually the lower one $V_0^{(-)}(\rho)$. With these basis functions, they construct the matrix representation $\boldsymbol{H}$ of their total Hamiltonian (describing the electronic motion, the bending vibration, and the rotation about the z axis) in the ℓ basis. They then use the eigenvectors of the 2×2 matrix with the elements given in equations (75) and (76) to construct a unitary matrix $\boldsymbol{S}$.[6] The matrix $\boldsymbol{S}$ is used in an orthogonal transformation to obtain the matrix representation of the Hamiltonian in a form with minimized off-diagonal elements. The transformed matrix is denoted $\boldsymbol{H}'$. Finally energies and numerical basis functions are obtained using one set of the ℓ basis functions for the second electronic state. These are used to set up the final matrix (called $\boldsymbol{H}''$) to be diagonalized. In practice, the pre-diagonalization procedure of Jungen and Merer is highly analogous to that of BDD, even though the mathematical formulations are somewhat different. Duxbury and Dixon (1981) have shown by numerical examples that the Jungen and Merer approach is equivalent to the BDD scheme. In both the Jungen and Merer and the BDD approaches, spin–orbit coupling is accounted for by using a Hund's case (a) basis set in which the electron spin is quantized along the molecule-fixed z axis.

The BDD approach, modified slightly to allow for resonant interactions between the two electronic states [Duxbury and Dixon (1981); Alijah and Duxbury (1990)] has been used by Duxbury and co-workers to analyze Renner perturbations in a number of triatomic molecules. For recent examples of this work, see Duxbury, Alijah and Trieling (1993), Alijah and Duxbury (1994), and Duxbury, McDonald and Alijah (1996). In recent years, two extensions of this approach have been made to account for all vibrational motions of a triatomic molecule. One is the work by Jensen and co-workers [Jensen *et al.* (1995a), Kolbuszewski *et al.* (1996) and Osmann *et al.* (1997)] described in the preceding sections, and the other is the 'stretch–bender' approach by Duxbury,

[6] The matrix $\boldsymbol{S}$ has nothing to do with the total electron spin.

Jungen, and co-workers [Duxbury *et al.* (1998a, 1998b)]. The 'stretch–bender' approach is very similar to the theory by Jensen and co-workers outlined here. One important difference is that in the 'stretch–bender' approach, the 'constant-bond-length' reference configuration introduced by HBJ is replaced by a reference configuration in which the bond lengths are functions of ρ. This is the philosophy of the *semirigid-bender* Hamiltonian [Bunker and Landsberg (1977)]. As the molecule bends, the bond lengths relax so that the molecule follows a 'minimum energy path' on the potential energy surface. It is to be expected that with this type of reference configuration, converged rovibronic energies can be obtained with fewer stretching basis functions than in the case of a reference configuration with fixed bond lengths.

An alternative method for the calculation of rovibronic energies of Renner-degenerate electronic states for triatomic molecules, which is not influenced by the work of BDD and of Jungen and Merer, has been developed by Carter, Handy, and co-workers [Carter and Handy (1984), Carter *et al.* (1990), and Green *et al.* (1991)]. Their nuclear kinetic energy operator is expressed in terms of geometrically defined vibrational coordinates: the instantaneous values of the bond lengths and the bond angle. No attempt is made to minimize the kinetic energy coupling between the various types of motion (rotation, bending, and stretching) by means of Eckart and Sayvetz conditions [see Hougen, Bunker and Johns (1970) and Chapter 15 of Bunker and Jensen (1998)]. The kinetic energy operator is extended to account for the effects of the electronic angular momentum by the terms given in equation (22), and the eigenvalues of the extended Hamiltonian are obtained in a variational calculation. The kinetic energy operator is not expanded in the manner indicated by equation (20). This type of approach to the calculation of rovibronic energies is discussed in Section 13.2.5 of Bunker and Jensen (1998).

Obviously, chain molecules with more than three nuclei can also exhibit the Renner effect. In this case, the theoretical description is much more complicated than in the case of a triatomic, chiefly because the zero-order treatment which describes the electronic motion, the rotation about the molecular axis, and the bending vibration, now involves more than one bending coordinate. Some work has been done towards the calculation of rovibronic energies for tetra-atomic molecules in Renner-degenerate electronic states [see Perić *et al.* (1997) and references cited therein].

15.8 References

Alijah, A., and Duxbury, G., 1990, *Mol. Phys.* **70**, 605–622.
Alijah, A., and Duxbury, G., 1994, *J. Opt. Soc. Am. B* **11**, 208–218.
Barrow, T., Dixon, R. N., and Duxbury, G., 1974, *Mol. Phys.* **27**, 1217–1234.
Brown, J. M., 1977, *J. Mol. Spectrosc.* **68**, 412–422.
Brown, J. M., and Howard, B. J., 1976, *Mol. Phys.* **31**, 1517–1525.
Brown, J. M., and Jørgensen, F., 1983, *Adv. Chem. Phys.* **LII**, 117–180.
Bunker, P. R., and Jensen, P., 1998, *Molecular Symmetry and Spectroscopy*, 2nd edition; NRC Research Press: Ottawa.
Bunker, P. R., and Landsberg, B. M., 1977, *J. Mol. Spectrosc.* **67**, 374–385.
Bunker, P. R., and Moss, R. E., 1980, *J. Mol. Spectrosc.* **80**, 217–228.
Carter, S., and Handy, N. C., 1984, *Mol. Phys.* **52**, 1367–1391.
Carter, S., Handy, N. C., Rosmus, P., and Chambaud, G., 1990, *Mol. Phys.* **71**, 605–622.
Dressler, K., and Ramsay, D. A., 1959, *Philos. Trans. R. Soc. London, Ser. A* **251**, 553–602.
Duxbury, G., and Dixon, R. N., 1981, *Mol. Phys.* **43**, 255–274.
Duxbury, G., and Jungen, C., 1988, *Mol. Phys.* **63**, 981–998.
Duxbury, G., McDonald, B. D., and Alijah, A., 1996, *Mol. Phys.* **89**, 767–790.
Duxbury, G., McDonald, B. D., Van Gogh, M., Alijah, A., Jungen, C., and Palivan, H., 1998a, *J. Chem. Phys.* **108**, 2336–2350.
Duxbury, G., Alijah, A., McDonald, B. D., and Jungen, C., 1998b, *J. Chem. Phys.* **108**, 2351–2360.
Duxbury, G., Alijah, A., and Trieling, R. R., 1993, *J. Chem. Phys.* **98**, 811–825.
Freed, K. F., and Lombardi, J. R., 1966, *J. Chem. Phys.* **45**, 591–598.
Green, Jr, W. H., Handy, N. C., Knowles, P. J., and Carter, S., 1991, *J. Chem. Phys.* **94**, 118–132.

Gu, J.-P., Hirsch, G., and Buenker, R. J., 1998, Personal communication.
Herzberg, G., and Johns, J. W. C., 1967, *Proc. R. Soc. London, Ser. A* **298**, 142–157.
Hougen, J. T., Bunker, P. R., and Johns, J. W. C., 1970, *J. Mol. Spectrosc.* **34**, 136–172.
Jensen, P., 1983, *Comput. Phys. Rep.* **1**, 1–55.
Jensen, P., 1988a, *J. Mol. Spectrosc.* **128**, 478–501.
Jensen, P., 1988b, *J. Chem. Soc., Faraday Trans. 2* **84**, 1315–1340.
Jensen, P., 1988c, *J. Mol. Spectrosc.* **132**, 429–457.
Jensen, P., 1992, Calculation of molecular rotation–vibration energies directly from the potential energy function, in *Methods in Computational Molecular Physics*, Wilson, S., and Diercksen, G. H. F., Eds; Plenum: New York.
Jensen, P., 1994, The MORBID method, in *Molecules in the Stellar Environment*, Lecture Notes in Physics no. 428, Jørgensen, U. G., Ed, Springer: Berlin.
Jensen, P., Brumm, M., Kraemer, W. P., and Bunker, P. R., 1995a, *J. Mol. Spectrosc.* **171**, 31–57.
Jensen, P., Brumm, M., Kraemer, W. P., and Bunker, P. R., 1995b, *J. Mol. Spectrosc.* **172**, 194–204.
Jungen, C., and Merer, A. J., 1976, in *Molecular Spectroscopy: Modern Research*, Vol. II, Narahari Rao, K., Ed; Academic Press: New York, chapter 3.
Jungen, C., and Merer, A. J., 1980, *Mol. Phys.* **40**, 1–23.
Kolbuszewski, M., Bunker, P. R., Kraemer, W. P., Osmann, G., and Jensen, P., 1996, *Mol. Phys.* **88**, 105–124.
Kraemer, W. P., Jensen, P., and Bunker, P. R., 1994, *Can. J. Phys.* **72**, 871–878.
Lee, T. J., Fox, D. J., Schaefer III, H. F., and Pitzer, R. S., 1984, *J. Chem. Phys.* **81**, 356–361.
Osmann, G., Bunker, P. R., Jensen, P., and Kraemer, W. P., 1997, *Chem. Phys.* **225**, 33–54.
Pople, J. A., 1960, *Mol. Phys.* **3**, 16–22.
Pople, J. A., and Longuet-Higgins, H. C., 1958, *Mol. Phys.* **1**, 372–383.
Perić, M., Ostojić, B., Schäfer, B., and Engels, B., 1997, *Chem. Phys.* **225**, 63–76.
Perić, M., Peyerimhoff, S. D., and Buenker, R. J., 1983, *Mol. Phys.* **49**, 379–400.
Perić, M., Peyerimhoff, S. D., and Buenker, R. J., 1985, *Int. Rev. Phys. Chem.* **4**, 85–124.
Renner, R., 1934, *Z. Phys.* **92**, 172–193.
Zare, R. N., 1988, *Angular Momentum*; Wiley: New York.

16 THE RENNER–TELLER EFFECT: THE EFFECTIVE HAMILTONIAN APPROACH

John M. Brown

University of Oxford, UK

16.1 Introduction

For a linear triatomic molecule in a degenerate electronic state, the potential energy curve as a function of the bond stretching coordinates is two-fold degenerate throughout the whole range because the cylindrical symmetry of the molecule is preserved by these vibrational coordinates. However, if the nuclei are displaced from the linear conformation by bending, the degeneracy is lifted. Two non-degenerate electronic states are formed, one with a wavefunction which is

Computational Molecular Spectroscopy. Edited by Per Jensen and Philip R. Bunker

symmetric with respect to reflection in the plane of the bent triatomic molecule (A′) and the other which is antisymmetric (A″)[1]. The separation between these two states and their relative energy order depend on the detailed electronic properties of the molecule. This behaviour of the potential energy curves as a function of the bending co-ordinate is shown in Figure 16.1.

The description of the vibrational levels for this pair of nearly degenerate potentials was first investigated in detail by Renner (1934) for a Π electronic state. However, the generalities of the problem had already been discussed by Herzberg and Teller in a preceding paper (1933). It is for this reason that some authors refer to vibronic coupling in a linear molecule as the Renner–Teller effect while others prefer to call it simply the Renner effect. Both names are used in the literature to describe a single phenomenon. In this chapter, we shall follow Herzberg (1991) and refer to it as the Renner–Teller or R–T effect. Renner introduced a dimensionless parameter ε which can be used to gauge the size of the R–T effect (in the harmonic approximation). We shall call the two curves in Figure 16.1 V' and V'' to indicate their symmetry under reflection the plane of the molecule. They can be written as

$$V' = \tfrac{1}{2}\lambda' Q_2^2, \tag{1a}$$

$$V'' = \tfrac{1}{2}\lambda'' Q_2^2, \tag{1b}$$

where Q_2 is the bending normal coordinate (i.e. it is mass weighted) and λ', λ'' are the appropriate harmonic force constants. The expressions in equation (1) can be rewritten as

$$V', V'' = \tfrac{1}{2}(1 \pm \varepsilon)\lambda_2 Q_2^2, \tag{2}$$

where

$$\varepsilon = \frac{\lambda' - \lambda''}{\lambda' + \lambda''}, \tag{3}$$

and λ_2 is the mean force constant,

$$\lambda_2 = \tfrac{1}{2}(\lambda' + \lambda''). \tag{4}$$

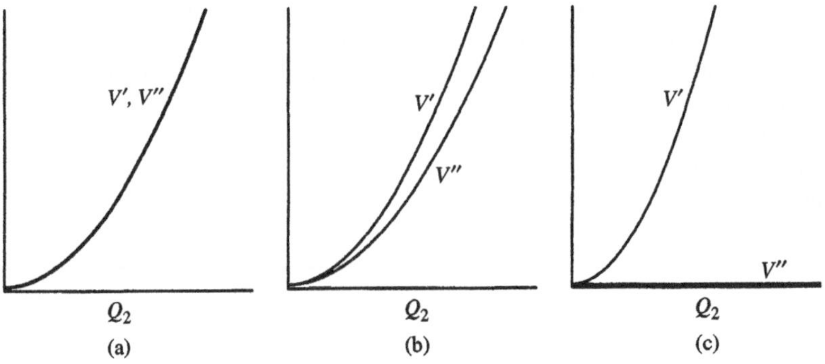

Figure 16.1. Some possible harmonic potential energy curves for the bending vibration of a linear triatomic molecule in a Π electronic state. In these diagrams, Q_2 is the bending coordinate. The curve V' is symmetric with respect to reflection in the plane formed by the bent molecule and the curve V'' is antisymmetric. In part (a), the Renner–Teller parameter ε is zero and the curves V' and V'' remain degenerate over the whole range of Q_2. For parts (b) and (c), ε equals 0.15 and 1.0, respectively. The energy order of the two curves is reversed for negative values of ε.

[1] For triatomic molecules of the type XY_2 with orthorhombic symmetry, the symmetric and antisymmetric wavefunctions have symmetry A_1 and B_1 (for Π electronic states in the linear configuration which arise from electrons in p orbitals on the X atom) and A_1 and B_1 or B_2 and A_2 (for Δ or Π states arising from d orbitals on the central atom).

We note that, when ε is positive, the symmetric curve V' lies *above* the antisymmetric curve V'' and vice versa. Figure 16.1 shows how the potential energy curves depend on the parameter ε. The parameter has a meaningful range $-1 \leq \varepsilon \leq +1$. At the limits of this range, the lower potential curve is a horizontal line, independent of Q_2 (see Figure 16.1(c)). The complication in solving the R–T problem is that the vibrational motion samples *both* potential energy curves in general and so depends on the separation between them.

In his discussion of the R–T effect, Renner used the molecule CO_2 as his example even though the molecule was not known to display such effects (this comment still holds true today). It was some 25 years before the first experimental example of the R–T effect was identified in the electronic spectrum of NH_2 by Dressler and Ramsay (1957, 1959). In fact, theirs was an example of a molecule which shows such a large effect in its ground state that the lower of the two potential energy curves has a minimum not at 180° but rather at 103°, as shown in Figure 16.2. Formally, it corresponds to a molecule with a value for ε larger in magnitude than unity although of course the lower potential curve is highly anharmonic. Both these curves correlate with a $^2\Pi$ state in the linear configuration. The ground state of NH_2 can be modelled reasonably well as an isolated doublet state of a bent molecule. The upper state, which displays some of the characteristics associated with a linear molecule, was described by Dressler and Ramsay as 'half a Renner state'. On this basis, Dressler and Ramsay arrived at the correct assignment of the vibronic quantum numbers. Soon afterwards, Pople and Longuet-Higgins (1958) showed that the observed pattern of energy levels could be reproduced with a model similar to that described by Renner (1934).

Many other examples of molecules which show the R–T effect have been identified since that time. Two classes of the phenomenon have emerged. The first class contains those molecules

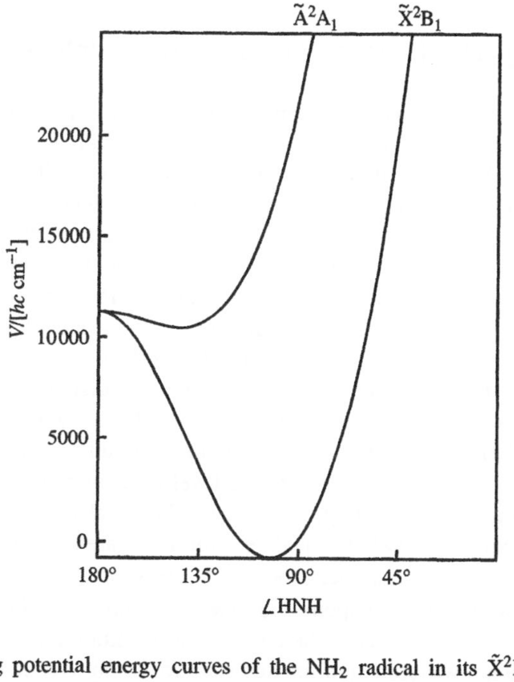

Figure 16.2. The bending potential energy curves of the NH_2 radical in its $\tilde{X}^2B_1$ and $\tilde{A}^2A_1$ electronic states, as determined by Jungen and Merer (1980). Note that the equilibrium configuration of this molecule is non-linear in both these states. In other words, both potential curves are highly anharmonic and the treatment described in this chapter is not appropriate.

which show a relatively small R–T effect, as envisaged originally by Renner. These can be described in terms of a model based on a linear molecule. The other class consists of molecules which show a large R–T effect for which either one or both components of the Π electronic state are non-linear at equilibrium; the radical NH_2 provides the prototype of this class of problem. Such molecules are usually described in terms of the detailed potential curves and are essentially approached from the standpoint of the bent molecule. (Note, however, that the solutions must encompass the possibility of the molecule achieving the linear configuration through vibrational excitation.) This approach has been developed fully by several authors including Barrow and Dixon (1973), Barrow, Dixon and Duxbury (1974), Jungen and Merer (1980), and Jensen, Osmann and Bunker in this volume.

The present chapter is concerned with the description of energy levels of triatomic molecules in Π states which are subject to a small R–T effect and so retain their essential linear character. The original approach of Renner (1934) can be used with advantage to develop an effective Hamiltonian which acts only within the degenerate state in question and yet still describes the energy levels quantitatively and in detail. It is therefore possible to fit the parameters of the model to experimental data in an unprejudiced way and with very little prior knowledge. Other, more complicated situations can also be envisaged. For example, molecules in Δ electronic states are also subject to the R–T effect, see for example Brown and Jørgensen (1982), Jensen *et al.* (1995), Duxbury *et al.* (1998), and Körsgen, Evenson and Brown (1998). In addition, linear molecules with four or more atoms also show R–T effects although in this case more than one R–T parameter is required [Petelin and Kiselev (1972), and Tang and Saito (1996)]. This article, however, is confined to linear triatomic molecules in Π electronic states. The complete variety of phenomena can be demonstrated by this, the simplest of examples.

16.2 Qualitative aspects of energy levels: symmetry labels

The energy levels of a two-dimensional harmonic oscillator (such as the bending vibration of a linear triatomic molecule) form an equally spaced ladder with a degeneracy $(v_2 + 1)$ for a level v_2. This pattern is distinctly altered by the R–T effect; each vibrational level splits into a number of sub-components. The nature of the resultant energy level scheme can be appreciated qualitatively from symmetry considerations. Neglecting nuclear hyperfine effects, there are four angular momenta involved in the description of an individual rotational level: the electronic orbital angular momentum $\boldsymbol{L}$, the electron spin angular momentum $\boldsymbol{S}$, the vibrational angular momentum $\boldsymbol{G}$ and the end-over-end rotational angular momentum $\boldsymbol{R}$. We postpone discussion of the rotational contribution to the energy until a later section. The R–T effect causes a coupling of two of the angular momenta, $\boldsymbol{L}$ and $\boldsymbol{G}$, so that we can no longer think of distinct electronic and vibrational motion or states. Rather, we must consider coupled or *vibronic* states. The symmetry of the vibronic states can be derived very simply from the direct product of the orbital and vibrational symmetries. Let us consider the $v_2 = 1$ vibrational level of a linear triatomic molecule in a $^1\Pi$ electronic state. In the absence of coupling, we can describe the electronic wavefunctions in terms of the orbital quantum number Λ (which equals ±1 for a Π state). The vibrational wavefunction is characterized by two quantum numbers, v_2 and l; the former labels the vibrational energy level and the latter describes the component of the vibrational angular momentum along the internuclear, z axis. For the level $v_2 = 1$, the component quantum number l takes values ±1. In other words, the vibrational wavefunction for this state forms an irreducible representation of Π symmetry also. Using

$$\Gamma_{\text{vibronic}} = \Gamma_{\text{electronic}} \otimes \Gamma_{\text{vibration}}, \tag{5}$$

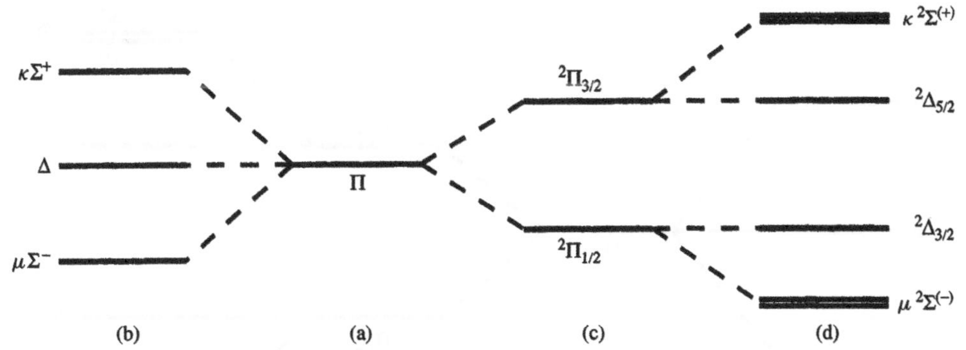

Figure 16.3. The spin–vibronic levels of a triatomic molecule in the $v_2 = 1$ level of a $^2\Pi$ electronic state. Part (b) shows the effect of the Renner–Teller interaction on the simple vibrational level shown in (a). Note that there is one unique level (Δ) and two non-unique levels (Σ). The two Σ levels are split apart by a first-order R–T effect. The energy ordering of these two levels is appropriate for a positive value for ε. The Σ^+ and Σ^- designation is swapped over if the sign of ε is reversed but the energy level pattern is unaffected by this change. Part (c) of this diagram shows the effect of electronic spin–orbit coupling on its own. The order of the levels in this case corresponds to a positive value of the spin–orbit coupling parameter A. Part (d) shows the combined effects of the R–T and spin–orbit coupling. It can be seen that the unique $^2\Delta$ component is not affected directly by R–T coupling and so shows a 'normal' spin–orbit splitting. The separation of the two Σ vibronic states depends on the R–T and spin–orbit coupling in quadrature. The symmetry labels Σ^+ and Σ^- are rigorously correct in part (a). However, when the effect of spin–orbit coupling is included as well, the $^2\Sigma^+$ and $^2\Sigma^-$ components are mixed and the $\pm$ characteristic is spoiled. For this reason, the states are labelled in some other way, such as the μ and κ energy ordering labels introduced by Hougen (1962a).

we deduce that three vibronic states are formed for the level $v_2 = 1$, with symmetry $\Sigma^+ \oplus \Sigma^- \oplus \Delta$. The single vibrational level is split into three components, separated typically by several tens of wavenumber units, as a result of the R–T effect; it is shown schematically in Figure 16.3(b).

For a large R–T effect, the quantum numbers for the two coupled angular momenta, Λ and l, cease to be good ones. However, the quantum number equal to their sum, $K = \Lambda + l$, is unaffected by the coupling and remains good; it is referred to as the vibronic quantum number. It can be seen in Figure 16.3 that, for the level $v_2 = 1$, there are two Σ states but only one Δ state. This is a general characteristic of the vibronic structure of any bending vibrational level of a linear triatomic molecule in a Π electronic state. There is always a single or *unique* level with $K = v_2 + 1$ (the largest possible value for K in a Π state); all the other vibronic components, with smaller values of K, occur in pairs. Thus for the zero point level, l can only take the value 0 so that $K = \pm 1$, i.e. there is only one unique level of Π symmetry. For $v_2 = 2$ on the other hand, l can take the values 0 and ± 2 so that K takes the values ± 3, ± 1 and ∓ 1. The resultant pair of Π levels and the unique Φ level are shown in Figure 16.4. We shall see later that, although the pairs of levels with the same values for K in a given vibrational manifold are mixed directly by R–T coupling, the unique level is not affected in first order by this interaction.

The R–T interaction is typically of the order of $100\,\text{cm}^{-1}$ in magnitude. If the linear triatomic molecule is in a $^{2S+1}\Pi$ state with S greater than zero, spin–orbit coupling effects of a similar magnitude are likely to occur. In this situation, a Hund's case (a) coupling scheme is appropriate and the spin angular momentum (quantum number S) is quantized along the linear z axis (component quantum number Σ). Let us consider the case of a $^2\Pi$ state. In the absence of R–T coupling, as for a diatomic molecule, the electronic state is split into two spin components characterized by a quantum number Ω where $\Omega = \Lambda + \Sigma$. For a $^2\Pi$ state, Ω takes the values $\pm\frac{1}{2}$ and $\pm\frac{3}{2}$

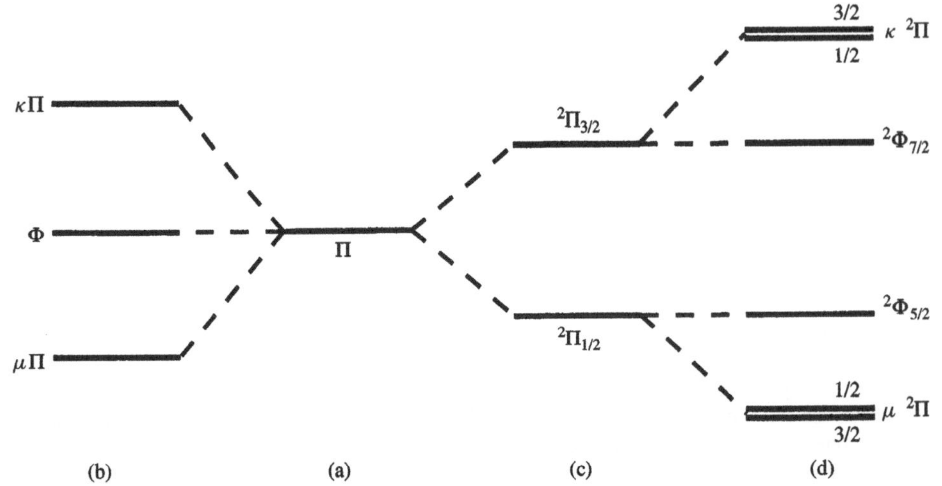

Figure 16.4. The spin–vibronic levels of a triatomic molecule in the $v_2 = 2$ level of a $^2\Pi$ electronic state. Part (b) shows the effect of Renner–Teller coupling alone on the simple vibrational level, shown in part (a). There is one unique level (Φ) and two non-unique levels (Π). Part (c) shows the effect of spin–orbit coupling on its own and part (d) shows the combined effects of R–T and spin–orbit couplings. The energy orderings shown correspond to positive values for ε, g_K and A. If the sign of ε is reversed, the parity labels of the μ and κ vibronic components are swapped over but the energy level pattern is unaltered. The values for the case (a) quantum number P for the μ and κ $^2\Pi$ vibronic states are given by the appropriate levels.

(the $\pm$ sign choices arise from the orbital degeneracy of this state). For a linear triatomic molecule, we must consider the effects of both R–T and spin–orbit couplings. The electron orbital angular momentum is involved in both interactions and the vibrational and spin angular momenta vie for its attention. There is a competition between the two effects and, as a result, neither K nor Ω is sure to be a good quantum number. Only the sum of all three component quantum number, $P = \Lambda + l + \Sigma$, remains good (and even then, only in the absence of rotational motion). The additional splittings caused by the spin–orbit interactions are shown in Figures 16.3 and 16.4 for the $v_2 = 1$ and 2 levels of a $^2\Pi$ state. Note in particular in Figure 16.3 that the $^2\Sigma^+$ and $^2\Sigma^-$ vibronic states, which are distinguished by the R–T effect, are directly mixed by spin–orbit coupling and, as a result, are pushed even further apart. The strict vibronic symmetry (superscript + or −, referring to behaviour of the vibronic coordinates under reflection in a plane containing the linear z axis) is thus spoiled by this effect. The two states retain $^2\Sigma$ character but are distinguished by another, less rigorous label such as 'upper' and 'lower' or κ and μ, as introduced by Hougen (1962a).

16.3 The Renner–Teller effect for Π states: the effective Hamiltonian

The basis of Renner's description of vibronic coupling (1934) is that the vibrational displacement generates an electric dipole moment which in turn is responsible for mixing the electronic states of the molecule. Since the mixing obeys dipole selection rules, the vibronic wavefunctions for a Π electronic state are perturbed to first order by Σ and Δ electronic states[2] and the resultant energy levels are affected in the second (and higher) order of perturbation theory. Such effects may be described as second order contributions to the R–T coupling. Pople and Longuet-Higgins

[2] For centro-symmetric molecules, the perturbations involved also obey the selection rule g $\leftrightarrow$ u. This is consistent with the breakdown of this symmetry as the molecule bends.

(1958), on the other hand, suggest that it is the quadrupolar term in the multipole expansion which is responsible for the vibronic coupling. This term makes a first-order contribution to the energy, that is, within the levels of the Π electronic state. In reality, the R–T effect is the aggregate of the first- and second-order contributions (and indeed of higher-order effects) as first pointed out by Carrington *et al.* (1971); see also Howard (1970). The dipolar term in second order mimics the quadrupolar term in first order and so can be represented by an operator of identical form in the effective Hamiltonian for the vibronic levels of the Π electronic state.

16.3.1 VIBRONIC COUPLING TREATED BY PERTURBATION THEORY

Relatively small vibronic couplings of electronic states can be reliably treated by perturbation theory; by 'small' in this context, we refer to molecules with a R–T parameter ε less than about 0.25 in magnitude. Such a treatment leads naturally to the construction of an effective Hamiltonian which can be used in turn to define the eigenstates of the molecule concerned.

The non-relativistic vibronic Hamiltonian for a molecule can be expressed in a molecule-fixed coordinate system as

$$H = T_e + T_n + V_{ee}(q) + V_{en}(q, Q) + V_{nn}(Q), \tag{6}$$

where T_e and T_n are the electronic and nuclear kinetic energies respectively (both measured in the molecule-fixed frame) and V is the Coulomb potential energy of the electron and/or nuclei according to the subscripts e or n. The electronic and nuclear (vibrational) coordinates are represented by q and Q, respectively. Anticipating a treatment by perturbation theory, it is convenient to define a zeroth-order Hamiltonian as

$$H_0 = T_e + V_{ee}(q) + V_{en}(q, Q_0) + V_{nn}(Q_0) + T_n + \tfrac{1}{2}\lambda_2\left(Q_{2x}^2 + Q_{2y}^2\right) \tag{7}$$

where Q_0 represents the equilibrium (linear) configuration of the nuclei, Q_{2x} and Q_{2y} are the normal coordinates for the degenerate bending mode (dimension : mass$^{1/2}$ length) and the potential energy parameter λ_2 is related to the bending harmonic wavenumber (in cm^{-1}) by

$$\lambda_2 = 4\pi c^2\omega_2^2. \tag{8}$$

The parameter λ_2 could be defined as the appropriate expectation value of the potential energy terms in equation (6) for each electronic state but, for simplicity, we shall take it to be some suitably chosen mean for the electronic states involved in the perturbation calculation. In other words, the bending wavenumber is assumed to be the same for all states in zeroth order. The form of the zeroth-order Hamiltonian allows us to write the basis functions as the product of an electronic factor $|\eta, \Lambda\rangle$ where η serves to identify the electronic state in question and a vibrational factor $|v_2, l\rangle$ where

$$\left\{T_n + \tfrac{1}{2}\lambda_2\left(Q_{2x}^2 + Q_{2y}^2\right)\right\}|v_2, l\rangle = (v_2 + 1)hc\omega_2|v_2, l\rangle. \tag{9}$$

Note that the vibrational stretching modes have not been explicitly included in this treatment because they are not relevant to the discussion. The choice of the zeroth-order Hamiltonian (7) has two advantages. First, it allows the complete separation of electronic and vibrational coordinates so that the zeroth-order wavefunction can be written as the product of an electronic wavefunction $\phi_e^0(q)$ and a nuclear wavefunction $\chi_v(Q)$; this is called the crude adiabatic approximation by Longuet-Higgins (1961). Second, the inclusion of the last two terms in equation (7) ensures that the vibrational levels of each electronic state are separated in zeroth order, at least as far as the bending vibration is concerned. This allows the terms which govern the vibronic energy level distribution in the effective Hamiltonian to appear in a lower order of perturbation theory.

The terms in the Hamiltonian (6) which are not included in H_0 are treated as a perturbation H':

$$H' = V_{en}(q, Q) - V_{en}(q, Q_0) + V_{nn}(Q) - V_{nn}(Q_0) - \tfrac{1}{2}\lambda_2\left(Q_{2x}^2 + Q_{2y}^2\right). \tag{10}$$

For small displacements of the nuclei from the linear configuration, it is valid to develop the potential energy as a Taylor series expansion about the equilibrium structure,

$$H' = V_{11}Q_2\cos(\theta - \phi) + V_{20}Q_2^2 + V_{22}Q_2^2\cos 2(\theta - \phi) + O\left(Q_2^3\right), \tag{11}$$

where Q_2 and ϕ are the alternative bending vibrational coordinates defined by

$$Q_2 = \left[Q_{2x}^2 + Q_{2y}^2\right]^{1/2}, \tag{12}$$

$$\phi = \tan^{-1}(Q_{2y}/Q_{2x}), \tag{13}$$

and θ is the electronic azimuthal angle, defined such that

$$\langle\Lambda'|\exp(\pm 2i\theta)|\Lambda\rangle = (-1)\delta_{\Lambda',\Lambda\pm 2}. \tag{14}$$

We need also to define the phase convention for the vibrational wavefunctions. It is convenient at this point to introduce the dimensionless vibrational coordinate q_2:

$$q_2 = \gamma_2^{1/2}Q_2, \tag{15}$$

where

$$\gamma_2 = hc\omega_2/\hbar^2. \tag{16}$$

Using the associated shift operators $q_\pm \equiv q_2\exp(\pm i\phi)$, the phase convention of di Lauro and Mills (1961) is given by

$$2^{1/2}q_\pm|v_2 l\rangle = [v_2 + 2 \pm l]^{1/2}|v_2 + 1, l \pm 1\rangle + [v_2 \mp l]^{1/2}|v_2 - 1, l \pm 1\rangle. \tag{17}$$

In other words the non-zero matrix elements are

$$\langle v_2 + 1, l \pm 1|q_\pm|v_2, l\rangle = \langle v_2, l|q_\mp|v_2 + 1, l \pm 1\rangle = (2)^{-1/2}[v_2 \pm l + 2]^{1/2}. \tag{18}$$

16.3.2 THE EFFECTIVE RENNER–TELLER HAMILTONIAN

We now incorporate the effects of the vibronic operator H' by standard degenerate perturbation theory [Messiah (1962)]. It is a relatively straightforward exercise to derive an effective Hamiltonian which describes these vibronic coupling effects to an adequate accuracy but which operates only within the manifold of vibrational (and rotational) states of the Π electronic state [Brown (1977)]. The result is

$$H_{RT}/hc = \tfrac{1}{2}\varepsilon\omega_2\left(q_+^2 e^{-2i\theta} + q_-^2 e^{2i\theta}\right) + g_K(G_z + L_z)L_z. \tag{19}$$

The first of the two terms in this Hamiltonian is essentially the operator first derived by Renner (1934). The magnitude of the effect is given by the parameter $\varepsilon\omega_2$ and it is easy to see from the operators involved that it has non-zero matrix elements with $\Delta v_2 = 0, \pm 2$ and $\Delta\Lambda = -\Delta l = \pm 2$, in other words, it mixes the two electronic components $\Lambda = \pm 1$ of the Π state and preserves the quantum number K. As anticipated, the R–T parameter has both first- and second-order contributions, the latter from the admixture of excited Σ states. Their explicit forms are given in Table 16.1. The sign of the second-order contribution to the R–T parameter depends on the

Table 16.1. Expressions for the parameters in the effective vibronic Hamiltonian for a linear molecule in a Π electronic state, derived by perturbation theory.

Parameter	Reduced mass dependence
$\varepsilon\omega_2 = \varepsilon^{(1)}\omega_2 + \varepsilon^{(2)}\omega_2$	
$\varepsilon^{(1)}\omega_2 = (hc\gamma_2)^{-1}\langle\eta\|V_{22}\|\eta'\rangle$	$\mu_2^{-1/2}$
$\varepsilon^{(2)}\omega_2 = -(2hc\gamma_2)^{-1}\sum^{\dagger}_{\Sigma\text{states}}(-1)^s\left\|\langle\eta\|V_{11}\,e^{i\theta}\|\eta\rangle\right\|^2 \times \left\{1+(hc\omega_2/\Delta E)^2\right\}/\Delta E$	$\sim\mu_2^{-1/2}$
$g_K = \omega_2(4\gamma_2)^{-1}\sum^{\dagger}_{\eta'}(-1)^p\|\langle\eta\|V_{11}\|\eta'\rangle\|^2/(\Delta E)^2$	μ_2^{-1}
$\Delta A = A_\eta(4\gamma_2)^{-1}\sum^{\dagger}_{\eta'}\|\langle\eta\|V_{11}\|\eta'\rangle\|^2 \times \left\{\left[1-(-1)^p\right]\frac{A_{\eta'}}{A_\eta}-1\right\}/(\Delta E)^2$	$\mu_2^{-1/2}$
$\Delta g_L = -(4\gamma_2)^{-1}\sum^{\dagger}_{\eta'}(-1)^p\|\langle\eta\|V_{11}\|\eta'\rangle\|^2/(\Delta E)^2$	$\mu_2^{-1/2}$

$$\gamma_2 = hc\omega_2/\hbar^2$$

η, η' are the labels of the Π and Σ or Δ electronic states respectively.
ΔE is the energy separation between the zeroth order electronic states, $\Delta E = E^0(\eta', \Lambda') - E^0(\eta, \Lambda)$.
In the summation over Σ electronic states, s is even or odd according as the state is Σ^+ or Σ^-.
In the summation over Σ and Δ states, p is even or odd for Σ or Δ states respectively.
μ_2 is the reduced mass for the bending vibration of the linear triatomic molecule; it is equal to $2m_Am_B/(m_A+2m_B)$ for a symmetrical AB_2 molecule.

sign of the energy denominator ΔE and on whether the Σ state has Σ^+ or Σ^- character. To a good approximation, the R–T parameter $\varepsilon\omega_2$ is proportional to $\mu_2^{-1/2}$ where μ_2 is the reduced mass for the bending vibrational coordinate. This is of course the same reduced mass dependence as the harmonic bending wavenumber ω_2 which implies that the dimensionless parameter ε is a property of the electronic state only, that is, it is isotope independent.

The second term in the effective Hamiltonian, equation (9), is a small but significant correction first derived by Brown (1977); it is typically about two orders of magnitude smaller than the leading term. The operators G_z and L_z represent the vibrational and electron orbital momentum about the linear axis ($-i\hbar\partial/\partial\phi$ and $-i\hbar\partial/\partial\theta$ respectively). The g_K parameter arises from the difference in the second order mixing of excited Σ and Δ electronic states. Its form is also given in Table 16.1. This parameter has a stronger dependence on the reduced mass than $\varepsilon\omega_2$, namely μ_2^{-1}. The effect of this term in the Hamiltonian is to shift levels with different K values, even those arising in the same v_2 manifold, relative to each other.

16.3.3 OTHER MANIFESTATIONS OF VIBRONIC COUPLING IN THE EFFECTIVE HAMILTONIAN

For a pure Π electronic state, the expectation value of L_z, is $\hbar\Lambda$ where $\Lambda = \pm 1$. When such a Π state is mixed with other Δ or Σ states through the bending dependence of the electrostatic potential, the expectation value of L_z is modified to an extent dependent on the excitation of the bending vibration (i.e. on v_2). All the terms in the effective Hamiltonian which involve the orbital angular momentum are modified in this way. In practice, however, the orbital angular momentum is taken to be unity in the effective Hamiltonian and the effects of vibronic coupling are absorbed into the parameters instead. Thus for example, the spin–orbit coupling term $A_{\text{eff}}L_zS_z$ is modified as follows:

$$A_{\text{eff}} = A + \Delta A(v_2 + 1). \quad (20)$$

The expression for ΔA, derived by second-order perturbation theory, is given in Table 16.1. Similarly, the orbital contribution to the magnetic dipole moment of the linear molecule $-g_{\text{L}}^{\text{eff}}\mu_{\text{B}}$ $\boldsymbol{L}$ varies with v_2 as

$$g_{\text{L}}^{\text{eff}} = g_{\text{L}} + \Delta g_{\text{L}}(v_2 + 1); \quad (21)$$

the expression for Δg_{L} is also given in Table 16.1. Finally, the nuclear spin–electron orbital hyperfine interaction $a_{\text{eff}}I_zL_z$ is modified in the same way:

$$a_{\text{eff}} = a + \Delta a(v_2 + 1). \quad (22)$$

The expression for Δa is the same as that for ΔA in Table 16.1 with the upper case As replaced by lower case as throughout.

Because these modifications of the terms in the effective Hamiltonian all have the same physical origin, they are related to each other, either rigorously

$$g_{\text{K}} = -\omega_2\, \Delta g_{\text{L}}, \quad (23)$$

or approximately

$$\Delta A \simeq A\, \Delta g_{\text{L}}, \quad (24)$$

$$\Delta a \simeq a\, \Delta g_{\text{L}}. \quad (25)$$

These relationships follow directly from the expressions given in Table 16.1.

16.4 Full description of molecular energy levels

16.4.1 THE FULL EFFECTIVE HAMILTONIAN

The general layout of the vibronic energy levels is described by the vibrational and R–T Hamiltonians. However, for the open-shell molecules which show the R–T effect, there are many other contributions to the energy which have to be taken into account for an accurate description. For molecules with $S \geq \frac{1}{2}$, the next largest interaction is spin–orbit coupling. Within a Π electronic state, it can be described by the simple operator

$$H_{\text{so}}/hc = A_{\text{eff}}L_zS_z, \quad (26)$$

where A_{eff} is the effective spin–orbit coupling constant for the $^2\Pi$ state. The incorporation of this term into the Renner–Teller problem was first tackled for the case of a $^2\Pi$ state by Pople (1960). He produced complicated energy level formulae which can nowadays be simply by-passed by computer diagonalization of the effective Hamiltonian.

The next most significant term is the rotational kinetic energy. The contribution of this term to the energy levels of a molecule in a $^2\Pi$ state was first discussed in a ground-breaking paper by Hougen (1962a). The Hamiltonian which operates within a $^{2S+1}\Pi$ state is written

$$H_{\text{rot}}/hc = B(\boldsymbol{J} - \boldsymbol{G} - \boldsymbol{S})^2, \quad (27)$$

where $B = \hbar/(2Ihc)$ is the rotational constant and $\boldsymbol{J}$, $\boldsymbol{G}$ and $\boldsymbol{S}$ are the total, vibrational and spin angular momenta respectively. No mention is made here of $\boldsymbol{L}$ because this operator mixes different electronic states and is therefore suppressed in the effective Hamiltonian. It is more convenient to write equation (27) in terms of the angular momentum N where $N = J - S$ in situations where Hund's case (b) coupling applies.

In addition to these two terms, there are several others which, though smaller in magnitude, must also be included to interpret modern experimental data adequately. They are the following:

Spin–spin coupling $$H_{ss}/hc = \tfrac{2}{3}\lambda(3S_z^2 - S^2), \tag{28}$$

Spin–rotation coupling $$H_{sr}/hc = \gamma(J - G - S)\cdot S, \tag{29}$$

Λ-type doubling $$\begin{aligned} H_{\Lambda\mathrm{doub}}/hc = &-\tfrac{1}{2}o(S_+^2\,\mathrm{e}^{-2i\theta} + S_-^2\,\mathrm{e}^{2i\theta}) \\ &+\tfrac{1}{2}p(N_+S_+\,\mathrm{e}^{-2i\theta} + N_-S_-\,\mathrm{e}^{2i\theta}) \\ &-\tfrac{1}{2}q(N_+^2\mathrm{e}^{-2i\theta} + N_-^2\,\mathrm{e}^{2i\theta}). \end{aligned} \tag{30}$$

We recall that θ is the electronic azimuthal angle.

l-type doubling: $$\begin{aligned} H_{l\mathrm{doub}}/hc = &-\tfrac{1}{2}o_G(S_+^2\,\mathrm{e}^{-2i\phi} + S_-^2\,\mathrm{e}^{2i\phi}), \\ &+\tfrac{1}{2}p_G(N_+S_+\,\mathrm{e}^{-2i\phi} + N_-S_-\,\mathrm{e}^{2i\phi}) \\ &-\tfrac{1}{2}q_G(N_+^2\,\mathrm{e}^{-2i\phi} + N_-^2\,\mathrm{e}^{2i\phi}). \end{aligned} \tag{31}$$

We recall that ϕ is the vibrational azimuthal angle.

The operators which represent the nuclear hyperfine interactions, both magnetic and electric, must also be included if the molecule contains a nuclear with non-zero nuclear spin I:

Magnetic hyperfine $$\begin{aligned} H_{\mathrm{mhf}}/hc = &\, aL_zI_z + bI\cdot S + cI_zS_z \\ &+\tfrac{1}{2}d(I_+S_+\,\mathrm{e}^{-2i\theta} + I_-S_-\,\mathrm{e}^{2i\theta}), \end{aligned} \tag{32}$$

Electric hyperfine $$H_Q/hc = -eT^2(Q)\cdot T^2(\nabla E). \tag{33}$$

Finally, there are operators which represent centrifugal distortion corrections to the various terms. The most important are the corrections to

(i) the rotational kinetic energy

$$H_{cd}/hc = -D(J - G - S)^4, \tag{34}$$

and (ii) the R–T coupling [Gillett (1994)]

$$H_{\mathrm{RTcd}}/hc = \tfrac{1}{4}\varepsilon\omega_{2D}\left[(q_+^2\,\mathrm{e}^{-2i\theta} + q_-^2\,\mathrm{e}^{2i\theta}), (J - G - S)^2\right]_+. \tag{35}$$

16.4.2 REPRESENTATION OF THE EFFECTIVE HAMILTONIAN: BASIS FUNCTIONS

The full effective Hamiltonian consists of the vibrational Hamiltonian, equation (9), the R–T Hamiltonian, equation (19) and all the contributions from equations (26)–(35). It gives all the various energy levels associated with the bending vibrational levels of a linear triatomic molecule. It can also be extended to include the two stretching vibrations although we ignore them in this article. This is tantamount to assuming that the stretching vibrations are decoupled from the other motions through some sort of perturbation treatment. The representation of the effective Hamiltonian is most conveniently constructed in a parity-conserving combination of Hund's case (a) basis functions of the form:

$$\tfrac{1}{\sqrt{2}}\{|\eta\Lambda; v_2 l; S\Sigma; JPM_J\rangle \pm (-1)^p|\eta - \Lambda; v_2, -l; S, -\Sigma; J, -P, M_J\rangle\}, \tag{36}$$

where the phase factor coefficient is given by

$$p = J - S + l, \tag{37}$$

and η labels the electronic state in question. The upper and lower sign choices correspond to states of positive and negative parity respectively (that is, the behaviour of the wavefunction under space-fixed inversion, see Brown *et al.* 1978).

For a $^{2S+1}\Pi$ electronic state, the basis functions which characterize a vibronic state with given (v_2, K) fall into two categories. The first category is denoted ψ_1 and is characterized by $l_1 = K - 1$, that is $\Lambda = 1$:

$$\begin{aligned}|\psi_1; v_2; S\Sigma; J\pm\rangle = \tfrac{1}{\sqrt{2}}\{&|1; v_2, l_1, K; S\Sigma; JK + \Sigma, M_J\rangle \\ &\pm (-1)^p|-1; v_2, -l_1, K; S, -\Sigma; J, -(K+\Sigma), M_J\rangle\}.\end{aligned} \tag{38}$$

The second category is denoted ψ_2 and is characterized by $l_2 = K + 1 = l_1 + 2$, that is $\Lambda = -1$:

$$\begin{aligned}|\psi_2; v_2; S\Sigma; J\pm\rangle = \tfrac{1}{\sqrt{2}}\{&|-1; v_2, l_2; K, S\Sigma; J, K + \Sigma, M_J\rangle \\ &\pm (-1)^p|1; v_2, -l_2; K, S - \Sigma; J, -(K+\Sigma), M_J\rangle\}.\end{aligned} \tag{39}$$

Each of the two categories ψ_1 and ψ_2 is spanned by the $(2S+1)$ permitted values for Σ from $-S$ to $+S$. Note that, in writing these basis functions, K has been taken as an unsigned quantum number.

There are two special cases for the above choice of basis functions.

(i) *Unique levels*. For the lowest level with a given value for K, that is the level with $v_2 = K - 1$, only the ψ_1 combination exists because the largest value which l can take is $l_{max} = v_2$ (or $K - 1$). Consequently, the value $l' = K + 1$ is not allowed. For all larger values for v_2, both the ψ_1 and ψ_2 forms exist. In other words, there are *two* vibronic levels with a given K-value for each v_2 manifold of levels.

(ii) Σ *vibronic levels* ($K = 0$). Σ vibronic levels require $l = \pm 1$ for a Π electronic state. In this case, the ψ_1 and ψ_2 wavefunctions are not independent. Either one form or the other can be used to model the $K = 0$ levels; we shall, by convention, use ψ_1. For each vibrational level with an even value for v_2, there are *two* Σ vibronic states. These two states have Σ^+ and Σ^- symmetry in the limit of vanishingly small spin–orbit coupling [Hougen (1962a)].

16.4.3 MATRIX ELEMENTS OF THE EFFECTIVE HAMILTONIAN

The matrix representation of the effective Hamiltonian can now be constructed using the basis functions in equations (38) and (39). This is a well-established procedure and we shall not go into details here.

The R–T Hamiltonian, equation (19), has matrix elements which obey the following selection rules:

$$\Delta\Lambda = -\Delta l = \pm 2; \Delta v_2 = 0; \pm 2; \Delta K = 0; \Delta\Sigma = 0. \tag{40}$$

Thus, for all states with $K > 0$, H_{RT} connects only ψ_1 and ψ_2 functions.

Diagonal in v_2:

$$\langle\psi_2; v_2; J\pm|H_{RT}|\psi_1; v_2; J\pm\rangle = -\tfrac{1}{2}\varepsilon\omega_2\left[n^2 - K^2\right]^{1/2},$$
$$\text{where} \quad n = v_2 + 1. \tag{41}$$

Off-diagonal in v_2:

$$\langle\psi_2; v_2+2; J\pm|H_{\rm RT}|\psi_1; v_2 J\pm\rangle = -\tfrac{1}{4}\varepsilon\omega_2[(n+K+2)(n+K)]^{1/2}, \tag{42a}$$

$$\langle\psi_1; v_2+2; J\pm|H_{\rm RT}|\psi_2; v_2 J\pm\rangle = -\tfrac{1}{4}\varepsilon\omega_2[(n-K+2)(n-K)]^{1/2}. \tag{42b}$$

For states with $K=0$, $H_{\rm RT}$ links the two parts of ψ_1 and so, in this case, the matrix elements contain the phase factor $(-1)^p$.

Diagonal in v_2:

$$\langle\psi_1; v_2; J\pm|H_{\rm RT}|\psi_1; v_2; J\pm\rangle = \mp(-1)^p\tfrac{1}{2}\varepsilon\omega_2 n. \tag{43}$$

Off-diagonal in v_2:

$$\langle\psi_1; v_2+2; J\pm|H_{\rm RT}|\psi_1; v_2; J\pm\rangle = \mp(-1)^p\tfrac{1}{4}\varepsilon\omega_2[n(n+2)]^{1/2}. \tag{44}$$

We note that the signs of the matrix elements of $H_{\rm RT}$ are different from those used by most previous workers [e.g. Hougen (1962a)]. This is a result of our choice of a different phase convention, equation (37).

An important characteristic of the R–T Hamiltonian is that it mixes bending vibrational levels differing in v_2 by 2. As a result, the matrix representation of a set of vibronic levels with a given value of K is infinite in dimension. For molecules which are subject only to a weak R–T effect, it is possible to model these off-diagonal terms by perturbation theory [see, e.g. Pople (1960)]. However, for larger R–T coupling or with modern data of increasing precision, it is necessary to diagonalize the matrix representation, truncated at some suitable finite dimension.

The other terms in the effective Hamiltonian are more familiar. Terms such as $H_{\rm so}$ and $H_{\rm rot}$ only have non-zero matrix elements which are diagonal in ψ_1 or ψ_2. The parity-dependent operators, the Λ-type or l-type doubling terms, are a little more complicated. $H_{\Lambda\rm doub}$ for example obeys selection rules

$$\Delta\Lambda = \pm 2, \Delta l = 0 \Rightarrow \Delta K = \pm 2.$$

Thus the matrix elements of this operator are diagonal for $K=1(\Pi)$ vibronic states but are otherwise off-diagonal (e.g. $K=0$ and 2 mixing). For $K=1$, l must equal zero and $H_{\Lambda\rm doub}$ connects the two parts of the ψ_1 function but not ψ_2. Similar comments can be made about $H_{l\rm doub}$ except that, in this case for $K=1$, the operator connects ψ_1 and ψ_2 and does not act within them.

Explicit expressions for the matrix elements of the various operator terms in the effective Hamiltonian have been given in various places in the literature; see for example Brown *et al.* (1978).

16.4.4 TREATMENT OF OFF-DIAGONAL RENNER–TELLER TERMS BY PERTURBATION THEORY

A complication in the quantum mechanical treatment of the R–T effect is that it mixes bending vibrational levels of the molecule with v_2 values differing by two. Provided such effects are small, they can be treated adequately by perturbation theory. This requires

$$\omega_2 \gg |\varepsilon\omega_2|, \tag{45a}$$

$$\text{i.e. } |\varepsilon| \ll 1. \tag{45b}$$

Renner–Teller effect for a molecule in a $^1\Pi$ state

Let us assume that the condition (45) applies for a molecule in a $^1\Pi$ state. The energy levels of the molecule in the bending vibrational level v are affected by R–T mixing with the levels $v+2$

and $v-2$ according to the second-order perturbation expression:

$$E^{(2)} = \sum_{v' \neq v} \langle v|H_{\mathrm{RT}}|v'\rangle\langle v'|H_{\mathrm{RT}}|v\rangle/(E_v - E_{v'}). \tag{46}$$

For compactness, we shall drop the subscript on the bending vibrational quantum number in this section. Whenever v occurs, it refers to the bending vibration. It is convenient to write the off-diagonal matrix elements of H_{RT}, given in equation (42), as

$$\langle v \pm 2; l \pm 2; -\Lambda|H_{\mathrm{RT}}|v, l, \Lambda\rangle = -\tfrac{1}{4}\varepsilon\omega_2[(n \pm \Lambda K + 2)(n \pm \Lambda K)]^{1/2}, \tag{47}$$

where the orbital quantum number Λ can take values ± 1, $n = v + 1$ and the matrix element links states with a value of K equal to $\Lambda + l$, as usual. Substituting in equation (46) and ignoring anharmonic corrections, we have

$$\begin{aligned} E^{(2)} &= \langle vl\Lambda|H_{\mathrm{RT}}|v+2, l', \Lambda'\rangle^2/(E_n - E_{n+2}) + \langle vl\Lambda|H_{\mathrm{RT}}|v-2, l', \Lambda'\rangle^2/(E_n - E_{n-2}) \\ &= \tfrac{1}{16}(\varepsilon\omega_2)^2[(n+\Lambda K)(n+\Lambda K+2) - (n-\Lambda K)(n-\Lambda K-2)]/(2\omega_2) \\ &= -\tfrac{1}{8}\varepsilon^2\omega_2(\Lambda K+1)n. \end{aligned} \tag{48}$$

For a unique level, $\Lambda K = n$ and so

$$E^{(2)}_{\mathrm{unique}} = -\tfrac{1}{8}\varepsilon^2\omega_2 n(n+1). \tag{49}$$

The values for this second-order correction for the lowest few vibrational levels are given in Table 16.2.

We recall that the vibronic sub-states of a non-unique level are mixed in first order by the R–T effect through the matrix element

$$\langle v, l+2\Lambda, -\Lambda|H_{\mathrm{RT}}|v, l, \Lambda\rangle = -\tfrac{1}{2}\varepsilon\omega_2\left[n^2 - K^2\right]^{1/2}. \tag{50}$$

A treatment similar to that above but using third-order degenerate perturbation theory [Messiah (1962)] shows that the effects of the matrix elements of H_{RT} off-diagonal in v can be represented by an additional element diagonal in v of the form:

$$\langle v, l+2\Lambda, -\Lambda|H_{\mathrm{RT}}|v, l, \Lambda\rangle^{(3)} = -\tfrac{1}{2}\varepsilon\omega_2\left[n^2 - K^2\right]^{1/2}\tfrac{1}{16}\varepsilon^2\left(2 - K^2\right). \tag{51}$$

Table 16.2. Second- and third-order contributions to the vibronic components of a level $|v_2\Lambda K\rangle$ in a $^1\Pi$ state of a linear triatomic molecule, treated by perturbation theory.

Second order	$n=1$	$(v_2=0)$	$K=1$	$E^{(2)} = -\frac{1}{4}\varepsilon^2\omega_2$
	$n=2$	$(v_2=1)$	$K=2$	$E^{(2)} = -\frac{3}{4}\varepsilon^2\omega_2$
			$K=0$	$E^{(2)} = -\frac{1}{4}\varepsilon^2\omega_2$
	$n=3$	$(v_2=2)$	$K=3$	$E^{(2)} = -\frac{3}{2}\varepsilon^2\omega_2$
			$K=1, \Lambda=1$	$E^{(2)} = -\frac{3}{4}\varepsilon^2\omega_2$
			$K=1, \Lambda=-1$	$E^{(2)} = 0$
Third order		$K=0$	$\varepsilon_{\mathrm{corr}} = \varepsilon\left(1+\frac{1}{8}\varepsilon^2\right)$	
		$K=1$	$\varepsilon_{\mathrm{corr}} = \varepsilon\left(1+\frac{1}{16}\varepsilon^2\right)$	
		$K=2$	$\varepsilon_{\mathrm{corr}} = \varepsilon\left(1-\frac{1}{8}\varepsilon^2\right)$	

Comparison of equations (50) and (51) shows that these third-order effects can be included with an off-diagonal element of the same form as equation (50) but with a corrected R–T parameter

$$\varepsilon_{\text{corr}} = \varepsilon\left[1 + \tfrac{1}{16}\varepsilon^2(2 - K^2)\right], \tag{52}$$

independent of v. Values for this corrected parameter for the first three values of K are also given in Table 16.2.

If the effects of the $\Delta v = 2$ matrix elements of H_{RT} are folded in by perturbation theory as described above, the vibronic levels for a given K value can be represented by a 2×2 matrix. Diagonalization of this matrix gives the two vibronic eigenvalues

$$E^{\pm}_{n\Lambda K} = n\omega_2 + g_{\text{K}}K\Lambda \pm \tfrac{1}{2}\varepsilon_{\text{corr}}\omega_2\left[n^2 - K^2\right]^{1/2} + E^{(2)}. \tag{53}$$

The contributions from the rotational kinetic energy, equation (27), and the parity-doubling terms, equations (30) and (31),

$$B\left[J(J+1) + \Lambda^2 - K^2\right] \pm \tfrac{1}{2}q_{\text{eff}}J(J+1)\delta_{K,1},$$

are added to these levels. The vibronic levels with $K = 0$ are a special case. Putting $K = 0$ in equation (53) and taking the value for $\varepsilon_{\text{corr}}$ from Table 16.2, we obtain to third order

$$E^{\pm(3)}_{n\Lambda 0} = n\left(1 \pm \tfrac{1}{2}\varepsilon - \tfrac{1}{8}\varepsilon^2 \pm \tfrac{1}{16}\varepsilon^3\right)\omega_2. \tag{54}$$

In the limit of infinite-order perturbation theory, this expression goes over to the exact result:

$$E^{\pm}_{n\Lambda 0} = n(1 \pm \varepsilon)^{1/2}\omega_2. \tag{55}$$

These are the eigenvalues of a harmonic oscillator within the potential functions V' and V'' given in equation (2). Each potential curve supports its own set of non-degenerate vibrational levels for the case $K = 0$ and there is no interaction between them.

Renner–Teller effect for a molecule in a $^2\Pi$ state

The description of a molecule in a $^2\Pi$ state subject to R–T effect is more complicated than that of a molecule in a $^1\Pi$ state because of the involvement of spin–orbit coupling. These two effects compete with each other for the orbital angular momentum as has been discussed qualitatively in Section 16.2. Two situations which can be treated by perturbation theory arise in practice. In the first, the spin–orbit splittings are much smaller than the separation between bending vibrational levels, that is

$$|A|, |\varepsilon\omega_2| \ll \omega_2. \tag{56}$$

This is the case treated by Pople (1960) in his early paper. In the second situation, the spin–orbit splitting is larger than or comparable with the vibrational spacing:

$$|A| \gtrsim \omega_2 \gg |\varepsilon\omega_2| \tag{57}$$

For the first of these two cases, we can determine the modification of the expectation value of L_z, the electronic orbital angular momentum, by a third-order perturbation treatment similar to that described in the previous sub-section. The general result is

$$\langle vl\Lambda|L_z|vl\Lambda\rangle^{(3)} = -\tfrac{1}{16}\Lambda\varepsilon^2\left[n^2 + 2\Lambda K + (\Lambda K)^2\right]. \tag{58}$$

For a unique level, this simplifies to

$$\langle vl\Lambda|L_z|vl\Lambda\rangle^{(3)}_{\text{unique}} = -\tfrac{1}{8}\varepsilon^2 n(n+1), \tag{59}$$

a result first given by Hougen (1962a). Using this result, we can write an effective spin–orbit coupling constant for a unique level $n = K$:

$$A_{\text{eff}} = (A + \Delta A n)\left[1 - \tfrac{1}{8}\varepsilon^2 n(n+1)\right], \tag{60}$$

see Table 16.1.

Using the perturbation corrections described above and in the previous subsection, we can represent the effective Hamiltonian operator for a bending level v_2 by a matrix of dimension $2n \times 2n$ (two states for each allowed K value from the spin degeneracy). To a good approximation, this factorizes into n 2×2 matrices because the only operators which link states of different K value are the Λ- and l-type doubling terms which are small in magnitude. The matrix representation of the spin-vibronic components of the $v_2 = 1$ level of a molecule in a $^2\Pi$ state is given in Table 16.3. This type of model has been applied successfully to description of the $v_2 = 1$ levels of several linear triatomic molecules, such as NCO in its ground $^2\Pi$ state [Bolman *et al.* (1975), and Gillett (1994)]. The major parameters for this molecule are $\omega_2 = 534.06\,\text{cm}^{-1}$, $\varepsilon = -0.1437$ and $A = -96.088\,\text{cm}^{-1}$.

When the spin–orbit splitting is comparable with or larger than the spacing between the bending vibrational levels, in accordance with equation (36), the perturbation treatment must be set up differently. The spin–orbit coupling operator, equation (26), is now included in the zeroth-order Hamiltonian so that the spin degeneracy is lifted in zeroth-order. A treatment along these lines has been given by Crozet *et al.* (1995). The matrix representation of the spin–vibronic components of the $v_2 = 1$ level of a molecule in a $^2\Pi$ state in this situation is given in Table 16.4. The representation is only slightly more complicated than that given for the same level but with $|A| \ll \omega_2$ in Table 16.3. This approach has been used successfully to describe the levels with $v_2 = 1$ and $v_2 = 2$ of $CuCl_2$ in its ground $^2\Pi_g$ state [Crozet *et al.* (1995)] where $\omega_2 = 94.55\,\text{cm}^{-1}$, $\varepsilon = -0.1704$ and $A = -251.5\,\text{cm}^{-1}$ (the latter value is not accurately determined).

Renner–Teller effect for a molecule in a $^3\Pi$ state

The general approach described above can be easily extended to describe the energy levels of molecules in Π electronic states of higher multiplicity [Hougen (1962b)]. The only examples of molecules in this class which have so far been observed experimentally are in $^3\Pi$ states, e.g. NCN [Herzberg and Travis (1964)] and CCO [Devillers and Ramsay (1971)]. As an example of the representation which can be used to describe the vibronic levels of such molecules, we give the matrix representation of the $K = 0(^3\Sigma)$ vibronic components of a linear triatomic molecule in a bending vibrational level with odd v_2 of a $^3\Pi$ electronic state in Table 16.5. This matrix is taken from a paper by Beaton and Brown (1997) on the NCN radical, following Hougen's approach. We note the important result that the order of the $P = 0^+$ and 0^- energy levels, which can be derived from the middle element of the 3×3 matrix, depends directly on the sign of $\varepsilon\omega_2$. If the order of these two levels can be determined from experimental observation, so too can the sign of the R–T parameter ε.

16.5 Other considerations

16.5.1 ANHARMONIC EFFECTS

The bending vibrational potential energy of a molecule in a non-degenerate electronic state is slightly anharmonic (typically to the extent of a few per cent). In the same way, so is the pair of potential curves for a linear triatomic molecule in a Π electronic state. Furthermore, there is no reason why the anharmonic corrections to two potential curves should be the same. As a result, there are small additional contributions to the R–T Hamiltonian in equation (19) which depend

Table 16.3. Matrix representation of the spin–vibronic components of the bending level $v_2 = 1$ of a molecule in a $^2\Pi$ state (for a given J value) for $\omega_2 \gg |A|, |\varepsilon\omega_2|$.

$\|\Lambda, l, \Sigma, \pm\rangle$	$^2\Delta_{5/2}\|1, 1, \frac{1}{2}\rangle$	$^2\Delta_{5/2}\|1, 1, -\frac{1}{2}\rangle$	$^2\Sigma_{1/2}\|1, -1, \frac{1}{2}\rangle$	$^2\Sigma_{1/2}\|1, -1, -\frac{1}{2}\rangle$
$\langle 1, 1, \frac{1}{2}\|$	$2\omega_2 + 2g_K - \frac{3}{4}\varepsilon^2\omega_2 + \frac{1}{2}A\left(1 - \frac{3}{4}\varepsilon^2\right) + B(z-4) - D\left(z^2 - 7z + 13\right)$	$-(z-3)^{1/2}\left[B^* - 2D(z-2)\right]$	$-\frac{1}{2}q_G[z(z-3)]^{1/2}$	$\pm(-1)^p\frac{1}{2}q[z(z-3)]^{1/2}$
$\langle 1, 1, -\frac{1}{2}\|$		$2\omega_2 + 2g_K - \frac{3}{4}\varepsilon^2\omega_2 - \frac{1}{2}A\left(1 - \frac{3}{4}\varepsilon^2\right) - \gamma + Bz - D\left(z^2 + z - 3\right)$	$\pm(-1)^p\frac{1}{2}q[z(z+1)]^{1/2} + \frac{1}{2}(p_G + 2q_G)z^{1/2}$	$\mp(-1)^p\frac{1}{2}(p+2q)z^{1/2} - \frac{1}{2}q_G[z(z+1)]^{1/2}$
$\langle 1, -1, \frac{1}{2}\|$			$2\omega_2 - \frac{1}{4}\varepsilon^2\omega_2 + \frac{1}{2}A\left(1 - \frac{1}{4}\varepsilon^2\right) \pm (-1)^p\varepsilon\omega_{2D}\left(J + \frac{1}{2}\right) + B(z+2) - D(z^2 + 5z + 5)$	$-(z+1)^{1/2}\left[B^* - 2D(z+2)\right] \mp (-1)^p\varepsilon\left(1 + \frac{1}{8}\varepsilon^2\right)\omega_2 \mp (-1)^p\varepsilon\omega_{2D}(z+2)$
$\langle 1, -1, -\frac{1}{2}\|$				$2\omega_2 - \frac{1}{4}\varepsilon^2\omega_2 - \frac{1}{2}A\left(1 - \frac{1}{4}\varepsilon^2\right) - \gamma \pm (-1)^p\varepsilon\omega_{2D}\left(J + \frac{1}{2}\right) + B(z+2) - D\left(z^2 + 5z + 5\right)$

The upper/lower sign choice refers to levels of +/– parity respectively.

$z = \left(J + \frac{1}{2}\right)^2 - 1$, $B^* = B - \frac{1}{2}\gamma$, the exponent $p = J - \frac{1}{2}$.

Table 16.4. Matrix representation of the spin–vibronic components of the bending level $v_2 = 1$ of a molecule in a $^2\Pi$ state for a given J value for $|A| \geq \omega_2 \gg |\varepsilon\omega_2|$.

$\vert\Lambda, l, \Sigma, \pm\rangle$	$^2\Delta_{5/2}\,\vert 1, 1, \frac{1}{2}\rangle$	$^2\Delta_{3/2}\,\vert 1, 1, -\frac{1}{2}\rangle$	$^2\Sigma_{1/2}\,\vert 1, -1, \frac{1}{2}\rangle$	$^2\Sigma_{-1/2}\,\vert 1, -1, -\frac{1}{2}\rangle$
$\langle 1, 1, \frac{1}{2}\vert$	$2\omega_2 + 2g_K - \frac{3(\varepsilon\omega_2)^2}{2(2\omega - A)} + \frac{1}{2}A + B(z-4) - D\left(z^2 - 7z + 13\right)$	$-(z-3)^{1/2}\left[B^* - 2D(z-2)\right]$	$-\frac{1}{2}q_G[z(z-1)]^{1/2}$	$\pm(-1)^p \frac{1}{2} q[z(z-3)]^{1/2}$
$\langle 1, 1, -\frac{1}{2}\vert$		$2\omega_2 + 2g_K - \frac{3(\varepsilon\omega_2)^2}{2(2\omega + A)} - \frac{1}{2}A - \gamma + Bz - D(z^2 + z - 3)$	$\pm(-1)^p q[z(z+1)]^{1/2} + \frac{1}{2}(p_G + 2q_G)z^{1/2}$	$\mp(-1)^p \frac{1}{2}(p + 2q)z^{1/2} - \frac{1}{2}q_G[z(z+1)]^{1/2}$
$\langle 1, -1, \frac{1}{2}\vert$			$2\omega_2 - \frac{(\varepsilon\omega_2)^2}{2(2\omega_2 - A)} + \frac{1}{2}A \pm (-1)^p \varepsilon\omega_{2D}\left(J + \frac{1}{2}\right) + B(z+2) - D(z^2 + 5z + 5)$	$-(z+1)^{1/2}\left[B^* - 2D(z+2)\right] \mp (-1)^p \varepsilon_{corr}\omega_2 \mp (-1)^p \varepsilon\omega_{2D}(z+2)$
$\langle 1, -1, -\frac{1}{2}\vert$				$2\omega_2 - \frac{(\varepsilon\omega_2)^2}{2(2\omega_2 + A)} - \frac{1}{2}A - \gamma \pm (-1)^p \varepsilon\omega_{2D}\left(J + \frac{1}{2}\right) + B(z+2) - D\left(z^2 + 5z + 5\right)$

The upper/lower sign choice refers to levels of +/− parity respectively.

$$z = \left(J + \tfrac{1}{2}\right)^2 - 1, \quad \text{exponent } p = J - \tfrac{1}{2}, \quad B^* = B - \tfrac{1}{2}\gamma,$$

$$\varepsilon_{corr} = \varepsilon\left[1 + \frac{(\varepsilon\omega_2)^2(8\omega_2^2 - 6A^2)}{4(4\omega_2^2 - A^2)}\right].$$

Table 16.5. Matrix representation of the $K = 0(0v_2 0)$ levels, with odd v_2, of a molecule in a $^3\Pi$ electronic state for a given J value.

$\|\Lambda, l, \Sigma, \pm\rangle$	$^3\Sigma_1\|1,-1,1\rangle$	$^3\Sigma_0\|1,-1,0\rangle$	$^3\Sigma_{-1}\|1,-1,-1\rangle$
$\langle 1,-1,1\|$	$(v_2+1)\omega_2 - \frac{1}{8}\varepsilon^2\omega_2 \times (v_2+1) + A + \frac{2}{3}\lambda - \gamma + Bx - Dx(2+x)$	$-(2x)^{1/2}\left[B^* - 2D(x+1)\right] \pm (-1)^p\frac{1}{2}\varepsilon\omega_{2D}(2x)^{1/2}(v_2+1)$	$\mp(-1)^p\frac{1}{2}\varepsilon\omega_2(v_2+1) - 2Dx \mp (-1)^p\frac{1}{2}\varepsilon\omega_{2D} \times (v_2+1)x$
$\langle 1,-1,0\|$		$(v_2+1)\omega_2 - \frac{1}{8}\varepsilon^2\omega_2(v_2+1) - \frac{4}{3}\lambda - 2\gamma \mp (-1)^p\frac{1}{2}\varepsilon\omega_2(v_2+1) + B(x+2) - D(x^2+8x+4)$	$-(2x)^{1/2}\left[B^* - 2D(x+1)\right] \pm (-1)^p\frac{1}{2}\varepsilon\omega_{2D}(2x)^{1/2} \times (v_2+1)$
$\langle 1,-1,-1\|$			$(v_2+1)\omega_2 - \frac{1}{8}\varepsilon^2\omega_2(v_2+1) + A + \frac{2}{3}\lambda - \gamma + Bx - Dx(2+x)$

The upper/lower sign choice refers to levels of +/− parity, respectively.
$x = J(J+1)$, exponent $p = J$, $B^* = B - \frac{1}{2}\gamma$.

on higher powers of the vibrational coordinates q_1, q_2 and q_3. For a linear triatomic molecule ABC in a $^1\Sigma^+$ state, there are six cubic terms in the anharmonic potential

$$V^{(3)} = \tfrac{1}{6}\phi_{111}q_1^3 + \tfrac{1}{2}\phi_{113}q_1^2q_3 + \tfrac{1}{2}\phi_{133}q_1q_3^2 + \tfrac{1}{6}\phi_{333}q_3^3 + \tfrac{1}{2}\phi_{122}q_1q_2^2 + \tfrac{1}{2}\phi_{322}q_3q_2^2, \quad (61)$$

and nine quartic terms in $V^{(4)}$ [see Mills (1972)]. However, a perturbation treatment based on the harmonic oscillator shows that the vibrational energy levels can be described to this order of accuracy by the expression

$$E\left(v_1, v_2^l, v_3\right) = E_{\text{harm}} + E_{\text{anharm}}, \quad (62)$$

where

$$E_{\text{anharm}} = x_{11}\left(v_1 + \tfrac{1}{2}\right)^2 + x_{22}(v_2+1)^2 + x_{33}\left(v_3 + \tfrac{1}{2}\right)^2 + x_{12}\left(v_1 + \tfrac{1}{2}\right)(v_2+1) + x_{13}\left(v_1 + \tfrac{1}{2}\right)\left(v_3 + \tfrac{1}{2}\right) + x_{23}(v_2+1)\left(v_3 + \tfrac{1}{2}\right) + g_{22}l^2. \quad (63)$$

There are seven anharmonic parameters in equation (63) which can be written explicitly in terms of the 15 anharmonic potential coefficients and the three harmonic vibrational frequencies. In some ways, this is a disappointing realization because even a complete determination of all the parameters in equation (63) is not sufficient to define the anharmonic force field to the quartic level exactly. However, it does mean that no more than seven anharmonic parameters are needed in practice to model a real set of energy levels.

Brown and Jørgensen (1982) have investigated the extension of these ideas to a triatomic molecule in a Π electronic state. It turns out that there are only six independent, anharmonic contributions to the difference between the potentials V' and V'' in equation (2) and not the 15 that might have been feared because they must all involve the factor q_2^2. However to formulate the R–T problem to this order of anharmonic effects, 17 parameters are required (including the parameter g_K which has been mentioned earlier). This is a depressingly large number of parameters required to define the model fully although not quite the worst possible scenario when $16 + 6$ or 21 parameters would be required. No experimental study exists which comes anywhere near providing sufficient information to define all these anharmonic parameters for a given molecule.

However, there are some studies which reveal the effects of specific anharmonic terms in the potential, through the observation of a Fermi resonance between the levels $v_1 = 1$ and $v_2 = 2$ which occurs when $\omega_1 \approx 2\omega_2$. This topic was first discussed for a molecule in a Π state by Hougen (1962c) and by Hougen and Jesson (1963). They showed that two parameters are needed to describe the coupling between the Π vibronic components of the levels (100) and (020), W_1 and W_2. The former corresponds to the usual Fermi resonance parameter

$$W_1 = \tfrac{1}{2}(\phi'_{122} + \phi''_{122}) \tag{64}$$

and the second is a new parameter which depends on the difference between the anharmonic contributions of this form to the two potential curves V' and V'',

$$W_2 = \tfrac{1}{2}(\phi'_{122} - \phi''_{122}). \tag{65}$$

This model has been implemented in a few cases where the Fermi resonance is significant, most notably for that of NCO [Woodward, Fletcher and Brown (1987)] where $\omega_1 = 1273.0\,\text{cm}^{-1}$ and $\omega_2 = 534.1\,\text{cm}^{-1}$. The matrix representation of the effective Hamiltonian operator in this case has to be extended to cover the Π vibronic components of both $v_1 = 1$ and $v_2 = 2$ levels.

16.5.2 *K*-TYPE DOUBLING EFFECTS

Each level of a non-rotating triatomic molecule in a Π electronic state with $K > 0$ is two-fold degenerate. When the molecule rotates, this degeneracy is lifted by the combined effects of Λ-type and l-type doubling [see equations (30) and (31)]. The phenomenon is known as K-type doubling. There is unfortunately a different sign convention used for the two contributing effects. For Λ-type doubling in a $^1\Pi$ state, a positive value for the parameter q implies that the level with parity $(-1)^J$ lies above that with parity $(-1)^{J+1}$. In the traditional treatment of l-type doubling however, a positive value for q implies the opposite order for the two parity doublets [Nielsen (1951), and Herzberg (1991)]. The confusion was compounded because the same symbol q was used for both parameters. Because of this, Brown (1975) suggested that, in the discussion of K-type doubling, the sign convention for Λ-doubling should be followed and that an alternative l-type doubling parameter q_G should be introduced which is defined with the opposite convention to the traditional l-type doubling constant. With this convention, which we have used in equation (31), the parameter q_G is usually negative.

The Λ-doubling operator obeys the selection rule $\Delta\Lambda = \pm 2$, $\Delta l = 0$ (see equation (30)) whereas the l-type doubling operator obeys $\Delta\Lambda = 0$, $\Delta l = \pm 2$ (equation (31)). Both of them therefore obey the selection rule $\Delta K = \pm 2$. Consequently, there is only a first order K-type doubling effect for $K = 1$ levels, because in this case $K = +1$ and $K = -1$ states are mixed directly. This produces a splitting of these levels

$$\Delta\nu_K = \pm\tfrac{1}{2} q_{\text{eff}} J(J+1) \tag{66}$$

where q_{eff} is a function of q and q_G. The explicit form for q_{eff} has been discussed for certain situations by Brown (1975) and Crozet *et al.* (1995).

Acknowledgement

I am very grateful to Dr. Sara Beaton for comments on an earlier draft of this chapter.

16.6 References

Barrow, T., and Dixon, R. N., 1973, *Mol. Phys.* **25**, 137–144.
Barrow, T., Dixon, R. N., and Duxbury, G., 1974, *Mol. Phys.* **27**, 1217–1234.
Beaton, S. A., and Brown, J. M., 1997, *J. Mol. Spectrosc.* **183**, 347–359.
Bolman, P. S. H., Brown, J. M., Carrington, A., Kopp, I., and Ramsay, D. A., 1975, *Proc. R. Soc. London, Ser. A*, **343**, 17–44.
Brown, J. M., 1975, *J. Mol. Spectrosc.* **56**, 159–162.
Brown, J. M., 1977, *J. Mol. Spectrosc.* **68**, 412–422.
Brown, J. M., Kaise, M., Kerr, C. M. L., and Milton, D. J., 1978, *Mol. Phys.* **36**, 553–583.
Brown, J. M., and Jørgensen, F., 1982, *Mol. Phys.* **47**, 1065–1086.
Carrington, A., Fabris, A. R., Howard, B. J., and Lucas, N. J. D., 1971, *Mol. Phys.* **20**, 961–980.
Crozet, P., Ross, A. J., Bacis, R., Barnes, M. P., and Brown, J. M., 1995, *J. Mol. Spectrosc.* **172**, 43–56.
Devillers, C., and Ramsay, D. A., 1971, *Can. J. Phys.* **49**, 2839–2858.
di Lauro, C., and Mills, I. M., 1961, *J. Mol. Spectrosc.* **21**, 386–413.
Dressler, K., and Ramsay, D. A., 1957, *J. Chem. Phys.* **27**, 971–972.
Dressler, K., and Ramsay, D. A., 1959, *Philos. Trans. R. Soc. London, Ser. A.* **251**, 553–581.
Duxbury, G., McDonald, B. D., van Goch, M., Alijah, A., Jungen, Ch., and Palivan, H., 1998, *J. Chem. Phys.* **108**, 2336–2350.
Gillett, D. A., 1994, D. Phil. Thesis (Oxford University).
Herzberg, G., 1991, *Molecular Spectra and Molecular Structure III. Electronic Spectra and Electronic Structure of Polyatomic Molecules*; Kreiger: Malabar, FL.
Herzberg, G., and Teller, E., 1933, *Z. Phys. Chem. B*, **21**, 410–446.
Herzberg, G., and Travis, D. N., 1964, *Can. J. Phys.* **42**, 1658–1675.
Hougen, J. T., 1962a, *J. Chem. Phys.* **36**, 519–534.
Hougen, J. T., 1962b, *J. Chem. Phys.* **36**, 1874–1881.
Hougen, J. T., 1962c, *J. Chem. Phys.* **37**, 403–408.
Hougen, J. T., and Jesson, J. P., 1963, *J. Chem. Phys.* **38**, 1524–1525.
Howard, B. J., 1970, Ph.D. Thesis, Southampton University.
Jensen, P., Brumm, M., Kraemer, W. P., and Bunker, P. R., 1995, *J. Mol. Spectrosc.* **171**, 31–57.
Jungen, Ch., and Merer, A. J., 1980, *Mol. Phys.* **40**, 1–23.
Körsgen, H., Evenson, K. M., and Brown, J. M., 1998, *J. Chem. Phys.* **107**, 1025–1027.
Longuet-Higgins, H. C., 1961, *Adv. Spectrosc.* **2**, 429–472.
Messiah, A., 1962, *Quantum Mechanics*; North-Holland: Amsterdam.
Mills, I. M., 1972, Vibration–rotation structure in asymmetric and symmetric top molecules, in *Molecular Spectroscopy: Modern Research*, Rao, K. N., and Matthews, C. W., Eds; Academic Press: New York.
Neilsen, H. H., 1951, *Rev. Mod. Phys.* **23**, 90–136.
Petelin, A. N., and Kiselev, A. A., 1972, *Int. J. Quant. Chem.* **6**, 701–716.
Pople, J. A., and Longuet-Higgins, H. C., 1958, *Mol. Phys.* **1**, 372–383.
Pople, J. A., 1960, *Mol. Phys.* **3**, 16–22.
Renner, R., 1934, *Z. Physik.* **92**, 172–193.
Tang, J., and Saito, S., 1996, *J. Chem. Phys.* **105**, 8020–8033.
Woodward, D. R., Fletcher, D. A., and Brown, J. M., 1987, *Mol. Phys.* **62**, 517–536.

17 SPIN–ORBIT COUPLING AND THE JAHN–TELLER EFFECT: VIBRONIC AND SPIN–VIBRONIC ANGULAR MOMENTA

Timothy A. Barckholtz and Terry A. Miller

The Ohio State University, Columbus OH, USA

17.1 Introduction

The concept of symmetry and its breaking holds a special and important place in the physical sciences. One of the best known examples of such symmetry breaking is the Jahn–Teller effect. In their original paper, Jahn and Teller (1937) demonstrated that a symmetry-dictated electronic degeneracy in a non-linear molecule will always be raised by a non-totally symmetric distortion linear in nuclear coordinates [Bunker and Jensen (1998)].

If the Jahn–Teller distortion is large it will result in a wavefunction localized around the minimum (minima) of the resulting potential energy surface (PES), which occurs (occur) at other than the symmetrical nuclear configuration. In this case, frequently referred to as the *static*

Computational Molecular Spectroscopy. Edited by Per Jensen and Philip R. Bunker

Jahn–Teller effect, the vibronic and spin–vibronic eigenvalues and eigenfunctions can be calculated much as for any other asymmetric molecule.

A rather more interesting case and the one we consider herein occurs when the Jahn–Teller distortion is not large enough to localize the molecule about a single minimum on the PES for the duration of the experimental measurement. This case is often called the *dynamical* Jahn–Teller effect. In this latter case the calculation of the vibronic and spin–vibronic eigenvalues and eigenfunctions is considerably more complex. Many of the principles of such calculations were laid out [Child (1963), Child and Longuet-Higgins (1961), and Longuet-Higgins *et al.* (1958)] in the late 1950s and early 1960s. While pioneering, these calculations were perturbative in nature and performed only for molecules with a single Jahn–Teller active vibrational mode. They were therefore of limited value to an experimentalist attempting to decipher the spectra of a molecule with more than three atoms and various strengths of Jahn–Teller coupling.

The next advancement in the calculation of the vibronic levels of a Jahn–Teller state came in the early 1980s, when the introduction of high speed computers made possible large calculations of the vibronic energy levels. The first applications of these calculations were the ground states of the halogen-substituted benzene ions, for which new jet-cooled laser-induced fluorescence spectra and related experiments offered the first reliable data [Cossart-Magos, Cossart and Leach (1979a, 1979b, 1979c), Cossart-Magos and Leach (1980a, 1980b), Cossart-Magos *et al.* (1983), Sears, Miller and Bondybey (1980, 1981a, 1981b), Yu *et al.* (1990), Bondybey *et al.* (1980), and Miller and Bondybey (1983)] on isolated molecules with several Jahn–Teller active modes, which could test quantitative Jahn–Teller calculations. Since these states exhibited both linear and quadratic Jahn–Teller coupling, in multiple active modes, these calculations were quite substantial in size, requiring the earliest versions of the Cray supercomputers to produce solutions.

One simplifying feature of the calculations on the substituted benzene ion ground states was that spin–orbit coupling was negligibly small. Because Jahn–Teller coupling and spin–orbit coupling both involve the electronic angular momentum of the state, a true solution for the vibronic structure of any non-singlet degenerate state must simultaneously include both of these effects. The most common approach has been first to solve the Jahn–Teller problem, and then to apply several approximations to deduce the spin–orbit splitting of the levels [Ham (1965, 1972), and Child (1963)].

During the course of the analysis of the vibronic structure of the ground states of the CF_3O and CF_3S radicals [Barckholtz, Yang and Miller (1999)], we realized that separating the Jahn–Teller and spin–orbit coupling into two separate problems has some inherent deficiencies. We therefore reinvestigated the Jahn–Teller problem, including spin–orbit coupling directly in the Hamiltonian, rather than as an 'afterthought' to the Jahn–Teller solutions. We have recently published this research [Barckholtz and Miller (1998)], which we will refer to as Paper I throughout this chapter. In that work, we described in some detail calculations of this kind and reported on several features of Jahn–Teller and spin–orbit coupling that had, for the most part, been neglected.

It has generally not been recognized that the operator that couples the vibrational and electronic angular momenta depends on the symmetry species of the electronic state and point group of the molecule. As we showed in Paper I, this has a profound effect on the definition of the conserved quantum number for linear Jahn–Teller coupling. As we will discuss quantitatively in this chapter, the Coriolis coupling is significantly different between these two general cases of linear Jahn–Teller coupling.

A second effect we will discuss in this chapter is a direct result of the combination of spin–orbit and Jahn–Teller coupling. The most noticeable effect of the addition of spin–orbit coupling to the Jahn–Teller problem is a splitting of the vibronic energy levels into two spin–vibronic energy levels. At higher resolution, though, as we showed in Paper I, the effective spin–rotation constant of the molecule can be drastically affected by the presence of both spin–orbit and Jahn–Teller coupling. The two spin components of a vibronic level have usually been assumed to have identical Coriolis coupling constants. However, they in fact can have quite different Coriolis constants, the difference of which is manifested in the observed spin–rotation constant of the state. In this chapter, we will present detailed calculations of this effect for an example state.

17.2 Hamiltonian, basis set, and eigenvectors

The appropriate Hamiltonian operator for the components of a degenerate electronic state is the sum of a number of terms,

$$\hat{\mathcal{H}} = \hat{\mathcal{H}}_{\mathrm{T}} + \hat{V} + \hat{\mathcal{H}}_{\mathrm{SO}} + \hat{\mathcal{H}}_{\mathrm{rot}}, \tag{1}$$

where $\hat{\mathcal{H}}_{\mathrm{T}}$ is the vibrational kinetic energy of the nuclei. The sum of $\hat{\mathcal{H}}_{\mathrm{SO}}$ and the potential $\hat{V}$ define the potential energy surface (PES). We refer the reader to our previous papers [Miller and Bondybey (1983), and Barckholtz and Miller (1998)] on this topic for the terms of the power series expansion of the potential $\hat{V}$ and for the details of the PES. For our purposes here, it suffices to say that the potential has, in addition to the standard harmonic oscillator terms for the degenerate modes, terms linear and quadratic in the vibrational coordinates of the degenerate modes that account for linear and quadratic Jahn–Teller coupling. In this chapter, we will consider only linear Jahn–Teller coupling in a single active mode with a harmonic vibrational frequency ω_e and a linear Jahn–Teller coupling constant D (see Paper I for precise definitions of these parameters). The spin–orbit Hamiltonian $\hat{\mathcal{H}}_{\mathrm{SO}}$ is parameterized by the product $a\zeta_e$, where ζ_e is the projection of the electronic orbital angular momentum on the principal axis and a is the spin–orbit coupling constant. The details of the rotational Hamiltonian have been presented elsewhere [Barckholtz and Miller (1998)].

The basis set used to compute the Hamiltonian matrix is the product of electronic, vibrational, and spin functions:

$$|\Lambda\rangle \prod_{i=1}^{p} |v_i, l_i\rangle \, |\Sigma\rangle \,, \tag{2}$$

where $\Lambda = \pm 1$; $v_i = 0, 1, 2, \ldots$; $l_i = v_i, v_i - 2, v_i - 4, \ldots, -v_i + 2, -v_i$; $\Sigma = -S, -S + 1, \ldots, S - 1, S$. We include in the basis set only those vibrational modes that are Jahn–Teller active (see Tables 2 and 3 of Paper I). Under the harmonic oscillator approximation, each of the v_i can take any positive integer value, which makes the basis set infinitely large. Therefore, the basis set for each Jahn–Teller active mode must be truncated to a manageable level. In general, the larger the Jahn–Teller coupling constants are for each mode, the larger the basis set must be for that mode. In practice, a relatively small basis set is typically used to do some initial calculations. The basis set is then expanded until additional basis functions have a negligible effect on the eigenvalues.

In Paper I, we developed a general 'selection rule' for which symmetries of vibrational modes of a point group will be Jahn–Teller active for a given electronic state's symmetry. In our notation, s_e and s_v are the numerical labels from the irreducible representations of the electronic and vibrational symmetries, respectively (see equations (3) and (5) of Paper I). As we showed, the following equality must hold true for the kth-order Jahn–Teller term in the Hamiltonian to be non-zero,

$$\left(2s_e + (-1)^{s_k} k s_v\right) \bmod n = 0, \tag{3}$$

where s_k must be either 0 or 1. The choice of s_k is dictated by the symmetry of the state and its point group, and has a direct impact on the relative phases of the electronic and vibrational terms in the Hamiltonian. As we show in the remainder of this chapter, the value of s_k has no effect on the eigenenergies of the vibronic or spin–vibronic levels, but has a direct impact on the Coriolis and spin–rotation parameters determined via their eigenfunctions.

Jahn–Teller coupling destroys the electronic and vibrational angular momenta quantum numbers Λ and l_i, respectively. (Strictly speaking, Λ is not a quantum number, but for most purposes can be treated as one.) However, a new, conserved quantum number, j, can be defined by the linear combination

$$j = l_t + \tfrac{1}{2}(-1)^{s_1}\Lambda, \tag{4}$$

where $l_t = \Sigma_i l_i$ and $s_1 = 0$ or 1 is determined from the symmetry of the state (see Table 2 of Paper I). The Jahn–Teller quantum number j is conserved for linear coupling ($k = 1$), though not for quadratic coupling.

We have used two general forms for the eigenvectors resulting from the diagonalization of the Hamiltonian matrix, depending on whether spin–orbit coupling is or is not included directly in the vibronic Hamiltonian. For the approximation when spin–orbit coupling is not included directly in the calculation, the form of the eigenvectors is

$$|j, n_j, \alpha\rangle\,|\Sigma\rangle = \sum_i \left(c_{i,n_j}, |\Lambda_i\rangle \prod_{m=1}^{p} |v_{m,i}, l_{m,i}\rangle \right) |\Sigma\rangle\,, \tag{5}$$

where α is the symmetry of the level under the molecule's nominal point group. When spin–orbit coupling is included in the Hamiltonian, the eigenvector notation is

$$|n_j, j, \Omega, \alpha'\rangle = \sum_i \left(c_{i,n_j,\Omega}|\Lambda_i\rangle \prod_{m=1}^{p} |v_{m,i}, l_{m,i}\rangle\,|\Sigma_i\rangle \right), \tag{6}$$

where $\Omega = \Lambda - l_t + \Sigma$ and the summation over i runs over all of the basis functions used in the calculation. For a coefficient $c_{i,n_j,\Omega}$ to be nonzero, the equality of equation (4) for j must be satisfied by the basis function. Each eigenvector $|n_j, j, \Omega, \alpha'\rangle$ has an associated eigenvalue $E_{j,n_j,\Omega}$. The notation $|n_j, j, \Omega, \alpha'\rangle$ indicates to which j-block the level corresponds and which eigenvector, n_j, it is from that symmetry block, with the lowest energy solution of each symmetry block being $n_j = 1$. In the strong spin–orbit coupling limit the vibrational quantum number v again becomes useful and it can be used in place of n_j. The symbol α' represents the symmetry of the level under the spin-double group appropriate for the molecule.

17.3 Computational details

A Fortran90 computer program, SOCJT (pronounced 'sock-it'), has been written to calculate the eigenenergies and eigenfunctions of a molecule described by the Hamiltonian of equation (1), excluding $\hat{\mathcal{H}}_{\mathrm{rot}}$. The program can handle an arbitrary number of active modes and the truncation of the basis set is limited only by available memory and computer time. The program has been written for the Cray T90 computer and for DOS on personal computers. It is this program that has provided the numerical results presented in the remainder of this chapter. The program is available upon request from the authors.

One of the features of this program that makes it amenable to realistic systems is its ability to handle an arbitrary number of Jahn–Teller active vibrational modes. The basis set, equation (2), is the set of all possible combinations of v and l for each mode. The generation of the basis set uses a recursive subroutine so that any number of vibrational modes can be included in the basis set. In this way, only one program had to be written, which had no limit on the number of active modes, a significant improvement over previous programs written for this purpose.

To make the diagonalization routine most efficient, the symmetry blocking of the Hamiltonian matrix is utilized to the fullest extent possible. For linear Jahn–Teller coupling, the basis set and Hamiltonian matrix are constructed in blocks diagonal in j (defined by equation (4)) and Σ, while for quadratic coupling the selection rule on j of $\Delta j = \pm 3$ is utilized.

While all of the examples discussed in Paper I were appropriate for doublet electronic states $S = 1/2$, we have written the program to be generalized for any half-integer or integer value of S. The program is not applicable to the cubic point groups, as the higher symmetry of these point groups would alter the structure of the program significantly. The program can be used for Renner calculations of linear molecules by setting the linear coupling constant to zero and treating the quadratic Jahn–Teller coupling constant as the Renner coupling constant ε.

The efficiency in the generation of the Hamiltonian matrix is maximized by the utilization of the selection rules on the basis set. Because the matrix is extremely sparse ($\approx$5 % of the matrix elements are non-zero for a single-mode linear Jahn–Teller coupling case; $\approx$0.001 % of the matrix elements are non-zero for a five-mode linear Jahn–Teller coupling case), only the non-zero matrix elements are stored in memory. Three one-dimensional arrays hold the value, row index, and column index of each of the non-zero matrix elements. In this way, the memory requirements of the calculation scale linearly rather than quadratically with basis set size.

The sparseness of the Hamiltonian matrix is also utilized in the diagonalization routines, which have been optimized for sparse matrices. Furthermore, the Hamiltonian matrix is a positive-definite symmetric Hermitian matrix, which allows even more highly optimized diagonalization routines to be used. Because in spectroscopic problems only the lowest energy levels are usually observed, diagonalization routines that find only the lowest eigenvalue solutions are used. The version of the program written for the Cray T90 uses algorithms from the NAG libraries, which are written and optimized specifically to take advantage of the Cray's vector capabilities. The DOS version uses the Underwood [Underwood (1975)] method of diagonalizing sparse matrices using code freely available over the Internet at http://www.netlib.org.

One limitation of the Cray version of the program is that it is not able to use the eigenvectors from a previous calculation to begin the iterations of the diagonalization routine. The DOS version does have this capability, so that even though the DOS version will not be as fast as the Cray version, it will be relatively more efficient for the calculation of the eigenvalues and eigenvectors when a parameter (such as D_i or $a\zeta_e$) has changed only slightly from one calculation to the next. One limitation on this capability is that the basis set must be exactly the same from one calculation to the next, i.e., that an additional mode has not been added to the calculation, or that the maximum values of v on each vibrational mode have not been changed.

Table 17.1 illustrates the size of the basis set, number of non-zero matrix elements, and approximate computational times for several different coupling combinations. As the first several rows indicate, increasing the basis set by an additional vibrational mode dramatically increases the size and computational cost of the calculation. Quadratic Jahn–Teller coupling also significantly increases the size of the calculation, though not as dramatically as does the addition of another vibrational mode. As can be seen from the table, the computational costs on the Cray T90 are minimal for even relatively large calculations. The computational time required for the DOS version, while approximately 25 times longer than on the Cray T90, is still reasonable for most situations. One of the advantages of the DOS version over the Cray version is that the DOS version is able to use as initial guesses to the iterative procedure the final results from a previous calculation.

The program also calculates several quantities directly related to the spectroscopy of the state. When spin–orbit coupling is not directly included in the Hamiltonian, the Ham quenching parameter d_{j,n_j} (see equations (80)–(83) of Paper I) is computed for each vibronic energy level. The Coriolis coupling constant (see later) is also calculated for each level. Perhaps most relevant to experimentalists, the relative transition intensities for E $\leftrightarrow$ A electronic transitions are calculated, both for cold- and for hot-band spectra. Lastly, to aid in the fitting of the energy levels of the

Table 17.1. Size of basis set, number of non-zero matrix elements, and approximate timings for several different types of spin–orbit/Jahn–Teller calculations.

Linear	Quadratic	Spin–orbit	No. modes	$[v_{max}]^a$	N^b	n^c	Time (Cray)[d]	Time (DOS)[e]
Y	N	N	1	[35]	36	71	1	6
Y	N	N	2	[7,5]	133	345	1	21
Y	N	N	3	[7,5,3]	1246	3983	23	400
Y	N	N	4	[7,5,3,2]	7236	26767	166	4100
Y	N	N	5	[7,5,3,2,2]	42098	176876	6000	≈150000
Y	N	Y	1	[35]	36	71	1	11
Y	N	Y	2	[7,5]	133	345	2	38
Y	N	Y	3	[7,5,3]	1246	3983	25	800
Y	Y	N	1	[35]	164	508	1	23
Y	Y	N	2	[7,5]	465	1498	2	50
Y	Y	N	3	[7,5,3]	4516	17247	35	770
Y	Y	Y	1	[35]	164	508	1	36
Y	Y	Y	2	[7,5]	465	1498	3	70
Y	Y	Y	3	[7,5,3]	4516	17247	50	2100

[a]Maximum values of v included in the basis set for each vibrational mode.
[b]Size of basis set for the $j = 1/2$ block.
[c]Number of non-zero matrix elements in the lower triangle of the Hamiltonian matrix.
[d]Approximate time, in seconds, for finding the 15 lowest energy solutions of each block up to $j = 15/2$, on a Cray T90.
[e]Same as footnote d except for execution under DOS on a 200 MHz Cyrix-686.

state, the derivatives of the energies with respect to the parameters of the coupling calculation (all $\omega_{e,i}$, D_i, K_i, and $a\zeta_e$) are computed.

17.4 Energies of the vibronic and spin–vibronic levels

In the remainder of this chapter, we will focus on the differences (or lack thereof) between the two general symmetry cases, $s_1 = 0$ (appropriate for the most well-studied case, a ^{2}E state of C_{3v} symmetry) and $s_1 = 1$ (appropriate for many states of molecules that belong to point groups with C_5 or higher axes). Three properties will be discussed: the energies of the vibronic levels under linear Jahn–Teller coupling, the Coriolis coupling constants under linear Jahn–Teller coupling, and the effective spin–rotation constant induced by the presence of spin–orbit coupling. For illustration, we present only the case of a single Jahn–Teller active mode; results for more than one active mode are easily calculated using our computer program.

17.4.1 ENERGIES OF THE VIBRONIC LEVELS UNDER LINEAR JAHN–TELLER COUPLING

We show in Figure 17.1 the energies, as a function of D, of the lowest several vibronic energy levels of a Jahn–Teller state. There is in fact *no* difference in the energies of the vibronic levels between the two general symmetry cases. Thus, if all one is interested in is the energies of the levels, the symmetry of the state can be completely neglected.

Figure 17.1 shows the absolute energies of the vibronic levels as a function of D, relative to the energy of the symmetric point of the PES, which is defined as the zero of energy for Figure 17.1. If, instead, the zero of energy is defined as the energy of the lowest-energy vibronic level, the energies of the levels are given by Figure 17.2. Figure 17.2 is a more useful diagram when examining the spectroscopy of these levels. Because the lowest energy vibronic level, denoted $|\frac{1}{2}, 1, e\rangle|\Sigma = \pm\frac{1}{2}\rangle$ in the notation of equation (6), is the zero-point energy level of the molecule,

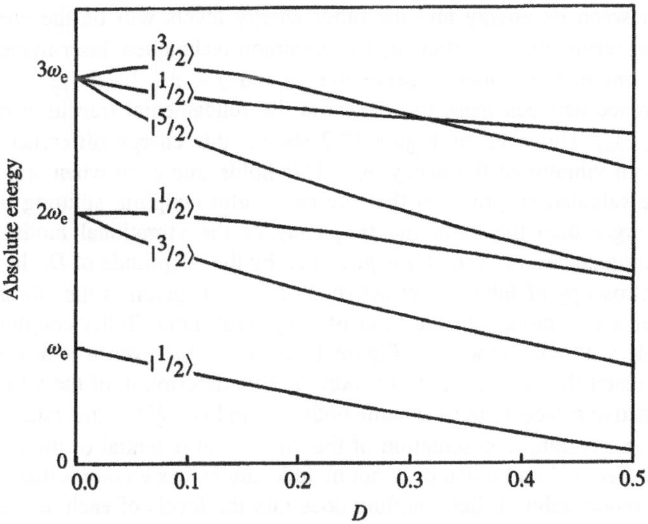

Figure 17.1. Absolute energies of the vibronic levels for linear coupling in a single Jahn–Teller active mode over the range of $D = 0$ to $D = 0.5$. The ordinate scale is in units of ω_e while the abscissa is the dimensionless linear Jahn–Teller coupling constant D. The levels are labeled by their value of j.

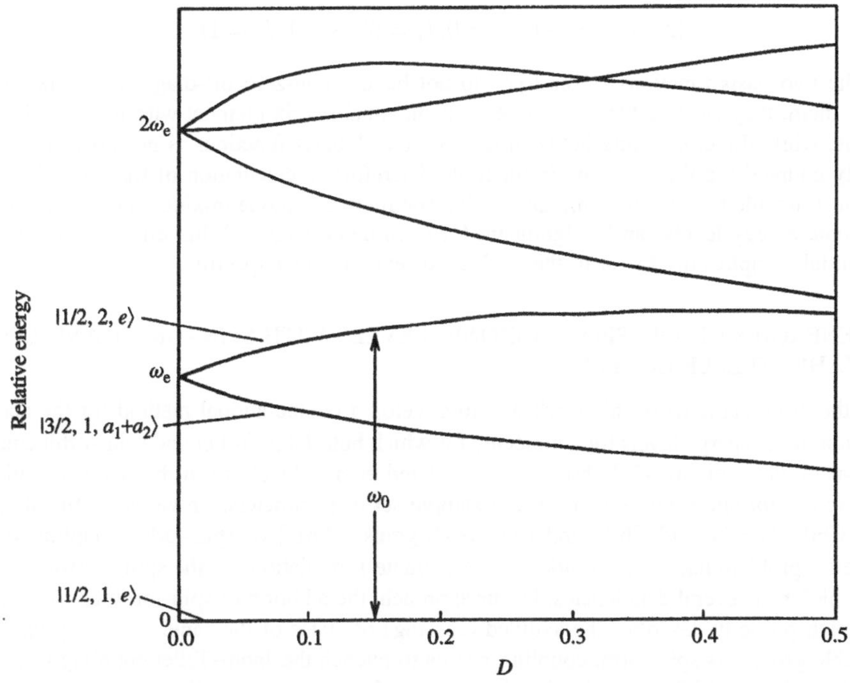

Figure 17.2. Relative energies of the vibronic levels, $|j, n_j, \alpha\rangle$, for linear Jahn–Teller coupling in a single active mode over the range of $D_i = 0$ to $D_i = 0.5$. The ordinate scale is in units of $\omega_{e,i}$ while the abscissa is the dimensionless linear Jahn–Teller coupling constant D_i, with the zero of energy set as the energy of the lowest level.

the differences between its energy and the other energy levels will be the energies measured in spectroscopic experiments. In particular, two common techniques, laser-induced fluorescence and infrared absorption, most often observe the second $j = \pm\frac{1}{2}$ level ($|\frac{1}{2}, 2, e\rangle$, Figure 17.2). The energy difference between these two levels is the fundamental transition frequency of the vibrational mode, $\omega_{0,i}$. However, as Figure 17.2 shows, this energy difference will be greater than the equilibrium vibrational frequency $\omega_{e,i}$. This holds true even when spin–orbit coupling is included in the calculation, provided that the spin–orbit coupling splitting of the electronic potential is not larger than the harmonic frequency of the vibrational mode. The difference between the two frequencies $\omega_{e,i}$ and $\omega_{0,i}$ is governed by the magnitude of D_i. In many published papers of the spectroscopy of Jahn–Teller active states, $\omega_{0,i}$ is given as the vibrational frequency of the Jahn–Teller active modes. In the limit of very weak Jahn–Teller coupling, $\omega_{0,i}$ and $\omega_{e,i}$ will be comparable with each other, but Figure 17.2 shows that even a relatively small D_i can cause $\omega_{0,i}$ to be greater than $\omega_{e,i}$ by 25 % or more. A true description of the vibrational structure of a Jahn–Teller active molecule must include both $\omega_{e,i}$ and D_i. While the value of $\omega_{0,i}$ is useful, by itself it is not an accurate representation of the vibrational potential of the molecule.

While the Jahn–Teller Hamiltonian does not include any matrix elements that are non-diagonal in more than one mode, Jahn–Teller coupling does mix the levels of each active mode with the others. To see how this can be the case, consider, for a two-mode calculation, the three basis functions

$$|0\rangle = |\Lambda = +1\rangle |v_1 = 0, l_1 = 0\rangle |v_2 = 0, l_2 = 0\rangle, \tag{7}$$

$$|1\rangle = |\Lambda = -1\rangle |v_1 = 1, l_1 = 1\rangle |v_2 = 0, l_2 = 0\rangle, \tag{8}$$

$$|2\rangle = |\Lambda = -1\rangle |v_1 = 0, l_1 = 0\rangle |v_2 = 1, l_2 = 1\rangle. \tag{9}$$

While the two basis functions $|1\rangle$ and $|2\rangle$ do not have a non-zero off-diagonal matrix element between them, they each do have a non-zero off-diagonal matrix element with the basis function $|0\rangle$. Thus, while direct coupling between the two $v = 1$ basis functions is not present, they are indirectly coupled via the $v = 0$ basis function. Therefore, a calculation of the vibronic energy levels must include both modes simultaneously. The impact of mode mixing on the calculation of the vibronic energy levels can be significant for even modest Jahn–Teller effects and represents a substantial complicating factor to the analysis of experimental spectra.

17.4.2 ENERGIES OF THE SPIN–VIBRONIC LEVELS WITH SPIN–ORBIT AND LINEAR JAHN–TELLER COUPLING

One of the principal aims of this work was to develop a computational method for the accurate calculation of the energy levels for a molecule for which both Jahn–Teller and spin–orbit coupling are non-negligible. Figure 17.3 shows the calculated energy levels from the diagonalization of the full spin–vibronic Hamiltonian for an example set of parameters. Previous treatments [Ham (1965), Child (1963), and Child and Longuet-Higgins (1961)] of spin–orbit coupling for the Jahn–Teller problem had only considered approximate formulations of the spin–vibronic energy levels, which had several deficiencies. In our approach, the addition of spin–orbit coupling to the Hamiltonian properly describes the avoided crossings of states of the same symmetry under the spin double group. As spin–orbit coupling begins to quench the Jahn–Teller coupling (the right-hand side of Figure 17.3), the vibrational spacings of a harmonic oscillator for a pair of $^2E_{3/2}$ and $^2E_{1/2}$ electronic states begin to be recovered. Notice, however, that Jahn–Teller coupling still has two significant effects on the energy levels: the vibrational spacing has yet to recover its harmonic value (100 cm^{-1} in this example) and the individual v_i levels are still split according to their l_i values, much as they are in the absence of spin–orbit coupling.

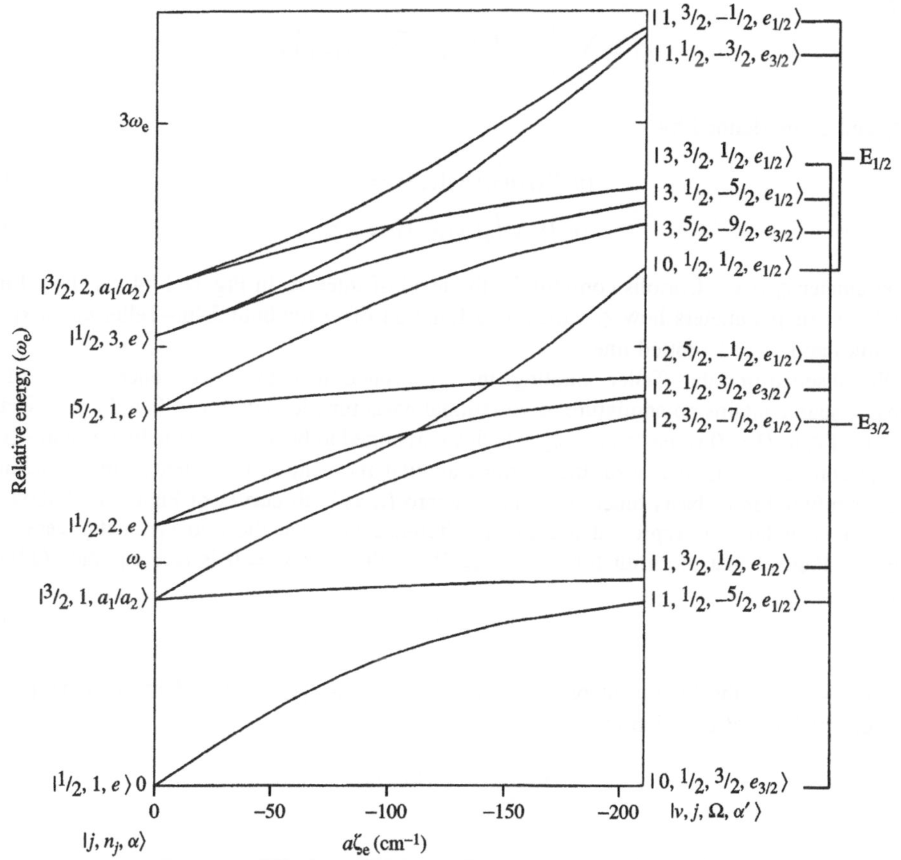

Figure 17.3. Calculated relative energy levels for a ^{2}E state of C_{3v} symmetry including linear Jahn–Teller coupling and spin–orbit coupling for a single Jahn–Teller active mode. Only those levels originating from $v = 0$, 1 and 2 are shown. The energy levels were calculated assuming $D_i = 0.125$ and $\omega_{e,i} = 100\,\text{cm}^{-1}$ and varying $a\zeta_e$. At $a\zeta_e = 0$, the lowest several energy levels are labeled by the ket $|j, n_j, \alpha\rangle$ and have components, $\Sigma = \pm\frac{1}{2}$. At $a\zeta_e = -200\,\text{cm}^{-1}$ the energy levels are labeled by the ket $|v, j, \Omega, \alpha'\rangle$. In the strong spin–orbit limit, the value of $\Omega = \Lambda - l_t + \Sigma$ determines [Barckholtz and Miller (1998)] the C_{3v} double group symmetry α'. The vibrational quantum number v is also included since it again becomes meaningful when spin–orbit coupling dominates Jahn–Teller coupling.

17.5 Coriolis coupling in the absence of spin–orbit coupling

Up to this point, we have considered only the eigenvalues of the vibronic and spin–vibronic levels of an orbitally degenerate state. However, as we show in this and the following section, the corresponding eigenfunctions are very important. They determine several experimentally observable quantities, such as the vibronic and spin–vibronic angular momenta of the state. These properties are drastically altered by the magnitude of the Jahn–Teller coupling and the symmetry of the state.

In Paper I, we gave the following expression the Coriolis coupling operator,

$$\hat{\mathcal{H}}_{\text{COR}} = -2A\langle j, n_j, \alpha = e|\hat{L}_z + \hat{G}_z|j, n_j, \alpha = e\rangle\hat{N}_z \tag{10}$$

$$= -2A\zeta_t\hat{N}_z, \tag{11}$$

where

$$\zeta_t = \sum_i \left[c_{i,nj}^2 \left(\Lambda_i \zeta_e + \sum_{m=1}^{p} l_{m,i} \zeta_m \right) \right], \tag{12}$$

and ζ_e and ζ_i are defined by

$$\langle v, l | \langle \Lambda | \hat{G}_z | \Lambda \rangle | v, l \rangle = l \zeta_i, \tag{13}$$

$$\langle v, l | \langle \Lambda | \hat{L}_z | \Lambda \rangle | v, l \rangle = \Lambda \zeta_e. \tag{14}$$

The parameter ζ_t is the Coriolis constant for the level of interest. In Figure 17.4 we show for an example set of parameters how ζ_t varies as a function of D for both Jahn–Teller cases $s_1 = 0$ (solid line) and $s_1 = 1$ (dotted line).

In the absence of Jahn–Teller coupling, the eigenfunction of the lowest energy level of the electronic state contains no admixture of vibrational basis functions with $v \neq 0$ and $l \neq 0$. Hence, the value of ζ_t at $D = 0$ is the same as ζ_e, which we assumed to be unity. As the linear Jahn–Teller coupling constant D_i is increased, the vibronic angular momentum is decreased by the addition to the eigenfunction of basis functions with non-zero l_i, as both curves of Figure 17.4 show.

The limits at large D represent a distinct difference between the two symmetry cases. We showed in Paper I that the limit for ζ_t at large D for the case $s_1 = 0$ is (see equation (110) of Paper I)

$$\lim_{D \to \infty} \zeta_t = j \zeta_i. \tag{15}$$

Because $j = \frac{1}{2}$ for the lowest energy level, ζ_t converges to $\frac{1}{2}\zeta_i$ ($= 0.125$ in our example). A more general form of equation (15) is

$$\lim_{D \to \infty} \zeta_t = (-1)^{s_1} j \zeta_k. \tag{16}$$

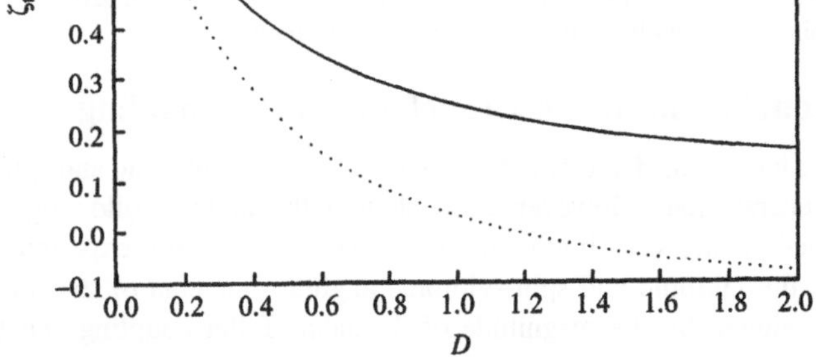

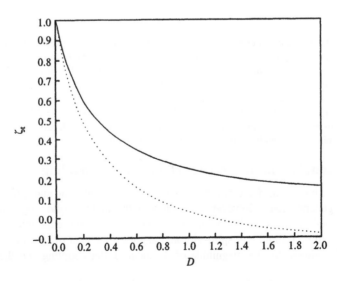

Figure 17.4. Coriolis coupling constant for the lowest energy level, calculated for a single Jahn–Teller active mode over the range of $D = 0$ to $D = 2.0$. The solid line is the result for the symmetry case $s_1 = 0$ while the dotted line is for $s_1 = 1$. The parameters used the calculation were $\omega_e = 500\,\mathrm{cm}^{-1}$, $\zeta_e = 1.0$, and $\zeta_i = 0.25$.

For the symmetry case of $s_1 = 1$, then, the limit at large D has the same magnitude as for $s_1 = 0$, but the opposite sign. Because in each case the value in the absence of linear Jahn–Teller coupling is the same ($\zeta_e = 1$), the two curves of Figure 17.4 have distinctly different shapes to arrive at their respective limits for large D.

17.6 The effective spin–rotation coupling constant induced by spin–orbit coupling

The Coriolis constant ζ_t is often of sufficient magnitude that only moderate resolution is needed to resolve Coriolis structure in experimental spectra. In nearly all rotationally resolved spectra analyzed to date, the analyses assumed that the two spin–orbit components of the level had identical Coriolis constants ζ_t^{Σ} (where the superscript $\Sigma = \pm\frac{1}{2}$ indicates a particular spin component). As we showed in Paper I, the two spin components *do not* necessarily have identical values of ζ_t^{Σ}.

The ramification of different values of ζ_t^{Σ} for a pair of $\Sigma = \pm\frac{1}{2}$ spin components is that if they are analyzed using a model that assumes $\zeta_t^{1/2} = \zeta_t^{-1/2}$, then a very large spin–rotation coupling constant must be used to fit the rotational levels satisfactorily. As we showed in Paper I, the effective spin–rotation constant $\varepsilon_{aa}^{\text{eff}}$ contains contributions from both the inherent spin–rotation term, ε_{aa}^{0}, of the state, as well as a second term, denoted ε_{aa}^{2v},

$$\varepsilon_{aa}^{\text{eff}} = \varepsilon_{aa}^{0} + \varepsilon_{aa}^{2v}. \tag{17}$$

We further showed that ε_{aa}^{2v} is proportional to the difference between the Coriolis coupling constants of the two spin components, $\Delta\zeta_t$,

$$\frac{\varepsilon_{aa}^{2v}}{2A} = -\Delta\zeta_t, \tag{18}$$

where

$$\Delta\zeta_t = \zeta_t^{+1/2} - \zeta_t^{-1/2}, \tag{19}$$

and A is the rotational constant along the symmetry axis. This formulation of the spin–rotation constant in terms of the Coriolis coupling constants of the two spin components, which can be calculated using the Jahn–Teller and spin–orbit computer program, was sufficient to explain the formerly anomalously large spin–rotation constants in the ground state of the methoxy radical ($-1.3533\,\text{cm}^{-1}$) [Liu *et al.* (1989), Liu *et al.* (1990), and Liu, Yu and Miller (1990)] and that observed in the excited state of the $CdCH_3$ radical ($15.8\,\text{cm}^{-1}$) [Cerny *et al.* (1993), and Pushkarsky, Barckholtz and Miller (1999)].

In Figure 17.5 we show accurate calculations of $\Delta\zeta_t$ as a function of $a\zeta_e$, for both symmetry cases for an example set of parameters. While the difference between the two symmetry cases for $\Delta\zeta_t$ is not as pronounced as for ζ_t, Figure 17.4, the difference is non-negligible. The graph clearly shows that the larger the spin–orbit coupling is for the system, the larger the effective spin–rotation constant will be.

17.7 Conclusions

In this chapter, we have presented accurate calculations of the energy, Coriolis coupling constant, and the effective spin–rotation constant induced by spin–orbit coupling for the vibronic and spin–vibronic levels of Jahn–Teller active states. While we chose to show results for various parameters for a molecule with only one active mode, the principals of these calculations can be extended to any other combination of active modes and coupling constants, using the computer

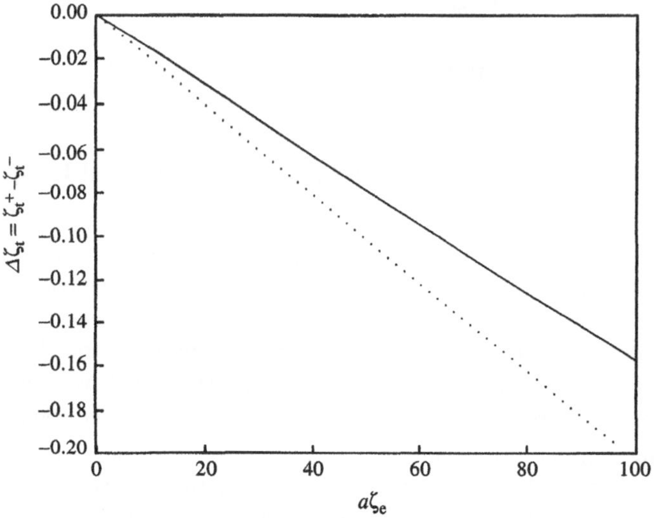

Figure 17.5. The difference, $\Delta\zeta_t$, of the Coriolis coupling constant of the two spin components of the lowest energy level, calculated for a single Jahn–Teller active mode over the range of $a\zeta_e = 0$ to $a\zeta_e = 100$. The solid line is the result for the symmetry case $s_1 = 0$ while the dotted line is for $s_1 = 1$. The parameters used the calculation were $\omega_e = 500\,\mathrm{cm}^{-1}$, $D = 0.25$, $\zeta_e = 1.0$, and $\zeta_i = 0.25$.

program, SOCJT. In the past, it has generally been the case that only the energies and electronic transition intensities have been used to deduce the Jahn–Teller and spin–orbit coupling constants of a Jahn–Teller active molecule. The results exemplified in Figures 17.4 and 17.5 show that Jahn–Teller and spin–orbit coupling also have significant effects on the expectation values of these vibronic and spin–vibronic levels. We therefore expect in the future that the coupling parameters of Jahn–Teller active states will be deduced not just from the energies of the vibronic and spin–vibronic levels, but also from the molecular parameters obtained from their rotationally resolved spectra.

Acknowledgments

We thank the National Science Foundation for support of this work, in the form of a grant to TAM (# 9320909) and a fellowship to TAB. We are also grateful to the Ohio Supercomputer Center (Grant # PAS540) for a generous allotment of computer time.

17.8 References

Barckholtz, T. A., and Miller, T. A., 1998, *Int. Rev. Phys. Chem.* **17**, 435–524.
Barckholtz, T. A., Yang, M. C., and Miller, T. A., 1999, *Mol. Phys.*, **97**, 239–254.
Bondybey, V. E., Sears, T. J., English, J. H., and Miller, T. A., 1980, *J. Chem. Phys.* **73**, 2063–2068.
Bunker, P., and Jensen, P., 1998, *Molecular Symmetry and Spectroscopy*, 2nd edition; NRC Press: Ottawa.
Cerny, T. M., Tan, X. Q., Williamson, J. M., Robles, E. S. J., Ellis, A. M., and Miller, T. A., 1993, *J. Chem. Phys.* **99**, 9376–9388.
Child, M. S., 1963, *J. Mol. Spectrosc.* **10**, 357–365.
Child, M. S., and Longuet-Higgins, H. C., 1961, *Proc. R. Soc. London, Ser. A* **245**, 259–294.
Cossart-Magos, C., and Leach, S., 1980a, *Chem. Phys.* **48**, 329–348.
Cossart-Magos, C., and Leach, S., 1980b, *Chem. Phys.* **48**, 349–358.
Cossart-Magos, C., Cossart, D., and Leach, S., 1979a, *Mol. Phys.* **37**, 793.
Cossart-Magos, C., Cossart, D., and Leach, S., 1979b, *Chem. Phys.* **41**, 345–362.

Cossart-Magos, C., Cossart, D., and Leach, S., 1979c, *Chem. Phys.* **41**, 363–372.
Cossart-Magos, C., Cossart, D., Leach, S., Maier, J. P., and Misev, L., 1983, *J. Chem. Phys.* **78**, 3673–3687.
Ham, F. S., 1965, *Phys. Rev. A* **138**, 1727–1740.
Ham, F. S. (1972), in *Electron Paramagnetic Resonance*; Plenum: New York, pp. 1–119.
Jahn, H. A., and Teller, E., 1937, *Proc. R. Soc. London, Ser. A* **161**, 220–235.
Liu, X., Damo, C., Lin, T.-Y. D., Foster, S. C., Misra, P., Yu, L., and Miller, T. A., 1989, *J. Phys. Chem.* **93**, 2266–2275.
Liu, X., Foster, S. C., Williamson, J. M., Yu, L., and Miller, T. A., 1990, *Mol. Phys.* **69**, 357–367.
Liu, X., Yu, L., and Miller, T. A., 1990, *J. Mol. Spectrosc.* **140**, 112–125.
Longuet-Higgins, H. C., Öpik, U., Pryce, M. H. L., and Sack, R. A., 1958, *Proc. R. Soc. London, Ser. A* **244**, 1–16.
Miller, T. A., and Bondybey, V. E., (1983), in *Molecular Ions: Spectroscopy, Structure, and Chemistry*; North-Holland: Amsterdam, pp. 201–230.
Pushkarsky, M., Barckholtz, T. A., and Miller, T. A., 1999, *J. Chem. Phys.* **110**, 2016–2028.
Sears, T., Miller, T. A., and Bondybey, V. E., 1980, *J. Chem. Phys.* **72**, 6070–6080.
Sears, T., Miller, T. A., and Bondybey, V. E., 1981a, *Discuss. Faraday Soc.* **71**, 175–180.
Sears, T., Miller, T. A., and Bondybey, V. E., 1981b, *J. Chem. Phys.* **74**, 3240–3248.
Underwood, R. R., 1975, PhD Thesis, Stanford University.
Yu, L., Foster, S. C., Williamson, J. M., and Miller, T. A., 1990, *J. Chem. Phys.* **92**, 5794–5800.

PART 5

DYNAMICS

18 SEMICLASSICAL RESONANCE APPROXIMATIONS AND WAVEPACKET TECHNIQUES

Mark S. Child
University of Oxford, UK

18.1 Introduction

In planning this chapter, it soon became apparent that it would be impossible to cover the wide variety of semiclassical quantization methods relevant to molecular spectroscopy [see for example Chapter 7 of Child (1991)]. The decision was therefore made to restrict attention to two techniques—one that is largely interpretative and the other predictive.

The interpretative approach covers the polyad spheres of Xiao and Kellman (1989), rotational energy surfaces of Harter and Patterson (1984), and related topics that fit naturally into the effective Hamiltonian schemes of molecular spectroscopy. The aim here is to show that the partitioning of the classical phase space, due to anharmonic interactions, is reflected in a partitioning

Computational Molecular Spectroscopy. Edited by Per Jensen and Philip R. Bunker

of the eigenvalue spectrum. Each zone of the phase space will be shown to be centered on a stable classical periodic orbit, which serves to organize the eigenvalues of states in the basin around it. Classical degeneracies between these orbits lead for example to clustering of the rotational eigenvalues [Harter and Patterson (1984), Sadovskii *et al.* (1990), and Kozin *et al.* (1992)] and vibrational eigenvalues [Lawton and Child (1980), Jaffé and Brumer (1980), and Child and Halonen (1984)]. Each such classical phase space zone is then separated from its neighbors by manifolds associated with an unstable orbit, whose energy in a given polyad (or with given angular momentum) may often be calculated analytically. Diagrams are given for a variety of systems to emphasize the systematic differences in the eigenvalue patterns on either side of such a separatrix, from one polyad to the next.

The second type of application concerns trajectory-based semiclassical wavepacket techniques, pioneered by Heller (1975) and his co-workers and more recently extended by Herman and Kluck (1984), Heller (1991), Kay (1994a,b), Walton and Manolopoulos (1996) and Brewer, Hulme and Manolopoulos (1997). Such methods are naturally adapted to the calculation of Franck–Condon spectra [Heller (1981)] and resonance Raman spectra [Lee and Heller (1979, 1982)], and the recent initial value methods may also be used for general semiclassical quantization purposes. The technique is to decompose the initial gaussian packet into a swarm of associated coherent states [Heller (1975)], each of which follows a classical trajectory defined by its center. The relative proximity of neighbors in the swarm is readily measured by use of gaussian overlap integrals, while interference effects are readily incorporated by accumulating the classical action along the trajectory. The resulting scheme is intuitively attractive and computationally undemanding with regard to storage, because the effort depends on the number of trajectories required for convergence, rather than the dimensionality of the system. The scaling properties with respect to system size remain to be fully investigated, but order N convergence has already been demonstrated for the low energy states of systems with up to 15 fully coupled vibrational degrees of freedom [Brewer, Hulme and Manolopoulos (1997)].

The chapter starts with a brief introduction to classical angle–action variables and correspondence principles, which play an important role in the semiclassical interpretation of effective Hamiltonian systems. Section 18.3.1 introduces the polyad sphere concept from the view point of a coupled-Morse local mode Hamiltonian, which is treated in some analytical detail. Extensions to more complicated systems are also indicated [Lu and Kellman (1997)], with particular reference to acetylene [Kellman and Xiao (1990), Svitak *et al.* (1995), Rose and Kellman (1995), and Jung and Taylor (1999)]. The mathematical connection between these vibrational problems and the semiclassical theory of rotational clustering [Harter and Patterson (1984), Sadovskii *et al.* (1990), and Kozin *et al.* (1992)] are brought out in Section 18.3.2. Computational details relevant to the estimation of clustering splittings are described in Section 18.3.3.

Section 18.4.1 introduces the semiclassical wavepacket method by reference to Heller's (1975) original frozen gaussian technique. Modern extensions of the theory using a so-called initial value representation, based on the Herman–Kluck (1984) propagator are described in Section 18.4.2. Finally computational questions are addressed in Section 18.4.3.

18.2 The classical framework

The aim of any semiclassical theory is to use trajectory information, derived by solution of Hamilton's equations of motion [see Section 8.1 of Goldstein (1980)],

$$\frac{\mathrm{d}q}{\mathrm{d}t} = \frac{\partial H}{\partial p}, \qquad \frac{\mathrm{d}p}{\mathrm{d}t} = -\frac{\partial H}{\partial q}, \tag{1}$$

to interpret and predict the spectrum. Semiclassical methods take particular advantage of the flexibility of this Hamiltonian formulation, because q and p are not restricted to the normal

Cartesian coordinates and momenta. There is a systematic *canonical transformation theory* to pass from one set of *canonical variables* (q, p) to another (Q, P) in which equations of the same form also apply. Details are given in Sections 9.1–9.2 of Goldstein (1980) and Appendix E of Child (1991). The interesting connection with unitary transformations in quantum mechanics is discussed in Miller's (1974) definitive article and in Appendix C of Child (1991). The special role of the transformation to *action–angle variables* is discussed at some length in the following section, in view of the Bohr–Sommerfeld connection between actions and quantum numbers.

Other important considerations concerns the *geometry of the classical phase space* spanned by the $2n$ conjugate variables of the system, and the *regularity* or *chaotic* nature of the classical motions (see Percival (1977)). A regular system has n constants of the motion, corresponding to n good quantum numbers, with the motions restricted to the n-dimensional surface of a torus, as illustrated in Figure 18.3 below. In separable or near separable situations the entire phase space is occupied a single set of concentric tori, and a single set of quantized actions is sufficient to label the all energy states. The spectroscopic analog would be described for example by a simple Dunham expansion. The argument that follows is that the introduction of anharmonic resonance terms serves to divide the classical phase space into separate regions (see Sections 4.2–4.4 of Tabor (1989) and Section 7.1 of Child (1991)). Each region contains its own set of concentric tori, centered on a *stable periodic orbit* of the system and its own set of angle–action variables (hence logically also its own quantum numbers) while the boundaries between the separate regions are each defined by a *separatrix* associated with an *unstable periodic orbit* of the system. Sections 18.3.1–18.3.4 below aim to show how the quantum mechanical spectrum shows clear imprints of these divisions.

The semiclassical wavepacket techniques discussed in Section 18.4 are also affected by the nature of the classical phase space, particularly in the presence of classical *chaos* which arises in the vicinity of the unstable orbits. The phase volume occupied by chaotic motions typically increases gradually as the total energy increases, leaving *islands of regularity* around the stable periodic orbits, which are assumed by RRKM theory to decrease to zero at the dissociation limit. The consequence for semiclassical wavepacket techniques is that wavepackets initiated in the regular regions remain tied to the stable periodic orbit, thereby regularly revisiting its initial location. Conversely the exponential divergence of trajectories in a chaotic region leads to rapid spreading of the wavepacket and much slower convergence of the quantization process.

18.2.1 CLASSICAL ANGLE–ACTION VARIABLES

Much of the elegance of classical mechanics lies in the freedom to transform from momentum p and coordinate q to other sets of canonical variables [Goldstein (1988), Chapter 9]. The transformation to action–angle variables (I, ϕ) plays a special role in the classical–quantal correspondence because it is designed to ensure that the energy depends only on the action,

$$E = H(I), \tag{2}$$

just as the quantum mechanical energy levels are specified only by the quantum numbers. The form in equation (2) ensures, via Hamilton's equations

$$\begin{aligned} \frac{\mathrm{d}I}{\mathrm{d}t} &= -\frac{\partial H}{\partial \phi} = 0, \\ \frac{\mathrm{d}\phi}{\mathrm{d}t} &= \frac{\partial H}{\partial I} = \omega(I), \end{aligned} \tag{3}$$

that I is a constant of the motion and that ϕ varies linearly with time at an action-dependent frequency $\omega(I)$, whose functional form depends on the system in question.

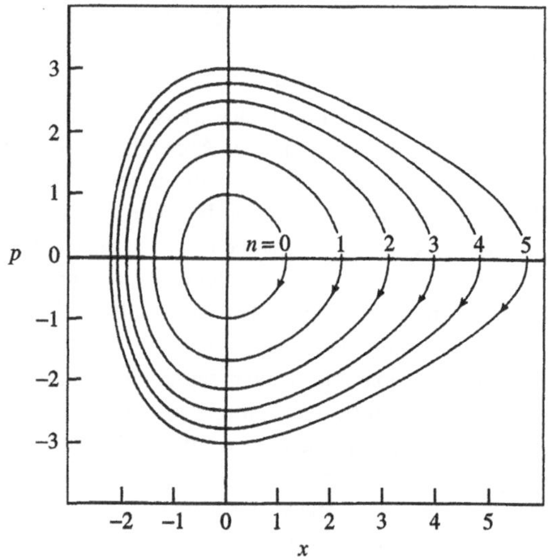

Figure 18.1. Phase space orbits of the Morse oscillator (taken from Child (1991) with permission).

To understand the systematic prescription for these special (I, ϕ) variables it is convenient to regard the classical orbits in the (q, p) plane in Figure 18.1 as energy contours. The associated action values specified by the enclosed areas

$$I = \frac{1}{2\pi} \oint p \, \mathrm{d}q, \tag{4}$$

are then clearly functions of the energy and vice versa. Here

$$p = \pm\sqrt{2m[E - V(q)]}, \tag{5}$$

with the upper and lower signs being taken for points in the upper and lower half-planes in Figure 18.1 respectively. The angle ϕ, which identifies a particular point on the orbit, is specified as

$$\phi = \omega(I) \int_{q_0}^{q} \frac{m}{p(q)} \, \mathrm{d}q, \tag{6}$$

where q_0 is any convenient reference point; one then readily verifies that

$$\frac{\mathrm{d}\phi}{\mathrm{d}t} = \omega(I) \frac{m}{p} \frac{\mathrm{d}q}{\mathrm{d}t} = \omega(I). \tag{7}$$

Equations (4)–(7) can frequently be manipulated to express (p, q) in terms of (I, ϕ). The functional relationship and the forms of $H(I)$ and or $\omega(I)$ in equations (2) and (3) depend on the nature of the potential function $V(q)$. The simplest case is that of the simple harmonic oscillator, with Hamiltonian

$$H = \frac{1}{2m} p^2 + \frac{1}{2} m\omega^2 q^2, \tag{8}$$

for which the transformation

$$\begin{aligned} q &= \sqrt{2I/m\omega} \cos\phi, \\ p &= -\sqrt{2Im\omega} \sin\phi, \end{aligned} \tag{9}$$

readily yields

$$H(I) = I\omega,$$
$$\omega(I) = \omega = \text{const.} \tag{10}$$

Equation (9) implies a choice of q_0 in equation (6) as the outer turning point on each orbit in Figure 18.1; the negative sign in the formula for p is required for consistency with Hamilton's equations; $(\mathrm{d}q/\mathrm{d}t) = (\partial H/\partial p)$ and $(\mathrm{d}p/\mathrm{d}t) = -(\partial H/\partial q)$.

Equations corresponding to equations (9) and (10) for a variety of other systems are collected in Appendix E of Child (1991). The following forms for the body-fixed components of angular momentum are relevant to what follows

$$\begin{aligned} J_x &= \sqrt{J^2 - K^2}\cos\phi_K, \\ J_y &= -\sqrt{J^2 - K^2}\sin\phi_K, \\ J_z &= K. \end{aligned} \tag{11}$$

They differ from other literature forms [Augustin and Miller (1974), and Gray and Davis (1989)] in the choice of origin for ϕ_K, which is taken in equation (11) to confirm with the normal angular momentum phase conventions. J in equation (11) is evaluated as $(j + \frac{1}{2})\hbar$ in semiclassical theory and K as $k\hbar$ [Child (1991), Section 3.1].

18.2.2 CORRESPONDENCE PRINCIPLES

The fundamental correspondence between the classical action I and the quantum number n is expressed by the Bohr–Sommerfeld quantization condition

$$I = \frac{1}{2\pi}\oint p\,\mathrm{d}q = (n + \delta)\hbar, \tag{12}$$

where the Maslov index δ is equal to the number of classical turning points divided by 4 [Child (1991)]. Thus $\delta = \frac{1}{2}$ for vibrational and orbital motion and $\delta = 0$ for rotation in a plane. The mathematical origin of these Maslov indices is discussed in Sections 2.2, 2.3 and 3.1 of Child (1991), and a pictorial explanation is available from our knowledge of one-dimensional wavefunctions. In the plane rotational case, motion around the circle is entirely classically accessible and the boundary conditions require an overall phase change of $2\pi n$ around the circle; the corresponding semiclassical phase change, $(1/\hbar)\oint p\,\mathrm{d}q = 2\pi n$, requires $\delta = 0$ in equation (12). The difference in a case restricted by turning points, a and b say, is that part of the wavefunction lies in the classically accessible regions, $q < a$ and $q > b$. Seen in sinusoidal terms the nth wavefunction has an overall phase change of $(n + 1)\pi$ of which two contributions of $\pi/4$ may be ascribed to the classically forbidden regions, because the wavefunction has a point of inflection at each turning point. Consequently in semiclassical terms $(1/\hbar)\int_a^b p\,\mathrm{d}q = (n + \frac{1}{2})$, which corresponds to equation (12) with $\delta = \frac{1}{2}$, bearing in mind that the integral above in taken around a complete cycle.

Other correspondences are obtained by introducing the semiclassical action operator

$$\hat{I} = \mathrm{i}\hbar\left(\frac{\partial}{\partial\phi}\right) + \delta\hbar, \tag{13}$$

which preserves the proper commutation relation

$$[\hat{I}, \phi] = -\mathrm{i}\hbar, \tag{14}$$

but is not in fact exact because the classical action is restricted to positive values [Carruthers and Nieto (1968), Leaf (1969), and Augustin and Rabitz (1979)]. Within the semiclassical approximation, however, the periodic eigenfunctions of $\hat{I}$ are given for all systems by

$$\psi_n(\phi) = \frac{1}{\sqrt{2\pi}} e^{in\phi}, \tag{15}$$

with eigenvalues corresponding to equation (13)

$$\hat{I}\psi_n(\phi) = (n+\delta)\hbar\psi_n(\phi). \tag{16}$$

Moreover, since $\hat{I}$ commutes with the obvious quantum mechanical analog of $H(I)$, namely

$$\hat{H} = H(\hat{I}), \tag{17}$$

it follows that the $\psi_n(\phi)$ are also eigenfunctions of $\hat{H}$ with eigenvalues

$$E_n = H[(n+\delta)\hbar]. \tag{18}$$

Consequently

$$\frac{1}{2}(E_{n+1} - E_{n-1}) \simeq \frac{\partial H}{\partial n} = \hbar\left(\frac{\partial H}{\partial I}\right) = \hbar\omega(I). \tag{19}$$

In other words the local quantum level spacing is equal to $\hbar$ times the classical frequency at the relevant mean action.

Equation (17) may be extended to more general functions $A(I, \phi)$ to produce equivalent operators

$$\hat{A} = A(\hat{I}, \phi), \tag{20}$$

with the proviso that any matrix elements $\langle n'|\hat{A}|n\rangle$ are evaluated at a mean quantum number $\overline{n}$ in order to ensure the hermicity of the matrix A; thus

$$\langle n'|\hat{A}|n\rangle = \frac{1}{2\pi}\int_0^{2\pi} A[(\overline{n}+\delta)\hbar, \phi]\, e^{i(n-n')\phi}\, d\phi. \tag{21}$$

Equation (21) gives the Heisenberg correspondence between quantal matrix elements and classical Fourier components. A simple example is provided by the harmonic oscillator creation (or raising) operator, which transforms in the light of equations (9) to

$$\hat{a}^\dagger = \frac{1}{\sqrt{2}}\left(\sqrt{m\omega}\hat{q} - i\hat{p}/\sqrt{m\omega}\right) = \sqrt{\hat{I}}\, e^{i\phi}. \tag{22}$$

It follows from equations (15) and (21) that $\langle n'|\hat{a}^\dagger|n\rangle = 0$ unless $n' = n+1$, while

$$\langle n+1|\hat{a}^\dagger|n\rangle = \sqrt{\left(\overline{n}+\tfrac{1}{2}\right)\hbar}. \tag{23}$$

Equation (23) agrees with the exact quantum mechanical result if $\overline{n}$ is taken as the arithmetic mean, $\overline{n} = n + \frac{1}{2}$.

A similarly exact result is also obtained for the angular momentum shift operators implied by equation (11). The $|jk\rangle$ wavefunction

$$\phi_{jk} = \frac{1}{2\pi}\exp(ij\phi_j + ik\phi_k) \tag{24}$$

is clearly shifted up in k by one unit, leaving j invariant, by the operator

$$\hat{J}_x - i\hat{J}_y = \sqrt{\hat{J}^2 - \hat{K}^2}\, e^{i\phi_k}, \tag{25}$$

where $\hat{J}$ and $\hat{K}$ have δ values, $\delta_J = \frac{1}{2}$ and $\delta_K = 0$ respectively. The matrix element $\langle j, k+1|\hat{J}_x - i\hat{J}_y|j, k\rangle$ is therefore readily deduced to be

$$\langle j, k+1|(\hat{J}_x - i\hat{J}_y)/\hbar|j, k\rangle = \sqrt{\left(j+\tfrac{1}{2}\right)^2 - \overline{k}^2} = \sqrt{j(j+1) - k(k+1)}, \tag{26}$$

if the arithmetic mean $\overline{k} = k + \frac{1}{2}$ is taken.

18.3 Geometry of the classical phase space

On moving from one to two degrees of freedom the phase space becomes four dimensional. Hence some method of projection is required to display the partitions between its component regions. The most useful is the Poincaré surface of section technique. Consider as an illustration the trajectories in Figure 18.2(a) for the coupled Morse oscillator Hamiltonian

$$H = \frac{1}{2m}(p_a^2 + p_b^2) + k_{ab}x_a x_b + D\left(1 - e^{-\beta x_a}\right)^2 + D\left(1 - e^{-\beta x_b}\right)^2. \tag{27}$$

Each of the three trajectories is seen to have the same energy because their corners lie on the same potential energy contour. Their roughly rectangular boundaries are termed the caustics of the classical motion. The Poincaré section of Figure 18.2(b) is obtained by plotting x_b and p_b whenever a trajectory crosses the sectioning line $x_a = X$ with $p_a > 0$. The three concentric dotted curves are maps of the three trajectories and the central point at $x_b = p_b = 0$, around which the others are organized, is a map of a special trajectory–the symmetry determined periodic orbit along the line $x_b = 0$ in Figure 18.2(a), which retraces itself indefinitely. A similar map could be used to display the (x_a, p_a) motion, by taking a section in the line $x_b = 0$ with $p_b > 0$. It would again show three concentric dotted curves, organized around a central point.

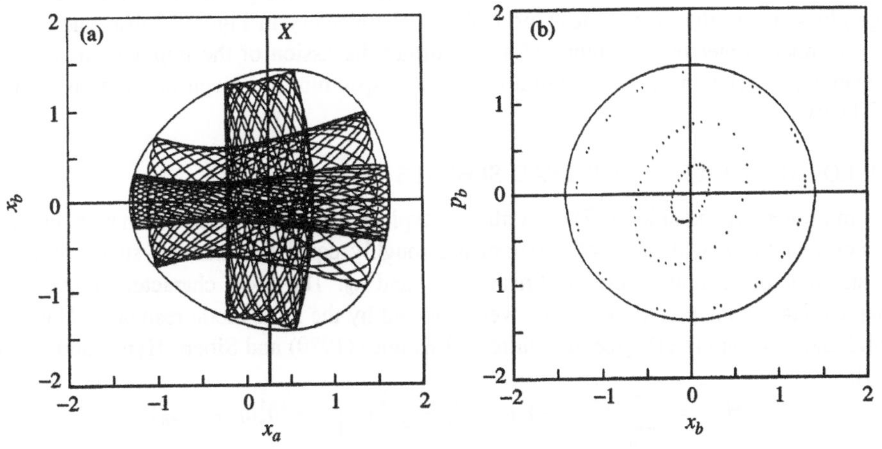

Figure 18.2. Normal motions of the Hamiltonian in equation (27) at $E = 2.0$, (a) as Lissajous figures, and (b) as a Poincaré section (adapted from Child (1991) with permission).

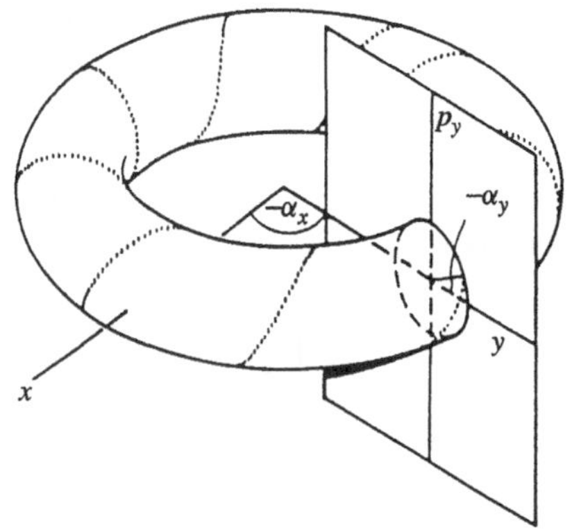

Figure 18.3. Quasi-periodic motion over the surface of an invariant torus. The cut indicates a Poincaré section (taken from Child (1991) with permission).

The picture that emerges whereby the representative phase point in any trajectory cycles around two independent periodic orbits may be visualized as motion on an invariant torus, as shown in Figure 18.3. It is termed invariant because the area enclosed by each of the dotted orbits in Figure 18.2(b) is invariant to the choice of sectioning plane X, provided that it cuts opposite caustics of the trajectory in question [Einstein (1917)]. The EBK [Einstein (1917), Brillouin (1926), and Keller (1958)] quantization procedure, practical details of which have been reviewed by Noid, Kosykowski and Marcus (1981), is based on adjusting the energy and trajectory starting conditions until the two enclosed areas simultaneously satisfy the Bohr–Sommerfeld conditions of equation (12).

Complications can occur, however, as a result of changes in the character of the periodic orbits. For example, above a certain threshold, dependent on the coefficients in equation (22) the simple patterns in Figure 18.2 go over to those in Figure 18.4. The same Hamiltonian gives rise to the two quite different trajectory types in Figures 18.4(a) and 18.4(b), and the Poincaré section in Figure 18.4(c) contains a central cross in place of the central dot in Figure 18.2(b). The explanation for these changes is that the symmetric stretching periodic orbit along the line $x_b = 0$ in Figures 18.4(a) and 18.4(b) has become unstable to small changes in its vicinity; one type of change gives rise to the symmetrical 'normal mode' trajectory in Figure 18.4(a) and another to the 'local mode' trajectory in Figure 18.4(b). Further discussion of the implications of periodic orbit stability in relation to the quantum mechanical spectrum are given in Sections 18.3.1 and 18.3.2 below.

18.3.1 LOCAL MODES AND POLYAD SPHERES

The Hamiltonian in equation (27) describes a typical 1:1 anharmonic resonance in that the frequencies of the two degrees of freedom are roughly equal (at least for small anharmonicities) and that the coupling term is linear in x_a and x_b. The local character of the quantum mechanical Hamiltonian is known to be well modeled by the local mode resonance Hamiltonian of Child and Lawton (1981), [see also Jaffé and Brumer (1980) and Sibert, Hynes and Reinhardt (1982)].

$$\hat{H}_{\text{res}} = \sum_{i=a,b} \left[\left(\hat{v}_i + \tfrac{1}{2}\right)\omega - \left(\hat{v}_i + \tfrac{1}{2}\right)^2 \omega x \right] - \lambda\left(\hat{a}_a^\dagger \hat{a}_b + \hat{a}_a \hat{a}_b^\dagger\right), \tag{28}$$

in which ω, ωx and λ depend on the parameters in equation (27) and the choice of creation and annihilation (or raising and lowering) operators restricts the coupling to a given overtone manifold

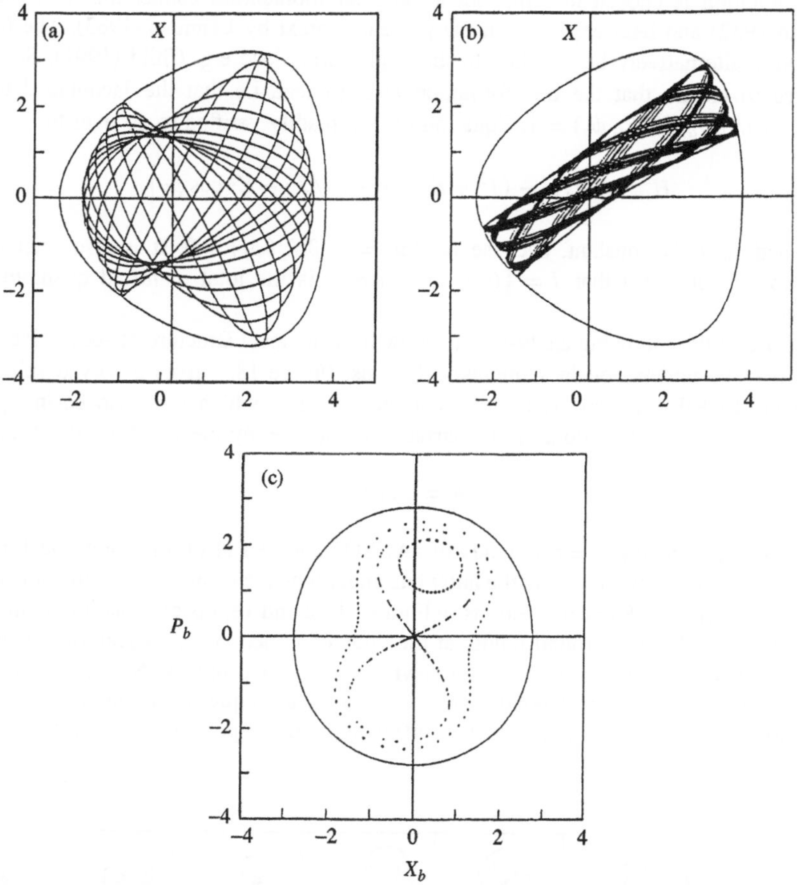

Figure 18.4. Bifurcated motions of the Hamiltonian in equation (61) at $E = 4.0$ showing (a) a normal trajectory, (b) a local trajectory and (c) the Poincarè section (adapted from Child (1991) with permission).

with $v_a + v_b = \text{const}$. A positive λ value in equation (28) ensures that $v_1 < v_3$ in normal mode notation. The polyad sphere is a device due to Xiao and Kellman (1989) for arbitrary resonances, designed to bring out the qualitative features of the quantal spectrum. The details given below for the 1:1 case illustrate the method, while also demonstrating the connection with the alternative Darling–Dennison formulation of local mode theory [Lehmann (1983), and Mills and Robiette (1985)].

The starting point is to derive the following classical equivalent of equation (28) by using equations (12) and (22) to substitute for $(\hat{v}_i + \frac{1}{2})$ and the creation/annihilation operators, to give

$$H_{\text{res}} = \sum_{i=a,b} \left[I_i \omega - I_i^2 \omega x \right] - 2\lambda \sqrt{I_a I_b} \cos(\phi_a - \phi_b), \tag{29}$$

where the actions are measured in units of $\hbar$. The next step is to transform to a new set of angle–action variables

$$\begin{aligned} J &= \tfrac{1}{2}(I_a + I_b) \qquad \phi_J = \phi_a + \phi_b, \\ K &= \tfrac{1}{2}(I_a - I_b) \qquad \phi_K = \phi_a - \phi_b, \end{aligned} \tag{30}$$

where the notation is chosen to bring out an angular momentum connection, first recognized by Kellman(1982) and later exploited in the present context by Lehmann(1983). The factors of $\frac{1}{2}$, which may alternatively be attached to the angle terms [see e.g. Child (1991), Section 7.3] are required to ensure that the transformation is canonical, i.e that the Jacobian determinant $\partial(J, K, \phi_i, \phi_K)/\ \partial(I_a, I_b, \phi_a, \phi_b) = 1$. Equation (28) is readily verified to go over to

$$H_{\mathrm{res}} = 2\Big[J\omega - (J^2 + K^2)\omega x - \lambda\sqrt{J^2 - K^2}\cos\phi_K\Big]. \tag{31}$$

The total action, J, is constant, because $(\mathrm{d}J/\mathrm{d}t) = -(\partial H/\partial\phi_J) = 0$. It is also seen from the definition in equation (30) that $J = \frac{1}{2}(v + 1)$, where v is the total manifold quantum number $v = v_a + v_b$.

The purpose of the following analysis is to show how the level structure of successive overtone manifolds may be understood in semiclassical terms. Figure 18.5 gives a polyad sphere [Xiao and Kellman (1989)] overview by plotting contours of H_{res}, which may also be interpreted as trajectories of the (K, ϕ_K) motion, on the surface of a sphere, by means of the substitution

$$K = J\cos\theta. \tag{32}$$

The diagrams relate to the $v = 6$ polyads of (A) SO_2 and (B) H_2O, of which the former still displays normal mode motions, while the latter has undergone the local mode bifurcation The qualitative similarity with the Poincaré sections in Figures 18.2 and 18.4 is obvious. In Figure 18.5(A) the trajectories simply circle round points at the ends of an axis in the equatorial plane, which actually correspond to the symmetric and antisymmetric normal modes. Note for example from equation (30) that the individual bond phases ϕ_a and ϕ_b are equal at the point $\phi_K = 0$ on the positive x axis in Figure 18.5, and that they differ by π at the point $\phi_K = \pi$ on the x axis. By

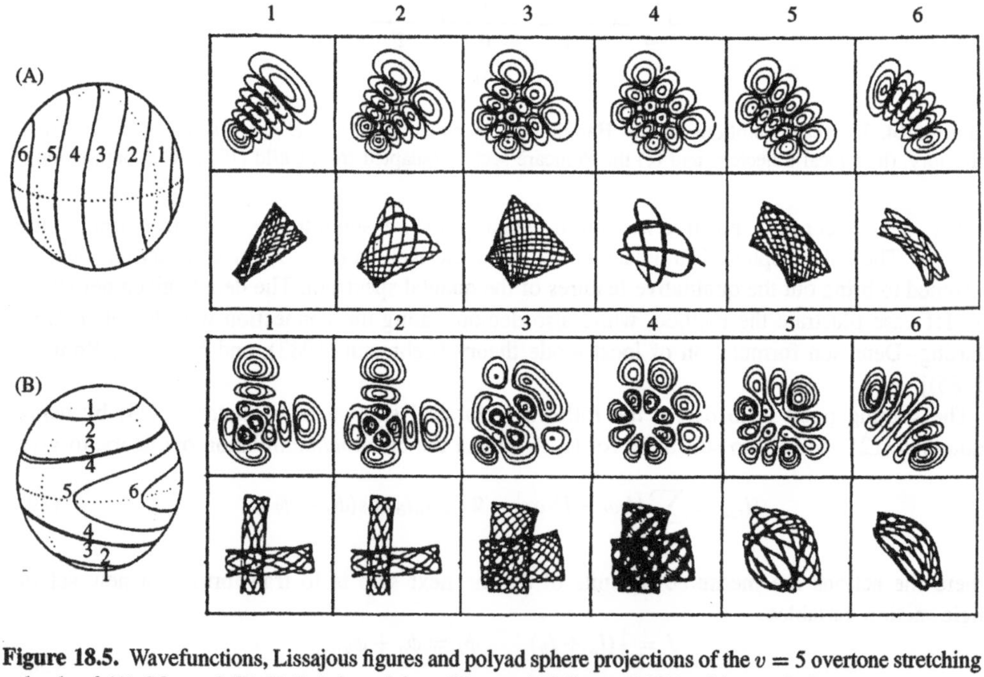

Figure 18.5. Wavefunctions, Lissajous figures and polyad sphere projections of the $v = 5$ overtone stretching polyads of (A) SO_2 and (B) H_2O (adapted from Xiao and Kellman (1992) with permission).

contrast Figure 18.5(B) shows a bifurcation into local and normal regions of the phase space. Equation (31) is designed to allow these changes to be followed analytically.

The nature of the fixed points, at which $(dK/dt) = (\partial H/\partial\phi_K) = 0$ and $(d\phi_K/dt) = -(\partial H/\partial K) = 0$, are of particular importance. The character of most such points (K_0, ϕ_0) may inferred from the local quadratic expansion

$$H_{res} \simeq H_{res}(K_0, \phi_0) + \frac{1}{2}(K - K_0, \phi - \phi_0)\begin{pmatrix} h_{kk} & h_{k\phi} \\ h_{\phi k} & h_{\phi\phi} \end{pmatrix}\begin{pmatrix} K - K_0 \\ \phi - \phi_0 \end{pmatrix}, \tag{33}$$

here h is the hessian matrix of second derivations. If $\det(h) > 0$, the eigenvalues of h are either both positive or both negative. Both of these so-called elliptic possibilities lead to stable local oscillations around (K_0, ϕ_0) at a frequency $\omega_0 = \sqrt{\det(h)}$ with the sense of oscillation (clockwise or anticlockwise) depending on the sign of the eigenvalues (positive or negative). A second possibility is that $\det(h) < 0$, in which case the point (K_0, ϕ_0) is said to be hyperbolic or unstable. Problems arise, however, if the fixed point lies at one of the poles of the polyad sphere, with $K = J$ or $K = -J$, at which the angle variable ϕ_K is undefined, because the hessian matrix h is undefined. The Hamiltonian in equation (31) does not allow such possibilities but later examples will occur in which the local time derivatives are given by

$$\frac{dK}{dt} \propto \left(J^2 - K^2\right)^{\alpha} \qquad \alpha > 0, \tag{34}$$

$$\frac{d\phi}{dt} = a + b\cos\phi. \tag{35}$$

K is then fixed at $K = \pm J$ by equation (34) and the character of the point is determined by the relative values of a and b in equation (35); it is stable if $d\phi/dt$ is finite (i.e if $|a/b| > 1$) and unstable otherwise, and the frequency in the stable case is given by

$$|\omega_0| = \frac{2\pi}{T} = 2\pi\Big/\int_0^{2\pi}\frac{d\phi}{a + b\cos\phi} = \sqrt{(a^2 - b^2)}. \tag{36}$$

The dimensions of ω_0 are T^{-1} in standard physical units or $(2\pi cT)^{-1}$ if H_{res} is converted to cm^{-1} and K is measured in units of $\hbar$. Its sign depends on whether the actual motion is vibrational or rotational in character, because the natural sense of rotation is anticlockwise while vibrations follow a clockwise path in the phase plane (see Figure 18.1). Consequently since ω_0 is used in the following applications to determine the 'vibrational' zero-point correction, the sign of ω_0 is that of a in equation (36) in vibrational applications, but of $-a$ in rotational ones. An alternative approach, which avoids this complication is to ensure that the orientation of the quantization axis eliminates any stable fixed points at the north and south poles; the sign of ω_0 is then unambiguously fixed by the signs of the eigenvalues of the hessian matrix.

Proceeding to details, the properties of the three possible fixed points of H_{res} in equation (31) are listed in Table 18.1, from which it is apparent that the transition between Figure 18.5(a) and 18.5(b) occurs when $\rho_v = 2J\omega x/\lambda = (v+1)\omega x/\lambda = 1$. Table 18.1 shows that there are only two real fixed points (of types I and II) for $\rho_v < 1$, both being stable. The type II point becomes unstable, however, for $\rho_v > 1$ and two new real stable points of type III appear.

The energies and local frequencies of the various fixed points may also be used to put the overall character of the quantum mechanical overtone spectrum in perspective, as illustrated for a model of H_2O stretching modes in Figure 18.6. The energy points are numerical eigenvalues of $\hat{H}_{res}$ in equation (28), modified by subtraction of

$$E_v^o = (v+1)\omega - \tfrac{1}{2}(v+1)^2\omega x, \tag{37}$$

Table 18.1. Fixed points of the local mode Hamiltonian.

Type	K_0	ϕ_0	$H'_{\text{res}}(K_0, \phi_0)$	ω_0
I	0	0	$-(v+1)\lambda$	$\sqrt{4\lambda^2(1-\rho_v)}$
II	0	π	$(v+1)\lambda$	$-\sqrt{4\lambda^2(1+\rho_v)}$
III	$\pm\left(\frac{\lambda}{2\omega x}\right)\sqrt{\rho_v^2-1}$	π	$-\frac{\lambda^2}{2\omega x}[1+\rho_v^2]$	$\sqrt{4\lambda^2(\rho_v^2-1)}$

$H'_{\text{res}} = H_{\text{res}} - (v+1)\omega + \frac{1}{2}(v+1)^2\omega x$.
$\rho_v = 2J\omega x/\lambda = (v+1)\omega x/\lambda$.
H_{res} is given by equation (31) with $J = \frac{1}{2}(v+1)$.
The local mode bifurcation occurs when ω_0 changes from real to imaginary at the type I point and vice versa for the type III point.

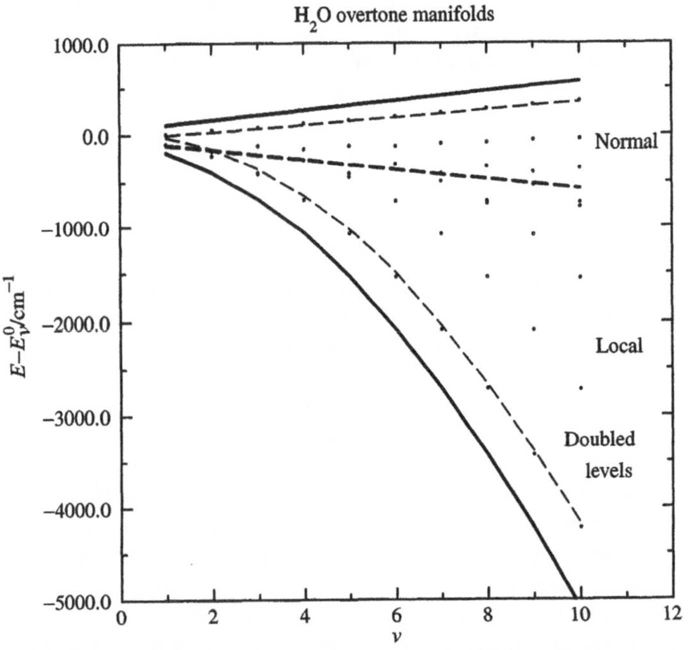

Figure 18.6. Periodic orbit-related partitions of the H_2O stretching eigenvalue spectrum. Solid lines give the energies of the quantized stable periodic orbits, with actions $J = (v+1)/2$ as listed in Table 18.1. The heavy dashed line follows the corresponding energy of the unstable symmetric stretching orbit, and the light dashed lines give approximate zero-point corrections according to the ω_0 entries in Table 18.1. The parameter values are $\omega = 3870.74\,\text{cm}^{-1}$, $\omega x = 84.17\,\text{cm}^{-1}$ and $\lambda = 49.5\,\text{cm}^{-1}$. Energies are plotted relative to $E_0 = (v+1)\omega - 0.5(v+1)^2\omega x$.

which is the purely J-dependent part of H_{res} in equation (30), expressed in terms of $v = 2J - 1$. The upper and lower continuous lines follow the modified energies of the type I and type II fixed points in Table 18.1 as a function of v, the lighter dashed lines give the harmonic zero-point energy corrections $\pm\frac{1}{2}\omega_0$ to these fixed points, where the sign depends on that of the eigenvalues of h. Finally the heavy dashed line in Figure 18.6 gives the energy of the unstable type II point. Energy levels in the local mode region below this line are close doublets arising from the classical double degeneracy of the type III points. This analytical fixed point analysis clearly gives an excellent semiquantitative picture of the overall level structure. Section 18.3.3 describes a more accurate quantization procedure that takes account of the dynamical tunneling, responsible for local mode splittings, as well as anharmonicities in the motion around the stable fixed points.

As a final note on the local mode story, it is interesting to establish the semiclassical connection between the coupled anharmonic oscillator picture in equation (28) and the commonly used equivalent Darling–Dennison model [Lehmann (1983), and Mills and Robiette (1985)]. The steps are to use equation (11) to perform the angular momentum axis rotation

$$\begin{aligned} K = J_z = J'_x &= \sqrt{J^2 - K'^2}\cos\phi_K, \\ \sqrt{J^2 - K^2}\cos\phi &= J_x = J'_z = K', \end{aligned} \tag{38}$$

with J and K' expressed in terms of normal mode analogs of equation (30), using I_1 and I_3 in place of I_a and I_b. The following substitution implied by equation (22)

$$\left(J^2 - K'^2\right)\cos 2\phi_{K'} = I_1 I_3 \cos 2(\phi_1 - \phi_3) \Longrightarrow a_1^{\dagger 2} a_3 + a_1 a_3^{\dagger 2} \tag{39}$$

is also required. The final transformation yields

$$\begin{aligned} H_{\text{res}} = {} & \left(v_1 + \tfrac{1}{2}\right)(\omega - \lambda) + \left(v_3 + \tfrac{1}{2}\right)(\omega + \lambda) \\ & - \tfrac{1}{2}\left[\left(v_1 + \tfrac{1}{2}\right)^2 + \left(v_3 + \tfrac{1}{2}\right)^2 + 4\left(v_1 + \tfrac{1}{2}\right)\left(v_3 + \tfrac{1}{2}\right) + a_1^{+2} a_3^2 + a_1^2 a_3^{+2}\right]\omega x, \end{aligned} \tag{40}$$

which is the Darling–Dennison Hamiltonian, within the $x - K$ relations of Mills and Robiette (1985).

This 1:1 local mode or 2:2 Darling–Dennison model has been analyzed in detail to illustrate a semiclassical technique that also applies to more complicated single resonance situations and to the rotational clustering which is discussed in Section 18.3.2. The same polyad sphere representation is applicable to any system coupled by a single resonance term–regardless of the powers to which the creation/annihilation operators are raised or the dimensionality of the system. The cases of 2:1 Fermi resonance and the 2345 multimode resonance in acetylene have been addressed in detail in the literature [Kellman and Xiao (1990), Svitak *et al.* (1995), and Rose and Kellman (1995)]. The general canonical transformation from zeroth-order (I_i, ϕ_i) to resonance-adapted (J_r, θ_r) angle–action variables, analogous to equation (30), may be systematized in the following way. Suppose that

$$\begin{aligned} H_{\text{res}} = {} & \sum_{i=1}^{n}\left(v_i + \frac{1}{2}\right)\omega_i - \sum_{i,j=1}^{n}\left(v_i + \frac{1}{2}\right)\left(v_j + \frac{1}{2}\right)x_{ij} \\ & + K\left[\Pi\left(a_i^{\dagger}\right)^{S_i}\Pi a_i^{T_i} + \text{c.c}\right], \end{aligned} \tag{41}$$

where the vector of frequencies $(\omega_1, \omega_2, \omega_3 \ldots \omega_n)$ is approximately parallel to an integer component vector $(\omega_1, \omega_2, \omega_3 \ldots \omega_n)$. The system will be in resonance if $\boldsymbol{w} \cdot \boldsymbol{V} = 0$, where $V_i = S_i - T_i$, and the classical resonance term involving $\cos(\phi_a - \phi_b)$ in equation (29) will generalize to one involving $\cos(\boldsymbol{V} \cdot \boldsymbol{\phi})$ where $\boldsymbol{\phi}$ is the vector of zeroth-order angles ϕ_i. The aim is then to perform a canonical transformation such that the large ω_i dependent terms are concentrated into a single term involving the first new action J_1, while the cosine term has the last new angle θ_n as its argument. In other words $J_1 \propto \boldsymbol{I} \cdot \boldsymbol{w}$ and $\theta_n = \boldsymbol{V} \cdot \boldsymbol{\phi}$.

A classical generator of the following form [Goldstein (1980)]

$$F_2(\phi_i J_r) = \tilde{J} R\phi \tag{42}$$

is used to perform the canonical transformation, where $\tilde{J}$ is the row vector of new actions, ϕ is a column vector of old angles and R is a matrix to be determined. The generating equations take the forms [Goldstein (1980), Section 9.1]

$$\begin{aligned} \tilde{I} &= \nabla_\phi F_2 = \tilde{J} R, \\ \theta &= \nabla_J F_2 = R\phi. \end{aligned} \tag{43}$$

The requirement $J_1 \propto \boldsymbol{I} \cdot \boldsymbol{w}$ means that rows 2 to n of R must be orthogonal to $\boldsymbol{w}$, with the final row equal to $\boldsymbol{V}$ (in order to ensure that $\theta_n = \boldsymbol{V} \cdot \phi$). The first row is conveniently taken proportional to $\boldsymbol{w}$. The reverse transformations to equation (43) are facilitated by choosing all rows of $\boldsymbol{V}$ to be mutually orthogonal, but this is not essential. There is in fact no unique canonical transformation even for a given angle specification by equation (43). The only firm requirement is that $[\partial(J_1, J_2 \ldots J_n)/\partial(I_1, I_2 \ldots I_n)] \times [\partial(\theta_1\theta_2 \ldots \theta_n)/\partial(\phi_1, \phi_2, \ldots \phi_n)] = 1$ and there are many different linear combinations for the $\{J_r\}$ with the same Jacobian determinant. The vital feature for present purposes is that the angle choice $\theta_n = \boldsymbol{V} \cdot \phi$ projects the entire motion onto the two-dimensional (J_n, θ_n) phase plane and hence onto the surface of the polyad sphere. The remaining actions J_r, $(r = 1, 2 \ldots n)$ which are constants of the motion, appear as parameters in H_{res}, whose changes can lead to bifurcations in the properties of the fixed points [Rose and Kellman (1985)]. Different choices for the definitions for the J_r simply affect the apparent origin of J_n, thereby causing a rotation of the sphere without altering the fixed point topology.

The extension of the Hamiltonian to include say p resonance terms gives rise to motion in a $2p$-dimensional phase space, which may even be chaotic [Lu and Kellman (1997)]. Nevertheless the scheme in equation (43) may be extended to set up the transformed classical resonance Hamiltonian and to identify conserved polyad numbers [Fried and Ezra (1987), and Kellman (1990)]. Each of the p resonance terms has its own resonance vector $\boldsymbol{V}_r$, $r = (n - p + 1), \ldots n$, which must appear as a row of R, so that $\theta_r = \boldsymbol{V}_r \cdot \phi$ in equation (43). The remaining conserved actions or polyad numbers J_r, $r = 1 \ldots (n - p)$, are defined by the columns of R^{-1}, because $\tilde{J} = \tilde{I} R^{-1}$; these columns must be orthogonal to the rows $\boldsymbol{V}_r$ and may be chosen to be orthogonal to each other. The remaining rows of R (columns of R^{-1}) may be taken as combinations of the known columns of R^{-1} (rows of R) subject to $RR^{-1} = I$.

As an example, the full five-mode Hamiltonian used by Smith and Winn (1988) to fit the acetylene spectrum includes three resonance terms and two constants of the motion. Kellman (1990) uses three resonance vectors (1, 0, −1, 0, 0), (0, −1, 1, −1, −1) and (1, −1, 2, −2, 0) in the five-dimensional space to identify two constants of the motion, expressed in our notation as

$$\begin{aligned} J_1 &= 5I_1 + 3I_2 + 5I_3 + I_4 + I_5, \\ J_2 &= 4I_1 - 22I_2 + 4I_3 + 13I_4 + 13I_5. \end{aligned} \tag{44}$$

J_1 is the expected total polyad number for a 5:3:5:1:1 resonance, but J_2 is a less obvious constant of the motion.

Extensions of the previous bifurcation (or fixed point) analysis to these multi-resonance situations have been discussed in some detail by Lu and Kellman (1997). Complications arise from the fact that each of the p resonance terms in the Hamiltonian has an associated action–angle pair. The search for fixed points must therefore be performed in a $2p$-dimensional space. Secondly the character of the resulting fixed points is not restricted to 'stable' or 'unstable' types; it depends on the eigenvalues of the hessian second derivative matrix, which necessarily occur in p pairs with product unity Lu and Kellman (1997). The most ambitious application to date is a four-mode study of the excited pure bending polyads of acetylene by Jacobson *et al* (1999), which demonstrated the existence of three distinct phase space zones in the higher polyads, each with its own assignable progressions of symmetry-labeled eigenvalues.

18.3.2 ROTATIONAL CLUSTERING

The existence of rotational clustering, to give higher close degeneracies than those imposed by symmetry, was first predicted by Dorney and Watson (1972). As illustrated below, the simplest example is actually the familiar high K and low K doubling of asymmetric top energy levels. More interesting cases include four-fold clustering in the asymmetric tops close to the oblate limit [Pavlichenkov and Zhilinski (1985), Makarewicz and Pyka (1989), Kozin *et al.* (1992), Kozin and Jensen (1994), and Jensen, Osmann and Kozin (1997)], high degeneracy clusters of SF_6 [Harter and Patterson (1984)] and a series of papers on vibrational–rotational clustering in tetrahedral species [see for example Sadovskii *et al.* (1990)]. Semiclassical quantization procedures are discussed in the following section. Here attention is restricted to the semi-quantitative divisions between different parts of a clustered spectrum, by extension of the foregoing fixed point analysis.

It is convenient to start by substituting from equation (11) in the rigid asymmetric top Hamiltonian to obtain

$$H_{\text{rot}} = AK^2 + \left(B\cos^2\phi + C\sin^2\phi\right)\left(J^2 - K^2\right), \tag{45}$$

bearing in mind that $J = j + \frac{1}{2}$ in semiclassical applications. Figure 18.7, taken from Harter and Patterson (1984), gives a rotational energy surface (RES) representation of this expression by substituting $K = J\cos\theta$ and plotting contours of H_{rot} as the radial distance in a spherical polar plot. The contours, or rotational energy trajectories, are seen to circle the J_a and J_c axes and to avoid J_b, each trajectory being doubly degenerate in classical mechanics as a consequence of time reversal symmetry. The most important qualitative point about the diagram is that the dotted line (or separatrix) through J_b divides the surface into distinct near prolate and near oblate regions around J_c and J_a respectively. In addition the classically degenerate (clockwise and anticlockwise) pairs of trajectories are more widely separated in phase space, the closer they

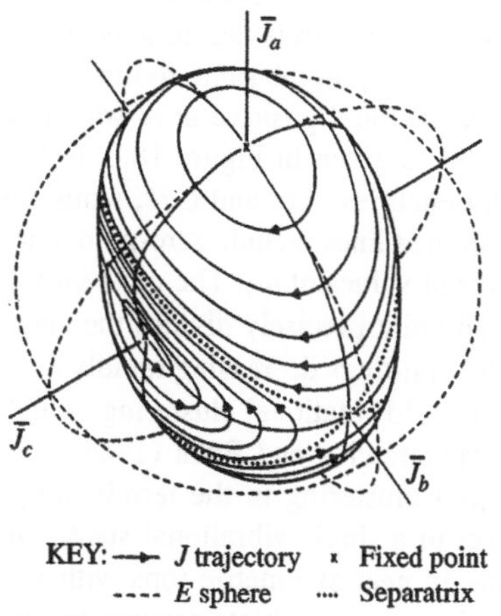

Figure 18.7. A rotational energy surface (RES) representation of the motions of a rigid asymmetric top (taken from Harter and Patterson (1984) with permission).

Table 18.2. Fixed points of the rotational Hamiltonians.

Asymmetric top [equation (45)]

Type	K_0	$\cos 2\phi_0$	Energy	ω_0
a	J	?	AJ^2	$-2J\sqrt{(A-B)(A-C)}$
b	0	1	BJ^2	$2J\sqrt{-(A-B)(B-C)}$
c	0	-1	CJ^2	$2J\sqrt{(A-C)(B-C)}$

Notes

(1) $J = j + \frac{1}{2}$ in semiclassical applications, with $K = 0, 1 \ldots j$.
(2) ω_0 is imaginary at the unstable type b fixed point.

Four-fold clustering model [equation (46)][b]

Type	K_0	$\cos 2\phi_0$	Energy	ω_0
a	J	?	$AJ^2 - (D_K + D_{JK})J^4$	$-2J\sqrt{(y-z-2D_K J^2)(y+z-2D_K J^2)}$
b	0	1	BJ^2	$2J\sqrt{-2z(y-z)}$
c	0	-1	CJ^2	$2J\sqrt{2z(y+z)}$
4-fold	$\pm\sqrt{\dfrac{y-z}{2D_K}}$	1	$BJ^2 + D_K K_0^4$	$-4\lvert K_0\rvert\sqrt{-z(y-z-2D_K J^2)}$

Notes

(1) $J = j + \frac{1}{2}$ in semiclassical applications, with $K = 0, 1 \ldots j$.
(2) ω_0 is complex at the unstable type b fixed point, and the four fold clustering occurs when ω_0 at a changes from real to imaginary and vice versa for the four-fold points.
[b] x, y and z are defined by equation (47).

are to the axis in question; consequently they are less readily split by dynamical tunneling (see below for a quantitative discussion).

The same message is obtained by applying the fixed point analysis of the previous section to equation (45), the results of which are collected in Table 18.2; note a non-physical fixed point at $K = J$ with $|\cos\phi| > 1$. The frequency ω_0 at the type a point, where ϕ is indeterminate, may be derived either from equation (36) or by a different choice of quantization axis in equation (45). The classical fixed point behavior must be irrelevant to axis labels and choice of the b axis as z, which puts the two stable points, a and c, on the same footing, clearly shows that the eigenvalues of the hessian matrix are negative at a and positive at c, which accounts for the signs of ω_0. A pictorial illustration of these results is given in Figure 18.8, in which the upper and lower solid lines follow the stable fixed point energies, AJ^2 and CJ^2, while the long dashed line follows the separatrix energy BJ^2. The light dashed lines include zero-point corrections for the motions around J_a and J_c, determined by the relevant values of ω_0. The most important point about Figure 18.8 is that the energy of the b-type fixed point precisely divides the energies into near prolate and near oblate groups, with the asymmetry splittings decreasing rapidly away from the central dashed line.

The semiclassical origin of four-fold rotational clustering, which was first discussed by Pavlichenkov and Zhilinski (1985), Makarewicz and Pyka (1989), and Kozin *et al.* (1992) is more interesting. We refer here to type I clustering in the terminology of Kozin and Jensen (1994), which is a purely rotational effect in a single vibrational state, whereas type II is a combination of the normal rotational doubling in near symmetric tops with vibrational local mode doubling. The type I four-fold clustering arises in near oblate asymmetric tops such as H_2Se [Kozin *et al.* (1992)] as a result of a sign change in the effective rotational constant difference $(A - B)_{\text{eff}}$, due to centrifugal distortion.

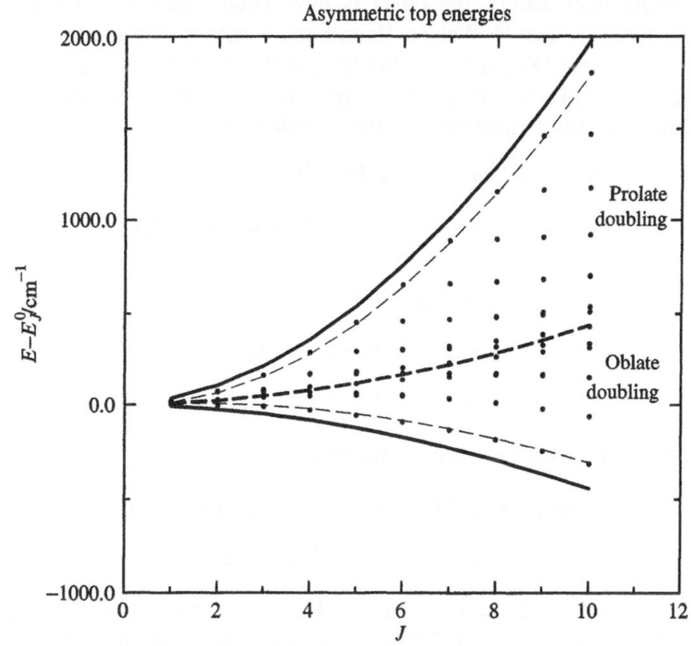

Figure 18.8. Periodic orbit related partitions of the rotational eigenvalues of an asymmetric top with $A = 30\,\mathrm{cm}^{-1}$, $B = 16\,\mathrm{cm}^{-1}$ and $C = 8\,\mathrm{cm}^{-1}$. Notations for the various lines are the same as in Figure 18.6. Energies are plotted relative to $E_J^0 = \frac{1}{2}(B + C)J^2$.

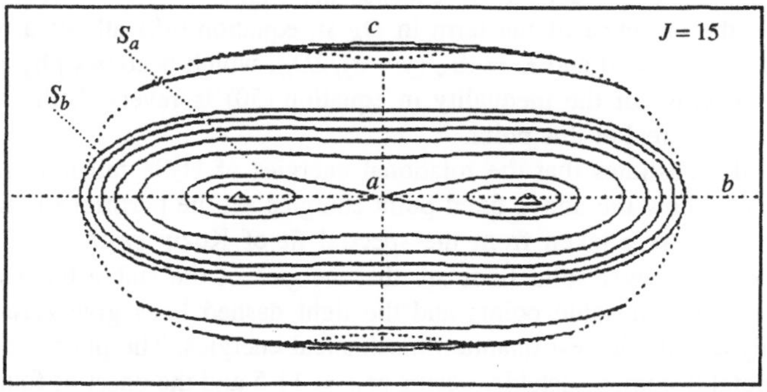

Figure 18.9. A four-fold clustering bifurcation on the rotational energy surface (RES) of H_2Se (adapted from Kozin *et al.* (1992) with permission).

Figure 18.9 which is taken from Kozin *et al.* (1992) shows that the resultant loss of stability of the a-axis fixed point on the RES is accompanied by a pitch-fork bifurcation, mathematically analogous to that responsible for the normal/local mode transition in Section 18.3.1, which leads in this case to two new stable points lying between the now unstable points on the a and b axes. An equivalent pair also appears at the opposite pole of the RES and the motions around these four fixed points are responsible for the four-fold clustering. Harter and Patterson (1984) give an idealized perspective view of the distorted RES, with a sketch of relevant atomic motions at the new stable angular momentum fixed points.

Turning to mathematical details, the effect is most readily described by including the D_{JK} and D_K distortion terms in the I^R Watson Hamiltonian; both contribute to the sign change in $(A - B)_{\text{eff}}$, as do other terms [Kozin *et al.* (1992)], and the D_K term is essential, as we shall see, in also providing the quartic restoring force responsible for the appearance of the new stable points. After substituting from equation (11) the Hamiltonian becomes

$$\begin{aligned} H_{\text{rot}} &= AJ_z^2 + BJ_x^2 + CJ_y^2 - D_{JK}J^2J_z^2 - D_KJ_z^4 \\ &= xJ^2 + yK^2 + z(J^2 - K^2)\cos 2\phi - D_KK^4, \end{aligned} \tag{46}$$

where

$$\begin{aligned} x &= \tfrac{1}{2}(B + C), \\ y &= A - \tfrac{1}{2}(B + C) - D_{JK}J^2, \\ z &= \tfrac{1}{2}(B - C). \end{aligned} \tag{47}$$

and again $J = j + \frac{1}{2}$. The fixed points must therefore satisfy

$$\partial H_{\text{rot}}/\partial K = 2K\left[y - z\cos 2\phi - 2D_KK^2\right] = 0, \tag{48}$$

$$\partial H_{\text{rot}}/\partial \phi = -2z\left(J^2 - K^2\right)\sin 2\phi = 0. \tag{49}$$

Solutions with $K = 0$ and $\cos 2\phi = \pm 1$ correspond to the relatively uninteresting points on the b and c axes with energies BJ^2 and CJ^2 respectively. The solution with $K = J$ and $\cos 2\phi$ indeterminate is the a axis point with energy $AJ^2 - (D_{JK} + D_K)J^4$, which is seen from a transition to complex frequency ω_0 in Table 18.2 to become unstable when

$$y - z - 2D_KJ^2 = A - B - (D_{JK} + 2D_K)J^2 < 0. \tag{50}$$

Each of these points is doubly degenerate because the sign of J may be reversed in classical mechanics. Finally the presence of the term in D_K in equation (48) allows a fourth, four-fold degenerate solution with $\phi = 0$ and $K = \pm\sqrt{(y - z)/2D_K}$, which becomes physically accessible $(K^2 < J^2)$ when the sense of the inequality in equation (50) is reversed; the frequency ω_0 in Table 18.2 also becomes real at this point.

Figure 18.10(a) demonstrates that the rotational energies of H_2Se fall into three groups and Figure 18.10(b) shows how the above fixed point energies set the partitions between them. The spectroscopic constants were taken from the spectral fit of Kozin *et al.* (1992). As in earlier diagrams, the upper and lower solid lines are the energies of the stable fixed points, the long dashed ones those of the unstable points and the light dashed lines give zero-point corrected estimates of the highest and lowest quantum mechanical energies. The pitch-fork bifurcation, at which the a-axis point becomes unstable, occurs at $J = 11.5$ and the onset of four-fold clustering to the right of the two unstable (a axis and b axis) lines is clearly indicated. Again the semiclassical fixed point analysis provides simple qualitative divisions between different parts of the quantum mechanical spectrum.

Even richer rotational clustering patterns appear in the spectra of spherical tops. Dorney and Watson (1972) started the discussion, but the fullest semiclassical analysis has been given by Harter and Patterson (1984) in relation to SF_6. The Hamiltonian was modeled to include quartic distortion terms in the form

$$H = BJ^2 + 10t_{044}\left(J_x^4 + J_y^4 + J_z^4 - \tfrac{3}{5}J^4\right), \tag{51}$$

which gives rise to a rotational energy surface of the form given in Figure 18.11. There are clearly six equivalent fixed points on the four-fold axes, eight equivalent points on the three-fold axes

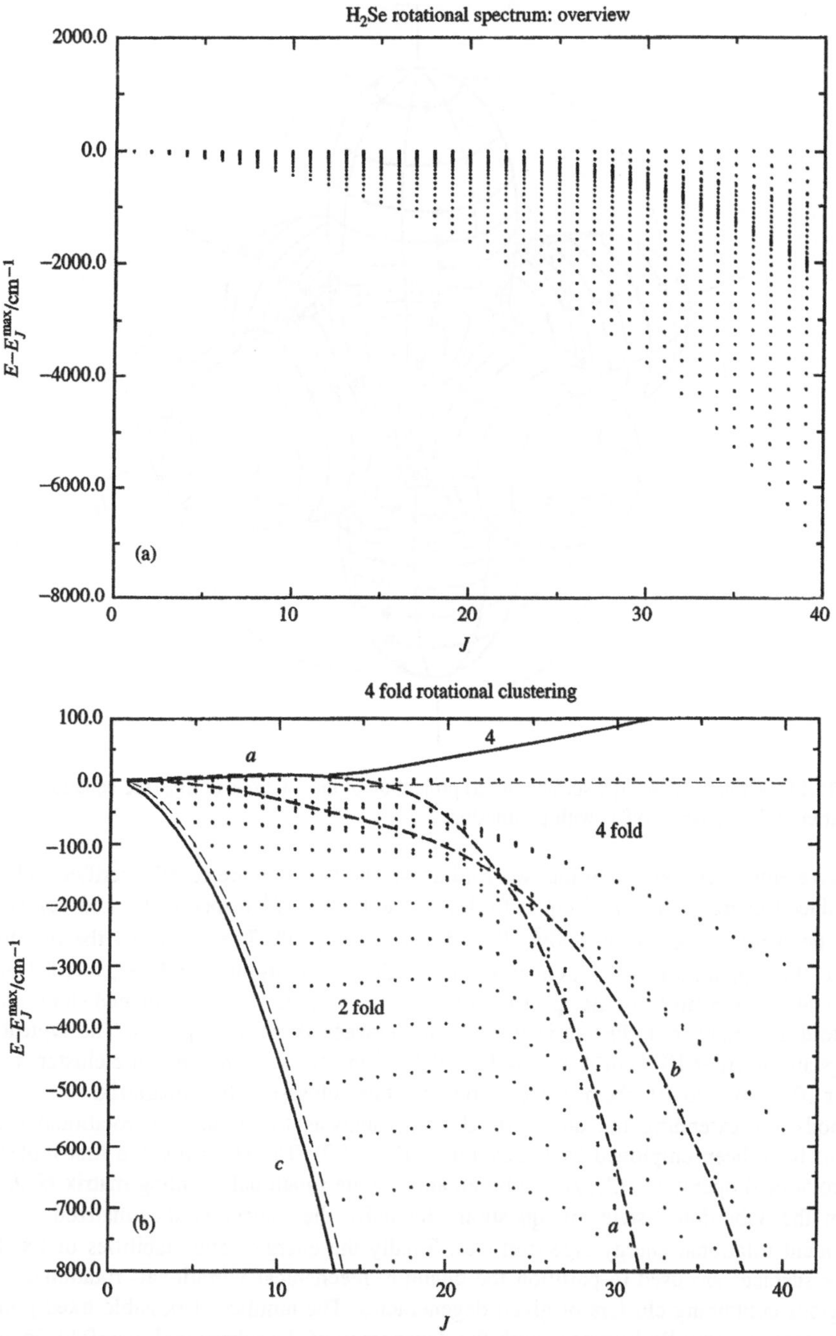

Figure 18.10. Rotational eigenvalues of the Hamiltonian in equation (51), with parameter values taken from Kozin *et al.* (1992). The overview in panel (a) indicates three distinct regions of the spectrum and panel (b) shows the periodic orbit-related partitions. The notations *a*, *b*, *c* and 4 on the heavy lines indicate the energies of the corresponding periodic orbits in Table 18.2. The light dashed lines are approximations to the highest and lowest energy levels for each *J* value. Energies are plotted with respect to the highest eigenvalue for given *J*.

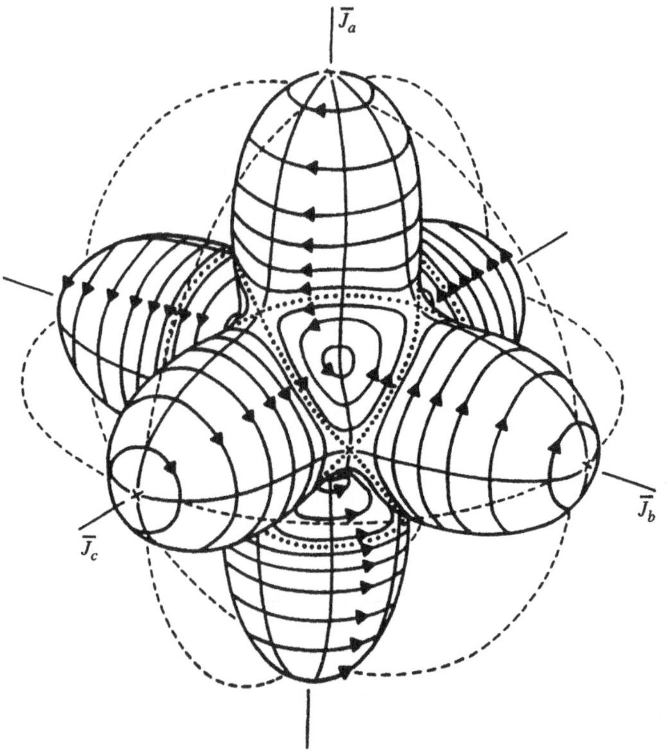

Figure 18.11. A rotational energy surface (RES) projection of the rotations of SF_6 (see equation (51)) (taken from Harter and Patterson (1984) with permission).

and twelve equivalent points on the two-fold axes. The first two are readily verified to be stable, with reduced energies $H - BJ^2$ equal to $4t_{044}J^4$ and $-\frac{8}{3}t_{044}J^4$ respectively; and corresponding hessian determinants given by $800t_{044}^2J^6$ and $(6400/9)t_{044}{}^2J^6$. The energy of the two fold axis points on the separatrix is $-\frac{1}{10}J^4$. Quantum mechanical calculations with $B = 0.091083\,\text{cm}^{-1}$ and $t_{044} = 5.44\,\text{Hz}$, derived from the spectrum of Bordé and Bordé (1982) do indeed show sequences of six-fold and eight-fold clusters in the regions between these fixed points. The typical cluster spacings are of order $10^{-4}\,\text{cm}^{-1}$ at $J = 30$ and $10^{-2}\,\text{cm}^{-1}$at $J = 88$, while the cluster widths are of order $10^{-6}\,\text{cm}^{-1}$ to $10^{-23}\,\text{cm}^{-1}$ except for the states closest to the separatrix.

Methods for extending the above fixed point analysis to vibrational–rotational clustering problems have been employed by Sadovskii *et al.* (1990). The steps involve first replacing the single rotational operator $H(J_a, J_b, J_c)$ by an appropriate rotational coupling matrix $\boldsymbol{H}(J_a, J_b, J_c)$ between the vibrational states in question. Secondly the matrix is diagonalized to produce J-dependent rotational energy eigensurfaces. Finally the energies and stabilities of fixed points on these surfaces are used to partition the quantum mechanical vibrational–rotational spectrum into regions containing clusters of given degeneracies. The number of possible fixed points of a given symmetry naturally increases with the degeneracy of the vibrational manifold. In addition, pairs of fixed point energy curves of the same symmetry may cross at a so-called diabolic point associated with a conical intersection between two rotational eigensurfaces. Figure 18.12, taken from Sadovskii *et al.* (1990) shows an example relevant to the v_2/v_4 dyad of SiH_4, which gives rise to 12-fold, 8-fold and 6-fold clusters associated with stable fixed points on the C_2, C_3 and C_4 axes respectively.

SILANE Nu2, Nu4 diad Parameters of G. Pierre *et al* - J. Phys. **43** (1982)

^{28}SiH$_4$ QC points: □ ◇ ×

23.10.1987.

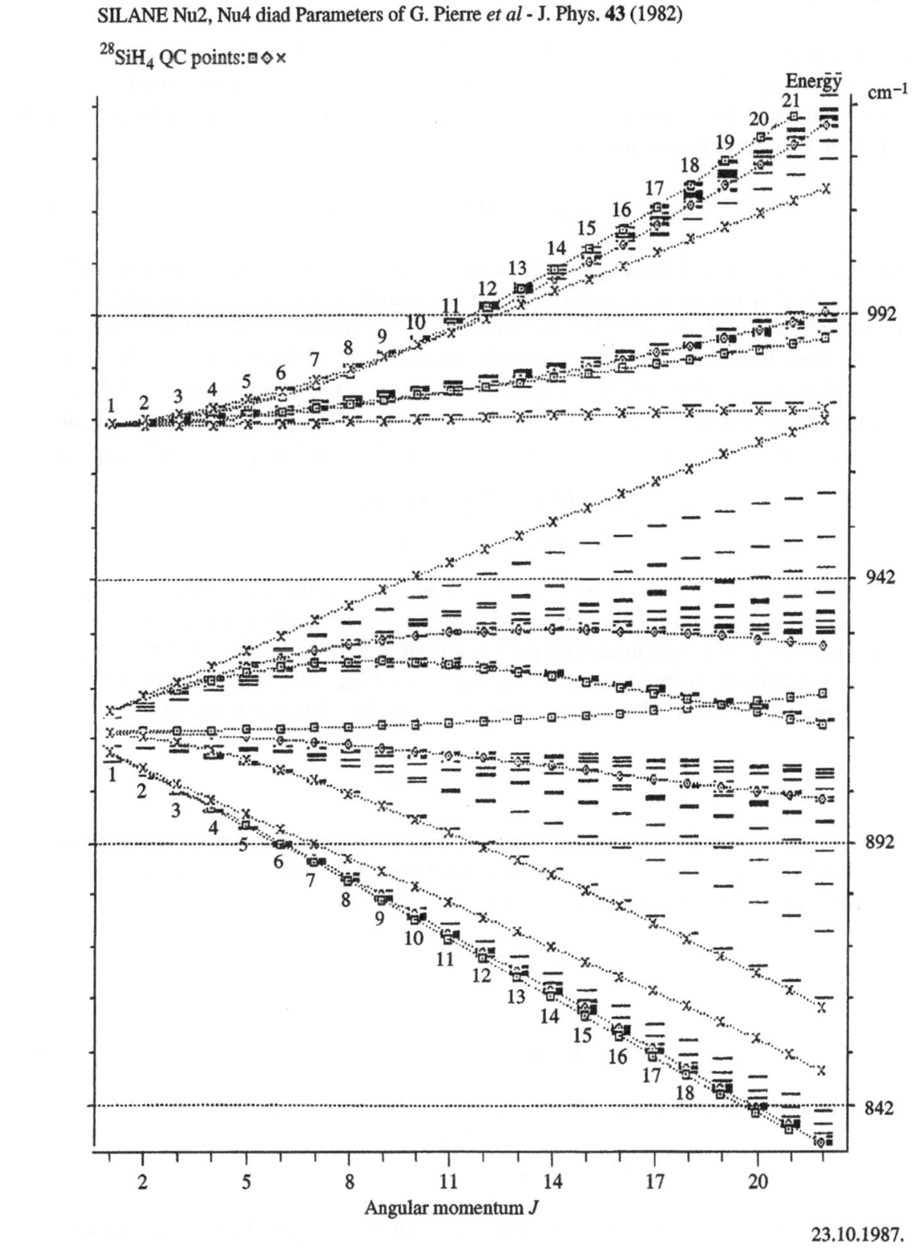

Figure 18.12. Periodic orbit related partitions of the rotational eigenvalues of the ν_2/ν_4 diad of SiH$_4$ (taken from Sadovskii *et al.* (1990) with permission).

18.3.3 TUNNELING-CORRECTED BOHR QUANTIZATION

Sections 18.3.1 and 18.3.2 concerned the qualitative organization of the quantum mechanical eigenvalue spectrum around the energies of the fixed points of a relevant semiclassical resonance Hamiltonian. Quantitative determination of the energy levels, including in particular the small clustering splittings, may be achieved by the following tunneling-corrected Bohr quantization procedure.

The assumption is that the polyad or clustering Hamiltonian depends on a single pair of conjugate variables, say (K, ϕ), plus various parameters. Hence for any given parameter set and type of motion one can always express K in terms of E and ϕ and adjust the energy until the Bohr quantization condition is satisfied; i.e.

$$\Phi_n = \int_a^b K(E_n, \phi)\,\mathrm{d}\phi = 2\pi(n+\delta), \tag{52}$$

where a and b are the classical turning points of the ϕ motion. It is usually most convenient to label the axes such that the motion occurs around the quantization axis [Harter and Patterson (1984)] in which case $a = 0$ and $b = 2\pi$; moreover $\delta = 0$ in equation (52), because the motion appears as a rotation. Equation (52) determines the mean energy of the nth eigenvalue cluster.

The cluster splittings are obtained by computing the phase integral analogous to (52) along a non-classical tunneling path from one equivalent orbit to another. The procedure may be illustrated by the example of the asymmetric rotor, for which the Hamiltonian in equation (45) implies that

$$K^2 = \frac{E - \left(B\cos^2\phi + C\sin^2\phi\right)J^2}{A - \left(B\cos^2\phi + C\sin^2\phi\right)}, \tag{53}$$

for the a axis motions with $AJ^2 > E > BJ^2$. The two equivalent orbits in this case are the positive and negative roots of this equation, and their separation fluctuates slowly for real ϕ values, with a minimum separation at $\phi = 0$. The tunneling path is therefore defined by the substitution $\phi = \mathrm{i}\phi'$, in order to take a classically forbidden path from $K_{\min}$ to $-K_{\min}$. The half-way point is reached when $\cosh 2\phi'_{E_n} = [2E - (B+C)J^2]/(B-C)$ and the tunneling integral is given by

$$\theta_n = 2\int_0^{\phi'_{E_n}} K(E_n, \mathrm{i}\phi')\,\mathrm{d}\phi', \tag{54}$$

because the two equivalent $K > 0$ and $K < 0$ orbits are separated by twice their individual separations from $(K, \phi) = (0, \mathrm{i}\phi'_{E_n})$. The tunneling splittings are then most conveniently estimated as the eigenvalues of the effective Hamiltonian [Harter and Patterson (1984)],

$$H_{\text{eff}} = \begin{pmatrix} E_n & 2S_n \\ 2S_n & E_n \end{pmatrix}, \tag{55}$$

where

$$S_n = \mathrm{e}^{-\theta_n}/T_n, \tag{56}$$

in which T_n is the classical time period of the equivalent classical trajectories

$$T_n = (\partial\Phi/\partial E)_{E=E_n} = \int_0^{2\pi} \left(\frac{\partial K}{\partial E}\right) \mathrm{d}\phi. \tag{57}$$

The factor 2 appears in equation (55) because there are two tunneling paths between each trajectory pair–one with $\phi = \mathrm{i}\phi'$ and the other with $\phi = \pi + \mathrm{i}\phi'$. Equation (55) is equivalent to an alternative procedures used by Colwell and Handy (1978), but it is more convenient for more intricate situations.

A more complicated example occurs for the six equivalent C_4 axis rotor states implied by the SF_6 rotational energy surface in Figure 18.11, each of which has a tunneling integral to its four nearest neighbors, which determines the overall splitting of the six-fold cluster. The detailed splitting pattern then depends on the nodal structure of the axis-localized rotor states. The details worked out by Patterson and Harter (1977) give a classification according to an index m_r where

Table 18.3. Octahedral rotational cluster splitting patterns[a].

Symmetry	0_4	1_4	2_4	3_4
A_1	$E_n + 4S_n$	—	—	—
A_2	—	—	$E_n - 4S_n$	—
E	$E_n - 2S_n$	—	$E_n + 2S_n$	—
T_1	E_n	$E_n - 2S_n$	—	$E_n - 2S_n$
T_2	—	$E_n + 2S_n$	E_n	$E_n + 2S_n$
	0_3	1_3	2_3	
A_1	$E_n + 3S_n$	—		
A_2	$E_n - 3S_n$	—		
E	—	E_n	E_n	
T_1	$E_n + S_n$	$E_n - 2S_n$	$E_n - 2S_n$	
T_2	$E_n - S_n$	$E_n + 2S_n$	$E_n + 2S_n$	

[a]Patterson and Harter (1975) give additional terms for tunneling to second and third nearest neighbors, which become important for energies close to the separatrix.

Table 18.4. Quantum mechanical and semiclassical $J = 30$ cluster splittings for SF_6[a].

n_4	$10^4 E_n$(QM)	$10^4 E_n$(SC)	S_n(QM)	S_n(SC)
30	5.31	5.29	2.63×10^{-11}	2.50×10^{-11}
29	3.54	3.53	9.55×10^{-10}	9.57×10^{-10}
28	2.04	2.03	1.53×10^{-8}	1.56×10^{-8}
27	0.80	0.79	1.42×10^{-7}	1.47×10^{-7}
26	−0.20	−0.22	8.32×10^{-7}	8.65×10^{-7}
25	−0.97	−1.00	2.34×10^{-6}	3.20×10^{-6}
n_3				
30	−3.57	−3.54	1.75×10^{-7}	1.68×10^{-7}
29	−2.48	−2.44	2.02×10^{-6}	2.15×10^{-6}
28	−1.68	−1.66	6.59×10^{-6}	7.40×10^{-6}

[a]Results in Hz are given relative to the term BJ^2 in equation (51), with $t_{044} = 5.44$ Hz [Harter and Patterson (1984)].

r is the periodicity of the axis and $m_r = n_{\text{mod } r}$, n being the quantum number in equation (52). Table 18.3 gives the symmetries and splitting patterns for the six-fold C_4 clusters and eight-fold C_3 clusters on the O_h rotational energy surfaces, and Table 18.4 gives a short extract of the results reported by Harter and Patterson (1984).

The necessary computations involve quadrature over a relatively coarse energy grid for the integrals Φ, θ and T in equations (52), (54) and (57) followed by interpolation for the energies such that $\Phi_n = 2\pi n$–provided that an explicit form for $K(E, \phi)$ is available. Alternatively the trajectories may be integrated by Hamilton's equations,

$$\dot{K} = -\partial H/\partial\phi, \qquad \dot{\phi} = \partial H/\partial K, \tag{58}$$

over a range of energies between the relevant fixed points. The integral Φ is evaluated as

$$\Phi = \int_0^T K(E, T)\dot{\phi}\, dt, \tag{59}$$

where T, which is required by equation (56), is the period of the trajectory. The tunneling integral θ in equation (54) is obtained by switching to an imaginary time increment at $\phi = 0$ (or $K = 0$) and accumulating a similar integral to equation (59) until K (or ϕ) reaches its value at the unstable fixed point between the relevant equivalent trajectories. It is instructive to follow the complex time discussions of Miller and George (1972) and Child (1991) to appreciate the details.

18.4 Semiclassical time-dependent wavepacket techniques

Any time-dependent study of a wavepacket $|\Psi(t_0)\rangle$ evolving to $|\Psi(t)\rangle$ under a bound time-independent Hamiltonian rests on the spectral decomposition

$$|\Psi(t)\rangle = \sum_n |\Psi_n\rangle\langle\Psi_n|\Psi(t_0)\rangle e^{-iE_n(t-t_0)/\hbar}. \tag{60}$$

The quantum mechanical eigenvalues are normally extracted by Fourier transformation of the auto-correlation function, which is conveniently expressed for what follows as

$$\begin{aligned} C(t) = \langle\Psi(t_0)|\Psi(t)\rangle &= \sum_n \langle\Psi(t_0)|\Psi_n\rangle e^{-iE_n(t-t_0)}\langle\Psi_n|\Psi(t_0)\rangle \\ &= \langle\Psi(t_0)|K(t,t_0)|\Psi(t_0)\rangle, \end{aligned} \tag{61}$$

where

$$K(t,t_0) = \sum_n |\Psi_n\rangle e^{-iH(t-t_0)/\hbar}\langle\Psi_n| \tag{62}$$

is the spectral representation of the time evolution operator. The half Fourier transform from t_0 to $t_0 + \infty$ may be expressed as

$$\begin{aligned} S(E) &= \frac{1}{T}\int_{t_0}^{t_0+T} \langle\Psi(t_0)|\Psi(t)\rangle e^{iE\cdot t/\hbar} \\ &= \sum_n |\langle\Psi(t_0)|\Psi_n\rangle|^2 \frac{\sin[(E-E_n)T/2\hbar]}{[(E-E_n)T/2\hbar]} \exp\left[\frac{i(E-E_n)T}{2\hbar}\right] \\ &= \frac{1}{T}\langle\Psi(t_0)|G^+(E)|\Psi(t_0)\rangle, \end{aligned} \tag{63}$$

where $G^{(+)}(E)$ is the Green's function

$$G^+(E) = -i\hbar \lim_{\varepsilon\to\infty}\left(\frac{1}{E-H+i\varepsilon}\right). \tag{64}$$

It is evident from equation (63) that $\mathrm{Re}\{S(E)\}$ peaks at the eigenvalues $E = E_n$, with a resolution of order $\Delta E = \hbar/T$, and that the relative intensities with respect to n follow the Franck–Condon decomposition of the initial state $\Psi(t_0)$.

Equations (60)–(64) apply to any time-dependent wavepacket. Many semiclassical applications rely on the properties of the gaussian wavepacket (or coherent state)

$$\Psi(x,t) = \left[\frac{2\alpha(0)}{\pi\hbar}\right]^{1/4} \exp\left[\frac{1}{\hbar}\left\{-\alpha(t)[x-q(t)]^2 + ip(t)[x-q(t)] + c(t)\right\}\right]. \tag{65}$$

The most important results apply to harmonic Hamiltonians, containing only, possibly time-dependent [Ge and Child (1997)], quadratic terms in the quantum mechanical operators $\hat{p}^2$, $\hat{q}^2$

and $\hat{p}\hat{q}+\hat{q}\hat{p}$, in which case $[q(t), p(t)]$ in equation (65) are readily shown to follow Hamilton's equations [Heller (1975)]

$$\frac{dq}{dt}=\frac{\partial H}{\partial p}, \qquad \frac{dp}{dt}=-\frac{\partial H}{\partial q}. \tag{66}$$

In addition the exponent $\alpha(t)$ also evolves 'classically' in the form [Lee and Heller (1982)]

$$\alpha=-\frac{i}{2}\left(\frac{P_z}{Z}\right) \tag{67}$$

$$\frac{dZ}{dt}=\frac{\partial H}{\partial P_z}, \qquad \frac{dP_z}{dt}=-\frac{\partial H}{\partial Z}, \tag{68}$$

and the phase term $c(t)$ is given by [Ge and Child (1997)]

$$c(t)=i\int_{t_0}^{t} L(t)\,dt-\frac{\hbar}{2}\ln\left[\frac{Z(t)}{Z(t_0)}\right], \tag{69}$$

where $L(t)$ is the classical Lagrangian

$$L(t)=p\dot{q}-H. \tag{70}$$

These results generalize immediately to many quadratic degrees of freedom.

18.4.1 SPECTRAL QUANTIZATION; FRANCK–CONDON SPECTRA AND RAMAN INTENSITIES

Although valid for any locally quadratic expansion, the time evolution of $\alpha(t)$ according to equations (67) and (68) is often inconvenient because $\alpha(t)$ becomes complex, although the sign of its real part is always preserved in a quadratic environment [Ge and Child (1997)]. The modeling of Franck–Condon spectra [Heller (1981)] and the spectral quantization procedure of DeLeon and Heller (1983, 1984) are therefore based on a frozen gaussian approximation, with α in equation (65) held constant. It is also assumed that $Z=Z_0e^{i\omega(t-t_0)}$, $P_Z=2i\alpha\, e^{i\omega(t-t_0)}$, by equation (69), so that

$$c(t)=i\int_{0}^{t} L(t)\,dt-\frac{i}{2}\hbar\omega t, \tag{71}$$

where ω is the frequency of the underlying oscillator.

We first consider the n-dimensional spectral quantization procedure of DeLeon and Heller (1983, 1984) which involves a frozen gaussian wavepacket of the form

$$\Psi(\boldsymbol{x},t)=\prod_{i=1}^{n}\left(\frac{2\alpha_i}{\pi\hbar}\right)^{1/4}\exp\left\{\frac{1}{\hbar}\left[\sum_{i=1}^{n}\phi_i(x_i,t)+i\int_{t_0}^{t}L(t)\,dt\right]\right\}, \tag{72}$$

where

$$\phi_i(x_i,t)=-\alpha_i[x_i-q_i(t)]^2+ip_i(t)[x_i-q_i(t)]-\frac{i\hbar\omega_i t}{2}, \tag{73}$$

$$L(t)=\sum_i p_i q_i-H, \tag{74}$$

in which $q_i(t)$ and $p_i(t)$ follow a classical trajectory from the initial center of the wavepacket. The simplicity of the method is that evaluation of the auto-correlation function $\langle\Psi(t_0)|\Psi(t)\rangle$ merely involves products of gaussian integrals. A typical term, generalized for the possibility that $\phi_{ik}(x_i, t)$ and $\phi_{ik'}(x_i, t)$ might follow different guiding trajectories is given by

$$\begin{aligned} I^i_{kk'}(t) &= \left(\frac{2\alpha_i}{\pi\hbar}\right)^{1/2} \int_{-\infty}^{\infty} \exp\left[\phi_{ik}(x_i, t) + \phi^*_{2k'}(x_i, t_o)\right] \mathrm{d}x \\ &= \exp\left\{\frac{1}{\hbar}\left[-\frac{\alpha_i}{2}\left(q_{ik} - q^0_{ik'}\right) - \frac{1}{2\alpha_i}\left(p_{ik} - p^0_{ik'}\right)\right.\right. \\ &\quad \left.\left. + \frac{\mathrm{i}}{2}\left(q_{ik} - q^0_{ik'}\right)\left(p_{ik} - p^0_{ik'}\right) - \frac{\mathrm{i}\hbar\omega_i}{2}(t - t_0)\right]\right\}, \end{aligned} \tag{75}$$

in which the classical time dependence of q_i and p_i has been suppressed to avoid undue congestion. The spectrum itself, in the present single trajectory approximation, is therefore a product of Fourier transforms of $I^i_{kk'}(t)$. A simple illustration is provided by the harmonic oscillator, with $q = \sqrt{2I/m\omega}\cos\omega t$, $p = \sqrt{2Im\omega}\sin\omega t$ and $\alpha = (m\omega/2)$, for which the auto-correlation function may be verified to become

$$\langle\Psi(t_0)|\Psi(t)\rangle = \exp\left\{\frac{1}{\hbar}\left[I(\cos\omega t + \mathrm{i}\sin\omega t - 1) - \frac{\mathrm{i}}{2}\hbar\omega(t - t_0)\right]\right\}. \tag{76}$$

It follows by expanding the exponential in powers of $(\cos\omega + \mathrm{i}\sin\omega t)$ that the Fourier transform $S(E)$ in equation (63) remains finite only for $E_n = (n + \frac{1}{2})\hbar\omega$ and that the intensity distribution with respect to n varies with the action I of the chosen classical trajectory.

Turning to the general n-dimensional case, it is necessary to make preliminary estimates of the frequencies ω_i which appear in equation (75), and the Fourier transform is evaluated as

$$S(E) = \frac{1}{T}\int_{t_0}^{t_0+T} [I_1(t)I_2(t)\ldots I_n(t)] \exp\left\{\frac{\mathrm{i}}{\hbar}\left[\int_{t_0}^{t} L(t)\,\mathrm{d}t - Et\right]\right\} \mathrm{d}t. \tag{77}$$

Assignment of the peaks in the resulting spectrum then leads to improved estimates of the ω_i.

In considering the calculation of Franck–Condon spectra, the above single trajectory frozen gaussian approach will often fail because no single trajectory will sample sufficient phase space on a strongly anharmonic upper state potential surface—the problem being that the spread of the wavepacket has been suppressed. An alternative type of spreading may, however, be introduced, while keeping α constant, by an ingenious seeded gaussian device, due to Heller (1975). It rests on the fact that any single initial gaussian function may be expanded as a swarm of others. The expansion may be performed in many ways, but the favored generic approximation for $\Psi(x, 0)$ given by equation (65), with $\alpha = \text{const}$ [Heller (1975)] would be

$$\begin{aligned} \Psi(x, 0) &= \left(\frac{2\mathrm{e}\alpha}{\pi\hbar}\right)^{1/4} \frac{1}{N}\sum_{k=0}^{N-1} \Psi_k(x), \\ \Psi_k(x) &= \exp\left\{\frac{1}{\hbar}\left[-\alpha\left(x - q^0_k\right)^2 + \mathrm{i}p^0_k\left(x - q^0_k\right) + \mathrm{i}\gamma_k\right]\right\}, \\ q^0_k &= q^0 + q_k, \qquad p^0_k = p^0 + p_k, \\ \gamma_k &= \frac{1}{2}p_k q_k + p^0 q_k - \frac{\pi k\hbar}{N}, \\ q_k &= \sqrt{\hbar/2\alpha}\cos(2\pi k/N), \qquad p_k = -\sqrt{2\alpha\hbar}\sin(2\pi k/N). \end{aligned} \tag{78}$$

The initial gaussian is seen to be distributed over a circle in phase space around its center at (q^0, p^0); and pictures of the evolution of the resulting swarm are given by Child (1991), where a similar approximation for higher harmonic oscillator states is given. The same general idea may also be applied to non-separable situations by distributing gaussians at equal time intervals along an appropriate quantizing trajectory [Davis and Heller (1981)] and following their time evolution on an upper potential surface.

Calculation of the electronic absorption intensity at frequency ω

$$\mathcal{E}(\omega) = \lim_{T\to\infty} \frac{1}{T} \int_0^T \langle \Psi_0(\boldsymbol{x}, 0)|\mu|\Psi'(\boldsymbol{x}, t)\rangle \exp\left[\frac{\mathrm{i}(\hbar\omega + E_0)t}{\hbar}\right] \mathrm{d}t \tag{79}$$

is now evaluated by distributing $\Psi_0(\boldsymbol{x}, 0)$ over the points (q_k^0, p_k^0) given by equation (78). Each gaussian evolves according to equation (72) along a classical trajectory from its initial center (q_k^0, p_k^0) and the transition moment at time t is given by combinations of the gaussian integrals $I_{kk'}^i(t)$ in equation (75) multiplied by the value of the dipole function μ at the gaussian product center, $x_i = [q_{ik}(t) + q_{ik'}^0(0)]/2$. Finally evaluation of the Fourier transform gives $\mathcal{E}(\omega)$. Figure 18.13 gives an example from Heller's(1981) paper. Numerical results from DeLeon and Heller (1984) are given in Table 18.5.

The same general approach may be extended to resonance Raman spectroscopy [Lee and Heller (1982)], for which the relevant scattering amplitude is given by

$$\mathcal{E}_{\mathrm{on}}(\omega) = \lim_{T\to\infty} \frac{1}{T} \int_0^T \langle \Psi_n(\boldsymbol{x}, t)|\Psi_0'(\boldsymbol{x}, t)\rangle \exp\left[\frac{\mathrm{i}(\hbar\omega + E_0)}{\hbar}\right] \mathrm{d}t, \tag{80}$$

where $\Psi_n(\boldsymbol{x}, t)$ is the time-dependent Raman excited stationary state and $\Psi_0'(\boldsymbol{x}, t)$ is the gaussian swarm projected from the $\Psi_0(\boldsymbol{x}, t)$ onto the upper potential energy surface. $\Psi_0'(\boldsymbol{x}, t)$ is treated

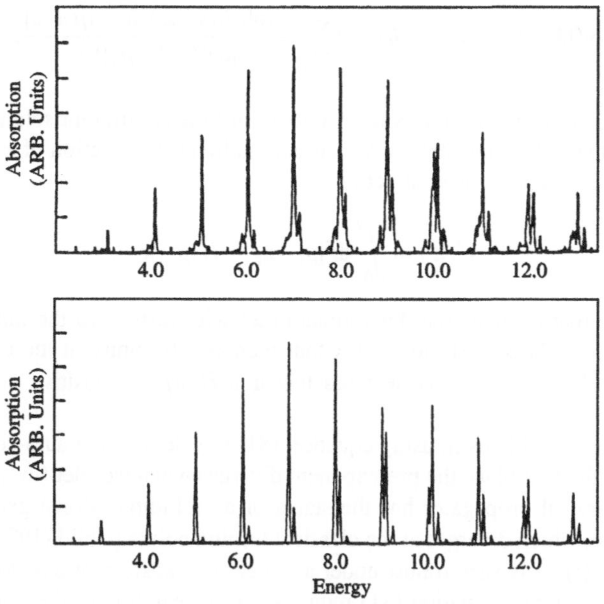

Figure 18.13. Franck–Condon spectra for a Fermi resonance model. The upper and lower panels show results from the frozen gaussian approximation and a converged variational calculation respectively (taken from Heller (1981) with permission).

Table 18.5. Spectral quantization for the model of DeLeon and Heller(1984).

Trajectory no.	n_x	n_y	E_q	E_{sq}
1	0	4	5.026	5.046
1	1	3	5.278	5.276
1	2	2	5.505	5.506
2	3	6	10.622	10.637
2	4	5	10.866	10.867
2	5	4	11.091	11.097
2	8	1	11.634	11.634

exactly as the Franck–Condon case and a gaussian representation of $\Psi_n(\boldsymbol{x}, t)$ is given in the harmonic approximation by the higher state generalization of equations (78) given by Child (1991), or more accurately by the seeded gaussian approach of Davis and Heller (1981).

18.4.2 INITIAL VALUE METHODS

A serious weakness of the previous frozen gaussian method is that the proper spreading of the wavepacket is suppressed. The deficiencies of this constraint may be alleviated by repetitions of the gaussian seeding equation (78), but the resulting wavepacket proliferation raises acute problems for long-time propagation on strongly anharmonic potentials [Littlejohn (1986)]. Attention has therefore turned in recent years to semiclassical approximations to the propagator in equation (62). The starting point is the Van Vleck (1928) semiclassical approximation to the coordinate representation.

$$\langle \boldsymbol{x}'|K_{\rm sc}(t)|\boldsymbol{x}\rangle = (2\pi i\hbar)^{-n/2} \sum_{\rm traj} \frac{\exp[iS_t(\boldsymbol{x}', \boldsymbol{x})/\hbar - i\mu\pi/2]}{|\det(\partial \boldsymbol{x}'/\partial \boldsymbol{p})|^{1/2}}, \tag{81}$$

where the sum is taken over trajectories from $\boldsymbol{x}$ to $\boldsymbol{x}'$ in time t; different solutions correspond to different values of the initial momentum, which is a multivalued function of $\boldsymbol{x}'$ and $\boldsymbol{x}$. The term $S_t(\boldsymbol{x}', \boldsymbol{x})$ is the corresponding classical action

$$S_t(\boldsymbol{x}', \boldsymbol{x}) = \int_0^t dt'[\boldsymbol{p}_{t'} \cdot \boldsymbol{x}_{t'} - H(\boldsymbol{p}_{t'}, \boldsymbol{x}_{t'})] \tag{82}$$

and $(\partial \boldsymbol{x}'/\partial \boldsymbol{p})$ is the matrix of partial derivations of $\boldsymbol{x}'$ with respect to the initial momentum $\boldsymbol{p}$. Finally μ in equation (81) is a Maslov index that increases by unity at the caustics, where det $(\partial \boldsymbol{x}'/\partial \boldsymbol{p}) = 0$, provided that the inverse mass tensor $\partial^2 H/\partial p^2$ is positive definite [Gutzwiller (1967)].

There are two major problems in using equation (81) to calculate the auto-correlation function in equation (61). In the first place the pre-exponent diverges whenever det $(\partial \boldsymbol{x}'/\partial \boldsymbol{p})$ changes sign, because the semiclassical propagator has the status of a multidimensional generalization of the JWKB wavefunction, which diverges at the classical turning points [Child (1991), Section 18.2.1] Nevertheless $\langle \boldsymbol{x}'|K_{\rm sc}(t)|\boldsymbol{x}\rangle$ is very robust under a variety of transformations [Miller (1974)]. The more serious problem is that equation (81) requires a root search over the initial momentum to discover the trajectories from $\boldsymbol{x}$ to $\boldsymbol{x}'$ in time t.

To avoid these difficulties various authors [Herman and Kluck (1984), Heller (1991), Kay (1994a), Sepulveda and Heller (1994a), and Walton and Manolopoulos (1996)] have replaced equation (81) by an integral over initial coordinates and momenta to obtain an initial value representation for $\langle x'|K_{sc}(t)|x\rangle$. Herman and Kluck (1984) initiated this development of the theory but the most general form was obtained by Kay (1994a), who considered conditions under which the following general gaussian convolution reduces to equation (81):

$$\langle x'|K_{sc}(t)|x\rangle = (2\pi\hbar)^{-n}\iint dp\, dq\, C_{pqt}\, e^{i\phi_{pqt}(x',x)/\hbar}, \tag{83}$$

where

$$\begin{aligned}\phi_{pqt}(x',x) = S_{pqt} + p_t\cdot(x'-q_t) - p\cdot(x-q) + i\hbar\big[(x'-q_t)\cdot\gamma^{(1)}\cdot(x'-q_t) \\ + (x-q)\cdot\gamma^{(2)}\cdot(x-q) + 2(x'-q_t)\cdot\gamma^{(3)}\cdot(x-q)\big].\end{aligned} \tag{84}$$

in which $\gamma^{(1)}$, $\gamma^{(2)}$ and $\gamma^{(3)}$ are constant n-dimensional matrices, such that the real part of $\begin{pmatrix}\gamma^{(1)} & \gamma^{(3)} \\ \gamma^{(3)} & \gamma^{(2)}\end{pmatrix}$ is positive definite, and $q_t(p,q)$ and $p_t(p,q)$ are the position and momentum at time t on a trajectory from (p, q) at $t = 0$. The reduction of equation (83) to equation (81) is too complicated to follow in detail. The necessary manipulations involve the properties of the monodromy (stability or symplectic) matrix [Goldstein (1980), Section 9.3], with elements defined by the $2n \times 2n$ partial derivatives of (p_t, q_t) with respect to (p, q). Two notations are employed in the literature.

$$\begin{pmatrix} M_{pp}, & M_{pq} \\ M_{qp}, & M_{qq}\end{pmatrix} = \begin{pmatrix}\partial P/\partial p, & \partial P/\partial q \\ \partial Q/\partial p, & \partial Q/\partial p\end{pmatrix}, \tag{85}$$

in which (P, Q) should be identified with (p_t, q_t). An important property is that $\det(M) = 1$, as a consequence of Liouville's Theorem [Goldstein (1980)]; in addition $M = I$ at time $t = 0$. Various relationships between the sub-blocks of M are also listed by Kay (1994a). A second important ingredient of the theory is given by the classical action identities [Kay (1994a)], which may be expressed in the row vector form:

$$\begin{aligned}\nabla_q^T S_{qpt} &= p_t^T M_{qq} - p^T, \\ \nabla_p^T S_{qpt} &= p_t^T M_{qp}.\end{aligned} \tag{86}$$

Kay's (1994a) major result is that equations (81) and (83) agree, within a quadratic approximation to $S_{qpt}(x', x)$ with $\gamma^{(3)} = 0$, provided that

$$C_{pqt} = \left(\frac{1}{2\pi i\hbar}\right)^{n/2}\left\{\det\left[-M_{pq} + 2i\hbar\gamma^{(1)}M_{qq} - 2i\hbar\left(-M_{pp} + 2i\hbar\gamma^{(1)}M_{qp}\right)\gamma^{(2)}\right]\right\}^{1/2} \tag{87}$$

A more complicated form is also given for the case when $\gamma^{(3)} \neq 0$. Equation (87) reduces to the form due to Herman and Kluck (1984) in the special case $\gamma^{(1)} = \gamma^{(2)} =$ diagonal. Another important practical consideration is that C_{pqt} may be shown to be continuous throughout the propagation [Kay (1994a)].

The fact that $\phi_{pqt}(x', x)$ is strictly quadratic in x and x' means that the passage from $\langle x'|K_{sc}(t)|x\rangle$ to the auto-correlation functions may be performed analytically for any initial gaussian wavepacket

$$\Psi(0) = N\exp\left\{\frac{1}{\hbar}\left[-\frac{1}{2}(x-x_0)\cdot\alpha\cdot(x-x_0) + ip_0\cdot(x-x_0)\right]\right\}, \tag{88}$$

where $N = \{\det(\alpha/\pi\hbar)\}^{1/4}$. The simplest case occurs for $\gamma^{(1)} = \gamma^{(2)} = \alpha/2\hbar =$ diagonal, $\gamma^{(3)} = 0$, when the combination of equations (61), (83), (84) and (88) yields the original form of Herman and Kluck(1984)

$$C(t) = \langle \Psi(0)|K_{sc}(t)|\Psi(0)\rangle = \frac{1}{(2\pi\hbar)}n \int\int \mathrm{d}p\, \mathrm{d}\boldsymbol{q} \mathrm{R}_{\boldsymbol{pq}t}\, \mathrm{e}^{\Phi^K_{\boldsymbol{pq}t}(p_0,x_0)/\hbar}, \tag{89}$$

where

$$R_{\boldsymbol{pq}t} = \left\{\det\left[\tfrac{1}{2}\left(\boldsymbol{M}_{pp} + \alpha\boldsymbol{M}_{qq}\alpha^{-1} - \mathrm{i}\alpha\boldsymbol{M}_{qp} + \mathrm{i}\boldsymbol{M}_{pq}\boldsymbol{\alpha}^{-1}\right)\right]\right\}^{1/2} \tag{90}$$

and

$$\begin{aligned}\Phi^K_{\boldsymbol{pq}t} = {} & \mathrm{i}S_{\boldsymbol{pq}t} - \frac{\mathrm{i}}{2}(\boldsymbol{p}_t + \boldsymbol{p}_0)\cdot(\boldsymbol{q}_t - \boldsymbol{x}_0) + \frac{\mathrm{i}}{2}(\boldsymbol{p} + \boldsymbol{p}_0)\cdot(\boldsymbol{q} - \boldsymbol{x}_0) \\ & - \frac{1}{4}[(\boldsymbol{q}_t - \boldsymbol{x}_0)\cdot\alpha\cdot(\boldsymbol{q}_t - \boldsymbol{x}_0) + (\boldsymbol{q} - \boldsymbol{x}_0)\cdot\alpha\cdot(\boldsymbol{q} - \boldsymbol{x}_0) \\ & + (\boldsymbol{p}_t - \boldsymbol{p}_0)\cdot\alpha^{-1}\cdot(\boldsymbol{p}_t - \boldsymbol{p}_0) + (\boldsymbol{p} - \boldsymbol{p}_0)\cdot\alpha^{-1}\cdot(\boldsymbol{p} - \boldsymbol{p}_0)].\end{aligned} \tag{91}$$

Sepulveda and Heller (1994a) use different arguments to reach an analogous, less symmetrical, form in which $\Phi^K_{\boldsymbol{pq}t}$ would be replaced in the present notation by

$$\begin{aligned}\Phi^{\mathrm{SH}}_{\boldsymbol{pq}t} = {} & \mathrm{i}S_{\boldsymbol{pq}t} - \mathrm{i}\boldsymbol{p}_t\cdot(\boldsymbol{x}_t - \boldsymbol{x}_0) + \mathrm{i}\boldsymbol{p}_0\cdot(\boldsymbol{q} - \boldsymbol{x}_0) \\ & - \tfrac{1}{2}(\boldsymbol{q} - \boldsymbol{x}_0)\cdot\alpha\cdot(\boldsymbol{q} - \boldsymbol{x}_0) - \tfrac{1}{2}(\boldsymbol{p}_t - \boldsymbol{p}_0)\cdot\alpha^{-1}\cdot(\boldsymbol{p}_t - \boldsymbol{p}_0),\end{aligned} \tag{92}$$

and $R_{\boldsymbol{pq}t}$ by $(\mathrm{i})^{-n/2}|\det(M_{pp})|\frac{1}{2}\,\mathrm{e}^{-\mathrm{i}\nu\pi/2}$ where ν is a modified Maslov index.

Turning to computational details, the time evolution of the initial wave-packet in equation (88) yields the associated spectrum by evaluation of $C(t)$ in equation (89), or the Sepulveda and Heller (1994a) variant, followed by the Fourier transformation in equation (63). Terms in the integrand of equation (89) are obtained by following trajectories initiated at $(\boldsymbol{p}, \boldsymbol{q})$ according to the equations

$$\begin{aligned}\dot{\boldsymbol{p}}_t &= -\nabla_{\boldsymbol{q}}H, & \boldsymbol{p}_t(0) &= \boldsymbol{p}, \\ \dot{\boldsymbol{q}}_t &= \nabla_p H, & \boldsymbol{q}_t(0) &= \boldsymbol{q}, \\ \dot{S} &= \boldsymbol{p}\cdot\dot{\boldsymbol{q}} - H, & S(0) &= 0 \\ \dot{\boldsymbol{M}} &= \boldsymbol{KM}, & \boldsymbol{M}(0) &= \boldsymbol{I},\end{aligned} \tag{93}$$

where

$$\boldsymbol{K} = \begin{pmatrix} -\boldsymbol{H}_{\mathrm{qp}} & -\boldsymbol{H}_{\mathrm{qq}} \\ \boldsymbol{H}_{\mathrm{pp}} & \boldsymbol{H}_{\mathrm{qp}} \end{pmatrix}. \tag{94}$$

in which for example $\boldsymbol{H}_{qp}$ is the matrix of partial derivatives of H with respect to the components first of $\boldsymbol{p}$ and second of $\boldsymbol{q}$. Brewer, Hulme and Manolopoulos (1997) give details of a symplectic integrator that preserves the symplectic symmetry [Goldstein (1980)] $\boldsymbol{M}^{\mathrm{T}}\boldsymbol{J}\boldsymbol{M} = \boldsymbol{J}$, where

$$\boldsymbol{J} = \begin{pmatrix} 0 & -I \\ I & 0 \end{pmatrix}.$$

Computational complications can arise in evaluation of the integral over $\boldsymbol{p}$ and $\boldsymbol{q}$ in equation (89) due to rapid oscillations of the integrand; but their effects may be diminished by the cellular dynamics algorithm of Heller (1991) and Sepulveda and Heller (1994a) or the equivalent Filinov (1986) transformation of Walton and Manolopoulos (1996). To see the effect, consider the integral

$$I = \int_{-\infty}^{\infty} \mathrm{e}^{\mathrm{i}\alpha q^2}\,\mathrm{d}q = \left(\frac{\pi}{\alpha}\right)^{1/2} \mathrm{e}^{\mathrm{i}\pi/4}, \qquad \alpha > 0, \tag{95}$$

which is readily evaluated analytically by displacing the integration contour into the complex plane [Arfken (1985)]. Accurate numerical computation over real q is, however, complicated by the increasingly rapid oscillations of $e^{i\alpha q^2}$ as $|q| \to \infty$. On the other hand, introduction of the identity in the form of a gaussian integral, followed by interchange of the integration variables casts equation (95) into the alternative form

$$I = \left(\frac{\gamma}{\pi}\right)^{1/2} \int_{-\infty}^{\infty} \int_{-\infty}^{\infty} \exp\left[-\gamma(u-q)^2 + i\alpha q^2\right] dq\, du$$
$$= \left(\frac{\gamma}{\gamma - i\alpha}\right)^{1/2} \int_{-\infty}^{\infty} \exp\left[\frac{-\alpha\gamma(\alpha - i\gamma)}{(\alpha^2 + \gamma^2)} u^2\right] du. \tag{96}$$

Equation (96) reverts to equation (95) in the limit $\gamma \to \infty$, but the oscillations of the integrand are rapidly damped by a gaussian envelope for $\gamma \ll \alpha$, thereby accelerating the rate of convergence. The choice of γ in relation to α is a matter of judgement and experience [Brewer, Hulme and Manolopoulos (1997)]. The cellular dynamics algorithm of Heller (1991) evaluates the multi-dimensional analog of equation (96) by quadrature [see Sepulveda and Heller (1994a) for details], whereas Kay (1994a,b), Walton and Manolopoulos (1996), and Brewer, Hulme and Manolopoulos (1997) use the Box–Müller Monte Carlo algorithm [Press *et al.* (1992)]. Both sets of authors make a quadratic approximation to Φ_{pqt} in equation (89) to perform the equivalent of the u integral in equation (96); the resulting working equations are collected in a convenient form by Brewer, Hulme and Manolopoulos (1997).

Tests of this initial value method have been reported by Kluck, Herman and Davis (1986), Heller (1991), Kay (1994b), Sepulveda and Heller (1994a,b), Walton and Manolopoulos (1996), and Brewer, Hulme and Manolopoulos (1997). The strengths and limitations as a black-box routine are well illustrated by the results of Kay (1994b) for a strongly chaotic two-dimensional quartic oscillator system. Serious problems were encountered because of the rapid growth of certain elements of the monodromy matrix $\boldsymbol{M}$ in the pre-exponent term R_{pqt} given by equation (90). The problem is that $\boldsymbol{M}$ has some exponentially large eigenvalues in a chaotic situation, which means that some terms in $\boldsymbol{M}$ must become exponentially large; and the resulting difficulties are aggravated by the long trajectory runs required, by the uncertainty principle, to resolve the spectrum. Kay (1994b) responded to this situation by simply discarding trajectories for which the equivalent of $|R_{pqt}|$ exceeded a threshold value, with a resultant rejection rate of up to 97 % for the longest trajectory runs. On the other hand, if the initial wavepacket was centered close to the most weakly unstable periodic orbit of the system, useful results were obtained from a small sample of only 500 trajectories. The latter results became excellent if the sample was increased to 10 000, but even larger samples were required for initial wavepackets localized in more chaotic regions of the phase space. The same general point is vividly illustrated in Figure 18.14, which is taken from Walton and Manolopoulos (1996). The system in question is the relatively weakly chaotic Henon–Hieles (1964) model. The striking point is that $|C(t)|$ oscillates in the expected more or less regular manner only for relatively short times, and thereafter begins to fluctuate wildly, up to values well in excess of unity, owing to an exponential increase of the pre-exponent R_{pqt}; see below, however, for corrections to this behavior.

The above comments are included as a warning for strongly chaotic systems, but it should be emphasized that Kay's (1994b) test is a very severe one. In addition the number of required trajectories is markedly reduced by the cellular dynamics or Filinov (1986) algorithms. Thus the cellular dynamics study of Sepulveda and Heller (1994a) reproduce the spectrum of the weakly chaotic Barbanis system to recognizable accuracy, with only 49 trajectories, and obtains an excellent description if the number is increased to 3800. Similarly Walton and Manolopoulos (1996) completely removed the wild oscillations in Figure 18.14, by means of the Filinov (1986)

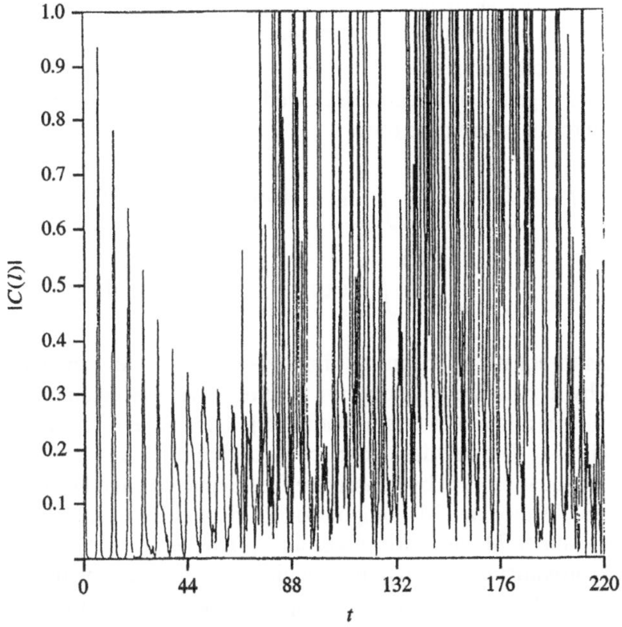

Figure 18.14. The auto-correlation function $C(t)$ for an initial value study of the chaotic Henon–Hieles system, before Filinov smoothing. Note that $|C(t)|$ incorrectly exceeds unity as time increases (taken from Walton and Manolopoulos (1996) with permission).

smoothing, and the subsequent Fourier transform gave an excellent picture of the spectrum with only 2047 trajectories. Finally Brewer, Hulme and Manolopoulos (1997) find that the number of required trajectories increases only linearly with the number of coupled degrees of freedom–up to 15 for the largest system investigated. It must be noted, however, that the relevant excitation states were very low. In general one can expect a linear increase if the underlying dynamics is regular [or quasiperiodic, Noid, Kosykowski and Marcus (1981)], because each trajectory moves over a fixed n-dimensional torus in the $2n$-dimensional phase space; in addition the eigenvalues of $\boldsymbol{M}$ have modulus unity, so that the magnitude of $|R_{pqt}|$ remains within reasonable bounds. The classically chaotic situation is much more demanding, because each trajectory will move more or less erratically over the entire phase volume, and the number required to sample this volume in a finite time will increase exponentially with the number of degrees of freedom.

An important computational advantage of these semiclassical initial value procedures is that the memory requirements are limited to those required to propagate the $2n$ coordinates and momenta and the $2n \times 2n$ monodromy matrix $\boldsymbol{M}$, while the retained information at each time step includes only the complex quantities R_{pqt} and Φ_{pqt} in equation (89). It is therefore quite feasible to perform a calculation in up to 15 degrees of freedom [Brewer, Hulme and Manolopoulos (1997)], provided that the underlying classical motion is regular or only weakly chaotic. In many systems persistent regular islands of phase space will be located around the stable periodic orbits (or anharmonic resonances) of the system, and these will probably be found to be the regions most favorable for initial value quantization. If so, the situation will be somewhat similar to that encountered in the first part of the chapter, in which the stable fixed points play the role of the stable periodic orbits. The difference is that the polyad resonance approximations simply ignore coupling to other degrees of freedom, whereas the wavepacket allows weak interactions with all degrees of freedom, but remains dynamically tied to the underlying periodic orbit.

18.5 Conclusions

This chapter had two objectives. The first was to show how divisions of the classical phase space are reflected in a partitioning of the quantum mechanical eigenvalue spectrum. Particular emphasis was given to single resonance vibrational polyad approximations and rotational clustering problems, for which much of the argument can be followed analytically. Specific applications were given to the polyad sphere representation [Xiao and Kellman (1989)] of local mode and normal mode dynamics [Sibert, Hynes and Reinhardt (1982)], two-fold and four-fold rotational clustering in asymmetric tops [Kozin (1992)] and high degeneracy rotational clusters in spherical tops [Harter and Patterson (1984)]. Extensions of the semiclassical theory to more complicated situations were also indicated.

The second intention was to give an overview of semiclassical wavepacket techniques for spectroscopic problems. The early frozen gaussian approach of Heller (1975) and co-workers was illustrated by application to spectral quantization [DeLeon and Heller (1983,1984)] and the calculation of Franck–Condon spectra [Heller (1981)] and resonance Raman spectra [Lee and Heller (1979, 1982)]. More recent work on 'initial value' approximations derived from the Herman–Kluck (1984) propagator was also discussed. The theory, which is given in its most general form by Kay (1994a), has attractive computational advantages for high dimensional systems [Brewer, Hulme and Manolopoulos (1997)], because memory requirements are minimal compared with any conventional variational approach. Difficulties arise from the number of trajectories required for convergence, especially in strongly chaotic situations, but there has already been substantial progress in smoothing out some of the most troublesome contributions to the auto-correlation function, upon which accurate computation of the spectrum rests. The cellular dynamics routine of Heller (1991) and the Filinov (1986) smoothing routine, suggested by Makri and Miller (1987) in a different context, and implemented by Walton and Manolopoulos (1996), have been particularly effective. The method remains to be fully tested but it seems likely to scale linearly with system size for systems with regular (rather than chaotic) classical dynamics, including in particular states that are semiclassically located close to stable or only weakly unstable classical periodic orbits.

Acknowledgements

It is a pleasure to acknowledge that the work was partially supported by the National Science Foundation through a grant for the Institute of Theoretical Atomic and Molecular Physics at Harvard University and Smithsonian Astrophysical Laboratory. The author is also grateful to Dr Xiaogang Wang for helpful comments on the manuscript.

18.6 References

Arfken, G., 1985, *Mathematical Methods for Physicists*, 3rd edition; Academic Press: New York.
Augustin, S. D., and Miller, W. H., 1974, *J. Chem. Phys.* **61**, 3155–3163.
Augustin, S. D., and Rabitz, H., 1979, *J. Chem. Phys.* **71**, 4956–4968.
Bordé, J., and Bordé, Ch., 1982, *Chem. Phys.* **71**, 417–441.
Brewer, M. L., Hulme, J. M. S., and Manolopoulos, D. E., 1997, *J. Chem. Phys.* **106**, 4832–4839.
Brillóuin, L., 1926, *J. Phys.* **7**, 353–338.
Carruthers, P., and Nieto, M. M., 1968, *Rev. Mod. Phys.* **40**, 411–440.
Child, M. S., 1991, *Semiclassical Mechanics with Molecular Applications*; Oxford University Press: Oxford.
Child, M. S., and Halonen, L., 1984, *Adv. Chem. Phys.* **57**, 1–58.
Child, M. S., and Lawton, R. T., 1981, *Far. Disc. Chem. Soc.* **71**, 273–285.
Colwell, S. M., and Handy, N. C., 1978, *Mol. Phys.* **35**, 1183–1190.
Davis, M. J., and Heller, E. J., 1981, *J. Chem. Phys.* **75**, 3916–3924.
DeLeon, N., and Heller, E. J., 1983, *J. Chem. Phys.* **78**, 4005–4017.
DeLeon, N., and Heller, E. J., 1984, *J. Chem. Phys.* **81**, 5957–5975.
Dorney, A. J., and Watson, J. K. G., 1972, *J. Mol. Spectrosc.* **42**, 135–148.

Einstein, A., 1917, *Verh. Dtsch. Phys. Ges.* **19**, 82–92.
Filinov, V. S., 1986, *Nucl. Phys.* B **271**, 717–725.
Fried, L. E., and Ezra, G. S., 1987, *J. Chem. Phys.* **86**, 6270–6282.
Ge, Y. C., and Child, M. S., 1997, *Phys. Rev. Lett.* **78**, 2507–2510.
Goldstein, H., 1980, *Classical Mechanics*, 2nd edition; Addison-Wesley: New York.
Gray, S. K., and Davis, M., 1989, *J. Chem. Phys.* **90**, 5420–5433.
Gutzwiller, M. C., 1967, *J. Math. Phys.* **8**, 1979–2000.
Harter, W. G., and Patterson, C. W., 1984, *J. Chem. Phys.* **80**, 4241–4261.
Heller, E. J., 1975, *J. Chem. Phys.* **62**, 1544–1555.
Heller, E. J., 1981, *J. Chem. Phys.* **75**, 2923–2931.
Heller, E. J., 1991, *J. Chem. Phys.* **94**, 2723–2729.
Henon, M., and Hieles, C., 1964, *Astron. J.* **69**, 73–79.
Herman, M. F., and Kluck, E., 1984, *Chem. Phys.* **91**, 27–34.
Jacobson, M. P., Jung, C., Taylor, H. S., and Field, R. W., 1999, *J. Chem. Phys.* **111**, 600–608.
Jaffé, C., and Brumer, P., 1980, *J. Chem. Phys.* **73**, 5646–5658.
Jensen, P., Osmann, G., and Kozin, I. N., 1997, in *Advanced Series in Physical Chemistry: Vibrational–Rotational Spectroscopy and Molecular Dynamics'*, Papoušek, D., Ed.; World Scientific: Singapore, pp. 298–351.
Kay, K. G., 1994a, *J. Chem. Phys.* **100**, 4377–4392.
Kay, K. G., 1994b, *J. Chem. Phys.* **101**, 2250–2260.
Keller, J. B., 1958, *Ann. Phys.* **4**, 180–184.
Kellman, M. E., 1982, *J. Chem. Phys.* **76**, 4528.
Kellman, M. E., 1990, *J. Chem. Phys.* **93**, 6630–6635.
Kellman, M. E., and Xiao, L., 1990, *J. Chem. Phys.* **93**, 5821–5825.
Kluck, E., Herman, M. F., and Davis, H. L., 1986, *J. Chem. Phys.* **84**, 326–334.
Kozin, J. N., and Jensen, P., 1994, *J. Mol. Spectrosc.* **163**, 483–509.
Kozin, I. N., Belov, S. P., Polyansky, O. L., and Tretyakov, M. Yu., 1992, *J. Mol. Spectrosc.* **152**, 13–28.
Lawton, R. T., and Child, M. S., 1980, *Mol. Phys.* **40**, 773–792.
Leaf, B., 1969, *J. Math. Phys.* **10**, 1980–1987.
Lee, S.-Y., and Heller, E. J., 1979, *J. Chem. Phys.* **71**, 4777–4788.
Lee, S.-Y., and Heller, E. J., 1982, *J. Chem. Phys.* **76**, 3035–3044.
Lehmann, K., 1983, *J. Chem. Phys.* **79**, 1098(L).
Littlejohn, R. G., 1986, *Phys. Rep.* **138**, 193–291.
Lu, Z., and Kellman, M. E., 1997, *J. Chem. Phys.* **107**, 1–15.
Makarewicz, J., and Pyka, J., 1989, *Mol. Phys.* **68**, 107–127.
Makri, N., and Miller, W. H., 1987, *Chem. Phys. Lett.* **139**, 10–14.
Miller, W. H., 1974, *Adv. Chem. Phys.* **25**, 69–177.
Miller, W. H., and George, T. F., 1972, *J. Chem. Phys.* **56**, 5637–5652, 5668–5681.
Mills, I. M., and Robiette, 1985, *Mol. Phys.* **56**, 743–765.
Noid, D. W., Kosykowski, M. L., and Marcus, R. A., 1981, *Annu. Rev. Phys. Chem.* **32**, 267–309.
Pavlichenkov, I. M., and Zhilinskii, 1985, *Chem. Phys.* **100**, 339–354.
Patterson, C. W., and Harter, W. G., 1977, *J. Chem. Phys.* **66**, 4886–4892.
Percival, I. C., 1977, *Adv. Chem. Phys.* **36**, 1–61.
Press, W. H., Flannery, B. P., Teukolsky, S. A., and Vetterling, W. T., 1992, *Numerical Recipes: The Art of Scientific Computing*, 2nd edition; Cambridge University Press: Cambridge.
Rose, J., and Kellman, M. E., 1995, *J. Chem. Phys.* **103**, 7255–7268.
Sadovskii, D. A., Zhilinskii, B. I., Champion, J. P. and Pierre, G., 1990, *J. Chem. Phys.* **92**, 1523–1537.
Sepulveda, M. A., and Heller, E. J., 1994a, *J. Chem. Phys.* **101**, 8004–8015.
Sepulveda, M. A., and Heller, E. J., 1994b, *J. Chem. Phys.* **101**, 8016–8027.
Sibert, E. L., Hynes, J. T., and Reinhardt, W. P., 1982, *J. Chem. Phys.* **77**, 3595–3604.
Smith, B. C., and Winn, J. S., 1988, *J. Chem. Phys.* **89**, 4638–4645.
Svitak, J., Lu. Z., Rose, J., and Kellman, M. E., 1995, *J. Chem. Phys.* **102**, 4340–4354.
Tabor, M., 1989, *Chaos and Integrability in Non-Linear Dynamics*; Wiley-Interscience: New York.
Van Vleck, J. H., 1928, *Proc. Natl. Acad. Sci.* **14**, 178–188.
Walton, A. R., and Manolopoulos, D. E., 1996, *Mol. Phys.* **87**, 961–978.
Xiao, L., and Kellman, M. E., 1989, *J. Chem. Phys.* **90**, 6086–6098.

19 FORMING SUPERPOSITION STATES

Tamar Seideman

National Research Council of Canada, Ottawa, Canada

Computational Molecular Spectroscopy. Edited by Per Jensen and Philip R. Bunker

19.1 Introduction

Most chapters in this book are concerned with the properties and the dynamics of highly energy-resolved states. Such states are typically prepared experimentally by excitation of a well-defined ground state with a narrow band source. Their theoretical study generally employs time-independent quantum mechanical techniques. More often than not calculations are carried out in close contact with energy-resolved spectroscopic experiments and aim at the elucidation of structural information from measured spectra. Since the resolution of current energy-domain spectroscopy approaches $0.001\,\mathrm{cm}^{-1}$ in the IR regime, it is clear that the numerical methods employed in transforming the data into structural information need be highly accurate.

The present chapter deals with the complementary field of time-resolved dynamics. Wavepackets, coherent superpositions of energy eigenstates, are prepared in the laboratory by excitation with short, precisely-phased, broad-band laser pulses. Time resolution translates into resolution in coordinate space. It allows the experimentalist to prepare a nonstationary, spatially well-defined state using a first (pump) pulse and follow its time-evolution using a second, time-delayed (probe) pulse. Such time-resolved pump–probe spectroscopy has been applied in recent years to the study of a broad range of phenomena in physics, chemistry, biology and surface science [Zewail (1994), Manz and Wöste (1995), Manz (1997), and Zhong and Zewail (1988)].

Since time and energy are related by Fourier transformation (for isolated systems), it is clear that time- and energy-resolved observables contain the same information about the probed system. It is nevertheless important to stress that time-resolved spectroscopy complements, rather than overlaps with, or replaces the traditional tools of energy-resolved spectroscopy. It is appropriate to different types of systems and advantageous when employed to address different types of problems. Typically less accurate structural information is sought for more complex systems. Often an intuitive view of the molecular motion and qualitative understanding of the complex processes it underlies are the goal.

Some of the general properties of wavepackets which make them a useful tool for achieving this goal are described in Section 19.2. The nature of time-resolved experiments and their objectives require the development and application of different theoretical methods than those employed to study energy-resolved spectroscopy. Such methods need to take into account the time dependence of the Hamiltonian as well as specific features of short pulse lasers. Several of the methods currently employed to study wavepacket dynamics are reviewed in Section 19.3. In Section 19.4 we use the computational methods described in Section 19.3 to illustrate some of the general wavepacket phenomena introduced in Section 19.2 by means of specific examples. In the final section we mention several of the applications of wavepacket technology in molecular dynamics. Although the sections of this chapter are related, each can be read independently of the others. In particular, the next section contains mostly elementary material. Readers familiar with time-dependent quantum mechanics are advised to proceed directly to Section 19.3.

A thorough and comprehensive review of theoretical studies of wavepackets, which includes also a survey of progress in time-independent methodology and many insightful observations, was recently given by Manz (1997). The availability of this up-to-date work allows us, in the present chapter, to omit a survey of the literature and to focus on fundamental concepts, new, partially unpublished ideas, expected to pave the way to future developments, and the connection with recent and ongoing experiments. In addition to the comprehensive work of Manz, we refer the reader to several more specific reviews: a two-volume book by Zewail (1994) contains a collection of mainly, but not exclusively, experimental studies of femtochemistry, starting with the first femtosecond studies of unimolecular fragmentation in the eighties and ending with electron diffraction studies reported in 1994. An up-to-date review, complementing the above reference was recently given by Zhong and Zewail (1998). Ultrashort pulse experiments are generally carried out at nonperturbative intensities, at least in the gas phase. Molecular dynamics in strong fields is thus intimately related to wavepacket phenomena. Intensity effects are not described in this chapter (save for a brief discussion of rotational excitation in moderately intense fields in Section 19.4.3). Readers interested in strong field effects are referred to the review of Corkum *et al.* (1997). An introduction to electronic nonadiabaticity and its spectroscopic detection using ultrashort pulses is given in the review of Stock and Domcke (1997). For time-dependent studies of femtosecond dynamics in solution see, for instance, the work of Benjamin and Wilson (1989) and the review articles of Mukamel (1990), Pollard and Mathies (1992) and Whitnell and Wilson (1993). Wavepacket dynamics of adsorbates on conducting and semiconducting surfaces has been recently reviewed by Guo, Saalfrank and Seideman (1999).

19.2 Why wavepackets? The quantum–classical correspondence and the short-time evolution of wavepackets

19.2.1 CONTINUUM WAVEPACKETS

Historically wavepackets emerged out of the quest for understanding quantum mechanics via the correspondence principle. Intuitively one seeks a quantum mechanical state of matter that in a certain limit will describe the motion of a classical particle whose position and momentum are well-defined at all times. A single energy eigenstate does not have such a limit. In the classical limit an eigenstate approaches a statistical mixture of classical particles [Messiah (1958)] rather than a single, spatially well-defined particle. This is most conveniently understood by exploiting the analogy to light waves.

Consider first the case of free motion (where the potential vanishes). A monochromatic wave of energy $E = \hbar\omega$ and wave vector $\boldsymbol{k}$,

$$\exp[\mathrm{i}(\boldsymbol{k} \cdot \boldsymbol{r} - \omega t)], \tag{1}$$

describes a fully delocalized state of matter, its probability density is isotropic. Envision, by contrast, a superposition of waves of the form (1) with different energies (for simplicity in one dimension),

$$\psi(q, t) = \int f(k' - k)\,\mathrm{e}^{\mathrm{i}(k'q - \omega' t)}\,\mathrm{d}k', \tag{2}$$

where $\omega' = \omega(k')$ etc. We express the complex coefficient function $f(k' - k)$ in terms of its phase and amplitude as $f = A\mathrm{e}^{\mathrm{i}\alpha}$ and group the various phases in equation (2) as

$$\Phi \equiv k'q - \omega' t + \alpha,$$

with which

$$\psi(q, t) = \int A(k' - k)\,\mathrm{e}^{\mathrm{i}\Phi}\,\mathrm{d}k'. \tag{3}$$

Clearly, the superposition state has significant amplitude only for q such that Φ does not vary strongly with k'. It peaks at

$$q_{\text{center}} = q\left(\left.\frac{\mathrm{d}\Phi}{\mathrm{d}k'}\right|_{k'=k} = 0\right) = t\frac{\mathrm{d}\omega}{\mathrm{d}k} - \frac{\mathrm{d}\alpha}{\mathrm{d}k}, \tag{4}$$

and is nonzero in a region of order

$$\Delta q \sim 1/\Delta k \tag{5}$$

about q_{center}. The order of magnitude estimation denoted $\sim$ in equation (5) is made more precise in Appendix A, where we use the theory of simultaneous measurements to derive Heisenberg's uncertainty relation between coordinates and momenta.

From equation (4) we see that the center of the wavepacket moves with velocity $\dot{q}_{\text{center}} = d\omega/dk$, which, using deBroglie's equation $p = \hbar k = h/\lambda$, is readily seen to be the classical velocity $\dot{q} = dE/dp$ of a single particle with momentum p.

Equation (2) is what we seek. It is a spatially defined quantum state of matter which, in a well-defined limit–the limit of an infinitely broad momentum distribution $f(k)$–describes the motion of a fully localized particle.

We note that $f(k)$ is the Fourier transform of the initial wavepacket,

$$\psi(q,0) = \frac{1}{\sqrt{2\pi}}\int dk\, f(k)\, e^{ikq} = \frac{1}{2\pi}\int dq'\,\psi(q',0)\int dk\, e^{ik(q-q')} = \psi(q,0). \tag{6}$$

Equation (2) is thus the simplest example of a continuum wavepacket, namely that of a free particle wavepacket.

19.2.1.1 Gaussian wavepackets

We proceed by using equation (6) to identify a particular (and useful) functional form of the superposition coefficient function $f(k)$. Specifically, we will show that the equality limit of the Schwarz inequality, when applied to suitably defined functions, defines a wavepacket in which the uncertainty product is minimized and which thus approaches as closely as possible the limit of a classical particle. The present derivation of minimum uncertainty wavepackets is not the simplest but is preferred here since it introduces a number of useful concepts and illustrates several of the lovely properties of wavepackets. The ideas developed here will serve in the introduction of a convenient wavepacket propagation technique in Section 19.3.3.2 and in the discussion of short pulses in Section 19.4.

First we note that the Schwarz inequality [Courant and Hilbert (1989)],

$$\left(\int |f|^2\, d\tau\right)\left(\int |g|^2\, d\tau\right) \geq \left|\int f^* g\, d\tau\right|^2 \tag{7}$$

becomes an equality if the functions g and f are proportional, $g = \lambda f$:

$$\left(\int |f|^2\, d\tau\right)\left(\int |\lambda f|^2 d\tau\right) = |\lambda|^2\left[\int |f|^2\, d\tau\right]^2 = \left|\int f^*\lambda f\, d\tau\right|^2. \tag{8}$$

For $f = (\hat{A} - \overline{A})\psi$, $g = (\hat{B} - \overline{B})\psi$ (see Appendix A), where $\hat{A}$ and $\hat{B}$ are arbitrary Hermitian operators with average values $\overline{A}$ and $\overline{B}$, we have from equation (80)

$$\int \psi^*(\hat{A} - \overline{A})(\hat{B} - \overline{B})\psi\, d\tau = \lambda(\Delta A)^2, \tag{9a}$$

$$\int \psi^*(\hat{B} - \overline{B})(\hat{A} - \overline{A})\psi\, d\tau = \frac{1}{\lambda}(\Delta B)^2, \tag{9b}$$

where ΔA and ΔB are the variances of A and B, respectively. Taking the sum and difference of the left hand sides of equations (9) and noting that $\overline{D}$ of equation (83) vanishes in the limit of Equation (8) one finds,

$$\lambda(\Delta A)^2 + \frac{1}{\lambda}(\Delta B)^2 = 0, \tag{10a}$$

$$\lambda(\Delta A)^2 - \frac{1}{\lambda}(\Delta B)^2 = i\overline{C}, \tag{10b}$$

and hence

$$\lambda = \frac{i\overline{C}}{2(\Delta A)^2}. \tag{10c}$$

Thus, for the strict equality limit of the Schwarz inequality to obtain in state ψ it should satisfy

$$(\hat{B} - \overline{B})\psi = \frac{i\overline{C}}{2(\Delta A)^2}(\hat{A} - \overline{A})\psi. \tag{11}$$

Of specific interest is the case of the position and momentum operators. From equation (84) $\overline{C} = \hbar$ and hence $\lambda = (i\hbar/2(\Delta q)^2)$. Equation (11) yields in that case the differential equation

$$\left(\frac{\hbar}{i}\frac{\hat{d}}{dq} - \overline{p}\right)\psi = \frac{i\hbar}{2(\Delta q)^2}(\hat{q} - \overline{q})\psi, \tag{12}$$

whose (normalized) solution is

$$\psi(q) = \left[2\pi(\Delta q)^2\right]^{-1/4}\exp\left[-\frac{(q-\overline{q})^2}{4(\Delta q)^2} + \frac{i\overline{p}q}{\hbar}\right]. \tag{13}$$

Equation (13) describes a a wavepacket of plane waves with Gaussian amplitudes,

$$\psi(q) = \psi(q, t = 0) = \frac{1}{\sqrt{2\pi}}\int f(p)\,e^{ipq/\hbar}\,dp, \tag{14}$$

where $f(p)$ is a Gaussian function of p, given as the Fourier transform of $\psi(q, 0)$,

$$\begin{aligned} f(p) &= \frac{1}{\sqrt{2\pi}}\int \psi(q,0)\,e^{-ipq/\hbar}\,dq \\ &= \left[\frac{2(\Delta q)^2}{\pi}\right]^{1/4}\exp\left[-\frac{(p-\overline{p})^2(\Delta q)^2}{\hbar^2} - i\overline{q}\frac{(p-\overline{p})}{\hbar}\right]. \end{aligned} \tag{15}$$

We have shown that the Gaussian wavepacket defined by $\psi(q, 0)$ [or, equivalently, by $f(p)$] is a minimum uncertainty wavepacket satisfying the strict equality limit of equation (85), $\Delta q\Delta p = \frac{1}{2}\hbar$. Another interesting property of the Gaussian form, utilized below, is that a wavepacket which is initially Gaussian will remain Gaussian in a quadratic (or lower order polynomial) potential. This is readily seen by applying the propagator $e^{-iHt/\hbar}$ to equation (13) where H is a harmonic oscillator Hamiltonian. A related feature of interest is that quantum mechanics and classical mechanics, regarded as theories of distributions, are indistinguishable for the case of Gaussian wavepackets. We will return to Gaussian wavepackets in the following sections.

19.2.1.2 Wavepacket spreading

Ehrenfest's theorem (1927) states that the expectation values of the coordinates and the momenta of a quantum mechanical system evolve in time following the classical equations of motion,

$$\frac{d}{dt}\overline{q}_i = \overline{\frac{\partial H}{\partial p_i}}, \quad (i = 1, 2, \ldots, F), \tag{16a}$$

$$\frac{d}{dt}\overline{p}_i = -\overline{\frac{\partial H}{\partial q_i}}, \quad (i = 1, 2, \ldots, F), \tag{16b}$$

where $\overline{A} = \overline{A}(t) \equiv \langle\psi(t)|A|\psi(t)\rangle$. Returning to the one-dimensional wavepacket model,

$$H = \frac{p^2}{2m} + V(q), \tag{17}$$

we denote by χ and η the variances of q and p, respectively [equation (79)],

$$\begin{aligned} \chi &\equiv (\Delta q)^2, \\ \eta &\equiv (\Delta p)^2. \end{aligned} \tag{18}$$

In the classical limit $\overline{q} \to q_{\text{classical}}$, $\overline{p} \to p_{\text{classical}}$, and

$$\epsilon \equiv \overline{H} - E_{\text{classical}} = \frac{\eta}{2m} + \overline{V} - V(\overline{q}) \tag{19}$$

is constant. With this limit in mind we assume that $\chi(t)$ is small at all times and expand the potential as

$$\begin{aligned} V(q) &= V_{\text{classical}} + (q-\overline{q})V'_{\text{classical}} + \tfrac{1}{2}(q-\overline{q})^2 V''_{\text{classical}} + \cdots, \\ V'(q) &= V'_{\text{classical}} + (q-\overline{q})V''_{\text{classical}} + \tfrac{1}{2}(q-\overline{q})^2 V'''_{\text{classical}} + \cdots. \end{aligned} \tag{20}$$

Hence,

$$\begin{aligned} \overline{V} &= V_{\text{classical}} + \tfrac{1}{2}\chi V''_{\text{classical}} + \cdots, \\ \overline{V'} &= V'_{\text{classical}} + \tfrac{1}{2}\chi V'''_{\text{classical}} + \cdots. \end{aligned} \tag{21}$$

and we have

$$\frac{\mathrm{d}}{\mathrm{d}t}\overline{q} = \frac{\overline{p}}{m}, \quad \frac{\mathrm{d}}{\mathrm{d}t}\overline{p} = -\overline{V'}. \tag{22}$$

In the limit

$$\overline{V'} \longrightarrow V'_{\text{classical}},$$

Ehrenfest's equations reduce to Hamilton's equations of motion. Recalling the time-derivative of a quantum mechanical expectation value,

$$i\hbar\frac{\mathrm{d}}{\mathrm{d}t}\overline{A} = \overline{[A, H]} + i\hbar\overline{\frac{\partial A}{\partial t}}, \tag{23}$$

we find

$$\frac{\mathrm{d}\chi}{\mathrm{d}t} = \frac{\mathrm{d}}{\mathrm{d}t}\overline{q^2 - \overline{q}^2} = \frac{1}{i\hbar}\overline{\left[q^2 - \overline{q}^2, H\right]} + \overline{\frac{\partial}{\partial t}\left(q^2 - \overline{q}^2\right)} = \frac{1}{m}(\overline{pq + qp} - 2\overline{p}\overline{q}), \tag{24}$$

where we noted the commutation relations

$$\left[q^2, p^2\right] = 2i\hbar(pq + qp), \quad \left[\overline{q}^2, p^2\right] = 0$$

and used Ehrenfest's equation for $(\mathrm{d}/\mathrm{d}t)\overline{q}$. Similarly

$$\frac{\mathrm{d}^2\chi}{\mathrm{d}t^2} = \frac{2\eta}{m^2} - \frac{1}{m}(\overline{V'q + qV'} - 2\overline{q}\overline{V'}). \tag{25}$$

In the limit of small χ we neglect the third terms of equation (20), $V'(q) \to V'_{\text{classical}} + (q - \overline{q})V''_{\text{classical}}$, with which

$$\frac{\mathrm{d}^2\chi}{\mathrm{d}t^2} \longrightarrow \frac{2}{m^2}\left(\eta - mV''_{\text{classical}}\chi\right) = \frac{4}{m}\left(\epsilon - V''_{\text{classical}}\chi\right). \tag{26}$$

Consider once again the free particle example, $V = 0$, $\bar{v} = (\bar{p}/m) =$ constant. From equation (19) $\eta(t) = \eta(t = 0) = 2m\epsilon$, and using equation (26)

$$\frac{d^2\chi}{dt^2} = \frac{2\eta}{m^2}. \tag{27}$$

Hence,

$$\chi = \chi_0 + \dot{\chi}_0 t + \frac{\eta_0}{m^2}t^2. \tag{28}$$

We find that the free particle wavepacket spreads indefinitely. The quantum mechanical evolution resembles that of a classical particle whose position is sharply defined only at very short times (while the expectation value of all coordinates follow the classical equations of motion). Although derived here for the simplest model for a continuum wavepacket, rapid spreading with time is general to scattering wavepackets.

19.2.2 DISCRETE WAVEPACKETS

Consider next the discrete analog of equation (2),

$$|\Psi(t)\rangle = \sum_n C_n |n\rangle e^{-iE_n t/\hbar} \tag{29}$$

where n is a given discrete quantum number (or a collection thereof), $|n\rangle$ and E_n are stationary eigenstates and the corresponding eigenvalues of an appropriately chosen zero-order Hamiltonian $H_0 = H - V(t)$,

$$H_0|n\rangle = E_n|n\rangle, \tag{30}$$

and C_n is a distribution of coefficients of width Δn centered about $\bar{n}$, determined by the mode of preparation of the wavepacket. Transition to the classical limit requires C_n to be such that the wavepacket will be initially strongly localized in space. That is, $\Delta q \ll L$ where q is the coordinate conjugate to n and L is the classical orbit corresponding to the peak of the coefficients distribution, $n \sim \bar{n}$. The uncertainty relation, equation (85), translates to

$$\Delta E \propto v\Delta p \sim \hbar\frac{\omega L}{\Delta q} \tag{31}$$

where ΔE is the energetic width of the wavepacket and ω is the fundamental frequency. For a quasi-harmonic vibrational mode with reduced mass m, for instance, $L \propto (E/m\omega^2)^{1/2}$ and $\Delta n \propto \bar{n}^{1/2}[\sqrt{\hbar/m\omega}/\Delta q]$. Hence the classical limit $\Delta q \ll L$ corresponds to a broad distribution in the conjugate quantum number space, $\Delta n \gg 1$. Large $\bar{n}$ alone is not sufficient, in general, for the wavepacket to behave quasi-classically.

19.2.2.1 Short time dynamics

Consider next the time evolution of the wavepacket subsequent to its preparation stage. At very short time, $t \ll T_{\mathrm{per}}$ (where $T_{\mathrm{per}} = 2\pi/\omega$ is the classical period), the discrete nature of the spectrum is not resolved and the initially localized wavepacket moves along a classical trajectory, spreading as it evolves, much like the free wavepacket of equation (2). By contrast to the continuum analog, however, the discrete wavepacket re-localizes periodically. This is readily seen from equation (29): In the harmonic limit,

$$|\Psi(t)\rangle = \sum_n C_n |n\rangle e^{-in\omega t} \tag{32}$$

(up to an overall phase). Clearly, all terms are in phase at times which are multiples of the classical period $T_{\rm per} = 2\pi/\omega$. Thus, the wavepacket exhibits a sequence of spikes at times $t = lT_{\rm per}$ where $\Psi(t=0)$ is reconstructed. To examine its shape between periods we return to Ehrenfest's theorem, equation (16) and consider again the harmonic case, $V = m\omega^2 q^2/2$. Using equation (26)

$$\frac{d^2\chi}{dt^2} \sim \frac{4}{m}\left(\epsilon - m\omega^2\chi\right) = -4\omega^2\left(\chi - \frac{\epsilon}{m\omega^2}\right), \tag{33}$$

or

$$\chi = \frac{\epsilon}{m\omega^2} + A\cos(2\omega t) + B\sin(2\omega t). \tag{34}$$

That is, the wavepacket spreads and relocalizes periodically. Thus, in the limit of equidistant levels the wavepacket oscillates indefinitely at the harmonic frequency ω, while its width oscillates at a frequency 2ω. The same behavior was predicted above based on equation (32). We will see in Section 19.4 that at short times the behavior of several types of realistic discrete wavepackets follows closely the periodic oscillations predicted by the simple harmonic oscillator model and expected on the basis of classical intuition. The classical picture, however, holds only at short times, before the anharmonicity of realistic molecular potentials is resolved. At longer times (where 'long' is quantified below) anharmonicities introduce purely quantum mechanical effects, as discussed in the next subsection.

19.2.2.2 Long time dynamics: dephasing, revivals and fractional revivals

In order to illustrate the effect of anharmonicities, it is convenient to rewrite $\Psi(t)$ of equation (29) as

$$|\Psi(t)\rangle = e^{-iE_{\overline{n}}t/\hbar} \sum_{m=-\infty}^{\infty} C_{\overline{n}+m}|\overline{n}+m\rangle e^{-i[E_{\overline{n}+m}-E_{\overline{n}}]t/\hbar}, \tag{35}$$

where $m = n - \overline{n}$, and expand the energy about $E_{\overline{n}}$ as

$$E(n) = E_{\overline{n}} + \left.\frac{dE(n)}{dn}\right|_{n=\overline{n}}(n-\overline{n}) + \frac{1}{2}\left.\frac{d^2E(n)}{dn^2}\right|_{n=\overline{n}}(n-\overline{n})^2 + \frac{1}{6}\left.\frac{d^3E(n)}{dn^3}\right|_{n=\overline{n}}(n-\overline{n})^3 + \cdots. \tag{36}$$

At short times the term linear in $n - \overline{n}$ dominates and the motion is quasi-harmonic, equation (32). At later times the third term in equation (36) introduces an additional phase shift between the wavepacket components and the equidistant peaks of equation (32) disappear; the wavepacket dephases. Nevertheless, the phase relationship between the wavepacket components is not lost. At times

$$t = T_{\rm rev} = \frac{2\pi\hbar}{\frac{1}{2}d^2E_n/dn^2|_{\overline{n}}} \tag{37}$$

the phases arising from the quadratic terms in equation (36) are exact multiples of 2π and, to the extent that the cubic and higher terms are negligible at $T_{\rm rev}$, the initial wavepacket is reconstructed. $T_{\rm rev}$ is called the revival time and is a direct observable, providing a measure of the anharmonicity of the underlying potential [Averbukh and Perel'man (1989), and Leichtle, Averbukh and Schleich (1996)]. It is an interesting property of molecular stretching modes that a quadratic expansion of the eigenenergy is often quite accurate. (It is exact in the limit of a Morse potential, where the energy includes only linear and quadratic terms, $E_n = \hbar\omega(n - \alpha n^2)$ giving $T_{\rm rev} = 2\pi/\omega\alpha$.) Hence realistic wavepackets of diatomic typically exhibit a sequence of revivals at times $kT_{\rm rev}$, where the initially prepared state revives. Detailed examination of the phase relationship between the wavepacket components [Averbukh and Perel'man (1989)] shows that at specific fractions of the revival time, $t = (q/r)T_{\text{rev}}$, where q/r is an arbitrary irrational number, the wavepacket splits into r (for odd r) or $r/2$ (for even r) spatially separated localized wavepackets, distributed along the classical orbit. For instance, at the half revival, $t = \frac{1}{2}T_{\text{rev}}$ the wavepacket is fully reconstructed but phase-shifted by π from $\Psi(t = 0)$. At the quarter revival it splits into two sub-wavepackets. The time dependence of the wavepacket (and hence the signals to which is gives rise) in the vicinity of fractional revivals was analyzed by Leichtle, Averbukh and Schleich (1996) within a fully quantum mechanical framework. A semiclassical discussion is given by Mallalieu and Stroud (1995). The studies of Leichtle, Averbukh and Schleich (1996) focus on the best studied case of vibrational or electronic wavepackets. The revival structure of rotational wavepackets is qualitatively different and quite fascinating, as shown by Seideman (1999). Revival structures have been observed experimentally in electronic [Yeazell and Stroud (1991)] and in vibrational [Bowman, Dantus and Zewail (1989), Baumert *et al.* (1992), and Vrakking, Villeneuve, and Stolow (1996)] wavepackets, as well as in one-atom masers [Rempe, Walther and Klein (1987)] and optical lattices [Raithel, Phillips and Rolston (1998)]. Examples are given in Section 19.4.

19.3 Predicting wavepacket motion

Time-dependent approaches in theoretical research predated time-resolved experiments of molecular vibrations [Zewail (1994)] by several decades. Their development was marked by a number of highlights, including the 1968 prediction of quantum beats by Jortner and Berry (1968) [see also the works of Bixon and Jortner (1969), Bixon, Jortner and Dothan (1969), Nitzan, Jortner and Rentzepis (1972) and references to related work therein], the introduction of time-dependent grid methods to follow the course of chemical reactions by Wyatt and coworkers [McCullough and Wyatt (1969)], Heller's formulation of CW spectra in terms of time-dependent correlation functions [Heller (1981a)] and his introduction of Gaussian wavepackets as a computational tool in reaction dynamics [Heller (1975)]. Several of these ideas had their origin in the yet earlier works of Schrödinger (1926), Ehrenfest (1927), Heisenberg (1927), Kubo (1952) and others.

Since the time-dependent and time-independent frameworks are related by Fourier transformations, it is clear that any problem [including problems involving explicitly time-dependent Hamiltonians; Shapiro (1993), and Seideman (1995)] can be formulated in either frame. Nevertheless, just as some phenomena are best observed using time-resolved experimental techniques while others are more conveniently observed using energy-resolved technology, some problems are most conveniently (or most efficiently) formulated in the time domain while others are best approached by the more traditional energy-resolved methods.

The most natural application of time-depdent methods is to the case of time-dependent Hamiltonians. While the most commonly studied such Hamiltonians are those that involve an external laser field, time dependence of the Hamiltonian is explicit also in the description of other transient phenomena [Seideman (1997a), and Seideman and Guo (1997)] and is introduced in a variety of system–bath approaches (see, e.g., Section 19.3.4).

In this section we discuss numerical and semi-analytical methods for solving the time-dependent Schrödinger equation,

$$i\hbar\frac{\partial}{\partial t}|\Psi(t)\rangle = H|\Psi(t)\rangle, \tag{38a}$$

or, in an integral form,

$$\Psi(t) = U(t, t_0)\Psi(t_0), \quad U(t, t_0) = \exp\left[-\frac{i}{\hbar}\int_{t_0}^{t} dt' H(t')\right]. \tag{38b}$$

These include fully quantum mechanical schemes, applicable to few mode systems, hybrid quantum–classical techniques, describing wavepacket dynamics coupled to a classical environment and semiclassical approximations, often introduced so as to allow physical insight, rather than for efficiency. We thus exclude a wide variety of time-dependent approaches for solving von Neuman's equation for evolution of the density matrix,

$$\dot{\rho} = -\frac{\mathrm{i}}{\hbar}[H, \rho] \tag{39}$$

where the density matrix, ρ, is analogous to the probability distribution of classical mechanics. Means of propagating equation (39), typically by partitioning the modes into a small system embedded in a large reservoir and taking the trace over the bath modes, have been extensively discussed in the literature and are reviewed, e.g., in the book by Blum (1981) and the article of Mukamel (1990). The reader is referred to the literature also for reviews of purely classical methods [Whitnell and Wilson (1993)] and statistical approaches [Levine (1985)].

We do not provide a complete survey of the literature but rather focus on a more detailed description of a few commonly employed approaches, whose discussion, in our personal judgment, should provide general insight as well as a useful computational tool. A detailed survey of numerical time-dependent methods is given in the review article of Manz (1997).

19.3.1 WAVEPACKET EXPANSIONS

A straightforward approach to solving equation (38) is based on the expansion of equation (29) in terms of the stationary eigenstates of a given reference Hamiltonian H_0. Perhaps the most familiar application of the expansion scheme is the discovery of quantum beats by Jortner and Berry (1968), who expanded an electronically excited wavepacket, coherently prepared by a short pulse (short compared with the level spacings) in terms of mixed states, superpositions of coupled Born–Oppenheimer eigenstates, and analyzed the subsequent nonradiative decay.

In the general case the expansion coefficients in equation (29) are solved for by substituting the expansion in the time-dependent Schrödinger equation, equation (38a), and employing the orthogonality of the $|n\rangle$ to obtain

$$\mathrm{i}\hbar\dot{C}_n(t) = \sum_{n' \neq n} C_{n'}(t)\langle n|H(t) - H_0|n'\rangle \mathrm{e}^{\mathrm{i}(E_n - E_{n'})t/\hbar}, \tag{40}$$

where $|n\rangle$ and E_n are defined by equation (30). To specify $\Psi(t)$ fully equation (40) needs to be supplemented by a set of initial conditions $C_n(t \to -\infty)$ appropriate to the mode of preparation of the system. The sum over the collective index n in equation (29) is replaced by summation over the discrete and integration over the continuum indices in case H_0 has both a discrete and a continuum spectrum.

The eigenstate expansion scheme is often also the most efficient one [Krause, Shapiro and Bersohn (1991)]. In the case of the pump–probe scenario, for instance, an expansion in terms of the eigenstates of the complete, field-free Hamiltonian $H_0 = H - V(t)$, allows the propagation during the field-free delay between the pulses to be performed analytically at no computational cost. The same approach was found useful in studies of photochemical events on conducting surfaces [Seideman (1997a), and Seideman and Guo (1997)], where, similar to the short-pulse, pump–probe case, long time propagation free of time-dependent perturbation follows a brief evolution under the time-dependent Hamiltonian. Conceptually equation (29) has the advantage of allowing a direct link between the wavepacket evolution and the stationary eigenstates of a Hamiltonian natural to the system and the studied problem.

In many cases the nature of the perturbation $V(t)$ allows for an analytical approximation of the C_n. In that event the computational effort reduces to solution of the linear algebra problem, equation (30). Consider, as a familiar example, electronic excitation from an initially prepared vibrational eigenstate $|n_i\rangle$ into a superposition of vibrational eigenstates $\{|n\rangle\}$ of a different electronic state. Using equation (40)

$$i\hbar\dot{C}_{n_i} = \sum_n C_n(t)\langle n_i|V(t)|n\rangle e^{i(E_{n_i}-E_n)t/\hbar}, \tag{41a}$$

$$i\hbar\dot{C}_n = C_{n_i}(t)\langle n|V(t)|n_i\rangle e^{i(E_n-E_{n_i})t/\hbar}, \quad n = 0, \ldots, \tag{41b}$$

with the initial condition $C_n(t=-\infty) = \delta_{n,n_i}$, and the field–matter interaction is given within the electric dipole approximation as

$$V(t) = H(t) - H_0 = -\boldsymbol{\mu}\cdot\boldsymbol{\varepsilon}(t), \tag{42}$$

$\boldsymbol{\mu}$ being the transition dipole operator and $\boldsymbol{\varepsilon}(t)$ the laser field. Assuming first that the laser field is weak, $|V(t)| \ll E_n$, $n = 0, \ldots$, we invoke first-order perturbation theory within which $C_{n_i}(t)$ is constant,

$$C_{n_i}(t) = C_{n_i}(t=-\infty) = 1. \tag{43}$$

Substituting equation (43) in equation (41b) we have

$$C_n = \frac{i\sqrt{2\pi}}{\hbar}\langle n|\boldsymbol{\mu}\cdot\hat{\boldsymbol{\varepsilon}}|n_i\rangle\tilde{\varepsilon}[(E_n - E_{n_i})/\hbar], \tag{44}$$

where $\hat{\boldsymbol{\varepsilon}}$ is a unit vector along the polarization direction,

$$\tilde{\varepsilon}(\omega) = \frac{1}{\sqrt{2\pi}}\int_{-\infty}^{\infty} dt\, \varepsilon(t)\, e^{i\omega t} \tag{45}$$

is the Fourier transform of the pulse and $\varepsilon(t)$ is a smooth envelope, $\vec{\varepsilon}(t) = \hat{\boldsymbol{\varepsilon}}\varepsilon(t)$.

A rather different, nonperturbative analytical approximation for C_n could be obtained in case the laser pulse spans a dense distribution of excited levels. It thus applies to the quasiclassical limit of short time and/or large mass, $\tau^{-1} \gg |E_n - E_{n-1}|$, τ being the pulse duration. Substituting equation (41b) in equation (41a) we have

$$\begin{aligned} C_{n_i} &= \frac{i}{\hbar}\sum_n \langle n_i|\boldsymbol{\mu}\cdot\hat{\boldsymbol{\varepsilon}}|n\rangle \int^t dt'\, C_n(t')\varepsilon(t')\, e^{i(E_{n_i}-E_n)t'/\hbar} + 1 \\ &= -\frac{1}{\hbar^2}\sum_n |\langle n_i|\boldsymbol{\mu}\cdot\hat{\boldsymbol{\varepsilon}}|n\rangle|^2 \int^t dt' \int^{t'} dt''\, C_{n_i}(t'')\varepsilon(t'')\, e^{i(E_{n_i}-E_n)(t'-t'')/\hbar}\varepsilon(t') + 1. \end{aligned} \tag{46}$$

We next use the assumption that the rapid excitation populates a dense manifold of excited levels to replace the n-summation by integration over E_n,

$$\sum_n \longrightarrow \rho\int_{-\infty}^{\infty} dE_n, \tag{47}$$

where ρ is the density of states, $\rho = dn/dE_n$, and energy is measured with respect to an intermediate excited state level. Assuming that $\langle n|\boldsymbol{\mu}\cdot\hat{\boldsymbol{\varepsilon}}|n_i\rangle$ vary slowly with energy as compared to the rapidly oscillating term, $\exp[iE_n(t''-t')/\hbar]$, the integral over E_n reduces to a δ-function,

$$\rho\int dE_n\, |\langle n_i|\boldsymbol{\mu}\cdot\hat{\boldsymbol{\varepsilon}}|n\rangle|^2\, e^{iE_n(t''-t')/\hbar} \approx 2\pi\rho\hbar|\langle n_i\rangle\boldsymbol{\mu}\cdot\hat{\boldsymbol{\varepsilon}}|\bar{n}\rangle|^2\delta(t''-t'), \tag{48}$$

where $\langle n_i|\boldsymbol{\mu}\cdot\hat{\boldsymbol{\varepsilon}}|\overline{n}\rangle$ is an n-averaged matrix element. Thus,

$$C_{n_i}(t) = -\frac{\pi\rho}{\hbar}\int_{-\infty}^{t} \mathrm{d}t' \, |\langle n_i|V(t')|\overline{n}\rangle|^2 C_{n_i}(t') + 1, \tag{49}$$

the solution of which is

$$C_{n_i}(t) = \mathrm{e}^{-\Gamma(t)/2}, \tag{50a}$$

where

$$\Gamma(t) = \frac{2\pi\rho}{\hbar}\int_{-\infty}^{t} \mathrm{d}t' \, |\langle n_i|V(t')|\overline{n}\rangle|^2. \tag{50b}$$

The C_n are now readily evaluated by substitution of equation (50) in equation (41b). Equation (50) shows that in the limit of a dense manifold, equation (47), the initially prepared state decays in a non-single-exponential manner during the field–matter interaction.

The expansion approach is natural to time-dependent problems where proper account of total angular momentum is essential. One such example is the problem of rotational wavepackets [Seideman (1995)] discussed in Section 19.4.3. Another example is pump–probe studies where the probe scheme is based on the detection of time-dependent alignment. This is the case, for instance, when photofragment or photoelectron [Reid (1993), Reid, Duxon and Towrie (1994), Seideman (1997b), Althorpe and Seideman (1999)] angular distributions are measured.

19.3.2 GRID METHODS

A second straightforward and increasingly popular approach for solving equation (38) is expansion of the Hamiltonian on a spatial grid. This scheme was employed as early as 1969 by McCullough and Wyatt, who used finite difference to evaluate the differential operators in equation (38) and applied the scheme to study the collinear $H + H_2$ reaction [McCullough and Wyatt (1969)]. This early study predated the development of efficient grid methods (see later) and hence presented a formidable computation task (while providing reaction probabilities that have been long known from higher accuracy time-independent studies). Nevertheless, it allowed new insights into the reaction dynamics through snapshots of the reactive flux, revealing transient phenomena which could not be uncovered by previous time-dependent studies.

The introduction of fast Fourier transform (FFT) techniques to wavepacket calculations, by Feit and co-workers (1982, 1983) and by Kosloff and Kosloff (1983) led to a breakthrough in the development of time-dependent grid methods. Here the actions of the potential and kinetic energy operators are computed, respectively, in the coordinate and momentum spaces,

$$V\Psi(q,t) = V(q)\Psi(q,t), \tag{51a}$$

$$T\tilde{\Psi}(q,t) = \frac{\hbar^2 k^2}{2m}\tilde{\Psi}(k,t), \tag{51b}$$

$\tilde{\Psi}$ being the transform of Ψ, and FFT and its inverse are used to transform the wavepacket efficiently between the two domains.

Significantly, roughly an order of magnitude larger time and coordinate step sizes as compared with the earlier finite difference method were found to give converged results with the FFT-based schemes [Feit and Fleck (1982), Feit, Fleck and Steiger (1983), and Kosloff and Kosloff (1983)]. Together with improved scaling of the computational effort in the number of grid points ($N \log N$ as compared with $\sim N^2$), the larger time and coordinate steps made the grid expansion scheme the method of choice for a large variety of applications, not only for conceptual but also for numerical reasons.

More recent progress in the development of time-dependent grid methods includes the introduction of polynomial expansions of the time evolution operator, e.g., Chebyshev expansions [Tal-Ezer and Kosloff (1984)], applications of the interaction representation [Das and Tannor (1990)] and development of schemes that treat time as an additional coordinate [Peskin and Moiseyev (1993)]. Different grid expansion approaches for the solution of the Schrödinger equation were recently reviewed by Kosloff (1996).

19.3.3 SEMICLASSICAL METHODS

As discussed in Section 19.2.1, wavepackets emerged historically to a large extent out of the quantum–classical correspondence and the quest for a quantum mechanical state of matter that would correlate, in a given limit, with the motion of a classical particle. This idea is nicely illustrated by a variety of time-dependent semiclassical approaches formulated during the past few decades, of which we outline two. Our choice of the semiclassical initial value representation and the Gaussian wavepacket approximation as examples is not accidental. While both methods emphasize the quantum–classical correspondence and take advantage of the dynamical information contained in classical trajectories, they make different approximations and are complementary in terms of applications.

19.3.3.1 Semiclassical initial value representation

One of the most successful and widely applicable time-dependent semiclassical approaches is the semiclassical initial value representation (SC-IVR), developed by Miller (1970, 1974). This approach is, at present, undergoing a rebirth of interest [Keshavamurthy and Miller (1994), Spath and Miller (1996), Sun and Miller (1997, 1998), Sun, Wang and Miller (1998a, 1998b), Batista *et al.* (1999), Kay (1994, 1997), Campolieti and Brumer (1994, 1997, 1998), Provost and Brumer (1995), and Walton and Manolopoulos (1995)] as a means of constructing the propagator from ordinary, real-valued classical trajectories, determined by specifying initial conditions.

A semiclassical expression for the transition probability from state $|\Psi_i\rangle$ to state $|\Psi_f\rangle$ is obtained by inserting the semiclassical approximation for the matrix element of the propagator, $e^{-iHt/\hbar}$ in the standard quantum mechanical expression for the transition amplitude,

$$
\begin{aligned}
T_{f,i}(t) &= \langle\Psi_f|\,e^{-iHt/\hbar}|\Psi_i\rangle \\
&= \int dq_1\,dq_2\,\Psi_f^*(q_2)\langle q_2|\,e^{-iHt/\hbar}|q_1\rangle\Psi_i(q_1) \\
&\simeq \sum\int dq_1\,dq_2\,\Psi_f^*(q_2)\left[2\pi\hbar\left|\frac{\partial q_2(q_1,p_1)}{\partial p_1}\right|\right]^{-1/2} \\
&\quad\times\exp\left[iS(q_2,q_1)-i\left(\nu+\frac{1}{2}\right)\pi/2\right]\Psi_i(q_1),
\end{aligned}
\tag{52}
$$

where $q_2(q_1, p_1)$ is the final position resulting from a classical trajectory with initial conditions (q_1, p_1) and $p_1 = p_1(q_2, q_1)$ is determined by the nonlinear boundary condition

$$q_2(q_1, p_1) = q_2. \tag{53}$$

The sum in the third line of equation (52) is over all roots of equation (53), $S(q_2, q_1)$ is the classical action integral,

$$S(q_2, q_1) = \int_0^t dt'\,\{p(t')\dot{q}(t') - H[p(t'), q(t')]\}, \tag{54}$$

and ν is the Maslov index, the number of zeros of the Jacobian $\partial q_2(q_1, p_1)/\partial p_1$ in the time interval $t = 0$ to t. In equations (52)–(54) we omit the dependence of q_2, p_2 and the classical action on time.

Equation (53) implies that for fixed q_1 one needs to search for p_1 such that the final position at time t is the specified q_2, clearly, a nontrivial numerical task. The IVR transformation of the semiclassical amplitude [Miller (1970)] eliminates the 'root search' by transforming from q_2 to p_1 as the integration variable,

$$\mathrm{d}q_2 = \mathrm{d}p_1 \left| \frac{\partial(q_1, p_1)}{\partial p_1} \right| \tag{55}$$

with which equation (52) takes the form

$$T_{f,i} \simeq \int \mathrm{d}q_1\, \mathrm{d}p_1\, \Psi_f^*[q_2(q_1, p_1)] \left[\left| \frac{\partial q_2(q_1, p_1)}{\partial p_1} \right| \Big/ 2\pi\hbar \right]^{1/2}$$
$$\times \exp\left[\mathrm{i}S(q_1, p_1) - \mathrm{i}\left(\nu + \frac{1}{2} \right) \pi/2 \right] \Psi_i(q_1). \tag{56}$$

The numerical advantages of equation (56) over equation (52) are clear. First, the Jacobian factor in the denominator of equation (52) (which has zeros at the classical turning points) has been eliminated, hence equation (56), by contrast to equation (52) is numerically stable. Second, equation (56) is based on classical trajectories specified by their initial conditions, rather than by double-ended boundary conditions. It thus circumvents the need to search for roots of equation (53) which complicates practical implementation of equation (52).

The SC-IVR has been successfully used in the study of a variety of quantum mechanical phenomena, including interference in chemical reactions [Keshavamurthy and Miller (1994)], tunneling [Kay (1997)], electronically nonadiabatic dynamics [Sun, Wang and Miller (1998a, 1998b)] and time-resolved photodetachment spectra [Batista *et al.* (1999)].

19.3.3.2 Gaussian wavepacket propagation

The Gaussian wavepacket propagation (GWP) technique was introduced by Heller in the seventies [Heller (1975); see also the works of Schrödinger (1926) and Ehrenfest (1927)] and was subsequently extended and utilized by a large number of authors [see, for instance, Heller (1976), Kulander and Heller (1978), Heller (1981b), Herman and Kluk (1984), Kolár, Ali and O'Shea, (1989), Heather and Metiu (1985), Braun, Metiu and Engel (1998), and Yan and Seideman (1999)]. Related approaches were recently developed by Billig (1997) and by Madhusoodanan and Kay (1998). Although the scheme and variants have been successfully employed in a variety of problems, it is described here primarily as an illustration of several of the ideas discussed in Section 19.2.

Similar to the SC-IVR, the GWP scheme is based on the assumption that much of the physics is contained in the classical method. To circumvent the violation of the uncertainty principle, inherent in classical mechanics, one surrounds the classical trajectories by 'quantum cells' of area h^F, where F is the number of degrees of freedom. This is naturally achieved by associating with each trajectory a Gaussian wavepacket, equation (13).

As noted in Section 19.2.1, wavepackets which are initially Gaussian remain Gaussian in quadratic (or lower order polynomial) potentials, although they may spread and acquire phase factors. This motivates expansion of the potential to second order about the instantaneous center of the wavepacket, $q_t = \overline{q}$ in equation (13),

$$V(q) \approx V_0 + \left.\frac{\mathrm{d}V}{\mathrm{d}q}\right|_{q=q_t} (q - q_t) + \frac{1}{2} \left.\frac{\mathrm{d}^2V}{\mathrm{d}q^2}\right|_{q=q_t} (q - q_t)^2. \tag{57}$$

Approximating the potential by the time-dependent expansion equation (57) and the wavefunction by the corresponding generalized Gaussian,

$$\Psi(q,t) = \exp\left[\frac{\mathrm{i}}{\hbar}\alpha_t(q-q_t)^2 + \frac{\mathrm{i}}{\hbar}p_t(q-q_t) + \frac{\mathrm{i}}{\hbar}\gamma_t\right], \tag{58}$$

we obtain from equation (38a) a set of first-order equations for the four parameters of equation (58),

$$\begin{aligned}
\dot{q}_t &= \partial H/\partial p_t,\\
\dot{p}_t &= -\partial H/\partial q_t,\\
\dot{\alpha}_t &= -\frac{2}{m}\alpha_t^2 - \frac{1}{2}\frac{\mathrm{d}^2 V}{\mathrm{d}q^2},\\
\dot{\gamma}_t &= \mathrm{i}\hbar\alpha_t/m + p_t\dot{q}_t - E.
\end{aligned} \tag{59}$$

The first two equations are Hamilton's equations of motion, as expected from Ehrenfest's Theorem, equations (16). The last equation identifies the phase of the wavepacket as the action integral along the trajectory, equation (54). Thus, the center of the wavepacket in position and momentum spaces follows a classical trajectory evolving on the exact potential energy surface. Its width varies in time subject to linear forces and the phase it acquires in the course of the trajectory is the familiar action integral of semiclassical mechanics. Heller (1976), Heather and Metiu (1985), Kolár, Ali and O'Shea (1989) and others introduced different variational approaches to improve the performance of the GWP scheme in application to anharmonic potentials.

Later studies [Heller (1981b)] replaced the distorting Gaussians by 'frozen Gaussians', Gaussian wavepackets with fixed form and centers that follow classical trajectories. Of major importance then is the representation of the initial and final states as properly phased *superpositions* of Gaussian wavepackets,

$$\Psi(q, t=0) = \sum_n C_n g_n(q), \tag{60}$$

where g_n are given by $\psi(q)$ of equation (13). From the discussion of Section 19.2.1 it is clear that the individual frozen Gaussians will generally do a rather poor job following the motion of the corresponding exact superposition components (except in the harmonic or linearly forced harmonic case, where Gaussian evolution is exact). Nevertheless [Heller (1981b)], the properly chosen superposition in equation (60) is expected to be more successful in reproducing the exact motion. In classical terms, this (coherent) averaging out of the error inherent in the propagation of the individual Gaussians arises from position–momentum correlation that develops between the superposition components in the course of their time propagation. Clearly, such 'collective correlation' depends crucially on the choice of the expansion coefficients, C_n in equation (60) which, in general, is neither trivial nor unique. Several physically motivated means of determining the C_n are discussed in [Heller (1981b)].

Herman and Kluk (1984) developed a Monte Carlo algorithm for selecting the initial superposition. This procedure removes the intuition from the choice of the basis set for the price of increasing the number of trajectories that need be integrated. We refer the reader to the review article of Gerber, Kosloff and Berman (1986) for description and comparison of different GWP-based methods.

The GWP approximation was introduced 25 years ago [Heller (1975)] as a means of evaluating reaction probabilities and CW spectra. At least at its simplest and most intuitively appealing form, it experienced serious difficulties in dealing with realistic (anharmonic) potentials. Recent work [Braun, Metiu and Engel (1998)] showed that the method provides an accurate and efficient means of describing femtosecond excitation in molecules, suggesting a range of new, better suited applications. The use of distorting Gaussians to study the evolution of molecular wavepackets in an optical trap, where the potential is nearly harmonic, is described by Yan and Seideman (1999).

19.3.4 TIME-DEPENDENT HARTREE APPROACHES

A widely employed class of time-dependent methods, which in principle can be made numerically exact but also lends itself to various approximations, is the time-dependent Hartree (TDH) (or time-dependent self-consistent field) scheme [Dirac (1930), Makri and Miller (1987), Alimi *et al.* (1990), Kotler, Neria and Nitzan (1991), Manthe, Meyer and Cederbaum (1992), Fang and Guo (1995), Raab *et al.* (1999) and others; see the review article of Gerber and Ratner (1988) and the work of Beck and Meyer (1997) for more complete lists of references]. At the simplest level TDH is an approximate method, based on the *anzatz* that the multidimensional wavefunction can be written as a product of single-mode functions

$$\Psi(q_1, q_2, \ldots, q_F, t) \approx \prod_{i=1}^{F} \phi_i(q_i, t) \exp\left[\mathrm{i}\int^t \left\langle \prod_i \phi_i(t') \left| V(t') \right| \prod_i \phi_i(t') \right\rangle \mathrm{d}t' \right]. \tag{61}$$

Each of the single mode functions evolves subject to a mean-field force, determined by the instantaneous dynamics of all other functions,

$$\mathrm{i}\hbar\dot{\phi}_i(q_i, t) \approx \langle \phi_1, \ldots, \phi_{i-1}, \phi_{i+1}, \ldots, \phi_F | H | \phi_1, \ldots, \phi_{i-1}, \phi_{i+1}, \ldots, \phi_F \rangle \phi_i. \tag{62}$$

The advantages and disadvantages of the TDH approach are evident from equations (61) and (62). Since the computational effort scales exponentially in the dimensionality of a quantum mechanical system, factorization of the multimode problem into single-mode problems drastically reduces the computational cost. On the other hand, only indirect interactions between the modes are accounted for through the mean field potentials. The neglect of correlations results in failure of the method when the wavepacket is delocalized or when it bifurcates.

A general remedy, borrowing from experience with similar problems in electronic structure theory, is the extension of equation (61) to include several configurations. A number of such multiconfiguration TDH (MCTDH) schemes have been introduced in recent years [Makri and Miller (1987), Kotler, Neria and Nitzan (1991), Manthe, Meyer and Cederbaum (1992), Fang and Guo (1995), and Beck and Meyer (1997)], illustrating the possibility of improving the accuracy at the cost of increased effort and vice versa, depending on the application. In particular it was shown that MCTDH can be formulated so as to converge to the exact solution with enough configurations [Manthe Meyer and Cederbaum (1992)]. Here

$$\Psi(q_1, q_2, \ldots, q_F, t) \approx \sum_{j_1=1}^{n_1} \cdots \sum_{j_F=1}^{n_F} A_{J_1 \ldots j_F}(t) \prod_{i=1}^{F} \phi_{j_i}^{(i)}(q_i, t), \tag{63}$$

where the sum runs over all combinations of one-dimensional basis functions and the equations of motion of the basis functions and the expansion coefficients are determined variationally. Thus, for the price of propagating a complicated set of nonlinear coupled equations, one is able to describe the dynamics to a desired accuracy with relatively few, optimally evolving functions. For detailed discussion of the computational scaling properties of equation (63) see the work of Beck and Meyer (1997). A recent application of MCTDH to calculate the absorption spectrum of pyrazine using a two electronic states, 24-mode model illustrates the power of the method [Raab *et al.* (1999)]

Importantly, TDH provides a natural framework for the development of hybrid quantum–classical [Alimi *et al.* (1990)] approximations, where certain modes are described quantum mechanically while others are approximated classically. Within quantum–classical TDH the classical modes evolve subject to Hamilton's equations and their influence on the quantum modes is approximated by averaging the potential over an ensemble of trajectories with different initial conditions. Since classical mechanics can be currently applied to thousands of coupled modes, such hybrid approaches are particularly useful for chemical reactions occurring in condensed phases, on surfaces and in large polyatomics or clusters, where a separation of the modes into classical and quantal, based on mass or time-scale arguments, can be often assumed. Similarly, semiclassical–classical and quantum–semiclassical TDH studies were found useful. Sun and Miller (1997) described a mixed classical-semiclassical TDSCF-like scheme within the SC-IVR formulation (Section 19.3.3.1). Alimi, Garcia-Vela and Gerber (1992) and Yan and Seideman (1999) employed the semiclassical GWA scheme (Section 19.3.3.2) within a semiclassical–classical and a quantum–semiclassical approach, respectively.

19.4 Modes of motion

In this section we discuss wavepacket dynamics in different degrees of freedom, including electronic, vibrational, rotational and (discrete) translational wavepackets. We illustrate several of the general wavepacket phenomena described in Section 19.2 by means of specific examples and employ several of the theoretical tools developed in Section 19.3 to analyze the observables. We start with the simplest and most commonly studied wavepackets and proceed with conceptually more complex and only recently explored wavepackets. Since the focus of this book is on molecular systems, we expand on vibrational and rotational wavepackets and discuss only briefly electronic and translational wavepackets.

19.4.1 ELECTRONIC WAVEPACKETS

Rydberg wavepackets, coherent superpositions of electronic Rydberg levels of atoms, have been studied extensively, theoretically [Parker and Stroud (1986), Krause *et al.* (1997)] and experimentally [Schumacher, Lyons and Gallagher (1997), Vrijen, Lankhuijzen and Noordam (1997), Weinacht, Ahn and Bucksbaum (1998), and Strehle, Weichmann and Gerber (1998)] during the past dozen years. Since the spectrum of atomic Rydberg levels in the quasi-classical, high-n regime is dense (the electronic level spacing scaling as n^{-3}), they provide the opportunity of generating experimentally a quantum mechanical state which exhibits many of the characteristics of a classical electron revolving in a Kepler orbit. Figure 19.1 illustrates, for instance, a calculated wavepacket of hydrogenic Rydberg levels at different times during its early evolution stages [Parker and Stroud (1986)]. The wavepacket was prepared by excitation of the ground state with a 10 ps pulse tuned to the vicinity of $\overline{n} = 85$. It first moves out to the classical turning point of the corresponding Kepler orbit while its spread in position space decreases. After reaching the turning point it reverses its direction and accelerates back toward the nucleus, spreading in coordinate space as it moves. It envelopes the nucleus and returns to the classical turning point. The periodic motion continues for several cycles, the period of motion being that expected for a classical particle revolving in the Kepler orbit corresponding to the level of excitation, $T_{\text{per}} \sim 2\pi\overline{n}^3 \sim 93.4$ ps.

Thus, the Rydberg wavepacket of Figure 19.1 nicely fulfills the premise of old quantum mechanics–only at short times, however. After about five cycles at the fundamental frequency, $\omega = 2\pi/T_{\text{per}}$, the wavepacket dephases and the periodic structure is lost. In agreement with the prediction of Section 19.2.2.2, it is reconstructed after about $T_{\text{rev}} = 2.6$ ns, at which point oscillation at T_{per} are renewed.

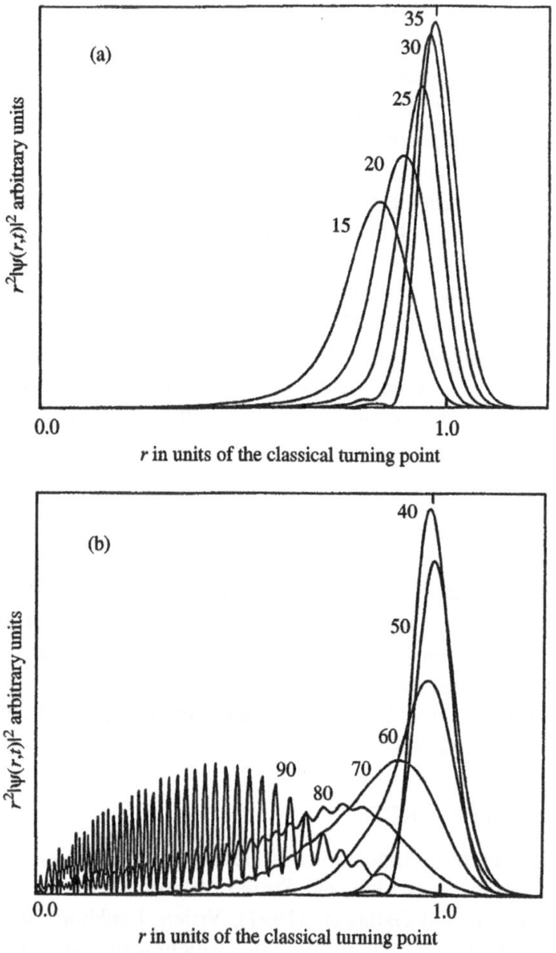

Figure 19.1. Electronic wavepacket dynamics: a calculated wavepacket of hydrogenic Rydberg levels during the first Kepler orbit. The pulse duration is 10 ps and its central frequency is tuned to the vicinity of $n = 85$. The curves are marked by time (in ps). Part (a) illustrates motion during the first half-period and part (b) during the second half-period. [Parker and Stroud (1986)].

Recent advances illustrated a variety of new applications of electronic wavepackets. Schumacher, Lyons and Gallagher (1997) showed the possibility of utilizing Rydberg wavepackets to study configuration interaction between electronic manifolds, in effect recording the flow of population back and forth between two configurations. Vrijen, Lankhuijzen and Noordam (1997) succeeded in imparting a momentum 'kick' to an electronic wavepacket using THz short pulse $\tau < 2\pi\bar{n}^3$ radiation and in monitoring the subsequent dynamics. Weinacht, Ahn and Bucksbaum (1998) reported shaping of electronic wavepackets using programmable pulses and determination of their amplitude and quantum phase, by analogy to optical spectral interference.

19.4.2 VIBRATIONAL WAVEPACKETS

While electronic wavepackets serve to study fundamental problems in wave mechanics and the quantum–classical correspondence, vibrational wavepackets serve primarily as a practical tool in molecular dynamics, capable of providing structural information about excited electronic states of polyatomic molecules and qualitative insight into complex reaction mechanisms which are difficult or impossible to determine by energy-resolved techniques.

This is typically achieved by preparing a localized superposition of vibrational states by broadband excitation from a well-defined eigenstate of a different electronic surface. Equilibrium mismatch between the initial and the probed potential energy surfaces ensures that the vibrational wavepacket is nonstationary in most molecular systems. It may vibrate in one [Bowman, Dantus and Zewail (1989), Baumert *et al.* (1995), Fischer *et al.* (1996)] or several coupled vibrational modes [Zewail (1994), Resch *et al.* (1999)] and may also probe different electronic surfaces [Stock and Domcke (1997), Blanchet *et al.* (1999)] or different arrangement channels [Zhong and Zewail (1998)]. In the course of the time evolution features of the potential energy surface are encoded in the wavepacket. This information can be extracted by using a second pulse to project the wavepacket onto a different electronic state. The final state can be another excited state of the neutral molecule [Zewail (1994)], an ion state [Baumert, Helbing and Gerber (1997), Assion *et al.* (1998), and Fischer *et al.* (1996)] or also the initial ground state. Serving as a template of the wavepacket, the final state should be a well-understood one. Ideally the probe stage would be rapid and simple, to avoid entangling the wavepacket dynamics with the probe dynamics. In reality, of course, this is rarely precisely the case but a sufficiently short pulse often allows nearing the ideal situation. Thus, the broad band of femtosecond sources, which makes them a useful tool for preparing vibrational wavepackets–transformed to the time domain–is responsible for their utility as a probe.

Figure 19.2(a) shows a prototypical example of a pump–probe experiment [Fischer, Vrakking, Villeneuve *et al.* (1996)] where a pump pulse prepares a wavepacket on the B-state of I_2 by excitation from the ground state and the probe pulse follows its time evolution by rapid ionization. The detected signal is shown in Figure 19.2(b). In this example the probe pulse detects the vibrating wavepacket each time it returns to the inner turning point, hence the signal oscillates at the vibrational period of B-state iodine, 340 fs at the level of excitation ($\overline{v} \sim 15$), as predicted by equation (32). Thus, the observed signal throughout the short time region of Figure 19.2(b) mimics well the harmonic oscillations of a classical particle rolling on the $I_2(B)$ potential energy surface, in agreement with the discussion of Section 19.2.2.1 and indeed with physical intuition. As is clearly seen in Figure 19.2(b), the classical picture fails as soon as the anharmonicity of realistic molecular potentials becomes noticeable, $t \sim 3$ ps, and wavepacket dephasing replaces periodic localization at $t = lT_{\text{per}}$, see Section 19.2.2.2. Nevertheless, coherence is not lost since spontaneous emission or collisions do not take place on the time scale of the experiment. In Figure 19.2(c) we show the wavepacket in the vicinity of the half revival, $T_{\text{rev}}/2 = \pi/\omega\alpha = 18$ ps for I_2, within a Morse approximation of the vibrational spectrum. As predicted in Section 19.2.2.2, the initial localized wavepacket is reconstructed and periodic motion at the fundamental frequency is renewed. More details of the experiment are given in the article of Fischer, Vrakking, Villeneuve *et al.* (1996).

While the I_2 experiment described in Figure 19.2 illustrates some of the fundamental concepts of wavepacket motion discussed in Section 19.2 and the potential of short pulse pump–probe techniques, it is clear that it does not reveal new structural information about the iodine molecule. In fact, the structural information about vibrational frequencies and anharmonicity constants of di- or triatomic systems that state-of-the-art time-domain experiments can provide does not near the accuracy of the information extractable from energy-resolved experiments.

The power of the time-domain approach to molecular dynamics lies in its ability to provide less precise information about more complex molecular motions and reaction mechanisms. The possibility of 'clocking' complex polyatomic reactions was illustrated, e.g., by Zewail and coworkers (1994). Resch *et al.* (1999) proposed a means of following in time the flow of energy between

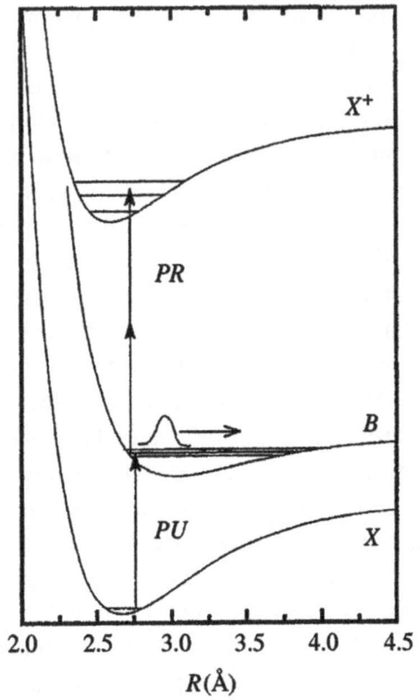

(a)

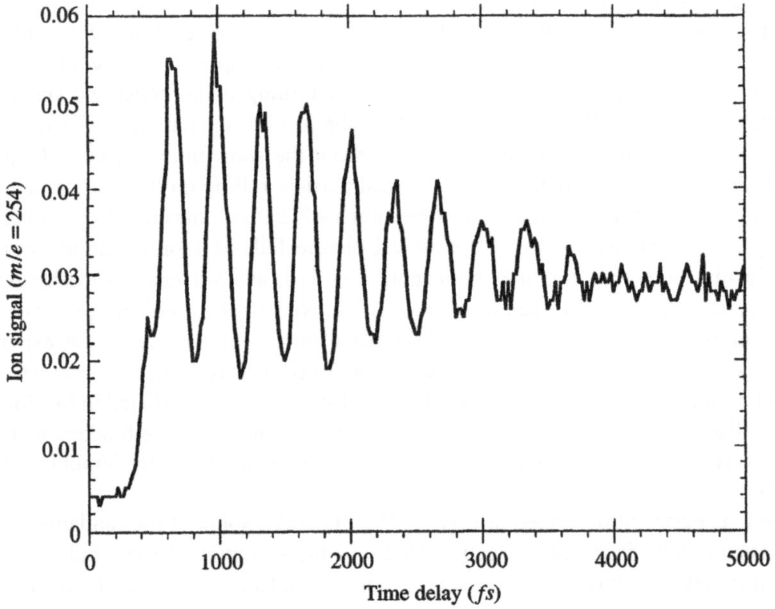

(b)

Figure 19.2. Vibrational wavepacket dynamics: a nonstationary wavepacket is prepared by short pulse excitation of X-state I_2 into the B state. The wavepacket evolution is probed via two-photon ionization into the ground state of the ion. (a) I_2 potential energy curves. (b) The molecular ion signal as a function of the pump–probe delay time during the first 5 ps of the wavepacket evolution. (c) The molecular ion signal during the late stages of the wavepacket evolution, illustrating its dephasing and revival. [Fischer *et al.* (1996)].

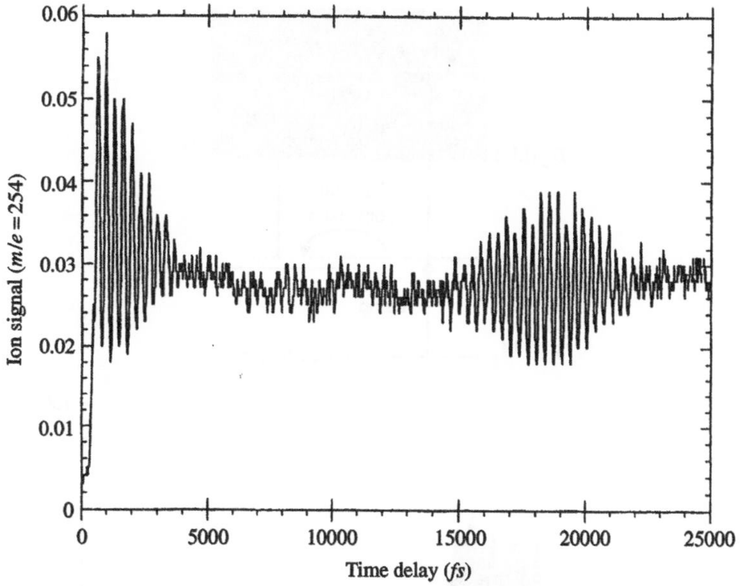

(c)

Figure 19.2. *(continued)*

coupled vibrational modes of a polyatomic molecule, a very general phenomena in multidimensional systems which is difficult (if possible) to quantify by other means. In a related study, Bingermann *et al.* (1997) observed the decay of excitation of a stretching vibration using time-resolved photofragmentation spectroscopy. The application of pump–probe spectroscopy to elucidate reaction pathways was illustrated through many examples by Zewail (1994), Zhong and Zewail (1998) and Bañares *et al.* (1998). The simulations of pseudorotations of Na_3 by Schön and Köppel (1995) demonstrated the sensitivity of the signal to Jahn–Teller-type coupling.

Figure 19.3 illustrates the possibility of following in time internal conversion and energy dissipation in polyatomic molecules through the example of all-*trans* decatetraene. Panel (a) gives the energy level scheme of decatetraene, showing the states relevant to the experiment [Blanchet *et al.* (1999)]. The pump field prepares a coherent superposition of the optically bright S_2 electronic state. The S_2 undergoes fast internal conversion into a ca 0.7 eV lower-lying S_1 state. The probe field projects the evolving, coupled wavepacket onto the ionization continuum. The arrows indicate the ionization propensity rules, based on Koopmans' Theorem [Eland (1984)]: the S_2 ionizes preferentially into the ground (D_0) state of the ion while the S_1 ionizes preferentially into the lowest excited (D_1) state. The kinetic energy of the electrons originating from S_1 neutrals is thus markedly different from that of electrons originating from S_2 neutrals.

Panel (b) shows the measured photoelectron kinetic energy spectra of decatetraene pumped at 287 nm and probed at 235 nm for several pump–probe delay times. At early times the energy distribution is structureless and peaked at high energy, ca 2.5 eV. These electrons originate from ionization of the initially prepared, low S_2 vibrational levels. As the S_2 state internally converts, on a time-scale of ca 400 fs, the energy distribution is shifted to a broad, structured band centered about 1.5 eV. The low energy band is due to ionization of vibrationally hot S_1 molecules into the D_1 state of the cation. The structure of the low energy band reflects the vibrational dynamics in the S_1 state, including mode coupling that is induced by the electronic nonadiabaticity. We note that the wavepacket does not return to the S_2 state, as would be expected in the absence

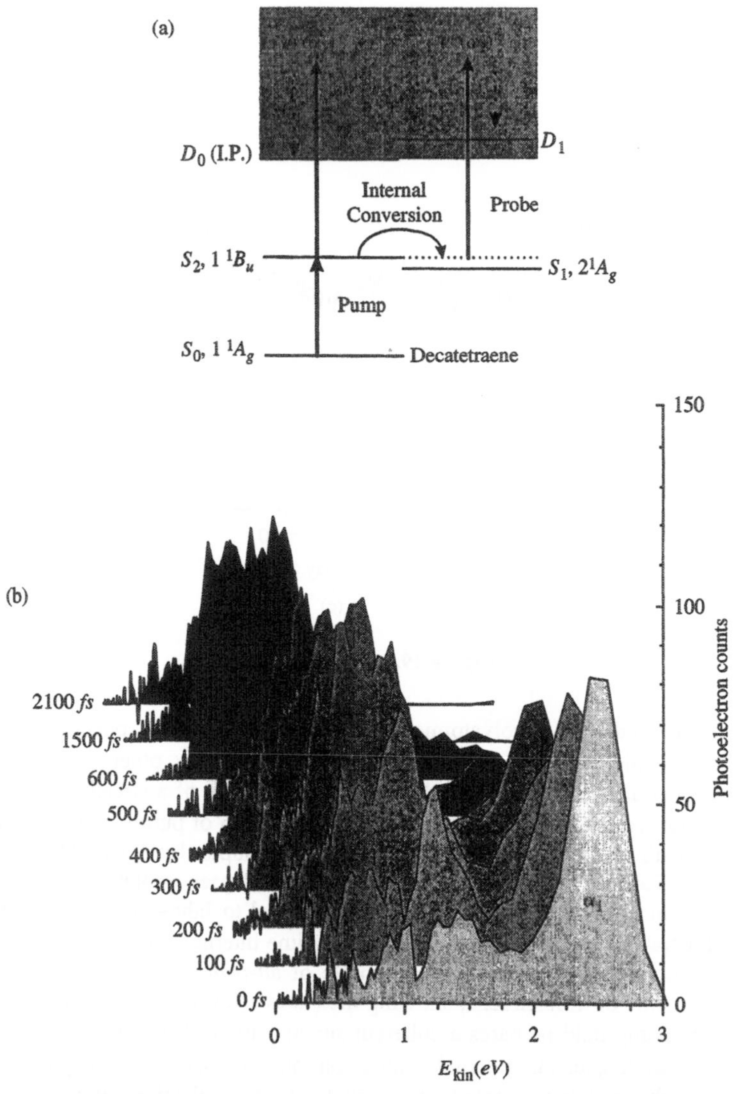

Figure 19.3. Time-resolved internal conversion. (a) Level scheme of decatetraene showing the neutral and ion states relevant to the experiment. A 100 fs pulse excites the ground state into the S_2 state. The S_2 wavepacket undergoes internal conversion into the lower lying S_1. The ionization propensity rules are indicated by solid arrows. (b) Time-resolved photoelectron kinetic energy spectra of decatetraene. The energy distribution changes rapidly, within ca 400 fs, from a structureless peak centered about 2.5 eV into a broad, structured peak centered about 1 eV. It reflects the internal conversion from the bright S_2 state into the S_1 as well as the vibrational dynamics that is induced by the electronic coupling. [Blanchet *et al.* (1999)].

of additional coupling mechanisms. Rather, the signal decays on a 3.5 ps time scale, as the S_1 population slowly converts to the ground state. Further discussions of photoelectron time-resolved spectroscopy as a route to understanding electronic nonadiabaticity are given in the review of Stock and Domcke (1997) and the works of Blanchet and Stolow (1998) and Blanchet *et al.* (1999).

19.4.3 ROTATIONAL WAVEPACKETS

Vibrational wavepackets have been studied extensively, both theoretically and experimentally during the past twelve years [Manz (1997), and Zewail (1994)]. Rotational wavepackets, coherent superpositions of total angular momentum levels, have been introduced only recently [Seideman (1995), Bandrauk and Aubanel (1995), Charron, Giusti-Suzor and Mies (1995), Vrakking and Stolte (1997), Seideman (1997b), Althorpe and Seideman (1999)] and have received so far little attention. [We distinguish below between rotational wavepackets and related fields such as rotational coherence spectroscopy, and the optical Kerr effect; see the works of Felker (1992) and of McMorrow, Lotshaw and Kenney-Wallace (1988).] Similar to the vibrational analog, a rotational wavepacket is a spatially well-defined superposition state which is correspondingly broad in the conjugate quantum number space, here $\Delta\theta\,\Delta J \gtrsim \hbar$, where J is the total angular momentum and θ is the angle between the molecular symmetry axis and a given axis fixed in space. Thus, rotational wavepackets are spatially *aligned*. However, the rotational and vibrational cases differ qualitatively in all other respects. One distinction follows trivially from the vast difference in level density between vibrational and rotational spectra. Vibrational level spacings are typically several orders of magnitude larger than rotational spacings. For the case of I_2 discussed above, for instance, the vibrational level spacing is $80-200\,\text{cm}^{-1}$ while the rotational spacing is $2JB_e$, where B_e, the rotational constant, is $0.037\,\text{cm}^{-1}$ in the ground state. A typical pulse duration of 50 fs allows simultaneous excitation of 5–6 vibrational levels of I_2 (and yet fewer with lighter systems). The same pulse spans several dozens of rotational levels and hence could, at least in principle, excite a much broader and hence better spatially defined rotational wavepacket.

A more qualitative distinction, however, appears to complicate issues: while the excitation of vibrational levels is limited only by energy considerations, the excitation of rotational levels is subject to the $\Delta J = 0, \pm 1$ electric dipole selection rules in a single photon process. It is here that the laser intensity serves for advantage. In a nonperturbative field, tuned near resonance between two electronic states, Rabi-type cycling between the states allows the exchange of an additional quantum of angular momentum between the field and the molecule on each transition, populating, after several cycles, coherent rotational wavepackets of opposite parity on both electronic states. This idea is quantified below [Seideman (1995), Vrakking and Stolte (1997), Seideman (1997b), Althorpe and Seideman (1999)]. At higher intensities Rabi cycles that populate several rotational levels are possible also between a bound and a dissociative state [Bandrauk and Aubel (1995), Charron, Giusti-Suzor and Mies (1995)]. Similarly, at nonresonant frequencies, far below electronic transitions, a rotational wavepacket on a single electronic state can be produced through two-photon, $\Delta J = 0, \pm 2$ cycles [Friedrich and Herschbach (1995), Seideman (1997c, 1997d), and Yan and Seideman (1999)].

A third distinction between the vibrational and rotational cases follows from the difference between the potential energy surfaces underlying the motion. In typical vibrational wavepacket experiments, equilibrium mismatch between ground and excited potential surfaces ensures that a wavepacket prepared through optical excitation of a ground state eigenfunction is nonstationary–analogous to a classical particle that is placed initially far from equilibrium. Rotational wavepackets evolve subject to a field-induced potential, which, as such, is subject to control. The ground and excited states correspond usually [not necessarily, Althorpe and Seideman (1999)] to the same equilibrium configuration and hence the centers of both wavepackets are essentially stationary. This results in a practical distinction between the two types of wavepackets. As noted above, vibrational wavepackets serve primarily to probe the electronic structure underlying the motion. Most applications of rotational wavepackets relate to their alignment property.

Aligning molecules has been one of the most important goals of modern reaction dynamics, because of the role of alignment in stereodynamical studies, gas–surface research and surface catalysis [Friedrich, Pullman and Herschbach (1991), and Cho and Bernstein (1991)]. Very recently it was shown that molecular alignment could be combined with focusing of the center-of-mass motion of molecules, opening a range of new applications of aligned molecules [Seideman (1997c, 1997d), Yan and Seideman (1999)]. Although other alignment techniques are available, alignment in intense laser fields, through excitation of a broad rotational wavepacket, is currently believed to provide significant advantages. These include the generality of the strong field scheme, the much sharper alignment it allows [Seideman (1995), Althorpe and Seideman (1999), and Yan and Seideman (1999)], as compared with conventional techniques and its ability to preserve alignment under field-free conditions (see later).

In order to appreciate the properties of rotational wavepackets it is useful to review briefly their theoretical description. Using equations (29)

$$|\Psi(t)\rangle = \sum_{\xi=0,1} \sum_{vJMK} C_{\xi}^{vJMK}(t)|\xi\, vJMK\rangle \exp\left(-\mathrm{i}E_{\xi}^{vJK}t/\hbar\right), \tag{64}$$

where we specified explicitly the quantum indices of equation (39), $n = \{\xi, v, J, M, K\}$, ξ is an electronic index, J is the total angular momentum, M and K are the projections of J onto the space-fixed and body-fixed z axes, respectively, and v is a collective vibrational index. In the case of linear or symmetric top molecules

$$\left\langle \hat{R}\boldsymbol{q} \middle| \xi vJMK\right\rangle = \overline{D}_{KM}^{J}(\hat{R})\psi_{\xi}^{vJK}(\boldsymbol{q}) \tag{65}$$

where $\hat{R}$ denotes the Euler angles of rotation of the body-fixed with respect to the space-fixed frame,

$$\overline{D}_{KM}^{J} = \sqrt{\frac{2J+1}{8\pi^2}} D_{KM}^{J}, \tag{66}$$

$D_{m'm}^{j}$ being standard rotation matrices [Edmonds (1960)], and ψ_{ξ}^{vJK} are vibronic eigenstates. For simplicity of notation we omitted the index ξ from the vibrational and rotational quantum numbers. Summation over vibrational indices allows for the possibility that the laser band spans several vibrational levels, producing a rovibrational or a rovibronic wavepacket.

The interaction term is given by equation (42) and we expand the dipole vector and the field polarization vector in spherical unit vectors finding,

$$\boldsymbol{\mu} \cdot \hat{\epsilon} = \sum_{qs} (-1)^q \epsilon_q \mu_s D_{s-q}^{1}(\hat{R}) \tag{67}$$

where s determines the nature of the transition and q the field polarization.

Using equations (65)–(67), the matrix elements in equations (40) are given as

$$\langle n|\boldsymbol{\mu} \cdot \epsilon(t)|n'\rangle = \epsilon(t) \sum_{qs} (-1)^q \epsilon_q W(JMK|qs|J'M'K')T(\xi vJK|s|\xi' v' J'K'), \tag{68}$$

where $W(JMK|qs|J'M'K')$ is an integral over the Euler angles,

$$W(JMK|qs|J'M'K') = \int \mathrm{d}\hat{R}\, \overline{D}_{KM}^{J*} D_{s-q}^{1} \overline{D}_{K'M'}^{J'} \tag{69}$$

and $T(\xi vJK|s|\xi' v' J'K')$ is an integral over the dynamical variables,

$$T(\xi vJK|s|\xi' v' J'K') = \int \mathrm{d}\boldsymbol{q}\, \psi_{\xi}^{vJK} \mu_s \psi_{\xi'}^{v'J'K'}. \tag{70}$$

Equation (69) for the orientational part of the matrix element is readily solved analytically [Bunker and Jensen (1998)],

$$W(JMK|qs|J'M'K') = (-1)^{K+M}\sqrt{(2J+1)(2J'+1)}\begin{pmatrix} J & 1 & J' \\ -K & s & K' \end{pmatrix}\begin{pmatrix} J & 1 & J' \\ -M & -q & M' \end{pmatrix} \tag{71}$$

The dynamical information is contained in the T matrix elements whose evaluation is generally a nontrivial numerical problem.

In the weak field limit, assuming that the system has been initially prepared in a single rovibrational eigenstate of the ground electronic Hamiltonian,

$$C_{\xi}^{vJMK}(t=0) = \delta_{\xi,0}\delta_{v,v_i}\delta_{J,J_i}\delta_{M,M_i}\delta_{K,K_i} \tag{72}$$

Equation (44) gives for the excited $\xi = 1$ amplitudes

$$C_1^{vJMK}(t) \approx \frac{i\sqrt{2\pi}}{\hbar}\langle 1vJMK|\boldsymbol{\mu}\cdot\hat{\epsilon}|0v_iJ_iM_iK_i\rangle\tilde{\epsilon}\left[\left(E_1^{vJK} - E_0^{v_iJ_iK_i}\right)/\hbar\right] \tag{73}$$

where $\tilde{\epsilon}(\omega)$ is given by equation (45). The properties of the 3-j symbols in equation (71) yield the well-known electric dipole selection rules in a single photon transition, $J = J_i$, $J_i \pm 1$, $M = M_i$, $M_i \pm 1$, $K = K_i$, $K_i \pm 1$. Under nonperturbative intensities the system Rabi-oscillates between the ground and excited electronic manifolds, exchanging an additional unit of angular momentum with the field on each transition. Very soon a coherent wavepacket of rotational states is populated in both electronic states.

The degree of rotational excitation is determined either by the pulse duration, τ or by the relative magnitudes of the detuning from resonance, Δ^J and the effective Rabi coupling, Ω_R^J. The detuning increases linearly with J, while the Rabi coupling is linear in the field strength and only weakly J dependent, through the J dependence of the Franck–Condon overlap. In case time is the limiting factor, i.e., in the case of a short and/or intense pulse, the number of levels accessed is roughly the number of cycles the system has time to make between the two electronic states,

$$J \lesssim \tau\Omega_R^J. \tag{74}$$

In case the intensity upper bounds the degree of rotational excitation we have that the maximum J accessed is determined through the approximate equality

$$\Omega_R^J \sim \Delta^J. \tag{75}$$

Consider, for instance, a linearly polarized field, $q = 0$ in equation (67), tuned to resonance with a parallel transition, $s = 0$. Equation (71) shows that both space- and body-fixed projections of the total angular momentum vector are conserved, $M = M_i$, $K = K_i$. If the system is initially rotationally cold (hence low M_i, $K_i \leq J_i$), $\Psi(t)$ describes a superposition of (properly phased) high total angular momentum states with close to zero projections onto the space-fixed and body-fixed axes. It is possible to show that, under rather general conditions, the phase relationship between the wavepacket components is such that the wavepacket is necessarily aligned.

The alignment property of intense-field-induced rotational wavepackets can be *qualitatively* understood classically. The field creates a potential for the angular motion, $V \propto \cos\theta$ in the case of linear polarization and a parallel transition. Provided that the well depth is large compared with the rotational temperature and that the field is applied long enough to impart sufficient angular momentum to the molecule, the free rotational motion transforms into small-amplitude librations about the polarization axis. It is worth restressing, however, that alignment is a coherent wavepacket effect. It relies on the phase relationship between the wavepacket components.

Figure 19.4(a) shows a theoretical simulation of the ground component of a Li_2 rotational wavepacket during a $\tau = 200$ fs Gaussian pulse. At the beginning of the pulse the system is naturally isotropic. As the field is turned on, higher rotational levels are excited and the wavepacket becomes gradually defined in space. Toward the end of the pulse it is peaked along the polarization direction. Figure 19.4(b) illustrates the evolution of the wavepacket after turn-off of the pulse, under field-free conditions. The initially aligned wavepacket undergoes a dense series of dephasings and rephasings, remaining on the average sharply peaked in the forward direction. Remarkably, the alignment is significantly enhanced after the pulse. An average measure of the time evolving alignment is the expectation value of $\cos^2\theta$ in the wavepacket. As shown in the insets of Figures 19.4(a) and 19.4(b), $\langle\cos^2\theta\rangle$ continues to grow during the short-time, classical-like evolution immediately following the pulse turn-off and is further enhanced at later times, where interference sets in. Similar alignment enhancement was found numerically also in other systems [Vrakking and Stolte (1997), and Althorpe and Seideman (1999)]. Its origin is illustrated by means of an analytical model in [Seideman (1999)].

Rotational wavepacket dynamics is a young but rapidly advancing field of research. Recent work showed the possibilities of confining molecules to a plane using circularly polarized nonresonant light, enriching the rotational level composition of the wavepacket by using chirped field [Karczmarek *et al.* (1999)], breaking its symmetry to produce an oriented state using two phase-related laser fields [Charron, Giusti-Suzor and Mies (1995), Vrakking and Stolte (1997)] and setting a wavepacket into oscillatory motion between alignment parallel and perpendicular to the field axis using linearly polarized field resonant with a perpendicular transition [Althorpe and Seideman (1999)].

19.4.4 DISCRETE CENTER-OF-MASS WAVEPACKETS

In Section 19.2.1 we described continuum wavepackets of the center-of-mass motion. Such wavepackets were studied since the early days of quantum mechanics and played an important role in the development of time-dependent scattering theory. Their bound-state analog, quantized systems describing periodic motion of the center of mass, were studied only recently. Their realization in the laboratory had to await the development of techniques for discretizing the translational mode of quantum mechanical systems and preparing nonstationary superpositions of the discrete levels.

In this section we describe two types of wavepackets of the center-of-mass motion, differing in their properties, mode of preparation and applications. Wavepackets of atoms confined in optical lattices [Rudy, Ejnisman and Bigelow (1997), Görlitz *et al.* (1997), and Raithel, Phillips and Rolston (1998)] and wavepackets of charge carriers in semiconductor structures [Leo *et al.* (1991), Feldmann *et al.* (1992), and Weiner, (1994)].

19.4.4.1 Wavepackets in optical lattices

Optical lattices are periodic potentials for atomic motion which resemble in some respects the familiar solid-state lattices but are based on the interaction of the atoms with laser fields, rather than on interatomic forces. Such potentials are formed by the interference of several laser beams tuned near resonance with an atomic transition to produce a standing wave intensity pattern. It is possible to show that (below saturation) the potential energy for the center-of-mass motion of atoms evolving subject to the light pattern is proportional to the laser intensity, $U(z) \sim I(z)/\Delta$ where z is the quantization direction of the lattice, parallel to the laser beam, $I(z)$ is the intensity and Δ is the detuning from resonance. Hence $U(z)$ is periodic in space with minima spaced by $\lambda/4$, λ being the wavelength of light. The amplitude of the periodic potential, U_0, is typically small, of order 10^{-8}–10^{-9} eV. Nevertheless, it is currently possible to cool atoms of simple level

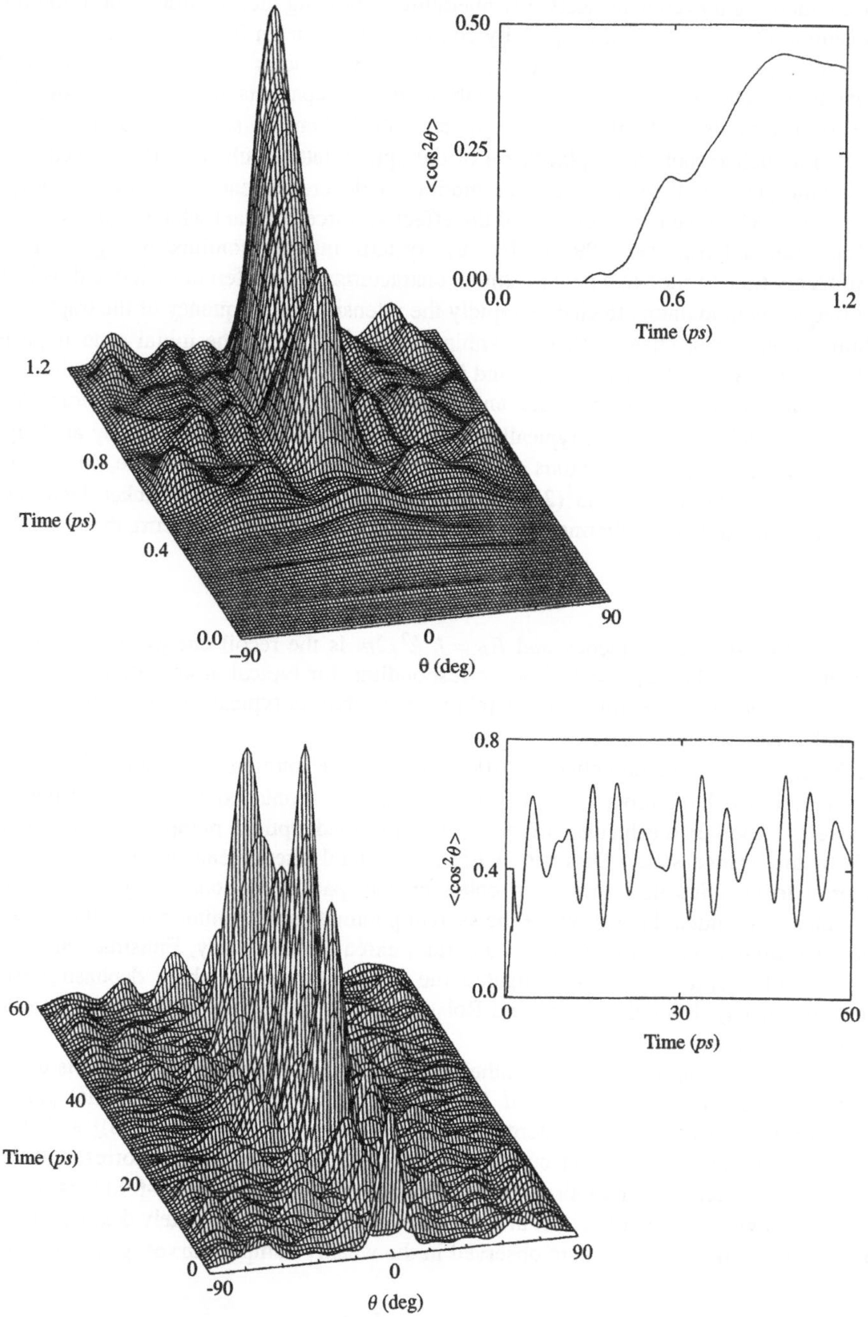

Figure 19.4. Rotational wavepacket dynamics. (a) A rotational wavepacket of Li_2 molecules versus the polar Euler angle θ and time during the laser pulse. The intensity is $10^{11}\,\mathrm{W\,cm^{-2}}$ and the pulse peaks at 600 fs with full width of 200 fs. As higher rotational levels are excited the wavepacket becomes better aligned. (b) The rotational wavepacket after turn-off of the laser pulse. The alignment is significantly enhanced under field-free conditions. The expectation value of $\cos^2\theta$, given as inset in both panels, quantifies the early and late alignment enhancement stages. [Seideman (1999)].

structure to micro- and even nano-Kelvin temperatures, allowing their confinement to the minima of the potential. Once the system is spatially confined its spectrum is discretized.

A coherent superposition of discrete center-of-mass levels can next be generated by suddenly shifting the lattice. By analogy to the familiar vibrational wavepackets discussed in Section 19.4.2, where a short laser pulse shifts the system to a new potential energy surface on a time scale short compared with nuclear motion, displacement of the optical lattice subjects the confined atoms to a new potential and sets them into periodic motion. A different means of inducing wavepacket motion in an optical lattice is modulation of the effective force constant which characterizes $U(z)$ [Rudy, Ejnisman and Bigelow (1997)]. Here the system, initially confined in a given potential $U(z)$ is suddenly transformed to a new potential, characterized by different period and well-depth, by using frequency modulators to change rapidly the intensity and frequency of the trapping light. In the limit of an abrupt change (that is, within a time $\delta_t \ll T_{\text{per}}$), the initial state is projected essentially unaltered into the second well and is set in oscillating breathing-type motion.

At short times the atoms oscillate back and forth in the potential wells with a common phase and harmonic period $T_{\text{per}} \propto U_0^{-1/2}$, typically in the microsecond regime. Again, by analogy with the wavepackets discussed in Sections 4.1–4.3, the potential is not harmonic, in the present problem it is sinusoidal, $U(z) \propto \sin^2(2kz)$ $(k = 2\pi/\lambda)$, and hence the wavepacket dephases on a time scale determined by the anharmonicity constants. Up to the quadratic term, the energy levels are given as

$$E_v \approx \hbar\omega\left(v + \tfrac{1}{2}\right) + \tfrac{1}{2}E_{\text{R}}v^2 \tag{76}$$

where ω is the oscillation frequency and $E_R = \hbar^2k^2/2m$ is the recoil energy. The revival time, [see equation (37)] is thus $T_{\text{rev}} = 4\pi/E_{\text{R}}$, corresponding, for typical atomic masses and optical wavelengths, to the 10^{-4} s regime, considerably longer than is typical of electronic, vibrational or rotational wavepackets.

The coherent revival structure characteristic of wavepacket motions is complicated in the case of optical lattices by dissipation, i.e., coherence loss due to spontaneous emission of the excited atomic levels. (Another mechanism of decay is so-termed optical pumping among magnetic sublevels of the ground state.) Dissipation being a general phenomena in complex systems, it is of interest to examine its imprints on coherent wavepacket motions using a simple model system, such as the optical lattice, where the system parameters determining the balance between dephasing and dissipation can be tuned to a certain extent. While Rudy, Ejnisman and Bigelow (1997) and Görlitz *et al.* (1997) investigated the coherent regime, where dephasing strongly dominates over decay, Raithel, Phillips and Rolston (1998) studied the long-time regime, where the two phenomena compete.

Figure 19.5, taken from the work of Raithel, Phillips and Rolston (1998), interprets the experimental wavepacket signals by means of Monte Carlo simulations. The potential parameters correspond to the experiment in Cs atoms [Raithel, Phillips and Rolston (1998)] and the solid and dashed curves [upper and lower curves in Figures 19.5(a) and 19.5(c)] correspond, respectively, to the actual excited state lifetime for spontaneous emission and to a 20-fold longer lifetime. Dissipation is shown to have little effect at short times, panel (b). It completely destroys, however, the full and fractional revivals that are observed in the weakly damped curves, panels (a) and (c).

19.4.4.2 Wavepackets of charge-carriers in semiconductors

Related to wavepacket motion in optical lattices, and also to wavepackets of atomic electronic levels, is the recently observed phenomenon of coherent charge oscillations in semiconductor double and multiple quantum well ('superlattice') structures [Leo *et al.* (1991), Feldmann *et al.* (1992), and Weiner (1994)]. A schematic energy level diagram of a superlattice in shown in Figure 19.6. In the absence of an electric field, coupling between neighboring wells results in

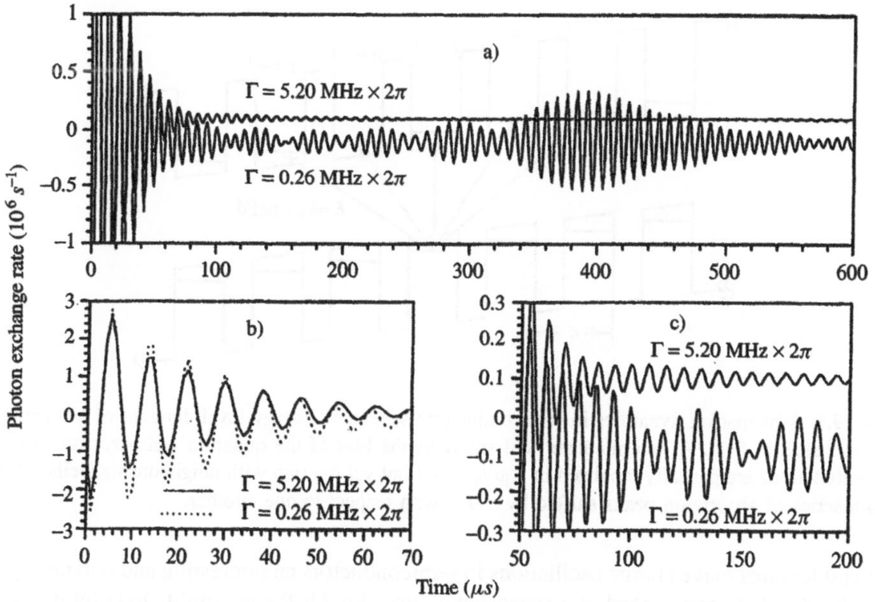

Figure 19.5. Wavepacket dynamics in an optical lattice: calculated wavepacket signals following a shift of the optical lattice by $dz = 0.14\lambda$ with a time constant of 3 μs. The lower curves in panels (a) and (c) and the dashed curve in panel (b) correspond to the limit of weak dissipation, where coherence is nearly fully conserved (excited state decay rate of 0.26 MHz × 2π). In that limit anharmonicity-induced dephasing and subsequent revivals are clearly observed. The upper curves in (a) and (c) and the solid one in (b) correspond to an excited state decay rate of 5.2 MHz × 2π, with which dissipation dominates. [Raithel, Phillips and Rolston (1998)].

delocalization of the electron and hole wavefunctions and the formation of energy minibands. When an external field is applied in the direction of growth of the quantum well, the wavefunctions localize and the energy is discretized.

At intermediate field strengths, partial overlap of the wavefunctions of different wells allows interband absorption of light to produce electrons with wavefunction centered in either the same well or neighboring ones. This gives rise to a series of absorption peaks, a so-termed Wannier–Stark ladder, with energies

$$E = E_0 + neFd \quad n = 0, \pm 1, \dots, \tag{77}$$

where E_0 is the vertical transition energy, e the elementary charge, F the applied field and d the superlattice period, see Figure 19.6 [Feldmann *et al.* (1992), and Weiner (1994)]. Viewed in the time domain, the quasi-momentum $\boldsymbol{k}$ of an electron in a periodic lattice increases at a rate proportional to the external field, $\dot{k} = eF$. When $\boldsymbol{k}$ reaches the edge of the Brillouin zone the electron is reflected, giving rise to periodic oscillations in $\boldsymbol{k}$ space, so-termed Bloch oscillations, with period $T_{\text{per}} = 2\pi/eFd$. Thus, Bloch oscillations are conveniently thought of as coherent motion of a wavepacket of Wannier–Stark levels.

The loss of coherence as a result of coupling with the solid environment qualitatively changes the dynamics and its theoretical formulation. Since, however, phase exchange with the environment takes place on a ca 10 ps time scale, it does not preclude the observation of a variety of wavepacket phenomena nor the application of wavepacket control schemes to manipulate the motion of charge carriers in semiconductor structures [Weiner (1994)].

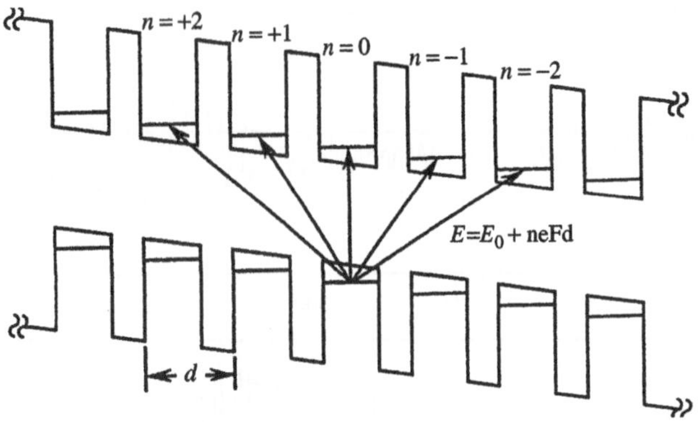

Figure 19.6. Wavepacket dynamics in semiconductors: schematic energy level diagram of a superlattice in an applied electric field. At intermediate field strengths the bias of the quantum well structure is such that the wavefunctions are partially localized in a given well but yet overlap with neighbouring wells. This gives rise to a series of absorption peaks shifted by $neFd$ with respect to one another.

Several features make charge oscillations in semiconductors an interesting and potentially useful system for the study and control of wavepacket motion. First is the possibility to tailor the potential through epitaxial growth and band-gap engineering techniques. This provides both understanding of and control over the Hamiltonian in many semiconductor structures. In molecular systems, by contrast, the Hamiltonian underlying the motion is usually complicated and in most cases unknown. Second is the application of compound semiconductor materials in advanced industries such as optical communication. Understanding and hence learning to control the motion of charge carriers is thus an interesting challenge which may also carry practical benefit.

19.5 Conclusions

The goal of this chapter was to complement earlier chapters in this book, which focused on energy-resolved spectroscopy and dynamics, with a discussion of the dynamics of time-resolved states, coherent superpositions of energy eigenstates called wavepackets. Subsequent to the introduction of several basic concepts in Section 19.2, we outlined, in Section 19.3, a number of the common methods for calculation of wavepacket dynamics. In Section 19.4 we employed these methods to illustrate some of the wavepacket phenomena introduced in Section 19.2 by means of specific examples.

Inevitably, only a small subset of the interesting questions in time-resolved research were explored, and, furthermore, a survey of the relevant literature was not provided. Several of the related areas of current intensive research which the reader may wish to explore in the recent literature are listed in the bibliographical notes below. These include wavepacket control, population transfer to a target state using a pulse sequence, the related problem of electronic population inversion with tailored short pulses, wavepacket interferometry and its application as a diagnostic tool, wavepacket molecular separation techniques, imaging of nonstationary wavefunctions and potentials using time-resolved data and short-pulse-induced reactions.

19.6 Bibliographical notes

19.6.1 WAVEPACKET CONTROL

Tannor and Rice (1985) proposed the possibility of controlling reactions through use of short-pulse excitation followed by short-pulse dump at a controlled instant. See the review article of

Rice (1997). The power of optimal control theory for the calculation of the temporal shape and spectral content of the pulse that would maximize the yield of a specified product was recognized by Rabitz and coworkers [Shi, Woody and Rabitz (1988)] and subsequently explored by several groups. See, for instance, the reviews of Rabitz (1992, 1997). Tracking control differs from optimal control in that the time-dependent tracking of an observable form of data is required, rather than physical results at a finite target time [Zhu, Smit and Rabitz (1999)].

Control of electronic branching ratios by variation of the timing between two transform-limited ultrashort pulses was illustrated by Herek, Materny and Zewail (1994). Experimental demonstration of branching ratio control by tailored pulses have been reported by Bardeen *et al.* (1997a) and by Assion *et al.* (1998). See the work of Bardeen *et al.* (1997b) for demonstration of quantum control of vibration localization.

19.6.2 POPULATION TRANSFER

The transfer of population from an initial state into a single final quantum state using a counter-intuitive sequence of pulses, named stimulated Raman adiabatic passage (STIRAP) was demonstrated in a large number of experiments by Bergmann and coworkers. See, for instance, the articles of Gaubatz *et al.* (1990) and Vitanov, Shore, and Bergmann (1998). A review was given by Bergmann and Shore (1995). Extensions of STIRAP were proposed by Unanyan *et al.* (1998), Kobrak and Rice (1998) and Davis and Warren (1999). Electronic population inversion using intense chirped pulses is discussed in the theoretical work of Cao, Bardeen and Wilson (1998). See also the experimental studies of Bardeen *et al.* (1997) and Bardeen *et al.* (1998).

19.6.3 RELATED TOPICS

Imaging of nonstationary wavefunctions and excited state potentials through the combination of time- and frequency-resolved data was discussed by Shapiro (1995, 1996b). Wavepacket interferometry using a sequence of femtosecond pulses is demonstrated in the works of Scherer *et al.* (1991) and Blanchet, Bouchène and Girard (1998). For an application of interferometry see the work of Leichtle *et al.* (1998). A wavepacket approach to isotope separation was described by Averbukh *et al.* (1996). Short-pulse-induced reactions are discussed by Backhaus and Schmidt (1997).

Acknowledgments

It is a pleasure to thank Dr. Albert Stolow for many stimulating conversations as well as for reading the manuscript and making helpful comments.

19.7 Appendix

SIMULTANEOUS MEASUREMENTS

A system can be assigned a definite value of an observable A only if it is in an eigenstate of the corresponding operator $\hat{A}$. For two observables A and B to be simultaneously assignable the corresponding operators $\hat{A}$ and $\hat{B}$ must have a common complete set,

$$\begin{aligned} \hat{A}\psi_{a_j,b_j} &= a_j\psi_{a_j,b_j}, \\ \hat{B}\psi_{a_j,b_j} &= b_j\psi_{a_j,b_j}. \end{aligned} \tag{78}$$

Thus, for all ψ in the set $(\hat{A}\hat{B} - \hat{B}\hat{A})\psi_{a_j,b_j} = 0$ and hence $\hat{A}\hat{B} = \hat{B}\hat{A}$.

Assume next that $\hat{A}$ and $\hat{B}$ are noncommuting and Hermitian and denote by $\overline{A}$ and $\overline{B}$ the expectation values of $\hat{A}$ and $\hat{B}$, respectively. The variances

$$(\Delta A)^2 = \int \psi^* \left(\hat{A} - \overline{A}\right)^2 \psi \, \mathrm{d}\tau$$
$$(\Delta B)^2 = \int \psi^* \left(\hat{B} - \overline{B}\right)^2 \psi \, \mathrm{d}\tau \tag{79}$$

obey $\Delta A \Delta B \geq \left|\frac{1}{2}\overline{C}\right|$ where $\overline{C}$ is the expectation value of a Hermitian operator $\hat{C}$ defined as $\hat{C} = (\hat{A}\hat{B} - \hat{B}\hat{A})/i$. This can be shown by noting that

$$(\Delta A)^2 = \int [(\hat{A} - \overline{A})\psi]^* (\hat{A} - \overline{A})\psi \, \mathrm{d}\tau = \int \left|(\hat{A} - \overline{A})\psi\right|^2 \mathrm{d}\tau, \tag{80}$$

and similarly for $(\Delta B)^2$. Using the Schwarz inequality,

$$\left(\int |f|^2 \, \mathrm{d}\tau\right) \left(\int |g|^2 \, \mathrm{d}\tau\right) \geq \left|\int f^* g \, \mathrm{d}\tau\right|^2,$$

with $f = (\hat{A} - \overline{A})\psi$ and $g = (\hat{B} - \overline{B})\psi$, we have

$$(\Delta A)^2 (\Delta B)^2 \geq \left|\int \psi^* (\hat{A} - \overline{A})(\hat{B} - \overline{B})\psi \, \mathrm{d}\tau\right|^2. \tag{81}$$

Consider next the operator $(\hat{A} - \overline{A})(\hat{B} - \overline{B})$ and express it in terms of its Hermitian and anti-Hermitian components as

$$\begin{aligned}(\hat{A} - \overline{A})(\hat{B} - \overline{B}) &= \frac{(\hat{A} - \overline{A})(\hat{B} - \overline{B}) + (\hat{B} - \overline{B})(\hat{A} - \overline{A})}{2} + \mathrm{i}\frac{(\hat{A} - \overline{A})(\hat{B} - \overline{B}) - (\hat{B} - \overline{B})(\hat{A} - \overline{A})}{2\mathrm{i}} \\ &= \hat{D} + \mathrm{i}\left(\frac{1}{2}\hat{C}\right).\end{aligned}$$

($\hat{D}$ and $\hat{C}$ are Hermitian since they are the sum and difference of an operator and its adjoint.) We have that

$$(\Delta A)^2 (\Delta B)^2 \geq \left|\int \psi^* \left(\hat{D} + \mathrm{i}\frac{1}{2}\hat{C}\right) \psi \, \mathrm{d}\tau\right|^2. \tag{82}$$

Since the expectation values of $\hat{C}$ and $\hat{D}$ are real,

$$(\Delta A)^2 (\Delta B)^2 \geq \overline{D}^2 + \tfrac{1}{4}\overline{C}^2 \geq \tfrac{1}{4}\overline{C}^2; \tag{83}$$

hence $\Delta A \Delta B \geq \frac{1}{2}\overline{C}$.

Of specific interest are the momentum and position operators, $\hat{A} = \hat{q}$, $\hat{B} = \hat{p}$ with which

$$(\hat{q}\hat{p} - \hat{p}\hat{q})\psi = \left(q\frac{\hbar}{\mathrm{i}}\frac{\partial}{\partial q} - \frac{\hbar}{\mathrm{i}}\frac{\partial}{\partial q}q\right)\psi = -\mathrm{i}\hbar\left[q\frac{\partial \psi}{\partial q} - \frac{\partial}{\partial q}(q\psi)\right] = \mathrm{i}\hbar\psi. \tag{84}$$

Thus, $\hat{q}\hat{p} - \hat{p}\hat{q} = \mathrm{i}\hbar$ giving the familiar result

$$\Delta q \Delta p \geq \tfrac{1}{2}\hbar. \tag{85}$$

19.8 References

Alimi, R., Garcia-Vela, A., and Gerber, R. B., 1992, *J. Chem. Phys.* **96**, 2034–2038.
Alimi, R., Gerber, R.B., Hammerich, A.D., Kosloff, R., and Ratner, M.A., 1990, *J. Chem. Phys.* **93**, 6484–6490.
Althorpe, S. C., and Seideman, T., 1999, *J. Chem. Phys.* **110**, 147–155
Assion, A., Baumert, T., Bergt, M., Brixner, T., Kiefer, B., Seyfried, V., Strehle, M., Gerber, G., 1998, *Science* **282**, 919–921.
Assion, A., Baumert, T., Helbing, J., Seyfried, V., Gerber, G., 1996, *Chem. Phys. Lett.* **259**, 488–494.
Averbukh, I. Sh., and Perel'man, N. F., 1989, *Sov. Phys. JETP* **69**, 464–469.
Averbukh, I. Sh., Vrakking, M. J. J., Villeneuve, D. M., and Stolow, A., 1996, *Phys. Rev. Lett.* **77**, 3518–3521.
Backhaus, P., and Schmidt, B., 1997 *Chem. Phys.* **217**, 131–144.
Bañares, L., Baumert, T., Bergt, M., Kiefe., B., and Gerber, G., 1998 *J. Chem. Phys.* **108**, 5799–5811.
Bandrauk, A. D., and Aubanel, E. E., 1995, *Chem. Phys.* **198**, 159.
Bardeen, C. J., Che, J., Wilson, K. R., Yakovlev, V. V., Apkarian, V. A., Martenes, C. C., Zadoyan, R., Koler, B., and Messina, M., 1997a, *J. Chem. Phys.* **106**, 8486–8503.
Bardeen, C. J., Che, J., Wilson, K. R., Yakovlev, V. V., Cong, P., Kohler, B., Krause, J. L., and Messina, M., 1997b, *J. Phys. Chem. A* **101**, 3815–3822.
Bardeen, C. J., Yakovlev, V. V., Squier, J. A., and Wilson, K. R., 1998, *J. Am. Chem. Soc.* **120**, 13023–13027.
Bardeen, C. J., Yakovlev, V. V., Wilson, K. R., Carpenter, S. D., Weber, P. M., and Warren, W. S., 1997, *Chem. Phys. Lett.* **280**, 11–158.
Batista, V. S., Zanni, M. T., Greenblatt, B. J., Neumark, D. M., and Miller, W. H., 1999 *J. Chem. Phys.* **110**, 3736–3747.
Baumart, T., Engel, V., Röttgermann, C., Strunz, W.T., and Gerber, G., 1992, *Chem. Phys. Lett.* **191**, 639–644.
Baumert, T., and Gerber, G., 1995, *Adv. At. Mol. Opt. Phys.* **35**, 163–208.
Baumert, T., Helbing, J., Gerber, G., 1997, *Adv. Chem. Phys.* **101**, 47–82.
Baumert, T., Thalweiser, V., Weiss, V., and Gerber, G., 1995, in *Femtosecond Chemistry*, Manz, J., and Wöste, L., Eds, VCH: Weinheim.
Beck, M. H., and Meyer, H.-D., 1997, *Z. Phys. D* **42**, 113–129.
Benjamin, I., and Wilson, K. R., 1989, *J. Chem. Phys.* **90**, 4176–4197.
Bergmann, K., and Shore, B. W., 1995, in *Molecular Dynamics and Spectroscopy by Stimulated Emision Pumping*, Dai, H.-L., and Field, R. W., Eds, Advanced Series in Physical Chemistry, Vol. 4; World Scientific: Singapore, pp. 315–373.
Billig, G. D., 1997, *J. Chem. Phys.* **107**, 4286–4294.
Bingermann, D., Gorman, M. P., King, A. M., and Crim, F., 1997, *J. Chem. Phys.* **107**, 661–664.
Bixon, M., and Jortner, J., 1969, *J. Chem. Phys.* **50**, 3284–3290.
Bixon, M., and Jortner, J., and Dothan, Y., 1969, *Mol. Phys.* **17**, 109–126.
Blanchet, V., Bouchène, M. A., and Girard, B., 1998, *J. Chem. Phys.* **108**, 4862–4876.
Blanchet, V., and Stolow, A., 1998, *J. Chem. Phys.* **108**, 4371–4374.
Blanchet, V., Zgierski, M., Seideman, T., and Stolow, A., 1999, *Nature* **401**, 52.
Blum, K., 1981, *Density Matrix Theory and Applications*; Plenum: New York.
Bowman, R. M., Dantus, M., and Zewail, A. H., 1989 *Chem. Phys. Lett.* **161**, 445–450.
Braun, M., Metiu, H., and Engel, V., 1998, *J. Chem. Phys.* **108**, 8983–8988.
Bunker, P. R., and Jensen, P., 1998, *Molecular Symmetry and Spectroscopy*; NRC Research Press: Ottawa.
Campolieti, G., and Brumer, P., 1994, *Phys. Rev. A* **50**, 997–1018.
Campolieti, G., and Brumer, P., 1997, *J. Chem. Phys.* **107**, 791–804.
Campolieti, G., and Brumer, P., 1998, *J. Chem. Phys.* **109**, 2999–3003.
Cao, J., Bardeen, C. J., Wilson, K. R., 1998, *Phys. Rev. Lett.* **80**, 1406–1409.
Charron, E., Giusti-Suzor, A., and Mies, F. H., 1995, *Phys. Rev. Lett.* **75**, 2815–2818.
Cho, V. A., and Bernstein, R. B., 1991, *J. Chem. Phys.* **93**, 8129–8136.
Corkum, P. B., Ivanov, M. Yu., and Wright, J. S., 1997, *Annu. Rev. Phys. Chem.* **48**, 387–406.
Courant, R., and Hilbert, D., (1989), *Methods of Mathematical Physics*; Wiley: New York.
Das, S., and Tannor, D. J., 1990, *J. Chem. Phys.* **92**, 3403–3409.
Davis, J. C., Warren, W. S., 1999, *J. Chem. Phys.* **110**, 4229–4237.
Dirac, P. A. M., 1930, *Proc. Cambridge Philos. Soc.* **26**, 376–385.
Edmonds, A. R., 1960, *Angular Momentum in Quantum Mechanics*; 2nd edition; Princeton University Press: Princeton, NJ.
Ehrenfest, P., 1927, *Z. Phys.* **45**, 455–457.
Eland, J. H. D., 1984, *Photoelectron Spectroscopy*; Butterworths: London.
Fang, J.-Y., and Guo, H., 1995, *J. Chem. Phys.* **102**, 2404–2412.
Feit, M. D., and Fleck, J. A. Jr., 1983, *J. Chem. Phys.* **78**, 301–308.
Feit, M. D., Fleck, J. A. Jr., and Steiger, A., 1982, *J. Comput. Phys.* **47**, 412–433.
Feldmann, J., Leo, K., Shah, J., Miller, D. A. B., Cunningham, J. E., Meier, T., von Plessen, G., Schulze, A., Thomas, P., and Schmitt-Rink, S., 1992, *Phys. Rev. A* **46**, 7252–7255.

Felker, P. M., 1992, *J. Chem. Phys.* **96**, 7844–7857.
Fischer, I., Vrakking, M. J. J., Villeneuve, D. M., and Stolow, A., 1996, *Chem. Phys.* **207**, 331.
Fried, L. E., and Mukamel, S., 1990, *J. Chem. Phys.* **93**, 3063–3071.
Friedrich, B., and Herschbach, D. R., 1995, *Phys. Rev. Lett.* **74**, 4623–4626.
Friedrich, B., Pullmann, D.P., and Herschbach, D. R., 1991, *J. Chem. Phys.* **95**, 8118–8129.
Garcia-Vela, A., and Gerber, R. B., 1993, *J. Chem. Phys.* **98**, 427–436.
Gaubatz, U., Rudecki, P., Schiemann, S., and Bergmann, K., 1990, *J. Chem. Phys.* **92**, 5363–5376.
Gerber, R. B., Kosloff, R., and Berman, M., 1986, *Comput. Phys. Rep.* **5**, 59.
Gerber, R. B., and Ratner, M. A., 1988, *J. Chem. Phys.* **92**, 3252–3260.
Gordon, R. J., Rice, S. A., 1997, *Annu. Rev. Phys. Chem.* **48**, 601–643.
Görlitz, A., Weidemüller, M., Hänsch, T. W., and Hemmerich, A., 1997, *Phys. Rev. Lett.* **78**, 2096–2099.
Gradshteyn, I. S., and Ryzhik, I. M., 1980, *Table of Integrals, Series, and Products*; Academic Press: Orlando, FL.
Guo, G., Saalfrank, P., and Seideman, T., 1999, *Prog. Surf. Sci.* **62**, 239.
Heather, R., and Metiu, H., 1985, *Chem. Phys. Lett.* **118**, 558–563.
Heisenberg, W., 1927, *Z. Phys.* **41**, 239–267.
Heller, E. J., 1975, *J. Chem. Phys.* **62**, 1544–1555.
Heller, E. J., 1976, **64**, 63–73.
Heller, E. J., 1981a, *Acc. Chem. Res.* **14**, 368–375.
Heller, E. J., 1981b, *J. Chem. Phys.* **75**, 2923–2932.
Herek, J. L., Materny, A., Zewail, A. H., 1994 *Chem. Phys. Lett.* **228**, 15–26.
Herman, M. F., and Kluk, E., 1984, *Chem. Phys.* **91**, 27–34.
Jortner, J., and Berry, R. S., 1968, *J. Chem. Phys.* **48**, 2757–2766.
Karczmarek, J., Wright, J., Corkum, P. B., and Ivanov, M. Yu., 1999, *Phys. Rev. Lett.* submitted.
Kay, K. G., 1994, *J. Chem. Phys.* **100**, 4377–4392.
Kay, K. G., 1997, *J. Chem. Phys.* **107**, 2313–2328.
Keshavamurthy, S., and Miller, W. H., 1994, *Chem. Phys. Lett.* **218**, 189–195.
Kobrak, M. N., and Rice, S. A., 1998, *Phys. Rev. A* **57**, 2885–2894.
Kohler, B., Krause, J. L., Raksi, F., Wilson, K. R., Yakovlev, V. V., Whitnell, R. M., Yan, Y., 1995, *Acc. Chem. Res.* **28**, 133–141.
Kolár, M., Ali, M. K., and O'Shea, S. F., 1989, *J. Chem. Phys.* **90**, 1036–1042.
Kosloff, D., and Kosloff, R., 1983, *J. Comp. Phys.* **52**, 35–53; *J. Chem. Phys.* **79**, 1823–1833.
Kosloff, R., 1996, Quantum molecular dynamics on grids, dynamics of molecules and chemical reactions, in *Dynamics of Molecules and Chemical Reactions*, Wyatt, R. E., and Zhang, J. Z., Eds; Marcel Dekker: New York, pp. 185–230.
Kotler, Z., Neria, E., and Nitzan, A., 1991, *Comput. Phys. Commun.* **63**, 43–258.
Krause, J. L., Schaefer, K. J., Ben-Nun, M., and Wilson, K. R., 1997, *Phys. Rev. Lett.* **79**, 4978–4981.
Krause, J. L., Shapiro, M., and Bersohn, R., 1991, *J. Chem. Phys.* **94**, 5499–5507.
Kubo, R., 1952, *Phys. Rev.* **86**, 929–937.
Kulander, K. C., and Heller, E. J., 1978, *J. Chem. Phys.* **69**, 2439–2449.
Leichtle, C., Averbukh, I. Sh., and Schleich, W. P., 1996, *Phys. Rev. Lett.* **77**, 3999–4002; *Phys. Rev. A* **54**, 5299–5312.
Leichtle, C., Schleich, W. P., Averbukh, I. Sh., Shapiro, M, 1998, *J. Chem. Phys.* **108**, 6057–6068.
Leo, K., Shah, J., Göbel, O., Damen, T. C., Schmitt-Rink, S., Schäfer, W., and Köhler, K., 1991, *Phys. Rev. Lett.* **66**, 201–204.
Levine, R. D., 1985, in *Theory of Chemical Reaction Dynamics*, Baer, M., Ed.; CRC: West Palm Beach, pp. 1–64.
Madhusoodanan, M., and Kay, K. G., 1998, *J. Chem. Phys.* **109**, 2644–2655.
Makri, N., and Miller, W. H., 1987, *J. Chem. Phys.* **87**, 5781–5787.
Mallalieu, M., and Stroud, Jr. C. R. , 1995, *Phys. Rev. A* **51**, 1827–1835.
Manthe, U., Meyer, D.-H., and Cederbaum, L. S., 1992, *J. Chem. Phys.* **97**, 3199–3213.
Manz, J., 1997, in *Femtochemistry and Femtobiology*, Sundström, V., Ed.; World Scientific: Singapore.
Manz, J., and Wöste, L., 1995, *Femtosecond Chemistry*; VCH: Weinheim.
McCullough Jr., E. A., and Wyatt, R. E., 1969, *J. Chem. Phys.* **51**, 1253–1255.
McMorrow, D., Lotshaw, W. T., and Kenney-Wallace, G. A., 1988, *IEEE J. Quat. Electron.* **24**, 443–454.
Messiah, A., 1958, *Quantum Mechanics*; Wiley: New York.
Miller, W. H., 1970, *J. Chem. Phys.* **53**, 1949–1959; 3578–3587.
Miller, W. H., 1974, *Adv. Chem. Phys.* **25**, 69–177 (see in particular pp. 164 and 171).
Mukamel, S., 1990, *Annu. Rev. Phys. Chem.* **41**, 647–681.
Nitzan, A., Jortner, J., and Retzepis, P. M., 1972, *Proc. R. Soc. London, Ser. A* **327**, 367–391.
Parker, J., and Stroud Jr. C. R., 1986, *Phys. Rev. Lett.* **56**, 716–719.
Peskin, U., and Moiseyev, N., 1993, *J. Chem. Phys.* **99**, 4590–5496.
Pollard, W. T., and Mathies, R. A., 1992, *Annu. Rev. Phys. Chem.* **43**, 497–523.
Provost, D., and Brumer, P., 1995, *Phys. Rev. Lett.* **74**, 250–253.
Raab, A., Worth, G. A., Meyer, H.-D., and Cederbaum, L. S., 1999 *J. Chem. Phys.* **110**, 936–946.

Rabitz, H., 1992, in *Coherence Phenomena in Atoms and Molecules in Laser Fields*, Bandrauk, A.D., and Wallace, S.C., Eds, Plenum: New York.
Rabitz, H., 1997, *Adv. Chem. Phys.* **101**, 315–326.
Raithel, G., Phillips, W. D., and Rolston, S. L., 1998, *Phys. Rev. Lett.* **81**, 3615–3618
Reid, K, 1993, *Chem. Phys. Lett.* **215**, 25.
Reid, K., Duxon, S. P., and Towrie, M., 1994, *Chem. Phys. Lett.* **228**, 351.
Rempe, G., Walther, H., and Klein, N., 1987, *Phys. Rev. Lett.* **58**, 353–356.
Resch, K., Blanchet, V., Stolow, A., and Seideman, T., 1999, *J. Chem. Phys.*, submitted.
Rice, S. A., 1997, *Adv. Chem. Phys.* **101**, 213–283.
Rudy, P., Ejnisman, R., and Bigelow, N. P., 1997, *Phys. Rev. Lett.* **78**, 4906–4909.
Scherer, N. F., Carlson, R. J., Matro, A., Du, M., Ruggiero, A. J., Romero-Rochin, V., Cina, J. A., Fleming, G. R., and Rice, S. A., 1991, *J. Chem. Phys.* **95**, 1487–1511.
Schön, J., and Köpel, H., 1995, *J. Chem. Phys.* **103**, 9292–9303.
Schrödinger, E., 1926, *Naturwiss.* **14**, 664–666.
Schumacher, D. W., Lyons, B. J., and Gallagher, T. F., 1997, *Phys. Rev. Lett.* **78**, 4359–4362.
Seideman, T., 1995, *J. Chem. Phys.* **103**, 7887–7896.
Seideman, T., 1997a, *J. Chem. Phys.* **106**, 417–431.
Seideman, T., 1997b, *J. Chem. Phys.* **107**, 7859–7868.
Seideman, T., 1997c, *Phys. Rev. A* **56**, R17–R20.
Seideman, T., 1997d, *J. Chem. Phys.* **106**, 2881–2892; *J. Chem. Phys.* **107**, 10420–10429.
Seideman, T., 1999, *Phys. Rev. Lett.* **82**, 4971.
Seideman, T., and Guo, H., 1997, *J. Chem. Phys.* **107**, 8627–8636.
Shapiro, M., 1993, *J. Chem. Phys.* **97**, 7396–7411.
Shapiro, M., 1995, *J. Chem. Phys.* **103**, 1748–1754.
Shapiro, M., 1996a, *Phys. Rev. A* **54**, 1504–1509.
Shapiro, M., 1996b, *J. Chem. Phys.* **100**, 7859–7866; *Chem Phys.* **207**, 317–329.
Shi, S., Woody, F. I., and Rabitz, H., 1988, *J. Chem. Phys.* **88**, 6870–6883.
Spath, B. W., and Miller, W. H., 1996, *Chem. Phys. Lett.* **262**, 486–494.
Stock, G., and Domcke, W., 1997, *Adv. Chem. Phys.* **100**, 1–170.
Strehle, M., Weichmann, U., and Gerber, G., 1998, *Phys. Rev. A* **58**, 450–455.
Sun, X., and Miller, W. H., 1997, *J. Chem. Phys.* **106**, 916–927.
Sun, X., and Miller, W. H., 1998, *J. Chem. Phys.* **108**, 8870.
Sun, X., Wang, H., and Miller, W. H., 1998a, *J. Chem. Phys.* **109**, 4190–4200.
Sun, X., Wang, H., and Miller, W. H., 1998b, *J. Chem. Phys.* **109**, 7064–7074; *Adv. Chem. Phys.* **100**, 1–169.
Tal-Ezer, H., and Kosloff, R., 1984, *J. Chem. Phys.* **81**, 3967–3971.
Tannor, D. J., and Rice, A., 1985, *J. Chem. Phys.* **83**, 5013.
Unanyan, R., Fleischhauser, M., Shore, B. W., and Bergmann, K., 1998, *Optics. Commun.* **155**, 144–154.
Vitanov, N. V., Shore, B. W., Bergmann, K., 1998, *Eur. Phys. J.D* **4**, 15–30.
Vrakking, M. J. J., and Stolte, S., 1997, *Chem. Phys. Lett.* **271**, 209–216.
Vrakking, M. J. J., Villeneuve, D. M., and Stolow, A., 1996, *Phys. Rev. A* **54**, 1–4.
Vrijen, R. B., Lankhuijzen, G. M., and Noordam, L. D., 1997, *Phys. Rev. Lett.* **79**, 617–620.
Walton, A. R., and Manolopoulos, D. E., 1995, *Chem. Phys. Lett.* **244**, 448–455.
Weinacht, T. C., Ahn, J., and Bucksbaum, P. H., 1998, *Phys. Rev. Lett.* **80**, 5508–5511.
Weiner, A. M., 1994, *J. Opt. Soc. Am. B* **11**, 2480–2491.
Whitnell, R. M., and Wilson, K. R., 1993, in *Reviews in Computational Chemistry*, Lipkowitz, K. B., and Boyd, D. B., Eds, Vol. IV, VCH: New York, pp. 67–148.
Yan, Z.-C., and Seideman, T., 1999, *J. Chem. Phys.* **111**, 4113.
Yeazell, J. A., and Stroud Jr. C. R., 1991, *Phys. Rev. A* **43**, 5153–5156.
Zewail, A. H., 1994, *Femtochemistry: Ultrafast Dynamics of the Chemical Bond*; World Scientific: Singapore.
Zewail, A. H., 1997, *Adv. Chem. Phys.* **101**, 3–46.
Zhong, D., and Zewail, A. H., 1998, *J. Chem. Phys. A* **102**, 4031–4058, and references cited therein.
Zhu, W., Smit, M., and Rabitz, H., 1999, *J. Chem. Phys.* **110**, 1905–1915.

20 *AB INITIO* MOLECULAR DYNAMICS

John S. Tse and Roger Rousseau
National Research Council of Canada, Ottawa, Canada

20.1 Introduction

In the last two decades, we have witnessed rapid growth in the capability of theoretical chemistry both in the development of new methodologies and in the seemingly unending advances in computer technologies. The availability of highly sophisticated quantum mechanical methods allows the accurate prediction of structures, energies and almost all physical and spectroscopic properties of molecules. One of the limitations of these conventional quantum chemistry methods is in the treatment of molecular dynamics. In the traditional approach, the dynamics can be treated by the calculation of the potential energy surface, which is used in classical trajectory calculations. The simultaneous calculation of the electronic potential energy surface and nuclear

Computational Molecular Spectroscopy. Edited by Per Jensen and Philip R. Bunker

dynamics is particularly important in the study of fluxional molecules where there are no well-defined static structures. In principle, if the Born–Oppenheimer approximation on the separation of the electronic and nuclei motions is valid and the motions of the atoms in a molecule can be described by classical Newtonian mechanics, the nuclear dynamics problem reduces to the solution of a set of equation-of-motion with the forces computed directly from the electronic wave function. This approach, although theoretically sound is practically inefficient. In 1985 Car and Parrinello proposed a revolutionary approach to the solution of the electronic problem of very large solid state systems [Car and Parrinello (1985)]. This approach, based on treating the parameters in the electronic wave function as dynamical variables, can be easily coupled with the classical Newtonian mechanics of the nuclei. This then allows both the electrons and nuclei to be treated dynamically in a unified manner. Thus, the nuclei can be given a kinetic energy which allows them to propagate through configuration space whilst the electrons are kept in their ground state without the need to optimize the electronic wave function at each time step of the molecular dynamics simulation. The Car–Parrinello (CP) or *ab initio* molecular dynamics method (AIMD) has made an important impact in the study of clusters and fluxional molecules and the dynamical study of chemical reaction mechanisms particularly in the condensed phase. In this article, the fundamentals of the CP method will be reviewed. This is followed by some technical details including a discussion of the pseudopotential plane wave method. The application of the CP method is illustrated through a series of examples. Finally, the extension of the CP method to quantum systems is presented.

20.2 Basic principles

20.2.1 CAR–PARRINELLO MOLECULAR DYNAMICS

Within the basis-set representation (localized or plane wave) of the molecular electronic wave function, the self-consistent solution to the Schrödinger equation in the independent electron approximation reduces to a pseudo-eigenvalue problem which can be solved by iterative diagonalization of the secular determinant. Since the size of the secular determinant increases as N^2 with the size, N, of the basis set eventually the diagonalization step will become a formidable numerical problem for very large systems such as solids. Car and Parrinello proposed an ingenious method to alleviate this obstacle [Car and Parrinello (1985)]. Central to this idea is the analogy between the spirit of the variational principle and minimization. It is recognized that the minimization of the total energy of a system by varying the electronic wave function can be achieved by a classical molecular dynamics (MD) like procedure if the coefficients of the basis set are treated as dynamical variables [Parrinello (1990), Remler and Madden (1990), Payne *et al.* (1992), and Wathelet, Meloni and André (1995)]. In classical mechanics, the dynamics of a physical system is given by the Lagrangian (L),

$$L = T - V, \tag{1}$$

where V is the potential energy and T is the kinetic energy. The equation of motion is given by

$$\frac{\mathrm{d}}{\mathrm{d}t}\left(\frac{\partial L}{\partial \dot{q}}\right) - \left(\frac{\partial L}{\partial q}\right) = 0, \tag{2}$$

where q is a generalized coordinate. In the CP formalism, if the wave function of the system (Ψ) is expanded in terms of the basis set (φ),

$$\Psi = \sum_i c_i \varphi_i. \tag{3}$$

Instead of the usual variation procedure, the energy minimization can be achieved using a molecular dynamics procedure. In this approach, the coefficient c_i are treated as dynamical variables and are varied until the total energy of the system is at a minimum. One can create a 'classical' Lagrangian for this system,

$$L = \mu \sum_i \left\langle \dot{\varphi}_i \middle| \dot{\varphi}_i \right\rangle - E(\Psi, R) + \sum_{ij} \Lambda_{ij} \left(\int \varphi_i^* \varphi_j \, \mathrm{d}^3 r - \delta_{ij} \right). \tag{4}$$

The first term corresponds to the fictitious kinetic energy of the wave function where μ is the 'fictitious' mass, which has no physical meaning or relationship to the real electron mass. The mass is fictitious and thus any value may be chosen and it in principle will provide identical dynamics. However, in practice the choice of mass does have an effect on the size of time step employed in a molecular dynamics simulations. This point will be addressed below when nuclear dynamics is also included within this scheme. The second term is simply the total energy of the system. The last term ensures the orthogonality between the electron orbitals $\langle \varphi_i | \varphi_j \rangle = \delta_{ij}$ and Λ_{ij} is the element of a matrix of Lagrangian multipliers. The total energy, $E(\Psi, R) = \langle \Psi | H | \Psi \rangle$ can be calculated from familiar methods such as the Hartree–Fock (HF) based methods [Szabo and Ostlund (1996)] or within the local density functional (LDA) formalism [Hohenberg and Kohn (1964), Kohn and Sham (1965)]. A molecular dynamics approach is used to optimize the coefficients $\{c_i\}$ where the equation of motion of each electron wave function can be obtained from L,

$$\frac{\mathrm{d}}{\mathrm{d}t}\left(\frac{\partial L}{\partial \dot{\varphi}_i{}^*}\right) = \frac{\partial L}{\partial \varphi_i{}^*},$$

which gives

$$\mu \ddot{\varphi}_i = -H\varphi_i + \sum_j \Lambda_{ij} \varphi_j. \tag{5}$$

This expression is analogous to the classical equation of motion subjected to holonomic constraints if $-H\varphi_i$ can be associated as the 'fictitious' electronic force. The equations of motions can be integrated readily by the well-known classical molecular dynamics Verlet scheme or its variants [Allen and Tildesley (1990)].

In principle, the extension of the CP electronic method to include the dynamics of the nuclei is straightforward. The nuclear dynamics obeys the Newtonian mechanics,

$$F_A = M_A \frac{\mathrm{d}^2 R_A}{\mathrm{d}t^2}, \tag{6}$$

where F_A is the force acting on nucleus A with mass M_A. As long as the stationary electronic wave function is completely the force can be computed from the derivative of the energy with respect to the nuclear coordinate via the application of the Hellmann–Feynman theorem.

$$F_A = \frac{-\partial E}{\partial R_A} = \left\langle \Psi(r) \middle| \frac{\partial H}{\partial R_A} \middle| \Psi(r) \right\rangle. \tag{7}$$

Therefore, the Lagrangian for the coupled electron–ion molecular dynamics becomes

$$L = \mu \sum_i \left\langle \dot{\varphi}_i \middle| \dot{\varphi}_i \right\rangle + \frac{1}{2} \sum_A M_A \dot{R}_A^2 - E(\Psi, R) + \sum_{ij} \Lambda_{ij} \left(\int \varphi_i^* \varphi_j \, \mathrm{d}^3 r - \delta_{ij} \right), \tag{8}$$

where the additional term simply describes the kinetic energy of the nuclei. With this classical Lagrangian, the fictitious equation of motion for the coefficients of the electronic wave function and the conventional classical nuclear dynamics are coupled together. With this Lagrangian it is possible to derive equations of motion which move the electrons and the nuclei at the same time. The CP method thus provides a very efficient scheme for the simultaneous calculation of the nuclear trajectory and updates of the electronic wave function corresponding to the new nuclear configurations. This innovative approach permits the *ab initio* calculation of the molecular dynamics of a system.

To initiate a CP molecular dynamics calculation, a starting optimized electronic wave function is needed. The electronic optimization can be performed with a simulated-annealing technique. Thus, the simplest method is to solve the following first-order differential equation of the electronic wave function in the fictitious time variable.

$$\varphi_i(r,t) = \frac{-\delta E}{\delta\varphi_i^*(r,t)} + \sum_j \Lambda_{ij}\varphi_j(r,t). \tag{9}$$

In this steepest descent method, the electronic degrees of freedom are driven to a minimum by moving along the gradient defined by

$$\frac{\delta E}{\delta\varphi_i^*(r,t)} = H\varphi_i(r,t). \tag{10}$$

The orthogonality of the electron orbitals can be guaranteed by the simple Gramm–Schmidt orthogonalization procedure. The electrons can then move with the nuclei and remain on the Born–Oppenheimer surface. This then introduces a constraint on the size of the time step which is employed in a molecular dynamics simulation. If the time step is too large the electrons cannot keep up with the nuclei and they are no longer in the electronic ground state. What is required is to tune the system with a proper choice of the fictitious mass which will allow the electrons to follow the nuclei. In practice this restricts the choice of $\mu = 300$–1000 a.u and the molecular dynamics time step to about 5–10 a.u. (i.e. 0.1–0.2 fs). Unfortunately, in some cases, the local steepest gradient does not necessarily provide the best direction for the minimum in the potential energy surface and may results in very slow convergence. Many other methods such as conjugate gradient, preconditioning, pseudo-second-order equation, direct inversion of iterative space and analytic-integration have been proposed and implemented to alleviate the convergence problem.

20.2.2 THE PSEUDOPOTENTIAL PLANE WAVE METHOD

The original formulation of the CP method was intended for the solution of the electronic problem for large extended systems. For these complex systems, the local density functional (LDA) approximation is the method of choice. In practice, the one-electron Kohn–Sham equations can be efficiently solved if the electronic wave function is expanded in a plane wave basis set. Although this approach is devised to treat periodic systems, it can also be used for molecular system by the use of a sufficiently large simulation box such that the interactions between the molecule and its periodic images are minimal. Plane waves behave like periodic sine and cosine functions. One serious drawback for using these functions is that a large number of plane waves are needed to describe accurately a wave function with steep curvature such as the core orbital and tightly bounded valence orbital of the first row main group and transition metal atoms. The curvature of an electron orbital is governed by the strength of the corresponding (attractive) atomic potential. Since only the valence electrons are responsible for the chemical bonding, a way to reduce the size of the numerical problem without adversely affecting the accuracy is to approximate the core potential by a smooth pseudopotential. A pseudopotential is constructed to replace the atomic all-electron potential such that core states are eliminated and the valence electrons are described by nodeless pseudo wave functions [Sutton (1993)]. The principle of the pseudopotential approximation is as follows. The pseudo-valence (plane wave) orbital $|k\rangle$ can be constructed to be orthogonalized to the core by projecting out the core states $|\sigma\rangle$,

$$|\phi_k\rangle = (1 - \hat{P})|k\rangle, \tag{11}$$

where the projection operator is

$$\hat{P} = \sum_{\sigma} |\sigma\rangle \langle\sigma|k\rangle , \tag{12}$$

since the total energy of the system is given by

$$H|k\rangle = (T + V)|k\rangle = E_k|k\rangle. \tag{13}$$

The solution to the pseudo-Hamiltonian corresponding to the pseudo-valence orbital is

$$H(1 - \hat{P})|\phi_k\rangle = E_k(1 - \hat{P})|\phi_k\rangle,$$

$$(H - H\hat{P} + E_k\hat{P})|\phi_k\rangle = E_k|\phi_k\rangle.$$

We define a pseudo-Hamiltonian H_{ps}

$$H_{\text{ps}} = H - H\hat{P} + E_k\hat{P} = H + (E_k - H)\hat{P} = T + V + (E_k - H)\hat{P} = T + V_{\text{ps}}, \tag{14}$$

with the pseudopotential V_{ps} given by

$$V_{\text{ps}} = V + (E_k - H)\hat{P}. \tag{15}$$

Therefore, the Schrödinger equation for the all-electron system reduces to a valence electron only equation.

$$H_{\text{ps}}|\phi_k\rangle = E_k|\phi_k\rangle. \tag{16}$$

The fundamental assumption of a pseudopotential is that the strength of the negative and attractive atomic potential V can be reduced by a large and positive term $(E_k - H)P$ so that the new eigenvalue problem is almost a free-electron equation and, thus, the plane wave basis set is the most appropriate solution.

In general, a pseudopotential is constructed subject to the following criteria:

(1) The all electron (ε_l) and pseudo-valence eigenvalues ($\varepsilon_{l,\text{ps}}$) must agree for a chosen atomic configuration

$$\varepsilon_l = \varepsilon_{l,\text{ps}}. \tag{17a}$$

(2) The nodeless pseudo wave function must reproduce the all-electron valence wave function,

$$\phi_i(r) = \phi_{i,\text{ps}}(r), \tag{17b}$$

where the radius $r \geq r_c$ and r_c defines the cutoff radius of each valence state.

(3) The integrated charge density for $r \geq r_c$ must be equivalent for the all-electron and pseudo-valence wave function.

$$\langle\phi_i(r)|\phi_i(r)\rangle_{r\geq r_c} = \langle\phi_{i,\text{ps}}(r)|\phi_{i,\text{ps}}(r)\rangle_{r\geq r_c}. \tag{17c}$$

(4) The logarithmic derivatives of the all-electron and pseudo wave function and their first energy derivatives agree for $r \geq r_c$

$$-\frac{1}{2}\left[(r\phi)^2 \frac{\mathrm{d}}{\mathrm{d}\varepsilon}\frac{\mathrm{d}}{\mathrm{d}r}\ln\phi\right]_{\mathrm{r}} = \int_0^r r^2|\phi|^2\,\mathrm{d}r. \tag{17d}$$

The atomic pseudopotential can be constructed as follows [Singh (1993), Fuchs and Scheffler (1999)]:

(1) Select a reference electron configuration for the atom and solve the Schrödinger equation to obtain the all-electron atomic potential, valence wave functions, eigenvalues and charge density.

(2) For each wave function with different azimuthal quantum number l, a core cutoff radius r_{cl} is selected. The r_{cl} must be chosen beyond the outermost node of the all-electron wave function.

(3) An approximate potential $V_1(r)$ is constructed by removing the singularity in the all electron atomic potential $V_{\mathrm{AE},l}(r)$,

$$V_l(r) = V_{\mathrm{AE},l}(r)\left[1 - f_1\left(\frac{r}{r_{cl}}\right)\right] + C_l f_2\left(\frac{r}{r_{cl}}\right), \tag{18}$$

where $f(x)$ is a monotonic function with $f(0) = 1$ and $f(\infty) = 0$ and it decays rapidly at $x = 1$. The wave function $\phi_l(r)$ corresponding to $V_l(r)$ is obtained by solving the atomic Schrödinger equation. The coefficient C_l is adjusted by such that the eigenvalue is the same as the all-electron value.

(4) Norm conservation condition is imposed to this wave function by adding a correction in the core region.

$$\phi^l_{\mathrm{ps}}(r) = \gamma_l[\phi_l(r) + \delta_l g_l(r)], \tag{19}$$

where γ_l is the ratio $\phi^l_{\mathrm{ps}}(r)/\phi_l(r)$ at r_{cl}, δ_l is the parameter chosen to ensure norm conservation and $g_l(r)$ is a regular function that vanishes rapidly for $r > r_{cl}$.

(5) The 'unscreened' pseudopotential is obtained by numerical inversion of the Schrödinger equation.

(6) The pseudopotential can then be extracted from the 'unscreened' potential by removing the contributions from valence Hartree (Coulomb) and exchange correlation potential.

The construction of highly accurate and efficient pseudopotential depends on the choice of the form for $f(r)$ and $g(r)$. For example, in the well-known Bachelet–Haman–Schluster (BHS) pseudopotential [Bachelet, Haman and Schluster (1982)], the functions are set to $f(r) = g_l(r) = \exp(-x^4)$ and r_{cl} is choosen to be between the outermost node and the outermost maximum in the wave function. Pseudopotentials constructed by other authors may adopt different functional forms. In the popular Troullier–Martins scheme [Troullier and Martins (1991)], additional constraints requiring the curvature of the unscreened pseudopotential to vanish at the origin and the first four derivatives of the pseudo and all-electron wave functions to agree at the core cutoff radius.

In passing, it should be noted that, apart from the simplicity in the form of the wave function, a plane wave basis set has a common Cartesian origin and is not atom centered. This unique feature significantly reduces the complexity in the solution of the Schrödinger equation. It can be shown that the total energy of a system can be written in analytical expressions where the components can be evaluated efficiency via fast Fourier transform (FFT) methods on a computer [Wathelet, Meloni and André (1995)].

20.3 Structure and dynamics of fluxional molecules

20.3.1 MOLECULAR DYNAMICS AND CALCULATION OF SPECTRAL PROPERTIES

One of the distinct advantages of CPMD over the traditional *ab initio* quantum chemistry approach of analyzing static molecular configurations is the ability to study the nuclear dynamics. The possibility of calculating the atom trajectory on-the-fly has been very useful in the determination of the equilibrium structure of a complex system with multiple minima using the method of simulated annealing. In this approach, a system is initially heated at high temperature. Periodically, the system is allowed to quench to remove the kinetic energy and settles to a local minimum. This process can then be repeated several times in the hope of locating the global minimum conformation. This method in combination of CPMD has been applied extensively to the study of the structures of metal clusters [Andreoni (1987), and Röthlisberger and Andreoni (1991)].

MD calculations are also very useful in extracting equilibrium thermodynamics, static and dynamical information from a system. One of the most important and useful quantities that can be evaluated from a MD simulation is the time correlation function C_{AB} that correlates two time dependent quantities A and B [Allen and Tildesley (1990)],

$$C_{AB}(t) = \langle A(t)B(t + t_0)\rangle. \tag{20}$$

This function is important because it provides detailed information on the dynamics of a system. Furthermore, the Fourier transform of a time correlation function is often related to relevant experimental observables. For example, the Fourier transform of the nuclear velocity–time autocorrelation function $v(t)$ is related to the spectral vibrational density of states $F(\omega)$.

$$F(\omega) = \int_0^\infty \mathrm{e}^{\mathrm{i}\omega t}\langle v(0)v(t)\rangle \mathrm{d}t. \tag{21}$$

The time correlation formalism was recently extended for the calculation of the infrared absorption coefficient $\alpha(\omega)$ for an extended system. The absolute absorption coefficient of an isotropic system can be obtained from [Silvestrelli, Bernasconi and Parrinello (1997), and Debernardi *et al.* (1998)]

$$\alpha(\omega) = \frac{4\pi\omega}{3nc}\mathrm{Im}\sum_\gamma \mathcal{X}_{\gamma\gamma}, \tag{22}$$

where n is the index of refraction and c is the speed of light. The imaginary part of the electronic susceptibility $\mathcal{X}_{\gamma\gamma}$ is given by

$$\mathrm{Im}\,\mathcal{X}_{\gamma\gamma} = \frac{V}{2\hbar}\left(1 - \mathrm{e}^{-\hbar\omega/k_\mathrm{B}T}\right)\int_0^\infty \left\langle \hat{P}_\gamma(0)\hat{P}_\gamma(t)\right\rangle \mathrm{e}^{\mathrm{i}\omega t}\,\mathrm{d}t, \tag{23}$$

where $P_\gamma(t)$ is the polarization vector along the γ direction. The total polarization can be evaluated from a CPMD simulation using the discrete Berry phase at the single-point limit [Resta (1994), and Yaschenko *et al.* (1998)]. Berry phase is an observable that cannot be cast as the expectation value of any operator being instead a gauge-invariant phase of the wave function. For a one-dimensional periodic system the discrete phase difference $\Delta\varphi_s$ between two Bloch wave functions at neighboring point k_s and k_{s+1} differs by an amount $2\pi/a$, where a is the length of the repeating unit. The phase shift is defined as,

$$\mathrm{e}^{-\mathrm{i}\Delta\varphi_s} = \frac{\left\langle \Psi(k_s)\middle|\mathrm{e}^{-\mathrm{i}(2\pi/a)\hat{X}}\middle|\Psi(k_{s+1})\right\rangle}{\left|\left\langle \Psi(k_s)\middle|\mathrm{e}^{-\mathrm{i}(2\pi/a)\hat{X}}\middle|\Psi(k_{s+1})\right\rangle\right|},$$

$$\Delta\varphi_s = -\mathrm{Im}\left(\ln\left\langle \Psi(k_s)\middle|\mathrm{e}^{-\mathrm{i}(2\pi/a)X}\middle|\Psi(k_{s+1})\right\rangle\right), \tag{24}$$

where X is the position operator. In the framework of the modern theory of macroscopic polarization, it can be shown that the electronic polarization ΔP_{el} is the time integral of the macroscopic current $J_{\mathrm{el}}(t)$, which is proportional to the time derivative of the Berry phase [Yaschenko *et al.* (1998)],

$$\Delta P_{\text{el}} = \frac{e}{2\pi}[\Delta\phi(\delta t) - \Delta\phi(0)], \tag{25a}$$

$$J_{\text{el}}(t) = \frac{e}{2\pi}\frac{\mathrm{d}}{\mathrm{d}t}[\Delta\phi(t)]. \tag{25b}$$

In a CPMD simulation, the single-point Berry phase can be calculated at each time step. Approximating the time derivative with a finite difference, the electronic current (generalized to three-dimensions) becomes

$$J_{\text{el}} \approx \frac{e}{(2\pi)^3 a^2}\frac{\Delta\Phi(t+\delta t) - \Delta\Phi(t)}{\delta t}, \tag{26}$$

where δt is the CP timestep. Therefore, the macroscopic polarization can be computed at each MD time step and the infrared absorption coefficient obtained from the Fourier transform of the polarization time autocorrelation function.

This approach is distinctly advantageous for large systems which have a relatively high number of thermally populated potential energy minima. For these types of molecule, the traditional *ab initio* quantum chemistry approach of analyzing static configurations is severely hindered by the large number of minima and more importantly finding them all. As opposed to methods based upon solving the nuclear dynamics problem via wavepacket dynamics or methods based upon the time-independent Schrödinger equation the spectroscopic parameters obtained from CP simulations are far less accurate. However, these more traditional spectroscopic tools are limited by the availability of good potential energy surfaces which are not readily obtainable for systems of larger then a few nuclei. The CP approach, on the other hand, allows one to study systems of up to a few hounded atoms but with the trade off of less accurate spectroscopic data. To appreciate the limitations and advantages of the CP approach the following four subsections are meant to illustrate the types of data that can be obtained from these simulations.

20.3.2 A SIMPLE EXAMPLE–AMMONIA

Here we illustrate the application of CPMD using a very simple example–the structure optimization and the calculation of the infrared spectrum of a NH_3 molecule. The very attractive potential for the 'p' channel in first row elements such as N poses a challenge to the construction of a 'soft' pseudopotential. We have used the Troullier–Martins method [Troullier and Martins (1991)] to generate a gradient-corrected (BLYP) pseudopotential using cutoff radii $r_s = r_d = 1.348$ a.u. and $r_p = 0.981$ a.u. The unscreened pseudopotential and the valence pseudo wave functions are shown in Figure 20.1. It can easily be seen that the pseudopotential (Figure 20.1(b)) for the 'p' channel is quite steep compared with both the s channels. The calculated valence pseudo wave functions are compared with the corresponding all-electron wave functions in Figure 20.1(a). The all-electron N 2s orbital which has a radial node is now replaced by a nodeless pseudo wave function and the two wave functions match exactly close to the cutoff radius $r_s = 1.348$ a.u. The hydrogen pseudopotential is generated in a similar manner. The starting wave function for NH_3 is computed from the pseudopotentials with a plane wave energy cutoff of 22.5 hartree. The geometry of the molecule is then optimized by the direct calculation of the forces. The optimized geometrical parameters r(N–H) = 1.0248 Å and < (H–N–H) = 105.66° can be compared with those obtained from conventional method of r(N–H) = 1.0254 Å and < (H–N–H) = 105.59° using the 6-311G** basis set with the BLYP functional. To compute the infrared spectrum, a molecular dynamics calculation was performed. To ensure proper sampling in the phase space, the temperature of the system is kept at 50 K via Nosé thermostats. A time step of 4 a.u. was used in the integration of the equation of motion and a trajectory consisted of 32 000 time steps was collected. The polarization autocorrelation function was calculated via the Berry phase method

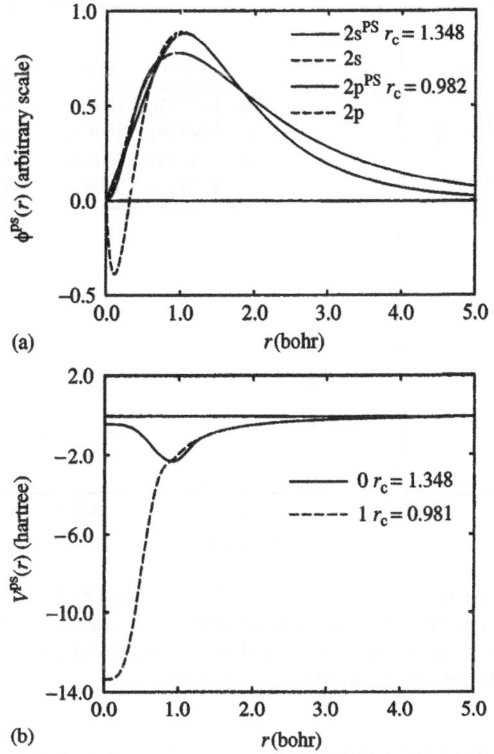

Figure 20.1. (a) The pseudo wave functions (solid lines) for the valence 2s and 2p channels of the nitrogen atoms are compared with the actual 2s and 2p radial wave functions (dashed lines) as obtained from an all-electron calculation with the same functional. (b) The pseudopotential for the s channel (solid line) and p channel (dashed line).

described above and the Fourier transform gives the infrared absorptivity. The results of the calculation are depicted in Figure 20.2. According to C_{3v} molecular symmetry, there are four infrared active modes in NH_3 which is in agreement with the calculated spectrum. The most intense vibration is the A_1 H–N–H bending mode at 1047 cm^{-1}. The degenerate bending *E* vibrations calculated at 1595 cm^{-1} has substantially lower intensity. The N–H stretch modes at 3167 cm^{-1} and 3270 cm^{-1} have the lowest intensities. The CPMD calculated vibrational frequencies and intensities for NH_3 are in reasonable agreement with those obtained from the conventional *ab initio* quantum chemistry method. Except for the *E* bending mode, the calculated relative intensities between the two different approaches are quite similar. The CPMD vibrational frequencies are generally lowered than the experimental values. This difference is probably because the functional used to calculate the potential energy surface is poor or because the plane wave expansion may not be completely converged.

20.3.3 THE STRUCTURES AND DYNAMICS OF PROTONATED HYDROCARBONS

The structure and dynamics of protonated acetylene ($C_2H_3^+$) has been the subject of special interest, see Tse, Klug and Laasonen (1995) for example. *Ab initio* quantum mechanical calculations have established that the classical vinyl cation and the nonclassical H-bridged structures are nearly equal in stability. A recent Coulomb explosion imaging (CEI) study suggests that the

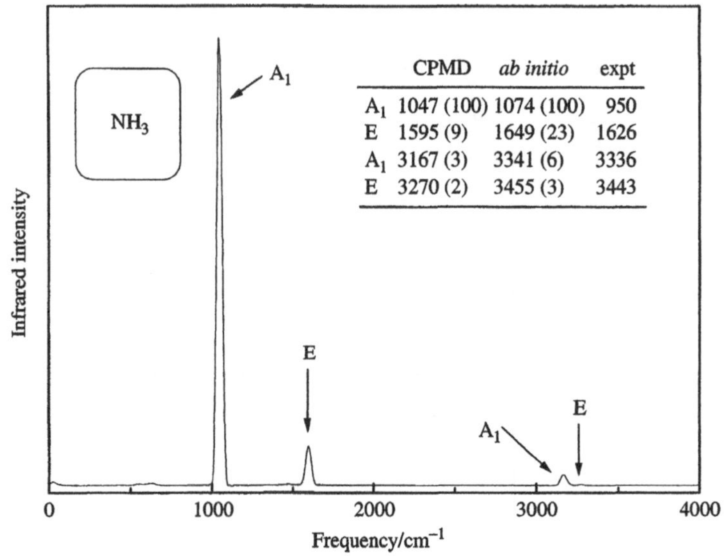

Figure 20.2. Simulated IR spectrum of ammonia molecule as obtained by Fourier transform of the dipole–dipole autocorrelation function. Intensity is in arbitrary units. The inset shows the relative intensities and frequencies as obtained dynamically from CPMD and from a static harmonic calculation with the same functional. Experimental data are also included for comparison.

bridging proton is delocalized and orbits around the C=C bond [Vager, Zajfman, Graber, *et al.* (1993)]. The tunneling splitting of the rotational spectrum has been analyzed in detail [Gomez and Bunker (1990), and Bogey *et al.* (1994)] indicating that the molecule visits several potential energy minima but is still sufficiently rigid to allow for full spectroscopic characterization. The fluxional nature of protonated acetylene is an ideal candidate for the application of the CP method. To this end, Vanderbilt ultrasoft pseudopotentials [Vanderbilt (1990)] were constructed for C and H atoms. All calculations were performed in a face-center cubic box of size 30 a.u. with a plane wave energy cutoff of 20 Ryd. For acetylene (C_2H_4), the calculated C–H and C–C bond lengths of 1.107 and 1.378 Å are to be compared with the observed values of 1.085 and 1.339 Å. CP calculation on $C_2H_3{}^+$ also revealed two nearly degenerate isomers with the nonclassical structure more stable by 1.2 kcal mol^{-1}. In the nonclassical structure, the distance of the bridging H to the C atom is 1.328 Å with a C–C bond of 1.240 Å. In the classical structure, the terminal H to C distance is 1.108 Å with a C–C bond of 1.216 Å. The calculated geometries of these isomers are in excellent agreement with previous *ab initio* calculations [Lee and Schaefer (1986), Pople (1987)].

After optimization of the electronic wave function and geometry of the ground state structure, dynamic calculations were performed at 300 K. The atomic trajectory was collected for 6.43 ps and used in the ensuing analysis. The results of the dynamical calculations using the temporal evolution of the separations of the three H atoms from the two acetylene C atoms as order parameters are summarized in Figure 20.3. The fluxional nature of $C_2H_3{}^+$ is clearly shown as the large and periodic fluctuations in the C–H distances. The bridging isomer is characterized by two equivalent C–H bonds at 1.27 Å and the vinyl structure by an H atom closer to one of the C (1.11 Å) and further (2.238 Å) from the other. The data collection starts with H_1 at the bridging position and H_2 and H_3 attached to C_a and C_b respectively. In the ensuing 1000 time steps, H_1 moves towards C_b and away from C_b (Figure 20.3(a)). Simultaneously, H_3 moves away from C_b and swing towards the middle of the C–C bond on the opposite side (Figure 20.3(c)) while H_2

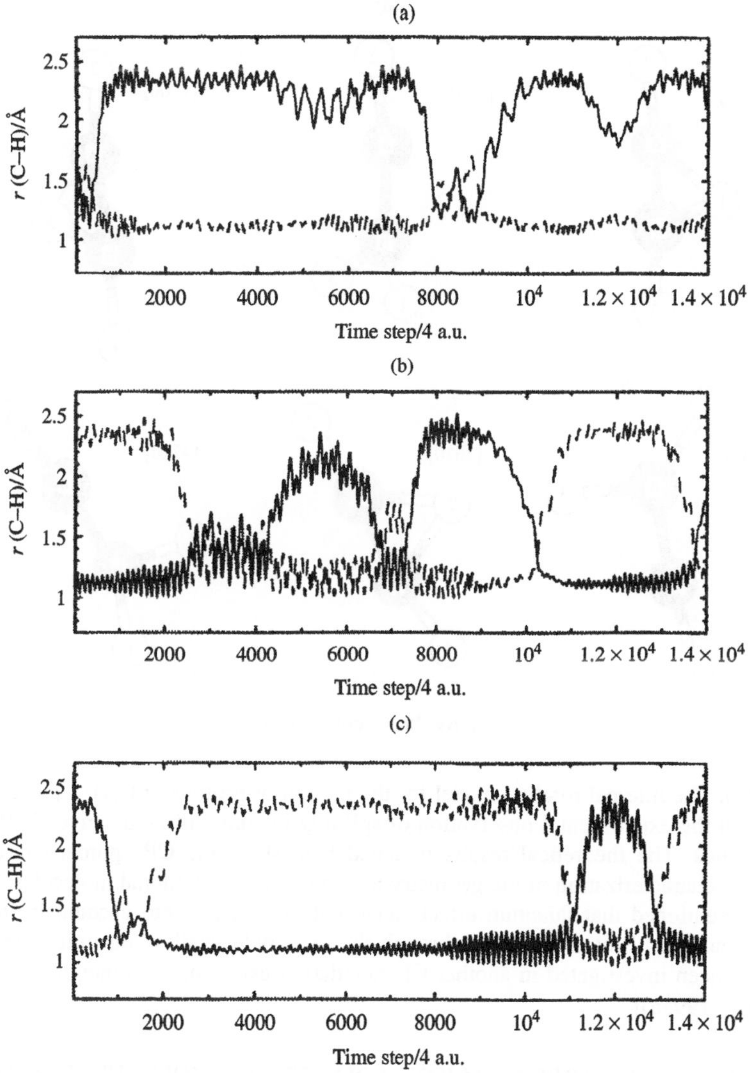

Figure 20.3. The dynamics of $C_2H_3{}^+$ may be understood by monitoring the C_a–H (dashed line) and C_b–H (solid line) for the protons H_1 (a), H_2 (b) and H_3 (c). (d) Snapshots of the configurations as obtained from an *ab initio* molecular dynamics trajectory.

remains attached to C_a. This leads to an exchange of H_1 and H_3. The H atoms then vibrate about their mean positions for another 600 time steps. After that, at time step 1600, the bridging H_3 moves towards C_a and H_2 shifts to the bridging position (Figures 20.3(b) and 20.3(c)). From time steps 7000 to 10 000 a similar exchange between H_1 and H_2 is observed, resulting H_1 occupying the bridging position. The internal rotation of the protons from one end of the C–C bond via the symmetric bridging position is repeated between H_2 and H_3 at time step 10.4000. 'Recoil' actions as shown at time steps 5000 and 11 700 occasionally disrupt this orderly proton migration. In both cases, H_1 (Figure 20.3(a)) attempted to move towards the central position but was 'pulled' back by C_b. The proton rearrangement is clearly depicted in the snapshots of the molecule in

(d)

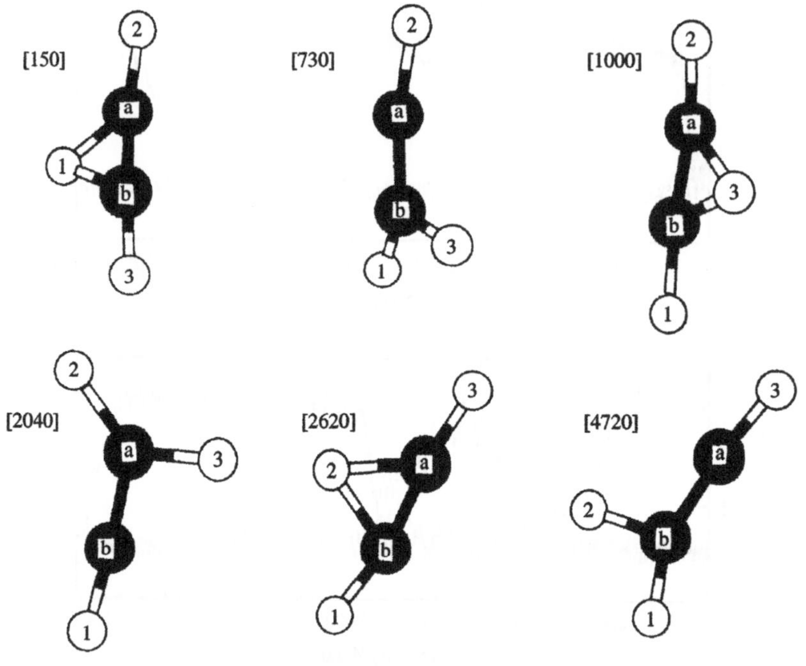

Figure 20.3. *(continued)*

Figure 20.3(d). The internal rotation model for the motion of protons in $C_2H_3^+$ presented here is consistent with the experimental observation of splitting in some infrared active C–H stretching vibrational bands. The theoretical results obtained here show the CP approach is a powerful method for the characterization of the geometry and dynamics of fluxional molecules. In passing, it should be mentioned that quantum effects have not been taken into account in the classical dynamical simulation. Proton tunneling through the extremely shallow potential energy surface of $C_2H_3^+$ has been investigated in another CP calculation using the path integral method [Marx and Parrinello (1995a)].

20.3.4 DYNAMICS OF CARBONIUM IONS SOLVATED BY MOLECULAR HYDROGEN

Protonated alkanes are highly reactive intermediates that form in the acid-catalyzed transformation of hydrocarbons. These carbonium ions are highly fluxional, and CH_5^+, for all practical purpose, does not have a unique, stable equilibrium geometry [Scuseria (1993), Schreiner, Kim, Schaefer, *et al.* (1993), Miller Noga *et al.* (1997)]. The hydrogen atoms are predicted to scramble almost freely among multiple equivalent minima. Studying the dynamics of such molecules is quite challenging. This has been studied by the experiment technique of solvating CH_5^+ with molecular hydrogen in order to slow down the dynamics of the scrambling motions. It is noted that for the CH_5^+ species the traditional methods based upon solving the time-independent Schrödinger equation are extremely accurate for explaining spectroscopic data [East, Kolbuszewski and Bunker (1997)]. However, this approach becomes severely hindered for these solvated species owing to the larger number of atoms. Since it is difficult to treat the floppy molecule as a static entity, CPMD method are an ideal theoretical method for this application [Tse, Klug and Laasonen (1995), Boo *et al.* (1995), Marx and Parrinello (1995b), and Marx and Parrinello (1999)].

The dynamics of $CH_5^+(H_2)_n$ ($n = 0, 1, 2, 3$) obtained from CP calculations are summarized in Figure 20.4 where the temporal evolution of H–H distances (~3 ps) are shown. At start of the simulation for CH_5^+, H_2, and H_3 is characterized by a short H–H distance (~1.1 Å) and three strong C–H bonds formed by H_1, H_2 and H_4. This structure can be loosely described as a $CH_3^+ \ldots H_2$ molecule with the H_2 forming a three-center two-electron (3c2e) bond with the C atom. As the simulation progresses, all the H–H distances fluctuated between 1.0 and 2.2 Å and the 3c2e bond could be formed between any pairs of H atoms.

For $CH_5^+(H_2)$ (Figure 20.4(b)), the CH_5^+ core remains floppy. For example, both H_3–H_4 and H_3–H_1 distances showed considerable fluctuations similar to those observed for CH_5^+. However, during the entire simulation, the H_2 molecule was always bound to the H_3 atom. Moreover, when min(H_3–H_n), which is the minimum values of the distances between H_3 and any one of the other H atoms in the CH_5^+ core, was compared with the corresponding min(H_1–H_n), the former fluctuated much less and had a smaller average value than the latter. This surprising observation indicates that the 3c2e bond is localized around H_3.

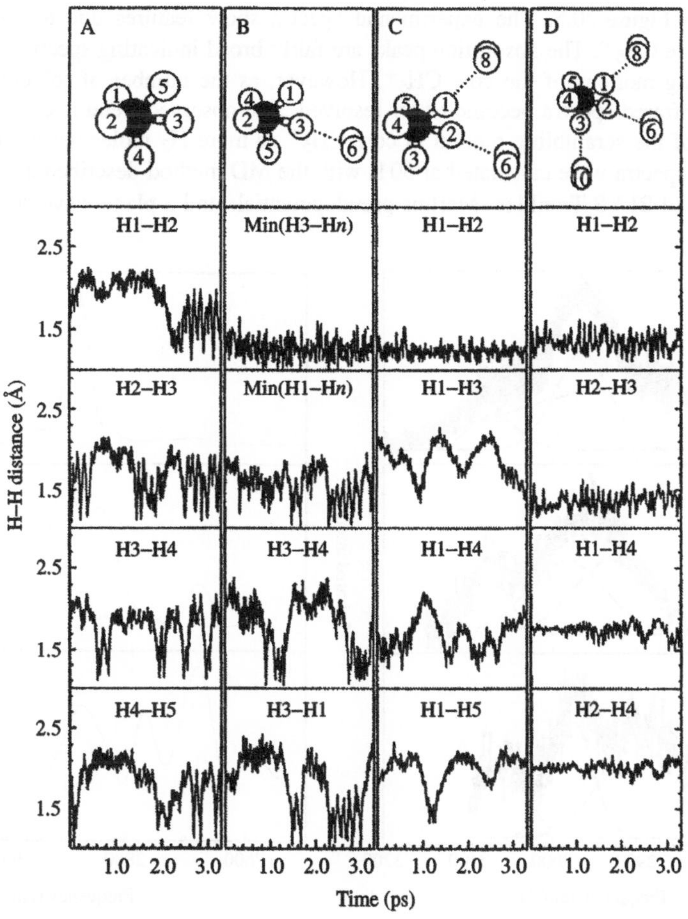

Figure 20.4. Evolution of the H–H distances in the CH_5^+ core during the molecular dynamics simulation: (a) CH_5^+, (b) $CH_5^+(H_2)$ (c) $CH_5^+(H_2)_2$ (d) $CH_5^+(H_2)_3$. The hydrogen atoms on CH_5^+ are numbered 1–5 and those on the solvent H_2 molecules 6–10.

For $CH_5^+(H_2)_2$ (Figure 20.4(c)), the stabilization effect of two weakly bound H_2 molecules becomes more prominent. The two H_2 were always bound to H_1 and H_2 of the core CH_5^+, respectively, and the H_1-H_2 distance stay around 1.25 Å with markedly small fluctuations (±0.05 Å root mean square). This result suggests that the 3c2e bond is localized to the C atom and the two H atoms weakly bonded to the two solvent H_2 molecules. The effect could be attributed to the electron deficiency in the 3c2e bond that attracts the two solvent H_2 molecules. In contrast, the remaining three H atoms belonging to the CH_5^+ core continued to exhibit large-amplitude motions as indicated by the large fluctuations in min(H_1-H_n) distance in Figure 20.4(c).

For $CH_5^+(H_2)_3$ (Figure 20.4(d)), the CH_5^+ core becomes semi-rigid. The three H_2 molecules were bound to H_1, H_2 and H_3 of the CH_5^+ core during the entire simulation. The in-plane wagging motion of H_2 between H_1 and H_3 contributed to the slightly larger fluctuations in the H_1-H_2 and H_2-H_3 distances than in the H_1-H_2 distance for CH_5^+ $(H_2)_2$. However, the fluctuations in the H_1-H_4 and H_2-H_4 distances were only 0.4 Å, indicating a considerable slowdown in the CH_5^+ scrambling motions.

The CPMD results corroborate the experimentally observed trend in the infrared spectra. The experimental infrared spectra for CH_5^+ $(H_2)_n$ $(n = 1, 2, 3)$ are compared with the CPMD simulated spectra in Figure 20.5. The experimental spectra show features due to C–H stretching modes in the core CH_5^+. The absorption peaks are fairly broad indicating spectra congestion due to the scrambling motions of the core CH_5^+. However, as the number of solvent H_2 molecule increases, the infrared spectra become better resolved. The observed trend is consistent with the slowing down of the scrambling motion in core CH_5^+ as more H_2 molecules are attached to it. The theoretical spectra were calculated at 50 K with the MD method described above employing gradient-corrected BLYP Troullier–Martins pseudopotentials and a plane wave cutoff of 35 Ryd.

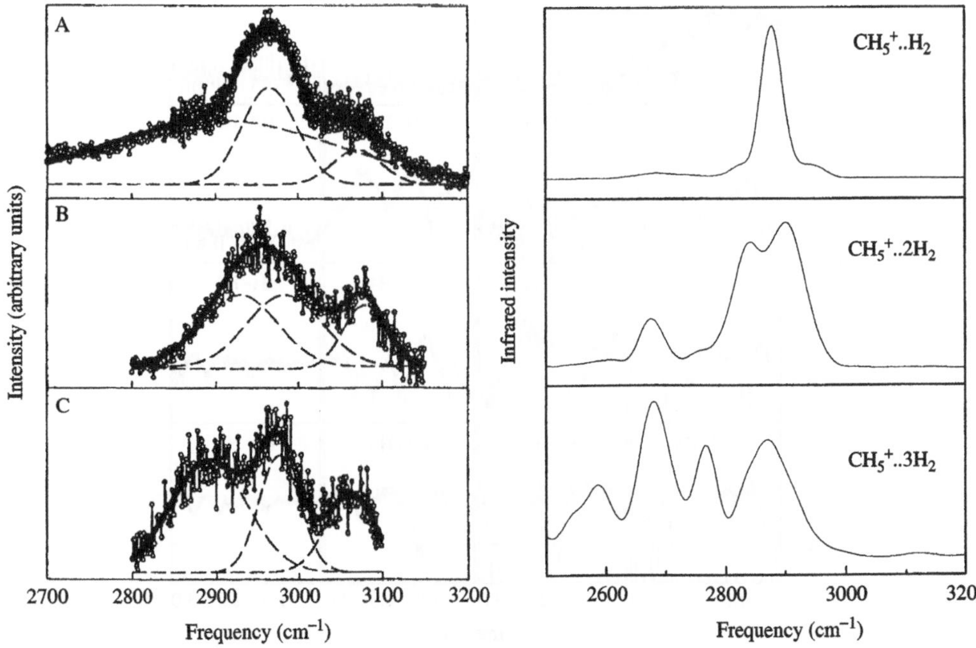

Figure 20.5. Infrared spectrum of C–H stretching modes for (a) $CH_5^+(H_2)$, (b) $CH_5^+((H_2)_2$, (c) $CH_5^+(H_2)_3$. Left sub-panels are experimental spectra right sub-panels are those obtained from *ab initio* molecular dynamics simulation.

Although the agreement with experiment is only qualitative, the theoretical infrared spectra reflect the essential features observed in the experimental spectra. The calculated frequencies are generally 100–180 cm^{-1} lower than the observed values [Boo *et al.* (1995)]. In $CH_5^+(H_2)$, a single strong absorption feature centered at 2880 cm^{-1} with weak shoulders on both sides of the main peak is predicted. A single peak at 2965 cm^{-1} with two side bands were observed in the experimental spectrum. The attachment of a second H_2 changes the infrared spectrum dramatically. The spectrum now shows three strong features at 2677, 2843 and 2901 cm^{-1}. This is in agreement with the observed infrared bands at 2926, 2957 and 3078 cm^{-1}. In particular, the experimental energy spread between the low and high energy peaks of 221 cm^{-1} is in good agreement with the calculated 224 cm^{-1}. The infrared spectrum of $CH_5^+(H_2)_3$ is the most complicated. Since the low end energy cutoff of the experimental spectrum is at 2800 cm^{-1}, it is difficult to make a direct comparison with the theoretical results. However, it is not unreasonable to correlate the calculated infrared adsorption peaks at 2689, 2770 and 2871 cm^{-1} to the observed bands at 2892, 2977 and 3062 cm^{-1}.

The advantage of CPMD method in the study for the structure and dynamics of floppy molecules such as protonated alkanes is clearly demonstrated above. It is difficult for conventional *ab initio* methods to compute the infrared spectra of molecules with large amplitude motions. The CPMD method takes into account of the anharmonic and temperature effects in a natural way and, in combination with accurate quantum chemistry calculations can provide very useful information in the assignment of the vibrational spectra of these molecules. A recent example is on the assignment of the vibrational spectra of $C_2H_7^+$ [East *et al.* (1998)]

20.3.5 CLUSTER CHEMISTRY

An emerging paradigm in cluster science is that of size selectivity and the tunability of cluster properties as the cluster evolves in size from a small molecule of a few atoms to a bulk solid [Haberland (1993), Brack (1993), and de Heer (1993)]. A fundamental first step in the understanding of these trends is to obtain a description of the cluster geometry which has been proven possible by direct comparison of experimental absorption spectra with theoretical ones [Bonnicic-Koutecký, Fantucci, and Koutecký (1991), Jarrold (1991), Blanc *et al.* (1991), Dugourd *et al.* (1991), Blanc *et al.* (1992), Honea *et al.* (1993), Giesen T. F. *et al.* (1994), and Rubio *et al.* (1996)]. By iteratively performing this procedure for a series of clusters one may systematically obtain descriptions of the structural, spectroscopic and thermodynamic properties as a function of cluster size. Another aspect to this problem, and subsequently one more difficult to address, is the question of the interaction between individual molecules and size-selected clusters as a function of particle size.

As an illustrative example of how one may employ *ab initio* molecular dynamics methods in questions of this nature we shall consider the example of the interaction of size-selected clusters of gold (ranging from 1 to 15 atoms) with methanol [Rousseau *et al.* (1998)]. One ideal probe for the detection of changes in molecular structure upon adsorption in the vibrational spectrum. In particular, the shift in the vibrational frequency of the C–O stretching mode, ν_{C-O}, has been used to investigate the adsorption process [Okumura *et al.* (1990), Huisken and Stemmler (1993), Pribble and Zwier (1994), Rayner *et al.* (1995), and Kirkwood, Winkel, and Stace (1995)]. It has been found that a single methanol adsorbed onto the surface of neutral Fe_{5-15}, Cu_{3-11}, Ag_{3-22} and Au_{3-13} exhibits a red shift of about 15–40 cm^{-1} for ν_{C-O} which is dependent of the type of metal atom but independent of the cluster size [Knickelbein (1996), and Knickelbein and Koretsky (1998)]. This is fundamentally different then the behavior found from the positively charged cluster Au_{1-15}^+, where there is a marked dependence upon cluster frequency [Dietrich *et al.* (1996), Rousseau *et al.* (1998), Rousseau and Marx (2000a), and Dietrich *et al.* (2000)]. The red shift $\Delta\nu_{C-O} = \nu_{C-O}$ (gas) $-\nu_{C-O}$ (adduct) is plotted in Figure 20.6. The largest red shift

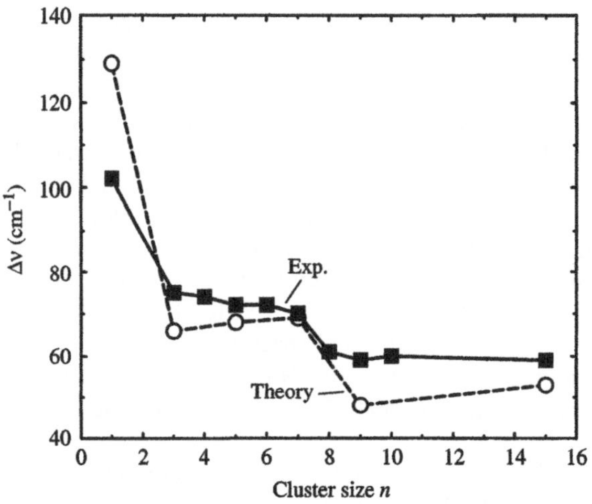

Figure 20.6. The shift in C–O stretching frequency, $\Delta\nu$, for methanol adsorbed onto the surface of size-selected Au_n^+ clusters as a function of cluster size n. The values of $\Delta\nu$ are relative to the C–O stretching frequency of the free methanol molecule. Dark squares are experiment and opaque circles are theory lines are draw as a guide to the eye.

$\Delta\nu$ exceeds $100\,cm^{-1}$ for the Au_1–CH_3OH^+ species. For the clusters adducts containing 3 to 7 gold atoms there is a shift of about $75\,cm^{-1}$. Finally, for the adducts with 8–15 gold atoms $\Delta\nu$ is about $60\,cm^{-1}$. Two fundamental concerns arise from these observations. First, the magnitude of $\Delta\nu$ is larger than the $30\,cm^{-1}$ found for neutral gold cluster–methanol adduct species and second the value of $\Delta\nu$ behaves in a stepwise fashion as a function of cluster size.

To understand the origin of these observations one must first consider the underlying cluster structure. This can be done in an analogous manner as described in previous sections; equilibration of an n-atom cluster at high temperature followed by simulated annealing to 0 K. This procedure may then be repeated systematically to obtain a suitable description of the low temperature cluster isomers. In a second step, the adduct molecule may then be placed within the simulation cell and allowed to react with the cluster during the course of a finite temperature simulation (300 K in the case of the Au_n–CH_3OH and Au_n–CH_3OH^+ adducts presented here). Once these species have equilibrated they should be again annealed to obtain the lowest energy configurations. After a suitably large number of adducts are obtained it is then possible to systematically construct viable low energy adducts without this laborious procedure, although it is advised that species generated in this fashion should be well equilibrated in a short MD trajectory in order to guarantee that they are in at least a local minimum on the potential energy surface. Vibrational spectra may be obtained from these species as outlined above by further finite temperature simulation or by harmonic frequencies as are routinely obtained from static configurations.

The cluster adducts obtained for Au_n–CH_3OH^+ ($n = 1, 3, 5, 7, 9, 15$) and Au_n–CH_3OH ($n = 4, 8$) via this prescription are shown in Figure 20.7 [Rousseau *et al.* (1998), and Rousseau and Marx (2000a)]. Theoretical values of $\Delta\nu$ as obtained from harmonic frequencies are plotted in Figure 20.6. It is important to note that to avoid inherent difficulties associated with treating open shell systems within DFT calculations only closed shell species are considered. First we examine the lowest energy species obtained for the positively charged cluster adducts. In general, all the cluster adducts are stabilized by a direct Au–O bond and the lowest energy species are invariably those where the gold atom attached to oxygen, Au*, has the lowest coordination number.

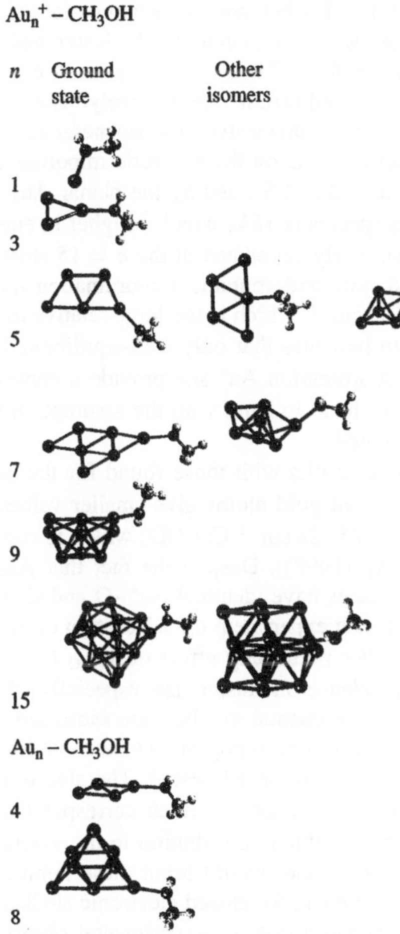

Figure 20.7. The structures of adducts formed between methanol CH_3OH and Au_n^+ ($n = 1, 3, 5, 7, 9$ and 15) clusters and Au_n ($n = 4$ and 8). The atoms are represented by: Au (black spheres), O (dark grey spheres), C (medium grey spheres) and H (light grey spheres).

The largest red shift is obtained from the positively charged $n = 1$ adduct (129 cm^{-1} and subsequently this species also exhibits the shortest Au*–O (2.04 Å) and longest C–O bonds (1.49 Å compared with the 1.43 Å calculated for an isolated methanol molecule). The lowest energy species Au_n–CH_3OH^+ ($n = 3$–7) are all composed of a two-dimensional Au_n cluster with the methanol attached to a two coordinate Au* binding site; $\Delta\nu = 66$–69 cm^{-1} with Au*–O and C–O bond lengths between 2.09 and 2.10 Å and between 1.47 and 1.48 Å respectively. For the larger clusters a planar geometry is no longer a viable energetic alternative and thus the methanol molecule must attach itself to the lowest coordination site of a nonplanar cluster. In the case the positively charged 9 and 15 clusters the molecule attaches to an Au* position with a coordination number of 4. This results in $\Delta\nu = 48$–53 cm^{-1} and Au*–O and C–O bond lengths of 2.12–2.14 and 1.47 Å respectively. Thus, the lowest energy species obtained from this approach appear to account for the observed phenomena and suggest that the cause is deeply connected to the underlying cluster structure.

To begin to unravel the reason for this behavior an examination of higher energy isomers is in order. For the $n = 5$ and 7 species a nonplanar $Au_n{}^+$ cluster and its corresponding adduct have also been generated which are about 7 kcal mol^{-1} higher in energy than the ground state species and provide $\Delta\nu$ values of 45 and 60 cm^{-1} respectively. This is surprising in light of the fact that the actual independent clusters themselves are isoenergetic (within 1 kcal mol^{-1}) with their planar analogs. Moreover, an estimate on the energetic importance of coordination number may be obtained by considering the adduct formed by the planar $Au_5{}^+$ cluster and methanol at the four-coordinate Au* site. This species is 16 kcal mol^{-1} higher in energy then the ground state with a red shift $\Delta\nu$ of 77 cm^{-1}. Similarly, an adduct of the $n = 15$ cluster which is 23 kcal mol^{-1} higher in energy then the ground state and contains a coordination number of 6 for Au*. This adduct provides a value of $\Delta\nu = 63$ cm^{-1}, which is too large relative to the other members of the nonplanar cluster series. This then indicates that only well-equilibrated low energy species with methanol attached to the lowest coordination Au* site provide a consistent interpretation of the experimental trend in $\Delta\nu$; this goes hand in hand with the assumption that the species observed in the experiment are well equilibrated.

It is interesting to compare these results with those found for the neutral species. Here both the cluster adducts with four and eight gold atoms give smaller values for the red shift ($\Delta\nu =$ 37–40 cm^{-1} for CH_3OH and $\Delta\nu = 25$–28 cm^{-1} CD_3OD) which is constant, $\Delta\nu = 30$ cm^{-1}, for CD_3OD [Knickelbein and Koretsky (1998)]. Despite the fact that Au_4 is a planar cluster and Au_8 is a nonplanar cluster these species have identical Au*–O and C–O bond lengths (2.17 and 1.46 Å respectively). In agreement with experiment the adsorption of the methanol molecule onto the neutral gold species has a smaller perturbing effect on ν_{C-O} than with the cationic species and does not exhibit a sizable dependence on cluster size; especially when one considers that the error bars on both theoretical and experimental numbers are estimated to be 3 cm^{-1}.

Now that a suitable structural model has been presented to account for the experimental observables the question of *why* this occurs may be addressed. The step-like behavior of $\Delta\nu$ for the cationic adducts occurs at values of $n = 1$ and 7 which correspond to the number of valence electrons of 0 and 6; assuming only s orbital contribution to the valence shell. Addition of two electrons from an oxygen atom of methanol would bring these numbers up to 2 and 8 respectively; which correspond to magic numbers for closed electronic shells as dictated by the jellium model [Brack (1993)]. This then suggests that the experimental observations may be attributed to an electronic shell closing effect as has been proposed to explain the interaction of carbon monoxide molecules with $Cu_n{}^+$ clusters [Nygren *et al.* (1991)]. This however, seems unlikely when one considers that only the ground state isomers account for the observed experimental frequency. If the step-like behavior of $\Delta\nu$ were due to an electron shell closing mechanism then a strong dependence upon cluster geometry does not seem likely. This then forces us to consider an alternative picture which relies the on molecular geometry.

A consideration of the structural aspects of this problem is aided by the observation that the Au*–O bond length is consistent with that observed in coordination compounds of gold and oxygen containing ligands. If this is a coordination bond then this suggests that Au* should have an appreciable positive charge relative to the remaining gold atoms and methanol should also acquire some net charge due to the donation of electrons into the cluster framework. This is indeed found via Mulliken population analysis. Methanol acquires a net positive charge of 0.20–0.35 (depending on cluster size) for the cationic species and 0.10 for the neutral species. For the cationic species Au* contains 30–50 % of the positive charge in the cluster fragment for the planar $n = 3$–7 species and 10–20 % for the nonplanar $n = 9$ and 15 species; in the isolated cationic clusters (without methanol) this distribution is more equal and the Au* site invariably becomes more charged upon adsorption of methanol. Moreover, Au* also acquires a small positive charge in the neutral adducts of about 0.10.

The picture which emerges from this analysis is that the presence of a direct Au*–O coordination bond polarizes the electrons such that Au* and methanol become slightly more positive and the remaining Au atoms become slightly more negative. Furthermore, this polarization of electron density is less effective as the dimension of the cluster increases from 0 to 3 and as such has less of a perturbing effect on the C–O vibrational frequency. Thus, for these cationic clusters methanol acts as a probe which effectively measures the dimension of the underlying cluster structure. For the neutral species this polarization is less effective because there is less positive charge to begin with. Thus, the coordination bond formed between neutral Au clusters and oxygen is subsequently weaker and not sensitive enough to act as a structural probe. This example illustrates that a detailed knowledge of the structure of clusters and adducts as obtained from CP simulations is indeed useful for unraveling the trends in spectroscopic data. Moreover, these results suggest that the structures of other clusters may be probed by picking a suitable sensor molecule. This species should form a sufficiently strong coordination bond with the nuclei within the cluster, yet not such a strong interaction that it perturbs the underlying cluster structure.

20.4 Quantum effects

20.4.1 WHEN AND WHY QUANTUM EFFECTS ARE IMPORTANT

In the previous sections the nuclear motion has been described implicitly by classical mechanics. Although, valid for heavy nuclei this description is less appropriate for the lighter elements. This is especially so in the case of clusters of light atoms where the bonding potential felt by the nuclei may be weak. For example, it is well known that the properties of helium or hydrogen clusters are dominated by the quantum mechanics of the nuclei at low temperatures, which leads to such phenomena as quantum-liquid behavior [Whaley (1994), and Scharf, Martyna, and Klein (1992)], superfluidity [Sindzingre, Klein and Ceperley (1989), and Sindzingre, Ceperley and Klein (1991)], binary phase separation [Chakravarty (1995)], and plasticity [Štich *et al.* (1997)]. These quantum effects are at heart governed by zero-point kinetic energy and tunneling which are not described within the regular *ab initio* molecular dynamics formalism. In this section an outline of a method by which such processes may be studied within a fully quantum mechanical framework for both nuclei and electrons will be outlined. In the latter portion of this section an example, for illustration, of the type of results and how to analyze them will be considered.

20.4.2 PATH INTEGRAL CAR–PARRINELLO MOLECULAR DYNAMICS

To begin to see how to model quantum effects it is necessary to start with some very simple ideas from quantum statistical mechanics [Feynman (1972), Feynman and Hibbs (1965), and Ceperley (1995)]. To aid in this discussion we introduce the concept of a density operator which is related to the partition function, Q, in the following manner,

$$Q = \sum_n \langle\psi_n|\exp(-\beta H)|\psi_n\rangle = \int \mathrm{d}r \sum_n \psi_n^*(r)\rho\psi_n(r) = \int \mathrm{d}r\rho(r, r, \beta) = \mathrm{Tr}[\rho], \qquad (27)$$

where $\beta = 1/k_\mathrm{b}T$ and ρ is the quantum mechanical density operator, which has the form,

$$\rho = \exp(-\beta H) = 1 - \beta H + \frac{\beta^2}{2}HH\ldots. \qquad (28)$$

The partition function is therefore the trace of the density matrix. In general, the density operator is not diagonal in a cartesian coordinate representation and thus depends on two spacial variables r_1 and r_2. However, for the purposes of the trace of this operator we are only interested in the diagonal elements $r_1 = r_2$. This concept at first may seem a bit cumbersome but in fact it holds the key to understanding just how quantum effects may be included in a molecular dynamics formalism.

Before diving into the idea of path integrals and so forth, it may be useful to review some things about these density operators and to try an example. First, it is important to note that this operator obeys the equation,

$$\frac{\partial \rho}{\partial \beta} = -H\rho, \tag{29}$$

which will serve as a useful tool in the following discussion. To illustrate how to obtain ρ, we examine the Hamiltonian for a free particle moving in one dimension. This Hamiltonian is described by the equation

$$H = \frac{-\hbar^2}{2m}\frac{\partial^2}{\partial x^2}. \tag{30}$$

To find the form of ρ in real space for a given temperature, $\rho(x, x', \beta)$, it is noted that equation (29) becomes,

$$\frac{\partial \rho(x, x', \beta)}{\partial \beta} = -\frac{\hbar^2}{2m}\frac{\partial^2 \rho(x, x', \beta)}{\partial x^2}, \tag{31}$$

which has the form of a diffusion equation. Furthermore, at higher temperatures we may invoke the correspondence principle, i.e. the particles will occupy states with high quantum number and thus act classically. This provides for a suitable boundary condition of $\rho(x, x', 0) = \delta(x - x')$ and leads to a solution,

$$\rho(x, x', \beta) \approx \left(\frac{m}{2\pi\beta\hbar^2}\right)^{1/2} \exp\left(\frac{-m(x - x')^2}{2\beta\hbar^2}\right). \tag{32}$$

This equation may be generalized in a straightforward fashion to give the expression for ρ for a free particle in three dimensions

$$\rho_{\text{free}}(r, r', \beta) \approx \left(\frac{m}{2\pi\beta\hbar^2}\right)^{3/2} \exp\left(\frac{-m(r - r')^2}{2\beta\hbar^2}\right). \tag{33}$$

To reconnect this back to the partition function we simply take the trace of this density matrix.

$$Q_{\text{free}} = \text{Tr}[\rho_{\text{free}}] = \int \mathrm{d}r\rho_{\text{free}}(r, r, \beta) = \left(\frac{m}{2\pi\beta\hbar^2}\right)^{3/2} V, \tag{34}$$

which provides the well-known result for the partition function of translation.

To take the first step towards how to develop a quantum mechanical description of nuclear motion in a CP-type simulation we re-partition the density operator into P equal components,

$$\mathrm{e}^{-\beta H} = \mathrm{e}^{-\beta H/P}\mathrm{e}^{-\beta H/P}\mathrm{e}^{-\beta H/P}\ldots(P \text{ times}). \tag{35}$$

This may then be inserted into the partition function to give,

$$Q = \int \mathrm{d}r_1 \langle r_1|\,\rho\,|r_1\rangle = \int \mathrm{d}r_1 \langle r_1|\,\mathrm{e}^{-\beta H/P}\mathrm{e}^{-\beta H/P}\mathrm{e}^{-\beta H/P}\ldots|r_1\rangle, \tag{36}$$

which may be rewritten with the aid of the completeness relation, $\int \mathrm{d}r|r\rangle\langle r| = 1$, to an expression which contains the matrix elements $\rho(r, r', \beta/P)$,

$$\begin{aligned} Q &= \int \ldots \int \mathrm{d}r_1\,\mathrm{d}r_2\,\mathrm{d}r_3 \ldots \mathrm{d}r_P \langle r_1|\,\rho\,|r_2\rangle \langle r_2|\,\rho\,|r_3\rangle \ldots \langle r_P|\,\rho\,|r_1\rangle \\ &= \int \ldots \int \mathrm{d}r_1\,\mathrm{d}r_2\,\mathrm{d}r_3 \ldots \mathrm{d}r_P\rho(r_1, r_2, \beta/P)\rho(r_2, r_3, \beta/P)\ldots\rho(r_P, r_1, \beta/P). \end{aligned} \tag{37}$$

Writing out Q in this form suggests an alternate way of thinking about the problem in terms of integration over a path starting at r_1 and passing through the points $r_2, r_3 \ldots r_P$ and finally returning back to r_1. This is illustrated in Figure 20.8. In the limit of an infinitely large number of steps along this path we effectively sample, upon integration, all paths which start and terminate at the same position. By integrating over all *paths* from r_1 back to itself and all points r_1 we obtain Q.

The reader may wonder what is the advantage of such an approach which on the surface seems to have complicated the matter by introducing the necessity of dealing with many integrations over off-diagonal elements of the density matrix. This in reality is not a problem as we shall now show because this path integral approach allows us to replace our Hamiltonian with an effective Hamiltonian which is much easier to deal with. To understand just how this is accomplished we consider the Hamiltonian of a particle which is subject to an external potential $V(r)$,

$$H = \frac{-\hbar^2}{2m}\nabla^2 + V(r). \tag{38}$$

Along an infinitely small segment of a path, ε, the particle may be described almost like a free particle and will feel only a small effect from the external potential $V(r)$. As a result the density matrix may be written down as a small variation from that for a free particle,

$$\rho = \rho_{\text{free}} + \delta\rho. \tag{39}$$

An expression $\delta\rho$ can be obtained from a first-order perturbation expansion [Feynman (1972)],

$$\delta\rho = -V(r)\varepsilon\rho_{\text{free}}. \tag{40}$$

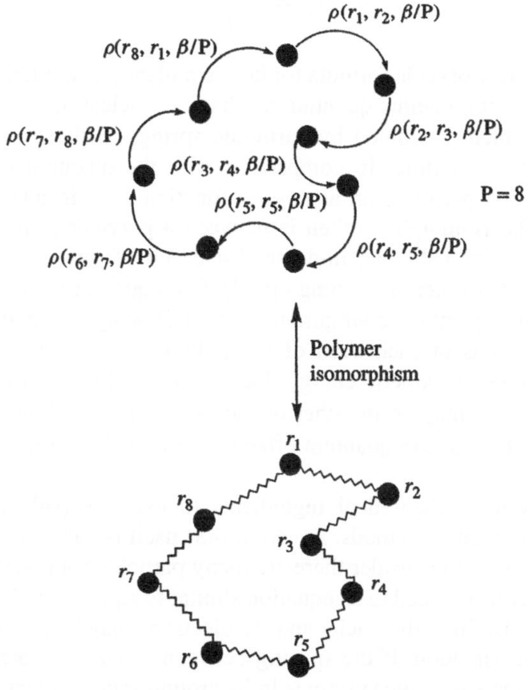

Figure 20.8. Graphical representation of polymer isomorphism.

This allows us to make the approximation for $\rho(r, r', \beta/P)$ in our path integral as

$$\rho(r, r', \beta/P) = \left(1 - \frac{V(r)\beta}{P}\right) \rho_{\text{free}}(r, r', \beta/P) \approx \rho_{\text{free}}(r, r', \beta/P) \exp\left(\frac{-V(r)\beta}{P}\right). \tag{41}$$

Using the expression which we derived above (equation (33)) we find

$$\begin{aligned} \rho(r, r', \beta/P) &\approx \left(\frac{Pm}{2\pi\beta\hbar^2}\right)^{3/2} \exp\left(\frac{-Pm(r-r')^2}{2\beta\hbar^2}\right) \exp\left(\frac{-\beta V(r)}{P}\right) \\ &\approx \left(\frac{Pm}{2\pi\beta\hbar^2}\right)^{3/2} \exp\left(\frac{-Pm(r-r')^2}{2\beta\hbar^2} - \frac{\beta V(r)}{P}\right). \end{aligned} \tag{42}$$

For $P \to \infty$ this expression is exact and for sufficiently large P it is a reliable approximation and allows us to use a much simpler form of the density operator where the exact Hamiltonian is replaced with an effective potential with harmonic terms.

This approximation allows us now to rewrite the partition function as

$$Q = \lim_{P\to\infty} \Pi_{s=1}^{P} \left(\frac{Pm}{2\pi\beta\hbar^2}\right)^{3/2} \int \mathrm{d}r^3 \exp\left[-\beta \sum_{s=1}^{P} \left(\frac{m\omega_P^2 (r^s - r^{s=1})^2}{2} + \frac{V(r^s)}{P}\right)\right], \tag{43}$$

where

$$r^{P+1} = r^1,$$
$$\omega_P^2 = \frac{P}{\beta^2\hbar^2}.$$

In this way we have a more workable formula for both simulation and interpretation. This formula actually maps the problem of modeling quantum mechanical nuclear motion onto that of a cyclical polymer with P beads which are joined by harmonic springs as illustrated in Figure 20.8. We note also that $\beta\hbar$ has units of time. In comparison with the quantum mechanical propagator $U(t) = U(0)\exp(-itH)$, $\beta\hbar$ performs an analogous function as, $-it$, and as such is referred to as imaginary time, τ. The isomorphism then is between a polymer in imaginary time and that of quantum mechanical motion. For an in-depth discussion of this isomorphism see Chandler and Wolynes (1981), and Chandler and Leung (1994). On a qualitative footing this isomorphism allows us to examine the quantum configurations by following along the polymer backbone and comparing what happens in each bead of the polymer chain. Consider, for example, the double well potential where the kinetic energy does not exceed the barrier height as illustrated in Figure 20.9. This polymer may be in either of the two minima or both due to tunneling; this formalism makes it easy to evaluate quantum effects by examining to see if all the beads are in the same or both minima.

We now have the essential theoretical ingredients to examine path integral CP molecular dynamics (PI-CPMD) and related methods. Equation (43) itself is valid for one particle; however, in the systems which we wish to consider there are many particles both nuclei and electrons. This complicated situation may be reduced to an equation similar to equation (43) by employing several simplifying approximations. First, the nuclei and the electrons may be decoupled by invoking the Born–Oppenheimer approximation. If the spacing between electronic states is much larger then kT we may then assume the electronic system is in its ground state and may therefore be modeled by density functional theory in the Kohn–Sham approximation. As such, the quantum nuclei move

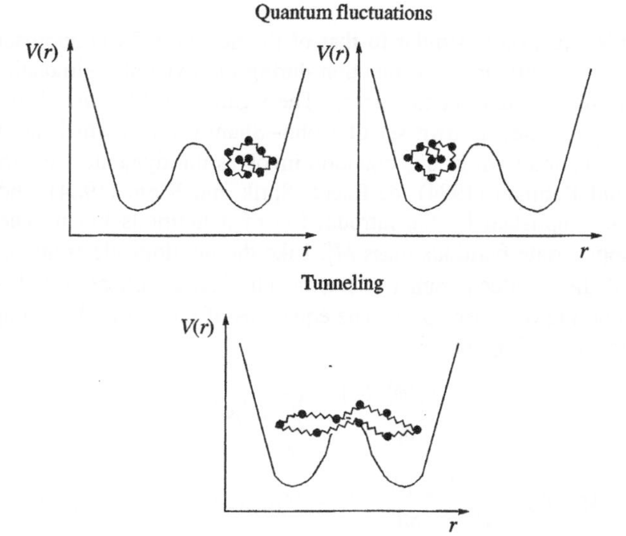

Figure 20.9. The polymer isomorphism in a double well potential. If all the beads of the polymer (black circles) are located within one potential well at a given time step but in another at a later time step this may be attributed to surmounting the barrier via the available kinetic energy. Tunneling, on the other hand, is expressed by finding some of the polymer beads in each minimum at a given time step.

on a ground state electronic potential energy surface, $E_{\text{elec}}[\phi_p^s, R_l^s]$, where the electronic energy for the configuration of polymer bead, s, is a functional of the atomic coordinates R_l^s and the Kohn–Sham orbital, ϕ_i^s of that bead. A further approximation is now made concerning the nuclei. These particles themselves are treated as distinguishable, i.e. neglecting nuclear exchange effects. This allows the density matrix to be written as a product function of the density matrices of the individual nuclei in the system. Under these restrictions the canonical partition function becomes:

$$Q = \lim_{P\to\infty} \Pi_{s=1}^{P} \Pi_{l=1}^{N} \left(\frac{Pm}{2\pi\beta\hbar^2}\right)^{3/2} \int \mathrm{d}R_l^s \exp$$
$$\left[-\beta \sum_{s=1}^{P}\sum_{l=1}^{N}\left(\frac{M_l \omega_P^2 (R_l^s - R_l^{s+1})^2}{2} + \frac{E_{\text{elec}}[\phi_i^s, R_l^s]}{P}\right)\right]. \tag{44}$$

This equation as written is the key to path integral *ab initio* simulations whether it is used in connection with Monte Carlo [Weht *et al.* (1998)], or molecular dynamics [Cheng, Barnett and Landman (1995), Barnett *et al.* (1995)] sampling of phase space. For the purpose of this presentation we shall concentrate on the CP-type implementation as first proposed by Marx and Parrinello [Marx and Parrinello (1994,1996), Tuckerman *et al.* (1996a,1997)]. The PI-CPMD Lagrangian has the form:

$$L = \frac{1}{P}\sum_{s=1}^{P}\left\{\sum_i \mu \left\langle \dot{\phi}_i^s \middle| \dot{\phi}_i^s \right\rangle - E_{\text{elec}}\left[\phi_i^s, R_l^s\right] + \sum_{ij} \Delta_{ij}^s \left(\left\langle \phi_i^s \middle| \phi_j^s \right\rangle - \delta_{ij}\right)\right\}$$
$$+ \sum_{s=1}^{P}\left\{\sum_{l=1}^{N} \frac{1}{2} M_l' \left(\dot{R}_l^s\right)^2 - \sum_{l=1}^{N} \frac{1}{2} M_l \omega_P^2 \left(R_l^s - R_l^{s+1}\right)^2\right\}. \tag{45}$$

The first sum in this equation is similar to that of the normal CPMD approach which involves the propagation of the electronic wave function during the dynamics simulation with the aid of a fictitious electron mass μ and kinetic energy. The significant difference here is that now each polymer bead, or replica, has its own set of Kohn–Sham orbitals which need to be moved at each time step. The second term in this equation involves the dynamical motion of the polymer chain [Parrinello and Rahman (1984), de Raedt, Sprik and Klein (1984), and Hall and Berne (1984)] which is accomplished by the introduction of a fictitious kinetic energy term for the polymer with an appropriate fictitious mass M'_I. Like the fictitious electronic kinetic energy, the time dependence of these latter parameters has no physical significance and is a tool to allow for the effective sampling of phase space. The equations of motion in their simplest form follow straightforwardly from this Lagrangian:

$$\mu\ddot{\phi}_i^s = -\frac{\delta E_{\text{elec}}[\phi_i^s, R_I^s]}{\delta\phi_i^{*s}} + \sum_j \Delta_{ij}^s \phi_j^s, \tag{46}$$

$$M'_I(\dot{R}_I^s) = \frac{1}{P}\frac{\delta E_{\text{elec}}[\phi_i^s, R_I^s]}{\delta R_I^s} - M_I\omega_P^2(2R_I^s - R_I^{s-1} - R_I^{s+1}). \tag{47}$$

However, these equations as written do not lead to an effective algorithm for two reasons. The first is the well-known problem with all path integral molecular dynamics methods which has to deal with the ergodicity of the trajectories. The problem arises from the harmonic term in the effective potential which corresponds to relatively stiff oscillations compared with the motion of the nuclei. As such, it is difficult to keep the energy flow between the inter- and intra-replica modes in equilibrium with the result of a poor sampling of the available phase space. To circumvent these difficulties, schemes based on an extended Lagrangians including Nosé–Hoover thermostat chains on the atomic degrees of freedom have been proposed [Tuckerman *et al.* (1996a,1996b)]. Another difficulty arises from the transfer of energy between the electrons and these same inter-replica harmonic modes. Although this harmonic term is not well coupled to the nuclear oscillation it is much closer in frequency to those of the electrons. This requires further extension of the Lagrangian where the electrons themselves are also coupled to Nosé–Hoover chains. The net result is that the equations of motion become more complicated and one must resort to more sophisticated integration schemes than the simple algorithms described in a previous section: these methods for solving the equations of motion are discussed elsewhere [Tuckerman *et al.* (1993,1996a)].

Another limitation on the PI-CPMD approach is that of the number of replicas which one may employ in a given simulation. In practice, the PI-CPMD simulation involves running P coupled CPMD calculations simultaneously. Thus, the practical demands on the computer resources generally limits P to 16–64 replicas, which means that the actual evaluation of quantum effects can only be quasi-quantitative.

With all these prohibiting factors one may ask why perform a PI-CPMD-type simulation? For systems with a large number of degrees of freedom or where bond breaking plays a central role in the dynamics reliable parameterized potentials are not available. This prohibits the use of other alternatives which rely on these types of potentials; for systems such as metal clusters there is no real alternative. As a final note to this section, a warning is given. Quantum effects are highly sensitive to the quality of the whole potential energy surface. Therefore, it becomes necessary to have stringent criteria for the choice of density functional method which is employed in the simulation. Comparison of the performance of a functional, relative to high quality correlated electron calculations, in terms of optimized molecular configurations, vibrational frequencies, energy differences between isomers and reaction barriers is strongly advised before an exhaustive PI-CPMD study is begun.

20.4.3 QUANTUM FLUCTUATIONS IN LITHIUM CLUSTERS

In recent years it has become clear that thermal fluctuations play an important role in determining the properties of metal clusters. Here we may mention thermal plasmon broadening [Bertsch and Tománek (1989), Wang *et al.* (1993), Pacheco and Broglia (1989), and Pacheco and Schöne (1997)] and the effects of a finite electronic temperature [Yannoules and Landman (1997)]. With the advent of atomistic *ab initio* molecular dynamics it has become possible to study thermal effects beyond jellium-type models [Brack (1993)] while preserving a similar accuracy as the static electronic structure calculations. Again, it was found that thermal fluctuations of the nuclear skeleton have sizable consequences at experimentally relevant temperatures. In the case of lithium, quantum effects were recognized to be crucial for the understanding of the solid, but very limited attention was given to its clusters [Gerber and Schumacher (1978), and Martins, Car and Buttet (1983)] despite the fact that the potential energy surfaces are known to be flat and to possess many minima of similar energy, an ideal condition for large quantum effects at low temperature. This property manifests itself in the interesting dynamics found in classical *ab initio* simulations of Li_n clusters at finite temperatures [Kawai, Tombrello and Weare (1994), Jellinek *et al.* (1994), Reichardt *et al.* (1997), Jones, Lichtenstein and Hutter (1997), and Gibson and Carter (1997)].

Recently, Li_n clusters with $n = 20$, 40, and 92 atoms were investigated at 30–600 K using path integral Monte Carlo simulations to include the quantum nature of the nuclei in conjunction with a finite version of the one–component plasma to take into account the electronic structure [Ballone and Milani (1992)]. The essential conclusions of this treatment are that 'zero-point motion and tunneling play a crucial role in the determination of the structural and thermodynamic properties' and that 'the clusters are fluid like at all temperatures'. However, this is in stark contrast with static electronic structure calculations: it is generally found that the electronic absorption of small lithium clusters with up to eight atoms can be understood in terms of well-defined static structures [Rubio *et al.* (1996), and Rao, Jena and Ray (1996)]. It should be mentioned that the temperature dependence of spectra down to the order of 10 K is now experimentally accessible for small alkali clusters and allows for the determination of their 'melting point' [Ellert *et al.* (1995), and Schmidt *et al.* (1997)], which means that such issues are of practical relevance in the assignment of cluster structures. We now demonstrate how the PI-CPMD approach has been used to resolve this controversy [Rousseau and Marx (1998), and Rousseau and Marx (1999)]. In order to make contact with both the electronic structure calculations using $n \leq 8$ and the path integral Monte Carlo simulations, $n \geq 20$ two cluster sizes Li_8 and Li_{20} have been examined within the context of low temperature simulations at 10 K with classical and quantum ^6Li nuclei using $P = 16$ replicas and as a convergence test $P = 32$ for Li_{20}.

A convenient place to start the discussion is to analyze the Li–Li bond length distribution, $P(r)$, in Figure 20.10 as obtained from equilibrated 10 K simulations. The $P(r)$ of the Li_8 cluster is characterized by two well-separated prominent peaks stemming from the first- and second-nearest neighbor pairs. In the classical low temperature simulation (dashed line), an additional splitting of both peaks is clearly discernible. This is a signature of the presence of short (approximately 2.8 Å) and long (approximately 3.1 Å) bond alternation as previously noted from static structure optimizations [Rousseau and Marx (1997)]. The same is true for the Li_{20} cluster, except that the distribution is much more structured because of the presence of a third-nearest neighbor distance. The situation changes *qualitatively* when quantum fluctuations are taken into account at 10 K (solid lines): the bond length alternation is destroyed. Thus, the pronounced fine structure of the peaks vanishes and one is left with the coarse structuring in terms of neighbor-shells only. In order to get a feeling for the influence of the quantum effects in terms of thermal excitations a comparison with the thermal fluctuations obtained in a simulation with classical Li nuclei at 100 K (circles) is included in Figure 20.10. These fluctuations *qualitatively* mimic the low-temperature

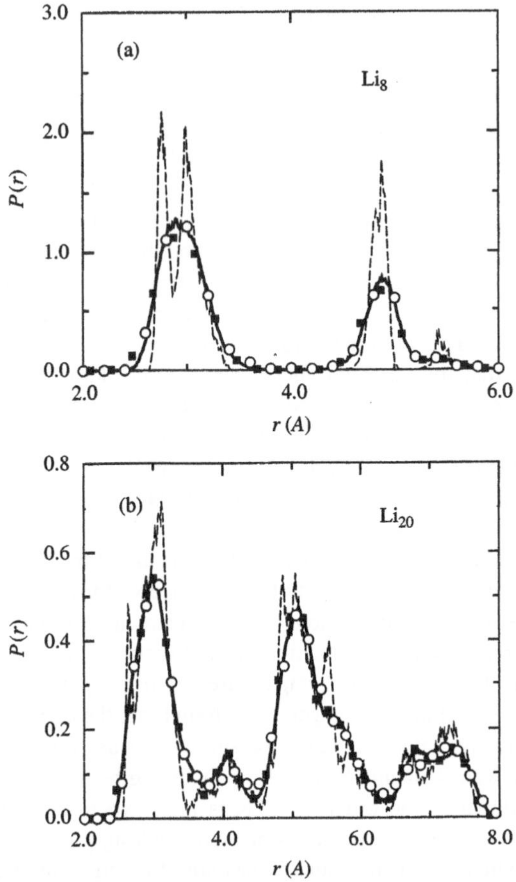

Figure 20.10. Bond length distribution spectra of Li_8 (a) and Li_{20} (b). The dashed line is obtained from a 10 K simulation with a classical description of the lithium nuclear motion, the solid line is obtained from a 10 K simulation with a quantum mechanical description of the lithium nuclear motion. The opaque circles correspond to a 100 K simulation with classical Li nuclei and black squares correspond to a quasi-harmonic broadening (see text for details) applied to the static optimized configuration of each cluster.

quantum fluctuations and suggest a 'matching temperature' of about 100 K; although this exact temperature will depend on the property under consideration.

An important thermodynamic measure to quantify the rigidity or floppiness of clusters is the r.m.s. relative bond length fluctuation [Beck, Doll and Freeman (1989), and Haberland (1994)]:

$$\delta = \frac{2}{N(N-1)} \sum_{ij} \frac{\sqrt{\langle r_{ij}^2 \rangle - \langle r_{ij} \rangle^2}}{\langle r_{ij} \rangle}, \tag{48}$$

where r_{ij} denotes the distance between the nuclei i and j. For the low temperature simulations with classical nuclei both clusters are characterized by $\delta \cong 1\,\%$; which is typical of frozen rigid structures. Quantum fluctuations at 10 K or classical fluctuations at 100 K lead to significantly larger values of about 4–5 % for both cluster sizes. However, this is still much lower than the value of typically $\delta \cong 10\text{–}20\,\%$ required to classify a cluster as liquid like in the sense that it visits several distinct minima corresponding to different structures. Thus, the cold quantum clusters are trapped around a minimum of the potential energy surface, but the nuclei exert a non-negligible zero-point motion.

A shell by shell analysis of the amplitude of the zero-point motion is provided by the r.m.s. position displacement correlation function:

$$R_i(\tau) = \sqrt{\langle |r_i(0) - r_i(\tau)|^2 \rangle} \tag{49}$$

in imaginary time

$$\tau \in [0, \beta\hbar],$$

which measures the 'size',

$$R_i(\beta\hbar/2),$$

and more importantly the localization properties of a particle; note that this quantity is strictly zero in the classical case. This function is displayed in Figure 20.11. For Li_{20} the innermost

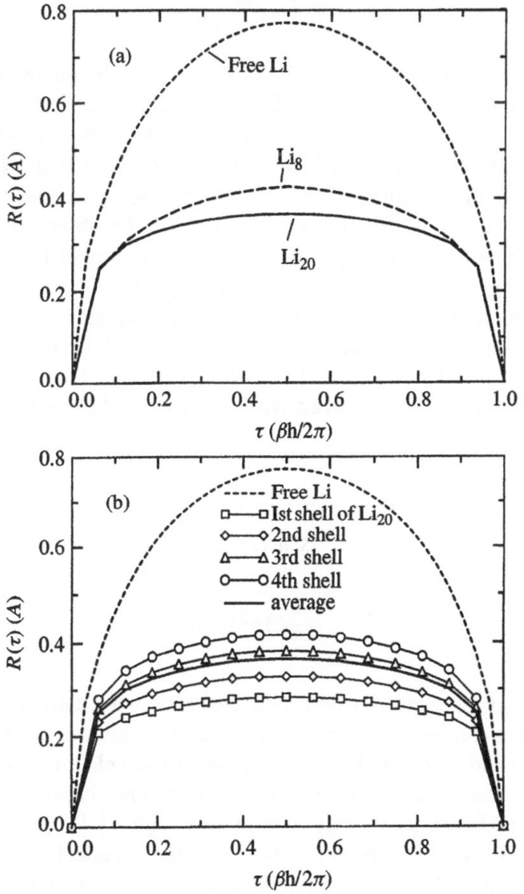

Figure 20.11. (a) The r.m.s. position displacement correlation function averaged over all lithium atoms for Li_8 and Li_{20}. (b) The r.m.s position displacement function averaged over individual shells of atoms as one moves out from the center of the Li_{20} cluster.

nuclei are most localized having a size of about 0.28 Å (compared with 0.78 Å for a free lithium nucleus at 10 K), but that the delocalization increases steadily to roughly 0.42 Å in the outermost shell. The latter value is similar to that obtained from the Li_8 cluster where all atoms can be considered on the surface. Thus, the most significant quantum fluctuations occur for the atoms on the *surface* of the clusters. It is this zero-point motion of the order of a quarter to half an ångström that ultimately destroys the distinction between the 2.8 and 3.1 Å short and long bonds. On the other hand, neither tunneling nor intershell particle exchange is observed.

In the absence of tunneling it is worth while to explore how well a simple (quasi-) harmonic approximation for the zero-point motion will fare. Towards this end, it is appropriate to consider the average nuclear radius of gyration,

$$\Delta = \sqrt{\left\langle \frac{1}{N} \sum_{i=1}^{N} \frac{1}{\beta\hbar} \int_0^{\beta\hbar} d\tau |r_i(\tau) - r_i^c|^2 \right\rangle}, \tag{50}$$

which is another measure of a particle's size relative to its centroid,

$$r_i^c = \int_0^{\beta\hbar} d\tau \frac{r_i(\tau)}{\beta\hbar}.$$

For the Li_8 and Li_{20} clusters at 10 K, $\Delta = 0.27$ and 0.23 Å. If the quantum nuclei behave like a collection of independent harmonic oscillators then one could emulate the quantum broadening of bond length distributions, $P(r)$, *exactly* by simply applying a Gaussian broadening $\propto_{\exp} (-r^2/2\sigma^2)$ with $\sigma^2 = \Delta^2/3$ to the underlying optimized static structures. The essentially perfect agreement of these curves (squares) with the full quantum simulation at 10 K (solid lines) in Figure 20.10 shows that (effective) harmonic zero-point motion can account for the observed quantum effects. As a further visualization of this motion, a sample quantum configuration of 16 replicas is superimposed on top of the static configuration for both clusters in Figure 20.12. For both Li_8 and Li_{20}, the lithium nuclei show a large spatial delocalization but in general the spread of the probability density for each nucleus is centered about the path centroid with no evidence of tunneling.

Is it possible to understand the uncovered behavior in terms of the chemical bonding of the underlying static structures? This is achieved with the help of the electron localization function (ELF) [Becke and Edgecombe (1990)]:

$$\eta(r) = \frac{1}{1 + \left[\frac{\sum_j |\nabla \phi_j(r)|^2 - \frac{1}{4} \frac{(\nabla \rho(r))^2}{\rho(r)}}{\frac{3}{5} (3\pi^2)^{5/3} \rho^{5/3}(r)} \right]^2}. \tag{52}$$

This function $\eta(r)$ is large in regions where two electrons with antiparallel spin are paired in space, whence its maxima can be associated with attractor basins due to electron pairing. ELF is normalized between zero and unity and its value for the uniform electron gas is 1/2. To understand this function in greater detail consider for example a core state in an atom (a typical *localized* state) where the local kinetic energy density will be very high $\sum_i |\nabla \phi_i|^2$ but so will the gradient of the electronic density $\nabla \rho(r)$. In such a case these two terms cancel each other and $\eta(r) = 1.0$. On the other hand, a valence electron in a conduction band of a metal (a typically *delocalized* state) will have a larger kinetic energy density relative to the gradient of $\rho(r)$ and result in lower values of $\eta(r)$. For a discussion and applications of this function see Silvi and Savin (1994), Savin *et al.* (1997) and Rousseau and Marx (2000b).

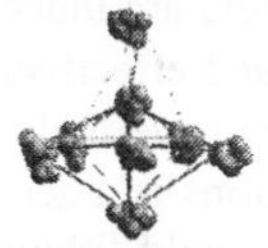

Li_8

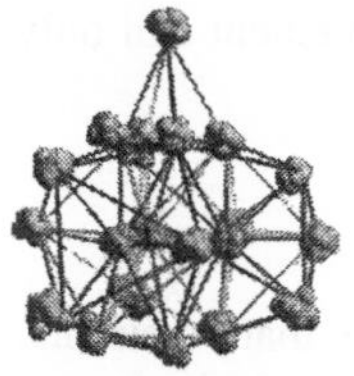

Li_{20}

Figure 20.12. Representative quantum configuration for Li_8 and Li_{20}.

Li_8

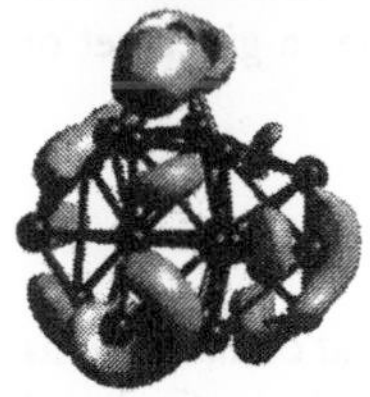

Li_{20}

Figure 20.13. Electron localization function (ELF) of Li_8 and Li_{20} as represented as an isosurface $\eta = 0.9$.

In Figure 20.13 this function for Li_8 and Li_{20} is depicted as an isosurface, $\eta = 0.9$, in order to show all the attractor basins. This function reveals that there is a strong and anisotropic electron pair localization at the *surfaces* of both clusters, whereas pairing in the cluster core is of similar magnitude as in the homogeneous electron gas reference. This suggests the picture of localized surface states with directional bonding and a more metallic cluster core. It also explains the rigidity of the clusters with respect to major rearrangements since these are expected to occur mainly at the cluster surface via its intrinsic defects. It is this inhomogeneity and directionality of the chemical bonding which is underestimated by the uniform background of negative charge in the framework of the one-component plasmon model, where fluid-like behavior was found mainly on the cluster surfaces. The ELF analysis in Figure 20.13 furthermore reveals that the electron pairs associated with the basins are not shared between only two or three atoms each, but rather distributed over several atoms 'multi-center bonding'. In addition, the present analysis shows that the long bonds are those that are contained within the basins of strong pairing, whereas the short bonds are preferentially located between such regions. Nevertheless, the topology of ELF does not change if the bond alternation of a particular isomer is distorted such that not all bonds have a similar length. Its topology, however, is in general clearly different for distinct isomers. This means that the observed quantum fluctuations do not induce a qualitative change of the electronic structure and thus of the associated excitation spectra relative to the underlying static isomer.

The overall picture that emerges from this analysis is that by taking properly into account both the electronic structure and the quantum motion of the nuclei the reported contradictory results may be reconciled. On the one hand, quantum effects in small lithium clusters are indeed significant as they destroy the distinction between short and long chemical bonds, and effectively correspond to thermal excitations of about 100 K. On the other hand, the clusters are nevertheless quasi-rigid at low temperatures and the main consequence of the quantum fluctuations is a zero-point motion broadening of a single underlying structure. Thus, optical spectra of cold lithium clusters can be *qualitatively* understood with the help of electronic structure calculations of a few representative static species as a discriminative tool. However, *quantitative* agreement can only be expected if quantum dispersion is taken into account.

20.5 Conclusions

Although CP molecular dynamics has its origins in solid state physics it, and its many variants, have had impact in many different areas of physics, chemistry and biology. The goal of this chapter is to provide the basic concepts which underlie the rapidly emerging field of *ab initio* molecular dynamics. We have purposely limited ourselves to a discussion of topics which seemed most relevant to molecular spectroscopy. This chapter is designed rather to serve as an introduction to the field and to indicate where these methods may be applied with maximum effectiveness. This general class of methodologies occupies an intermediate ground between the traditional quantum chemical approach which reveals aspects about the energy domain and wave packet dynamics which provides insight about the time domain. With CP one has the insights of both the static and the dynamic. The future will undoubtedly see many new advances where more and more molecular properties may be calculated on an ensemble of molecules under a given set of thermodynamic conditions as opposed to on just a single static configuration.

Acknowledgements

This chapter would not have been possible without the many useful discussions and collaborations which the authors have had with the members of the Car–Parrinello community. In particular, it is a pleasure to thanks W. Andeoni, P. Ballone, M. Bernasconi, M. Boero, J. Hutter, D. D. Klug, K. Laasonan, D. Marx, M. Parrinello, C. Rovira, P. L. Silvestrelli, and M. E. Tuckerman for all of their input over the past few years. All the calculations presented in this article have been performed with the CPMD *ab initio* molecular dynamics package written by J. Hutter, P. Ballone, M. Bernasconi, P. Focher, E. Fois, S. Goedecker, D. Marx, M. Parrinello and M. Tuckerman, Max Planck Institut für Festkörperforschung (Stuttgart) and IBM Research (1990–1999).

20.6 References

Allen, M. P., and Tildesley, D., 1990, *Computer Simulation of Liquids*; Oxford University Press: Oxford.
Andreoni, W., 1987, in *Elemental and Molecular Clusters*, Benedek, G., Martin, T. P., and Pacchiono, G., Eds; Springer: Berlin, pp. 2016–2213.
Bachelet, G. B., Harman, D. R., and Schlüter, 1982, *Phys. Rev.*, **B26**, 4199–1228.
Ballone, P., and Milani, P., 1992, *Phys. Rev. B*, **45**, 11 222–11 225.
Beck, T. L. Doll, J. D., and Freeman, D. L., 1989, *J. Chem. Phys.* **90**, 5651–5650.
Becke, A. D., and Edgecombe, K. E., 1990, *J. Chem. Phys.* **92**, 5397–5403.
Bertsch, G. F., and Tománek, D., (1989), *Phys. Rev. B*, **40**, 2749–2551.
Blanc, J., Broyer, M., Chevaleyer, J., Dugourd, Ph. Kübling, H. Labastie, P. Ulbrich, M. Wolf, J. P., and Wöste, L., 1991, *Z. Phys. D*, **19**, 7–24.
Blanc, J., Bonacic-Koutecký, V., Broyer, M., Chevaleyre, J., Dugourd, Ph., Koutecký, J., Scheuch, C., Wolf, J. P., and Wöste, L., 1992, *J. Chem. Phys.* **96**, 1793–1809.
Barnett, R. N., Cheng, H.-P., Hakkinen, H., and Landman, U., 1995, *J. Phys. Chem.* **99**, 7731–7752.
Bogey, M., Bolvin, H., Cordonnier, M., Demuynck, C., Destombes, J. L., Escribano, R., and Gomez, P. C., 1994, *Can. J. Phys.* **72**, 967–975.
Bonacic-Koutecký, V., Fantucci, P., and Koutecký, J., 1991, *Chem. Rev.* **91**, 1035–1108.

Boo, D. W., Liu, Z. F., Suits, A. G., Tse, J. S., and Lee, Y.T., 1995, *Sci.* **269**, 57–59.
Brack, M., 1993, *Rev. Mod. Phys.* **65**, 677–733.
Car, R., and Parrinello, M., 1985, *Phys. Rev. Lett.*, **55**, 2471–2474.
Chakravarty, C., 1995, *Phys. Rev. Lett.* **75**, 1727–1730.
Chandler, D., and Wolynes, P. G. J., 1981, *Chem. Phys.* **74**, 4078–4096.
Chandler, D., and Leung, K., 1994, *Ann. Rev. Phys. Chem.* **45**, 557–594.
Cheng, H. P., Barnett, R. N., and Landmman, U., 1995, *Chem. Phys. Lett.* **242**, 1–8.
Ceperley, D. M., 1995, *Rev. Mod. Phys.* **67**, 279–355.
Debermardi, A., Bernasconi, B., Cardona, M., and Parrinello, M., 1997, *Appl. Phys. Lett.* **71**, 2692–2694.
de Heer, W. A., 1993, *Rev. Mod. Phys.* **65**, 611–676.
de Raedt, B., Sprik, M., and Klein, M. L., 1984, *J. Chem. Phys.* **80**, 5719–5724.
Dietrich, G., Dasgupta, K., Krückeberg, S., Lützenkirchen, K., Schweikhard, L., Walther, C., and Ziegler, J., 1996, *Chem. Phys. Lett.* **259**, 397–402.
Dietrich, G., Krückeberg, S., Lützenkirchen, K., Schweikhard, L., and Walther, C., 2000, *J. Chem. Phys.* **112**, 752–760.
Dugourd, Ph., Blanc, J., Bonacic-Koutecký, V., Broyer, M., Chevaleyre, J., Koutecký, J., Pittner, J., Wolf, J. P., and Wöste, L., 1991, *Phys. Rev. Lett.* **67**, 2638–2642.
East, A. L. L., Kolbuszewski, M., and Bunker, P. R., 1997, *J. Phys. Chem. A*, **101**, 6746–6752.
East, A. L. L., Liu, Z. F., McCague, C., Cheng, K., and Tse, J. S., 1998, *J. Phys. Chem. A*, **102**, 10 903–10 911.
Ellert, C., Schmidt, M., Schmitt, C., Reiners, T., and Haberland, H., 1995, *Phys. Rev. Lett.* **75**, 1731–1734.
Feynman, R. P., 1972, *Statistical Mechanics A Set of Lectures*; Addisen-Wesley: Reading, MA.
Feynman, R. P., and Hibbs, A. R., 1965, *Quantum Mechanics and Path Integrals*; McGraw-Hill: New York.
Fuchs, M., and Scheffler, M., 1999, *Comput. Phys. Commun.*, (in press).
Gerber, W. H., and Schumacher, E., 1978, *J. Chem. Phys.* **69**, 1692–17 003.
Gibson, D. A., and Carter, E. A., 1997, *Chem. Phys. Lett.* **271**, 266–272.
Giesen, T. F., van Orden, A., Hwang, H. J., Fellers, R. S., Provencal, R. A., and Saykally, R., 1994, *Sci.* **265**, 756–758.
Gomez, P. C., and Bunker, P. R., 1990, *Chem. Phys. Lett.* **165**, 351–354.
Haberland, H., 1994, *Clusters of Atoms and Molecules I and II*; Springer: Berlin.
Hall, R. W., and Berne, B. J., 1984, *J. Chem. Phys.* **81**, 3641–3643.
Honea, E. C., Ogura, A., Murray, C. A., Raghavachari, K., Sprenger, S. O., Jarrold, M. F., and Brown, W. L., 1993, *Nature (London)*, **366**, 42–45.
Huisken, F., and Stemmler, M., 1993, *J. Chem. Phys.* **98**, 7680–7691.
Jarrold, M. F., 1991, *Sci.* **252**, 1085–1087.
Jellinek, J., Bonacic-Koutecký, V., Fantucci, P., and Wiechert, M., 1994, *J. Chem. Phys.* **101**, 10092–10 100.
Jones, R. O., Lichtenstein, A. L., and Hutter, J., 1997, *J. Chem. Phys.* **106**, 4566–4575.
Kawai, R., Tombrello, J. F., and Weare, J. H., 1994, *Phys. Rev. A*, **49**, 4236–4239.
Kirkwood, D. A., Winkel, J. F., and Stace, A. J., 1995, *Chem. Phys. Lett.* **247**, 332–338.
Knickelbein, M. B., 1996, *J. Chem. Phys.* **104**, 3517–3525.
Knickelbein, M. B., and Koretsky, G. M., 1998, *J. Phys. Chem. A*, **102**, 580–586.
Martins, J. L., Car, R., and Buttet, J., 1983, *J. Chem. Phys.* **78**, 5646–5655.
Marx, D., and Parrinello, P., 1994, *Z. Phys. D Rapid Note*, **94**, 143–144.
Marx, D., and Parrinello, M., 1995a, *Sci.* **271**, 179–181.
Marx, D., and Parrinello, M., 1995b, *Nature*, **375**, 216–218.
Marx, D., and Parrinello, M., 1996, *J. Chem. Phys.* **104**, 4077–4082.
Marx, D., and Parrinello, M., 1999, *Sci.* **284**, 59–61.
Miller, H., Kutzelnigg, W., Noga, J., and Klopper, W., 1997, *J. Chem. Phys.* **106**, 1863–1869.
Nygren, M. A., Siegbahn, P. E. M., Jin, C., Guo, T., and Smalley, R. E., 1991, *J. Chem. Phys.* **95**, 6181–6184.
Okumura, M., Yeh, L. I., Myers, J. D., and Lee, Y. T., 1990, *J. Phys. Chem.* **94**, 3416–3427.
Pacheco, J. M., and Broglia, R. A., 1989, *Phys. Rev. Lett.* **62**, 1400–1403.
Pacheco, J. M., and Schöne, W.-D., 1997, *Phys. Rev. Lett.* **79**, 4986–4999.
Parrinello, M., 1990, in *Modern Techniques in Computational Chemistry*, Clementi, Ed.; ESCOM: Leiden, pp. 731–744.
Parrinello, M., and Rahman, A., 1984, *J. Chem. Phys.* **80**, 860–867.
Payne, M. C., Teter, M. P., Allan, D. C., Arias, T. A., and Joannopoulos, J. D., (1992), *Rev. Mod. Phys.* **64**, 1046–1097.
Pribble, R. N., and Zwier, T. S., 1994, *Sci.* **265**, 75–77.
Rao, B. K., Jena, P., and Ray, A. J., 1997, *Phys. Rev. Lett.* **76**, 2878–2881.
Rayner, D. M., Lian, L., Fournier, R., Mitchell, S. A., and Hackett, P. A., 1995, *Phys. Rev. Lett.* **74**, 2070–2074.
Reichardt, D., Bonacic-Koutecký, V., Fantucci, P., and Jellinek, J., 1997, *Chem. Phys. Lett.* **279**, 129–139.
Remler, D. K., and Madden, P., 1990, *Mol. Phys.* **70**, 921–966.
Resta, R., 1994, *Rev. Mod. Phys.* **66**, 899–916.
Röthlisberger, U., and Andreoni, W., 1991, *J. Chem. Phys.* **94**, 8129–8151.

Rousseau, R., Dietrich, G., Krückeberg, S., Lützenkirchen, L., Marx, D., Schweikhard, L., and Walther, C., 1998, *Chem. Phys. Lett.* **295**, 41–46.
Rousseau, R., and Marx, D., 1997, *Phys. Rev. A*, **56**, 617–625.
Rousseau, R., and Marx, D., 1998, *Phys. Rev. Lett.* **80**, 2574–2577.
Rousseau, R., and Marx, D., 1999, *J. Chem. Phys.* **111**, 5091–5101.
Rousseau, R., and Marx, D., 2000a, *J. Chem. Phys.* **112**, 761–769.
Rousseau, R., and Marx, D., 2000b, *Chem. Eur. J.* in press.
Rubio, A., Alonso, J. A., Blasé, X., Balbas, L. C., and Louie, S. G., 1996, *Phys. Rev. Lett.* **77**, 247–251.
Savin, A., Nesper, R., Wengert, S., and Fässler, T. F., 1997, *Angew. Chem. Int. Ed. Engl.* **36**, 1808–1832.
Scharf, D., Martyna, G. J., and Klein, M. L., 1992, *Chem. Phys. Lett.* **197**, 236–241.
Schmidt, M., Kusche, R., Kronmüller, W., von Issendorff, B., and Haberland, H., 1997, *Phys. Rev. Lett.* **79**, 99–102.
Sindzingre, P., Klein, M. L., and Ceperley, D. M., 1989, *Phys. Rev. Lett.* **63**, 1601–1604.
Sindzingre, P., Ceperley, D. M., and Klein, M. L., 1991, *Phys. Rev. Lett.* **67**, 1871–1875.
Singh, D. J., 1993, *Planewaves, Pseudopotentials and the LAPW Method*; Kluwer: Dordrecht.
Silvestrelli, P. L., Bernasconi, M., and Parrinello, 1997, *Chem. Phys. Lett.* **277**, 478–482.
Silvi, B., and Savin, A., 1994, *Nature (London)*, **371**, 683–686.
Štich, I., Marx, D., Parrinello, M., and Terakura, K., 1997, *Phys. Rev. Lett.* **78**, 3669–3671.
Sutton, A. P., 1993, *Electronic Structure of Materials*; Clarendon Press: Oxford.
Szabo, A., and Ostlund, N., 1989, *Modern Quantum Chemistry*; McGraw-Hill: New York.
Troullier, N., and Martins, J. L., 1991, *Phys. Rev.* **B43**, 1993–2006.
Tse, J. S., Klug, D. D., and Laasonen, K., 1995, *Phys. Rev. Lett.* **74**, 876–879.
Tuckerman, M. E., Berne, B., Martyna, G. J., and Klein, M. L., 1993, *J. Chem. Phys.* **99**, 2796–2808.
Tuckerman, M. E., Marx, D., Klein, M. L., and Parrinello, M., 1996a, *J. Chem. Phys.* **104**, 5579–5588.
Tuckerman, M. E., Marx, D., Klein, M. L., and Parrinello, M., 1997, *Sci.* **275**, 817–819.
Tuckerman, M. E., Ungar, J. P., van Rosenvinge, T., and Klein, M., 1996b, *J. Phys. Chem.* **100**, 12 878–12 887.
Vanderbilt, D., 1990, *Phys. Rev.* **B41**, 7892–7895.
Wang, Y., Lewenkopf, C., Tománek, D., Bertsch, G., and Saito, S., 1993, *Chem. Phys. Lett.* **205**, 521–528.
Wathelet, V., Meloni, F., and André, J. M., 1995, in *Modern Techniques in Computational Chemistry*, Clementi, Ed.; ESCOM: Leiden, pp. 465–490.
Whaley, K. B., 1994, *Int. Rev. Phys. Chem.* **13**, 41–76.
Weht, R. O., Kohanoff, J., Estrin, D. A., and Chakravarty, C., 1998, *J. Phys. Chem.* **108**, 8848–8861.
Yannouleas, C., and Landman, U., 1997, *Phys. Rev. Lett.* **78**, 1424–1427.
Yaschenko, E., Fu, L., Resca, L., and Resta, R., 1998, *Phys. Rev.* **B58**, 1222–1229.

INDEX

Printed and bound by CPI Group (UK) Ltd, Croydon, CR0 4YY
27/10/2024
14580294-0004